Synthesis, Processing, Structure and Properties of Polymer Materials

Synthesis, Processing, Structure and Properties of Polymer Materials

Editors

Andrzej Puszka
Beata Podkościelna

MDPI • Basel • Beijing • Wuhan • Barcelona • Belgrade • Manchester • Tokyo • Cluj • Tianjin

Editors

Andrzej Puszka
Department of Polymer Chemistry, Institute of Chemical Sciences, Faculty of Chemistry
Maria Curie-Sklodowska University
Lublin
Poland

Beata Podkościelna
Department of Polymer Chemistry, Institute of Chemical Sciences, Faculty of Chemistry
Maria Curie-Sklodowska University
Lublin
Poland

Editorial Office
MDPI
St. Alban-Anlage 66
4052 Basel, Switzerland

This is a reprint of articles from the Special Issue published online in the open access journal *Polymers* (ISSN 2073-4360) (available at: www.mdpi.com/journal/polymers/special_issues/synthe_proces_struct_propert_polym_mater).

For citation purposes, cite each article independently as indicated on the article page online and as indicated below:

LastName, A.A.; LastName, B.B.; LastName, C.C. Article Title. *Journal Name* **Year**, *Volume Number*, Page Range.

ISBN 978-3-0365-5898-1 (Hbk)
ISBN 978-3-0365-5897-4 (PDF)

Contents

About the Editors

Andrzej Puszka

Andrzej Puszka graduated in Chemistry at Department of Polymer Chemistry of the Maria Curie-Sklodowska University (MCSU) in Lublin in 2008. After completing his master's degree, he began his doctoral studies at the Department of Polymer Chemistry of the MCSU. In 2015, he obtained his D.Sc. degree in Department of Polymer Chemistry, working on synthesis and characterization of new thermoplastic polyurethanes witch sulfur atoms. Since 2015 he works at the Department of Polymer Chemistry, Institute of Chemical Science, MCS University. The results of his research effort carried out between 2013 and 2022 were recorded in the publication of 22 articles in international journals, 1 patent and several contributions to national and international congresses.

The main scientific achievements of Dr Puszka focus on the synthesis of new monomers and polymers and their modifications, investigation of thermal, mechanical, physicochemical and viscoelastic properties of new polymeric materials.

Beata Podkościelna

Beata Podkościelna obtained her D.Sc. degree in 2009 at Maria Curie-Sklodowska University (MCSU) in Lublin, Poland. She completed her habilitation in 2016, and in 2018 she was employed at the position of university professor at the Department of Polymer Chemistry, Institute of Chemical Science, MCS University. She is the author and co-author of 92 original papers in international journals, 42 papers, published as conference materials, 3 patents, and a supervisor of 2 D.Sc. theses.

The major scientific interest of Professor Podkościelna is in polymer chemistry, with a special focus on the synthesis of new monomers and polymers, modification and application of lignin in polymeric materials, chemical modification of polymeric microspheres, photochemistry, physico-chemical and thermal studies of polymers, polymer composites, biocomposites, polymeric blends, hybrids materials and their degradation in different conditions.

polymers

MDPI

Editorial

Special Issue: Synthesis, Processing, Structure and Properties of Polymer Materials

Andrzej Puszka * and Beata Podkościelna *

Department of Polymer Chemistry, Faculty of Chemistry, Institute of Chemical Sciences, Maria Curie-Skłodowska University in Lublin, Gliniana 33, 20-614 Lublin, Poland
* Correspondence: andrzej.puszka@mail.umcs.pl (A.P.); beata.podkoscielna@mail.umcs.pl (B.P.)

1. Introduction

Polymeric materials are widely used in many different technical fields. Modern polymers due to their properties are increasingly replacing traditional construction materials. Polymers and plastics that are non-flammable and resistant to very high temperatures, electrically conductive polymers, fully biodegradable plastics, shape memory polymers and self-repairing polymers are also known. Therefore they can be used to make elements of machines and mechanisms or can be applied as functional advanced materials. The development of polymeric materials is constantly being driven by increasing market demand.

Oil and gas processing produces polymers that are used to make inexpensive, very versatile plastics that make everyday life much easier. But as fossil fuels become scarcer, new challenges arise for the plastics and polymers market.

Innovative plastics are providing ever higher performance, but creation of new materials must be in line with environmental policy. Sustainable thinking is important at every stage, from the design, synthesis, processing and disposal of plastics after use. Researchers and industry are increasingly turning to waste-free and low-emission technologies. As a result, it is possible to obtain materials that are fully functional and exhibit good performance properties.

Due to the growing interest in modern polymeric materials, scientific information related to specialized literature on the complex chemical structure of plastics, as well as the possibility of modifying their features and properties during processing may be useful for specialists from various fields of science and technology.

The aim of this Special Issue was to highlight the progress and fundamental aspects for the synthesis, characterization, properties, and application of novel polymeric materials, as well as their copolymers, composites, and nanocomposites. The development of polymers is the greatest of all materials and the articles presented in our book are perfectly in line with the latest discoveries in this topic.

Citation: Puszka, A.; Podkościelna, B. Special Issue: Synthesis, Processing, Structure and Properties of Polymer Materials. *Polymers* **2022**, *14*, 4550. https://doi.org/10.3390/polym14214550

Received: 24 October 2022
Accepted: 24 October 2022
Published: 27 October 2022

2. Short Description of the Articles presented in This Special Issue

In this Special Issue, 33 original articles have been published on various topics, including the preparation, characterization, and some examples of application of polymeric materials. This issue includes review articles [1–8] and research articles [9–33].

In the paper of Usri, Jamain and Makmud [1] methods for the synthesis of cyclotriphosphazene derivatives are outlined. The flame retardant and dielectric properties of various types of polymers doped with these compounds are also presented.

The relationship between the structure and properties of azo-compound-based liquid crystals was described by Razali and Jamain [2]. In the paper, they have demonstrated that few factors influenced on the formation of different liquid crystals: the length of the alkyl terminal chain, inter/intra-molecular interaction, presence of spacer, spacer length, polarization effects, odd-even effects, and the presence of an electron-withdrawing group or an electron-donating group.

Another 5 papers were devoted to organic-inorganic materials [3] and polymer composites [4–7]. In the review paper of Ikake, Hara and Shimizu [3], the authors discussed the organic-inorganic hybrid materials which have become indispensable high-performance and highly functional materials. This is owing to the improved dispersion control in hybrid materials and the emergence of functional ionic liquids. Harmonization of both these factors has enabled the utilization of functional 3D network structures and nanodispersions in composite materials. The authors summarized the historical development of hybrid materials prepared using the sol-gel method and the birth of ionic liquids.

In the paper of Mousavi et al. [4], summarized the recent advances in using bioactive Graphene Quantum Dots-based polymer composites in drug delivery, gene delivery, thermal therapy, thermodynamic therapy, bioimaging, tissue engineering, bioactive GQD synthesis, and GQD green resuscitation, in addition to examining GQD-based polymer composites.

The paper of Zaidi et al. [5] investigated the impact of carbon nanotubes influence on the properties of the polymer/CNT nanocomposies for photovoltaic application.

The paper of Ogunleye and Rusnakova [6], reviews some of the studies related to the techniques associated with fibre-prestressing in polymer matrix development. The reports from this study provide some basis for selecting a suitable technique for prestressing as well as measuring residual stresses in composite materials.

In the paper of Gómez-Gast et al. [7], the authors presents an overview of polyhydroxyalcanoate–vegetal fiber composites, the effects of the fiber type, and the production method's impact on the mechanical, thermal, barrier properties, and biodegradability, all relevant for biopackaging.

In the last review paper of Liu an Claesson [8], the authors summarize recent experimental investigations of the interfacial lubrication properties of surfaces coated with bottlebrush bio-lubricants and bioinspired bottlebrush polymers. The authors also discuss recent advances in understanding intermolecular synergy in aqueous lubrication including natural and synthetic polymers. Finally, opportunities and challenges in developing efficient aqueous boundary lubrication systems are outlined.

The scope of research articles of papers included a Special Issue is very wide. It includes papers on the synthesis and study of the properties of new thermoplastic polymers [9–13], obtaining of composites [14–16], nanocomposites [17–19], microspheres [20,21], polymeric blends [22–24], hydrophobic nanostructures [25], polymeric membranes [26,27], hydrogels [28] and others [29–33].

In the paper of Hernández-Fernández et al. [9], the effects of arsine on the synthesis and thermal degradation of virgin polypropylene and proposes reaction mechanisms that allow understanding of its behaviour are studied.

In another paper, Wnuczek et al. [10] proposed the synthesis of novel polycarbonates obtained without the use of toxic bisphenol A.

In the next two articles, the authors presented a method of obtaining new polyurethanes. Gomez et al. [11] described the preparation and properties of bio-polyurethanes obtained from residual palm oil without and with jatropha oil or algae oil as additions. Such polymers have a high potential for the production of environmentally friendly bio-based polyurethanes. Puszka and Sikora [12] presented the synthesis and detailed characterization of the structure and properties of poly(thiourethane-urethane) elastomers with a polycarbonate soft segment.

The paper of Serra-Aguila et al. [13] examined the development of a procedure to predict tensile moduli at different strain rates and temperatures, using experimental data from three-point-bending dynamic mechanical analysis. The method was validated by means of a prediction of tensile moduli of polyamide PA66 in the linear elastic range, over a temperature range that included the glass-transition temperature.

Several and important components of the Special Issue were papers describing polymer composites as well as nanocomposites. Ahmad and Abdul Rahman [14], used an electrospinning technique to obtain nanofibers of lignin-filled polyacrylonitrile composites.

Removal of lignin from the polymer matrix by selective dissolution yielded fibers with a porous structure was described in detail.

In the paper of Fouly et al. [15] the green composite from corn cob powder and epoxy resin was presented. For the materials obtained, they studied the mechanical and tribological properties, and compared the experimental results achieved with theoretical calculations.

In turn Wang [16] examined the effect of a halogen-free flame retardant on the kinetics of the thiol-ene reaction. The results indicated that flame retardant promoted cross-linking reactions, to form a compact char layer and retarded further the thermal degradation of the polymer matrix.

Díez-Rodríguez et al. [17] obtained poly(ε-caprolactone) composites with mesoporous SBA-15 nanosilica by extrusion. They studied the effect of the filler on the morphology, thermal and mechanical properties of the obtained compositions.

In the article of Merzah et al. [18], the authors presented the preparation of treated nanoclay composites with epoxy resin. The changes in mechanical and thermomechanical properties as well as morphology depending on the amount of filler in the composites were determined.

Chen et al. [19] used an in-situ polymerization method to prepare electrically conductive PANDB/γ-Al_2O_3 core-shell nanocomposites. For the resulting composites, the authors determined the changes in electrical conductivity and characterized the morphology and structure of the materials.

Wang's and others' paper [20] describes synthesis of novel crosslinked polyphosphaz ene-aromatic ether organic–inorganic hybrid microspheres with different structures. The materials were synthesized via precipitation polycondensation between hexachlorocyclotriphosphazene (HCCP) and bisphenol monomers. The changes in the structure, morphology and wettability of the films obtained by immersing a silicon wafer in ethanol-dispersed microspheres were studied.

In the paper of Rezić et al. [21], optimization of the microencapsulation process through two steps of simple and complex coacervation for further effective functionalization of polymers by dip-coating methodology was presented. The authors prepared functionalized different polymer surfaces with antimicrobial core-shell microcapsules and characterized of polymers after functionalization by several spectroscopic and microscopic techniques.

Masek et al. [22] have studied the influence of natural substances on the ageing properties of epoxidized natural rubber (ENR) and poly (lactic acid) (PLA) eco-friendly elastic blends. Additionally, the ENR/PLA blends were filled with natural pro-health substances of potentially antioxidative behavior, namely, δ-tocopherol (vitamin E), curcumin, δ-carotene and quercetin. Their finding paves new opportunities for bio-based and green anti-ageing systems employed in polymer technology.

In the paper of Lee et al. [23], high-density polyethylene (HDPE)/ethylene vinyl acetate (EVA)/polyurethane (PU) blends were prepared by radiation crosslinking to improve the thermal and mechanical properties of HDPE. This modification method is extremely important and can be applied to various polymer products requiring high heat resistance and flexibility, such as electric cables and industrial pipes.

The aim of the work by Chang et al. [24] was to obtain fully biodegradable polymer blends derived from thermoplastic starch and poly (butylene adipate-*co*-terephthalate). These blends were further foamed with carbon dioxide to improve the flexibility and reduce the brittleness of the finished materials. Due to the fact, that obtained materials exhibits improved flexibility, lightweight and environment-friendly nature, they could be used in electronics packaging materials and has medical equipment application potential in the future.

Hossain and Drmosh [25] reported a simple and inexpensive one-step process to fabricate polymer-templated hydrophobic nanostructures for H_2 sensing application. As a polymer matrix, they used commercial polycarbonate which they subjected to appropriate chemical treatment and then deposited palladium particles on it.

In the paper of Li et al. [26], alkali-induced grafting proton exchange membranes with *co*-grafting PVDF with α-methylstyrene and acrylonitrile were reported. The effects of the types and contents of the alkalis and solvents on the grafting polymerization were investigated and the obtained PVDF-based proton exchange membranes were characterized in terms of the morphology, liquid uptake capability, swelling, ion exchange capacity, conductivity, mechanical property and so on.

In the paper of Rosiles-González et al. [27], the authors reported the synthesis of block and random copolymers of 2-acrylamido-2-methyl-1-propane sulfonic acid (AMPS) and methyl methacrylate (MMA), with different AMPS feed ratios. These solution-processable copolymers with strongly sulfonated acid groups resulted in membranes with tunable ion exchange and water absorption capacities.

Chi et al. [28] obtained and characterized novel metal-supramolecular hydrogels. The hydrophilic polymers were based on acrylic acid and maleic anhydride, and in addition, the polymer chain was grafted with terpyridine units. The supramolecular hydrogels with different terpyridine contents were prepared via the Ni^{2+}–terpyridine coordination bonds formed in aqueous solution. Based on the results, the authors assume that such coordination hydrogels in the future can be used in environmental reactions, biosensors and the similar.

In the paper of Radwan et al. [29] are presented study of the mechanical properties of peat stabilized with different percentages of fillers (including PP fibers).

El-Sayed [30] described the preparation and degradation of geopolymer concrete samples during compression.

Pérez et al. [31], reported the effect of flocculation time, flocculant dose, and water quality on the shear strength of kaolin aggregates. The research contained in this work are useful for decision making in the design and operation of water thickeners and clarifiers.

In the paper of Lorente et al. [32], the relationship between the processing conditions of the solution blow spinning process used to prepare nonwoven mats of polyethylene oxide (PEO), and the structure and morphology of the resulting materials are studied. In this research, the influence of the solvent used to preparation the PEO solutions to be blow spun was considered. The authors concluded, that by changing the solvent composition used to dissolve, the different thermal, mechanical and dissolution behaviors of the material were found.

Pîrvu and Deleanu [33] have attempted to identify failure mechanisms and to measure a well-accepted characteristic, backface signature (BFS), for the designed ballistic protection package. The authors measured BFS and investigated failure mechanisms of this particular package in order to evaluate it as a future application in body armor.

Author Contributions: Conceptualization, A.P. and B.P.; writing—original draft preparation, A.P. and B.P.; writing—review and editing, A.P. and B.P.; supervision, A.P. and B.P.; All authors have read and agreed to the published version of the manuscript.

Funding: This research received no external funding.

Institutional Review Board Statement: Not applicable.

Data Availability Statement: Not applicable.

Acknowledgments: The Guest Editors would like to emphasise their immense gratitude to the Editors-in-Chief of Polymers journal for the fantastic opportunity to manage this Special Issue. Likewise, to all of the authors from many different countries who have contributed to the success of this Special Issue, we thank them for their very high-quality work. Mention should also be made of the reviewers who provided support with insightful comments that undoubtedly improved the quality of the submitted papers. Finally, thanks are due to the Section Managing Editor, for her thorough and constant guidance of the editorial process.

Conflicts of Interest: The authors declare no conflict of interest.

References

1. Usri, S.N.K.; Jamain, Z.; Makmud, M.Z.H. A Review on Synthesis, Structural, Flame Retardancy and Dielectric Properties of Hexasubstituted Cyclotriphosphazene. *Polymers* **2021**, *13*, 2916. [CrossRef] [PubMed]
2. Razali, N.A.; Jamain, Z. Liquid Crystals Investigation Behavior on Azo-Based Compounds: A Review. *Polymers* **2021**, *13*, 3462. [CrossRef] [PubMed]
3. Ikake, H.; Hara, S.; Shimizu, S. Skillful Control of Dispersion and 3D Network Structures: Advances in Functional Organic–Inorganic Nano-Hybrid Materials Prepared Using the Sol-Gel Method. *Polymers* **2022**, *14*, 3247. [CrossRef] [PubMed]
4. Mousavi, S.M.; Hashemi, S.A.; Kalashgrani, M.Y.; Omidifar, N.; Bahrani, S.; Vijayakameswara Rao, N.; Babapoor, A.; Gholami, A.; Chiang, W.-H. Bioactive Graphene Quantum Dots Based Polymer Composite for Biomedical Applications. *Polymers* **2022**, *14*, 617. [CrossRef] [PubMed]
5. Zaidi, B.; Smida, N.; Althobaiti, M.G.; Aldajani, A.G.; Almdhaibri, S.D. Polymer/Carbon Nanotube Based Nanocomposites for Photovoltaic Application: Functionalization, Structural, and Optical Properties. *Polymers* **2022**, *14*, 1093. [CrossRef]
6. Ogunleye, R.O.; Rusnakova, S. A Review of Prestressed Fibre-Reinforced Polymer Matrix Composites. *Polymers* **2022**, *14*, 60. [CrossRef]
7. Gómez-Gast, N.; López Cuellar, M.D.R.; Vergara-Porras, B.; Vieyra, H. Biopackaging Potential Alternatives: Bioplastic Composites of Polyhydroxyalkanoates and Vegetal Fibers. *Polymers* **2022**, *14*, 1114. [CrossRef]
8. Liu, X.; Claesson, P.M. Bioinspired Bottlebrush Polymers for Aqueous Boundary Lubrication. *Polymers* **2022**, *14*, 2724. [CrossRef]
9. Hernández-Fernández, J.; Guerra, Y.; Puello-Polo, E.; Marquez, E. Effects of Different Concentrations of Arsine on the Synthesis and Final Properties of Polypropylene. *Polymers* **2022**, *14*, 3123. [CrossRef]
10. Wnuczek, K.; Puszka, A.; Podkościelna, B. Synthesis and Spectroscopic Analyses of New Polycarbonates Based on Bisphenol A-Free Components. *Polymers* **2021**, *13*, 4437. [CrossRef]
11. Gomez, J.C.; Zakaria, R.; Aung, M.M.; Mokhtar, M.N.; Yunus, R. Synthesis and Characterization of Polyurethanes from Residual Palm Oil with High Poly-Unsaturated Fatty Acid Oils as Additive. *Polymers* **2021**, *13*, 4214. [CrossRef] [PubMed]
12. Puszka, A.; Sikora, J.W. Synthesis and Characterization of New Polycarbonate-Based Poly(thiourethane-urethane)s. *Polymers* **2022**, *14*, 2933. [CrossRef]
13. Serra-Aguila, A.; Puigoriol-Forcada, J.M.; Reyes, G.; Menacho, J. Estimation of Tensile Modulus of a Thermoplastic Material from Dynamic Mechanical Analysis: Application to Polyamide 66. *Polymers* **2022**, *14*, 1210. [CrossRef] [PubMed]
14. Ahmad, M.A.T.; Abdul Rahman, N. Preparation and Characterization of Highly Porous Polyacrylonitrile Electrospun Nanofibers Using Lignin as Soft Template via Selective Chemical Dissolution Technique. *Polymers* **2021**, *13*, 3938. [CrossRef] [PubMed]
15. Fouly, A.; Abdo, H.S.; Seikh, A.H.; Alluhydan, K.; Alkhammash, H.I.; Alnaser, I.A.; Abdo, M.S. Evaluation of Mechanical and Tribological Properties of Corn Cob-Reinforced Epoxy-Based Composites—Theoretical and Experimental Study. *Polymers* **2021**, *13*, 4407. [CrossRef] [PubMed]
16. Wang, J. Study on Fire Behavior, Thermal Stability and Degradation Kinetics of Thiol-Ene with Poly(aminopropyl/phenyl)silsesqui oxane. *Polymers* **2022**, *14*, 1142. [CrossRef] [PubMed]
17. Díez-Rodríguez, T.M.; Blázquez-Blázquez, E.; Antunes, N.L.C.; Ribeiro, M.R.; Pérez, E.; Cerrada, M.L. Nanocomposites of PCL and SBA-15 Particles Prepared by Extrusion: Structural Characteristics, Confinement of PCL Chains within SBA-15 Nanometric Channels and Mechanical Behavior. *Polymers* **2022**, *14*, 129. [CrossRef]
18. Merzah, Z.F.; Fakhry, S.; Allami, T.G.; Yuhana, N.Y.; Alamiery, A. Enhancement of the Properties of Hybridizing Epoxy and Nanoclay for Mechanical, Industrial, and Biomedical Applications. *Polymers* **2022**, *14*, 526. [CrossRef]
19. Chen, C.-H.; Lin, Y.-C.; Lin, H.-M. Surface Modification of γ-Al_2O_3 Nanoparticles Using Conductive Polyaniline Doped by Dodecylbenzene Sulfonic Acid. *Polymers* **2022**, *14*, 2232. [CrossRef]
20. Wang, Y.; Yu, Y.; Li, L.; Zhang, H.; Chen, Z.; Yang, Y.; Jiang, Z.; Mu, J. Preparation and Properties of Novel Crosslinked Polyphosphazene-Aromatic Ethers Organic–Inorganic Hybrid Microspheres. *Polymers* **2022**, *14*, 2411. [CrossRef]
21. Rezić, I.; Somogyi Škoc, M.; Majdak, M.; Jurić, S.; Stracenski, K.S.; Vinceković, M. Functionalization of Polymer Surface with Antimicrobial Microcapsules. *Polymers* **2022**, *14*, 1961. [CrossRef] [PubMed]
22. Masek, A.; Cichosz, S. Biocomposites of Epoxidized Natural Rubber/Poly(Lactic Acid) Modified with Natural Substances: Influence of Biomolecules on the Aging Properties (Part II). *Polymers* **2021**, *13*, 1677. [CrossRef] [PubMed]
23. Lee, J.-G.; Jeong, J.-O.; Jeong, S.-I.; Park, J.-S. Radiation-Based Crosslinking Technique for Enhanced Thermal and Mechanical Properties of HDPE/EVA/PU Blends. *Polymers* **2021**, *13*, 2832. [CrossRef]
24. Chang, C.-J.; Venkatesan, M.; Cho, C.-J.; Chung, P.-Y.; Chandrasekar, J.; Lee, C.-H.; Wang, H.-T.; Wong, C.-M.; Kuo, C.-C. Thermoplastic Starch with Poly(butylene adipate-*co*-terephthalate) Blends Foamed by Supercritical Carbon Dioxide. *Polymers* **2022**, *14*, 1952. [CrossRef] [PubMed]
25. Hossain, M.K.; Drmosh, Q.A. Polymer-Templated Durable and Hydrophobic Nanostructures for Hydrogen Gas Sensing Applications. *Polymers* **2021**, *13*, 4470. [CrossRef] [PubMed]
26. Li, S.; Li, X.; Fu, P.; Zhang, Y. Alkali-Grafting Proton Exchange Membranes Based on Co-Grafting of α-Methylstyrene and Acrylonitrile into PVDF. *Polymers* **2022**, *14*, 2424. [CrossRef]
27. Rosiles-González, V.; Le Lagadec, R.; Varguez-Catzim, P.; Loria-Bastarrachea, M.I.; González-Díaz, A.; Hernández-Núñez, E.; Aguilar-Vega, M.; González-Díaz, M.O. Preparation and Characterization of Strongly Sulfonated Acid Block and Random Copolymer Membranes for Acetic Acid Esterification with 2-Propanol. *Polymers* **2022**, *14*, 2595. [CrossRef]

28. Chi, Z.; Ma, C.; He, Z.; Ma, Z.; Chen, X.; Huang, Z. A Supramolecular Hydrogel Based on Copolymers of Acrylic Acid and Maleic Anhydride Derivatives with Terpyridine Motifs. *Polymers* **2022**, *14*, 2857. [CrossRef]
29. Radwan, M.K.H.; Lee, F.W.; Woon, Y.B.; Yew, M.K.; Mo, K.H.; Wai, S.H. A Study of the Strength Performance of Peat Soil: A Modified Cement-Based Stabilization Agent Using Fly Ash and Polypropylene Fiber. *Polymers* **2021**, *13*, 4059. [CrossRef]
30. El-Sayed, T.A. Axial Compression Behavior of Ferrocement Geopolymer HSC Columns. *Polymers* **2021**, *13*, 3789. [CrossRef]
31. Pérez, K.; Toro, N.; Jeldres, M.; Gálvez, E.; Robles, P.; Alvarado, O.; Toledo, P.G.; Jeldres, R.I. Estimating the Shear Resistance of Flocculated Kaolin Aggregates: Effect of Flocculation Time, Flocculant Dose, and Water Quality. *Polymers* **2022**, *14*, 1381. [CrossRef] [PubMed]
32. Lorente, M.Á.; González-Gaitano, G.; González-Benito, J. Preparation, Properties and Water Dissolution Behavior of Polyethylene Oxide Mats Prepared by Solution Blow Spinning. *Polymers* **2022**, *14*, 1299. [CrossRef] [PubMed]
33. Pîrvu, C.; Deleanu, L. Failure Investigation of Layered LFT SB1plus Package after Ballistic Tests for Level IIA. *Polymers* **2021**, *13*, 2912. [CrossRef] [PubMed]

polymers

MDPI

Article

Synthesis and Characterization of New Polycarbonate-Based Poly(thiourethane-urethane)s

Andrzej Puszka [1,*] and Janusz W. Sikora [2]

1 Department of Polymer Chemistry, Faculty of Chemistry, Institute of Chemical Sciences, Maria Curie-Skłodowska University in Lublin, Gliniana 33, 20-614 Lublin, Poland

2 Department of Technology and Polymer Processing, Faculty of Mechanical Engineering, Lublin University of Technology, Nadbystrzycka 36, 20-618 Lublin, Poland; janusz.sikora@pollub.pl

* Correspondence: andrzej.puszka@mail.umcs.pl

Abstract: The new segmented poly(thiourethane-urethane)s (PTURs) based on 1,1′-methanediylbis(4-isocyanatocyclohexane) (HMDI, *Desmodur W*®), polycarbonate diol (PCD, *Desmophen C2200*) and (methanediyldibenzene-4,1-diyl)dimethanethiol were synthesized by one-step melt polyaddition method. The obtained PTURs, with a content of 30–60 wt% of the hard segments (HS), were tested in which the influence of changes in the HS content on their properties was determined. The polymers were characterized by Fourier transform infrared spectroscopy (FTIR), gel permeation chromatography (GPC), thermal analysis (DSC, TGA) and thermomechanical analysis (DMTA). Additionally, tensile strength, optical (refractive index, UV-VIS and color) and surface properties of the obtained polymers (contact angle and surface free energy) and adhesion to copper were examined. FTIR analysis verified the supposed structure of the polymers obtained and showed a complete conversion of the isocyanate groups. TGA analysis confirmed the relatively good thermal stability of the polymers. On the other hand, after performing the DSC analysis, it was possible to state that the obtained materials were partially or completely amorphous, and the microphase separation decreased with increasing HS content in the polymer. Similar observations were made from the DMTA data. In addition, the hardness, tensile strength, modulus of elasticity, storage modulus, adhesion to copper, refractive index and total free surface energy increased with increasing HS content in the polymer.

Keywords: thermoplastic polyurethanes; surface free energy; dithiol; mechanical properties; differential scanning calorimetry; optical properties

Citation: Puszka, A.; Sikora, J.W. Synthesis and Characterization of New Polycarbonate-Based Poly(thiourethane-urethane)s. *Polymers* **2022**, *14*, 2933. https://doi.org/10.3390/polym14142933

Academic Editor: Sándor Kéki

Received: 26 June 2022
Accepted: 18 July 2022
Published: 20 July 2022

1. Introduction

Polyurethanes (PUs) are an important group of polymers that have a number of unique properties [1–4], which provides opportunities for their use in various industries, e.g., packaging, electronic, automotive, construction, building and medicine [5–11]. The various applications of polyurethanes also include the modification of textiles in order to provide them with increased mechanical strength and water resistance [12,13]. Among polyurethane materials, we can distinguish foams, adhesives, varnishes, fibers, elastomers, etc. [5]. The PUs can also form suitable blends [14–18] or composites [19–23].

PUs are generally formed by reacting diisocyanates with polyols or diamines [2]. Flexible polyol macro-chains have low glass transition temperatures (T_g) and form soft segments (SS), while hydrogen-bonded urethane linkages create hard segments (HS) of the polymer network structure [1–4,7,8,13]. These segments may be self-soluble with good mixing or may remain separate as segmented polyurethane (SPU)-forming domains. The quantities and separations of HS and SS determine the properties of the formed polymer. HS ensures a high modulus, while SS ensures the extensibility of the product [3,4]. Hydrogen bonding among hard and hard segments (HS-HS) and hard and soft segments (HS-SS) is created due to the presence of polar functional N-H and C=O groups on PU chains. The level of miscibility of HS and SS relies on the amount of hydrogen bonding between these segments [2,7,8,13]. In situations where there is a stronger hydrogen bond between HS and HS, phase separation is predominant, resulting in high performance products [1,3]. The phase intermixing is due to strong intra- or intermolecular interactions between HS and SS. Indeed, this is a key element highlighting the relevance of the 3D positioning of the functional groups [24], which has a significant influence on the mechanical properties of the final product. SPUs with different properties can be obtained by varying the type and the amount of initial components, by using different molecular weight diol component or by changing the preparation process. All these mechanisms change the amount and distribution of the HS and SS and lead to variation in the final properties [10,11]. This aspect of molar mass variation for polymeric network has been highlighted as well by De Keer et al. [25].

The soft segments consist of polyether, polyester or polycarbonate polyols, while the hard segments are formed by reactions between a diisocyanate and low molecular weight diol, dithiol or diamine chain extender (see Figure 1).

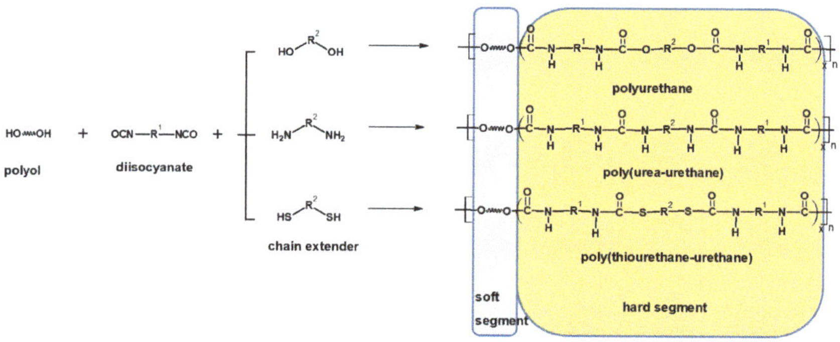

Figure 1. General scheme of reactions between polyol, diisocynate and different chain extenders.

Proper selection of the appropriate flexible segments in the case of biomedical materials is extremely important due to the scope of their potential application. This choice determines the hydrolytic stability of the SS. The issue of the hydrolytic stability of polyurethane materials has been the subject of many authors' research for many years [26–28]. The first PUs used in medicine were synthesized with the use of polyester diols, but the presence of ester bonds made them highly hydrolytic unstable. The chemical structure, and above all a highly developed surface, easily accessible to organic fluids, cells and enzymes, causes rapid hydrolytic degradation of these polymers. The resulting carboxyl groups reduce the pH value locally, additionally catalyzing their further destruction. For this reason, the polyester component in PUs has been replaced by a polyether component which is resistant to hydrolysis [28]. However, research has shown that poly(ether-urethane)s are in turn susceptible to so-called oxidative degradation, which is the result of the joint interaction of body fluids, oxidizing compounds and stresses. A mechanism for this degradation has been proposed, involving the initiation of the process by adhesion and activation of cells (phagocytes), followed by the release of oxidizing substances (HOO·) that attack the polymer in the polyether segment [28]. A mechanism that may additionally occur during the degradation of such PUs under physiological conditions is acid hydrolysis. Such slow hydrolysis can take place, for example, by the action of hydrogen ions secreted by leukocytes. In this reaction, the oxygen atom of the PUs ether group undergoes protonation, which leads to chain breakage. This course of the reaction is consistent with reports that PUs based on polyethers are much more easily degraded in an acidic environment than poly(carbonate-urethane)s. For these reasons, hydrolytically and oxidatively stable oligomeric polycarbonate diols are used as the polyol component of the newest generation of PUs, which in turn made it possible to obtain biologically stable poly(carbonate-urethane)s [28]. The hydrolytic stability of polycarbonates is associated with low water permeability, which is attributed to the rigidity of the polymer chains. Moreover, unlike polyesters, the hydrolysis of the carbonate bond leads to the formation of two groups, -OH and CO_2, i.e., products that do not further catalyze polycarbonate hydrolysis [26–28].

Another fragment of polyurethane materials that can be hydrolyzed is the HS fragment, which is much more resistant to degradation than the soft segment. As it results from the available literature, amine, alcohol and CO_2 are formed during the hydrolysis of urethane bonds. In turn, urea bonds are hydrolyzed to amines and CO_2.

For special applications, other chain extenders can be used that contain reactive hydrogen atoms in their structure, e.g., dithiols [29–34]. The appearance of sulfur atoms in the polymer structure, depending on the type of functional group, can enhance some key properties, e.g., thermal and chemical resistance and optical properties (refractive index). The use of chain extenders containing sulfur atoms in their structure is the subject of mine and my colleagues' research. Of the many systems available, we have synthesized derivatives of diphenylmethane [33–47], diphenylethane [48–52], benzophenone [53–55], diphenyl ether [56] and diphenyl sulphide [38,57–60]. In our previous studies, we have attempted to obtain and investigate the effects of such unconventional chain extenders on the structural, thermal, mechanical and optical properties of novel polyurethane materials. Polymers obtained with the use of unconventional chain extenders were characterized by hardness up to 81/61 Sh A/D and tensile strength up to 48.9 MPa, and these values are higher than those of the commercial Chronoflex® AL 80A biomaterial. Such materials were also characterized by better optical properties (refractive index 1.525 vs. 1.501, transmittance 88.5% vs. 82.1%) compared to polyurethanes obtained with butane-1,4-diol as the chain extender.

In this study, various poly(thiourethane-urethane)s (PTURs) were synthesized by using cycloaliphatic 1,1′-methanediylbis(4-isocyanatocyclohexane) (*Desmodur W®*, HMDI) and sulfur-containing chain extender (methanediyldibenzene-4,1-diyl)dimethanethiol (dithiol), forming hard segments in the obtained polymers. The soft segments were composed of PCD diol (*Desmophen® C2200*). By applying the one-step melt method, we obtained PTURs with HS contents of 30, 40, 50 and 60 wt%. The influence of the composition on the properties of the obtained PTURs was investigated. For the synthesized polymers, using Fourier transform infrared spectroscopy (FTIR), the physicochemical (reduced viscosities, gel permeation chromatography (GPC), contact angles (CAs) and surface free energy (SFE), optical (refractive index, transparency and color), thermal (differential scanning calorimetry (DSC) and thermogravimetric analysis (TGA)), thermomechanical (dynamic thermal mechanical analysis (DMTA)), mechanical (hardness and tensile test) and adhesive properties have been studied.

2. Materials and Methods

2.1. Materials

Dithiol (T_m = 85 °C) was obtained at the Department of Polymer Chemistry, Maria Curie-Skłodowska University [33,61]. HMDI (*Desmodur W®*) and PCD of $\overline{M_n}$ = 2000 g/mol (*Desmphen® C2200*) were kindly supplied by Covestro (Leverkusen, Germany). Before its use, the PCD was heated at 90 °C in vacuo for 10 h. Dibutyltin dilaurate (DBTDL) from Merck-Schuchardt (Hohenbrunn, Germany) was used as received. The analytical reagents were as follows: 1,1,2,2-tetrachloroethane (TChE), diiodomethane and tetrahydrofuran (THF) (Aldrich, St. Louis, MO, USA), 1–bromonaphthalene (POCh S.A., Gliwice, Poland), redistilled water (Millipore, UMCS, Lublin, Poland) and Optylite® physiological saline (KabiPac, Kutno, Poland) were used in purchased form.

2.2. Measurements Methods

2.2.1. Fourier Transform Infrared Spectroscopy (FTIR)

FTIR spectra were obtained using the attenuated total internal reflection (ATR/FTIR) method with a Bruker TENSOR 27 spectrophotometer (Ettlingen, Germany) equipped with a PIKE measuring cell with crystalline diamond embedded in zinc selenide. The FTIR spectra were recorded from 4000 to 600 cm^{-1}, at a resolution of 2 cm^{-1} and with 128 scans per sample in the absorption mode. The PTURs were in the form of a 1 mm-thick pressure-formed sheet.

2.2.2. Physicochemical Characterization

1. Gel permeation chromatography (GPC)

The number ($\overline{M_n}$) and mass ($\overline{M_w}$) average molar mass (Daltons (Da))), and the molar mass dispersity ($Đ_M$, $Đ_M = \overline{M_w}/\overline{M_n}$) [62] of the PTURs were determined by gel permeation chromatography (GPC) carried out on a Viscotek GPCMax (Westborough, MA, USA) equipped with Triple Detector Array TDA305. The eluent, with the flow rate 1 mL/min, was tetrahydrofuran (THF), and the operation temperature was set to be 35 °C. The molar mass was calibrated with polystyrene standards.

2. Reduced viscosities

Reduced viscosities (η_{red}, dL/g) of 0.5% the polymer solution in (TChE) were determined in an Ubbelohde viscometer (Gliwice, Poland) at 25 °C.

3. Contact angles (CAs) and Surface Free Energy (SFE)

Contact angles (CAs) of PTURs were measured at 20 °C with a contact angle goniometer (KRÜSS GmbH, Hamburg, Germany) with water and diiodomethane droplets. The volume of droplets was 2 µL. Each sample was analyzed five times and the average value of the contact angle was designated. For calculation of the surface free energy according to the method of Owens, Wendt, Rabel and Kaelble [63], the Krüss ADVANCE (KRÜSS GmbH, Hamburg, Germany) software was used. The sessile drop orientation and the ellipse fitting method were used for data analysis.

4. Hydrolytic resistance

The PTURs (weighing approximately 1 g) were placed in Optylite® saline. Its ion concentration was as follows (mmol/dm^3): 141 Na^+, 34 CH_3COO^-, 3 $C_6H_5O_7{}^{3-}$, 2 Ca^{2+}, 5 K^+, 1 Mg^{2+}, 109 Cl^-. The solution temperature was 37 °C and the immersion time was 8 weeks. After each week, the samples were taken from the solution, washed in distillated water and dried in vacuum at 60 °C. Then, changes in their mass (%) were determined.

2.2.3. Thermal and Thermomechanical Properties

1. Differential Scanning Calorimetry (DSC)

DSC analysis was carried out using a Netzsch 204 F1 Phoenix calorimeter (Günzbung, Germany), in accordance with the standard ISO 11357-1:2016 [64]. Each sample (weighing about 10 mg) was analyzed with the following program: cooling and isotherm for 3 min at −100 °C, heating to 200 °C, next cooling to −100 °C and then heating to 200 °C. The test was carried out at a heating rate of 10 °C/min, and under an argon atmosphere (flow 30 mL/min. Aluminum crucibles with a pierced lid were used as measuring cells, while an analogous empty crucible was used as reference. The glass transition temperatures (T_g) were read from the inflection point of the DSC curves, and the melting temperatures (T_m) from the maximums of the endothermic peaks.

2. Thermogravimetric Analysis (TGA)

TGA was carried out using a Netzsch STA 449 F1 Jupiter thermal analyzer (Selb, Germany) in a helium environment (gas flow = 20 cm^3/min), with the temperature ranging from 30 to 800 °C and a heating rate of 10 °C/min. Polymer samples weighing about 10 mg were placed in an aluminum(III) oxide (Al_2O_3) crucible and an analogous crucible was used as the reference.

3. Dynamic Mechanical Thermal Analysis (DMTA)

DMTA of PTURs was performed in tensile mode using DMA Q800 Analyzer TA Instruments (New Castle, DE, USA). Calibration was performed as per the manufacturer's recommendations included in Advantage Software, version 5.5.24 (TA Instruments, New Castle, DE, USA). The experiments were carried out on rectangular samples of dimensions close to 1 mm thick, 5 mm wide and 30 mm long. Polymer shapes were prepared using the Carver hydraulic press with heated plates. Press molding was performed at 100–120 °C under a 10–30 MPa pressure. Experimental conditions employed were a frequency of 1 Hz and static stain of 0.05%, with the scanning temperature ranging from −100 °C to 150 °C in the air conditions and a temperature ramp of 3 °C/min. The samples were cut from the pressed sheets. The variations in storage modulus (E'), loss modulus (E'') and tangent delta (tanδ) versus temperature were determined.

2.2.4. Mechanical Properties

Tensile testing was carried out using a Zwick/Roell Z010 tensile-testing machine (Ulm, Germany) agreeing to the International Standard ISO 527-2:2012 [65] at the speed of 100 mm/min at 23 °C. Polymeric samples (1 mm thick and 6 mm wide for the section measured) were cut from the pressed sheet. Polymer shapes were prepared using the Carver hydraulic press with heated plates. Press molding was performed at 100–120 °C under a 10–30 MPa pressure. The hardness of the PTURs was determined by the Shore A/D method using a Zwick 7206/H04 hardness tester (Ulm, Germany). The readings were taken after 15 s at the temperature of 23 °C [66].

The single lap-shear strength of polymers on a copper plate with dimensions of 100 mm × 25 mm × 1.5 mm was measured in accordance with the Polish Standard PN EN 1465: 2009 [67] using the Zwick/Roell Z010 device (Ulm, Germany). An adhesive joint from a polymer sample of 12.5 mm × 25 mm × 0.2 mm dimensions was prepared by pressing the polymer between two copper plates (prepared according to PN-EN-13887:2005 [68]). The strength of such an adhesive joint was measured at the tensile speed of 2 mm/min at 23 °C.

2.2.5. Optical Properties

1. Refractive index (RI)

RI was measured by using Conbest Abbe's Refractometer Type 325 (Krakow, Poland) instrument according to Method A of the European Standard EN ISO 489:2022 [69] at 23 °C. The contact liquid was 1–bromonaphthalene.

2. Transmittance

The ultraviolet-visible (UV/vis) spectra of the PTURs were obtained by a UV-2550 (Shimadzu, Kioto, Japan) UV spectrophotometer at a scanning rate of 200 nm/min and in the range of 200–900 nm.

3. Color

The color of the polymers was determined using a X-Rite Ci4200 spectrophotometer in accordance with the standard ASTM E308 [70]. The color is defined in the CIELab system, where it is specified in L*, a*, b* space. Parameter a* is the color from green (negative values) to red (positive values); parameter b* is the color from blue (negative values) to yellow (positive values); and parameter L* is the brightness, describing the grey scale from black to white (value 0 corresponds to black and 100 to white).

2.3. Polymer Synthesis

PTURs with hard-segment contents of 30, 40, 50 and 60 wt% were prepared by a one-step melt polymerization from dithiol, HMDI and at the NCO/(OH + SH) molar ratio of 1.07. The general procedure for the synthesis of PTURs using this method was as follows. Oligomer diol and dithiol (0.01 mol together) and HMDI (0.0107 mol) were heated with stirring under dry nitrogen to 90 °C in an oil bath. A catalytic amount of DBTDL (about 0.03 g) was added to the formed clear melt and polymerization rapidly began with vigorous stirring. The reaction temperature was gradually elevated to 130 °C and the formed rubber-like product was additionally heated at this temperature for 2 h. Schematic representation of the polymer synthesis is given in Figure 2.

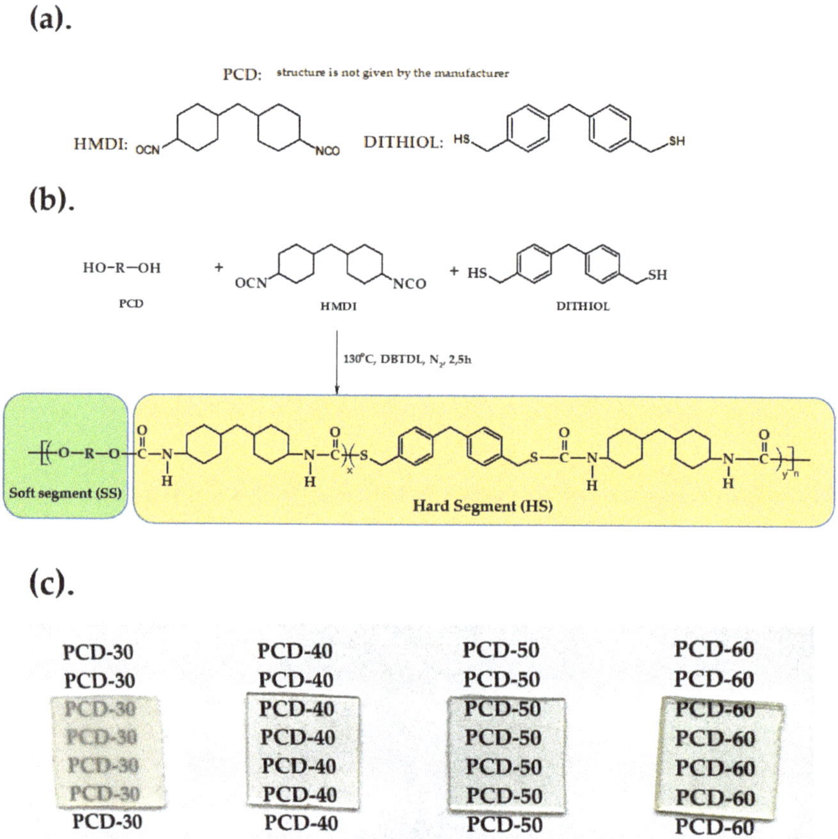

Figure 2. (**a**). Chemical structures of the reactants; (**b**) Schematic representation of the synthesis route of polymers; (**c**) Images of the synthesized polymers.

Designations and composition used to synthesize the PTURs are given in Table 1. The polymers were designated as X-Y, where X is the abbreviation of soft segment and Y represents the hard-segment content.

Table 1. Designations of the PTURs.

PTUR	Amount of Dithiol (mol %)	Amount of Soft Segment (mol %)	[1] Hard-Segment Content (wt%)
PCD-30	53	47	29.87
PCD-40	67	33	39.83
PCD-50	77	23	50.15
PCD-60	85	15	59.88

[1] The hard-segment content (wt%) is calculated by using the expression $\frac{W_{dithiol}+W_{HMDI}}{W_{dithiol}+W_{HMDI}+W_{SS}}$, where $W_{dithiol}$, W_{HMDI} and W_{SS} are the mass of dithiol, HMDI and soft segment, respectively.

Our use of a polycarbonate soft segment as a building block was extremely important in terms of the potential use of the newly obtained PTURs as biomedical materials. As is commonly known, depending on what the obtained material is to be used for, we can use different SS. Our goal was to obtain polyurethane materials that could be used in long-term implantation. Therefore, we chose polycarbonate among the different types of SS. Additional advantages of polyurethanes obtained with the use of PCD are their better resistance to hydrolysis than polymers obtained from polyester segments and more favorable strength properties than polymers with a polyether SS.

3. Results and Discussion

The polymers obtained were transparent or partially transparent (PCD-30) solids (see Figure 2c). All these polymers dissolved at room temperature in THF, TChE, chloroform, *N*,*N*-dimethylacetamide and *N*,*N*-dimethylformamide, but they were insoluble in *N*-methyl-2-pyrrolidone and dimethyl sulfoxide.

3.1. FTIR

The structures of all the polymers were verified by FTIR spectroscopy. Characteristic bands of different groups present in PTURs are visible in the FTIR spectra (see Figure 3).

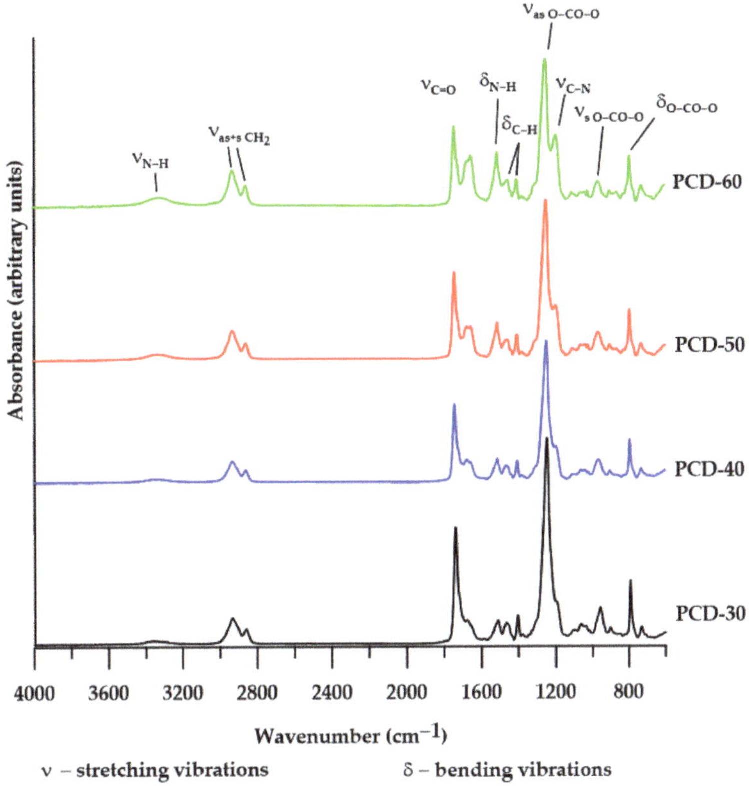

Figure 3. FTIR spectra of synthesized polymers.

On the FTIR spectra obtained for all the studied PTURs, a broad absorption band in the range 3351–3309 cm^{-1} was observed. It is typical for the N-H stretching vibrations of urethane and thiourethane groups [33,34]. For the same fragments of the above-mentioned groups, we also observe bending vibrations in the range 1511–1509 cm^{-1} [33,34]. The next clear bands visible in each spectrum are the peaks corresponding to C-H stretching vibrations of the CH_2 group; asymmetric ones are visible in the range 2933–2926 cm^{-1}, while symmetric ones in the range 2859–2854 cm^{-1}. Next, we can read the bands corresponding to the C-H bending vibration of the CH_2 group in the cyclohexane structure and in the aliphatic CH_2; the asymmetric ones are located in the range 1465–1450 cm^{-1}, while the symmetric ones are located in the range 1404–1403 cm^{-1} [33]. FTIR spectra of all the obtained polymers also indicated strong absorption in the range 1741–1738 cm^{-1} and 1676–1654 cm^{-1}. The former range corresponds to C=O stretching vibrations from carbonate groups while the latter corresponds to the same vibrations in the thiourethane groups [33,34]. In the case of C=O stretching vibrations occurring in the urethane group, the band corresponding to them can only be read for the structure containing 40% of the HS—1700 cm^{-1}. In the case of the other polymers, these bands are hardly visible due to the broad peak coming from the carbonate group. The next observed bands are those corresponding to the stretching vibrations of the carbonate group; for asymmetric ones—1249–1245 cm^{-1} and for symmetric ones—957–900 cm^{-1} [33]. On all spectra, we can also distinguish the bending vibration peaks of the carbonate group out-of-plane, which are in the range 792–791 cm^{-1}, while in-plane, they are in the range 731–727 cm^{-1}. For structures with a high content of HS (50 and 60 wt%), we can distinguish additional bands in the range 1195–1194 cm^{-1}. We attribute them to C-N stretching and N-H bending vibrations. These peaks are not observed for polymer structures with lower hard-segment content [33,34].

The absence of the –NCO band at about 2260 cm^{-1} shows that all –NCO groups were converted to urethane and thiourethane groups [34]. Table 2 presents major absorption bands appearing for the FTIR spectra of the obtained PTURs.

Table 2. Basic absorption bands appearing for the FTIR spectra of the obtained PTURs.

PTUR	Types of Vibrations [cm^{-1}]													
	ν_{N-H}	ν_{C-H} CH_2		$\nu_{C=O}$ Carbonyl	$\nu_{C=O}$ Thiour.		δ_{N-H}	δ_{C-H} 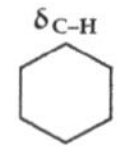 and CH_2		ν_{O-CO-O} Carbonyl		ν_{C-N} + δ_{N-H}	δ_{O-CO-O} Carbonyl	
		ν^{as}	ν^{sym}		H-Bonded	Free		ν^{as}	ν^{sym}	ν^{as}	ν^{sym}		Out of Plane	in pl.
PCD-30	3351	2933	2859	1738	-	1676	1511	1465	1404	1245	957	-	791	731
PCD-40	3351	2930	2857	1740	-	1676	1510	1458	1403	1245	901	-	792	730
PCD-50	3338	2928	2856	1740	1654	1672	1510	1458	1403	1246	900	1195	792	729
PCD-60	3309	2926	2854	1741	1654	-	1509	1450	1404	1249	900	1194	791	727

The changes in the intensity of the individual bands are closely related to the changes in the chemical composition of different polymers.

3.2. *Physicochemical Characterization*

3.2.1. Reduced Viscosities and GPC

Reduced viscosity values of the obtained polymers (given in Table 3) ranged from 0.62 to 3.63 dL/g, and in each series they decreased with increasing SS content in the sample. The obtained PTURs were materials which exhibit high molar masses ($\overline{M_n}$ in the range of 21,700–42,000 Da and $\overline{M_w}$ in the range of 33,000–62,000 Da) and relatively low dispersities ($Đ_M$ ranged from 1.43 to 1.66). The low molar mass dispersities of the obtained polymers results from the relatively long mixing time of the contents of the reaction flasks during the syntheses. This is due to the fact that at 135 °C, all polymers were in a plasticized state. This indicates the high homogeneity of the polymers obtained.

Table 3. η_{red} values and GPC data of the PTURs.

PTUR	η_{red} (dL/g)	$\overline{M_n}$ (Da)	$\overline{M_w}$ (Da)	$Đ_M$
PCD-30	3.63	41,900	60,100	1.43
PCD-40	1.90	32,000	53,200	1.66
PCD-50	1.54	42,000	62,000	1.48
PCD-60	0.62	21,700	33,000	1.52

3.2.2. CAs and SFE

The surface properties of the obtained polymers were determined on the basis of the CA and SFE values, and the obtained results are shown in Figure 4.

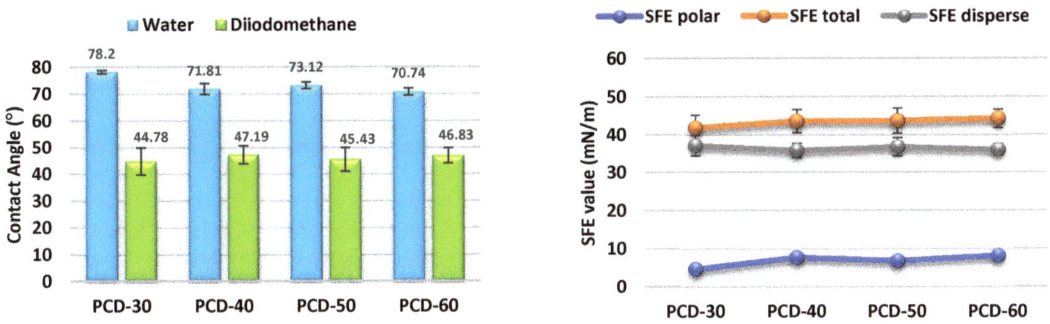

(**a**) (**b**)

Figure 4. Quantifying the contact angles of (**a**) and SFE (**b**) of synthesized polymers.

When using polymers as biomaterials, the chemical properties and surface properties are of key importance. Depending on the type of biomaterial used, both hydrophobic and hydrophilic properties may be advantageous. The CA and SFE values are significant in determining these properties. For polyurethanes, SFE values exceed 50 mN/m; therefore, they are classified as polar materials [63].

An important aspect for the researcher was the combination of biocompatibility with the value of critical surface energy. This problem was dealt with by Baier [71] in his research. According to the hypothesis proposed by him, the surface of the material with the so-called critical surface energy at the level of 20–30 mN/m is characterized by thrombogenicity and was defined by Baier as a hypothetical zone of biocompatibility, understood as the zone of minimal cell adhesion. On the other hand, materials with a critical surface energy above 40 mN/m promote cell adhesion. Therefore, they are good materials, especially for orthopedics, where the adhesion of cells to the implant surface and the overgrowth of the implant with tissues play a key role in the healing process.

It is commonly accepted that hydrophilic surfaces have CA values up to about 30°, and hydrophobic surfaces above 90° [63]. As it can be seen from Figure 3, all obtained PTURs were hydrophilic (CAs for water between 70.74° and 78.2°).

3.2.3. Hydrolytic Resistance

The hydrolytic resistance of polyurethane material in the human body environment is extremely important in biomedical applications. To preliminarily evaluate this property, PTURs with 50% wt. content were incubated in Opylite® salt solution for 8 weeks. The observed changes in the mass of the test sample are presented in Figure 5.

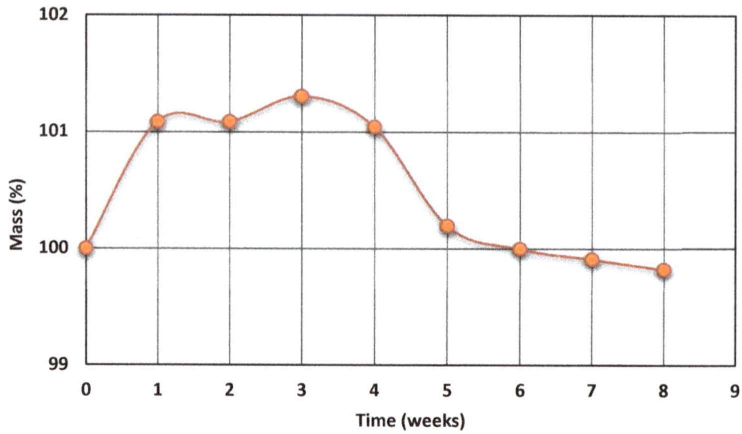

Figure 5. Diagram of sample mass change during Optylite® liquid incubation.

As can be seen in Figure 5, during the initial stage of incubation, the physiological liquid was absorbed by the polymer chain (in this stage the polymer was swollen). After appropriate saturation with the solution (and loosening of the polymer chain), the hydrolysis of carbonate bonds took place, which shows the small weight loss of the tested sample, which is visible on the curve. The conducted research shows that the PCD-50 polymer is hydrolytically stable in a multi-ion physiological fluid for about 6 weeks and its hydrolysis only takes place after this time.

It is noteworthy that after 8 weeks, the mass of the samples decreased by about 0.2%. According to the available literature, significant changes in the hydrolytic stability of polyurethanes only occur during the longer incubation period [72–74].

3.3. *Thermal and Thermomechanical Properties*

3.3.1. DSC

The changes in physical transformation of the obtained PTURs were determined by DSC analysis. In order to better interpret the obtained results, the compounds that constituted SS in the obtained polymers were also tested. The numerical data of the analyses are presented in Table 4, while the shapes of the DSC curves are presented in Figure 6.

Table 4. DSC data of the synthesized PTURs and SS.

PTUR	T_g [°C]		T_m [°C]		ΔH [J/g]	
	I [a]	II [b]	I [a]	II [b]	I [a]	II [b]
PCD-30	−2	−15	50	-	34.1	-
PCD-40	1	−12	47; 132	-	0.7; 0.6	-
PCD-50	32	20	52; 146	-	1.9; 2.2	-
PCD-60	44	33	55	-	2.6	-
PCD	−40	−50	21; 56	5.6; 40.1; 53.3	6.6; 64.8	3.6; 3.3; 53.0

[a, b] first and second heating scan, respectively.

The determined T_g values of the obtained PTURs were in the range of −2–44 °C in the first heating cycle and −15–33 °C in the second heating cycle. These values increased with the increase in the content of the HS in the polymers, with a significant increase being observed for the polymer with 50 wt% of hard segments. Considering the differences in the T_g values of polymers and pure soft segments (PCD: −40 °C), it can be concluded that with the increase in the content of the hard segment in PTURs, their microphase separation degree decreased. The T_g values in the second heating are much lower than in the first heating, i.e., repeated heating of the sample lowers its glass transition temperature. Polymers with a 50 and 60 wt% of the hard segment showed T_g values slightly higher than room temperature; therefore, they should be located on the border of elastomers and plastomers. More precise viscoelastic properties of the obtained materials were determined by the DMTA method and the description of the obtained results is presented in the rest of this work.

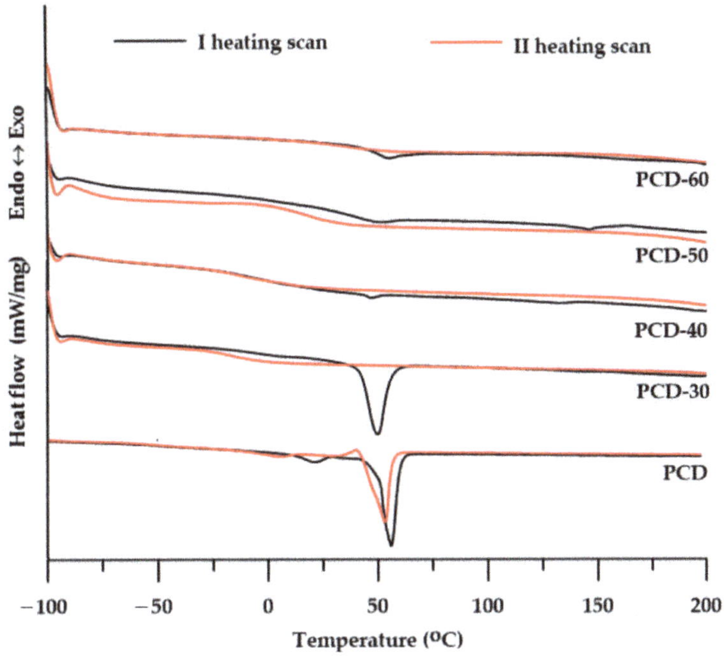

Figure 6. DSC curves of the synthesized polymers.

On the DSC curves of all polymers (see Figure 6) from the I heating cycle, one or two endothermic peaks with T_m values in the range of 47–146 °C were observed. The peaks with the lowest T_m values in the range of 47–55 °C and ΔH values in the range of 0.7–34.1 J/g may be responsible for the melting of the soft segment. In turn, small visible endothermic peaks with T_m values ranging from 132 to 146 °C should be attributed to the melting of more or less ordered hard segments [33]. The most intense peak of SS melting is visible for the PCD-30 polymer, while for the PCD-50 and PCD-60 polymers the heat of this transformation is very low, which makes it difficult to speak of a visible peak. The DSC curves of the second heat no longer reveal any melting energy effects of SS or HS. This indicates that the intrinsic capacity of the obtained polymers is very low.

The low ΔH values determined from the I heating cycle and the absence of endothermic peaks in the II heating cycle curves indicate a slight tendency of the obtained PTURs to form ordered structures. The polymer with the lowest content of hard segments (PCD-30) was characterized by the highest degree of ordering, but only within the domains of soft segments.

3.3.2. TGA

The thermal stability of the obtained PTURs was determined by means of thermogravimetric analysis carried out under an inert gas atmosphere.

As can be seen from the data in Table 5, the obtained polymers had higher thermal resistance compared to the regular polymer whose TG analysis was presented in a previous work [30]. Analyzing the effect of changing the HS content in PTURs, it can be seen that its increase in the polymer caused a decrease in T_5, T_{10} and T_{50}.

Table 5. TG data of the synthesized PTURs.

PTUR	T_5 [1] (°C)	T_{10} [2] (°C)	T_{50} [3] (°C)	T_{max} [4] (°C)
PCD-30	281	295	343	286; 333; 360; 429
PCD-40	279	289	341	292; 339; 362; 444
PCD-50	278	288	337	292; 334; 367; 446
PCD-60	275	285	333	295; 342; 451

[1, 2, 3] The temperatures of 5%, 10% and 50% mass loss, respectively; [4] The temperatures of maximum rate of mass loss.

Based on previous studies [33], and by analyzing the data in Table 4, it can be said that the decomposition process of the obtained PTURs took place in several stages. The DTG curves shown in Figure 7 revealed relatively intense peaks at the maximum in the range of 286–295 °C, which were associated with the decomposition of thiourethane bonds. The intensity of this peaks increased with increasing HS content in the polymer. The further peaks with maxima at 333–342 °C can be attributed to the decomposition of urethane bonds. A third degradation step for polymers with an HS content below 60 wt% (peaks at about 362–367 °C) is related to the degradation of the polycarbonate soft segment. The last decomposition step for all polymers (peaks at about 429–451 °C) can be attributed to the decomposition of the aromatic structure fragments of the obtained PTURs [33,34]. The small peaks seen at temperatures around 600 °C may be related to the decomposition of the solid products formed in the earlier stages.

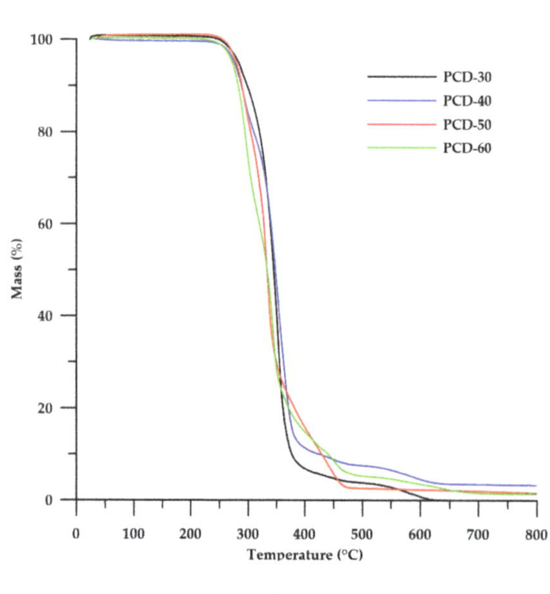

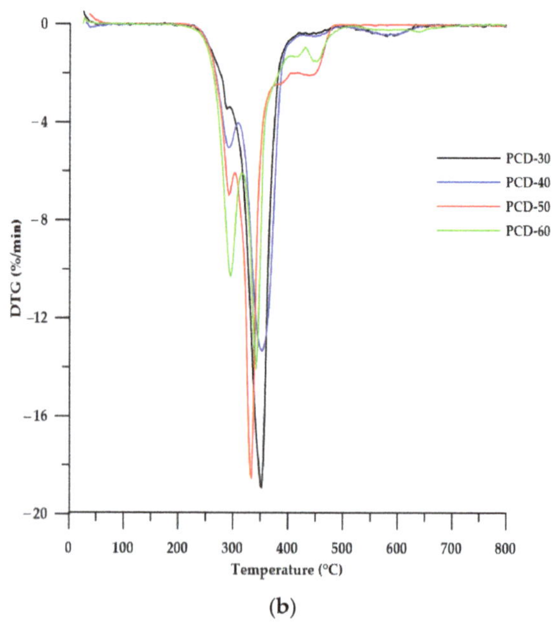

Figure 7. TG (**a**) and DTG (**b**) curves of the PTURs.

3.3.3. DMTA

To investigate the influence of the soft segments on the viscoelastic properties of PTURs, dynamic mechanical thermal analysis was performed. Changes in storage modulus (E') and mechanical loss factor (tanδ) with temperature are shown in Figure 8, while DMTA data are summarized in Table 6.

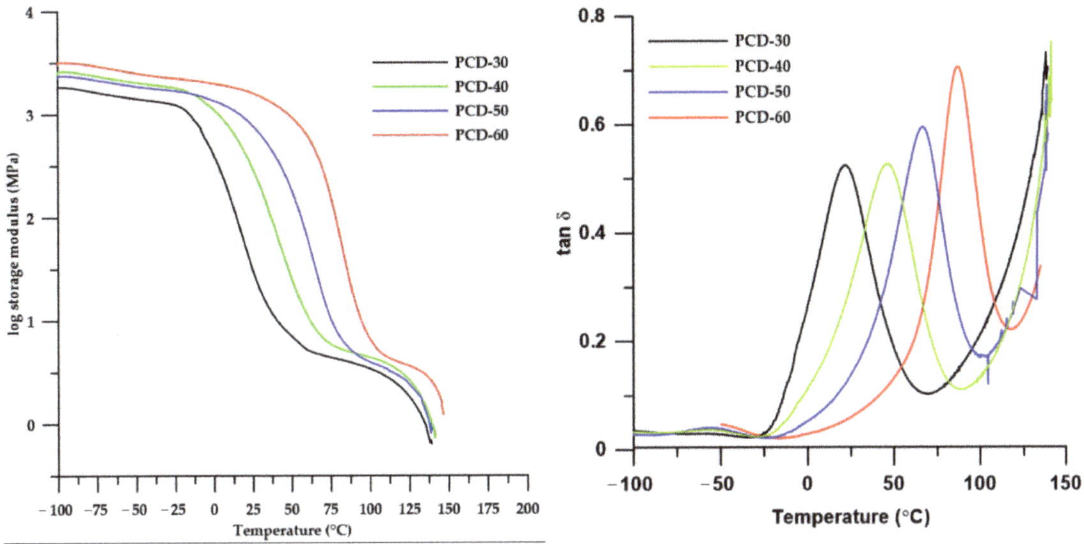

(a) (b)

Figure 8. Temperature dependence of storage modulus (**a**) and tanδ (**b**) of synthesized PTURs.

Table 6. DMTA results of the PTURs.

PTUR	E'_{onset} (°C)	E'_{20} (MPa)	E''_{max} (°C)	T $_{\tan\delta max}$ (°C)	$\tan\delta_{max}$	FWHM (°C)
PCD-30	−6.69	52	−5.81	21.07	0.459	48.45
PCD-40	10.55	418	5.81	46.19	0.448	40.23
PCD-50	36.98	910	27.42	66.37	0.461	31.43
PCD-60	61.41	1704	50.95	86.87	0.535	24.48

As can be seen from the data in Table 6, with the increase in the content of HS in the polymer, almost all parameters determined by the DMTA method increased. The exception was the damping value ($\tan\delta_{max}$) and the FWHM parameter (full width at half maximum), which is a degree of sample homogeneity.

On the basis of the results obtained, it can be stated that as the content of HS in the polymers increased, their glass transition temperatures increased, which is in agreement with the results of the DSC analysis. Increasing the value of the storage modulus in the series of polymers causes materials to become harder and stiffer. These conclusions were confirmed by the results of the strength test results described later in this paper. On the other hand, the decrease in the value of the FWHM parameter in the series confirms the results of the DSC, which found that as the HS content increased, hard and soft segments decreased, causing the polymer to have a more homogeneous structure.

Considering the curves of the dependence of the storage modulus on the temperature (Figure 8a), it can be concluded that, among all the materials, the polymers PCD-30 and PCD-40 exhibit the most favorable viscoelastic properties. This is due to the fact that these polymers possess rubber-like properties (i.e., they are in the rubber-elastic state) in the widest temperature range. The occurrence of this state over a wide temperature range is characteristic of elastomeric materials. The other polymers are characterized by the fact that the melting area begins almost immediately after the transition from the glassy state, in which the polymers transform into a melted amorphous state. In this state, polymers can be easily processed using conventional processing methods, such as injection molding, extrusion or 3D printing.

The tan delta vs. temperature curves show two peaks corresponding to two types of relaxations. The first one (with maxima between −60 and −50 °C) is connected to the main chain movements of the soft segment and local movements of thiourethane and urethane polar groups. The second type of relaxation (called primary relaxation) with maxima between 21 and 87 °C is related to the glass transition of the soft segment in the polymer. The intensity of the principal relaxation peak determines the material's ability to damp (absorb) vibrations; the smaller its value, the greater the material's damping capacity [32].

3.4. Mechanical Properties

The hardness, tensile strength, elongation at break, modulus of elasticity and lap shear strength are listed in Table 7.

Table 7. Hardness and mechanical properties of PTURs.

PTUR	Hardness (Sh)		Tensile Strength (MPa)	Elongation at Break (%)	Modulus of Elasticity (MPa)	Lap Shear Strength (MPa)
	A	D				
PCD-30	71.75 ± 1.50	25.00 ± 0.82	34.43 ± 0.23	350 ± 0	1.82 ± 0.13	4.32 ± 0.33
PCD-40	84.75 ± 0.50	34.25 ± 2.99	45.76 ± 2.71	275 ± 0	4.35 ± 0.56	14.70 ± 0.51
PCD-50	96.25 ± 0.50	53.00 ± 1.87	46.54 ± 4.06	223 ± 2.89	69.51 ± 2.19	15.93 ± 0.64
PCD-60	91.50 ± 1.15	66.00 ± 1.41	51.11 ± 2.37	75 ± 0	289.88 ± 2.88	18.05 ± 0.44

Shore hardness of the obtained PTURs was determined on two scales: A and D. On the A scale it was in the range of 71.75–96.25 ShA, while on the D scale it was in the range of 25.00–66.00 ShD. It depends on the composition of the polymer. In general, the higher the content of the hard segment, the higher the hardness values of the polymers defined in both scales. The exception is the PCD-60 polymer, the hardness of which on the A scale is lower than that of the PCD-50 polymer, while at such high values the results can be misleading due to the near maximum range of the measurement scale.

All of the obtained polymers exhibit relatively high tensile strengths ranging from 34.43 to 51.11 MPa, which generally increased with increasing hard segment content. The elongation at break is in the range of 75–350%. The value of this property decreased as the hard segment content increased. For the modulus of elasticity, the values are in the range of 1.82–289.88 MPa. The modulus values are in agreement with the determined T_g values. The PCD-30 and PCD-40 polymers that had T_g below room temperature also have moduli typical of elastomers. The other two polymers that had T_g values near room temperature also have significantly higher modulus values. The polymer with the highest content of the hard segment and the highest T_g value (PCD-60) is also characterized by the highest value of the modulus of elasticity and the highest hardness on the D scale.

The analysis of the influence of the hard segment content on the adhesion properties to copper showed that the more content there is, the greater the strength of the adhesive joint. This dependence confirms that with an increase in the amount of extender, and thus with an increase in the amount of sulfur atoms in the polymer, the affinity (adhesion) of such material to copper increases. This is in accordance with previous studies, and the values achieved for the obtained materials were higher than those previously described in the literature [33,34,38,49,58].

3.5. Optical Properties

3.5.1. Refractive Index and Transparency

The obtained polymers had refractive index values ranging from 1.5135 to 1.5615 (see Table 8), and these values increased with increasing HS content in the polymer. This relationship is in agreement with previous results, which confirm that the refractive index value depends on the presence and amount of sulfur atoms in the polymer; the more sulfur atoms there are, the higher the refractive index of the polymer.

Table 8. Refractive index and transmittance of PTURs.

PTUR	Refractive Index	Transmittance (%)	
		T_{500} [1]	T_{800} [2]
PCD-30	- [3]	63.41 ± 0.012	71.78 ± 0.009
PCD-40	1.5135 ± 0.002	76.52 ± 0.008	82.91 ± 0.014
PCD-50	1.5355 ± 0.003	76.05 ± 0.007	82.89 ± 0.011
PCD-60	1.5615 ± 0.003	72.67 ± 0.010	80.15 ± 0.007

[1,2] transmittance at 500 and 800 nm, respectively. [3] opaque.

In contrast to the refractive indexes, the transparency of the obtained polymers decreased with increasing HS content in the polymer. The exception was the polymer PCD-30, which was nontransparent (see Figure 2c). The polymer with the highest transparency was PCD-40, with little difference compared to the transparency of PCD-50. The PCD-30 polymer had the lowest transparency, which was due to the relatively high degree of ordering within the soft segment domains. To further characterize the color dependence of the HS content in the polymer, a color test according to ASTM E308 was performed.

3.5.2. Color

The results of the color tests of the obtained polymers are shown in Figure 9.

As can be seen from Figure 9a, the brightness L* of the polymers in the series generally decreases (except for the PCD-50 polymer), which indicates that the proportion of white color in the series decreases. On the other hand, the increase in parameter b* in the series causes that the proportion of yellow color to increase, which can be seen by observing the appearance of the samples in Figure 2c. The increase in parameter b* in the series can be related to the increase in the number of sulfur atoms in the polymer, i.e., the more sulfur there is, the more yellow the polymer. A similar relationship was observed for parameter a*, as the increase in parameter a* causes a shift in the color of the obtained polymer towards the green color.

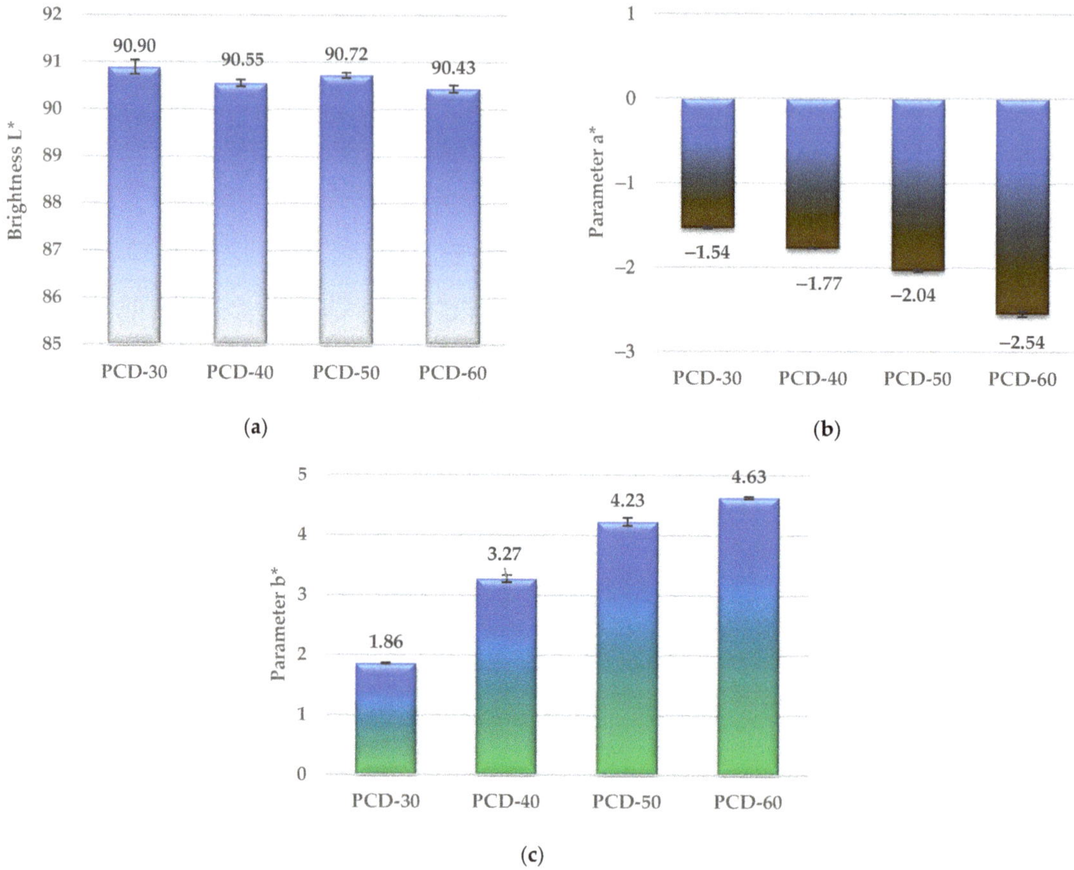

Figure 9. Brightness L* (**a**), parameter a* (**b**) and parameter b* (**c**) of synthesized PTURs.

4. Conclusions

The use of a one-step melt polyaddition method, carried out with a slight excess of isocyanate groups to hydroxyl and thiol groups (NCO/(OH + SH) = 1.07), enabled the preparation of high-molecular PTURs. The obtained polymers were colorless opaque rubber-like solids with relatively high transparency. PCD-40 and PCD-50 polymers exhibited similar and best transparency. The synthesized PTURs were soluble at room temperature in DMF, TChE and THF. All polymers of NMP and DMSO showed high resistance.

The η_{red} values determined were in the range of 0.62–3.63 dL/g. It was observed that as the hard segment content increased, the viscosities of the obtained PTURs decreased. The values determined $\overline{M_n}$ and $\overline{M_w}$ for THF-soluble polymers were in the range of 21,700–42,000 Da and 33,000–62,000 Da, respectively. The dispersity of molar masses contained in the range 1.46–1.66 was quite low for the polymers obtained by the one-step melt method. This indicates that their structures were highly homogeneous. After examining the refractive index, it was found that as the content of the hard segment in the polymer increased, the value of this index increased.

The TGA investigations allow us to conclude that the obtained PTURs were characterized by relatively good thermal stability. From the values obtained, it can be seen that the mass loss temperatures decreased with the increase in the hard segment content. DSC analysis showed that the obtained PTURs exhibit a partially ordered structure. The determined T_g values were in the range of −2–33 °C in the I heating cycle and −15–44 °C in the II heating cycle. These values increased with increasing hard segment content in the polymers. From the obtained T_g values, it was concluded that the obtained PTURs were elastomers or plastomers. DMTA analysis confirmed these observations.

Upon examination of the mechanical properties, it was found that all the obtained polymers exhibited relatively high tensile strengths (34.43–51.11 MPa), which increased with the increase in the hard segment content. The same relationship was exhibited by the shore hardness values. On the other hand, the elongation at break decreased with the increase in hard segment content. From the PTUR adhesion tests, it is evident that as the hard segment content in the polymer increases (and thus the sulfur content increases), the adhesion strength increases.

Author Contributions: Conceptualization, A.P.; methodology, A.P.; software, A.P. and J.W.S.; validation, A.P.; formal analysis, A.P. and J.W.S.; investigation, A.P.; resources, A.P.; data curation, A.P.; writing—original draft preparation, A.P.; writing—review and editing, A.P. and J.W.S.; visualization, A.P.; supervision, A.P. and J.W.S.; project administration, A.P. All authors have read and agreed to the published version of the manuscript.

Funding: This research received no external funding.

Institutional Review Board Statement: Not applicable.

Informed Consent Statement: Not applicable.

Data Availability Statement: The data presented in this study are available on request from the corresponding author.

Acknowledgments: The authors would like to thank Miroslaw Kluczka from the Covestro (formerly Bayer MaterialScience) for supplying the *Desmodur W®* and *Desmophen® C2200* used for polymer synthesis.

Conflicts of Interest: The authors declare no conflict of interest.

References

1. Janik, H.; Sienkiewicz, M.; Kucinska-Lipka, J. 9-Polyurethanes. In *Handbook of Thermoset Plastics*, 3rd ed.; Dodiuk, H., Goodman, S.H., Eds.; William Andrew Publishing: Norwich, NY, USA, 2014; pp. 253–295. [CrossRef]
2. Wirpsza, Z. *Polyurethanes—Chemistry, Technology, Application. Warszawa: WNT*; Ellis Horwood Ltd.: Chichester, UK, 1991.
3. Drobny, J.G. 9-Thermoplastic Polyurethane Elastomers. In *Plastics Design Library, Handbook of Thermoplastic Elastomers*, 2nd ed.; Drobny, J.G., Ed.; William Andrew Publishing: Norwich, NY, USA, 2014; pp. 233–253. [CrossRef]
4. Fink, J.K. Chapter 2-Poly(urethane)s. In *Plastics Design Library, Reactive Polymers Fundamentals and Applications*, 2nd ed.; Fink, J.K., Ed.; William Andrew Publishing: Norwich, NY, USA, 2013; pp. 49–93. [CrossRef]
5. Datta, J.; Kasprzyk, P.; Błażek, K.; Włoch, M. Synthesis, structure and properties of poly(ester-urethane)s obtained using bio-based and petrochemical 1,3-propanediol and 1,4-butanediol. *J. Therm. Anal. Calorim.* **2017**, *130*, 261–276. [CrossRef]
6. Lei, W.; Fang, C.; Zhou, X.; Li, J.; Yang, R.; Zhang, Z.; Liu, D. Thermal properties of polyurethane elastomer with different flexible molecular chain based on para-phenylene diisocyanate. *J. Mater. Sci. Technol.* **2017**, *33*, 1424–1432. [CrossRef]
7. Hou, X.; Sun, L.; Wei, W.; Taylor, D.K.; Su, S.; Yu, H. Structure and performance control of high-damping bio-based thermoplastic polyurethane. *J. Appl. Polym. Sci.* **2022**, *139*, e52059. [CrossRef]
8. Špírková, M.; Hodan, J.; Kobera, L.; Kredatusová, J.; Kubies, D.; Machová, L.; Poręba, R.; Serkis, M.; Zhigunov, A.; Kotek, J. The influence of the length of the degradable segment on the functional properties and hydrolytic stability of multi-component polyurethane elastomeric films. *Polym. Deg. Stab.* **2017**, *137*, 216–228. [CrossRef]
9. Zhang, C.; Jiang, X.; Zhao, Z.; Mao, L.; Zhang, L.; Coates, P. Effects of wide-range γ-irradiation doses on the structures and properties of 4,4′-dicyclohexyl methane diisocyanate based poly(carbonate urethane)s. *J. Appl. Polym. Sci.* **2014**, *131*, 41049. [CrossRef]
10. Kwiatkowski, K.; Nachman, M. The Abrasive Wear Resistance of the Segmented Linear Polyurethane Elastomers Based on a Variety of Polyols as Soft Segments. *Polymers* **2017**, *9*, 705. [CrossRef]
11. Kull, K.L.; Bass, R.W.; Craft, G.; Julien, T.; Marangon, E.; Marrouat, C.; Harmon, J.P. Synthesis and characterization of an ultra-soft poly(carbonate urethane). *Eur. Polym. J.* **2015**, *71*, 510–522. [CrossRef]
12. Patti, A.; Acierno, D. The Puncture and Water Resistance of Polyurethane- Impregnated Fabrics after UV Weathering. *Polymers* **2020**, *12*, 15. [CrossRef] [PubMed]
13. Patti, A.; Costa, F.; Perrotti, M.; Barbarino, D.; Acierno, D. Polyurethane Impregnation for Improving the Mechanical and the Water Resistance of Polypropylene-Based Textiles. *Materials* **2021**, *14*, 1951. [CrossRef]

14. Samaniego-Aguilar, K.; Sánchez-Safont, E.; Arrillaga, A.; Anakabe, J.; Gamez-Perez, J.; Cabedo, L. In Service Performance of Toughened PHBV/TPU Blends Obtained by Reactive Extrusion for Injected Parts. *Polymers* **2022**, *14*, 2337. [CrossRef]
15. Fang, H.; Zhang, L.; Chen, A.; Wu, F. Improvement of Mechanical Property for PLA/TPU Blend by Adding PLA-TPU Copolymers Prepared via In Situ Ring-Opening Polymerization. *Polymers* **2022**, *14*, 1530. [CrossRef] [PubMed]
16. Formela, K.; Kurańska, M.; Barczewski, M. Recent Advances in Development of Waste-Based Polymer Materials: A Review. *Polymers* **2022**, *14*, 1050. [CrossRef] [PubMed]
17. Lee, J.-G.; Jeong, J.-O.; Jeong, S.-I.; Park, J.-S. Radiation-Based Crosslinking Technique for Enhanced Thermal and Mechanical Properties of HDPE/EVA/PU Blends. *Polymers* **2021**, *13*, 2832. [CrossRef] [PubMed]
18. Tahir, M.; Heinrich, G.; Mahmood, N.; Boldt, R.; Wießner, S.; Stöckelhuber, K.W. Blending In Situ Polyurethane-Urea with Different Kinds of Rubber: Performance and Compatibility Aspects. *Materials* **2018**, *11*, 2175. [CrossRef]
19. Yang, Z.; Wu, H.; Zhang, R.; Deng, K.; Li, Y.; Liu, Z.; Zhong, Q.; Kang, Y. Effect of Graphene/Spherical Graphite Ratio on the Properties of PLA/TPU Composites. *Polymers* **2022**, *14*, 2538. [CrossRef]
20. Shiu, B.-C.; Hsu, P.-W.; Lin, J.-H.; Chien, L.-F.; Lin, J.-H.; Lou, C.-W. A Study on Preparation and Property Evaluations of Composites Consisting of TPU/Triclosan Membranes and Tencel®/LMPET Nonwoven Fabrics. *Polymers* **2022**, *14*, 2514. [CrossRef]
21. Luo, Z.; Li, X.; Zhao, S.; Xu, L.; Liu, L. Structure and Dielectric Properties of TPU Composite Filled with CNTs@PDA Nanofibers and MXene Nanosheets. *Polymers* **2022**, *14*, 2157. [CrossRef]
22. Zafar, K.; Zia, K.M.; Alzhrani, R.M.; Almalki, A.H.; Alshehri, S. Biocompatibility and Hemolytic Activity Studies of Synthesized Alginate-Based Polyurethanes. *Polymers* **2022**, *14*, 2091. [CrossRef]
23. Brzeska, J.; Jasik, G.; Sikorska, W.; Mendrek, B.; Karczewski, J.; Kowalczuk, M.; Rutkowska, M. Susceptibility to Degradation in Soil of Branched Polyesterurethane Blends with Polylactide and Starch. *Polymers* **2022**, *14*, 2086. [CrossRef]
24. De Keer, L.; Kilic, K.I.; Van Steenberge, P.H.M.; Daelemans, L.; Kodura, D.; Frisch, H.; De Clerck, K.; Reyniers, M.F.; Barner-Kowollik, C.; Dauskardt, R.H.; et al. Computational prediction of the molecular configuration of three-dimensional network polymers. *Nat. Mater.* **2021**, *20*, 1422–1430. [CrossRef]
25. De Keer, L.; Cavallli, F.; Estupii, D.; Krüger, A.J.D.; Rocha, S.; Van Steenberge, P.H.M.; Reyniers, M.-F.; De Laporte, L.; Hofkens, J.; Barner, L.; et al. Synergy of Advanced Experimental and Modeling Tools to Underpin the Synthesis of Static Step-Growth-Based Networks Involving Polymeric Precursor Building Blocks. *Macromolecules* **2021**, *54*, 9280–9298. [CrossRef]
26. Schollenberger, C.S.; Stewart, F.D. Thermoplastic Polyurethane Hydrolysis Stability. *J. Elastoplast.* **1971**, *3*, 28–56. [CrossRef]
27. Tanzi, M.C.; Mantovani, D.; Petrini, P.; Guidoin, R.; Laroche, G. Chemical stability of polyether urethanes versus polycarbonate urethanes. *J. Biomed. Mater. Res.* **1997**, *36*, 550–559. [CrossRef]
28. Xie, F.; Zhang, T.; Bryant, P.; Kurusingal, V.; Colwell, J.M.; Laycock, B. Degradation and stabilization of polyurethane elastomers. *Prog. Polym. Sci.* **2019**, *90*, 211–268. [CrossRef]
29. Shin, J.; Lee, J.; Jeong, H.M. Properties of polythiourethanes prepared by thiol–isocyanate click reaction. *J. Appl. Polym. Sci.* **2017**, *135*, 46070. [CrossRef]
30. Rahmawati, R.; Nozaki, S.; Kojio, K.; Takahara, A.; Shinohara, N.; Yamasaki, S. Microphase-separated structure and mechanical properties of cycloaliphatic diisocyanate-based thiourethane elastomers. *Polym. J.* **2019**, *51*, 265–273. [CrossRef]
31. Ellson, G.; Carrier, X.; Walton, J.; Mahmood, S.F.; Yang, K.; Salazar, J.; Voit, W.E. Tough thiourethane thermoplastics for fused filament fabrication. *J. Appl. Polym. Sci.* **2017**, *135*, 45574. [CrossRef]
32. Shin, J.; Matsushima, H.; Chan, J.W.; Hoyle, C.E. Segmented Polythiourethane Elastomers through Sequential Thiol−Ene and Thiol−Isocyanate Reactions. *Macromolecules* **2009**, *42*, 3294–3301. [CrossRef]
33. Kultys, A.; Puszka, A. Transparent poly(thiourethane-urethane)s based on dithiol chain extender. *J. Therm. Anal. Calorim.* **2014**, *117*, 1427–1439. [CrossRef]
34. Rogulska, M.; Kultys, A.; Olszewska, E. New thermoplastic poly(thiourethane-urethane) elastomers based on hexane-1,6-diyldiisocyanate (HDI). *J. Therm. Anal. Calorim.* **2013**, *114*, 903–916. [CrossRef]
35. Puszka, A. Thermal and Mechanical Behavior of New Transparent Thermoplastic Polyurethane Elastomers Derived from Cycloaliphatic Diisocyanate. *Polymers* **2018**, *10*, 537. [CrossRef]
36. Puszka, A.; Kultys, A.; Rogulska, M. Influence of DMPA content on the properties of new thermoplastic poly(ether-urethane) elastomers. *J. Elastom. Plast.* **2018**, *50*, 140–150. [CrossRef]
37. Puszka, A.; Kultys, A. The influence of soft segments on some properties of new transparent segmented polyurethanes. *Polym. Adv. Technol.* **2017**, *28*, 1937–1944. [CrossRef]
38. Rogulska, M.; Kultys, A.; Pikus, S. The effect of chain extender structure on the properties of new thermoplastic poly(carbonate–urethane)s derived from MDI. *J. Therm. Anal. Calorim.* **2017**, *127*, 2325–2339. [CrossRef]
39. Puszka, A.; Kultys, A. New thermoplastic polyurethane elastomers based on aliphatic diisocyanate. *J. Therm. Anal. Calorim.* **2017**, *128*, 407–416. [CrossRef]
40. Rogulska, M.; Kultys, A.; Puszka, A. New thermoplastic poly(carbonate-urethane)s based on chain extenders with sulfur atoms. *Chem. Pap.* **2017**, *71*, 1195–1204. [CrossRef]
41. Kultys, A.; Rogulska, M.; Pikus, S. New thermoplastic segmented polyurethanes with hard segment derived from 4,4′-diphenylmethane diisocyanate and methylenebis(1,4-phenylenemethylenethio)dialcanols. *J. Appl. Polym. Sci.* **2012**, *123*, 331–346. [CrossRef]

42. Kultys, A.; Rogulska, M.; Gluchowska, H. The effect of soft segment structure on the properties of novel thermoplastic polyurethane elastomers based on an unconventional chain extender. *Polym. Int.* **2011**, *60*, 652–659. [CrossRef]
43. Kultys, A.; Rogulska, M. New thermoplastic poly(carbonate-urethane) elastomers. *Pol. J. Chem. Technol.* **2011**, *13*, 23–30. [CrossRef]
44. Kultys, A.; Rogulska, M.; Pikus, S.; Skrzypiec, K. The synthesis and characterization of new thermoplastic poly(carbonate-urethane) elastomers derived from HDI and aliphatic-aromatic chain extenders. *Eur. Polym. J.* **2009**, *45*, 2629–2643. [CrossRef]
45. Kultys, A.; Pikus, S. Polyurethanes containing sulfur. III. New thermoplastic HDI-based segmented polyurethanes with diphenylmethane unit in their structure. *J. Polym. Sci. Pol. Chem.* **2001**, *39*, 1733–1742. [CrossRef]
46. Kultys, A.; Podkoscielny, W.; Majewski, W. Polyurethanes containing sulfur. II. New thermoplastic nonsegmented polyurethanes with diphenylmethane unit in their structure. *J. Polym. Sci. Part A Polym. Chem.* **2000**, *38*, 1767–1773. [CrossRef]
47. Kultys, A.; Puszka, A. New thermoplastic polyurethane elastomers based on sulfur-containing chain extenders. *Pol. J. Chem. Technol.* **2013**, *15*, 65–70. [CrossRef]
48. Rogulska, M. New thermoplastic poly(carbonate-urethane)s based on diphenylethane-derivative chain extenders—the effect of chain extender structure on thermal and mechanical properties. *J. Therm. Anal. Calorim.* **2020**, *139*, 3107–3121. [CrossRef]
49. Rogulska, M.; Maciejewska, M.; Olszewska, E. New thermoplastic poly(carbonate-urethane)s based on diphenylethane derivative chain extender. *J. Therm. Anal. Calorim.* **2020**, *139*, 1049–1068. [CrossRef]
50. Rogulska, M.; Podkoscielny, W.; Kultys, A.; Pikus, S.; Pozdzik, E. Studies on thermoplastic polyurethanes based on new diphenylethane-derivative diols. I. Synthesis and characterization of nonsegmented polyurethanes from HDI and MDI. *Eur. Polym. J.* **2006**, *42*, 1786–1797. [CrossRef]
51. Rogulska, M.; Kultys, A.; Podkoscielny, W. Studies on thermoplastic polyurethanes based on new diphenylethane-derivative diols. II. Synthesis and characterization of segmented polyurethanes from HDI and MDI. *Eur. Polym. J.* **2007**, *43*, 1402–1414. [CrossRef]
52. Rogulska, M.; Kultys, A.; Pikus, S. Studies on thermoplastic Polyurethanes based on new diphenylethane-derivative diols. III. The effect of molecular mass and structure of soft segment on some properties of segmented polyurethanes. *J. Appl. Polym. Sci.* **2008**, *110*, 1677–1689. [CrossRef]
53. Kultys, A.; Podkoscielny, W.; Pikus, S. Polyurethanes containing sulfur. I. New thermoplastic polyurethanes with benzophenone unit in their structure. *J. Polym. Sci. Pol. Chem.* **1999**, *37*, 4140–4150. [CrossRef]
54. Kultys, A. Polyesters containing sulfur. VIII. New benzophenone derivative thiopolyesterdiols synthesis and characterization. Their use for thermoplastic polyurethanes preparation. *J. Polym. Sci. Pol. Chem.* **2000**, *38*, 3977–3983. [CrossRef]
55. Kultys, A.; Rogulska, M.; Pikus, S. The synthesis and characterization of new thermoplastic poly(thiourethane-urethane)s. *J. Polym. Sci. Pol. Chem.* **2008**, *46*, 1770–1782. [CrossRef]
56. Wdowicka, D.; Podkoscielny, W.; Kultys, A. Polyurethanes (III): Polyaddition products of 4,4′-Bis(2-hydroxyethylenethio) diphenyl ether and 2,4-tolylene diisocyanate. *Iran. Polym. J.* **2000**, *9*, 97–103.
57. Rogulska, M.; Kultys, A.; Lubczak, J. New thermoplastic polyurethane elastomers based on aliphatic-aromatic chain extenders with different content of sulfur atoms. *J. Therm. Anal. Calorim.* **2015**, *121*, 397–410. [CrossRef]
58. Rogulska, M.; Kultys, A. Aliphatic polycarbonate-based thermoplastic polyurethane elastomers containing diphenyl sulfide units. *J. Therm. Anal. Calorim.* **2016**, *126*, 225–243. [CrossRef]
59. Rogulska, M. Transparent sulfur-containing thermoplastic polyurethanes with polyether and polycarbonate soft segments. *Polym. Bull.* **2018**, *75*, 1211–1235. [CrossRef]
60. Rogulska, M. Polycarbonate-based thermoplastic polyurethane elastomers modified by DMPA. *Polym. Bull.* **2019**, *76*, 4719–4733. [CrossRef]
61. Wnuczek, K.; Puszka, A.; Podkościelna, B. Synthesis and Spectroscopic Analyses of New Polycarbonates Based on Bisphenol A-Free Components. *Polymers* **2021**, *13*, 4437. [CrossRef] [PubMed]
62. Stepto, R.F.T. Dispersity in polymer science. *Pure Appl. Chem.* **2009**, *81*, 351–353. [CrossRef]
63. Król, P.; Uram, Ł.; Król, B.; Pielichowska, K.; Walczak, M. Study of chemical, physico-mechanical and biological properties of 4,4′-methylenebis(cyclohexyl isocyanate)-based polyurethane films. *Mater. Sci. Eng. C* **2018**, *93*, 483–494. [CrossRef] [PubMed]
64. *ISO 11357:2016*; Plastics—Differential Scanning Calorimetry (DSC). International Organization of Standardization: Geneva, Switzerland, 2016.
65. *ISO 527-2:2012*; Plastics—Determination of Tensile Properties—Part 2. International Organization of Standardization: Geneva, Switzerland, 2012.
66. *EN ISO. 868:2003*; Plastics—Plastics and Ebonite—Determination of Indentation Hardness by Means of a Durometer (ShoreHardness). International Organization of Standardization: Geneva, Switzerland, 2003.
67. *PN EN 1465:2009*; Adhesives. Determination of Tensile Lap-Shear Strength of Bonded Assemblies. European Committee for Standardization: Brussels, Belgium, 2009.
68. *PN-EN 13887:2005;* Structural Adhesives. Guidelines for Surface Preparation of Metals and Plastics Prior to Adhesive Bonding. European Committee for Standardization: Brussels, Belgium, 2005.
69. *ISO 489:2022;* Plastics—Determination of Refractive Index. International Organization of Standardization: Geneva, Switzerland, 2022.

70. *ASTM E308*; Standard Practice for Computing the Colour of Objects by Using the CIE System. ASTM International (ASTM): London, UK, 2018.
71. Baier, R.E. The role of surface energy in thrombogenesis. *Bull. N. Y. Acad. Med.* **1972**, *48*, 257–272. [PubMed]
72. Kim, H.W.; Knowles, J.C.; Kim, H.E. Hydroxyapatite/poly(ε-caprolactone) composite coatings on hydroxyapatite porous bone scaffold for drug delivery. *Biomaterials* **2004**, *25*, 1279–1287. [CrossRef] [PubMed]
73. Serrano, M.C.; Pagani, R.; Vallet-Regi, M.; Pena, J.; Ramila, A.; Izquierdo, I.; Portoles, M.T. In vitro biocompatibility assessment of poly(ε-caprolactone) films using L929 mouse fibroblasts. *Biomaterials* **2004**, *25*, 5603–5611. [CrossRef] [PubMed]
74. Williams, J.M.; Adewunmi, A.; Schek, R.M.; Flanagan, C.L.; Krebsbach, P.H.; Feinberg, S.E.; Hollister, S.J.; Das, S. Bone tissue engineering using polycaprolactone scaffolds fabricated via selective laser sintering. *Biomaterials* **2005**, *23*, 4817–4827. [CrossRef] [PubMed]

polymers

MDPI

Article

Synthesis and Spectroscopic Analyses of New Polycarbonates Based on Bisphenol A-Free Components

Krystyna Wnuczek *, Andrzej Puszka and Beata Podkościelna

Department of Polymer Chemistry, Faculty of Chemistry, Institute of Chemical Sciences, Maria Curie-Skłodowska University, Gliniana 33, 20-614 Lublin, Poland; andrzej.puszka@umcs.pl (A.P.); beata.podkoscielna@mail.umcs.pl (B.P.)

* Correspondence: krystyna.wnuczek@poczta.umcs.lublin.pl

Abstract: This paper discusses a new synthesis of bisphenol A-free polycarbonates based on four aliphatic–aromatic systems. In the first stage, different types of monomers (with/without sulfur) derived from diphenylmethane were synthesized. Then, new polycarbonates were prepared in the reactions with diphenyl carbonate (DPC) by transesterification and polycondensation reactions. Three different catalysts (zinc acetate, 4-(dimethylamino)pyridine and benzyltriethylammonium chloride) were tested. The structures of the compounds were confirmed by Nuclear Molecular Resonance spectroscopy (NMR) in each stage. The chemical structures of the obtained polycarbonates were verified by means of Attenuated Total Reflectance Fourier Transform infrared spectroscopy (ATR–FTIR). The presence of a carbonyl group in the infrared spectrum confirmed polycarbonate formation. Thermal studies by differential scanning calorimetry (DSC) were carried out to determine the melting temperatures of the monomers. A gel permeation chromatography analysis (GPC) of the polycarbonates was performed in order to investigate their molar masses. Thermal analysis proved the purity of the obtained monomers; the curves showed a characteristic signal of melting. The obtained polycarbonates were characterized as having high resistance to organic solvents, including tetrahydrofuran. The GPC analysis proved their relatively large molar masses and their low dispersity.

Keywords: polycarbonates; transesterification; polycondensation

Citation: Wnuczek, K.; Puszka, A.; Podkościelna, B. Synthesis and Spectroscopic Analyses of New Polycarbonates Based on Bisphenol A-Free Components. *Polymers* **2021**, *13*, 4437. https://doi.org/10.3390/polym13244437

Academic Editors: Andrzej Puszka, Beata Podkościelna and Dan Rosu

Received: 19 November 2021
Accepted: 13 December 2021
Published: 17 December 2021

1. Introduction

Polycarbonates are materials encountered in everyday life. They are a class of thermoplastic polymers that were formally esters of carbonic acid. They are characterized by numerous advantages such as good hardness, ductility, rigidity, transparency, and toughness and excellent mechanical properties [1]. Compared to other thermoplastics, the most advantageous properties are as follows: high impact strength, good dielectric properties, good dimensional stability, wide operating temperature range, high creep strength, small water absorption, and a self-extinguishing tendency. In addition, most of the polycarbonates are non-toxic, very hard, abrasion-resistant, and chemical-resistant materials [2,3].

Bisphenol A polycarbonate (BPA-PC), derived from petroleum, is one of the most important and widely commercialized polycarbonates. However, bisphenol A is a toxic compound that can induce chronic toxicity and environmental problems [4]. BPA-PC is the carbonic acid polyester derived from 2,2-bis(4-hydroxyphenyl) propane. This is the best known polycarbonic resin of this type because of its good mechanical, thermal, and electrical properties, as well as being made from available raw materials. Most of the commercially available polycarbonates have been synthesized using bisphenol A. This compound is a precursor of important plastics, primarily of some polycarbonates and epoxy resins [5]. The use of bisphenol A as a diol for the synthesis of polycarbonates is controversial. The detection of bisphenol A in the environment and food products has been the subject of much recent research. Some studies prove that the thermal treatment of any food packaging product containing bisphenol A causes it to be released into the food. Studies proved that at temperatures above 70 °C and at high humidity, polycarbonate is hydrolyzed to bisphenol A [6–10].

The conventional phosgene process for polycarbonate preparation has been eliminated because of the toxicity of phosgene. Historically, polycarbonates were obtained by the polycondensation of phosgene with aromatic diols [11]. More and more often, there are reports in the literature which describe other "phosgene-free" methods of synthesizing this group of polymers, based mainly on the transesterification of appropriate diols (mainly butane-1,4-diol and bisphenol A) and dimethyl or diphenyl carbonate.

In recent years a safe and environmentally favorable process for the synthesis of polycarbonates has been a research goal [12]. Diphenyl carbonate (DPC) and bisphenol A are the main raw materials for polycarbonate synthesis. The non-phosgene route became popular for polycarbonate synthesis through the melt transesterification of DPC and BPA [13–15]. Diphenyl carbonate is a sustainable and environmentally benign reagent mainly used as a phosgene substitute for the synthesis of polycarbonates. The transesterification of bisphenol A and diphenyl carbonate and ring-opening polymerization of macrocyclic oligomers are commonly known routes [16]. As the transesterification is achieved through melt polymerization, the solvent is not necessary. However, so far there is no reliable process to produce DPC from phenol without the use of phosgene. The synthesis of macrocyclic oligomers is essentially the same reaction as the production of PC from bisphenol A and phosgene. In this context, the wholly non-phosgene process to produce PC would be a very desirable alternative [17].

There are many entries in the literature regarding the transesterification of diols followed by polycondensation to polycarbonates. Kim and Lee compared the transesterification of bisphenol A with that of diphenyl carbonate or dimethyl carbonate to obtain polycarbonate precursors. They also used the direct oxidative carbonylation of bisphenol A (with carbon monoxide) to obtain polycarbonate precursors for the synthesis of phosgene-free polycarbonates. They concluded that the melt transesterification of bisphenol A and diphenyl carbonate occurred readily to produce reactive precursors without a significant equilibrium constraint [18]. On the other hand, these methods were used to obtain high-molecular-weight polymers. Park and co-workers have published a synthesis procedure to obtain polycarbonates with a molecular weight of 100,000−200,000 Da. In the first step, oligomers were formed bearing almost equal numbers of hydroxyl and methyl carbonate end-groups. In the second step, the condensation reaction was conducted at a high temperature to connect the −OH and –O–C(O)–OCH_3 chain-ends while removing the generated methanol under reduced pressure [19]. Sun and Kucling have described the synthesis of high-molecular-weight polycarbonates based on organo-catalysis. However, in most cases, the use of metal-based catalysts is required for the preparation of aliphatic polycarbonates by the polycondensation method, which are difficult to remove completely from the final polymer [20]. Their study was focused on the synthesis of high-molecular-weight aliphatic polycarbonates using organo-catalysts via a two-step polycondensation of dimethyl carbonate and a linear alkane diol as monomers. In our paper, three catalysts were used and all of them were organic compounds.

With the growing concern about environmental pollution and global warming, developing eco-friendly materials has become a key global necessity. The main goal of this study was to develop a new method of obtaining polycarbonate materials so that their syntheses fit into the ideals of green chemistry.

The aim of this research was to synthesize new polymeric compounds based on safe (BPA-free), non-toxic components. A method of synthesizing aliphatic–aromatic compounds based on diphenylmethane has been developed. Twelve new polycarbonates based on diols or dithiol with the DPC monomer were obtained. Polycarbonates were prepared by the polycondensation reaction and their chemical structures were confirmed by means of ATR–FTIR. Their properties were examined by DSC analysis. Briefly, our method applied the DPC reagent to introduce carbonate groups. The transesterification reaction of DPC with diols/dithiol in the melt phase was performed, resulting in the formation of PC precursors. In the next step, PC compounds are amenable to the condensation polymerization reaction. The polycondensation step occurs under vacuum conditions to remove the phenol produced as a recyclable byproduct. The precursors obtained by this route have phenyl carbonate and/or hydroxy end-groups [18,21]. As diols/dithiol components, the diphenylmethane derivatives were used. The chemical structures of monomers were confirmed by NMR (and the ATR–FTIR analysis in the Supplementary Materials). For the obtained polycarbonates, ATR–FTIR and GPC were performed. A detailed study concerning the thermal analysis of the obtained polycarbonates (thermogravimetry and DSC) will be discussed in the next paper.

2. Materials and Methods

2.1. Chemicals

Diphenyl carbonate, methane dichloride, and tetrahydrofuran were purchased from Merck (Merck, Darmstadt, Germany). Catalysts: zinc acetate, 4-(dimethylamino)pyridine (DMAP) and benzyltriethylammonium chloride also were obtained from Merck (Merck, Darmstand, Germany). Purified water was delivered by Millipore (Millipore, UMCS Lublin, Poland).

2.2. Methods

The ^{1}H and ^{13}C NMR spectra were recorded using a Bruker Avance 300 MSL instrument (Bruker, Coventry, United Kingdom) operating at 500 MHz for ^{1}H and 75 MHz for ^{13}C resonance frequency. Chemical shifts were referenced to deuterated chloroform ($CDCl_3$), which served as an internal standard. The coupling constants (J) are given in Hz. The abbreviations for signal patterns are as follows: s, singlet; d, doublet; t, triplet; q, quartet; m, multiplet; b, broad.

The attenuated total reflection (ATR) was recorded using infrared Fourier transform spectroscopy (ATR-FTIR) on a TENSOR 27, Bruker spectrometer, equipped with a diamond crystal (Germany). The spectra were recorded in the range of 600–4000 cm^{-1} with 64 scans per spectrum at a resolution of 4 cm^{-1}.

Differential scanning calorimetry (DSC) curves were obtained with the use of a DSC Netzsch 204 calorimeter (Netzsch, Günzbung, Germany). All DSC measurements were made using aluminum pans with pierced lids sample mass of 5–10 mg in a nitrogen atmosphere (30 mL/min). As the reference, an empty aluminum crucible was used. Dynamic scans were made at a heating rate of 10 K/min. The heating cycle was in the temperature range 0–200 °C. Parameters such as melting temperature (T_m) and enthalpy of melting (ΔH_m) were also determined.

The number ($\overline{M}_n$), weight ($\overline{M}_w$), average molar mass (g/mol), and molar mass dispersity (D_M) of the obtained polycarbonates were determined by gel permeation chromatography (GPC) performed on a Viscotek GPC max (Viscotek, Kennesaw, USA) equipped with the triple detector array TDA 305. The eluent was tetrahydrofuran (THF), the flow rate was 1 mL/min, the operation temperature was set to 35 °C, and the molar mass was calibrated with polystyrene standards.

2.3. Synthesis of Monomers

The syntheses of monomers were carried out according to the methods developed in the Department of Polymer Chemistry, UMCS. The reaction scheme is presented in Figure 1. The syntheses of diols E and H and dithiol have been reported in the literature [22,23]. The description of the syntheses of these three monomers can be found in the Supplementary Materials (S2.3). A detailed description of the synthesis of the diol M is provided below.

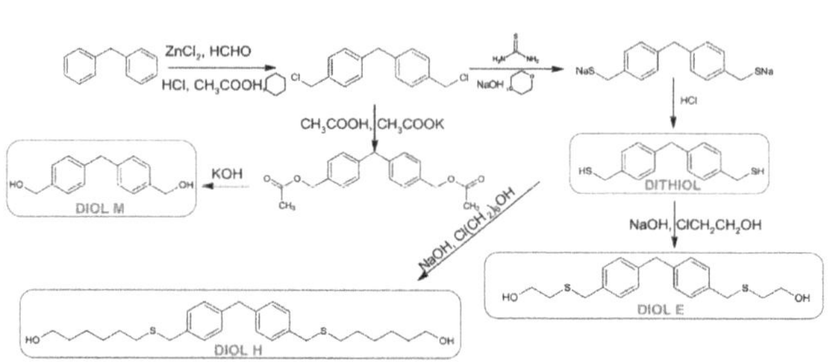

Figure 1. Scheme of monomers synthesis.

Synthesis of (methanediyldibenzene-4,1-diyl)dimethanol (diol M)

In total, 78 g (0.37 mol) of 1,1-methanediylbis[4-(chloromethyl)benzene] and 77 g (0.78 mol) of potassium acetate were weighed and added into a 1000 cm^3 round-bottom flask. Boiling pebbles and 600 cm^3 of pure acetic acid were added. The contents of the flask were boiled gently for 6 h. In order to avoid the sediment sticking to the flask, the whole construction was shaken. The solution was then decanted into a beaker and the KCl remained in the flask. An amount of 1 L of distilled water was added to the solution. An oil crystallized while stirring and was separated. The product was filtered off using a funnel and, after being transferred to a flask, 300 cm^3 of 10% aqueous Na_2CO_3 solution was poured over it, and then it was filtered again in the funnel. The resulting product was transferred to a 1000 cm^3 flask and treated with 5% aqueous KOH solution (100 cm^3) and 650 cm^3 of methanol. The flask was heated under reflux. The state of boiling was maintained for 3 h. Then, the alcohol was distilled off. The separated precipitate was filtered off using a funnel and then washed with distilled water until reaching neutral pH. The crude product was purified by crystallization from ethyl acetate (1 g per 7 cm^3 of the solvent).

2.4. Synthesis of Polymers

The synthesis method is based on two consecutive reactions. In the first stage, the transesterification reaction takes place (with simultaneous removal of the phenol byproduct from the reaction medium), and in the next stage, polycondensation takes place. For the production of polymers, diphenyl carbonate was selected as a monomer introducing carbonate groups (for the production of aliphatic–aromatic polycarbonates). Diphenyl carbonate and one of the monomers were placed in a three-necked flask equipped with a mechanical stirrer, gas inlet, thermometer and water pump. Substrates were added in a stoichiometric ratio of 1:1. The catalyst was added in an amount equivalent to 0.1 mol%. The reactions were conducted in a nitrogen atmosphere for 2 h after the complete melting of the components. The temperature was kept at 140–150 °C. As the reaction proceeded, phenol was released, which was fed by a water pump to the flask. After two hours, the gas supply and the water pump were disconnected. Reactions were carried out for another 2 h under a vacuum pump, maintaining the temperature at 140–150 °C. The resulting polycarbonates were placed into beakers, treated with dichloromethane (50 mL), and precipitated with methanol. Then, they were left to evaporate the solvent. For complete drying of polycarbonates, the beakers were placed in an oven for 12 h (50 °C). Figure 2 presents the laboratory glass diagram. Figures 3 and 4 show the schemes of reactions and probable reaction courses.

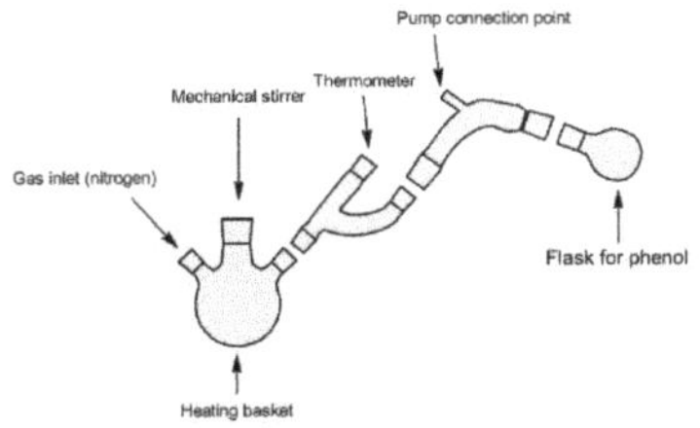

Figure 2. Diagram of laboratory glass used in the synthesis of polycarbonates.

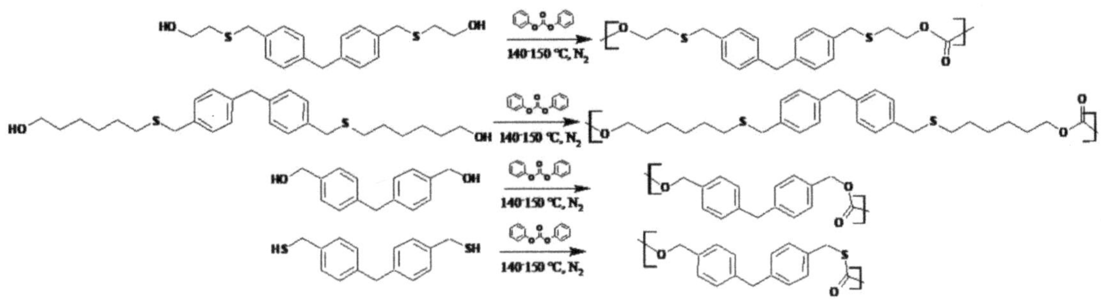

Figure 3. Scheme of syntheses of polycarbonates.

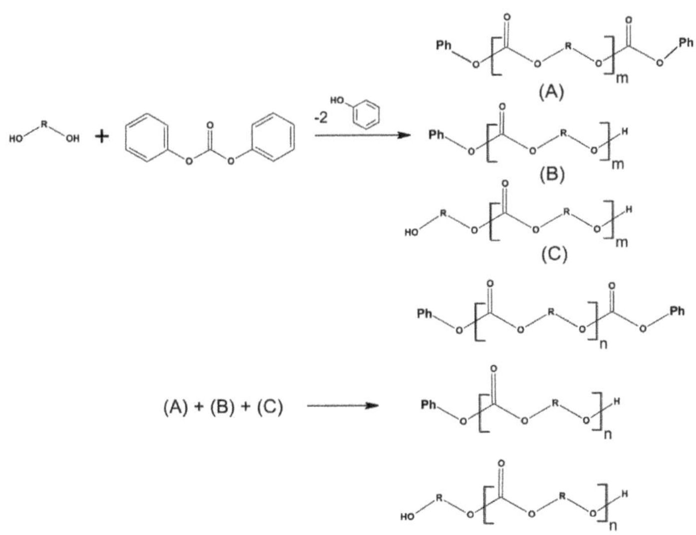

Figure 4. The probable course of the transesterification and polycondensation reactions with the use of diols and DPC [24]. (**A**), (**B**), (**C**): Oligomers.

Three types of catalysts were used in the reactions. Their structural formulae are listed in Figure 5. In total twelve parallel reactions were performed: each of the four monomers with three catalysts. Twelve products and twelve comparative materials (after precipitation with methanol) were obtained.

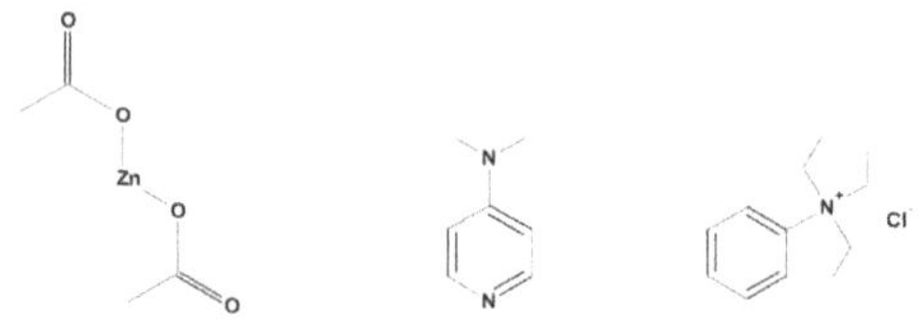

(a) (b) (c)

Figure 5. Catalysts: (**a**) zinc acetate, (**b**) DMAP, and (**c**) benzyltriethylammonium chloride.

3. Results and Discussion

3.1. ^{1}H and ^{13}C NMR Analysis

The ^{1}H and ^{13}C NMR analyses were performed for all synthesized monomers and the chlorine derivative of diphenylmethane (1,1′-methanediylbis[4-(chloromethyl)benzene]). NMR spectroscopy was applied in order to determine the monomers' structures. Deuterated chloroform was used as the solvent. The NMR analysis allowed confirmation of the structures of the obtained monomers. All NMR spectra are presented in Figures 6–15. The detailed information about the signals is presented below:

(a) 1,1′-methanediylbis[4-(chloromethyl)benzene]

^{1}HNMR (500 MHz, $CDCl_3$—d, δ ppm): 7.35 (d, J = 8.1 Hz, 4H), 7.21 (d, J = 8.21 Hz, 4H), 4.60 (s, 4H), 4.02 (s, 2H).
^{13}CNMR (126 MHz, $CDCl_3$—d, δ ppm): 141.13 (PhC), 135.47 (CPh), 129.31 (Ph), 128.87 (Ph), 46.13 (Cl-CH_2-Ph), 41.32 (Ph-CH_2-Ph).

(b) Diol M

^{1}HNMR (500 MHz, $CDCl_3$—d, δ ppm): 7.31 (d, J = 8.1 Hz, 4H), 7.21 (d, H = 8.21 Hz, 4H), 4.68 (s, 4H), 4.00 (s, 2H), 1.64 (s, 2H).
^{13}CNMR (126 MHz, $CDCl_3$—d, δ ppm): 140.59 (PhC), 138.71 (CPh), 129.11 (Ph), 127.32 (Ph), 65.20 (OH-CH_2-Ph), 41.34 (Ph-CH_2-Ph).

(c) Dithiol

^{1}HNMR (500 MHz, $CDCl_3$—d, δ ppm): 7.29 (d, J = 8.1 Hz, 4H), 7.22–7.00 (m, 4H), 3.99 (s, 2H), 3.75 (d, J = 7.5 Hz, 4H), 1.79 (t, J = 7.5 Hz, 2H).
^{13}CNMR (126 MHz, $CDCl_3$—d, δ ppm): 139.89 (PhC), 138.99 (CPh), 129.23 (Ph), 128.22 (Ph), 41.23 (-CH_2), 28.70 (SH-CH_2-Ph).

(d) Diol E

^{1}HNMR (500 MHz, $CDCl_3$—d, δ ppm): 7.25 (d, J = 8.1 Hz, 8H), 3.95 (s, 2H), 3.71 (s, 4H), 3.68 (t, J = 6.1 Hz, 4H), 2.64 (s, 4H), 2.44 (s, 2H).
^{13}CNMR (126 MHz, $CDCl_3$—d, δ ppm): 139.98 (PhC), 135.87 (CPh), 129.16 (Ph), 129.02 (Ph), 60.32 (-S-CH2-Ph), 41.21 (Ph-CH_2-Ph), 35.47 (OH-CH_2-), 34.33 (-CH_2-S-).

(e) Diol H

^{1}HNMR (500 MHz, $CDCl_3$—d, δ ppm): 7.24 (d, J = 8.1 Hz, 4H), 7.14 (d, J = 8.1 Hz, 4H), 3.96 (s, 2H), 3.69 (m, 4H), 3.64 (t, J = 6.6 Hz, 4H), 2.44 (s, 4H), 2.10 (s, 2H), 1.68–1.31 (m,16H).
^{13}CNMR (126 MHz, $CDCl_3$—d, δ ppm): 139.73 (PhC), 136.37 (CPh), 129.02 (Ph), 62.86 (-S-CH_2-Ph), 41.21 (Ph-CH_2-Ph), 36.00 (OH-CH_2-), 32.50 (-CH_2(II)-), 31.33 (-CH_2(III)-), 29.14 (-CH_2(IV)-), 28.58 (-CH_2(V)-), 25.33 (-CH_2-S-).

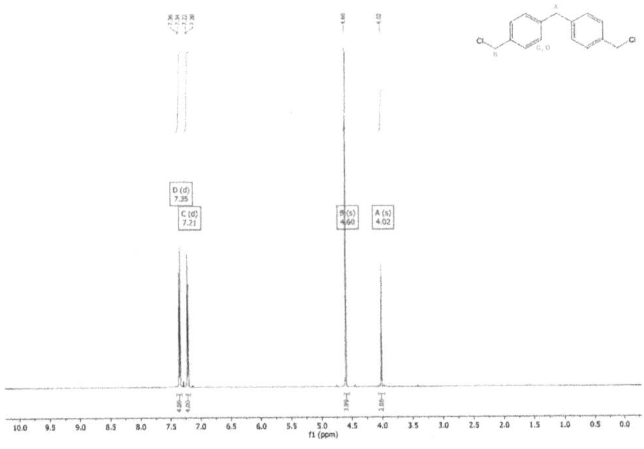

Figure 6. ^{1}HNMR spectrum of 1,1′-methanediylbis[4-(chloromethyl)benzene].

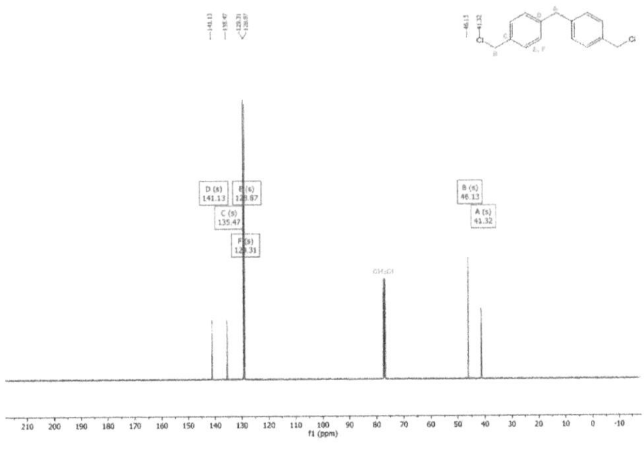

Figure 7. ^{13}CNMR spectrum of 1,1′-methanediylbis[4-(chloromethyl)benzene].

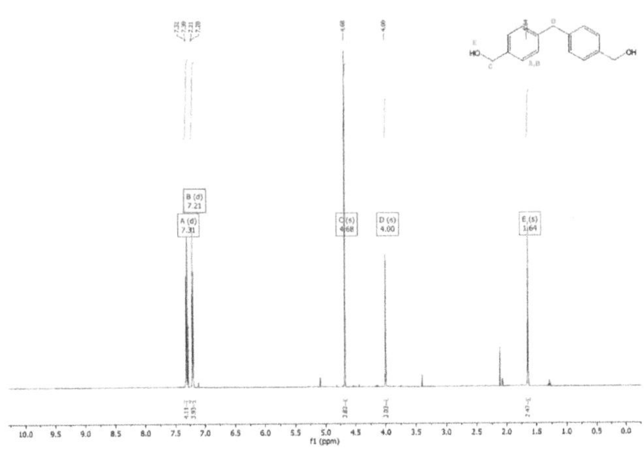

Figure 8. ^{1}HNMR spectrum of diol M.

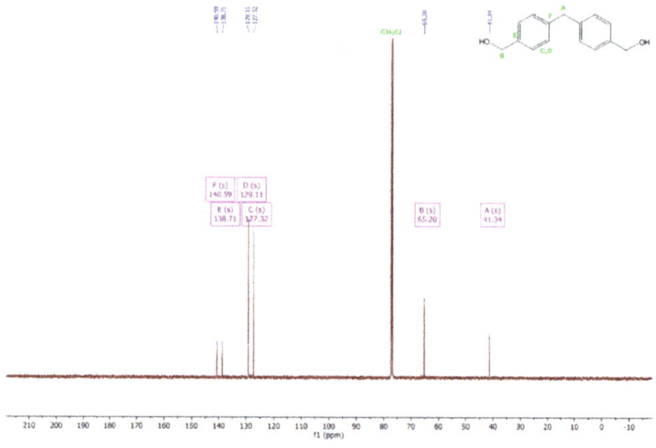

Figure 9. ^{13}CNMR spectrum of diol M.

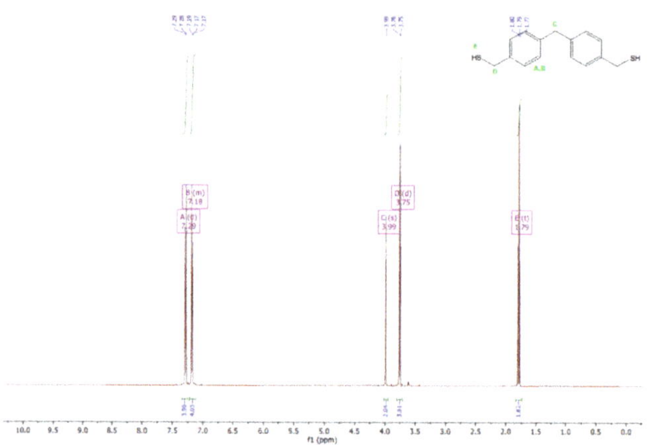

Figure 10. ^{1}HNMR spectrum of dithiol.

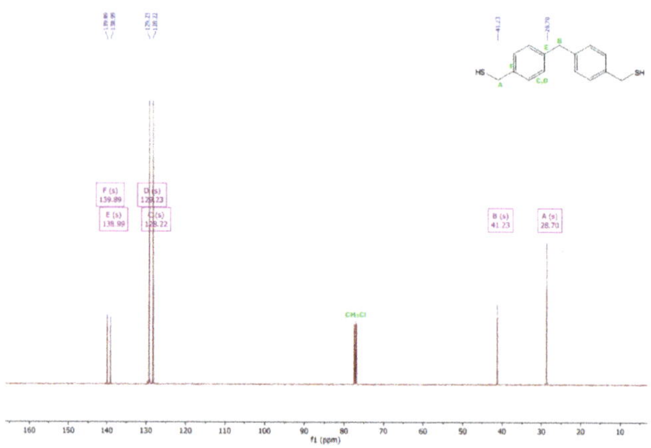

Figure 11. ^{13}CNMR spectrum of dithiol.

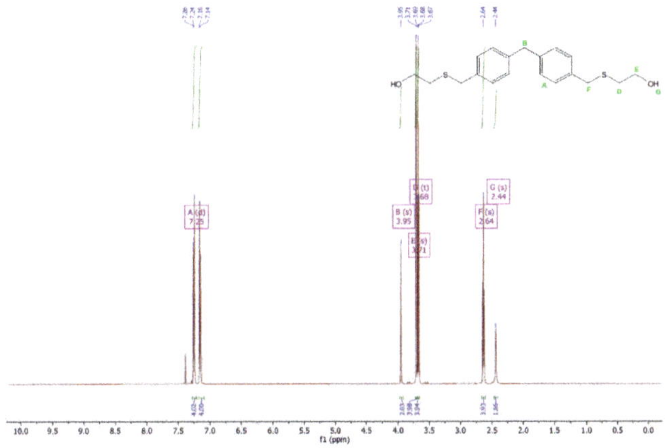

Figure 12. ^{1}HNMR spectrum of diol E.

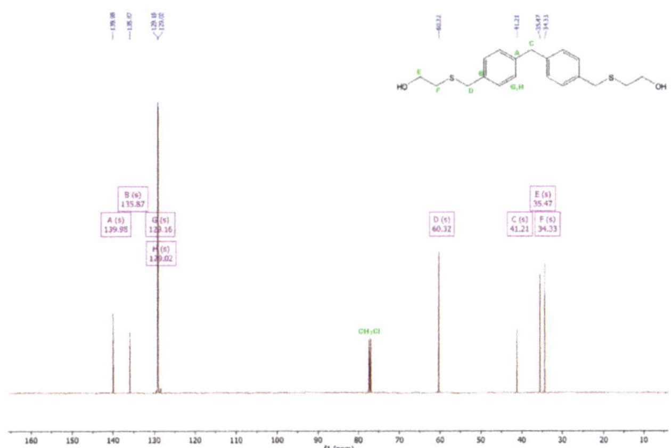

Figure 13. ^{13}CNMR spectrum of diol E.

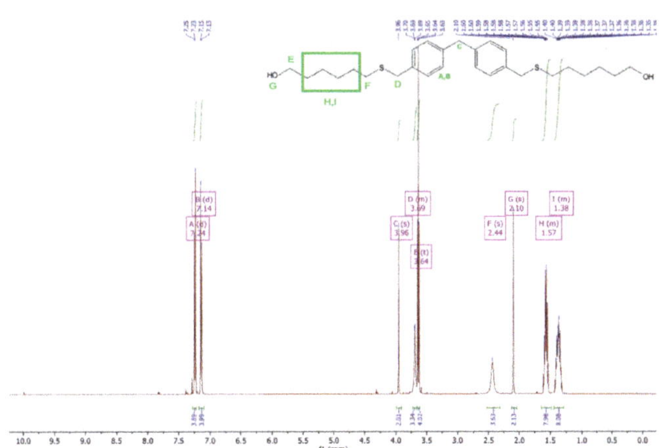

Figure 14. ^{1}HNMR spectrum of diol H.

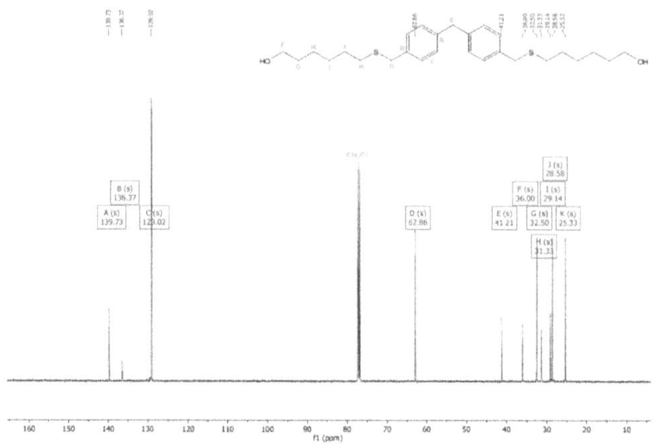

Figure 15. ^{13}CNMR spectrum of diol H.

3.2. ATR–FTIR Analysis

The chemical structure of polycarbonates was confirmed by attenuated total reflection–Fourier transform infrared spectroscopy. The study consisted of observing the changes in the positions of the absorption bands for the characteristic functional groups of the obtained polycarbonates. Figures 16–19 present the results of the polycarbonates analyses: the pure polycarbonate and the polymers obtained after precipitation from methane dichloride and methanol. The most important vibrations occurring in the spectra are presented in Table 1.

Table 1. Characteristic signals in ATR–FTIR spectra [cm^{-1}].

Polymer	C–H Aliph.	C–H Arom.	C=C Arom.	C–O	C=O	–OH
Diol E + DPC + Zinc acetate	2916	881	1596 1463	1239 1172	1743	3426
Diol E + DPC + DMAP	2915	816	1598	1239 1171	1744	-
Diol E + DPC + Benzyltriethylammonium chloride	2916	880	1597	1295 1238 1104	-	3458
Diol H + DPC + Zinc acetate	2923	943	1449	1257 1210	1742	-
Diol H + DPC + DMAP	2929	943	1598 1494	1259	1743	-
Diol H + DPC + Benzyltriethylammonium chloride	2931	988 819	1452	1239 1180 1100	1760	3408
Diol M + DPC + Zinc acetate	2961	924 854	1469	1228	1738	-
Diol M + DPC + DMAP	2917	915 860	1469	1235 1172 1109	-	3319
Diol M + DPC + Benzyltriethylammonium chloride	2929	915 860	1587 1490	1253 1230 1021	1740	-

Table 1. *Cont.*

Polymer	C–H Aliph.	C–H Arom.	C=C Arom.	C–O	C=O	–OH
Dithiol + DPC + Zinc acetate	2958	920 863	1593	1178 1106	1771	-
Dithiol + DPC + DMAP	2960	914 825	1595	1238 1172 1119	-	3315
Dithiol + DPC + Benzyltriethylammonium chloride	2962	925 811	1596	1256 1231 1174	1740	3370

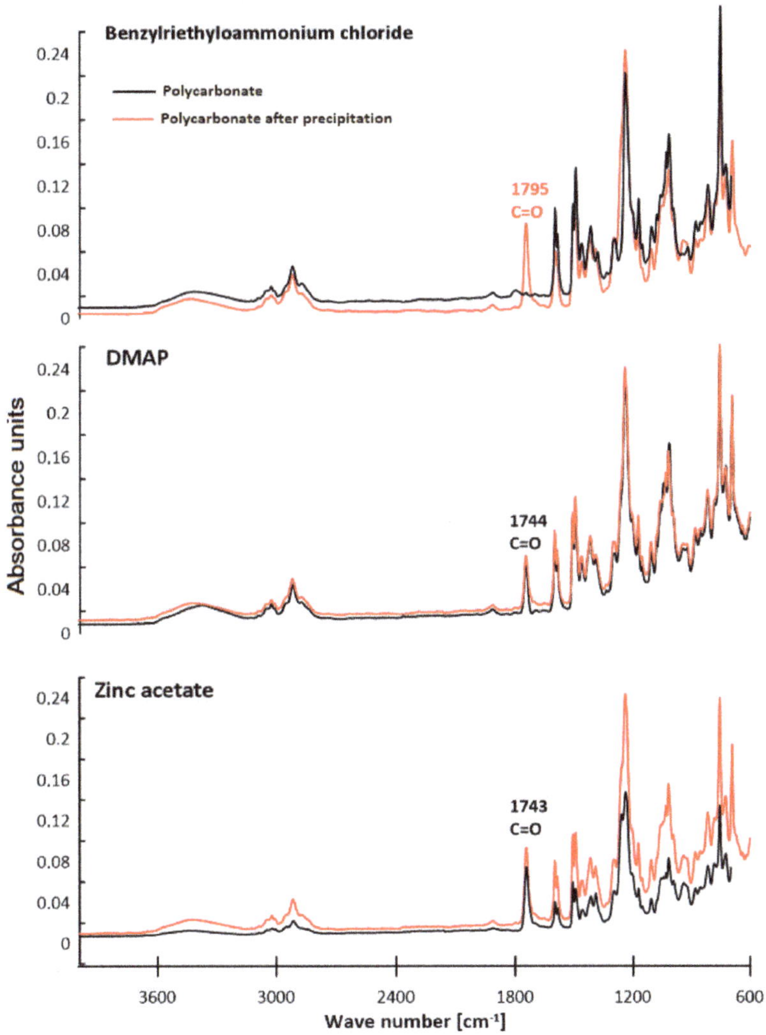

Figure 16. ATR–FTIR spectra of polycarbonates based on diol E.

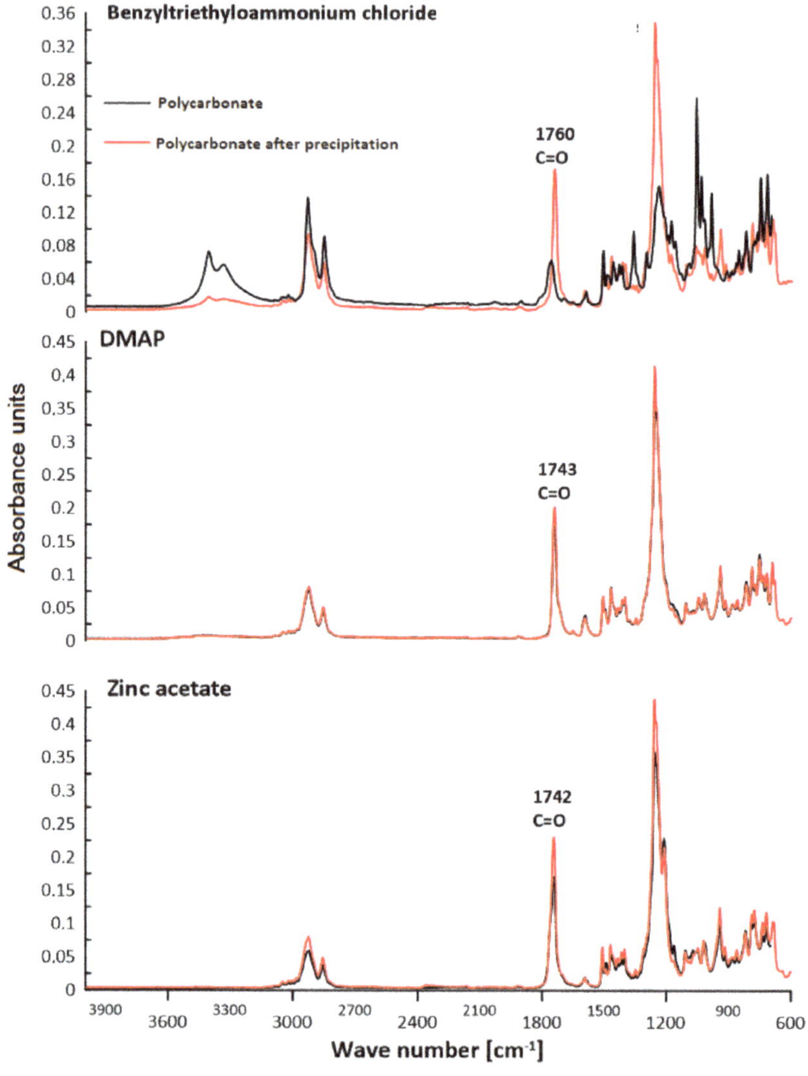

Figure 17. ATR–FTIR spectra of polycarbonates based on diol H.

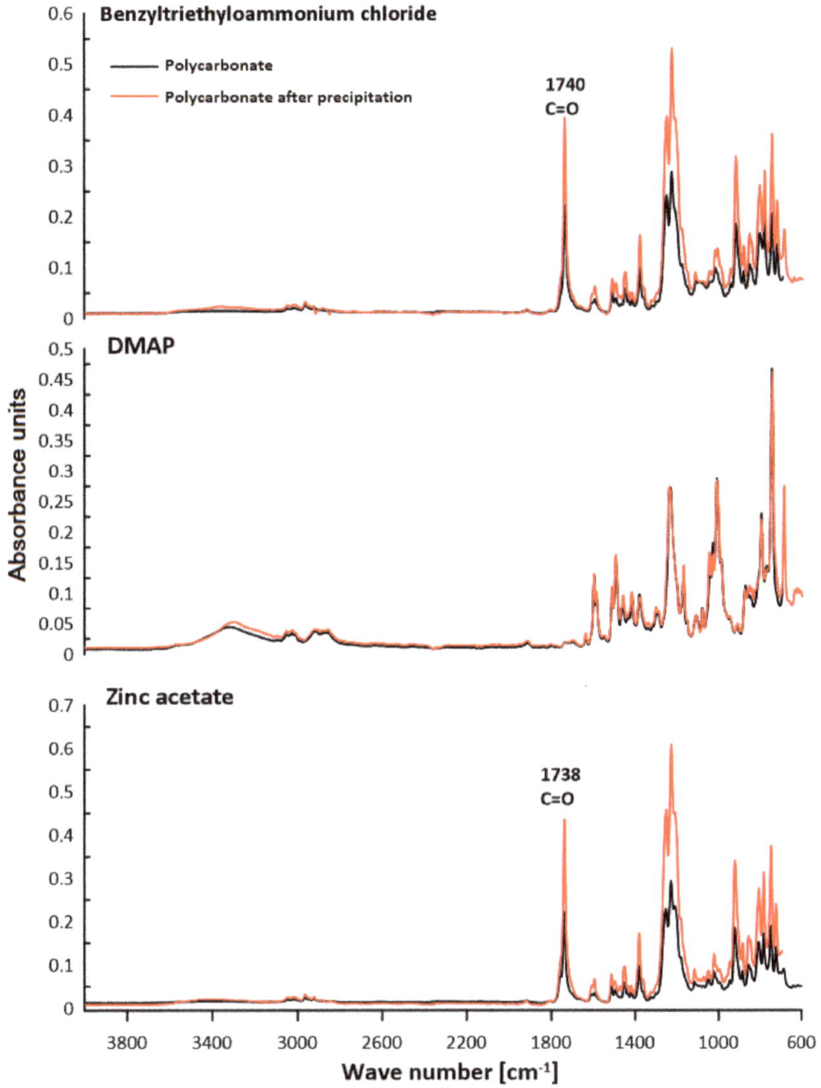

Figure 18. ATR–FTIR spectra of polycarbonates based on diol M.

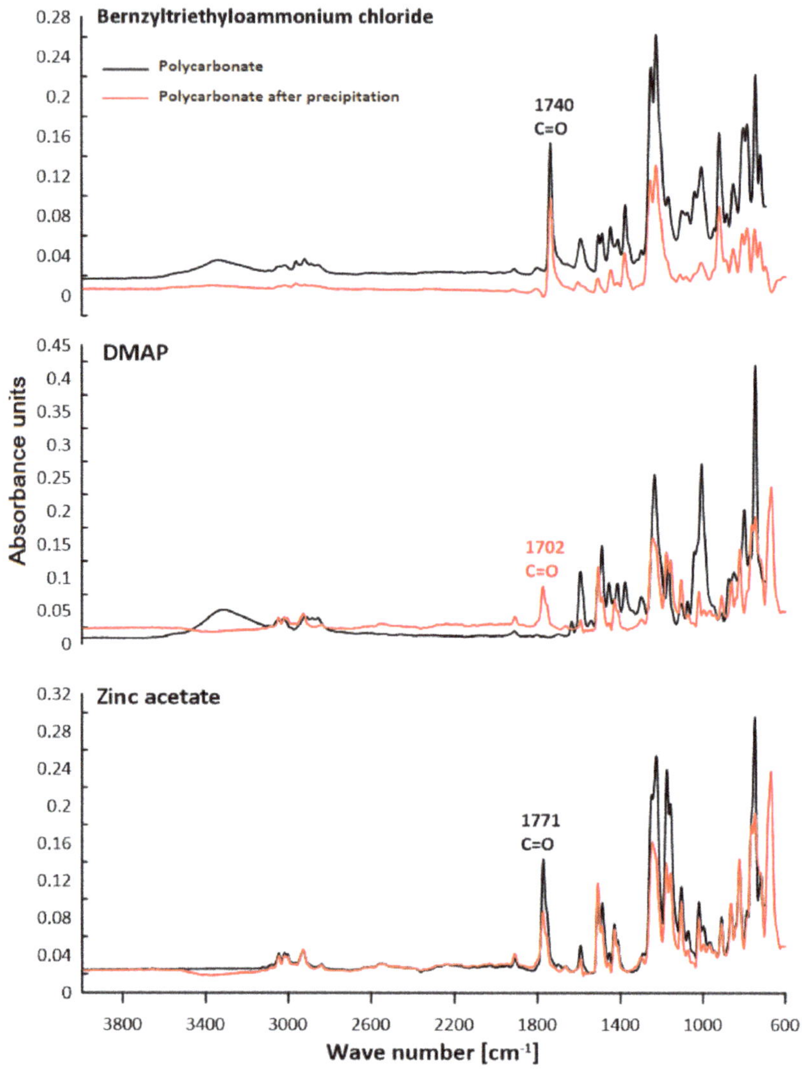

Figure 19. ATR–FTIR spectra of polycarbonates based on dithiol.

Due to the fact that the main components of the polymers are diphenyl carbonate and the diphenylmethane derivatives, the spectra are similar. The presence of the carbonyl group was considered as evidence of the polycarbonates' formation. The valence vibrations of the C=O group occur in the range of 1600–1900 cm^{-1}. In the aromatic esters this range is narrowed down to 1730–1780 cm^{-1}. In our previous studies of composites with polycarbonate as a filler, the carbonyl group peak was even narrower. The signal from the carbonyl group was in the range of 1730–1725 cm^{-1} for all materials [25]. In the present study, this effect was found for all samples except for the polymers: diol E + DPC + benzyltriethylammonium chloride, diol M + DPC + DMAP, dithiol + DPC + DMAP. In these cases, polycarbonates were not formed. The vibrations of the C–O group for saturated esters are visible in the range of 1050–1330 cm^{-1} for all samples. The signal for the hydroxyl group occurs for most samples. This could be related to the presence of the OH group in diols. Another characteristic absorption band in the range of 2962–2915 cm^{-1}

is derived from the stretching vibrations from the C–H aliphatic groups. This effect is visible for each composite. Multiple bands ranging from 1596 cm^{-1} to 1449 cm^{-1} can be associated with the vibrations of C–H and C=C bonds related to the benzene rings and aromatic skeletons. These come from both the aromatic diols and dithiol. The peaks around 1400–1450 cm^{-1} originate from the C–H deformation in the -CH_2- group. The bands around 1295–1012 cm^{-1} could be attributed to C–O stretching vibrations. The signal at 811–988 cm^{-1} was also associated with the C–C vibrations from the aromatic part. The spectra for the polymers obtained after precipitation from methane dichloride and methanol are characterized mostly by greater signal intensity. As for the carbonyl group, it was not visible in the spectra of diol M + DPC + DMAP. To sum up, the presence of the carbonyl group (C=O) appeared in 20 of 24 samples. Our previous studies discussed the effect of sulfur atoms on the infrared spectra [26,27]. The spectra of monomers are included in the Supplementary Materials (Figure S1 and Table S1).

3.3. DSC Analysis (Differential Scanning Calorimetry)

The thermal properties of the obtained monomers were studied by means of DSC analysis. Characteristic parameters were determined and collected in Table 2. Monomers were tested in the temperature range 0–200 °C (Figure 20). The DSC analysis allowed the determination of the minimum temperature of the transesterification reaction of the monomers used with DPC.

Table 2. DSC data of monomers.

Monomer	T_m [°C]	ΔH_m [J/g]
Diol E	85	143
Diol H	68	161
Diol M	119	154
Dithiol	84	101
DPC	84	134

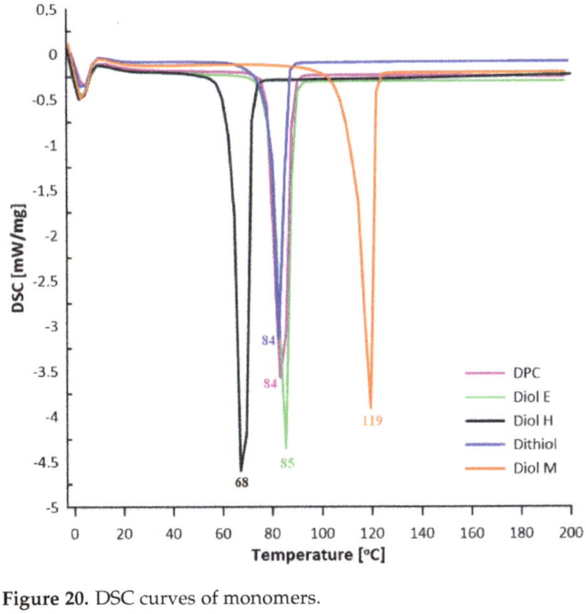

Figure 20. DSC curves of monomers.

3.4. GPC Analysis

The purpose of the GPC analysis was to determine the molar mass distribution of the obtained polycarbonates. Due to their high chemical resistance (only two polymers completely dissolved in tetrahydrofuran), it was not possible to perform the analysis: the remaining polycarbonates were insoluble or only partially dissolved in THF. The polymers that dissolved in THF (i.e., diol H + DPC + zinc acetate, diol H + DPC + DMAP) were characterized by $\overline{M}_n$ = 1570 and 6000 g/mol, $\overline{M}_w$ = 4346 and 11,387 g/mol, and molar mass dispersity ($Đ_M$) of 2.768 and 1.898, respectively. Figure 21 shows the chromatograms for these polymers.

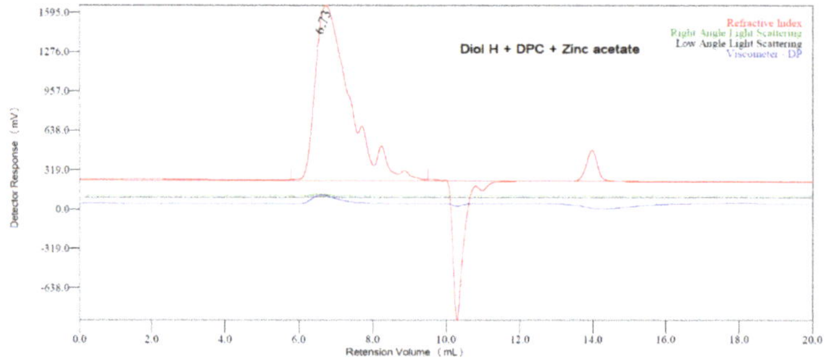

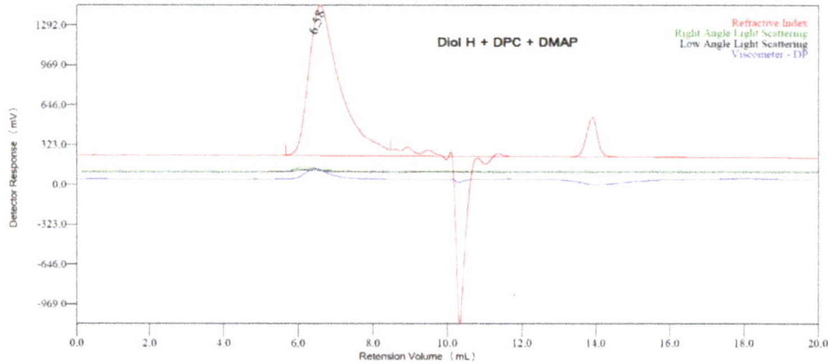

Figure 21. GPC chromatograms for polycarbonates dissolved in THF.

The asymmetry of the peaks in the chromatograms indicates the heterogeneity of the molar masses of the polymers (presence of fractions with different molar masses, including oligomers).

4. Conclusions

As a result of the multistage synthesis reactions, four aromatic–aliphatic compounds were obtained as diphenylmethane derivatives: diol M, diol E, diol H, and dithiol. The chemical structure of each compound was confirmed by the spectroscopic techniques ^{13}C and ^{1}H NMR. Additionally, the ATR–FTIR spectra of the compounds are shown. The new polycarbonates were obtained using transesterification reactions and the polycondensation process. Three catalysts were used. The presence of the carbonyl group in the infrared spectrum was used to confirm the formation of the polymer. Taking into account raw materials and comparative materials, the carbonyl group appeared in 20 samples. The polycarbonates were characterized by high resistance to THF. The obtained results of the GPC analysis indicated relatively large molar masses and small dispersity.

In summary, the use of unconventional compounds such as diols resulted in the production of polycarbonates free of toxic BPA.

Supplementary Materials: The following are available online at https://www.mdpi.com/article/10.3390/polym13244437/s1, Figure S1: ATR–FTIR analyses of monomers; Table S1: Characteristic signals in ATR–FTIR spectra of monomers [cm^{-1}].

Author Contributions: Conceptualization, A.P. and B.P.; methodology, A.P.; software, K.W.; validation, K.W., A.P. and B.P.; formal analysis, A.P. and B.P.; investigation, K.W.; resources, K.W.; data curation, K.W. and A.P.; writing—original draft preparation, K.W.; writing—review and editing, A.P. and B.P.; visualization, K.W.; supervision, B.P. and A.P.; project administration, K.W., A.P. and B.P. All authors have read and agreed to the published version of the manuscript.

Funding: This research received no external funding.

Institutional Review Board Statement: Not applicable.

Informed Consent Statement: Not applicable.

Data Availability Statement: The data presented in this study are available on request from the corresponding author.

Conflicts of Interest: The authors declare no conflict of interest.

References

1. Hammani, S.; Moulai-Mostefa, N.; Benyahia, L.; Tassin, J.F. Effects of composition and extrusion parameters on the morphological development and rheological properties of PP/PC blends. Co-continuity investigation. *J. Polym. Res.* **2012**, *19*, 994. [CrossRef]
2. Fleck, N.A.; Stronge, W.J.; Liu, J.H. High strain rate shear response of polycarbonate and polymethyl methacrylate. *Proc. R. Soc. Lond. A Math. Phys. Sci.* **1990**, *429*, 459–479.
3. Yan, Y.; Mao, Y.; Li, B.; Zhou, P. Machinability of the Thermoplastic Polymers: PEEK, PI, and PMMA. *Polymers* **2021**, *13*, 69. [CrossRef] [PubMed]
4. Shen, X.; Liu, S.; Wang, Q.; Zhang, H.; Wang, G. Synthesis of Poly (isosorbide carbonate) via Melt Polycondensation Catalyzed by a KF/MgO Catalyst. *Chem. Res. Chin. Univ.* **2019**, *35*, 721–728. [CrossRef]
5. Pivnenko, K.; Pedersen, G.A.; Eriksson, E.; Astrup, T.F. Waste material recycling: Assessment of contaminants limiting recycling. *Waste Manag.* **2015**, *44*, 39–47. [CrossRef]
6. Vandenberg, L.N.; Maffini, M.V.; Sonnenschein, C.; Rubin, B.S.; Soto, A.M. Bisphenol-A and the great divide: A review of controversies in the field of endocrine disruption. *Endocr. Rev.* **2009**, *30*, 75–95. [CrossRef]
7. Geens, T.; Aerts, D.; Berthot, C.; Bourguignon, J.P.; Goeyens, L.; Lecomte, P. A review of dietary and non-dietary exposure to bisphenol-A. *Food Chem. Toxicol.* **2012**, *50*, 3725–3740. [CrossRef]
8. Geens, T.; Goeyens, L.; Covaci, A. Are potential sources for human exposure to bisphenol-A overlooked? *Int. J. Hyg. Environ. Health* **2011**, *214*, 339–347. [CrossRef]
9. Sun, A.F.; Kang, L.; Xiang, X.; Lil, H.; Luol, C.; Luo1, R.; Lu1, C.; Peng, X. Recent advances and progress in the detection of bisphenol A. *Anal. Bioanal. Chem.* **2016**, *408*, 6913–6927. [CrossRef]
10. Bair, H.E.; Falcone, D.R.; Hellman, M.Y.; Johnson, G.E.; Kelleher, P.G. Hydrolysis of polycarbonate to yield BPA. *J. Appl. Polym. Sci.* **1981**, *26*, 1777. [CrossRef]
11. Wang, L.; Xiao, B.; Wang, G.Y.; Wu, J. Synthesis of polycarbonate diol catalyzed by metal-organic framework $Zn_4O[CO_2$-C_6H_4-$CO_2]_3$. *Sci. China Chem.* **2011**, *54*, 1468–1473. [CrossRef]
12. Hirotoshi, I.; Takeuchi, K.; Michihiko, A.; Mitsuru, U. Oxidative carbonylation of phenol to diphenyl carbonate catalyzed by Pd-pyridyl complexes tethered on polymer support. *Catal. Commun.* **2001**, *2*, 145–150.
13. Komiya, K.; Fukuoka, S.; Aminaka, M.; Hasegawa, K.; Hachiya, H.; Okamato, H.; Watanabe, T.; Yoneda, H.; Fukawa, I.; Dozono, T. New Process for Producing Polycarbonate Without Phosgene and Methylene Chloride. In *ACS Symposium Series*; Anastas, T., Williamson, T.C., Eds.; ACS: Washington, DC, USA, 1996; Volume 626, pp. 20–32.
14. Hsu, J.P.; Wong, J.J. Kinetic modeling of melt transesterification of diphenyl carbonate and bisphenol-A. *Polymer* **2003**, *44*, 5851–5857. [CrossRef]
15. Kreye, O.; Meier, M.A.R. Base catalyzed sustainable synthesis of phenyl esters from carboxylic acids using diphenyl carbonate. *RSC Adv.* **2015**, *5*, 53155–53160. [CrossRef]
16. Brunelle, D.J.; Sannon, T.G. Preparation and polymerization of bisphenol A cyclic oligomeric carbonates. *Macromolecules* **1991**, *24*, 3035. [CrossRef]
17. Haba, O.; Itakura, I.; Ueda, M.; Kuze, S. Synthesis of Polycarbonate from Dimethyl Carbonate and Bisphenol-A Through a Non-Phosgene Process. *Polym. Chem.* **1999**, *37*, 2087–2093. [CrossRef]
18. Kim, W.B.; Lee, J.S. Comparison of Polycarbonate Precursors Synthesized from Catalytic Reactions of Bisphenol-A with Diphenyl Carbonate, Dimethyl Carbonate, or Carbon Monoxide. *J. Appl. Polym. Sci.* **2002**, *86*, 937–947. [CrossRef]
19. Park, J.H.; Jeon, J.Y.; Lee, J.J.; Jang, Y.; Varghese, J.K.; Lee, B.Y. Preparation of High-Molecular-Weight Aliphatic Polycarbonates by Condensation Polymerization of Diols and Dimethyl Carbonate. *Macromolecules* **2013**, *46*, 3301–3308. [CrossRef]

20. Sun, J.; Kuckling, D. Synthesis of high-molecular-weight aliphatic polycarbonates by organo-catalysis. *Polym. Chem.* **2016**, *7*, 1642–1649.
21. Kim, W.B.; Joshi, U.A.; Lee, J.S. Making Polycarbonates without Employing Phosgene: An Overview on Catalytic Chemistry of Intermediate and Precursor Syntheses for Polycarbonate. *Ind. Eng. Chem. Res.* **2004**, *43*, 1897–1914. [CrossRef]
22. Puszka, A.; Kultys, A.; Rogulska, M. Influence of DMPA content on the properties of new thermoplastic poly(ether-urethane) elastomers. *J. Elastomers Plast.* **2018**, *50*, 140–150. [CrossRef]
23. Kultys, A.; Puszka, A. New thermoplastic polyurethane elastomers based on sulfur-containing chain extenders. *Pol. J. Chem. Tech.* **2013**, *15*, 4. [CrossRef]
24. Song, M.; Yang, X.; Wang, G. Synthesis of polycarbonate diols (PCDLs) via two-step process using CH_3COONa as an effective catalyst. *Chem. Res. Chin. Univ.* **2018**, *34*, 578–583. [CrossRef]
25. Wnuczek, K.; Puszka, A.; Klapiszewski, Ł.; Podkościelna, B. Preparation, thermal and thermo-mechanical characterization of polymeric blends based on di(meth)acrylate monomers. *Polymers* **2021**, *13*, 878. [CrossRef]
26. Fila, K.; Podkościelna, B.; Podgórski, M. Cross-linked polythiomethacrylate esters based on naphthalene—synthesis, properties and reprocessing. *Materials* **2020**, *13*, 3021. [CrossRef]
27. Fila, K.; Goliszek, M.; Podkościelna, B.; Podgórski, M. Polymer side-chain modification in methacrylate and styrene copolymers through thiol-thioester dynamic exchange. *Eur. Polym. J.* **2020**, *136*, 109918. [CrossRef]

polymers

Review

Skillful Control of Dispersion and 3D Network Structures: Advances in Functional Organic–Inorganic Nano-Hybrid Materials Prepared Using the Sol-Gel Method

Hiroki Ikake [1], Shuta Hara [2,*] and Shigeru Shimizu [1]

[1] Department of Materials and Applied Chemistry, College of Science and Technology, Nihon University, 1-8-14 Kandasurugadai, Chiyoda-ku, Tokyo 101-8308, Japan
[2] Department of Material and Life Chemistry, Faculty of Engineering, Kanagawa University, 3-6-1 Rokkakubashi, Kanagawa-ku, Yokohama 221-8686, Japan
* Correspondence: ft102160vg@kanagawa-u.ac.jp

Abstract: Organic–inorganic hybrid materials have become indispensable high-performance and highly functional materials. This is owing to the improved dispersion control in hybrid materials and emergence of functional ionic liquids. Harmonization of both these factors has enabled the utilization of functional 3D network structures and nanodispersions in composite materials. Polymeric materials endow materials with flexibility, toughness, and shape-memory properties, whereas inorganic materials provide materials with unique optical, electrical, and magnetic properties due to their nanosize. Organic–inorganic hybrid materials have evolved into novel materials that go beyond the composite rule. In this review, the historical development of hybrid materials prepared using the sol-gel method and the birth of ionic liquids have been summarized. In addition, the historical results leading to the development of functional 3D network structures and dispersion control have also been presented, as well as a review of the research on functional ionic liquids, which are of current interest. The authors also summarize the results of their research on functional ionic liquids. The design of new organic–inorganic hybrid materials has been discussed and the future prospects of new polymer composite materials provided.

Keywords: sol-gel process; 3D network hybrid materials; nanoparticles; nanodispersity; ionic liquids

Citation: Ikake, H.; Hara, S.; Shimizu, S. Skillful Control of Dispersion and 3D Network Structures: Advances in Functional Organic–Inorganic Nano-Hybrid Materials Prepared Using the Sol-Gel Method. *Polymers* **2022**, *14*, 3247. https://doi.org/10.3390/polym14163247

Academic Editor: Andrzej Puszka

Received: 15 July 2022
Accepted: 4 August 2022
Published: 9 August 2022

1. Birth of Organic–Inorganic Hybrid Materials

Organic polymers are low-weight, flexible, and supple, and have excellent formability and workability. Conversely, inorganic materials have excellent functionality, such as high elasticity and heat resistance, as well as optical, magnetic, and electrical properties. Consequently, organic–inorganic hybrid materials combine the excellent properties of both these materials. By appropriately designing the substances and elements to be combined as well as the way in which they are combined, they can be transformed into unique materials with new and diverse functionalities that go beyond their mere combination [1–8].

Hybrids between organic polymers and inorganic materials that exhibit excellent properties are found in nature. For example, shells, teeth, bones, siliceous algae, and rice are typical organic–inorganic hybrids created in nature via biomineralization, in which the inorganic substances are synthesized in vivo. Plants absorb silica from the soil through their roots, which is then accumulated as amorphous silica (opal) in their stems and leaves [9,10]. Strong wind-resistant and rain-resistant stems are supported by silica. In the leaves, the scattering effect of opal promotes plant photosynthesis, and opal precipitation near the leaf pores regulates the water content. Some plants are sensitive to salt damage. In regions with low rainfall, plants protect themselves from salt damage by actively forming hybrids with high silica contents. Animals too use their abundant calcium deposits. For example, *Trochus niloticus* (Figure 1), which inhabits the seas of Okinawa and other coral reefs around the world, has a densely stacked structure of polygonal aragonite ($CaCO_3$) (width ~8 mm; thickness ~0.9 mm; stacking interval ~40 nm) [11].

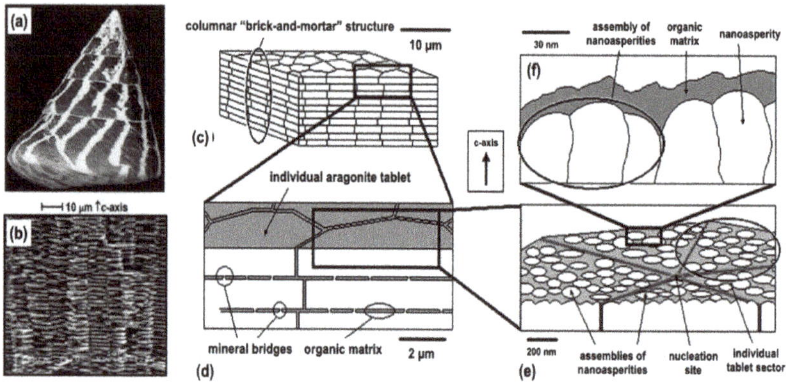

Figure 1. (**a**) Photograph of *Trochus niloticus* and (**b**) lateral-view SEM image of the internal stacked columnar nacreous layer microstructure of a fresh sample of *Trochus niloticus*. A schematic representation of the multi-scale hierarchical structure of the nacreous layer: (**c**) 10 μm, (**d**) 2 μm, and (**e**) 200 nm length scale of the individual tablet features, and (**f**) 30 nm length scale showing the nanoasperity structure. From ref. [11]. Adapted with permission from Ref. [11]. Copyright 2005, Materials Research Society.

The structure consists of a planar arrangement of ca. 95 wt.% aragonite tablets and ca. 5 wt.% biopolymers connecting the tablets. This regular arrangement allows one to see the beautiful structural color of the pearls, which is their characteristic luster. The high homogeneity and orderliness of the nacreous layer allows it to develop excellent mechanical properties. In other words, *Trochus niloticus* is protected by an excellent high-performance hybrid material.

The wings of the morpho butterfly emit a brilliant iridescent blue (Figure 2). Nanophotonic structures are built on wing scales in a structure so dense that it is almost mystical. Although the physics of structural color is well-established, replicating wing structures by conventional top-down lithography remains challenging.

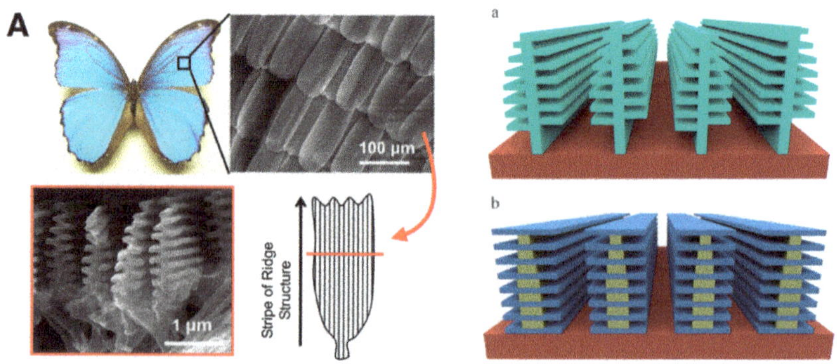

Figure 2. (**A**) Schematic representation of the scales of morpho butterfly wings. (**a**) Christmas tree structure of lamellar layers similar to actual wing scales. (**b**) Designed scales to be fabricated with aligned lamellae structures of PMMA/LOR (LOR; lift-of-resist supplied by MicroChem Corp.) alternate layers. From ref. [12].

Natural organic–inorganic hybrid materials are excellent composites with a dense structure consisting of an organic and an inorganic phase. Elaborate structures, such as homogeneous dispersions, phase separation, and dense structures, endow hybrid materials with functionalities beyond those predicted (composite law). Recently, considerable research attention has been directed toward the development of biomimetic hybrid materials that mimic excellent natural hybrid materials, which are rich in functionality [12–14]. In particular, organic–inorganic hybrid materials exhibiting ultrastructures with nanoscale dimensions in the dispersed phase have attracted considerable research interest [15–17]. The ultrafine structure of the dispersed phase gives rise to quantum effects that are not observed in the bulk material. The excellent properties of these materials, such as those observed for the aforementioned natural hybrid materials, have led to their widespread application in artificially prepared hybrid materials. In the field of optics, these materials satisfy the transparency, orientation, and wide viewing angles required for flat-screen displays.

Inorganic materials also exhibit properties that are different from their bulk counterparts, which can be hybridized with polymers to provide various properties, such as gas barrier [18], light modulation [19], and electrical conductivity [7,20]. These properties cannot be obtained from the bulk polymers and are new properties that can only be achieved via the formation of an ultrafine structure with inorganic materials and hybridization on a molecular level. However, the nanometer-sized dispersion of the inorganic materials in the polymer matrix generates extremely high interfacial free energy on the surface of the nanoparticles. This causes the nanoparticles to aggregate in the matrix, which not only causes the nanoparticles to lose their unique properties, but also reduces their suppleness, toughness, and formability. Therefore, the development of methods to produce highly dispersed inorganic nanoparticles in a polymer matrix is important for the future use of nanoparticles in devices.

2. Preparation of Hybrid Materials Using the Sol-Gel Method

The sol-gel method can be used to hybridize organic polymers and inorganic substances on a molecular level. The sol-gel method prepares these materials via the gelation of the precursor solution upon the hydrolysis and polycondensation reaction of an alkoxysilane. Because the starting materials are liquids, it can be applied to many organic polymers and various inorganic materials by adjusting the miscibility of the two components. The resulting hybrids show a variety of forms, from networked structures formed between the organic and inorganic phases to simply dispersed microparticles in a polymer matrix [21], which have a significant bearing on the overall functionality of the material.

The sol-gel method has a long history. Metal alkoxides were developed as raw materials to prepare hybrid materials by Ebelmen [22] in 1846. Research using the sol-gel method, which had stagnated for a long time, was triggered by the preparation of sintered polycrystals using metal alkoxides by Mazdiyasni et al. [23] in 1969 and Dislich [24] in 1971. Dislich developed the low-temperature hydrolysis of silicate esters, which is thought to be the beginning of the use of silicate ester hydrolysis as a glass production method. Silica gel was prepared via the hydrolysis of silicate esters. Dislich used the hydrolysis and polycondensation of metal alkoxides to prepare powdered gel particles, which were then hot-pressed to produce transparent PYREX® glass. Subsequently, research on the use of various metal alkoxides and the preparation of ceramics began to flourish. The sol-gel method has become a promising technique for hybridization using heat-sensitive organic materials because glasses and ceramics can be synthesized under relatively mild conditions. In particular, the development of hybrid materials comprising organic polymers and inorganic materials, and the development of hard contact lens materials by Philipp and Schmidt [25] in 1984, are regarded as epoch-making events. Mixed solutions containing epoxysilane, methacryloxysilane, and titanium alkoxide were subjected to condensation reactions, to which methacrylate (MMA) monomer was added to obtain an injectable homogeneous reaction solution. The viscous solution was heated and cured at temperatures up to 150 °C to obtain colorless, transparent hybrid materials in which the MMA monomer was polymerized into polymethyl methacrylate (PMMA). The PMMA hybrid has a network structure constructed from siloxane (Si-O-Si) bonds, which is also bonded to titanium (Si-O-Ti) via these bonds. The 3D network structure imparts high tensile strength, flexibility, and toughness to the hybrid. The dimethylsiloxane backbone and presence of alcohol groups in the side chains endow the hybrid material with good oxygen permeability, good wettability to tears, and excellent resistance to protein adhesion.

Organic–inorganic hybrid materials prepared using the sol-gel method have evolved from mere blends of organic polymers and inorganic substances to materials with new functional properties that are neither organic polymers nor inorganic substances due to their combination on a molecular level. Recently, the hybridization of many organic polymers and inorganic materials using the sol-gel method has been extensively carried out and the sol-gel method has attracted considerable interest as a technology used for the development of new materials. As mentioned beforehand, the 3D network structure is very important in regard to the physical properties of hybrid materials, and the silane coupling agents described in the next section significantly contribute to the creation of these 3D structures in a highly controlled manner.

The preparation of hybrid materials comprising organic polymers and metal oxides using the sol-gel method from the perspective of the silane coupling agents used is presented in the next section. The formation of cross-linking between the organic and inorganic phases by the silane coupling agents, the historical development of high-performance organic–inorganic hybrid materials, and the new adverse effects of hybridization are also described.

3. Creation of Nano-Hybrids Using Silane Coupling Agents

Recently, the preparation of organic–inorganic hybrid materials using the sol-gel method has attracted interest in the homogeneous dispersion or compositing of inorganic materials in nanometer size in a polymer matrix. However, the nanometer size of these particles results in high interfacial free energy on the surfaces of particles and the particles aggregate to reduce this free energy. In this process, small gaps are created between the agglomerating microparticles. As the polymer chains cannot penetrate these gaps, a difference in the osmotic pressure occurs between the inside and outside of the polymer, resulting in weak agglomeration [26,27] due to the solvent depletion effect [28]. This leads to microscopic phase separation and heterogeneous dispersion in many hybrid materials. The particle surface potential is important for the stable dispersion of the particles. This implies that the dispersion can be controlled using the appropriate design of the surfaces of particles.

The preparation of organic–inorganic hybrid materials using the sol-gel method has been extensively studied by Chujo et al. [29–33]. Using the sol-gel method, organic–inorganic hybrid materials are mainly prepared utilizing the following methods: **(i)** mixing and composite formation of the pre-synthesized organic polymers and inorganic materials, **(ii)** synthesis and composite formation of the organic polymers in the presence of the inorganic materials, and **(iii)** in situ preparation and composite formation of the inorganic nanoparticles in a polymer matrix depending on the type of organic polymer or inorganic material used and the expected material properties. These techniques have been used where appropriate. Recently, method **(iii)** has become the standard because hybridization on a molecular level is often achieved. When preparing silica via the sol-gel method, a silicon alkoxide, such as tetraethoxysilane [$Si(OC_2H_5)_4$; TEOS] or tetramethoxysilane [$Si(OCH_3)_4$; TMOS], is often used as the precursor. Hydrolysis and polycondensation of TEOS forms silica with a 3D network structure constructed via Si-O-Si bonds, as shown in Figure 3 [34].

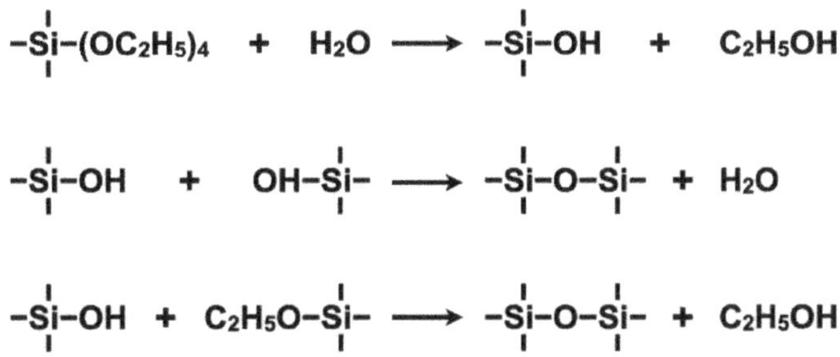

Figure 3. Key chemical reactions for the hydrolysis and polycondensation of tetraethyl orthosilicate (TEOS) under acidic or basic conditions. This is the central basic reaction in the sol-gel method, from which many functional hybrids are prepared.

Nanometer-sized silica can be dispersed in situ in a polymer matrix when TEOS is mixed with an organic polymer and subjected to the sol-gel reaction. Metal oxide nanoparticles with 3D network structures constructed by M-O-M bonds can be synthesized in situ by changing the starting material from Si to a metal (M) alkoxide [$M(OR)_n$; M = Ti, W, Cu, Mg, etc.; R = CH_3, C_2H_5, C_3H_7, and C_4H_9] [35,36]. According to Sanchez et al. [37], hybrids are classified as Class I or Class II based on the bonding state of the organic and inorganic materials. Class I refers to hybrids in which the inorganic substances are dispersed and combined in the polymer matrix via weak interactions, such as van der Waals forces, hydrogen bonds, and electrostatic forces, whereas Class II refers to hybrids in which the organic polymers and inorganic substances are combined via strong bonds, such as covalent and coordination bonds. Metal–carbon bonds are susceptible to hydrolysis and the bond between the two atoms is weak. Hence, silicon–carbon bonds, which are less susceptible to hydrolysis, are often utilized in Class II hybrids, and organic groups and silane coupling agents incorporating silicon–carbon bonds are often used in these organic–inorganic hybrid materials. In particular, the polymer side chains and termini are often modified, substituted, or copolymerized with silane coupling agents in hybrid materials comprising organic polymers and inorganic materials [38–40]. Furthermore, silane coupling agents not only improve the dispersibility of the nanoparticles, but also have excellent features that easily form Si-O-Si or Si-O-M bonds and 3D network structures [41] between the organic and inorganic phases.

The 3D network structure not only gives the hybrid material flexibility, but also contributes to the weather resistance of the material. One of the general-purpose polymers is polyurethane (PU), a generic term for polymers with urethane (-NHCOO-) groups in the main chain structure, obtained by the polyaddition reaction of polyhydric alcohols and isocyanates. Although PU exhibits excellent rubber elasticity due to the pseudo-network structure induced by isocyanates, aromatic isocyanates are vulnerable to light exposure, causing yellowing and oxidative degradation. Ikake et al. [42,43] focused on the network structure of PU and constructed 3D networks with high photostability by using a cross-linking agent instead of aromatic isocyanates. The properties of PU elastomers derive from this network structure, which also offers the advantage that the cross-linking density can be easily adjusted by adding appropriate amounts of metal alkoxides. Both ends of the polyol were modified with silane coupling agents, and titanium tetraisopropoxide was added to generate titania nanoparticles in situ, forming a 3D network within the PU matrix. The silane coupling agent enabled the titania nanoparticles to be highly dispersed in the PU matrix, resulting in hybrids with excellent transparency in the visible light range and UV-protection properties without yellowing. Furthermore, by changing the type of metal alkoxide used, various properties can be imparted to the PU. For example, copper(II) ethoxide (CuOEt) can be used to produce CuO nanoparticles, which do not contribute to the network structure because CuOEt is bifunctional. However, the PU matrix is reinforced by the Si-O-Si bonds originating from the coupling agent and the CuO nanoparticles dispersed in its network mesh (Class I hybrid). This effectively utilizes the band energy of CuO, making the PU hybrid material transparent in the visible light region, but able to block near-infrared (heat) radiation. In the case of tungsten trioxide (WO_3), which exhibits photochromic properties, WO_3 forms a tungsten bronze structure [44] upon UV irradiation, giving it a bright bronze-blue color. However, the resulting hybrid properties considerably depend on the method used to form the WO_3 composite. The choice of the WO_3 precursor is important because the use of tungsten alkoxide produces Class II hybrids, which do not exhibit photochromic properties. Conversely, Class I hybrids are produced when sodium tungstate(VI) dihydrate is used as the precursor and a WO_3 solution in which the Na-type is changed to H-type via a cation exchange resin is used. The resulting hybrids show good photochromic properties [45,46]. Leaustic et al. [47] showed that UV irradiation of an aqueous suspension of WO_3 formed tungsten bronze structures using electron spin resonance spectroscopy. Colton et al. [48] also showed that the d-orbital electrons of tungsten were excited by UV light causing a transition from W(VI) to W(V) with a decrease in energy of 1.2 eV using the X-ray photoelectron spectra of amorphous WO_3 powder samples. In Class II hybrids, the d-orbital of tungsten contributes to the bonding interactions, which may explain why the transition is less likely to occur and why photochromic properties were not exhibited. This is an example of the effect of the starting material used in the sol-gel method on the materials' properties.

Furthermore, the sol-gel method is useful for preparing thermosetting resins. The hybridization of polyimide (PI) with silica using the sol-gel method was reported by Morikawa et al. [49] in 1992. Tetraethoxysilane (TEOS) was added to polyamide acid (PAA) to obtain silica particles via in situ thermal cross-linking and dispersed in the PI matrix. The silica content was low because the hybrid is a Class I direct dispersion with a uniform dispersion achieved at < 8 wt.%; the dispersion was non-uniform at higher silica contents. The PI hybrid was completed upon cyclization of the PAA end during the thermal imidization reaction; the effect of modifying both ends of PAA with a silane coupling agent on the thermal imidization reaction was not determined and is unknown. In 1997, Srinivasan et al. [50] synthesized and characterized PAA oligomers bearing trimethoxysilyl groups introduced upon modification by a silane coupling agent. The oligomers were then thermally cured to form a cross-linked network with $T_g > 450$ °C and a thermal decomposition temperature of 500 °C. The PAA oligomers show excellent performance as low-viscosity precursors to cross-linked PI networks. Chan et al. [51] reported in 2002 that both ends of PAA can be modified with 3-aminopropyltrimethoxysilane (APrTMOS) and TEOS added to prepare PI-Silica hybrid materials. Silica was homogeneously dispersed by the silane coupling agent and the silica content in the hybrid was increased to 54.9 wt.%. At low silica contents, the hybrids form a 3D network structure constructed by Si-O-Si bonds, whereas, at higher silica contents, more Si-OH residues are formed upon hydrolysis of the alkoxy groups in APrTMOS or TMOS.

In the field of electronic materials, in addition to electrical properties, resins with excellent heat resistance and low coefficient of thermal expansion are used, especially thermosetting resins. Polyimide (PI), for example, is an example of such a resin, and we expect that the application range of PI will be further expanded by adding heat dissipation properties to PI. We have synthesized PI consisting of 1, 2, 3, 4-cyclobutane tetracarboxylic dianhydride (CBDA) and 2, 2-bis[4-(4-aminophenoxy)phenyl]propane (BAPP) [52]. The PI matrix is not heat-dissipating by itself, so new PI hybrids can be fabricated by adding heat-dissipating inorganic materials such as magnesia (MgO). In 2018, Hara et al. [53] prepared a PI/MgO hybrid material with MgO finely dispersed in a PI matrix cross-linked with APrTMOS. Magnesium ethoxide (MgOEt) was used as a precursor for MgO and the MgO particles prepared in situ using the sol-gel reaction of APrTMOS-terminated PAA and MgOEt. MgO particles have high thermal diffusivity. The dispersion of MgO particles in the matrix creates thermal conduction pathways in the direction of the thickness of the hybrid film. In the hybrids prepared with 20 wt.% of the MgO hybrids, the thermal diffusivity and thermal conductivity values were approximately twofold higher than as those of PI alone.

Currently, there is a wide variety of silane coupling agents; however, the polymer matrix and type of metal alkoxide that can be dispersed depend on the coupling agent used. In some cases, hybrid properties can be improved by promoting the formation of crosslinks in the hybrid material, whereas, in other cases, the hybrid properties can be utilized using nanodispersed inorganic materials. Therefore, it is important to select an appropriate structure, such as a 3D network, according to the desired functionality required for the desired application. In the sol-gel method, a combination of several metal alkoxides can impart various properties to the material. Conversely, an increase in the number of inorganic components in the polymer matrix can affect the conformation of the polymer chains, and although the material gains new functionalities, it also loses the original properties of the organic polymer, such as formability, flexibility, and toughness.

Subsequently, the interactions formed between the polymer matrix and inorganic materials in the hybrid material, the particulate dispersion, and composite formation in the hybrid material derived from these interactions need to be considered.

4. Birth of Ionic Liquids, from Designer Solvents to Functional Ionic Liquids

Recently, ionic liquids have attracted considerable attention as a "third solvent," similar to organic solvents and water. Ionic liquids have a long history and have been the subject of numerous review articles [54–60]. They are generally defined as salts with melting points below 100 °C. The first ionic liquids were molten salts reported by Gabriel and Weiner [61] in 1888. In 1914, Walden [62] reported that ethylammonium nitrate is a molten salt with a melting point of 12 °C. This is the original form of today's ionic liquids. They have subsequently attracted attention as functional liquids, mainly in the field of electrochemistry, and considerable research attention has been turned on ionic liquids with melting points near room temperature from the viewpoint of their convenience. Ionic liquids, which are also known as "designer solvents," are available in many varieties due to the number of cation and anion species available. As organic ions are also used, there is a high degree of freedom in terms of their molecular design. However, the complex interactions formed between ions and high ionic concentrations make it difficult to predict the properties of ionic liquids.

In 1992, Wilkes et al. [63] developed an ionic liquid using BF_4^- as the anion, which was found to be a room-temperature molten salt that is more stable to water and air than conventional ionic liquids. Since then, the research interest in ionic liquids and number of reports have increased. The excellent properties of ionic liquids as substances (salts) have been utilized in a wide range of practical applications, such as solar cell electrolytes [64], antistatic agents [65], lubricants [66], and antibacterial agents [67]. In addition, ionic liquids are not only used as solvents, but also as additives and functional liquids. Recently, ionic liquids have been used as functional materials, such as magnetic ionic liquids [68–71].

Ionic liquids can dissolve solutes that are not soluble in common organic solvents. A milestone was reported in 2002 by Swatloski et al. [72], who used an ionic liquid containing 1-butyl-3-methylimidazolium cations ($[C_{4mim}]^+$) to dissolve cellulose, a natural polymer. Fukaya et al. [73] clarified the correlation between the structure and polarity of ionic liquids and reported that a group of ionic liquids using various phosphoric acid derivatives as the anion have high polarity (evaluated using the Kamlet–Taft parameter [74]), which were stable and dissolved cellulose without heating. In particular, cellulose forms strong intra- and intermolecular hydrogen bonds. Therefore, it is advisable to design ionic liquids with high hydrogen bond-donating properties (donor type; the b-value used in the Kamlet-Taft parameter), which are effective for hydrogen bond breaking (relaxation) [75]. Abe et al. [76] reported that ionic liquids selectively dissolve cellulose when dissolved in an untreated biomass resource, such as bran. Bran is a macromolecular complex comprising cellulose, hemicellulose, and lignin, in which lignin forms a 3D network structure [75]. Therefore, lignin is not easily dissolved by the high polarity alone, but by appropriately controlling the polarity of the ionic liquid so that only cellulose can be isolated and extracted from bran. In other words, the design of the ionic liquid showed that it was highly selective for penetration into the bran matrix.

The affinity of ionic liquids toward polymeric materials is not limited to natural polymers such as cellulose; they also exhibit affinity toward synthetic polymers. The high compatibility and selectivity of ionic liquids with polymers can lead to the creation of new materials using ionic liquids as solvents and/or additives in the polymer matrix. Ueki et al. [77] reported the diverse compatibilities of various polymers using a hydrophobic ionic liquid ($[C_{2min}][NTf_2]$). PMMA is miscible with $[C_{2min}][NTf_2]$ in any ratio and the in situ polymerization of the ionic liquid with MMA and a crosslinker enables the preparation of gels loaded with the ionic liquid. Aqueous solutions of poly(N-isopropylacrylamide) (PNIPAm) have a lower critical solution temperature (LCST) [78] of ~304 K, which is also observed in the volume phase transition of PNIPAm gels. By changing the solvent used in this system to $[C_{2min}][NTf_2]$, the phase separation behavior follows a completely opposite trend, with PNIPAm + $[C_{2min}][NTf_2]$ exhibiting an upper critical solution temperature (UCST). Conversely, polybenzyl methacrylate (PBnMA) shows an LCST that is soluble at low temperature, but insoluble at higher temperature. Scott et al. [79] reported that the addition of an ionic liquid ($[C_{4min}][PF_6]$) to PMMA, a polymeric material with a high glass transition temperature (T_g), acts as a plasticizer to lower the T_g. It was reported that the addition of $[C_{4min}][PF_6]$) to the PMMA matrix improves the thermal stability of PMMA by imparting the properties of the ionic liquid to PMMA.

5. Ionic Liquids/Collaborations with Hybrid Materials

The addition of inorganic substances to polymers significantly reduces their polymer-specific properties, such as formability and toughness. This is because the dispersion of inorganic substances on a nanometer scale results in strong interactions (e.g., hydrogen bonds and coordination bonds) between the inorganic components and the polymer, which unintentionally increases the cross-linking density of the matrix. However, if the interaction between inorganic components and polymers is completely inhibited, the freely behaving inorganic components will aggregate in the matrix. Therefore, the cross-linking density of the inorganic components and polymers must be controlled.

Hara [80] synthesized several ionic liquids and investigated the dispersion and stability of inorganic particles by ionic liquids. In general, the Hildebrand solubility parameter (SP value) is used as a guide when selecting a solvent for a polymer. However, the SP value concept is only valid for systems with weak interactions, such as London dispersion forces, and is often not applicable in highly polar solvents. In the case of polymers, the solubility differs depending on the polymer concentration and molecular weight distribution. In addition, the estimation of the SP values of ionic liquids is difficult because they are mainly based on Coulombic interactions, which often do not match the actual system. In such a situation, we found an ionic liquid, tetrabutylphosphonium chloride (TBPC; P[4,4,4,4]Cl), which specifically inhibited the hydrogen bonds formed between the polymer and inorganic components. When nanoparticles, such as titania, are added to PMMA, the hydrogen bonding between the C=O groups in PMMA and the OH groups present on the titania surface is enhanced. Hara et al. [81] reported that upon loading TBPC onto a hybrid material, only the hydrogen bonds were inhibited, and the hybrid film became flexible (Figure 4).

Due to its high affinity for PMMA, TBPC has no effect on the transparency in the visible light region. The cross-linking density of the PMMA hybrid material can also be adjusted by simply adjusting the MSi content; the addition of TBPC reduced the cross-linking density due to hydrogen bonding, resulting in a significant increase in the shape-memory performance (fourfold higher deformation rate) and toughness (28-fold higher) when compared to the hybrid prepared without TBPC. TBPC also reduced the cross-linking density of the material. Furthermore, the inhibitory effect of TBPC on the hydrogen bonding and thermal dissociation of the matrix can be controlled in the presence of TBPC [82].

The incorporation of titania and thermally dissociable carboxyl groups into the network structure of the hybrid can be used to prepare hybrid materials with reversible cross-linking, in which the cross-linked structure is retained at low temperature and dissociated upon increasing temperature; the cross-linked structure is reformed upon decreasing the temperature (Figure 5). In other words, a melt-moldable network polymer was created for the first time using organic–inorganic hybrid materials prepared via the sol-gel method.

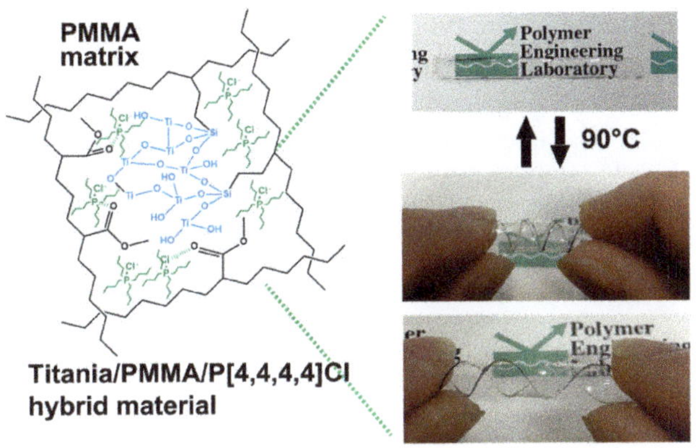

Figure 4. Schematic representation of the hybrid material composed of poly(methyl methacrylate) (PMMA), titania, and tetrabutylphosphonium chloride (TBPC). The transparency, toughness, and shape memory of the PMMA/titania hybrid are improved.

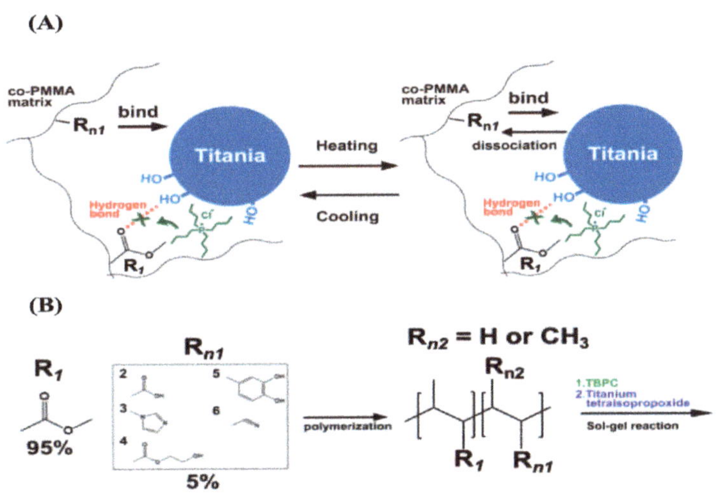

Figure 5. (**A**) TBPC inhibits hydrogen bonding between the carbonyl group of MMA (R1) and the hydroxy groups on the surface of titania particles; the functional group (Rn) of the other candidate monomer is thermally reversibly bound and unbound in the PMMA copolymer/titania hybrid material. (**B**) Monomers with functional groups capable of bonding to the surface of titania particles; the functional group (Rn) of the other candidate monomer is thermally reversibly bound and unbound in the PMMA copolymer/titania hybrid material. Monomers with functional groups capable of radical polymerization: R2 is MA, R3 is VI, R4 is HE, R5 is DMA, and R6 is AN. From ref. [82]. Reprinted with permission from Ref. [82]. Copyright 2020 American Chemical Society.

Notably, the selectivity of the ionic liquid to act on the polar sites in the PMMA matrix enables the hybrid material to be melt-molded. Furthermore, the original properties of the matrix, such as the toughness and high transparency of PMMA, were retained. Titania did not aggregate before and after melt-molding and retained its UV-shielding ability, and the shape-memory properties of PMMA were maintained due to the hydrogen bond-inhibiting and plasticizing effects of TBPC. In other words, ionic liquids have the effect of balancing the basic properties of the constituents. This has enabled hybrid materials to evolve into highly functional materials that combine the superior properties of their incorporated components.

6. Creation of Functional Hybrids in the Presence of Ionic Liquids

The functional manifestations of hybrid materials prepared using the sol-gel method have been described. It is clear that the hybrid properties are highly related to the network structure of the matrix and dispersibility of the inorganic material. Here, we discuss the future prospects for the creation of functional organic–inorganic hybrid materials from the viewpoint of their dispersion in the matrix. The durability of functional nanoparticles in the matrix is particularly important for the preparation of functional hybrid materials and, from the viewpoint of their functional properties, the nature of the composite materials will evolve from the homogeneous dispersion systems currently used to dispersion-controlled systems in the future. Aida et al. [83] reported that UV irradiation of titania photoinduced a switch from Ti(IV) to Ti(III), and that the magnetic properties of titania switched from antimagnetic to paramagnetic. Despite its excellent magnetic properties, Ti(III) is easily oxidized in the presence of oxygen and cannot withstand long-term use, which is a new problem. Hara et al. [84] exploited the packing effect of the cyclohexane skeleton to control the oxygen permeability. The lifetime of Ti(III) in polycarbonate (PC), which does not have a cyclohexane skeleton, was ~2 h, but by nanodispersing titania in PC with a cyclohexane skeleton, they succeeded in extending the lifetime to ~10 d, which is 120 times longer than previously obtained. Furthermore, by covering the PC/titania hybrid film with a photomask and irradiating it with ultraviolet light at 250 W for 2 h, only the areas exposed to light were photoinduced to change to Ti(III) and become paramagnetic. The film is expected to be used as a flexible magnetic lithography substrate because it enables flexible patterning of the magnetic materials and magnetization (Figure 6).

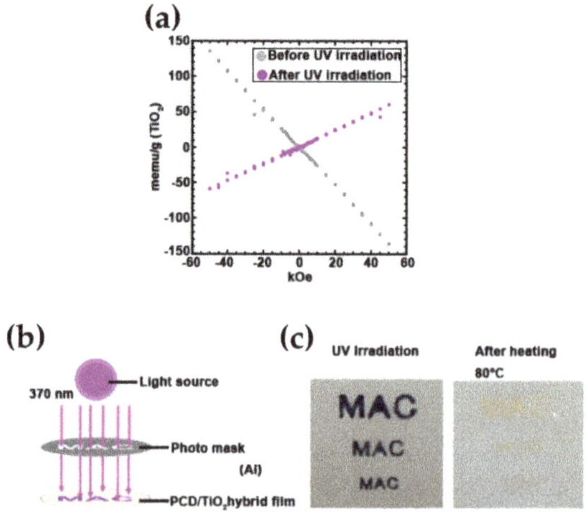

Figure 6. (**a**) Superconducting quantum interference device (SQUID) magnetometry measurements of PC100: before (grey plot) and after (purple plot) UV irradiation. (**b**) Image of the photolithography of PC100 using a UV light source. (**c**) Photograph of PC100 upon photolithography and heat treatment.

However, the sol-gel method also faced the problem that the high dispersion and random orientation of titania did not allow a sufficient magnetic susceptibility to be achieved for use as a magnetic lithography substrate. Therefore, it is necessary to develop new substrates with aligned Ti(III) magnetic domains. In other words, the magnetic nanoparticles require the development of technology to control the phase separation process during the dispersion of the nanoparticles, i.e., the magnetic properties need to be improved by clustering the nanoparticles together. Ikake et al. [85] have developed a method to control the phase separation by combining the photoinduced phase transition and shape-memory processes (Figure 7).

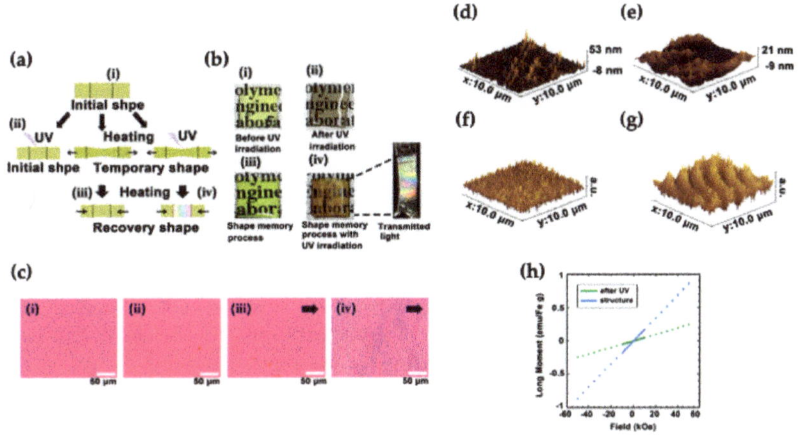

Figure 7. (**a**) Preparation of $FeCl_3$-oriented hybrid films. (**b**) Photographs of Ti15TB40Fe5 after the (iii) shape memory and (iv) UV irradiation and shape-memory processes. (**c**) Photographs of Ti15TB40Fe5 after each process and state, with the image (iv) expanded to show a photograph of the transmitted light. The white bar represents 50 mM. AFM images of each state of Ti15TB40Fe5 after (**d**) UV irradiation and (**e**) UV irradiation and the shape-memory process. MFM images of each state of Ti15TB40Fe5 after (**f**) UV irradiation and (**g**) UV irradiation and the shape-memory process. (**h**) Magnetometry measurements of Ti15TB40Fe5 after UV irradiation (green plot) and the shape-memory process (blue plot). From ref. [85]. Adapted with permission from Ref. [85]. Copyright 2022, The Royal Society of Chemistry.

We found that UV irradiation of the highly dispersed magnetic ionic liquid TBP [$FeCl_4$] in a polymer matrix causes phase separation of the $FeCl_3$ particles, which coincidentally align along the polymer chain when the polymer chain is uniaxially stretched. Hybrids with aligned particles exhibited a vivid color. When the hybrid material was irradiated with UV light without the shape-memory process, the hybrid material was colored brown by the $FeCl_3$, but the magnetic susceptibility did not increase. By incorporating the aforementioned shape-memory property, the magnetic nanoparticles were not only aligned inside the hybrid material, but the magnetic domains were also aligned, and the magnetic susceptibility increased threefold when compared with the hybrid material in which $FeCl_3$ was uniformly dispersed. Furthermore, partial patterning was possible when the hybrid film was covered with a photomask and irradiated with ultraviolet light and the shape-memory process carried out with only the UV-irradiated areas showing paramagnetism. This is the first reported attempt toward the preparation and morphological control of organic–inorganic hybrid materials via the combination of a photoinduced phase transition and shape-memory process. This knowledge is of great importance in the creation of organic–inorganic hybrid materials. In contrast to conventional hybrid materials, which focus on the highly homogeneous dispersion of inorganic particles in a polymer matrix, in this system, the hybrid material is prepared by promoting the microphase-separated structure of the particles. In particular, the use of the shape-memory property, which is also a characteristic of network polymers, and the use of the hybrid structure itself as a reaction field, are very unique, even though they are the product of chance.

7. Conclusions

This paper summarizes the creation of organic–inorganic hybrid materials as a historical fact. Various techniques and dispersion types were discussed, including in situ formation of inorganic particles using metal alkoxides, physical dispersion and dispersion by chemical bonding, but all have in common the creation of hybrid materials with a highly homogeneous dispersion in a polymer matrix. It is precisely for this reason that organic–inorganic hybrid materials have developed into essential materials with high performance and functionality. From a simple blend of organic and inorganic materials, a completely new material has emerged that is neither organic nor inorganic. Recently, the incorporation of ionic liquids into the field of organic–inorganic hybrid materials has revealed that ionic liquids are deeply involved in the control of the microstructure of hybrid materials, from their role as solvents to their selective action on polar sites. The use of functional ionic liquids as hybrid raw materials has also dramatically improved the dispersibility of inorganic nanoparticles down to the single-particle level, enabling the creation of highly nanodispersed hybrid materials. Functional ionic liquids have made it possible to functionally control 3D network structures and nanodispersion of fine particles. However, there are still unexplored areas. Biomimetic hybrid materials such as those described above are still some way off. Further research is awaited into the formation of dense structures corresponding to their functions and the control of their structure. Biomimetic hybrid materials are fabricated using only natural energy. Methods that actively utilize natural energy for the creation of new materials will be increasingly needed in the future. Only by solving these problems will organic–inorganic hybrid materials be transformed into novel functional materials that go beyond the composite rule.

Author Contributions: Conceptualization, S.H.; writing—original draft preparation, H.I.; writing—review and editing, S.H.; visualization, S.S.; supervision, S.S.; project administration, S.H.; funding acquisition, S.S. All authors have read and agreed to the published version of the manuscript.

Funding: This research received no external funding.

Institutional Review Board Statement: Not applicable.

Informed Consent Statement: Not applicable.

Data Availability Statement: Not applicable.

Acknowledgments: The technical and financial support provided by Nihon University is gratefully acknowledged.

Conflicts of Interest: The authors declare no conflict of interest.

References

1. Schubert, U.; Hüsing, N.; Lorenz, A. Hybrid inorganic-organic materials by sol-gel processing of organofunctional metal alkoxides. *Chem. Mater.* **1995**, *7*, 2010–2027. [CrossRef]
2. Judeinstein, P.; Sanchez, C. Hybrid organic-inorganic materials: A land of multidisciplinarity. *J. Mater. Chem.* **1996**, *6*, 511–525. [CrossRef]
3. Wen, J.; Wilkes, G.L. Organic/inorganic hybrid network materials by the sol-gel approach. *Chem. Mater.* **1996**, *8*, 1667–1681. [CrossRef]
4. Schottner, G. Hybrid sol-gel derived polymers: Applications of multifunctional materials. *Chem. Mater.* **2001**, *13*, 3422–3435. [CrossRef]
5. Sanchez, C.; Julián, B.; Belleville, P.; Popall, M. Applications of hybrid organic-inorganic nanocomposites. *J. Mater. Chem.* **2005**, *15*, 3559–3592. [CrossRef]
6. Mammeri, F.; Bourhis, E.L.; Rozesa, L.; Sanchez, C. Mechanical properties of hybrid organic-inorganic materials. *J. Mater. Chem.* **2005**, *15*, 3787–3811. [CrossRef]
7. Caseri, W.R. Nanocomposites of polymers and inorganic particles: Preparation, structure, and properties. *Mater. Sci. Technol.* **2006**, *22*, 807–817. [CrossRef]

8. Mir, S.H.; Nagahara, L.A.; Thundat, T.; Mokarian-Tabari, P.; Furukawa, H.; Khosla, A. Organic-inorganic hybrid functional materials: An integrated platform for applied technologies. *J. Electrochem. Soc.* **2018**, *165*, B3137–B3156. [CrossRef]
9. Ma, J.F.; Tamai, K.; Yamaji, N.; Mitani, N.; Konishi, S.; Katsuhara, M.; Ishiguro, M.; Murata, Y.; Yano, M. A silicon transporter in rice. *Nature* **2006**, *440*, 688–691. [CrossRef]
10. Ma, J.F.; Yamaji, N.; Mitani, N.; Tamai, K.; Konishi, S.; Fujiwara, T.; Katsuhara, M.; Yano, M. An efflux transporter of silicon in rice. *Nature* **2007**, *448*, 209–212. [CrossRef]
11. Bruet, B.J.F.; Qi, H.J.; Boyce, M.C.; Panas, R.; Tai, K.; Frick, L.; Ortiz, C. Nanoscale morphology and indentation of individual nacre tablets from gastropod mollusk *Trochus niloticus*. *J. Mater. Res.* **2005**, *20*, 2400–2419. [CrossRef]
12. Zhang, S.; Chen, Y. Nanofabrication and coloration study of artificial Morpho butterfly wings with aligned lamellae layers. *Sci. Rep.* **2015**, *5*, 16637. [CrossRef] [PubMed]
13. Dumanli, A.G.; Savin, T. Recent advances in the biomimicry of structural colors. *Chem. Soc. Rev.* **2016**, *45*, 6698–6724. [CrossRef]
14. Yin, Z.; Hannard, F.; Barthelat, F. Impact-resistant nacre-like transparent materials. *Science* **2019**, *364*, 1260–1263. [CrossRef]
15. Sugawara, A.; Ishii, T.; Kato, T. Self-organized calcium carbonate with regular surface-relief structures. *Angew. Chem. Int. Ed.* **2003**, *42*, 5299–5303. [CrossRef] [PubMed]
16. Nakayama, M.; Kajiyama, S.; Kumamoto, A.; Nishimura, T.; Ikuhara, Y.; Yamato, M.; Kato, T. Stimuli-responsive hydroxyapatite liquid crystal with macroscopically controllable ordering and magneto-optical functions. *Nat. Commun.* **2018**, *9*, 1–9. [CrossRef] [PubMed]
17. Sun, N.; Jia, Y.; Wang, C.; Xia, J.; Dai, L.; Li, J. Dopamine-mediated biomineralization of calcium phosphate as a strategy to facilely synthesize functionalized hybrids. *J. Phys. Chem. Lett.* **2021**, *12*, 10235–10241. [CrossRef] [PubMed]
18. Yeh, J.M.; Huang, K.Y.; Dai, C.F.; Chand, B.G.; Weng, C.J. Organic-acid-catalyzed sol-gel route for preparing poly(methyl methacrylate)-silica methacrylate)-silica hybrid materials. *J. Appl. Polym. Sci.* **2008**, *110*, 2108–2114. [CrossRef]
19. Chen, C.; Sun, X.; Wang, F.; Zhang, F.; Wang, H.; Shi, Z.; Cui, Z.; Zhang, D. Electro-optic modulator based on novel organic-inorganic hybrid nonlinear optical materials. *IEEE J. Quantum Electron.* **2012**, *48*, 61–66. [CrossRef]
20. Gangopadhyay, R.; De, A. Conducting polymer nanocomposites: A brief overview. *Chem. Mater.* **2000**, *12*, 608–622. [CrossRef]
21. Mackenzie, J.D. Structures and properties of ormosils. *J. Sol-Gel Sci. Technol.* **1994**, *2*, 81–86. [CrossRef]
22. Ebelmen, M. Sur les produits de la decomposition des espèces minérales de la famille des silicates. *Ann. Des Mines* **1845**, *7*, 3–66.
23. Mazdiyasni, K.S.; Dolloff, R.T.; Smith II, J.S. Preparation of high-purity submicron barium titanate powders. *J. Am. Ceram. Soc.* **1969**, *52*, 523–526. [CrossRef]
24. Dislich, H. New routes to multicomponent oxide glasses. *Angew. Chem. Int. Ed.* **1971**, *10*, 363–370. [CrossRef]
25. Philipp, G.; Schmidt, H. New materials for contact lenses prepared from Si- and Ti-alkoxides by the sol-gel process. *J. Non-Cryst. Solids* **1984**, *63*, 283–292. [CrossRef]
26. Scheutjens, J.M.; Fleer, G.J. Interaction between two adsorbed polymer layers. *Macromolecules* **1985**, *18*, 1882–1900. [CrossRef]
27. Fleer, G.J.; Scheutjens, J.M.H.M. Interaction between adsorbed layers of macromolecules. *J. Colloid Interface Sci.* **1986**, *111*, 504–515. [CrossRef]
28. Oosawa, F.; Asakura, S. Surface tension of high polymer solutions. *J. Chem. Phys.* **1954**, *22*, 1255–1256. [CrossRef]
29. Chujo, Y.; Saegusa, T. Organic polymer hybrids with silica gel formed by means of the sol-gel method. *Adv. Polym. Sci.* **1992**, *100*, 11–29.
30. Chujo, Y. Organic-inorganic polymer hybrids. *Polym. Mater.* **1996**, *6*, 4793–4798.
31. Chujo, Y. Organic-inorganic hybrid materials. *Curr. Opin. Solid State Mater. Sci.* **1996**, *1*, 806–811. [CrossRef]
32. Chujo, Y.; Tamaki, R. New preparation methods for organic-inorganic polymer hybrids. *MRS Bull.* **2001**, *26*, 389–392. [CrossRef]
33. Ogoshi, T.; Chujo, Y. Organic-inorganic polymer hybrids prepared by the sol-gel method. *Compos. Interfaces* **2005**, *11*, 539–566. [CrossRef]
34. Aelion, R.; Loebel, A.; Eirich, F. Hydrolysis of ethyl silicate. *J. Am. Chem. Soc.* **1950**, *72*, 5705–5712. [CrossRef]
35. Boyd, T. Preparation and properties of esters of polyorthotitanic acid. *J. Polym. Sci.* **1951**, *7*, 591–602. [CrossRef]
36. Bradley, D.C.; Carter, D.G. Metal oxide alkoxide polymers, part I. The hydrolysis of some primary alkoxides of zirconium. *Can. J. Chem.* **1961**, *39*, 1434–1443. [CrossRef]
37. Sanchez, C.; Ribot, F. Design of hybrid organic-inorganic materials synthesized via sol-gel chemistry. *New J. Chem.* **1994**, *18*, 1007–1047.
38. Schmidt, H.K. Organically modified silicates as inorganic-organic polymers. *ACS Symp. Ser.* **1988**, *360*, 333–344.
39. Schmidt, H. Inorganic-organic composites by sol-gel techniques. *Mater. Res. Soc. Symp. Proc.* **1989**, *171*, 3–13. [CrossRef]
40. Schmidt, H.K. Aspects of chemistry and chemical processing of organically modified ceramics. *Mater. Res. Soc. Symp. Proc.* **1990**, *180*, 961–973. [CrossRef]
41. Schmidt, H.; Krug, H. Sol-gel-based inorganic-organic composite materials. *ACS Symp. Ser.* **1994**, *572*, 183–194.
42. Ikake, H.; Hashimoto, W.; Kawai, M.; Kamiyama, N.; Shimizu, S.; Kurita, K.; Yano, S. Optical properties and microstructures of poly(tetramethylene)/titania hybrid materials with protection against UV-rays. *Kobunshi Ronbunshu* **2002**, *59*, 539–547. [CrossRef]
43. Yada, K.; Tanaka, S.; Ikake, H.; Shimizu, S.; Kurita, K. Preparation and physical properties of poly(propylene glycol)/titania hybrid materials. *Kobunshi Ronbunshu* **2006**, *63*, 727–736. [CrossRef]
44. Deb, S.K. Optical and photoelectric properties and color centers in thin films of tungsten oxide. *Philos. Mag.* **1973**, *27*, 801–822.

45. Ikake, H.; Hashimoto, W.; Obara, T.; Kurita, K.; Yano, S. Photochromic properties and microstructures of poly(tetramethylene oxide)/tungsten trioxide hybrid materials. *Kobunshi Ronbunshu* **2000**, *57*, 376–382. [CrossRef]
46. Ikake, H.; Fukuda, Y.; Shimizu, S.; Kurita, K.; Yano, S. Microstructures of poly(ethylene glycol)/tungsten oxide hybrid films exhibiting photochromic properties. *Kobunshi Ronbunshu* **2002**, *59*, 608–615. [CrossRef]
47. Leaustic, A.; Babonneau, F.; Livage, J. Photoreactivity of WO_3 dispersions: Spin trapping and electron spin resonance detection of radical intermediates. *J. Phys. Chem.* **1986**, *90*, 4193–4198. [CrossRef]
48. Colton, R.J.; Guzman, A.M.; Rabalais, J.W. Photochromism and electrochromism in amorphous transition metal oxide films. *Acc. Chem. Res.* **1978**, *11*, 170–176. [CrossRef]
49. Morikawa, A.; Iyoko, Y.; Kakimoto, M.; Imai, Y. Preparation of a new class of polyimide-silica hybrid films by the sol-gel process. *Polym. J.* **1992**, *24*, 107–113. [CrossRef]
50. Srinivasan, S.A.; Hedrick, J.L.; Miller, R.D.; Pietro, R.D. Crosslinked networks based on trimethoxysilyl functionalized poly(amic ethyl ester) chain extendable oligomers. *Polymer* **1997**, *38*, 3129–3133. [CrossRef]
51. Chang, C.C.; Chen, W.C. Synthesis and optical properties of polyimide-silica hybrid thin films. *Chem. Mater.* **2002**, *14*, 4242–4248. [CrossRef]
52. Yamashiro, R.; Ikake, H.; Shimizu, S.; Kurita, K. Preparation of polyimide/silica hybrid materials. *Polym. Prepr. Jpn.* **2008**, *57*, 688.
53. Hara, S.; Ikake, H.; Shimizu, S. Features and applications of polyimide/magnesia hybrid films. *Jpn. Plast.* **2018**, *69*, 33–36.
54. Welton, T. Room-temperature ionic liquids: Solvents for synthesis and catalysis. *Chem. Rev.* **1999**, *99*, 2071–2083. [CrossRef]
55. Holbrey, J.D.; Seddon, K.R. Ionic Liquids. *Clean. Prod. Process.* **1999**, *1*, 223–236. [CrossRef]
56. Wasserscheid, P.; Keim, W. Ionic liquids—New solutions for transition metal catalysis. *Angew. Chem. Int. Ed.* **2000**, *39*, 3772–3789. [CrossRef]
57. Wilkes, J.S. A short history of ionic liquids—From molten salts to neoteric solvents. *Green Chem.* **2002**, *4*, 73–80. [CrossRef]
58. Rogers, R.D.; Seddon, K.R. Ionic liquids—Solvents of the future? *Science* **2003**, *302*, 792–793. [CrossRef]
59. Seddon, K.R. Ionic liquids—A taste of the future. *Nat. Mater.* **2003**, *2*, 363–365. [CrossRef]
60. Plechkova, N.V.; Seddon, K.R. Applications of ionic liquids in the chemical industry. *Chem. Soc. Rev.* **2008**, *37*, 123–150. [CrossRef]
61. Gabriel, S.; Weiner, J. Ueber einige Abkömmlinge des Propylamins. *Eur. J. Inorg. Chem.* **1888**, *21*, 2669–2679. [CrossRef]
62. Walden, V.P. Ueber die Molekulargrösse und elektrisele Leitfähigkeit einiger gesehmolzenen Salze. *Bull. l'Acad. Impér. Sci.* **1914**, *8*, 405–422.
63. Wilkes, J.S.; Zaworotko, M.J. Air and water stable 1-ethyl-3-methylimidazolium-based ionic liquids. *J. Chem. Soc. Chem. Commun.* **1992**, *13*, 965–967. [CrossRef]
64. Papageorgiou, N.; Athanassov, Y.; Armand, M.; Bonhôte, P.; Pettersson, H.; Azam, A.; Grätzel, M. The performance and stability of ambient temperature molten salts for solar cell applications. *J. Electrochem. Soc.* **1996**, *143*, 3099–3108. [CrossRef]
65. Cieniecka-Rosłonkiewicz, A.; Pernak, J.; Kubis-Feder, J.; Ramani, A.; Robertson, A.J.; Seddon, K.R. Synthesis, anti-microbial activities, and anti-electrostatic properties of phosphonium-based ionic liquids. *Green Chem.* **2005**, *7*, 855–862. [CrossRef]
66. Liu, X.Q.; Zhou, F.; Liang, Y.M.; Liu, W.M. Tribological performance of phosphonium-based ionic liquids for an aluminum-on-steel system and opinions on lubrication mechanism. *Wear* **2006**, *261*, 1174–1179. [CrossRef]
67. Qin, J.; Guo, J.; Xu, Q.; Zheng, Z.; Mao, H.; Yan, F. Synthesis of pyrrolidinium-type poly(ionic liquid) membranes for antibacterial applications. *ACS Appl. Mater. Interfaces* **2017**, *9*, 10504–10511. [CrossRef]
68. Hayashi, S.; Hamaguchi, H. Discovery of a magnetic ionic liquid [bmim]$FeCl_4$. *Chem. Lett.* **2004**, *33*, 1590–1591. [CrossRef]
69. Yoshida, Y.; Otsuka, A.; Saito, G.; Natsume, S.; Nishibori, E.; Takata, M.; Sakata, M.; Takahashi, M.; Yoko, T. Conducting and magnetic properties of 1-ethyl-3-methylimidazolium (EMI) salts containing paramagnetic irons: Liquids [EMI][M^{III} Cl_4] (M = Fe and $Fe_{0:5}Ga_{0:5}$) and solid $[EMI]_2[Fe^{II}Cl_4]$. *Bull. Chem. Soc. Jpn.* **2005**, *78*, 1921–1928. [CrossRef]
70. Xu, W.; Dai, Q.; Wang, Y.; Hu, X.; Xu, P.; Ni, R.; Meng, J. Creating magnetic ionic liquid-molecularly imprinted polymers for selective extraction of lysozyme. *RSC Adv.* **2018**, *8*, 21850–21856. [CrossRef] [PubMed]
71. Correia, D.M.; Fernandes, L.C.; García-Astrain, C.; Tariq, M.; Esperança, J.M.S.; Bermudez, V.Z.; Méndez, S.L. Magnetic ionic liquid/polymer composites: Tailoring physico-chemical properties by ionic liquid content and solvent evaporation temperature. *Compos. B Eng.* **2019**, *178*, 107516. [CrossRef]
72. Swatloski, R.P.; Spear, S.K.; Holbrey, J.D.; Rogers, R.D. Dissolution of cellulose with ionic liquids. *J. Am. Chem. Soc.* **2002**, *124*, 4974–4975. [CrossRef]
73. Fukaya, Y.; Hayashi, K.; Wada, M.; Ohno, H. Cellulose dissolution with polar ionic liquids under mild conditions: Required factors for anions. *Green Chem.* **2008**, *10*, 44–46. [CrossRef]
74. Kamlet, M.J.; Abboud, J.M.; Abraham, M.H.; Taft, R.W. Linear solvation energy relationships. 23. A comprehensive collection of the solvatochromic parameters, p*, α, and β, and some methods for simplifying the generalized solvatochromic equation. *J. Org. Chem.* **1983**, *48*, 2877–2887. [CrossRef]
75. Brandt, A.; Gräsvik, J.; Hallett, J.P.; Welton, T. Deconstruction of lignocellulosic biomass with ionic liquids. *Green Chem.* **2013**, *15*, 550–583. [CrossRef]
76. Abe, M.; Fukaya, Y.; Ohno, H. Extraction of polysaccharides from bran with phosphonate or phosphinate-derived ionic liquids under short mixing time and low temperature. *Green Chem.* **2010**, *12*, 1274–1280. [CrossRef]
77. Ueki, T.; Watanabe, M. Macromolecules in ionic liquids: Progress, challenges, and opportunities. *Macromolecules* **2008**, *41*, 3739–3749. [CrossRef]

78. Shimizu, S.; Kurita, K.; Furusaka, M. Segment-segment interactions of poly(N-isopropylacrylamide) in aqueous methanol solutions by using small-angle scattering. *Appl. Phys. A Mater. Sci. Process.* **2002**, *74*, S389–S391. [CrossRef]
79. Scott, M.P.; Brazel, C.S.; Benton, M.G.; Mays, J.W.; Holbrey, J.D.; Rogers, R.D. Application of ionic liquids as plasticizers for poly(methyl methacrylate)methacrylate). *Chem. Commun.* **2002**, *13*, 1370–1371. [CrossRef]
80. Hara, S.; Ikake, H.; Shimizu, S. Compound, Dispersant, Complex, Dispersion, and Method for Producing Complex. JP. Patent No. JP6736121, 5 August 2020.
81. Hara, S.; Ishizu, M.; Watanabe, S.; Kaneko, T.; Toyama, T.; Shimizu, S.; Ikake, H. Improvement of the transparency, mechanical, and shape memory properties of polymethylmethacrylate/titania hybrid films using tetrabutylphosphonium chloride. *Polym. Chem.* **2019**, *10*, 4761–4896. [CrossRef]
82. Hara, S.; Tomono, M.; Fukumoto, K.; Kubodera, M.; Kato, N.; Kaneko, T.; Toyama, T.; Shimizu, S.; Ikake, H. Melt-moldable copolymethacrylate/titania thermoreversible polymer networks with shape memory. *ACS Appl. Polym. Mater.* **2020**, *2*, 5654–5663. [CrossRef]
83. Wang, X.; Li, X.; Aya, S.; Araoka, F.; Ishida, Y.; Kikkawa, A.; Kriener, M.; Taguchi, Y.; Ebina, Y.; Sasaki, T.; et al. Reversible switching of the magnetic orientation of titanate nanosheets by photochemical reduction and autoxidation. *J. Am. Chem. Soc.* **2018**, *140*, 16396–16401. [CrossRef] [PubMed]
84. Hara, S.; Kurebayashi, S.; Sanae, G.; Watanabe, S.; Kaneko, T.; Toyama, T.; Shimizu, S.; Ikake, H. Polycarbonate/titania hybrid films with localized photo-induced magnetic-phase transition. *Nanomaterials* **2021**, *11*, 5. [CrossRef] [PubMed]
85. Ikake, H.; Hara, S.; Kurebayashi, S.; Kubodera, M.; Watanabe, S.; Hamada, K.; Shimizu, S. Development of a magnetic hybrid material capable of photoinduced phase separation of iron chloride by shape memory and photolithography. *J. Mater. Chem. C* **2022**, *10*, 7849–7856. [CrossRef]

polymers MDPI

Article

Effects of Different Concentrations of Arsine on the Synthesis and Final Properties of Polypropylene

Joaquín Hernández-Fernández [1,*], Yoleima Guerra [2], Esneyder Puello-Polo [3] and Edgar Marquez [4,*]

1 Chemistry Program, Department of Natural and Exact Sciences, San Pablo Campus, University of Cartagena, Cartagena 130015, Colombia
2 Centro de Investigación en Ciencias e Ingeniería, CECOPAT&A, Cartagena 131001, Colombia; yoleima.guerra@cecopat.com
3 Grupo de Investigación en Oxi/Hidrotratamiento Catalítico Y Nuevos Materiales, Programa de Química-Ciencias Básicas, Universidad del Atlántico, Puerto Colombia 081001, Colombia; esneyderpuello@mail.uniatlantico.edu.co
4 Grupo de Investigaciones en Química Y Biología, Departamento de Química Y Biología, Facultad de Ciencias Básicas, Universidad del Norte, Carrera 51B, Km 5, Vía Puerto Colombia, Barranquilla 081007, Colombia
* Correspondence: hernandez548@hotmail.com (J.H.-F.); ebrazon@uninorte.edu.co (E.M.); Tel.: +57-301-5624990 (J.H.-F.)

Abstract: This article studies the effects of arsine on the synthesis and thermal degradation of 4 samples of virgin polypropylene (PP-virgin) and proposes reaction mechanisms that allow understanding of its behaviour. Different points are monitored during the polypropylene synthesis to perform TGA, DSC, FT-IR, RDX, and MFI analyses later. The content of AsH_3 in polypropylene varies between 0.05 and 4.73 ppm, and of arsenic in virgin PP residues between 0.001 and 4.32 ppm for PP0 and PP10, increasing in fluidity index from 3.0 to 24.51. The origin of thermo-oxidative degradation is explained by the reaction mechanisms of the Molecule AsH_3 with the active titanium center of the ZN catalyst and the subsequent oxidation to form radical complexes. OO-AsH-$TiCl_4$-$MgCl_2$ and $(OO\text{-as-}OO)_2$ -$TiCl_4$-$MgCl_2$, which, by radical reactions, give rise to the formation of functional groups aldehyde, ketone, alcohol, carboxylic acid, CO, CO_2, PP-Polyol, PP-Polyether, and PP-Isopropylethers. These species caused the TG and DTG curves to increase degradation peaks in pp samples.

Keywords: arsine; ligands; polypropylene; catalyst; degradation

Citation: Hernández-Fernández, J.; Guerra, Y.; Puello-Polo, E.; Marquez, E. Effects of Different Concentrations of Arsine on the Synthesis and Final Properties of Polypropylene. *Polymers* **2022**, *14*, 3123. https://doi.org/10.3390/polym14153123

Academic Editors: Beata Podkościelna and Andrzej Puszka

Received: 20 June 2022
Accepted: 29 July 2022
Published: 31 July 2022

1. Introduction

Arsine (AsH_3) is a chemical compound of inorganic nature formed by an arsenic atom and three hydrogen atoms, and has a general structure ER_3 (E = As, N, P, Sb, Bi, and R = H, alkyl, aryl, halogens) and a tetrahedral geometry, with the pair of non-shared electrons located in one of its vertices, which can act as a soft Lewis base [1,2]. In addition, it can bind to transition metals (M), specifically those that have partially filled d-orbitals and with vacant s and p orbitals, and form M-ER_3 complexes using a covalent bond between the arsine and the metal [1,3]. The coordination of the metals of the d-block of these compounds is given two bonds at once: the donor bond σ and acceptor π. The donor bond σ is established between the pair of nonbonding electrons of the ER_3 ligand, belonging to the HOMO orbital of the ligand (Highest Occupied Molecular Orbital) and the vacant orbitals of the metal (LUMO, Lowest Unoccupied Molecular Orbital). On the other hand, for the π bond, an action called retro bonding occurs between the filled or half-filled orbitals of the metal (HOMO of the metal) and the anti-bonding σ orbitals of the E-R ligand and/or the empty d orbitals of the heteroatom (E) [4,5]. Arsine complexes for use in catalytic processes are of interest for their organometallic characteristics [6–10], such as their oxidation state, metal coordination number, and the ease by which they can form vacant coordination sites, among others, thereby allowing us to understand its reactivity and behaviour in catalytic systems [11,12]. The use of analysis techniques such as FTIR, X-ray fluorescence, and NMR are necessary for studying these interactions between ligands and metals so that the oxidation of compounds can be evaluated [4,13–15].

The effect of AsH_3 on the synthesis of PP is of importance since, in the polymerization stage of propylene, fundamental raw materials are required, such as Propylene, Ziegler catalyst Natta (ZN), triethylaluminium co-catalyst (TEAL), selectivity control people, hydrogen (H_2), and nitrogen (N_2) [16,17], where at least one can act as Lewis acid or have transition metals is its structure. PP synthesis begins with the initiation reaction, which is the activation of the double bond and the active site, followed by a polymerization stage and a polymerization completion stage [18]. The catalyst is titanium (IV) chloride ($TiCl_4$), which is a compound with a tendency to form different complexes with a wide diversity of ligands as a result of being a strong Lewis acid [19,20]. In the catalytic system used for the synthesis of PP, the effect of impurities such as this and other components, such as inhibitors of PP polymerization, have been examined and quantified [15,21–28]. In this sense, AsH_3 is an inhibitor of polymerization reactions. It is observed that its effect is irremediable to the catalytic system and its coordination reaction with the active center of Ti in the catalyst, as well as to the different surfaces of $MgCl_2$ and with the alkyl co-catalyst [29–31].

In this research, the synthesis of 4 PP is conducted, and specific concentrations of AsH_3 are dosed at each stage [32] to evaluate its impact on the efficiency of the catalyst. Regarding the PP production rate during the polymerization stage, the effect on the onset of thermal degradation of PP will be evaluated. The significant alterations in the physicochemical properties of PP are of great importance, and have not been observed in previous research. These must be studied so that adequate reaction mechanisms can be explained and proposed in order to help understand the effect of arsine.

2. Materials and Methods

2.1. Standards and Reagents

For the elaboration of this work, a fourth-generation spherical Ziegler-Natta catalyst with $MgCl_2$ support with 3.6% by weight of Ti was used; Diisobutyl phthalate (DIBP) as an in-house donor supplied by Sudchemie, Germany. It was used as a triethylaluminium co-catalyst (98% purity TEAL) from Merck, Germany, diluted in n-heptane. As an external donor, cyclohexyl methyl dimethoxysilane (CMDS) acquired from Merck, Germany, was used. Shazand Petrochemical, Iran, provided polymer-grade propylene. The hydrogen and nitrogen used had a purity of 99.999% [33,34].

We worked with AsH_3 at 99.999% purity and to guarantee the concentrations of 0.1 and 3.0 ppm in LPG balance, which were required for this research. A dilusor in line was used, which allowed for the creation of the mixtures of AsH_3 and LPG in the required proportions.

2.2. Polymerization Process

The PP samples were synthesized by polymerization of propylene in the gas phase with the use of a ZN catalyst [35]. The process is shown in Figure 1, and this process consists of a fluidized bed reactor, where nitrogen was initially fed to purge the equipment; subsequently, the propylene and hydrogen that provide the fluidization and absorb the heat of the reaction were fed the catalyst, TEAL, the selectivity control agent, and the nitrogen, in order to carry out the polymerization in discontinuous mode; the quantities are shown in Table 1. This process was conducted at 70 °C and 27 bar pressure. Gases leaving the reactor passed through a compressor to be transported to a heat exchanger where they were cooled and recirculated to the reactor. The resin obtained passed to a purge tower fed with nitrogen and steam to remove traces of hydrocarbons that left the system to obtain a virgin resin [17].

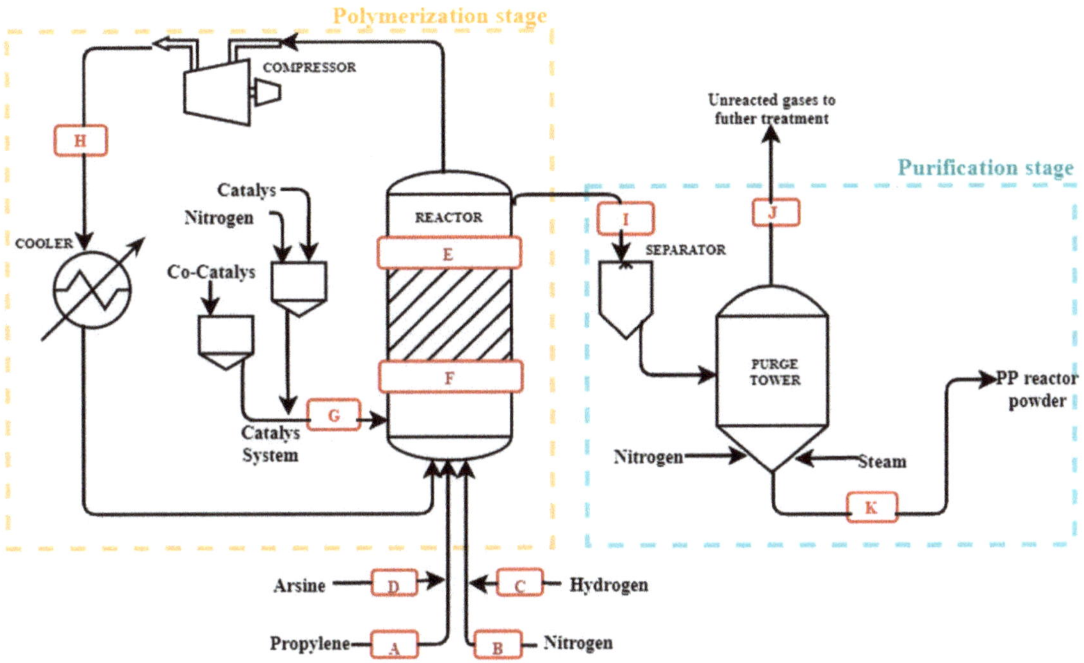

Figure 1. Diagram of the polymerization process and sampling points.

Table 1. Reagents for polymerization.

Run	Catalyst [kg/h]	Propylene [TM/h]	TEAl [1] [kg/h]	Hydrogen [g/h]	Nitrogen	Arsine [ppm]	Selectivity Control Agent [mol/h]	T [°C]	Pressure [bar]
0	5	1.2	0.25	30	.	0	1	70	27
1	5	1.2	0.25	30	.	0.05	1	70	27
5	5	1.2	0.25	30	.	0.84	1	70	27
10	5	1.2	0.25	30	.	4.73	1	70	27

[1] triethylaluminium co-catalyst.

During the process, sampling points were established, as shown in Figure 1. The first points correspond to the feed where samples of propylene were taken (A) that came from liquefied petroleum gases (LPG), nitrogen (B), hydrogen (C), and arsine (D), which came from the same as the propylene from LPG. The following points are those taken from within the reactor; these are composed of gas samples inside the reactor (E), catalyst inside the reactor (F), a sample of the catalytic system (G), and recirculated propylene (H). The last points were taken in the degassing stage, the resin sample that came out of the reactor (I), the gases that were synthesized and removed by virgin PP desorption (J), and the PP obtained from the process (K). Because this process was repeated several times with different concentrations of arsine, a number was assigned to each of them, with zero as the target.

Throughout the process, sampling points were established, as shown in Figure 1. These were found in the feed, in the reaction system, and in the degassing stage. Table 2 describes the sampling points and the identification of the samples; because this process was repeated several times with different concentrations of arsine, a number was assigned to each of them, with zero as the target.

Table 2. Identification and sampling points.

Stage	Point	Substance	State of Origin	Identification
Feeding	A	Propylene	LPG	A0, A1, A5, A10
	B	Nitrogen	Gas	B0, B1, B5, B10
	C	Hydrogen	Gas	C0, C1, C5, C10
	D	Arsine	LPG	D0, D1, D5, D10
Reaction	E	Gases	Inside the reactor	E0, E1, E5, E10
	F	Catalyst	Inside the reactor	F0, F1, F5, F10
	G	Catalyst	Catalytic system	G0, G1, G5, G10
	H	Propylene	Recovered	H0, H1, H5, H10
Degassing	I	Resin	Exits the reactor	I0, I1, I5, I10
	J	Gases	Retired in the purge	J0, J1, J5, J10
	K	Polypropylene	Dust	K0, K1, K5, K10

2.3. *Analysis*

2.3.1. Thermogravimetric Analysis-TGA

To determine the effects of arsine on the thermal degradation of the PP samples obtained, the Perkin Elmer TGA7 equipment performed a thermogravimetric analysis (TGA) at a heating rate of 20 °C min^{-1} and under an N_2 flow rate of 60 mL min^{-1}. For this analysis, 10 mg samples were used to obtain the TG and DTG curves [36,37].

2.3.2. Differential Scanning Calorimetry Analysis-DSC

A differential heat scanning (DSC) analysis was performed on a DSC Standard Cell RC. To ensure an identical thermal history, the sample was heated from 0 °C to 230 °C at a rate of 10 °C min^{-1} and subsequently cooled from 230 °C to 0 °C at the same speed.

2.3.3. Melt Flow Index-MFI

The melt flow index (MFI) was measured using a Tinius Olsen MP1200 plastometer. The temperature inside the cylinder of the plastometer was 230 °C and a 2.16 kg piston was used to displace the melt. After the MFI data was obtained, the average molecular weight for each of the PP samples was evaluated using the Bremner approximation [38], which was obtained by correlations that were formulated by the study of commercial polypropylene samples with MI values between 0.7 and 12, measuring the variation in the molecular weight distribution by SEC, using differential refractive index and low-angle laser light scattering detectors [39].

$$M_W{}^{3.7} = 1675/((\text{MFI230 °C 2.16 kg}) \times 10^{-21}) \quad (1)$$

2.3.4. X-ray Fluorescence

This analysis was performed on Malvern's Axios FAST analytical equipment, which allowed us to see the concentration of arsenic and arsine present at the sample points.

2.3.5. Infrared Fourier Transform Analysis-FT-IR

Fourier transform infrared spectroscopy (FT-IR) was used to determine the most significant structural changes in the PP matrix and the ZN catalyst due to the reaction of ZN Ti with AsH_3, and to evaluate how this interaction affected the thermal stability of PP. FT-IR analysis was performed using a Nicolet 6700 (Thermo Scientific) infrared spectrometer using the attenuated total reflectance (ATR) method. The samples were analyzed in the form of films; these films were obtained by compression moulding of the PP in a CARVER 3895 hot press to obtain 300 mm diameter films of ≈100 μm thickness.

2.3.6. Gas Chromatography Analysis with Selective Mass Detector (GC-MS)

A GC-MS 7890B from Agilent technologies was used for arsine analysis [36]. Figure 2 shows the chromatograms of the standards for AsH_3.

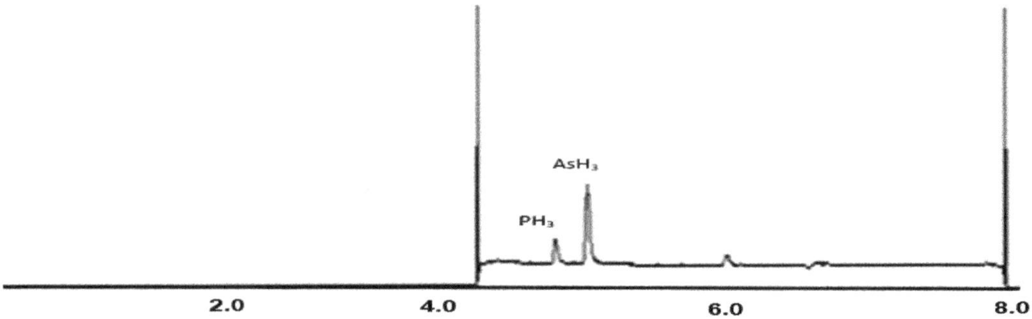

Figure 2. Arsine chromatogram.

2.3.7. Headspace-GC-MS

PP samples were analyzed using an Agilent 7694E Headspace Sampler with the following conditions, GC cycle time: 60 min, furnace: 150 °C, transfer line: 175 °C, loop: 175 °C, vial equilibrium time: 60 min, stirring speed: off, loop size: 1.0 mL, loop filling time: 0.2 min, loop equilibrium time: 0.02 min, injection time: 0.3 min, pressurization time: 0.4.

3. Results

3.1. Polymerization Process

Polymerization occurs in three stages. The first is the formation of the active site consisting of the alkylation of a pentacoordinate Ti (III) ion for the titanium atom that is on the surface of the layers of the α-$TiCl_3$ [40,41] and the activation of the double bond of the monomer. This is given by the coordination of the monomer when establishing a giving bond of electrons p of the double bond with the empty orbital of the transition metal. During the reaction, an intermediate is formed where the unsubstituted carbon binds to the transition metal and the other to the alkyl group [18]. The second stage is the propagation of the monomer, where the active center is linked to the transition metal-alkyl, formed by the constitution of a coordination complex between the growth chain, the added monomer, and the active center of the catalyst [41]. The third stage is the culmination of polymerization, which can be achieved with the separation of hydrogen β, with the removal of a hydride, or by deactivation of the complex using H_2 [41,42]. The addition of arsine during polymerization can affect this process due to its ability to bind to transition metals such as Ti present in the catalyst, inhibiting its effectiveness. Thus, in this work, samples are taken at strategic points that allow for the measurement of the content of this compound and for the proposition of a reaction mechanism, so that the behavior of the PP-virgin arsine and its effects can be determined.

3.1.1. Quantification of Arsine at the Polymerization Stage

The variation of the concentrations of arsine presented in the samples taken at the different selected points of the polymerization process is shown in Figure 1, where the points detailing the presence of these components can be observed. Point D is the initial concentration of arsine introduced to the process, which varies from 0.05 to 4.73 ppm, referring to the presence of impurities not removed during the purification stage of the refinery grade propylene, which translates to a failure of the petrochemical plants that supply this compound. Point H refers to the recovered propylene; the values obtained are equal to 0, which means that AsH_3 was adsorbed or absorbed during the polymerization reaction. Point E shows the gaseous propylene found in the reactor; the presence of AsH_3 at this point indicates that the compound is part of the reaction or interacts in the formation of the polymer being adsorbed, which had been confirmed at point H. Therefore, there must be a reaction mechanism between the catalyst and the arsine that allows the latter to act as a component of the polymerization reaction, which is possible due to the arsine's ability to form M-ER_3 bonds, binding to the transition metal of the active center of the catalyst affecting the polymerization reaction, and due to the efficiency of the catalyst.

3.1.2. Reaction Mechanisms of AsH_3 Proposed during Polymerization: Reaction of Residual AsH_3 with $TiCl_4/MgCl_2$

In the mechanism proposed in Figure 3, AsH_3 competes with the propylene monomer for the active site of Ti. First, a complex π is formed by coordinating the AsH_3 with the Ti of the $TiCl_4/MgCl_2$ complex. The AsH_3-Ti interaction is conducted via the interaction of the electropositive Ti with the free electron pair of arsenic, which predominates the electrons of the π bond of propylene. The formation of a complex π in Ti-propylene has no barriers and is accompanied by an energy gain less than that of AsH_3-Ti, so the latter reaction is predominant. The propylene insertion barrier varies between 6–12 kcal mol^{-1} [43–46]. The high probability of occurrence of the insertion reaction is supported by thermodynamics, since a favoring of approximately 20 kcal mol^{-1} is observed [30,47,48]. In the PP synthesis process, the polymer chain's propagation takes place by moving propylene into Ti-PP, where the PP-alkyl chain has the olefin inserted [47,49]. The PP chain's growth is affected when Ti's active centre reacts with inhibitors of different polarities. The occurrence of these reactions depends on energetic factors. The coordination of the different inhibitors to the Ti Center on the surface of $MgCl_2$ is favored by 27.2 kcal mol^{-1} for Ti-H_2O, 15.1 kcal mol^{-1} for Ti-H_2S, 8.4 kcal mol−1 for Ti-CO_2, 13.1 kcal mol^{-1} for Ti-O_2, and 30.6 kcal mol^{-1} for Ti-CH_3OH interactions [26]. Since AsH_3 exhibits polarities intermediate to the poisons mentioned above, it is expected that its values are within the ranges discussed above. As shown in Figure 3, AsH_3 interferes with the formation of propylene complexes and their insertion. It is worth mentioning that the interactions of these impurities, including AsH3 with the active centers of Ti, are reversible when these inhibitors are removed from the system, and thereafter, the active center of Ti resumes polymerization [50].

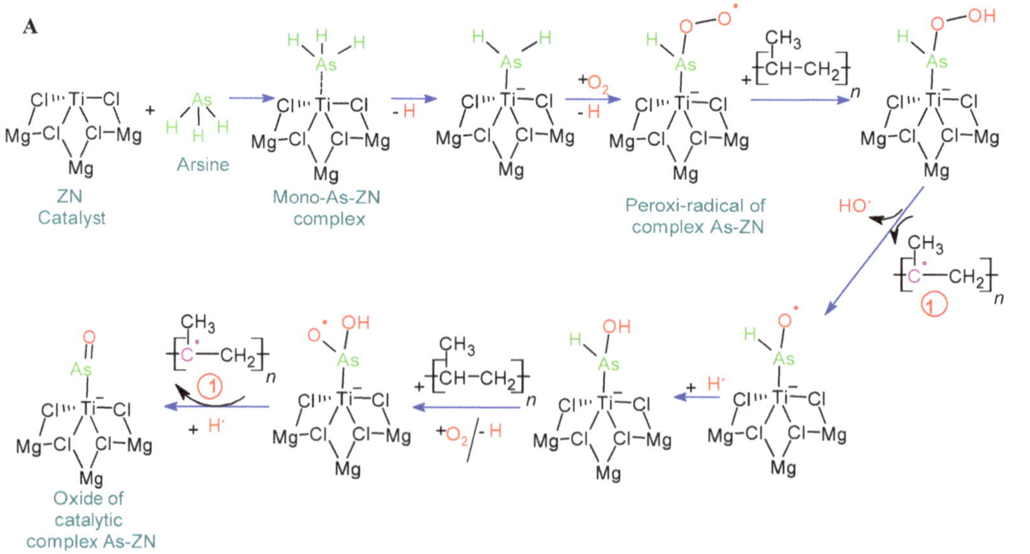

Figure 3. *Cont.*

B

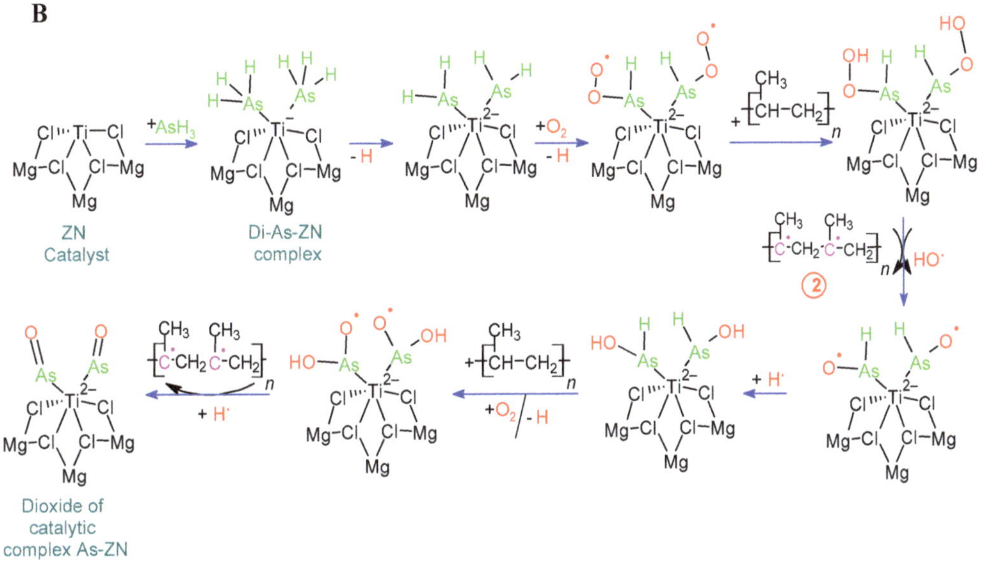

Figure 3. Reaction mechanism of residual AsH_3 with $TiCl_4/MgCl_2$.

3.1.3. Discussion of FTIR Analysis

Figure 4 shows the catalyst's infrared spectra with various arsine levels used during the synthesis of PP. The above identifies species of oxidized arsenic in the As-ZN complex, thus supporting the mechanisms prepositioned in Figure 3. Concerning the fundamental vibration As-O and As = O in FTIR, the force constants of an As-O bond estimated by the use of Gordy's rule are 8.7 × 105 and 4.6 × 105 cm^{-1} dinas for a single bond and double bond, respectively [51]. When substituted in the fundamental equation for a harmonic oscillator, these values give VAs-O 964 and 776 cm^{-1} for a "pure" single bond and a double bond, respectively. The values for the fundamental vibration As-0 and As = O were, in most cases, within 10–40 cm^{-1} of 920 cm^{-1} the average of the frequencies calculated for a single and double bond [51,52]. The observed frequency suggests a higher bond order than one that may well result from the subsequent donation of the pπ electrons of the oxygen atom using the empty dπ orbitals of the arsenic atom. Figure 4 shows that the band's growth between 800 and 970 cm^{-1} is associated with the increase in the concentration of oxidized arsenic atoms in the As-ZN complex, and an increase in the area of these peaks is observed, being greater for the K10 sample. This last catalyst sample was the one that was in contact with the highest levels of AsH_3. The spectra indicate that the oxidized arsenic that is part of a chemical complex within the catalyst was higher when AsH_3 levels increased.

Arsenic of AsH_3-$TiCl_4$-$MgCl_2$ is oxidized in the presence of oxygen atoms to form a complex of radicals O-O-AsH_2-$TiCl_4$-$MgCl_2$, which are the initiators of the degradation of wasting PP. In Figure 5, the effect of this complex of radicals O-O-AsH-$TiCl_4$-$MgCl_2$ is shown on the beginning of the degradation of PP, which consists of the abstraction of hydrogen in the PP-virgin, generating a stable tertiary radical, which, when reacting with a molecule O_2, will produce a Peroxy-PP-virgin radical, which extracts hydrogen from a neighboring virgin-PP chain, thereby forming the virgin-PP-Hydroperoxide. The latter, after a homolytic rupture, will give rise to the Alkoxy-PP-Virgin radical [53–56]. The radical complex of O-O-ash-tiCl_4-$MgCl_2$, after the abstraction of hydrogen, forms the peroxy complex H-O-O-AsH-$TiCl_4$-$MgCl_2$ that, by hemolysis, forms the radical O-AsH-$TiCl_4$-$MgCl_2$. The latter can perform successive abstractions of hydrogen from the tertiary carbon of the PP-virgin chain to continue increasing its degradation or reaction with hydrogen radicals to end the radicalization process and form the stable complex HO-

AsH-$TiCl_4$-$MgCl_2$. This last complex also participates in the radical reaction with PP-virgin and with oxygen atoms to form the radical tertiary polymer and the intermediate complex HO-As-OH-$TiCl_4$-$MgCl_2$ that, by dehydration, forms the new adduct O = As-$TiCl_4$-$MgCl_2$. Figure 3b shows the reaction of two molecules of AsH_3 with the active center of Ti of the Catalyst ZN, to form the complex $(AsH_3)_2$-$TiCl_4$-$MgCl_2$ that subsequently oxidizes in the presence of oxygen atoms to form the radical complex (O-O-As-O-O-)$_2$-$TiCl_4$-$MgCl_2$, which are the initiators of the degradation of PP-virgin residues by abstraction of multiple tertiary carbon protons, by the formation of multiple tertiary radicals of the PP-virgin chain and the intermediate complex HO-As-OH-$TiCl_4$-$MgCl_2$ that, by dehydration, forms the new adduct O = As-$TiCl_4$-$MgCl_2$.

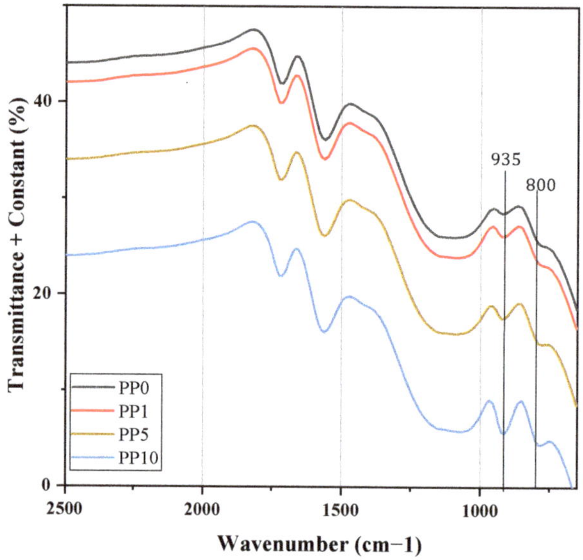

Figure 4. Infrared spectra of the catalyst with distinct levels of arsine.

3.2. *Effects of Arsine on the Properties of PP*

3.2.1. Effects of Arsenic Content on Virgin PP MFI

The effects of the arsenic content in the samples are reflected in the change of the MFI obtained for each of them, presenting a linear correlation with an R^2 of 0.99786; where the concentration of income of AsH_3 increases, so does the MFI, as shown in Figure 5. This means that the arsine from impurities (point A) participates in the polymerization reaction, forming a stable complex present in the polymer chain that changes some properties of the resin obtained, including the M_w of the samples, which was calculated using the formula proposed by Brenner. Its variation is shown in Figure 5, demonstrating that as the arsenic content present in the resins increases, the MFI decreases.

Consequently, small chain scissors and fractures of the polymer structure during degradation are very frequent, occurring as the average molecular weight decreases and the MFI increases. It should be noted that the residue contents of the catalyst, such as Ti, Al, Cl, and Fe, do not present significant differences, thus the changes presented in this paragraph are not generated by residues other than arsine, as has been presented in other studies [36,57].

From these data, it can be inferred that, during the reaction, the arsine reacts with the ZN catalyst. During the development of this work, a reaction mechanism was proposed where it was evident that the arsine competed with the propylene molecule in the polymerization stage, reacting with the active center of the ZN that influences the growth of the polymer chain and presenting evidence of the high reactivity of the catalyst.

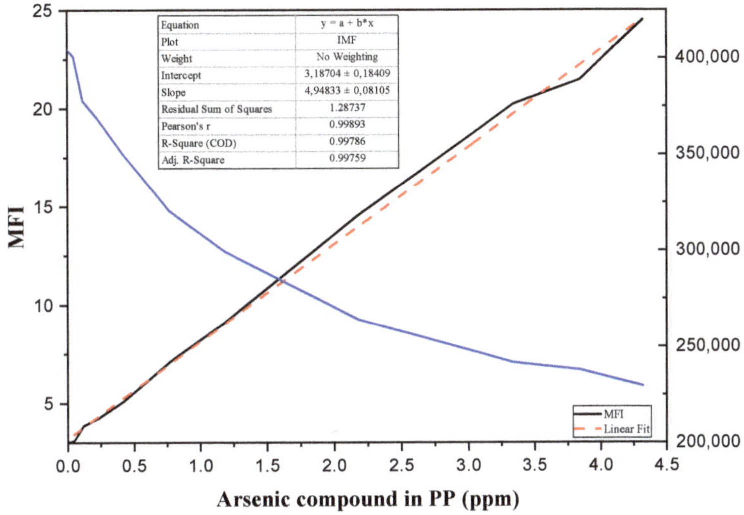

Figure 5. Relationship between the arsenic content in the PP and the MFI, and the variation of the M_W.

3.2.2. Degradation of Residual PP

The thermal degradation of the samples was evaluated by TGA and DSC, the data obtained show a decrease of the thermal stability of the material as the arsenic content increases. These results are presented in Figure 6. The thermal behavior of the K0 sample does not have a significant variation, as presented in the literature [19]. K0 and K1, which have a zero and very low content of arsenic, respectively, in their matrix, have a peak rupture at 390 and 430 °C with a mass loss of 22%.

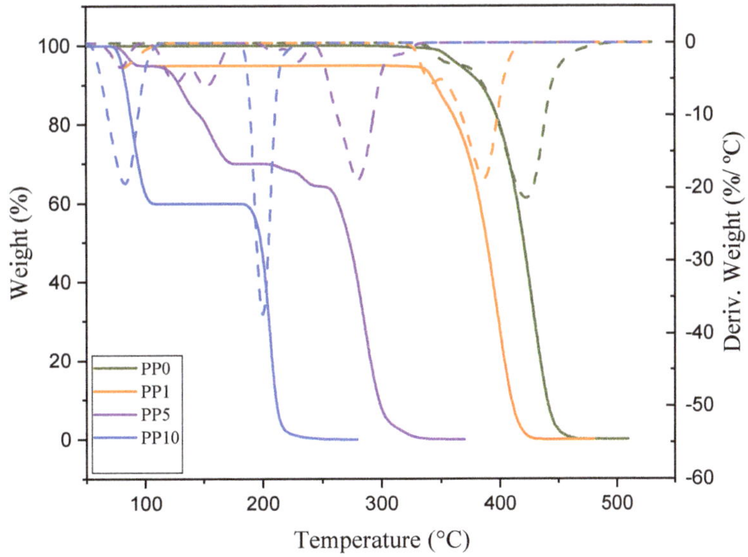

Figure 6. TGA and DTG of PP samples.

The addition of AsH_3 in the polymerization implies marked changes in the performance and macromolecular structure of the final material obtained, demonstrating that their molecular weight and thermal stability are too low to have functional groups different from the hydrocarbon. Previous research has demonstrated a link between the polymer's low molecular weight and photooxidative degradation, as indicated by the carbonyl index in the 1712 cm^{-1} band of FTIR (Figure 7), which is impacted when the concentration of arsine and arsenic in the sample increases [57–59]. To understand this behavior, a reaction mechanism is proposed to explain the reduction in the physicochemical properties of PP when it has arsine content.

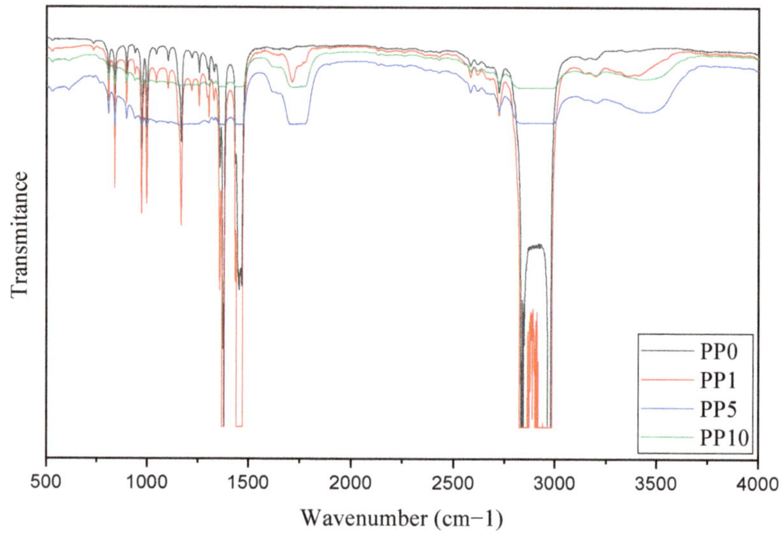

Figure 7. Carbonyl index of PP samples.

3.3. Degassing Process

The degassing process is the second stage of the process, in which the resin passes when leaving the reactor. This consists of the purge tower, where nitrogen and value are fed to remove the hydrocarbons, and the adsorption or absorption of arsine by the resin obtained is confirmed [35].

3.3.1. Identification and Quantification of Arsine in the Degassing Stage of PP Resin

Table 3 shows the concentrations obtained from the different compounds during the degassing stage taken at the available sampling points together with the input data for the overall process. Point I: the samples obtained from this point are of newly polymerized resin, which contains arsine either absorbed or adsorbed. Point J: volatile compounds are removed, having values equal to 0 for the arsine content, which allows us to rule out the absorption of both arsine and arsenic at the interface of the polymer. At this point, the other compounds formed during polymerization that are purged are also determined, which allows us to determine the effect of the arsine on their composition. Finally, point K: showing that the resin obtained has the content of the two compounds, therefore, the products obtained from these processes are considered waste and may present variations concerning virgin resins that can contribute to the degradation of the material, thermal stability, and fluidity index, among others. The study of the effects on these properties allows us to determine applications for these generated residues.

Table 3. Compounds determined in the degassing stage.

Compound (ppm)	Point I				Point J				Point K			
	I0	I1	I5	I10	J0	J1	J5	J10	K0	K1	K5	K10
AsH_3	0	0	0	0	0	0	0	0	0	0	0	0
Alcohol	10.1	50.2	81.5	180.1	15.2	80.2	100	310.1	5.1	7.5	20.1	50.5
Ketone	19.5	80.2	160.1	220.1	30.2	130.1	220	280.7	4.5	10.6	31.5	45.5
Aldehydes	13.2	60.1	121.5	242.7	25.2	91.5	172	270.1	7.1	11.7	22.3	21.2
Acid	20.1	97.1	210.1	351.7	32.1	161.2	308	425.1	4.5	23.4	41.5	33.3
CO	0	0	0.2	1.1	0	0	0.4	3.2	0	0	0	0.1
CO_2	0.1	1.5	5.5	11.3	1.2	2.1	8	20.2	0	0.1	1.2	1.9

The compounds determined during this stage present a significant increase concerning the target; this increase may be related to the increase in the content of arsine fed in the polymerization stage. If we take point J as a reference, where most of these gaseous compounds are removed, it can be seen that the values are 3.5 to 5.27 times higher for the sample J1 concerning the white (J0); in the case of the J10 sample, the values are between 9.29 and 20.4 times higher with respect to the white, being alcohol the compound, where the greatest increase is shown. In addition, as the arsine content in the feed increases, the formation of CO production is promoted.

3.3.2. Mechanisms of Reaction of Peroxy Radicals in Arsenic Complexes to Form the VOCs Identified in PP

The Alkoxy-PP-virgin radical that has formed undergoes a homolytic rupture of the simple bond adjacent to the carbon of the oxygenated radical. This homolysis forms the methyl-PP-virgin ketone and the radical PP, as seen in Figure 8. The radical O-AsH-$TiCl_4$-$MgCl_2$ attacks the tertiary carbon of the methyl-PP-virgin ketone, to abstract the hydrogen atom and thus propagate the oxidation of the more stable carbon [58]. This new radical reacts with the oxygen atom to form the β-Hydroxyperoxy Methyl-PP-virgin Ketone, 2,4-pentanedione, 1-Hydroxy-2-propanone, and acetone. The latter two are formed after homolytic cleavage of another tertiary carbon bond, the formation of the acetone radical, and subsequent reactions with the hydroxyl and hydrogen radicals. On the other hand, the PP radical at the end of the chain reacts with O_2, giving rise to PP-hydroperoxide, to the PP-Alkoxy radical, and then to methanol-PP-virgin, which, by abstraction of the tertiary carbon proton by the radical O-AsH-$TiCl_4$-$MgCl_2$, forms an intermediate radical that, when reacting with O_2, decomposes into methanol and formaldehyde. Methanol-PP-virgin gives rise to Aldehyde-PP-virgin that, by homolysis of the H-C bond of the terminal carbonyl group, forms the Carbonyl-PP-virgin radical, which in turn, by reactions with hydroxyl radicals and hydrogen, gives rise to CO, CO_2 and formic acid, as shown in Figure 8.

The strong relationship between the concentration of AsH_3 residues and the arsenic content within the PP-virgin matrix with the MFI, TG, and DTG shows that increasing the concentration of AsH_3 drastically decreases the stability of PP, allowing us to propose an alternative reaction mechanism to that shown in Figure 3; a mechanism that comes even closer to the behavior of K10. Figure 8 shows the reaction of two Molecules AsH_3 with the active center of Ti of the catalyst ZN, to form the complex $(AsH_3)_2$-$TiCl_4$-$MgCl_2$ that is subsequently oxidized in the presence of oxygen atoms to form the radical complex $(O\text{-}O\text{-}As\text{-}O\text{-}O\text{-})_2$-TiCl4-$MgCl_2$, which are the initiators of the degradation of PP-virgin waste by abstraction of multiple tertiary carbon protons, by the formation of multiple tertiary radicals of the virgin-PP chain. These will react with the oxygen atoms by forming PP-Virgin poly peroxide that supports increased PP fluidity and low TG and DTG values. The poly peroxide of PP-Virgin, due to its chemical structure, has the appropriate stereochemistry to form hydrogen bonds with another poly peroxide of the PP-Virgin to give rise to a macromolecule of greater polarity, and in an acid medium, creates polyols of PP-virgin. Figure 8 also shows that PP-virgin polyoxide, by the effect of hydrogen and hydroxyl radicals, forms PP-virgin polyols with a carboxyl group at the end of the chain. The latter experiences homolytic ruptures and dehydration. In homolysis, it gives rise to acetone, 1-Hydroxy-2-propanone, isopropyl alcohol, and 1,2-Propanediol, and in dehydration, it forms stable tertiary carbocations, which, when reacting with different alcohols and polyols, gives rise to macromolecules of PP-virgin ether type.

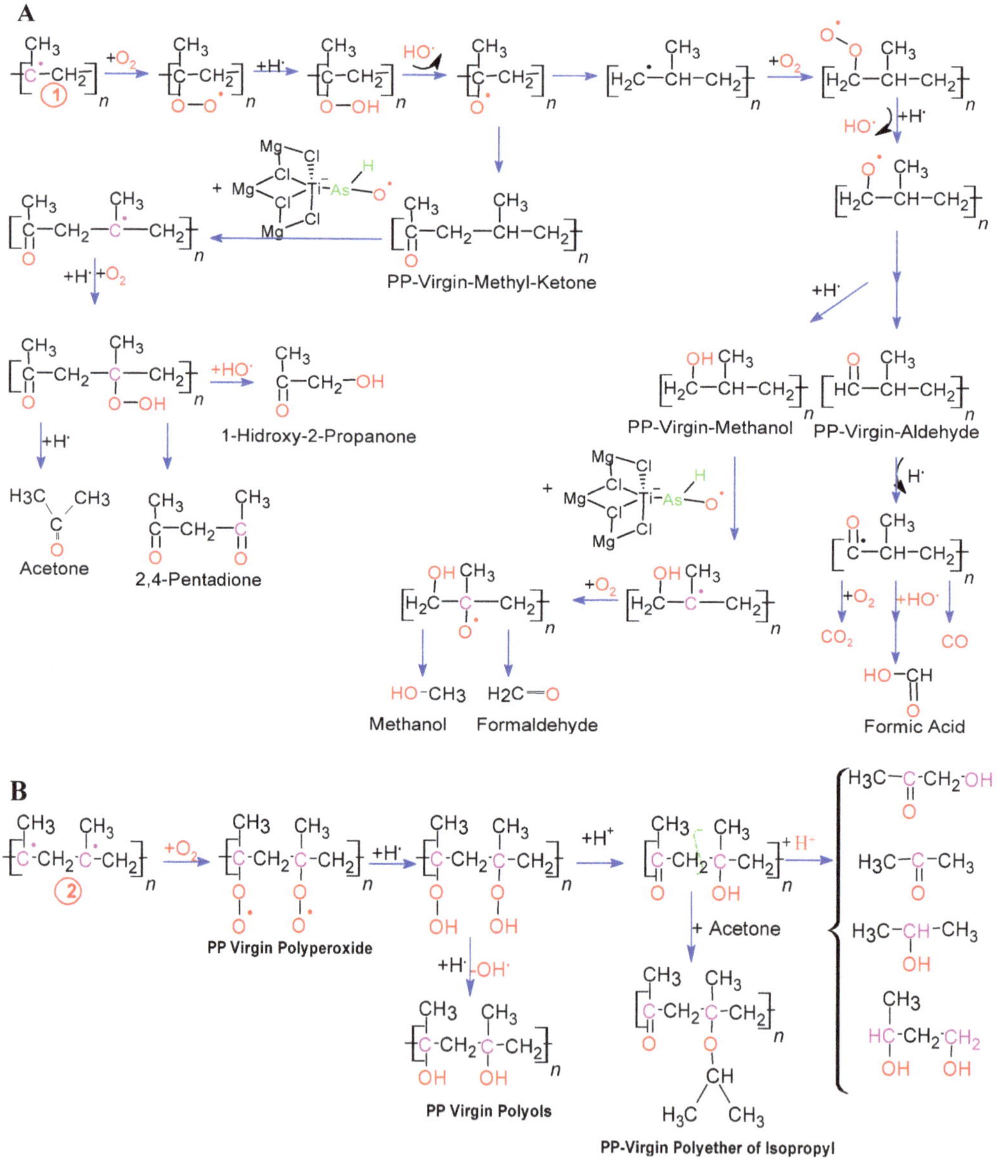

Figure 8. Mechanisms of reaction of peroxy radicals in arsenic complexes to form VOCs.

This study showed the impact that certain contaminants may have on the manufacture of polymers such as polypropylene. The selection and analysis of raw materials is critical to preventing this sort of reaction from happening during the industrial synthesis of PP. With this research, we expect that the purity of any raw materials will be thoroughly examined before application.

4. Conclusions

After the analysis of the samples obtained, it is possible to define the effect that the arsine has on thermal stability and other properties of PP, in addition to defining how an increase of the content of this chemical compound can further enlarge the difference in thermal stability, molecular weight, and MFI between the samples and standard; the first two decreasing and the last increasing. Likewise, the possible reaction mechanisms that occur between the arsine with the active titanium center of the ZN catalyst and the subsequent oxidation to form radical complexes of O-O-AsH-TiCl4-$MgCl_2$ and (O-O-As-O-O-)$_2$-$TiCl_4$-$MgCl_2$, which, by radical reactions, give formation to aldehydes, ketones, alcohols, carboxylic acids of functional groups, CO, CO_2, PP-Polyol, PP-ethers, and PP-Isopropyl ethers, can be defined.

Author Contributions: Conceptualization, J.H.-F.; methodology, J.H.-F. and E.P.-P.; validation, E.P.-P. and E.M.; formal analysis, J.H.-F. and Y.G.; investigation, Y.G. and J.H.-F.; writing—original draft preparation, Y.G.; writing—review and editing, E.M.; E.P.-P. and Y.G.; supervision, J.H.-F. and E.M.; project administration, J.H.-F. All authors have read and agreed to the published version of the manuscript.

Funding: This research received no external funding.

Institutional Review Board Statement: Not applicable.

Informed Consent Statement: Not applicable.

Data Availability Statement: The data presented in this study are available on request from the corresponding author.

Conflicts of Interest: The authors declare no conflict of interest.

References

1. Quinteros, J.G. *Sintesis y Aplicaciones de Ligandos Arsinas. Estudios de Sistemas Catalíticos de Pd y Au*; Universidad Nacional de Córdoba: Córdoba, Spain, 2016.
2. Gras, R.; Luong, J.; Hawryluk, M.; Monagle, M. Analysis of part-per-billion level of arsine and phosphine in light hydrocarbons by capillary flow technology and dielectric barrier discharge detector. *J. Chromatogr. A* **2010**, *1217*, 348–352. [CrossRef] [PubMed]
3. Zhang, J.; Li, X. Hydrogen bonding in the complexes formed by arsine and H-X molecules: A theoretical study. *Chem. Phys. Lett.* **2019**, *735*, 136767. [CrossRef]
4. Burt, J.; Levason, W.; Reid, G. Coordination chemistry of the main group elements with phosphine, arsine and stibine ligands. *Coord. Chem. Rev.* **2014**, *260*, 65–115. [CrossRef]
5. Green, J. Transition Metals in the Synthesis of Complex Organic Molecules. *J. Am. Chem. Soc.* **2010**, *132*, 1443. [CrossRef]
6. Orpen, A.G.; Connelly, N.G. Structural systematics: The role of PA. sigma.* orbitals in metal-phosphorus .pi.-bonding in redox-related pairs of M-PA3 complexes (A = R, Ar, OR; R = alkyl). *Organometallics* **1990**, *9*, 1206–1210. [CrossRef]
7. Crabtree, R.H. *The Organometallic Chemistry of the Transition Metals*; John Wiley & Sons, Inc.: Hoboken, NJ, USA, 2014. [CrossRef]
8. McCleverty, J.A.; Meyer, T.J. Comprehensive Coordination Chemistry II: From Biology to Nanotechnology. *J. Am. Chem. Soc.* **2004**, *126*. [CrossRef]
9. Treichel, P.M. *Phosphine, Arsine, and Stibine Complexes of the Transition Elementes*; Elsevier Scientific Publishing Co.: Amsterdam, The Netherlands, 1979; Volume 9. [CrossRef]
10. Wilkinson, G.; Gillard, R.D.; McCleverty, J.A. *Comprehensive Coordination Chemistry: The Synthesis, Reactions, Properties and Applications of Coordination Compounds*; Pergamon Press: Oxford, UK, 1987; Available online: https://inis.iaea.org/search/search.aspx?orig_q=RN:19039703 (accessed on 1 November 2021).
11. Rodríguez, M.E. *Funcionalización de Ligandos Coordinados*; Universidad de Coruña: Coruña, Spain, 2017.
12. Hernández-Fernández, J. Quantification of arsine and phosphine in industrial atmospheric emissions in Spain and Colombia. *Implementation of modified zeolites to reduce the environmental impact of emissions. Atmos. Pollut. Res.* **2021**, *12*, 167–176. [CrossRef]
13. Carrizo, D.S. *Estudios Computacionales en Catálisis Homogénea con Metales de Transición: Reacción de Stille, Activación de Enlaces C-H y β-Eliminación*; Universidad Nacional de Córdoba: Córdoba, Spain, 2016.
14. Arlinghaus, R.T.; Andrews, L. Infrared spectra of the PH3, AsH3, and SbH 3-HX hydrogen bonded complexes in solid argon J. *Chem. Phys.* **1984**, *81*, 4341–4351. [CrossRef]
15. Joaquin, H.-F.; Juan, L. Quantification of poisons for Ziegler Natta catalysts and effects on the production of polypropylene by gas chromatographic with simultaneous detection: Pulsed discharge helium ionization, mass spectrometry and flame ionization. *J. Chromatogr. A* **2020**, *1614*, 460736. [CrossRef]

16. Kurahashi, E.; Wada, T.; Nagai, T.; Chammingkwan, P.; Terano, M.; Taniike, T. Synthesis of polypropylene functionalized with a trace amount of reactive functional groups and its utilization in graft-type nanocomposites. *Polymer* **2018**, *158*, 46–52. [CrossRef]
17. Karol, F.J.; Jacobson, F.I. Catalysis and the Unipol Process. In *Studies in Surface Science and Catalysis*; Elsevier: Amsterdam, The Netherlands, 1986; Volume 25. [CrossRef]
18. Mier, J.; Artiaga, R.; Soto, L.G. *Síntesis de Polímeros. Pesos moleculares. Conformación y configuración. Elementos Estructurales con Materiales Polímeros: Ferrol*; Universidad de Coruña: Coruña, Spain, 1997; pp. 11–48.
19. Bailar, J.C.; Emeléus, H.J.; Nyholm, R.; Trotman-Dickenson, A.F. *Comprehensive Inorganic Chemistry*; Elsevier: Amsterdam, The Netherlands, 1973; Volume 3. [CrossRef]
20. Nikolaeva, M.; Mikenas, T.; Matsko, M.; Zakharov, V. Effect of $AlEt_3$ and an External Donor on the Distribution of Active Sites According to Their Stereospecificity in Propylene Polymerization over $TiCl_4/MgCl_2$ Catalysts with Different Titanium Content. *Macromol. Chem. Phys.* **2016**, *217*, 1384–1395. [CrossRef]
21. Panchenko, V.N.; Vorontsova, L.V.; Zakharov, V.A. Ziegler-Natta catalysts for propylene polymerization—Interaction of an external donor with the catalyst. *Polyolefins J.* **2017**, *4*, 87–97. [CrossRef]
22. Vizen, E.I.; Rishina, L.A.; Sosnovskaja, L.N.; Dyachkovsky, F.S.; Dubnikova, I.L.; Ladygina, T.A. Study of hydrogen effect in propylene polymerization on (with) the MgC12- Supported Ziegler-Natta catalyst-Part 2. Effect of Cs2 on polymerization centres. *Eur. Polym. J* **1994**, *30*, 1315–1318.
23. Kallio, K.; Wartmann, A.; Reichert, K.-H. Reactivation of a Poisoned Metallocene Catalyst by Irradiation with Visible Light. *Macromol. Rapid Commun.* **2002**, *23*, 187–190. [CrossRef]
24. Bahri-Laleh, N. Interaction of different poisons with $MgCl_2/TiCl_4$ based Ziegler-Natta catalysts. *Appl. Surf. Sci.* **2016**, *379*, 395–401. [CrossRef]
25. Asynkiewicz, S.P. Reactions of organoaluminium compounds with electron donors. *Pure Appl. Chem.* **1972**, *30*, 509–522. [CrossRef]
26. Hernández-Fernández, J. Quantification of oxygenates, sulphides, thiols and permanent gases in propylene. A multiple linear regression model to predict the loss of efficiency in polypropylene production on an industrial scale. *J. Chromatogr. A* **2020**, *1628*, 461478. [CrossRef] [PubMed]
27. Li, Z.; Yin, Y.; Wang, X.; Tu, D.M.; Kao, K.C. Formation and inhibition of free radicals in electrically stressed and aged insulating polymers. *J. Appl. Polym. Sci.* **2003**, *89*, 3416–3425. [CrossRef]
28. Cavallo, L.; Del Piero, S.; Ducéré, J.M.; Fedele, R.; Melchior, A.; Morini, G.; Piemontesi, F.; Tolazzi, M. Key interactions in heterogeneous Ziegler-Natta catalytic systems: Structure and energetics of TiCL4-lewis base complexes. *J. Phys. Chem. C* **2007**, *111*, 4412–4419. [CrossRef]
29. Bahri-Laleh, N.; Nekoomanesh-Haghighi, M.; Mirmohammadi, S.A. A DFT study on the effect of hydrogen in ethylene and propylene polymerization using a Ti-based heterogeneous Ziegler-Natta catalyst. *J. Organomet. Chem.* **2012**, *719*, 74–79. [CrossRef]
30. Correa, A.; Bahri-Laleh, N.; Cavallo, L. How well can DFT reproduce key interactions in Ziegler-Natta systems? *Macromol. Chem. Phys.* **2013**, *214*, 1980–1989. [CrossRef]
31. Joaquin, H.-F.; Juan, L.-M. Autocatalytic influence of different levels of arsine on the thermal stability and pyrolysis of polypropylene. *J. Anal. Appl. Pyrolysis* **2022**, *161*, 105385. [CrossRef]
32. Arshouee, G.H.; Zarand, S.M.G. Determination of Catalyst Residence Time, Heat Generation, and Monomer Conversion via Process Variables during Propylene Polymerization by a Mathematical Model. *Theor. Found. Chem. Eng.* **2021**, *55*, 140–152. [CrossRef]
33. Hernández-Fernandez, J.; Rodríguez, E. Determination of phenolic antioxidants additives in industrial wastewater from polypropylene production using solid phase extraction with high-performance liquid chromatography. *J. Chromatogr. A* **2019**, *1607*, 460442. [CrossRef] [PubMed]
34. Hernández-Fernández, J.; Lopez-Martinez, J.; Barceló, D. Quantification and elimination of substituted synthetic phenols and volatile organic compounds in the wastewater treatment plant during the production of industrial scale polypropylene. *Chemosphere* **2021**, *263*, 128027. [CrossRef] [PubMed]
35. Hernández-Fernández, J.; López-Martínez, J. Experimental study of the auto-catalytic effect of triethylaluminum and TiCl4 residuals at the onset of non-additive polypropylene degradation and their impact on thermo-oxidative degradation and pyrolysis. *J. Anal. Appl. Pyrolysis* **2021**, *155*, 105052. [CrossRef]
36. Zhang, S.; Li, B.; Lin, M.; Li, Q.; Gao, S.; Yi, W. Effect of a novel phosphorus-containing compound on the flame retardancy and thermal degradation of intumescent flame retardant polypropylene. *J. Appl. Polym. Sci.* **2011**, *122*, 3430–3439. [CrossRef]
37. Hernández-Fernández, J.; Rayón, E.; López, J.; Arrieta, M.P. Enhancing the Thermal Stability of Polypropylene by Blending with Low Amounts of Natural Antioxidants. *Macromol. Mater. Eng.* **2019**, *304*, 1900379. [CrossRef]
38. Bremner, T.; Rudin, A.; Cook, D.G. Melt flow index values and molecular weight distributions of commercial thermoplastics. *J. Appl. Polym. Sci.* **1990**, *41*, 1617–1627. [CrossRef]
39. Ivin, K.J.; Rooney, J.J.; Stewart, C.D.; Green, M.L.H.; Mahtab, R. Mechanism for the stereospecific polymerization of olefins by Ziegler–Natta catalysts. *J. Chem. Soc. Chem. Commun.* **1978**, *14*, 604–606. [CrossRef]
40. Padilla Paz, R.M. *Síntesis y Estudio de Cpmplejos Organometálicos de Iridio con N-Aril-4,5-dimetilen-1,3-oxazolidin-2-onas y Complejos de Cobre con Furoiltioureas*; Universidad Autónoma del estado de Hidalgo: Hidalgo, Mexico, 2006.
41. Castro, C.A.; Eguren, L.; Korswagen, R.P. Catalizadores Ziegler-Natta utilizados para polimerizar propileno y etileno. *Rev. De Química* **1987**, *1*, 5–13.

42. Natta, G.; Pasquon, I.; Giachetti, E. Kinetics of the stereospecific polymerization of polypropylene to isotactic polymers. In *Stereoregular Polymers and Stereospecific Polymerizations*; Elsevier: Amsterdam, The Netherlands, 1967. [CrossRef]
43. Chien, J.C.W.; Bres, P. Magnesium Chloride Supported High Mileage Catalysts for Olefin Polymerization. XIII. Effect of External Lewis Base on Ethylene Polymerization. *J. Polym. Sci. Part A Polym. Chem.* **1986**, *24*, 1967–1988. [CrossRef]
44. Bhaduri, S.; Mukhopadhyay, S.; Kulkarni, S.A. Role of titanium oxidation states in polymerization activity of Ziegler-Natta catalyst: A density functional study. *J. Organomet. Chem.* **2006**, *691*, 2810–2820. [CrossRef]
45. Mukhopadhyay, S.; Kulkarni, S.A.; Bhaduri, S. Density functional study on the role of electron donors in propylene polymerization using Ziegler-Natta catalyst. *J. Organomet. Chem.* **2005**, *690*, 1356–1365. [CrossRef]
46. Jensen, V.R.; Borve, K.J.; Ystenesg, M. Ziegler-Natta Ethylene Insertion Reaction for a Five-Coordinate Titanium Chloride Complex Bridged to an Aluminum Hydride Cocatalyst. *J. Am. Chem. Soc.* **1995**, *117*, 4109–4117. [CrossRef]
47. Zakharov, I.I.; Zakharov, V.A.; Zhidomirov, G.M. Quantum chemical studies of propene, ethylene, acetylene and dihydrogen reactivity in the insertion reaction into the titanium-alkyl bond. *Macromol. Theory Simul.* **1996**, *5*, 837–843. [CrossRef]
48. Cavallo, L.; Guerra, G.; Corradini, P. Mechanisms of Propagation and Termination Reactions in Classical Heterogeneous Ziegler−Natta Catalytic Systems: A Nonlocal Density Functional Study. *J. Am. Chem. Soc.* **1998**, *120*, 2428–2436. [CrossRef]
49. Cheremisinoff, N.P. *Handbook of Polymer Science and Technology: Synthesis and Properties*; John Wiley & Sons Ltd.: London, UK, 1989.
50. Gordy, W. A Relation between Bond Force Constants, Bond Orders, Bond Lengths, and the Electronegativities of the Bonded Atoms. *J. Chem. Phys.* **1946**, *14*, 305–320. [CrossRef]
51. Myneni, S.C.B.; Traina, S.J.; Waychunas, G.A.; Logan, T.J. Vibrational spectroscopy of functional group chemistry and arsenate coordination in ettringite. *Geochim. Cosmochim. Acta* **1998**, *62*, 3499–3514. [CrossRef]
52. Mowery, D.M.; Assink, R.A.; Derzon, D.K.; Klamo, S.B.; Clough, R.L.; Bernstein, R. Solid-State 13C NMR Investigation of the Oxidative Degradation of Selectively Labeled Polypropylene by Thermal Aging and γ-Irradiation. *Macromolecules* **2005**, *38*, 5035–5046. [CrossRef]
53. Hamid, S.H. *Handbook of Polymer Degradation*, 2nd ed.; Marcel Dekker: New York, NY, USA, 2000.
54. Carlsson, D.J.; Wiles, D.M. The Photooxidative Degradation of Polypropylene. Part I. Photooxidation and Photoinitiation Processes. *J. Macromol. Sci. Part C Polym. Rev.* **1976**, *14*, 65–106. [CrossRef]
55. Rangaraj, V.M.; Singh, S.; Devaraju, S.; Wadi, V.S.; Alhassan, S.; Anjum, D.H.; Mittal, V. Polypropylene/phosphazene nanotube nanocomposites: Thermal, mechanical, and flame retardation studies. *J. Appl. Polym. Sci.* **2020**, *137*, 49525. [CrossRef]
56. Kurt, G.; Kasgoz, A. Effects of molecular weight and molecular weight distribution on creep properties of polypropylene homopolymer. *J. Appl. Polym. Sci.* **2021**, *138*, 50722. [CrossRef]
57. Schoolenberg, G.E.; Vink, P. Ultra-violet degradation of polypropylene: 1. Degradation profile and thickness of the embrittled surface layer. *Polymer* **1991**, *32*, 432–437. [CrossRef]
58. Rouillon, C.; Bussiere, P.O.; Desnoux, E.; Collin, S.; Vial, C.; Therias, S.; Gardette, J.L. Is carbonyl index a quantitative probe to monitor polypropylene photodegradation? *Polym. Degrad. Stab.* **2016**, *128*, 200–208. [CrossRef]
59. Zhang, T.; Chen, K.; Zhang, Z.; Shi, R. Quantitative relationship between melting peak temperature and carbonyl index of polypropylene during UV aging based on date fitting. *IOP Conf. Ser. Mater. Sci. Eng.* **2018**, *322*, 022016. [CrossRef]

polymers MDPI

Article

A Supramolecular Hydrogel Based on Copolymers of Acrylic Acid and Maleic Anhydride Derivatives with Terpyridine Motifs

Zheng Chi, Chenchen Ma, Ziyuan He, Zihan Ma, Xuegang Chen * and Zhaoge Huang *

Key Laboratory of Rubber-Plastic of Ministry of Education (QUST), School of Polymer Science and Engineering, Qingdao University of Science and Technology, Qingdao 266042, China; cz17854239116@126.com (Z.C.); 19511618633@163.com (C.M.); hzy15309081778@126.com (Z.H.); mazihan20212021@163.com (Z.M.)
* Correspondence: xgchen@qust.edu.cn (X.C.); hzg@qust.edu.cn (Z.H.)

Abstract: A kind of terpyridine derivative (**NH_2-Tpy**) in which the amino was incorporated by a short alkyl chain was synthesized. Through grafting of terpyridine units into the hydrophilic copolymers of maleic anhydride and acrylic acid **PAAMa** via the reaction of the amino groups in **NH_2-Tpy** and the maleic anhydride units, a series of gelator polymers—**P1**, **P2**, and **P3**—containing different contents of terpyridine units was synthesized. Under coordination of Ni^{2+} and terpyridine ligands in linear polymers, the supramolecular hydrogels **H1**, **H2**, and **H3** with different cross-linking degrees were prepared. The linear polymers **P1–P3** had a strong absorption peak at about 290 nm in the UV-vis spectra which was attributed to π–π* transition, and there was a new peak at about 335 nm led by the metal-to-ligands charge transfer (MLCT) when coordinated with Ni^{2+} ions. According to the rheological behaviors, the storage modulus (G′) was larger than the loss modulus (G″). These hydrogels showed typical gel-like characteristics when the terpyridine content of the hydrogels exceeded 10%, and the hydrogels showed liquid-like characteristics when the terpyridine content of the hydrogels was less than 7%. The results of the micromorphological investigation of the xerogels from SEM illustrated the metal–terpyridine coordination cross-linking could have an important influence on the microstructures of the resulting hydrogels. Furthermore, these hydrogels based on supramolecular cross-links exhibited reversible solution–gel transition at different environmental temperatures. At the same time, the equilibrium swelling of the supramolecular hydrogels was 8.0–12.3 g/g, which increased with the decrease in the content of the terpyridine units in the resulting hydrogels.

Keywords: supramolecular hydrogel; acrylic acid; maleic anhydride; terpyridine; coordination interaction

Citation: Chi, Z.; Ma, C.; He, Z.; Ma, Z.; Chen, X.; Huang, Z. A Supramolecular Hydrogel Based on Copolymers of Acrylic Acid and Maleic Anhydride Derivatives with Terpyridine Motifs. *Polymers* **2022**, *14*, 2857. https://doi.org/10.3390/polym14142857

Academic Editors: Andrzej Puszka and Beata Podkościelna

Received: 30 May 2022
Accepted: 12 July 2022
Published: 13 July 2022

1. Introduction

Hydrogels are polymeric materials with 3D networks of cross-linked structures that can retain a large amount of water in their swollen state through capillary, osmotic, and hydration forces [1,2]. Great advances have been made in supramolecular polymers over the past several decades [3,4]. As one of the most important subfields, supramolecular hydrogels refer to the soft materials that are constructed via non-covalent interactions such as hydrogen bonds, π–π interactions, van der Waals interactions, hydrophobic interactions, ionic bonds, host–guest interactions, metal–ligands coordination interactions, etc., and have attracted great enthusiasm for these fields [5–9]. Supramolecular assembly systems based on multiple, orthogonal non-covalent motifs allow for the preparation of controlled, reversible, selective, multi-stimuli-responsive hydrogel materials, and supramolecular hydrogels play a vital role in biological science, drug transport, tissue engineering, wound dressing, and so on [10–12].

Generally, most covalent chemistry is irreversible, whereas most supramolecular non-covalent chemistry is reversible, and thus some thermodynamic behaviors could occur in these systems [13]. Many supramolecular hydrogels have stimulus–response characteristics, especially thermal responsiveness, and metal–ligand-coordination-based supramolecular hydrogels with thermo-responsive properties are promising [9,14,15]. The gel–sol transition temperature of supramolecular hydrogels could be adjusted by controlling the amount and strength of the non-covalent coordination bonds, which allows hydrogels to be more accurately applied in various fields [14]. Furthermore, since the metal–ligands coordination bond can usually be rebuilt after being destroyed in the appropriate environment, a supramolecular hydrogel based on metal–ligands has sometimes been used as a tunable thermosensitive material with reversibility [16].

It is known that 2,2′:6′,2′′-terpyridine (tpy)-based ligands show versatile binding with different divalent and trivalent metal ions to form bis-terpyridine complexes, whereas linear coordination polymers or network polymers are obtained from the coordination of ditopic or polytopic ligands with appropriate metals [17–19]. Therefore, the tpy-based building blocks applied in supramolecular chemistry and materials have attracted considerable interest, and the ease of modification in the ligands or polymer chains and the introduction of different metals into molecular structures have endowed these building blocks and resulting supramolecular materials with many interesting functional properties [20,21].

There have been some novel supramolecular hydrogels based on terpyridine-metal coordination in recent years [22–25]. Yu et al. reported a ditopic ligand comprising terpyridine and acetylene segments, and the fluorescence enhancement in the blue emission in solution or gel could occur when coordinated with zinc ions [22]. Yang et al. designed interpenetrating polymer network hydrogels based on metal–terpyridine coordination and dynamic covalent bonds, and the properties of these hydrogels can be adjusted via tunable metal–ligand interaction [23]. The modification of building blocks and the introduction of different metals supply more ways to optimize the properties of these novel supramolecular hydrogels.

Hydrogels originating from the hydrophilic group have good hydrophilicity. Obviously, terpyridine and some functional segments generally show hydrophobic properties, and it is difficult to form hydrogel systems for terpyridine alone. Linking with hydrophilic structures is a possible way to obtain hydrogels for terpyridine–metal complexes. Due to large steric hindrance, it is difficult to form the homo-polymer of maleic anhydride; however, copolymerization based on maleic anhydride is very common, and many copolymers containing maleic anhydride have been prepared and applied in scale inhibition and dispersing agents. As one of the hydrophilic monomers, maleic anhydride (MA) is often used in the preparation of hydrogels [26–28]. Furthermore, MA is easy to modify and functionalize for the copolymers containing maleic anhydride through the reaction of acid anhydride groups and other functional groups [29–31].

Herein, a terpyridine derivative with an amino group (NH_2-Tpy) was designed and synthesized (Scheme 1). Through the reaction of the amino in NH_2-Tpy and the anhydride group in a hydrophilic copolymer (PAAMa) of acrylic acid and maleic anhydride prepared from free radical copolymerization, the terpyridine segments were introduced in the copolymers and a series of hydrophilic target gelator polymers (PAAMa-Tpy) were obtained (Scheme 1). The supramolecular hydrogels with different terpyridine contents were prepared via the Ni^{2+}–terpyridine coordination bonds formed in aqueous solution. The photophysical properties, morphological characteristics, rheological properties, thermal responsiveness, and the equilibrium swelling ratio of the linear polymers and resulting metallo-supramolecular hydrogels based on non-covalent coordination were studied.

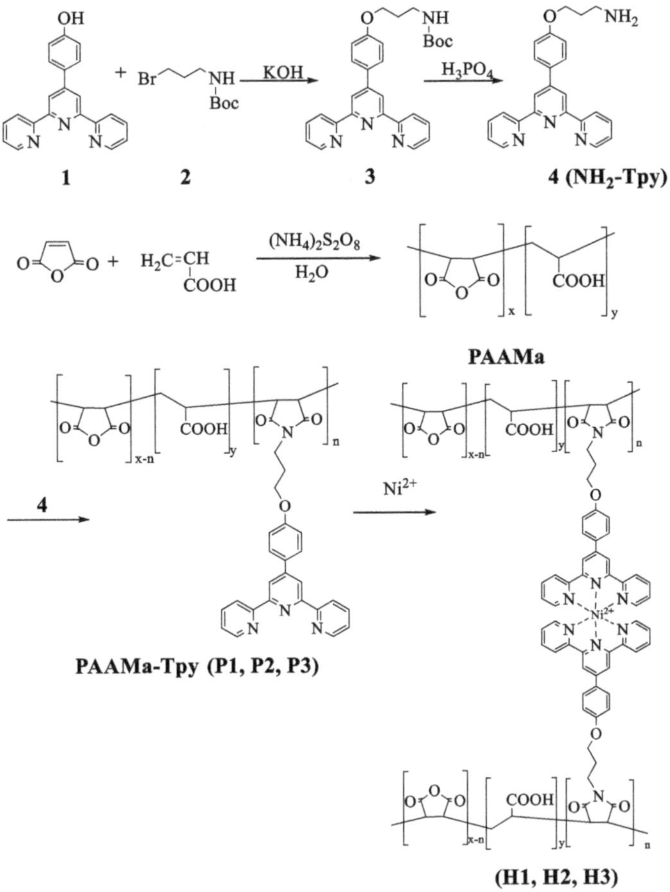

Scheme 1. Synthetic routes of the copolymer **PAAMa** and the terpyridine-containing polymers **P1**, **P2**, and **P3**, and the preparation of metallo-supramolecular hydrogels **H1**, **H2**, and **H3**.

2. Materials and Methods

All the chemicals were purchased from commercial sources and used without further purification unless otherwise specified. The 4′-(p-hydrophenyl)-2,2′:6′,2″-terpyridine (**1**) was synthesized via the Kröhnke method according to instructions in [32], 4′-(p-(3-(boc-amine) propoxy)-phenyl)- 2,2′:6′,2″-terpyridine (**3**) was synthesized according to the previous method in [33], and 4′-(p-(3-amine propoxy)-phenyl)- 2,2′:6′,2″-terpyridine (**NH₂-Tpy**) (4) was synthesized according to the reported method in [34]. The copolymer **PAAMa** was synthesized as indicated in the literature [35]. The gelator polymer (**PAAMa-Tpy**) was synthesized according to the instructions in [36]. The ^{1}H NMR spectra and ^{13}C NMR spectra were obtained with a Bruker Advance 500 spectrometer. IR spectra were recorded using a Bruker vertex 70 TGA-IR instrument. The UV spectra were recorded with a Hitachi U-4100 instrument. The morphological analysis of all hydrogels was carried out using a JSM-7500F instrument. The viscosity average molecular weight (M_η) was studied using an Ubbelohde viscometer. The swelling ratio of the resulting supramolecular hydrogel was measured at room temperature. The swelling ratio (g/g) of the sample was calculated: equilibrium swelling ratio (ESR) = $(m_t - m_0)/m_0$, where m_0 and m_t are the weights of the xerogel and hydrogel samples at time t, respectively. When a hydrogel reaches its swelling equilibrium state, its swelling ratio is called the equilibrium swelling ratio (ESR). In measurement, the free water should be removed carefully so that precise data can be obtained.

2.1. Synthesis of 4′-(p-(3-(boc-amine) propoxy)-phenyl)-2,2′:6′,2″-terpyridine (3)

Compound **1** (3.5 g, 10.7 mmol) and (3-(boc-amino)-propyl bromide) (**2**) (3.2 g, 13.4 mmol) were added to the solution of potassium hydroxide (0.7 g, 12.5 mmol) in ethyl alcohol (60 mL) under the protection of nitrogen and stirred for 48 h at 90 °C. After the reaction was completed, the solvent was evaporated, and the residue was separated using column chromatography with PE as eluent. Following this, 4′-(p-(3-(boc-amine) propoxy)-phenyl)-2,2′:6′,2″-terpyridine (**3**) (4.5 g, 9.3 mmol) was obtained as a white solid with a yield of 87%. ^{1}H NMR ($CDCl_3$): δ 8.74–8.66 (m, 6H), 7.90–7.88 (m, 4H), 7.37–7.34 (m, 2H), 7.03–7.01 (d, 2H, J = 8.8 Hz), 4.11–4.09 (t, 2H, J = 6.0 Hz), 3.38–3.36 (m, 2H), 2.04–2.01 (m, 2H), 1.46 (s, 9H), 0.98–0.94 (t, 1H, J = 7.4 Hz).

2.2. Synthesis of 4′-(p-(3-amine propoxy)-phenyl)-2,2′:6′,2″-terpyridine (NH_2-Tpy) (4)

To the solution of compound **3** (2.0 g, 4.1 mmol) in dichloromethane (15 mL) was added phosphoric acid aqueous solution (85 wt%) (1.5 mL, 7.0 mmol) dropwise; then, the mixture was vigorously stirred at room temperature for 2 h. Sodium hydroxide (0.9 g, 21.0 mmol) aqueous solution (10 mL) was added dropwise at 0 °C until the solvent became weakly alkaline (PH ≈ 8). The solvent was extracted with dichloromethane, and the organic phase was dried with magnesium sulfate and evaporated to obtain **NH_2-Tpy** as a white solid with a yield of 88%. FT-IR (KBr): 3960.26, 3734.43, 2928.74, 1696.56, 1659.53, 1600.96, 1516.05, 1438.43, 1389.36, 1247.81, 1178.37, 1093.31, 833.34, 788.26, 660.52 cm^{-1}. ^{1}H NMR ($CDCl_3$): δ 8.74–8.66 (m, 6H), 7.90–7.88 (m, 4H), 7.37–7.34 (m, 2H), 7.03–7.01 (d, 2H, J = 8.8 Hz), 4.11–4.08 (t, 2H, J = 6.0 Hz), 3.38–3.36 (m, 2H), 2.96–2.93 (t, 2H, J = 6.6 Hz), 2.04–2.01 (m, 2H). ^{13}C NMR ($CDCl_3$): 156.4, 155.8, 149.092, 136.8, 128.5, 123.7, 121.4, 118.2, 114.8, 77.3, 77.0, 76.8, 66.0, 39.2, 33.0.

2.3. Synthesis of Copolymer of Acrylic Acid and Maleic Anhydride PAAMa

To the solution of maleic anhydride (10.0 g, 102.0 mmol) in deionized water (30 mL) was added acrylic acid (18.2 g, 252.7 mmol), and the ammonium persulfate water solution (10 mL, 10 wt%). The solvent was stirred for 5 h at 85 °C, and was then cooled to room temperature. The solution was evaporated in a vacuum environment, and the copolymer **PAAMa** was obtained as a white solid with a yield of 96%. M_η = 21,000. ^{1}H NMR (D_2O): δ 4.40 (br, 2H), 3.92 (br, 2H), 1.79–1.96 (br, 5H), 1.49–1.50 (br, 10H).

2.4. General Synthetic and Purification Procedure of Gelator Polymer

To the three-neck flask was added **NH_2-Tpy** (2.4 g,6.3 mmol), **PAAMa** (4.5 g), and N,N-dimethylformamide (100 mL) under a nitrogen atmosphere. The mixture was stirred for 48 h at 130 °C, and then the solution was cooled to room temperature. The target gelator polymer **P2** (6.0 g) was obtained as a yellowish solid via precipitation with diethyl ether twice. The other polymers, **P1** and **P3**, with different contents of the terpyridine unit, were prepared with a similar method.

2.5. Preparation of Supramolecular Hydrogels

To the 0.14 mL aqueous solution of NaOH (0.1 g/mL) was added 0.1 g gelator polymer (**P1** or **P2** or **P3**) and 0.86 mL H_2O. After the polymer was dissolved completely, the 0.5 mL aqueous solution of $NiCl_2 \cdot 6H_2O$ with a concentration of 0.024 g/mL (for **P1**), 0.034 g/mL (for **P2**), and 0.042 g/mL (for **P3**) was added into the polymer solution to obtain supramolecular hydrogels **H1**, **H2**, and **H3**, respectively.

3. Results and Discussion

3.1. Synthesis of the Hydrophilic Polymers Containing Terpyridine Segments

Compound **1** was synthesized via the classic Kröhnke method according to the literature [32]. In an alkaline condition, amino-substituted bromide **2** reacted with the hydroxyl group in compound **1**, and the protected amino functional group could be incorporated on it with a high yield (Scheme 1). Through a deprotection process under an acid condition, the important intermediate **NH_2-Tpy** can be obtained. The copolymer of acrylic acid and maleic anhydride, **PAAMa**, was simply synthesized via free radical copolymerization of acrylic acid and maleic anhydride initiated by ammonium persulfate. The gelator polymers (**PAAMa-Tpy**) were synthesized via a reaction of different molar ratios of **NH_2-Tpy** with the polymer **PAAMa** in DMF. Through an amidation process of the terminal amino group in **NH_2-Tpy** with the anhydride unit in **PAAMa**, the terpyridine segments can be introduced onto **PAAMa** successfully [36]. The resulting linear polymers exhibited good solubility in water due to the large numbers of hydrophilic carboxyl groups from acrylic acid and maleic anhydride. The FTIR spectra of **PAAMa**, **P1**, **P2**, and **P3** are given in Figure 1. It can be seen that the strong stretching vibration absorption peaks of C=O at 1717 cm^{-1} and C-O at 1176 cm^{-1} for **PAAMa** appeared and there were not any aromatic structure signals. The characteristic bands of aromatic skeleton were detected at 1518 cm^{-1}, 1595 cm^{-1}, and 785 cm^{-1}, which were assigned to the aromatic structures in phenyl-substituted 2,2′:6′,2′′-terpyridine in polymers **P1**, **P2**, and **P3**. As seen in Figure 1, the absorption intensity of the aromatic skeleton structures became more and more strong from **P1** to **P3**, which is in agreement with the result of an increase in the terpyridine segments grafted onto the polymer for **P1**-**P3**. The NMR spectroscopy of **NH_2-Tpy**, **PAAMa**, and three gelator polymers with different terpyridine contents (**P1**, **P2**, and **P3**) are shown in Figure 2. The characteristic peaks of the aromatic part in **NH_2-Tpy** appeared between 7 and 9 ppm, whereas the **PAAMa** had no signal in this range, and the three polymers had the characteristic peaks of the terpyridine units at the same position as **NH_2-Tpy**, which indicated that the terpyridine moieties were incorporated onto **PAAMa** successfully.

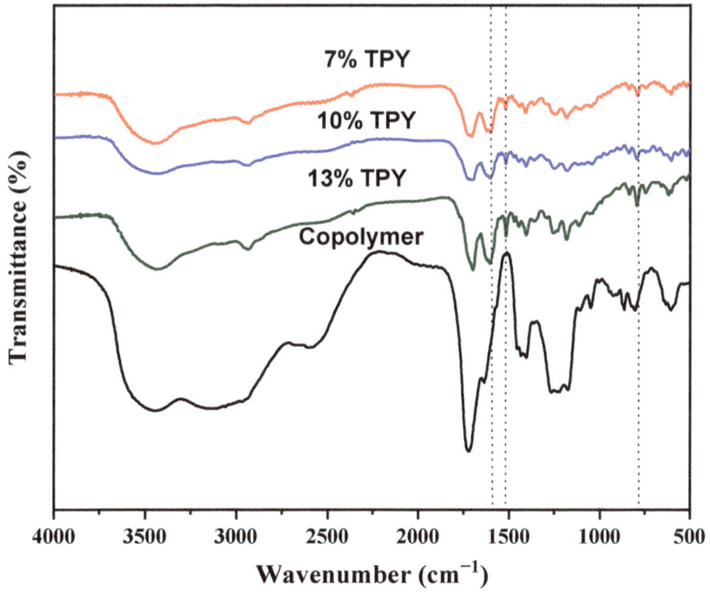

Figure 1. FTIR spectra of **PAAMa**, **P1**, **P2**, and **P3**.

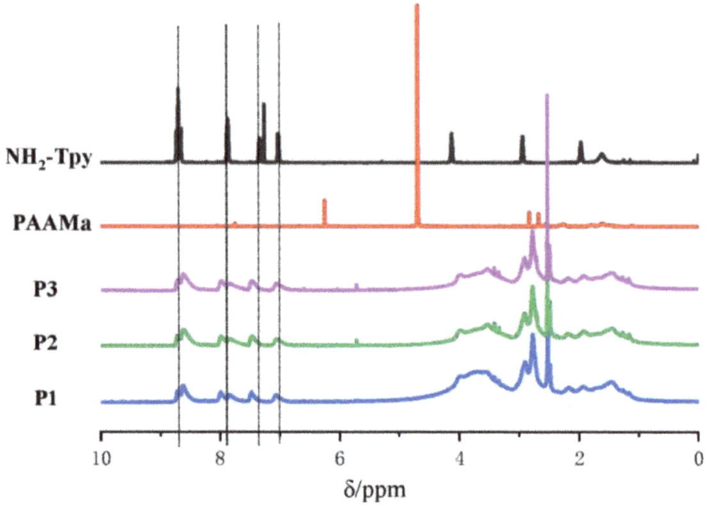

Figure 2. The NMR spectra of **NH_2-Tpy**, **PAAMa**, and the polymers **P1**, **P2**, and **P3**.

According to the integrating area of the NMR spectra, the content of the terpyridine segments in the resulting polymers (**PAAMa**) can be calculated, and the results obtained from NMR peaks and the feed percentage of the terpyridine-containing monomer **NH_2-Tpy** are summarized in Table 1. As shown in Table 1, the feed percentage of **NH_2-Tpy** was close to the content of terpyridine units calculated from NMR.

Table 1. The feed content of monomer **NH_2-Tpy**, and the content of terpyridine segments calculated from the NMR integrating area and the conversion rate of the radical polymerization for **P1**, **P2**, and **P3**.

Polymer	Feed Percentage of NH_2-Tpy (%)	Terpyridine Content from NMR (%)	Conversion Rate (%)
P1	7.2	6.9	95.8
P2	10.4	10.1	97.1
P3	13.5	12.7	94.1

3.2. Photophysical Properties of the Linear Polymers Containing Terpyridine Segments

Ultraviolet absorption spectroscopy was used to study the photophysical properties of **PAAMa-Tpy** and the interaction with Ni^{2+} in water solution. As shown in Figure 3, the UV spectra clearly showed that there was an absorption peak at about 290 nm caused by the π–π^* transition of the terpyridine-based conjugated structures, and the absorption intensity increased with the increase in the terpyridine content incorporated in the main chains of the resulting polymers [37,38]. An aqueous solution of $NiCl_2 \cdot 6H_2O$ (0.5 mL) with a concentration of 0.0024 g/mL, 0.0034 g/mL, and 0.0042 g/mL was added to an aqueous solution of **P1**, **P2**, and **P3** (9.5 mL) with a concentration of 0.0010 g/mL, respectively. A total of 0.1 mL of the above solution was diluted to 3 mL and tested. In addition to the shorter wavelength peak at about 290 nm similar to that of the uncoordinated polymers, a new absorption peak at about 335 nm appeared, which was attributed to the metal-to-ligand charge transfer (MLCT) after coordination with Ni^{2+}, which supports the formation of metal–ligand coordination between terpyridine and Ni^{2+} [24,37,39].

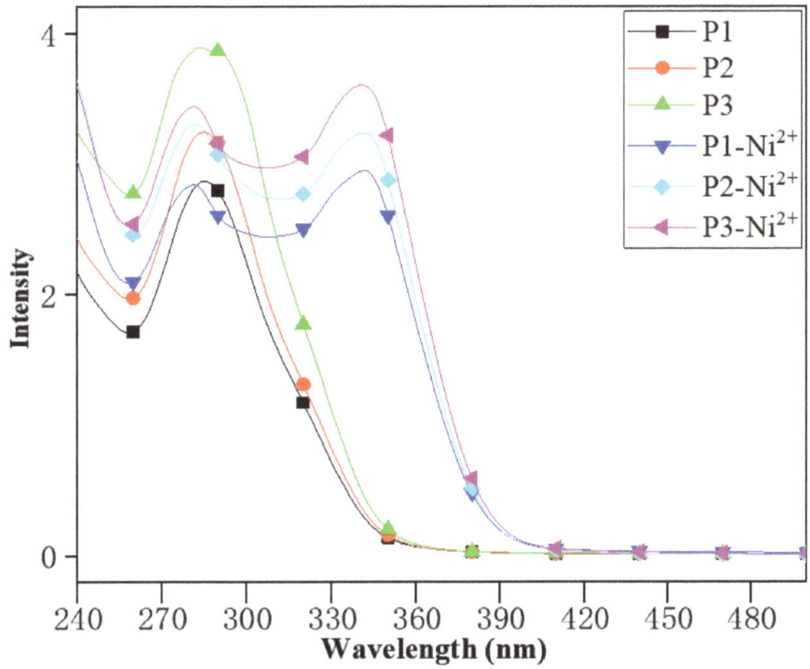

Figure 3. UV spectra of 33 ppm **P1** (7% terpyridine content), **P2** (10% terpyridine content), and **P3** (13% terpyridine content); and 33 ppm **P1**, **P2**, and **P3** with 5 ppm, 7 ppm, and 9 ppm Ni^{2+} in aqueous solution, respectively.

3.3. Preparation of the Metallo-Supramolecular Hydrogels

As shown in Figure 4, the Ni^{2+} was added to the aqueous solution of **PAAMa-Tpy**. The white complex was first produced as the complex of terpyridine-Ni^{2+} spread in the water; the transparency of the hydrogel then became higher, and finally, a uniform hydrogel was formed. After Ni^{2+} was added, the Ni^{2+} complexed with two terpyridines successively to generate mono-complex $Ni(terpyridine)^{2+}$ and bis-complex $Ni(terpyridine)_2{}^{2+}$ to obtain a twisted octahedral structure in which Ni^{2+} was coordinated with six nitrogen atoms [40,41]. The stability constant of bis(terpyridine)-complex $Ni(terpyridine)_2{}^{2+}$ can be estimated using the formation rate constant and dissociative rate constant between terpyridine and $Ni(terpyridine)^{2+}$. At 25 °C, the formation rate constant (k_f) of terpyridine and $Ni(terpyridine)^{2+}$ in water is $10^{5.3}$ $M^{-1}{\cdot}s^{-1}$, and the dissociative rate constant (k_d) of $Ni(terpyridine)_2{}^{2+}$ is $10^{-5.8}$ s^{-1}. Since the stability constant of $Ni(terpyridine)_2{}^{2+}$ $K_2 = k_f/k_d = 10^{11.1}$ M^{-1} and the stability constant of $Ni(terpyridine)^{2+}$ $K_1 = 10^{10.7}$ M^{-1}, the $Ni(terpyridine)_2{}^{2+}$ can exist stably at room temperature [42].

3.4. Micromorphological Study

To study the three-dimensional cross-linked networks inside the hydrogels, the microstructures of the xerogels prepared from the corresponding hydrogels via freeze drying were investigated through SEM. As shown in Figure 5a, the pores of the xerogel with 7% terpyridine content were irregular, and there was no molded cross-linked structure inside the hydrogel for **H1**. The partial collapse of the pores probably resulted from the fact that there was no effective or strong cross-linking structure for the low content of the coordination bonds. Figure 5b shows that the xerogel with 10% terpyridine content (**H2**) had uniform pore distribution and a regular structure, and the average diameter of the pores inside the xerogel was about 50–60 μm. Compared with **H1**, the increase in the content of terpyridine-unit-coordinated structures consequently led to a higher cross-linking degree and strength. The results indicate that there was a critical value of about 7–10% for the content of terpyridine units in forming effective cross-linking structures for these coordination-leading supramolecular hydrogels. As Figure 5c shows, the average diameter of the pores in the xerogel with 13% terpyridine content (**H3**) was slightly smaller than that of **H2** (about 50 μm). There were many nanofibers in xerogels **H2** and **H3**, whereas there were almost no nanofiber structures in **H1**. With increasing terpyridine content, the number of Ni^{2+}–terpyridine coordination bonds increased; these play a vital role in the formation of 3D network structures and their strength [43,44].

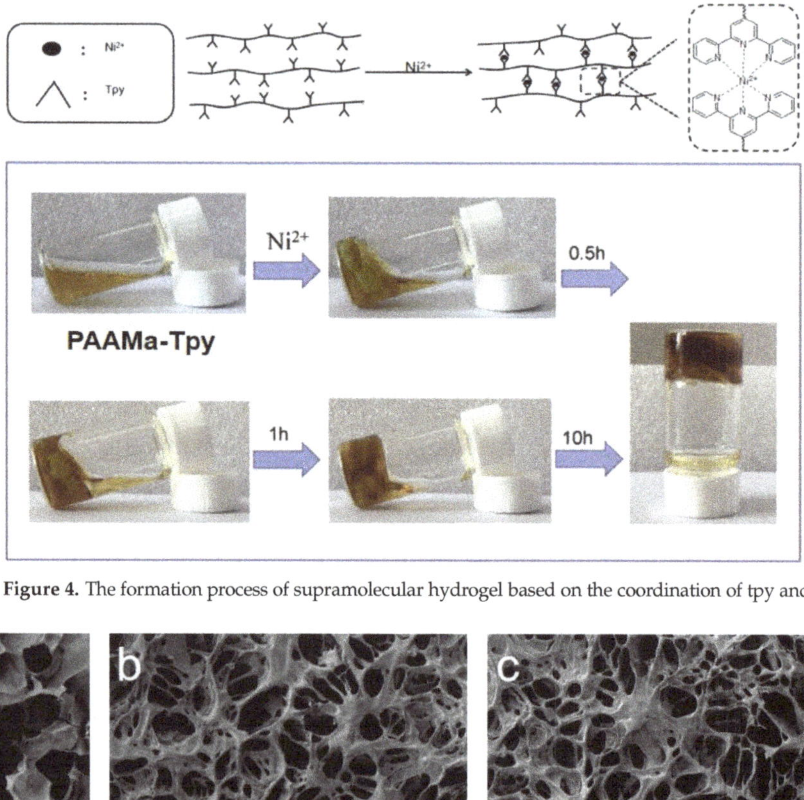

Figure 4. The formation process of supramolecular hydrogel based on the coordination of tpy and Ni^{2+}.

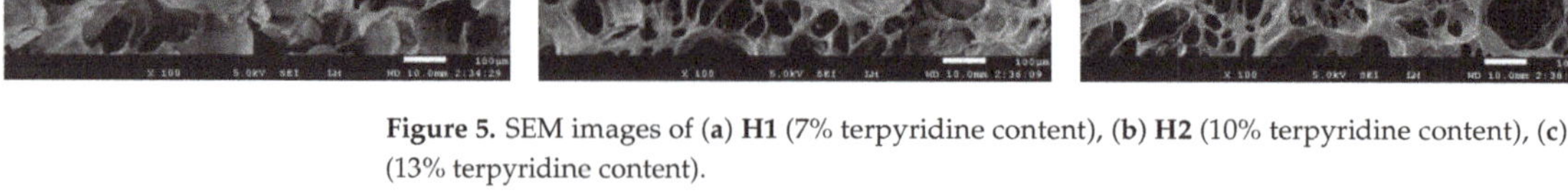

Figure 5. SEM images of (**a**) **H1** (7% terpyridine content), (**b**) **H2** (10% terpyridine content), (**c**) **H3** (13% terpyridine content).

3.5. Rheological Properties

The rheological properties of the three resulting metallo-supramolecular hydrogels with different terpyridine contents—**H1**, **H2**, and **H3**—with a water content of 90% at 25 °C were studied (Figure 6). As shown in Figure 6a, the changes in G′ (storage moduli) and G″ (loss moduli) of the hydrogels **H1**, **H2**, and **H3** as a function of applied frequency (0–100 Hz) were monitored. Both G′ and G″ increased obviously with increasing terpyridine content in hydrogels, from **H1**, **H2**, to **H3**, which was due to the greater number of metallo-supramolecular cross-link structures that could be formed with the higher content of terpyridine units in hydrogels [45]. At the same time, from 0 to 100 rad/s, both G′ and G″ increased with the increase in frequency and exhibited shear-thinning behaviors, which was attributed to the deformation of the material under shearing forces and the reorganization of the internal structures of the supramolecular hydrogels that appeared [46]. For hydrogel **H1**, with a lower terpyridine content, however, the G′ increased slightly with the increase in frequency and the differentials of G′ and G″ were rather lower at different frequency ranges, most likely because the supramolecular cross-link degree was very low and the networks had not been formed effectively. For all hydrogels, the G′ values were much larger than the corresponding G″, and the gaps between G′ and G″ of hydrogels with 7% (**H1**), 10% (**H2**), and 13% (**H3**) terpyridine content were about 700, 2000, and 2600 Pa, respectively, which were characteristic of an effective cross-linked network. Obviously, the coordination interaction of terpyridine ligands incorporated into the hydrophilic polymer chains with Ni^{2+} drove the formation of a viscoelastic 3D network, and consequently as the terpyridine content in hydrogels increased and then the supramolecular cross-link degree increased, the hydrogels became more solid-like.

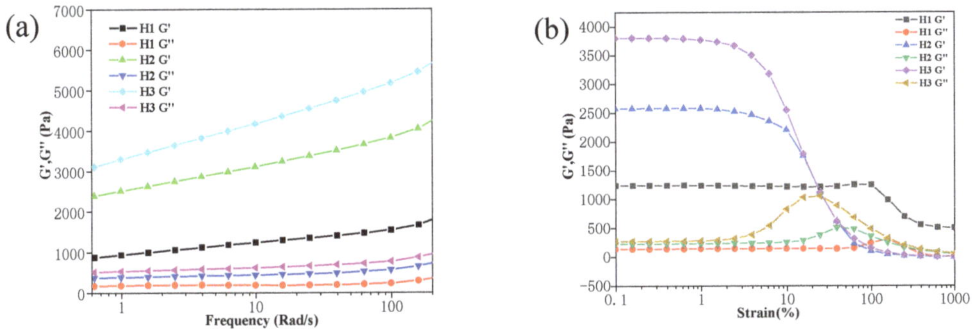

Figure 6. (**a**) Frequency sweep of three hydrogels with **H1** (7% terpyridine content), **H2** (10% terpyridine content), and **H3** (13% terpyridine content); (**b**) strain sweep of **H1**, **H2**, and **H3**.

Strain sweep measurements for **H1**, **H2**, and **H3** at a frequency of 6.28 Hz were carried out and the results were as shown in Figure 6b. As the strain increased, the G′ of **H3** decreased sharply, which was caused by the destruction of the Ni^{2+}–terpyridine coordination interactions in the hydrogel, and when the strain reached 20%, the curve of G″ appeared at a peak and was crossed over with G′ at this point, which showed that the internal cross-linked networks of the hydrogel were completely destroyed here [47,48]. The shapes of the G′ and G″ curves of the hydrogel **H2** were similar to those of the hydrogel **H3**, but the intersection point of the G′ and G″ curves for **H3** was at 1000 Pa, whereas the intersection point of the hydrogel **H2** was at 500 Pa, which was due to **H3** possessing more terpyridine and more coordination bonds [49]. The G″ peak of **H1** appeared at 250 Pa, which was lower than that of **H2**. By contrast, there was no intersection point between the G′ and G″ of **H1**, mostly due to the low content of terpyridine in the hydrogel **H1**, and there was not enough effective coordination for interaction-based cross-linked structures to be formed in this hydrogel.

3.6. Thermo-Responsive Properties

These polymer hydrogels based on metallo-supramolecular cross-linking exhibited interesting reversible sol–gel transition properties at different temperatures. To elucidate the relationship of the sol–gel transition temperature $T_{gel–sol}$ and the internal structure of the hydrogels, the $T_{gel–sol}$ of the hydrogels with different terpyridine contents and different water contents was investigated. As shown in Figure 7, the hydrogel **H1** gradually entered into a homogeneous solution state at about 55 °C for 1 h, and when the environmental temperature decreased to room temperature, the solution turned back to a gel state again slowly. This was probably caused by the disintegration of the Ni^{2+}–terpyridine coordination bonds in polymer hydrogels at the temperature above $T_{gel-sol}$, which were rebuilt at room temperature [50]. As shown in Table 2, the $T_{gel-sol}$ of the hydrogels **H1**, **H2**, and **H3** were 55 °C, 60 °C, and 75 °C, respectively. Although there existed equilibrium of free Tpy and $Ni(II)(Tpy)_2$ complexes, the equilibrium constant is related to temperature. When the temperature increased, the content of $Ni(II)(Tpy)_2$ was reduced to the critical value, and the number of cross-linking bonds was not high enough to keep hydrogels, the gel-to-sol transition occurred. Therefore, the hydrogel with a higher content of terpyridine-Ni(II) coordination bonds (H3) reduced to the critical content at higher temperature. Obviously, the increased cross-linking degree of the polymer hydrogels resulting from the greater number of Ni^{2+}–terpyridine coordination bonds could have increased their thermal stability and exhibited a higher $T_{gel-sol}$ [9]. Furthermore, the $T_{gel-sol}$ values of the hydrogel **H3** with water contents of 91%, 90%, and 85% were 70 °C, 75 °C, and 93 °C, respectively. The thermal stability of the hydrogels increased with the increase in the gelator content because of the increased intermolecular interactions and network strength [51,52]. After heating and cooling the same sample three times, the hydrogel still showed good phase transition characteristics, which indicates that the supramolecular hydrogels have an excellent dynamic reversible character.

Figure 7. The gel–sol–gel transition recycle of **H1**.

Table 2. The transition temperature of **H1**, **H2**, and **H3** with 90% water content.

Hydrogel	Transition Temperature
H1	55 °C
H2	60 °C
H3	75 °C

3.7. Swelling Behavior

These hydrogels contain a large number of hydrophilic groups in the main chains and exhibited water absorption properties. The swelling ratios of the metallo-supramolecular hydrogels **H1**, **H2**, and **H3** were measured at room temperature when a hydrogel reached its swelling equilibrium state. The results are shown in Figure 8. Generally, the water absorption ratio of the cross-linking polymer hydrogel decreased with increasing cross-linking degree of the hydrogel. As shown in Figure 8, the equilibrium swelling ratio of the resulting supramolecular hydrogels **H1**, **H2**, and **H3** were 12.3 g/g, 10.3 g/g, and 8.0 g/g, respectively. These results were caused by the different cross-linking degrees for different hydrogels. The cross-linking of these metallo-supramolecular hydrogels was obtained via coordination interactions between metal ions and terpyridine units incorporated in the linear polymer backbones. The content of terpyridine units in **P1**, **P2**, and **P3** were 7.0%, 10.0%, and 13.0%, respectively, and consequently the cross-linking degrees for the resulting hydrogels are in the following order: **H1** < **H2** < **H3**. Fewer terpyridine units in hydrophilic polymers led to fewer coordination bonds and a lower cross-linking degree in the hydrogels. The cross-linking network can swell more sufficiently, and more water can enter into the network and bind with the hydrophilic polymer structures. Consequently, the swelling ratio order for these supramolecular hydrogels was: **H1** > **H2** > **H3**. These supramolecular hydrogels showed high water-absorption speed, and the times to reach the swelling equilibrium state were 110 min, 130 min, and 140 min for **H1**, **H2**, and **H3**, respectively. Obviously, the low cross-linking degree was beneficial for the diffusion into the 3D hydrophilic network.

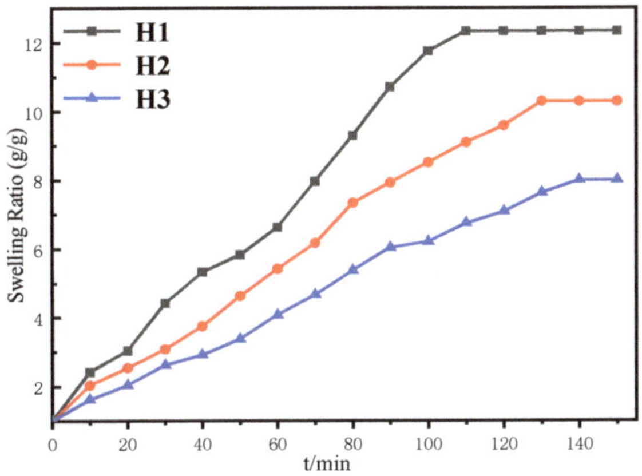

Figure 8. The equilibrium swelling ratio of the supramolecular polymer hydrogels **H1**, **H2**, and **H3**.

4. Conclusions

A novel metallo-supramolecular hydrogel was developed. The hydrophilic polymers were based on acrylic acid and maleic anhydride, and used radical polymerization. The gelator polymers containing different terpyridine units were obtained through grafting reactions and the NMR results indicate that the synthesis was successful. The metallo-supramolecular hydrogels were prepared via a coordination interaction of the terpyridine units and metal ions, and the photophysical properties, rheological properties, micromorphology, thermo-responsive properties, and equilibrium swelling behaviors were fully studied. The graft ratio of the terpyridine unit onto the main chains of the hydrophilic copolymers of acrylic acid and maleic anhydride can be tuned easily. The content of the ligand unit terpyridine played a vital role in the microstructures, mechanical performances, and equilibrium swelling ratio due to the fact that the cross-linking degree can be controlled conveniently via changing the content of terpyridine units in raw linear hydrophilic polymers. The interesting thermal-responsive characteristics originated from the non-covalent cross-linking bonds, and these can be further used as intelligent polymer hydrogels. In general, these hydrogels based on coordination show good promise for environmental responses, biosensors, and similar. In the future, adhesives, biocompatibility, photo-function, and so on based on different back-bone systems should be investigated and their application widened.

Author Contributions: Z.C. and C.M. contributed equally to this work. Conceptualization, Z.C., C.M., Z.H. (Zhaoge Huang), and X.C.; methodology, Z.C., C.M., and X.C.; validation, Z.H. (Ziyuan He) and Z.M.; formal analysis, Z.C.; investigation, Z.C. and C.M.; resources, Z.H. (Ziyuan He); data curation, Z.H. (Ziyuan He) and Z.M.; writing—original draft preparation, Z.C. and C.M.; writing—review and editing, X.C. and Z.H. (Zhaoge Huang); visualization, C.M.; supervision, X.C.; project administration, X.C. and Z.H. (Zhaoge Huang). All authors have read and agreed to the published version of the manuscript.

Funding: This research was funded by the National Natural Science Foundation of China (NSFC) (No.20804022) and the Natural Science Foundation of Shandong Province (No. ZR2011BM001).

Institutional Review Board Statement: Not applicable.

Informed Consent Statement: Not applicable.

Data Availability Statement: No statement.

Conflicts of Interest: The authors declare no conflict of interest.

References

1. Ahmed, E.M. Hydrogel: Preparation, Characterization, and Applications: A review. *J. Adv. Res.* **2015**, *6*, 105–121. [CrossRef] [PubMed]
2. Hu, W.K.; Wang, Z.J.; Xiao, Y.; Zhang, S.M.; Wang, J.L. Advances in Crosslinking Strategies of Biomedical Hydrogels. *Biomater. Sci.* **2019**, *7*, 843–855. [CrossRef] [PubMed]
3. Brunsveld, L.; Folmer, B.J.B.E.; Meijer, W.; Sijbesma, R.P. Supramolecular Polymers. *Chem. Rev.* **2001**, *101*, 4071–4098. [CrossRef]
4. Li, S.L.; Xiao, T.X.; Lin, C.; Wang, L.Y. Advanced Supramolecular Polymers Constructed by Orthogonal Self-Assembly. *Chem. Soc. Rev.* **2012**, *41*, 5950–5968. [CrossRef]
5. Hapiot, F.; Menuel, S.; Monflier, E. Thermoresponsive Hydrogels in Catalysis. *ACS Catal.* **2013**, *3*, 1006–1010. [CrossRef]
6. Ma, X.; Tian, H. Stimuli-Responsive Supramolecular Polymers in Aqueous Solution. *Acc. Chem. Res.* **2014**, *47*, 1971–1981. [CrossRef] [PubMed]
7. Shigemitsu, H.; Hamachi, I. Design Strategies of Stimuli-Responsive Supramolecular Hydrogels Relying on Structural Analyses and Cell-Mimicking Approaches. *Acc. Chem. Res.* **2017**, *50*, 740–750. [CrossRef]
8. Shao, T.; Falcone, N.; Kraatz, H.B. Supramolecular Peptide Gels: Influencing Properties by Metal Ion Coordination and Their Wide-Ranging Applications. *ACS Omega* **2020**, *5*, 1312–1317. [CrossRef]
9. Nandi, N.; Gayen, K.; Ghosh, S.; Bhunia, D.; Kirkham, S.; Sen, S.K.; Ghosh, S.; Hamley, I.W.; Banerjee, A. Amphiphilic Peptide-based Supramolecular, Non-Cytotoxic Stimuli-Responsive Hydrogels with Antibacterial Activity. *Biomacromolecules* **2017**, *18*, 3621–3629. [CrossRef]
10. Hymes, J.; Wolf, B. Biotinidase and Its Role in Biotin Metabolism. *Clin. Chim. Acta* **1996**, *255*, 1–11. [CrossRef]
11. Daigo, M.; Fumito, A.; Hideo, T.; Jungeun, K.; Katomasaki, T.; Takuzo, A. Ferroelectric Columnar Liquid Crystal Featuring Confined Polar Groups Within Core-Shell Architecture. *Science* **2012**, *336*, 209–213.
12. Santanu, B.; Yamuna, K.G. First Report of Phase Selective Gelation of Oil from Oil/Water Mixtures. Possible Implications Toward Containing Oil Spills. *Chem. Commun.* **2001**, *2*, 185–186.
13. Rowan, S.J.; Cantril, S.J.; Cousins, G.R.L.; Jeremy, K.M.; Sanders, J.K.M.; Stoddart, J.F. Dynamic Covalent Chemistry. *Angew. Chem. Int. Ed. Engl.* **2010**, *41*, 898–952. [CrossRef]
14. Sun, Z.F.; Li, Z.Y.; He, Y.H.; Shen, R.J.; Deng, L.; Yang, M.H.; Liang, Y.Z.; Zhang, Y. Ferrocenoyl Phenylalanine: A New Strategy Toward Supramolecular Hydrogels with Multistimuli Responsive Properties. *J. Am. Chem. Soc.* **2013**, *135*, 13379–13386. [CrossRef] [PubMed]
15. Zhao, X.J.; Wang, Q.; Du, M. Metal-Directed Assembly of 1-D and 2-D Coordination Polymers with Fluconazole and Dicyanamide Co-Ligand. *Inorg. Chim. Acta* **2007**, *360*, 1970–1976. [CrossRef]
16. Fullenkamp, D.E.; He, L.H.; Barrett, D.G.; Burghardt, W.R.; Messersmith, P.B. Mussel-Inspired Histidine-Based Transient Network Metal Coordination Hydrogels. *Macromolecules* **2013**, *46*, 1167–1174. [CrossRef]
17. Wang, C.; Flynn, N.T.; Langer, R. Controlled Structure and Properties of Thermoresponsive Nanoparticle-Hydrogel Composites. *Adv. Mater.* **2010**, *16*, 1074–1079. [CrossRef]
18. Khatyr, A.; Ziessel, R. Stepwise Construction of Pyrene Bridged Polytopic Ligands Carrying Acetylenic Tethers. *Tetrahedron Lett.* **2002**, *43*, 7431–7434. [CrossRef]
19. Stadler, A.M.; Puntoriero, F.; Campagna, S.; Kyritsakas, N.; Welter, L.; Lehn, J.M. Synthesis, Structural Features, Absorption Spectra, Redox Behaviour and Luminescence Properties of Ruthenium(Ii) Rack-Type Dinuclear Complexes with Ditopic, Hydrazone-Based Ligands. *Chemistry* **2005**, *11*, 3997–4009. [CrossRef]
20. Liu, P.; Shi, G.H.; Chen, X.G. Terpyridine-Containing π-Conjugated Polymers for Light-Emitting and Photovoltaic Materials. *Front. Chem.* **2020**, *8*, 592055. [CrossRef]
21. Andreas, W.; Andreas, W.; Florian, S.; Ulrich, S.S. Advances in The Field of π-conjugated 2,2′:6′,2″-Terpyridines. *Chem. Soc. Rev.* **2011**, *40*, 1459–1511.
22. Yang, H.; Ghiassinejad, S.; Ruymbeke, E.V.; Fustin, C.A. Tunable Interpenetrating Polymer Network Hydrogels Based on Dynamic Covalent Bonds and Metal–Ligand Bonds. *Macromolecules* **2020**, *53*, 6956–6967. [CrossRef]
23. Yu, X.; Wang, Z.; Li, Y.; Geng, L.; Ren, J.; Feng, G. Fluorescent and Electrochemical Supramolecular Coordination Polymer Hydrogels Formed from Ion-Tuned Self-Assembly of Small Bis-Terpyridine Monomer. *Inorg. Chem.* **2017**, *56*, 7512–7518. [CrossRef]
24. Xu, Y.; Jerca, F.A.; Jerca, V.V.; Richard, H. Self-Healing and Moldable Poly (2-isopropenyl-2-oxazoline) Supramolecular Hydrogels Based on a Transient Metal Coordination Network. *Macromolecules* **2020**, *53*, 6566–6575. [CrossRef]
25. Bekiari, V.; Lianos, P. Photophysical Behavior of Terpyridine-Lanthanide Ion Complexes Incorporated in A Poly (N,N-Dimethylacrylamide) Hydrogel. *Langmuir* **2006**, *22*, 8602–8606. [CrossRef] [PubMed]
26. Filho, C.M.C.; Bueno, P.V.A.; Matsushita, A.F.Y.; Vilsinski, B.H.; Valente, A.J.M. Uncommon Sorption Mechanism of Aromatic Compounds onto Poly (Vinyl Alcohol)/Chitosan/Maleic Anhydride-β-Cyclodextrin Hydrogels. *Polymers* **2020**, *12*, 877. [CrossRef] [PubMed]
27. Hennig, T.L.R.; Kascholke, C.; Hoetzel, R.; Hacker, M.C. Reactive and Atimuli-Responsive Maleic Anhydride Containing Macromers–Multi-Functional Cross-Linkers and Building Blocks for Hydrogel Fabrication. *React. Funct. Polym.* **2013**, *73*, 1480–1492.

28. Jakub, H.; Eva, C.K.; Miroslava, D.S.; Radka, H.; Jakub, S.; Martin, H.; Jiri, M.; Jiri, H.; Petr, L.; Roman, S. Hydrogel Tissue Expanders for Stomatology. Part II. Poly (styrene-maleic anhydride) Hydrogels. *Polymers* **2019**, *11*, 1087.
29. Schmidt, U.; Zschoche, S.; Werner, C. Modification of Poly (Octadecene-alt-Maleic Anhydride) Films by Reaction with Functional Amines. *J. Appl. Polym. Sci.* **2010**, *8*, 1255–1266. [CrossRef]
30. Uppu, D.S.S.M.; Akkapeddi, P.; Manjunath, G.B.; Yarlagadda, V.; Hoque, J.; Haldar, J. Polymers with Tunable Side-Chain Amphiphilicity as Non-Hemolytic Antibacterial Agents. *Chem. Commun.* **2013**, *49*, 9389–9391. [CrossRef]
31. Koning, C.; Ikker, A.; Borggreve, R.; Leemans, L.; Miller, M. Reactive Blending of Poly (Styrene-co-Maleic Anhydride) with Poly (Phenylene Oxide) by Addition of (Alpha)-Amino-Polystyrene. *Polymer* **1993**, *34*, 4410–4416. [CrossRef]
32. Hanabusa, K.; Nakamura, A.; Koyama, T.; Shirai, H. Synthesis, Polymerization, Copolymerization, and Transition-Metal Coordination of 4-(2,2′:6′,2″-terpyridin-4′-yl) styrene and its Polymers and Copolymers. *Macromol. Chem. Phys.* **1992**, *193*, 1309–1319. [CrossRef]
33. Chen, X.G.; Zhou, Q.G.; Cheng, Y.X.; Geng, Y.H.; Ma, D.G. Synthesis, Structure and Luminescence Properties of Zinc (II) Complexes with Terpyridine Derivatives as Ligand. *J. Lumin.* **2007**, *126*, 81–90. [CrossRef]
34. Li, B.; Berliner, M.; Buzon, R.; Chiu, C.K.F.; Colgan, S.T.; Kaneko, T.; Keene, N.; Kissel, W.; Le, T.; Leeman, K.R.; et al. Aqueous Phosphoric Acid as a Mild Reagent for Deprotection of Tert-Butyl Carbamates, Esters, and Ethers. *J. Org. Chem.* **2006**, *71*, 9045–9050. [CrossRef] [PubMed]
35. Yang, M.; Liu, B.T.; Gao, G.; Liu, X.L.; Liu, F.Q. Poly (Maleic Anhydride-co-Acrylic acid)/Poly (Ethylene Glycol) Hydrogels with pH- and Ionic-Strength-Responses. *Chin. J. Polym. Sci.* **2010**, *28*, 951–959. [CrossRef]
36. Barman, S.; Konai, M.M.; Samaddar, S.; Haldar, J. Amino-Acid Conjugated Polymers: Antibacterial Agents Effective against Drug-resistant A. baumannii with no Detectable Resistance. *ACS Appl. Mater. Interfaces* **2019**, *11*, 33559–33572. [CrossRef]
37. Zhou, G.C.; Harruna, I.I. Synthesis and Characterization of Bis (2,2′:6′,2″-Terpyridine) Ruthenium(II)-Connected Diblock Polymers via RAFT Polymerization. *Macromolecules* **2005**, *38*, 4114–4123. [CrossRef]
38. Holyer, R.H.; Hubbard, C.D.; Kettle, S.F.A.; Wilkins, R.G. The Kinetics of Replacement Reactions of Complexes of the Transition Metals with 1,10-Phenanthroline and 2,2′-Bipyridine. *Inorg. Chem.* **1965**, *4*, 929–935. [CrossRef]
39. Carlos, G.S.; Lohmeijer, B.G.G.; Meier, M.A.R.; Schubert, U.S. Synthesis of Terpyridine-Terminated Polymers by Anionic Polymerization. *Macromolecules* **2005**, *38*, 10388–10396.
40. Zhang, A.J.; Chen, C.J.; Lou, D.Y.; Miao, Y.H.; Qiao, C.X.; Hu, J.B. Nickel (II), Cadmium (II), and Copper (II) Complexes Based on Ditopic Terpyridine Derivative Ligand: Syntheses, Crystal Structures, and Luminescent Properties: Nickel (II), Cadmium (II), and Copper (II). *Z Anorg. Allg. Chem.* **2016**, *642*, 817–822. [CrossRef]
41. Naik, R.M.; Rai, J.; Srivastava, A.; Ratan, S.; Singh, I. Kinetic and Mechanistic Studies of Uncatalyzed Substitution of Coordinated Cyanide in Hexacyanoferrate (II) by 2,2′-Bipyridine. *Int. J. Chem. Phys. Sci.* **2015**, *4*, 105–114.
42. Holyer, R.H.; Hubbard, C.D.; Kettle, S.F.A.; Wilkins, R.G. The Kinetics of Replacement Reactions of Complexes of the Transition Metals with 2,2′,2″-Terpyridine. *Inorg. Chem.* **1966**, *5*, 622–625. [CrossRef]
43. Zhang, Y.; Gu, H.; Yang, Z.; Xu, B. Supramolecular Hydrogels Respond to Ligand-Receptor Interaction. *J. Am. Chem. Soc.* **2003**, *125*, 13680–13681. [CrossRef]
44. Reddy, S.M.M.; Augustine, G.; Ayyadurai, N.; Shanmugam, G. Biocytin-Based pH-Stimuli Responsive Supramolecular Multivariant Hydrogelator for Potential Applications. *ACS Appl. Bio. Mater.* **2018**, *1*, 1382–1388. [CrossRef] [PubMed]
45. Li, N.; Liu, C.J.; Chen, W. Facile Access to Guar Gum Based Supramolecular Hydrogels with Rapid Self-Healing Ability and Multi-Stimuli Responsive Gel-Sol Transitions. *J. Agric. Food Chem.* **2019**, *11*, 33559–33572.
46. Appel, E.A.; Loh, X.J.; Jones, S.T.; Biedermann, F.; Dreiss, C.A.; Scherman, O.A. Ultrahigh-Water-Content Supramolecular Hydrogels Exhibiting Multistimuli Responsiveness. *J. Am. Chem. Soc.* **2012**, *134*, 11767–11773. [CrossRef]
47. Song, G.S.; Zhao, Z.Y.; Peng, X.; He, C.C.; Weiss, R.A.; Wang, H.L. Rheological Behavior of Tough PVP-In Situ-PAAm Hydrogels Physically Cross-Linked by Cooperative Hydrogen Bonding. *Macromolecules* **2016**, *49*, 8265–8273. [CrossRef]
48. Kitazawa, Y.; Iwata, K.; Imaizumi, S.; Ahn, H.; Sung, Y.K.; Ueno, K.; Moon, J.P.; Watanabe, M. Gelation of Solvate Ionic Liquid by Self-Assembly of Block Copolymer and Characterization as Polymer Electrolyte. *Macromolecules* **2014**, *47*, 6009–6016. [CrossRef]
49. Miao, T.X.; Fenn, S.L.; Charron, P.N.; Floreani, R.A. Self-Healing and Thermoresponsive Dual-Cross-Linked Alginate Hydrogels Based on Supramolecular Inclusion Complexes. *Biomacromolecules* **2015**, *16*, 3740–3750. [CrossRef]
50. Liu, J.; Chen, G.; Guo, M.; Jiang, M. Dual Stimuli-Responsive Supramolecular Hydrogel Based on Hybrid Inclusion Complex (HIC). *Macromolecules* **2010**, *43*, 8086–8093. [CrossRef]
51. Sarkar, K.; Dastidar, P. Supramolecular Hydrogel Derived from a C3-Symmetric Boronic Acid Derivative for Stimuli Responsive Release of Insulin and Doxorubicin. *Langmuir* **2018**, *34*, 685–692. [CrossRef] [PubMed]
52. Raghavan, S.R.; Cipriano, B.H. Gel Formation: Phase Diagrams Using Tabletop Rheology and Calorimetry. In *Molecular Gels*; Springer: Berlin/Heidelberg, Germany, 2006; pp. 241–252.

polymers

Review

Bioinspired Bottlebrush Polymers for Aqueous Boundary Lubrication

Xiaoyan Liu [1,*] and Per M. Claesson [2]

1 School of Chemistry and Chemical Engineering, Shaanxi Normal University, Xi'an 710062, China
2 Division of Surface and Corrosion Science, Department of Chemistry, School of Engineering Sciences in Chemistry, Biotechnology and Health, KTH Royal Institute of Technology, SE-100 44 Stockholm, Sweden; percl@kth.se
* Correspondence: xiaoyanl@kth.se

Abstract: An extremely efficient lubrication system is achieved in synovial joints by means of bio-lubricants and sophisticated nanostructured surfaces that work together. Molecular bottlebrush structures play crucial roles for this superior tribosystem. For example, lubricin is an important bio-lubricant, and aggrecan associated with hyaluronan is important for the mechanical response of cartilage. Inspired by nature, synthetic bottlebrush polymers have been developed and excellent aqueous boundary lubrication has been achieved. In this review, we summarize recent experimental investigations of the interfacial lubrication properties of surfaces coated with bottlebrush bio-lubricants and bioinspired bottlebrush polymers. We also discuss recent advances in understanding intermolecular synergy in aqueous lubrication including natural and synthetic polymers. Finally, opportunities and challenges in developing efficient aqueous boundary lubrication systems are outlined.

Keywords: bioinspired bottlebrush polymers; aqueous boundary lubrication; friction; wear resistance

Citation: Liu, X.; Claesson, P.M. Bioinspired Bottlebrush Polymers for Aqueous Boundary Lubrication. *Polymers* **2022**, *14*, 2724. https://doi.org/10.3390/polym14132724

Academic Editors: Andrzej Puszka and Beata Podkościelna

Received: 16 June 2022
Accepted: 1 July 2022
Published: 3 July 2022

1. Introduction

The origin of friction is the energy dissipation processes that occur as two surfaces slide against each other [1]. Friction forces between two surfaces are often characterized by the effective friction coefficient, μ_{eff}, that is, calculated by taking the ratio of friction force (F_{Friction}) and applied load (F_{Load}). The friction coefficient in synovial joints in mammals was found to be as low as 0.001 as measured by hip function simulator machines, even though values reported in different studies vary significantly [2]. It can be safely concluded that the friction coefficient is well below 0.01. For example, Gale et al. have reported values in the range of 0.002~0.006 [3]. This extremely efficient aqueous lubrication ability of the synovial joint is achieved by an association of lubricin, hyaluronan, phospholipids, and aggrecan [4–6]. Lubricin and aggrecan, bottlebrush-like biomacromolecules that have densely grafted pendant chains are abundant in synovial joints [7,8]. It is believed that these bottlebrush structured biomacromolecules play key roles for a number of critical biological functions, for example, hydration, aqueous boundary lubrication, wear resistance as well as mediating the rheological and mechanical properties under pressure [8,9].

Inspired by nature, synthetic bottlebrush polymers consisting of a linear polymeric backbone on which side chains are densely grafted have been designed and synthesized [10–12]. Theoretical studies and computer simulations have been used to predict the interfacial properties of bottlebrush polymer layers. The adsorbed amount of different bottlebrush polymers at the mica/silica surfaces has been predicted by lattice mean-field theory and compared with experimental data [13]. The frictional forces between polymer brush-like layers formed by linear polymers have been predicted by molecular dynamic simulations and scaling theory [14,15]. In addition, the interactions between bottlebrush polyelectrolyte layers have been predicted by molecular dynamic simulations [16]. In recent years, considerable progress has been made in theory and computer simulation to investigate the interfacial properties of polymer brush layers. For more detailed discussions, we recommend the valuable reviews on the theoretical studies of interaction forces and the lubrication properties of polymer brush-like layers [17,18].

The adsorption and interfacial lubrication properties between surfaces coated with a range of different bioinspired bottlebrush polymers have been extensively investigated by various surface analytical techniques [11–13,19–51]. Optical reflectometry [27–29], ellipsometry [52], neutron reflectometry [28,52], and quartz crystal microbalance with dissipation (QCM-D) [27,29] have been used to study the adsorption and desorption of bottlebrush polymers on surfaces in aqueous media. Direct force measurement techniques including surface force apparatus (SFA) [26,43,44,46], atomic force microscopy-colloidal probe technique (AFM-CP) [12,30], and mini traction machine [11,31] have been used to determine surface forces and/or friction forces between the bottlebrush polymer layers. Such measurements have shown that very efficient lubrication properties can be achieved by bottlebrush structured polymers [12,19–23,30,31,48,53]. Such excellent performance and highly promising lubrication systems are of great general interest, not least for relieving arthritis pain [54].

To introduce the subject of this review, we start with a short introduction of the molecular structures and functions of several important bio-lubricants found in the synovial fluid. Next, we discuss recent experimental investigations of the interfacial lubrication properties of bottlebrush polymer layers including friction forces and wear resistance. In addition, synergistic aqueous lubrication mediated by natural and bottlebrush polymers and small amphiphilic molecules is also discussed. Finally, outlooks for future research in and use of aqueous boundary lubrication systems are provided.

2. Bio-Lubricants

Bio-lubricants including lubricin, hyaluronan, and phospholipids as well as aggrecan, which is important for the mechanical response of cartilage, play important roles in reducing friction forces and providing shock absorbing mechanical responses to articular cartilage surfaces [8]. These molecular structures are very different, but they share a common feature that they contain highly hydrophilic groups that bind water and allow for the hydration lubrication mechanism [55] to be operative. Lubricin is a glycoprotein with a bottlebrush structure. It is composed of the N-terminal by 2-somatomedin B (SMB)-like domains, the C-terminal with a hemopexin (PEX)-like domain, chondroitin sulfate (CS) side chain, and a densely glycosylated and mucin-like domain in the middle (Figure 1) [56,57]. It has been suggested that lubricin is able to counteract damages of the superficial zone of cartilage, and contributes to the preservation of chondrocytes in the joints [58]. The bottlebrush structure in lubricin and mucins also reduces the interpenetration zone, that is, the region where polymer chains on two opposing surfaces carrying such molecules overlap. This suggests that the bottlebrush structured polymers are promising candidates for reducing friction forces as the energy dissipation related to dragging polymer chains through the interpenetration zone is minimized. The interfacial lubrication studies of various types of mucins have shown that friction forces between surfaces can be reduced by mucins, which can be attributed to their extensive hydration, which arises from the large amount of oligomeric carbohydrate side chains [59–64]. Aggrecan also has a bottlebrush structure domain in the middle (Figure 1) [65,66]. The friction forces between surfaces with covalently attached cartilage aggrecan was found to be rather low ($\mu_{eff} = 0.03 \pm 0.01 \sim 0.11 \pm 0.01$, depending on the ionic strength of the solution) [67], even though this molecule's most important function in the synovial joint area is found within the cartilage where it is associated with hyaluronan, and together with collagen, builds the intricate nanostructure of cartilage. For instance, the compressive resistance of articular cartilage has important contributions from hierarchical brush-on-brush structures formed by one hyaluronan associated with as many as 100 aggrecan molecules [68].

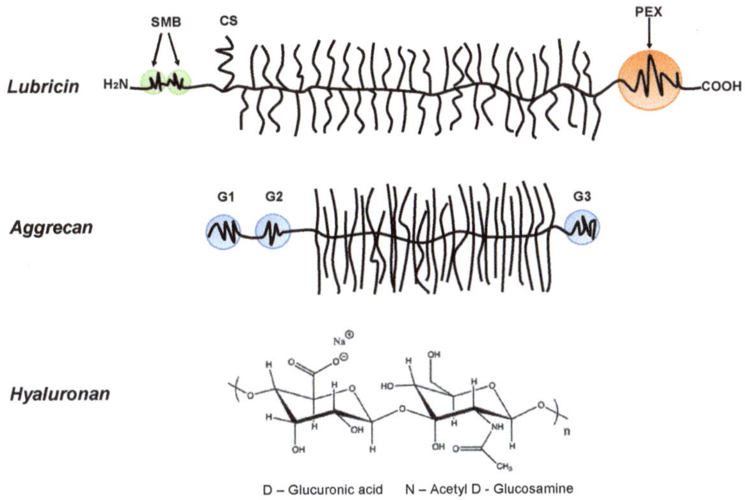

Figure 1. ***Lubricin:*** containing the N-terminal 2-somatomedin B (SMB)-like domains and the C-terminal hemopexin (PEX)-like domain. Lubricin also contains a chondroitin sulfate (CS) side chain, and in the middle region, it has a densely glycosylated and mucin-like domain. ***Aggrecan***: containing three globular domains (G1, G2, and G3), and in the middle, it has a large extended domain heavily modified with glycosaminoglycans. ***Hyaluronan (HA)***: the anionic disaccharide building unit of hyaluronan.

Hyaluronan and phospholipids also play important roles to achieve efficient aqueous lubrication systems [5,69–75]. Hyaluronan is a linear anionic polysaccharide (Figure 1). In fact, hyaluronan alone is not enough to reduce friction forces between surfaces [76,77]. However, hyaluronan is responsible for the viscous and elastic properties of the synovial fluid, which is very important for reducing the friction forces between articular cartilage [78,79]. It has also been reported that hyaluronan/aggrecan aggregates can achieve better lubrication than hyaluronan alone, which is due to the highly charged glycosaminoglycan segments on the aggrecan core protein [79]. In addition, the friction forces between damaged human cartilage can be reduced by self-assembled structures formed by hyaluronan and phospholipids [80]. The favorable interfacial lubrication properties of hyaluronan associated with phospholipids have also been observed in model systems [81–83]. The results of the friction force measurements between model silica surfaces coated with supported DPPC bilayers in the presence of hyaluronan have clearly indicated that the aggregates of hyaluronan/phospholipids are able to achieve low friction up to the pressure of 56 MPa, [81] which is higher than the pressure (25 $\pm$ 5.2 MPa) [84] that leads to the damage of the hyaline cartilage. Recently, multilayers have been prepared by the co-adsorption of HA and DPPC vesicles by Raj et al. [82], and the investigation showed that the friction coefficient between the layers was below 0.01 up to the pressure of 20 MPa. Over the last decade, Dedinaite and co-workers have investigated the intermolecular synergistic mechanism of the bio-lubricants in synovial fluid such as synergy pairs of hyaluronan and phospholipids [77,81–87]. Their studies indicate that synergistic lubrication can be achieved by the bio-lubricants working together. We summarize some of the observed synergistic effects of bio-lubricants regarding their interfacial lubricating properties in Table 1.

Table 1. The lubrication performance of the bio-lubricant aggregates. The table provides data for the effective friction coefficient, μ_{eff}, and the highest pressure, P, investigated.

Compositions	Substrate	Interfacial Lubricating Properties	Ref.
HA + DPPC liposomes	Damaged human cartilage	The reduction in friction was 69.5%, P = 1.3 MPa	[79]
HA + Aggrecan	Mica	μ_{eff} = 0.01, P = 1.6 MPa	[78]
HA + DPPC vesicles	Macroscopic glass surfaces	μ_{eff} = 0.1, P = 210 MPa	[77]
HA + DPPC bilayer	Silica	μ_{eff} = 0.03, P = 56 MPa	[80]
HA + DPPC vesicles	Silica	μ_{eff} < 0.01, P = 23 MPa	[81]
COMP + lubricin	PMMA	μ_{eff} = 0.06, P = 7 MPa	[85]
cross-linked HA + DOPC	Mica	μ_{eff} > 0.5, P = 2 MPa	[88]
HA + Lubricin	Mica	μ_{eff} = 0.09–0.4, P = 4 MPa	[89]
HA + Lubricin + Type II collagen	Gold versus SiO_2	μ_{eff} = 0.01, P = 0.013 MPa	[90]

3. Bottlebrush Polymers

3.1. Synthesis of Bottlebrush Polymers

Graft copolymers are composed of a polymeric backbone with densely grafted side chains, of which bottlebrush polymers are considered as a subset characterized by a very high grafting density. With increasing packing density of the side chains, a comb structure changes to a bottlebrush structure. This forces the backbone to adopt an extended conformation, and the molecule forms an overall cylindrical morphology. The conformation of the backbone and the length and the density of the side chains in bottlebrush polymers play significant roles in the lubrication properties. Inspired by bottlebrush bio-lubricants, synthetic bottlebrush polymers have been designed and prepared [91–97]. Conventional radical polymerization methods including atom transfer radical polymerization (ATRP), reversible addition-fragmentation chain transfer (RAFT), and nitroxide-mediated radical polymerization (NMP)) have been used to prepare precisely designed bottlebrush polymers [98–103]. According to how the side chains of the bottlebrush polymers are formed, the synthesis of bottlebrush polymers can be categorized into three approaches: (i) "grafting-from" (the polymerization of monomers from polyinitiators in the backbone) [99,100]; (ii) "grafting-through" (polymerization of macromonomers) [101,102]; and (iii) "grafting-to" (attachment of pre-formed side chains to the backbone) [104] (Figure 2). Novel bottlebrush polymers with advanced molecular structures such as double-brush [105], Janus [102], and core-shell [103] bottlebrush polymers have been prepared recently. Bottlebrush polymers can be applied in various fields such as templates of novel nanostructures, drug delivery, aqueous lubrication, and super-soft elastomers. In addition, novel cylindrical nanostructures such as segmented nanofibers [106] and anisotropic polymer nanostructures [107] have been developed. These emerging materials of precisely defined dimensions can be applied in many fields such as, sensors, bioimaging as well as lubrication [108,109]. Here, only a general overview of bottlebrush polymer synthesis is provided; for interested readers, we recommend a comprehensive review of the design and synthesis of bottlebrush polymers [10,92,93].

3.2. Adsorption of Bottlebrush Polyelectrolytes at Surfaces

The adsorption properties of bottlebrush polymers at the mica and silica surfaces have been predicted by the lattice mean-field theory by Linse and Claesson [13], Figure 3. The bottlebrush polymers were considered to consist of charged segments without side chains and uncharged segments with an attached side chain. The composition variable X stands for the percentage of charged segments ranging from X = 0 (uncharged bottlebrush polymer) to X = 100 (linear polyelectrolyte). The results of the theoretical modeling indicate that there is a large difference in the surface excess of bottlebrush polymers at the mica and silica surfaces. The difference in the surface excess between regular and random distributions is at most 20%, while a 5-fold difference in the surface excess between regular and diblock distributions has been predicted. These results indicate that the adsorption properties of bottlebrush polymers can be influenced by the polymer segment sequence, the non-electrostatic interactions as well as the surface charge density. To be able to tune the layer structure is important as the structure of the adsorbed polymer layers directly influences the interfacial lubrication properties of the polymer bearing surfaces.

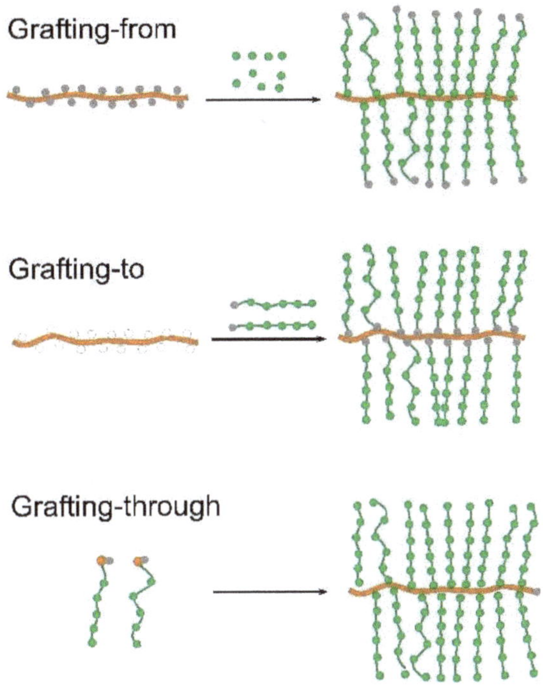

Figure 2. An illustration of three main synthetic approaches of bottlebrush polymer synthesis: grafting-from, grafting-to, and grafting-through.

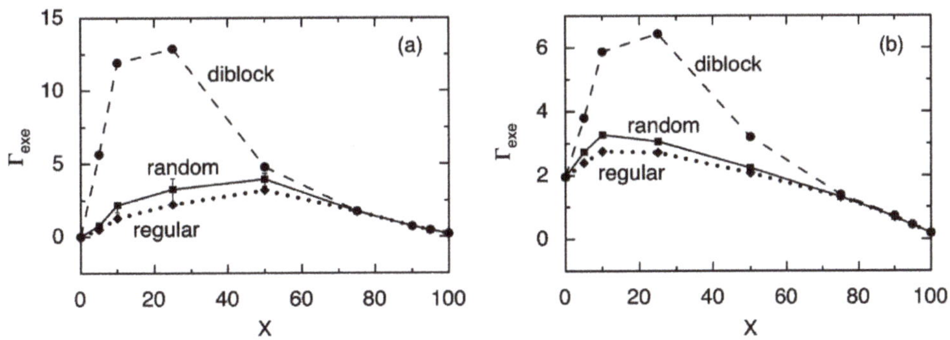

Figure 3. The theoretical surface excess (Γ_{exe}) as a function of the percentage of charged segments, X, on the mica (**a**) and silica (**b**) surfaces with random (squares), regular (diamonds), and diblock (circles) distributions of the uncharged and charged main-chain segments. In panel a, the fluctuations of Γ_{exe} given as twice the standard deviation for 100 realizations of random distributions with the fixed fraction of charged segments are given as error bars. The figures were adopted with permission from [13]. Copyright © 2022 American Chemical Society.

Polymer bottlebrushes attached to surfaces can be prepared by either the "grafting-to" or the "grafting-from" method. The "grafting-to" method includes physical adsorption as well as the chemical covalent attachment of pre-prepared polymers. The "grafting-from" method is a bottom-up approach where polymer chains are grown by surface-initiated polymerization from a substrate [110]. Surfaces with anchored polymer bottlebrushes have shown great potential in lubrication applications [12,20,30,111–116]. Notably, the branched brush configuration and layers expressing polymer loop structures have been formed on surfaces by the "grafting-to" method through electrostatic interactions and using diblock bottlebrush polymers [30] and triblock bottlebrush polymers [116], respectively. The friction coefficients between surfaces coated with the bottlebrush polymer were as low as 10^{-3} up to the pressure of 2.1 MPa [116]. The excellent lubrication properties of these polymers arise from the very limited chain interpenetration of two opposing compressed bottlebrushes due to the strong steric repulsive interactions between the densely grafted side chains of the polymers. Furthermore, the side chains of the polymers are highly hydrated, allowing for strongly bound but easily sheared water layers (i.e., the hydration lubrication mechanism). In this context, it should be noted that shear properties of thin aqueous layers outside hydrophilic surfaces are very similar to that of bulk water down to thicknesses in the nanometer range as measured using a vertically oriented force sensor and AFM [117]. Polymer brushes on the surfaces can also be formed by linear polymers, and in many cases, favorable lubrication can be achieved. Interested readers are referred to the review of Kreer [18] and the references therein.

3.3. *Interfacial Lubrication Properties of Bottlebrush Polymer Layers*

The interfacial lubrication properties of the surfaces coated with a polymer layer depend on the chemical structure of the polymer, the structure of the polymer layer, and the water content of the polymer layer. These properties are influenced by several factors including the polymer–surface affinity as well as the solvent quality. The basis of aqueous boundary lubrication is the presence of a strongly bound yet easily sheared water layer, therefore, water-based lubricants always contain strongly hydrophilic regions. Thus, synthetic water-based lubricants can be designed following two criteria: (a) they shall be strongly hydrated to provide a low friction force; and (b) they shall adsorb strongly to the surface being lubricated to provide a high load bearing capacity. The latter criteria means that the polymers should remain on the surface under the action of a high load even during sliding motion. Considering this criterium, one is tempted to draw the conclusion that covalently attached polymers are to preferred over adsorbed ones. However, one should also consider that covalently attached layers do not self-heal when worn. In contrast, adsorbed layers can easily self-heal when the lubricating polymer is present in the surrounding bulk solution.

3.4. *Random Bottlebrush Polymers*

Bottlebrush polymers containing cationic segments with hydrophilic side chains have been developed, and investigations of interfacial lubrication have shown that such polymers can achieve a low friction force [11,12,29,31,36,48]. For instance, Pettersson et al. studied the effect of the side chain and charge density of random bottlebrush polyelectrolytes, poly(ethylene oxide)-methyl ether methacrylate:methacryloxyethyl trimethylammonium chloride-X (PEO_{45}MEMA:METAC-*X*) by the AFM colloidal probe technique [12]. Here, X stands for the percentage of charged segments and 100-X is thus the percentage of segments carrying a 45 unit long poly(ethylene oxide) side chain. The effective friction coefficient (μ_{eff}) of PEO_{45}MEMA:METAC-*X* polyelectrolytes for the different systems is provided in Figure 4. Here, μ_{eff} was calculated by taking the ratio of friction force (F_{Friction}) and applied load (F_{Load}), at $F_{\text{Load}} \approx 7$ nN, which was in the vicinity of the highest measured F_{Load}. The results show that the PEO_{45}MEMA:METAC-X polyelectrolytes with a high percentage of PEO_{45} segments, $100 \leq 100 - X \leq 25$ ($0 < X < 75$), can achieve the lowest μ_{eff} at the mica–silica surface, implying that the frictional properties depend, to a significant degree, on which of the surfaces has the highest concentration of extended PEO_{45} segments.

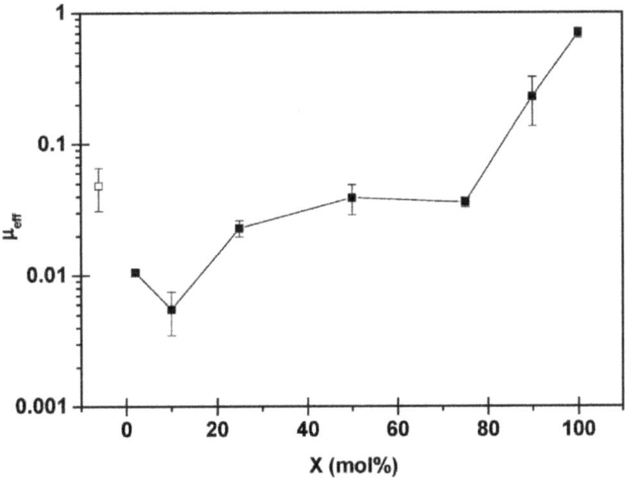

Figure 4. The effective friction coefficient (μ_{eff}) of PEO_{45}MEMA:METAC-*X* systems (filled squares), μ_{eff} of the bare mica–silica surface pair (unfilled square). The figures were adopted with permission from [12]. Copyright © 2022 American Chemical Society.

The effects of the architectural parameters of poly(L-lysine)-*g*-poly(ethylene glycol) (PLL-*g*-PEG) including side chain (PEG) length, Lys/PEG grafting ratio, and backbone chain (PLL) length on the lubrication properties were studied by Spencer and co-workers [31,48]. These studies have shown how the interfacial lubrication properties of PLL-*g*-PEG can be optimized by varying the length of the side chains and the grafting ratio. The efficient aqueous boundary lubrication ability of PLL-*g*-PEG as well as PEO_{45}MEMA:METAC-*X* was due to the densely grafted PEG side chains counteracting the interpenetration between the opposing layers under compression. The investigation of solvent quality on the lubricating ability of the surfaces coated with PLL-*g*-PEG has shown that the friction coefficient of the surfaces greatly increased with the decreasing solvent quality as the polymer layer structure changed from an expanded brush structure to a more random-coil-like structure [34,35]. In addition, the Scheutjens–Fleer self-consistent field theory has been employed to reveal how the architecture parameters, graft ratio, and graft length of random bottlebrush polymers influence the interaction between such polymers with a polyelectrolyte backbone and neutral hydrophilic side chains as well as an oppositely charged surface [118]. The modeling results indicated that the adsorption of random bottlebrush polymers is determined by the electrostatic and steric forces in the system, which can be affected by the graft ratio and graft length of the bottlebrush polymer. Particularly, it was demonstrated that the entropic penalty of adsorption increases as the side chains become longer.

3.5. Diblock Bottlebrush Polymers

Diblock bottlebrush polymers typically consist of one strongly adsorbing block and a second block with a bottlebrush structure with no or low tendency for adsorption. This polymer design is expected to provide an adsorbed layer structure with bottlebrush tails. Liu et al. studied the adsorption and frictional properties of diblock bottlebrush copolymers, $(METAC)_m$-*b*-$(PEO_{45}MEMA)_n$, at the silica surfaces [29]. This diblock copolymer was composed of a cationic block and an uncharged bottlebrush block. The adsorbed amount of the diblock polymer on silica was high, and the thickness of the adsorbed polymer layer was found to be around 46 nm with a water content of around 90%. This result is in line with the prediction of the theoretical modeling in which diblock bottlebrush copolymers have a higher adsorbed amount than random and regular bottlebrush polymers at the silica surface [13] (Figure 3b). A change in the adsorbed layer structure was observed after a few hundred seconds of adsorption. This change can be illustrated by the QCM-D data by plotting the change in dissipation (ΔD) as a function of the change in frequency (Δf) (left panel in Figure 5). The $\Delta D - \Delta f$ curve for (METAC)$_m$-*b*-(PEO$_{45}$MEMA)$_n$ initially followed that observed for the uncharged block ((PEO$_{45}$MEMA)$_n$), implying that initially, the diblock copolymer adsorbs parallel to the surface. However, for (METAC)m-*b*-(PEO$_{45}$MEMA)$_n$, the dissipation increased rapidly as the magnitude of the frequency change exceeded about 25 Hz. This means that (METAC)$_m$-*b*-(PEO$_{45}$MEMA)$_n$ changes its orientation from preferentially parallel to the silica surface at low coverage (low dissipation) to significantly more extended conformations at high coverage (high dissipation).

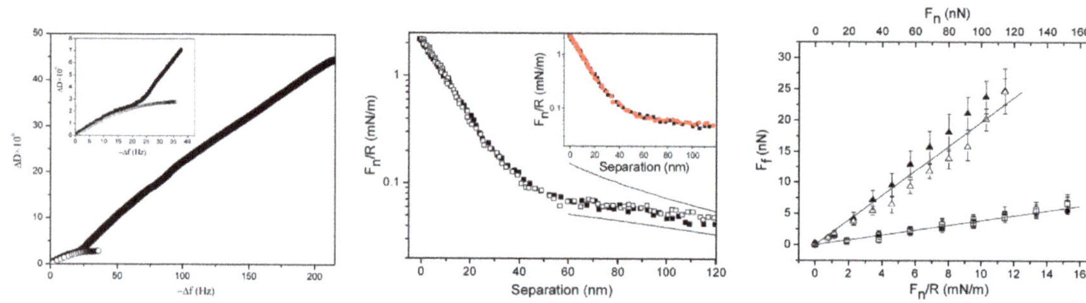

Figure 5. (**Left panel**) Dissipation change (ΔD) as a function of frequency change ($-\Delta f$) upon adsorption of (METAC)$_m$-*b*-(PEO$_{45}$MEMA)$_n$ (upper curve) and (PEO$_{45}$MEMA)$_n$ (lower curve) on silica. The inset shows the data in the range of $-\Delta f$ up to 40 Hz in more detail. The figures were adopted with permission from [29]. Copyright © 2022 American Chemical Society. (**Middle panel**) F_n/R as a function of separation between the silica surfaces coated with (METAC)$_m$-*b*-(PEO$_{45}$MEMA)$_n$. Fitted DLVO forces were obtained by using constant charge (upper line) and constant potential (lower line) boundary conditions. The inset shows the forces between the polymer layers prior to (black squares) and after (red circles) rinsing with water. (**Right panel**) Friction force (F_f) as a function of load (F_n/R and F_n) of the bare silica surfaces in water (triangles) and after the adsorption of (METAC)$_m$-*b*-(PEO$_{45}$MEMA)$_n$ in a 50 ppm polymer solution (the first cycle (squares) and the subsequent one (circles)). The straight lines were fitted to the data points obtained at low loads. The error bars corresponded to multiple friction force measurements. Filled and unfilled symbols represent the data points obtained on loading and unloading, respectively. The figures were adopted with permission from [30]. Copyright © 2022 The Royal Society of the Chemistry. In all cases, the polymer concentration of the aqueous solution was 50 ppm and the surface was silica.

The surface and friction forces of the silica surface coated with the (METAC)$_m$-*b*-(PEO$_{45}$MEMA)$_n$ polymer were studied by the AFM colloidal probe technique, and some results are shown in the middle and right panels in Figure 5. The force curves between the silica surfaces coated with the diblock bottlebrush polymer were consistent with a double-layer force at separation beyond 40 nm. However, the force increased more steeply at separations below 40 nm, indicating the presence of a long-range steric repulsion. The calculated DLVO force curves indicated that the apparent double-layer potential of the adsorbed layer was found to be 33 mV at the onset of the steric interaction at a separation of 40 nm. In addition, no hysteresis was observed between the forces measured on approach and retraction. The friction coefficient was found to be 0.03 up to a pressure of 50 MPa, and thus the layer remained intact up to this high pressure [30]. The low friction force can be attributed to the strong steric repulsion and very limited interpenetration between the branched brush layers, combined with the strong hydration of the PEO side chains. Moreover, the friction forces between the adsorbed bottlebrush diblock polymer layers increased slightly more than predicted by Amontons' rule at high applied loads, particularly in solutions with a high ionic strength. This suggests that another energy dissipative mechanism comes into play (e.g., some shear-induced lateral motion of the adsorbed molecules along the surface). However, the low friction forces were recovered when the applied load was reduced, indicating that attachments by electrostatic interaction may provide self-healing properties by re-adsorption or repositioning of the molecules desorbed or moved along the surface at high loads.

3.6. Triblock Bottlebrush Polymers

Triblock bottlebrush polymers typically consist of two adsorbing end blocks and a middle block with a bottlebrush structure with no or low tendency for adsorption. Such a polymer design is interesting as it will facilitate formations of bottlebrush loops. Recently, an ABA triblock bottlebrush polymer, poly[(quaternized 2-(dimethylaminoethyl) methacrylate)-*co*-methyl methacrylate] (P(DMAEMA-*co*-MMA)), has been prepared by combining ATRP and post-modification techniques [116]. The triblock bottlebrush polymer was built by two positively charged domains with P(DMAEMA-*co*-MMA) chains (A blocks) and a bottlebrush region (B block), decorated with poly(2-methacryloyloxyethyl phosphorylcholine) (PMPC). The surface and friction forces between the surfaces coated with the triblock bottlebrush polymers were studied and the results showed that μ_{eff} of the adsorbed polymer layer was as low as 0.0025 up to a pressure of 2.1 MPa in pure water (Figure 6). The friction coefficient increased to 0.0115 $\pm$ 0.0003 in phosphate buffered saline (PBS) solution at 2.1 MPa due to the conformational changes in the outer part of the adsorbed polymer in the presence of salt, which was attributed to the increased affinity of the bottlebrush block to the surface and the screening of the electrostatic interaction between blocks adsorbed on the surface. A single loop adsorbed structure was formed by the triblock bottlebrush polymer in pure water, while in the presence of PBS, the polymer adopted to conformations with multiple loops and trains. The experimental data suggest that polymer loops are efficient in reducing friction forces as it reduces the interpenetration zone. This extremely low friction is additionally due to strong osmotic repulsion between the densely grafted side chains of the triblock bottlebrush polymers and the sliding of the surfaces is favored by the flow of water molecules during shearing.

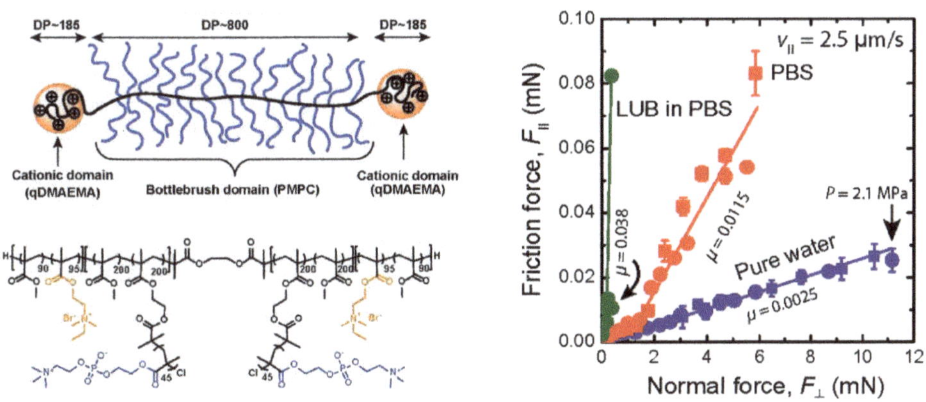

Figure 6. (**Left panel**) Schematic representation of the ABA triblock bottlebrush copolymer and the molecular structure of the triblock copolymer. (**Right panel**) Experimental results of the friction force vs. the normal force for the triblock bottlebrush adsorbed layer. The figures were adopted with permission from the study by Banquy et al. [116]. Copyright © 2022 American Chemical Society. The friction coefficient values of lubricin (LUB) in PBS in the right panel were adopted with permission from [119]. Copyright © 2022 Elsevier Inc.

3.7. Wear Resistance Influenced by Molecular Adsorption Strength

The molecular adsorption strength cannot be judged by the polymer adsorbed amount. For example, polyelectrolytes typically attach more strongly to oppositely charged surfaces the higher their charge density. In contrast, the adsorbed amount typically decreases with polyelectrolyte charge density. When surfaces are modified with thin adsorbed polymer layers to form a well-lubricated surface, it is important that the adsorption strength is high to promote a long lifetime and function of the layer. The adsorption of the polymers onto surfaces may take place by non-specific electrostatic and hydrophobic interactions, hydrogen bonding, or specific anchoring groups. Particular attention has recently been given to mussel-inspired anchoring of polymers where catechol groups serve as anchors on the surfaces [120,121]. Catechol groups adsorb at most types of surfaces under wet condition, even on wet organic surfaces [122], and this provides a new approach to form efficient aqueous boundary lubrication layers with strong anchors. Self-healing properties of the adsorbed polymer layers have been observed in electrostatically anchored polymers systems [29,30,123]. Claesson et al. studied how different functional attachment mechanisms influence the adsorption strength by investigating molecular wear using statistical and diblock copolymers using different anchoring mechanisms, electrostatic (NH_3^+), hydrogen bonding/dispersion interaction (catechol), or both NH_3^+ and catechol [124,125]. It is difficult to observe wear scars of a sub-nanometer depth by using AFM. An alternative for soft adsorbed layers on hard surfaces is to determine the change in surface stiffness with increasing wear. An increase in stiffness implies that the thickness of the adsorbed polymer layer is reduced due to wear, and this effect originates from a larger contribution to the stiffness from the underlying substrate. It has been found that the adsorption strength varied as NH_3^+/catechol > catechol > NH_3^+ anchoring on silica surfaces in water [124] (Figure 7). This result implies that the anchoring of polymers containing catechol groups have a better ability to protect the layer from abrasion compared to electrostatic interactions alone. We note that anchoring the lubricant to the surface with the aid of electrostatic interactions has the disadvantage that the load-bearing capacity decreases with the increasing ionic strength of the solution. This situation can be improved by using non-electrostatically anchoring groups such as catechol.

3.8. Synergistic Aqueous Lubrication Mediated by Aggregates of Natural Molecules and Polymers

The interfacial frictional behavior of bottlebrush polymers associated with surfactants/phospholipids has been reported [50,126]. The interactions between a co-adsorbed poly(acrylic acid)-poly(acrylamide) diblock copolymer with cationic surfactant layers on mica surfaces were investigated by Drummond et al., who found that the lubrication ability of the surface could be improved by the associated structure, which was attributed to the hemifusion instability of the adsorbed surfactant layers being inhibited by the copolymer [126]. Furthermore, the aqueous boundary lubrication properties of complexes of the anionic surfactant (sodium dodecyl sulfate) with positively charged polymers (poly[3-(2-methyl propionamido)-propyl]trimethylammonium chloride (PMAPTAC)) have been investigated [50]. The study showed that the friction force was very low, even at pressures up to 20 MPa, except for some friction peaks that occurred due to the load and shear-induced structural changes in the layer. The data clearly demonstrated that the association of the polymer and surfactant had a much better aqueous boundary lubrication performance than the polymer or surfactant alone. This result implies that the internal organization of the aggregate is essential for developing efficient polyelectrolyte–surfactant systems for aqueous lubrication. This study provides a new approach to develop efficient water-based lubrication systems formed by synergistic self-assembly at solid/aqueous interfaces by utilizing the self-assembly structures of polyelectrolytes and surfactants.

Faivre et al. studied the lubrication and wear protection of a mixture of bottlebrush polymers and linear polymers (hyaluronan, HA) and found that this mixture provided wear protection both in water and saline [43]. This was attributed to the synergy of the two polymers by which a boundary film was formed by entanglements of the bottlebrush polymer and HA. Faivre et al. also studied the effects of polyzwitterionic bottlebrush polymer architecture such as monoblock, diblock, and triblock on the wear resistance, where both individual bottlebrush polymers and mixture of the bottlebrush polymers with HA were investigated [46]. The results demonstrated that the rupture pressure (P^*) (i.e., the load bearing capacity) of the thin film formed by the bottlebrush polymers alone increased with the increasing content of the adhesive blocks on the bottlebrush polymers, $P^*_{triblock} > P^*_{diblock} > P^*_{monoblock}$. They suggested that this is because the monoblock polymer is rather peculiar since it has no terminal adhesive end blocks, the diblock polymer has similar conformation as the monoblock polymer, but also electrostatic interactions with the surface, while the triblock polymer tends to form loop structures that provide stronger anchoring to the surface and lower chain interpretation. In addition, the rupture pressure greatly increased when adding high-molecular weight HA into the bottlebrush polymer layers. This high protective performance was attributed to the strong intermolecular interaction between the bottlebrush polymers and HA. This study provides a new approach to enhance the interfacial lubrication and wear resistance of surfaces by using bottlebrush polymers and linear bio-lubricants, opening new routes toward the development of efficient aqueous boundary lubrication systems.

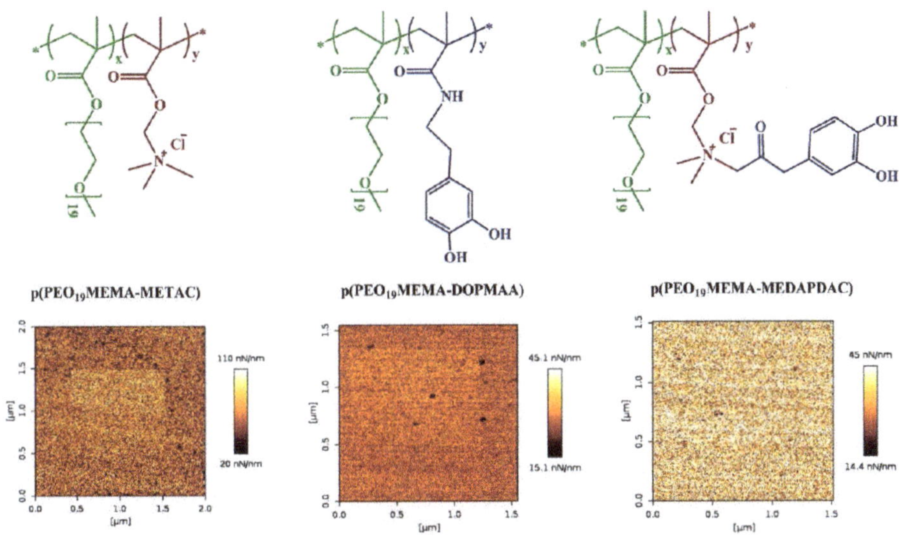

Figure 7. (**Top row**) The molecular structure of the catechol-based polymers. (**Bottom row**) Surface stiffness maps show the worn area in the middle of the AFM image. From left to right, at high loads (top of the worn area), wear was readily seen for electrostatic anchoring, less clearly seen when catechol groups were used, and not observable when both electrostatic anchoring and catechol groups were used. The figures were adopted with permission from [124]. Copyright © 2022 American Chemical Society.

4. Outlook

An excellent aqueous boundary lubrication system will have a low energy dissipation when the opposing layers slide against each other. This means that the interpenetration zone should be small and the outer layer should be highly hydrated to allow for hydration lubrication. The wear resistance of the lubricating layer should also be considered as it is important to have a high load bearing capacity. This means that the lubricant should be strongly anchored to the surface, and here, catechol groups are a promising alternative to electrostatic anchoring. It is also important to have an efficient self-healing ability, which can be achieved by physical adsorption and by having a reservoir of the lubricants available in bulk solution. Several studies have demonstrated that very efficient aqueous boundary lubrication systems can be formed by tuning the structure of bottlebrush polymers, the layer conformation of bottlebrush polymers, and the molecular adsorption strength. Furthermore, highly efficient intermolecular synergistic lubrication has been achieved by aggregates of natural or synthetical molecules, for instance, HA/aggrecan, HA/phospholipids, COMP/lubricin, HA/lubricin/Type II collagen, cross-linked HA/DOPC, polyelectrolyte/surfactant, and HA/bottlebrush polymer. These results have shown that efficient lubrication can be achieved as different types of molecules work together. However, more observations are needed to reveal the internal structure of the aggregates, leading to synergistic lubrication in polymer–surfactant, polymer–polymer, and polymer–bio-lubricant systems. It is thus important to understand the intermolecular synergy mechanisms in lubrication and we hope that future studies of water-based lubrication will pay increasing attention to intermolecular synergies. Moreover, the wear resistance of the lubricating layer is greatly influenced by the molecular adsorption strength. Thus, future studies focusing on the wear resistance of both the lubricating layer and the underlying surfaces should be of high value. In addition, modeling studies are needed to optimize the polymer design to achieve excellent lubrication for surfaces of different chemical composition and surface roughness, which would provide an additional molecular understanding of the aqueous lubrication mechanism [55,117]. These related studies may not only provide new approaches to aqueous lubrication processes in technical systems but also lead to the development of new treatments and technologies to alleviate pain and prevent cartilage degeneration. The review discusses the significance and effects of the bottlebrush polymer design for excellent lubrication properties and wear resistance and summarizes the recent observations of the lubrication properties of bottlebrush polymers. We predict that future studies will provide answers to the fundamental questions and lead to the development of new technologies for aqueous lubrication systems, for instance, smart lubricating coatings with anti-fouling and anti-corrosion properties as well as tissue-engineering scaffolds.

Author Contributions: Conceptualization, X.L. and P.M.C.; writing—original draft preparation, X.L. and P.M.C.; writing—review and editing, X.L. and P.M.C. All authors have read and agreed to the published version of the manuscript.

Funding: The research was funded by the National Natural Science Foundation of China (NSFC) (21902098).

Institutional Review Board Statement: Not applicable.

Informed Consent Statement: Not applicable.

Data Availability Statement: The data presented in this study are available on request from the corresponding author.

Acknowledgments: X.L. acknowledges financial support from the National Natural Science Foundation of China (NSFC) (21902098).

Conflicts of Interest: The authors declare no conflict of interest.

References

1. Singer, I.L.; Pollock, H. *Fundamentals of Friction: Macroscopic and Microscopic Processes*; Springer: Berlin/Heidelberg, Germany, 2012.
2. Forster, H.; Fisher, J. The Influence of Loading Time and Lubricant on the Friction of Articular Cartilage. *Proc. Inst. Mech. Eng. Part H J. Eng. Med.* **1996**, *210*, 109–119. [CrossRef] [PubMed]
3. Gale, L.R.; Coller, R.; Hargreaves, D.; Hills, B.A.; Crawford, R. The role of SAPL as a boundary lubricant in prosthetic joints. *Tribol. Int.* **2007**, *40*, 601–606. [CrossRef]
4. Wright, V.; Dowson, D. Lubrication and cartilage. *J. Anat.* **1976**, *121*, 107–118.
5. Hills, B.A. Surface-active phospholipid: A Pandora's box of clinical applications. Part II. Barrier and lubricating properties. *Intern. Med. J.* **2002**, *32*, 242–251. [CrossRef] [PubMed]
6. Klein, J. Molecular mechanisms of synovial joint lubrication. *Proc. Inst. Mech. Eng. Part J J. Eng. Tribol.* **2006**, *220*, 691–710. [CrossRef]

7. Levangie, P.K.; Norkin, C.C. *Joint Structure and Function: A Comprehensive Analysis*, 5th ed.; F. A. Davis Company: Philadelphia, PA, USA, 2011.
8. Dėdinaitė, A. Biomimetic lubrication. *Soft Matter* **2012**, *8*, 273–284. [CrossRef]
9. Abbasi, M.; Faust, L.; Wilhelm, M. Comb and Bottlebrush Polymers with Superior Rheological and Mechanical Properties. *Adv. Mater.* **2019**, *31*, 1806484. [CrossRef]
10. Verduzco, R.; Li, X.Y.; Pesek, S.L.; Stein, G.E. Structure, function, self-assembly, and applications of bottlebrush copolymers. *Chem. Soc. Rev.* **2015**, *44*, 2405–2420. [CrossRef]
11. Lee, S.; Muller, M.; Ratoi-Salagean, M.; Vörös, J.; Pasche, S.; De Paul, S.M.; Spikes, H.A.; Textor, M.; Spencer, N.D. Boundary Lubrication of Oxide Surfaces by Poly(L-lysine)-g-poly(ethylene glycol) (PLL-g-PEG) in Aqueous Media. *Tribol. Lett.* **2003**, *15*, 231–239. [CrossRef]
12. Pettersson, T.; Naderi, A.; Makuška, R.; Claesson, P.M. Lubrication Properties of Bottle-Brush Polyelectrolytes: An AFM Study on the Effect of Side Chain and Charge Density. *Langmuir* **2008**, *24*, 3336–3347. [CrossRef]
13. Linse, P.; Claesson, P.M. Modeling of Bottle-Brush Polymer Adsorption onto Mica and Silica Surfaces. *Macromolecules* **2009**, *42*, 6310–6318. [CrossRef]
14. Galuschko, A.; Spirin, L.; Kreer, T.; Johner, A.; Pastorino, C.; Wittmer, J.; Baschnagel, J. Frictional Forces between Strongly Compressed, Nonentangled Polymer Brushes: Molecular Dynamics Simulations and Scaling Theory. *Langmuir* **2010**, *26*, 6418–6429. [CrossRef] [PubMed]
15. Spirin, L.; Galuschko, A.; Kreer, T.; Johner, A.; Baschnagel, J.; Binder, K. Polymer-brush lubrication in the limit of strong compression. *Eur. Phys. J. E* **2010**, *33*, 307–311. [CrossRef] [PubMed]
16. Russano, D.; Carrillo, J.-M.Y.; Dobrynin, A.V. Interaction between Brush Layers of Bottle-Brush Polyelectrolytes: Molecular Dynamics Simulations. *Langmuir* **2011**, *27*, 11044–11051. [CrossRef] [PubMed]
17. Leermakers, F.; Zhulina, E.; Borisov, O. Interaction forces and lubrication of dendronized surfaces. *Curr. Opin. Colloid Interface Sci.* **2017**, *27*, 50–56. [CrossRef]
18. Kreer, T. Polymer-brush lubrication: A review of recent theoretical advances. *Soft Matter* **2016**, *12*, 3479–3501. [CrossRef]
19. Sakakibara, K.; Maeda, K.; Yoshikawa, C.; Tsujii, Y. Water Lubricating and Biocompatible Films of Bacterial Cellulose Nanofibers Surface-Modified with Densely Grafted, Concentrated Polymer Brushes. *ACS Appl. Nano Mater.* **2021**, *4*, 1503–1511. [CrossRef]
20. Yan, W.; Ramakrishna, S.N.; Spencer, N.D.; Benetti, E.M. Brushes, Graft Copolymers, or Bottlebrushes? The Effect of Polymer Architecture on the Nanotribological Properties of Grafted-from Assemblies. *Langmuir* **2019**, *35*, 11255–11264. [CrossRef]
21. Sun, Z.; Feeney, E.; Guan, Y.; Cook, S.G.; Gourdon, D.; Bonassar, L.J.; Putnam, D. Boundary mode lubrication of articular cartilage with a biomimetic diblock copolymer. *Proc. Natl. Acad. Sci. USA* **2019**, *116*, 12437–12441. [CrossRef]
22. Adibnia, V.; Olszewski, M.; De Crescenzo, G.; Matyjaszewski, K.; Banquy, X. Superlubricity of Zwitterionic Bottlebrush Polymers in the Presence of Multivalent Ions. *J. Am. Chem. Soc.* **2020**, *142*, 14843–14847. [CrossRef]
23. Jia, W.; Tian, J.; Bai, P.; Li, S.; Zeng, H.; Zhang, W.; Tian, Y. A novel comb-typed poly(oligo(ethylene glycol) methylether acrylate) as an excellent aqueous lubricant. *J. Colloid Interface Sci.* **2018**, *539*, 342–350. [CrossRef] [PubMed]
24. Wei, Q.; Fu, T.; Yue, Q.; Liu, H.; Ma, S.; Cai, M.; Zhou, F. Graphene oxide/brush-like polysaccharide copolymer nanohybrids as eco-friendly additives for water-based lubrication. *Tribol. Int.* **2021**, *157*, 106895. [CrossRef]
25. Claesson, P.; Makuska, R.; Varga, I.; Meszaros, R.; Titmuss, S.; Linse, P.; Pedersen, J.S.; Stubenrauch, C. Bottle-brush polymers: Adsorption at surfaces and interactions with surfactants. *Adv. Colloid Interface Sci.* **2010**, *155*, 50–57. [CrossRef]
26. Naderi, A.; Makuška, R.; Claesson, P.M. Interactions between bottle-brush polyelectrolyte layers: Effects of ionic strength and oppositely charged surfactant. *J. Colloid Interface Sci.* **2008**, *323*, 191–202. [CrossRef] [PubMed]
27. Olanya, G.; Iruthayaraj, J.; Poptoshev, E.; Makuska, R.; Vareikis, A.; Claesson, P.M. Adsorption Characteristics of Bottle-Brush Polymers on Silica: Effect of Side Chain and Charge Density. *Langmuir* **2008**, *24*, 5341–5349. [CrossRef]
28. Liu, X.; Dedinaite, A.; Nylander, T.; Dabkowska, A.P.; Skoda, M.; Makuska, R.; Claesson, P.M. Association of anionic surfactant and physisorbed branched brush layers probed by neutron and optical reflectometry. *J. Colloid Interface Sci.* **2014**, *440*, 245–252. [CrossRef]
29. Liu, X.; Dedinaite, A.; Rutland, M.; Thormann, E.; Visnevskij, C.; Makuska, R.; Claesson, P.M. Electrostatically Anchored Branched Brush Layers. *Langmuir* **2012**, *28*, 15537–15547. [CrossRef]
30. Liu, X.; Thormann, E.; Dedinaite, A.; Rutland, M.; Visnevskij, C.; Makuska, R.; Claesson, P.M. Low friction and high load bearing capacity layers formed by cationic-block-non-ionic bottle-brush copolymers in aqueous media. *Soft Matter* **2013**, *9*, 5361–5371. [CrossRef]
31. Müller, M.; Lee, S.; Spikes, H.A.; Spencer, N.D. The Influence of Molecular Architecture on the Macroscopic Lubrication Properties of the Brush-Like Co-polyelectrolyte Poly(L-lysine)-*g*-poly(ethylene glycol) (PLL-*g*-PEG) Adsorbed on Oxide Surfaces. *Tribol. Lett.* **2003**, *15*, 395–405. [CrossRef]
32. Dedinaite, A.; Thormann, E.; Olanya, G.; Claesson, P.M.; Nyström, B.; Kjøniksen, A.-L.; Zhu, K. Friction in aqueous media tuned by temperature-responsive polymer layers. *Soft Matter* **2010**, *6*, 2489–2498. [CrossRef]
33. Yan, X.; Perry, S.S.; Spencer, N.; Pasche, S.; De Paul, S.M.; Textor, M.; Lim, M.S. Reduction of Friction at Oxide Interfaces upon Polymer Adsorption from Aqueous Solutions. *Langmuir* **2003**, *20*, 423–428. [CrossRef] [PubMed]
34. Müller, M.T.; Yan, X.; Lee, S.; Perry, S.S.; Spencer, N.D. Preferential Solvation and Its Effect on the Lubrication Properties of a Surface-Bound, Brushlike Copolymer. *Macromolecules* **2005**, *38*, 3861–3866. [CrossRef]

35. Müller, M.T.; Yan, X.; Lee, S.; Perry, S.S.; Spencer, N.D. Lubrication Properties of a Brushlike Copolymer as a Function of the Amount of Solvent Absorbed within the Brush. *Macromolecules* **2005**, *38*, 5706–5713. [CrossRef]
36. Lee, S.; Spencer, N.D. Poly(l-lysine)-graft-poly(ethylene glycol): A versatile aqueous lubricant additive for tribosystems involving thermoplastics. *Lubr. Sci.* **2007**, *20*, 21–34. [CrossRef]
37. Hartung, W.; Drobek, T.; Lee, S.; Zürcher, S.; Spencer, N.D. The Influence of Anchoring-Group Structure on the Lubricating Properties of Brush-Forming Graft Copolymers in an Aqueous Medium. *Tribol. Lett.* **2008**, *31*, 119–128. [CrossRef]
38. Lee, S.; Spencer, N.D. Adsorption Properties of Poly(l-lysine)-*graft*-poly(ethylene glycol) (PLL-*g*-PEG) at a Hydrophobic Interface: Influence of Tribological Stress, pH, Salt Concentration, and Polymer Molecular Weight. *Langmuir* **2008**, *24*, 9479–9488. [CrossRef] [PubMed]
39. Perrino, C.; Lee, S.; Choi, S.W.; Maruyama, A.; Spencer, N.D. A Biomimetic Alternative to Poly(ethylene glycol) as an Antifouling Coating: Resistance to Nonspecific Protein Adsorption of Poly(l-lysine)-*graft*-dextran. *Langmuir* **2008**, *24*, 8850–8856. [CrossRef]
40. Iruthayaraj, J.; Olanya, G.; Claesson, P.M. Viscoelastic Properties of Adsorbed Bottle-brush Polymer Layers Studied by Quartz Crystal Microbalance—Dissipation Measurements. *J. Phys. Chem. C* **2008**, *112*, 15028–15036. [CrossRef]
41. Perrino, C.; Lee, S.; Spencer, N.D. End-grafted Sugar Chains as Aqueous Lubricant Additives: Synthesis and Macrotribological Tests of Poly(l-lysine)-graft-Dextran (PLL-g-dex) Copolymers. *Tribol. Lett.* **2008**, *33*, 83–96. [CrossRef]
42. Bijelic, G.; Shovsky, A.; Varga, I.; Makuska, R.; Claesson, P.M. Adsorption characteristics of brush polyelectrolytes on silicon oxynitride revealed by dual polarization interferometry. *J. Colloid Interface Sci.* **2010**, *348*, 189–197. [CrossRef]
43. Faivre, J.; Shrestha, B.R.; Burdynska, J.; Xie, G.; Moldovan, F.; Delair, T.; Benayoun, S.; David, L.; Matyjaszewski, K.; Banquy, X. Wear Protection without Surface Modification Using a Synergistic Mixture of Molecular Brushes and Linear Polymers. *ACS Nano* **2017**, *11*, 1762–1769. [CrossRef] [PubMed]
44. Faivre, J.; Shrestha, B.R.; Xie, G.; Delair, T.; David, L.; Matyjaszewski, K.; Banquy, X. Unraveling the Correlations between Conformation, Lubrication, and Chemical Stability of Bottlebrush Polymers at Interfaces. *Biomacromolecules* **2017**, *18*, 4002–4010. [CrossRef] [PubMed]
45. Faivre, J.; Montembault, A.; Sudre, G.; Shrestha, B.R.; Xie, G.; Matyjaszewski, K.; Benayoun, S.; Banquy, X.; Delair, T.; David, L. Lubrication and Wear Protection of Micro-Structured Hydrogels Using Bioinspired Fluids. *Biomacromolecules* **2018**, *20*, 326–335. [CrossRef] [PubMed]
46. Faivre, J.; Shrestha, B.R.; Xie, G.; Olszewski, M.; Adibnia, V.; Moldovan, F.; Montembault, A.; Sudre, G.; Delair, T.; David, L.; et al. Intermolecular Interactions between Bottlebrush Polymers Boost the Protection of Surfaces against Frictional Wear. *Chem. Mater.* **2018**, *30*, 4140–4149. [CrossRef]
47. Xia, Y.; Adibnia, V.; Huang, R.; Murschel, F.; Faivre, J.; Xie, G.; Olszewski, M.; De Crescenzo, G.; Qi, W.; He, Z.; et al. Biomimetic Bottlebrush Polymer Coatings for Fabrication of Ultralow Fouling Surfaces. *Angew. Chem. Int. Ed.* **2019**, *58*, 1308–1314. [CrossRef]
48. Perry, S.S.; Yan, X.; Limpoco, F.T.; Lee, S.; Müller, M.; Spencer, N.D. Tribological Properties of Poly(l-lysine)-*graft*-poly(ethylene glycol) Films: Influence of Polymer Architecture and Adsorbed Conformation. *ACS Appl. Mater. Interfaces* **2009**, *1*, 1224–1230. [CrossRef]
49. Hartung, W.; Rossi, A.; Lee, S.; Spencer, N.D. Aqueous Lubrication of SiC and Si3N4 Ceramics Aided by a Brush-like Copolymer Additive, Poly(l-lysine)-graft-poly(ethylene glycol). *Tribol. Lett.* **2009**, *34*, 201–210. [CrossRef]
50. Dedinaite, A.; Pettersson, T.; Mohanty, B.; Claesson, P.M. Lubrication by organized soft matter. *Soft Matter* **2010**, *6*, 1520–1526. [CrossRef]
51. Krivorotova, T.; Makuška, R.; Naderi, A.; Claesson, P.; Dėdinaitė, A. Synthesis and interfacial properties of novel cationic polyelectrolytes with brush-on-brush structure of poly(ethylene oxide) side chains. *Eur. Polym. J.* **2010**, *46*, 171–180. [CrossRef]
52. Moglianetti, M.; Campbell, R.A.; Nylander, T.; Varga, I.; Mohanty, B.; Claesson, P.M.; Makuška, R.; Titmuss, S. Interaction of sodium dodecyl sulfate and high charge density comb polymers at the silica/water interface. *Soft Matter* **2009**, *5*, 3646–3656. [CrossRef]
53. Xu, X.; Billing, M.; Ruths, M.; Klok, H.; Yu, J. Structure and Functionality of Polyelectrolyte Brushes: A Surface Force Perspective. *Chem. Asian J.* **2018**, *13*, 3411–3436. [CrossRef] [PubMed]
54. Mocny, P.; Klok, H.-A. Tribology of surface-grafted polymer brushes. *Mol. Syst. Des. Eng.* **2016**, *1*, 141–154. [CrossRef]
55. Ma, L.; Gaisinskaya-Kipnis, A.; Kampf, N.; Klein, J. Origins of hydration lubrication. *Nat. Commun.* **2015**, *6*, 6060. [CrossRef] [PubMed]
56. Estrella, R.P.; Whitelock, J.M.; Packer, N.H.; Karlsson, N.G. The glycosylation of human synovial lubricin: Implications for its role in inflammation. *Biochem. J.* **2010**, *429*, 359–367. [CrossRef]
57. Swann, D.A.; Sotman, S.; Dixon, M.; Brooks, C. The isolation and partial characterization of the major glycoprotein (LGP-I) from the articular lubricating fraction from bovine synovial fluid. *Biochem. J.* **1977**, *161*, 473–485. [CrossRef]
58. Jay, G.D.; Waller, K.A. The biology of Lubricin: Near frictionless joint motion. *Matrix Biol.* **2014**, *39*, 17–24. [CrossRef]
59. Pettersson, T.; Dėdinaitė, A. Normal and friction forces between mucin and mucin–chitosan layers in absence and presence of SDS. *J. Colloid Interface Sci.* **2008**, *324*, 246–256. [CrossRef]
60. Efremova, N.V.; Huang, Y.; Peppas, N.A.; Leckband, D.E. Direct Measurement of Interactions between Tethered Poly(ethylene glycol) Chains and Adsorbed Mucin Layers. *Langmuir* **2002**, *18*, 836–845. [CrossRef]
61. Harvey, N.M.; Yakubov, G.E.; Stokes, J.R.; Klein, J. Normal and Shear Forces between Surfaces Bearing Porcine Gastric Mucin, a High-Molecular-Weight Glycoprotein. *Biomacromolecules* **2011**, *12*, 1041–1050. [CrossRef]

62. Zappone, B.; Patil, N.J.; Madsen, J.B.; Pakkanen, K.I.; Lee, S. Molecular Structure and Equilibrium Forces of Bovine Submaxillary Mucin Adsorbed at a Solid–Liquid Interface. *Langmuir* **2015**, *31*, 4524–4533. [CrossRef]
63. An, J.; Dėdinaitė, A.; Nilsson, A.; Holgersson, J.; Claesson, P.M. Comparison of a Brush-with-Anchor and a Train-of-Brushes Mucin on Poly(methyl methacrylate) Surfaces: Adsorption, Surface Forces, and Friction. *Biomacromolecules* **2014**, *15*, 1515–1525. [CrossRef]
64. An, J.; Jin, C.; Dėdinaitė, A.; Holgersson, J.; Karlsson, N.G.; Claesson, P.M. Influence of Glycosylation on Interfacial Properties of Recombinant Mucins: Adsorption, Surface Forces, and Friction. *Langmuir* **2017**, *33*, 4386–4395. [CrossRef] [PubMed]
65. Seog, J.; Dean, D.; Plaas, A.H.K.; Wong-Palms, S.; Grodzinsky, A.J.; Ortiz, C. Direct Measurement of Glycosaminoglycan Intermolecular Interactions via High-Resolution Force Spectroscopy. *Macromolecules* **2002**, *35*, 5601–5615. [CrossRef]
66. Hardingham, T.E.; Fosang, A.J. Proteoglycans: Many forms and many functions. *FASEB J.* **1992**, *6*, 861–870. [CrossRef] [PubMed]
67. Han, L.; Dean, D.; Ortiz, C.; Grodzinsky, A.J. Lateral Nanomechanics of Cartilage Aggrecan Macromolecules. *Biophys. J.* **2007**, *92*, 1384–1398. [CrossRef] [PubMed]
68. Horkay, F.; Basser, P.J.; Hecht, A.-M.; Geissler, E. Gel-like behavior in aggrecan assemblies. *J. Chem. Phys.* **2008**, *128*, 135103. [CrossRef] [PubMed]
69. Goldberg, R.; Schroeder, A.; Silbert, G.; Turjeman, K.; Barenholz, Y.; Klein, J. Boundary Lubricants with Exceptionally Low Friction Coefficients Based on 2D Close-Packed Phosphatidylcholine Liposomes. *Adv. Mater.* **2011**, *23*, 3517–3521. [CrossRef]
70. Goldberg, R.; Schroeder, A.; Barenholz, Y.; Klein, J. Interactions between Adsorbed Hydrogenated Soy Phosphatidylcholine (HSPC) Vesicles at Physiologically High Pressures and Salt Concentrations. *Biophys. J.* **2011**, *100*, 2403–2411. [CrossRef]
71. Goldberg, R.; Klein, J. Liposomes as lubricants: Beyond drug delivery. *Chem. Phys. Lipids* **2012**, *165*, 374–381. [CrossRef]
72. Sivan, S.; Schroeder, A.; Verberne, G.; Merkher, Y.; Diminsky, D.; Priev, A.; Maroudas, A.; Halperin, G.; Nitzan, D.; Etsion, I.; et al. Liposomes Act as Effective Biolubricants for Friction Reduction in Human Synovial Joints. *Langmuir* **2009**, *26*, 1107–1116. [CrossRef]
73. Dėdinaitė, A.; Wieland, D.C.F.; Bełdowski, P.; Claesson, P.M. Biolubrication synergy: Hyaluronan–Phospholipid interactions at interfaces. *Adv. Colloid Interface Sci.* **2019**, *274*, 102050. [CrossRef] [PubMed]
74. Cao, Y.; Kampf, N.; Klein, J. Boundary Lubrication, Hemifusion, and Self-Healing of Binary Saturated and Monounsaturated Phosphatidylcholine Mixtures. *Langmuir* **2019**, *35*, 15459–15468. [CrossRef] [PubMed]
75. Lin, W.; Kluzek, M.; Iuster, N.; Shimoni, E.; Kampf, N.; Goldberg, R.; Klein, J. Cartilage-inspired, lipid-based boundary-lubricated hydrogels. *Science* **2020**, *370*, 335–338. [CrossRef] [PubMed]
76. Tadmor, R.; Chen, N.; Israelachvili, J. Normal and Shear Forces between Mica and Model Membrane Surfaces with Adsorbed Hyaluronan. *Macromolecules* **2003**, *36*, 9519–9526. [CrossRef]
77. Li, S.; Macakova, L.; Bełdowski, P.; Claesson, P.M.; Dėdinaitė, A. Phospholipids and Hyaluronan: From Molecular Interactions to Nano- and Macroscale Friction. *Colloids Interfaces* **2022**, *6*, 38. [CrossRef]
78. Lapcik, L., Jr.; Lapcik, L.; De Smedt, S.; Demeester, J.; Chabreček, P. Hyaluronan: Preparation, Structure, Properties, and Applications. *Chem. Rev.* **1998**, *98*, 2663–2684. [CrossRef]
79. Seror, J.; Merkher, Y.; Kampf, N.; Collinson, L.; Day, A.J.; Maroudas, A.; Klein, J. Articular Cartilage Proteoglycans As Boundary Lubricants: Structure and Frictional Interaction of Surface-Attached Hyaluronan and Hyaluronan–Aggrecan Complexes. *Biomacromolecules* **2011**, *12*, 3432–3443. [CrossRef]
80. Forsey, R.W.; Fisher, J.; Thompson, J.; Stone, M.H.; Bell, C.; Ingham, E. The effect of hyaluronic acid and phospholipid based lubricants on friction within a human cartilage damage model. *Biomaterials* **2006**, *27*, 4581–4590. [CrossRef]
81. Wang, M.; Liu, C.; Thormann, E.; Dėdinaitė, A. Hyaluronan and Phospholipid Association in Biolubrication. *Biomacromolecules* **2013**, *14*, 4198–4206. [CrossRef]
82. Raj, A.; Wang, M.; Zander, T.; Wieland, D.F.; Liu, X.; An, J.; Garamus, V.M.; Willumeit-Römer, R.; Fielden, M.; Claesson, P.M.; et al. Lubrication synergy: Mixture of hyaluronan and dipalmitoylphosphatidylcholine (DPPC) vesicles. *J. Colloid Interface Sci.* **2017**, *488*, 225–233. [CrossRef]
83. Liu, C.; Wang, M.; An, J.; Thormann, E.; Dėdinaitė, A. Hyaluronan and phospholipids in boundary lubrication. *Soft Matter* **2012**, *8*, 10241–10244. [CrossRef]
84. Spahn, G.; Wittig, R. Spannungs-und Bruchverhalten des gesunden Gelenkknorpels unter axialer Belastung. Eine biomechanische Untersuchung. *Zent. Für Chir.* **2003**, *128*, 78–82. [CrossRef] [PubMed]
85. Bełdowski, P.; Weber, P.; Dėdinaitė, A.; Claesson, P.M.; Gadomski, A. Physical crosslinking of hyaluronic acid in the presence of phospholipids in an aqueous nano-environment. *Soft Matter* **2018**, *14*, 8997–9004. [CrossRef]
86. Raj, A.; Wang, M.; Liu, C.; Ali, L.; Karlsson, N.G.; Claesson, P.M.; Dėdinaitė, A. Molecular synergy in biolubrication: The role of cartilage oligomeric matrix protein (COMP) in surface-structuring of lubricin. *J. Colloid Interface Sci.* **2017**, *495*, 200–206. [CrossRef] [PubMed]
87. Dėdinaitė, A.; Claesson, P.M. Synergies in lubrication. *Phys. Chem. Chem. Phys.* **2017**, *19*, 23677–23689. [CrossRef]
88. Yu, J.; Banquy, X.; Greene, G.; Lowrey, D.D.; Israelachvili, J.N. The Boundary Lubrication of Chemically Grafted and Cross-Linked Hyaluronic Acid in Phosphate Buffered Saline and Lipid Solutions Measured by the Surface Forces Apparatus. *Langmuir* **2011**, *28*, 2244–2250. [CrossRef]
89. Das, S.; Banquy, X.; Zappone, B.; Greene, G.W.; Jay, G.; Israelachvili, J.N. Synergistic Interactions between Grafted Hyaluronic Acid and Lubricin Provide Enhanced Wear Protection and Lubrication. *Biomacromolecules* **2013**, *14*, 1669–1677. [CrossRef]

90. Majd, S.E.; Kuijer, R.; Köwitsch, A.; Groth, T.; Schmidt, T.A.; Sharma, P.K. Both Hyaluronan and Collagen Type II Keep Proteoglycan 4 (Lubricin) at the Cartilage Surface in a Condition That Provides Low Friction during Boundary Lubrication. *Langmuir* **2014**, *30*, 14566–14572. [CrossRef]
91. Kelly, M.T.; Kent, E.W.; Zhao, B. Stepwise Conformational Transitions of Stimuli-Responsive Linear Ternary Heterografted Bottlebrush Polymers in Aqueous Solution. *Macromolecules* **2022**, *55*, 1629–1641. [CrossRef]
92. Li, Z.; Tang, M.; Liang, S.; Zhang, M.; Biesold, G.M.; He, Y.; Hao, S.-M.; Choi, W.; Liu, Y.; Peng, J.; et al. Bottlebrush polymers: From controlled synthesis, self-assembly, properties to applications. *Prog. Polym. Sci.* **2021**, *116*, 101387. [CrossRef]
93. Yin, L.; Liu, L.; Zhang, N. Brush-like polymers: Design, synthesis and applications. *Chem. Commun.* **2021**, *57*, 10484–10499. [CrossRef]
94. Ribeiro, J.P.M.; Mendonça, P.V.; Coelho, J.F.J.; Matyjaszewski, K.; Serra, A.C. Glycopolymer Brushes by Reversible Deactivation Radical Polymerization: Preparation, Applications, and Future Challenges. *Polymers* **2020**, *12*, 1268. [CrossRef]
95. Clarke, B.R.; Tew, G.N. Synthesis and characterization of poly(ethylene glycol) bottlebrush networks via ring-opening metathesis polymerization. *J. Appl. Polym. Sci.* **2022**, *60*, 1501–1510. [CrossRef]
96. Wu, Y.; Tang, Q.; Zhang, M.; Li, Z.; Zhu, W.; Liu, Z. Synthesis of bottlebrush polymers with v-shaped side chains. *Polymer* **2018**, *143*, 190–199. [CrossRef]
97. Matyjaszewski, K. Advanced Materials by Atom Transfer Radical Polymerization. *Adv. Mater.* **2018**, *30*, e1706441. [CrossRef]
98. Beers, K.L.; Gaynor, S.G.; Matyjaszewski, K.; Sheiko, S.S.; Möller, M. The Synthesis of Densely Grafted Copolymers by Atom Transfer Radical Polymerization. *Macromolecules* **1998**, *31*, 9413–9415. [CrossRef]
99. Lee, H.-I.; Boyce, J.R.; Nese, A.; Sheiko, S.S.; Matyjaszewski, K. pH-induced conformational changes of loosely grafted molecular brushes containing poly(acrylic acid) side chains. *Polymer* **2008**, *49*, 5490–5496. [CrossRef]
100. Chen, Y.; Zhou, H.; Sun, Z.; Li, H.; Huang, H.; Liu, L.; Chen, Y. Shell of amphiphilic molecular bottlebrush matters as unimolecular micelle. *Polymer* **2018**, *149*, 316–324. [CrossRef]
101. Cheng, C.; Khoshdel, E.; Wooley, K.L. Facile One-Pot Synthesis of Brush Polymers through Tandem Catalysis Using Grubbs' Catalyst for Both Ring-Opening Metathesis and Atom Transfer Radical Polymerizations. *Nano Lett.* **2006**, *6*, 1741–1746. [CrossRef]
102. Li, Y.; Themistou, E.; Zou, J.; Das, B.P.; Tsianou, M.; Cheng, C. Facile Synthesis and Visualization of Janus Double-Brush Copolymers. *ACS Macro Lett.* **2012**, *1*, 52–56. [CrossRef]
103. Börner, H.G.; Beers, A.K.; Matyjaszewski, K.; Sheiko, A.S.S.; Möller, M. Synthesis of Molecular Brushes with Block Copolymer Side Chains Using Atom Transfer Radical Polymerization. *Macromolecules* **2001**, *34*, 4375–4383. [CrossRef]
104. Gao, H.; Matyjaszewski, K. Synthesis of Molecular Brushes by "Grafting onto" Method: Combination of ATRP and Click Reactions. *J. Am. Chem. Soc.* **2007**, *129*, 6633–6639. [CrossRef]
105. Huang, K.; Canterbury, D.P.; Rzayev, J. Organosoluble polypyrrole nanotubes from core–shell bottlebrush copolymers. *Chem. Commun.* **2010**, *46*, 6326–6328. [CrossRef]
106. Jin, X.-H.; Price, M.B.; Finnegan, J.R.; Boott, C.E.; Richter, J.M.; Rao, A.; Menke, S.M.; Friend, R.H.; Whittell, G.R.; Manners, I. Long-range exciton transport in conjugated polymer nanofibers prepared by seeded growth. *Science* **2018**, *360*, 897–900. [CrossRef]
107. Hua, Z.; Jones, J.R.; Thomas, M.; Arno, M.C.; Souslov, A.; Wilks, T.R.; O'Reilly, R.K. Anisotropic polymer nanoparticles with controlled dimensions from the morphological transformation of isotropic seeds. *Nat. Commun.* **2019**, *10*, 5406. [CrossRef]
108. Foster, J.C.; Varlas, S.; Couturaud, B.; Coe, Z.; O'Reilly, R.K. Getting into Shape: Reflections on a New Generation of Cylindrical Nanostructures' Self-Assembly Using Polymer Building Blocks. *J. Am. Chem. Soc.* **2019**, *141*, 2742–2753. [CrossRef]
109. Pearce, A.K.; Wilks, T.R.; Arno, M.C.; O'Reilly, R.K. Synthesis and applications of anisotropic nanoparticles with precisely defined dimensions. *Nat. Rev. Chem.* **2020**, *5*, 21–45. [CrossRef]
110. Ma, S.; Zhang, X.; Yu, B.; Zhou, F. Brushing up functional materials. *NPG Asia Mater.* **2019**, *11*, 24. [CrossRef]
111. Yang, W.; Zhou, F. Polymer brushes for antibiofouling and lubrication. *Biosurf. Biotribol.* **2017**, *3*, 97–114. [CrossRef]
112. Lin, W.; Klein, J. Recent Progress in Cartilage Lubrication. *Adv. Mater.* **2021**, *33*, 2005513. [CrossRef]
113. Chen, M.; Briscoe, W.H.; Armes, S.P.; Klein, J. Lubrication at Physiological Pressures by Polyzwitterionic Brushes. *Science* **2009**, *323*, 1698–1701. [CrossRef]
114. Wei, Q.; Cai, M.; Zhou, F.; Liu, W. Dramatically Tuning Friction Using Responsive Polyelectrolyte Brushes. *Macromolecules* **2013**, *46*, 9368–9379. [CrossRef]
115. Yu, J.; Jackson, N.E.; Xu, X.; Morgenstern, Y.; Kaufman, Y.; Ruths, M.; de Pablo, J.J.; Tirrell, M. Multivalent counterions diminish the lubricity of polyelectrolyte brushes. *Science* **2018**, *360*, 1434–1438. [CrossRef]
116. Banquy, X.; Burdyńska, J.; Lee, D.W.; Matyjaszewski, K.; Israelachvili, J. Bioinspired Bottle-Brush Polymer Exhibits Low Friction and Amontons-like Behavior. *J. Am. Chem. Soc.* **2014**, *136*, 6199–6202. [CrossRef]
117. Ulcinas, A.; Valdre, G.; Snitka, V.; Miles, M.J.; Claesson, P.M.; Antognozzi, M. Shear Response of Nanoconfined Water on Muscovite Mica: Role of Cations. *Langmuir* **2011**, *27*, 10351–10355. [CrossRef]
118. Feuz, L.; Leermakers, F.A.M.; Textor, M.; Borisov, O. Adsorption of Molecular Brushes with Polyelectrolyte Backbones onto Oppositely Charged Surfaces: A Self-Consistent Field Theory. *Langmuir* **2008**, *24*, 7232–7244. [CrossRef]
119. Zappone, B.; Ruths, M.; Greene, G.; Jay, G.; Israelachvili, J.N. Adsorption, Lubrication, and Wear of Lubricin on Model Surfaces: Polymer Brush-Like Behavior of a Glycoprotein. *Biophys. J.* **2007**, *92*, 1693–1708. [CrossRef]

120. Kang, T.; Banquy, X.; Heo, J.; Lim, C.; Lynd, N.A.; Lundberg, P.; Oh, D.X.; Lee, H.-K.; Hong, Y.-K.; Hwang, D.S.; et al. Mussel-Inspired Anchoring of Polymer Loops That Provide Superior Surface Lubrication and Antifouling Properties. *ACS Nano* **2016**, *10*, 930–937. [CrossRef]
121. Sedó, J.; Saiz-Poseu, J.; Busqué, F.; Ruiz-Molina, D. Catechol-Based Biomimetic Functional Materials. *Adv. Mater.* **2013**, *25*, 653–701. [CrossRef]
122. Levine, Z.A.; Rapp, M.V.; Wei, W.; Mullen, R.G.; Wu, C.; Zerze, G.H.; Mittal, J.; Waite, J.H.; Israelachvili, J.N.; Shea, J.-E. Surface force measurements and simulations of mussel-derived peptide adhesives on wet organic surfaces. *Proc. Natl. Acad. Sci. USA* **2016**, *113*, 4332–4337. [CrossRef]
123. Lee, S.; Müller, M.; Heeb, R.; Zürcher, S.; Tosatti, S.; Heinrich, M.; Amstad, F.; Pechmann, S.; Spencer, N.D. Self-healing behavior of a polyelectrolyte-based lubricant additive for aqueous lubrication of oxide materials. *Tribol. Lett.* **2006**, *24*, 217–223. [CrossRef]
124. Dobryden, I.; Steponavičiūtė, M.; Klimkevičius, V.; Makuška, R.; Dėdinaitė, A.; Liu, X.; Corkery, R.W.; Claesson, P.M. Bioinspired Adhesion Polymers: Wear Resistance of Adsorption Layers. *Langmuir* **2019**, *35*, 15515–15525. [CrossRef]
125. Dobryden, I.; Steponavičiūtė, M.; Hedman, D.; Klimkevičius, V.; Makuška, R.; Dėdinaitė, A.; Liu, X.; Corkery, R.W.; Claesson, P.M. Local Wear of Catechol-Containing Diblock Copolymer Layers: Wear Volume, Stick–Slip, and Nanomechanical Changes. *J. Phys. Chem. C* **2021**, *125*, 21277–21292. [CrossRef]
126. Drummond, C.; Marinov, G.; Richetti, P. Reinforcement of a Surfactant Boundary Lubricant Film by a Hydrophilic−Hydrophilic Diblock Copolymer. *Langmuir* **2008**, *24*, 1560–1565. [CrossRef]

polymers

MDPI

Article

Preparation and Characterization of Strongly Sulfonated Acid Block and Random Copolymer Membranes for Acetic Acid Esterification with 2-Propanol

Verónica Rosiles-González [1], Ronan Le Lagadec [2], Paulina Varguez-Catzim [1], María I. Loria-Bastarrachea [1], Abigail González-Díaz [3], Emanuel Hernández-Núñez [4], Manuel Aguilar-Vega [1] and María Ortencia González-Díaz [5,*]

[1] Laboratorio de Membranas, Unidad de Materiales, Centro de Investigación Científica de Yucatán, A.C., Calle 43 No. 130, Chuburná de Hidalgo, Mérida 97205, Mexico; very_1994@hotmail.com (V.R.-G.); paulivar41@gmail.com (P.V.-C.); marisa@cicy.mx (M.I.L.-B.); mjav@cicy.mx (M.A.-V.)

[2] Instituto de Química, Universidad Nacional Autónoma de México, Circuito Exterior s/n, Ciudad Universitaria, Ciudad de México 04510, Mexico; ronan@unam.mx

[3] Instituto Nacional de Electricidad y Energías Limpias, Reforma 113, Col. Palmira, Cuernavaca 62490, Mexico; abigail.gonzalez@ineel.mx

[4] Centro de Investigación y de Estudios Avanzados del IPN, Unidad Mérida, Mérida 97310, Mexico; emanuel.hernandez@cinvestav.mx

[5] CONACYT—Centro de Investigación Científica de Yucatán, A.C., Calle 43 No. 130, Chuburná de Hidalgo, Mérida 97205, Mexico

* Correspondence: maria.gonzalez@cicy.mx

Citation: Rosiles-González, V.; Le Lagadec, R.; Varguez-Catzim, P.; Loria-Bastarrachea, M.I.; González-Díaz, A.; Hernández-Núñez, E.; Aguilar-Vega, M.; González-Díaz, M.O. Preparation and Characterization of Strongly Sulfonated Acid Block and Random Copolymer Membranes for Acetic Acid Esterification with 2-Propanol. *Polymers* **2022**, *14*, 2595. https://doi.org/10.3390/polym14132595

Academic Editors: Andrzej Puszka and Beata Podkościelna

Received: 23 May 2022
Accepted: 20 June 2022
Published: 27 June 2022

Abstract: In this paper, we report the synthesis of block and random copolymers of 2-acrylamido-2-methyl-1-propane sulfonic acid (AMPS) and methyl methacrylate (MMA), with different AMPS feed ratios. These solution-processable copolymers with strongly sulfonated acid groups resulted in membranes with tunable ion exchange (IEC) and water absorption capacities. AFM images confirmed the microphase separation of **PAMPS-*b*-PMMA-1:1** block copolymer membrane, annealed under the appropriate conditions. The resulting copolymers from the random combination of a 1:1 molar ratio of AMPS and MMA monomers are effective at enhancing the esterification conversion of acetic acid, when compared with a reaction catalyzed by PAMPS-*b*-PMMA block copolymers and the previously studied catalytic membranes. With the **PAMPS-*co*-PMMA-1:1** membrane, the esterification reaction using acetic acid achieved 85% isopropyl acetate. These results are closely correlated with the increase in IEC (2.63 mmol H^+g^{-1}) and the relationship between weight loss (20.3%) and swelling degree (68%) in 2-propanol.

Keywords: block copolymers; random copolymers; catalytic membranes; esterification; isopropyl acetate

1. Introduction

Over the years, polymeric membrane technology has been successfully applied in fuel cells, natural gas purification, gas water treatment, hydrogen recovery, and drug delivery systems that involve separation or transport processes [1–3]. Recently, catalytic membranes have been used in processes that involve chemical reactions wherein separation of the reaction products may or may not take place, such as in dehydrogenation reactions, selective oxidation, methane-steam reforming, the Fischer-Tropsch reaction, dimerization, and esterification and transesterification reactions [4–7]. In the field of esterification and transesterification reactions, catalytic membranes present the advantage of combining chemical conversion and molecular separation in one step, thereby reducing the separation cost [8] and presenting a low environmental impact [9]. Several solid acids and ionic polymers, such as zirconium sulfate ($Zr(SO_4)_2$), sulfonated polyethersulfone (SPES)/phosphotungstic acid (PWA), poly(styrene sulfonic acid) (PSSA), poly(2-acrylamido-2-methyl-1-propane sulfonic acid) (PAMPS), sulfonated polyphenyl sulfone (SPPS) or sulfonated polyether sulfone (SPES) with different sulfonation degrees (SD), embedded or blended in a polymeric matrix of poly(vinyl alcohol) (PVA), polyacrylonitrile (PAN), polyether sulfone (PES) or polyphenyl sulfone (PPS), have been explored as catalytic membranes for esterification and transesterification reactions [10–18]. Most of these catalytic membranes are blends of at least two different polymers that present a decrease in their catalytic stability on reusing the materials, which is attributed to the loss of solid acid or ionic polymer from the polymeric matrix [14]. Thus, the development of functional membranes with strong acid groups that allow higher catalytic efficiency without catalytic activity loss due to reuse is now required, to prepare new heterogeneous acid catalysts. Within this context, copolymers (either block or random) with rationally designed chemical structures could be considered new alternatives that are able to overcome the shortcomings of blends to form a catalytic membrane via these reactions. Both random and block copolymers are expected to prevent the loss of ionic polymers, which occurs with blending in the polymeric matrix due to the strong covalent bonds that link together the monomers in the macromolecule, increasing the membrane's useful life and reusability in esterification/transesterification reactions. Moreover, in the case of block copolymers, it seems plausible that they are able to form micro-channels within the polymeric structure that will promote more effective transport of the reagents through the membrane.

Isopropyl acetate (IPOAc) is commercially produced for use in a wide range of chemical applications. It is used as an industrial solvent for synthetic resins and polymers such as vinyl copolymers, acrylics, polyesters, nitrocellulose, ethyl cellulose, epoxy resins, methacrylic resins, polyamides, and alkyds, and as a solvent in adhesive production, as well as in printing inks and as an aroma compound in perfumes [19,20]. Isopropyl acetate is produced in the liquid-phase esterification reaction of acetic acid with 2-propanol, in the presence of homogeneous or heterogeneous acidic catalysts [20]. In an industrial process, the unreacted acetic acid is recirculated in order to increase the overall yield, and the separation of the reaction products is carried out by reactive distillation, which involves high energy consumption [21]. The use of catalytic membranes as heterogeneous catalysts in the acetic acid (HOAc) esterification reaction emerges as an attractive alternative that reduces the production cost due to energy savings, the elimination of the catalyst separation step, and because it is an environmentally friendly process. For example, Zhang et al. [22] used a catalytically active membrane with a composite structure of PVA/PES and ion-exchange resins in the esterification of acetic acid and *n*-butanol, reaching 71.9% and 91.4% conversion in a batch and pervaporation reactors at 85 °C, respectively. Caetano et al. [23] reached ~90% biodiesel conversion in the esterification of palmitic acid with methanol, catalyzed by a PVA membrane crosslinked with sulfosuccinic acid (PVA_SSA40) using a 1:63 alcohol:acid molar ratio at 60 °C. Recently, Wang et al. [24] elaborated a novel, highly porous membrane by crosslinking PVA with 5-sulfosalicylic acid. This membrane was used as a catalyst in the esterification of acetic acid and ethanol in a pervaporation membrane reactor (PVMR), and an acetic acid conversion of 98.4% was reached at 75 °C in 12 h. A new type of catalytic composite membrane, containing a support layer of commercial porous PAN, a middle layer of a dense sodium alginate (SA)/MoS_2, and an upper layer of a PVA/p-hydroxybenzene sulfonic acid (Pha) blend was reported by Liu et al. [25]. The authors reported that the acetic acid conversion rate reached 94.3% under assisted pervaporation at 75 °C in 12 h, which is 29% higher than that of the batch reaction.

Thus, in the present work, we synthesized a series of block and random copolymers of 2-acrylamido-2-methyl-1-propane sulfonic acid (AMPS) and methyl methacrylate (MMA) with different AMPS feed ratios. In these copolymers, strongly acidic poly(2-acrylamido-2-methyl-1-propanesulfonic acid) (PAMPS) is expected to catalyze the esterification reaction, whereas poly(methyl methacrylate) (PMMA), due to its hydrophobic nature should control water swelling of PAMPS [20]. Physicochemical characterization and tests of the ion exchange capacity (IEC) and absorption capacity of PAMPS-*b*-PMMA and PAMPS-*co*-PMMA as copolymers membranes were carried out, and their performance as heterogeneous catalysts on the acetic acid esterification with 2-propanol in a batch reactor are discussed.

2. Materials and Methods

2.1. Materials and Reagents

First, 2-acrylamido-2-methylpropane sulfonic acid (AMPS, 99%), sodium 2-acrylamido-2-methylpropanesulfonate (AMPSA, 50 wt% in H_2O), methyl methacrylate (MMA, 99%), water (HPLC), potassium persulfate (KPS, ≥99.0%), methanol (MeOH, 99.9%), N,N-dimethylformamide (DMF, 99.8%), ethyl 2-bromoisobutyrate (EBiB, 98%), 2-propanol (99.8%) and 1,10-phenanthroline (phen) were purchased from Aldrich, Toluca (Edo. México) México. Acetic acid (HOAc, 99.9%) was supplied by Fermont, Monterrey (Nuevo León) México. MMA was purified by passing it through an inhibitor remover column. The [Ru(*o*-C_6H_4-2-py)(phen)$(MeCN)_2$]PF_6 catalyst was synthesized as previously reported elsewhere [26].

2.2. Synthesis of Block and Random Copolymers

For both copolymers (block and random), the $[AMPS]_o$: $[MMA]_o$ feed molar ratios were 1:1 and 1:2.

2.2.1. Synthesis of PAMPS-*b*-PMMA

The block copolymers were synthesized by radical polymerization, catalyzed by a ruthenium complex (Scheme 1), as reported previously by Martínez-Cornejo et al. [27]. In a typical synthesis, a 25-mL Schlenk flask was charged with AMPS (0.39 g, 1.88 mmol), [Ru(*o*-C_6H_4-2-py)(phen)$(MeCN)_2$]PF_6 (12.5 mg, 0.0185 mmol) and EBiB (2.75 µL, 0.0185 mmol), in 0.5 mL DMF at 80 °C for 16 h. A known amount of previously degassed MMA (1.88 or 3.76 mmol) was then added to the reaction mixture using a syringe under a N_2 purge and the solution was heated again at 80 °C for 20 h. The resultant copolymer was poured into diethyl ether and the product obtained thereby was filtered. After dissolving in dichloromethane, the polymeric solution was purified through a silica gel column.

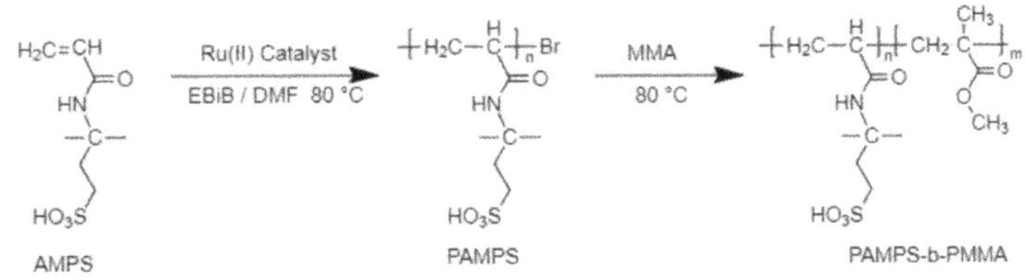

Scheme 1. Synthesis of PAMPS-*b*-PMMA by ATRP, catalyzed by a Ru(II) complex.

2.2.2. Synthesis of PAMPS-*co*-PMMA

The random copolymers were synthesized by free radical polymerization, using potassium persulfate (KPS) as an initiator. As reported by Shen et al. [28], AMPS-Na (3.68 mL, 9.68 mmol), KPS (0.3 mol %), a known amount of MMA (9.68 or 19.3 mmol), and previously deoxygenated water (0.5 mL) were introduced to a 50-mL Schlenk tube. The homogeneous mixture was degassed and the reaction tube was immersed in an oil bath at 60 °C. Polymerization was stopped after 2 h, and the reaction mixture was precipitated into ethanol. Finally, the copolymers were dried at 80 °C under vacuum for 24 h (see Scheme 2).

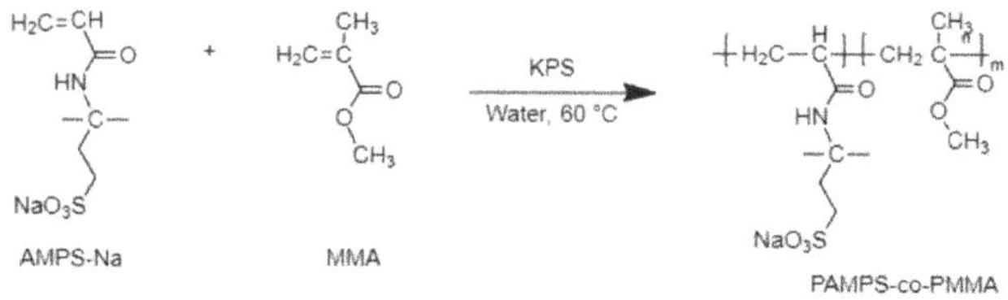

Scheme 2. Synthesis of PAMPS-*co*-PMMA by free radical polymerization.

2.3. Preparation of Dense Membranes

The dense membranes were prepared by the casting method, using 5% solutions of the copolymers. Two different solvents were used: methanol to dissolve PAMPS-*b*-PMMA and N,*N*-dimethylformamide (DMF) for PAMPS-*co*-PMMA. The solutions were poured into an aluminum ring, then the solvent was slowly evaporated at 30 and 80 °C, respectively. The membranes were dried under vacuum at 80 °C for 24 h. Finally, the PAMPS-*co*-PMMA membranes were immersed in a 1 M HCl methanol solution for 24 h to convert the sulfonate groups to acid.

2.4. Characterization

The ^{1}H-NMR analyses were performed on a Varian NMR-600 spectrometer (Agilent technologies, Santa Clara, CA, USA), using CD_3OD for PAMPS-*b*-PMMA and DMF-d_7 for PAMPS-*co*-PMMA. An Agilent 1100 series GPC-SEC (Agilent technologies, Santa Clara, CA, USA), equipped with two columns (Zorbax 60-S and 1000-S) and a RI detector, was used to measure the molecular weights. A mobile phase (flow rate of 0.7 mL min^{-1}) of DMF with 0.05 % lithium bromide (LiBr) and PMMA standards was used for the analysis. A TA Instrument DSC 2920 (Columbus, OH, USA) calorimeter was used to evaluate the glass transition temperatures of the polymers at a heating rate of 10 °C/min, in a nitrogen atmosphere, between 50 °C and 200 °C. The surface morphology of the block copolymer membrane was analyzed with a tapping-mode atomic force microscope (AFM) using a Bruker Multimode 8 instrument (Bruker, Billerica, MA, USA).

The ion exchange capacity (IEC, mmol H^+g^{-1}) was measured using a titration method. First, 100 mg of membrane samples that had previously been dried were immersed in NaOH solution (0.1 M, 5 mL) at room temperature for 24 h [28]. Thereafter, the remaining solution was titrated with 0.02 N HCl, using phenolphthalein as an indicator. The IEC value was calculated by:

$$\text{IEC}\left(\text{mmolg}^{-1}\right) = \frac{M_{NaOH}V_{NaOH} - M_{HCl}V_{HCl}}{W_s} \quad (1)$$

where V_{NaOH} and M_{NaOH} are the NaOH volume consumed and the respective molarity; V_{HCl} and M_{HCl} are the volume and molarity of the HCl consumed in the titration, respectively; and W_s is the membrane sample weight.

Solvent uptake and weight loss: Dry membrane samples were immersed in 2-propanol or methanol (5 mL) for 5 days at 60 °C. The samples were blotted with filter paper and then weighed every 24 h. Finally, after 5 days, the samples were dried at 60 °C under vacuum for 24 h and weighed. The 2-propanol or methanol uptake (%) and % weight loss (WL) were calculated by:

$$\text{Solvent uptake}\ (\%) = \frac{m - m_o}{m_o} \times 100 \quad (2)$$

$$\text{WL} = \frac{m_o - m_s}{m_o} \times 100 \quad (3)$$

where $m - m_0$ is the weight difference between the swollen mass and initial mass of the membrane sample, and m_s is the dry sample mass after 2-propanol uptake.

2.5. Esterification Reaction Performance of Copolymer Membranes

The esterification reaction of acetic acid was carried out in a series of 12 mL screw-cap vials under stirring at 60 °C. In each vial, 20 mg of membrane, cut into small pieces, was immersed in 3 mL of 2-propanol for 24 h. The esterification reaction was then initiated with the addition of 0.25 mL of acetic acid; it was continued up to a reaction time of 29 h. Vials were removed at specified intervals and the titration technique with 1 M NaOH was used to quantify the esterification reaction kinetics. The conversion (C) to isopropyl acetate (IPOAc) was calculated using Equation (4) [22,29]:

$$C = \left(1 - \frac{n_t}{n_o}\right) \times 100 \quad (4)$$

where n_o is the initial number of CH_3COOH moles and n_t is the number of CH_3COOH moles at time t.

3. Results and Discussion

3.1. Synthesis and Characterization of PAMPS-b-PMMA

The PAMPS-*b*-PMMA copolymers were synthesized by atom transfer radical polymerization (ATRP), catalyzed by a ruthenium (II) complex. The chemical structure and composition of the block copolymers were confirmed by ^{1}H-NMR, where the spectra (Figure 1a) exhibit the characteristic signals of each block (PMMA and PAMPS). Four signals at 1.54, 1.68, 1.93, and 3.10 ppm correspond to polymer backbone protons (a, b), methyl protons (c), and to methylene protons (d) adjacent to the acid group, respectively [27,30]. Moreover, with the incorporation of PMMA as a block, the copolymer shows the presence of broad peaks between 1.10 and 0.80 ppm (f) and around 3.62 ppm (g) due to main-chain protons [31]. The PAMPS-*b*-PMMA molar composition was calculated by the ^{1}H-NMR integration ratio of the peaks, at 3.10 ppm from $–CH_2$- (d) of PAMPS and at 3.62 ppm from $–CH_3$ (g) of PMMA. In this case, the molar compositions calculated by ^{1}H-NMR were lower than the theoretical ones. For block copolymers with monomer concentrations and feed $[AMPS]_o$: $[MMA]_o$ molar ratios of 1:1 (50/50 mol %) and 1:2 (33/67 mol %), the copolymer molar compositions found by NMR were 24/76 and 12/88 mol %, respectively, which is attributed to the fact that a low concentration of PAMPS macroinitiator chains allows the growth of a second PMMA block, due to the incompatibility of a purely hydrophilic ionic block (PAMPS) with the purely hydrophobic nature of PMMA [32].

Two different glass transition temperatures (T_g), associated with individual polymer PAMPS (124 °C) and PMMA (114 °C) blocks, were expected to be observed in the DSC curves, due to their inherent immiscibility [33,34]. However, as can be seen in Figure 2, no clear T_g was presented by the PAMPS-*b*-PMMA copolymers, which may be due to the large number of water molecules bound to the copolymer chains [34,35], the presence of chain interactions or crosslinking between the different functional groups, [28] and/or the close values of their T_gs.

The immiscibility of the amphiphilic block copolymers leads to microphase separation, where well-defined nanostructures can be obtained under the appropriate conditions [36,37]. In this context, the resultant morphologies of the block copolymer membrane (with 24% PAMPS and roughly 90 μm membrane thickness) were investigated after a prolonged or fast annealing step on silicon surfaces. Figure 3 shows the tapping-mode AFM images of **PAMPS-*b*-PMMA-1:1** (24/76 mol%) when thermally annealed under vacuum at 120 °C for 72 h (Figure 3a) and when annealed using a microwave reactor for 180 s at 120 °C in a toluene environment (Figure 3b,c), as reported by B. Xiaojiang et al. [37].

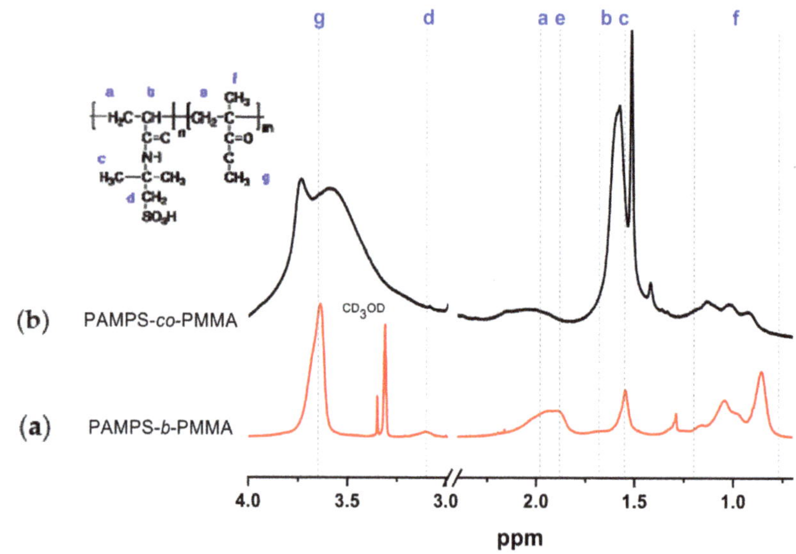

Figure 1. (**a**) ^{1}H-NMR spectrum of PAMPS-*b*-PMMA in CD_3OD and (**b**) PAMPS-*co*-PMMA in DMF-d_7. Molar composition $[AMPS]_o/[MMA]_o = 1/2$.

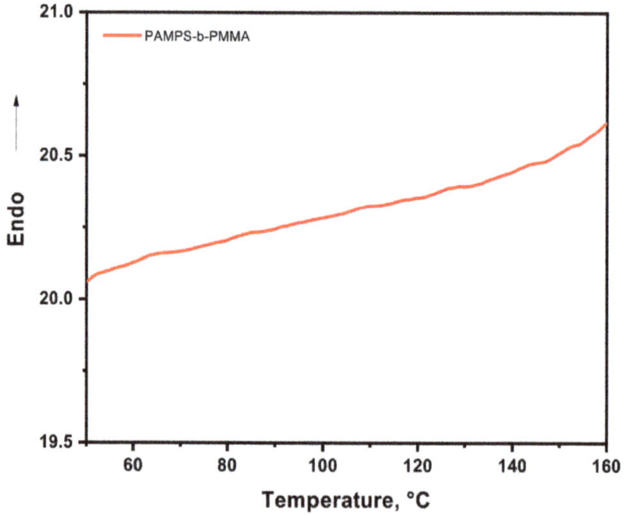

Figure 2. The DSC second-heating curve of PAMPS-*b*-PMMA copolymer.

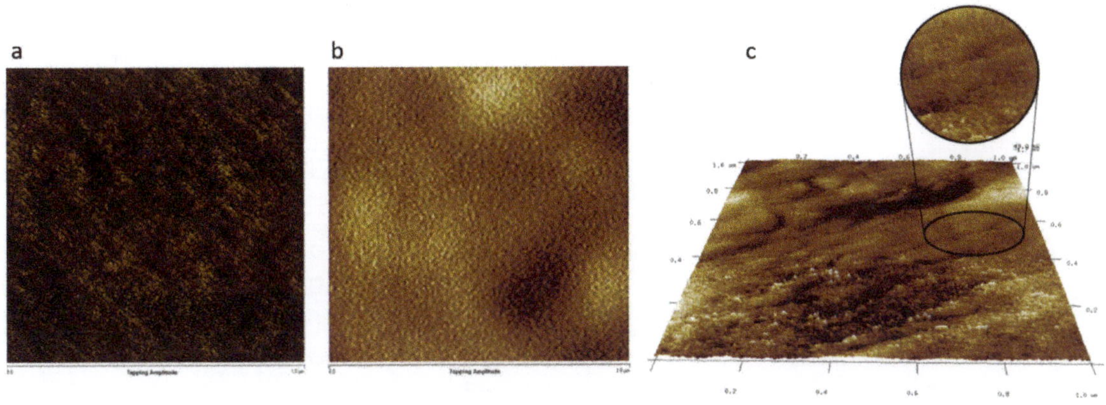

Figure 3. AFM images of the **PAMPS-*b*-PMMA-1:1** membrane: (**a**) thermally annealed at 120 °C for 72 h, (**b**) microwave-annealed for 180 s in the presence of toluene solvent, and (**c**) the 3-dimensional texture of the image in (**b**).

The membrane surface (Figure 3a) exhibits a microphase separation structure, where the paler regions are due to the hydrophobic component, while the darker regions represent the hydrophilic PAMPS domains, which appear dark because they contain a high water fraction [38]. However, a well-defined morphology was not obtained in the **PAMPS-*b*-PMMA-1:1** membrane annealed at 120 °C. Only segregated domains were observed, which was to be expected since it is difficult to obtain a completely ordered state by thermal annealing [39]. Meanwhile, the microwave-annealed **PAMPS-*b*-PMMA-1:1** membrane (Figure 3b) clearly shows an array of almost hexagonally perforated layer (HPL) morphology. This ordered morphology is obtained due to the blocks' composition (f_{PAMPS} = 0.25), as well as the tendency of sulfonated block copolymers to produce hexagonally packed cylinders when exposed to a toluene environment [40,41]. These results confirm that PAMPS-*b*-PMMA block copolymers were obtained.

3.2. Synthesis and Characterization of PAMPS-co-PMMA

On the other hand, it is well known that polymers must have the appropriate properties for the fabrication of polymeric membranes for a particular application. For example, in the case of catalytic membranes for esterification or transesterification reactions, it is desirable for the polymers to present a high concentration of active sites (acid or base), adequate swelling properties, and membrane-forming ability, as well as high molecular weight, among other properties [18,20,27,42,43]. Thus, for comparison, random copolymers with the same theoretical molar composition were synthesized using the free radical polymerization technique; Figure 1 provides a representative ^{1}H-NMR spectrum for each block and random copolymer with the highest PMMA concentration. The free-radical polymerization of AMPS with MMA using potassium persulfate (KPS) as an initiator in an aqueous medium was not successful. Therefore, it was necessary to use AMPS as its deprotonated sodium salt (AMPSA) and, subsequently, to protonate the formed copolymer to obtain PAMPS-*co*-PMMA. The ^{1}H-NMR signals corresponding to each proton in the PAMPSA-*co*-PMMA were observed in the spectra (Figure 1b), which became broader and shifted slightly downfield when compared with PAMPS-*b*-PMMA. The molar composition could not be estimated by the integration ratio because of the overlapping between the aforementioned signals at 3.10 ppm (d) of PAMPS and at 3.62 ppm of (g) of PMMA. However, according to the IEC values, the theoretical composition in the feed agrees with that of the experimental one, as will be discussed later. In comparison with PAMPS-*b*-PMMA, PAMPS-*co*-PMMA random copolymers with higher PAMPS concentrations turn out to be insoluble in methanol. For this reason, DMF-d_7 or DMF was used for their characterization and membrane preparation, respectively.

An important difference between random and block copolymers was found in their molecular weights and polydispersity index (*Đ*) values (see Table 1). The number-average molecular weights (M_n) and *Đ* values for block copolymers PAMPS-*b*-PMMA were in the range of 68,000–78,500 and 1.45–1.48, respectively, whereas for PAMPS-*co*-PMMA random copolymers, these values increased by a factor of 5.4 and 5.8, with *Đ* values ranging from 1.87 to 1.91. The reason for this behavior is that high molecular-weight polymers are formed immediately in free-radical polymerization and remain unchanged during the polymerization process, whereas in a living polymerization, such as ATRP, the number-average molecular weight increases linearly with conversion, leading to narrow molecular weight distribution [44].

Table 1. Composition and molecular weights of block and random copolymers.

Copolymer	PAMPS (%mmol) [a]	PMMA (%mmol) [a]	$M_n \times 10^{-4}$ ($gmol^{-1}$) [b]	$M_w \times 10^{-4}$ ($gmol^{-1}$) [b]	Đ [b]	IEC [c] (mmol H^+g^{-1})
PAMPS-*b*-PMMA-1:1	24	76	7.85	11.6	1.48	1.40
PAMPS-*b*-PMMA-1:1	12	88	6.80	9.86	1.45	0.55
PAMPS-*co*-PMMA-1:1	-	-	42.5	81.2	1.91	2.63
PAMPS-*co*-PMMA-1:1	-	-	39.7	74.2	1.87	1.57

[a] Molar composition, estimated by ^{1}H-NMR; [b] IEC values, expressed as mmol H^+g^{-1} of membrane; [c] molecular weights.

However, while both copolymers are capable of membrane-forming behavior, more flexible and practical membranes were obtained with high-molecular-weight PAMPS-*co*-PMMA copolymers.

3.3. Ion Exchange Capacity (IEC) and Swelling Degree

Ion exchange capacity (IEC), which indicates the total number of active sites in the membrane that may be able to catalyze a reaction, and swelling degree, which allows reactants to access sulfonic acid groups, are important parameters for evaluating the catalytic performance of a membrane [18,20,42]. The IEC value depends on the amount of PAMPS in the copolymer. The block copolymers with 12 and 25% PAMPS exhibited IEC values of 0.55 and 1.40 mmol H^+g^{-1}, respectively, whereas in the random copolymers, the IEC values were 2.63 and 1.57 mmol H^+g^{-1} (see Table 1). These latter values corresponded to a PAMPS composition of 50 and 34% in the copolymers, which agrees with the theoretical feed molar ratios of $[AMPS]_o$: $[MMA]_o$ of 1:1 and 1:2 respectively. The IEC values of **PAMPS-*b*-PMMA-1:1**, **PAMPS-*co*-PMMA-1:1** and **PAMPS-*co*-PMMA-1:2** membranes were similar to those reported for a transesterification catalytic reaction for methyl-ester formation, ranging between 1.26 and 3.80 mmol H^+g^{-1}, and were higher than those for a Nafion-117 membrane (0.92 mmol H^+g^{-1}) [7,12,18,20,42].

Figures 4 and 5 show the swelling degree and weight loss values for PAMPS-*b*-PMMA and PAMPS-*co*-PMMA membranes. In addition, 2-propanol uptake values of 80% and 50.2% were found for membranes **PAMPS-*b*-PMMA-1:1** and **PAMPS-*b*-PMMA-1:2**, respectively, due to the high hydrophilicity of PAMPS. These results indicate that 2-propanol uptake increases with the increase of PAMPS concentration in the copolymer. Both the PAMPS-*b*-PMMA membranes were completely soluble in methanol. It can also be seen that the stability of the membranes cast from PAMPS-*co*-PMMA increased when they were immersed in 2-propanol or methanol (see Figure 3), which is due to the high content of random sulfonic groups in the polymeric chain, which allows a certain degree of physical crosslinking via hydrogen bonding between the sulfonic functional groups present. In fact, the swelling degree of **PAMPS-*co*-PMMA-1:1** and **PAMPS-*co*-PMMA-1:2** diminishes considerably in 2-propanol, in comparison to membranes prepared with block copolymers. Moreover, the PAMPS-*co*-PMMA membranes were insoluble in methanol, presenting 100.5% and 81.3% methanol uptake, respectively. In comparison, the crosslinked PVA/PAMPS catalytic membranes reported by Corzo-González et al., prepared using poly(vinyl alcohol) (PVA) and PAMPS blends that are similar to PAMPS-*co*-PMMA random copolymer membranes, presented a methanol uptake increase of 50–60% with increasing PAMPS concentration from 10 to 30% in the blend [20]. This result confirms that the presence of the random sulfonic group in the material allows a certain degree of physical crosslinking that increases membrane stability in solution [45].

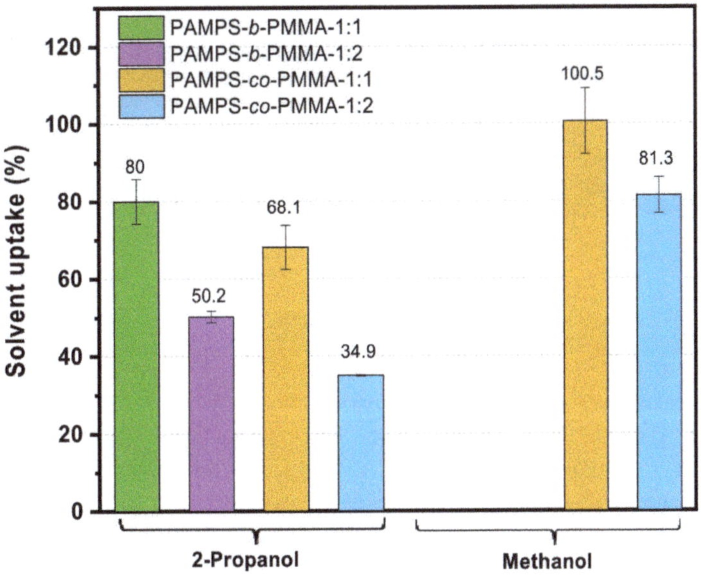

Figure 4. The 2-propanol and methanol uptake of membranes made from PAMPS-*b*-PMMA and PAMPS-*co*-PMMA copolymers at different PAMPS concentrations.

Moreover, as can be seen in Figure 5, the weight losses (WL, %), after 2-propanol immersion for 5 days, were 20.17 ± 1.7 and 19.7 ± 0.6 for the **PAMPS-*co*-PMMA-1:1** and **PAMPS-*co*-PMMA-1:2** membranes, respectively, whereas membranes made from PAMPS-*b*-PMMA copolymers presented WL values of 35.1 ± 2.5 and 31.3 ± 5.8, with feed $[AMPS]_o$: $[MMA]_o$ molar ratios of 1:1 and 1:2, respectively. The high weight-loss values in block copolymer membranes are attributed to their lower molecular weight, in comparison with those obtained in random copolymers. PAMPS-*b*-PMMA turned out to create fragile and slightly brittle membranes when submerged in 2-propanol, making it difficult to extract samples from the solvent for weighing in the WL experiment. Finally, the PAMPS-*co*-PMMA membranes presented WL values of around 50% after methanol immersion.

Due to their high IEC and degree of swelling in 2-propanol, as well as an optimal relationship between weight loss in 2-propanol and IEC, the **PAMPS-*co*-PMMA-1:1** and **PAMPS-*b*-PMMA-1:1** membranes were chosen to study their catalytic activity in the esterification reaction of acetic acid with 2-propanol.

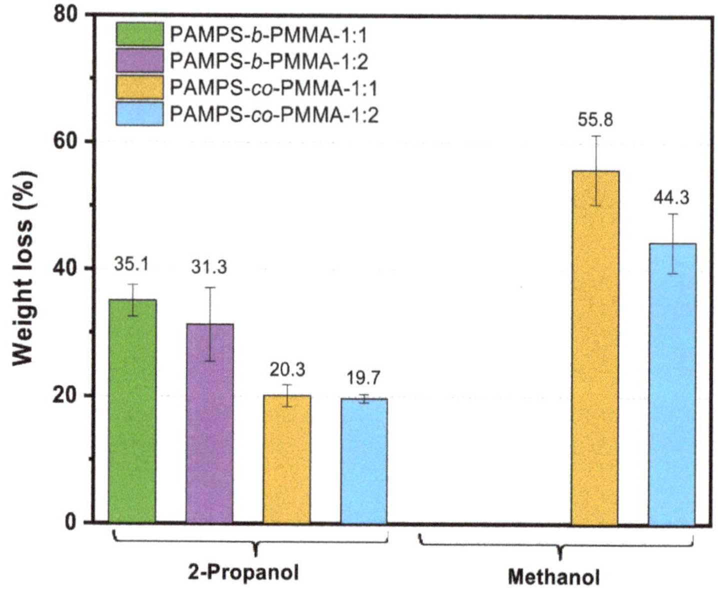

Figure 5. Weight loss percentages for membranes PAMPS-*b*-PMMA and PAMPS-*co*-PMMA in 2-propanol and methanol.

3.4. Catalytic Performance in the Esterification Reaction

Figure 6 shows the kinetic reaction of the acetic acid esterification catalyzed by **PAMPS-*co*-PMMA-1:1** or **PAMPS-*b*-PMMA-1:1**. In addition, the reaction was performed without a catalyst. It was observed that the esterification reaction proceeds even in the absence of the catalyst, due to the weak acidity of acetic acid. However, it barely reaches a 9% conversion in 29 h. Conversely, 17% isopropyl acetate conversion was obtained with the **PAMPS-*b*-PMMA-1:1** membrane in the same period of time, despite its high IEC value (1.40 mmol $H^{+}g^{-1}$) and swelling degree (80%) in 2-propanol that allow the diffusion of the reactants to access sulfonic groups for catalyzing the esterification reaction. The low isopropyl acetate conversion of the **PAMPS-*b*-PMMA-1:1** membrane could be attributed to its poor stability in solution and its high weight loss in 2-propanol (35.1%), making it difficult for it to absorb the moisture produced by the esterification reaction [46].

On the other hand, in the presence of a **PAMPS-*co*-PMMA-1:1** catalytic membrane, the conversion is greatly accelerated compared to the **PAMPS-*b*-PMMA-1:1** membrane, reaching 85% in 29 h. The result can be ascribed to the high IEC value (2.63 mmol $H^{+}g^{-1}$) and adequate 2-propanol uptake (68%), which allow reactants to diffuse through the swollen membrane and access the acid groups. Figure 6 shows a rapid isopropyl acetate conversion in the first 15 h, up to 75%. After that, there is a stabilization period in the following 14 h, with isopropyl acetate conversion at between 75 and 85%. The efficiency of **PAMPS-*b*-PMMA-1:1** as a catalytic membrane in acetic acid esterification is expected to be enhanced in a hybrid process involving a catalytic reaction and pervaporation for reaction product extraction [22].

Table 2 shows the **PAMPS-*co*-PMMA-1:1** catalytic performance for esterification, in comparison with Amberlyst and various PVA-based catalytic membranes [22,23,46,47]. Amberlyst 15 presented a slightly lower conversion rate (78%) at 75 °C in 29 h [46] than was observed for **PAMPS-*co*-PMMA-1:1** at 60 °C. The membrane used by Neguyen et al. [47] was based on poly(styrene sulfonic acid) in a PVA matrix, whereas the second membrane used PVA as a polymeric matrix crosslinked with sulfosuccinic acid (SSA) [23]. Both PVA membranes also presented a slightly lower conversion rate (70% and 83% respectively) than **PAMPS-*co*-PMMA-1:1**. If we compare these with a PVA/PES membrane embedded with ion-exchange resin, this clearly shows a lower conversion rate, even though the reaction was carried out at 75 °C.

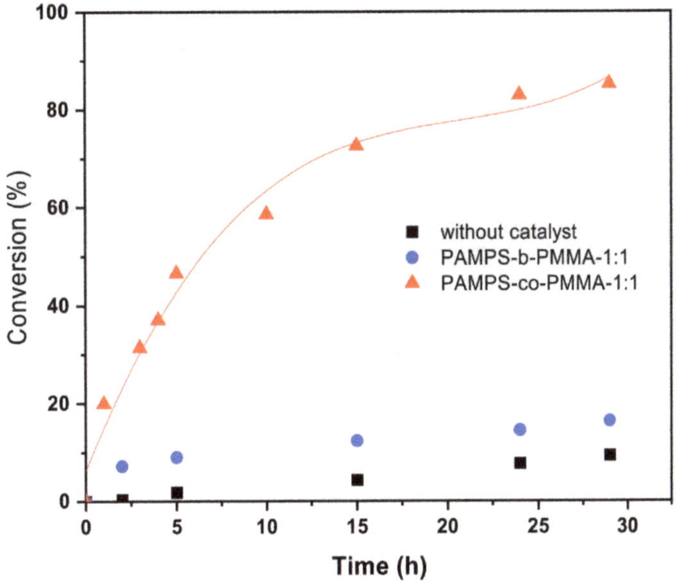

Figure 6. Isopropyl acetate conversion versus reaction time for PAMPS-*b*-PMMA-1:1 (▲), PAMPS-*co*-PMMA-1:1 (●), and without catalyst (■), at 60 °C for 29 h.

Table 2. Esterification performance by Amberlyst and PVA-based acid catalytic membranes, in comparison with the **PAMPS-*co*-PMMA-1:1** membrane reported in this work.

Material	Ref.	T (°C)	IEC mmol H^+g^{-1}	Raw Material	Conversion (%)	Time(h)
Amberlyst 15	46	75	4.75	Acetic acid/2-propanol	78	29
PVA/PSSH	47	50	-	Propanoic acid/propanol	70	25
PVA_SSA40	23	60	-	Oleic acid/methanol	~83	25
PVA/PES/Ion-Exchange resin	22	75	-	Acetic acid/n-butanol	68.2	20
PAMPS-*co*-PMMA-1:1	This work	60	2.63	Acetic acid/2-propanol	85.1	29

PVA: polyvinyl alcohol; SSA: sulfosuccinic acid; PSSH: poly(styrene sulfonic acid); PES: polyethersulfone.

Overall, the **PAMPS-*co*-PMMA-1:1** catalytic membrane evinces a more efficient performance than a PVA/PSSH catalytic blend membrane [47], a PVA/PES membrane embedded with ion-exchange resin [22], and PVA cross-linked with sulfosuccinic acid, (SSA) [23] in terms of esterification reactions.

4. Conclusions

A series of block and random copolymers of MMA with different AMPS feed ratios were successfully synthesized via atom transfer radical polymerization (ATRP) catalyzed by a ruthenium(II) complex and free radical polymerization, respectively. Although these copolymers presented good film-forming ability, the PAMPS-*co*-PMMA random copolymers were more flexible and versatile membranes, due to their high molecular weight in comparison with block copolymers. PAMPS-*co*-PMMA membranes presented the highest IEC values (1.57–2.63 mmol H^+g^{-1}), with 2-propanol uptake of between 34 and 68%, whereas PAMPS-*b*-PMMA exhibited IEC values and 2-propanol uptake in the range of

0.55–1.40 mmol $H^{+}g^{-1}$ and 50–80%, respectively. These results are well correlated with the increase in the molar composition of PAMPS in the copolymer. It was observed that the catalytic activity performance of the **PAMPS-*co*-PMMA-1:1** membrane, as a heterogeneous catalyst in acetic acid esterification with 2-propanol at 60 °C, is more effective than in the membranes obtained with block copolymers. A conversion rate of 85% was obtained with this **PAMPS-*co*-PMMA-1:1** membrane, which is expected to be enhanced in a hybrid process involving reaction and separation in one single unit.

Author Contributions: Data curation, V.R.-G. and P.V.-C.; funding acquisition, M.O.G.-D.; investigation, V.R.-G. and P.V.-C.; methodology, E.H.-N.; supervision, M.O.G.-D.; validation, M.I.L.-B.; writing—original draft, M.O.G.-D.; writing—review and editing, R.L.L., A.G.-D. and M.A.-V. All authors have read and agreed to the published version of the manuscript.

Funding: This research was funded by CONACYT México CB 286973.

Acknowledgments: The authors thank LANNBIO CINVESTAV-Merida (projects FOMIX-Yucatan 2008-108160, CONACyT LAB-2009-01 No. 123913, and CB20121 178947) for NMR spectroscopic analysis and the CONACyT grant 268595-2016 for Raman-AFM acquisition. V.R.G. acknowledges the support of CONACYT through grant 929315, and R.L.L. thanks the DGAPA—UNAM (PAPIIT project IN- 211522).

Conflicts of Interest: The authors declare that there is no conflict of interest.

References

1. Luo, S.; Zhang, Q.; Zhu, L.; Lin, H.; Kazanowska, B.; Doherty, C.; Hill, A.; Gao, P.; Guo, R. Highly selective and permeable microporous polymer membranes for hydrogen purification and CO_2 removal from natural gas. *Chem. Mater.* **2018**, *30*, 5322–5332. [CrossRef]
2. Aguilar-Vega, M.; Perez-Padilla, Y.; Loria-Bastarrachea, M.I. Sulfonated membranes from random aramide copolyisophtalamides with increasing sulfonation degree: Characterization for possible use as solid electrolyte in fuel cell. *Polym. Plast. Technol. Eng.* **2015**, *54*, 711–718. [CrossRef]
3. Sequeira, R.S.; Miguel, S.P.; Cabral, C.S.D.; Moreira, A.F.; Ferreira, A.F.; Correia, I.J. Development of a poly(vinyl alcohol)/lysine electrospun membrane-based drug delivery system for improved skin regeneration. *Int. J. Pharm.* **2019**, *570*, 118640. [CrossRef] [PubMed]
4. Shelepova, E.V.; Vedyagin, A.A. Intensification of the dehydrogenation process of different hydrocarbons in a catalytic membrane reactor. *Chem. Eng. Process. Process Intensif.* **2020**, *155*, 108072. [CrossRef]
5. Li, Y.L.; Zhang, W.; Li, W.; Xing, W. Optimization of dual-functional membrane and application for esterification enhancement. *Chem. Eng. Process. Process Intensif.* **2019**, *139*, 103–112. [CrossRef]
6. Zhang, J.; Li, X.; Song, Y.; Ji, Y.; Cui, Z.; Li, J.; Younas, M. Biodiesel production through heterogeneous catalysis using a novel poly(phenylene sulfide) catalytic membrane. *Energy Fuels* **2020**, *34*, 7422–7429. [CrossRef]
7. Heijnen, J.H.M.; De Bruijn, V.G.; Van den Broeke, L.J.P.; Keurentjes, J.T.F. Micellar catalysis for selective epoxidations of linear alkenes. *Chem. Eng. Process.* **2003**, *42*, 223–230. [CrossRef]
8. Abdallah, H. A review on catalytic membranes production and applications. *Bull. Chem. React. Eng. Catal.* **2017**, *12*, 136–156. [CrossRef]
9. Zhang, H.; Ding, J.; Qiu, Y.; Zhao, Z. Kinetics of esterification of acidified oil with different alcohols by a cation ion-exchange resin/polyethersulfone hybrid catalytic membrane. *Bioresour. Technol.* **2012**, *112*, 28–33. [CrossRef]
10. Shi, W.; He, B.; Cao, Y.; Li, J.; Yan, F.; Cui, Z.; Zou, Z.; Guo, S.; Qian, X. Continuous esterification to produce biodiesel by SPES/PES/NWF composite catalytic membrane in flow-through membrane reactor: Experimental and kinetic studies. *Bioresour. Technol.* **2013**, *129*, 100–107. [CrossRef]
11. Zhu, M.; He, B.; Shi, W.; Feng, Y.; Ding, J.; Li, J.; Zeng, F. Preparation and characterization of PSSA/PVA catalytic membrane for biodiesel production. *Fuel* **2010**, *89*, 2299–2304. [CrossRef]
12. Guerreiro, L.; Pereira, P.M.; Fonseca, I.M.; Martin-Aranda, R.M.; Ramos, A.M.; Diaz, J.M.L.; Oliveira, R.; Vital, J. PVA embedded hydrotalcite membranes as basic catalysts for biodiesel synthesis by soybean oil methanolysis. *Catal. Today* **2010**, *156*, 191–197. [CrossRef]
13. Casimiro, M.H.; Silva, A.G.; Alvarez, R.; Ferreira, L.M.; Ramos, A.M.; Vital, J. PVA supported catalytic membranes obtained by γ-irradiation for biodiesel production. *Radiat. Phys. Chem.* **2014**, *94*, 171–175. [CrossRef]
14. Shi, W.; He, B.; Li, J. Esterification of acidified oil with methanol by SPES/PES catalytic membrane. *Bioresour. Technol.* **2011**, *102*, 5389–5393. [CrossRef] [PubMed]
15. Shi, W.; Li, H.; Zhou, R.; Qin, X.; Zhang, H.; Su, Y.; Du, Q. Preparation and characterization of phosphotungstic acid/PVA nanofiber composite catalytyc membranes via electrospinning for biodiesel production. *Fuel* **2016**, *180*, 759–766. [CrossRef]

16. Gómez-Trejo-López, E.; González-Díaz, M.O.; Aguilar-Vega, M. Waste cooking oil transesterification by sulfonated polyphenylsulfone catalytic membrane: Characterization and biodiesel production yield. *Renew. Energy* **2022**, *182*, 1219–1227. [CrossRef]
17. Shi, W.; Yang, M.; Li, H.; Zhou, R.; Zhang, H. Preparation and characterization of sulfonated poly(ether sulfone) (SPES)/phosphotungstic acid (PWA) hybrid membrane for biodiesel production. *Catal. Lett.* **2015**, *145*, 1581–1590. [CrossRef]
18. Corzo-González, Z.; Loria-Bastarrachea, M.I.; Hernández-Nuñez, E.; Aguilar-Vega, M.; González-Díaz, M.O. Preparation and characterization of crosslinked PVA/PAMPS blends catalytic membranes for biodiesel production. *Polym. Bull.* **2017**, *74*, 2741–2754. [CrossRef]
19. Kuan, L.; Shih-Bo, H.; Wan-Jen, H.; Cheng-Ching, Y.; Ming-Jer, L.; Hsiao-Ping, H. Design and control of reactive distillation for ethyl and isopropyl acetates Production with Azeotropic Feeds. *Chem. Eng.* **2007**, *62*, 878.
20. Qi, W.; Malone, M.F. Semibatch reactive distillation for isopropyl acetate symthesis. *Ind. Eng. Chem. Res.* **2011**, *50*, 1272–1277. [CrossRef]
21. Ohyama, K.O.; Shimada, G.I.; Tokumoto, Y.; Sakamoto, K.Z. Process for the Preparation of Isopropyl Acetate. U.S. Patent 5384426, 7 December 1993.
22. Zhang, W.; Qing, W.; Chen, N.; Ren, Z.; Chen, J.; Sun, W. Enhancement of esterification conversion using novel composite catalytically active pervaporation membranes. *J. Memb. Sci.* **2014**, *451*, 285–292. [CrossRef]
23. Caetano, C.S.; Guerreiro, L.; Fonseca, I.M.; Ramos, A.M.; Vital, J.; Castanheiro, J.E. Esterification of fatty acids to biodiesel over polymers with sulfonic acid groups. *App. Catal. A-Gen.* **2009**, *359*, 41–46. [CrossRef]
24. Wang, T.; Shi, J.; Liang, Y.; Han, J.; Tong, Y.; Li, W. Novel SPVA/g-C3N4-SA/PAN Pervaporation Membranes with Porous Catalytic Layers for Esterification Enhancement. *Ind. Eng. Chem. Res.* **2021**, *60*, 6089–6100. [CrossRef]
25. Liu, Q.; Shi, J.; Wang, T.; Dong, W.; Li, W.; Xing, W. A novel catalytic composite membrane with anti-swelling for enhancing esterification of acetic acid with ethanol. *Chem. Eng. J. Adv.* **2021**, *6*, 100088. [CrossRef]
26. Ryabov, D.A.; Le Lagadec, R.; Estevez, H.; Toscano, A.R.; Hernandez, S.; Alexandrova, L.; Kurova, S.V.; Fischer, A.; Sirlin, C.; Pfeffer, M. Synthesis, characterization, and electrochemistry of biorelevant photosensitive low potential orthometalated ruthenium complexes. *Inorg. Chem.* **2005**, *44*, 222–230. [CrossRef]
27. Martínez-Cornejo, V.; Velázquez-Roblero, J.; Rosiles-González, V.; Correa-Duran, M.; Avila-Ortega, A.; Hernández-Núñez, E.; Le Lagadec, R.; González-Díaz, M.O. Synthesis of poly(2-acrylamido-2-methylpropane sulfonic acid) and its block copolymers with methyl methacrylate and 2-hydroxyethyl methacrylate by quasiliving radical polymerization catalyzed by a cyclometalated ruthenium (II) complex. *Polymers* **2020**, *12*, 1663. [CrossRef]
28. Shen, Y.; Xi, J.; Qiu, X.; Zhu, W. A new proton conducting membrane based on copolymer of methyl methacrylate and 2-acrylamido-2-methyl-1-propanesulfonic acid for direct methanol fuel cells. *Electrochim. Acta* **2007**, *52*, 6956–6961. [CrossRef]
29. Bagnell, L.; Cavell, K.; Hodges, A.M.; Mau, A.W.H.; Seen, A.J. The use of catalytically active pervaporation membranes in esterification reactions to simultaneously increase product yield, membrane permselectivity and flux. *J. Memb. Sci.* **1993**, *85*, 291–299. [CrossRef]
30. Nadim, E.; Bouhendi, H.; Ziaee, F.; Nouri, A. Kinetic study of the aqueous free-radical polymerization of 2-acrylamido-2-methyl-1-propanesulfonic acid via an online proton nuclear magnetic resonance technique. *J. Appl. Polym. Sci.* **2012**, *126*, 156–161. [CrossRef]
31. Vargas, N.A.; Espinosa, J.N.; Lopez, M.S.; Ryabov, D.A.; Le Lagadec, R.; Alexandrova, L. Light-driven living/controlled radical polymerization of hydrophobic monomers catalyzed by ruthenium (II) metalacycles. *Macromolecules* **2012**, *45*, 8135–8146.
32. Evdokia, K.O.; Aikaterini, B.; Georgios, B.; Joannis, K.K. Poly (sodium styrene sulfonate)-b-poly(methyl methacrylate) diblock copolymers through direct atom transfer radical polymerization: Influence of hydrophilic–hydrophobic balance on self-organization in aqueous solution. *Eur. Polym. J.* **2011**, *47*, 752–761.
33. Pal, S.; Mondal, R.; Guha, S.; Chatterjee, U.; Jewrajka, S.K. Homogeneous phase crosslinked poly(acrylonitrile-co-2-acrylamido-2methyl-1-propanesulfonic acid) conetwork cation exchange membranes showing high electrochemical properties and electrodialysis performance. *Polymer* **2019**, *180*, 121680. [CrossRef]
34. Lowe, A.B.; McCormick, C.L. Reversible addition–fragmentation chain transfer (RAFT) radical polymerization and the synthesis of water-soluble (co)polymers under homogeneous conditions in organic and aqueous media. *Prog. Polym. Sci.* **2007**, *32*, 283–351. [CrossRef]
35. Mincheva, R.; Paneva, D.; Mespouille, L.; Manolova, N.; Rashkov, I.; Dubois, P. Optimized water-based ATRP of an anionic monomer: Comprehension and properties characterization. *J. Polym. Sci. Part A Polym. Chem.* **2009**, *47*, 1108–1119. [CrossRef]
36. Vukovic, I.; Brinke, G.T.; Loos, K. Hexagonally perforated layer morphology in PS-b-P4VP(DPP) supramolecules. *Macromolecules* **2012**, *45*, 9409–9418. [CrossRef]
37. Zhang, X.; Harris, K.D.; Wu, N.L.Y.; Murphy, J.N.; Buriak, J.M. Fast assembly of ordered block copolymer nanostructures through microwave annealing. *ACS Nano.* **2010**, *4*, 7021–7029. [CrossRef]
38. Jung, B.; Kim, B.; Yang, J.M. Transport of methanol and protons through partially sulfonated polymer blend membranes for direct methanol fuel cell. *J. Membr. Sci.* **2004**, *245*, 61–69. [CrossRef]
39. Park, C.; Yoon, J.; Thomas, E.L. Enabling nanotechnology with self assembled block copolymer patterns. *Polymer* **2003**, *44*, 6725–6760. [CrossRef]
40. Truong, P.V.; Black, R.L.; Coote, J.P.; Lee, B.; Ardebili, H.; Stein, G.E. Systematic approaches to tailor the morphologies and transport properties of solution-cast sulfonatetd pentablock copolymers. *ACS Appl. Polym. Mater.* **2019**, *1*, 8–17. [CrossRef]

41. Park, M.J.; Kim, S.Y. Ion transport in sulfonated polymers. *J. Polym. Sci. Part B Polym. Phys.* **2013**, *51*, 481–493. [CrossRef]
42. Aca-Aca, G.; Loría-Bastarrachea, M.I.; Ruiz-Treviño, F.A.; Aguilar-Vega, M. Transesterification of soybean oil by PAAc catalytic membrane: Sorption properties and reactive performance for biodiesel production. *Renew. Energy* **2018**, *116*, 250–257. [CrossRef]
43. Talebian-Kiakalaieh, A.; Amin, N.A.S.; Mazaheri, H. A review on novel processes of biodiesel production from waste cooking oil. *Appl. Energy* **2013**, *104*, 683–717. [CrossRef]
44. Odian, G. *Principles of Polymerization*, 4th ed.; John Wiley & Sons, Inc.: Hoboken, NJ, USA, 2004.
45. Chi-An, D.; Chun-Jie, C.; An-Cheng, K.; Wei-Bor, T.; Wen-Shiang, C.; Wei-Ming, L.; Wen-Pin, S.; Chien-Ching, M. Polymer Actuator Based on PVA/PAMPS Ionic Membrane: Optimization of Ionic Transport Properties. *Sens. Actuator A Phys.* **2009**, *155*, 152–162.
46. Sanz, M.T.; Gmehling, J. Esterification of fatty acids to biodiesel over polymers with sulfonic pervaporation part (I): Kinetics and pervaporation studies. *Chem. Eng. J.* **2006**, *123*, 1–8. [CrossRef]
47. Nguyen, Q.T.; M'Bareck, C.O.; David, M.O.; Métayer, M.; Alexandre, S. Ion exchange membranes made of semi-interpenetrating polymer networks, used for pervaporation-assited esterification and ion transport. *Mat. Res. Innov.* **2003**, *7*, 212–219. [CrossRef]

polymers

MDPI

Article

Alkali-Grafting Proton Exchange Membranes Based on Co-Grafting of α-Methylstyrene and Acrylonitrile into PVDF

Shufeng Li *, Xuelin Li, Pengfei Fu and Yao Zhang

Key Laboratory of Advanced Textile Composites, College of Textile Science and Engineering, Tiangong University, Tianjin 300387, China; 2130010014@tiangong.edu.cn (X.L.); 2030010010@tiangong.edu.cn (P.F.); 2031010096@tiangong.edu.cn (Y.Z.)
* Correspondence: lishufeng@tiangong.edu.cn; Tel.: +86-022-8395-5287

Abstract: A novel alkali-induced grafting polymerization was designed to synthesize a PFGPA proton exchange membrane based on the co-grafting of α-methyl styrene (AMS) and acrylonitrile (AN) into the poly(vinylidenedifluoride) (PVDF) membrane. Three kinds of alkali treatments were used: by immersing the PVDF membranes into a 1 M NaOH solution and mixing the PVDF powders with 16% or 20% Na_4SiO_4. Then, AMS with AN could be co-grafted into the PVDF backbones in two grafting solvents, THF or IPA/water. Finally, the grafted membranes were sulfonated to provide the PFGPA membranes. In the experiments, the Na_4SiO_4 treatments showed a greater grafting degree than the NaOH treatment. The grafting degree increased with the increasing amount of Na_4SiO_4. The grafting solvent also influenced the grafting degree. A 40–50 percent grafting degree was obtained in either the THF or IPA/water solvent after the Na_4SiO_4 treatment and the THF resulted in a greater grafting degree. FTIR and XPS testified that the PFGPA membranes had been prepared and a partial hydrolysis of the cyano group from AN occurred. The PFGPA membranes with the grafting degree of about 40–50 percent showed a better dimensional stability in methanol, greater water uptake capabilities, and lower ion exchange capacities and conductivities than the Nafion 117 membranes. The PFGPA membrane with the 16% Na_4SiO_4 treatment and THF as the grafting solvent exhibited a better chemical stability. The obtained experimental results will provide a guide for the synthesis of alkali-grafted PFGPA membranes in practical use.

Keywords: PVDF; alkali-grafting; α-methyl styrene; acrylonitrile; proton exchange membrane

Citation: Li, S.; Li, X.; Fu, P.; Zhang, Y. Alkali-Grafting Proton Exchange Membranes Based on Co-Grafting of α-Methylstyrene and Acrylonitrile into PVDF. *Polymers* **2022**, *14*, 2424. https://doi.org/10.3390/polym14122424

Academic Editors: Andrzej Puszka and Beata Podkościelna

Received: 15 May 2022
Accepted: 10 June 2022
Published: 15 June 2022

1. Introduction

Recently, proton exchange membrane fuel cells (PEMFCs) have attracted more and more attention as portable, clean, and efficient electrochemical energy conversion systems due to their low operation temperature, long lifetime, and high power density [1]. The proton exchange membrane (PEM) is one of the key components of the PEMFCs, which provide the channel for the proton transportation [2]. Nafion, a perfluorosulfonic acid polymer with high ion conductivity, thermal, and mechanical stability, is the most successfully commercialized PEM [3]. However, Nafion membranes exhibit the drawbacks of high cost, low performance under the low relative humidity condition, which limits the commercial application on a large-scale [4]. Hence, many efforts have been made to develop alternative membranes as substitutes for Nafion [5] such as the organic/inorganic hybrid membranes [6,7], sulfonated aromatic polymers [8], blended Nafion composite membranes [9], and fluorine containing polymers by grafting polymerization [10]. Sulfonated aromatic polymers exhibit excellent water uptake capability, ion exchange capacity, and conductivity. Their shortcomings are the great swelling and bad dimensional stabilities, which further deteriorate the mechanical properties of the membranes in application. Fluorine containing polymers [10] including the fluorinated (PTFE, FEP, and PFA) or partially fluorinated (PVDF and ETFE) membranes possess excellent chemical and thermal stability, but never have conductive groups. Therefore, how to introduce ion exchange groups into the fluorinated polymeric matrix is the key to successfully preparing the fluorinated PEMs.

Radiation grafting is an effective way to prepare the PEMs for fuel cells [11–13]. The properties of the membranes can be easily controlled by modifying the grafting polymerization parameters. Styrene (St) [14], due to its low cost and adjustable ion exchange capacity by sulfonation, is an appropriate material to be radiation-grafted into the fluorinated membranes. In addition, acrylonitrile (AN), methylacrylonitrile (MAN) [15], and dimethylaminoethyl methacrylate (DMAEMA) [16] are also applied. It is noted that radiation grafting is strict to the polymerizing instrumentation due to the use of high energy γ- and X-ray, which is relatively toxic and expensive.

Recently, Qiu [17] proposed an alkali (NaOH) solution-induced grafting polymerization to synthesize the PVDF-*g*-PSSA membrane, which displayed a higher conductivity of 0.1119 S/cm at 25 °C and lower methanol permeability than Nafion. Guo [18] investigated the tetrasodium orthosilicate (Na_4SiO_4)-induced grafting polymerization of PVDF with a polystyrene sulfonated acid membrane. The experimental results indicated that polystyrene was easily grafted into the PVDF by an alkali solution. With the increasing content of NaOH or Na_4SiO_4, the proton conductivity of the PVDF-*g*-PSSA membranes were gradually increased.

Styrene, or its sulfonated derivatives, is the most commonly utilized radiation grafting monomer into fluorinated polymers. However, the main shortcoming of polystyrene derivatives is that the benzylic C–H bond is easily broken to form benzyl radicals, resulting in a subsequent chain scission by β-fragmentation [19,20]. Using fluorinated monomers (e.g., α, α, β-trifluorostyrene (TFS)) can increase the chemical stability. Nevertheless, TFS possesses low grafting kinetics [21] and sulfonation degree [22], leading to a too long reaction time and unacceptable ion exchange capacity. Using the pre-sulfonated monomers (e.g., sulfonyl fluoride containing TFS derivatives [21]) can improve the grafting kinetics, but they are too expensive to be utilized on a large-scale. Ring-substituted styrenic monomers such as *p*-methylstyrene or *p*-tert-butylstyrene have also been studied as alternative monomers in order to increase the chemical stability [23]. Theoretically, α-methyl styrene (AMS), a α-substituted styrenic derivative that is commercially available, is the most chemically stable, and the polymerized AMS showed superior stability to polystyrene under oxidized conditions [24]. However, the homopolymerization of AMS is difficult due to the low steric hindrance of the α-methyl group and low ceiling temperature [25]. Grafting of AMS with other monomers such as AN or methacrylonitrile (MAN) [26–28] could enhance the polymerization conversion. However, few studies have been found on the alkali grafted PVDF with AMS and its co-monomers.

In this paper, alkali-induced grafting proton exchange membranes with *co*-grafting PVDF with α-methylstyrene and acrylonitrile were reported. The effects of the types and contents of the alkalis and solvents on the grafting polymerization were investigated and the obtained PVDF-based proton exchange membranes (PFGPA) were characterized in terms of the morphology, liquid uptake capability, swelling, ion exchange capacity (IEC), conductivity, mechanical property and so on. The PFGPA membranes possess excellent dimensional stability and low swelling in methanol, indicating a potential application prospect. Due to the advantages of low cost, ease of preparation, and relative environmental-friendliness, the alkali grafting polymerization is attractive and exhibits a new approach to synthesizing PFGPA membranes.

2. Materials and Methods

2.1. Materials

Kynar® polyvinylidene fluoride (PVDF) resins for batteries (Mw = 400,000~600,000) were purchased from Arkema, France. Alpha-methylstyrene (AMS) was purchased from Futian Chemical Com., Zhaoqing, China. Acrylonitrile (AN), N-methyl pyrrolidone (NMP), dibenzoyl peroxide (BPO), tetrasodium orthosilicate (Na_4SiO_4), 1,2-dichloroethane, tetrahydrofuran (THF), isopropanol (IPA), hydrogen peroxide, and methanol were purchased from

Guangfu Fine Chemical Com., Tianjin, China. All reagents were used as received except that AN and AMS were distilled prior to use.

2.2. Preparation of the PFGPA Membranes

Preparation of the PFGPA membranes is shown in Scheme 1. Two types of alkalis, NaOH and Na_4SiO_4, were used to treat the PVDF membranes.

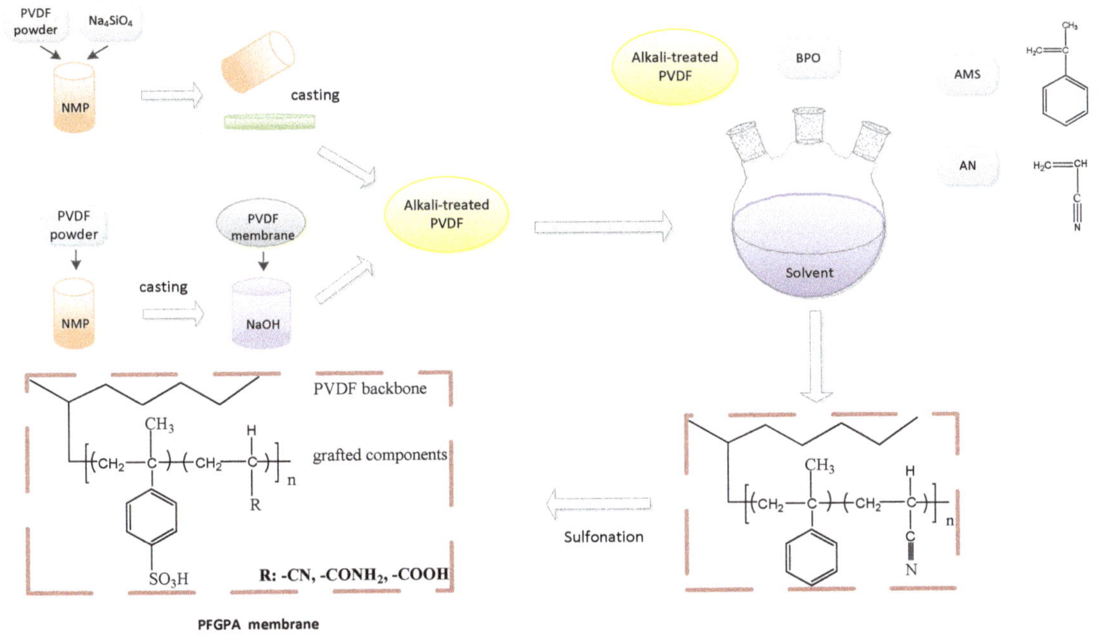

Scheme 1. The preparation of the PFGPA membranes.

NaOH treatment process: Certain amounts of PVDF powders and NMP solution were mixed, stirred for 2 h, and poured onto the glass plate. The obtained PVDF membranes were dried under vacuum at 60 °C for 10 h. Then, they were immersed into a 50 mL 1.0 M NaOH ethanol solution with 150 mg tetrabutylammonium bromide (TBAB) being added at 60 °C for 20 min. The treated PVDF membranes were washed by deionized water to neutral.

Na_4SiO_4 treatment process: Certain amounts of PVDF powders were dissolved in NMP solution containing 3 wt.% deionized water, then the Na_4SiO_4 powders were added. The mixture was stirred for 2 h, poured onto the glass plate and dried under vacuum to afford the Na_4SiO_4-treated PVDF membranes, expressed as "PVDF_Na_4SiO_4 percent".

Before grafting, the alkali-treated PVDF membranes were swelled in 1,2-dichloromethane for 2 h. These were respectively immersed into a solvent, IPA/water (I/W) or THF, for grafting polymerization with AN and AMS as the *co*-monomers and BPO as the initiator. The grafting polymerization was performed at 50 °C for 60 h. The obtained PVDF-*g*-P(AMS-AN) were washed with acetone and 1,2-dichloromethane three times, 2 h each, to remove the residue reactants on the surface and dried under vacuum at 60 °C for 12 h. The grafting degree (*GD*) of the PVDF-*g*-P(AMS-AN) was calculated according to Equation (1).

$$GD = \frac{m_1 - m_0}{m_0} \times 100\% \tag{1}$$

where GD is the grafting degree; m_0 and m_1 are the masses of the PVDF membranes before and after the grafting polymerization, respectively.

Afterward, the grafted PVDF membranes were swelled in 1,2-dichloromethane for 2 h. Then, they were sulfonated in a mixture of chlorosulfonic acid/1,2-dichloromethane (5:100, v:v) at room temperature for 2 h [26] to obtain the PFGPA membranes. The PFGPA membranes were washed with deionized water to neutral and dried under vacuum. The obtained PFGPA membranes were expressed as "PFGPA_X_Y", where X represents the NaOH or percent of Na_4SiO_4 and Y represents the grafting solvent, respectively.

2.3. SEM

Morphologies of the samples were observed on a S-4800 field emission scanning electron microscope from Hitachi, Japan. Prior to the measurement, the samples were dried in the vacuum oven at 60 °C for 24 h and sprayed with Au powders on the surfaces.

2.4. IR Analysis

FTIR was performed on a Nicolet iS50 spectrometer from ThermoFisher Scientific Co., Ltd, Shanghai, China. The spectra were measured in transmittance mode in a wavenumber range of 4000–600 cm^{-1} with a resolution of 8 cm^{-1}.

2.5. XPS

The X-ray photoelectron spectroscopy (XPS) was performed with a D8 DISCOVER instrument from BRUKER, Germany, with monochromatic Al Kα radiation (1600 W, 40 KV, 40 mA) and 2θ range of 5–45°. The XPS data were analyzed and curve fitted by the XPS peak software.

2.6. Liquid Uptake Capability and Swelling Ability

The membranes were cut into the dimensions of about 2 × 2 cm^2 and dried at 120 °C for 2 h to completely remove the absorbed water to afford the dry weight, length, and width. The liquid uptake capabilities were measured in three types of liquids: deionized water, 1 mol/L methanol aqueous water, and pure methanol, respectively. The membranes were immersed into the liquid at room temperature, taken out periodically, and the water on the surface was quickly removed using filter paper to provide the wet weight, length, and width.

The liquid uptake capability and area swelling were calculated according to Equations (2) and (3), respectively,

$$Liquid\ uptake = \frac{W_{wet} - W_{dry}}{W_{dry}} \times 100\% \quad (2)$$

$$Area\ swelling = \frac{A_{wet} - A_{dry}}{A_{dry}} \times 100\% \quad (3)$$

where W_{dry} and A_{dry} are the mass and surface area in the dry, respectively; W_{wet} and A_{wet} are the mass and surface area in the wet, respectively.

2.7. Chemical Stability

The PFGPA membranes were tailored into the dimensions of 2 × 2 cm^2. Chemical stability tests were evaluated by the degradation experiment of immersing the membranes into 30 wt.% H_2O_2 aqueous solution at 60 °C for 6 h. Then, the membranes were intermittently taken out from the H_2O_2 solution and weighed after wiping off the water on the surface.

2.8. Ionic Exchange Capacity

Ionic exchange capacities (IECs) of the PFGPA membranes were determined by acid–base titration. Before the test, the dried PFGPA membrane was immersed into a 0.05 mol/L

NaOH solution for 24 h. Afterward, the membrane was taken out and washed with deionized water several times. The washed water was combined and poured into the NaOH solution. Then, the NaOH solution was neutralized with a 0.06 mol/L HCl solution. IEC ($mmol \cdot g^{-1}$) was calculated according to Equation (4).

$$IEC = \frac{V_{NaOH} \times C_{NaOH} - V_{HCl} \times C_{HCl}}{W} \tag{4}$$

where V_{NaOH} and V_{HCl} are the volume of NaOH and HCl (mL), respectively; C_{NaOH} and C_{HCl} are the concentration of NaOH and HCl ($mol \cdot L^{-1}$), respectively; W is the weight of the membrane in the dry.

2.9. Proton Conductivity

The proton conductivity measurement was performed by a double probe method in SI 1287 AC impedance, UK. The membranes were placed between two Pt electrodes with the fixed distance under the AC voltage of 10 mV. Conductivities σ of the membranes were examined at room temperature according to Equation (5).

$$\sigma = \frac{L}{R \times W \times T} \tag{5}$$

where R is the resistance (Ω) of the sample; T and W are the thickness and the width, respectively; L of 1.5 cm represents the distance between two Pt electrodes.

2.10. Mechanical Measurement

The mechanical properties of the modified PVDF membranes were measured by a strong electronic universal material testing machine at room temperature. The membranes were fixed with clamps with the distance of 25 mm. The clamps were moved at the speed of 20.0 mm/min until the membrane was broken.

3. Results and Discussion

3.1. Preparation of the PFGPA Membranes

PVDF possesses excellent chemical and thermal stability and it is hard to directly introduce the other structural units into the polymeric backbone. In order to facilitate the grafting polymerization, two types of alkalis, NaOH and Na_4SiO_4, were used to generate the active reaction points in PVDF. Alkali treatments affect the positions and amounts of the active points in PVDF. The NaOH treatment was performed by immersing the PVDF membranes into a 1 M NaOH solution at 60 °C for 20 min and the alkali-corrosion generally happened on the surfaces of the PVDF membranes. The Na_4SiO_4 treatment was performed by mixing the PVDF powders with Na_4SiO_4 in NMP, so that the active points were distributed evenly on the cross-section [29] of the PVDF membranes, maximizing the likelihood of the grafting polymerization.

Afterward, the active PVDF membranes were immersed into a mixture containing the *co*-monomers and the initiators to perform the grafting polymerization. The effects of the alkali, the solvents, and amounts of the monomers on the grafting degree were discussed. In the grafting polymerization, two types of grafting solvent, THF and IPA/water, were used, which influenced the solubility of the monomers and compositions, structures, and properties of the resulted polymers. In the IPA/water solvent, AMS and AN are hydrophobic and immiscible. Gubler [28] proposed that due to the poor polarity of AMS, using the IPA/water solvent (5:2, *v*/*v*) could increase the effective AMS concentration at the grafting front, resulting in a higher AMS incorporation into the graft copolymer. When an AMS molar fraction of 0.6 in the feed composition was required, the AMS:AN molar ratio of 1:1 in the graft component was obtained. In our experiments, an AMS molar fraction of 0.6 by Gubler's method was used.

Table 1 lists the parameters of the alkali, monomers, and solvents in the grafting polymerization. When the amounts of AMS and AN were lower, the grafting degree by the

NaOH treatment was about 2 percent in either the IPA/water or THF solvent (No. 1, 4). The grafting degree by the Na_4SiO_4 treatment increased with the increasing amount of Na_4SiO_4 in either the IPA/water (No. 2, 3) or THF (No. 5, 6). For the 20% Na_4SiO_4 treatment, the IPA/water solvent exhibited a greater grafting degree of 20 percent than 13.8 percent in THF. When the volume ratio of AMS/AN to the solvent increased to 4:1, the IPA/water solvent resulted in a greater grafting degree of about 30–50 percent (No. 7–9). For the NaOH treatment, the grafting degree exceeded 30 percent. The treatment of 16% Na_4SiO_4 reached a grafting degree of 49 percent, greater than 43.61 percent of the 20% Na_4SiO_4, manifesting a potential application prospect. When THF was used as the solvent, in order to improve the grafting degree, $AlCl_3$, a Lewis acid, which had successfully improved the radiation-induced grafting polymerization of PVDF [30,31], was used to effectively catalyze the alkali-induced grafting polymerization in our experiments. The obtained grafting degree was slightly greater than that in IPA/water (No. 10–12). These experimental results indicate that the PVDF-*g*-P(AMS-*co*-AN) with a grafting degree of 40–50 percent had been successfully achieved in either the IPA/water or THF solvent system.

Table 1. The preparation of the PFGPA membranes.

No.	Sample	Alkali Treatments		Grafting Polymerization Factors					GD	IEC	Cond.
		NaOH mol/L	Na_4SiO_4 %	BPO /g	AMS/mL	AN/ mL	THF/mL	I/W/ mL	%	mmol/g	$\times 10^2$ S/cm
1		1		0.35	16	5		50/20	2.91	-	-
2			16	0.35	16	5		50/20	8.65	-	-
3			20	0.35	16	5		50/20	20.03	-	-
4		1		0.35	16	5	70		2.33	-	-
5			16	0.35	16	5	70		8.61	-	-
6			20	0.35	16	5	70		13.83	-	-
7	PFGPA_NaOH_I/W	1		0.46	71	24		17/7	35.87	0.56	-
8	PFGPA_16%_I/W		16	0.46	71	24		17/7	49.02	0.89	1.33
9	PFGPA_20%_I/W		20	0.46	71	24		17/7	43.61	0.74	1.33
10	PFGPA_NaOH_THF	1		0.46	71	24	24		33.74	0.41	0.72
11	PFGPA_16%_THF		16	0.46	71	24	24		49.64	0.57	1.13
12	PFGPA_20%_THF		20	0.46	71	24	24		50.83	0.78	1.51

Finally, the PVDF-*g*-P(AMS-*co*-AN) membranes were sulfonated according to the literature [26] to obtain the resulted PFGPA membranes.

3.2. *Characterization of the PFGPA Membranes*

3.2.1. Morphologies

The PFGPA membranes were obtained sequentially by alkali treatment, grafting polymerization, and sulfonation. Figure 1 reveals the SEM photographs of the PVDF membranes after alkali treatment. It was found that the pristine PVDF membrane possessed a smooth surface. After the PVDF membrane was treated by 1 M NaOH solution, they still remained smooth, but several tiny holes and cracks appeared. The PVDF membranes by the 16% Na_4SiO_4 treatment manifested a slightly uneven thickness and some shallow strips emerged. Treatment with 20% Na_4SiO_4 further magnified the nonuniformity of the thicknesses and several defects caused by mixing the Na_4SiO_4 were clearly observed. It was concluded that whenever NaOH or Na_4SiO_4 were utilized, the alkali treatments led to a structural damage of the PVDF membranes. This is because the alkali treatment destroyed the PVDF backbones, generating the active points to perform the grafting polymerization.

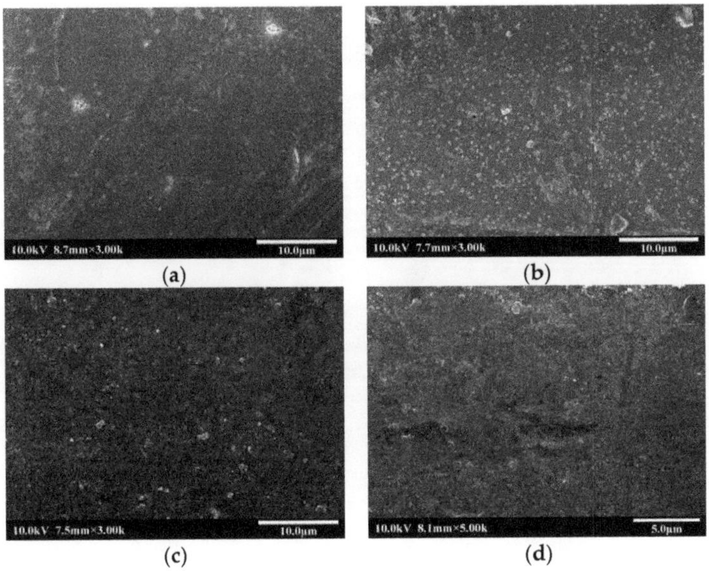

Figure 1. The morphologies of the PVDF membranes by various alkali treatments. (**a**) pristine PVDF; (**b**) treatment by NaOH; (**c**) treatment by 16% Na_4SiO_4; (**d**) treatment by 20% Na_4SiO_4.

After alkali treatment, the PVDF membranes were graft polymerized, respectively, in two solvents, IPA/water or THF. Figure 2 shows the morphologies of the grafted PVDF membranes. In the IPA/water solvent, the grafted PVDF membranes with NaOH treatment showed a rough, uneven surface with a few holes. In comparison, the THF solvent resulted in more and greater holes, indicating a worse mechanical property. For the treatment by 16% Na_4SiO_4, the grafted PVDF membranes in the IPA/water manifested many thin and scaly stripes, along which a few tiny cracks were found. Treatment by 20% Na_4SiO_4 led to folded scaly stripes, rougher surfaces, and greater cracks. Distinctly, when THF was used as the solvent, the grafted PVDF membrane with the treatment of 16% Na_4SiO_4 showed a relatively smooth surface and a few holes were evidently formed. For the treatment by 20% Na_4SiO_4, the membranes possessed more and greater holes, but still remained smooth.

Our experimental results showed that the solvents distinctly influenced the morphologies of the grafted PVDF membranes. These could be explained by the differences in the solubilities of AN and AMS in IPA/water and THF. When the grafting polymerization was performed in IPA/water, AN and AM tended to concentrate on the side of the IPA and were grafted into the PVDF backbones. On the side of water, due to the insolubility, less AMS and AN were grafted. Concentration differences in AN and AMS caused by the heterogeneity of the IPA/water solvent brought about a nonuniform surface of the grafted PVDF membranes. When THF was used, the AN and AMS were dissolved well and the grafting polymerization could be performed evenly, which led to a smooth surface.

Finally, the grafted PVDF membranes were sulfonated by chlorosulfonic acid to afford the PFGPA membranes (Figure 3). The PFGPA surfaces in the grafting solvent of IPA/water with three alkali treatments of 1 M NaOH, 16% and 20% Na_4SiO_4, respectively, all showed scaly surfaces, along with longer and deeper cracks, and somewhere the scales fell. For THF as the grafting solvent, the surfaces of the PFGPA membranes presented as being more even and the holes became greater.

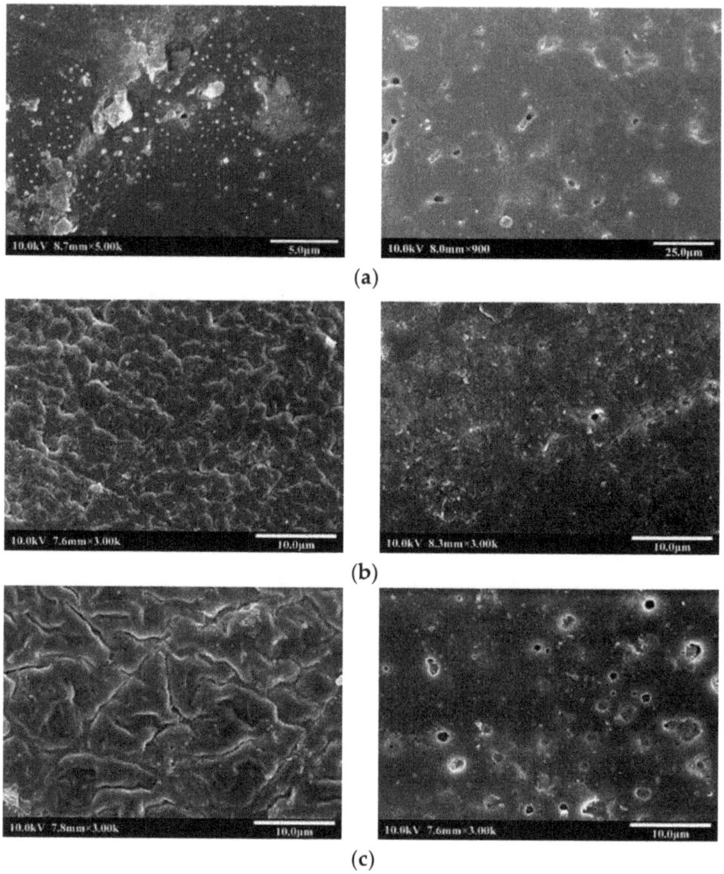

Figure 2. The morphologies of the grafted PVDF membranes with various alkali treatments in IPA/water (**left**) and THF (**right**). (**a**) 1 M NaOH, (**b**) 16% Na_4SiO_4, (**c**) 20% Na_4SiO_4.

3.2.2. IR Analysis

The IR spectrograms of the pristine, 16% Na_4SiO_4-treated and grafted PVDF, and the PFGPA membranes are presented in Figure 4. In the pristine PVDF membranes, absorption bands at 1400 cm^{-1}, 1164 cm^{-1}, and 879 cm^{-1} were the characteristic peaks of PVDF, representing the C–C, C–F, and –CH_2 vibrations, respectively. After the treatment of 16% Na_4SiO_4, the absorption band of the C=C bond at 1564 cm^{-1} appeared, indicating the formation of the active points. The grafted PVDF membranes showed the appearance of absorption bands at 2235 cm^{-1}, which was attributed to the C≡N bond of AN [26] and 700 cm^{-1} assigned to the aromatic C–H deformation of the mono-substituted benzene ring of AMS [16], whereas the absorption band of the C=C bond at 1564 cm^{-1} disappeared, showing that the AMS and AN had been successively grafted into the PVDF. After the grafted PVDF was sulfonated, the membranes manifested new absorption bands at 1035 cm^{-1} and 1006 cm^{-1}, which were the characteristic peaks of the –SO_3H group [16]. The absorption band at 700 cm^{-1} disappeared, confirming the sulfonation of AMS. It was also noted that in the PFGPA membranes, a lower absorption peak at 1660 cm^{-1} emerged, indicating a partial hydrolysis of the CN bond to the carboxyl group [32].

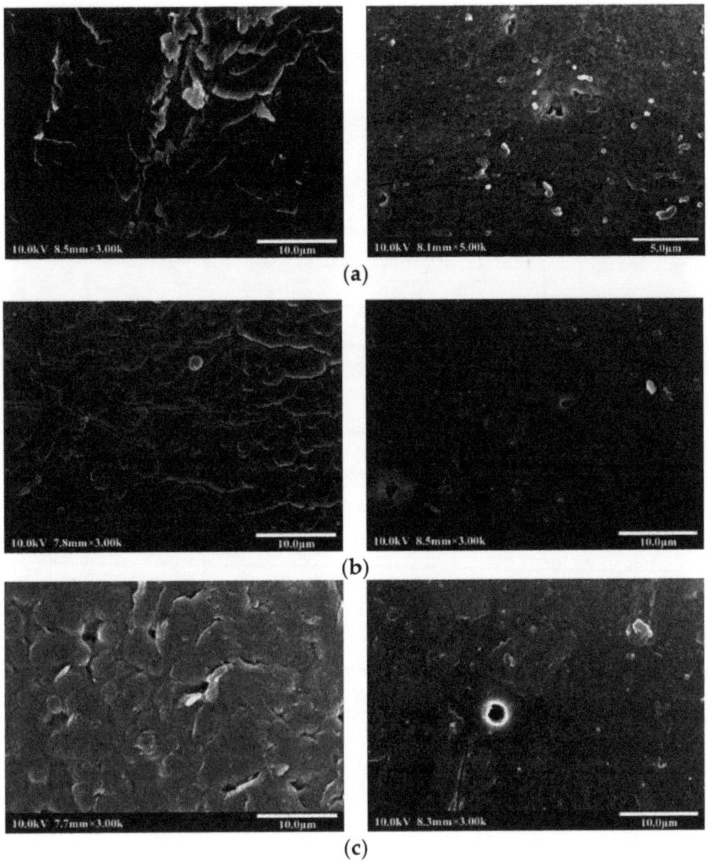

Figure 3. The PFGPA membranes in IPA/water (**left**) and THF (**right**) with various alkali treatments. (**a**) 1 M NaOH, (**b**) 16% Na_4SiO_4, (**c**) 20% Na_4SiO_4.

3.2.3. XPS

The XPS analysis of the PFGPA membranes with the treatment of 16% Na_4SiO_4 is shown in Figure 5. The PFGPA membranes showed peaks of C 1s and F 1s at about 285 eV and 688 eV, attributed to the PVDF backbone [17]. The peaks of N 1s, O 1s, and S 2p at about 400 eV, 532 eV, and 168 eV, respectively, confirmed the grafting of AMS and AN into the PVDF membranes and sulfonation of the grafted PVDF membranes. Si 2p at 182 eV was also observed, indicating the residue of silicon oxide from Na_4SiO_4. From the N 1s core-level spectrum (Figure 5b), the N 1s spectrum of the PFGPA membrane could be curve-fitted with two peaks at the BEs of 399.89 eV and 401.55 eV due to the N≡C– and –NH–CO– groups, respectively, which demonstrated that the N≡C units had been partly protonated during the hydrolysis process to generate the amide groups [32]. The peak of O 1s at around 532 eV was curve-fitted with four peaks at the BEs of 531.65 eV, 532.10 eV, 532.90 eV, and 533.90 eV, corresponding to $–SO_3H$, silicon oxide, $–CONH_2$, and –COOH, respectively. This showed that a few N≡C groups in AMS were hydrolyzed into amides and carboxylic acid.

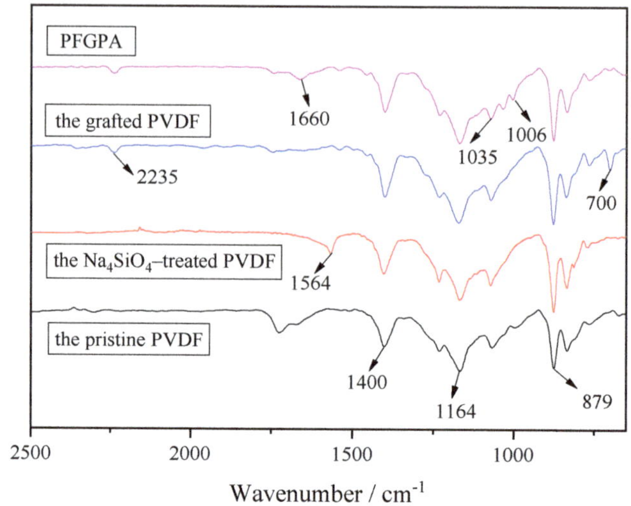

Figure 4. The IR spectrum of the modified PVDF membranes.

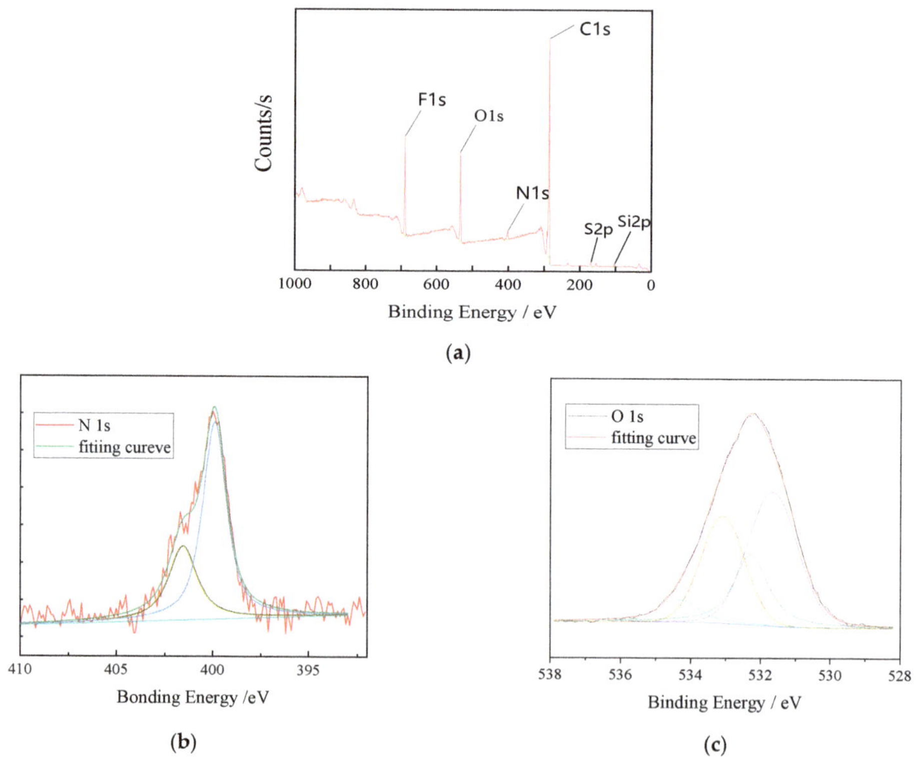

Figure 5. The XPS (**a**), N1s (**b**), and O1s (**c**) fitting curves of the PFGPA membranes.

Therefore, it was concluded that PVDF-g-P(AMS-co-AN) had been successfully prepared and the sulfonic grafted PVDF membranes (PFGPA) were partly hydrolyzed. The hydrolysis of the cyano groups was due to the acid sulfonation condition, under which the

N≡C groups were first hydrolyzed into the amides and then converted into the carboxylic acid [26]. The hydrolysis of the cyano groups was unexpected [26] and the sulfonation of the PFGPA membranes will be systematically investigated later.

3.3. *Performances of the PFPGA Membranes*

In our experiments, due to the higher grafting degree of 40–50 percent by 16% or 20% Na_4SiO_4 treatment, four types of the PFGPA membranes in either THF or IPA/water (I/W) were selected to measure their properties.

3.3.1. Liquid Uptake Capability and Area Swelling

In this section, three liquids, deionized water, 1 M methanol aqueous solution, and pure methanol, were used to evaluate the liquid uptake capability and area swelling.

Figure 6 shows the water uptake capabilities and area swelling of various PFGPA membranes for 192 h. For the water uptake capabilities, the four PFGPA membranes and Nafion all peaked and then decreased during the 192 h. The PFGPA membranes obtained in the grafting solvent of IPA/water showed a greater water uptake capability than those in THF. It was noticeable that PFGPA_16%_I/W demonstrated the quickest and greatest water uptake. It reached 75 percent at 24 h, peaked to 90 percent at 96 h, and finally decreased to 50 percent at 192 h. PFGPA_16%_THF, PFGPA_20%_THF, and PFGPA_20%_I/W showed the maximum water uptake of 82, 40, and 54 percent, respectively, much greater than the 17 percent of Nafion.

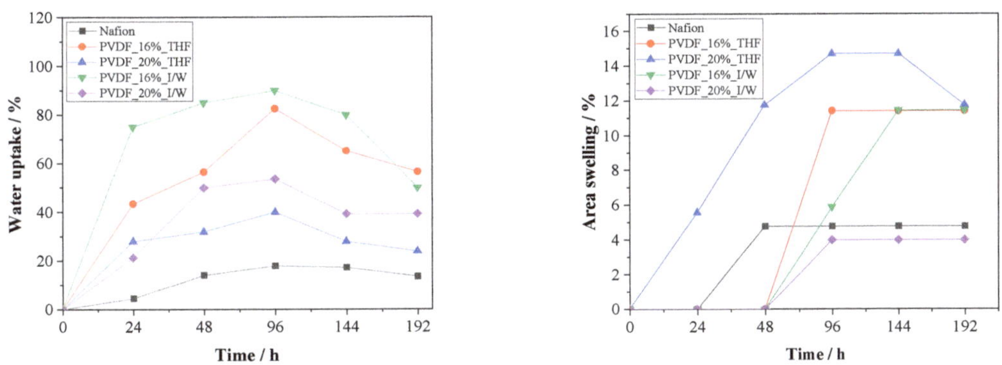

(a) (b)

Figure 6. The water uptake capability (**a**) and area swelling (**b**) of the PFGPA membranes during 192 h.

With respect to the area swelling in deionized water, the four PFGPA membranes peaked and then remained almost unchangeable for 192 h except that PVDF_20%_THF showed the greatest area swelling of 15 percent at 96 h and then declined. Nafion displayed a maximum area swelling of 4.76 percent at 24 h. The PFGPA membranes grafted in THF showed a greater area swelling than Nafion. The PFGPA membranes grafted in the IPA/water solvent showed a similar area swelling with Nafion at 96 h. Afterward, the PFPGA_16%_I/W manifested an increasing area, swelling to 11.5 percent. However, the PFPGA_20%_I/W maintained a lower swelling of about 4 percent. These experimental results indicated that the IPA/water solvent provided a better dimensional stability in water than THF.

Figure 7 displays the liquid uptake capabilities and area swelling of various PFGPA membranes in deionized water, 1 M methanol aqueous solution, and methanol for 48 h. Nafion showed the greatest methanol uptake capability, then the 1 M methanol aqueous solution and the water uptake was the lowest. In contrast, the four PFGPA membranes all exhibited the greatest water uptake capabilities, then the 1 M methanol aqueous solution. The methanol uptake capabilities were the lowest, no more than half that of Nafion, indicating a significant methanol resistance.

For the area swelling, similar to the liquid uptake capability, Nafion revealed a remarkable area swelling in methanol of about 50 percent, almost 10 times as much as that in water and 5 times in the 1 M methanol aqueous solution. However, the four PFGPA membranes exhibited an area swelling in methanol of no more than 20 percent, much lower than that of Nafion. Especially for PFGPA_16%_I/W and PFGPA_20%_I/W, the area swellings were about 5 percent, indicating a better dimensional stability and promising application in direct methanol fuel cells.

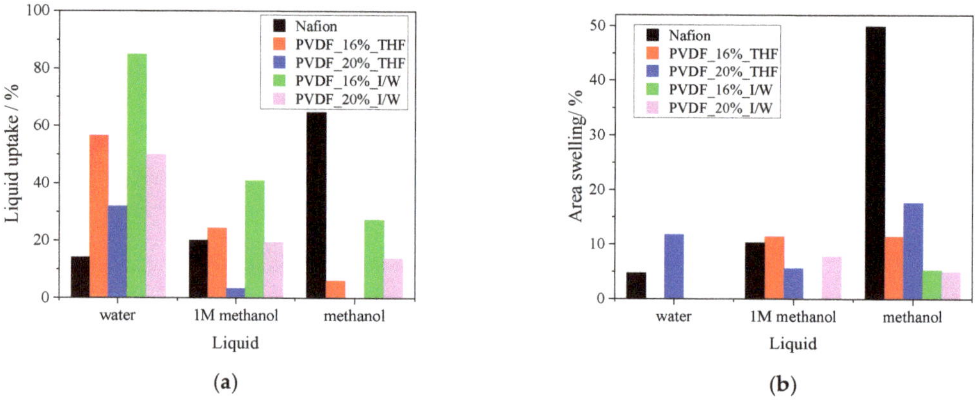

Figure 7. The liquid uptake capabilities (**a**) and area swelling (**b**) of the PFGPA membranes for 48 h.

3.3.2. Chemical Stability

It is essential that the proton exchange membranes should maintain excellent chemical properties during the operation of the fuel cell. In this paper, the chemical properties of the four PFGPA membranes were evaluated by soaking in 30% H_2O_2 at 60 °C and the weight changes were determined. The radicals generated by H_2O_2 resulted in the degradation of the PVDF membranes. Figure 8 shows the weight changes of the four PFGPA membranes. In comparison, Nafion was also tested under the same experimental conditions. It was observed that Nafion showed a weight loss of less than 10 percent, indicating the best chemical stability. Except for PFGPA_16%_THF, the other three PFGPA membranes quickly descended to about 50 percent at 6 h and then leveled off. The PFGPA_16%_THF declined to about 80 percent at 18 h, then slowed down to about 75 percent, slightly lower than Nafion, but greater than 60 percent of AIEM-1 at 6% H_2O_2 solution by Hu [16].

On the other hand, the PFGPA_16%_THF and PFGPA_16%_I/W revealed different chemical stabilities. The grafted chains accounted for 33 percent of the weight due to their similar grafting degree of about 49 percent in Section 3.1. After degradation in H_2O_2, the weight of the PFGPA_16%_I/W decreased to about 40 percent, meaning that the grafted chains were almost completely degraded. However, for the PFGPA_16%_THF, the residue weight was still 75 percent. This indicated that only part of the grafted chains decomposed from the PVDF base and about 27 percent of the grafted chains were still retained, proving a better chemical stability.

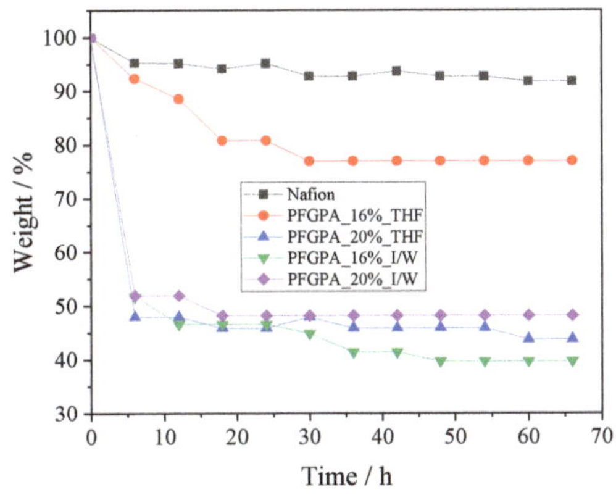

Figure 8. The chemical stabilities of the PFGPA membranes.

3.3.3. Ion Exchange Capability and Conductivity

The PFGPA membranes have sulfonic acid groups and possess the ion exchange capacities (IEC). The measured IEC values of the four PFGPA membranes are listed in Table 1. It was noted that the IPA/water solvent in grafting polymerization displayed slightly greater IEC values of above 0.7 mmol/g than THF. In particular, PFGPA_16%_I/W displayed the greatest IEC value of 0.89 mmol/g, close to 0.91 mmol/g of Nafion measured in our experiment, exhibiting a valuable application prospect. Whenever the grafting solvent was either THF or IPA/water, the PFGPA membranes with 20% of Na_4SiO_4 treatment showed an IEC value of about 0.75 mmol/g. The PFGPA_16%_THF showed the lowest IEC value of 0.56 mmol/g.

Figure 9 shows the impedance plots of the PFGPA membranes and the conductivities are listed in Table 1. The impedances of Nafion and the PFGPA membranes were mainly influenced by the charge transfer process. In our experiment, Nafion showed the measured conductivity of 0.0390 S/cm with the lower solution and charge transfer impedances than the four PFGPA membranes. In comparison, the four PFGPA membranes possessed a lower conductivity of about 0.0113–0.0151 S/cm. As far as the effects of the grafting solvents were concerned, the THF resulted in a lower solution impedance than that of IPA/water. PFGPA_16%_THF and PFGPA_20%_THF possessed a similar conductivity of 0.0133 S/cm, showing that the content of Na_4SiO_4 never greatly influenced the proton transfer capability. For the grafting solvent of IPA/water, the conductivity of the PFGPA membranes increased with the Na_4SiO_4 content.

3.3.4. Mechanical Properties

So far, many of the PVDF-based proton exchange membranes by radiation grafting have been synthesized as substituents for Nafion [15–18], but few of their mechanical properties have been reported. In general, the radiation grafting destroyed the backbones of the PVDF and thus decreased the mechanical properties. The greater the grafting degree, the greater the decrease in the mechanical property. In this paper, the mechanical properties of the obtained PFGPA membranes were measured and the data are listed in Table 2. The effects of the alkali treatments and grafting solvents are discussed below.

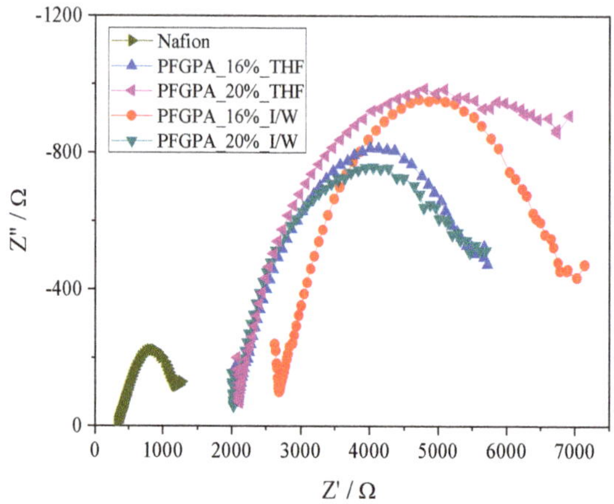

Figure 9. The impedance plots of the PFGPA membranes.

Table 2. The mechanical properties of the PFGPA membranes.

Sample	Max. Load/N	Elastic Modulus/MPa	Stress/Mpa	Strain/%
Pristine PVDF	68.14	1771.37	34.07	17.6
PVDF-16%	16.53	554.89	8.27	4.8
PVDF-20%	9.24	536.78	4.62	5.1
PFGPA-16%-THF	7.828	0.92	3.92	4.3
PFGPA-16%-I/W	10.79	0.043	5.40	2.8
PFGPA_NaOH-THF	25.07	639.40	12.53	4.9
PFGPA_NaOH-I/W	14.29	366.63	7.14	4.0

The PEMs have to possess the excellent mechanical properties to withstand the tough operating conditions. To investigate the mechanical properties of the PFGPA membranes, the pristine and Na_4SiO_4-treated PVDF membranes and the NaOH-treated PFGPA membranes were used as a contrast. The pristine PVDF membranes showed an obvious plastic deformation with the broken stress of 34.07 MPa, a strain of 17.6%, and elastic modulus of 1771.37 MPa. After the PVDF was treated by Na_4SiO_4, the mechanical properties declined dramatically. Treatment by 16% Na_4SiO_4 showed a stress of 8.27 MPa and strain of 4.8 percent, a quarter of the pristine PVDF membranes. Treatment by 20% Na_4SiO_4 intensified the decrease with the stress of 4.62 MPa and the strain of 5.1 percent. After the grafting polymerization and sulfonation, compared with the PVDF_16%, the PFGPA_16%_I/W and PFGPA_16%_THF decreased in broken strain to about a half, indicating a worse mechanical property. The IPA/water brought about a lower decline in strain than the THF. For the PVDF_20%_THF and PVDF_20%_I/W, the membranes were too fragile to be measured under the experimental conditions.

Comparatively, the PFGPA_NaOH_THF and PFGPA_NaOH_I/W showed the broken stresses of 12.53 MPa and 7.14 MPa, and strains of 4.9% and 4.0%, respectively, which were close to PVDF_16% and greater than PFGPA_16%_THF and PFGPA_16%_I/W. These could be explained by the fact that PFGPA_NaOH_THF and PFGPA_NaOH_I/W possessed the grafting degree of about 35 percent, lower than 49 percent of PFGPA_16%_THF and PFGPA_16%_I/W, which resulted in a lower decrease in the mechanical property.

Our experiments demonstrated that the alkali treatment influenced the mechanical property of the PVDF membranes. The alkali treatment destroyed the PVDF backbone more than the grafting polymerization and sulfonation. The Na_4SiO_4 treatment led to a greater decrease in mechanical property than the NaOH treatment and the obtained PFGPA membranes became fragile [33].

4. Conclusions

The PFGPA membranes were prepared by alkali-induced grafting polymerization and subsequent sulfonation. The AMS could be grafted with AN into the PVDF base and IR and XPS analysis testified that AMS and AN were successfully grafted into the PVDF backbone and a partial hydrolysis of the CN group in AN occurred. A higher alkali content resulted in a higher grafting degree. THF as the grafting solvent could dissolve the AMS and AN, so that a smoother membrane and higher grafting degree were obtained than that in the IPA/water solvent. The PFGPA membranes exhibited a greater water uptake capability and better dimensional stability in methanol than Nafion, though they possessed lower IEC values and conductivities. The PFGPA_16%_THF membrane with a grafting degree of 49 percent manifested a slightly lower chemical stability than Nafion, indicating a promising application. It was also noted that the alkali treatment brought about a decrease in the mechanical property of the PVDF membranes and limits their practical application in fuel cells. The next research will focus on the enhancement of the conductivities of the PFGPA membranes and the improvement in the mechanical properties. The obtained experimental conclusions would be instructive to optimize the PFGPA membranes prepared by an alkali-grafted polymerization and promote their practical applications in PEMs.

Author Contributions: Conceptualization, S.L.; Data curation, formal analysis, investigation, and resources, X.L.; Software, P.F. and Y.Z.; Writing—original draft and writing—review & editing, S.L. All authors have read and agreed to the published version of the manuscript.

Funding: This research was funded by the National Natural Science Fund of China (grant number 51603144).

Acknowledgments: The authors would like to thank the Analytical & Testing Center of Tiangong University for the SEM, IR, and XPS analysis.

Conflicts of Interest: The authors declare no conflict of interest.

References

1. Ke, X.; Zhang, Y.; Gohs, U.; Drache, M.; Beuermann, S. Polymer electrolyte membranes prepared by graft copolymerization of 2-acrylamido-2-methylpropane sulfonic acid and acrylic acid on PVDF and ETFE activated by electron beam treatment. *Polymers* **2019**, *11*, 1175. [CrossRef] [PubMed]
2. Mojarrad, N.R.; Iskandarani, B.; Taşdemir, A.; Yürüm, A.; Gürsel, S.A.; Kaplan, B.Y. Nanofiber based hybrid sulfonated silica/P(VDF-TrFE) membranes for PEM fuel cells. *Int. J. Hydrogen Energy* **2021**, *46*, 13583–13593. [CrossRef]
3. Vijayalekshmi, V.; Khastgir, D. Fabrication and comprehensive investigation of physicochemical and electrochemical properties of chitosan-silica supported silico tungstic acid nanocomposite membranes for fuel cell applications. *Energy* **2018**, *142*, 313–330.
4. Heitner-Wirguin, C. Recent advances in perfluorinated ionomer membranes: Structure, properties and applications. *J. Membr. Sci.* **1996**, *120*, 1–33. [CrossRef]
5. Ketpang, K.; Lee, K.; Shanmugam, S. Facile synthesis of porous metal oxide nanotubes and modified Nafion composite membranes for polymer electrolyte fuel cells operated under low relative humidity. *ACS Appl. Mater. Interfaces* **2014**, *6*, 16734–16744. [CrossRef]
6. Nagarale, R.K.; Gohil, G.S.; Shahi, V.K.; Rangarajan, R. Organic/inorganic hybrid membrane: Thermally stable cation-exchange membrane prepared by the sol–gel method. *Macromolecules* **2003**, *37*, 10023–10030. [CrossRef]
7. Rikukawa, M.; Sanui, K. Proton-conducting polymer electrolyte membranes based on hydrocarbon polymers. *Prog. Polym. Sci.* **2000**, *25*, 1463–1502. [CrossRef]
8. Hickner, M.A.; Ghassemi, H.; Kim, Y.S.; Einsla, B.R.; McGrath, J.E. Alternative polymer systems for proton exchange membranes (PEMs). *Chem. Rev.* **2004**, *104*, 4587–4612. [CrossRef]
9. Oh, K.; Kwon, O.; Son, B.; Lee, D.H.; Shanmugam, S. Nafion-sulfonated silica composite membrane for proton exchange membrane fuel cells under operating low humidity condition. *J. Membr. Sci.* **2019**, *583*, 103–109. [CrossRef]
10. Gubler, L.; Gürsel, S.A.; Scherer, G.G. Radiation grafted membranes for polymer electrolyte fuel cells. *Fuel Cells* **2005**, *5*, 317–335. [CrossRef]
11. Chen, J.; Asano, M.; Maekawa, Y.; Sakamura, T.; Kubota, H.; Yoshida, M. Preparation of ETFE-based fuel cell membranes using UV-induced photografting and electron beam-induced crosslinking techniques. *J. Membr. Sci.* **2006**, *283*, 373–379. [CrossRef]
12. Hasegawa, S.; Suzuki, Y.; Maekawa, Y. Preparation of poly(ether ether ketone)-based polymer electrolytes for fuel cell membranes using grafting technique. *Radiat. Phys. Chem.* **2008**, *77*, 617–621. [CrossRef]
13. Chen, J.; Asano, M.; Yamaki, T.; Yoshida, M. Preparation and characterization of chemically stable polymer electrolyte membranes by radiation-induced graft copolymerization of four monomers into ETFE films. *J. Membr. Sci.* **2006**, *269*, 194–204. [CrossRef]

14. Fu, R.Q.; Woo, J.J.; Seo, S.J.; Lee, J.S.; Moon, S.H. Sulfonated polystyrene/polyvinyl chloride composite membranes for PEMFC applications. *J. Membr. Sci.* **2008**, *309*, 156–164. [CrossRef]
15. Zhang, Z.; Jetsrisuparb, K.; Wokaun, A.; Gubler, L. Study of nitrile-containing proton exchange membranes prepared by radiation grafting: Performance and degradation in the polymer electrolyte fuel cell. *J. Power Sources* **2013**, *243*, 306–316. [CrossRef]
16. Hu, G.; Wang, Y.; Ma, J.; Qiu, J.; Peng, J.; Li, J.; Zhai, M. A novel amphoteric ion exchange membrane synthesized by radiation-induced grafting α-methylstyrene and N, N-dimethylaminoethyl methacrylate for vanadium redox flow battery application. *J. Membr. Sci.* **2012**, *407–408*, 184–192. [CrossRef]
17. Qiu, X.; Li, W.; Zhang, S.; Liang, H.; Zhu, W. The Microstructure and character of the PVDF-*g*-PSSA membrane prepared by solution grafting. *J. Electrochem. Soc.* **2003**, *150*, 917–921. [CrossRef]
18. Guo, G.B.; An, S.L.; Kou, S.S. Preparation and properties of modified poly(vinylidene fluoride) grafted with polystyrene sulfonated acid membrane. *Polym. Mater. Sci. Eng.* **2010**, *26*, 112–115.
19. Hübner, G.; Roduner, E. EPR investigation of HO/radical initiated degradation reactions of sulfonated aromatics as model compounds for fuel cell proton conducting membranes. *J. Mater. Chem.* **1999**, *9*, 409–418. [CrossRef]
20. Dockheer, S.M.; Gubler, L.; Bounds, P.L.; Domazou, A.S.; Scherer, G.G.; Wokaun, A.; Koppenol, W.H. Damage to fuel cell membranes. Reaction of HO with an oligomer of poly(sodium styrene sulfonate) and subsequent reaction with O_2. *Phys. Chem. Chem. Phys.* **2010**, *12*, 11609–11616. [CrossRef]
21. Gursel, S.A.; Yang, Z.; Choudhury, B.; Roelofs, M.G.; Scherer, G.G. Radiation grafted membranes using a trifluorostyrene derivative. *J. Electrochem. Soc.* **2006**, *10*, A1964–A1970. [CrossRef]
22. Hodgdon, R.B. Sulfonated Polymers of α, α, β-Trifluorostyrene, with Applications to Structures and Cells. U.S. Patent 3341366, 12 September 1967.
23. Chen, J.; Asano, M.; Yamaki, T.; Yoshida, M. Improvement of chemical stability of polymer electrolyte fuel cell membranes by grafting of new substituted styrene monomers into ETFE films. *J. Mater. Sci.* **2006**, *41*, 1289–1292. [CrossRef]
24. Assink, R.A.; Arnold, C.; Hollandsworth, R.P. Preparation of oxidatively stable cation-exchange membranes by the elimination of tertiary hydrogens. *J. Membr. Sci.* **1991**, *56*, 143–151. [CrossRef]
25. Sharma, K.R. Thermal terpolymerization of alpha-methylstyrene, acrylonitrile and styrene. *Polymer* **2000**, *41*, 1305–1308. [CrossRef]
26. Woo, J.J.; Seo, S.J.; Yun, S.H.; Fu, R.Q.; Yang, T.H.; Moon, S.H. Enhanced stability and proton conductivity of sulfonated polystyrene/PVC composite membranes through proper copolymerization of styrene with alpha-methylstyrene and acrylonitrile. *J. Membr. Sci.* **2010**, *363*, 80–86. [CrossRef]
27. Li, J.; Muto, F.; Miura, T.; Oshima, A.; Washio, M.; Ikeda, S.; Katsumura, Y. Improving the properties of the proton exchange membranes by introducing alpha-methylstyrene in the pre-irradiation induced graft polymerization. *Eur. Polym. J.* **2006**, *42*, 1222–1228. [CrossRef]
28. Gubler, L.; Slaski, M.; Wallasch, F.; Wokaun, A.; Scherer, G.G. Radiation grafted fuel cell membranes based on co-grafting of alpha-methylstyrene and methacrylonitrile into a fluoropolymer base film. *J. Membr. Sci.* **2009**, *339*, 68–77. [CrossRef]
29. Guo, G.B.; An, S.L.; Kou, S.S. Preparation and performance of modified poly(vinylidene fluoride) grafted onto a blended polystyrene sulfonated acid membrane. *Acta Phys.-Chim. Sin.* **2009**, *25*, 2161–2166.
30. Riberio, M.R.; Portela, M.F.; Deffieux, A.; Cramail, H.; Rocha, M.F. Isospecific homo and copolymerization of styrene with ethylene in the presence of VCl_3, $AlCl_3$ as catalyst. *Macromol. Rapid Commun.* **1996**, *17*, 461–469. [CrossRef]
31. Deng, Y.; Peng, C.; Liu, P.; Lu, J.; Zeng, L. Studies on the cationic copolymerization of alpha-pinene and styrene with complex $SbCl_3/AlCl_3$ catalyst systems. 1. Effects of the polymerization conditions on the copolymerization products. *J.M.S. Pure Appl. Chem.* **1996**, *A33*, 995–1004.
32. Henkensmeier, D.; Wallasch, F.; Gubler, L. Radiation grafted ETFE-graft-poly(α-methyl styrenesulfonic acid-co-methacrylonitrile) membranes for fuel cell applications. *J. Membr. Sci.* **2013**, *447*, 228–235. [CrossRef]
33. Dias, A.J.; McCarthy, T.J. Dehydrofluorination of poly(vinylidene Fluoride) in dimethylformamide solution: Synthesis of an operationally soluble semiconducting polymer. *J. Polym. Sci. Polym. Chem. E* **1985**, *23*, 1057–1061. [CrossRef]

polymers

Article

Preparation and Properties of Novel Crosslinked Polyphosphazene-Aromatic Ethers Organic–Inorganic Hybrid Microspheres

Yan Wang [1], Yunwu Yu [2], Long Li [3], Hongbo Zhang [1], Zheng Chen [1], Yanchao Yang [1], Zhenhua Jiang [1] and Jianxin Mu [1,*]

[1] Key Laboratory of High-Performance Plastics, Ministry of Education, National & Local Joint Engineering Laboratory for Synthesis Technology of High Performance Ploymer, College of Chemistry, Jilin University, 2699 Qianjin Street, Changchun 130012, China; wyan2012@jlu.edu.cn (Y.W.); zhanghb20@mails.jlu.edu.cn (H.Z.); chenzheng2013@jlu.edu.cn (Z.C.); yangyanchao@jlu.edu.cn (Y.Y.); jiangzhenhua@jlu.edu.cn (Z.J.)

[2] School of Materials Science and Engineering, Shenyang Jianzhu University, Shenyang 110168, China; yuyunwu@sjzu.edu.cn

[3] College of Materials Science and Engineering, Shenyang University of Chemical Technology, Shenyang 110142, China; lilong@syuct.edu.cn

* Correspondence: jianxin_mu@jlu.edu.cn

Abstract: A series of novel crosslinked polyphosphazene-aromatic ether organic–inorganic hybrid microspheres with different structures were prepared via precipitation polycondensation between hexachlorocyclotriphosphazene (HCCP) and bisphenol monomers. The bisphenol monomers have different numbers of $–CF_3$ in the side group, which correspond to distinct oligomeric species-absorbing mechanisms. The wetting behavior of the microsphere surface was evaluated using a water contact angle (CA) measurement, which increased with the increase in the content of $–CF_3$ in polyphosphazene. We also investigated the effects of HCCP concentration and ultrasonic power on the morphology of the microspheres.

Keywords: polyphosphazene; micro-nanospheres; species-absorbing mechanisms; hydrophobicity; thermochemical

Citation: Wang, Y.; Yu, Y.; Li, L.; Zhang, H.; Chen, Z.; Yang, Y.; Jiang, Z.; Mu, J. Preparation and Properties of Novel Crosslinked Polyphosphazene-Aromatic Ethers Organic–Inorganic Hybrid Microspheres. *Polymers* **2022**, *14*, 2411. https://doi.org/10.3390/polym14122411

Academic Editors: Andrzej Puszka and Beata Podkościelna

Received: 1 May 2022
Accepted: 9 June 2022
Published: 14 June 2022

1. Introduction

In recent years, people have shown great interest in the research into and application of the polyphosphazene family of materials, because they not only have excellent thermal and chemical stability [1,2], but also can be surface-modified through interface reactions that connect functional groups and bioactive molecules to the surface. The traditional phosphonitrile material is a linear polymer with an alternating P-N main chain synthesized by means of ring-opening polymerization of hexachlorocyclophosphazene and triphosphazene. It has been widely used in solid polyelectrolytes, biomedical and organometallic polymers and flame-retardant materials [3]. However, to expand its application further, linear polyphosphazenes have the disadvantage of low yield and high cost which must be overcome. For a long time, other kinds of phosphonitrile materials have attracted attention. They are ring-linear and ring-matrix polymers, in which phosphonitrile rings are connected by outer ring groups to form linear chains or crosslinked matrices. Because the complexity of synthesis seriously limits the design ability of cyclic phosphine materials, more and more studies are focusing on the preparation of cyclic matrix materials.

In the past few decades, various phosphonitrile polymers have been reported. Because of their high thermal stability, they are used widely in adhesives [4], flame-retardant polymer additives [5–7] and thermosetting resins [8,9]. Most cyclic polymers are synthesized through many complex procedures: the phosphonitrile ring is first joined to bifunctional compounds such as hydroquinone or aminophenol, and then the precursors are crosslinked with other bifunctional compounds such as anhydride and carbonyl chloride. It is difficult to totally avoid the chain tensioning reaction in this procedure, which results in a faulty structure in the final ring matrix product. To generate a unique cyclic trimer, an active group of bifunctional components must first be protected, then deprotected, which complicates the reaction process and limits its use [6,10,11].

Ion-crosslinked polyphosphazene hydrogel microspheres have previously been reported as being made using the ion complexation process and have the potential to be used for protein encapsulation. Polycrystalline poly[cyclotriphosphazene co-(4,4′-sulfonyldiphenol)] nanotubes were synthesized by means of the in situ template method [12,13], and silver nanocables coated with hybrid poly[cyclotriphosphazene co-(4,4′-sulfonyldiphenol)] were obtained by means of the hard template method [14,15]. These results show that polyphosphazenes are inorganic–organic hybrid micro-nano materials with potential value at the molecular level for the preparation of nanomaterials. However, there are few studies on the functionalization of these polyphosphazenes [16–20]. Therefore, it is necessary to investigate the functionalization of inorganic–organic hybrid materials and provide new opportunities for their further application.

We describe a simple, broad, and successful one-pot synthesis method for surface-functionalized novel crosslinked polyphosphazene-aromatic ether organic–inorganic hybrid microspheres by means of a single-step precipitation polymerization. Our approach allows the direct one-pot synthesis of a series of functional polyphosphazene microspheres, poly[cyclotriphosphazene-co-(3,5-ditrifluoromethyl)phenylhydroquinone] (6FPZF), poly[cyclotriphosphazene-co-(3-trifluoromethyl)phenylhydroquinone] (3FPZF) and poly[cyclotriphosphazene-co-2-phenylhydroquinone] (TPZF). To the best of our knowledge, this is the first report on these polyphosphazene-aromatic ether microspheres. In contrast to a previous report [21], we investigate the morphology of polyphosphazene-aromatic ether microspheres with different side group structures on the polymer chain. Additionally, we focus on the influence of different side groups of the bisphenol monomers on the oligomeric species-absorbing mechanism. The morphology, particle sizes, chemical structure and hydrophobic properties of the resulting microspheres are characterized.

2. Materials and Methods

2.1. Materials

Hexachlorocyclotriphosphazene (HCCP) (synthesized as described in the literature [3]) was recrystallized from petroleum ether. The melting point of the purified HCCP was 113–116 °C. (3,5-ditrifluoromethyl)phenylhydroquinone (6FPH) and (3-trifluoromethyl) phenylhydroquinone (3FPH) were synthesized and characterized by the method reported in [22,23]. 2-phenylhydroquinone (TPH) was purchased from Alfa Aesar Chemical Co. Ltd., Shanghai, China. Acetonitrile, acetone, tetrahydrofuran (THF) and triethylamine (TEA) were purchased from Beijing Chemical works Co. Ltd., Beijing, China and used without further purification.

2.2. Characterization

The ultrasonic bath used in the experiment was produced by Kunshan Ultrasonic Instrument Company, and the model was KQ-300GV (Kunshan, China). The samples were tested for infrared spectra. The infrared spectra (KBr particles or films) were measured on a Nicolet 410 Fourier (Madison, WI, USA) transform infrared spectrometer at room temperature (25.8 °C). The scanning range of the sample was 450–4000 cm^{-1}. Scanning electron microscopy (SEM) measurements were performed on an SHIMADZUSSX-550 electron microscope (Kyoto, Japan) at an accelerating voltage of 15 kV. Each sample (4 mg) for SEM observations was dispersed in ethanol (1 mL) and then deposited onto the surface of aluminum foil. The specimens were coated with gold before the SEM observations. Transmission electron microscopy (TEM) measurements were obtainedusing a JEOL JEM1200-EX transmission electron microscope (Tokyo, Japan). The pressed polyphosphazene microspheres were analyzed by X-ray photoelectron spectroscopy (XPS) (ESCALAB.MK II, Thermo Scientific, Horsham, UK). The ESCALAB.MK II scanning ESCA microprobe instrument adopted a monochrome Al KαX-ray source (1486.60 eV). Wettability experiments were characterized by the water contact angle (CA) test. The CA was measured on the JC2000C2 contact angle system (POWEREACH/JC2002C2 CA meter, Shanghai, China). At room temperature, water droplets (approx. 4.0 μL) were dropped onto the coating surface of polyphosphazene microspheres carefully. The contact angle value of each sample was obtained by measuring five different positions of the same sample and taking the average value. Thermogravimetric analysis (TGA) was used for evaluating the thermal stability of the polyphosphazene microspheres by the PE Pyris 1 TGA thermal analyzer system (Perkin Elmer, Waltham, MA, USA). Before analysis, we kept the sample at a constant temperature for 10 min at 100 °C in a nitrogen atmosphere to ensure that it was dry and free of moisture. Then, we cooled the sample to 80 °C, then raised the temperature to 800 °C at a heating rate of 10 °C min^{-1} and recorded the temperature of each sample under 5% and 10% weight loss.

2.3. Synthesis

The preparation of microspheres was carried out as follows (Scheme 1): 4 mL of TEA was added to 100 mL of acetonitrile solution containing 0.2 g (0.575 mmol) HCCP. Then, 6FPH, 3FPH or TPH was added into the above solution. Polycondensation was carried out in an ultrasonic bath (150 W, 45 kHz) at 20 °C for 3 h. After the reaction, the precipitated product was filtered and washed three times with 100mL of THF and 100mL of deionized water. Then, the product was dried in a vacuum (−0.1MPa) at 60 °C for 12 h to obtain white powdered 6FPZF, 3FPZF and TPZF micro-nanospheres. Synthesis yield was about 80–90% wt%, calculated from HCCP.

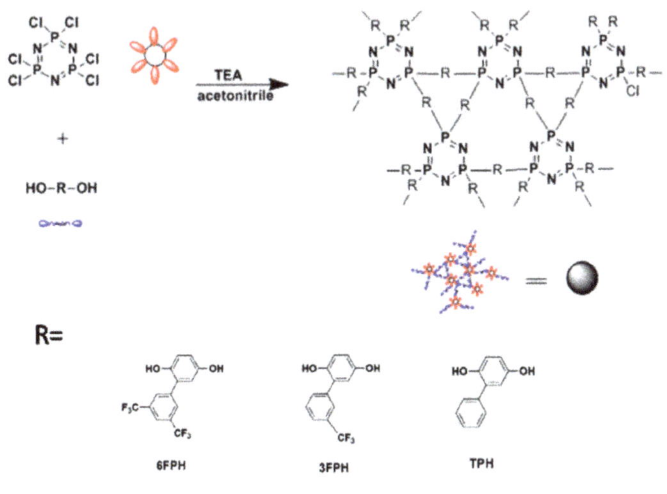

Scheme 1. Synthetic route and suggested chemical structures of polymeric micro-nanospheres.

2.4. Film Preparation for CA Measurement

Polyphosphazene hybrid organic–inorganic micro-nanospheres were prepared in a 0.5 mg mL^{-1} ethanol dispersion. Before use, the dispersion was treated with ultrasound (240 W, 45 kHz) for 30 min. A clean silicon wafer was immersed in the microsphere dispersion, after which it was removed, and the excess liquid was quickly evaporated under infrared light. Finally, the silicon wafer impregnated with polyphosphazene microspheres was dried in a vacuum at 60 °C for 1 h to form a hydrophobic surface layer.

3. Results and Discussion

In the acetonitrile solution, the reaction was carried out in a hypertonic bath with excess TEA as the acid receptor. The polycondensation of HCCP with equimolar phenol produced prepolymer and hydrogen chloride. The reaction of hydrogen chloride with triethylamine accelerated the polycondensation reaction. Therefore, fully crosslinked polymers were prepared. The functional monomer initially formed a complex with the template. After polymerization, their functional groups were fixed by highly crosslinked polymer structures.

Figure 1 shows the chemical structure of the prepared polyphosphazene micro-nanospheres, which was tested by Fourier transform infrared spectroscopy (FT-IR). The stretching vibration of the P-O-(PH) group has a strong absorption peak at 961 cm^{-1}. The absorption at 3200–3300 cm^{-1} corresponds to the phenolic hydroxyl groups of 6FPH, 3FPH and TPH, which no longer exist after the polycondensation. Therefore, the polycondensation between HCCP and hydroxybenzene was successfully completed.

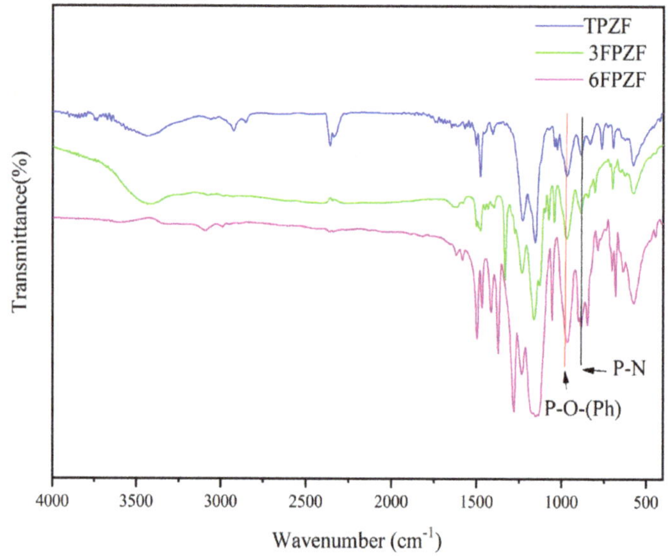

Figure 1. FT−IR spectra of 6FPZF, 3FPZF and TPZF micro-nanospheres.

According to a previous study, in which Choe and his colleagues proposed the mechanism of oligomer species absorption, the formation process of polyphosphazene micro-nanospheres also follows a similar mechanism [24]. In the initial stage of precipitation polymerization, primary micro-nanospheres are generated through the continuous aggregation of primary nuclear particles. Once stable particles are produced, these particles will grow by continuously absorbing oligomers instead of forming particles at one time. Therefore, there are almost no pores in the micro-nanospheres obtained at the end of polymerization. It is pointed out that only when the molecular weight of the oligomer exceeded the critical value (critical molecular weight $\overline{M}_{W.C}$) did the precipitation of the crosslinked oligomer produce the initial nucleus. This critical value $\overline{M}_{W.C}$ was determined by the solubility of oligomers.

The morphology of as-synthesized microspheres was investigated by means of transmission electron microscopy (TEM). Figure 2 shows that all three of the species of obtained polyphosphazene microsphere exhibit a clean surface, and there are no discernable nanostructures on the microspheres.

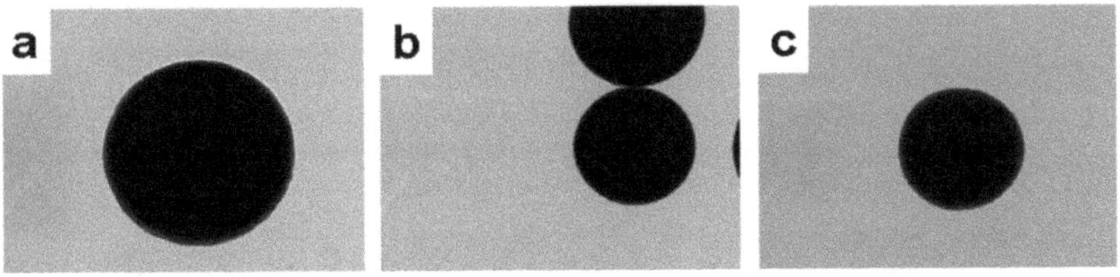

Figure 2. TEM images of polyphosphazene micro-nanospheres: (**a**) 6FPZF micro-nanospheres; (**b**) 3FPZF micro-nanospheres; (**c**) TPZF micro-nanospheres.

The morphology and particle sizes of the obtained microspheres were characterized with SEM. As shown in Figure 3, it can be seen clearly that 6FPZF and 3FPZF provide visible microspheres and there are no TPZF microspheres when the molar ratio of HCCP to bisphenol is 1:2. The reason for this phenomenon is that the $-CF_3$ groups with strong polarityaccelerate the nucleophilic substitution reactions. Therefore, the 6FPH containing two $-CF_3$ groups possessed higher reactivity with HCCP to form stable microspheres compared with 3FPH and TPH. The reaction degree between TPH and HCCP was too low to reach the critical molecular weight $\overline{M}_{W.C}$, and thus the oligomers could not precipitate from the solution.

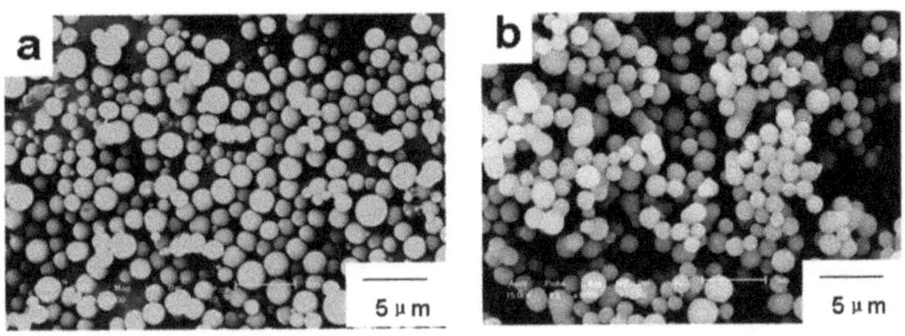

Figure 3. SEM images of polymer microspheres prepared when the molar ratio of HCCP to bisphenol was 1:2: (**a**) 6FPZF micro-nanospheres; (**b**) 3FPZF micro-nanospheres.

Figure 4 shows the SEM microphotographs of crosslinked polymer micro-nanospheres prepared when the molar ratio of HCCP to bisphenol was 1:3. The 6FPZF and 3FPZF microspheres exhibit clean surfaces and no adhesion, indicating the independence between microspheres. The diameter of the microspheres was less than 2 μm and about 1.5 μm on average. Under the same reaction conditions, the size of TPZF microspheres was relatively smaller, which showed that the $-CF_3$ group had a great influence on the growth of particles in the process of precipitation polymerization.

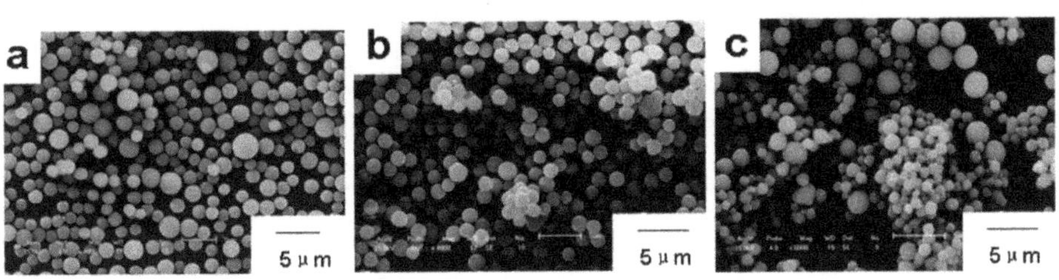

Figure 4. SEM images of polymer microspheres prepared when the molar ratioof HCCP to bisphenol was 1:3: (**a**) 6FPZF micro-nanospheres [21] Copyright © 2013, Springer-Verlag Berlin Heidelberg; (**b**) 3FPZF micro-nanospheres; (**c**) TPZF micro-nanospheres.

The particle morphology and sizes of the obtained 6FPZF, 3FPZF and TPZF microspheres prepared when the molar ratio of HCCP to bisphenol was 1:6 were characterized with SEM (Figure 5). It was obvious that the polymer microspheres with narrower dispersion and smooth surfaces were obtained with decreasing molar ratios of HCCP to 6FPH. Similar phenomena were also observed in 3FPZF and TPZF microspheres, due to the similar structures of their polymer chains.

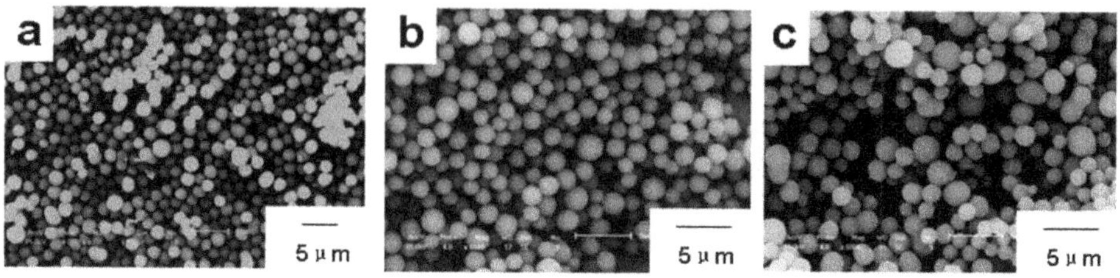

Figure 5. SEM images of polymer microspheres prepared when the molar ratio of HCCP to bisphenol was 1:6: (**a**) 6FPZF micro-nanospheres; (**b**) 3FPZF micro-nanospheres; (**c**) TPZF micro-nanospheres.

In the process of precipitation polymerization, temperature and ultrasonic power have a great influence on the growth of particles. Figure 6 shows the number–average sphere diameter (Dn) of the 6FPZF, 3FPZF and TPZF microspheres in various reaction conditions. The particle size of 3FPZF and TPZF microspheres increased with the increase of ultrasonic power, and compared with 6FPZF microspheres, the particle size always adhered to a certain law, namely 3FPZF > TPZF > 6FPZF, as shown in Figure 6.

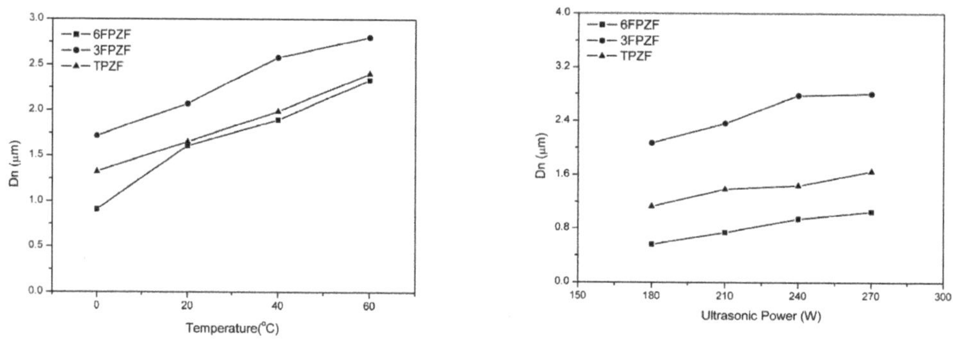

(**a**) (**b**)

Figure 6. (**a**) Dependence of number–average particle size (Dn) of polymer microspheres on temperature of the 6FPZF, 3FPZF and TPZF microspheres; (**b**) effect of ultrasonic power on number-average particle size (Dn) of the 6FPZF, 3FPZF and TPZF microspheres.

Figure 7 shows the SEM microphotographs of crosslinked polymer micro-nanospheres prepared by means of a single-step precipitation polymerization. In this experiment, the microsphere morphology differed with the varying molar ratios of HCCP to hydroxybenzene. The 6FPZF microspheres exhibited clean surfaces and almost complete spheres when the molar ratio of HCCP to 6FPH ranged from 1:2 to 1:7 (Figure 7(a1–f1)). The microspheres were independent of each other, and no microsphere adhesion was observed. The diameter was less than 2 μm and about 1.5 μm on average. The average particle sizes of 3FPZF micro-nanospheres were1.57, 2.03, 2.12, 2.02, 1.76, and 1.74 μm. Firstly, the particle size of as-synthesized 3FPZF microspheres increased as the molar ratios of HCCP to 3FPH decreased(Figure 7(a2–c2)).The maximum particle size was achieved when the molar ratio of HCCP to 3FPH was 1:4 (Figure 7(c2)).Then, the particle size of the as-synthesized

3FPZF microspheres increased with the increase in the molar ratio of HCCP to 3FPH (Figure 7(d2–f2)). This phenomenon was due to the low level of reaction at high molar ratios of HCCP to 3FPH. At the initial stage of precipitation polymerization, the oligomeric species precipitated from the solution when the molecular weight was not high enough. Thus, the particle size of obtained 3FPZF microspheres was smaller. However, when the molar ratio was lower, more oligomers of 3FPZF formed; these oligomers exceeded the critical molecular weight $M_{W.C}$ and precipitated from the solution. The lower the molar ratio of HCCP to 3FPH, the higher the number of nucleus particles, thus resulting in a smaller particle size of 3FPZF microspheres. As shown in Figure 7(a3–e3), when the molar ratio of HCCP to TPH was 1:2, there were no TPZF microspheres. The TPZF microspheres had non-uniform particle sizes when the molar ratio of HCCP to TPH was 1:3 (Figure 7(a3)). Then, with the decrease in the molar ratio of HCCP to 3FPH, stable particles were generated. The average particle size was 1.97μm when the molar ratio of HCCP to TPH was 1:4. However, doublet and triplet particles were observed at the higher molar ratio of HCCP to TPH. The particle sizes of 6FPZF, 3FPZF and TPZF microspheres with varying molar ratios are shown in Table 1. The differing morphologies of the resultant microspheres are attributed to the $–CF_3$ group with strong polarity. It was easier to perform a nucleophilic substitution reaction because of the existence of $–CF_3$. Otherwise, the existence of $–CF_3$ also led to the addition to the geometry and free volume in the crosslinked system. The TPZF microspheres did not have $–CF_3$ in the side group, and thus there was a more flexible P-N skeleton on the sphere surface. This was also confirmed by XPS measurement.

Table 1. Particle sizes of 6FPZF, 3FPZF and TPZF microspheres with varying molar ratios.

	1:2	1:3	1:4	1:5	1:6	1:7
6FPZF	1.43 μm	1.53 μm	1.5 μm	1.54 μm	1.51 μm	1.42 μm
3FPZF	1.57 μm	2.03 μm	2.12 μm	2.02 μm	1.76 μm	1.74 μm
TPZF	/	non-uniform	1.97 μm	non-uniform	non-uniform	non-uniform

X-ray photoelectron spectra of polyphosphazene microspheres were obtained and are shown in Figure 8 and Table 2. The concentration of phosphorus atoms on the TPZF sphere surface was 5.29%, higher than that on the 6FPZF sphere surface (2.86%). The concentrations of fluorine atoms and phosphorus atoms on the 3FPZF sphere surface were both higher than the others. This result is consistent with the formation mechanism and chemical structure of the above microspheres. These results also have some impact on their wetting behavior.

Table 2. Atomic concentrations of the polyphosphazene microsphere surfaces.

	P%	Cl%	C%	N%	O%	F%
6FPZF	2.86	1.82	64.92	2.7	20.48	7.22
3FPZF	5.97	4.7	63.65	5.38	11.61	8.7
TPZF	5.29	2.64	72.97	4.69	14.41	0

The wettability of the prepared coating surface was tested using the water contact angle (CA). Figure 9 shows the shape of water droplets on the silicon wafer after dip coating with polyphosphazene micro-nanospheres. It was found that the silicon wafers dip-coated with the prepared 6FPZF, 3FPZF and TPZF micro-nanospheres have a water CA of 137°, 114° and 95°, respectively. This could be mainly attributed to the fluorine atom content on the spheres' surfaces. The surface of 3FPZF not only had a high fluoride content, but also hydrophilic phosphorus and a nitrogen skeleton.

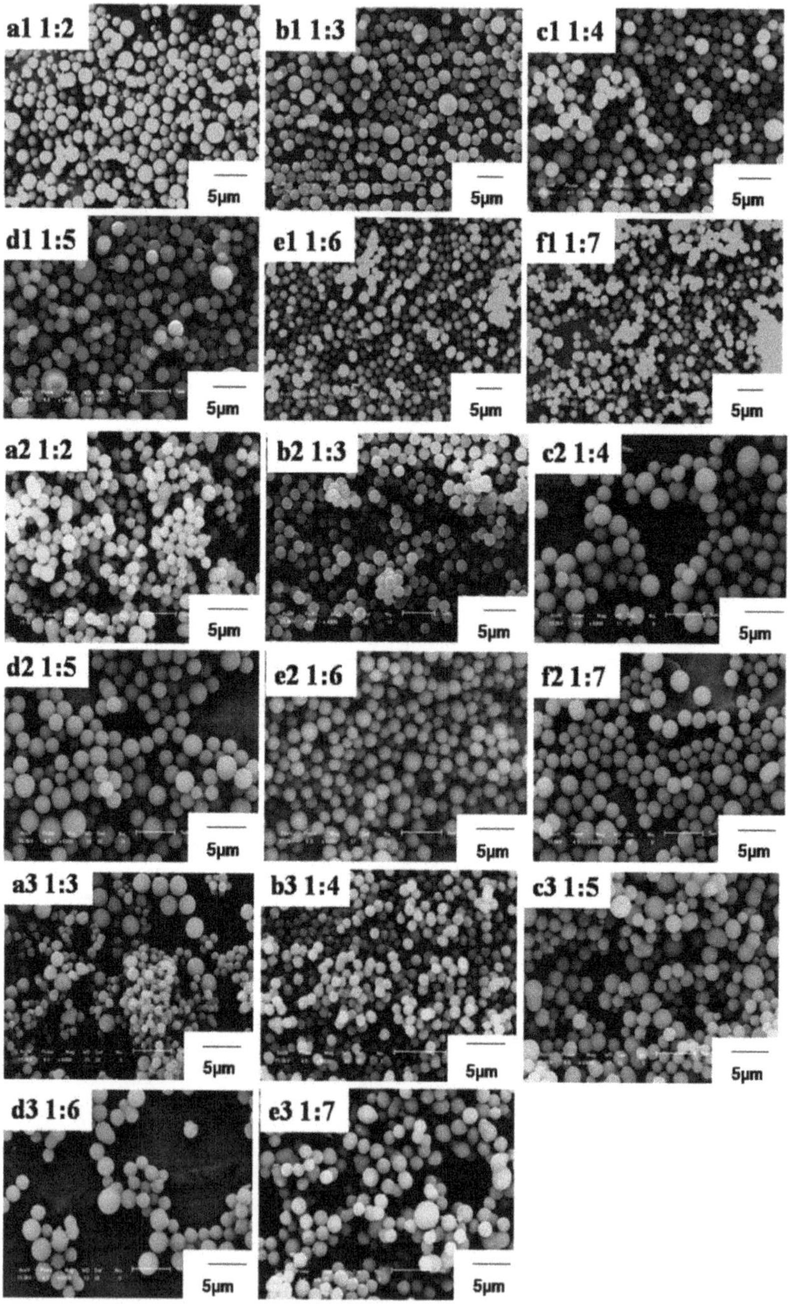

Figure 7. SEM micrographs of the polymeric particles prepared by precipitation polymerization with various molar ratios of HCCP to hydroxybenzene: (**a1**–**f1**) 6FPZF microspheres with molar ratios of HCCP to 6FPH of 1:2 to 1:7; (**a2**–**f2**) 3FPZF microspheres with molar ratios of HCCP to 3FPH of 1:2 to 1:7; (**a3**–**e3**) TPZF microspheres with molar ratios of HCCP to TPH of 1:2 to 1:7.

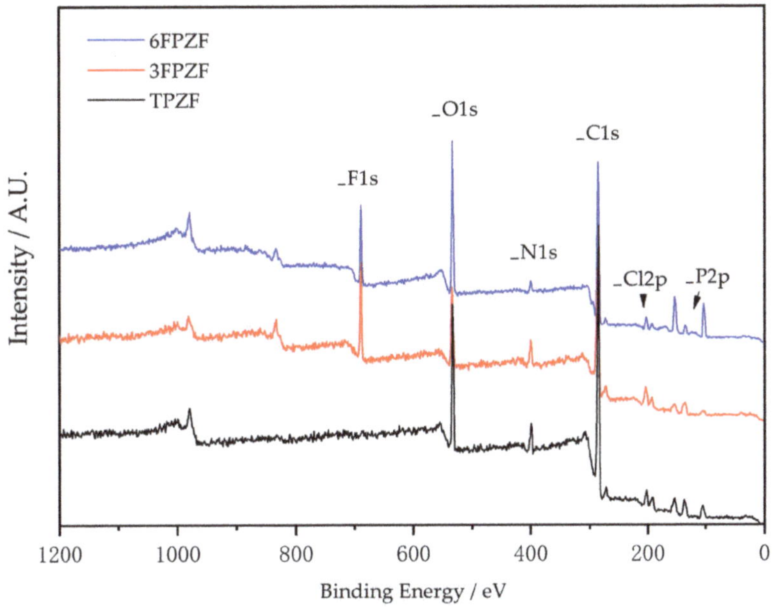

Figure 8. XPS spectra of the polyphosphazene microsphere surfaces.

Figure 9. Shape of a water droplet on a silicon wafer dip-coated with polyphosphazene micro-nanospheres: (**a**) 6FPZF, (**b**) 3FPZF, (**c**) TPZF.

The thermal properties of polyphosphazene micro-nanospheres prepared by precipitation polymerization were characterized by means of thermogravimetric analysis, as shown in Figure 10. The temperatures at a 5% weight loss (T_d) of 6FPZF, 3FPZF and TPZF were 366 °C, 339 °C and 244 °C, respectively, as demonstrated in the weight loss curves. Obviously, the T_ds of the polymers increased in the following order: TPZF, 3FPZF and 6FPZF. This is explained by the introduction of aromatic chains and strong C–F bonds. The residues after the decomposition of 6FPZF, 3FPZF and TPZF in nitrogen at a temperature of 800 °C were 35.5%, 42.9% and 70.8% of the original weight, respectively. The reason for the high residue rate of TPZF is the high content of phosphorus and nitrogen in the main chain of phosphonitrile. This is also in line with the chemical structure and formation mechanism of the above microspheres.

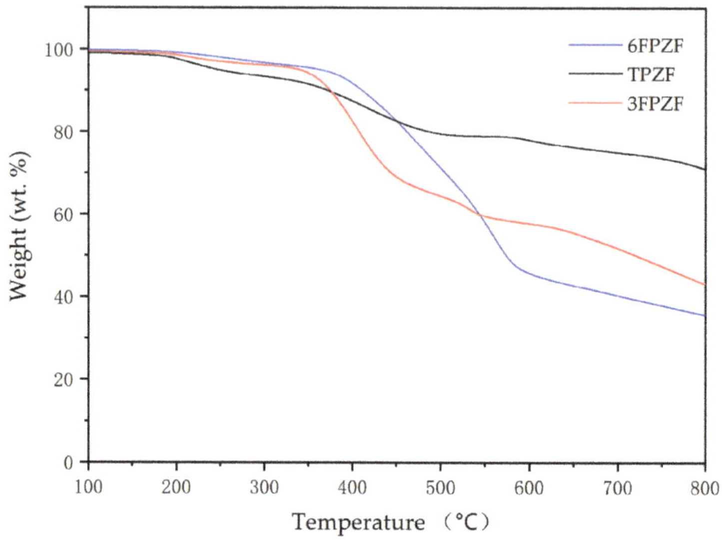

Figure 10. TGA curves of polyphosphazene micro-nanospheres.

4. Conclusions

In conclusion, novel crosslinked polyphosphazene-aromatic ether organic–inorganic hybrid microspheres with different structures were prepared by means of the precipitation polymerization of a hexachlorocyclotriphosphazene (HCCP) monomer. The different formation mechanisms of polyphosphazene microspheres with different structures are caused by the absorption mechanisms of different oligomers. The resulting as-synthesized polyphosphazene microspheres exhibited clean surfaces, and there is no discernable nanostructure on the microsphere. These polyphosphazene micro-nanospheres exhibit outstanding thermochemical stability. The wettability of spheres with different structures is caused by the difference in the atomic composition on the surface of spheres. This result indicates that the spheres might have potential applications as hydrophobic materials.

Author Contributions: Conceptualization, Z.J. and J.M.; formal analysis, H.Z.; resources, L.L. and Y.Y. (Yanchao Yang); data curation, Y.W.; writing—original draft preparation, Y.W.; writing—review and editing, Y.W. and Y.Y. (Yunwu Yu); visualization, Z.C., Z.J. and J.M.; funding acquisition, Y.Y. (Yunwu Yu) and Z.C. All authors have read and agreed to the published version of the manuscript.

Funding: This work was funded by the Foundation of Liaoning Province Education Administration (No. lnqn202019) and the Jilin Province Science and Technology Department Science Fund (No. 20190701009GH and 20200404177YY and Jilin University Fund No. 415010300062).

Institutional Review Board Statement: Not applicable.

Informed Consent Statement: Not applicable.

Data Availability Statement: Not applicable.

Conflicts of Interest: The authors declare no conflict of interest.

References

1. Singler, R.E.; Schneider, N.S.; Hagnauer, G.L. Polyphosphazenes—Synthesis-Properties-Applications. *Polym. Eng. Sci.* **1975**, *15*, 321–338. [CrossRef]
2. Allcock, H.R. Inorganic-Organic Polymers. *Adv. Mater.* **1994**, *6*, 106–115. [CrossRef]
3. De Jaeger, R.; Gleria, M. Poly(organophosphazene)s and related compounds: Synthesis, properties and applications. *Prog. Polym. Sci.* **1998**, *23*, 179–276. [CrossRef]

4. Barrett, E.W.; Phelps, M.V.B.; Silva, R.J.; Gaumond, R.P.; Allcock, H.R. Patterning poly(organophosphazenes) for selective cell adhesion applications. *Biomacromolecules* **2005**, *6*, 1689–1697. [CrossRef]
5. Chen-Yang, Y.W.; Lee, H.F.; Yuan, C.Y. A flame-retardant phosphate and cyclotriphosphazene-containing epoxy resin: Synthesis and properties. *J. Polym. Sci. Part A Polym. Chem.* **2000**, *38*, 972–981. [CrossRef]
6. Medici, A.; Fantin, G.; Pedrini, P.; Gleria, M.; Minto, F. Functionalization of Phosphazenes. 1. Synthesis of phosphazene materials containing hydroxyl-groups. *Macromolecules* **1992**, *25*, 2569–2574. [CrossRef]
7. Gleria, M.; Bolognesi, A.; Porzio, W.; Catellani, M.; Destri, S.; Audisio, G. Grafting reactions onto poly(organophosphazenes). 1. The case of poly bis(4-isopropylphenoxy)phosphazene-g-polystyrene copolymers. *Macromolecules* **1987**, *20*, 469–473. [CrossRef]
8. Luther, T.A.; Stewart, F.F.; Lash, R.P.; Wey, J.E.; Harrup, M.K. Synthesis and characterization of poly{hexakis (methyl)(4-hydroxyphenoxy) cyclotriphosphazene}. *J. Appl. Polym. Sci.* **2001**, *82*, 3439–3446. [CrossRef]
9. Mathew, D.; Nair, C.P.R.; Ninan, K.N. Phosphazene-triazine cyclomatrix network polymers: Some aspects of synthesis, thermal- and flame-retardant characteristics. *Polym. Int.* **2000**, *49*, 48–56. [CrossRef]
10. Femec, D.A.; McCaffrey, R.R. Solvent replacement in the synthesis of persubstituted phosphonitrilic hydroquinone prepolymer materials. *J. Appl. Polym. Sci.* **1994**, *52*, 501–505. [CrossRef]
11. Zhang, T.; Cai, Q.; Wu, D.Z.; Jin, R.G. Phosphazene cyclomatrix network polymers: Some aspects of the synthesis, characterization, and flame-retardant mechanisms of polymer. *J. Appl. Polym. Sci.* **2005**, *95*, 880–889. [CrossRef]
12. Zhu, L.; Xu, Y.Y.; Yuan, W.Z.; Xi, J.Y.; Huang, X.B.; Tang, X.Z.; Zheng, S.X. One-pot synthesis of poly(cyclotriphosphazene-co-4,4′-sulfonyldiphenol) nanotubes via an in situ template approach. *Adv. Mater.* **2006**, *18*, 2997–3000. [CrossRef]
13. Fu, H.W.; Huang, X.B.; Huang, Y.W.; Zhu, L.; Zhu, Y.; Tang, X.Z. Facile preparation of branched phosphazene-containing nanotubes via an in situ template approach. *Macromol. Mater. Eng.* **2008**, *293*, 173–177. [CrossRef]
14. Fu, J.; Huang, X.; Huang, Y.; Pan, Y.; Zhu, Y.; Tang, X. Preparation of Silver Nanocables Wrapped with Highly Cross-Linked Organic-Inorganic Hybrid Polyphosphazenes via a Hard-Template Approach. *J. Phys. Chem. C* **2008**, *112*, 16840–16844. [CrossRef]
15. Zhang, P.; Huang, X.; Fu, H.; Huang, Y.; Zhu, Y.; Tang, X. A One-Pot Approach to Novel Cross-Linked Polyphosphazene Microspheres with Active Amino Groups. *Macromol. Chem. Phys.* **2009**, *210*, 792–798. [CrossRef]
16. Huang, Y.W.; Fu, J.W.; Pan, Y.; Huang, X.B.; Tang, X.Z. Pervaporation of ethanol aqueous solution by polyphosphazene membranes: Effect of pendant groups. *Sep. Purif. Technol.* **2009**, *66*, 504–509. [CrossRef]
17. Zhang, X.Y.; Dai, Q.Y.; Huang, X.B.; Tang, X.Z. Synthesis and characterization of novel magnetic Fe_3O_4/polyphosphazene nanofibers. *Solid State Sci.* **2009**, *11*, 1861–1865. [CrossRef]
18. Chen, K.Y.; Liu, Y.; Hu, Y.J.; Yuan, M.Y.; Zheng, X.W.; Huang, X.B. Facile synthesis of amino-functionalized polyphosphazene microspheres and their application for highly sensitive fluorescence detection of Fe^{3+}. *J. Appl. Polym. Sci.* **2020**, *137*, 48937. [CrossRef]
19. Chen, C.; Zhu, X.-Y.; Gao, Q.-L.; Fang, F.; Huang, X.-J. Surface modification of cyclomatrix polyphosphazene microsphere by thiol-ene chemistry and lectin recognition. *Appl. Surf. Sci.* **2016**, *387*, 1029–1036. [CrossRef]
20. Ding, G.J.; Wang, A.; Shi, X.M.; Li, J.M.; You, L.J.; Wang, S.Y. Preparation of multiple-spectra encoded polyphosphazene microspheres and application for antibody detection. *Polym. Bull.* **2021**, 1–21. [CrossRef]
21. Wang, Y.; Shi, L.L.; Zhang, W.H.; Jiang, Z.H.; Mu, J.X. A comparative structure-property study of polyphosphazene micro-nano spheres. *Polym. Bull.* **2014**, *71*, 275–285. [CrossRef]
22. Wang, G.B.; Wang, D.; Jiang, Z.H.; Chen, C.H.; Zhang, W.J.; Wu, Z.W. The synthesis and properties of fluoropoly(aryl ether ketone)s. *Chem. J. Chin. Univ. Chin.* **2001**, *22*, 1053–1056.
23. Liu, B.J.; Hu, W.; Chen, C.H.; Jiang, Z.H.; Zhang, W.J.; Wu, Z.W.; Matsumoto, T. Soluble aromatic poly(ether ketone)s with a pendant 3,5-ditrifluoromethylphenyl group. *Polymer* **2004**, *45*, 3241–3247. [CrossRef]
24. Shim, S.E.; Yang, S.Y.; Choe, S.J. Mechanism of the formation of stable microspheres by precipitation copolymerization of styrene and divinylbenzene. *J. Polym. Sci. Part A Polym. Chem.* **2004**, *42*, 3967–3974. [CrossRef]

polymers

MDPI

Article

Surface Modification of γ-Al_2O_3 Nanoparticles Using Conductive Polyaniline Doped by Dodecylbenzene Sulfonic Acid

Cheng-Ho Chen *, Ying-Chen Lin and Hung-Mao Lin

Department of Chemical and Materials Engineering, Southern Taiwan University of Science and Technology, Tainan City 710, Taiwan; ma340101@stust.edu.tw (Y.-C.L.); hmlin@stust.edu.tw (H.-M.L.)
* Correspondence: chchen@stust.edu.tw; Tel.: +886-6-2430522

Abstract: In this study, electrically conductive PANDB/γ-Al_2O_3 core–shell nanocomposites were synthesized by surface modification of γ-Al_2O_3 nanoparticles using polyaniline doped with dodecylbenzene sulfonic acid. The PANDB/γ-Al_2O_3 core–shell nanocomposites were synthesized by in situ polymerization. Pure PANDB and the PANDB/γ-Al_2O_3 core–shell nanocomposites were characterized using Fourier transform infrared spectroscopy, ultraviolet–visible spectroscopy, transmission electron microscopy, field emission scanning electron microscopy, and measurement of a four-point probe. The conductivity of the PANDB/γ-Al_2O_3 core–shell nanocomposite was about 0.72 S/cm when the weight ratio of aniline/γ-Al_2O_3 was 3/1. The results showed that the conductivity of the PANDB/γ-Al_2O_3 core–shell nanocomposite decreased with increasing amounts of γ-Al_2O_3 nanoparticles. The transmission electron microscopy results indicated that the γ-Al_2O_3 nanoparticles were thoroughly coated with PANDB to form a core–shell structure. Transmission electron microscopy and field emission scanning electron microscopy images of the conductive PANDB/γ-Al_2O_3 core–shell nanocomposites also showed that the thickness of the PANDB layer decreased as the amount of γ-Al_2O_3 was increased.

Keywords: polyaniline; dodecylbenzene sulfonic acid; γ-Al_2O_3; in situ polymerization; core–shell nanocomposite

Citation: Chen, C.-H.; Lin, Y.-C.; Lin, H.-M. Surface Modification of γ-Al_2O_3 Nanoparticles Using Conductive Polyaniline Doped by Dodecylbenzene Sulfonic Acid. *Polymers* **2022**, *14*, 2232. https://doi.org/10.3390/polym14112232

Academic Editors: Andrzej Puszka and Beata Podkościelna

Received: 8 April 2022
Accepted: 30 May 2022
Published: 31 May 2022

1. Introduction

Most polymers are insulators because they are made of covalent bonds without free-moving electrons or ions. Inherently conductive polymers (ICPs) are a special class of synthetic polymers with unique electro-optical characteristics. ICPs possess conjugated chains with alternating single and double bonds [1]. Polyaniline (PANI) is an important member of the intrinsically conductive polymer (ICP) family. Since PANI is easy to synthesize and exhibits a wide range of conductivity, low operational voltage, unique electrochemical properties, and environmental stability, numerous researchers have studied it extensively [2–6] and have used it in many applications, such as secondary batteries [7,8], biosensors [9,10], corrosion protectors [11,12], antistatic packaging materials [13], and light-emitting diodes (LEDs) [14].

The encapsulation of inorganic materials inside a PANI shell has become the most popular and interesting aspect of nanocomposites in recent years. The combination of inorganic components with electrically conductive PANI (emeraldine salt (ES)-type PANI) has attracted considerable attention due to the novel physical and chemical properties of the resulting nanocomposite and potential applications. It has great potential applications in the fields of medication delivery, biosensors, chemical assembly, materials science, etc. These core–shell conductive nanocomposites can provide new synergistic properties that cannot be obtained from individual materials alone. Many groups have reported PANI/inorganic core–shell nanocomposites such as PANI/bagasse fiber (BF) [15], PANI/Y_2O_3 [16], PANI/$NiCo_2O_4$ [17], PANI/T-ZnOw [18], PANI/TiO_2 [19], PANI/MnO_2 [20], PANI/alumina [21–24], and PANI/clay [25,26].

However, PANI has some processing disadvantages, such as low or no solubility in most common organic solvents and poor processing properties [1]. Many researchers have attempted to synthesize soluble conductive PANI doped with various dopants. The most promising and attractive approach is to synthesize PANI doped with dodecylbenzene sulfonic acid (DBSA) (PANDB) in aqueous solution [27] or emulsion polymerization [28]. In these synthetic methods, DBSA can act as both a surfactant and a dopant during the synthesis. Since the molecular chain of DBSA has a lipophilic group ($-C_{12}H_{25}$), PANDB-based materials are soluble in common organic solvents. Therefore, PANDB-based materials can have more diverse applications.

Aluminum oxide (Al_2O_3) has good physical properties, such as thermal stability, corrosion resistance, abrasion resistance, electrical insulation, and high mechanical strength. Therefore, Al_2O_3 is the material most commonly applied in industrial applications. γ-Al_2O_3 is one of the metastable polymorphs of transition alumina, which has higher thermal stability compared with other polymorphs of Al_2O_3.

To our knowledge, no studies have been reported on the preparation of conductive PANDB/γ-Al_2O_3 core–shell nanocomposites via in situ polymerization of aniline (AN) in the presence of DBSA. In this study, γ-Al_2O_3 nanoparticles are embedded by PANDB via in situ polymerization to form conductive PANDB/γ-Al_2O_3 core–shell nanocomposites. Furthermore, the influences of the weight ratio of AN/γ-Al_2O_3 on the electrical conductivity, chemical structure, and morphology of the synthesized conductive PANDB/γ-Al_2O_3 core–shell nanocomposites are systemically examined using a four-point probe, Fourier transform infrared spectroscopy (FTIR), ultraviolet–visible spectroscopy (UV-Vis), transmission electron microscopy (TEM), and field emission scanning electron microscopy (FE-SEM).

2. Experimental

2.1. Materials

Aniline (AN), dodecylbenzene sulfonic acid (DBSA), and ammonium persulfate (APS) were purchased from Merck Co., Darmstadt, Germany. The γ-Al_2O_3 nanoparticles were purchased from Degussa Co., Frankfurt, Germany. The diameter of an individual γ-Al_2O_3 nanoparticle is about 10~30 nm.

2.1.1. Synthesis of Pure PANDB

PANDB was directly synthesized by chemical oxidative polymerization according to the modified procedure described by Cao et al. [29] and our previous study [30]. First, 8 g of aniline was mixed with 23 g of DBSA and 400 mL of distilled water to form a uniform milky white dispersion of aniline–DBSA complex in a 1000-milliliter four-neck flat-bottom reactor at room temperature with appropriate stirring. Then, APS solution (20 g of APS dissolved in 200 mL of distilled water) was slowly added to the reactor. The aniline:DBSA:APS molar ratio was 1:0.8:1. After 2 h synthesis, the dark green PANDB dispersion was precipitated by the addition of 600 mL of acetone. The reaction mixture was filtered and washed several times with deionized water until the filtrate was colorless. Finally, the resulting precipitate was collected and dried in an oven at 60 °C for 24 h.

2.1.2. Synthesis of Conductive PANDB/γ-Al_2O_3 Core–Shell Nanocomposites

Conductive PANDB/γ-Al_2O_3 core–shell nanocomposites were synthesized by in situ chemical oxidative polymerization. Two grams of γ-Al_2O_3 nanoparticles were dispersed in 100 mL of an aqueous solution containing 1 g of aniline, and the resulting dispersion was stirred at room temperature for 10 min. The weight ratio of AN/γ-Al_2O_3 was 3/1. Then, 50 mL of aqueous DBSA was added to the solution. After stirring for 10 min, 30 mL of APS aqueous solution was added dropwise to the dispersion with constant stirring. The molar ratio of aniline:DBSA:APS was 1:0.8:1. The resulting mixture was reacted at room temperature for 2 h. Subsequently, the product was washed with deionized water until the filtrate became colorless. Afterwards, the product was dried in a vacuum oven at 60 °C for 24 h. To determine the effect of the weight ratio of AN/γ-Al_2O_3 on the properties of the

PANDB/γ-Al_2O_3 core–shell nanocomposites, products with different weight ratios were also applied, namely 3/2, 3/3, 3/4, and 3/5.

2.2. Characterization

2.2.1. Electrical Conductivity Analysis

A sample of 0.1 g was weighed and then pressed at 3.0×10^5 psi for 2 min at room temperature. Four-point probe measurement (model: LSR4-KHT200, KeithLink Technology Co., Ltd., Taipei, Taiwan) was used to determine the electrical conductivity σ (S/cm) of the sample at room temperature.

2.2.2. FTIR Analysis

The chemical structure of the samples was detected by Fourier transform infrared spectroscopy (FTIR) (model Spectrum One; Perkin Elmer, Waltham, MA, USA) at 32 scans/s in the wavenumber range of 4000–400 cm^{-1}. The powdered sample was ground together with potassium bromide (KBr) (approximately 1:99 by weight) to a fine powder, and the homogeneous mixture was pressed into a pellet for analysis.

2.2.3. UV-Vis Analysis

The synthesized product was dispersed in absolute ethanol with ultrasonic stirring for 1 h at room temperature. A UV-Vis spectrophotometer (UV-Vis; Shimadzu, model UV-2401 PC, Kyoto, Japan) was used to measure the absorption of the sample solution in the wavelength range of 300–900 nm.

2.2.4. TEM Examination

The synthesized PANDB and PANDB/γ-Al_2O_3 nanocomposite samples were diluted and dispersed uniformly. They were then cast on carbon-coated copper grids to prepare for transmission electron microscopy (TEM) (JEM-1230, JEOL, Ltd., Tokyo, Japan) analysis. The microscope was operated at an accelerating voltage of 80 kV.

2.2.5. FE-SEM Examination

The samples were coated with a gold-palladium film. The surface morphology of the samples was observed using a field emission scanning electron microscope (JSM 6700F model; JEOL, Ltd., Tokyo, Japan).

3. Results and Discussion

Figure 1 clearly shows the color change of the polymerization solution at the weight ratio of AN/γ-Al_2O_3 = 3/1 after a reaction time of 3, 27, and 30 min. The color of the initial reaction solution was milky white. After adding the APS solution, the color of the reaction solution remained unchanged for the first 15 min and then turned light blue, blue, and finally turquoise within 15–27 min. This color change indicated the formation of the pernigraniline oxidation state of PANI [31]. After 27 to 30 min, the reaction solution became green to dark green quickly. This showed that the oxidation state of pernigraniline was converted to the oxidation and reduction states of emeraldine of PANI. The color change from green to dark green is due to the doping of DBSA onto the PANI backbone to form conductive PANDB (ES type).

The color change during the reaction can be used as an indicator of the polymerization rate of aniline. The experimental results show that the time required for color change increased (i.e., the polymerization rate decreased) as the AN/γ-Al_2O_3 weight ratio decreased. This observation was due to the fact that the probability of collision between AN and initiator (that is, APS) molecules was decreased while the amount of γ-Al_2O_3 was increased in the reaction solution. Therefore, the AN polymerization rate decreased as the weight of γ-Al_2O_3 was increased.

Figure 1. Color changes of the polymerization solution at the weight ratio of AN/γ-Al_2O_3 of 3/1 after (**a**) 3, (**b**) 27, and (**c**) 30 min reaction time.

Table 1 exhibits the electrical conductivities of pure γ-Al_2O_3, PANDB, and the PANDB/γ-Al_2O_3 core–shell nanocomposites. Since γ-Al_2O_3 has good electrical insulation, its electrical conductivity cannot be detected by four-point probe measurement. Meanwhile, the electrical conductivity of pure PANDB was 0.82 S/cm according to the test results. The results in Table 1 indicate that the electrical conductivity of the PANDB/γ-Al_2O_3 core–shell nanocomposites decreased as the weight ratio of AN/γ-Al_2O_3 was decreased. When the weight ratio of AN/γ-Al_2O_3 was 3/1, the electrical conductivity of the PANDB/γ-Al_2O_3 core–shell nanocomposite was about 0.72 S/cm. This result implies that PANDB could be successfully coated onto the surface of γ-Al_2O_3 nanoparticles, and the conductivity of γ-Al_2O_3 could be improved by forming electrically conductive PANDB/γ-Al_2O_3 core–shell nanocomposites. When the weight ratio of AN/γ-Al_2O_3 was decreased from 3/1 to 3/5, the conductivity of the PANDB/γ-Al_2O_3 core–shell nanocomposite decreased from 0.72 to 0.53 S/cm. This is due to the fact that the thickness of the conductive PANDB on γ-Al_2O_3 was relatively decreased with the increasing weight of γ-Al_2O_3.

Table 1. Influence of the weight ratio of AN/γ-Al_2O_3 on the electrical conductivity of PANDB/γ-Al_2O_3 core–shell nanocomposites.

Weight Ratio of AN/γ-Al_2O_3	Conductivity of PANDB and PANDB/γ-Al_2O_3 Nanocomposites (S/cm)
Pure γ-Al_2O_3	-
Pure PANDB	0.82
3/1	0.72
3/2	0.58
3/3	0.57
3/4	0.55
3/5	0.53

Figure 2 presents photographs of the γ-Al_2O_3 nanoparticles (Figure 2a) and the conductive PANDB/γ-Al_2O_3 core–shell nanocomposite synthesized at the weight ratio of AN/γ-Al_2O_3 = 3/1 through in situ polymerization (Figure 2b). The appearance of the γ-Al_2O_3 nanoparticles was white, whereas the appearance of the conductive PANDB/γ-Al_2O_3 core–shell nanocomposite was dark green. This is because PANDB is an emeraldine salt (ES) type with good electrical conductivity. Therefore, the color of the conductive PANDB/γ-Al_2O_3 core–shell nanocomposite was dark green.

Figure 2. Photographs of (**a**) γ-Al_2O_3 nanoparticles and (**b**) conductive PANDB/γ-Al_2O_3 core–shell nanocomposite synthesized via in situ polymerization (weight ratio of AN/γ-Al_2O_3 = 3/1).

Figure 3 shows the FTIR spectra of pure PANDB and the conductive PANDB/γ-Al_2O_3 core–shell nanocomposites synthesized at the weight ratios of AN/γ-Al_2O_3 = 3/2 and 3/5. Peaks at about 2853 and 2953 cm^{-1} were observed due to the stretching vibration mode of the -CH (-CH_3 or -CH_2-) for all samples. For pure PANDB (Figure 3a), the characteristic peaks at 1567 and 1491 cm^{-1} were due to the stretching vibrations of the N=Q=N and the N-B-N ring, respectively. The characteristic peak at 1300 cm^{-1} was attributed to the C-N stretching vibrations of the secondary amine in the main chain of PANDB. The peaks at 998–1040 cm^{-1} were due to the asymmetric and symmetric O=S=O stretching vibrations of DBSA. Characteristic peaks at 1100–1200 cm^{-1} were due to the B-NH-Q bond or the B-NH-B bond and the in-plane bending vibration of benzenoid or quinonoid C-H bonds (where B represents benzenic-type rings and Q represents quinonic-type rings). As these absorption peaks overlapped in the range of 1000 to 1200 cm^{-1}, a broad peak was observed. The peaks at 800–700 cm^{-1} were attributed to the characteristic feature of the B-NH-Q bond or the B-NH-B bond and the out-of-plane bending vibration of benzenoid or quinonoid -CH and –N–H bonds. These characteristic peaks are in good agreement with those reported in the literature [32,33]. The bands in the range of 400~600 cm^{-1} were associated with γ-Al_2O_3 nanoparticles. Note that there was no significant interaction between PANDB molecules and γ-Al_2O_3 nanoparticles from the FTIR results. The good adhesion of PANDB to γ-Al_2O_3 nanoparticles was assumed to be physicochemical in nature [21].

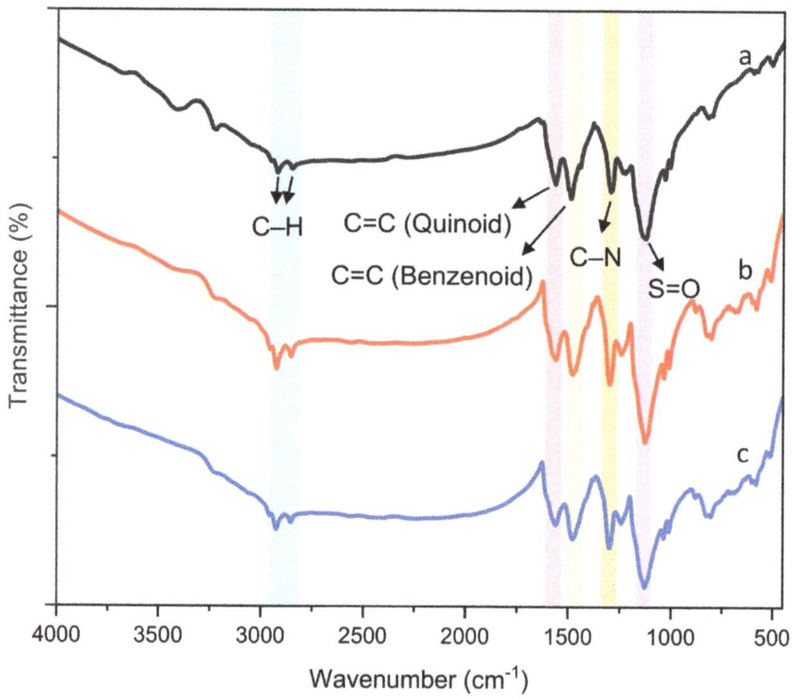

Figure 3. FTIR spectra of pure PANDB (**a**) and conductive PANDB/γ-Al_2O_3 core-shell nanocomposites synthesized at the weight ratios of AN/γ-Al_2O_3 = (**b**) 3/2 and (**c**) 3/5.

Figure 4 shows the UV-Vis absorption spectra of pure PANDB and the PANDB/γ-Al_2O_3 core–shell nanocomposites. For pure PANDB, three characteristic absorption peaks can be clearly observed in the UV-Vis spectrum at ~340, ~430, and ~830 nm (Figure 4a). The absorption peak at ~340 nm was due to the π–π^* transition of the benzenoid rings, while the peaks at ~430 and ~830 nm were attributed to the polaron–π^* transition and π–polaron transition, respectively [34,35]. The analytical results show that the synthesized PANDB was an emeraldine salt (ES) form. Three characteristic absorption peaks also appeared in the spectrum of PANDB/γ-Al_2O_3 core–shell nanocomposites owing to the presence of PANDB (Figure 4b–f). However, the absorption peak related to the π–polaron transition was shifted to a lower wavelength (from 840 to 820 nm) by decreasing the weight ratio of

AN/γ-Al_2O_3 from 3/1 to 3/5. This result was attributed to less PANDB being coated on the surface of the γ-Al_2O_3 nanoparticles when the amount of γ-Al_2O_3 was increased. This result is consistent with the decrease in conductivity.

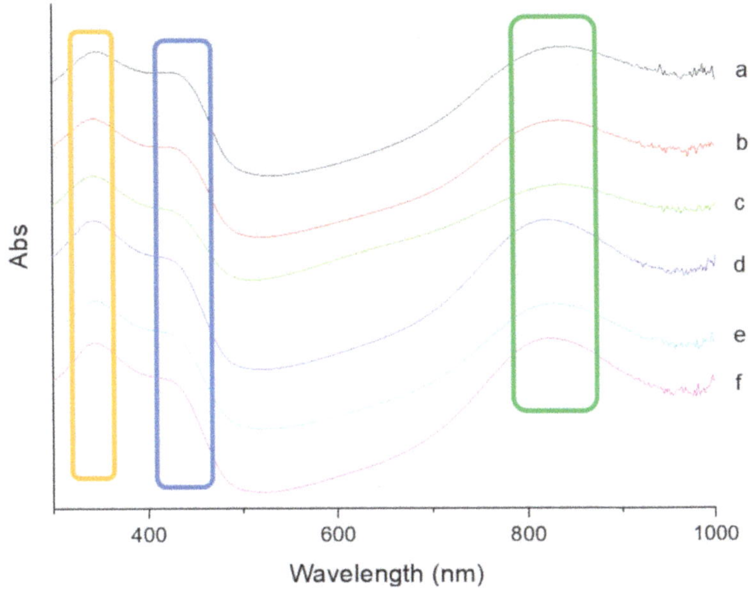

Figure 4. UV-Vis spectra of (**a**) pure PANDB and conductive PANDB/γ-Al_2O_3 core–shell nanocomposites synthesized at the weight ratios of AN/γ-Al_2O_3 = (**b**) 3/1, (**c**) 3/2, (**d**) 3/3, (**e**) 3/4, and (**f**) 3/5.

Figures 5 and 6 show the TEM and FE-SEM images of pure PANDB (Figures 5a and 6a), γ-Al_2O_3 nanoparticles (Figures 5b and 6b), and the PANDB/γ-Al_2O_3 core–shell nanocomposites synthesized at the weight ratios of AN/γ-Al_2O_3 = 3/1 and 3/3 (Figure 5c,d and Figure 6c,d). Figures 5a and 6a demonstrate that pure PANDB was aggregated by irregular particles and rod-like PANDB. Figure 5b shows an image of independent γ-Al_2O_3 nanoparticles and aggregates of γ-Al_2O_3 nanoparticles with irregular shapes due to the high surface energy of the nanoparticles. Figures 5b and 6b show that the diameter of individual γ-Al_2O_3 nanoparticles was about 10~30 nm and the diameter of the clusters of γ-Al_2O_3 nanoparticles ranged from 30 to 100 nm. Comparing Figure 5c with Figure 5d or Figure 6c with Figure 6d, the surface of the γ-Al_2O_3 nanoparticle was coated with less PANDB as the amount of γ-Al_2O_3 nanoparticle was increased. Therefore, the electrical conductivity of the PANDB/γ-Al_2O_3 core–shell nanocomposites decreased as the amount of γ-Al_2O_3 nanoparticle was increased.

Based on the result of the TEM images, the synthesis of conductive PANDB/γ-Al_2O_3 core–shell nanocomposites via an in situ polymerization process is shown in Figure 7. First, the TEM results indicated that the γ-Al_2O_3 nanoparticles were coated with an anilinium–DBSA complex. Then, by adding APS for polymerization, conductive PANDB/γ-Al_2O_3 core–shell nanocomposites could be obtained. Furthermore, the independent γ-Al_2O_3 nanoparticles and the aggregates of γ-Al_2O_3 nanoparticles were simultaneously coated by PANDB to form conductive PANDB/γ-Al_2O_3 core–shell nanocomposites.

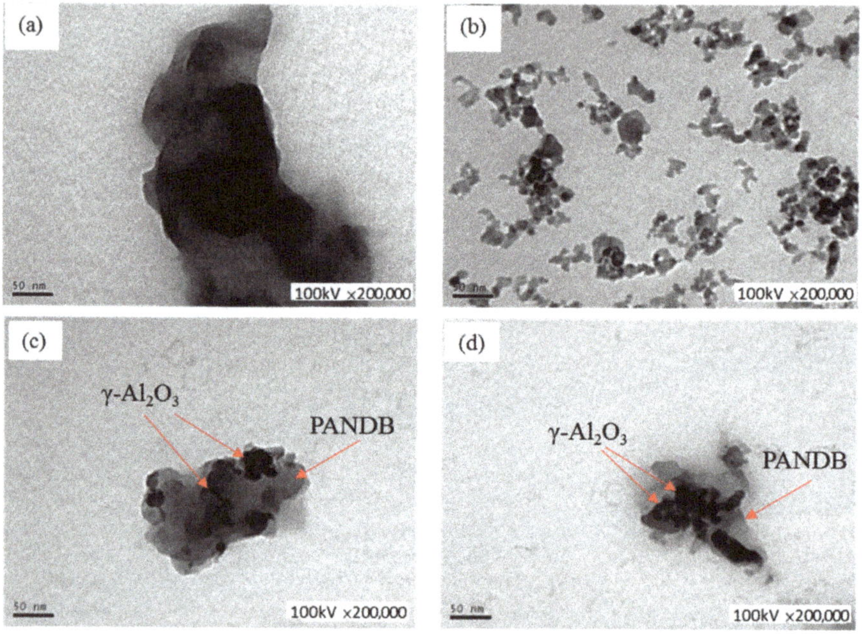

Figure 5. TEM images of (**a**) pure PANDB, (**b**) γ-Al_2O_3, and conductive PANDB/γ-Al_2O_3 core–shell nanocomposites synthesized at weight ratios of AN/γ-Al_2O_3 = (**c**) 3/1 and (**d**) 3/3.

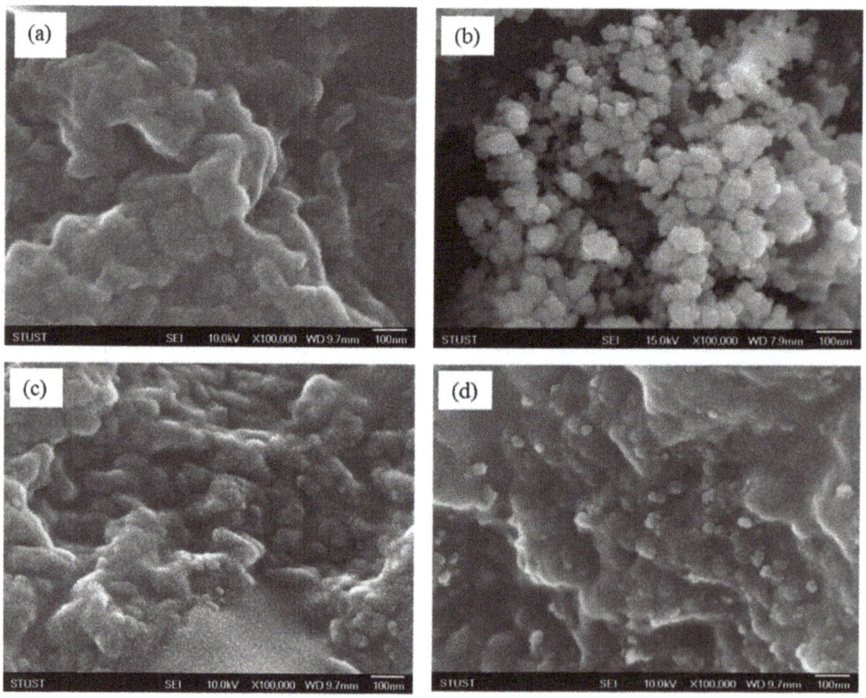

Figure 6. FE-SEM images of (**a**) pure PANDB, (**b**) γ-Al_2O_3, and conductive PANDB/γ-Al_2O_3 core–shell nanocomposites synthesized at weight ratios of AN/γ-Al_2O_3 = (**c**) 3/1 and (**d**) 3/3.

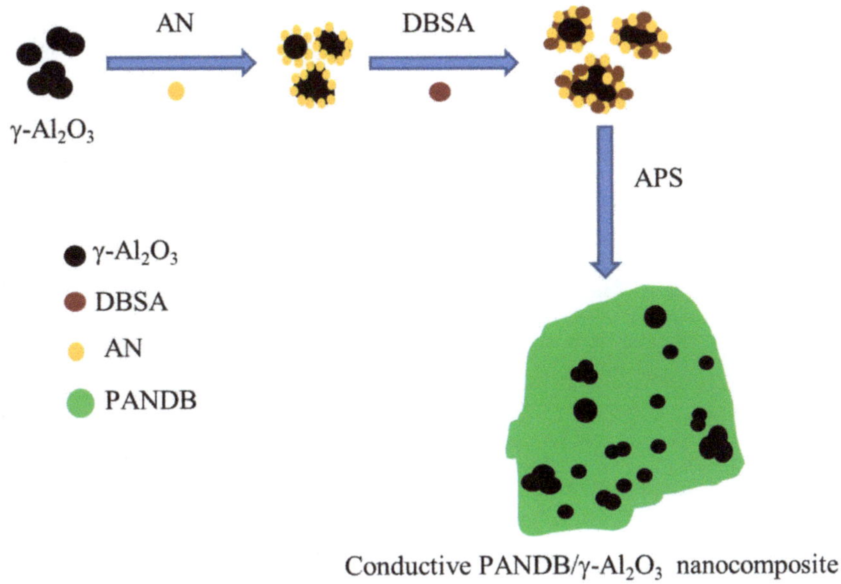

Figure 7. Synthesis process of conductive PANDB/γ-Al_2O_3 core–shell nanocomposites through in situ polymerization.

4. Conclusions

In this study, an in situ polymerization method was used to prepare electrically conductive PANDB/γ-Al_2O_3 core–shell nanocomposites. The core is γ-Al_2O_3, and the shell is PANDB. The electrical conductivity of pure PANDB was 0.82 S/cm. The electrical conductivity of the PANDB/γ-Al_2O_3 core–shell nanocomposite decreased with increasing the amount of γ-Al_2O_3. When the weight ratio of AN/γ-Al_2O_3 was decreased from 3/1 to 3/5, the conductivity of the PANDB/γ-Al_2O_3 core–shell nanocomposite decreased from 0.72 to 0.53 S/cm. Both the TEM and FE-SEM images showed that PANDB could be successfully coated on the surface of γ-Al_2O_3 nanoparticles. Therefore, the electrical conductivity of γ-Al_2O_3 could be improved by forming a conductive PANDB/γ-Al_2O_3 core–shell nanocomposite.

Author Contributions: C.-H.C. and H.-M.L. conceptualized and supervised the study. Y.-C.L. performed the experimental work. All authors contributed to the writing, editing, and analysis. All authors have read and agreed to the published version of the manuscript.

Funding: This research was funded by Southern Taiwan University of Science and Technology.

Institutional Review Board Statement: Not applicable.

Informed Consent Statement: Not applicable.

Data Availability Statement: Not applicable.

Acknowledgments: We acknowledge the financial support from Southern Taiwan University of Science and Technology. We also acknowledge C.J. Ko for her kind support for the FE-SEM.

Conflicts of Interest: The authors declare no conflict of interest.

References

1. Zare, E.N.; Makvandi, P.; Ashtari, B.; Rossi, F.; Motahari, A.; Perale, G. Progress in Conductive Polyaniline-Based Nanocomposites for Biomedical Applications: A Review. *J. Med. Chem.* **2020**, *63*, 1–22. [CrossRef] [PubMed]
2. Baker, C.O.; Huang, X.W.; Nelson, W.; Kaner, R.B. Polyaniline nanofibers: Broadening applications for conducting polymers. *Chem. Soc. Rev.* **2017**, *46*, 1510–1525. [CrossRef] [PubMed]

3. Reddy, K.R.; Karthik, K.V.; Prasad, S.B.B.; Soni, S.K.; Jeong, H.M.; Raghu, A.V. Enhanced photocatalytic activity of nanostructured titanium dioxide/polyaniline hybrid photocatalysts. *Polyhedron* **2016**, *120*, 169–174. [CrossRef]
4. Li, B.; Li, X.P.; Li, W.S.; Wang, Y.Q.; Uchaker, E.; Pei, Y.; Cao, X.; Li, S.; Huang, B.; Cao, G.Z. Mesoporous Tungsten Trioxide Polyaniline Nanocomposite as an Anode Material for High-Performance Lithium-Ion Batteries. *Chem. Nano Mat.* **2016**, *2*, 281–289. [CrossRef]
5. Gao, F.J.; Mu, J.; Bi, Z.X.; Wang, S.; Li, Z.L. Recent advances of polyaniline composites in anticorrosive coatings: A review. *Prog. Org. Coat.* **2021**, *151*, 106071. [CrossRef]
6. Yang, L.Y.; Xu, X.R.; Liu, M.D.; Chen, C.; Cui, J.; Chen, X.; Wu, K.; Sun, D.P. Wearable and flexible bacterial cellulose/polyaniline ammonia sensor based on a synergistic doping strategy. *Sens. Actuators B Chem.* **2021**, *334*, 129647. [CrossRef]
7. Venkatachalam, S.; Prabhakaran, P.V. Oligomeric phthalocyanine modified polyaniline—An electrode material for use in aqueous secondary batteries. *Synth. Met.* **1998**, *97*, 141–146. [CrossRef]
8. Ryu, K.S.; Kim, K.M.; Kang, S.G.; Lee, G.J.; Joo, J.S.; Chang, S.H. Electrochemical and physical characterization of lithium ionic salt doped polyaniline as a polymer electrode of lithium secondary battery. *Synth. Met.* **2000**, *110*, 213–217. [CrossRef]
9. Parente, A.H.; Marques, E.T., Jr.; Azevedo, W.M.; Diniz, F.B.; Melo, E.H.; Filro, J.L.L. Glucose biosensor using glucose oxidase immobilized in polyaniline. *Appl. Biochem. Biotechnol.* **1992**, *37*, 267–273. [CrossRef]
10. Leite, V.; Dasilva, V.L.; Azevedo, W.M.; Melo, E.H.; Filro, J.L.L. Increasing glucose determination range by flow injection analysis (FIA) using glucose oxidase immobilised on polyaniline. *Biotechnol. Tech.* **1994**, *8*, 133–136. [CrossRef]
11. Li, P.; Tan, T.C.; Lee, J.Y. Corrosion protection of mild steel by electroactive polyaniline coatings. *Synth. Met.* **1997**, *88*, 237–242. [CrossRef]
12. Pud, A.A.; Shapoval, G.S.; Kamarchik, P.; Ogurtsov, N.A.; Gromovaya, V.F.; Myronyuk, I.E.; Kontsur, Y.V. Electrochemical behavior of mild steel coated by polyaniline doped with organic sulfonic acids. *Synth. Met.* **1999**, *107*, 111–115. [CrossRef]
13. Subramaniam, C.K.; Kaiser, A.B.; Gilberd, P.W.; Wessling, B. Electronic transport properties of polyaniline/PVC blends. *J. Polym. Sci. Pol. Phys.* **1993**, *31*, 1425–1430. [CrossRef]
14. Gustafsson, G.; Cao, Y.; Treacy, G.M.; Klavetter, F.; Colaneri, N.; Heeger, A.J. Flexible light-emitting diodes made from soluble conducting polymers. *Nature* **1992**, *357*, 477–479. [CrossRef]
15. Yang, Z.; Qiu, M.N.; Yu, Y.; Wen, B.Y.; Cheng, L.L. A Novel Polyaniline-Coated Bagasse Fiber Composite with Core–Shell Heterostructure Provides Effective Electromagnetic Shielding Performance. *ACS Appl. Mater. Interf.* **2017**, *9*, 809–818.
16. Kowsari, E.; Faraghi, G. Ultrasound and ionic-liquid-assisted synthesis and characterization of polyaniline/Y_2O_3 nanocomposite with controlled conductivity. *Ultrason. Sonochem.* **2010**, *17*, 718–725. [CrossRef]
17. Yu, Z.Y.; Li, H.J.; Zhang, X.M.; Liu, N.K.; Tan, W.L.; Zhang, X.; Zhang, L.L. Facile synthesis of $NiCo_2O_4$@Polyaniline core–shell nanocomposite for sensitive determination of glucose. *Biosens. Bioelectron.* **2016**, *75*, 161–165. [CrossRef] [PubMed]
18. Chen, X.; Zhou, Z.; Lu, W.; Huang, T.; Hu, S. Preparation of core-shell structure T-ZnOw/polyaniline composites via graft polymerization. *Mater. Chem. Phys.* **2009**, *115*, 258–262. [CrossRef]
19. Gai, L.; Du, G.; Zuo, Z.; Wang, Y.; Liu, D.; Liu, H. Controlled synthesis of hydrogen Titanate-Polyaniline composite nanowires and their resistance-temperature characteristics. *J. Phys. Chem. C* **2009**, *113*, 7610–7615. [CrossRef]
20. Pan, L.; Pu, L.; Shi, Y.; Song, S.; Xu, Z.; Zhang, R.; Zheng, Y. Synthesis of polyaniline nanotubes with a reactive template of manganese oxide. *Adv. Mater.* **2007**, *19*, 461–464. [CrossRef]
21. Zhang, D.H. Preparation of Core–Shell Structured Alumina–Polyaniline Particles and Their Application for Corrosion Protection. *J. Appl. Polym. Sci.* **2006**, *101*, 4372–4377. [CrossRef]
22. Resan, S.A.; Essa, A.F. Preparation and study of the optical properties for polyaniline-Al_2O_3 nanocomposite. *Mater. Today Proc.* **2021**, *45*, 5819–5822. [CrossRef]
23. Bekhti, M.A.; Belardja, M.S.E.; Lafiah, M.; Chouli, F.; Benyoucef, A. Enhanced Tailored of Thermal Stability, Optical and Electrochemical Properties of PANI Matrix Containing Al_2O_3 Hybrid Materials Synthesized through In Situ Polymerization. *Polym. Compos.* **2021**, *42*, 6–14. [CrossRef]
24. Chen, C.H.; Lin, Y.C.; Yen, F.S. Synthesis and Characterization of Conducting PANDB/χ-Al_2O_3 Core-Shell Nanocomposites by In Situ Polymerization. *Polymers* **2021**, *13*, 2787. [CrossRef]
25. Yoshimoto, S.; Ohashi, F.; Kameyama, T. Characterization and thermal degradation studies on polyaniline-intercalated montmorillonite nanocomposites prepared by a solvent-free mechanochemical route. *J. Polym. Sci. Polym. Phys.* **2005**, *43*, 2705–2714. [CrossRef]
26. De León-Almazan, C.M.; Estrada-Moreno, I.A.; Páramo-García, U.; Rivera-Armenta, J.L. Polyaniline/clay nanocomposites. A comparative approach on the doping acid and the clay spacing technique. *Synth. Met.* **2018**, *236*, 61–67. [CrossRef]
27. Yin, W.; Ruckenstein, E. Soluble polyaniline co-doped with dodecyl benzene sulfonic acid and hydrochloric acid. *Synth. Met.* **2000**, *108*, 39–46. [CrossRef]
28. Qi, L.; Xie, A.; Chen, T.; Han, J.; Yu, L.; Zhang, M. Direct access to xylene solution of polyanilines via emulsion polymerization-extraction method facilitating the preparation of conductive film materials. *Mater. Lett.* **2019**, *254*, 361–363. [CrossRef]
29. Cao, Y.; Andreatta, A.; Heeger, A.J.; Smith, P. Influence of chemical polymerization conditions on the properties of polyaniline. *Polymers* **1989**, *30*, 2305–2311. [CrossRef]
30. Chen, C.H. Thermal studies of polyaniline doped with dodecyl benzene sulfonic acid directly prepared via aqueous dispersions. *J. Polym. Res.* **2002**, *9*, 195–200. [CrossRef]

31. Haba, Y.; Segal, E.; Narkis, M.; Titelman, G.I.; Siegmann, A. Polyaniline–DBSA/polymer blends prepared via aqueous dispersions. *Synth. Met.* **2000**, *110*, 189–193. [CrossRef]
32. Zhang, Z.; Wan, M. Composite films of nanostructured polyaniline with poly (vinyl alcohol). *Synth. Met.* **2002**, *128*, 83–89. [CrossRef]
33. Zhou, S.; Wu, T.; Kan, J.Q. Effect of methanol on morphology of polyaniline. *Eur. Polym. J.* **2007**, *43*, 395–402. [CrossRef]
34. Abdiryim, T.; Zhang, X.G.; Jamal, R. Comparative studies of solid-state synthesized polyaniline doped with inorganic acids. *Mater. Chem. Phys.* **2005**, *90*, 367–372. [CrossRef]
35. Gao, Y.; Shan, D.; Cao, F.; Gong, J.; Li, X.; Ma, H.Y.; Su, Z.M.; Qu, L.Y. Silver/polyaniline composite nanotubes: One-step synthesis and electrocatalytic activity for neurotransmitter dopamine. *J. Phys. Chem. C* **2009**, *113*, 15175–15181. [CrossRef]

polymers

MDPI

Article

Thermoplastic Starch with Poly(butylene adipate-*co*-terephthalate) Blends Foamed by Supercritical Carbon Dioxide

Chih-Jen Chang [1,†], Manikandan Venkatesan [1,†], Chia-Jung Cho [2,*], Ping-Yu Chung [2], Jayashree Chandrasekar [1], Chen-Hung Lee [3,*], Hsin-Ta Wang [1], Chang-Ming Wong [4] and Chi-Ching Kuo [1,*]

1 Institute of Organic and Polymeric Materials, Research and Development Center of Smart Textile Technology, National Taipei University of Technology, Taipei 10608, Taiwan; allan6730@gmail.com (C.-J.C.); manikandanchemist1093@gmail.com (M.V.); jayashreechem7@gmail.com (J.C.); htwang@mail.ntut.edu.tw (H.-T.W.)
2 Institute of Biotechnology and Chemical Engineering, I-Shou University, Kaohsiung 84001, Taiwan; isu10605060a@cloud.isu.edu.tw
3 Division of Cardiology, Department of Internal Medicine, Chang Gung Memorial Hospital-Linkou, Chang Gung University College of Medicine, Taoyuan 33305, Taiwan
4 CoreTech System Co., Ltd., Hsinchu 30265, Taiwan; cmwong@moldex3d.com
* Correspondence: ppaul288@isu.edu.tw (C.-J.C.); chl5265@gmail.com (C.-H.L.); kuocc@mail.ntut.edu.tw (C.-C.K.)
† These authors contributed equally to this work.

Abstract: Starch-based biodegradable foams with a high starch content are developed using industrial starch as the base material and supercritical CO_2 as blowing or foaming agents. The superior cushioning properties of these foams can lead to competitiveness in the market. Despite this, a weak melting strength property of starch is not sufficient to hold the foaming agents within it. Due to the rapid diffusion of foaming gas into the environment, it is difficult for starch to maintain pore structure in starch foams. Therefore, producing starch foam by using supercritical CO_2 foaming gas faces severe challenges. To overcome this, we have synthesized thermoplastic starch (TPS) by dispersing starch into water or glycerin. Consecutively, the TPS surface was modified by compatibilizer silane A (SA) to improve the dispersion with poly(butylene adipate-co-terephthalate) (PBAT) to become (TPS with SA)/PBAT composite foam. Furthermore, the foam-forming process was optimized by varying the ratios of TPS and PBAT under different forming temperatures of 85 °C to 105 °C, and two different pressures, 17 Mpa and 23 Mpa were studied in detail. The obtained results indicate that the SA surface modification on TPS can influence the great compatibility with PBAT blended foams (foam density: 0.16 g/cm^3); whereas unmodified TPS and PBAT (foam density: 0.349 g/cm^3) exhibit high foam density, rigid foam structure, and poor tensile properties. In addition, we have found that the 80% TPS/20% PBAT foam can be achieved with good flexible properties. Because of this flexibility, lightweight and environment-friendly nature, we have the opportunity to resolve the strong demands from the packing market.

Keywords: thermoplastic starch; silane; foam; carbon dioxide

Citation: Chang, C.-J.; Venkatesan, M.; Cho, C.-J.; Chung, P.-Y.; Chandrasekar, J.; Lee, C.-H.; Wang, H.-T.; Wong, C.-M.; Kuo, C.-C. Thermoplastic Starch with Poly(butylene adipate-*co*-terephthalate) Blends Foamed by Supercritical Carbon Dioxide. *Polymers* **2022**, *14*, 1952. https://doi.org/10.3390/polym14101952

Academic Editor: Andrzej Puszka

Received: 11 April 2022
Accepted: 7 May 2022
Published: 11 May 2022

1. Introduction

In the past few decades, many industries were developed to manufacture conventional plastics for daily products and polymer-based energy and AI technologies. Particularly, these included multifunctional movement and pressure sensors [1,2], optoelectronic devices [3–5], and wearable electronic devices [6–8]. However, these plastic products required several hundred years to fully decay into the soil [9]. Furthermore, their collection and storage for recycling is an undesirable and time-consuming process [10]. A viable strategy is to develop bio-based polymer materials as alternatives to these materials to minimize the use of non-biodegradable polymers [11–15]. Considering these, the key to biodegradable polymer-based foams can be used in a wide range of applications, owing to their biocompatibility [16,17], biodegradability, and renewability process [18].

Starch is an eco-friendly biopolymer, with low costs and is easily obtained, although the processing is complicated and difficult to purify. Thermoplastic starch (TPS) can be mixed with water or plasticizers to improve its processability. TPS has an opportunity to be processed by extrusion, injection, or molding to make films, sheets, foams, etc.

There are many works associated with TPS, such as the study of starch-based films [19–28], the production of innovative biodegradable products [29], the additives used to strengthen TPS [30], water or azodicarbonamide used as foaming agents to develop starch-based foams [18,26,31–38], and evaluating the performance of starch loose-fill foams and expanded polystyrene [39]. Foam products are often used in packing goods, which can easily become plastic garbage. Loose-fill foams made with biodegradable materials, such as TPS or TPS mixed with other plastics, have received attention recently.

Starches with high amylose content are able to produce TPS foams [18,26,31,32,35,36,38]. Starch grafting by poly(methyl acrylate) [34], or grafting polystyrene [37] using water as a foaming agent to make foams, has been studied. Due to the poor mechanical properties of TPS, it is not suitable for producing foam materials, whereas TPS can be blended with other materials to improve properties in order to make the composite foams. However, TPS is a hydrophilic material, which is not suitable for hydrophobic polymeric resins. Therefore, surface modification is necessary to enhance the compatibility between two materials. In general, fatty acids are preferred for generating the peroxide or hydroperoxide reactive species to attack the carbon linkages in the starch to blended with polystyrene (PS), polyolefin, thermoplastic polyurethane (TPU), and polylactide (PLA) with free-radical grafting of maleic anhydride mixed with TPS, etc., [40,41]. Maleated TPS with PBAT [42–45] or TPS with maleated PBAT [46] are the materials for research.

PBAT is a soft and biodegradable co-polyester synthesized from fossil resources, which is eco-friendly, easily processable, and has a high elongation at break in tensile tests. PBAT can also be used for goods packaging applications, even though the price is not affordable for daily usage. PBAT blended with TPS can lower the material price to possibly manufacture acceptable products and vice versa, TPS mixed with PBAT can improve TPS mechanical properties. In the aforementioned works regarding foaming, with the emergence of various social and environmental concepts, water is the major foaming agent for the TPS foaming process. Although it is difficult for TPS to be foamed under supercritical CO_2 as a foaming agent, we have addressed these disadvantages by optimizing the ratio between PBAT and foaming agents by using surface modification techniques.

In this study, we used an inexpensive industrial starch with water and glycerol added to become thermoplastic starch (thermoplastic starch, TPS), which was the main foaming material. Further, combined with innovative chemical modification technology, the melt strength was improved to solve the problem of the insufficient structural melt strength of thermoplastic starch. By expanding the molecular chains of TPS/biodegradable polyester composites and generating intermolecular entanglement with each other, the composites have a certain elasticity and buffering properties, which can support the cell structure during the cell growth process. It also has the advantages of energy saving and carbon emission reduction because of its bio-plastic foam material, which can be decomposed and recycled, thus solving the problem of white pollution caused by petrochemical plastics. This starch-based material and processing technique can bring about a new generation of materials that will inevitably develop eco-friendly applications.

2. Experimental Preparations

2.1. Materials and Procedure

There are three materials, i.e., tapioca starch/TPS (thermoplastic starch: from Roi Et Group, Yannawa, Thailand), PBAT (Poly(butylene adipate-co-terephthalate) (Ecoflex: from BASF, Lemförde, Germany), and Silane A 6040 (SA: from Ya-Hu-Chi industrial Co.,

Zhubei, Taiwan) that were used in the study. SA has three methoxy groups and one epoxy group; the molecular weight of SA is 0.236 kg/mole and the boiling point is 190 °C. Starch was first mixed with water and glycerin to form TPS and then the TPS surface was modified using SA for the feasible blending with PBAT. The blending ratios with symbols are displayed in Table 1. The sheet of PBAT, TPS/PBAT, and (SA/TPS)/PBAT blends were produced by a hot press at a temperature of 140 °C. The sheet area was 75 mm × 75 mm × 3 mm (thickness) and these sheets were used for foaming by CO_2 under supercritical conditions. In the above, the mixing, preparation, and foaming of this experiment were repeated five times to ensure a high degree of accuracy in the results, and it has excellent reproducibility.

Table 1. The composite material composition ratio and symbols corresponding list.

Symbols	Blends (by Weight)
[N-1]	50% TPS/50% PBAT
[N-2]	60% TPS/40% PBAT
[N-3]	70% TPS/30% PBAT
[S-0.5]	50% (TPS with 5PHR SA)/50% PBAT
[S-1]	50% (TPS with 10PHR SA)/50% PBAT
[S-2]	60% (TPS with 10PHR SA)/40% PBAT
[S-3]	70% (TPS with 10PHR SA)/30% PBAT
[S-4]	80% (TPS with 10PHR SA)/20% PBAT

2.2. Functional Group Chemical Modification TPS Experiment

The functional group chemical modification of the TPS experiment: First, the commercially available starch raw materials, water, and appropriate glycerin were kneaded in a mixing machine plastic spectrometer (Barbender MIX, Kulturstraße, Germany) to make the starch uniformly thermally plasticized into TPS. Then, a functional group modifier-coupling agent (SA) for TPS modification was added. The reaction temperature was 70~55 °C, and the reaction time was 30~60 min. The pelletizer was modified and pelletized to produce surface-modified SA/TPS blends.

2.3. The (SA/TPS)/PBAT Biodegradable Polyester Composite Mixing

The modified (SA/TPS) weight ratio of 50, 60, 70, and 80% was added with different proportions of 50, 40, 30, and 20% biodegradable polyester (PBAT) and then kneaded by the mixing machine plastic spectrometer. The SA/TPS and the biodegradable polyester are uniformly mixed and dispersed into the composite. The mixing machine plastic spectrometer temperature is 90 °C to 145 °C, and the granulation screw speed is 50~100 rpm. Similarly, a control sample of unmodified TPS was blended with PBAT.

2.4. The (SA/TPS)/PBAT Composite Foam Test Piece and Supercritical Foaming Experiment

Primarily, the (SA/TPS)/PBAT composite particles entered the hot press (DAKE, Grand Haven, MI, USA), and were hot-pressed at a temperature of 140 °C to 165 °C, then, a square test piece of 75 mm × 75 mm × 3 mm size was retrieved.

As the prepared composite square test piece was placed into the mold of supercritical foaming equipment (200-ton capacity, Tainan City, Jing Day Machinery Industrial Co., Ltd., Taiwan) and obtained incompatible foams according to the different foaming parameter settings, the manufactured parameters were as follows: The foaming temperature was 80~105 °C, the foaming pressure was 17 MPa and 23.8 MPa, and the time of CO_2 impregnation was for 60 min. This impregnation time can make the CO_2 in the material reach a saturated state.

2.5. The (SA/TPS)/PBAT Composite Foam Appearance and Internal Structure SEM Analysis

The foam samples after supercritical foaming were plated with gold on the cross-sectional surface by vacuum evaporation, and then the structure of the foam cells, the size of the bubbles, and the dispersion of TPS in the composite were observed by scanning electron microscopy (SEM, using a Hitachi TM4000 Plus, Hitachi High-Tech Fielding Corporation, Nagano, Japan), and the WD parameters were 15.6 mm, the HV voltage was at 5.0 kV.

2.6. Chemical Modification and Analysis of TPS Functional Group

At the interface of TPS and PBAT polymer substrate, adhesion is not good due to the viscous surface of TPS and the lipophilic polymer base. For improving the adhesion between TPS and PBAT polymer, surface treatment becomes essential. The addition of SA coupling agents provides the required adhesion property for their conjunction and lowers the interfacial shear strength of TPS to polymer substrates [47].

SA chain consists of the organic function and the alkoxy function at the terminals which interact with PBAT polymer and TPS to become more hydrophilic and improve the affinity with polymer [48].

$$X - (CH_2)_n - Si(OR')_3, \ n = 0 \sim 3 \tag{1}$$

The OR' is a hydrolyzable alkoxy group, and the X is an epoxy-functional group. Silanes OR' end form a hydrogen bond with the surface of TPS containing hydroxyl groups. Moreover, the polymer treated with silane can improve wetting, and the organic functional group X in the coupling agent reacts with the polymer to form an interpenetrating network. Eventually, a formed network renders reinforcement between the TPS and the polymer interfaces. As shown in [49,50], the silyl group is first hydrolyzed and then condensed, forming a bond of silanol during acid-base condensation. In addition, the hydroxyl groups on the substrate surface of the TPS, and silanol form a polysiloxane structure. The melt strength of the TPS structure did not prevent the foaming gas from escaping into the environment and maintaining the structure of the pores. Therefore, using supercritical CO_2 foam foaming technology in the development of starch foaming materials is facing challenges.

In this study, we synthesized the reactive functional groups at the ends of the thermoplastic starch molecular chain to produce a three-dimensional structure, so that the molecular chains can be entangled and intertwined with each other, and their viscoelastic and plasticizing behaviors achieve the strength requirements of the foam material structure. The process is shown in [49,50].

2.7. Supercritical Batch Foaming Process

Figure 1a schematically describes the processing system used for CO_2 batch foaming in the study. An upper and under part of a batch die is mounted on the upper and under the platform of the press, respectively. A small tunnel in the batch die was connected with a CO_2 injection unit by a stainless steel tube. The press can provide heat to raise the temperature of the batch die and force to seal foaming agent CO_2 in the die. The inside dimensions for the batch die are 250 mm in diameter and 13 mm in height. The foaming procedure was as follows: (a.) PBAT with thermoplastic starch and silane were blended to form a large sheet of 3 mm in thickness. Next, the large sheet was cut into several small square samples of 25 mm in length and 3 mm in width thickness for foaming. (b.)The batch die was heated to reach setting temperatures ranging from 80 °C to 105 °C first; then, five square samples were placed in the under part of the die, and next, we sealed the upper and under part of the dying by the force from the piston of the press. (c.) Foaming agent CO_2 was injected into the batch die by using a CO_2 injection unit through a stainless steel tube until the setting pressure, which is 17 MPa or 23.8 MPa. (d.) The five square samples were soaked with CO_2 under a supercritical condition for an hour to allow the samples to reach saturation conditions. (e.) The batch die was opened after an hour and the open rate of the die was instant less than 1 s by lowering the piston of the press. (f.) The five foam samples were taken out immediately from dying to prevent shrinkage from a high temperature. (g.) The foam samples were cooled at room temperature to let foam samples be stable. (h.)

The properties and structures of five foam samples at the same foaming conditions were evaluated when foam samples were at room temperature after 24 h.

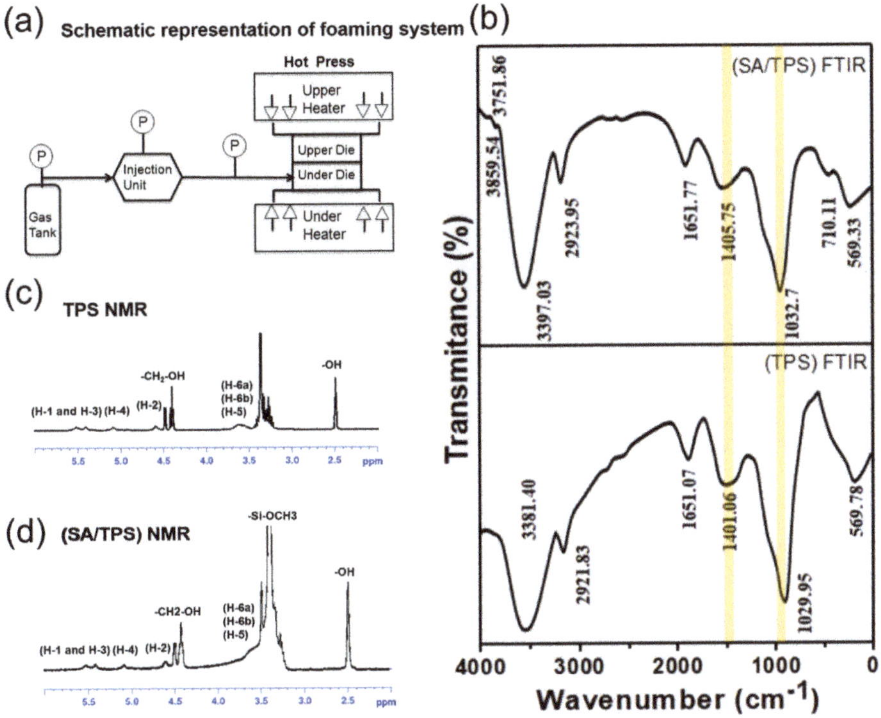

Figure 1. Schematic representation of foaming system and chemical modification for analysis of TPS functional group. (**a**) The processing system is used for CO_2 batch foaming. (**b**) FT-IR of TPS and (SA/TPS). (**c**,**d**) NMR of TPS and (SA/TPS).

3. Results and Discussions

3.1. FT-IR Analysis of TPS and SA/TPS

Figure 1b shows the FT-IR spectra of the TPS polymer functional groups before and after the surface modification. The main IR peaks at 3381 cm^{-1}, 2922 cm^{-1}, 1651 cm^{-1}, and 1401 cm^{-1} represent the O–H stretching, C–H and -CH_2 asymmetric stretching, C=O stretching, and -CH_2-deformation, while the peak position at 1030 cm^{-1} of starch may be caused by C–O–H stretching. The modified SA/TPS IR peaks are slightly shifted to 3397 cm^{-1}, 2924 cm^{-1}, 1652 cm^{-1}, 1401 $cm^{-1,}$ and 1033 cm^{-1}, respectively, which represented the SA interaction with the TPS functional groups. Especially, the peak position is shifted from 1405 cm^{-1} to 1115 cm^{-1}. The position of the corresponding peak of the SA modifier was with relatively low frequency. The SA modifier and this frequency are close to the corresponding starch where C–H is the -CH_2- asymmetric expansion and C–O–C expansion and contraction.

H NMR Analysis of TPS and SA/TPS

The TPS NMR analysis spectrum is shown in Figure 1c. The sample is dissolved in DMSO-d_6. Compared with literature data [48], there are small signals at 3.6 (H-6a, H-6b, and H-5), 4.6 (H-2), 5.08 (H-4), and 5.41–5.52 (H-1 and H-3) ppm. This indicates the resonance of starch. The small signals at 2.49 and 3.38 ppm are from the -OH (hydrogen bond and water) in DMSO and starch, respectively, and the resonances at 4.41 and 4.48 ppm are from the terminal proton -CH_2-OH group, respectively.

The (SA/TPS) NMR analysis spectrum is shown in Figure 1d, where (PEG/TPS) is dissolved in DMSO-d_6. This small signal is at 3.6 (H-6a, H-6b, and H-5), 4.6 (H-2), 5.08 (H-4), and 5.41–5.52 (H-1 and H-3) ppm, representing starch resonance [48]. The peaks at 2.49 and 3.38 ppm come from DMSO and the -OH group, and -OH (hydrogen bond and water) in starch. The resonances at 4.41 and 4.48 ppm come from the proton at the terminal functional group of -CH_2-OH, and the 3.49 ppm signal comes from the resonance of the SA modifier proton at the -Si-OCH_3 functional group.

3.2. The Composite Foam Density of TPS/PBAT

3.2.1. The Density of Foamed Material of PBAT

Figure 2 shows the density changes of PBAT foams impregnated with CO_2 at different foaming temperatures (80, 85, 90, 95, 100, and 105 °C) with two different foaming pressures (17 MPa and 23.8 MPa). The obtained results show that the foam density of PBAT was higher at a lower pressure than the high-pressure conditions. However, the pressure changes in PBAT foaming processes were not brought to any noticeable density differences. Consecutively, the temperature rises along with the above pressure condition and affects the PBAT foam density proportionally; furthermore, the obtained foam density was moved from 350 kg/m^3 to 160 kg/m^3.

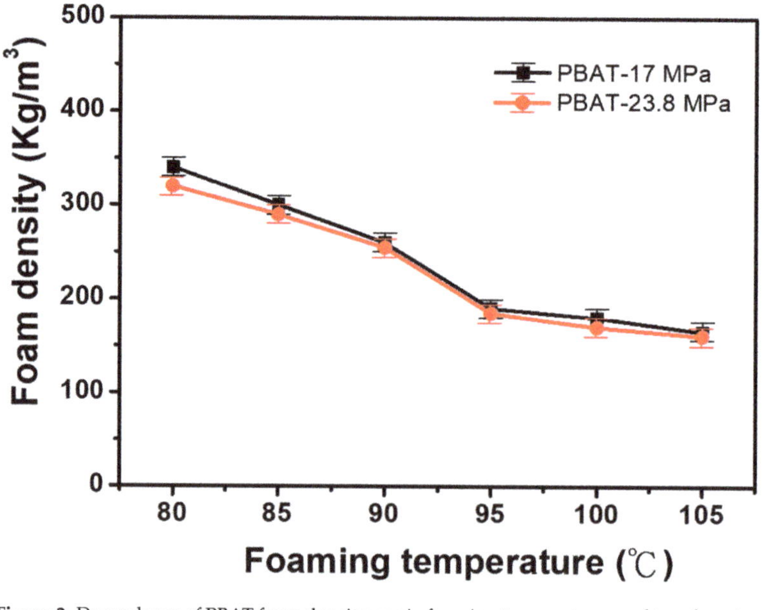

Figure 2. Dependence of PBAT foam density on six foaming temperatures and two foaming pressures.

3.2.2. The Density of TPS/PBAT Composite Foam

Figure 3a represents the foam density of TPS/PBAT blends by impregnating CO_2 at the foaming pressure of 17 Mpa and foaming temperatures between 80 °C and 105 °C. The three curves in Figure 3a indicate the trends of foam density for three blends, i.e., 70% TPS/30% PBAT [N-3], 60% TPS/40% PBAT [N-2], and 50% TPS/50% PBAT [N-1]. The form density of PBAT increases as the concentration of TPS increases in the TPS/PBAT composite. In particular, the forming density of [N-3] reaches its maximum at the forming temperature of 80 °C, which is approximately 1120 kg/m^3. However, the increasing temperature was not ideal for the foam-forming process. Because of the high temperature, the microcellular structure of the composite form starts to shrink, which leads to poor elongation properties. It can be found that 50% of TSP shows a 20% elongation with the lowest foam density of 340 kg/m^3. A low foam density of the three blends occurs around the foaming temperature of 100 °C, however, the foam density rises again when foaming temperatures are higher than 100 °C. TPS is a material that is not easy to foam, especially CO_2 as a foaming gas, which is mainly used as a filler for biological materials.

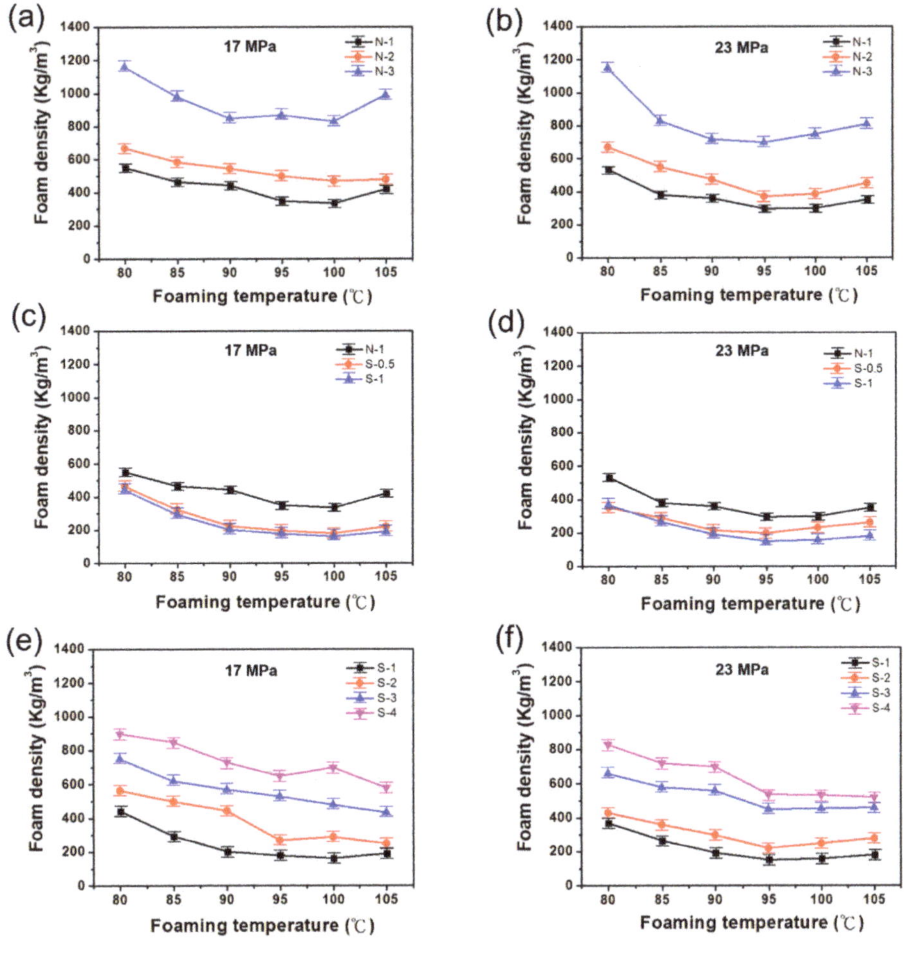

Figure 3. Dependence of TPS/PBAT and (TPS with SA)/PBAT composite foam density under six foaming temperatures and different foaming pressures. (**a**,**c**,**e**) Variation of composite foam density at 17 MPa. (**b**,**d**,**f**) Variation of composite foam density at 23 MPa.

Figure 3b shows the density change of TPS/PBAT composite foam obtained by impregnating CO_2 with TPS/PBAT composite at a foaming temperature range of 80 °C to 105 °C and a foaming pressure of 23.8 MPa. In Figure 3b, curves are representing the foam densities of the [N-1], [N-2], and [N-3] composites. From the obtained results, the maximum foam density of the composites at 80 °C was reduced from 1120 to 700 kg/m^3 because of the high-pressure environment; when the foaming temperature is 80 °C, compared with other foaming temperatures. Similarly, as the foaming temperature continues to rise, the density of all foams will continue to decrease until the foaming temperature is 95 °C. If the foaming temperature rises again, the density of the foams will also increase slowly.

The bubble structure cannot be supported, causing shrinkage (as shown in Figure S1). It was found that the lower foaming density of this foaming pressure of 23.8 Mpa for the composite [N-1] is about 300 kg/m^3, occurring at a foaming temperature of 95 °C, and the same foaming temperature varies as the TPS content increases. The foam density of the composite will also increase with the increasing TPS content. The TPS/PBAT composite system is more difficult to foam, however, the increase in foaming pressure will help the density of the composite foam to decrease. However, the bonding between TPS and PBAT was very poor, especially since the 80% TPS/20% PBAT blend is difficult to foam.

3.2.3. The Density of (SA/TPS)/PBAT Composite Foam

Figure 3c shows the density change of the 50% (SA/TPS)/50% PBAT composite under the foaming temperature range of 80°C to 105°C, and the foaming pressure of 17 MPa impregnated with CO_2. This experiment uses an SA modifier to modify the terminal functional groups of TPS, to verify whether it can promote the interface between TPS and PBAT to produce better compatibility. In Figure 3c, there were three curves representing the foam density of unmodified composite [N-1]: 50% TPS containing modifier SA addition and is 5 PHR (PHR: per hundred resin (TPS)) combined with 50% PBAT [S-0.5], and 50% TPS containing 10 PHR SA in TPS combined with 50% PBAT, [S-1]. Among them, [N-1] foam has the highest foam density, while [S-0.5] and [S-1] show a foaming temperature of 100 °C and have a lower foaming density of 180 kg/m^3 and 160 kg/m^3, respectively. It was proved that the SA concentration in TPS significantly modifies and improves physiological modification, such as tensile strength, elongation, and foam density. While the foaming temperature is higher than 100 °C, the density of the three kinds of composite foams all have an upward trend, however, the density of [N-1] foam is comparatively high, which also confirms the internal structure of the foam strength as poor.

Figure 3c,d used three types of the same composite and the same foaming temperature and impregnation gas, while Figure 3d used a higher foaming pressure of 23.8 MPa. The density of the three kinds of foam is under the same foaming temperature range, and the changing trend of the density of the composite foam is similar to Figure 3c. However, under higher foaming pressures, the density of the three types of composite foams is relatively low; the [N-1] compound and [S-0.5] compound have the same lower foaming. The low density of 300 kg/m^3 and 200 kg/m^3 occurred at the foaming temperature of 95 °C, whereas [S-1] shows the lowest foam density of 150 kg/m^3 at a foaming temperature of 100 °C.

3.2.4. The Influence of Foaming Temperature on the Density of Composite Foams with PBAT in Different Ratios (SA/TPS)

Figure 3 shows the foam density of SA (10 PHR) surface-modified TPS mixed with a PBAT polymer composite blend. At fixed pressure 17 MPa, the foam density of various mixing ratios of SA/TPS and PBAT was investigated under changing temperatures. Mixed weight ratios of 50%, 60%, 70%, and 80% SA/TPS were blended with 50%, 40%, 30% and 20% PBAT, as the mixed composites named [S-1], [S-2], [S-3] and [S-4], respectively. It can be found that the addition of 50% (SA/TPS) can obtain the lowest foam density of about 160 kg/m^3 at a foaming temperature of 100 °C. When (SA/TPS) is added at 60%, 70%, and 80%, the lowest foam density appears at the foaming temperature of 105 °C, and the foam density is 250 kg/m^3, 420 kg/m^3, and 580 kg/m^3, respectively. We found that increasing the concentration of the SA/TPS composite ratio in PBAT polymer was unfavorable to the foam composites, whereas the relative stability of the composites is also better, and therefore, increases the temperature of the lowest foaming density.

A similar procedure was followed in terms of the high pressure 23.8 MPa conditions, where it renders the four composites foam density changes values. Figure 3f shows that the ratio of (SA/TPS) in the composites is 50%, 60%, and 70%, and the lowest foam density is about 180 kg/m^3, 220 kg/m^3, and 450 kg/m^3. When the ratio of 80% (SA/TPS) is mixed, the lower foam density occurs when the foaming temperature is 105 °C, and the foam density is 520 kg/m^3. It can be found from the research, however, that the (SA/TPS) ratio is 70% and 80% of the composites if the foaming temperature of this experiment is 100 °C. There is no obvious change in the foam density at a higher foaming temperature. When the foaming pressure is 17 MPa, the density of the two-component composites foam shows a downward trend rather than an upward trend. This means that the more the proportion of (SA/TPS) in the composites is, the higher the heat resistance of the composites and the smaller the change in foam density. However, the density of the foam is relatively high, which means that (SA/TPS) is similarly filled in the composites. It is of little help to the foaming process of the composites.

3.3. *Appearance and Internal Structure of TPS/PBAT Composites Foam*

3.3.1. Appearance and Internal Microstructure of PBAT Foam

Figure 4 shows the change of the foam impregnated with CO_2 at the foaming temperature of 95 °C and the foaming pressure (a) 17 MP and (b) 23.8 MPa.

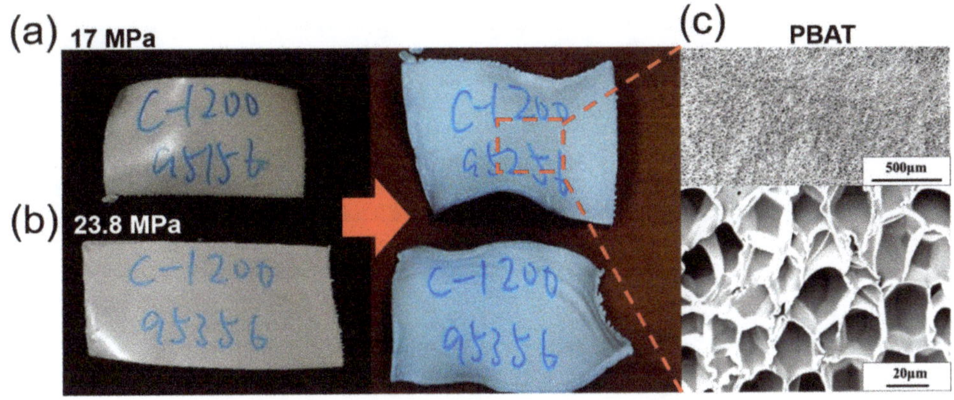

Figure 4. The changes in the foam impregnated with CO_2 at the foaming temperature of 95 °C and the foaming pressure (**a**) 17 MP and (**b**) 23.8 MPa. (**c**) The SEM image of the PBAT foams was magnified by 50 times and 1000 times, respectively, at a foaming temperature of 95 °C and a foaming pressure of 17 MPa.

For the appearance of the samples after foaming for 1 h, and stored at room temperature for 24 h, we were found that at the foaming temperature of 95 °C, the softer PBAT was found to be smooth and the foam surface was flat and elastic, however, the foam volume becomes larger as the foaming pressure increases.

After supercritical foaming, the foams were obtained from foaming pressures of 17 MPa and 23.8 MPa, and the initial appearance of the foam is without cracks. After 24 h of storage, the foam became deformed and smaller in volume, which caused the material damage. The reason, however, might be due to the carbon dioxide gas in the foam being released into the atmosphere faster than the air in the atmosphere penetrates into the foam, resulting in negative pressure created inside the foam during the cooling process; however, the foam structure cannot support this negative pressure, which causes the foam to shrink. Moreover, thermal expansion and contraction are also the reason. There was a small residual temperature inside the foam even after 2 h. Although, after more than 20 h of cooling, the temperature of the foam continues to decrease, causing the volume to continue to shrink.

Figure 4c is a SEM image of the internal microstructure of the CO_2 foam impregnated with PBAT at a foaming temperature of 95 °C and a foaming pressure of 17 MPa. The depicted figures of PBAT foams were magnified by 50 times and 1000 times, respectively, and it was found that PBAT foam was achieved with uniformly distributed bubbles under this foaming condition. Most of the bubbles have a pentagonal or hexagonal cross-section and each bubble has obvious bubble walls, which means that the foam has a closed-cell structure. Since the internal bubbles of the foam have a complete structure, the surface is smooth, and the average bubble diameter is about 25 μm.

3.3.2. The Appearance and Internal Structure of TPS/PBAT Composites Foam

The photographic images in Figure S2a render the internal structure of the unmodified [N-1] composite after the CO_2 foaming process at a temperature range of 80–105 °C and a fixed pressure of 17 MPa. We observed that the composite foams expand slightly, however, the surface properties remain smooth and hard due to their high foam density value at the foaming temperature of 80 °C. As the foaming temperature rises, the composite foam continues to expand. However, as the foaming temperature rises to 100 °C, the surface of the foam becomes uneven shapes that are relatively softer; therefore, raising the foaming temperature results in the lowest foam density. However, at the foaming temperature of 105 °C, the shape of the foam is completely distorted and rigid, the appearance of the foam surface becomes more rough and uneven, the flexibility of the foam deteriorates, and the volume is severely collapsed.

Figure S2b,c shows the photographic images of the unmodified composites [N-2] and [N-3], respectively. It was found that at the low foam temperatures of 85 °C and 90 °C, the composite foams only slightly expand, whereas the surface properties remained unchanged. When the foaming temperature rises, the expansion of the [N-2] foam is slightly larger than that of the [N-3] foam. The appearance of the two foams is distorted and deformed while the foaming temperature rises to 100 °C and 105 °C. Among them, the [N-3] foam has no flexibility, and the surface of the foam is rough, which is close to no foaming. Whereas, performing the same process under 23.8 MPa pressure conditions, the [N-1] foam volume is comparatively larger than at 17 MPa pressure conditions. Similarly, the surface of the [N-1] foam becomes a more rough and uneven shape. The photographic image was included in the supporting information (Figure S3a). The other two composites, [N-2] and [N-3], also produced a similar trend, as seen in Figure S3b,c. Although, at this higher foaming pressure, the softness of the foams was slightly better than the composites at the low foaming pressure of 17 MPa. However, the overall softness of the foam is still very poor, the surface is rough, and the density of the foam is still relatively high.

Figure 5a–c renders the cell structures of the [N-1], [N-2], and [N-3] foams made at the foaming temperature of 95 °C and the fixed foaming pressure of 17 MPa. Many large cells or pores and small TPS particles scattered in the foam were observed in Figure 5a. Large cells, more than 100 μm, and small cells of less than 20 μm coexist in the foams; furthermore, the cell wall is not intact and has tiny holes. Unreacted TPS particles can be seen clearly and the boundary between the TPS particles and cells is also very distinct. Moreover, the bubble size distribution of this foam is not uniform, where the diameter of the large bubbles was 10 times bigger than the small bubbles. Using the higher magnification SEM image, the bubble structure of the foam cannot show complete bubbles like the foam of PBAT in Figure 4c can. The bubble structure of this foam is inconsistent and the bubble wall is broken or merged with the neighboring bubbles, resulting in a large size pore volume. This is called the open-cell structure of the bubbles. Furthermore, the bubbles of the PBAT composites foam were affected by the undissolved TPS particles. Therefore, the degree of open-cell structure of the foam was much higher than that of the PBAT foam.

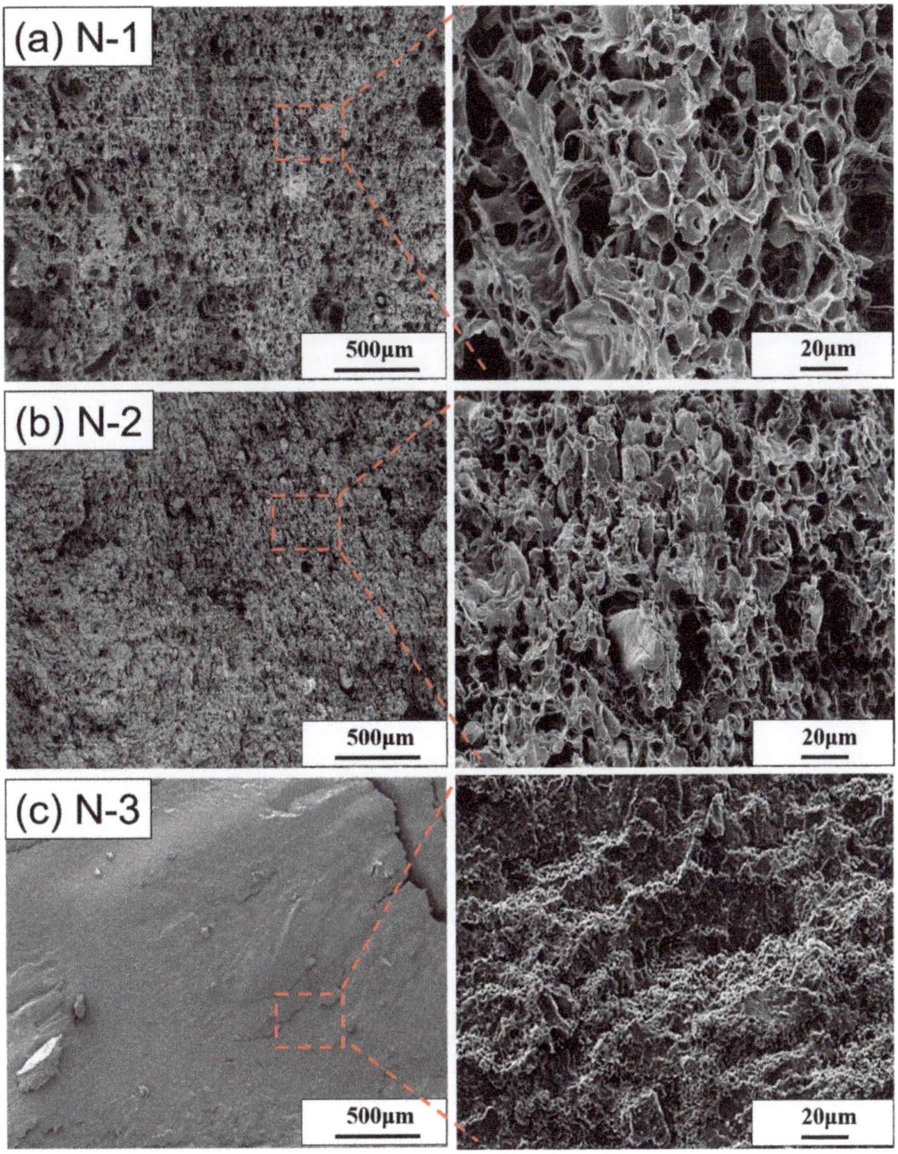

Figure 5. The SEM images of cell structure (**a**) [N-1], (**b**) [N-2], and (**c**) [N-3] at the foaming temperature of 95 °C and the foaming pressure of 17 MPa.

The cell structures for the [N-2] and [N-3] foams are shown in Figure 5b,c, respectively. It is not surprising that the addition of untreated TPS into PBAT improves the foam density. However, an increased concentration of TPS results in the undissolved particle agglomeration in the PBAT composite, such as the [N-2] and [N-3] composites showing 50 μm and 10 μm large and small pore volumes, respectively. Meanwhile, there are fewer incomplete bubbles in the foam were observed. Therefore, the obtained SEM image shows the uneven surface with a high open-cell structure, which causes the hardness and rigidity of the foams. It was figured that in the TPS/PBAT foam-forming process under supercritical

CO_2, the formed foam cells were particularly dependent on the PBAT rather than TPS. In Figure 5, it is proved that in the [N-1], [N-2], and [N-3] composites formed by unmodified TPS with PBAT, TPS seems to be used as a filler and has poor compatibility with PBAT. The mechanical properties of the composite will also deteriorate.

3.4. Appearance and Internal Structure of (SA/TPS)/PBAT Composites Foam

3.4.1. The (SA/TPS)/PBAT Composite Foam Appearance and Internal Structure

To improve the interfacial diffusion between the TSP and PBAT polymers, a surface grafting regent (SA) was introduced. Initially, two various fractions of SA were used to modify the surface of 50% TPS, such as [S-0.5] and [S-1]. Figure S4's photographic images of [S-0.5], and [S-1] are shown. The composite foams were produced under the same operating temperature and CO_2 foaming gas was passed at a fixed pressure of 17 MPa. The foam sample become softer and improved the elastic behavior. Among the series of foams, foaming temperatures of 95 °C and 100 °C formed samples exhibiting a large pore volume with the lowest foam density. However, we have noticed a slight surface roughness on these foams. When the foaming temperature continuously rises to 105 °C, the structures' appearance of foams for (a) [S-0.5] and (b) [S-1] are slightly deformed. However, when compared to the unmodified TPS/PBAT foams, this SA-modified TPS/PBAT provides better tensile strength and elongation properties.

At the same operational temperature, the foaming process was elevated to 23.8 MPa. The obtained results reveal the uneven surface as per the temperature raise (Figure S5). Furthermore, the large volume of foams occurred between a 90 °C to 100 °C temperature range; however, this resulted in foams becoming soft with slight surface roughness. When the foaming temperature is 105 °C, the volume of the foam tends to shrink. Thus, the foaming pressure seems to tend to increase the volume of the composites foam at the same foaming temperature, however, it is not conducive to high-temperature foaming at 105 °C. Figure S6a,b shows the cell structure of [S-0.5] and [S-1] composite foams at fixed parameters of 95 °C and 17 MPa foaming temperature and pressure. The surface of the TPS particles in the [S-0.5] foam (as shown in Figure S6a) is rougher than the surface of the TPS particles in the [N-1] foam (as shown in Figure 5), such as the surface of the unmodified TPS mixed foam. Although the TPS particles were modified with 5 PHR SA, the interface between this particle and PBAT was not clear. The TPS particles are uniformly dispersed inside the foam, however, the particle size is not uniform. The diameter of the TPS particles presented is less than 25 μm or greater than about 200 μm. Although the 5 PHR SA can increase the compatibility of TPS and PBAT, it limits the dispersion of TPS in PBAT. Most of the bubbles have a closed-cell structure; therefore, the bubble wall between the bubbles is obvious, and consecutively the bubble size distribution is relatively consistent. The SEM image shows that the average bubble diameter is about 10 μm, which is also the performance of good compatibility between TPS and PBAT.

The cell structure in Figure S6b shows that the boundary between TPS and PBAT becomes fuzzier when compared with Figure S6a; especially with the distinct small (TPS with SA) particles from PBAT being difficult, and the surface of the (TPS with SA) particles is surrounded by many small cells that are less than 10 μm in diameter. Most of the cells in Figure S6b's foam are closed cells. The phenomenon can be explained by the compatibility and miscibility in [S-1] having dramatically enhanced. However, large pores are also observed in Figure S6b, which indicates that compatibility decreases between large (TPS with SA) particles and PBAT. Therefore, the dispersion of (TPS with SA) in PBAT is also an important factor.

3.4.2. The Appearance and Internal Structure of (SA/TPS)/PBAT Foams with Different (SA/TPS) Ratios

Figure 6 shows the appearance of foams with different ratios (10 PHR SA/TPS) and PBAT composite at three foaming temperatures (at 95 °C, 100 °C, and 105 °C) and a foaming pressure of 23.8 MPa impregnated with CO_2. After the surface-modified 10 PHR

SA/TPS mixture was mixed with different ratios of PBAT to form the composite foams, such as [S-2], [S-3], and [S-4], it was observed that the volume of the foam decreased as the proportion of 10 PHR SA/TPS increased. This study was carried out at three foaming temperatures. As the foaming temperature rises, the volume of the foam increases. The volume expansion changes of [S-2] were the most obvious. However, for [S-4], the volume of the foam only slightly increased. This phenomenon proves that, when the TPS content reaches 80%, the interface between TPS and PBAT is still intact from the internal structure of the foam. Therefore, it is further confirmed that the addition of a modifier provides more compatibility between TPS and PBAT. When the proportion of TPS in the composite material is higher, it is easier to disperse in the composite material. Because the more reactive functional groups, PBAT and TPS are easier to connect with each other, the reaction makes the compatibility better.

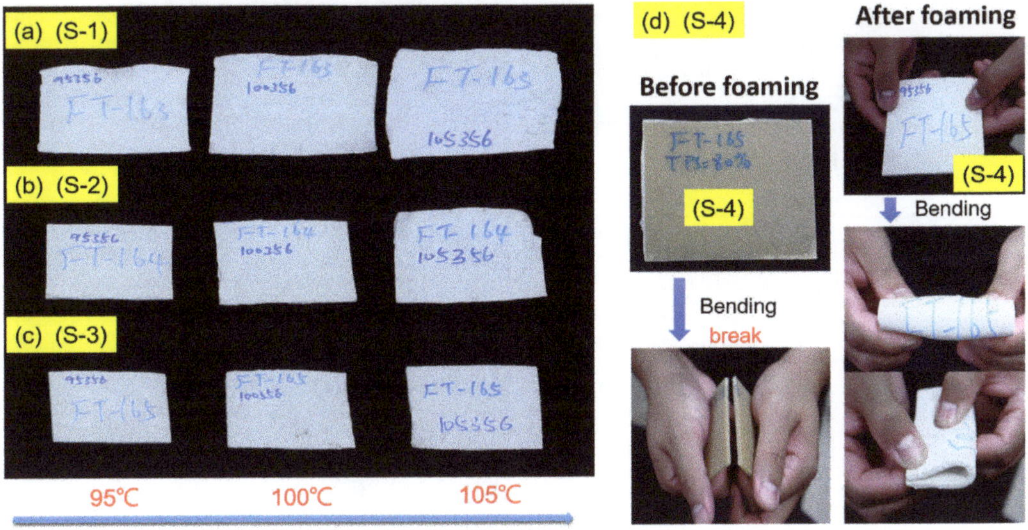

Figure 6. The appearance of foams with different ratios (SA/TPS) and PBAT composite (**a**) [S-1], (**b**) [S-2], and (**c**) [S-3] at three foaming temperatures and a foaming pressure of 23.8 MPa impregnated with CO_2. (**d**) Comparing the foam appearance characteristics and flexural limitations of the [S-4] before and after.

The SEM image shows that the TPS in the composite foams becomes a continuous phase, the large pores become less, and more TPS particles with extended shapes are found, which is proof of the better compatibility between the two. However, the unmodified TPS (such as N-3) composite material with 70% TPS content exhibits poor mechanical properties, whereas the SA-modified TPS foam has excellent flexibility after being folded even at higher concentrations. The very different material properties before and after foaming are illustrated.

Moreover, Figure S7 shows the appearance of the foam under the foaming temperature of 105 °C and different foaming pressures of 17 and 23.8 MPa impregnated with CO_2, as shown in Figure S7a–c, for [S-2], [S-3], and [S-4], respectively. At the same foaming temperature under 23.8 MPa, the three types of composites have a larger foam volume than the foaming pressure of 17 MPa. However, the surface of the foam deteriorates and large bubbles were generated. Whereas, the high proportion [S-4] foam provides a smaller volume than flat surface foam; although [S-4] has the least amount of foaming volumes, and the foam has flexural limitations, as shown in Figure 6d. The un-foamed composite board is easy to break, but the foamed board is not easy to break and has advantageous flexibility after bending, which also illustrates the different material properties after foaming.

The surface morphology change of the [S-1], [S-2], [S-3], and [S-4] composites at a low foaming pressure of 17 MPa with a fixed foaming temperature of 95 °C were depicted in the SEM image (Figure 7). Figure 7a shows that the composites [S-1] and [S-2] foams have similar internal structures with large pores, but the [S-1] foam contains smaller bubbles. Although these two foam shows a closed-cell structure without particle agglomeration, the bubble diameter of the [S-2] foam is about 15~20 μm. In addition, compared with the unmodified [N-2] foam (as shown in Figure 5b), the existence of complete bubbles cannot be seen.

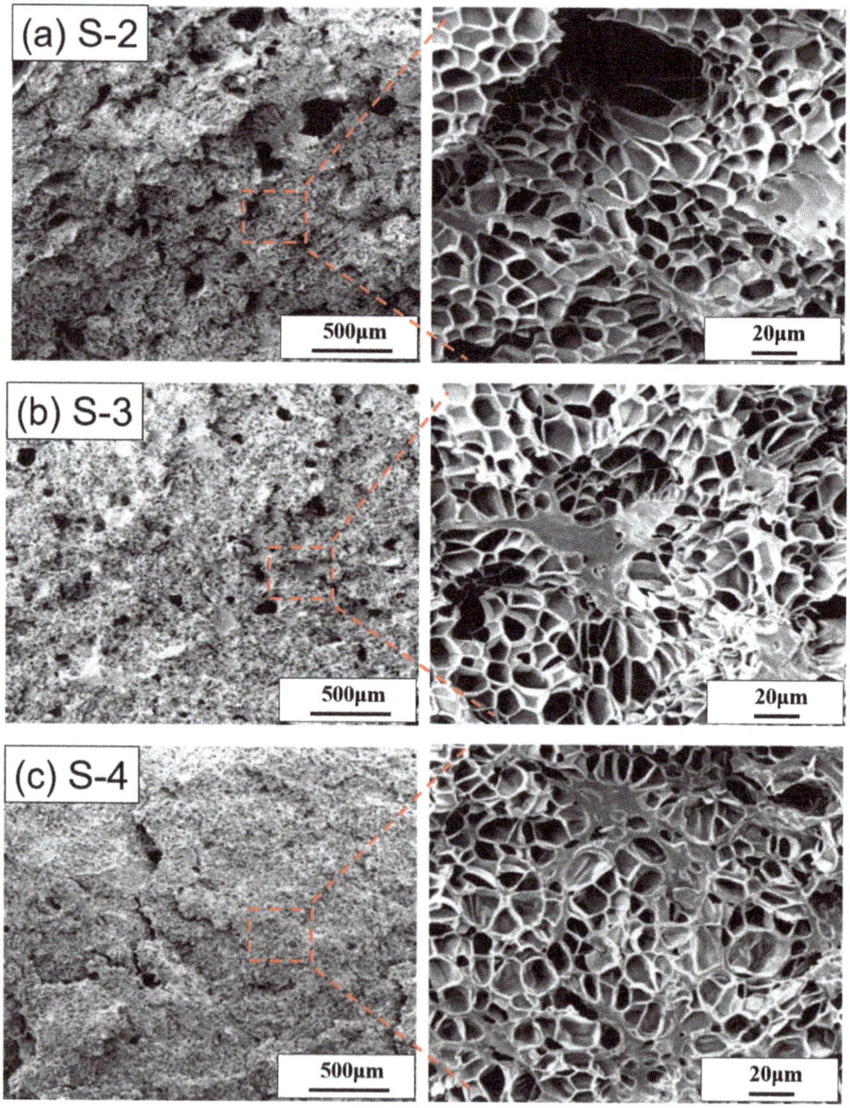

Figure 7. The SEM images of cell structure at the foaming temperature of 95 °C and the foaming pressure of 17 MPa. (**a**) [S-2], (**b**) [S-3], and (**c**) [S-4].

Figure 7b shows that [S-3] has large pores inside the foam and tends to decrease. Besides the large holes, the bubble size is still uniform. However, the shape of the bubble is slightly deformed, and it is no longer the pentagonal or hexagonal shape of normal foam. The bubbles are mostly closed-cell structures, and the interface with PBAT remains intact without gaps, which can be judged as good compatibility. Because of the high fraction of surface-modified SA/TPS, the foam expanded with more bubbles, and the foam bubble diameter is about 15–20 μm, even though the un-foamed areas were also found.

When combining the SA/TPS with PBAT, the functional group of SA forms a cross-link with PBAT and reduces the pore size, whereas the unmodified TPS with PBAT foam exhibits large pores with a broken cell structure.

Many (SA/TPS) large particles are not spherical (instead, it is a special shape, as shown in Figure 7b). Because of the compatibility of TPS with PBAT after modification; it appears to be one phase by SEM. However, surface unmodified TPS is incompatible with PBAT and obtained SEM images appeared with two phases. Many (SA/TPS) large particles are not spherical and are typically incompatible.

Figure 7c shows that the surface of [S-4] foaming still occurs in the composite; here, the size of the bubbles inside the foam is more delicate, and compared with the other foams, it has fewer larger pores. However, the shape of the bubbles is deformed and the bubbles are mostly closed-cell structures, and the bubble diameter is about 10~15 μm. The [S-4] composites solution shows a complete dispersity of SA/TPS due to the high SA content. As the (SA/TPS) material has more reactive functional groups, it is easier for PBAT and (SA/TPS) to contact each other to cause a reaction and better compatibility. Figure 7c shows more (SA/TPS) particles of special shape and fewer large pores in the foam, which is proof that PBAT and (SA/TPS) have better compatibility.

3.5. Mechanical Properties of the TPS/PBAT and the (SA/TPS)/PBAT Composites Foam

Table 2 show the ultimate tensile strength and elongation break and foam density for the unmodified foams [N-1], [N-2], and [N-3] and the surface-modified foams [S-1], [S-2], [S-3], and [S-4], respectively. These various composite foams are produced at the foaming temperature of 95 °C and the foaming pressure of 17 MPa. The unmodified TPS/PBAT composites have approximately 4% of the ultimate tensile strength and elongation at the break for these three TPS/PBAT foams and become much shorter. However, the surface-modified foams show increased ultimate tensile strength, such as the [S-1], [S-2], and [S-3] blended foams as being about 4 times, 2.5 times, and 2 times larger than that of the [N-1], [N-2], and [N-3] blended foams, respectively, although, the elongation at break becomes decreased. The foam density of the (TPS with SA)/PBAT blend has the same trend as that of the TPS/PBAT blend. In general, the ultimate tensile strength of the (TPS with SA)/PBAT blends is basically better than that of the TPS/PBAT blends; however, the two types of blends generally have a similar elongation at break (shown in Table 3).

Table 2. The mechanical properties of the (SA/TPS)/PBAT composites foam.

	[N-1] (F)	[N-2] (F)	[N-3] (F)	[S-1] (F)	[S-2] (F)	[S-3] (F)	[S-4] (F)
Tensile strength (kPa)	226 ±35	392 ±44	539 ±48	883 ±45	1030 ± 52	1157 ± 40	1236 ± 55
Elongation (%)	20 ± 3	9 ± 1	2 ± 1	37 ± 3	22 ± 2	13 ± 2	10 ± 1
Foam density (kg/m^3)	349 ± 2	540 ± 1	897 ± 1	187 ± 3	287 ± 2	515 ± 2	663 ± 1

Table 3. The mechanical properties of the TPS/PBAT composites.

	[N-1]	[N-2]	[N-3]	[S-1]	[S-2]	[S-3]	[S-4]
Tensile strength (Breaking point) (kPa)	226 ± 35	392 ± 44	539 ± 48	883 ± 45	1030 ± 52	1157 ± 40	1236 ± 55
Elongation (%)	20 ± 3	9 ± 1	2 ± 1	37 ± 3	22 ± 2	13 ± 2	10 ± 1

4. Conclusions

This research has successfully developed a highly flexible and compatible starch-based foam using supercritical carbon dioxide. This research completely covers the foam-forming optimization steps, including various foaming temperatures under two different pressures. Furthermore, blending ratios of the surface-modified (SA/TPS) and unmodified (TPS) starch were optimized with the PBAT foam. The results show that with the addition of surface-modified SA/TPS starch to PBAT, the composite foam turns into a soft altered foam density with improved elongation strength and tensile properties for flexible application. We also observed that the unmodified (TPS/PBAT) composite is not suitable for the foam to form due to its poor interface formation with PBAT. However, when compared to the internal microstructure, the SA/TPS/PBAT composite foam shows a uniform bubble with an average diameter of 25 μm and a closed-cell structure. Therefore, (TPS with SA)/PBAT blended foams exhibit lower foam density and better tensile properties than those of TPS/PBAT blended foams. In addition, this research can be applied to thermoplastic starch biodegradable foaming materials, combined with supercritical foaming green process technology, to develop starch composite foaming materials. This foam-forming processing technique could be used in electronics packaging materials and has medical equipment application potential in the future.

Supplementary Materials: The following supporting information can be downloaded at: https://www.mdpi.com/article/10.3390/polym14101952/s1, Figure S1: The SEM images of cell structure at the foaming temperature of 95 °C and the foaming pressure of 23.8 MPa, Figure S2. Dependence of TPS/PBAT foam appearance on six foaming temperatures and the foaming pressure of 17 MPa, Figure S3. Dependence of TPS/PBAT foam appearance, Figure S4. Dependence of (TPS with SA)/PBAT foam appearance, Figure S5. Dependence of TPS/PBAT foam appearance on six foaming temperatures and the foaming pressure of 23.8 MPa, Figure S6. The SEM images of cell structure, Figure S7. The appearance change of the foam under the different foaming pressure of 17 and 23.8 MPa.

Author Contributions: Conceptualization, C.-C.K., C.-H.L. and C.-J.C. (Chia-Jung Cho). conceived and supervised the project; C.-J.C. (Chih-Jen Chang), M.V. and H.-T.W. conceived and designed the experiments; C.-J.C. (Chih-Jen Chang), M.V. and C.-M.W. performed and analyzed the experiments; J.C., P.-Y.C. and C.-M.W. assisted with sample preparation, analyzing the SEM morphology and the other measurements; C.-J.C. (Chih-Jen Chang) wrote the paper; C.-J.C. (Chia-Jung Cho), C.-C.K. and C.-J.C. (Chih-Jen Chang) discussed the results and revised or commented on the manuscript. All authors have read and agreed to the published version of the manuscript.

Funding: This work was supported by (1) the Ministry of Science and Technology, Taiwan (Contracts: MOST 109-2221-E-027-114-MY3, MOST 110-2222-E-214-001-MY2, MOST 110-2622-E-027-019-) (2) I-Shou University (Contracts: ISU 111-S-01). The authors also thank the National Science Council of Taiwan (NSC-108-2221-E-182A-002-), Chang Gung Memorial Hospital (CMRPG3J1281, CMRPG3K1331, CMRPG3L1301, and CORPG3M0151), and the National Taipei University of Technology and Chang Gung Memorial Hospital Joint Research Program (NTUT-CGMH-111-04) for financially supporting this research.

Institutional Review Board Statement: Not applicable.

Informed Consent Statement: Not applicable.

Data Availability Statement: Not applicable.

Conflicts of Interest: The authors declare no conflict of interest.

References

1. Cho, C.-J.; Lu, S.-T.; Kuo, C.-C.; Liang, F.-C.; Chen, B.-Y.; Chu, C.-C. Pyrene or rhodamine derivative–modified surfaces of electrospun nanofibrous chemosensors for colorimetric and fluorescent determination of Cu^{2+}, Hg^{2+}, and pH. *React. Funct. Polym.* **2016**, *108*, 137–147. [CrossRef]
2. Venkatesan, M.; Veeramuthu, L.; Liang, F.-C.; Chen, W.-C.; Cho, C.-J.; Chen, C.-W.; Chen, J.-Y.; Yan, Y.; Chang, S.-H.; Kuo, C.-C. Evolution of electrospun nanofibers fluorescent and colorimetric sensors for environmental toxicants, pH, temperature, and cancer cells—A review with insights on applications. *Chem. Eng. J.* **2020**, *397*, 125431. [CrossRef]
3. Cho, C.-J.; Chen, S.-Y.; Kuo, C.-C.; Veeramuthu, L.; Au-Duong, A.-N.; Chiu, Y.-C.; Chang, S.-H. Morphology and optoelectronic characteristics of organic field-effect transistors based on blends of polylactic acid and poly(3-hexylthiophene). *Polym. J.* **2018**, *50*, 975–987. [CrossRef]
4. He, C.; Liang, F.; Veeramuthu, L.; Cho, C.; Benas, J.; Tzeng, Y.; Tseng, Y.; Chen, W.; Rwei, A.; Kuo, C. Super Tough and Spontaneous Water-Assisted Autonomous Self-Healing Elastomer for Underwater Wearable Electronics. *Adv. Sci.* **2021**, *8*, 2102275. [CrossRef]
5. Lin, C.C.; Jiang, D.-H.; Kuo, C.-C.; Cho, C.-J.; Tsai, Y.-H.; Satoh, T.; Su, C. Water-Resistant Efficient Stretchable Perovskite-Embedded Fiber Membranes for Light-Emitting Diodes. *ACS Appl. Mater. Interfaces* **2018**, *10*, 2210–2215. [CrossRef]
6. Jao, C.; Chang, J.; Ya, C.; Chen, W.; Cho, C.; Lin, J.; Chiu, Y.; Zhou, Y.; Kuo, C. Novel stretchable light-emitting diodes based on conjugated-rod block elastic-coil copolymers. *Polym. Int.* **2020**, *70*, 426–431. [CrossRef]
7. Veeramuthu, L.; Venkatesan, M.; Benas, J.-S.; Cho, C.-J.; Lee, C.-C.; Lieu, F.-K.; Lin, J.-H.; Lee, R.-H.; Kuo, C.-C. Recent Progress in Conducting Polymer Composite/Nanofiber-Based Strain and Pressure Sensors. *Polymers* **2021**, *13*, 4281. [CrossRef]
8. Venkatesan, M.; Chen, W.-C.; Cho, C.-J.; Veeramuthu, L.; Chen, L.-G.; Li, K.-Y.; Tsai, M.-L.; Lai, Y.-C.; Lee, W.-Y.; Chen, W.-C.; et al. Enhanced piezoelectric and photocatalytic performance of flexible energy harvester based on CsZn0.75Pb0.25I3/CNC–PVDF composite nanofibers. *Chem. Eng. J.* **2022**, *433*, 133620. [CrossRef]
9. Mao, H.-I.; Wang, L.-Y.; Chen, C.-W.; Hsu, K.-H.; Tsai, C.-H.; Cho, C.-J.; Yu, Y.-Y.; Rwei, S.-P.; Kuo, C.-C. Enhanced crystallization rate of bio-based poly(butylene succinate-co-propylene succinate) copolymers motivated by glycerol. *J. Polym. Res.* **2021**, *28*, 92. [CrossRef]
10. Chen, C.-W.; Liu, P.-H.; Lin, F.-J.; Cho, C.-J.; Wang, L.-Y.; Mao, H.-I.; Chiu, Y.-C.; Chang, S.-H.; Rwei, S.-P.; Kuo, C.-C. Influence of Different Molecular Weights and Concentrations of Poly(glycidyl methacrylate) on Recycled Poly(ethylene terephthalate): A Thermal, Mechanical, and Rheological Study. *J. Polym. Environ.* **2020**, *28*, 2880–2892. [CrossRef]
11. Chan, H.; Cho, C.; Hsu, K.; He, C.; Kuo, C.; Chu, C.; Chen, Y.; Chen, C.; Rwei, S. Smart Wearable Textiles with Breathable Properties and Repeatable Shaping in In Vitro Orthopedic Support from a Novel Biomass Thermoplastic Copolyester. *Macromol. Mater. Eng.* **2019**, *304*, 1900103. [CrossRef]
12. Cho, C.-J.; Chang, Y.-S.; Lin, Y.-Z.; Jiang, D.-H.; Chen, W.-H.; Lin, W.-Y.; Chen, C.-W.; Rwei, S.-P.; Kuo, C.-C. Green electrospun nanofiber membranes filter prepared from novel biomass thermoplastic copolyester: Morphologies and filtration properties. *J. Taiwan Inst. Chem. Eng.* **2020**, *106*, 206–214. [CrossRef]
13. Hsu, K.-H.; Chen, C.-W.; Wang, L.-Y.; Chan, H.-W.; He, C.-L.; Cho, C.-J.; Rwei, S.-P.; Kuo, C.-C. Bio-based thermoplastic poly(butylene succinate-co-propylene succinate) copolyesters: Effect of glycerol on thermal and mechanical properties. *Soft Matter* **2019**, *15*, 9710–9720. [CrossRef]
14. Lu, W.-C.; Chen, C.-Y.; Cho, C.-J.; Venkatesan, M.; Chiang, W.-H.; Yu, Y.-Y.; Lee, C.-H.; Lee, R.-H.; Rwei, S.-P.; Kuo, C.-C. Antibacterial Activity and Protection Efficiency of Polyvinyl Butyral Nanofibrous Membrane Containing Thymol Prepared through Vertical Electrospinning. *Polymers* **2021**, *13*, 1122. [CrossRef]
15. Lu, W.-C.; Chuang, F.-S.; Venkatesan, M.; Cho, C.-J.; Chen, P.-Y.; Tzeng, Y.-R.; Yu, Y.-Y.; Rwei, S.-P.; Kuo, C.-C. Synthesis of Water Resistance and Moisture-Permeable Nanofiber Using Sodium Alginate–Functionalized Waterborne Polyurethane. *Polymers* **2020**, *12*, 2882. [CrossRef]
16. Benas, J.-S.; Veeramuthu, L.; Chuang, Y.-Y.; Chuang, S.-Y.; Liang, F.-C.; Cho, C.-J.; Lee, W.-Y.; Yan, Y.; Zhou, Y.; Kuo, C.-C. Eco-friendly collagen-based bio-organic field effect transistor with improved memory characteristics. *Org. Electron.* **2020**, *86*, 105925. [CrossRef]
17. Chen, P.-Y.; Hsu, C.; Venkatesan, M.; Tseng, Y.-L.; Cho, C.-J.; Han, S.-T.; Zhou, Y.; Chiang, W.-H.; Kuo, C.-C. Enhanced electrical and thermal properties of semi-conductive PANI-CNCs with surface modified CNCs. *RSC Adv.* **2021**, *11*, 11444–11456. [CrossRef]
18. Motloung, M.P.; Ojijo, V.; Bandyopadhyay, J.; Ray, S.S. Cellulose Nanostructure-Based Biodegradable Nanocomposite Foams: A Brief Overview on the Recent Advancements and Perspectives. *Polymers* **2019**, *11*, 1270. [CrossRef]
19. Nanda, S.; Patra, B.R.; Patel, R.; Bakos, J.; Dalai, A.K. Innovations in applications and prospects of bioplastics and biopolymers: A review. *Environ. Chem. Lett.* **2021**, *20*, 379–395. [CrossRef]
20. Rai, P.; Mehrotra, S.; Priya, S.; Gnansounou, E.; Sharma, S.K. Recent advances in the sustainable design and applications of biodegradable polymers. *Bioresour. Technol.* **2021**, *325*, 124739. [CrossRef]

21. Abe, M.; Branciforti, M.; Brienzo, M. Biodegradation of Hemicellulose-Cellulose-Starch-Based Bioplastics and Microbial Polyesters. *Recycling* **2021**, *6*, 22. [CrossRef]
22. Arvanitoyannis, I.; Biliaderis, C.; Ogawa, H.; Kawasaki, N. Biodegradable films made from low-density polyethylene (LDPE), rice starch and potato starch for food packaging applications: Part *Carbohydr. Polym.* **1998**, *36*, 89–104. [CrossRef]
23. Arvanitoyannis, I.; Nakayama, A.; Aiba, S.-I. Edible films made from hydroxypropyl starch and gelatin and plasticized by polyols and water. *Carbohydr. Polym.* **1998**, *36*, 105–119. [CrossRef]
24. Bilck, A.P.; Grossmann, M.V.; Yamashita, F. Biodegradable mulch films for strawberry production. *Polym. Test.* **2010**, *29*, 471–476. [CrossRef]
25. Huneault, M.A.; Li, H. Morphology and properties of compatibilized polylactide/thermoplastic starch blends. *Polymer* **2007**, *48*, 270–280. [CrossRef]
26. Lourdin, D.; Della Valle, G.; Colonna, P. Influence of amylose content on starch films and foams. *Carbohydr. Polym.* **1995**, *27*, 261–270. [CrossRef]
27. Phothisarattana, D.; Wongphan, P.; Promhuad, K.; Promsorn, J.; Harnkarnsujarit, N. Biodegradable Poly(Butylene Adipate-Co-Terephthalate) and Thermoplastic Starch-Blended TiO_2 Nanocomposite Blown Films as Functional Active Packaging of Fresh Fruit. *Polymers* **2021**, *13*, 4192. [CrossRef]
28. Raquez, J.-M.; Nabar, Y.; Narayan, R.; Dubois, P. In situ compatibilization of maleated thermoplastic starch/polyester melt-blends by reactive extrusion. *Polym. Eng. Sci.* **2008**, *48*, 1747–1754. [CrossRef]
29. Ekielski, A.; Żelaziński, T.; Mishra, P.K.; Skudlarski, J. Properties of Biocomposites Produced with Thermoplastic Starch and Digestate: Physicochemical and Mechanical Characteristics. *Materials* **2021**, *14*, 6092. [CrossRef]
30. Gutiérrez, T.J.; Álvarez, V. Bionanocomposite films developed from corn starch and natural and modified nano-clays with or without added blueberry extract. *Food Hydrocoll.* **2018**, *77*, 407–420. [CrossRef]
31. Haurat, M.; Dumon, M. Amorphous Polymers' Foaming and Blends with Organic Foaming-Aid Structured Additives in Supercritical CO_2, a Way to Fabricate Porous Polymers from Macro to Nano Porosities in Batch or Continuous Processes. *Molecules* **2020**, *25*, 5320. [CrossRef] [PubMed]
32. Song, J.; Mi, J.; Zhou, H.; Wang, X.; Zhang, Y. Chain extension of poly (butylene adipate-co-terephthalate) and its microcellular foaming behaviors. *Polym. Degrad. Stab.* **2018**, *157*, 143–152. [CrossRef]
33. Bhatnagar, S.; Hanna, M.A. Starch-based plastic foams from various starch sources. *Cereal Chemistry* **1996**, *73*, 601–604.
34. Chen, L.; Gordon, S.H.; Imam, S.H. Starch Graft Poly(methyl acrylate) Loose-Fill Foam: Preparation, Properties and Degradation. *Biomacromolecules* **2004**, *5*, 238–244. [CrossRef]
35. Diyana, Z.; Jumaidin, R.; Selamat, M.; Ghazali, I.; Julmohammad, N.; Huda, N.; Ilyas, R. Physical Properties of Thermoplastic Starch Derived from Natural Resources and Its Blends: A Review. *Polymers* **2021**, *13*, 1396. [CrossRef]
36. Nabar, Y.; Raquez, J.M.; Dubois, A.P.; Narayan, R. Production of Starch Foams by Twin-Screw Extrusion: Effect of Maleated Poly(butylene adipate-*co*-terephthalate) as a Compatibilizer. *Biomacromolecules* **2005**, *6*, 807–817. [CrossRef] [PubMed]
37. Thanpitcha, T.; Kritchayanon, N.; Pentrakoon, D.; Pimpan, V. An Initial Study of Starch-g-polystyrene Foam Prepared by a Steaming Process. *J. Met. Mater. Miner.* **2003**, *12*, 1–6.
38. Willett, J.; Shogren, R. Processing and properties of extruded starch/polymer foams. *Polymer* **2002**, *43*, 5935–5947. [CrossRef]
39. Tatarka, P.D.; Cunningham, R.L. Properties of protective loose-fill foams. *J. Appl. Polym. Sci.* **1998**, *67*, 1157–1176. [CrossRef]
40. Fazeli, M.; Florez, J.P.; Simão, R.A. Improvement in adhesion of cellulose fibers to the thermoplastic starch matrix by plasma treatment modification. *Compos. Part B Eng.* **2019**, *163*, 207–216. [CrossRef]
41. Nazrin, A.; Sapuan, S.M.; Zuhri, M.Y.M.; Ilyas, R.; Syafiq, R.; Sherwani, S.F.K. Nanocellulose Reinforced Thermoplastic Starch (TPS), Polylactic Acid (PLA), and Polybutylene Succinate (PBS) for Food Packaging Applications. *Front. Chem.* **2020**, *8*, 213. [CrossRef] [PubMed]
42. Boonprasertpoh, A.; Pentrakoon, D.; Junkasem, J. Effect of PBAT on physical, morphological, and mechanical properties of PBS/PBAT foam. *Cell. Polym.* **2019**, *39*, 31–41. [CrossRef]
43. Fourati, Y.; Tarrés, Q.; Mutjé, P.; Boufi, S. PBAT/thermoplastic starch blends: Effect of compatibilizers on the rheological, mechanical and morphological properties. *Carbohydr. Polym.* **2018**, *199*, 51–57. [CrossRef]
44. Shi, X.; Qin, J.; Wang, L.; Ren, L.; Rong, F.; Li, D.; Wang, R.; Zhang, G. Introduction of stereocomplex crystallites of PLA for the solid and microcellular poly(lactide)/poly(butylene adipate-co-terephthalate) blends. *RSC Adv.* **2018**, *8*, 11850–11861. [CrossRef]
45. Tian, H.-L.; Wang, Z.-P.; Jia, S.-L.; Pan, H.-W.; Han, L.-J.; Bian, J.-J.; Li, Y.; Yang, H.-L.; Zhang, H.-L. Biodegradable Foaming Material of Poly(butylene adipate-co-terephthalate) (PBAT)/Poly(propylene carbonate) (PPC). *Chin. J. Polym. Sci.* **2021**, *40*, 208–219. [CrossRef]
46. Lackner, M.; Ivanič, F.; Kováčová, M.; Chodák, I. Mechanical properties and structure of mixtures of poly(butylene-adipate-co-terephthalate) (PBAT) with thermoplastic starch (TPS). *Int. J. Biobased Plast.* **2021**, *3*, 126–138. [CrossRef]
47. Masina, N.; Choonara, Y.E.; Kumar, P.; du Toit, L.C.; Govender, M.; Indermun, S.; Pillay, V. A review of the chemical modification techniques of starch. *Carbohydr. Polym.* **2017**, *157*, 1226–1236. [CrossRef]

48. Dehouche, N.; Idres, C.; Kaci, M.; Zembouai, I.; Bruzaud, S. Effects of various surface treatments on Aloe Vera fibers used as reinforcement in poly(3-hydroxybutyrate-co-3-hydroxyhexanoate) (PHBHHx) biocomposites. *Polym. Degrad. Stab.* **2020**, *175*, 109131. [CrossRef]
49. Petzold, K.; Einfeldt, L.; Günther, W.; Stein, A.; Klemm, D. Regioselective Functionalization of Starch: Synthesis and 1H NMR Characterization of 6-O-Silyl Ethers. *Biomacromolecules* **2001**, *2*, 965–969. [CrossRef]
50. Pothan, L.A.; Simon, F.; Spange, A.S.; Thomas, S. XPS Studies of Chemically Modified Banana Fibers. *Biomacromolecules* **2006**, *7*, 892–898. [CrossRef]

polymers

MDPI

Article

Functionalization of Polymer Surface with Antimicrobial Microcapsules

Iva Rezić [1,*], Maja Somogyi Škoc [2], Mislav Majdak [1], Slaven Jurić [3], Katarina Sopko Stracenski [3] and Marko Vinceković [3]

[1] Department of Applied Chemistry, Faculty of Textile Technology, University of Zagreb, 10000 Zagreb, Croatia; mislav.majdak@ttf.hr
[2] Department of Materials, Fibers and Textile Testing, Faculty of Textile Technology, University of Zagreb, 10000 Zagreb, Croatia; maja.somogyi@ttf.hr
[3] Faculty of Agriculture, University of Zagreb, 10000 Zagreb, Croatia; sjuric@agr.hr (S.J.); ksstracenski@agr.hr (K.S.S.); mvincekovic@agr.hr (M.V.)
* Correspondence: iva_rezic@net.hr

Abstract: The development of antimicrobial polymers is a priority for engineers fighting microbial resistant strains. Silver ions and silver nanoparticles can assist in enhancing the antimicrobial properties of microcapsules that release such substances in time which prolongs the efficiency of antimicrobial effects. Therefore, this study aimed to functionalize different polymer surfaces with antimicrobial core/shell microcapsules. Microcapsules were made of sodium alginate in shell and filled with antimicrobial silver in their core prior to application on the surface of polymer materials by dip-coating methodology. Characterization of polymers after functionalization was performed by several spectroscopic and microscopic techniques. After the characterization of polymers before and after the functionalization, the release of the active substances was monitored in time. The obtained test results can help with the calculation on the minimal concentration of antimicrobial silver that is encapsulated to achieve the desired amounts of release over time.

Keywords: microcapsules; dip coating; encapsulation; spectroscopy; microscopy; antibacterial silver

Citation: Rezić, I.; Somogyi Škoc, M.; Majdak, M.; Jurić, S.; Stracenski, K.S.; Vinceković, M. Functionalization of Polymer Surface with Antimicrobial Microcapsules. *Polymers* **2022**, *14*, 1961. https://doi.org/10.3390/polym14101961

Academic Editor: Andrzej Puszka

Received: 24 March 2022
Accepted: 5 May 2022
Published: 11 May 2022

1. Introduction

Microorganism strains that are resistant to a wide range of antibiotic-based penicillin drugs are known as the members of the "*super-bugs*" group. Conventional drugs that are often used cannot harm them, so they (e.g., methicillin, dicloxacillin, nafcillin or oxacillin) are not efficient against special strains of *Staphylococcus aureus*. The problems with antimicrobial resistance become even more important during the COVID-19 pandemic, since more people were hospitalized and suffered from compromised immune systems. The size of the nanoparticles in the antimicrobial coatings is much smaller than the size of the bacteria which is crucial for effective penetration through the cell wall and causes undesirable effects on microorganisms [1,2].

Silver is a powerful antimicrobial reagent [3–6]. During polymer surface modification, different techniques are used for characterization [7–13]. The most frequently used methods are different varieties of spectroscopic techniques [14–18]. In addition, liquid or ion chromatography helps with efficient sample preparation and the analysis of particular metal ions [19–23]. However, for complex samples such as encapsulated nanoparticles, the combination of different methodologies is needed.

Although the process of microencapsulation technology originates from the 1940s, during the last 10 years, encapsulated products have been extensively used in many industries, such as the agricultural, food, cosmetic and textile industries. In medical applications in particular, microencapsulated materials are showing an increase in the market since they can offer the prolonged release of the active compounds [3]. The encapsulation process is a technique in which the small particles are surrounded by the desired coating, resulting in a typical diameter of around 2 to 2000 µm. The average thickness of the walls is 0.5–150 µm and the core can contain 20 to 95% of the total weight. The choice of the methodological encapsulation technique depends on the desired purpose of the microcapsules.

There are many advantages of microencapsulation, and the most prominent are: the protection of the inner active species, controlled release, practical handling and targeted delivery. There are different varieties of encapsulation procedures, such are physical processes (air suspension coating, coextrusion with submerged or stationary nozzle, rotating disk atomization, spray drying or spray cooling), but also a variety of chemical encapsulation processes (matrix polymerization, liposome technology, simple or complex coacervation, solvent extraction, evaporation of solvents and others. Encapsulation by coacervation is a technique that involves production of core–shell microcapsules by using the interaction of oppositely charged polyelectrolytes in aqueous form. It is characterized by high payload and high encapsulation efficiency (higher than 99%), relatively low costs, and ability to use ambient room temperature and pressures [24]. Coacervation is a process in which an electrostatic interaction between two aqueous media occurs. There are many different types of hydrophilic polymers such as sodium alginate, gelatin, and chitosan that are usually used for the creation of polymeric micro-or nano-capsules [25].

Coacervation includes phase-separation for creation of two separated liquid phases. The first of them consists of a polymer-rich phase (coacervate) and the second one of polymer-depleted phase (equilibrium solution). Therefore, the process of coacervation can be used for both polar and non-polar species. In the case of more complex coacervation systems, an active ingredient can be encapsulated by the combination of two species, such as gelatin and gum arabic or sodium alginate and chitosan [26]. Chitosan is insoluble in alkaline pH medium, but it precipitates in contact with an alkaline solution. During this reaction, the particles are produced by adding the chitosan into an alkali solution using a compressed air nozzle. Lastly, the separation of encapsulated particles is achieved by filtration [27].

Coacervation is one of the oldest and most widely used techniques of microencapsulation. Based on the process, it can be divided into simple coacervation (using colloidal gelatin or chitosan, and complex coacervation in which the polymeric solution is prepared from two oppositely charged colloids, such as gelatin with gum arabic or chitosan (Table 1) [28].

Table 1. Shell and core materials used in simple and complex coacervation for textile applications.

	Simple Coacervation	
Polymer	**Coacervation Agent**	**Active Substance**
Chitosan	NaOH	Antimicrobial silver, essential oil
Chitosan	Sodium dodecyl sulphate	Antimicrobial silver, linseed oil
Chitosan	Sodium tripolyphosphate	Antimicrobial silver, honey, vitamins
	Complex coacervation	
A Polymer	**B Polymer**	**Active substance**
Gum arabic	Gelatin	Antimicrobial silver ions
Gum arabic	Chitosan	Antimicrobial silver, oils, perfumes, dyes
Chitosan	Gelatin	Antibacterial silver, essential oils

After creating antimicrobial microcapsules filled with silver active substances, such microcapsules need to be fixed at the surface of the polymer materials [28–31]. Dip-coating immersion is a process that is ideally fitted for such purpose since in it the substrate is immersed in a liquid and then extracted at a certain speed at a controlled temperature and atmospheric conditions. Antimicrobial components are present in the liquid form [32–34]. Recent focus is on the application of gold or silver nanoparticles [35–37] and for the characterization of such materials particular methodology and combination of spectroscopic and microscopic techniques is needed [38–40]. The process has three stages, which include immersion of the substrate, the forming of a wet layer and drying of the coating by solvent evaporation. Immersion is very slow, allowing the coating to be oriented into a more favorable, denser structure. Evaporation of the solvent can cause destabilization if it is not performed under favorable conditions. During this process the cross-linking occurs, in which the coating becomes stiffer and the resulting gel is glued by heat, while the coagulation temperature itself depends on the composition [41–43]. Many different parameters influencing the outcome and the thickness of the film during immersion exist; these are the viscosity of salts, surface tension, and pressure. By optimizing these parameters, the efficient adhesion of microcapsules is fixed to the surface of the polymer. Moreover, the antimicrobial effects are enhanced with silver nanoparticles [44–46]. Colloidal nanoparticles have many benefits [47–50], and their characterization is achieved with sophisticated methodology [51,52].

The antimicrobial core of microcapsules contains silver. Interesting work from Pandey et al. [6] has shown that the method of stabilization of silver nanoparticles strongly influences antimicrobial properties. In their research, the silver nanoparticles (AgNPs) were even more toxic to bacteria than to other microorganisms like fungi [6]. Thus, the stabilization of silver nanoparticles has broad antibacterial applications [7–11]. Encapsulated active species need to be glued to the surface of the polymers, and this is achieved by dip-coating methodology which uses the sol–gel mechanism of binding microcapsules to the textile carrier [53–56].

The sol-gel is one of the oldest procedures used on thin films. It results in a uniform coating that is very suitable for antimicrobial protection [57]. Due to their properties, microcapsules can be used in various branches of industry [58]. It is therefore not surprising that several studies are being conducted aimed at developing formulations intended for healthcare, especially as an integral part of textiles [59,60]. This work was focused on the functionalization of polymer surfaces by microcapsules filled with antimicrobial silver. The novelty and significance of this work included functionalization of the polymer surface not directly with nanoparticles, as is conventionally performed, but with shell/core microcapsules filled with antimicrobial silver compounds that enable release of antimicrobial silver in time and by this enhances the efficiency of the antimicrobial properties of the functionalized polymer. More precisely, this work is in optimization of the microencapsulation process through two steps of simple and complex coacervation for further effective functionalization of polymers by dip-coating methodology.

The main hypotheses of our work are the following: (i) antimicrobial microcapsules can be prepared with antimicrobial silver in different sizes and chemical compositions; (ii) the microcapsules can be easily applied on the surface of the polymers by dip-coating methodology; and (iii) novel materials with microcapsules release Ag in time, so they can be used as medical materials.

2. Materials and Methods

2.1. Materials and Instrumentation

In this investigation, the following reagents of highest purity were used: the dithizone, 1,5-diphenylthiocarbazone ($C_{13}H_{12}N_4S$) manufactured by Merck (Darmstadt, Germany), Ethanol 96% (C_2H_5OH) manufactured by Gram-Mol (Zagreb, Croatia), silver nitrate ($AgNO_3$) manufactured by Gram-Mol (Croatia), zinc sulfate 7-hydrate ($ZnSO_4 \times 7H_2O$) manufacturer Gram-Mol (Zagreb, Croatia), 3-glycidyloxypropyltrimethoxysilane (GLYMO, 98%) manufactured by Aldrich Chemicals, sodium alginate ($C_6H_9NaO_7$) manufacturer Sigma-Aldrich for testing.

Buchi Encapsulator B-390 manufacturer Buchi Labortechnik AG (Flawil, Switzerland) was used for encapsulation. Heidolph Unimax 1010, manufactured by Heidolph Instruments (Schwabach, Germany), was used to mix the aqueous medium with the microcapsules. A two-beam UV/VIS Lambda 20 spectrometer manufactured by Perkin Elmer (Waltham, MA, USA) was used to perform spectroscopic analysis. During the measurements, 10 mm quartz cuvettes manufactured by Perkin Elmer (Waltham, MA, USA) were used.

The samples after microcapsulation were tested using high-resolution optical microscopy (Tescan Vega company, Brno, Czech Republic) and afterwards a scanning electronic microscope "TESCAN VEGA TS5136LS" with an EDS detector. A microphotograph obtained from SEM microphotograph of encapsulated antimicrobial silver, as well as EXD analysis of the samples core and shell parts, are presented in the Results section.

2.2. Microencapsulation of Antimicrobial Silver

In this work, the encapsulation of the silver-filled microcapsule samples was performed by simple coacervation, within the sodium alginate samples that contained zinc sulfate 7-hydrate in its outer shell. Microcapsulation is a process that involves the separation of a macromolecular solution into two immiscible liquid phases: the first is a denser coacervate phase and the second one is a diluted equilibrium solution. Four basic steps are carried out under stirring and are presented in Figure 1 including: (1) dispersion of the active substance into a solution of a surface-active colloid; (2) precipitation of the colloid onto the dispersed droplets; (3) addition of a second colloid to induce the polymer–polymer complex coacervation; and (4) stabilization by adding a cross-linking agent chitosan (Figure 1) [28].

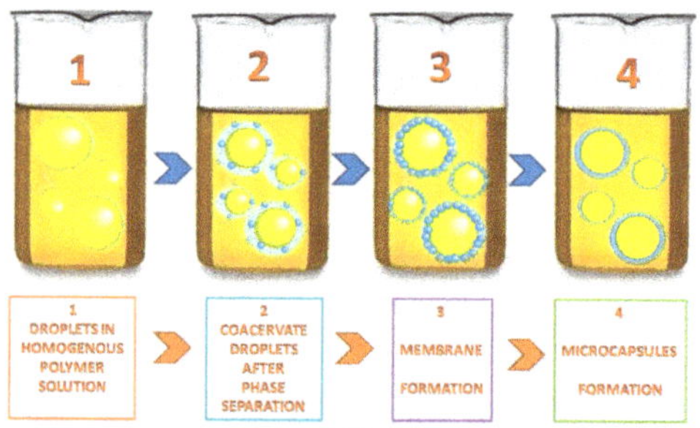

(A)

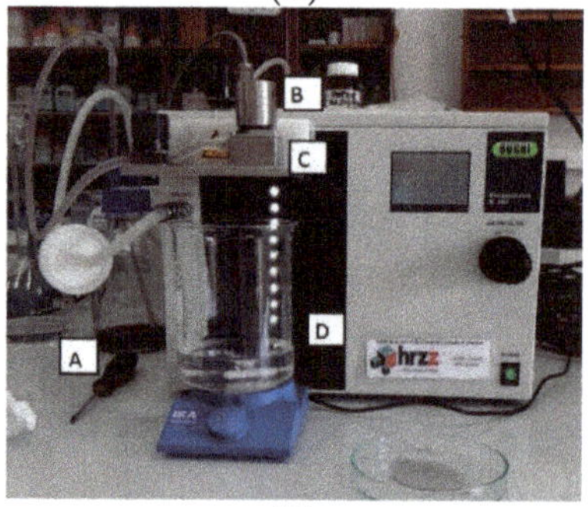

(B)

Figure 1. (**A**) Steps in encapsulation by coacervation prior the functionalization, (**B**) encapsulation apparature Buchi Encapsulator B-390 manufacturer Buchi Labortechnik AG (Flawil, Switzerland) (A—500 mL bottle with encapsulating agent; B—magnetic vibrator; C—heater; nozzle at the bottom; D—solidifying agent).

In this work, the encapsulation of the silver-filled microcapsule samples was performed by reaction in which polymer anions react with polyvalent cations, whereupon an outer shell is formed. The process is most commonly done with the aid of external ionotropic method who's steps are presented in Figure 1, including: (1) polymer droplets containing an active ingredient are immersed in a homogenous solution containing the crosslinker; (2) crosslinker interacts with the polymer; (3) formation of the outer shell [28]. Encapsulation was performed under the following conditions: pressure 530 mbar, amplitude 5, frequency 2000 Hz and heating 50 °C. During the encapsulation, the occurring droplets of the solution were falling into the zinc sulfate 7-hydrate solution. After completion of encapsulation, the resulting microcapsules were stirred for 60 min, filtered and washed with distilled water. The resulting microcapsules had a core filled with silver ions, and a shell composed of sodium alginate and zinc sulfate 7-hydrate. The encapsulation procedure was performed using a Buchi Encapsulator B-390. To perform the encapsulation, it was first necessary to prepare a solution of sodium alginate with a concentration of 1.5%, and a solution of zinc sulfate 7-hydrate with a concentration of 1M. A 1.5% sodium alginate solution was prepared by dissolving 7.50 g of sodium alginate in a 500 mL volumetric flask with distilled water.

Zinc sulphate 7-hydrate, on the other hand, was prepared by weighing 143.77 g of zinc sulphate 7-hydrate into a 500 mL volumetric flask, then adding water to the mark after which the contents of the volumetric flask had to be shaken well. After the sodium alginate was dissolved, a solution of sodium alginate and $AgNO_3$ was prepared. For this purpose, $AgNO_3$ weighing 0.0332 g was dissolved in a solution of sodium alginate with a volume of 100 mL, with stirring on a magnetic stirrer. The dissolution process was performed for 30 min. After dissolution, the solution had to be filtered using vacuum filtration. Prior to the encapsulation procedure, 100 mL of zinc sulfate 7-hydrate solution was poured into a 600 mL beaker. Zinc sulfate 7-hydrate served as a solidifying agent in the encapsulation process. In a 500 mL bottle, 100 mL of the filtered solution was poured, which, thanks to nitrogen pressure, passed through a silicone tube all the way to a 20 μm diameter nozzle. When passing all the way to the nozzle, a silicone tube passes by a heater that heats the solution, thus preventing the solution from clotting in the nozzle. At the same time, a magnetic vibrator was used to ensure the formation of as many microcapsules as possible. The encapsulator with the indicated parts is shown in Figure 1.

During encapsulation, droplets of the solution fall into the solidification solution, i.e., zinc sulfate 7-hydrate solution. After completion of encapsulation, the resulting microcapsules were stirred for 60 min to ensure solidification. After solidification, the microcapsules were filtered and washed with distilled water. The resulting microcapsules had a core composed of $AgNO_3$, and a shell with zinc sulfate 7-hydrate. The microcapsules were purple, but staining caused by the presence of $AgNO_3$ occurred.

After 30 min, the spheres were filtered, rinsed and stored in a cool and dark place prior to the characterization. Part of the microcapsules was dispersed in chitosan solution (0.5% CS in 1.0% CH_3COOH) under constant stirring (magnetic stirrer). The contact time between microspheres and chitosan solution was about 30 min to give chitosan time to form a layer around the microspheres and formation of microcapsules. Microcapsules were filtered, rinsed and stored in a cool and dark place prior to the characterization. A number of the microcapsules were allowed to air-dry at room temperature to reach their equilibrium moisture content.

2.3. *Monitoring of the Release of Silver Ions*

The active substance is released from the microcapsule by diffusion, rupture or dissolution of the coating. In this case the active substance, Ag^+ cations, is released by diffusion. To perform diffusion, it was first necessary to weigh, into a 100 mL beaker, 1 ± 0.0001 g of microcapsules. Random sampling was used to select microcapsules. After the weighing, 20 mL of distilled water was added via a pipette. A total of 23 aqueous solutions prepared in this way were used for testing. Afterwards, the solutions were shaken. The Unimax 1010 (manufacturer Heidolph instruments, Schwabach Germany) instrument was used for shaking. The shaking conditions were: default shaking speed 4 (the instrument has 10 possible speeds); shaking time 60 min; no heating. After 60 min, the solutions were filtered to separate the microcapsules, as microcapsules can disrupt spectrometric analysis.

2.4. *UV-VIS Characterization*

To monitor the course of the analysis, and to determine the concentration of Ag^+ cation, it is necessary to make a calibration diagram, i.e., the calibration curve. To perform quantitative and qualitative analysis, a two-beam UV/VIS Lambda 20 spectrometer was used. The test was performed in the UV and visible range, i.e., in the wavelength range 190 to 900 nm. Since the test was performed in the UV range, cuvettes made of quartz were used.

Before the test, the dithizone solution had to be diluted 1:9 with distilled water in a 10 mL volumetric flask. At the same time, the microcapsules contain zinc sulfate 7-hydrate, so it is possible that Zn interferes with the determination of silver. In order to reduce interference, i.e., to better detect the signal from the Ag^+ cation, the standard addition method was used. The standard addition method involves adding a standard solution of known concentration to a volumetric flask containing the test solution. In order to perform the UV-VIS spectroscopical investigation of colorless silver solution, first it was necessary to create a colored complex with dithizone (diphenylthicocarbazone), Sodium diethyldithiocarbamate, 4-(2-Pyridylazo) resorcinol monosodium salt, and Methyl N-[α-(8-hydroxy-7-quinolyl) benzyl] anthranilate. Dithizone, a complex compound that can be dissolved in carbon tetrachloride, chloroform, and 96% ethanol, was used for testing [61]. To avoid the use of toxic chemicals such as carbon tetrachloride and chloroform [62,63], 0.07% dithizone solution was used for testing purposes, which had to be filtered before use to remove residual dithizone residues. In the analysis, it is important to keep in mind that dithizone itself is not selective and reacts with all metals present in the test solution [63]. Before performing the analysis, it was necessary to make a calibration diagram. For the purpose of making the calibration diagram, a titron of AgNO3 with a concentration of 0.1 M was used as a standard solution. The calibration diagram was made in the following concentrations of standard solution $c_1 = 4.444 \times 10^{-3}$ M; $c_2 = 8.888 \times 10^{-3}$ M; $c_3 = 0.0178$ M; $c_4 = 0.0222$ M; $c_5 = 0.0444$ M. Standard solutions were added with dithizone to 25 mL volumetric flasks in a 1:1 volume ratio, achieving an excellent correlation of the calculated absorbances.

2.5. Chromatographic Characterization

The TLC method was developed for the analysis of metal ions that can be extracted from the microcapsules. During preliminary experiments, more than 30 stationary phases were investigated with different mobile phases. Hydrochloric acid and water were added to create optimal solvent mixtures. The chromatographic method used contained ACN, HCl, and H_2O in 60.00:19.17:20.83 *v*/*v*%, respectively.

2.6. Dip Coating of Polymers with Antimicrobial Microcapsules

The dip-coating process was performed in a following manner: firstly the sol was stirred magnetically to obtain optimal molar ratio of GLYMO to water, based on our previous experience to retain textile character (hand value, softness, etc.) [42,47,55]. Important parameters for the sol–gel properties (precursor, types of catalysts and solvent, molar ratio of precursor to solvent, pH value, and temperature) were also based on physical and chemical character of the chosen precursor GLYMO and catalyst, HCl.

Secondly, the solution of distilled water and ethanol was thoroughly mixed, since volatile substances can be removed from sol by evaporation before or during condensation. Then, GLYMO was added and the process was carried out under continuous magnetic stirring until a homogeneous solution was obtained. Thirdly, the samples were cut as to be in 5×5 cm^2 in size in order to go through the dip-coating process on a custom-made apparatus with the predetermined drawing speed of 1 mm/s. Lastly, such modified samples were left to gel at room temperature for 24 h, and were then dried at 100 °C for 1 h. For the antimicrobial treatment effects, silver ions originating from silver nitrate were added to the sol carefully and in small amounts, under higher mixing speed. The process was carried out under continuous magnetic stirring until a homogeneous solution was obtained.

The detailed description of the surface modification by using the sol-gel procedure was optimized in our previous work [47] where we introduced the technology of polymer functionalization by 3-glycidyloxypropyltrimethoxy-silane (GLYMO, Sigma Aldrich, Darmstadt, Germany, Europe). Nanoparticles in microcapsules were left to gel at room temperature and then heated at 100 °C for 60 min. Such modified products were characterized by different instrumental techniques, including thin layer chromatography (TLC), scanning electron microscopy with an EDX detector (SEM-EDX, Tescan Vega, Brno, Czech Republic), and FTIR-ATR (Perkin Elmer, Waltham, MA, USA) spectroscopy [47]. In this work, the results of modification by microcapsules filled with silver compounds will be shown. There are many crucial factors that influence the efficiency of the functionalization process: the speed of immersion, concentration and related viscosity of solutions, the surface tension, and the sample dimensions [43]. Dip coating was applied on both woven and nonwoven polymers. The important working parameter was the molar ratio of GLYMO to water, which was chosen based on the experimental results obtained. We observed that 1:1.5 is the stoichiometric ratio for total condensation of the alkoxide groups and 1:3 is the stoichiometric ratio for total hydrolysis of the alkoxide groups. The optimal molar ration was chosen from the obtained resulting surface: for example, the goal to sustain the homogenous distribution of antibacterial silver on the coating, smooth touch, optimal shine and the most flexible surface coating that resists force without cracking, was associated with optimized system parameters.

The last working parameter that influenced the dip-coating process was the composition of the sample. It was determined in accordance with the standard norm procedure ISO 1833.22. The sample was composed of cellulose fibers. Mass per unit was determined in accordance with the ISO 2286-2 and found to be 186.7 g m^2. No preparatory process on the textile substrate before dip-coating was performed (scouring, bleaching, or others).

2.7. FTIR Spectroscopical Characterization

FTIR characterization of polymers functionalized with antimicrobial microcapsules, was done by Fourier transform infrared spectrometer (Spectrum 100 FTIR, Perkin Elmer, Waltham, MA, USA), which applies KBr and Attenuated Total Reflectance (ATR) techniques. The spectra of samples were recorded in a frequency range from 400 to 4000 cm^{-1} in diffuse reflectance mode, at a resolution of 4 cm^{-1}. This instrument with this particular configuration enables the recording of spectra of the sample solid-state.

3. Results and Discussion

The results of this investigation are a step forward in finding the solutions against powerful resistant microorganisms. Microencapsulation is a fascinating technique that enables slow release of antimicrobial agents, can be applied on different materials without affecting the existing polymer, and has a valid shelf life. The focus of this investigation was on antibacterial silver ion that was the core compound of all prepared microcapsules. This material was chosen since it is known for its strong antimicrobial effects against a wide range of Gram-positive and Gram-negative bacteria, including antibiotic-resistant strains. Encapsulation methods enhance the applications of antimicrobial silver. The availability of microcapsules to slowly release silver has enabled potential storage, stabilization, and application in many different products. In this work, microcapsules were prepared as antimicrobial materials for medical applications (for example, on bandages, wound materials, sheets in hospitals, and others. The antimicrobial silver was chosen due to the reported literature references, and based on our testing. We reported the results of the antimicrobial investigation of the silver nanoparticles in our previous investigations [42,47], in which we determined the antimicrobial activity on *S. aureus* model microorganisms.

3.1. Characterization of Microcapsules with Antimicrobial Silver

Figure 2 shows the microcapsules prepared by simple coacervation process, filled with silver ions in the core, and the alginate shell with zinc sulfate 7 hydrate in the shell.

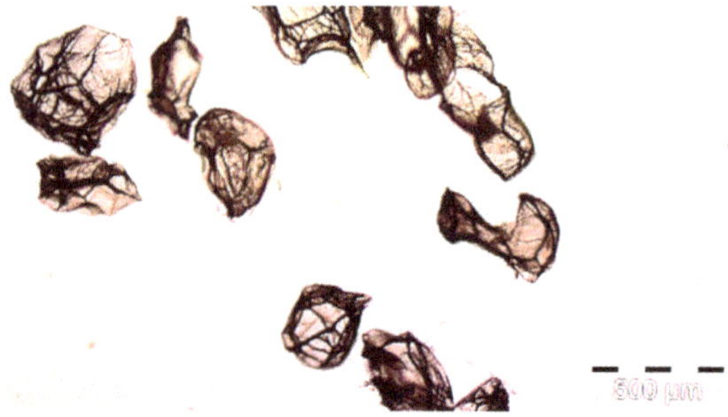

Figure 2. Microphotograph of prepared microcapsules filled with silver ion of 450 µm in diameter, prior functionalization of polymers recorded by optical microscopy with high resolution.

Some microcapsules were also encapsulated by additional adding of chitosan, during complete coacervation process. No other physical, chemical or morphological property of the antimicrobial silver filling of the microcapsules was varied during their production. Particularly, the focus was not to change any step of modification and functionalization on the polymer surface. To conclude, there was no difference among the final filling product of microcapsules.

After recording microphotographs of microcapsules after encapsulation using high-resolution optical microscopy, a scanning electron microscope was used in order to determine the dimensions of pores and shells of particular microcapsules after their lyophilisation. The combination of both techniques has been widely used for the characterization of microcapsules. The SEM results are presented in Figure 3A. As can be seen from this Figure, the average diameter of microcapsules was in the range of 140 to 285 µm in diameter. When comparing those results to the results presented in Figure 2, it can be concluded that the reduction in size of microcapsules after their lyophilization was in range of 37% to 69%. The most important result is the proof that there were no cracks on microcapsules after the lyophilization process. This points to the conclusion that the lyophilization process was adequate, not causing any damage to the shell of microcapsules. Similar results were obtained and discussed by Glaucia et al., 2013 [56], who discussed that these characteristics are important to ensure greater protection and retention of the encapsulated material. In their research the encapsulated sucralose by double emulsion followed by complex coacervation was cross-linked by solid bridges.

The results followed to conclusion that such solid bridges can be attributed to the lyophilization process, which was responsible for clustering of the microcapsules. [56]. Moreover, their results are in agreement to ours in which it was shown that the average particle size was not influenced by varying the shell material. However, Glaucia et al. observed that the concentration of encapsulating agent has influence on the average particle size [56]. In contrast, He et al. [57] did not use lyophilization but spray drying of the microcapsules prior to the SEM investigation.

Their results showed that the particle sizes of spray-dried microcapsules were distributed in much broader range, between 1 and 100 µm, while the most of them were distributed around 10–30 µm [57]. As can be seen from Figure 3A, in which the results of our research obtained by SEM are shown, the determined size of microcapsules was in range of 150 to 400 nm, while the size of the pores among them microcapsules was in range of 50 to 100 nm.

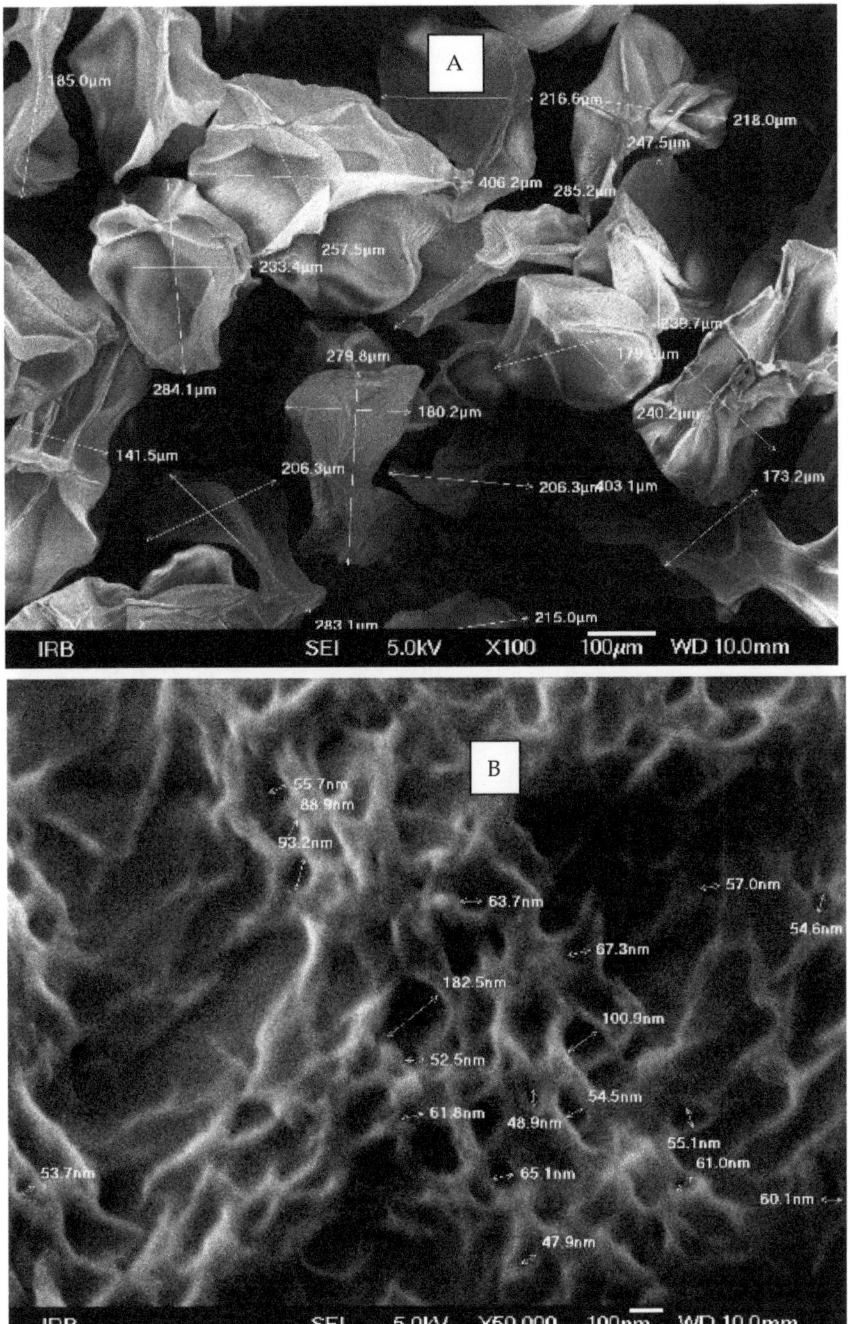

Figure 3. SEM microphotograph of prepared microcapsules filled with silver ions: (**A**) determination of dimensions of the liophilizated microcapsules and (**B**) calculated dimensions of the pores on their surface.

The results of SEM-EDX presented in Figure 4 show that the microcapsules contain zinc and silver. In order to investigate the release of silver from the core of the microcapsules, further investigation by chromatographical screening and quantitative UV-VIS spectroscopy was performed.

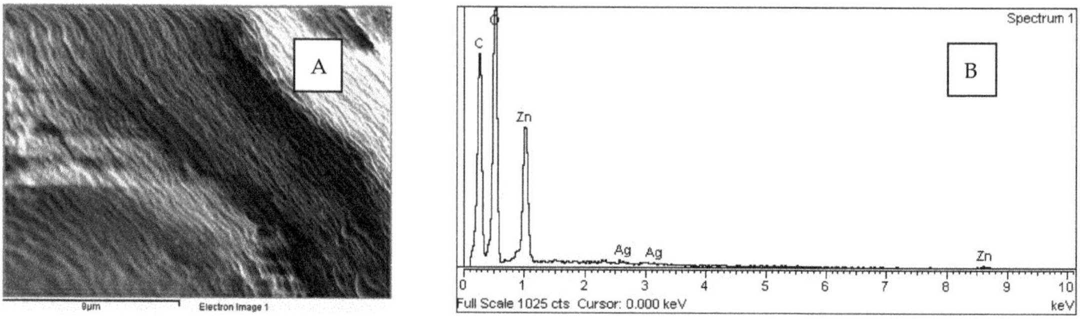

Figure 4. SEM-EDX chemical characterization: (**A**) the surface of the sample that was used for characterization; (**B**) SEM-EDX spectra showing the Ag in core and Zn in shell of antimicrobial microcapsules.

3.2. The Results of Chromatographic Prescreening

Silver from microcapsules were qualitatively identified by using thin-layer chromatography. The chromatographic investigation enables monitoring of silver among other metals, due to its distinguished R_F value. In the chromatographic system that has ACN: HCl: H_2O in 60.00: 19.17: 20.83 $v/v\%$, respectively, silver was detected by different indicators (Table 2).

Table 2. Thin-layer chromatography results of detected silver after using different visualization reagents; development by ACN: HCl:H_2O in 60.00:19.17:20.83 $v/v\%$, on cellulosic precoated plates 20 × 20 cm.

Detection of Silver through Its R_F Value and Characteristical Colors of Spots			
Without Indicator	**Dimethylglioxime**	**Quercetin**	**Alizarin Red S**
+	+	+	+ Zn can be detected as pink spot

3.3. The Results of UV-VIS Characterization

The UV-VIS characterization was performed in a range of 400 to 600 nm. Figure 5A presents the UV-VIS curves obtained during investigation, and Figure 5B the calibration curve calculated based on the results. In order to make the calibration curve, five $AgNO_3$ solutions of known concentrations were used. All samples were prepared in the same manner: firstly 1 mL of dithizone was added to five 25 mL volumetric flasks with $AgNO_3$ solutions. Secondly, the same volume of dithizone as $AgNO_3$ solutions was added to the samples. All absorption values were recorded at a maximal wavelength of 512 nm. The values of the recorded absorptions are shown in Figure 5A,B.

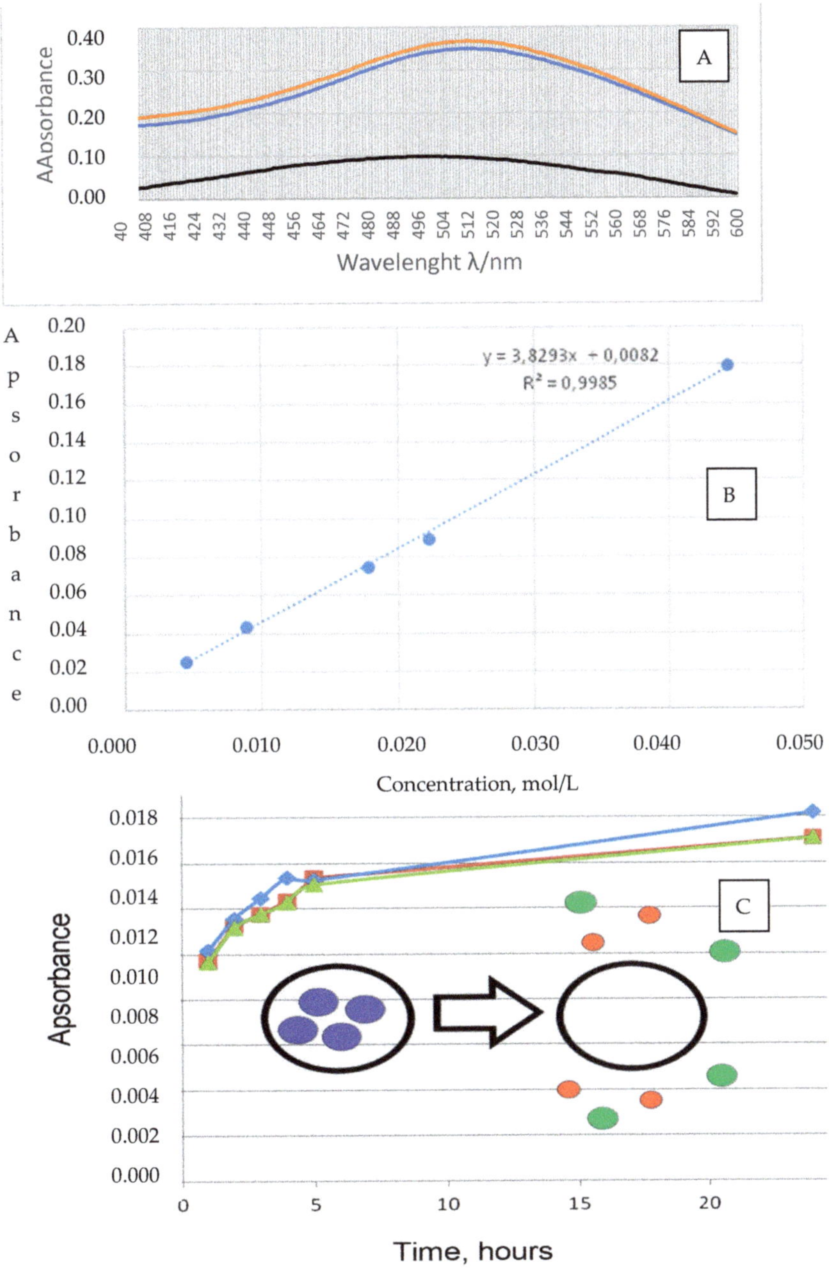

Figure 5. (**A**) UV-VIS curves obtained during measurements (black—silver from microcapsules; blue—standard reference solution with silver ions; and orange—mixture of silver from microcapsules and standard solution, spike solution); (**B**) calibration diagram of silver ion obtained by UV-VIS spectrometry; and (**C**) graphic representation of the silver release over 25 h from microcapsules (blue—450 µm microcapsules without chitosan, red—120 µm microcapsules without chitosan, green 120 µm microcapsules with chitosan). As can be seen in details, the release starts in the first 4 h and then it reaches a meta-stabile phase.

Before the analysis of the release of silver from antimicrobial microcapsules, the samples had to be determined. The calibration curve presented in Figure 5B was used for the determination of release of the silver (Figure 5C). During the diffusion in an aqueous medium, Ag^{+} cations "exit" from the microcapsule into the medium itself. The diffusion procedure itself is shown in Figure 5C, together with the obtained spectral curves. Quantitative analysis of silver in microcapsules of different diameters (120 or 450 μm) showed that samples contained between 100 and 350 μg/g of silver, respectively. Those amounts are valid for protective medical materials.

3.4. The Results of Functionalization of Woven and Non-Woven Polymers by Microcapsules

Table 3 presents the microphotograph of the samples after dip-coating functionalization recorded under the optical microscopy with specified chemical composition (woven and non-woven cotton and viscose polymers).

Table 3. Microphotograph of the samples after dip-coating functionalization (woven and non-woven cotton and viscose polymers).

	Before Modification, without Microcapsule	After Modification with Yellow Microcapsules
Woven and cotton	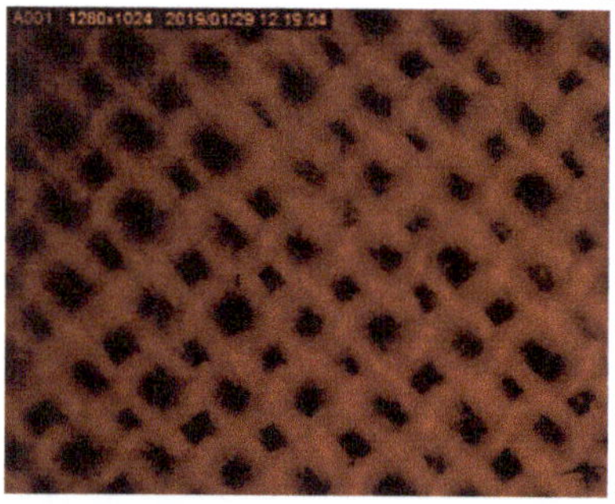	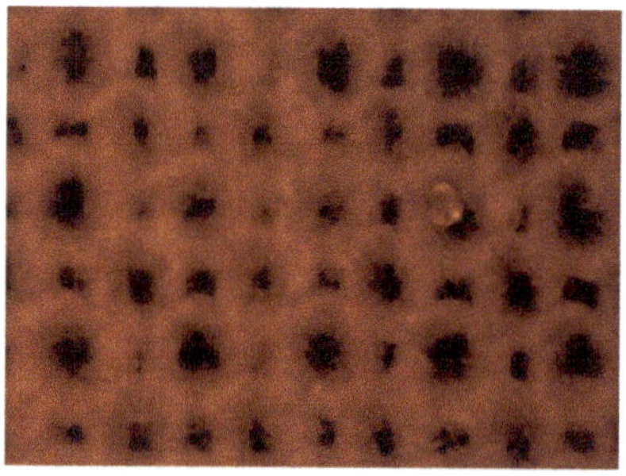
Woven cotton		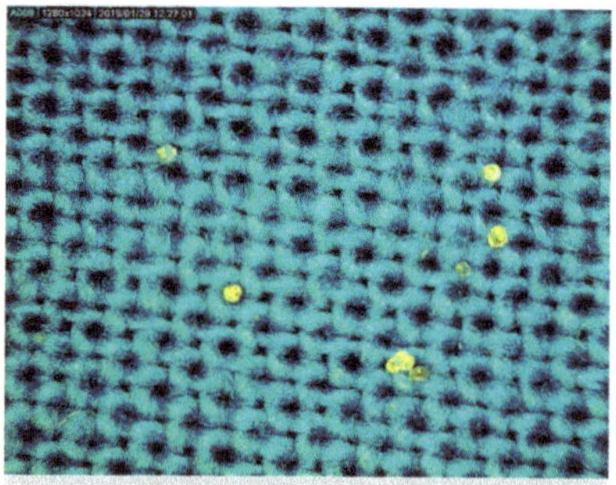
Non woven, viscose	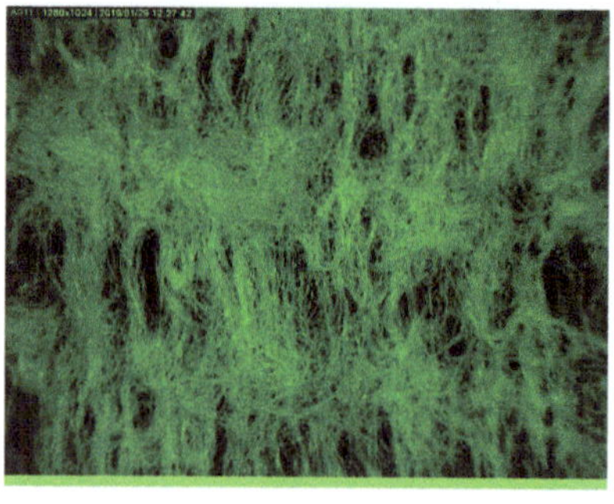	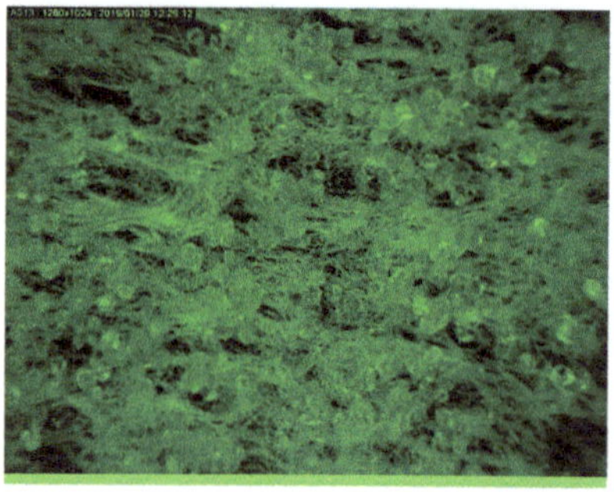

3.5. The Results of Spectroscopical and Microscopical Characterization

For the spectroscopic characterization of the samples, the Fourier-transform infrared spectrometry (Spectrum 100 FTIR, Perkin Elmer, Waltham, Massachusetts) was used, and the result is shown in Figure 6. The newly obtained microcapsules were then impregnated by dip-coating process on 100% viscose nonwoven fabric, and woven viscose or viscose/cellulose polymers. As can be seen from Figure 6, the dip coating process can be monitored through the formation of particular functional groups that are present after the formation in a sol-gel process.

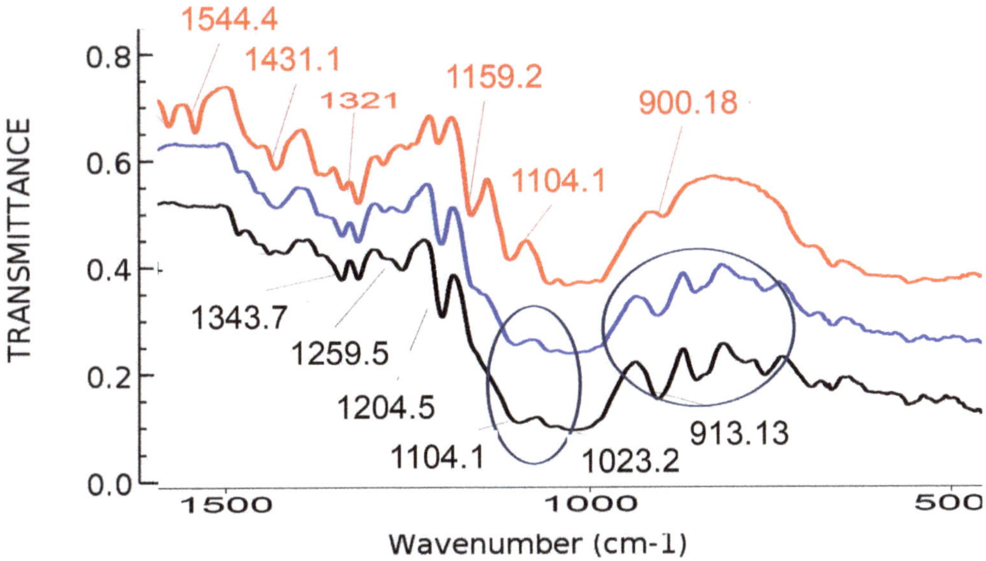

Figure 6. Fourier-transform infrared spectra of the samples before and after the modification: red line—FTIR-Attenuated Total Reflectance (ATR) spectra of cotton woven polymer sample before modification with GLYMO precursor; blue line—FTIR ATR spectra of cellulose sample after modification with organic GLYMO precursor; black line—FTIR ATR spectra of cellulose sample modified with GLYMO precursor and silver. Two blue circles define area of the epoxy groups determined around ~905 and 911 cm^{-1}, and the area around 1100 cm^{-1} is linked to Si–O groups.

ATR-FTIR is a spectroscopic methodology that allows analysis of the microcapsules without complicated sample preparation which can significantly increase the speed of analysis and the number of sample throughput. In determination of the shell thickness of antimicrobial microcapsules, the ATR-FTIR technique has a capability to penetrate in the micrometer range. Therefore, it is generally used to provide information on the functional groups, characterizing the microcapsules' shell properties.

Microcapsules were prepared in this research as a career of antimicrobial agents, particularly antimicrobial silver. In our previous research we have investigated the effects of different metal and metal-oxide nanoparticles, and our results have shown that small metal silver nanoparticles have very strong antimicrobial effects, not only against conventional Gram positive and Gram negative bacteria, but also against fungi. In addition, their application on the surface of polymers that were used in this research have resulted in activity against drug resistant methyl resistant staphylococcus aureus MRSA and methyl sensitive staphylococcus aureus MSSA strains.

In this work, the antibacterial effect of our antimicrobial polymer coated with microcapsules was based on previous research on antimicrobial effects of silver that started after 6h of exposure [42,47]. Therefore, the concentrations used for the research were chosen based on the parameters of the maximal antimicrobial efficiency of silver. The application of antimicrobial silver as potential material for bandages and open wounds was the main focus of this work (Table 1).

The potential strong efficiency of the newly formed materials is a result of functionalization that is shown in figures and tables, but particularly, at the FTIR-ATR spectra of samples after modification. Therefore, the antibacterial microcapsules were efficiently applied to the polymer surface and were filled with antimicrobial silver.

3.6. The Results of Antimicrobial Testing

The results of this investigation have shown that such surfaces might possess an antimicrobial effect due to the presence of silver particles, but this is a matter for future studies.

In our previous papers [42,47] we have reported that silver is a potential strong antimicrobial reagent. Its efficiency is enhanced when it is present in a form of nanoparticles. Such samples in size of 10 nm showed even potential antimicrobial investigation of metal and metal oxide nanoparticles on model microorganism *Staphylococcus aureus* ATCC 29213 by agar-well diffusion assay and reported zone of inhibition was 10 mm for 20 ppm of silver.

In their work, Özyildiz and co-workers [53] sought to develop antimicrobial textiles by using encapsulated ozonated red pepper seed oil in a wrapper made of gelatin and Arabic rubber. Our results correlate with the literature data [53], where the authors concluded that the processed textile materials show good antimicrobial properties against the bacteria *Staphylococcus aureus*, *Candida albicans*, and *Escherichia coli* [53]. In order to have enough data on comparison of microcapsules made by two different approaches—simple and complex coacervation—much more investigation needs to be performed, and this will be the focus of our further research. We have reported the results of our previous work, although there are many different sources that specify different antimicrobial activity based on the chemical and physical characteristics of particular nanoparticles. As is emphasized in this work, much more data should be collected in future work in order to clearly specify that new material is antimicrobial efficient, but for work focused on the preparation and fixation of newly prepared microcapsules on woven and non-woven materials.

In contrast to our research, Ramya and Maheshwari [54] applied microcapsules with a piper betel leaf extracts encapsulated inside the microcapsules. Unlike Özyildiz et al., they used fabrics composed of 50:50% (W/W) of cotton and bamboo blend fibers. At the same time, they used the method of exhaustion to apply microcapsules to textile material. Moreover, for the encapsulation purposes, the authors used spray encapsulation in which calcium chloride was sprayed on the drops of extract and sodium alginate. The authors stated that their result was a textile material with antimicrobial properties effective against bacterial species that are often found on foot wounds: *Staphylococcus*, *Bacillus*, *Klebsiella*, *Pseudomonas*, and *Proteus*. For this purpose, the antimicrobial testing was performed on materials containing piper betel extract, and on materials containing microcapsules with the encapsulated extract. Materials treated with microcapsules showed better results, especially after washing [54].

The antibacterial activity and mechanism of silver on *S. aureus* strain ATCC 6538P were investigated by Li and coauthors [44]. Their results have shown that the minimal bactericidal concentration (MBC) is 20 μg/mL. Those results can be compared to the results of other researchers; such is the paper by Mirzajani and his coworkers [45]. In this paper the Mirzajani et al. showed that 4 μg/mL of silver completely inhibits bacterial growth. This interaction was confirmed by Grigoreva et al. [46]. Shevtsova et al. [58] incorporated antimicrobial silver nanoparticles inside temperature-responsive hybrid nanomaterials based on modified halloysite nanotubes, an approach that offers many significant improvements to the application of antimicrobial materials. Moreover, Raczkowska et al. [59] presented results of "smart" antibacterial surfaces based on silver nanoparticles (AgNPs) embedded in temperature-responsive poly(di(ethylene glycol)methyl ether methacrylate)—(POEGMA188) as well as poly(4-vinylpyridine)—(P4VP) coatings attached to a glass surface. They managed to provoke a temperature-switched killing of the bacteria against *Escherichia coli* ATCC 25922 and *Staphylococcus aureus* ATCC 25923 [59]. Konnova et al. reported a simple procedure for fabrication of 50 nm silver nanoparticles and application of these nanoparticles against yeast or bacteria. They have applied cationic polymer-stabilized nanoparticles that were electrostatically adhered to microbial cells. By such forms, they were producing an even monolayer on the cell walls that resulted in fast delivery of silver nanoparticles into *C. elegans* micro worms. The fabrication of "cyborg" biological cells that contain surfaces functionalized with nanomaterials is today a fascinating new area in cell-surface engineering that is promising for many new applications. Therefore, it can be concluded that the results of our research correspond to the relevant and recent literature data. Moreover, we have shown that silver is one of the most prominent candidates for application on antibacterial medical materials produced by encapsulation.

4. Conclusions

Due to the growing resistance of harmful bacteria to conventional drugs, it is necessary to develop new solutions and new formulations intended to suppress their growth on medical textile materials. Encapsulation was proposed in this research as a fascinating technique that enables effective new ways of partial release of antimicrobial agents. We produced microcapsules of 120 and 450 µm from sodium alginate and zinc sulfate 7-hydrate, filled with silver, and characterized them by different microscopic and spectrometric methodologies. Afterwards the microcapsules were efficiently bonded to the surface of woven and nonwoven polymers (made of viscose and cellulose), for preparation of antimicrobial materials. Lastly, the release of active antimicrobial silver from microcapsules was monitored over 25 h in an aqueous medium in order to predict the antimicrobial potential of microcapsules, as potential filler for medical materials such are bandages and wound dressings. By this, it can be concluded that the main hypothesis of this work were proven: (i) antimicrobial microcapsules were efficiently prepared with antimicrobial silver in different sizes (120 and 450 µm) and chemical compositions (with and without chitosan) (ii) the microcapsules were efficiently applied on the surface of the polymers by dip-coating methodology; and (iii) novel materials with microcapsules released Ag in time so this is the proof that such items can be used as medical materials. Finally, based on the obtained results, the antimicrobial efficiency can be predicted to inhibit the growth of certain strains of microorganisms on open wounds.

Author Contributions: Conceptualization, I.R.; formal analysis, M.M., M.S.Š., S.J. and K.S.S.; methodology, M.S.Š. and I.R.; project administration, I.R.; resources, I.R.; software, I.R.; supervision I.R. and M.V.; writing—original draft, I.R.; writing—review and editing, I.R. and M.V. All authors have read and agreed to the published version of the manuscript.

Funding: This research was funded by the Croatian Science Foundation grant number IP-2019-04-1381 (project under title 'Antibacterial coating for biodegradable medicine materials ABBAMEDICA'). Any opinions, findings and conclusions or recommendations expressed in this material are those of the authors and do not necessarily reflect the views of the Croatian Science Foundation.

Data Availability Statement: Not applicable.

Conflicts of Interest: The authors declare no conflict of interest. The funders had no role in the design of the study; in the collection, analyses, or interpretation of data; in the writing of the manuscript, or in the decision to publish the results.

References

1. World Health Organization. Antimicrobial Resistance Global Report on Surveillance 2014. Available online: https//www.who.int/drugresistance/documents/surveillancereport/en/ (accessed on 7 January 2020).
2. Van Kourtis, A.P.; Hatfield, K.; Baggs, J.; Mu, Y.; See, I.; Epson, E.; Nadle, J.; Kainer, M.A.; Dumyati, G.; Petit, S.; et al. Vital Signs: Epidemiology and Recent Trends in Methicillin-Resistant and in Methicillin-Susceptible *Staphylococcus aureus* Bloodstream Infections—United States. *Morb. Mortal. Wkly. Rep.* **2019**, *68*, 214–219. [CrossRef] [PubMed]
3. Kirin, P. The Analysis of Antimicrobial Surface by Thin Layer Chromatography. Master's Thesis, University of Zagreb, Faculty of Textile Technology, Zagreb, Croatia, 2019.

4. Delaney, J.A.; Schneider-Lindner, V.; Brassard, P.; Suissa, S. Mortality after infection with methicillin-resistant *Staphylococcus aureus* (MRSA) diagnosed in the community. *BMC Med.* **2008**, *6*, 2. [CrossRef] [PubMed]
5. Padmavathy, N.; Vijayaraghavan, R. Enhanced bioactivity of ZnO nanoparticles-an antimicrobial study. *Sci. Technol. Adv. Mater.* **2008**, *9*, 035004. [CrossRef] [PubMed]
6. Pandey, S.; De Klerk, C.; Kim, J.; Kang, M.; Fosso-Kankeu, E. Eco Friendly Approach for Synthesis, Characterization and Biological Activities of Milk Protein Stabilized Silver Nanoparticles. *Polymers* **2020**, *12*, 1418. [CrossRef] [PubMed]
7. Pandey, S.; Ramontja, J. Sodium alginate stabilized silver nanoparticles-silica nanohybrid and their antibacterial characteristics. *Int. J. Biol. Macromol.* **2016**, *93*, 712–723. [CrossRef] [PubMed]
8. Beyth, N.; Houri-Haddad, Y.; Domb, A.; Khan, W.; Hazan, R. Alternative Antimicrobial Approach Nano-Antimicrobial Materials. *Complement. Altern. Med.* **2015**, *2015*, 246012. [CrossRef]
9. Dizaj, S.M.; Lotpoura, F.; Barzegar-Jalali, M.; Zarrintana, M.H.; Adibkiab, K. Antimicrobial activity of the metals and metal oxide nanoparticles. *Mater. Sci. Eng. C* **2014**, *44*, 278–284. [CrossRef]
10. Emami-Karvani, Z.; Chehrazi, P. Antibacterial activity of ZnO nanoparticle on Gram-positive and Gram-negative bacteria. *Afr. J. Microbiol. Res.* **2011**, *5*, 1368–1373.
11. Rasmussen, J.W.; Martinez, E.; Louka, P.; Wingett, D.G. Zinc oxide nanoparticles for selective destruction of tumor cells and potential for drug delivery applications. *Expert Opin. Drug Deliv.* **2010**, *7*, 1063–1077. [CrossRef]
12. Rastin, H.; Ormsby, R.T.; Atkins, G.J.; Losic, D. 3D Bioprinting of Methylcellulose/Gelatin-Methacryloyl (MC/GelMA) Bioink with High Shape Integrity. *ACS Appl. Bio Mater.* **2020**, *3*, 1815–1826. [CrossRef]
13. Peak, C.V.; Singh, K.A.; Adlouni, M.; Chen, J.; Gaharwar, A.K. Printing Therapeutic Proteins in 3D using Nanoengineered Bioink to Control and Direct Cell Migration. *Adv. Health Mater.* **2019**, *8*, e1801553. [CrossRef] [PubMed]
14. Mohammad, J.; Hajipour, M.J.; Fromm, K.M.; Ashkarran, A.A.; Aberasturi, D.J.; Larramendi, I.R.; Rojo, T.; Serpooshan, V.; Parak, W.J.; Mahmoud, M. Erratum: Antibacterial properties of nanoparticles. *Trends Biotech.* **2012**, *30*, 499–511.
15. Raghupathi, K.R.; Koodali, R.T.; Manna, A.C. Size-dependent bacterial growth inhibition and mechanism of antibacterial activity of zinc oxide nanoparticles. *Langmuir* **2011**, *27*, 4020–4028. [CrossRef] [PubMed]
16. Varner, K. *State of the Science Literature Review: Everything Nanosilver and More*; Scientific, Technical, Research, Engineering and Modeling Support Final Report; EPA/600/R-10/084; U.S. Environmental Protection Agency, Office of Research and Development: Washington, DC, USA, 2010; pp. 15–18.
17. Rastin, H.; Zhang, B.; Bi, J.; Hassan, K.; Tung, T.T.; Losic, D. 3D printing of cell-laden electroconductive bioinks for tissue engineering applications. *J. Mater. Chem. B* **2020**, *8*, 5862–5876. [CrossRef] [PubMed]
18. Rastin, H.; Zhang, B.; Mazinani, A.; Hassan, K.; Bi, J.; Tung, T.T.; Losic, D. 3D bioprinting of cell-laden electroconductive MXene nanocomposite bioinks. *Nanoscale* **2020**, *12*, 16069–16080. [CrossRef] [PubMed]
19. Abramova, A.V.; Abramova, V.O.; Bayazitova, V.M.; Voitova, Y.; Straumal, E.A.; Lermontov, S.A.; Cherdyntseva, T.A.; Braeutigam, P.; Weiße, M.; Günther, K. A sol-gel method for applying nanosized antibacterial particles to the surface of textile materials in an ultrasonic field. *Ultrason. Sonochem.* **2020**, *60*, 104788. [CrossRef]
20. Akhavan, F.; Majid Montazer, S. In situ sonosynthesis of nano TiO_2 on cotton fabric. *Ultrason. Sonochem.* **2014**, *21*, 681–691. [CrossRef]
21. Behzadnia, A.; Montazer, M.; MahmoudiRad, M. In situ photo sonosynthesis and characterize nonmetal/metal dual doped honeycomb-like ZnO nanocomposites on wool fabric. *Ultrason. Sonochem.* **2015**, *27*, 200–209. [CrossRef]
22. Ahmadizadegan, H.; Esmaielzadeha, S.; Ranjbara, M.; Marzban, Z.; Ghavas, F. Synthesis and characterization of polyester bionanocomposite membrane with ultrasonic irradiation process for gas permeation and antibacterial activity. *Ultrason. Sonochem.* **2018**, *41*, 538–550. [CrossRef]
23. Ghayempour, S.; Montazer, M. Ultrasound irradiation based *in-situ* synthesis of star-like *Tragacanth gum*/zinc oxide nanoparticles on cotton fabric. *Ultrason. Sonochem.* **2017**, *34*, 458–465. [CrossRef]
24. Timilsena, Y.P.; Akanbi, T.O.; Khalid, N.; Adhikari, B.; Barrow, C.J. Complex coacervation: Principles, mechanisms and applications in microencapsulation. *Int. J. Biol. Macromol.* **2019**, *121*, 1276–1286. [CrossRef] [PubMed]
25. Jana, S. Natural polymeric biodegradable nanoblend for macromolecules delivery. In *Recent Developments in Polymer Macro, Micro and Nano Blends*; Elsevier: Amsterdam, The Netherlands, 2017; pp. 289–312.
26. Chadha, S. Recent advances in nano-encapsulation technologies for controlled release of biostimulants and antimicrobial agents. In *Advances in Nano-Fertilizers and Nano-Pesticides in Agriculture*; Elsevier: Amsterdam, The Netherlands, 2021; pp. 111–146.
27. Giri, T.K. Nanoarchitectured Polysaccharide-Based Drug Carrier for Ocular Therapeutics. In *Nanoarchitectonics for Smart Delivery and Drug Targeting*; Elsevier: Amsterdam, The Netherlands, 2016; pp. 29–55.
28. Salaün, F. Microencapsulation technology for smart textile coatings. In *Textiles, Active Coatings for Smart Textiles*; Woodhead Publishing Series; Elsevier: Amsterdam, The Netherlands, 2016; pp. 179–220.
29. Yadav, A.; Prasad, V.; Kathe, A.A.; Raj, S.; Yadav, D.; Sundaramoorthy, C.; Vigneshwaran, N. Functional finishing in cotton fabrics using zinc oxide nanoparticles. *Bull. Mater. Sci.* **2006**, *29*, 641–645. [CrossRef]
30. Vigneshwaran, N.; Kumar, S.; Kathe, A.A.; Varadarajan, P.V.; Prasad, V. Functional finishing of cotton fabrics using zinc oxide soluble starch nanocomposites. *Nanotechnology* **2006**, *17*, 5087–5095. [CrossRef]

31. Wang, R.Y.; Zhang, W.; Zhang, L.Y.; Hua, T.; Tang, G.; Peng, X.Q.; Hao, M.H.; Zuo, Q.T. Adsorption characteristics of Cu(II) and Zn(II) by nano-alumina material synthesized by the sol-gel method in batch mode. *Environ. Sci Pollut. Res. Int.* **2019**, *26*, 1595–1605. [CrossRef]
32. Council of Europe. *European Pharmacopoeia Commission, European Pharmacopoeia*, 5th ed.; Council of Europe: Strasbourg, France, 2005; pp. 188–191.
33. EUCAST. European Committee for Antimicrobial Susceptibility Testing (EUCAST) of the European Society for Clinical Microbiology and Infectious Diseases (ESCMID). EUCAST Discussion Document, E.Def.5.1. *Clin. Microbiol. Infec.* **2003**, *9*, 1–10.
34. Peran, J.; Ercegović Ražić, S.; Kosalec, I.; Ziberi, F. Antimicrobial Effectiveness of Cellulose based Fabrics treated with Silver Nitrate Solution using Plasma Processes. *Tekstilec* **2017**, *60*, 247–253. [CrossRef]
35. Yan, N.; Zhu, Z.; Jin, L.; Guo, W.; Gan, Y.; Hu, S. Quantitative Characterization of Gold Nanoparticles by Coupling Thin Layer Chromatography with Laser Ablation Inductively Coupled Plasma Mass Spectrometry. *Anal. Chem.* **2015**, *16*, 6079–6087. [CrossRef]
36. Yan, N.; Zhu, Z.; He, D.; Jin, L.; Zheng, H.; Hu, S. Simultaneous Determination of Size and Quantification of Gold Nanoparticles by Direct Coupling Thin Layer Chromatography with Catalyzed Luminol Chemiluminescence. *Sci. Rep.* **2016**, *6*, 24577. [CrossRef]
37. Elci, S.G.; Yan, B.; Kim, S.T.; Saha, K.; Jiang, Y.; Klemmer, G.A.; Moyano, D.F.; Tonga, G.Y.; Rotello, V.M.; Vachet, R.W. Quantitative imaging of 2 nm monolayer-protected gold nanoparticle distributions in tissues using laser ablation inductively-coupled plasma mass spectrometry (LA-ICP-MS). *Analyst* **2016**, *21*, 2418–2425. [CrossRef]
38. Rezić, I.; Ćurković, L.; Ujević, M. Simple methods for characterization of metals in historical textile threads. *Talanta* **2010**, *82*, 237–244. [CrossRef]
39. Rezić, I.; Špehar, M.; Jakovljević, S. Characterization of Ag and Au nanolayers on Cu alloys by TLC, SEM-EDS, and ICP-OES. *Mater. Corr.* **2017**, *68*, 560–565. [CrossRef]
40. Hodoroaba, V.D.; Motzkus, C.; Macé, T.; Vaslin-Reimann, S. Performance of high-resolution SEM/EDX systems equipped with transmission mode (TSEM) for imaging and measurement of size and size distribution of spherical nanoparticles. *Microsc. Microanal.* **2014**, *20*, 602–612. [CrossRef]
41. Reza Mahdavi, S.; Talesh, T.A. The effect of ultrasonic irradiation on the structure, morphology and photocatalytic performance of ZnO nanoparticles by sol-gel method. *Ultrason. Sonochem.* **2017**, *39*, 504–510. [CrossRef]
42. Rezić, I.; Majdak, M.; Ljoljić Bilić, V.; Pokrovac, I.; Martinaga, L.; Somogyi Škoc, M.; Kosalec, I. Development of Antibacterial Protective Coatings Active Against MSSA and MRSA on Biodegradable Polymers. *Polymers* **2021**, *13*, 659. [CrossRef]
43. Tang, S.; Zheng, J. Antibacterial Activity of Silver Nanoparticles: Structural Effects. *Adv. Health Mater.* **2018**, *7*, 1701503. [CrossRef]
44. Li, W.R.; Xie, X.B.; Shi, Q.S.; Duan, S.S.; Ouyang, Y.S.; Chen, Y.B. Antibacterial effect of silver nanoparticles on *Staphylococcus aureus*. *Biometals* **2011**, *24*, 135–141. [CrossRef]
45. Mirzajani, F.; Ghassempour, A.; Aliahmadi, A.; Esmaeili, M.A. Antibacterial effect of silver nanoparticles on *Staphylococcus aureus*. *Res. Microbiol.* **2011**, *162*, 542–549. [CrossRef]
46. Grigor'eva, A.; Saranina, I.; Tikunova, N.; Safonov, A.; Timoshenko, N.; Rebrov, A.; Ryabchikova, E. Fine mechanisms of the interaction of silver nanoparticles with the cells of *Salmonella typhimurium* and *Staphylococcus aureus*. *Biometals* **2013**, *26*, 479–488. [CrossRef]
47. Martinaga Pintarić, L.; Somogy Škoc, M.; Ljoljić Bilić, V.; Pokrovac, I.; Kosalec, I.; Rezić, I. Synthesis, Modification and Characterization of Antimicrobial Textile Surface Containing ZnO Nanoparticles. *Polymers* **2020**, *12*, 1210. [CrossRef]
48. Moore, T.L.; Rodriguez-Lorenzo, L.; Hirsch, V.; Balog, S.; Urban, D.; Jud, C.; Rothen-Rutishauser, B.; Lattuada, M.; Petri-Fink, A. Nanoparticle colloidal stability in cell culture media and impact on cellular interactions. *Chem. Soc. Rev.* **2015**, *44*, 6287–6305. [CrossRef]
49. Panáček, A.; Kvítek, L.; Prucek, R.; Kolář, M.; Večeřová, R.; Pizúrová, N.; Zbořil, R. Silver Colloid Nanoparticles: Synthesis, Characterization, and Their Antibacterial Activity. *J. Phys. Chem. B* **2006**, *110*, 16248–16253. [CrossRef] [PubMed]
50. Ansari, M.; Khan, H.; Khan, A.; Sultan, A.; Azam, A. Characterization of clinical strains of MSSA, MRSA and MRSE isolated from skin and soft tissue infections and the antibacterial activity of ZnO nanoparticles. *World, J. Microb. Biotechnol.* **2012**, *28*, 1605–1613. [CrossRef] [PubMed]
51. Rezić, I.; Steffan, I. ICP-OES determination of metals present in textile materials. *Microchem. J.* **2007**, *85*, 46–51. [CrossRef]
52. Rezić, I.; Zeiner, M.; Steffan, I. Determination of 28 selected elements in textiles by axially viewed inductively coupled plasma optical emission spectrometry. *Talanta* **2011**, *83*, 865–871. [CrossRef]
53. Özyildiz, F.; Karagönlü, S.; Basal, G.; Uzel, A.; Bayraktar, O. Micro-encapsulation of ozonated red pepper seed oil with antimicrobial activity and application to nonwoven fabric. *Lett. Appl. Microbiol.* **2013**, *56*, 168–179. [CrossRef]
54. Ramya, K.; Maheshwari, V. Preparation of a naturally driven cotton wound dressing via honey, Tragacanth and Sumac. *Indian, J. Fibre Text. Res.* **2015**, *40*, 213–216.
55. Somogyi Škoc, M.; Macan, J.; Pezelj, E. Modification of polyurethane-coated fabrics by sol-gel thin films. *J. Appl. Pol. Sci.* **2014**, *131*, 39914.
56. Rocha-Selmi, G.A.; Theodoro, A.C.; Thomazini, M.; Bolini, H.M.A.; Favaro-Trindade, C.S. Double emulsion stage prior to complex coacervation process for microencapsulation of sweetener sucralose. *J. Food Eng.* **2013**, *119*, 28–32. [CrossRef]
57. He, R.; Ye, J.; Wang, L.; Su, P. Preparation and Evaluation of Microcapsules Encapsulating Royal Jelly Sieve Residue: Flavor and Release Profile. *Appl. Sci.* **2020**, *10*, 8126. [CrossRef]

58. Shevtsova, T.; Cavallaro, G.; Lazzara, G.; Milioto, S.; Donchak, V.; Harhay, K.; Korolko, S.; Budkowski, A.; Stetsyshyn, Y. Temperature-responsive hybrid nanomaterials based on modified halloysite nanotubes uploaded with silver nanoparticles. *Coll. Surf. A Physicochem. Eng. Asp.* **2022**, *641*, 128525. [CrossRef]
59. Raczkowska, J.; Stetsyshyn, Y.; Awsiuk, K.; Brzychczy-Włoch, M.; Gosiewski, T.; Jany, B.; Lishchynskyi, O.; Shymborska, Y.; Nastyshyn, S.; Bernasik, A.; et al. "Command" surfaces with thermo-switchable antibacterial activity. *Mater. Sci. Eng. C* **2019**, *103*, 109806. [CrossRef] [PubMed]
60. Konnova, S.A.; Danilushkina, A.A.; Fakhrullina, G.I.; Akhatova, F.S.; Badrutdinov, A.R.; Fakhrullin, R.F. Silver nanoparticle-coated "cyborg" microorganisms: Rapid assembly of polymer-stabilised nanoparticles on microbial cells. *RSC Adv.* **2015**, *5*, 13530–13537. [CrossRef]
61. Fries, J.; Getrost, H. *Organic Reagents for Trace Analysis*, 1st ed.; MERCK: Darmstadt, Germany, 1977.
62. Rood, A.S.; McGavran, P.D.; Aanenson, J.W.; Till, J.E. Stochastic estimates of exposure and cancer risk from carbon tetrachloride released to the air from the rocky flats plant. *Risk Anal.* **2001**, *21*, 675–695. [CrossRef] [PubMed]
63. Lévesque, B.; Ayotte, P.; Tardif, R.; Ferron, L.; Gingras, S.; Schlouch, E.; Gingras, G.; Levallois, P.; Dewailly, E. Cancer risk associated with household exposure to chloroform. *J. Toxicol. Environ. Health Part. A* **2002**, *65*, 489–502. [CrossRef] [PubMed]

polymers

MDPI

Article

Estimating the Shear Resistance of Flocculated Kaolin Aggregates: Effect of Flocculation Time, Flocculant Dose, and Water Quality

Kevin Pérez [1], Norman Toro [2], Matías Jeldres [1], Edelmira Gálvez [3], Pedro Robles [4], Omar Alvarado [5], Pedro G. Toledo [6] and Ricardo I. Jeldres [1,*]

1 Departamento de Ingeniería Química y Procesos de Minerales, Facultad de Ingeniería, Universidad de Antofagasta, Av. Angamos 601, Antofagasta 1240000, Chile; kevin.perez.salinas@ua.cl (K.P.); hugo.jeldres.valenzuela@ua.cl (M.J.)
2 Faculty of Engineering and Architecture, Universidad Arturo Prat, Almirante Juan José Latorre 2901, Antofagasta 1244260, Chile; notoro@unap.cl
3 Department of Metallurgical and Mining Engineering, North Catholic University, Angamos Av. 0610, Antofagasta 1270709, Chile; egalvez@ucn.cl
4 Escuela de Ingeniería Química, Pontificia Universidad Católica de Valparaíso, Valparaíso 2340000, Chile; pedro.robles@pucv.cl
5 Departamento de Química, Facultad de Ciencias, Universidad del Bio Bio, Av. Collao 1202, Concepción 4030000, Chile; oalvarado@ubiobio.cl
6 Department of Chemical Engineering and Laboratory of Surface Analysis (ASIF), Universidad de Concepcion, P.O. Box 160-C, Correo 3, Concepción 4030000, Chile; petoledo@udec.cl
* Correspondence: ricardo.jeldres@uantof.cl

Citation: Pérez, K.; Toro, N.; Jeldres, M.; Gálvez, E.; Robles, P.; Alvarado, O.; Toledo, P.G.; Jeldres, R.I. Estimating the Shear Resistance of Flocculated Kaolin Aggregates: Effect of Flocculation Time, Flocculant Dose, and Water Quality. *Polymers* **2022**, *14*, 1381. https://doi.org/10.3390/polym14071381

Academic Editors: Beata Podkościelna and Andrzej Puszka

Received: 21 February 2022
Accepted: 15 March 2022
Published: 29 March 2022

Abstract: The resistance of kaolin aggregates to shearing in water clarification and recovery operations is a critical input in designing thickener feed wells. A recently formulated but already available criterion is used to determine the shear strength of flocculated kaolin aggregates. The flocculant is a high molecular weight anionic polyelectrolyte. The resistance of the aggregates is evaluated as a function of flocculation time, flocculant dosage, and water quality. The determination is based on a standardized experimental method. First, the time evolution of the average size of kaolin flocs is measured when aggregates are exposed to incremental shear rates from a predetermined base value. Then, the results are fitted to a pseudo-first-order model that allows deriving a characteristic value of the shear rate of rupture associated with the upper limit of the strength of the aggregates. In seawater, at a given dose of flocculant, the strength of the aggregates increases with time up to a maximum; however, at longer times, the resistance decreases until it settles at a stable value corresponding to stable aggregates in size and structure. A higher flocculant dosage leads to stronger aggregates due to more bridges between particles and polymers, leading to a more intricate and resistant particle network. In industrial water with very low salt content, the resistance of the kaolin aggregates is higher than in seawater for the same dose of flocculant. The salt weakens the resistance of the aggregates and works against the efficiency of the flocculant. The study should be of practical interest to concentration plants that use seawater in their operations.

Keywords: kaolin flocculation; aggregate resistance; salinity; flocculation kinetic; shear rate

1. Introduction

The separation of solids from liquids in a thickener is driven by gravity leading to a dense slurry containing most solids under a zone of clarified liquid. The dense slurry is swept into the thickener discharge by rotating rakes from where it is transported to tailings deposits. The clarified liquid is the thickener overflow that is recirculated to the process. The liquid is generally water, which is recovered by thickening through sedimentation applied to product and tailings streams. When the particle sizes are small, <80 μm, the gravitational effects are negligible, and the sedimentation is poor. Thus, the application of water-soluble polymers is key to the formation of massive aggregates that accelerate the rate of sedimentation. The most widely used reagents to destabilize tailings particles and aggregates are high molecular weight anionic polyacrylamides (A-PAM) due to high sedimentation rates that are achieved at relatively low doses and costs [1,2].

Flocculation performance is susceptible to shear applied in the thickener feedwell. A low shear rate limits the dispersion and contact between particles and flocculant chains. Therefore, the flocs formed do not reach an adequate settlement. At the same time, an excessive shear causes the detachment of the polymer chains from the particles and breakage of particle–flocculant–particle bonds, impacting the size and structure of the aggregates. Heath et al. [3] used a population balance (PB) model to describe the aggregation/breakage kinetics in calcite flocculation in turbulent pipe flow, demonstrating that the aggregates decrease in size once a maximum size is reached due to fragmentation by shear and polymer degradation, both irreversible events.

Their size and structure characterize aggregates through properties such as size distribution, strength, and fractal dimension. These properties directly impact tailings management indicators such as the water recovery rate, the quality of the recovered water, and the rheological properties of the thickened slurries. The relationship between the mass m in a sphere of radius R of a self-similar fractal aggregate obeys $m \propto R^{D_f}$, with D_f the fractal dimension [4]. In a space of d Euclidean dimensions $D_f \leq d$, and if $d = 3$ then $1 \leq D_f \leq 3$. The largest values of D_f correspond to dense and resistant spherical flocs, whereas the smallest ones correspond to ultra-light and brittle rod-like aggregates. The prefactor represents the structure of the aggregate; however, it is not generally considered even though it is needed in conjunction with D_f to determine the mass of an aggregate from its size.

The fractal dimension of aggregates can be estimated using image analysis and light scattering techniques and hindered sedimentation models [5–7]. Image analysis is the oldest and most popular technique despite the immense data processing load involved and the resolution that may be insufficient for compact fractals [6]. Light scattering is the most appropriate technique if the flocs are small, optically translucent, and diluted. The technique has been used successfully in colloidal systems [8–11] but not in aggregates with high clay content. Estimates of D_f from hindered sedimentation data require the average size of the aggregates, which may be non-trivial if the slurry is strongly sheared before sedimentation because then the model can yield fractal dimensions that decrease with agitation [12]. The question is whether the aggregates are effectively no longer self-similar or the size definition should be adjusted for highly-sheared aggregates. The advantage of this last methodology is that it provides structural properties of aggregates under conditions that mimic flocculation in the feedwell of industrial thickeners.

Deng and Davè [13] analyzed the effect of impact velocity and surface energies on the structure of spherical particle agglomerates using the discrete element method (DEM). The authors found a significant reduction in the fractal dimension with increasing shear, stating that they grow irregularly when the aggregates are produced with a greater mixing intensity. Additionally, they consider that aggressive hydrodynamics deteriorates the structure of the aggregates, which promotes the reduction of the fractal dimension. Finally, Leiva et al. [12] analyzed the impact of flocculation time and shear on the structural characteristics of HPAM-flocculated clay tailings aggregates in seawater. The authors found that the more an aggregate is exposed to hydrodynamic disturbances, the more significant the deterioration of its structure, which is reflected in lower fractal dimension values.

The structural properties of the flocs are determined by the characteristics of the mineral, mineralogy, and size distribution; characteristics of the flocculant, structure, molecular weight, charge density and dosage; conditions of the environment, salinity, and pH; and shear rate during mixing [14–16]. Although various studies relate the control variables in the feedwell of a thickener with the properties of the aggregates and the flocculation performance [17–20], few do so when the medium is seawater. Using molecular dynamics, Quezada et al. [21] studied the impact of salinity on the conformation of an anionic polyelectrolyte on quartz surfaces. These authors demonstrated that increasing the salinity improves the adsorption of polyelectrolytes on quartz surfaces. However, at the same time, it reduces the radius of gyration of the polyelectrolyte, limiting the size of the aggregation achieved. Jeldres et al. [22] carried out flocculation and sedimentation experiments on suspensions of quartz and kaolinite mixtures, finding a direct relationship between flocs microscopic properties, fractal dimension, and some macroscopic observables such as sedimentation speed and yield stress of the sediment. Flocs with a larger fractal dimension settled at higher speeds, implying an increase in the yield stress of the sediment. This last parameter presented an exponential growth with the fractal dimension for all the shear conditions studied. The authors explain this behavior because all three parameters are dependent on both the size and structure of the aggregate.

Floc resistance is directly related to the number and strength of interparticle bonds [23,24]; it is considered a key parameter in solid/liquid separation since it determines the degree of floc breakage. A floc breaks if the applied stress is greater than the force that holds the particles together. Operationally, this is important since the small particles of a fragmented floc have a lower sedimentation rate. However, a high floc resistance could increase the sediment's rheological properties, which implies a higher energy consumption for its transport to the tailings deposit. Therefore, the degree of resistance must remain in a range that guarantees industrially acceptable rates of sedimentation and magnitudes of rheological properties.

Recently, Jeldres et al. [25] proposed a simple criterion to quantitatively estimate the strength of aggregates subjected to increasing mechanical shear disturbances. The monotonic relationship between the shear rate increments (ΔG) and the final size of the aggregates is used for a more quantitative estimate of the resistance of the target aggregates since a pseudo-first-order model can describe this relationship. The main result is a parameter that does not depend on the hydrodynamic conditions chosen to establish the disturbances unlike models available in the literature, which, although they work adequately within the experimental conditions carried out for their validation, are not necessarily extrapolated and comparable in different systems. The methodology was used to evaluate the impact of salinity on the resistance of clay aggregates. The results show that floc breakage is achieved with considerably lower agitation increments as the salinity of the system increases.

In this work, the resistance of kaolin flocs is related to its structural properties characterized by the fractal dimension. The kaolin suspension is flocculated with a high molecular weight anionic polyelectrolyte. The objective is to compare the strength of the aggregates and their fractal dimension both in seawater and in industrial water with low ionic charge. Additionally, the effect of the flocculation time is studied, being one of the variables handled in the thickeners' feedwell. Kaolinite is a typical clay in tailings of copper mining operations. The particles have a colloidal size and platelet shape that represent major challenges in thickening. The challenges relate to a decrease in sedimentation rate [26,27] and an increase in rheological parameters [28,29], which are closely related to the mechanical resistance of the suspension aggregate structure. The study is of practical interest to the copper industry, especially for seawater plants in their operations. The demand for seawater has been increasing in recent years [30–32] as a result of the search for alternatives to face the scarcity of water resources due to climate change.

2. Materials and Methods

2.1. Materials

Seawater from Bahía San Jorge in Antofagasta (Chile) was filtered at 1 μm using a UV filter system to remove all bacterial activity. The conductivity was 50.4 mS/cm. The ionic composition and the methods used are specified in Table 1. Industrial water was prepared with 0.01 M analytical grade NaCl from Merck (Santiago, Chile).

Table 1. Ionic concentration of seawater and methods of analysis.

Ion	Concentration (g/L)	Method
Na^+	10.9	Atomic absorption spectrometry
Mg^{2+}	1.38	Atomic absorption spectrometry
Ca^{2+}	0.4	Atomic absorption spectrometry
K^+	0.39	Atomic absorption spectrometry
Cl^-	19.6	Argentometric method
HCO^{3-}	0.15	Acid–base volumetry

Ward's Science kaolin particles were used. Its composition was determined by XRD analysis on a Bruker X-ray diffractometer, D8 Advance (Bruker, Billerica, MA, USA), using the 2020 ICDD (International Center for Diffraction Data) Powder Diffraction File database. The spectra showed that kaolin is mainly composed of halite and kaolinite (>10%) and some SiO2 (1–10%) (Figure 1). Volume-weighted particle size distribution (PSD) was obtained using a Microtrac S3500 laser diffraction analyzer (Verder Scientific, Newtown, PA, USA) and published in Jeldres et al. [25]. The data showed that 10% of the kaolin particles were smaller than 1.8 μm.

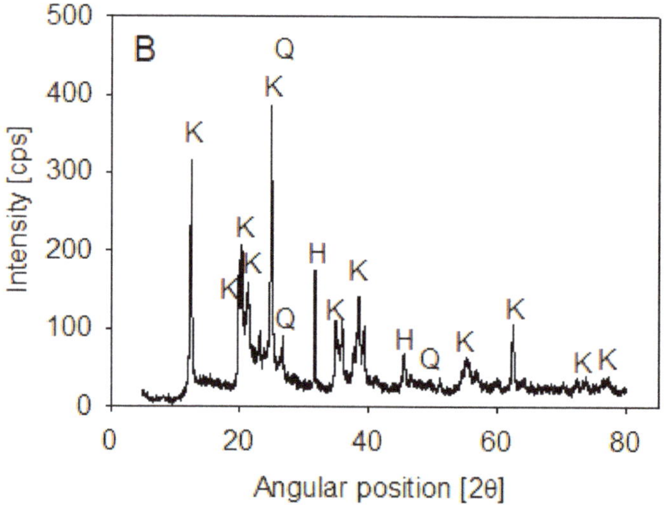

Figure 1. XRD of kaolin particles.

The flocculant used was commercial polyacrylamide SNF704, an anionic acrylate copolymer supplied by SNF-Chile with 98% purity. NaOH was used to reach pH 8.

2.2. *Aggregate Characterization*

The flocculation of kaolin particles was carried out in a cylindrical container with a diameter of 100 mm and a capacity of 1 L. An 80 mm diameter turbine impeller, attached to the end of a vertical shaft (4 mm diameter), was used to keep the particles suspended. The impeller base was located 20 mm above the bottom of the vessel.

The minerals were mixed with water (seawater and industrial water with 0.01 M NaCl) to give a total mass of 270 g. The mixture was stirred at 600 rpm (G = 1400 s^{-1} approximately) for 30 min at pH 8 to keep the particles dispersed before flocculation. Stirring was then reduced to 200 rpm (G = 273 s^{-1}). After 1 min, the flocculant was added in doses of 8 to 89 g/t. The evolution of the average size of flocs was analyzed for 6 min. The strength of the aggregates was evaluated by increasing the mixing intensity. For this, the flocculation tests were repeated at the standard agitation of 200 rpm, but after a determined flocculation time, which was considered to be in the range of 30–120 s, the mixing intensity was increased to promote shear fragmentation. Independent tests were performed by applying shear rate increments from the initial $G = 273\ s^{-1}$ (0 to 1516 s^{-1}).

The Focused Beam Reflectance Measurement (FBRM) system (Particle Track E25, Mettler Toledo, OH, USA) was used to measure the aggregate size evolution. The system consists of a processing unit with a 19 mm diameter probe tip positioned vertically in the flocculation vessel 10 mm above the stirrer and 20 mm off the axis of rotation. The technology is based on the backscattering of light using a laser beam rotating at 2 m/s and projected onto the sample suspension through a sapphire window (14 mm in diameter) at the end of the probe. The chord length, which correlates with the actual size of the particles, is obtained by multiplying the light backscatter time of each particle by the scanning speed. The instrument's software allows recording data processing into histograms of counts corresponding to chord lengths in selected channel sizes ranging from 1 μm to 1 mm as quickly as every 2 s. In this case, the chord length distributions (CLDs) represent 100 channels over the entire range, but the histograms are presented as line plots for easy comparison. The equipment offers two types of distributions, the unweighted CLD that provides enhanced resolution to fine particle changes and the squared-weighted CLD that provides enhanced resolution to coarse particle changes. The former is useful in studies of primary nucleation, particle growth, and aggregation, and the latter is useful in studies of breakage, attrition, secondary nucleation, and particle dissolution. The average size of the aggregates was obtained considering the squared-weighted chord distribution, which has been documented to be more similar to the distribution obtained with laser diffraction.

2.3. Shear Rate

The mean shear rate required by the aggregation and breakage of flocs was calculated from:

$$G = \left(\frac{\varepsilon\rho_{sus}}{\mu_{sus}}\right)^{1/2} \quad (1)$$

where μ_{sus} is the viscosity (4 mPas), ρ_{sus} is the pulp density, and ε is the average energy dissipation rate:

$$\varepsilon = \frac{N_p N^3 D^5}{V} \quad (2)$$

N_p is the impeller power number (0.6 in our case for a plane disk with gentle agitation [2]), N is the rotation speed, D and V are, respectively, the diameter of the impeller and the working volume of the impeller vessel. The density of the suspension ρ_{sus} was calculated from:

$$\rho_{sus} = \left(\frac{w}{\rho_s} + \frac{1-w}{\rho_w}\right)^{-1} \quad (3)$$

where w is the mass fraction of solids in suspension, and ρ_s and ρ_w are the density of solid and water, respectively. The parameters are indicated in Table 2.

Table 2. Data to calculate the mean shear rate.

Parameter	Value	Unit of Measurement
u_{sus}	0.004	kg/(m·s)
N_p	0.6	
D	0.08	m
W	0.04	
ρ_s	2600	kg/m^3
ρ_w	1000	kg/m^3
V	0.25	L

2.4. Sedimentation Tests

The setting rate was determined by interrupting the flocculation tests at specific preset times, 10, 60, 90, and 120 s. The resulting suspensions were gently poured into 300 cm^3 cylinders (35 mm internal diameter). Then, each cylinder with its content was slowly inverted three times (each cylinder inversion took ~4 s) and then placed on a flat surface to determine the sedimentation rate classically (the change in the liquid–solid interface was recorded over 10 min).

2.5. Fractal Dimension

The fractal dimension D_f of the aggregates was calculated from the model of Heath et al. [3], for the hindered settlement rate as a function of aggregate size, as

$$U_h = \frac{\overline{d_{agg}^2} g(\rho_s - \rho_l)\left(\frac{\overline{d_{agg}}}{d_p}\right)^{D_f-3}}{18\mu}\left(1 - \varphi_s\left(\frac{d_{agg}}{d_p}\right)^{3-D_f}\right)^{4.65} \quad (4)$$

where U_h is the hindered settling rate in m/s, d_p is the average size of primitive particles, d_{agg} is the average size of aggregates after some flocculation time, in this work both d_p and d_{agg} are approximated by the squared-weighted mean chord length, ρ_s and ρ_l are, respectively, the densities of solid and liquid phases, g is the acceleration of gravity, μ is the fluid viscosity, φ_s is the solid fraction, and D_f is the mass fractal dimension. Many of these parameters remain constant across experiments, for example, ρ_s, ρ_l, μ, φ_s, and d_p. While the hindered sedimentation rate was experimentally determined after flocculation as described above. D_f was calculated from Equation (4) using the standard method of least squares.

3. Results

3.1. Aggregation with Flocculant

Figure 2 shows the flocculation kinetics of clay suspensions in seawater (high salt) at polyacrylamide doses of 13, 34, and 89 g/t and industrial water (low salt) at 89 g/t at a shear rate of $G = 273\ s^{-1}$. Once the flocculant is added, the growth of the aggregates is practically instantaneous, and it only takes a few seconds to reach a maximum size. Subsequently, shear fragmentation causes a continuous reduction in the size of the aggregates until reaching a stable size. The average size of aggregates shows an increasing trend with the doses studied; for example, at 13 g/t in seawater, a size peak of 115 µm is reached, while at 89 g/t the peak is at 300 µm. A saline system can favor or harm flocculation. The final result depends on the competition of two mechanisms propitiated by the presence of cations, that is, (i) better flocculant adsorption via cationic bridges and (ii) electrostatic shielding of the active sites in the flocculant, leading to polymer curling. The results in Figure 2 show better flocculation in industrial water than in seawater, indicating that the high concentration of cations in seawater causes a significant increase in polymer curling, limiting its ability to form hydrogen bonds and thus its effectiveness as a flocculant. Figure 3 shows unweighted and squared-weighted chord length distributions (CLD) for the kaolin suspension at 4 wt%, using seawater at polyacrylamide doses of 13, 34, and 89 g/t and industrial water at 89 g/t. The chosen mixing time is 55 s, just before applying an increase in shear rate.

In seawater and as the flocculant dose increases, the squared weighted distribution shows that the particles form larger and larger aggregates due to a combined process of growth (which does not change the number of particles) and agglomeration (which reduces the number of particles). The size peak with 13 g/t of flocculant occurs at 100 µm with a count of just over 40 s^{-1}, while with 89 g/t the peak occurs at 200 µm with a frequency of just over 160 s^{-1}. The unweighted distribution is illustrative of the particle growth process when the flocculant dose increases. At any of the flocculant doses tested, the distribution is markedly bimodal. At 13 g/t of flocculant, the distribution shows similar size frequencies of fine and coarse particles, centered, respectively, at 4 and 40 µm. The flocculant captures only a few fines at this low dosage to form larger aggregates. At the highest rate of 34 g/t, the flocculant can capture many fines (about 4 μm) to form aggregates of just over 100 μm. Then, the frequency of fine particles drops dramatically to just under 300 s^{-1}. The frequency of coarse agglomerates also decreases to about 300 s^{-1}. The coarse ones consume fines, and the more they consume, the lower their number. At the highest dose tested of 89 g/t, the effectiveness of the flocculant is very high because it captures practically all the fines, whose frequency decreases to less than 50 s^{-1}, forming a few agglomerates that reach a peak size of 110 μm although with the lowest frequency, just over 200 s^{-1}. The coiling effect of the polymer chains at high salt in seawater ceases to be a limiting factor for flocculation when the flocculant dose is high enough.

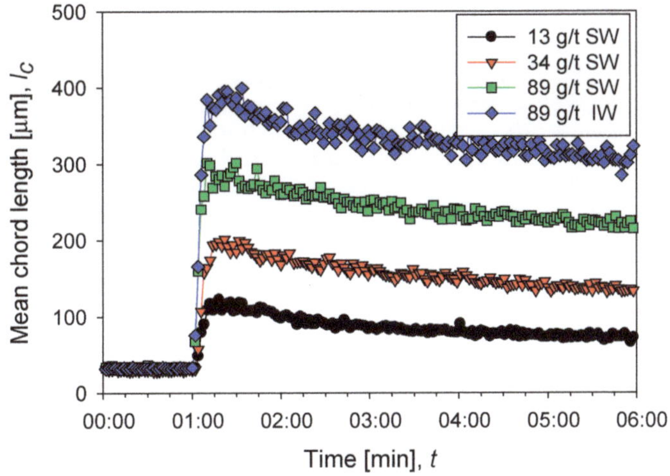

Figure 2. Evolution of the average size of aggregates of kaolin particles (chord length) before and after the addition of flocculant in different doses in seawater (SW) and industrial water (IW).

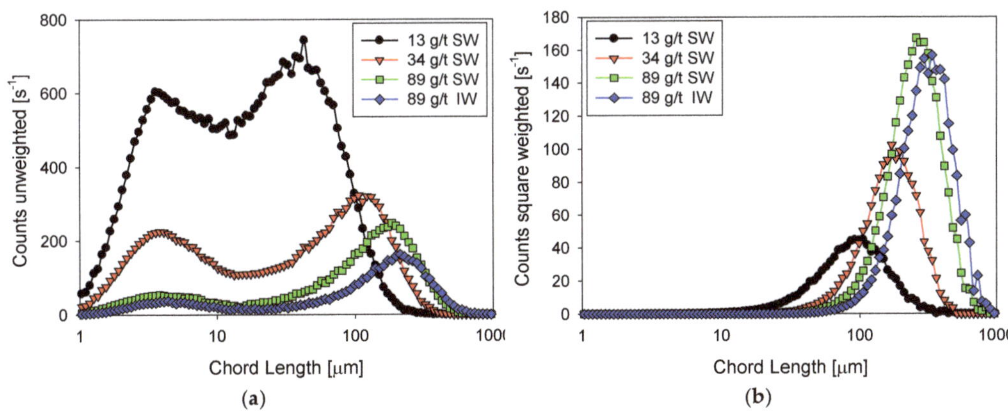

Figure 3. Unweighted (**a**) and squared-weighted (**b**) size distribution of kaolin suspensions with different doses of flocculant in SW and IW after three minutes of flocculation at 200 rpm immediately before increasing shear rate.

In industrial water with very low salt content and at the highest flocculant dosage of 89 g/t, the unweighted and square weighted distributions show little difference from the distributions in seawater. The flocculant forms fewer aggregates in industrial water, although larger than in seawater. The aggregate frequency in industrial water is only 150 s^{-1} (vs. 250 s^{-1} in seawater) with a peak size of 115 μm (vs. 110 μm in seawater). This result reveals that the polymer coiling effect in industrial water is less than in seawater. This effect on flocculation may be more dominant than the salt bridges that are more frequent in seawater.

3.2. Settling Rate

The sedimentation rate in seawater increases rapidly as the flocculant dose increases from 13 g/t to 89 g/t. Remarkable is that at low doses, 13 and 34 g/t, the rate does not change significantly and remains relatively constant as a function of mixing time. However, at 89 g/t the rate is more than double at low doses. This result is consistent with the flocculation kinetics shown in Figure 2. At higher doses, larger flocs are obtained, and therefore the population of fines is reduced (Figure 3), thus increasing the sedimentation rate. At the high flocculant dose of 89 g/t, the sedimentation rate shows a non-monotonic behavior as a function of the mixing time. At 30 s, a smooth but distinguishable maximum is observed, then the rate decreases with mixing time.

The sedimentation rate in industrial water is notoriously higher than in seawater, as shown in Figure 4 at 89 g/t of flocculant in both cases. This result is also consistent with the flocculation kinetics in Figure 2, which show aggregates with larger average sizes in industrial water. The particle size distributions in Figure 4 show that coarser aggregates are larger in industrial water than in seawater. In industrial water, the sedimentation rate shows a more pronounced maximum at 30 s.

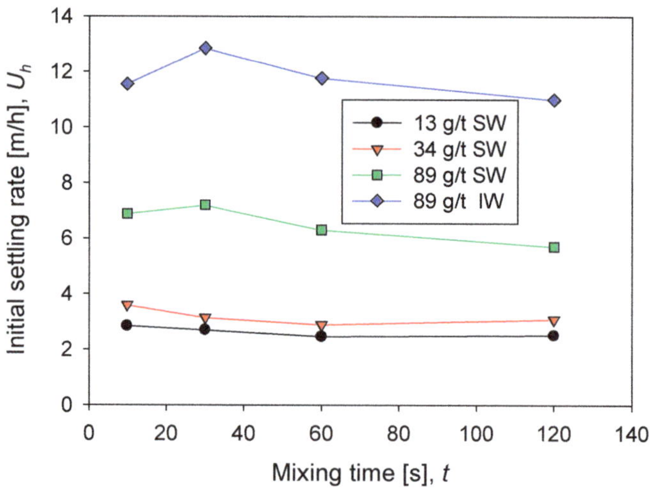

Figure 4. Sedimentation rate of kaolin suspensions with different doses of flocculant in SW and IW at a shear rate of 200 rpm.

3.3. Fractal Dimension

The aggregation of particles in the presence of flocculant occurs through hydrogen bonds and salt bridges if the medium is saline, leading to highly irregular porous structures that cannot be described by impermeable sphere models [33,34]. There is consensus that the structure of flocculated aggregates follows a scale-invariant growth that generates fractal structures [35]. The fractal dimension directly relates to macroscopic observables of industrial interest such as rheological and settling [22].

Figure 5 shows the behavior of the fractal dimension of kaolin aggregates as a function of the flocculation time at different doses of flocculant in seawater and in industrial water at the same shear rate. The fractal dimension of kaolin aggregates is not very sensitive to flocculation time; only at low doses of flocculant in seawater is a small decrease associated with a slight restructuring of the aggregates observed. This result is different in aggregates that come from mineral mixtures. For example, the fractal dimension of kaolin and quartz aggregates decreases with flocculation time, according to Leiva et al. [12]. At low flocculant doses in seawater, the flocs that form are small according to the flocculation kinetics (Figure 2) and the size distribution (Figure 3). Now according to their low fractal dimension (between 1.5 and 2), it is inferred that they are very fragile structures. This result is expected since the flocculant molecules are scarce, and therefore the probability of formation of bridges of any type is low. On the contrary, at high doses of flocculant in seawater, the flocs are large, and, in addition, according to their high fractal dimension (between 2 and 2.5), they have a resistant structure. This result is also expected because the greater the number of flocculant molecules, the greater the probability of forming interparticle bridges.

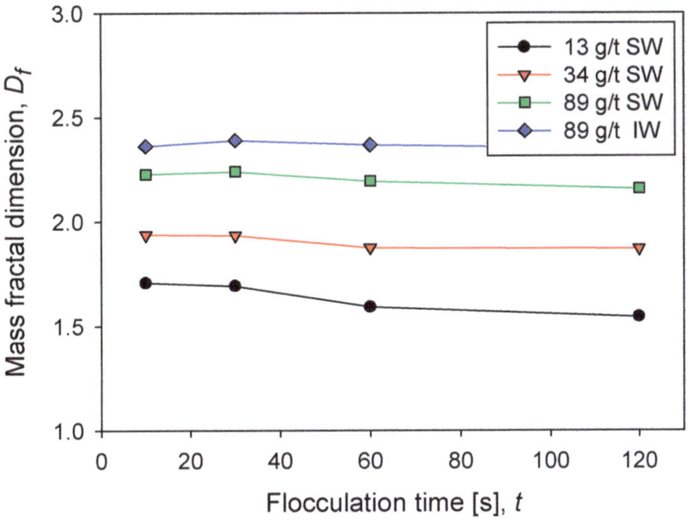

Figure 5. Fractal dimension of kaolin suspensions with different doses of flocculant in SW and IW at a shear rate of 200 rpm.

3.4. Resistance of Aggregates

The strength of the aggregates is obtained from the temporal evolution of the average size of the flocs when they face changes in the intensity of mixing, following the methodology recently proposed by Jeldres et al. [25]. Here we analyze the impact of flocculation time, flocculant dose, and type of water on the resistance of kaolin aggregates (4% by weight) flocculated with polyacrylamide.

Figure 6 shows the change in flocculation kinetics when the system is subjected to increased agitation intensity. The responses are described by fitting the experimental data to Equation (5):

$$y(t) = y_0 - A\left(1 - e^{-bt}\right) \quad (5)$$

where $y(t)$ is the average size of the aggregates as a function of time, y_0 is the average size of the aggregates just before the increment in agitation, b is the decay constant of the aggregate size, and A represents the amplitude of the size reduction of the aggregates, as shown in Equation (6).

$$A = y_0 - y_\infty \quad (6)$$

where y_∞ corresponds to the size reached after applying a specific increase in shear rate during an infinite time.

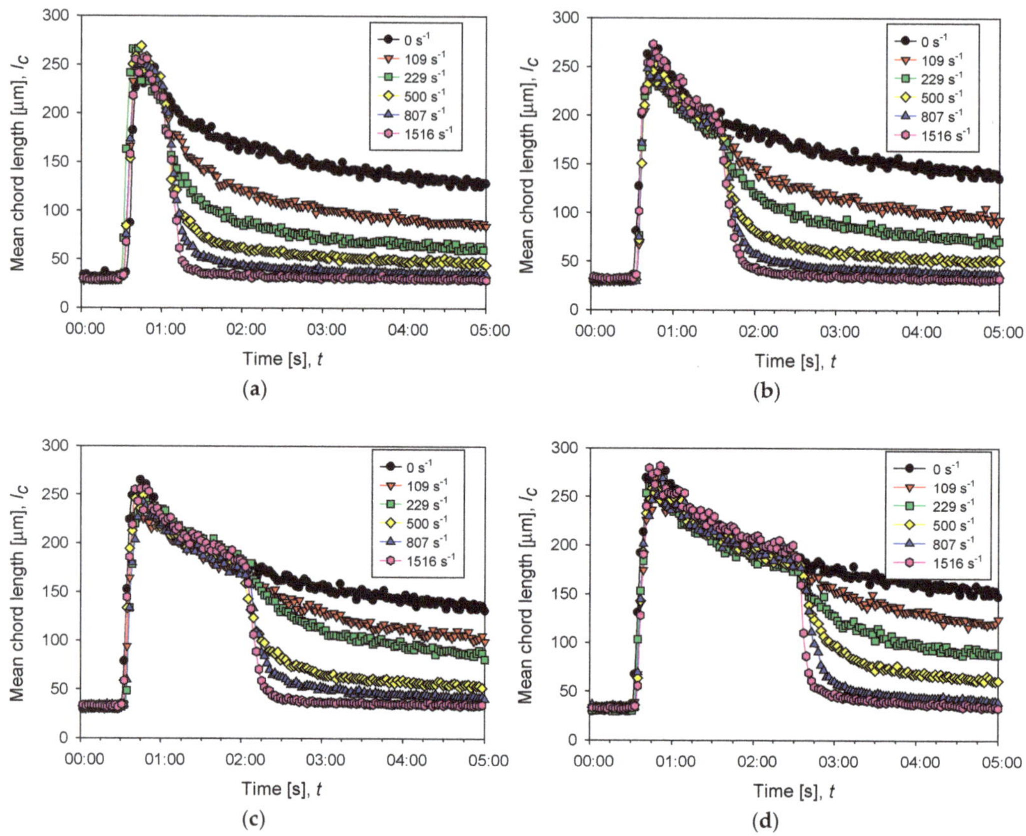

Figure 6. Evolution of the average size of aggregates of kaolin in suspensions with different doses of flocculant in SW and IW at a shear rate of 200 rpm (black dots) and after several increments of shear rate (color dots). The curves with increased shear show three stages, pre-flocculation (primary particles remain free), flocculation (particle size first increases sharply by aggregation and then decreases smoothly with time), and disaggregation (particle size decreases abruptly as shear is increased). (**a–c**) correspond to 13, 34, and 89 g/t of flocculant in seawater, respectively, while (**d**) corresponds to 89 g/t of flocculant in industrial water.

The maximum aggregate rupture at an infinite time for each increment in shear rate (ΔG) is estimated from Equation (7).

$$R_{max} = 100 \times A/(y_0 - d_0) \tag{7}$$

where R_{max} is the maximum aggregate rupture percent and d_0 is the average size of the primary particles. The R_{max} values calculated for aggregates at different conditions can be represented by Equation (8), which allows estimating the degree of rupture as a function of the agitation increments (ΔG). The results are shown in Figure 7.

$$R_{max} = R_{max,0} + U\left(1 - e^{-B\cdot\Delta G}\right) \tag{8}$$

where $R_{max,0}$ is the maximum aggregate rupture percent when there is no disturbance of agitation (constant mean shear rate of 273 s^{-1}), U is the difference between the maximum aggregate rupture at an infinite shear rate ($\Delta G \to \infty$) and the rupture degree at the base agitation of 273 s^{-1} (ΔG = 0), and B is the constant of flocs rupture due to the increase in agitation, whose inverse is considered a characteristic shear rate for the disintegration that should be useful as an indicator of the resistance of aggregates.

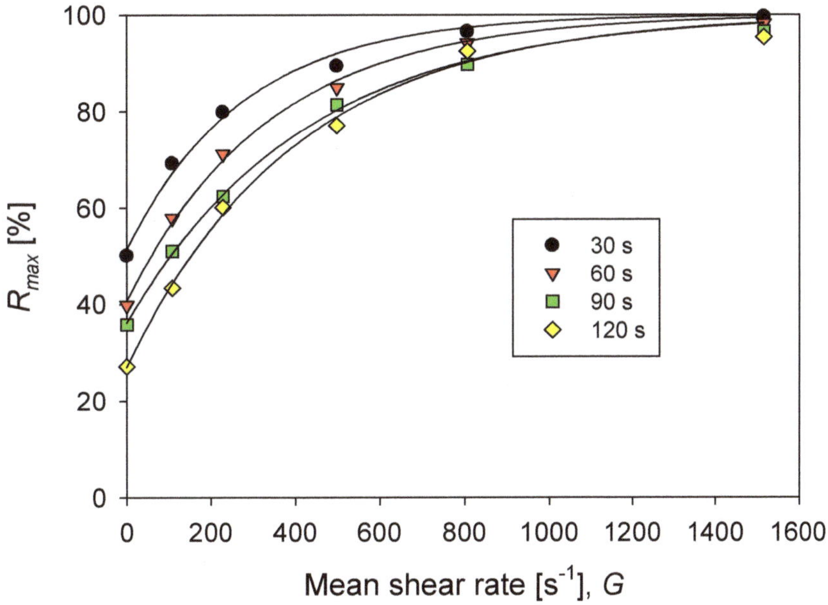

Figure 7. Percentage of maximum rupture (at an infinite time) of aggregates of kaolin in solutions with different doses of flocculant in SW and IW for each increase in agitation from the base of $G = 273\ s^{-1}$ (200 rpm).

After fitting the data to Equation (4), the restriction indicated in Equation (9) is added, which corresponds to the system response when the mixing intensity is very high ($\Delta G \to \infty$). In this case, a total rupture of the aggregate is considered, giving rise to a suspension of dispersed primary particles.

$$R_{max,0} + U = 100\% \tag{9}$$

The value of $R = 1/B$ required to obtain $R_{max} - R_{max,0} = 63.2\%$ defines the characteristic shear rate or aggregate resistance. The higher the characteristic shear rate, the greater the resistance of the aggregates. In a recent work [25], we show that this simple criterion offers advantages over current methods of estimating the strength of aggregates under increased shearing. We use it to analyze the strength of kaolin aggregates in pulps with different salinities. The strength of aggregates is their opposition to disintegration when shear is increased. The more resistant the aggregate, the lower the fragmentation. The practical impacts are varied, for example, in the design of thickener feedwells in clarification and water recovery operations.

Figure 8 shows the impact of flocculation time (prior to application of increased agitation) on the strength of kaolin aggregates in seawater at a particular dosage of flocculant. The minimum resistance R for the system and conditions studied is obtained after 30 s of mixing and reaches $R = 276\ s^{-1}$. The resistance increases steadily up to 90 s of mixing, reaching a maximum value of $R = 424\ s^{-1}$. After this maximum, a longer mixing time leads to a decrease in aggregate strength, $R = 400\ s^{-1}$. The smaller structures that are obtained with extended mixing are less prone to breakage and in the case studied here, they dominate the behavior above 100 s of flocculation, a condition in which the aggregates present stable size and structure.

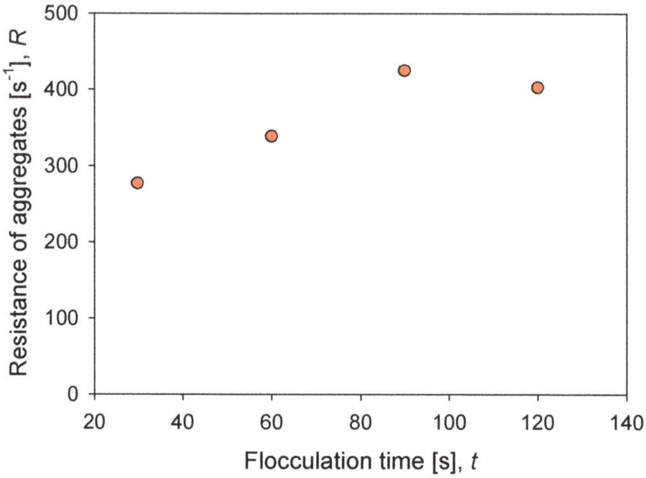

Figure 8. Effect of flocculation time on the resistance of kaolin aggregates (R) flocculated in SW with 89 g/t of flocculant.

Figure 9 shows the impact of the flocculant dosage on the strength of kaolin aggregates in seawater after a flocculation time of 60 s. With 8 g/t of flocculant, a minimum resistance R of 203 s^{-1} is obtained, which increases monotonically with the dose until it reaches a value slightly higher than R = 400 s^{-1} with 55 g/t of flocculant, and tends to settle after R = 445 s^{-1} with 89 g/t of flocculant. The flocculant dose is a critical aspect in the formation of the aggregates since it determines the number of polymer molecules available in the suspension to join the particles. Therefore, for a fixed surface area, an increase in polymer molecules is expected to generate a greater number of particle–flocculant bonds or a more intricate particle network. This explains the increasing aggregate strength in Figure 9 with flocculant dosages increasing from 8 to 55 g/t. At higher doses, 89 g/t, the resistance does not change significantly despite having a greater number of polymer molecules. In this case, it should be considered that the large aggregates are more prone to breakdown, which limits the value of the resistance.

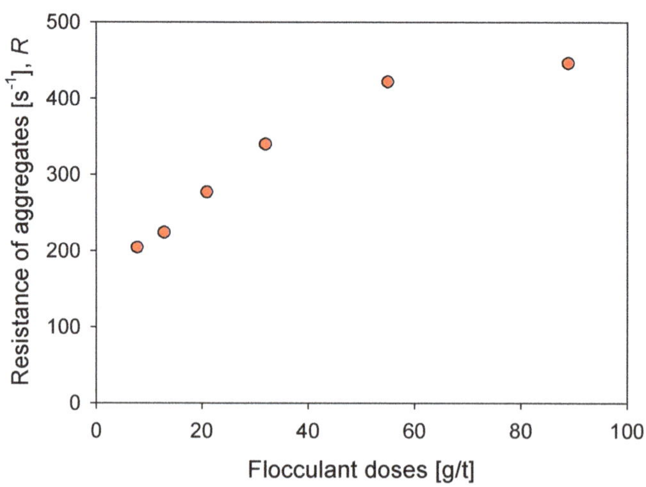

Figure 9. Effect of the flocculant dose on the resistance of kaolin aggregates (R) flocculated in SW after 60 s flocculation time.

Finally, the resistance of flocs in two types of water, industrial water (low salt load) and seawater (high salt load), is compared, considering two doses of flocculant and 60 s of flocculation time. The results are summarized in Figure 10. At any of the flocculant doses tested, the aggregates are more resistant in industrial water than in saltwater. At 21 g/t of flocculant, the resistance is $R = 350\ s^{-1}$ in industrial water and only $R = 276\ s^{-1}$ in seawater. At 89 g/t of flocculant, the resistance is $R = 552\ s^{-1}$ in industrial water and only $R = 446\ s^{-1}$ in seawater. Furthermore, the increase of flocculant dose in the same type of water leads to a greater resistance of the aggregates in industrial water than in seawater. When the dose increases from 21 to 89 g/t in industrial water, the resistance increases in 202 s^{-1}, while in seawater, it increases only in 170 s^{-1}. These results of the effect of salinity on the strength of kaolin aggregates were anticipated by Jeldres et al. recently [25]. The resistance of kaolin aggregates in freshwater without salt reaches a characteristic shear of $R = 530\ s^{-1}$, while in saltwater 0.1 M NaCl, the resistance reaches a modest $R = 320\ s^{-1}$. These authors observed that an increase in salinity generates aggregates that are more prone to fragmentation due to several factors. For example, changes in the aggregation modes of the primary particles, coiling of the flocculant chains due to a reduced repulsion between anionic functional groups, and changes in the particle–flocculant interactions due to a prevalence of the train configuration in the adsorption mode of the flocculant.

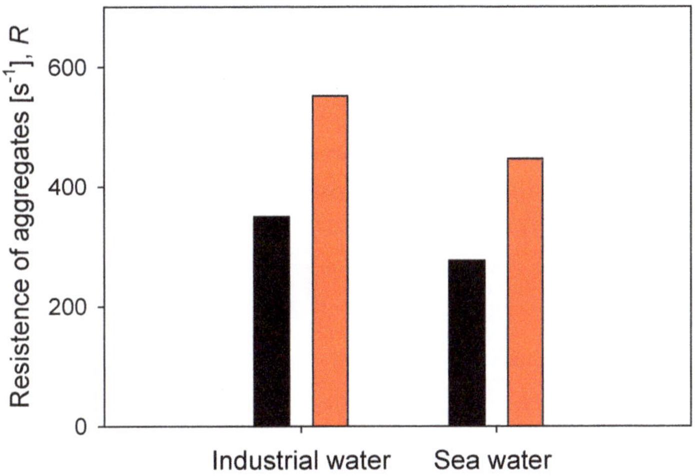

Figure 10. Effect of water type on the resistance of kaolin aggregates (R) flocculated at two doses of flocculant after 60 s flocculation time.

The resistance response of kaolin aggregates to different salinities was somewhat greater in the study by Jeldres et al. [25], whose experiments were carried out at pH 5.5. In contrast, in the present study, the experiments were carried out at pH 8, at which the kaolin particles are mostly anionic. At pH 5.5, in the work of Jeldres et al., kaolin particles in freshwater form aggregates by electrostatic attraction between cationic edges and anionic faces (giving rise to the well-known house-of-cards structure). However, if the ion concentration increases, the electrostatic shielding of the mineral surface reduces the number of edge–face bonds, and face–face type organizations prevail, which are more fragile, and also the efficiency of the flocculant is critically reduced by coiling. A pH of 8 does not favor edge–face bonds that provide resistance to the kaolin aggregates, and thus their resistance is somewhat lower than at pH 5.5. In the presence of seawater and at a pH of 8, on the one hand, salt bridges are activated that contribute to the resistance of the aggregates, and on the other, the flocculant chains are deactivated due to coiling; the resistance of the aggregates is the net effect of these two factors.

4. Conclusions

The time evolution of the average size of the kaolin flocs measured when the particle suspension is subjected to incremental shear rates from a predetermined base value, reveals a shear rupture rate characteristic of these aggregates and their formulation. We associate this shear rate with the upper strength limit of the structure of these aggregates. In this way, the effect of flocculation time, flocculant dose, and water quality on the shear strength of kaolin aggregates are studied. The kaolin aggregates are flocculated with a high molecular weight anionic polyelectrolyte at a pH of about 8. In seawater, at a given dose of flocculant, the strength of the aggregates increases with time up to a maximum value; however, at longer times, the resistance decreases until settling at a stable value corresponding to stable aggregates in size and structure. A higher flocculant dosage leads to stronger aggregates due to more particle–polymer bridges, leading to a more resistant particle network. In industrial water with low salt content, the resistance of the kaolin aggregates is higher than in seawater at the same dose of flocculant. Thus, the salt content weakens the resistance of the aggregates and works against the efficiency of the flocculant, corroborating previous studies. The kaolin aggregate rupture strength criterion used in this work is simple, obeys standardized tests that are relatively easy to replicate in the laboratory, and is useful for decision making in the design and operation of water thickeners and clarifiers.

Author Contributions: Conceptualization, M.J. and R.I.J.; methodology, K.P., O.A. and N.T.; formal analysis, P.G.T., N.T., E.G., P.R. and R.I.J.; resources, P.G.T. and R.I.J.; data curation, K.P. and E.G.; writing—original draft preparation, M.J.; writing—review and editing, P.R., O.A., P.G.T. and R.I.J. All authors have read and agreed to the published version of the manuscript.

Funding: This research was funded by the CRHIAM Project ANID/FONDAP/15130015. PGT and O.A. also thank project ANID/FSEQ/210002. RIJ thanks the Project ANID/FONDECYT/1211606.

Institutional Review Board Statement: Not applicable.

Informed Consent Statement: Not applicable.

Data Availability Statement: The data presented in this study are available on request from authors K. Pérez, M. Jeldres, and R.I. Jeldres.

Acknowledgments: K.P., O.A., R.I.J. and P.G.T. are grateful for the financial support from the CRHIAM Project ANID/FONDAP/15130015. R.I.J thanks Project ANID/FONDECYT/1211606. K.P. and M.J. acknowledge the infrastructure and support of the Programa de Doctorado en Ingeniería de Procesos de Minerales of the Universidad de Antofagasta. O.A. and P.G.T. are grateful for the support from the project ANID/FSEQ210002. P.R. thanks the Pontificia Universidad Católica de Valparaíso for the support provided.

Conflicts of Interest: The authors declare no conflict of interest.

References

1. Lee, L.T.; Rahbari, R.; Lecourtier, J.; Chauveteau, G. Adsorption of polyacrylamides on the different faces of kaolinites. *J. Colloid Interface Sci.* **1991**, *147*, 351–357. [CrossRef]
2. Nabzar, L.; Pefferkorn, E.; Varoqui, R. Polyacrylamide-sodium kaolinite interactions: Flocculation behavior of polymer clay suspensions. *J. Colloid Interface Sci.* **1984**, *102*, 380–388. [CrossRef]
3. Heath, A.R.; Bahri, P.A.; Fawell, P.D.; Farrow, J.B. Polymer flocculation of calcite: Population balance model. *AIChE J.* **2006**, *52*, 1641–1653. [CrossRef]
4. Meakin, P. Fractal aggregates. *Adv. Colloid Interface Sci.* **1987**, *28*, 249–331. [CrossRef]
5. Liang, L.; Peng, Y.; Tan, J.; Xie, G. A review of the modern characterization techniques for flocs in mineral processing. *Miner. Eng.* **2015**, *84*, 130–144. [CrossRef]
6. Bushell, G.C.; Yan, Y.D.; Woodfield, D.; Raper, J.; Amal, R. On techniques for the measurement of the mass fractal dimension of aggregates. *Adv. Colloid Interface Sci.* **2002**, *95*, 1–50. [CrossRef] [PubMed]
7. Florio, B.J.; Fawell, P.D.; Small, M. The use of the perimeter-area method to calculate the fractal dimension of aggregates. *Powder Technol.* **2019**, *343*, 551–559. [CrossRef]
8. Zhou, Z.; Franks, G.V. Flocculation mechanism induced by cationic polymers investigated by light scattering. *Langmuir* **2006**, *22*, 6775–6786. [CrossRef]

9. Rasteiro, M.G.; Garcia, F.A.P.; Ferreira, P.; Blanco, A.; Negro, C.; Antunes, E. The use of LDS as a tool to evaluate flocculation mechanisms. *Chem. Eng. Process. Process Intensif.* **2008**, *47*, 1323–1332. [CrossRef]
10. Kätzel, U.; Vorbau, M.; Stintz, M.; Gottschalk-Gaudig, T.; Barthel, H. Dynamic light scattering for the characterization of polydisperse fractal systems: II. Relation between structure and DLS results. *Part. Part. Syst. Charact.* **2008**, *25*, 19–30. [CrossRef]
11. Flesch, J.C.; Spicer, P.T.; Pratsinis, S.E. Laminar and turbulent shear-induced flocculation of fractal aggregates. *AIChE J.* **1999**, *45*, 1114–1124. [CrossRef]
12. Leiva, W.H.; Fawell, P.D.; Goñi, C.; Toro, N.; Jeldres, R.I. Temporal evolution of the structure of tailings aggregates flocculated in seawater. *Miner. Eng.* **2021**, *160*, 106708. [CrossRef]
13. Deng, X.; Davé, R.N. Breakage of fractal agglomerates. *Chem. Eng. Sci.* **2017**, *161*, 117–126. [CrossRef]
14. Grabsch, A.F.; Yahyaei, M.; Fawell, P.D. Number-sensitive particle size measurements for monitoring flocculation responses to different grinding conditions. *Miner. Eng.* **2020**, *145*, 106088. [CrossRef]
15. Dao, V.H.; Mohanarangam, K.; Fawell, P.D.; Simic, K.; Iyer, R.; Cameron, N.R.; Saito, K. Enhanced flocculation efficiency in a high-ionic-strength environment by the aid of anionic ABA triblock copolymers. *Langmuir* **2020**, *36*, 1538–1551. [CrossRef] [PubMed]
16. Stephens, D.W.; Fawell, P.D. Process equipment optimisation using CFD and surrogate models. *Prog. Comput. Fluid Dyn. Int. J.* **2015**, *15*, 102. [CrossRef]
17. Nguyen, T.; Farrow, J.; Smith, J.; Fawell, P. Design and development of a novel thickener feedwell using computational fluid dynamics. *J. S. Afr. Inst. Min. Metall.* **2012**, *112*, 939–948.
18. Fawell, P.D.; Nguyen, T.V.; Solnordal, C.B.; Stephens, D.W. Enhancing gravity thickener feedwell design and operation for optimal flocculation through the application of computational fluid dynamics. *Miner. Process. Extr. Metall. Rev.* **2019**, *42*, 496–510. [CrossRef]
19. Tanguay, M.; Fawell, P.; Adkins, S. Modelling the impact of two different flocculants on the performance of a thickener feedwell. *Appl. Math. Model.* **2014**, *38*, 4262–4276. [CrossRef]
20. Zaki, S.; Etarahony, M.; Elkady, M.; Abd-El-Haleem, D. The use of bioflocculant and bioflocculant-producing Bacillus mojavensis strain 32A to synthesize silver nanoparticles. *J. Nanomater.* **2014**, *2014*, 8. [CrossRef]
21. Quezada, G.R.; Jeldres, R.I.; Fawell, P.D.; Toledo, P.G. Use of molecular dynamics to study the conformation of an anionic polyelectrolyte in saline medium and its adsorption on a quartz surface. *Miner. Eng.* **2018**, *129*, 102–105. [CrossRef]
22. Jeldres, M.; Piceros, E.C.; Toro, N.; Torres, D.; Robles, P.; Leiva, W.H.; Jeldres, R.I. Copper tailing flocculation in seawater: Relating the yield stress with fractal aggregates at varied mixing conditions. *Metals* **2019**, *9*, 1295. [CrossRef]
23. Bache, D.H.; Johnson, C.; McGilligan, J.F.; Rasool, E. A conceptual view of floc structure in the sweep floc domain. In *Water Science and Technology*; Elsevier Science Ltd.: Amsterdam, The Netherlands, 1997; Volume 36, pp. 49–56.
24. Jarvis, P.; Jefferson, B.; Gregory, J.; Parsons, S.A. A review of floc strength and breakage. *Water Res.* **2005**, *39*, 3121–3137. [CrossRef]
25. Jeldres, M.; Ayala, L.; Robles, P.; Gálvez, E.; Leiva, W.H.; Toledo, P.G.; Jeldres, R.I. A criterion for estimating the strength of flocculated aggregates in salt solutions. *Minerals* **2021**, *11*, 713. [CrossRef]
26. Teh, E.J.; Leong, Y.K.; Liu, Y.; Fourie, A.B.; Fahey, M. Differences in the rheology and surface chemistry of kaolin clay slurries: The source of the variations. *Chem. Eng. Sci.* **2009**, *64*, 3817–3825. [CrossRef]
27. Liu, D.; Edraki, M.; Berry, L. Investigating the settling behaviour of saline tailing suspensions using kaolinite, bentonite, and illite clay minerals. *Powder Technol.* **2018**, *326*, 228–236. [CrossRef]
28. Basnayaka, L.; Subasinghe, N.; Albijanic, B. Influence of clays on the slurry rheology and flotation of a pyritic gold ore. *Appl. Clay Sci.* **2017**, *136*, 230–238. [CrossRef]
29. Forbes, E.; Chryss, A. Fundamentals of Clays: Surface and Colloid Science, and Rheology. In *Clays in the Minerals Processing Value Chain*; Grafe, M., Klauber, C., McFarlane, A.J., Robinson, D.J., Eds.; Cambridge University Press: Cambridge, UK, 2017; pp. 81–110.
30. Arancibia-Bravo, M.P.; López-Valdivieso, A.; Flores, L.F.; Cisternas, L.A. Effects of potassium propyl xanthate collector and sodium sulfite depressant on the floatability of chalcopyrite in seawater and KCl solutions. *Minerals* **2020**, *10*, 991. [CrossRef]
31. Hernández, P.; Dorador, A.; Martínez, M.; Toro, N.; Castillo, J.; Ghorbani, Y. Use of seawater/brine and caliche's calts as clean and environmentally friendly sources of chloride and nitrate ions for chalcopyrite concentrate leaching. *Minerals* **2020**, *10*, 477. [CrossRef]
32. Cisternas, L.A.; Gálvez, E.D. The use of seawater in mining. *Miner. Process. Extr. Metall. Rev.* **2018**, *39*, 18–33. [CrossRef]
33. Hawley, N. Settling velocity distribution of natural aggregates. *J. Geophys. Res.* **1982**, *87*, 9489. [CrossRef]
34. Gibbs, R.J. Estuarine flocs: Their size, settling velocity and density. *J. Geophys. Res.* **1985**, *90*, 3249–3251. [CrossRef]
35. Li, D.H.; Ganczarczyk, J. Fractal geometry of particle aggregates generated in water and wastewater treatment processes. *Environ. Sci. Technol.* **1989**, *23*, 1385–1389. [CrossRef]

polymers

MDPI

Article

Preparation, Properties and Water Dissolution Behavior of Polyethylene Oxide Mats Prepared by Solution Blow Spinning

Miguel Ángel Lorente [1], Gustavo González-Gaitano [2] and Javier González-Benito [1,*]

[1] Department of Materials Science and Engineering and Chemical Engineering and Institute ITQMAAB, Universidad Carlos III de Madrid, Avda. Universidad 30, 28911 Leganés, Spain; milorent@ing.uc3m.es
[2] Department of Chemistry, Universidad de Navarra, 31080 Pamplona, Spain; gaitano@unav.es
* Correspondence: javid@ing.uc3m.es; Tel.: +34-916248870

Abstract: The relationship between processing conditions, structure and morphology are key issues to understanding the final properties of materials. For instance, in the case of polymers to be used as scaffolds in tissue engineering, wound dressings and membranes, morphology tuning is essential to control mechanical and wettability behaviors. In this work, the relationship between the processing conditions of the solution blow spinning process (SBS) used to prepare nonwoven mats of polyethylene oxide (PEO), and the structure and morphology of the resulting materials are studied systematically, to account for the thermal and mechanical behaviors and dissolution in water. After finding the optimal SBS processing conditions (air pressure, feed rate, working distance and polymer concentration), the effect of the solvent composition has been considered. The structure and morphology of the blow spun fibers are studied as well as their thermal, mechanical behaviors and dissolution in water. We demonstrate that the morphology of the fibers (size and porosity) changes with the solvent composition, which is reflected in different thermal and mechanical responses and in the dissolution rates of the materials in water.

Keywords: solution blow spinning; polyethylene oxide; morphology; materials characterization; polymer dissolution

Citation: Lorente, M.Á.; González-Gaitano, G.; González-Benito, J. Preparation, Properties and Water Dissolution Behavior of Polyethylene Oxide Mats Prepared by Solution Blow Spinning. *Polymers* **2022**, *14*, 1299. https://doi.org/10.3390/polym14071299

Academic Editors: Andrzej Puszka and Beata Podkościelna

Received: 25 February 2022
Accepted: 22 March 2022
Published: 23 March 2022

1. Introduction

Polyethylene oxide, PEO, is a biocompatible, water-soluble polymer used extensively for the release of active agents upon its dissolution in an aqueous environment [1–3]. The dissolution rate mainly depends on the molar mass of the polymer but also on its crystallinity and morphology, factors that condition the surface accessible to the solvent in the dissolution process. In turn, these depend on the processing method used to produce the material. This is the case of polymeric fibers, whose size, shape, and entanglement degree can be controlled to produce different release profiles. Polymeric fibers are lately receiving much interest for their applications as scaffolds in tissue engineering [4,5], wound dressings [6] and membrane design [7]. Among the different methods to prepare fibrous materials in the form of films or mats, electrospinning, ES, and solution blow spinning, SBS, are the most frequently used. Fiber diameters ranging from a few microns to tens of nanometers can be produced with both methods, depending on the processing conditions [8,9]. In the case of ES, the polymeric solution is injected through a nozzle containing a capillary. The drop that forms at the tip of the capillary is then drawn by the action of an electrical potential between the nozzle and a substrate (collector) [10–12] and transforms into dry fibers as the solvent quickly evaporates. In SBS, the drop formed at the tip of the capillary is drawn by the action of a pressurized gas that passes through a concentric nozzle [13–15] and favors the evaporation and orientation of fibers towards the collector, usually a rotating cylinder. Although ES is more widespread, the interest in SBS over the last few years has considerably increased, due to its simplicity of handling, no need for the application of large electric fields and in-situ application. On the other hand, there is another technique named forcespinning that has great interest in preparing polymer nanofibers [16,17], however, like ES, it does not seem feasible when the material is expected to be dispensed on specific sites, such as for instance a wound to be treated. Up to now, it can be considered that there are very few articles published on the preparation of polymeric materials from SBS compared to those published where the materials are prepared by ES. For this reason, taking into account the potentiality of SBS to prepare materials for multiple applications, even more work should be conducted, since there are still many aspects to improve, for instance, control of final morphology, less loss of material, among others.

Preparation of fibrous materials by SBS requires a careful choice of the processing conditions. The parameters that can affect the final morphology [18–20] can be grouped as shown in Table 1.

Table 1. Parameters that affect final morphology of SBS polymer systems.

Parameters Associated to the Solution	Parameters Associated to the Processing	Environment Conditions
Polymer concentration	Working distance, WD (distance from the tip of the capillary to the collector)	Temperature
Type of solvent and composition	Injection or feeding rate, FR (velocity at which the polymer solution is injected)	Humidity
Viscosity	Gas pressure, Ap	
Surface tension	Rotational speed of the collector, RSC	
	Diameter of the capillary	
	Capillary protrusion from the nozzle exit	
Polymer concentration	Working distance, WD (distance from the tip of the capillary to the collector)	Temperature

The fine adjustment of these parameters can lead to different morphologies [21–23] and, consequently, to different mechanical, thermal and wettability properties. In fact, several works about the use of SBS are simply focused on the study of the influence of processing conditions, represented by the parameters shown in Table 1, on the final morphology of the materials prepared [12]. In addition, if active agents, such as therapeutic molecules, such as antimicrobials or anti-inflammatories are incorporated in the fiber structure, with the aim of providing multifunctionality to the material, they must be compatible with the polymer and soluble in the solvent mixture.

Although there are several very interesting biocompatible and biodegradable biomaterials, including BSA, chitosan, etc. [24,25]. PEO is a very simple and low-cost polymer to start exploring how SBS processing conditions affect final morphology. PEO is a linear hydrophilic polymer that is synthesized from ethylene oxide. It is easy to produce, water soluble, non-toxic, or sensitive to pH or physiological fluids. After contact with water, the PEO is hydrated and swollen, resulting in a layer of hydrogel that regulates the subsequent inlet of water and the corresponding dissolution of the active agents that may be inside the polymer. These special properties of hydration and swelling of the PEO make it a remarkably interesting candidate in the pharmaceutical industry for the manufacture of drug release systems [26].

Optimization of SBS processing conditions is thus of great importance in order to achieve the desired properties, which constitutes the objective of the present work, focused on the preparation of fibrous PEO-based mats. The influence of the processing conditions on the morphology and the thermal and mechanical response has been evaluated in terms of the ability of the polymer to be dissolved in water, an essential aspect for the design of PEO mats as a controlled drug delivery system. In this study, the influence of the solvent used to prepare the PEO solutions to be blow spun was considered. To change the solvent properties, a simple variation on the composition in a mixture of acetone and chloroform was chosen in this work.

2. Materials and Methods

2.1. Materials

Powder PEO (Mv = 100,000, CAS: 25322-68-3, density = 1.13 g/mL) was provided by Sigma-Aldrich (Sant Louis, MO, USA). Solutions of the polymer were prepared in a mixture of acetone, Ac $\geq$ 99.9% purity (CAS: 67-64-1, M = 58.08 g/mol, bp = 56 °C, density = 0.791 g/mL) and chloroform, Chl $\geq$ 99% purity (CAS: 67-66-3, M = 119.38 g/mol, bp = 61 °C, density = 1.48 g/mL). Both solvents were supplied by Sigma-Aldrich (Sant Louis, MO, USA).

2.2. Preparation of Materials

According to previous studies [27,28] preliminary working conditions for the SBS process were set (capillary diameter 0.6 mm, working distance, WD = 15 cm; injection rate, Fr = 0.5 mL/min; air pressure, Ap = 2 bar; rotational speed of the collector, RSC = 200 rpm; volume of injected solution 10 mL). The process was carried out at 23 °C and 40% relative humidity. For a given composition of the solvent mixture, PEO concentrations were 6, 8, 10, 12 and 14% wt/v. For a selected concentration of PEO, the solvent compositions considered were 4:6, 5:5, 6:4, 7:3, 8:2, 9:1 and 10:0 v/v in Chl:Ac, in such a way that the final materials obtained were named as PEO-46, PEO-55, PEO-64, PEO-73; PEO-82 and PEO-100, respectively. PEO concentrations lower than 6% led to very thin mats with no fibrous morphology. Likewise, polymer concentrations higher than 14%, and acetone proportion higher than 40% were not considered either due to poor polymer solubility.

2.3. Characterization Methods

Viscosity measurements were performed using a Haake iQ viscometer. The sample was placed between two circular plates at constant temperature (25 °C), and subjected to an oscillatory shear force, in the frequency range 100–900 s^{-1}, with an acquisition time of 30 s per measurement (100 points).

The structure of the materials was investigated by attenuated total reflectance Fourier transform infrared spectroscopy, ATR-FTIR, and X-ray diffraction, XRD. A Shimadzu IRAffinity-1S Spectrometer, equipped with a Golden Gate ATR accessory (diamond window) was used to collect the IR spectra (32 scans per interferogram), in the wavelength range 600–4000 cm^{-1}. XRD patterns were recorded with an X-ray Powder Diffractometer Bruker ECO D8 Advance (Bruker, Karslruhe, Germany) with a Bragg-Brentano configuration and a Lynxeye XE-T coupled detector, using the Cu-K_{a1}, λ = 1.5418 Å. Blow spun samples were placed on an amorphous glass holder and the diffractograms were recorded from 5° and 35° in 2θ at 2 s per step with 0.02° of step size.

The thermal behavior of the materials was studied by differential scanning calorimetry, DSC, using a Mettler Toledo 822e calorimeter under N_2 atmosphere, applying the following thermal cycles: (i) dynamic heating from 30 to 90 °C at 10 °C/min (materials prepared by SBS) and (ii) dynamic cooling from 90 to 30 °C at 10 °C/min (materials after erasing their processing and thermal histories).

Temperatures of melting, T_m, and crystallization, T_c, were obtained from the endothermic and exothermic peaks in the heating and cooling scans, respectively. Fusion, ΔH_m, and crystallization, ΔH_c, enthalpies were determined from the areas under the endothermic and exothermic peaks, respectively. The crystalline fraction was calculated for the samples produced by SBS before and after erasing their processing and thermal histories, χ_m and χ_c, using Equations (1) and (2), respectively:

$$\chi_m = \frac{\Delta H_m}{\Delta H_\infty} \quad (1)$$

$$\chi_c = \frac{\Delta H_c}{\Delta H_\infty} \quad (2)$$

where ΔH_∞ = 197 J/g [29–31] is the melting enthalpy of 100% crystalline PEO.

The morphology of the SBS materials was studied by field emission scanning electron microscopy (FESEM) using a TENEO-FEI microscope. The signal from the secondary electrons in the ETD detector was used to generate the FSEM images, applying an acceleration voltage of 5 kV. Samples were carbon-coated to avoid charge accumulation using a Leica EM ACE200 low-vacuum coater. Images were acquired using different tilt angles in order to perform a more complete morphological analysis, which was carried out with ImageJ V.1.52a (National Institute of Mental Health, Bethesda, MD, USA).

The specimens of the blow spun materials were detached from the collector considering its direction of rotation (the longest side of the specimen was parallel to the rotation direction). The length and width of the specimens were measured with a caliper (±0.01 mm accuracy), and the thickness with a Digimatic micrometer (Mitutoyo Corporation, Barcelona, Spain), ±1 μm accuracy. At least six specimens per sample were considered, and the dimensions were the average of five measurements.

The mechanical behavior of the materials was studied using a universal testing machine Microtest TPF-1D at a crosshead speed of 1 mm/min and a load cell of 5 kp. Hard rubber pieces were located on the inner side of the clamps to avoid the specimen sliding during the tests. Six well-dimensioned specimens per material were tested. From the analysis of the resulting stress-strain curves, the following parameters were obtained as averaged values: Young's modulus, E, from the slope of the stress-strain curve; the tensile stress, σ_r, representing the maximum stress value at which the specimen breaks, determined as the point at which the maximum stress is reduced by 5% [32]; the yield stress, σ_y, determined as the stress at which plastic deformation starts (stress-strain plot gives up being linear) [32]; and maximum elongation at failure.

For monitoring the dissolution process of the materials, an optical method was proposed, consisting in recording a video (Optika Microscope 4083.13 camera, exposure time of 39 milliseconds) from the output signal of an optical microscope (Olympus SZ-CTV with a light source Olympus Highlight 2100 and an SZ-40 objective with 40× magnification). The quantification of the fraction of dissolved material was carried out from the loss of transparency of the films by image analysis using video-image file converter software, Prism (V.7.14). A MatLab (R2018b) script was used to transform the raw frames images from the video to 8-bit greyscale and then to black and white, using an intensity threshold of 0.7 for each pixel (1 if the intensity is higher than 0.7, and 0 otherwise). The fraction of pixels with an intensity equal to 1 is taken as the fraction of specimen dissolved, which is then plotted versus time.

3. Results and Discussion

Processability by SBS is highly dependent on the viscosity of the solution [33,34]. High viscosities can produce nozzle obstruction or overpressure, while low viscous solutions, mainly associated with low polymer concentrations, usually lead to poor yields of material collected. Additionally, when using solvents of relatively high boiling point, the remaining solvent can be found in the blow-spun material, which favors spreading on the collector and formation of films instead of fibers. Therefore, a preliminary viscosity study using different empirical and semi empirical methods has been carried out in order to optimize the solvent composition for a given PEO concentration.

Viscosity as a function of the polymer concentration for different solvent mixtures is plotted in Figure 1. Different color bars correspond to each composition used (volume ratio) of solvent mixture Chl:Ac. The Huggins-Martins method [35,36] provided in all cases the best fits. Regardless of the solvent composition, there is a general trend where viscosity increases with the concentration of PEO and as expected, the higher the proportion of chloroform the higher the viscosity at a particular concentration of PEO, given the higher viscosity of chloroform (0.542 cP at 25 °C versus 0.316 cP for acetone at 25 °C). At concentrations above 10%, the SBS process became unstable due to overpressure in the nozzle, causing heterogeneous solution ejection. Therefore, the polymer concentration was set at 10% wt in all the SBS experiments since it was the highest concentration that produced a stable jet of solution.

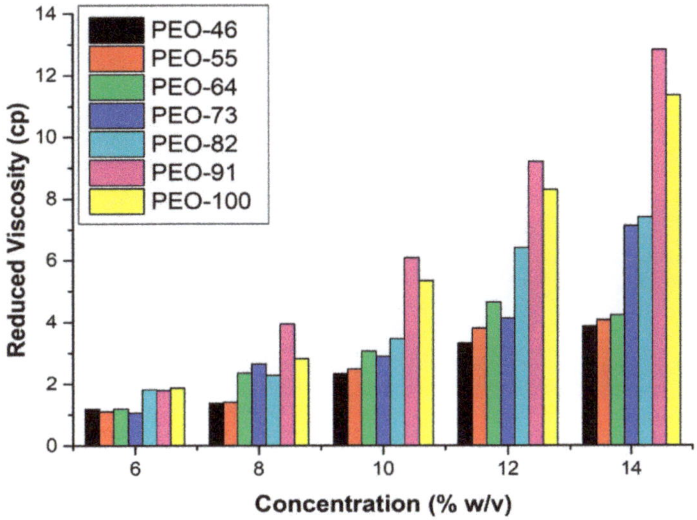

Figure 1. Viscosity of the PEO solutions as a function of concentration and composition of the solvent mixture.

The structure of the prepared materials was tested by FTIR-ATR and XRD. The infrared spectra of the blow spun materials are shown in Figure 2. The characteristic peaks of the solvents used (acetone and chloroform) are not observed in the spectra, for example, the absorption band due to the C-Cl stretching in chloroform at 665 cm^{-1} or the C=O stretching band in acetone at 1710 cm^{-1}, which reinforces the idea of the complete evaporation of the solvent during the SBS process.

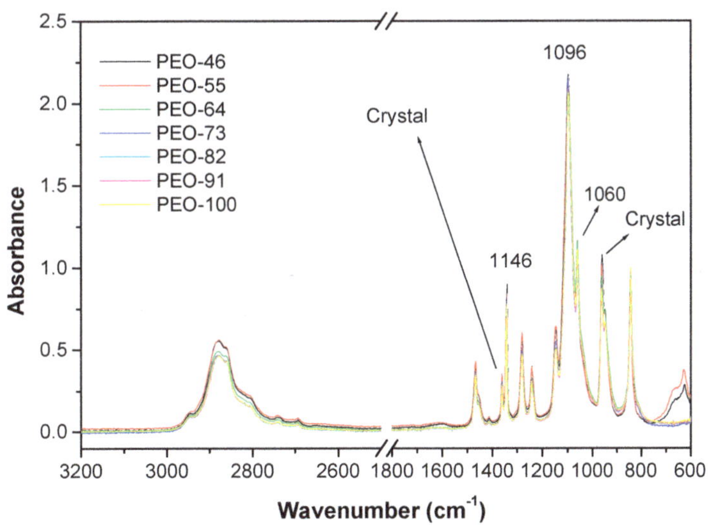

Figure 2. ATR-FTIR spectra of the blow spun PEO-based obtained from 10% wt solutions at different compositions Chl:Ac.

The band assignation of the PEO was conducted by Yosihara et al. [37]. The methylene C-H stretching appears in the range of 2675 cm^{-1} to 3010 cm^{-1}. The bands at 1146 cm^{-1}, 1095 cm^{-1} and 1060 cm^{-1} are assigned to a triplet stretching of ether group –C–O–C– [38–41] and can be used to study structural variations in the PEO since they depend on the chain conformation of semi-crystalline PEO [38–41]. To this effect, the spectra have been normalized to the absorption at 845 cm^{-1}, since that band is usually the least affected by changes in the structure, according to I. Pucic et al. [38]. Although the most intense signal is the one at 1095 cm^{-1}, the others follow a different order, since the peak at 1146 cm^{-1} is more intense than the one centered at 1060 cm^{-1} [38]. On the other hand, the presence of a double peak at 1359 cm^{-1} and 1340 cm^{-1} is clear evidence of crystalline PEO [41]. In general, a slight decrease of the relative intensities of the peaks at 1359 cm^{-1}, 1146 cm^{-1}, 1060 cm^{-1} and 960 cm^{-1}, would indicate a lower crystallinity degree with the content in chloroform [41].

Figure 3 shows the baseline corrected and normalized XRD diffractograms of the SBS materials. Characteristics peaks of crystalline PEO [42,43] can be observed at 19.3°, corresponding to the reflections due to the planes (120), and at 23.5°, associated with the planes (112) [23,44–47]. There are no significant differences among the diffraction patterns but a slight shift to low angles of the peak for the (120) planes and a small variation of the peak half-bandwidths and crystallinity degrees with the proportion of chloroform.

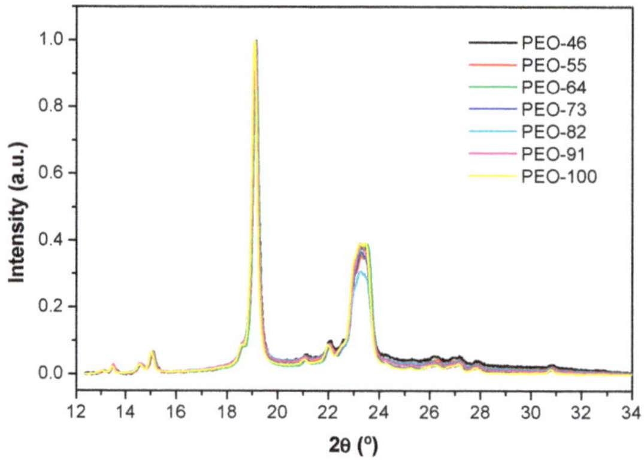

Figure 3. X-ray diffractograms of the blow spun PEO-based obtained from 10% wt solutions at different compositions Chl:Ac.

The crystallinity can be estimated semiquantitatively from the amorphous and crystalline contributions of the XRD pattern, deduced from deconvolution by Gaussian functions of the reflections. The fraction of crystalline polymer can be calculated by:

$$X_{XRD} = \frac{\sum_i A_i}{A_T} \tag{3}$$

where A_i represents the area of the deconvoluted peak i and A_T is the total area of the diffractogram. Additional parameters can be extracted from the diffractograms. The full width at half maximum, FWHM, of the diffraction peaks, is related to the size of the crystallites, D (Equation (4)), as well as the micro-strain, ε (Equation (5)), due to the presence of defects or the processing conditions.

$$D = \frac{k\lambda}{\beta_s \cos\theta} \tag{4}$$

$$\varepsilon = \frac{\beta_s}{4\tan\theta} \tag{5}$$

where k is a dimensionless shape factor whose value is near to 1; λ is the X-ray wavelength; β_s is the FWHM in radians after subtracting the instrumental line broadening (considered negligible with respect to FWHM of a peak in a polymer) and θ is the Bragg angle in radians.

The results of these calculations are shown in Table 2. It can be seen how, as the proportion of chloroform increases, the fraction of the crystalline phase slightly decreases. Besides, although among the solvent mixtures there is not a clear correlation between viscosity and D or ε, when pure chloroform is used to dissolve the polymer, the lowest micro-strain and highest crystallite size and crystallinity fraction are obtained. This result can be ascribed to the higher viscosity of the polymer solution. It is reasonable to think that higher viscosity makes it more difficult to drag the solution and the subsequent polymer chain preorientation. This would imply fewer constrictions for the polymer to crystallize leading to a more relaxed crystalline structure, with larger crystallite sizes and a higher degree of crystallinity.

Table 2. Estimated crystalline fraction, crystallite size and micro-strain generated for all materials studied.

	PEO-46	PEO-55	PEO-64	PEO-73	PEO-82	PEO-91	PEO-100
χ_{XDR}	0.56	0.63	0.74	0.70	0.65	0.70	0.76
D (nm)	34.4	35.2	34.0	33.5	34.0	34.5	35.8
$\varepsilon \times 10^3$	6.74	6.58	6.81	6.93	6.83	6.72	6.50

The thermal behavior of the SBS materials was studied by differential scanning calorimetry, DSC (Figure 4). In the case of the as-received commercial sample, a single homogeneous endothermic peak at 68.9 °C is observed, corresponding to the melting point of PEO reported by other authors [48] and by the supplier. On the other hand, highly heterogeneous endothermic transitions are observed for the blow spun polymer, occurring in a wider range of temperatures and even showing distinct overlapped peaks. This peculiar melting behavior of PEO could be ascribed to an instantaneous fractionation process according to the molar mass. Yan Kuo et al. showed that melting and crystallization points of PEO are highly dependent on its molar mass [49]. In the SBS process, solvent evaporation enriches the vapor phase in the most volatile component (acetone, in this case). Since PEO is less soluble in this solvent, higher proportions of acetone will induce the precipitation of polymer, which begins with the longer PEO chains. In other words, as the evaporation proceeds, fibers will form from polymers of different molar masses, from highest to lowest, in a sort of "time of flight" phase separation, which, leads to a heterogeneous thermal behavior, as detected in our DSC experiments. In general, when SBS PEO samples are prepared from chloroform proportions higher than 70%, a narrower endothermic transition is observed, although it is highly heterogeneous (one peak and two shoulders). These results suggest that the SBS process causes a very heterogeneous crystallization (crystallites and spherulites of different sizes and densities) that strongly depends on the nature of the solvent, tending to be more homogeneous the higher the proportion of chloroform in the mixture is. According to this reasoning, when the proportion of chloroform is high enough, all fractions of PEO can remain together, since chloroform is a better solvent of PEO than acetone, leading to a more homogeneous molar mass distribution in the fibers.

Crystallinity degrees, χ_m and χ_c, obtained from DSC thermograms are collected in Table 3. As can be seen, there are no big differences between the materials prepared with values of crystallinity degree in the range 66–71%, coincident with the range of crystallinities obtained by the XRD results (Table 1), however, no correlations with SBS conditions can be extracted.

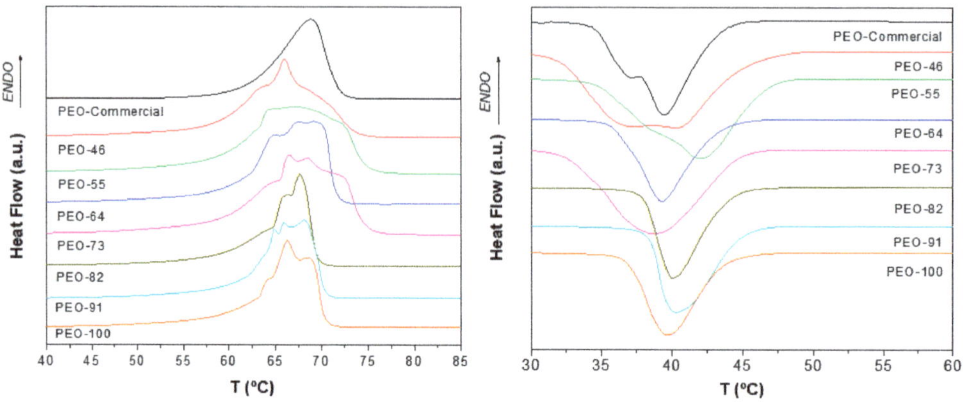

Figure 4. DSC thermograms of blow spun PEO at different compositions Chl:Ac. First heating scan (**left**) and subsequent cooling scan (**right**).

Table 3. Data obtained from DSC curves corresponding to: (a) melting transition and (b) crystallization process.

(a)	PEO 46	PEO 55	PEO 64	PEO 73	PEO 82	PEO 91	PEO 100	PEO Com
T_m (°C)	61.9	61.1	61.2	59.0	63.6	55.4	63.1	62.1
ΔH_m (J/g)	131.6	130.3	126.0	122.4	137.8	145.5	143.1	131.1
χ_m (%)	66.8	66.1	64.0	62.2	69.9	73.8	72.6	66.5
(b)	**PEO 46**	**PEO 55**	**PEO 64**	**PEO 73**	**PEO 82**	**PEO 91**	**PEO 100**	**PEO Com**
T_c (°C)	41.1	46.6	43.1	44.7	43.8	44.9	43.7	43.1
ΔH_c (J/g)	121.1	126.5	117.8	129.3	123.4	133.4	130.8	88.0
χ_c (%)	61.5	64.2	59.8	65.6	62.7	67.7	66.4	44.6

It is worth mentioning that, after erasing the processing and thermal history of the samples, one would expect very similar DSC thermograms in the cooling scan, and yet the DSC profiles strongly depend on the conditions of sample preparation. This only can be explained if the thermal treatment chosen to erase the processing history is not as effective as it should to completely disrupt the solid-state structure with complete diffusion of macromolecules, in order to perfectly mix the different molar masses fractions. In other words, although the interactions that maintain macromolecules in an ordered structure are overcome, the complete macromolecular flow was not achieved, keeping in some way a patterned seed for subsequent crystallization. It is clear, therefore, that to fully erase the thermal and processing history, higher temperatures or longer times would be necessary.

The materials morphology was studied by FSEM using the signal from secondary electrons, SE. Figure 5 shows images at different magnifications (500×; 4000× and 8000×) of the blow spun solutions at different solvent compositions.

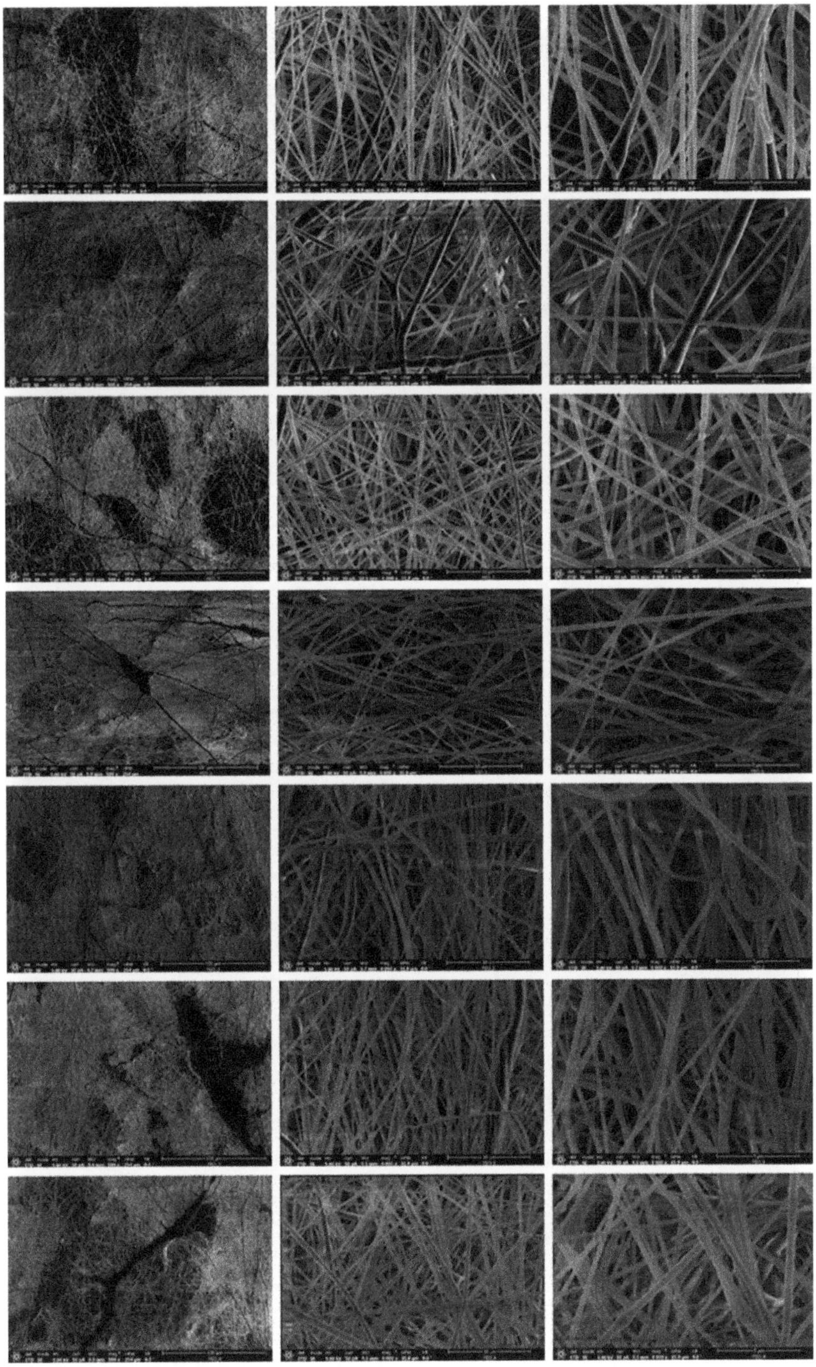

Figure 5. FESEM images of the blow spun fibers at different magnifications. Columns (left to right) represent the magnification, 500×, 4000× and 8000×. Rows (top to bottom) correspond to PEO-46, PEO-55, PEO-64, PEO-73, PEO-82, PEO-91AND PEO-100.

A highly homogeneous fibrous morphology can be visualized in all cases, with little presence of material accumulation in the typical form of bead-on-a-string shape or as heterogeneous corpuscles. By using different tilt angles of observation, morphological differences were not observed (Figure S1). In terms of heterogeneity, regions with different relative densities and entanglements of fibers exist, regardless of the solvent composition. The results of the fibers diameter analysis (Figures S2–S8) are shown in Table 4 (average diameter, <D>; and standard deviation, σ). It can be observed that the mean diameters do not depend on either the tilt angle of observation or the sample type. Likewise, the width and shape of the distributions are also very similar (Figures S2–S8).

Table 4. Morphological parameters of the blow spun fibers extracted from the SEM images analysis.

Sample	Tilt Angle (°)	<D> (nm)	Mean Diameter (nm)	σ (nm)
PEO-46	−5	308	302	240
	0	311		259
	5	302		205
	10	287		187
PEO-55	−5	346	301	302
	0	232		268
	5	316		232
	10	309		193
PEO-64	−5	293	294	237
	0	307		259
	5	289		253
	10	285		235
PEO-73	−5	290	304	180
	0	315		239
	5	324		227
	10	329		246
PEO-82	−5	363	339	290
	0	346		280
	5	295		145
	10	351		187
PEO-91	−5	280	294	225
	0	306		212
	5	274		201
	10	315		238
PEO-100	−5	247	288	169
	0	323		238
	5	319		299
	10	264		187

The porosity of the materials is a key parameter regarding the potential applications of the material. Two procedures have been used to estimate the porosity:

(a) Image analysis. Permits obtaining the mean pore area, the number of pores and air area (Table 5). The results can be extrapolated to a 3D network of fibers assuming a homogeneous porosity throughout the sample.

$$A_{\text{Air}} = \text{mean pore area} \times \text{number of pores} \quad (6)$$

$$\text{Porosity } (\%) = \frac{A_{\text{Air}}}{A_{\text{image}}} 100 \quad (7)$$

Table 5. Mean pore area estimated from image analysis.

	PEO-46	PEO-55	PEO-64	PEO-73	PEO-82	PEO-91	PEO-100
Mean Pore Area (μm)	0.88	1.28	0.78	0.96	1.21	0.97	0.87

(b) Gravimetry. The void space or volume of air, V_{Air}, can be determined by weighing, using a correction parameter (K) as the ratio between the apparent density (weight of the specimen, m, divided by its volume, $V_{specimen}$), ρ_A, and the density of the bulk material, ρ_R [50,51]:

$$V_{Air} = \frac{m\left(1 - K^2\right)}{\rho_A} \tag{8}$$

$$\text{Porosity } (\%) = \frac{V_{Air}}{V_{especimen}} 100 \tag{9}$$

The porosity values, given in percentage of air, are shown in Table 6. As expected, image analysis yields lower values of porosity since a projection in a plane or 2D images is considered. However, the important thing is that both methods do not show any particular trend with respect to the solvent composition used to blow spun the PEO.

Table 6. Estimated porosities deduced from image analysis and gravimetry.

		PEO-46	PEO-55	PEO-64	PEO-73	PEO-82	PEO-91	PEO-100
Porosity (%)	Image Analysis	39	31	37	38	38	38	36
	Gravimetry	62	61	61	68	73	63	64

Another important morphological feature that may influence the performance of the material is the preferential orientation of the fibers. This calculation was carried out on the SEM images by using the OrientationJ plugin in ImageJ software, the results were plotted as number distribution (fraction of fiber segments) as a function of the orientation angle, φ (Figure 6). In these plots, the number of peaks in the distribution gives information about the number of preferential orientations for the fibers, while the homogeneity of a particular orientation is deduced from the peak width. As can be seen, all the samples present a wide monomodal distribution, indicating no preferential orientation of the fibers.

The mechanical behavior of the materials was studied by tensile tests. Since these materials are highly porous, it is convenient to estimate the parameters that correspond to the bulk material, in addition to those that can be extracted directly from the tensile curves. In the first case, all factors affecting the mechanical behavior must be considered (fiber size, preferential orientation, porosity and crystallinity) but, in the second, consideration of fiber size and porosity would not make any sense.

The mathematical treatment applied to evaluate the porosity contribution from the tensile tests data has been described elsewhere [50]. In Figure 7, values of Young's modulus are plotted as a function of the solvent composition, with and without porosity correction.

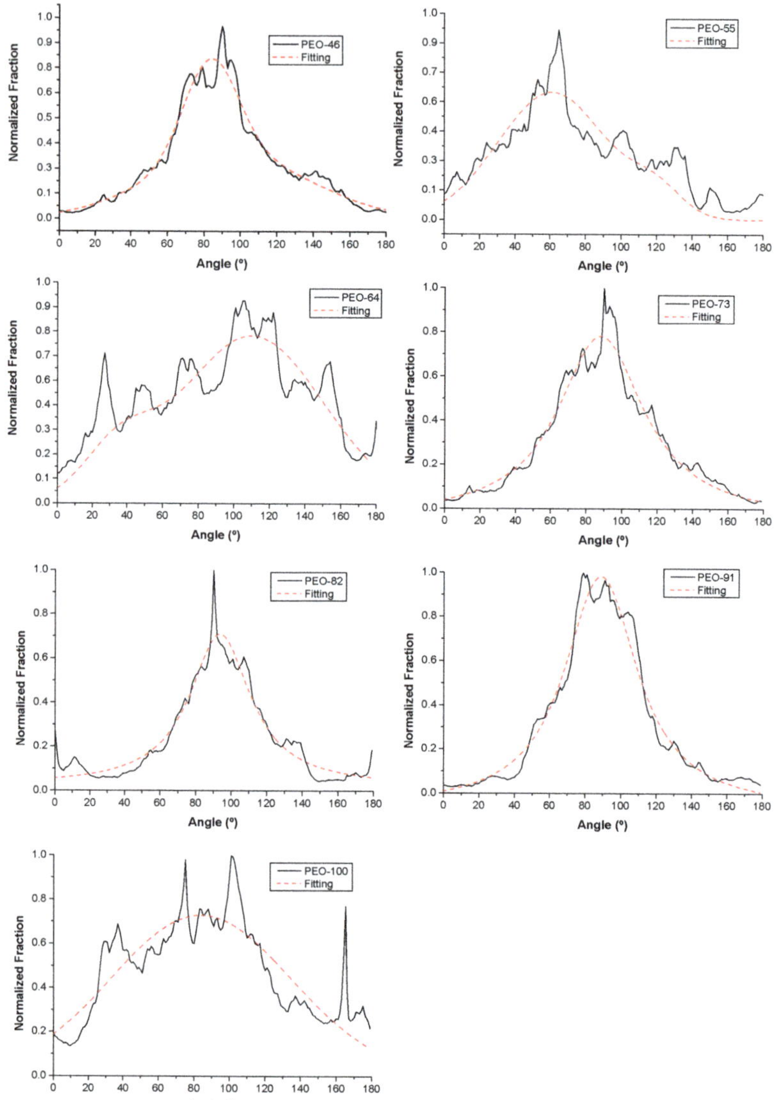

Figure 6. Normalized orientation distributions of blow spun PEO fibers at different compositions Chl:Ac.

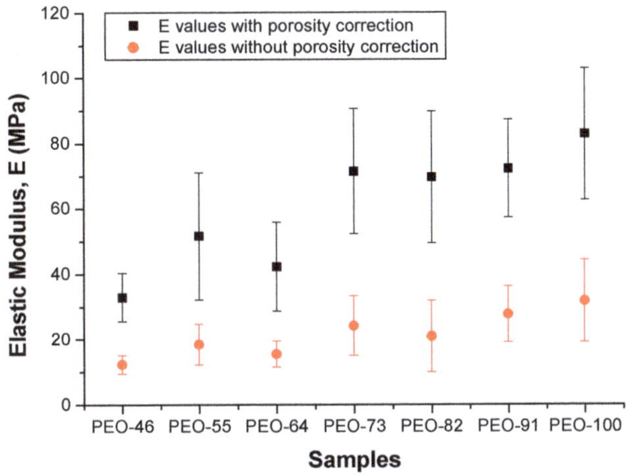

Figure 7. Young's modulus of the blow spun PEO mats, with and without porosity correction.

As expected, there is a clear difference between the moduli obtained for the as-prepared samples (red values) and after removing the contribution of the porosity (black values), being in this case around 60% higher, and close to the reported Young's modulus for PEO [42–55]. On the other hand, Young's moduli increase when the proportion of chloroform in the solvent is used for SBS regardless of the porosity. Other mechanical parameters were also extracted from the stress-strain plots, maximum stress, σ_{max} (Figure S9), stress at failure, σ_f (Figure S10), and yield strength, σ_y (Figure S11). All of them show similar tendencies, pointing out that mechanical parameters referring to the material strength increase as the proportion of chloroform in the solvent used for SBS increases. These results indicate that variations of structure and/or morphology are affecting mechanical behavior. As shown above, the porosity is practically constant. Likewise, negligible variations in crystallinity as well as in the preferential orientation of the constituent fibers are observed. However, a parameter that may account for the observed mechanical behavior is the observed heterogeneity of the distributions of molar masses of PEO. When the proportion of chloroform increases, a higher homogeneity in the molar mass distribution is expected and, a more homogeneous distribution of molar masses must improve the transmission of loads because of more effective intermolecular interactions.

Regarding the elastic and plastic behavior, they have been studied from the relative areas obtained from the initial strain value and the yield strain value, A_E, and from the yield strain and the rupture strain, A_P, respectively (Figure 8). As can be seen, the higher the chloroform content in the solvent used for SBS, the higher the plastic behavior of the material. This result is typical of semi-crystalline thermoplastic polymers, in which a wider plastic region is found when the interactions between macromolecules are stronger.

The dissolution process of the PEO blow spun materials in water was monitored by optical microscopy. Square specimens were placed between two glass slides and the sample was visualized in an optical microscope (Figure 9) [56,57]. Then, 0.25 mL of deionized water was added in between the glass slides, reaching the specimen by capillarity action. The evolution of the material was then recorded as a function of time using a video camera.

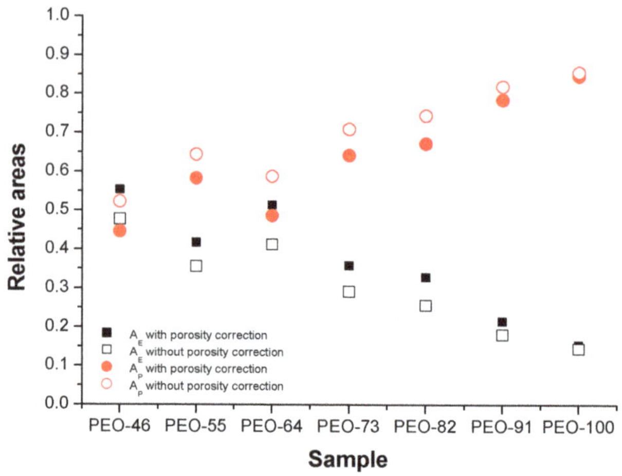

Figure 8. Elastic-plastic behavior of the different PEO-based materials before (circles) and after (squares) considering porosity.

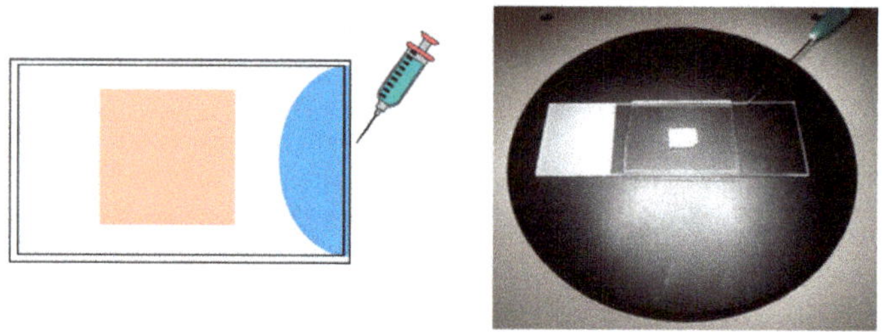

Figure 9. Set up for the dissolution tests.

As the dissolution occurs, light reflected due to the opaque solid reduces progressively as the medium becomes transparent. The quantitative monitoring of the process was performed by image analysis from video frames taken at different times (details of the data analysis in Section 2), as shown in Figures 10 and 11.

The fraction of specimen dissolved has been plotted as a function of time in Figure 11. All the dissolution plots show a sigmoidal profile: after an induction time, the dissolution process begins, until the curve levels off, marking the finalization of the dissolution. Under our experimental conditions, the sample is fully dissolved in less than 20 s. Depending on the sample type, the slope of the intermediate region of the plot differs. The PEO dissolution observed in this work is faster than in studies reported [58], since the materials here produced are nonwoven fibrous mats and highly porous. This wide contact surface makes the interaction with the solvent more effective and consequently the dissolution. Following the same reasoning, one would expect higher dissolution rates the thinner the fibers are (and the lower the density of fibers or the higher the porosity). As can be seen in Table 5, the average diameter is similar for all the blow spun material, but with certain differences in terms of porosity parameters (Tables 5 and 6). For instance, all the samples present similar porosities except PEO-55, which has a slightly lower value. On the other hand, PEO-55 and PEO-82 show a smaller number of pores with a higher mean pore area (Table 5), with these the samples showing different dissolution patterns and a slower dissolution rate.

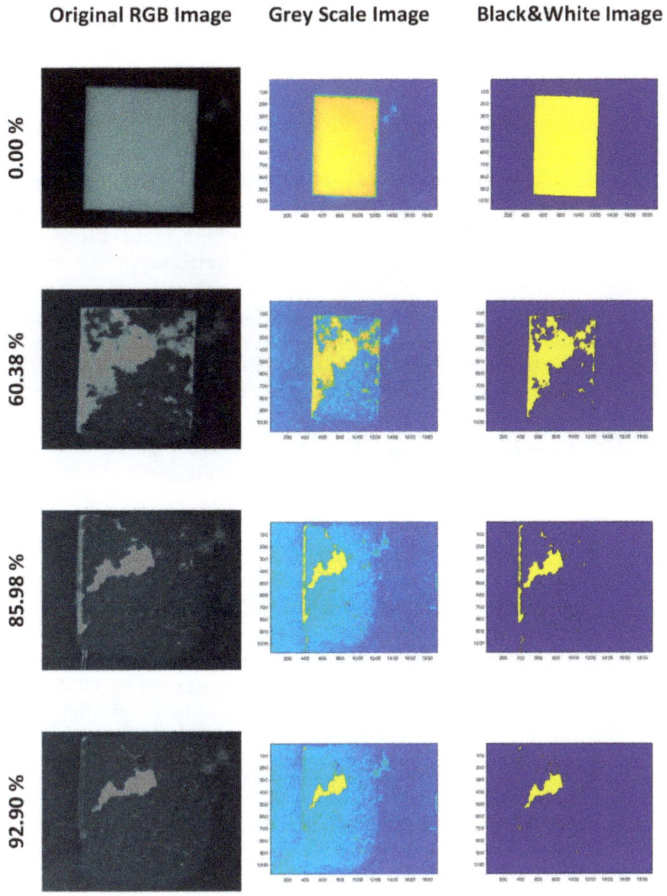

Figure 10. Image analysis to monitor dissolution process of the PEO-based specimens.

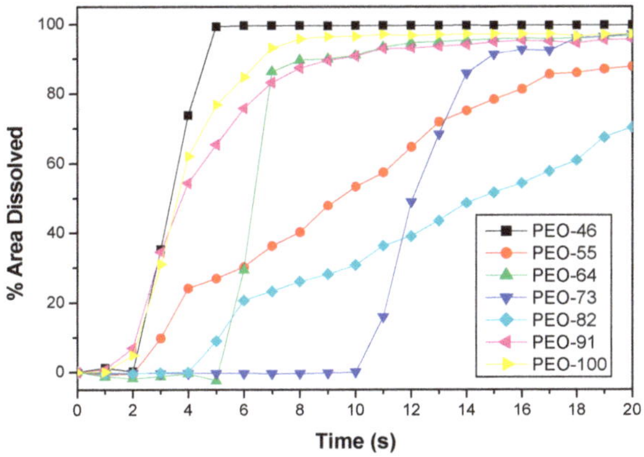

Figure 11. Dissolution profiles of blow spun PEO mats at different solvent compositions Chl:Ac.

4. Conclusions

The solution blow spinning method, SBS, was used to prepare PEO-based materials in the form of nonwoven mats constituted by submicrometric fibers. By changing the solvent composition used to dissolve the polymer (mixture of acetone and chloroform), different thermal, mechanical and dissolution behaviors of the material were found. From the study of the thermal and mechanical behavior, it is concluded that the composition of the solvent conditions, the evaporation rate, and therefore, the way the solid PEO fibers are formed, in an instantaneous or "time of flight" fractionation of the polymer, is according to the different molar masses. A higher proportion of chloroform favors the homogeneity of the molar mass distribution, which is reflected in better-defined thermal transitions in DSC and higher mechanical strength since the specific interactions occurring between the polymer chains are favored. Finally, the dissolution process of the mats is mainly conditioned by the global morphology of the fibrous material, higher porosity, and small fiber diameter leading to a faster dissolution rate in water, as a consequence of the increased polymer surface available to the solvent.

Supplementary Materials: The following supporting information can be downloaded at: https://www.mdpi.com/article/10.3390/polym14071299/s1. Figure S1. SEM images of the different PEO developed materials with a 4000× magnification at different tilts. Columns (from left to right) represent the different tilt angles: −5, 0, 5 and 10 degrees; whereas rows (from top to button) represent the different materials: PEO-46, PEO-55, PEO-64, PEO-73, PEO-82, PEO-91AND PEO-100. Figure S2. Fiber size distributions of PEO-46 as a function of the tilt. Figure S3. Fiber size distributions of PEO-55 as a function of the tilt. Figure S4. Fiber size distributions of PEO-64 as a function of the tilt. Figure S5. Fiber size distributions of PEO-73 as a function of the tilt. Figure S6. Fiber size distributions of PEO-82 as a function of the tilt. Figure S7. Fiber size distributions of PEO-91 as a function of the tilt. Figure S8. Fiber size distributions of PEO-100 as a function of the tilt. Figure S9. Maximum stress as a function of PEO based material. Figure S10. Stress at failure as a function of PEO based material. Figure S11. Yield stress as a function of PEO based material.

Author Contributions: M.Á.L. carried out all experimental tests except ATR-FTIR and DRX, data curation; formal analysis and discussion of the whole paper and editing. G.G.-G. was the responsible of carried out the ATR-FTIR and DRX experiments, formal analysis of the data obtained, funding acquisition, and writing and discussion of the article. J.G.-B. was the responsible of conceptualization; data curation; formal analysis; funding acquisition; methodology; project administration; resources; Software; supervision; validation; visualization; writing, review & editing. All authors have read and agreed to the published version of the manuscript.

Funding: This work was financially supported by AEI (Ministerio de Ciencia e Innovación of Spain, PID2020-112713RB-C22 and -C21]; the Universidad Carlos III de Madrid, Fondos de Investigación of Fco. Javier González Benito [2012/00130/004] and the strategic Action in Multifunctional Nanocomposite Materials [Code: 2011/00287/003].

Institutional Review Board Statement: Not applicable.

Informed Consent Statement: Not applicable.

Data Availability Statement: Not applicable.

Acknowledgments: Authors greatly appreciate help received form the technicians of the Department of Materials Science and Engineering and Chemical engineering of the Universidad Carlos III de Madrid and of the Department of Chemistry of the Universidad de Navarra.

Conflicts of Interest: The authors declare no conflict of interest.

References

1. Niazmand, R.; Razavizadeh, B.M. Active Polyethylene Films Incorporated with β-Cyclodextrin/Ferula Asafoetida Extract Inclusion Complexes: Sustained Release of Bioactive Agents. *Polym. Test.* **2021**, *95*, 107113. [CrossRef]
2. Zaitoon, A.; Lim, L.T.; Scott-Dupree, C. Activated Release of Ethyl Formate Vapor from Its Precursor Encapsulated in Ethyl Cellulose/Poly(Ethylene Oxide) Electrospun Nonwovens Intended for Active Packaging of Fresh Produce. *Food Hydrocoll.* **2021**, *112*, 106313. [CrossRef]

3. Zhao, X.; Chen, Z.; Lin, N.; Ma, J. Water Redispersion and Cytotoxicity of Reducing End-Modified Cellulose Nanocrystals by Grafting Long-Chain Poly(Ethylene Oxide). *Int. J. Biol. Macromol.* **2021**, *180*, 143–151. [CrossRef] [PubMed]
4. El Fawal, G.; Abu-Serie, M.M.; Mo, X.; Wang, H. Diethyldithiocarbamate/Silk Fibroin/Polyethylene Oxide Nanofibrous for Cancer Therapy: Fabrication, Characterization and in Vitro Evaluation. *Int. J. Biol. Macromol.* **2021**, *193*, 293–299. [CrossRef]
5. Eldurini, S.; Abd El-Hady, B.M.; Shafaa, M.W.; Gad, A.A.M.; Tolba, E. A Multicompartment Vascular Implant of Electrospun Wintergreen Oil/ Polycaprolactone Fibers Coated with Poly(Ethylene Oxide). *Biomed. J.* **2021**, *44*, 589–597. [CrossRef]
6. Ionescu, O.M.; Iacob, A.T.; Mignon, A.; van Vlierberghe, S.; Baican, M.; Danu, M.; Ibănescu, C.; Simionescu, N.; Profire, L. Design, Preparation and In Vitro Characterization of Biomimetic and Bioactive Chitosan/Polyethylene Oxide Based Nanofibers as Wound Dressings. *Int. J. Biol. Macromol.* **2021**, *193*, 996–1008. [CrossRef]
7. Bandehali, S.; Moghadassi, A.; Parvizian, F.; Hosseini, S.M.; Matsuura, T.; Joudaki, E. Advances in High Carbon Dioxide Separation Performance of Poly (Ethylene Oxide)-Based Membranes. *J. Energy Chem.* **2020**, *46*, 30–52. [CrossRef]
8. Wojasiński, M.; Pilarek, M.; Ciach, T. Comparative Studies of Electrospinning and Solution Blow Spinning Processes for the Production of Nanofibrous Poly(L-Lactic Acid) Materials for Biomedical Engineering. *Pol. J. Chem. Technol.* **2014**, *16*, 43–50. [CrossRef]
9. Oliveira, J.E.; Mattoso, L.H.C.; Orts, W.J.; Medeiros, E.S. Structural and Morphological Characterization of Micro and Nanofibers Produced by Electrospinning and Solution Blow Spinning: A Comparative Study. *Adv. Mater. Sci. Eng.* **2013**, *2013*, 409572. [CrossRef]
10. Zhang, C.; Yuan, X.; Wu, L.; Han, Y.; Sheng, J. Study on Morphology of Electrospun Poly(Vinyl Alcohol) Mats. *Eur. Polym. J.* **2005**, *41*, 423–432. [CrossRef]
11. Tan, S.H.; Inai, R.; Kotaki, M.; Ramakrishna, S. Systematic Parameter Study for Ultra-Fine Fiber Fabrication via Electrospinning Process. *Polymer* **2005**, *46*, 6128–6134. [CrossRef]
12. Kasiri, A.; Domínguez, J.E.; González-Benito, J. Morphology Optimization of Solution Blow Spun Polystyrene to Obtain Superhydrophobic Materials with High Ability of Oil Absorption. *Polym. Test.* **2020**, *91*, 106859. [CrossRef]
13. Medeiros, E.S.; Glenn, G.M.; Klamczynski, A.P.; Orts, W.J.; Mattoso, L.H.C. Solution Blow Spinning: A New Method to Produce Micro- and Nanofibers from Polymer Solutions. *J. Appl. Polym. Sci.* **2009**, *113*, 2322–2330. [CrossRef]
14. Zhang, L.; Kopperstad, P.; West, M.; Hedin, N.; Fong, H. Generation of Polymer Ultrafine Fibers through Solution (Air-) Blowing. *J. Appl. Polym. Sci.* **2009**, *114*, 3479–3486. [CrossRef]
15. Oliveira, J.E.; Moraes, E.A.; Costa, R.G.F.; Afonso, A.S.; Mattoso, L.H.C.; Orts, W.J.; Medeiros, E.S. Nano and Submicrometric Fibers of Poly(D, L -Lactide) Obtained by Solution Blow Spinning: Process and Solution Variables. *J. Appl. Polym. Sci.* **2011**, *122*, 3396–3405. [CrossRef]
16. Sarkar, K.; Gomez, C.; Zambrano, S.; Ramirez, M.; Hoyos, E.; Vasquez, H.; Lozan, K. Electrospinning to Forcespinning^{TM}. *Mater. Today* **2010**, *13*, 12–14. [CrossRef]
17. Mamidi, N.; Zuníga, A.E.; Villela-Castrejón, J. Engineering and evaluation of forcespun functionalized carbon nano-onions reinforced poly (ε-caprolactone) composite nanofibers for pH-responsive drug release. *Mater. Sci. Eng. C* **2020**, *112*, 110928. [CrossRef]
18. Lou, H.; Li, W.; Li, C.; Wang, X. Systematic Investigation on Parameters of Solution Blown Micro/Nanofibers Using Response Surface Methodology Based on Box-Behnken Design. *J. Appl. Polym. Sci.* **2013**, *130*, 1383–1391. [CrossRef]
19. Daristotle, J.L.; Behrens, A.M.; Sandler, A.D.; Kofinas, P. A Review of the Fundamental Principles and Applications of Solution Blow Spinning. *ACS Appl. Mater. Interfaces* **2016**, *8*, 34951–34963. [CrossRef]
20. Tutak, W.; Gelven, G.; Markle, C.; Palmer, X.L. Rapid Polymer Fiber Airbrushing: Impact of a Device Design on the Fiber Fabrication and Matrix Quality. *J. Appl. Polym. Sci.* **2015**, *132*, 42813. [CrossRef]
21. Li, J.; Song, G.; Yu, J.; Wang, Y.; Zhu, J.; Hu, Z. Preparation of Solution Blown Polyamic Acid Nanofibers and Their Imidization into Polyimide Nanofiber Mats. *Nanomaterials* **2017**, *7*, 395. [CrossRef] [PubMed]
22. da Silva Parize, D.D.; Foschini, M.M.; de Oliveira, J.E.; Klamczynski, A.P.; Glenn, G.M.; Marconcini, J.M.; Mattoso, L.H.C. Solution Blow Spinning: Parameters Optimization and Effects on the Properties of Nanofibers from Poly(Lactic Acid)/Dimethyl Carbonate Solutions. *J. Mater. Sci.* **2016**, *51*, 4627–4638. [CrossRef]
23. Polaskova, M.; Peer, P.; Cermak, R.; Ponizil, P. Effect of Thermal Treatment on Crystallinity of Poly(Ethylene Oxide) Electrospun Fibers. *Polymers* **2019**, *11*, 1384. [CrossRef]
24. Mamidi, N.; Manuel, R.; Delgadillo, V.; Gonzalez-Ortiz, A. Engineering of carbon nano-onion bioconjugates for biomedical applications. *Mater. Sci. Eng. C* **2021**, *120*, 111698. [CrossRef]
25. Mamidi, N.; Manuel, R.; Delgadillo, V. Design, fabrication and drug release potential of dual stimuli-responsive composite hydrogel nanoparticle interfaces. *Colloids Surf. B Biointerfaces* **2021**, *204*, 111819. [CrossRef] [PubMed]
26. Ma, L.; Deng, L.; Chen, J. Applications of poly(ethylene oxide) in controlled release tablet systems: A review. *Drug Dev. Ind. Pharm.* **2014**, *40*, 845–851. [CrossRef] [PubMed]
27. Oliveira, J.E.; Moraes, E.A.; Marconcini, J.M.; Mattoso, L.H.C.; Glenn, G.M.; Medeiros, E.S. Properties of Poly(Lactic Acid) and Poly(Ethylene Oxide) Solvent Polymer Mixtures and Nanofibers Made by Solution Blow Spinning. *J. Appl. Polym. Sci.* **2013**, *129*, 3672–3681. [CrossRef]
28. González Benito, J.; Teno, J.; Torres, D.; Diaz, M. Solution Blow Spinning and Obtaining Submicrometric Fibers of Different Polymers. *Int. J. Nanopart. Nanotechnol.* **2017**, *3*, 007. [CrossRef]

29. Silva, V.P.R.; Silva, G.G.; Caliman, V.; Rieumont, J.; de Miranda-Pinto, C.O.B.; Archanjo, B.S.; Neves, B.R.A. Morphology, Crystalline Structure and Thermal Properties of PEO/MEEP Blends. *Eur. Polym. J.* **2007**, *43*, 3283–3291. [CrossRef]
30. Chrissopoulou, K.; Andrikopoulos, K.S.; Fotiadou, S.; Bollas, S.; Karageorgaki, C.; Christofilos, D.; Voyiatzis, G.A.; Anastasiadis, S.H. Crystallinity and Chain Conformation in PEO/Layered Silicate Nanocomposites. *Macromolecules* **2011**, *44*, 9710–9722. [CrossRef]
31. Blaine, R.L. *Polymer Heats of Fusion*; Thermal Applications Note of TA Instruments: New Castle, DE, USA. Available online: http://www.tainstruments.com/pdf/literature/TN048.pdf (accessed on 1 March 2022).
32. *ASTM D882-10*; Standard Test Method for Tensile Properties of Thin Plastic Sheeting. ASTM International: West Conshohocken, PA, USA, 2018.
33. Dadol, G.C.; Kilic, A.; Tijing, L.D.; Lim, K.J.A.; Cabatingan, L.K.; Tan, N.P.B.; Stojanovska, E.; Polat, Y. Solution Blow Spinning (SBS) and SBS-Spun Nanofibers: Materials, Methods, and Applications. *Mater. Today Commun.* **2020**, *25*, 101656. [CrossRef]
34. Dias, F.T.G.; Rempel, S.P.; Agnol, L.D.; Bianchi, O. The Main Blow Spun Polymer Systems: Processing Conditions and Applications. *J. Polym. Res.* **2009**, *27*, 205. [CrossRef]
35. Huggins, M.L. The Viscosity of Dilute Solutions of Long-Chain Molecules. IV. Dependence on Concentration. *J. Am. Chem. Soc.* **1942**, *64*, 2716–2718. [CrossRef]
36. Ma, X.; Pawlik, M. Intrinsic Viscosities and Huggins Constants of Guar Gum in Alkali Metal Chloride Solutions. *Carbohydr. Polym.* **2007**, *70*, 15–24. [CrossRef]
37. Yoshihara, T.; Tadokoro, H.; Murahashi, S. Normal Vibrations of the Polymer Molecules of Helical Conformation. IV. Polyethylene Oxide and Polyethylene-d_4 Oxide. *J. Chem. Phys.* **1964**, *41*, 2902–2911. [CrossRef]
38. Pucić, I.; Jurkin, T. FTIR Assessment of Poly(Ethylene Oxide) Irradiated in Solid State, Melt and Aqeuous Solution. *Radiat. Phys. Chem.* **2012**, *81*, 1426–1429. [CrossRef]
39. Da Rosa, C.G.; Sganzerla, W.G.; de Oliveira Brisola Maciel, M.V.; de Melo, A.P.Z.; da Rosa Almeida, A.; Ramos Nunes, M.; Bertoldi, F.C.; Manique Barreto, P.L. Development of Poly (Ethylene Oxide) Bioactive Nanocomposite Films Functionalized with Zein Nanoparticles. *Colloids Surf. A Physicochem. Eng. Asp.* **2020**, *586*, 124268. [CrossRef]
40. Qin, Y.; Chen, K.; Zhang, H.; Luo, X.; Liang, S.; Tian, C.; Wang, J.; Zhang, L. Structure-Property Correlation of Poly(Ethylene Glycol) Based Form Stable Phase Change Materials with Different Crosslinking Structure. *Sol. Energy Mater. Sol. Cells* **2019**, *203*, 110192. [CrossRef]
41. Li, X.; Hsu, S.L. An Analysis of the Crystallization Behavior of Poly(Ethylene Oxide)/Poly(Methyl Methacrylate) Blends by Spectroscopic and Calorimetric Techniques. *J. Polym. Sci. Polym. Phys. Ed.* **1984**, *22*, 1331–1342. [CrossRef]
42. Wang, M.; Jin, H.J.; Kaplan, D.L.; Rutledge, G.C. Mechanical Properties of Electrospun Silk Fibers. *Macromolecules* **2004**, *37*, 6856–6864. [CrossRef]
43. Delgado, I. Analisis de Electrolitos Solidos Polimericos PEO/CF3CO2Li Por Difraccion de Rayos-X y SEM. *Rev. Timbaga* **2006**, *1*, 69–74.
44. Deitzel, J.M.; Kleinmeyer, J.D.; Hirvonen, J.K.; Beck Tan, N.C. Controlled Deposition of Electrospun Poly(Ethylene Oxide). *Polymer* **2001**, *42*, 8163–8170. [CrossRef]
45. Malathi, M.; Tamilarasan, K. Synthesis and Characterization of Polyethylene Oxide Based Nano Composite Electrolyte. *Sadhana* **2014**, *39*, 999–1007. [CrossRef]
46. Zhou, C.; Chu, R.; Wu, R.; Wu, Q. Electrospun Polyethylene Oxide/Cellulose Nanocrystal Composite Nanofibrous Mats with Homogeneous and Heterogeneous Microstructures. *Biomacromolecules* **2011**, *12*, 2617–2625. [CrossRef] [PubMed]
47. Tong, Y.; Lin, Y.; Wang, S.; Song, M. A Study of Crystallisation of Poly (Ethylene Oxide) and Polypropylene on Graphene Surface. *Polymer* **2015**, *73*, 52–61. [CrossRef]
48. Sánchez-Soto, P.J.; Ginés, J.M.; Arias, M.J.; Novák, C.; Ruiz-Conde, A.; García González, P. Effect of molecular mass on the melting temperature, enthalpy and entropy of hydroxy-terminated PEO. *J. Therm. Anal. Calorim.* **2002**, *67*, 189–197. [CrossRef]
49. Kou, Y.; Wang, S.; Luo, J.; Sun, K.; Zhang, J.; Tan, Z.; Shi, Q. Thermal Analysis and Heat Capacity Study of Polyethylene Glycol (PEG) Phase Change Materials for Thermal Energy Storage Applications. *J. Chem. Thermodyn.* **2019**, *128*, 259–274. [CrossRef]
50. Teno, J.; Corral, A.; Gorrasi, G.; Sorrentino, A.; Benito, J.G. Fibrous Nanocomposites Based on EVA40 Filled with Cu Nanoparticles and Their Potential Antibacterial Action. *Mater. Today Commun.* **2019**, *20*, 100581. [CrossRef]
51. Morais, S. Porosity of Fiber Mats from Different Polymer Solutions: Evaluating the Influence of Process and Solution Parameters. Master's Thesis, Leibniz Universität Hannover, Hannover, Germany, 2011.
52. Ichikawa, Y.; Washiyama, J.; Moteki, Y.; Noguchi, K.; Okuyama, K. Crystal Modification in Poly(Ethylene Succinate). *Polym. J.* **1995**, *27*, 1264–1266. [CrossRef]
53. Bellan, L.M.; Kameoka, J.; Craighead, H.G. Measurement of the Young's Moduli of Individual Polyethylene Oxide and Glass Nanofibres. *Nanotechnology* **2005**, *16*, 1095–1099. [CrossRef]
54. Jee, A.Y.; Lee, H.; Lee, Y.; Lee, M. Determination of the Elastic Modulus of Poly(Ethylene Oxide) Using a Photoisomerizing Dye. *Chem. Phys.* **2013**, *422*, 246–250. [CrossRef]
55. Warfield, R.W.; Barnet, F.R. Elastic Constanto of Bulk Polymers. *Angew. Makromol. Chem.* **1972**, *27*, 215–217. [CrossRef]
56. Yu, D.G.; Shen, X.X.; Branford-White, C.; White, K.; Zhu, L.M.; Annie Bligh, S.W. Oral Fast-Dissolving Drug Delivery Membranes Prepared from Electrospun Polyvinylpyrrolidone Ultrafine Fibers. *Nanotechnology* **2009**, *20*, 055104. [CrossRef]

57. Bai, Y.; Wang, D.; Zhang, Z.; Pan, J.; Cui, Z.; Yu, D.G.; Annie Bligh, S.W. Testing of Fast Dissolution of Ibuprofen from Its Electrospun Hydrophilic Polymer Nanocomposites. *Polym. Test.* **2021**, *93*, 106872. [CrossRef]
58. Körner, A.; Larsson, A.; Andersson, Å.; Piculell, L. Swelling and Polymer Erosion for Poly(Ethylene Oxide) Tablets of Different Molecular Weights Polydispersities. *J. Pharm. Sci.* **2010**, *99*, 1225–1238. [CrossRef] [PubMed]

polymers

Article

Estimation of Tensile Modulus of a Thermoplastic Material from Dynamic Mechanical Analysis: Application to Polyamide 66

Albert Serra-Aguila [1], Josep Maria Puigoriol-Forcada [2], Guillermo Reyes [2] and Joaquin Menacho [2,*]

[1] Passive Safety Department, Applus + IDIADA HQ, Santa Oliva, L'Albornar, P.O. Box 20, 43710 Tarragona, Spain; albert.serra@idiada.com
[2] IQS School of Engineering, Universitat Ramon Llull, Via Augusta 390, 08017 Barcelona, Spain; josep.puigoriol@iqs.url.edu (J.M.P.-F.); guillermo.reyes@iqs.url.edu (G.R.)
* Correspondence: joaquin.menacho@iqs.url.edu

Abstract: The mechanical properties of thermoplastic materials depend on temperature and strain rate. This study examined the development of a procedure to predict tensile moduli at different strain rates and temperatures, using experimental data from three-point-bending dynamic mechanical analysis (DMA). The method integrated different classical concepts of rheology to establish a closed formulation that will allow researchers save an important amount of time. Furthermore, it implied a significant decrease in the number of tests when compared to the commonly used procedure with a universal testing machine (UTM). The method was validated by means of a prediction of tensile moduli of polyamide PA66 in the linear elastic range, over a temperature range that included the glass-transition temperature. The method was applicable to thermo-rheologically simple materials under the hypotheses of isotropy, homogeneity, small deformations, and linear viscoelasticity. This method could be applicable to other thermoplastic materials, although it must be tested using these other materials to determine to what extent it can be applied reliably.

Keywords: polymer characterization; mechanical properties; viscoelasticity; DMA

Citation: Serra-Aguila, A.; Puigoriol-Forcada, J.M.; Reyes, G.; Menacho, J. Estimation of Tensile Modulus of a Thermoplastic Material from Dynamic Mechanical Analysis: Application to Polyamide 66. *Polymers* **2022**, *14*, 1210. https://doi.org/10.3390/polym14061210

Academic Editors: Andrzej Puszka and Beata Podkościelna

Received: 21 January 2022
Accepted: 11 March 2022
Published: 17 March 2022

1. Introduction

There are a variety of engineering applications involving viscoelastic properties of materials, ranging from the more classic structural applications [1,2], through nanocomposites and biopolymers [3], to biomechanical models [4,5] and biomedical engineering [6,7]. There is also a variety of mathematical models to fit the performance of materials that have viscoelastic behavior [8,9]. In spite of this variety of models, such as quasilinear models [10,11], fractional models [12,13], and other nonlinear models [14,15], linear models in the form of a Prony series are the simplest and therefore the most widely used in industrial practice [8,16]. In order to create mathematical models that allow for good calculations, the identification of the parameters of viscoelastic material models is a permanent issue in engineering research [17,18]. The limits of an accurate identification imply limitations to the applications in many engineering fields [5,7,19]. This task is more difficult, since these models must take into account the variation in the behavior of materials at different frequencies and temperatures [2,6,20]. Although dynamic mechanical analysis (DMA) is a well-established option for obtaining material-response data over a broad spectrum of frequencies and temperatures [21], new methods are currently being tested and developed to identify the material parameters: for applications at the micro- and nanoscale by atomic force microscopy [22–24] or optical tweezer [25]; for biological materials (challenging because of their intrinsic softness and labile nature) [26]; and for noncontact methods such as magnetic mechanical testing for biomaterials [27] or ultrasonic DMA for nondestructive inspection [28]. On the other hand, a number of studies have examined simpler methods to be able to estimate the properties of materials, reducing the experimental cost and complexity [18,29,30].

The use of a universal testing machine (UTM) to characterize a material is widespread. In industry, it is common to have the tensile modulus at room temperature and at a low strain rate (given by the supplier), but data on other temperatures and strain rates are of interest as well. However, using a UTM to determine viscoelastic properties requires, in general, a large number of tests. So, it takes a long time for a complete characterization. Therefore, the progress and challenge of this field is to estimate a large amount of information in a shorter time, by using only dynamic mechanical analysis (DMA) [18,21]. After the experimentation with DMA, if the information is treated through a series of mathematical models, it is possible to obtain the same information as in the use of a UTM, without the necessity of using this device. This procedure allows researchers to save time and money.

To avoid the time-consuming experimentation in UTM tests, the present article deals with experimental data from DMA tests to obtain tensile moduli of thermoplastic materials. Three-point-bending DMA experimental data were measured at certain values of amplitude, frequency, and temperature. With these input data, a mathematical model was created and its parameters were found, in order to predict the tensile moduli in the linear elastic zone for a range of temperatures and strain rates.

Consequently, the objective of this work was to predict the tensile modulus of polyamide PA66 as a function of temperature and strain rate from three-point-bending DMA data without using a UTM. The validation of the results was carried out by comparing them with experimental data.

Three hypotheses to be validated were assumed:

1. The relationships between the flexural modulus and frequency, and also between tensile modulus and strain rate, follow a logarithmic tendency;
2. There is an equivalence between the strain rate of a tensile test and the frequency and amplitude of the sinusoidal strain from the flexural tests;
3. There is a factor that permits a relationship between the tensile modulus and flexural modulus. This factor is not dependent on strain rate nor on temperature.

From these assumptions, the mathematical model was established. Then, it was applied to a polyamide as an example. At the end of the work, the hypotheses will be verified when contrasted with the experimentation.

The procedure began with the DMA experimental data (storage and loss moduli) over input parameters (amplitude, frequency, and temperature). Once these data were measured, with the use of the proposed mathematical model, tensile moduli were estimated at different temperatures and frequencies. Finally, these estimated values were compared to tensile tests in order to validate the material model.

2. Materials and Methods

The diagram in Figure 1 shows the concept of the present work. Starting from DMA data: on one hand, when comparing the flexural modulus (E_F) for different temperatures and frequencies, it was possible to state the factors for interconverting between temperatures and frequencies (a_T); on the other hand, a relationship between frequency and strain rate was determined ($\dot{\varepsilon}(f)$). As a result, the flexural modulus was calculated as a function of the temperature and strain rate. Assuming that the tensile modulus of the material was known from the literature, just for a single reference temperature and strain rate, the flexural modulus could then be calculated in the same reference conditions. This allowed us to state the relationship (*MR*) between the tensile and flexural moduli (third hypothesis of this work). As a result, a model could be built to predict the tensile modulus for every temperature and strain rate (in a certain range). Finally, it was possible to define a loop check (tensile loop) by using data from the UTM to validate the model.

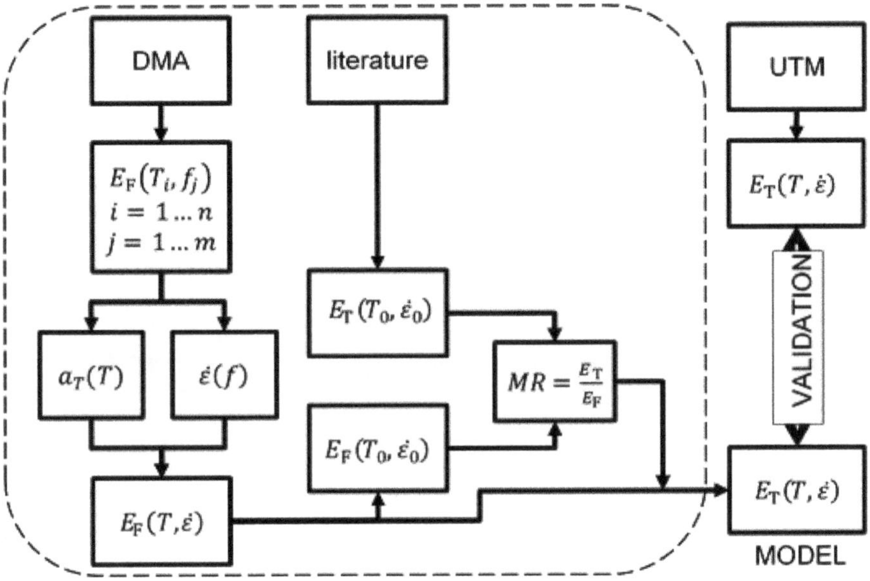

Figure 1. Diagram of the "tensile loop".

2.1. Mathematical Material Model

In order to state the mathematical model of a viscoelastic material, two different types of models were checked in the literature:

- Models depending on the strain rate: in which the mechanical property under study was defined as a function of the strain rate;
- Models depending on the temperature: in which the mechanical property under study was defined as a function of temperature.

2.1.1. Strain Rate Dependence

In polymers and composites, the modulus and the strength increase with an increase in the strain rate [16,31]. The relationship between these mechanical properties (modulus or strength) follow a logarithmic relationship as a function of the strain rate (first hypothesis of this work). One of the most well-known models that describes mechanical properties as a function of strain rate is the Johnson–Cook model [32,33].

The Johnson–Cook model is a mechanical model that was developed to describe the plastic behavior that a material suffers under impact conditions. The behavior of materials subjected to high strains, high temperatures, and high strain rates is reproduced. This model is defined using parameters measured in experimental procedures at different strain rates and temperatures. The yield stress is defined as [32]:

$$\sigma = \left(\sigma_0 + JC_1 \cdot \varepsilon_p{}^{JC_2}\right) \cdot \left(1 + JC_3 \cdot \ln \dot{\varepsilon}_p^*\right) \cdot \left(1 - T_H{}^{JC_4}\right) \tag{1}$$

where σ is the stress, σ_0 the yield stress at the reference temperature and at a low strain rate; ε_p the effective plastic strain; $\dot{\varepsilon}_p^*$ the effective plastic strain rate, normalized with the reference strain rate $\dot{\varepsilon_0} = 1\ s^{-1}$; JC_i $(i = 1 \ldots 4)$ are four coefficients of the Johnson–Cook model; and T_H is a dimensionless coefficient. This last coefficient is defined as [32,33]:

$$T_H = \frac{T - T_{ref}}{T_m - T_{ref}} \tag{2}$$

where T is the temperature, T_{ref} is the reference temperature (the ambient temperature, for instance), and T_m is the melting temperature of the material.

Although the Johnson–Cook model is related to the development of plastic strain, and the analysis of this work was related to thermoplastic materials within the linear viscoelastic region, this model was considered important in this study due to the logarithmic relationship between the modulus or strength and the strain rate.

Relying on this model, the present work proposed the first hypothesis mentioned above: the flexural modulus (or the tensile one) has, in general, a relationship with the strain rate that can be expressed by means of a logarithmic relationship of the form:

$$E_F(\dot{\varepsilon}) = E_F(\dot{\varepsilon}_{ref}) + SE_F \cdot \ln \dot{\varepsilon} \tag{3}$$

where $\dot{\varepsilon}$ is the strain rate, $E_F(\dot{\varepsilon})$ is the flexural modulus of the material for a strain rate $\dot{\varepsilon}$, and the reference strain rate $\dot{\varepsilon}_{ref} = 1$ mm/min. SE_F is the slope of the linearized expression, and can be calculated by fitting a line to the experimental data.

As will be seen below, a completely analogous expression was proposed for the tensile modulus.

The units for strain rate will always be mm/min in this work, in order to compare the different strain rates measured in tensile tests with the UTM results. This input parameter (strain rate) for these tests is given in mm/min as well.

2.1.2. Temperature Dependence

It is well known that in polymers and composites, the modulus decreases with an increase in temperature [16,34]. The relationships between mechanical properties as a function of temperature have been widely studied. Several constitutive properties (such as relaxation modulus or creep compliance) depend on temperature and time. The time–temperature superposition principle, as found in the literature [16,35,36], states that there is a certain equivalence between a variation in temperature and the time scale in which the processes of relaxation, creep, or deformation take place. For example, relaxation time varies in a certain (inverse) proportion with temperature. This means that the curves of a certain constitutive property with respect to time (or frequency) obtained at a higher temperature can be superimposed on those obtained at a lower temperature by applying a suitable change of scale to the time axis. Since the temporal (or frequency) scales used in this type of experiment are logarithmic, the change in time scale corresponds to a shift in the (logarithmic) time axis. The materials in which this principle of time–temperature superposition applies are called thermo-rheologically simple. In the present study, in order to deal with the influence of the temperature, a shift factor was used, a_T. The shift factor could be calculated with a shifting procedure.

Knowing how this shift factor affects the temperature is of great interest. The shift factor is defined as [16]:

$$a_T = \frac{t_T}{t_{T_{ref}}} \tag{4}$$

where a_T is the shift factor, t_T is the time to reach a certain value in a mechanical property (relaxation modulus or creep compliance) at temperature T, and t_{Tref} is the time to reach a certain value in a mechanical property at reference temperature T_{ref}. This factor accounts for the time–temperature superposition of the relaxation modulus (or creep compliance), so that t units of time at temperature T are equivalent to t/a_T units at temperature T_{ref}.

The shift factor also applies to constitutive functions expressed in the frequency domain. For the complex modulus $G^* = G' + j \cdot G''$, it leads to [16]:

$$G^*(f, T) = G^*(a_T \cdot f, T_{ref}) \tag{5}$$

Various empirical equations have been proposed to determine the temperature-dependent behavior of the shift. The Williams–Landel–Ferry (WLF) equation is useful when the material has a temperature above the glass transition T_g, as well as in the T_g region. This equation uses the following expression [16,34]:

$$\log a_T = \frac{-C_1(T - T_{\text{ref}})}{C_2 + (T - T_{\text{ref}})} \tag{6}$$

where T_{ref} is the reference temperature, and C_1 and C_2 the coefficients of the WLF equation. The values of the constants initially were thought to be universal constants, taking values of 17.44 and 51.6, respectively, when T_{ref} was considered T_g. Today, they are instead considered as parameters to curve-fit in each case [16].

When the temperature is below T_g, the relationship between the shift factor and temperature is given by the Arrhenius equation [37], which is defined using the following expression:

$$\log a_T = \frac{E_a}{2.303 \cdot R} \cdot \left(\frac{1}{T} - \frac{1}{T_{\text{ref}}} \right) \tag{7}$$

where E_a is the activation energy, T is the (absolute) temperature, and R the thermodynamic constant.

When working at different temperatures with WLF or Arrhenius shift functions, the number of temperatures to be analyzed in order to achieve a reliable adjustment of the shift factor over temperature depends on how much time the experiment lasts. In most cases, 4 or 5 temperatures can be enough [38]. The coefficients for the equations were found through the least-squares error method, after the equations were linearized [39].

In order to determine the shift coefficient of a material, this work began with the curves of the storage and loss moduli over a range of frequencies, provided by DMA tests at different temperatures. The curves given by these tests were shifted horizontally until they overlapped (at least in a certain range). The values of these shifts were then adjusted to the WLF or Arrhenius models.

However, according to what was previously exposed, two considerations were taken into account:

(a) In order to calculate the coefficients of shift functions, it was necessary to consider the change in behavior of the material when the glass-transition temperature was exceeded. Thus, either one temperature range or two temperature subranges needed to be used, depending on the case. The WLF equation presented a discontinuity when the test temperature was a certain temperature below T_{ref}: when $T_{\text{ref}} - T = C_2$. However, it was useful above T_{ref}.

The method of finding the coefficients of the equation was through linearization of Equation (6), resulting in:

$$\underbrace{\frac{1}{\log a_T}}_{y} = -\frac{C_2}{C_1} \cdot \underbrace{\frac{1}{T - T_{\text{ref}}}}_{x} - \frac{1}{C_1} \tag{8}$$

On the other hand, the Arrhenius equation presented the energy activation, the factor of which was nonsense in temperatures above T_g. However, it could be used for the subrange $T \leq T_g$. In this case, to find the coefficients, the linearization of (7) showed the form:

$$\underbrace{\log a_T}_{y} = \frac{E_a}{2.303 \cdot R} \cdot \underbrace{\frac{1}{T}}_{x} - \frac{E_a}{2.303 \cdot R} \cdot \frac{1}{T_{\text{ref}}} \tag{9}$$

(b) In the case in which there were two subranges, T_{ref} was required to be the same for every subrange in order to work within the entire temperature range indistinctly. In this way, the continuity of the values of the shift coefficient was preserved: the last temperature for the first subrange coincided with the first temperature for the second subrange; thus, the shift factor also had to coincide. Otherwise, there would be a discontinuity in the shift factor over temperature.

2.2. Moduli Ratio

The DMA tests carried out in this study were of a three-point-bending nature. They allowed us to determine the flexural modulus of the material. However, the goal of this work was to state the tensile modulus of the material. Therefore, it was necessary to establish a relationship between both modules. This relationship constituted the third hypothesis of this work. The proposed relationship was a simple proportion that remained constant for the considered range of temperatures and deformation rates. Hence, the moduli ratio (*MR*) was defined as the ratio of tensile modulus E_T to flexural modulus E_F:

$$MR = \frac{E_T}{E_F} \tag{10}$$

By definition, this moduli ratio was valid for tensile and flexural moduli at the same temperature and at the same strain rate.

This ratio was considered to be independent of the temperature and strain rate (third hypothesis of this work). It meant that this parameter was assumed to be constant in the ranges of temperature and strain rate under study. With this assumption, at the end, a comparison between flexural values and experimental tensile data was always necessary. If the differences were enough low, the assumption of the existence of this constant value along the ranges of the variables under study could be considered correct.

Moreover, if the moduli ratio is applied to (3), one finds:

$$E_T(\dot{\varepsilon}) = E_T(\dot{\varepsilon}_{\text{ref}}) + \underbrace{MR \cdot SE_F}_{SE_T} \cdot \ln\dot{\varepsilon} \tag{11}$$

So, a logarithmic line related the tensile modulus to the strain rate, and the slope of such line was that of the flexural times the moduli ratio.

2.3. Mathematical Model for Tensile Loop Validation

By combining the ideas outlined in the previous sections, a mathematical model for the tensile module could be constructed. That mathematical expression allowed us to calculate the tensile modulus as a function of the temperature and strain rate. This expression had two parts. One part represented the variation of tensile modulus as a function of temperature, and the other part as a function of the strain rate. The combination of both parts was the resultant tensile modulus:

$$E_T(T,\dot{\varepsilon}) = \underbrace{MR \cdot E_F(T,\dot{\varepsilon}_{\text{ref}})}_{E_T(T,\dot{\varepsilon}_{\text{ref}})} + MR \cdot SE_F \cdot \ln\dot{\varepsilon} \tag{12}$$

The tensile modulus at a given temperature can be found from the tensile modulus at a reference temperature. Therefore, the tensile modulus presents a logarithmic adjustment as a function of the strain rate, in which the independent term is the tensile modulus at a reference temperature.

For the mathematical model of a tensile loop, the tensile modulus can be determined as a function of temperature and strain rate, as shown in (12).

2.4. Material for Experiments

The experimental work was carried out with polyamide PA66 with no glass fiber. It was a lubricated and stabilized polyamide 66. It was a nylon resin with good mechanical properties, good high temperature performance, and chemical resistance. It is used in demanding applications such as in electrical and electronic devices, the automotive industry, furniture and domestic appliances, sporting goods, and packaging applications. The material used in this work was delivered by DuPont™ under the tradename ZYTEL® 103HSL NC010.

3. Experimental Application of the Method

The purpose of the method was to be able to predict the tensile modulus of a thermoplastic material (PA66) as a function of the strain rate and temperature from DMA data.

In order to apply this method, a series of steps were followed:

1. DMA experiments (sweeps);
2. Determination of the shift factor;
3. Moduli ratio;
4. Material model: temperature- and strain-rate-dependent;
5. Validation.

3.1. DMA Experiments

Several amplitude, temperature, and frequency sweeps were carried out by DMA [21]. The experiments were three-point-bending tests, carried out under the ASTM D5023-07 standard. With these sweeps, storage and loss moduli were measured over these three mentioned variables.

3.1.1. Amplitude Sweep

The amplitude sweep was the first experimental test with DMA. This test was performed at 30 °C and at 1 Hz. The measured storage modulus $G'(= E_F)$ for PA66 is presented in Figure 2.

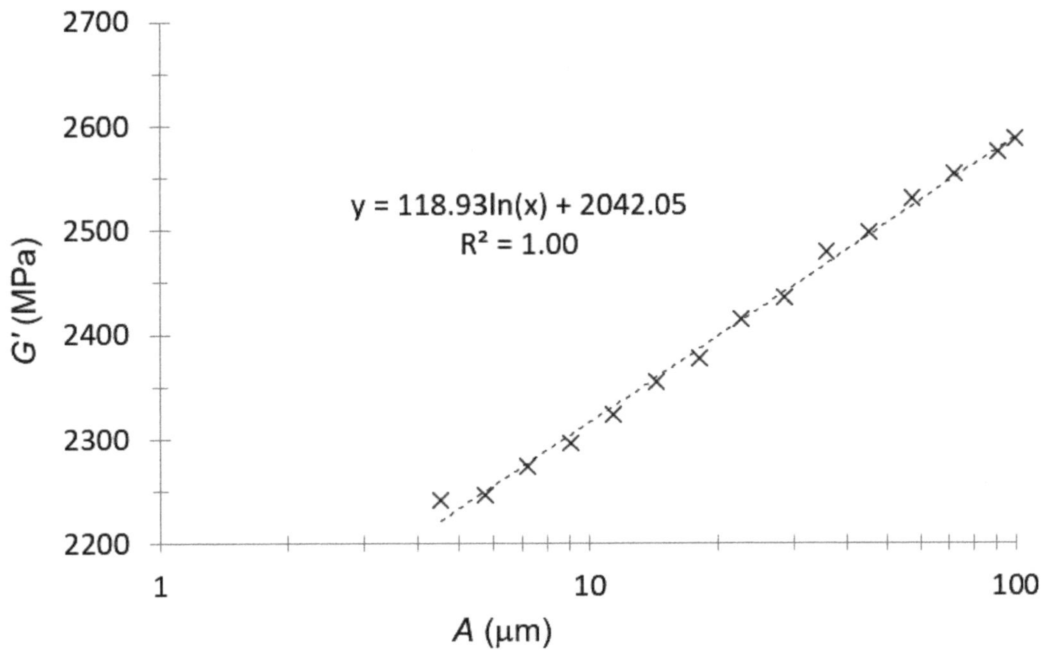

Figure 2. Storage modulus–amplitude plot for PA66.

For PA66, the amplitude to perform subsequent tests was 30 µm. This amplitude, considering the data in Figure 2, was located within the linear viscoelastic range.

The plot in Figure 3 was only to observe if the tendency between the flexural modulus and strain rate was the expected one (a logarithmic tendency). Since the strain in the DMA tests was sinusoidal, the average strain rate was taken as the length traveled in a complete oscillation divided by a period; i.e. four times the amplitude times the frequency (second hypothesis in this work):

$$\dot{\varepsilon} = 4 \cdot A \cdot f \tag{13}$$

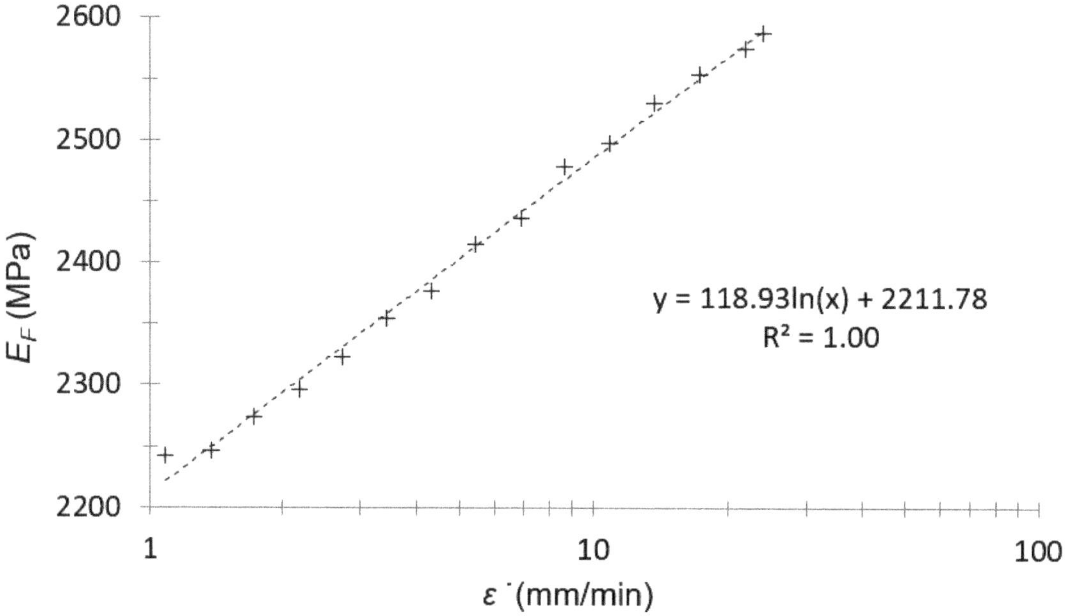

Figure 3. Flexural modulus–strain rate plot of amplitude sweep for PA66.

As can be seen in Figure 3, the logarithmic tendency was confirmed between the flexural modulus and the strain rate for this material (first hypothesis of this work).

3.1.2. Temperature Sweep

The temperature sweep was the second experimental test with DMA. This test was performed at 30 μm of amplitude and 1 Hz of frequency. The loss modulus for PA66 in these conditions is presented in Figure 4.

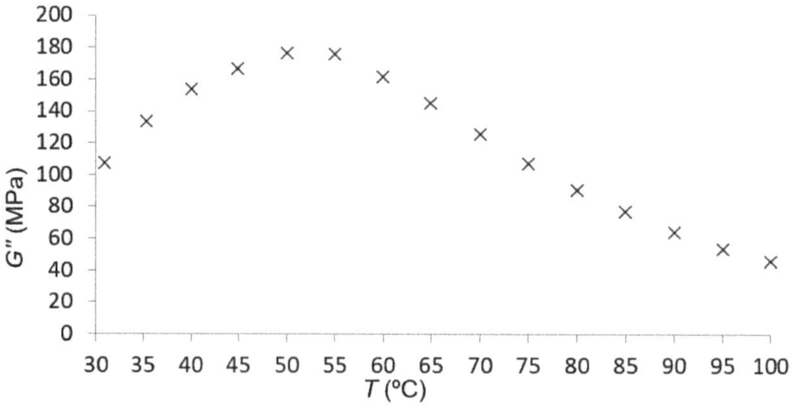

Figure 4. Loss modulus–temperature plot for PA66.

This second test was mostly used to observe if there was a maximum for the loss modulus at a specific temperature in the given conditions of amplitude and frequency.

According to the data in Figure 4, for the material under study, there was a maximum of the loss modulus at approximately 50 °C. This indicated that the subrange change, explained in Section 2.1.2, had to be considered at this temperature.

3.1.3. Frequency Sweep

The frequency sweep was the third and last experimental test with DMA. This test was performed at 30 μm of amplitude at different temperatures, from 30 to 100 °C, with an increment of 10 °C between the sweeps.

Frequency sweeps for the storage modulus and loss modulus for PA66 are presented in Figures 5 and 6, respectively. Three repetitions of each test were performed, and we took the average of these results.

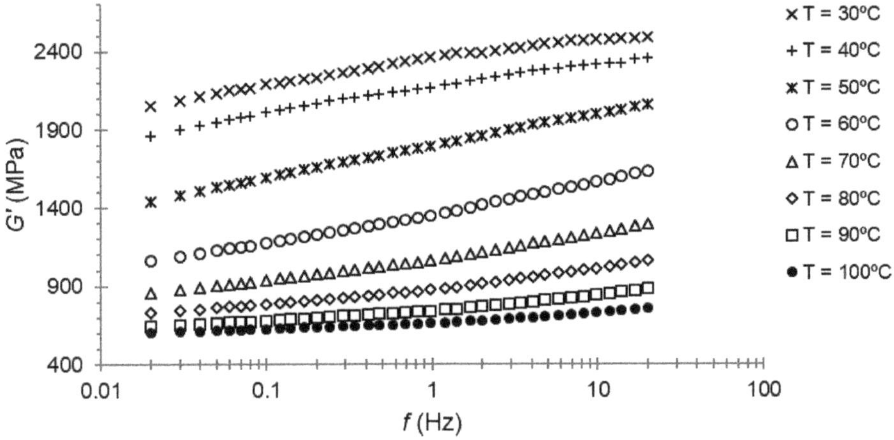

Figure 5. Storage modulus–frequency plot for PA66.

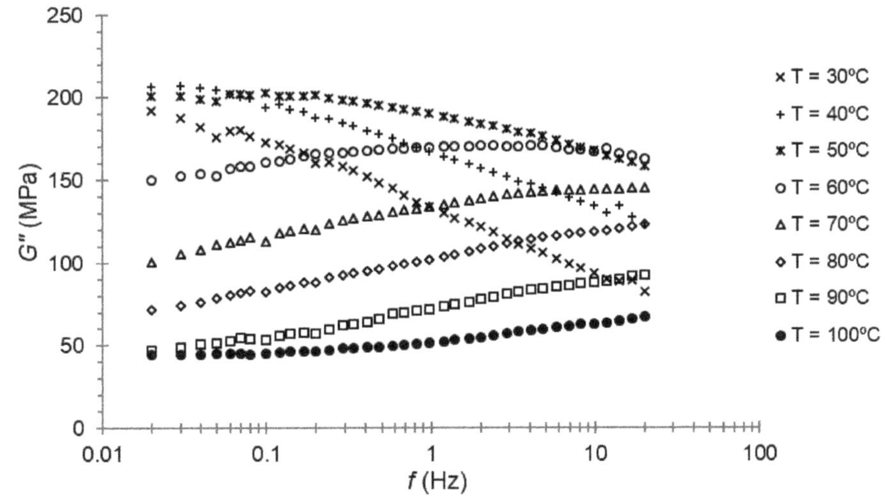

Figure 6. Loss modulus–frequency plot for PA66.

At first, the maximum of the loss modulus for 1 Hz of frequency was seen, as expected, at a temperature close to 50 °C; that is, the temperature at which the maximum of the loss modulus was found during the temperature sweep. When examining Figure 6, it can be seen that the maximum was reached between 40 and 50 °C for a lower frequency.

Once these frequency sweeps were performed, the experimental part of the method was finished.

The next step was to plot the flexural (or storage) modulus over the strain rate at the different temperatures tested. In this way, it was possible to find, for every temperature, the parameter SE_F, as well as the flexural modulus at a reference strain rate of 1 mm/min.

For PA66, the plot of the flexural modulus over the strain rate for every temperature is shown in Figure 7.

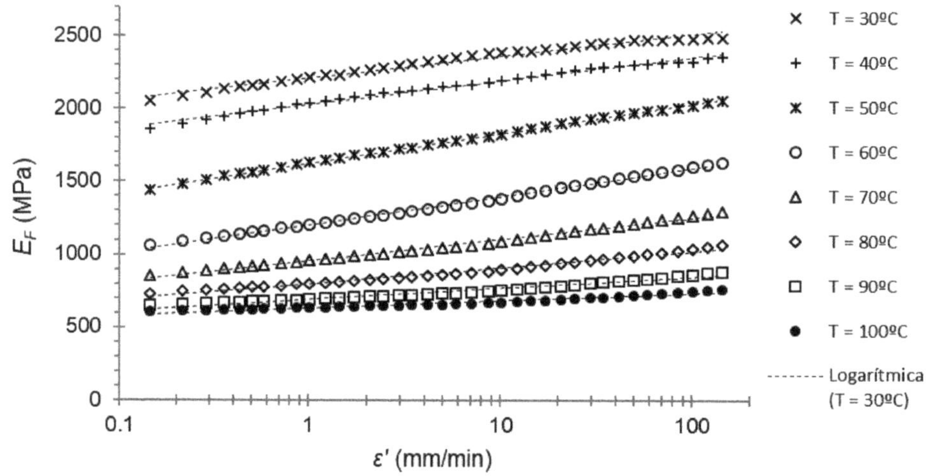

Figure 7. Flexural modulus–strain rate plot of the frequency sweep for PA66. Dotted lines are the mean-square-error fitting lines.

A logarithmic Equation (3) for every temperature was adjusted with a high coefficient of determination R^2. The resulting slope of the equation was the parameter SE_F in (3). These results are shown in Table 1.

Table 1. Slope and independent terms of the logarithmic adjustment between flexural modulus and strain rate for PA66.

T (°C)	SE_F (MPa)	E_F (MPa) 1 mm/min	R^2
30	65.0	2211	0.981
40	70.1	2029	0.992
50	88.7	1624	0.999
60	83.8	1203	0.997
70	63.6	957	0.992
80	48.0	799	0.982
90	33.7	690	0.974
100	21.6	632	0.963

At first, the maximum of the loss modulus for 1 Hz of frequency was found, as expected, at a temperature close to 50 °C; that is, the temperature at which the maximum of the loss modulus was found at the temperature sweep. When examining Figure 6, it can be seen that the maximum was found between 40 and 50 °C for a lower frequency.

Moreover, the logarithmic tendency was confirmed again between the flexural modulus and the strain rate for this material (first hypothesis of this work).

3.2. Determination of the Shift Factor

In the frequency sweep of the DMA tests, every curve was determined at one specific temperature (Figures 5 and 6). According the hypotheses, these curves could be horizontally shifted until they were superposed. This frequency–temperature superposition procedure was carried out by minimizing the squares-error between every couple of curves. The procedure resulted in just one curve with a larger frequency range.

As explained in Section 2.1.2, two cases were possible:

(a) If there was a maximum of the loss modulus over the frequency, two temperature subranges were necessary. On one hand, there was a first subrange that ranged from the minimum temperature to the temperature at which the maximum is found, adjusted with the Arrhenius equation. On the other hand, a second subrange that ranged from this last temperature to the maximum temperature, adjusted with the WLF equation.

(b) If there was no maximum of the loss modulus, just one range was applied, from the minimum to the maximum temperature. The choice of the shift function depended on the curve shape: if the loss modulus increased with temperature (a maximum was expected at a higher temperature), the Arrhenius equation was recommended; conversely, if the loss modulus decreased when the temperature increased, the WLF equation was recommended.

For PA66, two temperature subranges were defined: a first subrange from 30 to 50 °C, and a second one from 50 to 100 °C. The first subrange was adjusted with the Arrhenius Equation (7), and the second one with the WLF Equation (6). The frequency–temperature superposition returned the value of the shift coefficient a_T for each temperature, defined as the relationship between frequencies:

$$a_T = \frac{f_T}{f_{T_{\mathrm{ref}}}} \tag{14}$$

where f_T and $f_{T_{\mathrm{ref}}}$ are the corresponding frequencies of the curves T_{ref}. In the present case, T_{ref} was assumed to be 30 °C. For this reason, the shift factor at this specific temperature was 1. For the upper range, the shared value at 50 °C ensured the continuity of the shift coefficient.

The determination of the values of the shift coefficient was carried out by means of an algorithm of minimization of an objective function. A shift coefficient value was applied to the abscissae of each curve, calculated using the Arrhenius (from 30 to 50 °C) and WLF (from 50 °C) equations with the parameters undetermined. The objective function was calculated as the sum of squared errors between the junction points of the curve sections. This objective function was minimized, varying the parameters of the Arrhenius and WLF equations, using a nonlinear generalized reduced gradient algorithm. The Arrhenius Equation (7) gave:

$$\log a_T = \frac{7388.73}{T} - 24.39 \tag{15}$$

where T is absolute temperature (K). In addition, the linearization of the WLF Equation (6) gave:

$$\log a_T = \frac{-1}{0.17 + \frac{9.91}{T - T_{\mathrm{ref}}}} \tag{16}$$

Figure 8 shows the results of these formulas.

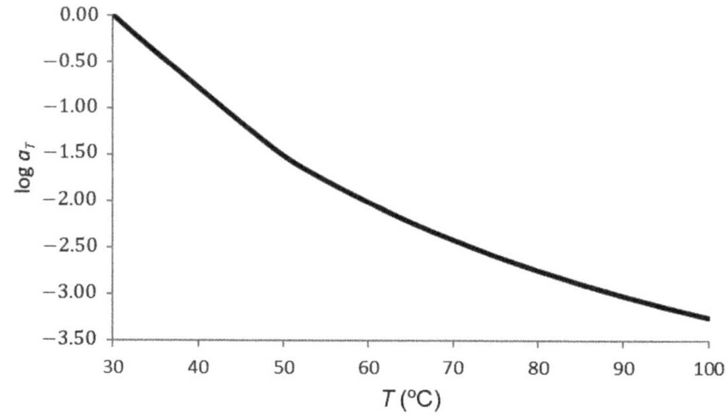

Figure 8. Shift factor versus temperature: a slight change can be seen in the glass-transition zone.

By applying the shift factor, the frequency range was extended. When applying this shift factor to the values of the frequencies at the different temperatures T, new shifted frequencies were obtained for T_{ref} = 30 °C (5): $f_{30} = a_T \cdot f_T$. Since the values of the shift coefficient were less than one, the points obtained at $T > T_{\mathrm{ref}}$ were shifted to the left (toward lower frequencies). By joining the curves thus displaced, the master curve could be drawn at T_{ref} (Figure 9).

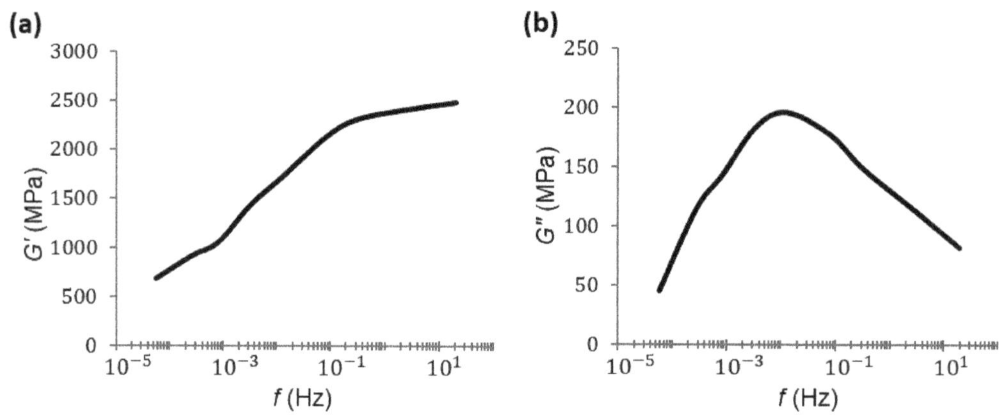

Figure 9. Master curve for the storage modulus (**a**) and loss modulus (**b**).

3.3. Moduli Ratio

To calculate the moduli ratio, it was necessary to know the tensile modulus at least at one temperature. In this case, there were published data of tensile tests of PA66 at 23 °C [40]. The tensile modulus was calculated according to ISO 527-1/-2:

$$E_T = \frac{\sigma_{0.0025} - \sigma_{0.0005}}{0.0025 - 0.0005} \quad (17)$$

The experimental value was $E_T(23\ ^\circ\mathrm{C}, 1\mathrm{mm/min}) = 2780$ MPa for a strain rate of 1 mm/min.

Due to the fact that the tensile modulus was known only at a given temperature (23 °C), the flexural modulus had to be shifted at 23 °C using the shift factor. As explained in Section 2.1.2, the shift factor had an influence on temperature and time or frequency due to the time–temperature superposition or frequency–temperature superposition, respectively. So, using the shift factor, a value of a mechanical property, such as the tensile modulus, measured at T_{ref}, could be predicted at a different temperatures T.

Therefore, an extrapolation to 23 °C through mathematical formula (15) had to be applied in order to find this value. The extrapolation was considered acceptable, because shift functions were found in the temperature range of 30 to 100 °C. This extrapolation gave us a theoretically calculated (or "synthetic") value of $E_F(23\ °C, 1mm/min)$.

From an experimental value (at 30 °C and a 30 μm amplitude in a range of frequencies), the model could find a "synthetic" value (for 23 °C and 1.00 mm/min of strain rate). This was performed using the following procedure:

a. The known flexural modulus at 30 °C was the independent term (corresponding to 1 mm/min) of the adjustment of this modulus as a function of the strain rate (Table 1): $E_F(30\ °C, 1mm/min) = 2211$ MPa;
b. The frequency that corresponded to $\dot{\varepsilon}_{ref} = 1$ mm/min was calculated by applying (13): this gave $f_{ref} = 0.139$ Hz;
c. The shift factor at 23 °C (a_{23}) was calculated (15) and then applied in (14) to calculate the corresponding frequency at that temperature: $f = a_{23} \cdot f_{ref} = 3.624 \cdot 0.139 = 0.504$ Hz;
d. The flexural modulus $E_F(23\ °C)$ could be calculated on the logarithmic expression at $f = a_{23} \cdot f_{ref}$: this was formula (3) taking the parameters of $T_{ref} = 30$ °C, but shifting the strain rate to that corresponding to the shifted frequency; that is, $\dot{\varepsilon} = 4 \cdot 0.030 \cdot 0.504 \cdot 60 = 3.625$ mm/min.

Then:

$$E_F(23\ °C, 1mm/min) = E_F(30\ °C, 1mm/min) + SE_F \cdot \ln(3.625) \tag{18}$$

According to Table 1, $SE_F = 65.0$. This resulted in a "synthetic" value of $E_F(23\ °C, 1mm/min) = 2295$ MPa.

Table 2 gives a summary of this procedure.

Table 2. Flexural and tensile moduli at 23 °C for PA66.

T (°C)	A (mm)	$\dot{\varepsilon}$ (mm/min)	f (Hz)	E_F (MPa)	E_T (MPa)
30	0.03	1.000	0.139	2211	-
30	0.03	3.625	0.504	2295	-
23	0.03	1.000	0.139	2295	2780

Finally, the moduli ratio was calculated as:

$$MR = \frac{E_T(23\ °C, 1\ mm/min)}{E_F(23\ °C, 1\ mm/min)} = 1.211 \tag{19}$$

As a reminder, the units for strain rate are in mm/min in this work in order to compare the results with those found in tensile tests with a UTM, in which the strain rate was entered in mm/min, as well.

3.4. Material Model: Temperature and Strain-Rate Dependence

In order to calculate the tensile modulus at every temperature, once the flexural modulus at every temperature and at 1 mm/min was measured, the tensile moduli at every temperature could be calculated by multiplying by moduli ratio:

$$E_T(T, 1\ mm/min) = MR \cdot E_F(T, 1\ mm/min) \tag{20}$$

The first step was the calculation of the flexural modulus at one specific temperature for the reference strain ratio $E_F(T, \dot{\varepsilon}_{ref})$. In our case, $\dot{\varepsilon}_{ref} = 1mm/min$. This could be calculated through linear interpolation from experimental data of the storage modulus. However, in order to construct the general model from the DMA tests, we calculated the slope of the logarithmic adjustment of the flexural modulus as a function of the strain rate (SE_F) at a reference temperature (in our case, at 30 °C).

Starting from this value, the flexural modulus (at $\dot{\varepsilon}_{ref}$) for any other temperature could be calculated using the shift coefficient, as was done in the previous section.

$$E_F(T,\ \dot{\varepsilon}_{ref}) = E_F(T_{ref}\ ,\ a_T \cdot \dot{\varepsilon}_{ref}) = E_F(T_{ref}, \dot{\varepsilon}_{ref}) + SE_F(T_{ref}) \cdot \ln(a_T \cdot \dot{\varepsilon}_{ref}) \quad (21)$$

The tensile modulus at the reference conditions could be automatically calculated:

$$E_T(T, \dot{\varepsilon}_{ref}) = MR \cdot E_F(T_{ref}, a_T \cdot \dot{\varepsilon}_{ref}) \quad (22)$$

Then, from (21) and (22), the general model (12) could be written for the material under study:

$$E_T(T, \dot{\varepsilon}) = MR \cdot [E_F(T_{ref} \cdot \dot{\varepsilon}_{ref}) + SE_F(T_{ref}) \cdot \ln(a_T \cdot \dot{\varepsilon}_{ref})] + MR \cdot SE_F(T) \cdot \ln \dot{\varepsilon} \quad (23)$$

In the studied example, for PA66 at 23 °C, $E_F(30\ °C, 1\ mm/min)$ was found to be 2211 MPa, $a_T = 3.625$, $MR = 1.211$, and $SE_F(30\ °C) = 65.0$. So, the model found:

$$E_T(23°C, \dot{\varepsilon}) = 1.211 \cdot [2211 + 65 \cdot \ln(3.625 \cdot 1) + SE_F(T) \cdot \ln(\dot{\varepsilon})] \quad (24)$$

with $\dot{\varepsilon}$ in mm/min. The value of the SE slope had to be obtained from the experimental curves (Table 1), taking the value of the closest temperature or interpolating. In the present case, we took $SE_F(23\ °C) \approx SE_F(30\ °C) = 65.0$, and a more specific expression could be written:

$$E_T(23\ °C, \dot{\varepsilon}) = 2780 + 78.7 \cdot \ln \dot{\varepsilon} \quad (25)$$

4. Validation

Some validations were conducted to close the loop. In order to do this, several values of the tensile modulus at different temperatures and strain rates, calculated using the presented method, were compared to experimental ones from the UTM tests.

4.1. Influence of Temperature

Once MR was known, the tensile modulus could be calculated at every temperature (for $\dot{\varepsilon} = 1$ mm/min) from the experimental data (Figure 7). The resulting plot of the tensile modulus over temperature for the material under study is shown in Figure 10.

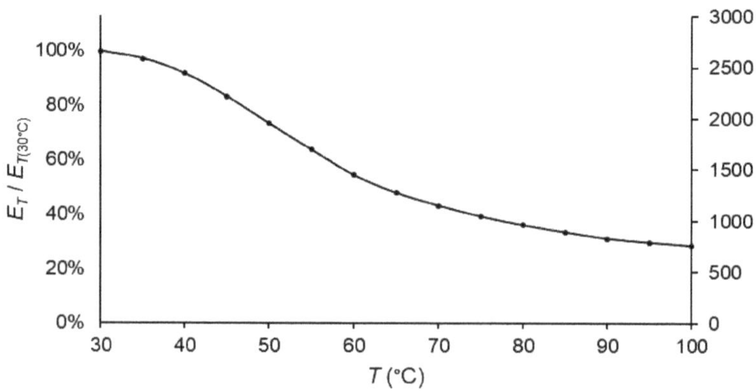

Figure 10. Tensile modulus vs. temperature plot for PA66, taking T = 30 °C as a reference.

As expected, the slope of the curve was higher if there was a subrange change; in other words, if there was a maximum of the loss modulus over the temperature. Moreover, in the same way as occurred for the storage/flexural moduli, the tensile modulus was lower at higher temperatures.

4.2. Validation by Temperature Dependence

In order to verify the values of the tensile modulus as a function of temperature, published stress–strain curves [40] at different temperatures for the same ZYTEL® 103HSL material from DuPont™ were compared to the prediction of the DMA results.

Taking the resulting values at 1 mm/min of the logarithmic fitting of the DMA results (Table 1) and applying the moduli ratio (1.211), this work assumed that it was possible to predict the tensile modulus. The resulting values were compared to the experimental ones for 40, 60, and 90 °C, as shown in Table 3.

Table 3. Tensile modulus: experimental and predicted by the moduli ratio.

T (°C)	E_T (MPa) Experimental	E_F (T,1 mm/min) Experimental (MPa)	$E_T = MR \cdot E_F$ (MPa)	Error (%)
23	2780	2295	2780	0.00
40	2538	2029	2457	3.19
60	1450	1203	1457	0.48
90	846	690	836	1.18

As defined in the experimentation in DMA, the temperature range under study was from 30 to 100 °C. For this reason, the last temperature published, 120 °C, could not be compared, as it was out of range in our study.

The three hypotheses of this work were applied here: the logarithmic tendency of the moduli related to frequency, the equivalence between the flexural and tensile moduli (constant moduli ratio), and the equivalence between the frequency and strain rate. The error introduced by those assumptions remained under 5% for PA66.

In addition, these results were compared with those of the study by Mhanna et al. [2] of a PET–FRP laminate material. The decrease in the modulus in the elastic zone in both studies were of the same order.

4.3. Validation by Strain Rate Influence

Stress–strain curves at different strain rates for PA66 also were determined with a UTM at different strain rates, in order to verify the values given by the model of the tensile modulus (22, 23). Three repetitions were conducted for each strain rate. The resulting values of the tensile modulus are shown in Table 4.

Table 4. Tensile modulus by tensile test in UTM compared to "synthetic" tensile modulus of PA66 for different strain rates at 23 °C.

	UTM Tests					Model	
	E_T (MPa)						
Strain Rate (mm/min)	1	2	3	SD	E_T (MPa)	E_T (MPa)	Error (%)
1.000	2732	2795	2813	42	2780	2780	0.00
3.466	2777	2800	2871	49	2816	2878	2.20
12.011	2812	2885	2958	73	2885	2976	3.15
41.628	2838	3012	3036	108	2962	3073	3.75
144.270	3013	3072	3095	43	3060	3171	3.63
499.000	3045	3081	3222	93	3116	3269	4.91

The "synthesized" values of the tensile modulus were calculated by the model (25). Once we performed the experiment and the calculations, it was possible to compare both values for the different strain rates tested. For the PA66, the mathematical model for the tensile loop as a function of the strain rate also was well fitted: errors remained within the 5% range, as shown in Table 4.

5. Conclusions

As stated in the objectives of this work, the developed method, with the use of the mathematical model proposed in this paper, predicted tensile moduli of thermoplastic materials as a function of the temperature and strain rate from DMA data without using a UTM.

The use of the described development involved an important saving of time, as well as a significant decrease in the number of tests, when compared to the procedure with a UTM.

In order to verify the tensile moduli, a "tensile loop" was developed, and the following conclusions were drawn:

- The parameter defined as the moduli ratio presented a constant relationship between the tensile modulus and flexural modulus, with good results;
- Following the sinusoidal tendency of flexural tests, the equation between the strain rate, frequency, and amplitude was confirmed;
- There was a logarithmic tendency between the flexural modulus and frequency, as well as between the tensile modulus and strain rate;
- The divergence of the experimental results given by the tensile tests and the predicted values using this model as a function of these two variables was small (<5%) for PA66.

The method was applicable under the hypotheses of isotropy, homogeneity, small deformations, and linear viscoelasticity. Therefore, the estimated values could be used for a linear elastic behavior. This excluded the mechanical behavior under high stress or strain. In principle, this method could be applicable to other thermoplastic materials, under the stated hypotheses, as long as they are thermo-rheologically simple materials; that is, wherever the principle of time–temperature superposition is valid. Naturally, this should be validated in each case.

Furthermore, the model was limited to the working range of the variables under study (strain rate and temperature).

Special attention must be paid to the glass-transition temperature. For the correct application of the model, it is necessary to determine this reference temperature. Significant variations occurred in the experimentation near this temperature. In the same way, the equation that related the shift factor with the temperature also was different below or above this temperature. Therefore, it is important to conduct an accurate experimental study around the glass-transition temperature, since important changes in the structural behavior of the material were observed experimentally, and there was a significant drop in the moduli. If that drop cannot be determined accurately, the model cannot be fed well enough to correctly estimate the tensile moduli.

Finally, the method presented here, verified for a polyamide 66, must be tested with other thermoplastic materials to determine to what extent it can be applied reliably. In addition, a future work will expand the range of the variables (temperature and strain rate) in order to determine the limits of the method's applicability.

Author Contributions: Conceptualization, A.S.-A., J.M.P.-F. and J.M.; Data curation, A.S.-A.; Formal analysis, A.S.-A. and J.M.; Investigation, A.S.-A. and J.M.P.-F.; Methodology, A.S.-A. and J.M.P.-F.; Project administration, J.M.P.-F.; Resources, A.S.-A. and J.M.P.-F.; Software, A.S.-A.; Supervision, J.M.P.-F., G.R. and J.M.; Validation, A.S.-A., J.M.P.-F. and G.R.; Visualization, G.R. and J.M.; Writing—original draft, A.S.-A.; Writing—review and editing, J.M.P.-F., G.R. and J.M. All authors have read and agreed to the published version of the manuscript.

Funding: This research received no external funding.

Institutional Review Board Statement: Not applicable.

Informed Consent Statement: Not applicable.

Conflicts of Interest: The authors declare no conflict of interest.

References

1. Matter, Y.S.; Darabseh, T.T.; Mourad, A.-H.I. Flutter analysis of a viscoelastic tapered wing under bending–torsion loading. *Meccanica* **2018**, *53*, 3673–3691. [CrossRef]
2. Mhanna, H.H.; Hawileh, R.A.; Abuzaid, W.; Naser, M.Z.; Abdalla, J.A. Experimental Investigation and Modeling of the Thermal Effect on the Mechanical Properties of Polyethylene-Terephthalate FRP Laminates. *J. Mater. Civ. Eng.* **2020**, *32*, 04020296. [CrossRef]
3. Sadasivuni, K.K.; Saha, P.; Adhikari, J.; Deshmukh, K.; Ahamed, M.B.; Cabibihan, J.J. Recent advances in mechanical properties of biopolymer composites: A review. *Polym. Compos.* **2020**, *41*, 32–59. [CrossRef]
4. Menacho, J.; Rotllant, L.; Molins, J.J.; Reyes, G.; García-Granada, A.A.; Balcells, M.; Martorell, J. Arterial pulse attenuation prediction using the decaying rate of a pressure wave in a viscoelastic material model. *Biomech. Model. Mechanobiol.* **2018**, *17*, 589–603. [CrossRef] [PubMed]
5. Bertaglia, G.; Caleffi, V.; Valiani, A. Modeling blood flow in viscoelastic vessels: The 1D augmented fluid–structure interaction system. *Comput. Methods Appl. Mech. Eng.* **2020**, *360*, 112772. [CrossRef]
6. Liao, Z.; Hossain, M.; Yao, X.; Navaratne, R.; Chagnon, G. A comprehensive thermo-viscoelastic experimental investigation of Ecoflex polymer. *Polym. Test.* **2020**, *86*, 106478. [CrossRef]
7. Rus, G.; Faris, I.H.; Torres, J.; Callejas, A.; Melchor, J. Why are viscosity and nonlinearity bound to make an impact in clinical elastographic diagnosis? *Sensors* **2020**, *20*, 2379. [CrossRef]
8. Serra-Aguila, A.; Puigoriol-Forcada, J.M.; Reyes, G.; Menacho, J. Viscoelastic models revisited: Characteristics and interconversion formulas for generalized Kelvin–Voigt and Maxwell models. *Acta Mech. Sin. Xuebao* **2019**, *35*, 1191–1209. [CrossRef]
9. Drozdov, A.D. *Finite Elasticity and Viscoelasticity*; World Scientific: Singapore, 1996; ISBN 9810224338.
10. García-Vilana, S.; Sánchez-Molina, D.; Llumà, J.; Galtés, I.; Velázquez-Ameijide, J.; Rebollo-Soria, M.C.; Arregui-Dalmases, C. Viscoelastic Characterization of Parasagittal Bridging Veins and Implications for Traumatic Brain Injury: A Pilot Study. *Bioengineering* **2021**, *8*, 145. [CrossRef]
11. Aryeetey, O.J.; Frank, M.; Lorenz, A.; Estermann, S.-J.; Reisinger, A.G.; Pahr, D.H. A parameter reduced adaptive quasi-linear viscoelastic model for soft biological tissue in uniaxial tension. *J. Mech. Behav. Biomed. Mater.* **2022**, *126*, 104999. [CrossRef] [PubMed]
12. Su, X.; Xu, W.; Chen, W.; Yang, H. Fractional creep and relaxation models of viscoelastic materials via a non-Newtonian time-varying viscosity: Physical interpretation. *Mech. Mater.* **2020**, *140*, 103222. [CrossRef]
13. Bonfanti, A.; Kaplan, J.L.; Charras, G.; Kabla, A. Fractional viscoelastic models for power-law materials. *Soft Matter* **2020**, *16*, 6002–6020. [CrossRef] [PubMed]
14. Soldatos, K.P. The generalized viscoelastic spring. *Proc. R. Soc. A Math. Phys. Eng. Sci.* **2020**, *476*, 20190881. [CrossRef]
15. Xu, Q.; Engquist, B.; Solaimanian, M.; Yan, K. A new nonlinear viscoelastic model and mathematical solution of solids for improving prediction accuracy. *Sci. Rep.* **2020**, *10*, 2202. [CrossRef] [PubMed]
16. Gutierrez-Lemini, D. *Engineering Viscoelasticity*; Springer: New York, NY, USA, 2014; ISBN 978-1-4614-8138-6.
17. Huayamares, S.; Grund, D.; Taha, I. Comparison between 3-point bending and torsion methods for determining the viscoelastic properties of fiber-reinforced epoxy. *Polym. Test.* **2020**, *85*, 106428. [CrossRef]
18. Chen, D.-L.; Chiu, T.-C.; Chen, T.-C.; Chung, M.-H.; Yang, P.-F.; Lai, Y.-S. Using DMA to Simultaneously Acquire Young's Relaxation Modulus and Time-dependent Poisson's Ratio of a Viscoelastic Material. *Procedia Eng.* **2014**, *79*, 153–159. [CrossRef]
19. Parveen, N.; Jana, P.K.; Schönhoff, M. Viscoelastic properties of polyelectrolyte multilayers swollen with ionic liquid solutions. *Polymers* **2019**, *11*, 1285. [CrossRef] [PubMed]
20. Obada, D.O.; Kuburi, L.S.; Dauda, M.; Umaru, S.; Dodoo-Arhin, D.; Balogun, M.B.; Iliyasu, I.; Iorpenda, M.J. Effect of variation in frequencies on the viscoelastic properties of coir and coconut husk powder reinforced polymer composites. *J. King Saud Univ. -Eng. Sci.* **2020**, *32*, 148–157. [CrossRef]
21. Menard, K.P.; Peter, K. *Dynamic Mechanical Analysis: A Practical Introduction*; CRC Press: Boca Raton, FL, USA; London, UK; New York, NY, USA; Washington, DC, USA, 1999; ISBN 0849386888.
22. Benaglia, S.; Amo, C.A.; Garcia, R. Fast, quantitative and high resolution mapping of viscoelastic properties with bimodal AFM. *Nanoscale* **2019**, *11*, 15289–15297. [CrossRef]
23. Chakravartula, A.M.; Pruitt, L.A.; Komvopoulos, K. Viscoelastic properties of plasma-treated low-density polyethylene surfaces determined by nanoscale dynamic mechanical analysis. *Mater. Res. Lett.* **2019**, *7*, 320–326. [CrossRef]
24. Uluutku, B.; López-Guerra, E.A.; Solares, S.D. A new method for obtaining model-free viscoelastic material properties from atomic force microscopy experiments using discrete integral transform techniques. *Beilstein J. Nanotechnol.* **2021**, *12*, 1063–1077. [CrossRef]
25. Kubacková, J.; Slabý, C.; Horvath, D.; Hovan, A.; Iványi, G.T.; Vizsnyiczai, G.; Kelemen, L.; Žoldák, G.; Tomori, Z.; Bánó, G. Assessing the Viscoelasticity of Photopolymer Nanowires Using a Three-Parameter Solid Model for Bending Recovery Motion. *Nanomaterials* **2021**, *11*, 2961. [CrossRef]
26. Tirella, A.; Mattei, G.; Ahluwalia, A. Strain rate viscoelastic analysis of soft and highly hydrated biomaterials. *J. Biomed. Mater. Res. Part A* **2014**, *102*, 3352–3360. [CrossRef] [PubMed]

27. Li, Y.; Hong, Y.; Xu, G.-K.; Liu, S.; Shi, Q.; Tang, D.; Yang, H.; Genin, G.M.; Lu, T.J.; Xu, F. Non-contact tensile viscoelastic characterization of microscale biological materials. *Acta Mech. Sin.* **2018**, *34*, 589–599. [CrossRef]
28. Lionetto, F.; Montagna, F.; Maffezzoli, A. Ultrasonic Dynamic Mechanical Analysis of Polymers. *Appl. Rheol.* **2005**, *15*, 326–335. [CrossRef]
29. Pérez-Peña, A.; García-Granada, A.A.; Menacho, J.; Molins, J.J.; Reyes, G. A methodology for damping measurement of engineering materials: Application to a structure under bending and torsion loading. *JVC J. Vib. Control* **2016**, *22*, 2471–2481. [CrossRef]
30. Tassieri, M.; Laurati, M.; Curtis, D.J.; Auhl, D.W.; Coppola, S.; Scalfati, A.; Hawkins, K.; Williams, P.R.; Cooper, J.M. i-Rheo: Measuring the materials' linear viscoelastic properties "in a step"! *J. Rheol.* **2016**, *60*, 649–660. [CrossRef]
31. Arao, Y.; Taniguchi, N.; Nishiwaki, T.; Hirayama, N.; Kawada, H. Strain-rate dependence of the tensile strength of glass fibers. *J. Mater. Sci.* **2012**, *47*, 4895–4903. [CrossRef]
32. Johnson, G.R.; Cook, W.H. A constitutive model and data for metals subjected to large strains, high strain rates and high temperatures. In Proceedings of the 7th International of Symposium on Ballistics, The Hague, The Netherlands, 19–21 April 1983; pp. 541–547.
33. Zhang, Y.; Outeiro, J.C.; Mabrouki, T. On the Selection of Johnson-cook Constitutive Model Parameters for Ti-6Al-4V Using Three Types of Numerical Models of Orthogonal Cutting. *Procedia CIRP* **2015**, *31*, 112–117. [CrossRef]
34. Plazek, D.J. Temperature Dependence of the Viscoelastic Behavior of Polystyrene. *J. Phys. Chem.* **1965**, *69*, 3480–3487. [CrossRef]
35. Álvarez-Vázquez, A.; Fernández-Canteli, A.; Castillo Ron, E.; Fernández Fernández, P.; Muñiz-Calvente, M.; Lamela Rey, M.J. A Novel Approach to Describe the Time–Temperature Conversion among Relaxation Curves of Viscoelastic Materials. *Materials* **2020**, *13*, 1809. [CrossRef] [PubMed]
36. Gergesova, M.; Zupančič, B.; Saprunov, I.; Emri, I. The closed form t-T-P shifting (CFS) algorithm. *J. Rheol.* **2011**, *55*, 1–16. [CrossRef]
37. Nantasetphong, W.; Jia, Z.; Amirkhizi, A.; Nemat-Nasser, S. Dynamic properties of polyurea-milled glass composites Part I: Experimental characterization. *Mech. Mater.* **2016**, *98*, 142–153. [CrossRef]
38. Mottahedi, M.; Dadalau, A.; Hafla, A.; Verl, A. Numerical Analysis of Relaxation Test Based on Prony Series Material Model. In *Integrated Systems, Design and Technology 2010*; Springer: Berlin/Heidelberg, Germany, 2011; pp. 79–91.
39. Knauss, W.G. The sensitivity of the time-temperature shift process to thermal variations-A note. *Mech. Time-Depend. Mater.* **2008**, *12*, 179–188. [CrossRef]
40. Dupont Zytel103 HSLNC010 Datasheet. Available online: https://www.campusplastics.com (accessed on 7 February 2022).

polymers MDPI

Article

Study on Fire Behavior, Thermal Stability and Degradation Kinetics of Thiol-Ene with Poly(aminopropyl/phenyl)silsesquioxane

Jiangbo Wang

School of Materials and Chemical Engineering, Ningbo University of Technology, Ningbo 315211, China; jiangbowang@nbut.edu.cn; Tel.: +86-0574-87081240

Abstract: In this article, the flame retardant poly(aminopropyl/phenyl)silsesquioxane (PA) was incorporated into thiol-ene (TE), to obtain a flame-retardant thiol-ene (FRTE) composite. The cone calorimeter (CONE) measurement results showed that, compared with neat TE, the peak of heat release rate (PHRR) and total heat release (THR) of FRTE have decreased by almost 23.7% and 14.5%, respectively. Thermogravimetric analysis (TGA) results further confirmed that the flame retardant PA could induce the initial thermal degradation of TE, and increased the amounts of residual char. Moreover, the activation energies of FRTE were calculated through the Kissinger and Flynn–Wall–Ozawa methods. Compared with the neat TE, the activation energies of FRTE were raised by the addition of PA. It indicated that the flame retardant PA promoted cross-linking reactions of TE, to form a compact char layer and retarded further the thermal degradation of the polymer matrix.

Keywords: flame retardancy; poly(aminopropyl/phenyl)silsesquioxane; thiol-ene; kinetics; activation energy

Citation: Wang, J. Study on Fire Behavior, Thermal Stability and Degradation Kinetics of Thiol-Ene with Poly(aminopropyl/phenyl)silsesquioxane. *Polymers* **2022**, *14*, 1142. https://doi.org/10.3390/polym14061142

Academic Editors: Andrzej Puszka and Beata Podkościelna

Received: 20 February 2022
Accepted: 10 March 2022
Published: 12 March 2022

1. Introduction

UV-photopolymerization is a simple and efficient way of generating cross-linked networks. Due to the advantages of solvent-free, environment-friendly, all active ingredients, and rapid curing under UV irradiation, UV-curing film has strong potential application in the field of coatings. A wide variety of monomers (including multifunctional acrylate and methacrylate monomers) have been found to undergo rapid photopolymerization under UV light, with the right amount of photoinitiator [1–3]. However, there are still many problems in the above monomer systems, such as the fact that it is unstable in oxygen, has uneven crosslinking, large internal stress in polymerization, is easy to cause volume shrinkage, and so on [4–8].

Thiol-ene (TE) photopolymerization is a novel photopolymerization, based on click chemistry, which is different from the step-growth reaction mechanism of an acrylate-based photopolymerization system. It has the characteristics of uniform cross-linking network, gel point delay, low volume shrinkage and low stress, which overcome the defects of previous conventional photopolymerization systems. In addition, the rate of the thiol-ene addition reaction is very fast, which is almost equal to the photopolymerization of acrylate under inert conditions. Conventional radical addition polymerization is difficult to carry out in the presence of oxygen, but the thiol-ene reaction is different from this. It can occur in the presence of oxygen and will not be affected. [9–13]. However, the thiol-ene polymer, like most organic polymers, is deficient in flame retardancy. Therefore, adding some additional substances into thiol-ene is necessary to reduce its flammability [14,15].

Halogen flame retardance, the earliest used flame retardant, is an important kind of organic flame retardant at present. With low price and additional excellent stability and compatibility, it has become one of the most used flame retardants in the world [16,17].

However, the halogen gas released from the combustion of halogen containing polymers will generate corrosive, harmful gas (hydrogen halide), when combined with water vapor, causing corrosion to some equipment and buildings. Halogen flame retardants will release strong carcinogens, such as dioxin and benzofuran, after combustion, affecting normal human metabolism and seriously damaging the environment [18–20].

The use of halogen-free flame retardants has become the development trend of polymer flame retardants. As a high-efficiency, smokeless, low-toxicity and pollution-free flame retardant, the phosphate flame retardant has attracted the interest of many researchers. At present, remarkable achievements have been made in synthesis and application. However, most phosphate flame retardants also have some disadvantages, such as high volatility, poor heat resistance, poor compatibility, and dripping during combustion. Inorganic flame retardants mainly include hydroxide (aluminum hydroxide, magnesium hydroxide), red phosphorus, tin series and borate (zinc borate) [21–24]. Inorganic flame retardants not only have a flame retardant effect, but also have a smoke suppression effect, and can inhibit the formation of hydrogen chloride. Inorganic flame retardants are widely used because they are non-toxic and non-corrosive. Today, with the increasing requirements of environmental protection, inorganic flame retardants show strong competitiveness and development potential. The disadvantage is that inorganic flame retardants generally have relatively large addition and low flame-retardant efficiency, which seriously damage other properties of the polymer matrix. As reported in the literature, polysiloxane has been demonstrated as an effective and 'environment friendly' flame retardant, for various polymers. Silicon, due to its low surface energy, migrates easily to the surface of the polymer matrix during combustion. Thus, the thermal degradation of the polymer can be effectively prevented, by forming a protective layer with excellent heat resistance [25–28]. However, as far as we know, no one has studied the effect of polysiloxane to enhance the fire behavior and thermal property of the thiol-ene system.

Thus, in this paper, poly(aminopropyl/phenyl)silsesquioxane (PA) was incorporated into thiol-ene to enhance the flame retardancy of the composites. We chose PA because the phenyl groups, in their structure, have excellent char-forming properties. Additionally, the amino group forms nitrogen during combustion, which also has a flame-retardant effect. Then, the fire behavior and thermal degradation behavior of siliconized-modified thiol-ene were investigated by cone calorimeter measurement and thermogravimetric analysis (TGA), respectively.

2. Materials and Methods

2.1. Materials

Trimethylolpropane tris(3-mercaptopropionate) (3T) was supplied by Bruno Bock Chemische Fabrik GmbH & Co. (Marschacht, Germany). Tetramethylammonium hydroxide (TMAOH) and phenyltriethoxysilane (PTES) were supplied by Alfa Aesar Chemical Reagent Co. Ltd. (Tewksbury, MA, USA). Sigma-Aldrich Reagent Co. Ltd. (St. Louis, MO, USA) supplied 2,2-Dimethoxy-2-phenylacetophenone (DMPA), pentaerythritol allyl ether (TAE), (3-aminopropyl)triethoxysilane (APS) and ethyl alcohol (EtOH) were all used as received.

2.2. Synthesis of Poly(aminopropyl/phenyl)silsesquioxane (PA)

As shown in Figure 1, the synthesis of PA was based on previous publications and the specific method was as follows [29,30]: EtOH (75 mL), distilled water (25 mL) and TMAOH (1 mL) were added into a 250 mL flask. Then, PTES and APS at different molar ratios were mixed in the above solution, accounting for 10 wt% of the total. Stirring was stopped after 8 h and left overnight. The supernatant was removed and the precipitate condensate was collected. It was then pumped and filtered with EtOH/distilled water (3/1 by volume) and washed with anhydrous EtOH. The product was dried in vacuum for 20 h at room temperature to obtain PA.

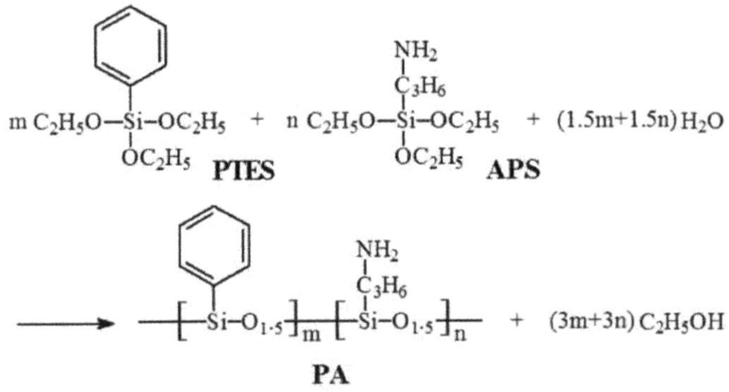

Figure 1. Synthesis of PA.

2.3. Preparation of TE Composites

For the composite preparation, 1 wt% DMPA was first dissolved in 3T and ultrasound was performed for 30 min. Then, equal amounts of TAE with 3T and PA (5 wt% of the total amount) were added to the mixture and stirred evenly. The mixture was further mixed with an ultrasonic device and the bubbles were removed (30 min). TE/PA (FRTE) composites were prepared by UV curing after pouring the mixture onto the glass substrate. For comparison, TE was prepared under the same technological conditions.

2.4. Characterization and Measurement

Cone calorimeter measurement was carried out using an FTT Conical Calorimeter (Fire Testing Technology Ltd., East Grinstead, West Sussex, UK) according to ASTM E1354. The heat flux was 50 kW/m^2 and the specimen size was $100 \times 100 \times 3$ mm^3. All specimens were measured in three groups and then averaged. Thermogravimetric analysis (TGA) was performed on the Q5000 TA Thermogravimetric Analyzer (TA Instrument Corp., New Castle, DE, USA). In a nitrogen atmosphere, about 10 mg of the sample was heated from 50 °C to 600 °C at 10 °C/min heating rate.

2.5. Thermal Degradation Theory

When studying the thermal transformation kinetics of solid chemical reactions, it is generally based on the following reaction rate [31,32]:

$$r = \frac{da}{dt} = kf(a) \tag{1}$$

where, *r* is the degradation rate, *a* is the conversion degree, *t* is the time, *k* is the rate constant, *f(a)* is the reaction model. It is generally assumed that *k* obeys the Arrhenius equation:

$$k = A\exp(-E/RT) \tag{2}$$

where, *A* is the pre-exponential factor, *E* is the activation energy, *R* is the universal gas constant and *T* is the temperature.

The influence relationship between degradation rate and temperature and sample weight change can be expressed as:

$$\frac{da}{dt} = Af(a)\exp(-E/RT) \tag{3}$$

Equation (3) can also be used in its integral form. Under isothermal conditions, the integral form is:

$$\ln t = E/RT - \ln[A/g(x)] \tag{4}$$

For non-isothermal degradation, Equation (3) becomes:

$$\frac{da}{dT} = (A/\beta)f(a)\exp(-E/RT) \tag{5}$$

where, β is the heating rate ($\beta = \frac{dT}{dt}$), $g(x)$ is the mechanism integrated forms ($g(x) = \int_0^a \frac{da}{f(a)}$).

(1) Kissinger method [33]

The equation of the Kissinger method can be expressed as follows:

$$\ln(\frac{\beta}{T_{\max}^2}) = \ln(\frac{AR}{E}) - \frac{E}{RT_{\max}} \tag{6}$$

where, $T_{\max}$ is the temperature of the peak rate.

The temperature of peak rate is determined by the DTG curves at different heating rates. Then draw with $1/T_{\max}$ as the abscissa and $\ln(\beta/T_{\max}^2)$ as the ordinate and fitting a straight line. The activation energy can be calculated from the slope of the line by the Kissinger equation.

(2) Flynn–Wall–Ozawa method [34,35]

The equation of the Flynn–Wall–Ozawa method is as follows:

$$\lg(\beta) = \lg AE/g(a)R - 2.315 - 0.457\frac{E}{RT} \tag{7}$$

As can be seen from the above equation, variable $\lg(\beta)$ is linearly proportional to variable $1/T$. The activation energy for any particular degree of degradation can be obtained by calculating the slope of the $\lg(\beta) - 1/T$ plots.

3. Results and Discussion

3.1. Flame Retardancy

There are many traditional fire hazard testing methods, but most of them use small instruments to test the performance, which is far from the actual situation when a fire occurs. The cone calorimeter is mainly based on the principle of oxygen consumption for testing. It provides a way to measure multiple different parameters in the same experiment. It has been shown that the cone calorimeter test results have a very good correlation with the parameters obtained from large-scale fire tests. Thus, it can be used to predict the burning behavior of materials in real fires [36]. The cone calorimeter of TE composites is presented in Figure 2. It could be obtained that the peak of heat release rate (PHRR) for the neat TE reached 2152.4 kW/m^2, which presented a very sharp heat release rate (HRR) curve and the combustion was complete after 321 s. Compared with neat TE, the incorporation of 5 wt% PA led to a strong reduction in PHRR, which reached a value of 1642.8 kW/m^2 and the PHRR was reduced by nearly 23.7%. The reduction in HRR was accompanied by a prolongation of burning time (from 321 to 409 s). Moreover, it was clear that the total heat release (THR) evidently decreased (from 188.0 to 160.7 MJ/m^2) for the FRTE composite, compared with the neat TE matrix.

The Flame Retardancy Index (FRI) was always used to evaluate the flame retardancy of resin systems [37,38]. The calculation equation of FRI is as follows:

$$\mathrm{FRI} = \frac{\left[\mathrm{THR} \times \left(\frac{\mathrm{PHRR}}{\mathrm{TTI}}\right)\right]_{\mathrm{TE}}}{\left[\mathrm{THR} \times \left(\frac{\mathrm{PHRR}}{\mathrm{TTI}}\right)\right]_{\mathrm{FRTE}}} \tag{8}$$

As shown in Table 1, the FRI value of PA was 2.31 after calculating the relevant parameters, which exhibited "good" flame retardancy performance.

Table 1. The parameters obtained from the cone calorimeter measurement.

Sample	PHRR (kW/m^2)	THR (MJ/m^2)	TTI (s)	FRI	Flame Retardancy Performance
TE	2152.4	188.0	102.5	-	-
FRTE	1642.8	160.7	154.5	2.31	good

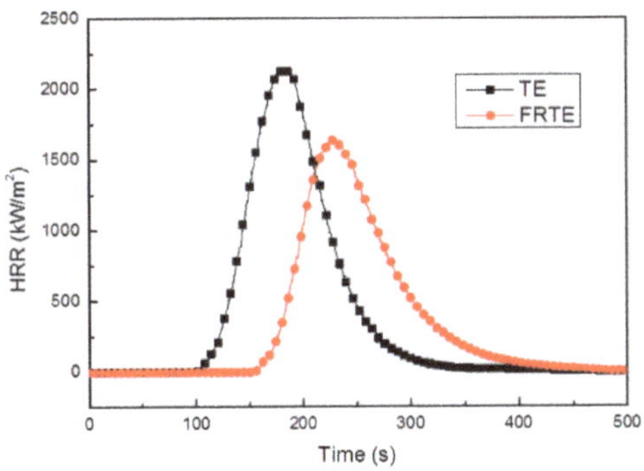

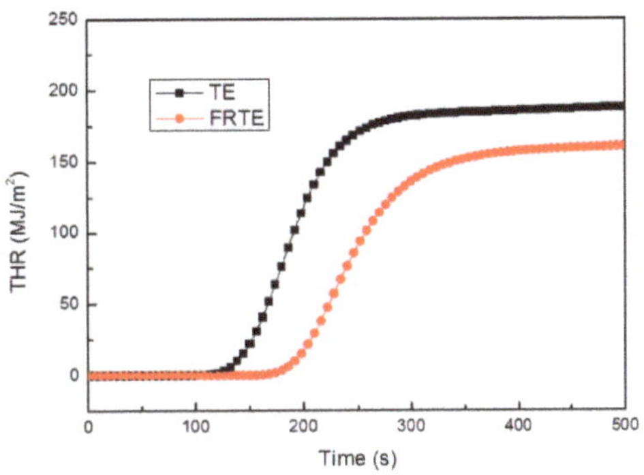

Figure 2. HRR (**a**) and THR (**b**) curves of TE composites.

The morphology of the char residue in the cone calorimeter after combustion is presented in Figure 3. It can be seen that the neat TE was fully burned and there was practically no residue (Figure 3a). In contrast, the amount of FRTE residue was large and the color was light yellow (Figure 3b). It was particularly important that the residue formed a compact and continual char layer. This obviously helps prevent the passage of heat and combustible substances in the fire and, finally, the flame retardancy of the polymer was raised.

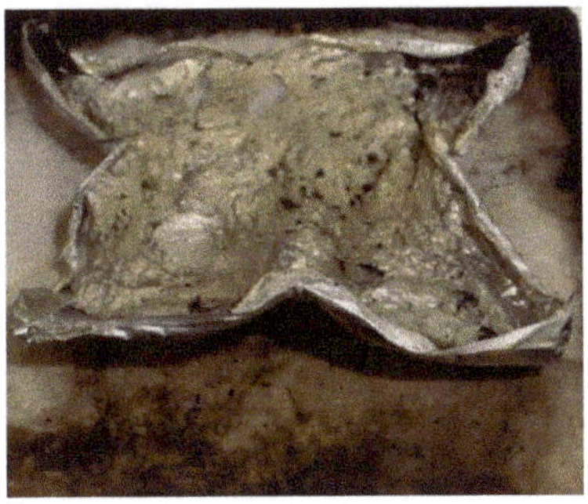

(**a**)

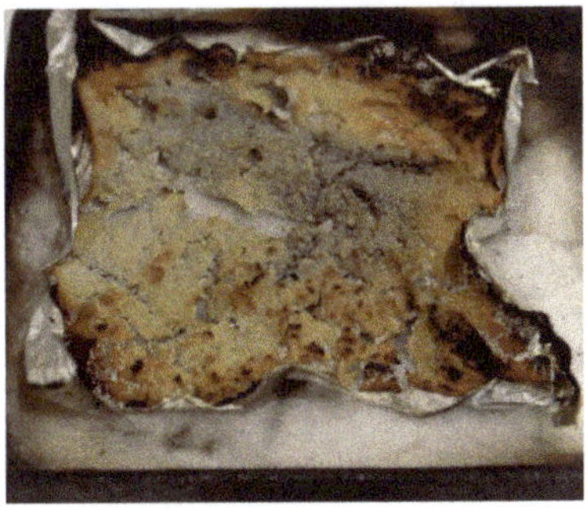

(**b**)

Figure 3. Residual char images of TE (**a**) and FRTE (**b**) after CONE measurement.

3.2. Thermal Stability

The TGA/DTG curves for the degradation of the TE composites, at a heating ramp rate of 10 °C/min in nitrogen, are presented in Figure 4, and the data are listed in Table 2. It revealed that the onset degradation temperature ($T_{5wt\%}$) of FRTE was much lower than that of neat TE. The $T_{5wt\%}$ for TE was 345.7 °C, but 327.3 °C for the FRTE, indicating that the onset degradation temperature of TE tended to decrease with the incorporation of PA. The major degradation in the FRTE occurred between 300–500 °C, which was similar to that of TE.

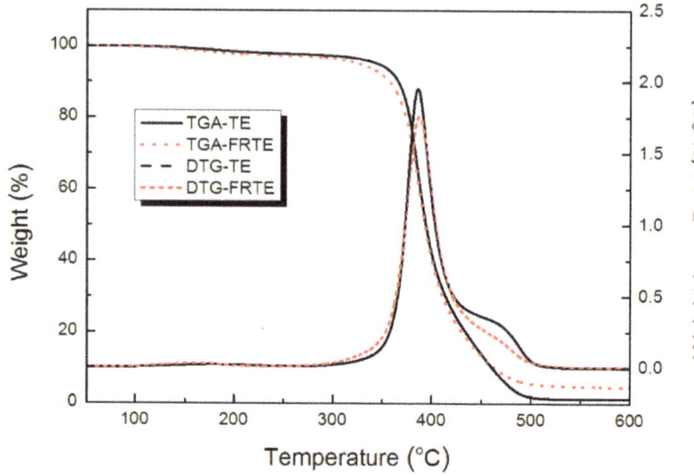

Figure 4. Thermal stability of TE composites.

Table 2. TGA data of TE composites.

Sample	Temperature (°C)		Peak Rate (wt%/°C)	Residue Char (wt%)
	$T_{5wt\%}$	T_{max}		
TE	345.7	385.9	1.95	1.33
FRTE	327.3	387.8	1.76	4.60

The char residues of the FRTE at 600 °C were higher than that of TE, and the amount of solid residue shifted from 1.33 wt% (for TE) to 4.60 wt% (for FRTE) of the initial weight. The phenyl groups in the structure had excellent char-forming properties and aminopropyl groups could promote the crosslinking reactions during the thermal degradation process. In addition to that, nitrogen could be formed from the amino group during combustion, which also has a flame-retardant effect. This result further confirmed that the branched silicone with aminopropyl and phenyl could induce the formation of the char layer, which might play an important role for the flame retardancy of the FRTE composite [39,40].

3.3. Thermal Degradation Kinetics

The TGA and DTG curves of the TE composites, at the heating rates of 5, 10, 20 and 40 °C/min, are shown in Figures 5 and 6. The curves revealed the different profiles, depending on heating rate, and two weight-loss stages occurred during degradation, which was consistent with the literature report [41]. The first major degradation in the TE composites emerged in a temperature range of 300–450 °C, whereas the second stage degradation was observed above 450 °C. The temperature of the peak rate (T_{max}) of the TE composites increased progressively as the heating rate increased. Generally, with the increase in heating rate, the time required for the sample to reach a certain temperature is shortened. Therefore, it could be seen from Figure 5 that, when the heating rate gradually increased from 5 °C/min to 40 °C/min, the TGA curve of the sample also moved to a higher temperature.

Figure 7 presents the Kissinger plots of $\ln(\beta/T_{\max}^2)$ versus $1000/T_{\max}$ for TE composites. The kinetic parameters of the first stage in thermal degradation, calculated by the Kissinger method, are summarized and compared in Table 3.

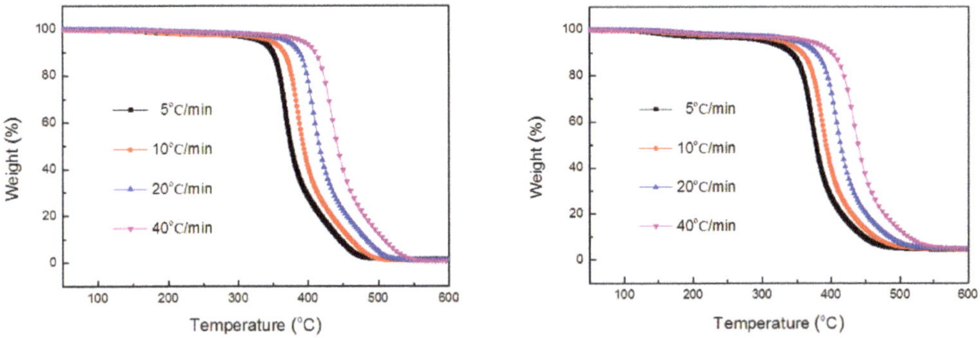

(a) (b)

Figure 5. TGA curves of TE (**a**) and FRTE (**b**) composites.

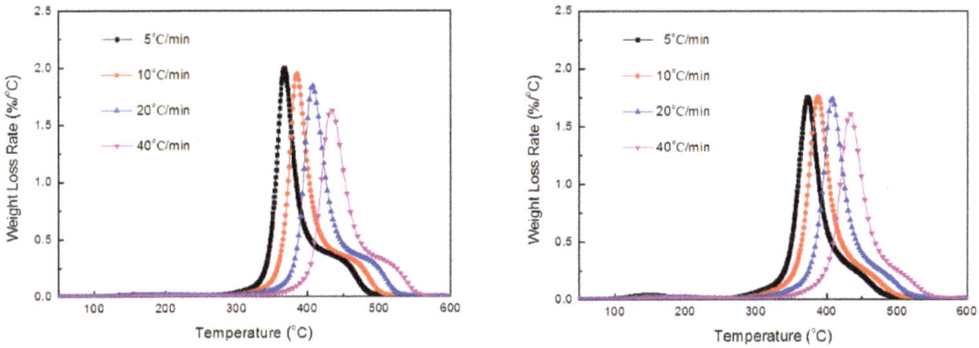

(a) (b)

Figure 6. DTG curves of TE (**a**) and FRTE (**b**) composites.

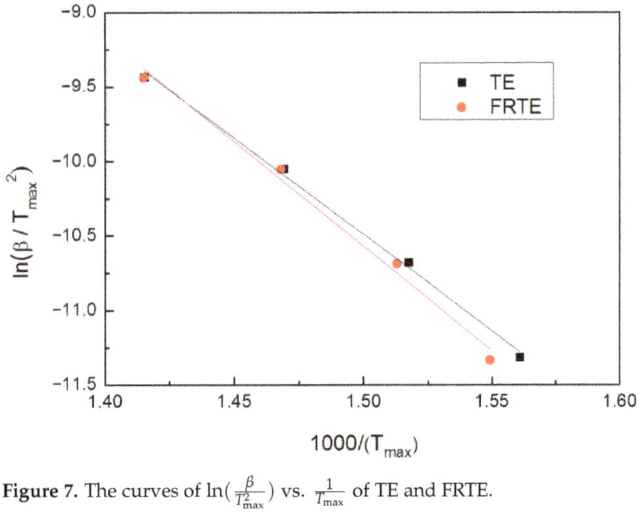

Figure 7. The curves of $\ln(\frac{\beta}{T_{max}^2})$ vs. $\frac{1}{T_{max}}$ of TE and FRTE.

Table 3. Kinetic data using the Kissinger method.

Sample	Temperature (°C)				E (kJ/mol)	lnA (1/min)
	5 °C/min	10 °C/min	20 °C/min	40 °C/min		
TE	367.5	385.9	407.5	433.4	107.4	11.4
FRTE	372.3	387.8	408.0	433.5	116.7	13.1

The kinetic parameters of TE changed with the incorporation of PA. For FRTE, the values of activation energy and *lnA* were 116.7 kJ/mol and 13.1/min, respectively, which were significantly higher than those of neat TE. In general, the incorporation of flame-retardant PA enhanced the thermal stability of TE [42].

The Flynn–Wall–Ozawa method is another kinetics analysis method and was used in this study. Compared with the Kissinger method, the Flynn–Wall–Ozawa method can analyze the change in activation energy of a flame retardant system in the whole thermal degradation process, through simple TGA data processing, and can, therefore, obtain more comprehensive and complete kinetic data.

Based on the data in Figure 4, and the equation of $a = \frac{w_0 - w_t}{w_0 - w_\infty}$ (w_0 is the initial weight of the sample, w_t is the sample weight at any temperature t, w_∞ is the final sample weight), the conversion degree as a function of temperature, relative to the decomposition of the TE and FRTE systems, can be calculated, as exhibited in Figure 8.

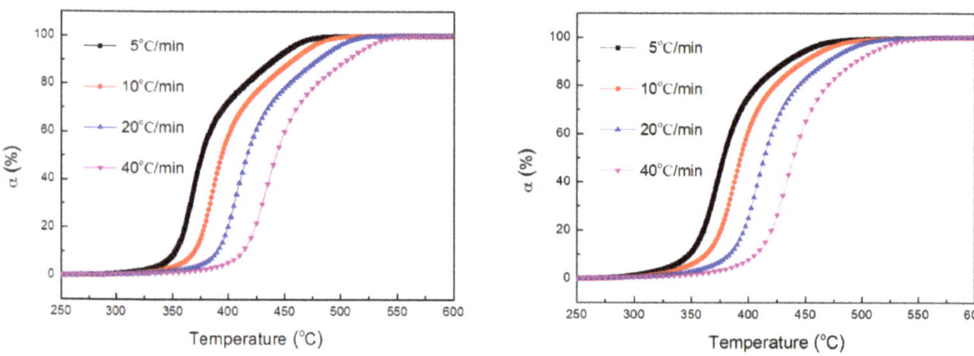

(**a**) (**b**)

Figure 8. Conversion of TE (**a**) and FRTE (**b**) as a function of temperature.

The activation energies of the thermal degradation for the TE composites could be calculated through Equation (7). The conversion values were 0.02, 0.05, 0.10, 0.20, 0.30, 0.40, 0.50, 0.60, 0.70, 0.80, 0.90, 0.95 and 0.98.

For the fitting straight lines, obtained in Figure 9, their R2 values were both higher than 99%. This means that the Flynn–Wall–Ozawa method was suitable for this research system. Moreover, the fitting lines, corresponding to TE and FRTE, were relatively parallel, which indicated that the research system should correspond to a single reaction mechanism. The activation energy curves are presented in Figure 10.

As seen, the same tendency as for the results from the Flynn–Wall–Ozawa method was obtained. A decrease in the activation energy, with the increasing conversion in the initial degradation stage (2~10%), was found. The activation energy of FRTE at 5% conversion was around 116.0 kJ/mol, whereas that of neat TE was 117.8 kJ/mol. It was reported that the earlier thermal degradation of polymer always occurred, due to the degradation of polysiloxane at a lower temperature [43]. Then, the activation energies of neat TE and FRTE both increased with increasing conversion. With the increase in activation energy, the thermal stability of the polymer was improved, and the degradation became difficult, which indicated that the flame retardancy of the polymer was improved. However, for FRTE, the incorporation of PA led to activation energy greater than that of neat TE ($\alpha \geq 10\%$). From these values, mean values of 124.4 kJ/mol and 129.1 kJ/mol were found for neat TE and FRTE, respectively, which was in best agreement with those obtained using the Kissinger method [44]. The above results indicated that the Si-$C_3H_6NH_2$ bond and Si-Ph bond of flame retardant PA may form some silyl radicals or siloxane derivatives, which could react with TE or the evolved products of TE. Thus, the cross-linking reactions were promoted to form a compact char layer in the FRTE composite and further pyrolysis during the thermal degradation process was retarded.

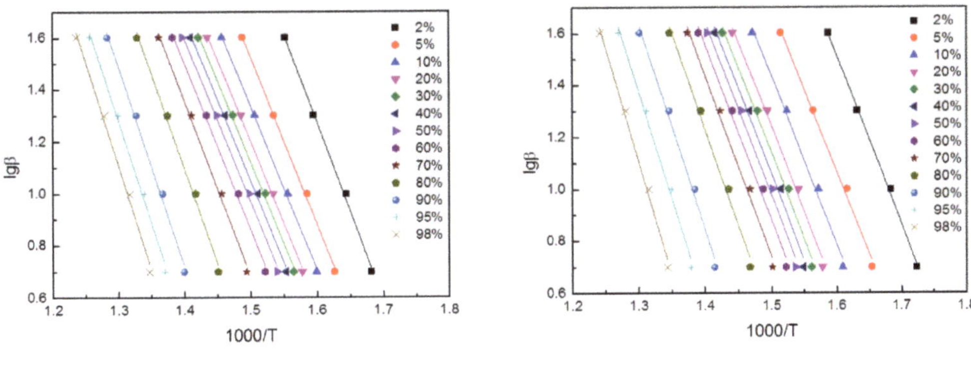

Figure 9. The curves of lg(β) vs. 1000/T of TE (**a**) and FRTE (**b**).

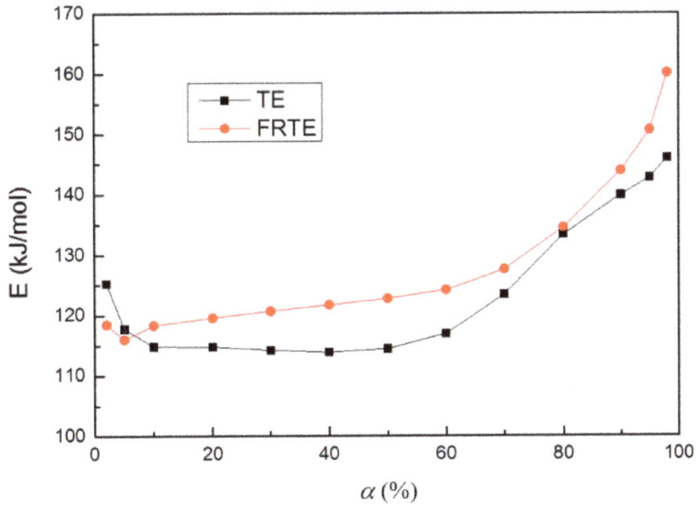

Figure 10. Activation energy curves by Flynn–Wall–Ozawa method.

4. Conclusions

A flame-retardant composite FRTE has been successfully prepared by the incorporation of poly(aminopropyl/phenyl)silsesquioxane into a thiol-ene matrix. The results of the cone calorimeter and TGA measurements showed that, compared with neat TE, the flame retardancy and thermal stability of FRTE were improved. Specifically, the PHRR and THR of FRTE were reduced by almost 23.7% and 14.5%, and the amounts of residual char were increased. Furthermore, the results from both the Kissinger and Flynn–Wall–Ozawa methods showed that the activation energies of FRTE were enhanced by the incorporation of PA, which indicated that the branched silicone with aminopropyl and phenyl promoted cross-linking reactions of TE, to form a compact char layer, and retarded further pyrolysis during the thermal degradation process of the polymer matrix.

Funding: This research was funded by the Ningbo Natural Science Foundation (2019A610032). This work was also supported by the Foundation (No. GZKF202127) of State Key Laboratory of Biobased Material and Green Papermaking, Qilu University of Technology, Shandong Academy of Sciences.

Institutional Review Board Statement: Not applicable.

Informed Consent Statement: Not applicable.

Data Availability Statement: The data used to support the findings of this study are available from the corresponding author upon request.

Acknowledgments: We gratefully acknowledge the financial support of the above funds and the researchers of all reports cited in our paper.

Conflicts of Interest: The authors declare no conflict of interest.

References

1. Yao, H.-Y.; Lin, H.-R.; Sue, G.-P.; Lin, Y.-J. Chitosan-based hydrogels prepared by UV polymerization for wound dressing. *Polym. Polym. Compos.* **2019**, *27*, 155–167. [CrossRef]
2. Wang, X.; Wang, X.; Song, L.; Xing, W.; Tang, G.; Hua, W.; Hu, Y. Preparation and thermal stability of UV-cured epoxy-based coatings modified with octamercaptopropyl POSS. *Thermochim. Acta* **2013**, *568*, 130–139. [CrossRef]
3. Qian, J.; Li, Z.; Huang, N.; Lu, Q.; Xia, J. UV-Irradiation Polymerization of Bis-EDOT Methane Derivatives and Their Application for Br_2 Detection. *Polymer* **2021**, *226*, 123808. [CrossRef]
4. Kalaithong, W.; Molloy, R.; Nalampang, K.; Somsunan, R. Design and optimization of polymerization parameters of carboxymethyl chitosan and sodium 2-acrylamido-2-methylpropane sulfonate hydrogels as wound dressing materials. *Eur. Polym. J.* **2021**, *143*, 110186. [CrossRef]
5. Wang, J.; Tsai, C.; Tsai, N.; Chiang, C.; Li, Y. An injectable, dual crosslinkable hybrid pectin methacrylate (PECMA)/gelatin methacryloyl (GelMA) hydrogel for skin hemostasis applications. *Int. J. Biol. Macromol.* **2021**, *185*, 441–450. [CrossRef] [PubMed]
6. Manjunath, M.; Naveen, K.; Vinod, P.; Balashanmugam, N.; Shankar, M. Mechanical characterization of UV photopolymerized PMMA with different photo-initiator concentration. *Appl. Mech. Mater.* **2021**, *903*, 11–16.
7. Kwisnek, L.; Heinz, S.; Wiggins, J.; Nazarenko, S. Multifunctional thiols as additives in UV-cured PEG-diacrylate membranes for CO_2 separation. *J. Membr. Sci.* **2011**, *369*, 429–436. [CrossRef]
8. Sparks, B.; Kuchera, T.; Jungman, M.; Richardson, A.; Savin, D.; Hait, S.; Lichtenhan, J.; Striegel, M.; Patton, D. Cyclic Tetravinylsiloxanetetraols as Hybrid Inorganic-Organic Thiol-Ene Networks. *J. Mater. Chem.* **2012**, *22*, 3817–3824. [CrossRef]
9. Hoyle, C.; Bowman, C. Thiol-ene click chemistry. *Angew. Chem. Int. Ed.* **2010**, *49*, 1540–1573. [CrossRef] [PubMed]
10. Ma, S.; Wang, Y.; Zhang, H.; Li, Y.; Ou, J.; Wei, Y.; Ye, M. One-step fabrication of cinchona-based hybrid monolithic chiral stationary phases via photo-initiated thiol-ene polymerization for cLC enantioseparation. *Talanta* **2019**, *198*, 432–439. [CrossRef] [PubMed]
11. Romera, C.; Oliveira, D.; Araújo, P.; Sayer, C. Biobased ester 2-(10-undecenoyloxy)ethyl methacrylate as asymmetrical diene monomer in thiol-ene polymerization. *Ind. Eng. Chem. Res.* **2019**, *58*, 21044–21055. [CrossRef]
12. Le, C.M.Q.; Morlet-Savary, F.; Chemtob, A. Role of thiol oxidation by air in the mechanism of the self-initiated thermal thiol–ene polymerization. *Polym. Chem.* **2021**, *12*, 6594–6605. [CrossRef]
13. Lee, J.; Lee, Y.; Park, S.; Ha, K. Preparation and Properties of Thiol-ene UV-cured Nanocomposites with Methacrylate-grafted Cellulose Nanocrystals as Fillers. *Polym. Korea* **2019**, *43*, 612–620. [CrossRef]
14. Çakmakçı, E.; Mülazim, Y.; Kahraman, M.; Apohan, N. Flame retardant thiol-ene photocured coatings. *React. Funct. Polym.* **2011**, *71*, 36–41. [CrossRef]
15. Çakmakçı, E.; Mülazim, Y.; Kahraman, M.; Apohan, N. Preparation and characterization of boron containing thiol-ene photocured. *Prog. Org. Coat.* **2012**, *75*, 28–32. [CrossRef]
16. Covaci, A.; Harrad, S.; Abdallah, M.; Ali, N.; Law, R.; Herzke, D.; Wit, C. Novel brominated flame retardants: A review of their analysis, environmental fate and behaviour. *Environ. Int.* **2011**, *37*, 532–556. [CrossRef] [PubMed]
17. Wit, C.; Herzke, D.; Vorkamp, K. Brominated flame retardants in the Arctic environment—Trends and new candidates. *Sci. Total Environ.* **2010**, *15*, 2885–2888. [CrossRef]
18. Rakotomalala, M.; Wagner, S.; Döring, M. Recent developments in halogen free flame retardants for epoxy resins for electrical and electronic applications. *Materials* **2010**, *3*, 4300–4327. [CrossRef] [PubMed]

19. Levchik, S.; Weil, E. Thermal decomposition, combustion and flame-retardancy of epoxy resins—A review of the recent literature. *Polym. Int.* **2004**, *53*, 1901–1929. [CrossRef]
20. Zhao, X.; Vignesh, H.; Llorca, J.; Wang, D. Impact of halogen-free flame retardant with varied phosphorus chemical surrounding on the properties of diglycidyl ether of bisphenol—A type epoxy resin: Synthesis, fire behaviour, flame-retardant mechanism and mechanical properties. *RSC Adv.* **2016**, *6*, 59226–59236. [CrossRef]
21. Wang, Y.; Feng, W.; Dong, Q.; Xie, M.; Peng, L.; Ding, Y.; Zhang, S.; Yang, M.; Zheng, G. Core-shell expandable graphite @ aluminum hydroxide as a flame-retardant for rigid polyurethane foams. *Polym. Degrad. Stab.* **2017**, *146*, 267–276. [CrossRef]
22. Chen, S.; Bao, X.; Wu, F.; Wang, J. Fire Retardancy and Thermogravimetric Kinetics of Thiol-Ene Degradation as Affected by Added 9,10-Dihydro-9-Oxa-10-Phosphaphenanthrene-10-Oxide (DOPO). *J. Macromol. Sci. Part B* **2021**, *61*, 1–9. [CrossRef]
23. Chen, X.; Yu, J.; He, M.; Guo, S.; Luo, L. Effects of zinc borate and microcapsulated red phosphorus on mechanical properties and flame retardancy of polypropylene/magnesium hydroxide composites. *J. Polym. Res.* **2009**, *16*, 357–362. [CrossRef]
24. Vahabi, H.; Laoutid, F.; Mehrpouya, M.; Saeb, S.; Dubois, S. Flame retardant polymer materials: An update and the future for 3D printing developments. *Mater. Sci. Eng. R Rep.* **2021**, *144*, 100604. [CrossRef]
25. Wang, W.; Peng, L.; Hsiue, G. Characterization and properties of new silicone-containing epoxy resin. *Polymer* **2000**, *41*, 6113–6120. [CrossRef]
26. Liu, S.; Ye, H.; Zhou, Y.; He, J.; Jiang, Z.; Zhao, J.; Huang, X. Study on flame-retardant mechanism of polycarbonate containing sulfonate-silsesquioxane-fluoro retardants by TGA and FTIR. *Polym. Degrad. Stab.* **2006**, *91*, 1808–1814. [CrossRef]
27. Wang, J.; Xin, Z. Flame retardancy, thermal, rheological and mechanical properties of polycarbonate/polysilsesquioxane system. *J. Appl. Polym. Sci.* **2010**, *115*, 330–337. [CrossRef]
28. Wu, K.; Song, L.; Hu, Y.; Lu, H.; Kandola, B.; Kandare, E. Synthesis and characterization of a functional polyhedral oligomeric silsesquioxane and its flame retardancy in epoxy resin. *Prog. Org. Coat.* **2009**, *65*, 490–497. [CrossRef]
29. Liu, S.; Lang, X.; Ye, H.; Zhang, S.; Zhao, J. Preparation and characterization of copolymerized aminopropyl/phenylsilsesquioxane microparticles. *Eur. Polym. J.* **2005**, *41*, 996–1001. [CrossRef]
30. Wu, F.; Bao, X.; Xu, H.; Kong, D.; Wang, J. Functionalization of graphene oxide with polysilicone: Synthesis, characterization and fire retardancy in thiol-ene systems. *J. Macromol. Sci. Part B* **2021**, *60*, 339–349. [CrossRef]
31. Bao, X.; Wu, F.; Wang, J. Thermal degradation behavior of epoxy resin containing modified carbon nanotubes. *Polymers* **2021**, *13*, 3332. [CrossRef] [PubMed]
32. Wang, J. Silicon-aluminum synergistic mechanism in flame retardancy of epoxy resin. *Polym. Compos.* **2014**, *35*, 1553–1558. [CrossRef]
33. Kissinger, H. Reaction kinetics in differential thermal analysis. *Anal. Chem.* **1957**, *29*, 1702–1706. [CrossRef]
34. Flynn, J. A quick, direct method for the determination of activation energy from thermogravimetric data. *J. Polym. Sci. Polym. Lett.* **1966**, *4*, 323–328. [CrossRef]
35. Flynn, J. Initial kinetic parameters from thermogravimetric rate and conversion data. *J. Polym. Sci. Polym. Lett.* **1967**, *5*, 191–196. [CrossRef]
36. Wang, L.; Yu, J.; Tang, Z.; Jiang, P. Synthesis, characteristic, and flammability of modified carbon nanotube/poly(ethylene-co-vinyl acetate) nanocomposites containing phosphorus and silicon. *J. Mater. Sci.* **2010**, *45*, 6668–6676. [CrossRef]
37. Vahabi, H.; Kandola, B.; Saeb, M. Flame Retardancy Index for thermoplastic composites. *Polymers* **2019**, *11*, 407. [CrossRef]
38. Qi, Y.; Weng, Z.; Zhang, K.; Wang, J.; Zhang, S.; Liu, C.; Jian, X. Magnolol-based bio-epoxy resin with acceptable glass transition temperature, processability and flame retardancy. *Chem. Eng. J.* **2020**, *387*, 124115. [CrossRef]
39. Wang, S.; Yang, X.; Li, Z.; Xu, X.; Liu, H.; Wang, D.; Min, H.; Shang, S. Novel eco-friendly maleopimaric acid based polysiloxane flame retardant and application in rigid polyurethane foam. *Compos. Sci. Technol.* **2020**, *198*, 108272. [CrossRef]
40. Wang, Q.; Xiong, L.; Liang, H.; Chen, L.; Huang, S. Synthesis of a novel polysiloxane containing phosphorus, and boron and its effect on flame retardancy, mechanical, and thermal properties of epoxy resin. *Polym. Compos.* **2018**, *39*, 807–814. [CrossRef]
41. Emre, B.; Oktay, B.; Kahraman, M.; Apohan, N. UV cured thiol-ene flame retardant hybrid coatings. *Prog. Org. Coat.* **2013**, *76*, 936–943.
42. Ahuja, D.; Kumar, L.; Kaushik, A. Thermal stability of starch bionanocomposites films: Exploring the role of esterified cellulose nanofibers isolated from crop residue. *Carbohydr. Polym.* **2021**, *255*, 117466. [CrossRef] [PubMed]
43. Zhou, W.; Yang, H. Flame retarding mechanism of polycarbonate containing methylphenyl-silicone. *Thermochim. Acta* **2007**, *452*, 43–48. [CrossRef]
44. Liu, B.; Wang, H.; Guo, X.; Yang, R.; Li, X. Effects of an Organic-Inorganic Hybrid Containing Allyl Benzoxazine and POSS on Thermal Properties and Flame Retardancy of Epoxy Resin. *Polymers* **2019**, *11*, 770. [CrossRef] [PubMed]

polymers

MDPI

Review

Biopackaging Potential Alternatives: Bioplastic Composites of Polyhydroxyalkanoates and Vegetal Fibers

Natalia Gómez-Gast [1], Ma Del Rocío López Cuellar [2], Berenice Vergara-Porras [1] and Horacio Vieyra [3,*]

1 Tecnologico de Monterrey, Escuela de Ingeniería y Ciencias, Carretera Lago de Guadalupe 3.5, Colonia Margarita Maza de Juárez, Atizapán de Zaragoza 52926, Mexico; a00354363@tec.mx (N.G.-G.); vergarabp@gmail.com (B.V.-P.)
2 Cuerpo Académico de Biotecnología Agroalimentaria (CABA), Institute of Food and Agricultural Sciences (ICAp), Autonomous University of Hidalgo State (UAEH), Av. Universidad Km. 1, Ex-Hda. De Aquetzalpa AP 32, Tulancingo de Bravo 43600, Mexico; marocio_lopez@uaeh.edu.mx
3 Tecnologico de Monterrey, Escuela de Ingeniería y Ciencias, Eduardo Monroy Cardenas 2000, San Antonio Buenavista, Toluca de Lerdo 50110, Mexico
* Correspondence: h.vieyra@tec.mx

Abstract: Initiatives to reduce plastic waste are currently under development worldwide. As a part of it, the European Union and private and public organizations in several countries are designing and implementing regulations for single-use plastics. For example, by 2030, plastic packaging and food containers must be reusable or recyclable. In another approach, researchers are developing biopolymers using biodegradable thermoplastics, such as polyhydroxyalkanoates (PHAs), to replace fossil derivatives. However, their production capacity, high production costs, and poor mechanical properties hinder the usability of these biopolymers. To overcome these limitations, biomaterials reinforced with natural fibers are acquiring more relevance as the world of bioplastics production is increasing. This review presents an overview of PHA–vegetal fiber composites, the effects of the fiber type, and the production method's impact on the mechanical, thermal, barrier properties, and biodegradability, all relevant for biopackaging. To acknowledge the behaviors and trends of the biomaterials reinforcement field, we searched for granted patents focusing on bio-packaging applications and gained insight into current industry developments and contributions.

Keywords: polyhydroxyalkanoates; fibers; mechanical properties; biodegradability; packaging; patents

Citation: Gómez-Gast, N.; López Cuellar, M.D.R.; Vergara-Porras, B.; Vieyra, H. Biopackaging Potential Alternatives: Bioplastic Composites of Polyhydroxyalkanoates and Vegetal Fibers. *Polymers* **2022**, *14*, 1114. https://doi.org/10.3390/polym14061114

Academic Editors: Andrzej Puszka and Beata Podkościelna

Received: 24 January 2022
Accepted: 7 March 2022
Published: 10 March 2022

1. Introduction

In 2020, the European Parliament approved a strategy for a circular economy in plastics. Some of the main challenges are increasing plastic reuse and recycling rates, achieving a competitive and efficient economy in the use of resources, and undertaking the effort to reduce marine litter [1,2]. The Commission proposal urged to ensure that by 2030 all plastic packaging placed in the European Union (EU) market will be reusable and recyclable in a cost-effective manner, improving the design and collection process and reducing single-use plastic, restricting the use of oxo-degradable products, and defining rules for labeling compostable and biodegradable plastics. It also suggested a list of single-use plastic items to be banned or restricted, such as food and drink containers, drink cups, cutlery, plates, stirrers, bottles, beakers, lids, lightweight plastic carrier bags, and oxo-degradable plastics, among others [3,4]. Oxo-degradable or oxo-biodegradable polymers are mainly petroleum-based products combined with additives to promote fragmentation. Their macromolecular chains split into small chains when these polymers come into contact with heat, oxygen, or light through either biotic or abiotic mechanisms. Examples include low-density polyethylene and polypropylene films combined with metal oxides (Fe_2O_3, Cu_xO, and ZnO) [5–7].

Several international organizations such as the Nations Convention on the Law of Seas (UNCLOS) and the International Maritime Organization (IMO) have contributed strategies for preventing, reducing, and controlling pollution from land-based sources, waste from vessels, pollution from the exploitation process, and marine plastic litter from ships [8,9]. At the same time, some countries are leading initiatives to reduce plastic waste. More than 100 governments affirmed their disposition to launch negotiations for a new global plastics agreement in recent years, and almost 127 countries approved legislation to regulate plastic bags and stimulate circular plastic economies. Germany has been named "the champion recycling country" for achieving a 67% recycling rate of household solid waste in 2017, followed by Austria and South Korea with 53%. In addition, China, Malaysia, Vietnam, Thailand, Indonesia, South Korea, Taiwan, India, and Turkey banned plastic waste importation [10–13].

Another factor that increases the urgency of finding solutions to reduce plastic waste is the Chinese prohibition of plastic waste importation. China used to import a large percentage of the global plastic waste for a manually recyclable process. In 2016, China imported approximately 8 million tons of waste from developed countries, and now those governments must recycle their waste or export it to other Asian countries [14,15]. In this context, bioplastics such as polyhydroxyalkanoates (PHAs) have emerged as part of the solution to plastic waste.

Biopackaging, in eco-conscious packaging, is any biodegradable packaging conceived for sustainability. It involves natural and synthetic biodegradable polymers, called biopolymers, that can include by-products of the agro-industry, such as fibers and inorganic or bioactive compounds, to be more respectful to the environment. These materials wrap or contain products temporarily for handling, transport, and storage [16–18]. Because biodegradability and biocompatibility are remarkable properties of polyhydroxyalkanoates (PHAs), these polymers are more suited for biopackaging than are synthetic plastics [19].

PHAs are linear and thermoplastic polymers. They can be produced by plants and bacteria such as *Delftia acidovorans*, *Pseudomonas mosselii*, *P. oleovorans*, *P. putida*, *Halomonas* sp., and *Escherichia coli* LS5218 when subjected to stress by lack of nutrients such as nitrogen, phosphorus, and others [20–22]. These bacteria store carbon in PHAs granules as an energy reserve [23,24]. PHAs are composed of hydroxyalkanoate monomers susceptible to other bacteria and fungi degradation. For example, microorganisms such as *Paucimonas lemoignei* release depolymerases to degrade the PHA into water-soluble monomers and oligomers [23,25,26]. For example, a PHA bottle can take less than three and a half years to degrade in a marine environment, a short time compared to that required for fossil-derived plastics [25], and a high degradation rate of PHA is achievable in soil (109 days) depending on the soil composition, bacterial population, and crystallinity degree of PHA [27,28].

There are bioplastics made with PHAs already on the market, but despite low-cost applications granted by recently developed extruders and molding machines, the production capacity of PHAs is lower than the current consumption of plastic [29,30]. According to Our World in Data, a project of the Global Change Data Lab, a non-profit organization based in the United Kingdom (Charity Number 1186433), in 2015, approximately 11% of PHA production was used for packaging, although other reports believe it to be 40% [31,32]. While the global capacity for bioplastics production in 2019 was 2.11 million tons, barely 0.57% of the total production of plastics, only 25 thousand tons were PHA (around 1% of total production). The European Bioplastics and Nova-Institute forecasted that by 2024, bioplastic production will be approximately 2.43 million tons, and PHA production will triple the current production [33,34], which is a significant growth but still lower than the current demand for plastics.

The high production cost of PHAs is another drawback to replacing conventional plastics. This cost comes from the complex production process that includes several steps such as selection of the raw material, bioreaction, separation and drying of the biomass, PHA extraction, and processing [35,36]. The raw material accounts for more than 50% of the production cost [37], whereas the price of PHA can be approximately 300 times the price of a polymer such as polypropylene. Thus, research has focused on selecting cheap raw materials and developing new strategies for PHA production. One of the alternatives to overcome this economic and technical barrier is the elaboration of composites with natural fibers. The vegetal fibers used as fillers are essential in reducing cost, increasing biodegradability, and tailoring mechanical and thermal properties. However, using natural fibers with PHA has limitations, such as low interfacial adhesion to the biopolymer matrix, poor matrix dispersion, and hydrophilic characteristics [38–41].

Some alternatives implemented to surpass the technical disadvantages include making composites with two or more fibers that complement each other, different fiber pretreatments, and compatibilizers [42].

This review aimed to analyze the composites made with PHAs and natural fibers as a realistic alternative for biopackaging.

2. PHA–Vegetal Fiber Composites

Biomaterials reinforced with natural fibers are acquiring more relevance as the world of bioplastics production is increasing. The keywords biodegradable, PHAs, fibers, natural, and vegetables were searched in the ISI Web of Knowledge and Scopus, resulting in 87 articles. Below, we discuss the main characteristics required for biocomposites meant for bio-packaging and the impact of the fiber and the preparation method on the composite properties.

2.1. Polyhydroxyalkanoate and Fiber Composites

Natural fibers are mainly composed of cellulose, hemicellulose, and lignin, which have different physical and chemical properties [43,44]. Different natural fibers have been used to reinforce PHAs, but their inclusion also presents some issues. For example, cellulose provides the strength of the fiber but has poor compatibility with hydrophobic polymers such as PHA [45,46]. Hemicellulose is amorphous and hydrophilic due to hydroxyl and acetyl groups; therefore, its mechanical properties are poor, and it retains moisture [47,48]. The lining is aromatic and amorphous but less hydrophilic than other components [46,47]. These characteristics cause low interfacial adhesion between fiber and matrix and the generation of polar groups, which generate poor dispersion in the matrix [49,50]. Thus, different pretreatments have been used to reduce the polarity and water absorption of fibers to improve the affinity between the fillers and the matrix and enhance the efficient stress transfer from matrix to fibers. Some examples of the primary pretreatments found in the literature, and the main changes observed in the fibers after the treatment are shown in Table 1.

With pretreatments of chopping or grinding, fibers are cut by mechanical methods and sieving, or micronized, to obtain smaller particle sizes to improve fiber–matrix adhesion and promote crystallization [51]. Grinding methods include cutting milling, impact milling, or ball milling [52]. Enzymatic pretreatment immerses the fibers in a pectinase, laccase, or cellulase solution to modify the fillers' surfaces and remove impurities [53]. In grafting, powder cellulose fibers undergo an esterification process and are subsequently dried. These fibers are further treated to increase their hydrophobicity [54]. In mechanical–hydrothermal pretreatment, fibers are immersed in warm water for surface modification and bacterial degradation [55,56]. Mechanical–chemical pretreatment consists of grinding the vegetal fillers followed by an alkali or solvent treatment to remove impurities and to improve the fiber–matrix adhesion [57]. Lastly, an argon plasma jet induces new functional groups on the cellulose surface in the plasma pretreatment, allowing the fibers to ultrasonicate and lyophilize [58]. Additional reports on pretreatment of the fiber with waxes suggest performance improvements without hindering the biodegradability of the composite. For instance, in a blend of PHBV, wheat, ATBC, and calcium carbonate, the pretreatment of the fiber with bio-based waxes improves the mechanical performance of the blend in terms of impact resistance, and in a composite of PHBV–potato–ATBC–calcium carbonate, the wax pretreatment of the fiber enhances the fiber–matrix adhesion and the mechanical properties of the composite [59,60].

Table 1. Examples of fiber pretreatments used for biodegradable composite production.

Pretreatment	Natural Fiber	Additive	Treatment Effect	Reference
Chopped grinding	Wood F		The fibers were too short to hinder the brittle fracture.	[51]
	Basalt F		The fiber–matrix adhesion improved.	[51]
	Rice husk		Irregular fibers in morphology or size	[42]
	Posidonia oceanica		Small size; Poor fiber–matrix interaction	[52]
	Cotton		Uniform fiber dispersion due to a grafting process of the polymer matrix.	[53]
Enzymatic	Bamboo	Pectinase, cellulase, laccase	Rough surface. Decreased fiber polarity. Better compatibility. Slight increase in tensile and impact strength of composites. Delay in the thermal decomposition of composites. Increased melting peak and crystallization rate. Lower water absorption.	[44]
Grafted	Cellulose	Palmitoyl chloride	Increased hydrophobicity. Better fiber distribution in the matrix. Improved elongation of fibers.	[45]
Mechanical/Hydrothermal	Ceiba	Immerse in water for bacterial degradation	*	[47]
	Piassava	*	Hemicellulose and lignin partially removed. Increased smooth surface area. Decreased average diameter. Better composites thermal stability.	[54]
Mechanical/Hydrothermal	Microcrystalline cellulose	*	Size and shape changed to nanocrystals and nanofibrils. CNC contributed to high and perfect PLLA crystal formation. Composites with CNC have better elongation than composites with MCC.	[55]
Mechanical–chemical	Luffa	Alkaline treatment (NaOH)	Hemicelluloses and lignin partially removed. Reduced diameter. Improved adhesion.	[56]
	Olive husk	Acetone–ethanol, NaOH	Hemicellulose and lignin eliminated. Smooth surface. Improved interfacial adhesion.	[57]
	Wheat straw	Alkaline treatment (NaOH)	Non-cellulosic components removed. Straight and smooth surface. Better fiber–matrix interface. Improved tensile strength.	[58]
	Vine shoots	Acetone	Reduction of lignin and resveratrol content. Improved biodegradability of the fiber.	[59]
	Seagrass	Acetic acid to wash, and alkaline treatment (NaOH)	Wax, hemicellulose, lignin, and calcium carbonate impurities removed. More reactive −OH groups in the fiber surface promoted better fiber–matrix interaction.	[60]
	Rice husk	Alkaline treatment (NaOH)	Wax, hemicellulose, and lignin removed. Increased crystallinity.	[60]

Table 1. *Cont.*

Pretreatment	Natural Fiber	Additive	Treatment Effect	Reference
	Almond shell	Alkaline treatment (NaOH)	Wax, hemicellulose, and lignin removed. Increased crystallinity.	[60]
	Radiata pine	Sigmacote, hexane, heptane	*	[61]
	Coconut	Oregan essential oil	Fiber length decreased with increasing screw speed. Antibacterial activity (bacteriostatic effect against *S. aureus*).	[48]
	Microcrystalline cellulose		The size was reduced to nano dimensions, and the shape changed to spherical and fibril.	[55]
	Kenaf	Alkaline treatment (NaOH) and silane (triethoxysilyl propylamine)	Improved interfacial fiber–matrix bond but not improved mechanical properties.	[33]
	Palm brunches (Efb)	Alkaline treatment (NaOH) and silane (triethoxysilyl propylamine)	Improved interfacial bond between fibers and matrix but not improved mechanical properties.	[33]
Plasma	Microcrystalline cellulose		Surface modification for a better matrix–fiber interface. Decreased thermal stability.	[49]

* Information not provided. Basalt F: basalt fiber; CNC: nanocrystalline cellulose; Efb: empty-fruit palm brunches; MCC: microcrystalline cellulose; NaOH: sodium hydroxide; PLLA: polylactic acid; Wood F: wood fiber.

2.2. Mechanical Properties of PHA–Vegetal Fiber Composites

In general, the composite polymers use less than 30 wt% of reinforcements, probably due to the difficulty in achieving a homogeneous dispersion of the vegetable fibers, and melt flow index also decreases, which hinders composites' processability. Young's modulus, tensile strength, and elongation at break are the most commonly reported mechanical properties, typically measured using the ASTM D638 Standard Test Method for Tensile Properties of Plastics. In addition, ASTM D790 for Flexural Test of Plastics, ASTM D256 2018 Standard Test Methods for Determining the Impact Resistance of Plastic Izod Pendulum, and ISO 527 Determination of Tensile Properties in Plastic Films are used. After analyzing the literature and organizing the information according to the matrix materials and fibers used as reinforcement in composites, we plotted the reported data, regardless of the preparation method, to acknowledge behaviors and trends of the biomaterial reinforcement using Tableau (Salesforce company) for visual analysis (Figures 1–3).

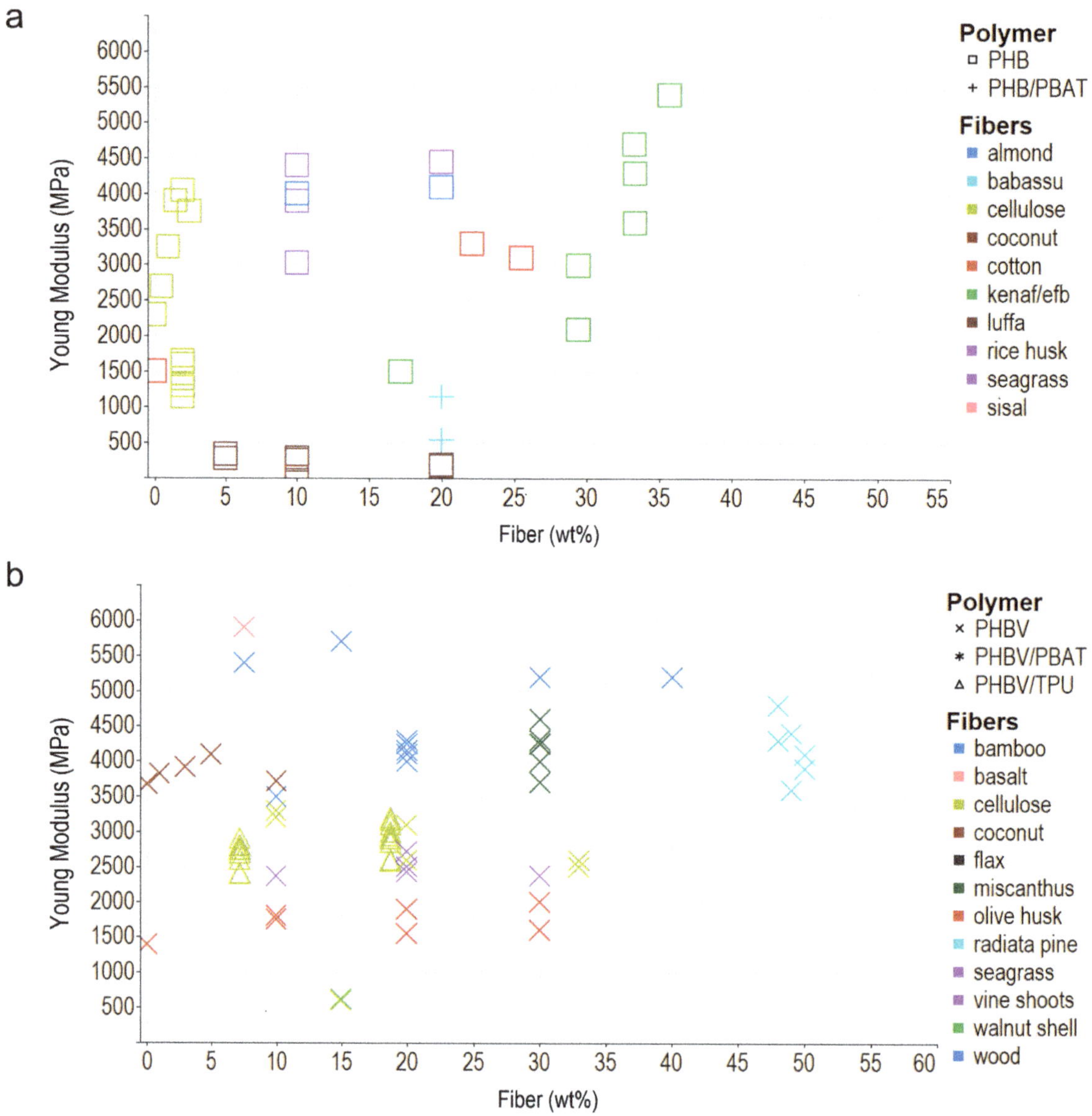

Figure 1. Young modulus of different PHA–fiber composites. (**a**) PHB and PHB/PBAT blends. (**b**) PHBV and its blends with PBAT and TPU. The colors refer to the fiber type (fillers), and the shapes refer to polymer types (matrix).

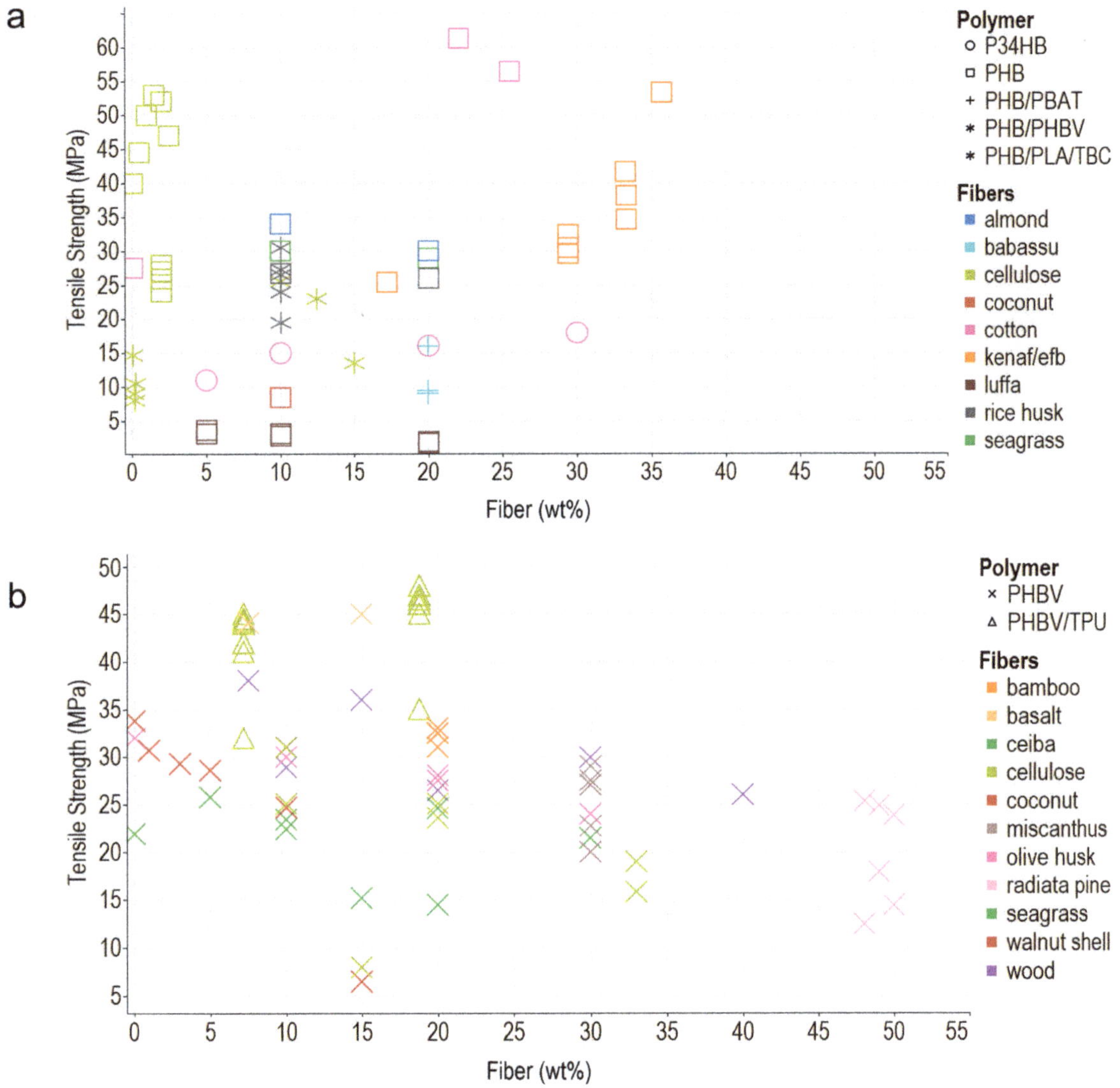

Figure 2. Tensile strength. (**a**) PHB and PHB/PBAT blends. (**b**) PHBV and its blends with PBAT and TPU. The colors refer to the fiber type (fillers), and the shapes refer to polymer types (matrix).

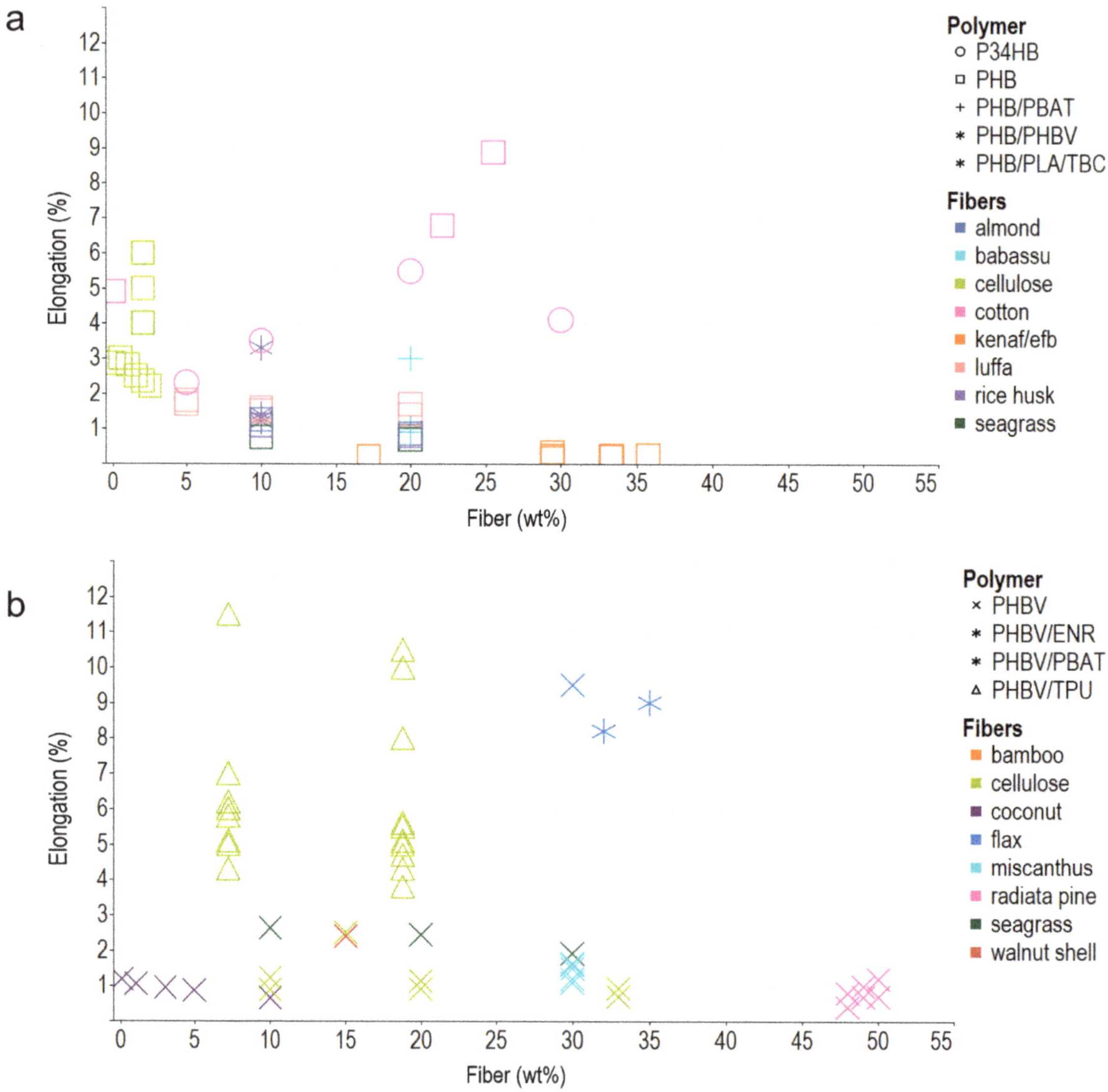

Figure 3. Elongation at break of PHA–fiber composites. (**a**) PHB and its blends with PBAT, PHBV, PLA, and TBC. (**b**) PHBV and its blends with ENR, PBAT, and TPU. The colors refer to the fiber type (fillers), and the shapes refer to polymer types (matrix).

2.2.1. Young's Modulus

The addition of fillers to the polymeric matrix enhances its resistance to stretching or deformation, a property related to the Young's modulus, independently of the poor filler matrix adhesion (Figure 1). Fibers modify the Young's modulus, measured in the elastic zone of the material's stress–strain curve, and the inclusions act as nucleating agents to promote crystallization [55,61]. The Young's modulus values of some traditional plastics range between 67 and 3100 MPa, and the Young's modulus of the composites studied in this review exhibited a wide range between 10 and 24,000 MPa [36,62,63]. The fiber type, fiber content, and pretreatment determine the mechanical properties. PHB (polyhydroxybutyrate) and PHBV (polyhydroxybutyrate-co-valerate) fiber composites, PHB/fiber (80/20 wt%) with almond fiber, and rice husk range between 2200 and 2500 MPa. Blends of PHBV bamboo, miscanthus, flax, or wood (70/30 wt%) had Young's modulus values between 3000 and 5000 MPa. This same range of Young's modulus values is observed using 50 wt% of radiata fiber or 10 wt% of coconut or cellulose. Nevertheless, incorporating plasticizers into the PHAs matrix reduces Young's modulus. In a composite of PHBV/TPU(thermoplastic polyurethane)–cellulose, where three different additives were used, the compatibilization between fiber and matrix enhanced and incremented the viscosity of the films, but Young's modulus decreased compared with the PHBV controls [64,65].

Fiber length impacts the crystal structure. For example, in PLA–PHB–cellulose composites, nanofibers promote crystal size reduction, perfect crystal formation, and homogeneous dispersion, resulting in a stiffer composite [66]. In another example, the addition of short coconut fibers (542–1100 μm) improves crystallinity; however, when fiber content increases, some aggregates are formed in the matrix due to poor dispersion of fibers, which can generate microfractures in the composites [67,68]. In a blend of PHB/PBAT(polybutylene adipate-*co*-terephthalate)–babassu, PHB and PBAT mutually inhibited crystallization, and the resulting composite was amorphous. At room temperature, the increase in PBAT content reduced Young's modulus compared with PHB film, but when samples were tested at -40 °C, the polymer chain movement was restricted, and the Young's modulus of the blend increased [69].

The method of obtaining the composites also improves the mechanical properties, providing alternatives to tailoring the elastic modulus of the matrix. Higher Young's modulus values are obtained when preparing the composite by stacking, casting, extrusion, and injection. For example, the Young's modulus of a composite of PHBV–wood (85/15 wt%) elaborated by extrusion–injection increased from 4500 MPa (neat PHBV) to 5667 MPa. In a composite with kenaf/Efb, the number of layers and the stiffness of the external layer contribute to increasing the Young's modulus. Moreover, a PHBV–flax (70/30 wt%) composite made by stacking reached 16.69 GPa, representing a 320% increase compared to the PHBV matrix due to fibers' properties (flax Young's modulus is 60–80 GPa). PHA–flax (70/30 wt%), also made by stacking, reached a Young's modulus of 10.27 GPa, probably due to a reduction of 2% in the porosity. However, when other polymers such as polybutylene adipate *co*-terephthalate (PBAT) or epoxidized natural rubber (ENR) were added to the matrix, the Young's modulus of the composites was reduced by 15% [48,70]. Although the method of obtaining the composites modifies their mechanical properties, the selection of the method depends on the intended usage of the produced material. For example, casting is ideal for laboratory testing, extrusion is ideal for profile generation, and injection for end products such as containers. The most common methods of composite preparation are listed in Table 2.

Table 2. Common methods to produce biodegradable composites.

Process	Matrix	Fiber	Additive	Applications	Reference
Casting	PHB	Luffa	*	Packaging	[71]
	PHBV	Ceiba	*	Fresh fruit packaging	[56]
	PLLA, PHB	MCC	Tributyl citrate	Food packaging	[72]
	P34HB	Cotton	Benzoyl peroxide and maleic anhydride.	Paper-based packing	[73]
Compression	PHB	Seagrass, almond shell, rice husk		Food packaging	[74]
	PHB	Microcrystalline cellulose	*	Biomedical and engineering uses	[58]

Table 2. *Cont.*

Process	Matrix	Fiber	Additive	Applications	Reference
	PHBV	Radiata pine	Polymethylene diphenyl diisocyanate	*	[75]
	PHB	Coconut, sisal	Glycerol	Small tubes and plastic bags for planting	[76]
Extrusion–compression	PHB	Wheat straw	*	Biomedical and food packaging, biodegradable polymer	[77]
	PHBV	Bamboo, luffa	*	*	[78]
	PHBV	Rice husk	TGIC, DCP	Food packaging	[51]
	PHBV	Coconut		Food packaging	[57]
Extrusion–injection	PHBV	Bamboo	*	*	[53]
	PHBV	Olive husk	*	Environmentally-friendly material	[79]
	PHBV	Radiata pine	*	Improved mechanical properties of PHBV composites	[80]
	PHBV	Wood/basalt	*	Long-life material products	[81]
	PHBV	Cellulose	*	Biocomposites with tailored properties	[54]
	PHB	Piassaba	*	*	[66]
	PHBV	Posidonia oceanica	*	Bio-container for plants	[82]
	PHBV, TPU	Cellulose	Hexamethylene diisocyanate, joncryl, TGIC.	*	[64]
	PHB, PBAT	Babassu	*	Several applications	[69]
	PHBV	Posidonia oceanica	ATBC	Seawater applications.	[83]
Extrusion–injection	PHBV	Miscanthus	DCP	*	[65]
Injection	PHBV	Nanocellulose, walnut, eggshell, tuff	*	Packaging for airline cosmetics food	[84]
Micro-compounding	PHBV	Vine shoots		Biodegradable materials	[85]
Stacking–compression	PHA, PLA, PBS, PP	Flax	*	Adjustable mechanical properties for industrial products	[86]
	PHBV, PBAT, ENR	Flax	Epoxy sizing	*	[70]
	PHB, PLA	Cotton	*	Building, furniture, or automotive products	[87]
	PHB	Efb, kenaf bast fiber	Triethyl citrate	Replacement of wood products	[42]

Processing conditions such as temperature, time, speed, or pressure were different for each report. * Information not provided. ATBC: acetyl tributyl citrate; DCP: dicumyl peroxide; Efb: empty-fruit palm brunches; ENR: epoxidized natural rubber; GMA: glycidyl methacrylate; MCC: microcrystalline cellulose; P34HB: poly-3-hydroxybutyrate-*co*-4-hydroxybutyrate; PBAT: polybutylene adipate-*co*-terephthalate; PBS: polybutylene succinate; PHA: polyhydroxyalkanoate; PHB or P3HB: poly(3-hydroxybutyrate); PHBV: polyhydroxyburytrate co-valerate); PLA: polylactic acid; PLLA: poly-L-lactide; PP: polypropylene; TGIC: triglycidyl isocyanurate; TPU: thermoplastic polyurethane.

2.2.2. Tensile Strength

Tensile strength (UTS) is the maximum point in the stress–strain curve that materials reach without fracture when a load is applied. In composites, the average UTS values range between 20 and 30 MPa, very close or below the UTS of the matrix, mainly due to poor adhesion (Figure 2). The PHBV/ENR–flax (65/35 wt%) composite made by staking achieved the highest UTS value, 188 MPa; PHA–flax reported an 82 MPa, despite the differences in the stiffness of the matrix [70,77]. Fibers properties modulate mechanical behavior. Composite's UTS depends on many factors such as fiber properties, size, weight, percent of content, processing temperature, and fiber–matrix adhesion [49,88,89].

As depicted in Figure 2, the amount of fiber affects the UTS. Some studies report a better performance when adding 1–10 wt% fiber, but the UTS tends to decrease when the content is higher, probably due to the formation of agglomerates [56]. In general, poor compatibility is more evident when filler weight is 30% because fibers act as stress concentrators, and the composites turn into a brittle material, and UTS decreases [80]. Hence, good interfacial adhesion and uniform fiber dispersion are required to improve UTS. Composites can withstand more significant stress before breaking when fibers efficiently transfer stress, which requires pretreatment before the compounding process [79]. Pretreatments aim to enhance fiber–matrix adhesion. For example, the grafting process of PHA and maleic anhydride (MA) used to reduce the hydrophilicity of composites had a positive effect on mechanical properties [73]. In addition, in a PHBV–bamboo composite, the enzymatic treatment leads to better fiber bonding and a UTS increase of 4% [53]. Likewise, plasma treatment in PHB–cellulose (98–2 wt%) increases the UTS by 15% [58].

The fiber size also affects the UTS. For example, in a PHBV–coconut fiber composite made by extrusion–compression molding, high long fiber content hinders chain-folding and the formation of the crystals, but high short fiber content leads to a nucleating effect, which results in a lower UTS for the blend [57]. Thus, fiber length is critical for the efficiency of the reinforcement. In a composite of PHB–sisal fibers, long fibers increase the blend's moisture content, affecting the interfacial interaction, which causes a reduction in the UTS from 22.3 to 11.9 and 18.2 MPa [76]. When fiber size appears not to affect the UTS, it is probably due to variability, which masks the fiber type and size effect [90].

2.2.3. Elongation at Break

Elongation at break is a differential between the initial longitude and the final longitude of a specimen after the breakage test, generally expressed as a percentage. The elongation at break is poor in fragile materials due to large spherulites generating processing gaps, and the addition of vegetal fillers to a matrix impacts the net's elongation. Figure 3 shows the relationship between fiber percentage and elongation at break. The elongation at break is lower than 7% in most composites. Contrary to Young's modulus, fiber addition leads to a reduction in elongation at break. The PHA–chestnut (90/10 wt%) composite made by compression molding has an elongation at break of 580%, the highest value reported (not depicted in Figure 3). In this composite, grafting with glycidyl methacrylate (GMA) contributed to more uniformly dispersed fibers, better wetting, and improved interfacial adhesion due to the similar hydrophilicity of the phases [91].

Typically, fibers compromise the elongation at break. Long fibers promote the formation of large crystals and, therefore, a more fragile material, whereas the crystals formed are smaller when short fibers are used [72]. In addition, fiber orientation may cause stiffness and reduce the flexibility of biopolymer nets [77]. However, incorporating plasticizers into the blends promotes fiber–matrix bonding. For instance, in the PHBV/TPU–cellulose blend produced by extrusion–injection, incorporating 1 phr of hexamethylene diisocyanate (HMDI) enhances the interfacial adhesion and improves the composite elongation by 150% [64]. The addition of ATBC increased the elongation at break to 6.4%, triple that of PHBV film [52]. Nonetheless, in some cases, the fiber content's effect exceeds the plasticizer's effect, as in the composite PHBV–posidonia (70/30 wt%), where the ATBC could not compensate the fiber's stiff effect [83]. In a composite made by lamination, the joncryl additive partially fills the gap between fiber and matrix but does not affect crystallization or elongation at break [87].

2.3. *Thermal Properties of PHA–Fiber Composites*

Two main techniques determine transition temperatures to characterize polymers. In differential scanning calorimetry (DSC), the specimens are subjected to two heating cycles. The first heating process has the objective to erase the thermal history of the polymer matrix and remove moisture because water acts as a plasticizer and modifies the properties of the polymers, and the second cycle identifies melting and crystallinity temperature, and, in some cases, the generation of crystals of different sizes [92,93]. Thermogravimetric analysis (TGA) measures the mass variation when the temperature changes [94,95].

Figure 4 shows the relationship between fiber percentage and melting and degradation temperatures. Fiber addition does not significantly modify the material's melting temperature, as seen in the compounds PHB–pissaba, PHB–rice husk, and PHBV–cellulose. This behavior is desirable since inexpensive vegetable fillers can be used in composites to reduce the cost without significantly affecting processability. Thermal stability is critical for packaging applications because some containers are exposed to high or low temperatures during shipping and storage. A biocomposite must endure heating or cooling processes [96–98].

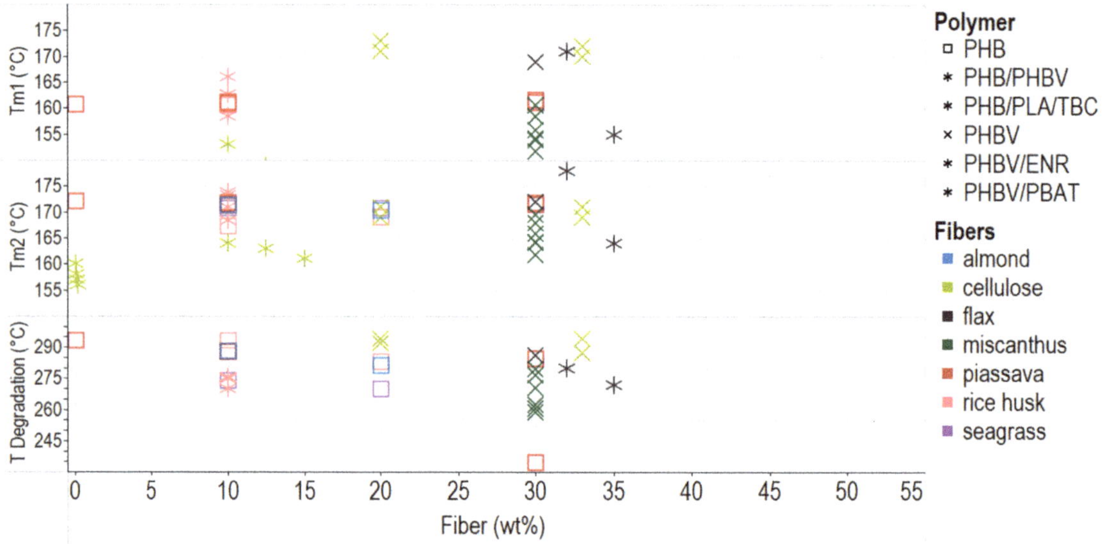

Figure 4. Thermal behavior of PHA–fiber composites. The colors refer to the fiber type (fillers), and the shapes refer to polymer types (matrix). Tm = temperature of melting.

The TGA shows that the degradation of composites PHA-vegetal fibers occurs in two main steps: first, the initial fiber and PHA degradation by hydrolysis, and second, the lining and residue degradation at 350 °C or more [99,100]. The addition of fibers also implies the addition of impurities, and the initial temperature of degradation (Tdeg) decreases; also, some interaction between fibers and PHBV matrix results in lower Tdeg of composites [101]. Fiber addition also causes differential melting temperatures in the first heating (Tm1) and the second cycle (Tm2) due to perfect crystal formation because more giant and more ordered crystals need more energy to melt again [51,84]. This behavior has been reported for PHB–flax, PHBV/ENR–flax, or PHBV–miscanthus composites (Figure 4). The degradation temperature of the composites decreases when using more significant amounts of the filler and additional treatments to reinforce, which implies less thermal stability.

Some treatments improve the thermal stability of composites due to the removal of pectin, cellulose, and other substances of the filler [78]; when fiber improves the interaction with the matrix, the thermal degradation is retarded [54]. Likewise, reactive agents impact the thermal behavior of composites. For example, DCP (>0.1 phr), the additive in a blend of PHBV–miscanthus (70–30 wt%). reduces the temperature of melting (T_m) of the blend by reducing crystallinity [65]. In some cases, additives mask the nucleating effect of the vegetable fiber in the polymeric matrix. In addition, plasticizers, typically used for internal lubrication, increase mobility and decrease the temperature of glass transition (T_g) [72].

The loss of the plasticizer usually appears in the first part of a TGA curve. This behavior is typical in PHB composites with glycerol and triethyl citrate (TEC), among others [76].

2.4. Barrier Properties of PHA–Fiber Composites

The barrier properties are essential for packaging materials, especially in food and shelf applications [102,103]. If a composite absorbs water or oils or has a high gas and vapor permeability, it is unsuitable for preserving the organoleptic properties of the package content [104]. The addition of vegetal fibers to a polymer matrix increases the porosity and the number of polar groups that result in absorbing water. Moreover, poor interfacial adhesion creates zones that efficiently uptake water [85,105]. The moisture reduces the mechanical properties of the blend and increases biodegradation because the water migrates to the amorphous zones and leads to polymer chain scission. Pretreatment of the vegetable fibers would reduce moisture absorption. Pretreatments such as esterification, use of NaOH, or enzymatic reduction achieve better dispersion and adhesion of the filler into the matrix, reducing the hygroscopicity [78,106].

Noteworthily, the water vapor transmission (WVP) of PHA films is similar to polyethyleneterephtalate (PET) films [104], but it increases with the addition of vegetal fibers because of the crystallinity changes generated by the fillers [107]. The WVP of a composite, as a measure of water vapor uptake, depends on fiber amount, crystallinity decrement, and changes in the molecular weight of the matrix [54]. The type of fiber and its hygroscopicity also affects the water vapor permeability [74]. For example, adding low amounts of fiber (2%) in a PHB–cellulose composite improves the crystallinity, reducing the diffusion process.

2.5. Biodegradability

A biodegradable polymer undergoes biodegradation, a chemical process during which microorganisms that are available in the environment decompose materials into natural substances such as water, carbon dioxide, and methane [108]. As per the ASTM D6400 definition, compostable plastics must demonstrate proper disintegration during the composting, an adequate level of inherent biodegradation, and no adverse impacts to support plant growth [109]. Most materials are biodegradable, but their biodegradation process might take hundreds of years [110]. Therefore, one of the objectives of biodegradable plastic developers is to achieve this process within an appropriate life span according to the use of the material and its subsequent disposal.

There are different methods to measure a polymer's biodegradability. The test selection depends on the organization (ASTM, EPA, ODEC, ISO), external conditions (aerobic, anaerobic, UV exposure), the environment (soil, marine water, compost), or their purpose (biodeterioration, assimilation, biofragmentation) [111,112]. Among the most common tests used to monitor biodegradation are weight loss, abiotic degradation, CO_2 measurement, macromolecular weight loss, oxygen consumption rate, and anaerobic digestion (biogas production–weight loss) in compliance with ISO 15814:1999, ISO 17556 (2019), ASTM G160–12, and ASTM D6691 standards [113–117]. CO_2 measurements with a respirometric test also help identify the material's shelf life according to ASTM D5988-96. These standards require that the material biodegrades in a certain period and leaves no toxic residue in the soil. Exposure to the environment (i.e., temperature, moisture, microbial population, pH, oxygen content) affects the biodegradation of a polymer; thus, a material that degrades by microbial activity under industrial composting conditions may not degrade in other conditions [118].

Reinforcement with vegetal fibers is expected to improve the biodegradability of the already biodegradable PHAs. Instead, biodegradation depends on soil composition, fiber amount and pretreatment, material stiffness, and processing [85]. For instance, the PHBV–shoot vine composite biodegradation rate is 83%. The content of lignin and polyphenols makes biodegradation difficult, but pretreatment of the fibers raises the composite's biodegradation to 97%. In a different example, PHB degradation ranged from 60 to 98%, depending on the method. PHB showed 64.3% degradation in 6 months using microbial fermentation in soil tests [119]. In a quasi-steady state, co-digestion of synthetic municipal primary sludge (SMWS) and PHB, after 45 days, exhibited approximately 80–98% conversion of PHB to biomethane [120]. In soil, P(3HB) specimens had been 60% degraded in 21 days, but they continued to degrade to 93% by day 35 [121]. For comparison, blending PHB with wood fiber yields conflicting results. The biodegradation of [P(3HB-*co*-3HHx)]/Kenaf during 48 days in mineral medium-soil reached 13%, whereas, in an aqueous-nutrient medium, it barely reached 2.4% [113]. Biodegradation of a composite of P(3HB-*co*-4HB)/wood under laboratory composting for 21 days could not be detected, but in an aqueous medium, it was 0.35% in two months. However, biodegradation was 35% per year in soil [114,116].

Additives also impact biodegradability. TPU (18–21 wt%) added to a PHBV–cellulose composite reduced the disintegration of the blend because the TPU covered the filler and interfered with the microorganism's access. When using HMDI, the degradation rate increased because the plasticizer blocked the effect of the TPU on the fibers [74]. Likewise, bio-based plastics with additives tested in soil media for 660 days did not show significant biodegradation. Instead, the PHA film reported 70% mineralization, very similar to the cellulose control film measured under the same conditions [122]. In a PHBV–posidonia composite, using ATBC as a plasticizer increases the polymer chain mobility and accelerates the disintegration of the blend [83]. This research measured the specimen's degradation in marine mesocosms and found degradation in warm seawater conditions. The specimens used in the test maintained their tensile properties by ten months, suggesting possible applications in marine ítems.

Disposal in landfills raises additional concerns for biodegradable plastics. A food packaging study observed that packaging made with biodegradable materials releases methane, more harmful than CO_2 [123]. Thus, alternate bioremediation strategies are needed, such as using methanotrophic microorganisms to reduce methane emissions in landfills or using the methane for energy generation [124–126] to take full advantage of biopackaging waste biodegradation.

2.6. Theoretical Modeling to Evaluate Performance and Applications of Polymer–Vegetal Fiber Composites

Although blending with fibers improves the polymers' mechanical properties, the extent of the fiber contribution is unknown. Theoretical modeling has helped infer the performance and modifications expected for these composites. PHA/hemp and PHA/jute (30 wt%) modeling showed damage on the matrix due to the different physical properties of the fillers. In this study, hemp was the best filler and achieved a better interface that supports higher mechanical loads [127]. A PHBV/oak wood flour composite was modeled using a modified Fickian diffusion law and the Halping-Tsai and Nicolas and Nicodemo model to predict composite properties, such as moisture absorption, stiffness, and strength. Despite the improved mechanical properties of the composite, the blend is susceptible to deterioration by the hygrothermal behavior of the fiber [128].

Further modeling and experimental testing are essential to make better predictions. A 3D model of PHBV–wheat straw designed to predict water vapor permeability using the finite element method (FEM) to the 3D structures permitted a better prediction of water vapor permeability dynamics of the composite [129]. Numerical homogenization and representative volume elements (RVEs) are used to model composites' effective elastic, thermal, and thermoelastic properties. This methodology allows for the preservation of fiber–matrix interactions and the predicted effective properties of blends, which can be further validated with experimental data [130].

3. PHAs Composite Applications in Packaging: Contributions of the Industry

Researchers have arduously worked on designing and preparing biocomposites, and one of the leading applications proposed for PHA–vegetal fiber composites is bio-packaging. Nevertheless, not all of the composites have possible industrial applicability. To determine the extent of industrial applicability of PHA–vegetal fiber composites, we searched for patents in the Lens database and Web of Science. After eliminating medical, veterinary, processing methods, machines, or equipment, we found 141 patents. The patents we found belong to the subgroup C08L6731, according to the Cooperative Patent Classification (CPC) system, comprising compositions of polyesters obtained by reactions that form a carboxylic ester bond in the main chain of polyester-amides, the subgroup CO8L2666, comprising polymers characterized by an additional compound in the mixture, and the subgroup Y10T428, comprising stock materials or miscellaneous.

The patents' owners are developing materials to adjust to the new normativity demanding reduction of the environmental impact. The inventions are summarized in Table 3. The patents offer materials with a wide range of applications, and biodegradability is paramount in these composites. Most patents related to food packaging acknowledge that PHAs are safe for food and contribute to preserving their organoleptic properties. Some materials can preserve and protect against contamination with bacteria and fungi [131], and others keep the product's aroma, flavor, and texture [132]. Most of the products are designed to serve as films for molding or coating. Others are ideal for stacking, where different layers complement each other. PHAs are brittle and stiff, but a layer of PLA, PLC, or PBS and a plasticizer reinforce the processability of the blend.

The relevance of biocomposites, particularly PHA–vegetal fiber composites, in many applications has attracted the industry's attention, mainly because fillers also reduce composite costs. Independently of their business sector, important companies invest in developing blends with PHAs. Some chemical companies such as DuPont, BASF, Cargill-Dow Polymers, Union Carbide, Bayer, Monsanto, Mitsui, and Eastman Chemical developed and currently sell biodegradable products. Among them are blends of PLA, PET, or PBAT, additives to enhance PLA properties, or even products ideal for blending with fibers such as ECOFLEX/ECOVIO and EASTAR BIO aliphatic–aromatic polymers, BAK (a polyester amide_, and BIOMAX for modified PET and PLA [133,134].

Table 3. Patents of materials with PHAs and natural fibers for packaging or similar applications.

Year	Publication Title	Owner	Matrix/Base	Filler/Additives	Applications	Publication Number	Group	Ref
2021	Composite materials, methods of making, methods of use, and articles incorporating the composite materials	NIKE INC			Articles that undergo water contact	US 10919257 B2	B32B *	[135]
2020	Film packaging for oral biologics	CEVA SANTE ANIMALE S A	Biodegradable polymers and petroleum-based polymers		Films for packaging oral biologics such as vaccines.	EP 2775986 B1	A61J, A61K, B32B, B65D *	[136]
	Compositions containing new polyesters	NOVAMONT SPA	Polyester, PHAs, aliphatic, and/or aromatic polyesters.		Mass-produced articles	US 10738149 B2	C08K, C08G, C08J, C08K, C08K, C08L *	[137]
	Biodegradable sheets	TIPA CORP LTD	PBS, PBAT, PHA, PLA		Biodegradable sheets	US 10675845 B2	B32B, C08K, C08L, C09D *	[138]
2019	Biodegradable fabric and methods of making and using the same	SANCTUARY SYSTEM LLC	PLA		Packaging material, health care articles, and household products	WO2019070633	D06M	
2016	Biodegradable sheet	TIPA CORP LTD	PBS, PBAT, PHA, PLA		Biodegradable sheets	US 10239292 B2	C08L, B32B, C08J, C08K,	[139]
	Bio-based modifiers for polyvinylchloride blends	METABOLIX INC	Polyvinylchloride (PVC) and PHA		Packaging	US 9505927 B2	B32B, B29C, Y10T	[140]
	Biodegradable polymer films and sheets suitable for use as laminate coatings as well as wraps and other packaging materials	BIOTEC BIOLOG NATURVERPACK	Polyester amides and other polyesters, and natural polymers	Inorganic fillers and plasticizers	Packaging, coating, and wrapping for fast food	EP 2357209 B1	D21H, B29C, B29K, B42D, C08J, C08L, D21D, D21H, Y10T	[141]
	Multilayer article comprising a biodegradable polymer-based layer and a cellulose–fiber-based support; method of manufacturing multilayer article and food accessory comprising a multilayer article	AHLSTROM OY	PHAs, PLA, polybutylene succinate (PBS), biopolyesters	Non-woven fiber layer, kraft, and parchment, food-safe adhesive	Food molds resistant to moisture, food accessory agro-food industry.	EP 2841263 B1	B32B, Y10T	[142]

Table 3. *Cont.*

Year	Publication Title	Owner	Matrix/Base	Filler/Additives	Applications	Publication Number	Group	Ref
	Process for manufacturing a composite article comprising cellulose pulp fibers and a thermoplastic matrix	SÖDRA SKOGSÄGARNA EKONOMISK FÖRENING	Polyolefins, PHAs, PLA, polycarbonates, polyvinyl, and mixtures thereof	Cellulose pulp fibers and lubricant	Food containers and packaging	EP 2847382 B1	D21H, B32B, B226, Y10T	
2015	Film with compostable heat seal layer	FRITO-LAY NORTH AMERICA INC/MOUNT III ELDRIDGE M; PALTA DEEPALI	PHBV, PHA		Flexible packaging film with heat seal layer	US 9162421 B2	C09J, C08K, C08L, C09J	[132]

* Information not provided. CPC classes: A61L, methods or utensils to sterilize material or objects, B, performing operations; C08J, composites processed after treatment; C08K, use of inorganic substances as composite ingredients; C08L, organic macromolecular compounds and their preparation; C09J, use of materials as adhesives; D21D, treatment of materials for papermaking; Y10T, technical subjects covered by former US classification.

4. Conclusions

Research on biodegradable polymers and vegetal inclusions has grown significantly in the last ten tears. Furthermore, the increasing patents of materials using PHAs highlight these biocomposites' roles in replacing fossil-derived plastics. This review identified different alternatives to tailoring composite properties. It is possible to use vegetal fibers to enhance Young's modulus, but that generates poor elongation at break and less thermal stability. Plasticizers enhance the composite elongation, but the degradation rate tends to decrease. Therefore, it is imperative to define objectives and applications before selecting the methodology to produce the blends. Designing a material that meets usage requirements without sacrificing quality standards is essential to compete with conventional plastics.

Most of the publications reviewed above claim that their new composite is ideal for bio-packaging, but the usability of the material is rarely evaluated. Usability tests should be implemented in the industrial environment to evaluate the interactions of the polymers with different food types and conditions. In addition, it would be interesting to incorporate antimicrobial compounds that prolong the product's life. Researchers have a worthy challenge in designing biopolymers and compounds suitable for packaging applications and exploring other possible uses such as tissue regeneration, plant growth, and automotive applications. Prospects also include designing and manufacturing materials composed of PHA and local agricultural by-products, characterizing them, and evaluating the possibility of using them in food packaging, potentializing at the same time the local circular economy.

Author Contributions: Original draft preparation, investigation, visualization, N.G.-G.; reviewing and editing, B.V.-P. and M.D.R.L.C.; resources, supervision, writing—reviewing and editing, H.V. All authors have read and agreed to the published version of the manuscript.

Funding: This research did not receive any specific grant from funding agencies in the public, commercial, or not-for-profit sectors.

Institutional Review Board Statement: Not applicable.

Informed Consent Statement: Not applicable.

Data Availability Statement: The data used to support the findings of this study are included within the article.

Acknowledgments: Natalia Gómez is grateful to the Mexican National Council of Science and Technology (CONACYT) and the Instituto Tecnológico y de Estudios Superiores de Monterrey (ITESM) for the scholarships received.

Conflicts of Interest: The authors declared that they have no conflicts of interest to this study. We declare that we do not have any commercial or associative interest that represents a conflict of interest in connection with the study submitted.

References

1. Cerrar el Círculo: Un Plan de Acción de la UE para la Economía Circular. *Comision Europea*. Available online: https://eur-lex.europa.eu/resource.html?uri=cellar:8a8ef5e8-99a0-11e5-b3b7-01aa75ed71a1.0011.02/DOC_1&format=PDF/ (accessed on 24 February 2022).
2. Erbach, G. Legislative Train The European Green Deal. *European Parliament*. 2021. Available online: https://www.europarl.europa.eu/legislative-train/theme-a-european-green-deal/file-european-green-deal/ (accessed on 20 February 2022).
3. A European Strategy for Plastics in a Circular Economy. *European Comission*. Available online: https://ec.europa.eu/environment/strategy/plastics-strategy_en/ (accessed on 20 February 2022).
4. Bourguignon, D. Single-use plastics and fishing gear. *European Parliament*. 2019, p. 12. Available online: https://www.europarl.europa.eu/RegData/etudes/BRIE/2018/625115/EPRS_BRI(2018)625115_EN.pdf (accessed on 20 February 2022).
5. Rizzarelli, P.; Rapisarda, M.; Ascione, L.; Innocenti, F.D.; La Mantia, F.P. Influence of photo-oxidation on the performance and soil degradation of oxo- and biodegradable polymer-based items for agricultural applications. *Polym. Degrad. Stab.* **2021**, *188*, 109578. [CrossRef]
6. Zaborowska, M.; Bernat, K.; Pszczółkowski, B.; Wojnowska-Baryła, I.; Kulikowska, D. Anaerobic Degradability of commercially available bio-based and oxo-degradable packaging materials in the context of their end of life in the waste management strategy. *Sustainability* **2021**, *13*, 6818. [CrossRef]

7. EU Monitor COM(2018)35—Impact of the Use of Oxo-Degradable Plastic, Including Oxo-Degradable Plastic Carrier Bags, on the Environment—EU Monitor. Available online: https://www.eumonitor.eu/9353000/1/j9vvik7m1c3gyxp/vkl2iurn2lzy (accessed on 24 February 2022).
8. United Nations United Nations Convention on the Law of the Sea Oceans and Law of the Sea. Division for Ocean Affairs and the Law of the Sea. 1982. Available online: https://www.un.org/depts/los/convention_agreements/convention_overview_convention.htm (accessed on 24 February 2022).
9. International Marine Organization Action Plan to Address Marine Plastic Litter from Ships. Available online: https://wwwcdn.imo.org/localresources/en/KnowledgeCentre/IndexofIMOResolutions/MEPCDocuments/MEPC.310(73).pdf (accessed on 20 February 2022).
10. Macintosh, A.; Simpson, A.; Neeman, T.; Dickson, K. Plastic bag bans: Lessons from the Australian capital territory. *Resour. Conserv. Recycl.* **2020**, *154*, 104638. [CrossRef]
11. Barrowclough, D.; Birkbeck, C. Transforming the global plastics economy: The role of economic policies in the global governance of plastic pollution. *Soc. Sci.* **2022**, *11*, 26. [CrossRef]
12. World Econcomic Forum Germany Recycles More than Any Other Country. World Economic Forum. Available online: https://www.weforum.org/agenda/2017/12/germany-recycles-more-than-any-other-country/ (accessed on 24 February 2022).
13. Sun, J.; Fang, C.; Chen, Z.; Chen, G. Regional cooperation in marine plastic waste cleanup in the south China sea region. *Sustainability* **2021**, *13*, 9221. [CrossRef]
14. UN Environment Programme How China's Ban of Plastic Waste Imports Can Help Us Beat Pollution. Available online: https://www.unep.org/news-and-stories/story/how-chinas-ban-plastic-waste-imports-can-help-us-beat-pollution (accessed on 24 February 2022).
15. Wang, C.; Zhao, L.; Lim, M.K.; Chen, W.Q.; Sutherland, J.W. Structure of the global plastic waste trade network and the impact of China's import Ban. *Resour. Conserv. Recycl.* **2020**, *153*, 104591. [CrossRef]
16. Masuelli, M.A. *Biopackaging*; CRC Press: Boca Raton, FL, USA, 2017; ISBN 1351648381.
17. Cuq, B.; Gontardt, N.; Guilbertt, S. Thermoplastic properties of fish myofibrillar proteins: Application to biopackaging fabrication. *Polymer* **1997**, *38*, 4071–4078. [CrossRef]
18. Debeaufort, F. Active biopackaging produced from by-products and waste from food and marine industries. *FEBS Open Bio* **2021**, *11*, 984–998. [CrossRef]
19. Muhamad, I.I.; Sabbagh, F.; Karim, N.A. *Chapter 7 Polyhydroxyalkanoates: A Valuable Secondary Metabolite Produced in Microorganisms and Plants*; Apple Academic Press: Palm Bay, FL, USA, 2016; Volume 3, ISBN 9781315366302.
20. Albuquerque, P.B.S.; Malafaia, C.B. Perspectives on the production, structural characteristics and potential applications of bioplastics derived from polyhydroxyalkanoates. *Int. J. Biol. Macromol.* **2018**, *107*, 615–625. [CrossRef]
21. Sabbagh, F.; Muhamad, I.I. Production of poly-hydroxyalkanoate as secondary metabolite with main focus on sustainable energy. *Renew. Sustain. Energy Rev.* **2017**, *72*, 95–104. [CrossRef]
22. Mojaveryazdi, F.S.; Muhamad, I.I.; Rezania, S.; Benham, H. Importance of glucose and pseudomonas in producing degradable plastics. *J. Teknol. Sci. Eng.* **2014**, *69*, 7–10. [CrossRef]
23. Pérez-Rivero, C.; López-Gómez, J.P.; Roy, I. A sustainable approach for the downstream processing of bacterial polyhydroxyalkanoates: State-of-the-art and latest developments. *Biochem. Eng. J.* **2019**, *150*, 107283. [CrossRef]
24. Wu, C.S. Mechanical properties, biocompatibility, and biodegradation of cross-linked cellulose acetate-reinforced polyester composites. *Carbohydr. Polym.* **2014**, *105*, 41–48. [CrossRef] [PubMed]
25. Dilkes-Hoffman, L.S.; Lant, P.A.; Laycock, B.; Pratt, S. The rate of biodegradation of PHA bioplastics in the marine environment: A meta-study. *Mar. Pollut. Bull.* **2019**, *142*, 15–24. [CrossRef]
26. Jendrossek, D. Peculiarities of PHA granules preparation and PHA depolymerase activity determination. *Appl. Microbiol. Biotechnol.* **2007**, *74*, 1186–1196. [CrossRef]
27. Ong, S.Y.; Sudesh, K. Effects of polyhydroxyalkanoate degradation on soil microbial community. *Polym. Degrad. Stab.* **2016**, *131*, 9–19. [CrossRef]
28. Boyandin, A.N.; Prudnikova, S.V.; Filipenko, M.L.; Khrapov, E.A.; Vasil'ev, A.D.; Volova, T.G. Biodegradation of polyhydroxyalkanoates by soil microbial communities of different structures and detection of PHA degrading microorganisms. *Appl. Biochem. Microbiol.* **2012**, *48*, 28–36. [CrossRef]
29. Mojaveryazdi, F.S.; Azimah, N.; Mohd, B.; Rezania, S.; Universiti, B.; Biotechnology, I.; Faculty, E.E. Production of biodegradable polymers (PHA) through low cost carbon sources: Green Chemistry. *Int. J. Chem. Environ. Eng.* **2012**, *4*, 184–188.
30. Carus, M. Bio-based polymers-Production capacity will triple from 3.5 million tonnes in 2011 to nearly 12 million tonnes in 2020. *Ind. Biotechnol.* **2013**, *9*, 81–84. [CrossRef]
31. Chen, Y.; Fan, D.; Han, Y.; Lyu, S.; Lu, Y.; Li, G.; Jiang, F.; Wang, S. Effect of high residual lignin on the properties of cellulose nanofibrils/films. *Cellulose* **2018**, *25*, 6421–6431. [CrossRef]
32. Hamilton, A.; Scheer, R.; Stakes, T.; Allan, S. Solving Plastic pollution through accountability. *WWF Plast. Rep.* **2019**, *25*, 6–46.
33. Bourguignon, D. New boost for jobs, growth and investment. *Legis. Train Eur. Parliam.* **2020**, 2016–2018.
34. Bioplastics Market Data–European Bioplastics. Available online: https://www.european-bioplastics.org/market/ (accessed on 16 January 2022).

35. Dietrich, K.; Dumont, M.J.; Del Rio, L.F.; Orsat, V. Sustainable PHA production in integrated lignocellulose biorefineries. *N. Biotechnol.* **2019**, *49*, 161–168. [CrossRef] [PubMed]
36. Keskin, G.; Klzll, G.; Bechelany, M.; Pochat-Bohatier, C.; Öner, M. Potential of polyhydroxyalkanoate (PHA) polymers family as substitutes of petroleum based polymers for packaging applications and solutions brought by their composites to form barrier materials. *Pure Appl. Chem.* **2017**, *89*, 1841–1848. [CrossRef]
37. Pérez-Arauz, A.O.; Aguilar-Rabiela, A.E.; Vargas-Torres, A.; Rodríguez-Hernández, A.-I.; Chavarría-Hernández, N.; Vergara-Porras, B.; López-Cuellar, M.R. Production and characterization of biodegradable films of a novel polyhydroxyalkanoate (PHA) synthesized from peanut oil. *Food Packag. Shelf Life* **2019**, *20*, 100297. [CrossRef]
38. Hassaini, L.; Kaci, M.; Benhamida, A.; Bruzaud, S.; Pillin, I.; Grohens, Y. The effects of PHBV-g-MA compatibilizer on morphology and properties of poly(3-hydroxybutyrate-Co-3-hydroxyvalerate)/olive husk flour composites. *J. Adhes. Sci. Technol.* **2016**, *30*, 2061–2080. [CrossRef]
39. Sood, M.; Dwivedi, G. Effect of fiber treatment on flexural properties of natural fiber reinforced composites: A review. *Egypt. J. Pet.* **2018**, *27*, 775–783. [CrossRef]
40. Michel, A.T.; Billington, S.L. Characterization of poly-hydroxybutyrate films and hemp fiber reinforced composites exposed to accelerated weathering. *Polym. Degrad. Stab.* **2012**, *97*, 870–878. [CrossRef]
41. Wu, C.S.; Wu, D.Y.; Wang, S.S. Preparation, characterization, and functionality of bio-based polyhydroxyalkanoate and renewable natural fiber with waste oyster shell composites. *Polym. Bull.* **2021**, *78*, 4817–4834. [CrossRef]
42. Khoshnava, S.M.; Rostami, R.; Ismail, M.; Rahmat, A.R.; Ogunbode, B.E. Woven hybrid Biocomposite: Mechanical properties of woven kenaf bast fibre/oil palm empty fruit bunches hybrid reinforced poly hydroxybutyrate biocomposite as non-structural building materials. *Constr. Build. Mater.* **2017**, *154*, 155–166. [CrossRef]
43. Pasangulapati, V.; Ramachandriya, K.D.; Kumar, A.; Wilkins, M.R.; Jones, C.L.; Huhnke, R.L. Effects of cellulose, hemicellulose and lignin on thermochemical conversion characteristics of the selected biomass. *Bioresour. Technol.* **2012**, *114*, 663–669. [CrossRef] [PubMed]
44. Pérez, J.; Muñoz-Dorado, J.; De La Rubia, T.; Martínez, J. Biodegradation and biological treatments of cellulose, hemicellulose and lignin: An overview. *Int. Microbiol.* **2002**, *5*, 53–63. [CrossRef] [PubMed]
45. Cazón, P.; Velázquez, G.; Vázquez, M. Regenerated cellulose films combined with glycerol and polyvinyl alcohol: Effect of moisture content on the physical properties. *Food Hydrocoll.* **2020**, *103*, 105657. [CrossRef]
46. Motaung, T.E.; Linganiso, L.Z. *Critical Review on Agrowaste Cellulose Applications for Biopolymers*; Springer India: New Delhi, India, 2018; Volume 22, ISBN 1258801892.
47. Melo, J.D.D.; Carvalho, L.F.M.; Medeiros, A.M.; Souto, C.R.O.; Paskocimas, C.A. A biodegradable composite material based on polyhydroxybutyrate (PHB) and carnauba fibers. *Compos. Part B Eng.* **2012**, *43*, 2827–2835. [CrossRef]
48. Bousfield, G.; Morin, S.; Jacquet, N.; Richel, A. Extraction and refinement of agricultural plant fibers for composites manufacturing. *Comptes Rendus Chim.* **2018**, *21*, 897–906. [CrossRef]
49. Chilali, A.; Assarar, M.; Zouari, W.; Kebir, H.; Ayad, R. Mechanical characterization and damage events of flax fabric-reinforced biopolymer composites. *Polym. Polym. Compos.* **2020**, *28*, 631–644. [CrossRef]
50. Meng, D.C.; Shen, R.; Yao, H.; Chen, J.C.; Wu, Q.; Chen, G.Q. Engineering the diversity of polyesters. *Curr. Opin. Biotechnol.* **2014**, *29*, 24–33. [CrossRef]
51. Melendez Rodriguez, B.; Torres Giner, S.; Aldureid, A.; Cabedo, L.; Lagaron, J.M. Reactive melt mixing of poly(3-hydroxybutyrate)/rice husk flour composites with purified biosustainably produced poly(3-hydroxybutyrate-co-3-hydroxyvalerate) beatriz. *Materials* **2019**, *12*, 2152. [CrossRef]
52. Martino, L.; Berthet, M.A.; Angellier-Coussy, H.; Gontard, N. Understanding external plasticization of melt extruded PHBV-wheat straw fibers biodegradable composites for food packaging. *J. Appl. Polym. Sci.* **2015**, *132*, 1–11. [CrossRef]
53. Zhuo, G.; Zhang, X.; Jin, X.; Wang, M.; Yang, X.; Li, S. Effect of different enzymatic treatment on mechanical, water absorption and thermal properties of bamboo fibers reinforced poly(hydroxybutyrate-co-valerate) biocomposites. *J. Polym. Environ.* **2020**, *28*, 2377–2385. [CrossRef]
54. David, G.; Gontard, N.; Angellier-Coussy, H. Mitigating the impact of cellulose particles on the performance of biopolyester-based composites by gas-phase esterification. *Polymers* **2019**, *11*, 200. [CrossRef] [PubMed]
55. Santos, E.B.C.; Barros, J.J.P.; de Moura, D.A.; Moreno, C.G.; de Carvalho Fim, F.; da Silva, L.B. Da Rheological and thermal behavior of PHB/piassava fiber residue-based green composites modified with warm water. *J. Mater. Res. Technol.* **2019**, *8*, 531–540. [CrossRef]
56. Varghese, S.A.; Pulikkalparambil, H.; Rangappa, S.M.; Siengchin, S.; Parameswaranpillai, J. Novel biodegradable polymer films based on poly(3-hydroxybutyrate-co-3-hydroxyvalerate) and Ceiba pentandra natural fibers for packaging applications. *Food Packag. Shelf Life* **2020**, *25*, 100538. [CrossRef]
57. Torres-Giner, S.; Hilliou, L.; Melendez-Rodriguez, B.; Figueroa-Lopez, K.J.; Madalena, D.; Cabedo, L.; Covas, J.A.; Vicente, A.A.; Lagaron, J.M. Melt processability, characterization, and antibacterial activity of compression-molded green composite sheets made of poly(3-hydroxybutyrate-co-3-hydroxyvalerate) reinforced with coconut fibers impregnated with oregano essential oil. *Food Packag. Shelf Life* **2018**, *17*, 39–49. [CrossRef]
58. Vizireanu, S.; Panaitescu, D.M.; Nicolae, C.A.; Frone, A.N.; Chiulan, I.; Ionita, M.D.; Satulu, V.; Carpen, L.G.; Petrescu, S.; Birjega, R.; et al. Cellulose defibrillation and functionalization by plasma in liquid treatment. *Sci. Rep.* **2018**, *8*, 15473. [CrossRef]

59. Gigante, V.; Cinelli, P.; Righetti, M.C.; Sandroni, M.; Polacco, G.; Seggiani, M.; Lazzeri, A. On the use of biobased waxes to tune thermal and mechanical properties of polyhydroxyalkanoates-bran biocomposites. *Polymers* **2020**, *12*, 2615. [CrossRef]
60. Righetti, M.C.; Cinelli, P.; Mallegni, N.; Stäbler, A.; Lazzeri, A. Thermal and mechanical properties of biocomposites made of poly(3-hydroxybutyrate-co-3-hydroxyvalerate) and potato pulp powder. *Polymers* **2019**, *11*, 308. [CrossRef]
61. Wong, S.; Shanks, R.; Hodzic, A. Properties of poly(3-hydroxybutyric acid) composites with flax fibres modified by plasticiser absorption. *Macromol. Mater. Eng.* **2002**, *287*, 647–655. [CrossRef]
62. Nakajima, H.; Dijkstra, P.; Loos, K. The recent developments in biobased polymers toward general and engineering applications: Polymers that are upgraded from biodegradable polymers, analogous to petroleum-derived polymers, and newly developed. *Polymers* **2017**, *9*, 523. [CrossRef]
63. Russo, P.; Carfagna, C.; Cimino, F.; Acierno, D.; Persico, P. Biodegradable composites reinforced with kenaf fibers: Thermal, mechanical, and morphological issues. *Adv. Polym. Technol.* **2012**, *32*, 474–485. [CrossRef]
64. Sánchez-Safont, E.L.; Arrillaga, A.; Anakabe, J.; Cabedo, L.; Gamez-Perez, J. Toughness enhancement of PHBV/TPU/cellulose compounds with reactive additives for compostable injected parts in industrial applications. *Int. J. Mol. Sci.* **2018**, *19*, 2102. [CrossRef] [PubMed]
65. Muthuraj, R.; Misra, M.; Mohanty, A.K. Reactive compatibilization and performance evaluation of miscanthus biofiber reinforced poly(hydroxybutyrate-co-hydroxyvalerate) biocomposites. *J. Appl. Polym. Sci.* **2017**, *134*, 1–10. [CrossRef]
66. Rojas dos Santos, N.R.; Middea, A.; da Costa Gonçalves, L.F.; Pereira, I.C.S.; Oliveira, R.N.; Lima, A.M.; Lima, E.M.B.; Minguita, A.P.S.; Moreira, A.P.D.; Tavares, M.I.B.; et al. Poly(lactic acid) biocomposites with mango waste and organo-montmorillonite for packaging. *J. Appl. Polym. Sci.* **2019**, *136*, 47512. [CrossRef]
67. Fiore, V.; Scalici, T.; Nicoletti, F.; Vitale, G.; Prestipino, M.; Valenza, A. A new eco-friendly chemical treatment of natural fibres: Effect of sodium bicarbonate on properties of sisal fibre and its epoxy composites. *Compos. Part B Eng.* **2016**, *85*, 150–160. [CrossRef]
68. Zhang, B.; Huang, C.; Zhao, H.; Wang, J.; Yin, C.; Zhang, L.; Zhao, Y. Eects of cellulose nanocrystals and cellulose nanofibers on the structure and properties of polyhydroxybutyrate nanocomposites. *Polymers* **2019**, *11*, 2063. [CrossRef]
69. Beber, V.C.; de Barros, S.; Banea, M.D.; Brede, M.; de Carvalho, L.H.; Hoffmann, R.; Costa, A.R.M.; Bezerra, E.B.; Silva, I.D.S.; Haag, K.; et al. Effect of Babassu natural filler on PBAT/PHB biodegradable blends: An investigation of thermal, mechanical, and morphological behavior. *Materials* **2018**, *11*, 820. [CrossRef]
70. Zaidi, Z.; Crosky, A. Unidirectional rubber-toughened green composites based on PHBV. *Sustainability* **2019**, *11*, 2411. [CrossRef]
71. Avecilla-RamIrez, A.M.; del Rocio Lopez-Cuellar, M.; Vergara-Porras, B.; Rodriguez-Hernandez, A.I.; Vazquez-Nunez, E. Characterization of poly-hydroxybutyrate/luffa fibers composite material. *BioResources* **2020**, *15*, 7159–7177. [CrossRef]
72. El-Hadi, A.M. Increase the elongation at break of poly (lactic acid) composites for use in food packaging films. *Sci. Rep.* **2017**, *7*, 1–14. [CrossRef]
73. Zhao, C.; Li, J.; He, B.; Zhao, L. Fabrication of hydrophobic biocomposite by combining cellulosic fibers with polyhydroxyalkanoate. *Cellulose* **2017**, *24*, 2265–2274. [CrossRef]
74. Sánchez-Safont, E.L.; Aldureid, A.; Lagarón, J.M.; Gámez-Pérez, J.; Cabedo, L. Biocomposites of different lignocellulosic wastes for sustainable food packaging applications. *Compos. Part B Eng.* **2018**, *145*, 215–225. [CrossRef]
75. Chan, C.M.; Vandi, L.J.; Pratt, S.; Halley, P.; Richardson, D.; Werker, A.; Laycock, B. Composites of Wood and biodegradable thermoplastics: A review. *Polym. Rev.* **2018**, *58*, 444–494. [CrossRef]
76. Hosokawa, M.N.; Darros, A.B.; Moris, V.A.D.S.; De Paiva, J.M.F. Polyhydroxybutyrate composites with random mats of sisal and coconut fibers. *Mater. Res.* **2017**, *20*, 279–290. [CrossRef]
77. Qasim, U.; Fatima, R.; Usman, M. Efficient barrier properties of mechanically enhanced agro-extracted cellulosic biocomposites. *Mater. Today Chem.* **2020**, *18*, 100378. [CrossRef]
78. Guo, Y.; Wang, L.; Chen, Y.; Luo, P.; Chen, T. Properties of luffa fiber reinforced PHBV biodegradable composites. *Polymers* **2019**, *11*, 1765. [CrossRef]
79. Hammiche, D.; Boukerrou, A.; Grohens, Y.; Guermazi, N.; Arrakhiz, F.E. Mechanical properties and biodegradation of biocomposites based on poly (hydroxybutyrate-co-valerate) and alfa fibers. *J. Polym. Res.* **2020**, *27*, 1–10. [CrossRef]
80. Vandi, L.-J.; Chan, C.M.; Werke, A.; Richardson, D.; Laycock, B.; Pratt, S. Extrusion of wood fibre reinforced poly(hydroxybutyrate-co-hydroxyvalerate) (PHBV) biocomposites: Statistical analysis of the effect of processing conditions on mechanical performance. *Polym. Degrad. Stab.* **2019**, *159*, 1–14. [CrossRef]
81. Mazur, K.; Kuciel, S. Mechanical and hydrothermal aging behaviour of polyhydroxybutyrate-co-valerate (PHBV) composites reinforced by natural fibres. *Molecules* **2019**, *24*, 3538. [CrossRef]
82. Balestri, E.; Vallerini, F.; Seggiani, M.; Cinelli, P.; Menicagli, V.; Vannini, C.; Lardicci, C. Use of bio-containers from seagrass wrack with nursery planting to improve the eco-sustainability of coastal habitat restoration. *J. Environ. Manag.* **2019**, *251*, 109604. [CrossRef]
83. Seggiani, M.; Cinelli, P.; Balestri, E.; Mallegni, N.; Stefanelli, E.; Rossi, A.; Lardicci, C.; Lazzeri, A. Novel sustainable composites based on poly(hydroxybutyrate-co-hydroxyvalerate) and seagrass beach-CAST fibers: Performance and degradability in marine environments. *Materials* **2018**, *11*, 772. [CrossRef]
84. Kuciel, S.; Mazur, K.; Jakubowska, P. Novel biorenewable composites based on poly (3-hydroxybutyrate-co-3-hydroxyvalerate) with natural fillers. *J. Polym. Environ.* **2019**, *27*, 803–815. [CrossRef]

85. David, G.; Michel, J.; Gastaldi, E.; Gontard, N.; Angellier-Coussy, H. How vine shoots as fillers impact the biodegradation of PHBV-based composites. *Int. J. Mol. Sci.* **2020**, *21*, 228. [CrossRef] [PubMed]
86. Pantaloni, D.; Shah, D.; Baley, C.; Bourmaud, A. Monitoring of mechanical performances of flax non-woven biocomposites during a home compost degradation. *Polym. Degrad. Stab.* **2020**, *177*, 109166. [CrossRef]
87. Battegazzore, D.; Frache, A.; Abt, T.; Maspoch, M.L. Epoxy coupling agent for PLA and PHB copolymer-based cotton fabric bio-composites. *Compos. Part B Eng.* **2018**, *148*, 188–197. [CrossRef]
88. Graupner, N.; Müssig, J. A comparison of the mechanical characteristics of kenaf and lyocell fibre reinforced poly(lactic acid) (PLA) and poly(3-hydroxybutyrate) (PHB) composites. *Compos. Part A Appl. Sci. Manuf.* **2011**, *42*, 2010–2019. [CrossRef]
89. Awaja, F.; Gilbert, M.; Kelly, G.; Fox, B.; Pigram, P.J. Adhesion of polymers. *Prog. Polym. Sci.* **2009**, *34*, 948–968. [CrossRef]
90. Berthet, M.-A.; Angellier-Coussy, H.; Guillard, V.; Gontard, N. Vegetal fiber-based biocomposites: Which stakes for food packaging applications? *J. Appl. Polym. Sci.* **2016**, *133*, 1–18. [CrossRef]
91. Wu, C.S.; Liao, H.T. The mechanical properties, biocompatibility and biodegradability of chestnut shell fibre and polyhydroxyalkanoate composites. *Polym. Degrad. Stab.* **2014**, *99*, 274–282. [CrossRef]
92. Iggui, K.; Le Moigne, N.; Kaci, M.; Cambe, S.; Degorce-Dumas, J.R.; Bergeret, A. A biodegradation study of poly(3-hydroxybutyrate-co-3-hydroxyvalerate)/organoclay nanocomposites in various environmental conditions. *Polym. Degrad. Stab.* **2015**, *119*, 77–86. [CrossRef]
93. Tian, J.; Zhang, R.; Wu, Y.; Xue, P. Additive manufacturing of wood flour/polyhydroxyalkanoates (PHA) fully bio-based composites based on micro-screw extrusion system. *Mater. Des.* **2021**, *199*, 109418. [CrossRef]
94. Vilchez, A.; Acevedo, F.; Cea, M.; Seeger, M.; Navia, R. Development and thermochemical characterization of an antioxidant material based on polyhydroxybutyrate electrospun microfibers. *Int. J. Biol. Macromol.* **2021**, *183*, 772–780. [CrossRef] [PubMed]
95. Briassoulis, D.; Tserotas, P.; Athanasoulia, I.G. Alternative optimization routes for improving the performance of poly(3-hydroxybutyrate) (PHB) based plastics. *J. Clean. Prod.* **2021**, *318*, 128555. [CrossRef]
96. Othman, S.H. Bio-nanocomposite materials for food packaging applications: Types of biopolymer and nano-sized filler. *Agric. Agric. Sci. Proc.* **2014**, *2*, 296–303. [CrossRef]
97. Tarrahi, R.; Fathi, Z.; Seydibeyoğlu, M.Ö.; Doustkhah, E.; Khataee, A. Polyhydroxyalkanoates (PHA): From production to nanoarchitecture. *Int. J. Biol. Macromol.* **2020**, *146*, 596–619. [CrossRef] [PubMed]
98. Olayil, R.; Arumuga Prabu, V.; DayaPrasad, S.; Naresh, K.; Rama Sreekanth, P.S. A review on the application of bio-nanocomposites for food packaging. *Mater. Today Proc.* **2021**. *in Press*. [CrossRef]
99. Camargo, F.A.; Innocentini-Mei, L.H.; Lemes, A.P.; Moraes, S.G.; Durán, N. Processing and characterization of composites of poly(3-hydroxybutyrate-co- hydroxyvalerate) and lignin from sugar cane bagasse. *J. Compos. Mater.* **2012**, *46*, 417–425. [CrossRef]
100. Nair, S.S.; Kuo, P.Y.; Chen, H.; Yan, N. Investigating the effect of lignin on the mechanical, thermal, and barrier properties of cellulose nanofibril reinforced epoxy composite. *Ind. Crops Prod.* **2017**, *100*, 208–217. [CrossRef]
101. Vandi, L.J.; Chan, C.M.; Werker, A.; Richardson, D.; Laycock, B.; Pratt, S. Wood-PHA composites: Mapping opportunities. *Polymers* **2018**, *10*, 751. [CrossRef]
102. Cherpinski, A.; Torres-Giner, S.; Vartiainen, J.; Peresin, M.S.; Lahtinen, P.; Lagaron, J.M. Improving the water resistance of nanocellulose-based films with polyhydroxyalkanoates processed by the electrospinning coating technique. *Cellulose* **2018**, *25*, 1291–1307. [CrossRef]
103. Seoane, I.T.; Luzi, F.; Puglia, D.; Cyras, V.P.; Manfredi, L.B. Enhancement of paperboard performance as packaging material by layering with plasticized polyhydroxybutyrate/nanocellulose coatings. *J. Appl. Polym. Sci.* **2018**, *135*, 46872. [CrossRef]
104. Cherpinski, A.; Torres-Giner, S.; Cabedo, L.; Méndez, J.A.; Lagaron, J.M. Multilayer structures based on annealed electrospun biopolymer coatings of interest in water and aroma barrier fiber-based food packaging applications. *J. Appl. Polym. Sci.* **2018**, *135*. [CrossRef]
105. Wei, L.; Liang, S.; McDonald, A.G. Thermophysical properties and biodegradation behavior of green composites made from polyhydroxybutyrate and potato peel waste fermentation residue. *Ind. Crops Prod.* **2015**, *69*, 91–103. [CrossRef]
106. Quiles-Carrillo, L.; Montanes, N.; Garcia-Garcia, D.; Carbonell-Verdu, A.; Balart, R.; Torres-Giner, S. Effect of different compatibilizers on injection-molded green composite pieces based on polylactide filled with almond shell flour. *Compos. Part B Eng.* **2018**, *147*, 76–85. [CrossRef]
107. Tănase, E.E.; Popa, M.E.; Râpă, M.; Popa, O. PHB/Cellulose fibers based materials: Physical, mechanical and barrier properties. *Agric. Agric. Sci. Proc.* **2015**, *6*, 608–615. [CrossRef]
108. Babu, R.P.; O'Connor, K.; Seeram, R. Current progress on bio-based polymers and their future trends. *Prog. Biomater.* **2013**, *2*, 8. [CrossRef]
109. A European Strategy for Plastics in a Circular Economy. *European Parliament*. Available online: https://ec.europa.eu/transparency/documents-register/detail?ref=COM(2018)28&lang=en/ (accessed on 10 January 2022).
110. Single-Use Plastics: New EU Rules to Reduce Marine Litter. *Delegation of the European Union to Montenegro*. Available online: https://eeas.europa.eu/delegations/montenegro_fa/45931/Single-use%20plastics:%20New%20EU%20rules%20to%20reduce%20marine%20litter/ (accessed on 10 January 2022).
111. Lucas, N.; Bienaime, C.; Belloy, C.; Queneudec, M.; Silvestre, F.; Nava-Saucedo, J.E. Polymer biodegradation: Mechanisms and estimation techniques—A review. *Chemosphere* **2008**, *73*, 429–442. [CrossRef]

112. Eubeler, J.P.; Zok, S.; Bernhard, M.; Knepper, T.P. Environmental biodegradation of synthetic polymers I. Test methodologies and procedures. *TrAC Trends Anal. Chem.* **2009**, *28*, 1057–1072. [CrossRef]
113. Joyyi, L.; Ahmad Thirmizir, M.Z.; Salim, M.S.; Han, L.; Murugan, P.; Kasuya, K.; Maurer, F.H.J.; Zainal Arifin, M.I.; Sudesh, K. Composite properties and biodegradation of biologically recovered P(3HB- co -3HHx) reinforced with short kenaf fibers. *Polym. Degrad. Stab.* **2017**, *137*, 100–108. [CrossRef]
114. Musioł, M.; Jurczyk, S.; Sobota, M.; Klim, M.; Sikorska, W.; Zięba, M.; Janeczek, H.; Rydz, J.; Kurcok, P.; Johnston, B.; et al. (Bio)Degradable Polymeric Materials for Sustainable Future—Part 3: Degradation Studies of the PHA/Wood Flour-Based Composites and Preliminary Tests of Antimicrobial Activity. *Materials* **2020**, *13*, 2200. [CrossRef]
115. Šerá, J.; Serbruyns, L.; De Wilde, B.; Koutný, M. Accelerated biodegradation testing of slowly degradable polyesters in soil. *Polym. Degrad. Stab.* **2020**, *171*, 109031. [CrossRef]
116. Chan, C.M.; Vandi, L.-J.; Pratt, S.; Halley, P.; Richardson, D.; Werker, A.; Laycock, B. Insights into the biodegradation of PHA / wood composites: Micro- and macroscopic changes. *Sustain. Mater. Technol.* **2019**, *21*, e00099. [CrossRef]
117. Iwańczuk, A.; Kozłowski, M.; Łukaszewicz, M.; Jabłoński, S. Anaerobic biodegradation of polymer composites filled with natural fibers. *J. Polym. Environ.* **2015**, *23*, 277–282. [CrossRef]
118. Narancic, T.; Verstichel, S.; Reddy Chaganti, S.; Morales-Gamez, L.; Kenny, S.T.; De Wilde, B.; Babu Padamati, R.; O'Connor, K.E. Biodegradable plastic blends create new possibilities for end-of-life management of plastics but they are not a panacea for plastic pollution. *Environ. Sci. Technol.* **2018**, *52*, 10441–10452. [CrossRef] [PubMed]
119. Jain, R.; Tiwari, A. Biosynthesis of planet friendly bioplastics using renewable carbon source. *J. Environ. Heal. Sci. Eng.* **2015**, *13*, 11. [CrossRef] [PubMed]
120. Benn, N.; Zitomer, D. Pretreatment and anaerobic co-digestion of selected PHB and PLA bioplastics. *Front. Environ. Sci.* **2018**, *5*, 93. [CrossRef]
121. Volova, T.G.; Prudnikova, S.V.; Vinogradova, O.N.; Syrvacheva, D.A.; Shishatskaya, E.I. Microbial degradation of polyhydroxyalkanoates with different chemical compositions and their biodegradability. *Microb. Ecol.* **2017**, *73*, 353–367. [CrossRef]
122. Gómez, E.F.; Michel, F.C. Biodegradability of conventional and bio-based plastics and natural fiber composites during composting, anaerobic digestion and long-term soil incubation. *Polym. Degrad. Stab.* **2013**, *98*, 2583–2591. [CrossRef]
123. Dilkes-Hoffman, L.S.; Lane, J.L.; Grant, T.; Pratt, S.; Lant, P.A.; Laycock, B. Environmental impact of biodegradable food packaging when considering food waste. *J. Clean. Prod.* **2018**, *180*, 325–334. [CrossRef]
124. Chiemchaisri, W.; Chiemchaisri, C.; Boonchaiyuttasak, J. Utilization of stabilized wastes for reducing methane emission from municipal solid waste disposal. *Bioresour. Technol.* **2013**, *141*, 199–204. [CrossRef]
125. Behera, S.K.; Park, J.M.; Kim, K.H.; Park, H.-S. Methane production from food waste leachate in laboratory-scale simulated landfill. *Waste Manag.* **2010**, *30*, 1502–1508. [CrossRef]
126. Ghosh, P.; Shah, G.; Chandra, R.; Sahota, S.; Kumar, H.; Vijay, V.K.; Thakur, I.S. Assessment of methane emissions and energy recovery potential from the municipal solid waste landfills of Delhi, India. *Bioresour. Technol.* **2019**, *272*, 611–615. [CrossRef] [PubMed]
127. Kaouche, N.; Mebrek, M.; Mokaddem, A.; Doumi, B.; Belkheir, M.; Boutaous, A. Theoretical study of the effect of the plant and synthetic fibers on the fiber-matrix interface damage of biocomposite materials based on PHAs (polyhydroxyalkanoates) biodegradable matrix. *Polym. Bull.* **2021**, 1–21. [CrossRef]
128. Srubar, W.V.; Billington, S.L. Nonlonear micromechanical modleing of hygrothermal effect on structural biobased composite materials. In *Multiscale and Multiphysics Processes in Geomechanics*; Springer: Berlin/Heidelberg, Germany, 2011; pp. 189–192.
129. Kabbej, M.; Guillard, V.; Angellier-Coussy, H.; Wolf, C.; Gontard, N.; Gaucel, S. 3D Modelling of mass transfer into bio-composite. *Polymers* **2021**, *13*, 2257. [CrossRef] [PubMed]
130. Deierling, P.E.; Zhupanska, O.I. Computational modeling of the effective properties of spatially graded composites. *Int. J. Mech. Sci.* **2018**, *145*, 145–157. [CrossRef]
131. Villada Castillo, H.S.; Navia Porras, D.P.; Castañedad Niño, J.P. Biodegradable Films Obtained from Cassava Starch and Their Manufacture Process. U.S. Patent No. 9,416,275, 16 August 2016.
132. Mount, E.; Palta, D. Film with compostable heat seal. U.S. Patent 9,162,421 B2, 20 October 2015.
133. Khemani, K.; Schmidt, H. Biodegradable Polymer Films and Sheets Suitable for Use as Laminate Coatings as well as Wraps and Other Packaging Materials. EP Patent 0879946A2, 28 December 2011.
134. Mohanty, A.; Khan, M.; Hinrichsen, G. Influence of chemical surface modification on the properties of biodegradable jute fabrics—polyester amide composites. *Compos. Part A Appl. Sci. Manuf.* **2000**, *31*, 143–150. [CrossRef]
135. Constantinou, J. Composite Materials, Methods of Making, Methods of Use, and Articles Incorporating the Composites Materials. U.S. Patent 10,91,9257 B2, 16 February 2021.
136. Wilke, J.; Kaiser, C.; Schuster, P. Film Packaging for Oral Biologics. EP Patent 277598 B1, 7 November 2012.
137. Bastioli, C.; Capuzzi, L.; Milizia, T.; Rallis, A.; Vallero, R. Compositions Containing New Polyester. U.S. Patent 10,738,149, 11 August 2020.
138. Neuman, T.; Konieezny, N.; Pelled, I. Biodegradable Sheets. U.S. Patent 10,675,845 B2, 9 June 2020.
139. Nissenbaum, D.; Neuman, T.; Pelled, D.; Garty, S.; Konieezny, N. Biodegradable Sheet. U.S. Patent 10,239,292 B2, 36 March 2019.
140. Kann, Y. Biobased Modifiers for Polyvinylchloride Blends. U.S. Patent 9,464,187, 11 October 2016.

141. Khemani, K. Biodegradable Polymer Films and Sheet Suitable for Use as Laminate Coatings as well as Wraps and Other Packaging Materials. EP Patent 2 357 209 B1, 16 March 2016.
142. Planchard, H. Multilayer Article Comprising a Biodegradable Polymer-Based Layer and a Cellulose-Fiber Based Support; Method of Manufacturinga Multilayer Article and Food Accessory Comprising a Multilayer Article. U.S. Patent 9,956,741, 1 May 2018.

polymers

MDPI

Review

Polymer/Carbon Nanotube Based Nanocomposites for Photovoltaic Application: Functionalization, Structural, and Optical Properties

Boubaker Zaidi [1,2,*], Nejmeddine Smida [3,4], Mohammed G. Althobaiti [5], Atheer G. Aldajani [1] and Saif D. Almdhaibri [1]

1 Department of Physics, College of Science and Humanities, Shaqra University, Dawadmi 11911, Saudi Arabia; s442481193@std.su.edu.sa (A.G.A.); s442500416@std.su.edu.sa (S.D.A.)
2 Laboratoire de Synthèse Asymétrique et Ingénierie Moléculaire de Matériaux Organiques Pour L'électronique Organique LR 18ES19, Department of Physics, Faculty of Science, University of Monastir, Monastir 5019, Tunisia
3 Department of Chemistry, College of Science and Humanities, Shaqra University, AlQuwaiiyah 19257, Saudi Arabia; nejm.smida@su.edu.sa
4 Laboratory of Interfaces and Advanced Materials, Faculty of Science, University of Monastir, Monastir 5019, Tunisia
5 Department of Physics, Faculty of Science, Taif University, Taif 21974, Saudi Arabia; m.althobaiti@tu.edu.sa
* Correspondence: boubaker@su.edu.sa

Abstract: We present a systematic review of nanostructured organic materials, including synthesis methods, functionalization, and applications. First, we report the chemical and physical procedures used for preparing the polymer/carbon nanotube composites described in the literature over the last decade. We compare the properties of different polymer-based prototypes of organic nanocomposites functionalized with carbon nanotubes. Theoretical and experimental vibrational investigations provide evidence of the molecular structure describing the interaction between both components, showing that the allowed amount of carbon nanotubes and their dispersion states differ across polymers. Moreover, the nature of the solvent used in the preparation has a significant impact on the dispersion process. The integration of these materials in photovoltaic applications is discussed, where the impact of nanoparticles is evidenced through the correlation between experimental analyses and theoretical approaches based on density functional theory. Alterations in optical properties, evaluated from the absorption and luminescence process, are coherent with the solar spectrum, and a good distribution of donor/acceptor interpenetration was observed. In all cases, it was demonstrated that the performance improvement is physically related to the charge transfer from the organic matrix to the nanoparticles.

Keywords: PVK; hexylthiophene; PANI; nanocomposite; photovoltaic cells; DFT

Citation: Zaidi, B.; Smida, N.; Althobaiti, M.G.; Aldajani, A.G.; Almdhaibri, S.D. Polymer/Carbon Nanotube Based Nanocomposites for Photovoltaic Application: Functionalization, Structural, and Optical Properties. *Polymers* **2022**, *14*, 1093. https://doi.org/10.3390/polym14061093

Academic Editors: Andrzej Puszka and Beata Podkościelna

Received: 22 January 2022
Accepted: 25 February 2022
Published: 9 March 2022

1. Introduction

Scientific communities are continuously exploring innovative properties of π-conjugated conducting polymers, used as active layers in various organic electronic devices [1–4]. Hence, diverse molecular engineering procedures have been reported, such as doping conjugated polymers with conductive elements to ameliorate their conductivities [5,6]. Compared to inorganic semiconductors, these materials offer good flexibility and easier processing, leading to easier modulation of electronic, optical, and mechanical properties [7–13]. Handicaps related to fragility and a lower operating lifetime can be overcome by adding a small amount of carbon nanotubes (CNTs) [14,15]. Recently, Subramanyam et al. demonstrated that the addition of CNTs helps achieve an increase in lifetime by 67% in an open atmosphere [16]. According to Francis et al., CNTs also improve charge transport and provide good thermal stability, allowing π-conjugated conducting polymers to be good candidates for electrical, thermal, sensing, and actuating applications [17]. If CNTs are well dispersed and the exciton diffusion length is sufficient, the configuration leads to charge and energy transfer in the organic matrix [18–20]. This transfer is a major parameter for good photovoltaic conversion. Hence, the separation process of electron–hole pairs should dominate if the diffusion length is sufficiently higher than the mean distances between nanoparticles. The resulting interpenetrating (bulk P–N hetero-nanojunctions) network results in a good overlap between the optical absorption (OA) spectral region and the solar spectrum [21]. Indeed, the decrease in the optical gap involves additional absorption features in the visible spectrum and leads to a higher charge separation and good collection efficiencies [22–25]. Hybrid materials, including inexpensive flexible polymers and their nanostructures, represent the next generation of photovoltaic solar cells [26]. Continuous progress in and improvement of fabrication methods is a major focus of research [27]. The aim is to ensure a uniform distribution and arrangement of nanoparticles in the polymer matrix. The performance of organic photovoltaic solar cells is comparable to that of monocrystalline silicon, where PCE reaches 24% [16].

CNTs exist in a bundled form in the solid state. The dispersion process requires mechanical or vibrational energy higher than van der Waals cohesive forces [28]. The direct mixing methods are based on mechanical agitation or vibrational forces using homogenizers with different velocities [29,30]. CNTs with better homogeneity and alignment can be achieved by applying to the nanocomposite an ultrasound wave and centrifugal forces of appropriate power for an appropriate amount of time [31,32]. Chemical or physical functionalization, through covalent or noncovalent bonding between CNTs and diverse active molecules, has also been reported [33,34]. For chemical functionalization of the CNT surface (Scheme 1a,b), the bonding can be via defect groups or by side wall bonding, as well as by other special reactive groups, such as lithium alkyl nitrenes and fluorine radicals [35–37]. These chemical modifications facilitate the grafting of other functional groups and give rise to covalent bonding with the polymers/molecules [38]. However, noncovalent functionalization can be hexahedral (Scheme 1c,d) or endohedral [39].

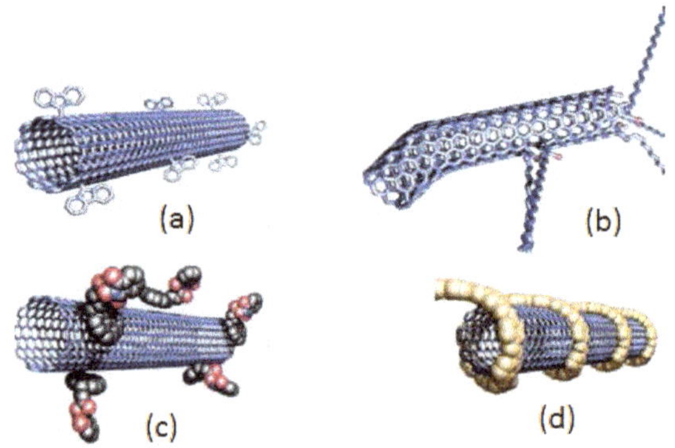

Scheme 1. Interaction between CNTs and functional molecules covalently (**a**,**b**) and noncovalently (**c**,**d**) [39].

For a polymer moiety, the polymerization process can be carried out after chemical functionalization, where first the monomers are linked to the tube and then the polymerization takes place [40,41]. If the direct mixing of polymers or in situ polymerized fragments with CNTs results in noncovalent linking, the process is called physical functionalization (Scheme 2) [42–46].

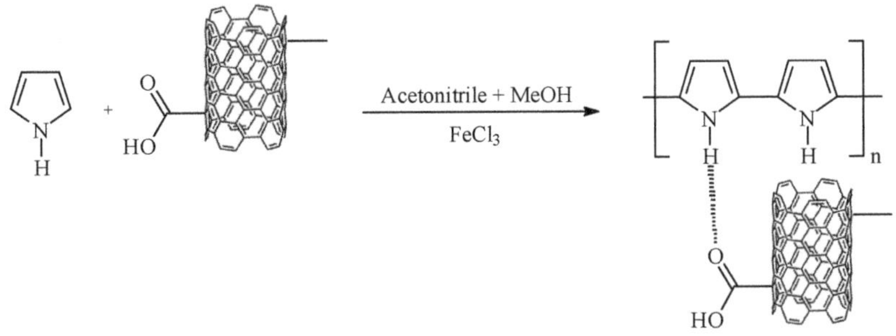

Scheme 2. Interaction between polypyrrole and multiwalled CNTs.

The dispersion process of nanoparticles is often of short duration. These nanoparticles have the tendency to agglomerate on the surface. To separate aggregates, the obtained solution should be exposed to ultrasonic baths and centrifuge forces with controlled frequencies and power [47]. In this work, we present various methods used over the last two decades for nanocomposite elaboration for a specific class of polymeric materials and their impact on the amelioration of photovoltaic properties. The study is restricted to polymer materials containing nitrogen in the aliphatic or cyclic sequences, such as poly(N-vinyl-carbazole) (PVK), poly(3-hexylthiophene (P3HT), and polyaniline (PANI). We report the results of our investigations in the last decade of the different polymeric nanostructures based on a large variety of polymers. As indicated for nanostructured materials, photoexcitations are followed by the establishment of charge transfer. The electronic structure modification, including the highest occupied molecular orbital (HOMO) and the lowest unoccupied molecular orbital (LUMO) levels, localized states, their energies, and the corresponding skin depth, should be evaluated in order to reduce injection barriers. Density functional theory (DFT) calculations are additional alternatives applied to a justified model structure to support experimental characterizations and physical interpretations [48–53]. Properties of organic materials that are experimentally inaccessible are in general described by DFT-based quantum calculations [50,53,54].

2. Organic Nanocomposite Synthesis and Characterization

2.1. CNT Treatment and Dispersion

CNTs (single-walled or multiwalled forms) need some treatment before functionalization with organic materials. Purification can be ensured either chemically, using a selective oxidation processor, or physically, based on filtration and/or centrifugation [55–58]. Metallic and semiconducting nanotubes can be separated by gradient-of-density ultracentrifugation techniques. The difference in their diameters is also exploited in the separation process [59,60]. Selective interaction with a conducting polymer is an alternative procedure that is efficient for separation in an acidic medium [61].

Good dispersion (homogeneous distribution) of CNTs in the polymer matrix is a major step that influences their properties [62]. CNTs largely exist in the bundled form due to strong van der Waals cohesive forces. It is difficult to achieve the isolated form, and the functionalization process leads to aggregate formation (Figure 1). Concerning the composite preparation, purified single-walled carbon nanotubes (SWCNTs) are dispersed in the same solvent with a polymer for an appropriate amount of time. The transitory dispersed phase of SWCNTs is immediately added in the desired amount to the polymer solution. Then, the sonication process and centrifuge forces are applied.

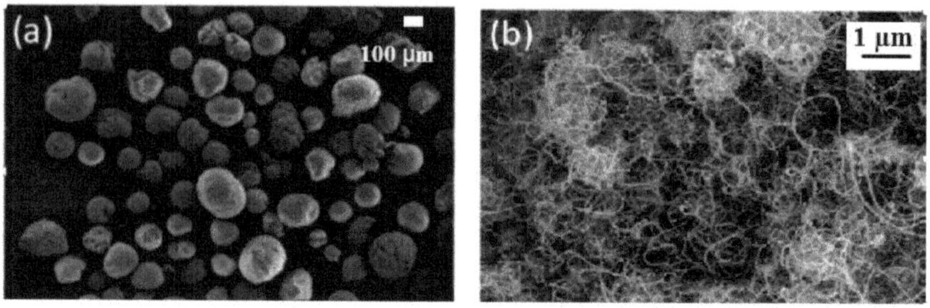

Figure 1. Structure of the bundled CNT (**a**); a zoom-in photo of the aggregates (**b**).

2.2. Materials and Methods

Composite solutions were deposited at room temperature with nearly uniform thickness on silica substrates for OA measurements. However, glass slides were used for FTIR and photoluminescence (PL) measurements. All substrates were already cleaned with deionized water and ethanol in an ultrasonic bath. All samples thus obtained were introduced into a Pyrex tube and heated under vacuum at a moderate temperature based on the thermal stability of the resulting material. Infrared absorption measurements were recorded with a resolution of 2 cm^{-1}. The OA spectra were recorded in the absorption mode with the wavelength varying from 200 to 2000 nm. For the PL experiment, depending on the material emission spectral range, different excitation wavelengths were used (280, 325, and 540 nm). Ultrafast time-resolved PL measurements were carried out with the regenerative amplified femtosecond laser system. The time-resolved emission spectra were spatiotemporally detected with a high-dynamic-range Hamamatsu C7700 streak camera with a temporal resolution <20 ps. The PL intensity and decay times were simulated with two coupled exponential decays. The TEM images were observed after adding liquid nitrogen to the Cu grid and then were taken at temperatures varying from 100 to 104 K. Theoretical data were procured from quantum calculations based on density functional theory (DFT). The modeling structure was fully optimized using the most popular Becke's three-parameter hybrid functional, B_3 [63], with the basis set 6-31G(d). Calculations applying to the proposed modeling structures were carried out with nonlocal correlation of Lee–Yang–Parr (LYP), abbreviated as the B_3LYP method. Theoretical OA spectra were calculated using the time-dependent density functional theory (TD-DFT) method with the 6-31G(d) basis.

2.3. Alignment and Orientation of CNTs

CNT orientation can be accomplished in solution during or after the polymer synthesis process. Chemical vapor deposition synthesis of SWCNTs results in a massive growth of superdense and vertically aligned nanotubes with enhanced catalytic activity of catalysts (Figure 2) [64]. CNTs can also be aligned under a magnetic or an electric field during the growth step, but some dynamic structural changes can occur due to the complexity of the process [65–68].

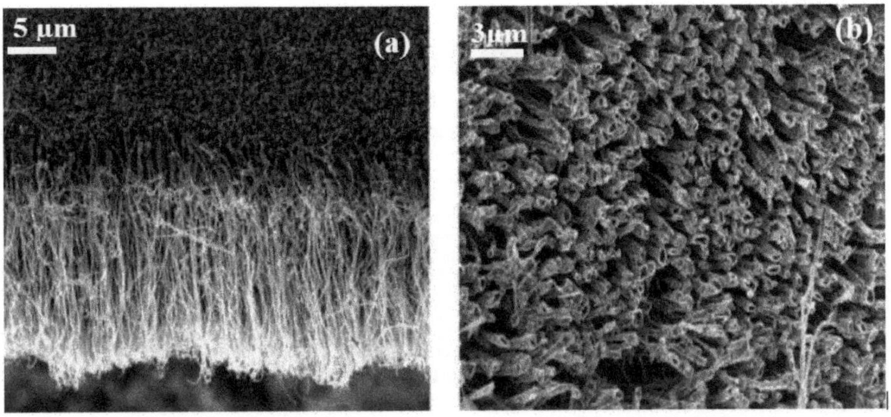

Figure 2. TEM pictures illustrating CNT orientation observed in (**a**) the lateral direction and (**b**) the longitudinal direction [69].

After CNT dispersion, the application of external electrical or magnetic perturbations allows the orientation of CNTs in the preferred direction [70–75]. This network formation results in three steps: the orientation of nanotubes, interaction, and network establishment (Scheme 3). Due to dipolar interaction, the aligned CNT configuration leads to tail-to-tail or head-to-head connections.

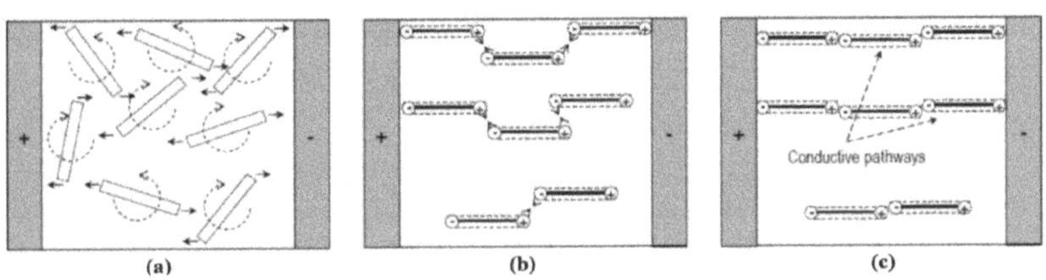

Scheme 3. Network formation: (**a**) orientation, (**b**) interaction between oriented CNTs, and (**c**) the resultant network [76].

In both cases, the oriented CNTs permit the reinforcing of optical and electronic transport properties in the preferred direction. This process helps enhance the performances of the electronic and optical devices. Moreover, annealing at moderate temperatures leads to a good repartition of CNTs into the organic polymeric matrix, which results in higher photovoltaic performances [77–79].

3. Applications of Organic Nanocomposites for Photovoltaic Cells

Organic solar cells represent new alternatives to those based on inorganic semiconductors [16]. Organic materials are characterized by flexibility, lightness, low cost, and easy processing. However, their intrinsic charge carrier mobility is lower (at max 10^{-1} cm^2 V^{-1} s^{-1}) compared to that of their homolog inorganic semiconductors, where it can reach 900 cm^2 V^{-1} s^{-1} for a single silicon crystal. Otherwise, handicaps related to lower mobility and fragility can be partially resolved when adding CNTs or phenyl C61 butyric acid methyl ester (PCBM) because carbon structures exhibit better mobility and rigidity than silicon [80]. Therefore, adding a small quantity of CNTs (single-walled or multiwalled forms) may improve the mechanical properties and lead to a good compatibility with the solar spectrum (UV–visible range). If, additionally, there is coherence between diffusion length (~20 nm) and the distances between CNTs, the separation process of electron–hole pairs will be improved. In fact, when adding CNTs, the entire network will be regular nano-P–N junctions, which represent photogeneration sites in all points. The charges transferred from the organic matrix to CNTs may be simultaneously collected and carried by CNTs. In the literature, the open-circuit current for some elaborate photovoltaic devices is in the range of a few tens of milliamperes per square centimeter [81–84]. This nanometric configuration is different from the conventional inorganic P–N photovoltaic cell. Indeed, for an inorganic device, a single junction separates the P and N regions, while in the nanometric interpenetrating network, junctions are distributed in the entire bulk, giving rise to more photoexcitations, leading to higher efficiency.

For organic photovoltaic conversion, efficiency started with values lower than 1%. However, the conversion efficiency was considerably improved beginning from the year 2000, reaching 2.5% and then 3.6% when PCBM was inserted in the organic matrix [85,86]. The relatively high conversion efficiency is obtained with a double heterostructure of fullerene (C_{60}) and copper phthalocyanine [87]. In 2011, a new record of 5% was attained using hybrid CNT/polymer materials and then ameliorated to reach 12% [10]. Compared to a silicon monocrystal, the power conversion efficiency (PCE) remains lower, but the easy processing and low cost make the material attractive [10]. The actual records were achieved by Cui et al. in 2020 and Subramanyam et al. in 2021, reaching, respectively, 17% and 24%, comparable to the PCE of porous and crystalline silicon [16,88].

4. Nanocomposites of Oligo-N-vinyl-carbazole/Single- and Multiwalled CNTs

Infrared, Raman, OA, and PL spectroscopies were used to describe the interaction between OVK and SWCNTs in polar and apolar solvents. The effect of the solvents' polarities on the dispersion process, i.e., repartition of SWCNTs on the organic matrix and its relationship with optical and vibrational properties, is reported. The observed changes in characteristics and their interpretations are discussed according to the literature [89–92]. Experimental results are correlated to those theoretically obtained from DFT; the functional and the method are described elsewhere [63,93–95].

4.1. Vibrational Properties

OVK/SWCNT composites were prepared at various concentrations in different solvents, such as chloroform and chlorobenzene, and were annealed to ameliorate the dispersion process [21,96]. FTIR spectra depicted in Figure 3 reveal that the most relevant vibrational modes of native OVK and SWCNTs are found in the composite spectrum. Moreover, in both cases, adding SWCNTs leads to some significant changes. Some bands with a relatively higher intensity ascribed to OVK vinyl groups disappear, which is strictly related to the linking between SWCNTs and OVK. This grafting is spontaneously established in the case of chloroform. However, when dispersing in chlorobenzene, certain spectra can be recorded after annealing treatment at 393 K (as in the case of nanocomposite dispersed in chloroform). Indeed, a better modification of the spectrum is clearly seen after annealing, and both spectra (a) and (c) present higher similarity. In both cases, the spectra reveal the appearance of a new feature at 1297 cm^{-1} that indicates the covalent interaction between organic OVK matrix and CNTs, as described for electrosynthesized PVK/SWCNTs [97].

Raman analyses (Figure 3(2)) reveal the presence of SWCNT bands, but those of OVK are severely attenuated. Radial and tangential modes depend on the used solvents and thermal treatment, which reflect the dispersion state of CNTs in the organic matrix (decrease in the band attributed to the bundled form at 181 cm^{-1}).

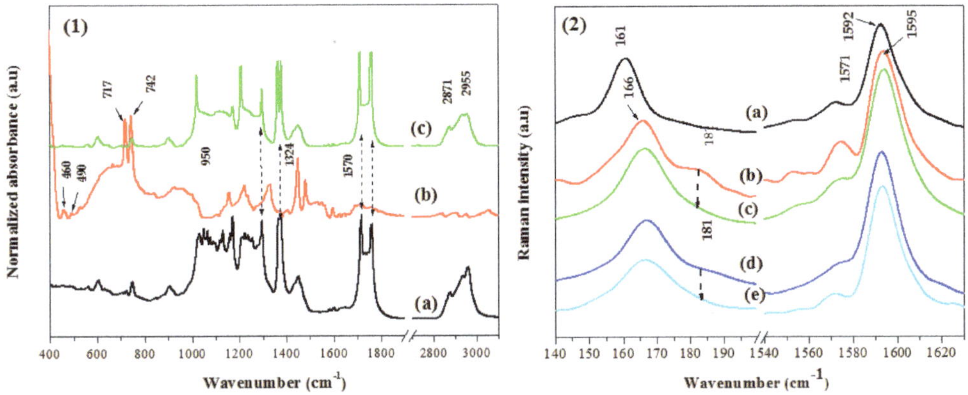

Figure 3. (**1**) Infrared spectra of the OVK/SWCNT composite dispersed in a chloroform solvent (**a**) and a chlorobenzene solvent (**b**) not annealed and (**c**) annealed at 333 K. (**2**) Raman radial and tangential modes: (**a**) SWCNTs; OVK/SWCNT in chloroform in the not-annealed (**b**) and annealed (**c**) states; OVK/SWCNT in chlorobenzene in the not-annealed (**d**) and annealed (**e**) states.

Both radial mode bands at 161 and 181 cm^{-1} are associated, respectively, with isolated and bundled SWCNTs [98,99]. The mean diameter of the tube can be evaluated from the empirical relation and varies from 1.34 to 1.38 nm [96,100]. The shift in radial mode toward higher wavenumbers has been interpreted as a consequence of inserting organic molecules into bundled SWCNTs. The disappearance of radial mode (bundled CNTs) in the case of the chlorobenzene solvent leads to the conclusion that the polar solvent leads to more dispersed nanotubes. The change from 2D to 3D symmetry is illustrated by the tangential mode narrowing and the decrease in intensity for the mode at 1571 cm^{-1} [101]. The spectrum of annealed states reveals shifting and band intensity modifications, promoting a better functionalization process.

4.2. Changes in Optical Properties

The most relevant information on the excitation and de-excitation process can be described by the complementarity of OA and PL measurements (Figure 4).

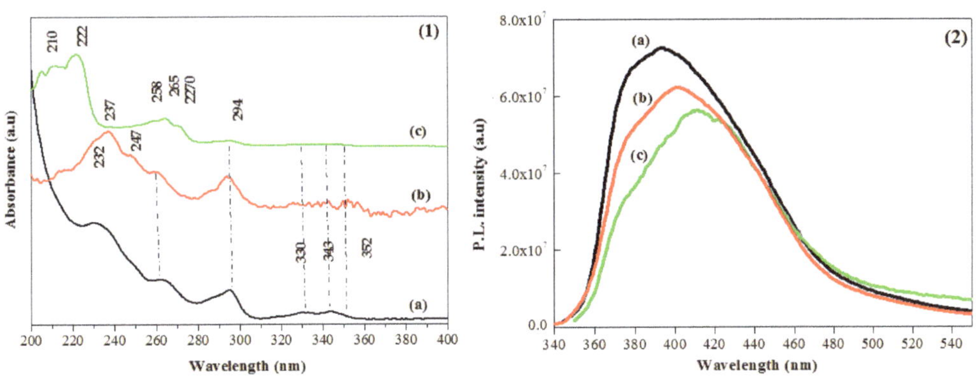

Figure 4. Changes in OA (**1**) and PL (**2**) spectra. Before SWCNT addition (**a**) and after SWCNT addition in chloroform (**b**) and in chlorobenzene (**c**).

Absorption spectra are affected by adding SWCNTs. For both solvents (chloroform and chlorobenzene), we see a redshift and a new band apparition, implying an interaction between both components. When chloroform is the solvent, in the spectral region of 400 to 2000 nm, there is a gradual decrease in the absorbance, similar to the case of other polymers when functionalized with SWCNTs [21,102]. Bands that appear in this spectral region (λ > 400 nm) correspond to the optical transitions for semiconducting SWCNTs. In chlorobenzene, the hypsochromic shift of the blue-side transition implies a decrease in the oligomer chain length and/or a decrease in the interchain interaction due to the good dispersion of SWCNTs in the organic matrix [81].

Chlorobenzene as a solvent quenches PL intensity after SWCNT insertion more efficiently (Figure 4(2)). This result is in good agreement with Raman results supporting good SWCNT dispersion. Indeed, photoexcitations are followed by a separation process rather than a recombination, because of which the observed PL is quenched [89,103,104]. We think that the photoexcitation produced in the OVK matrix can reach SWCNTs where the dissociation process is more preponderant rather than the recombination process. After annealing, the quenching effect becomes more pronounced and a new feature emission is created at around 499 nm (2.48 eV) when chloroform is the solvent (Figure 5(1)). As at a lower temperature (T = 87 K) and for both annealed samples (with and without SWCNTs), this weak emission band appears only in the case of the composite. It is, therefore, the consequence of the insertion of SWCNTs (Figure 5(2)).

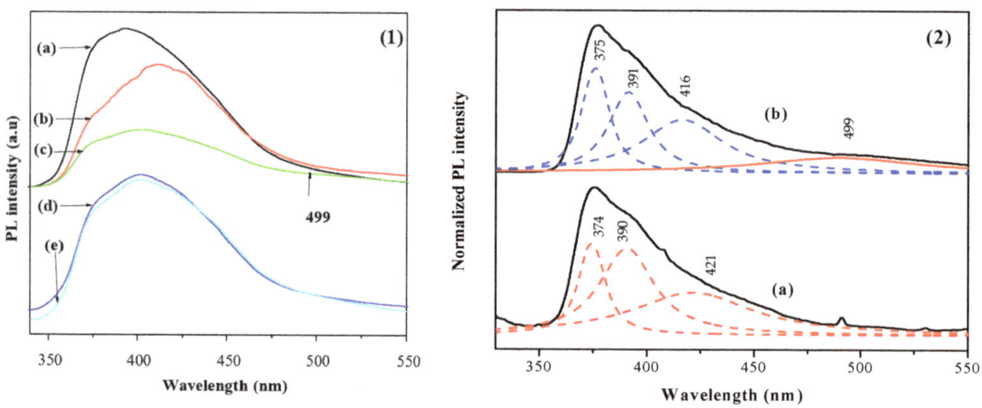

Figure 5. (**1**) Stationary PL spectra of (**a**) OVK and of nanocomposite in chloroform (**b**) not annealed and (**c**) annealed at 333 K and in chlorobenzene (**d**) not annealed and (**e**) annealed at 333 K. (**2**) PL spectra at T = 87 K of both annealed samples (**a**) OVK and with the OVK/SWCNT composite (**b**).

4.3. Interaction between SWCNTs and an Organic Matrix

The correlation of experimental and theoretical results evidences the covalent functionalization process between SWCNTs and oligo-N-vinyl carbazole [21]. The process does not require any thermal treatment when the dispersion is in chloroform as the solvent. The most affected chemical environments are vinylidene groups because, we assume, grafting occurs between these groups and the side wall (Scheme 4), similar to the case of PVK/MWCNT and for PVK/C60 composites [98,101]. However, after annealing, the similarity of both spectra supports the same grafting process.

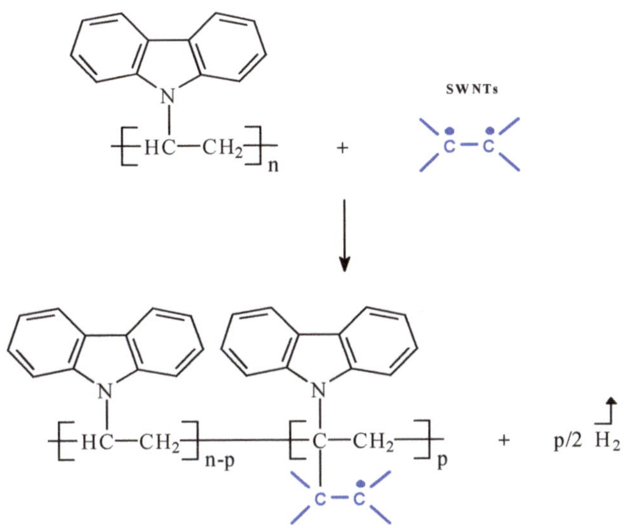

Scheme 4. Grafting mechanism of OVK and CNTs.

The proposed grafting mechanism, confirmed in our previous work, is consolidated by theoretical vibrational frequencies obtained either for OVK or for OVK/SWCNT composites [21]. First, experimental results are reproduced to confirm the modeling structures of primary materials. Then, the most probable grafting sites with higher reactivity are supported by force constant calculations in both neutral and oxidized states. In the resulting model structure of the nanocomposite, therefore, OVK is grafted on the side walls via vinylidene groups (Figure 6(1)).

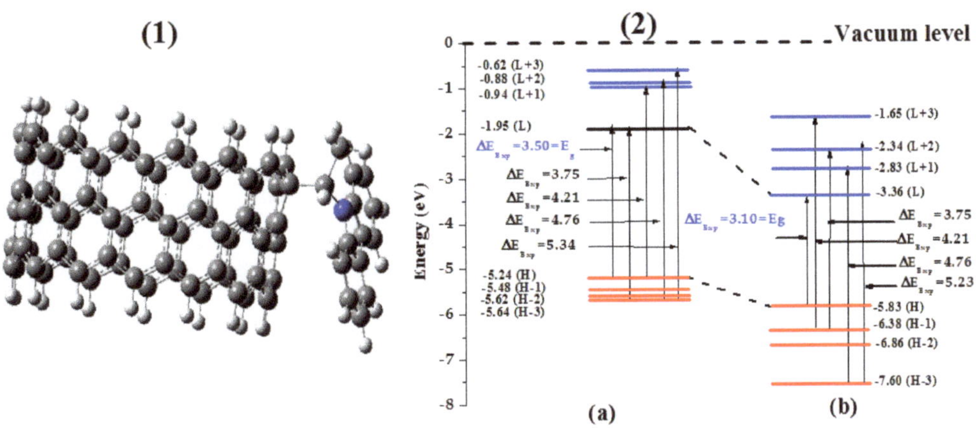

Figure 6. (**1**) Nanocomposite modeling structure used for DFT calculation and (**2**) change in the electronic structure from OVK (**a**) to the nanocomposite (**b**). H: HOMO; L: LUMO; Exp: experimental.

Based on the above modeling structure, DFT has also been calculated with the B3LYP/3-21G* method to carry out changes in the electronic structure. All the theoretical changes are depicted in Figure 6(2), which are in good accordance with experimental absorption transitions.

4.4. Structural Properties in Relationship with Transient Photoluminescence

Transmission electron microscope analyses were carried out to have an idea about the SWCNT distribution and arrangement (Figure 7) [18].

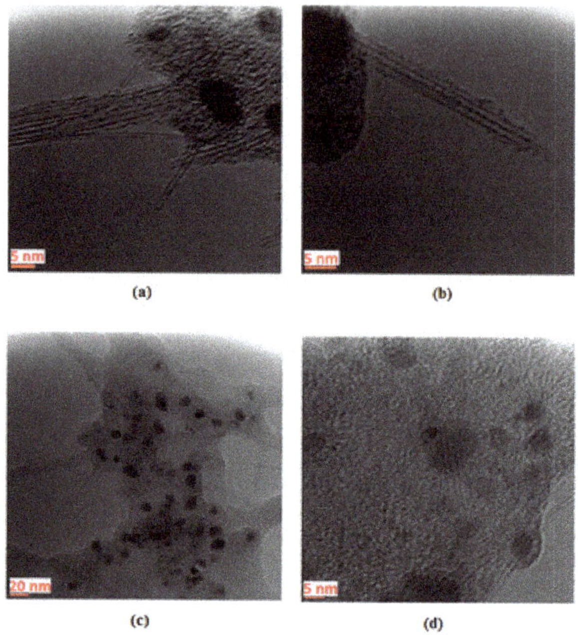

Figure 7. TEM pictures of OVK/SWCNT composites dispersed in chloroform (**a**,**b**) and in chlorobenzene (**c**,**d**) [18].

When chloroform is the solvent, bundled and isolated SWCNTs are detected, but with a lower concentration for the isolated form, supporting a good SWCNT dispersion process. Figure 7c shows a good uniformity and orientation of SWCNTs having an interdistance of nearly 10 nm, probably due to the thermal treatment. In chlorobenzene (Figure 7d), this distance varies from 3 to 20 nm and the structure contains bundled SWCNTs connected by individual ones. In fact, these structural differences can influence the dynamical properties of the excited states.

Results of PL decay (obtained by pumping at 330 and probing at 390 nm) show a decrease in lifetimes and imply exciton diffusion to the SWCNTs [18,105,106] (Table 1).

Table 1. Decay parameters of OVK/SWCNT nanocomposites [18].

Samples	P_1 (%)	P_2 (%)	τ_1 (ns)	τ_2 (ns)	A_1	A_2	τ_m (ns)
OVK	18.8	81.10	0.983	6.223	8.911	6.042	5.23
OVK/SWCNTs in chloroform	32.73	67.26	0.753	6.996	15.575	3.445	4.95
OVK/SWCNTs in chlorobenzene	32.98	67.01	0.662	3.848	13.747	4.804	2.796

The existence of two populations weighing P_1 and P_2, having lifetimes of τ_1 and τ_2, and having a mean lifetime τ_m has been interpreted to be because of shorter and longer segments, respectively. A similar decay evolution was reported in the case of other polymers functionalized with various nanoparticles [107,108], supporting a charge transfer between the organic matrix and SWCNTs [105,109]. Lifetimes evaluated from Figure 8(1) are shorter than those of PVK, which is coherent with the fact that faster decay is a signature of shorter segments [110,111]. The photoexcitations that occur in the organic matrix will diffuse and be transferred into the SWCNTs, and the electron–hole pairs will be separated due to the internal electric field created by hetero-nanojunctions [81,83]. The obtained decay times are coherent with PL cartography (Figure 8(2)), in which we show a more intense luminescence when chlorobenzene is the solvent. Indeed, a faster lifetime results in the condensation of excitations in the radiative pathways.

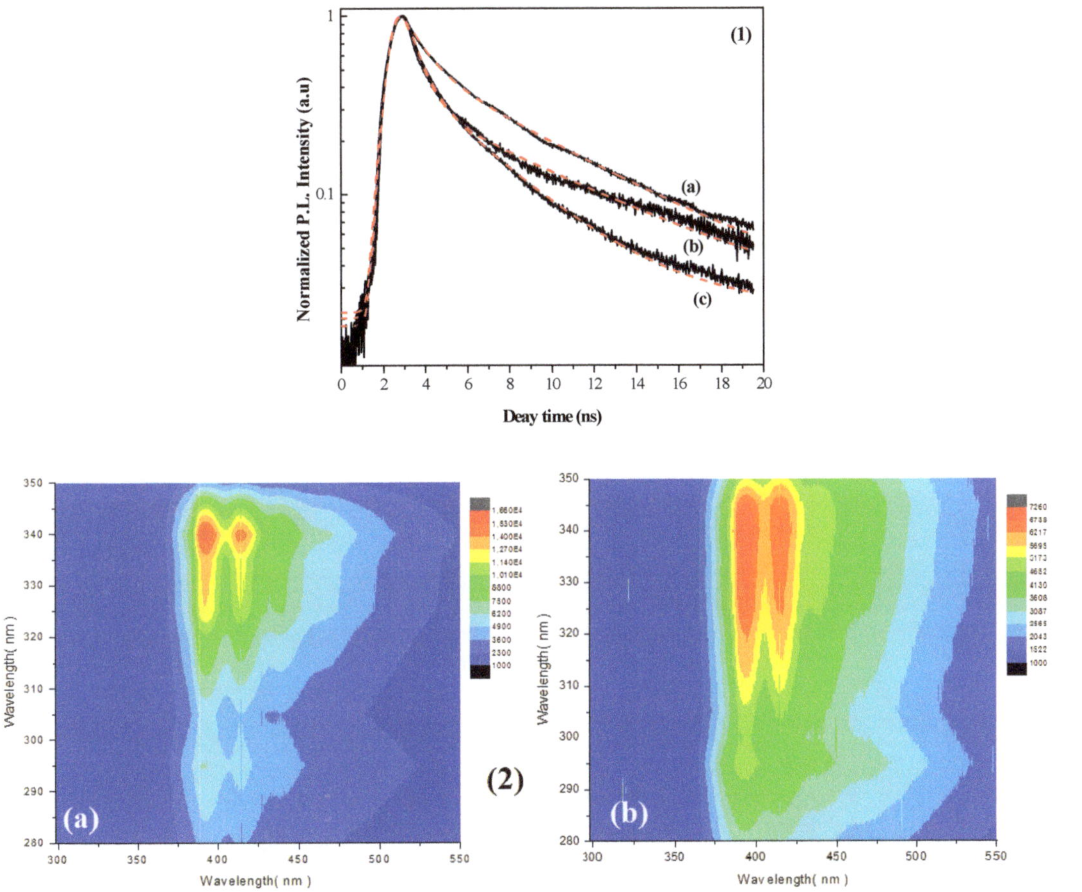

Figure 8. (**1**) PL decay OVK (**a**), OVK/SWCNTs in chloroform (**b**), and OVK/SWCNTs in chlorobenzene (**c**) (—: experimental; (. . . .): theoretical). (**2**) PL cartography of OVK/SWCNTs in chloroform (**a**) and in chlorobenzene (**b**).

5. Properties of PVK-3HT/SWCNT Nanocomposites

To reach a better dispersion process with a higher CNT concentration and to obtain good optical properties, another matrix containing PVK was synthetized [29,112]. Hexylthiophene, known by a large optical spectrum, was bridged with PVK, where the resulting matrix is called PVK-3HT. PVK-3HT/SWCNT nanocomposites were studied by FTIR and Raman spectroscopy, where the SWCNT weight concentration varied from 0% to 60%. It was shown that in all cases, SWCNTs are bundled and isolated. However, the D-band and the G-band, situated at 1250 and 1595 cm^{-1}, respectively, are nearly the same in all cases, similarly to the case of purified nanotubes.

5.1. Vibrational Study

On adding SWCNTs to PVK-3HT, there is no apparent change in the band positions and shapes of both compounds. However, the band intensities change and the appearance of new vibrational features at 1049 and 1384 cm^{-1} (Figure 9(2)) lead to the conclusion that there is an interaction between both components.

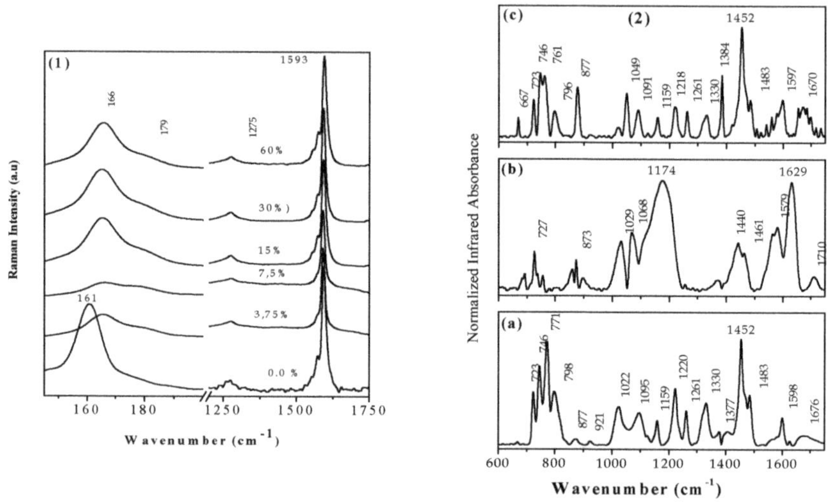

Figure 9. (**1**) Raman spectra of PVK-HT/SWCNTs at different CNT weight concentrations (0%, 3.75%, 7.5%, 15%, 30%, and 60%). (**2**) FTIR spectra of PVK-HT (**a**), SWCNTs (**b**), and PVK-HT/SWCNT composite (60%) (**c**).

5.2. Optical Properties Changes

UV–visible absorption analyses (Figure 10(1)) demonstrate that PVK-HT absorbs in the large spectral domain. The absorbance shape observed in the spectral domain ranging from 240 to 300 nm is the signature of the grafting of bicarbazole units as a part of the resulting copolymer [113]. The extension of the absorbance in the visible range indicates extended conjugated fragments of P3HT in the obtained copolymer [113,114]. On adding SWCNTs, no apparent change is observed for $\lambda < 400$. However, the band near 430 nm decreases and blue-shifts, which agrees well with the hypothesis of the diminution of interchain interaction and/or the decrease in the π-conjugation length [115].

The continuous PL quenching (Figure 10(2)) shows that the addition of CNTs has a direct impact on the excited-state dynamics of the prepared composite (exciting at 400 nm using a seep range of 1 ns) [111]. As in the case of the OVK/SWCNT composite, we think that the exciton diffusion length becomes progressively higher than the distance between neighboring nanotubes. In this case, the separation process is the most dominant behavior that can occur.

These results are in full accordance with the progressive decrease in emitted energy intensity as a function of time from 0 to 1 ns obtained in 3D maps (Figure 11).

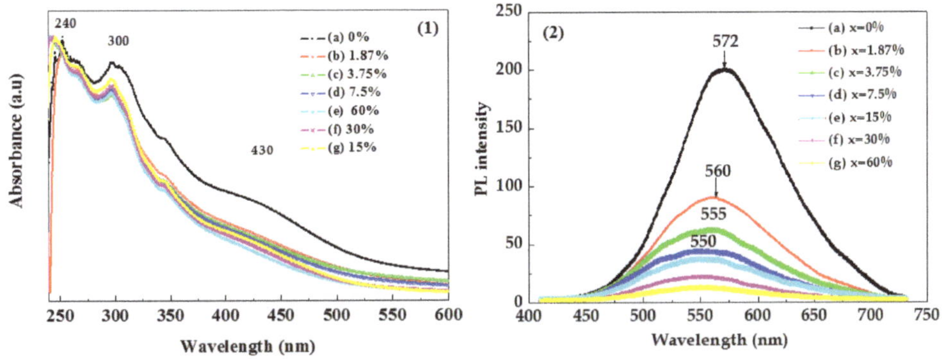

Figure 10. (**1**) OA and (**2**) PL spectra in chloroform solvent of pristine PVK-HeT (**a**) and those of PVK-HeT/SWCNT composites at various SWCNT weight concentrations: (**b**) 1.87%, (**c**) 3.75%, (**d**) 7.5%, (**e**) 15%, (**f**) 30%, and (**g**) 60%.

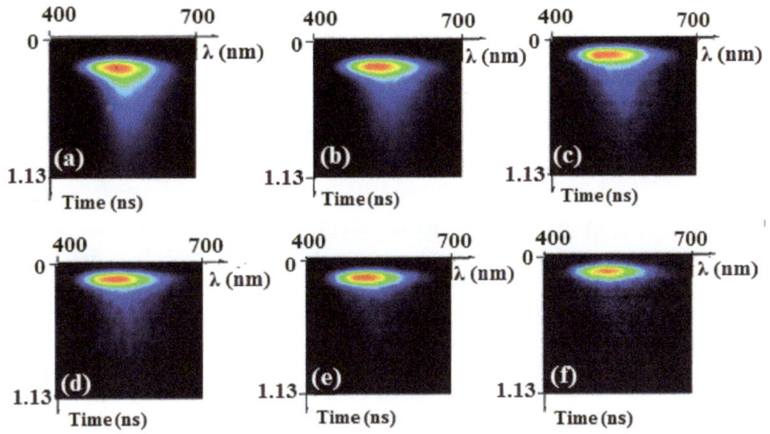

Figure 11. Contour plot of transient PL for PVK-HTs functionalized with different SWCNT concentrations: (**a**) 0%, (**b**) 1.87%, (**c**) 7.5%, (**d**) 15%, (**e**) 30%, and (**f**) 60%.

Time-resolved PL spectra are presented in Figure 12. A progressively faster decay time is obtained when increasing the SWCNT weight concentration, as in the case of the PPV/SWCNT composite [111]. This behavior confirms that SWCNTs play a role in charge separation after exciton diffusion. Indeed, a shorter lifetime is an indicator of the decrease in the distance traveled by the exciton. The increase in the SWCNT concentration will, therefore, progressively reduce these distances. If these distances become lower than the intrinsic diffusion length, charge transfer from the organic matrix to CNTs can occur. This transfer leads to a separation of electron–hole pairs rather than their recombination, because of which there is a progressive quenching of stationary PL.

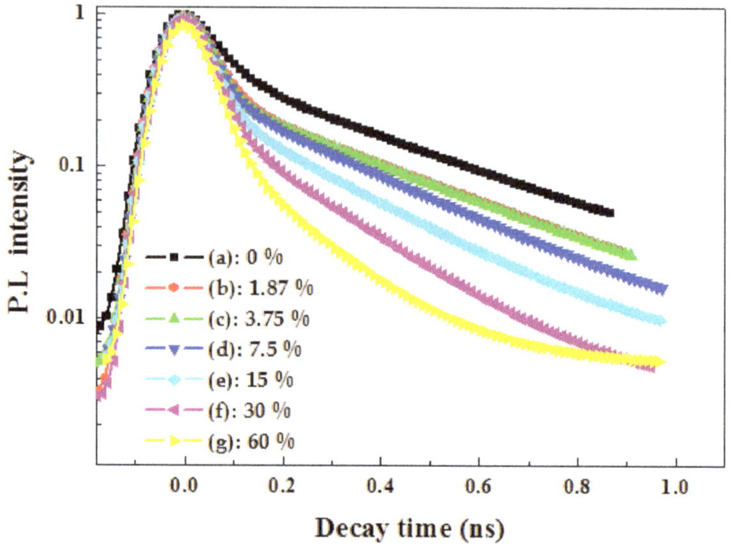

Figure 12. Time-resolved PL decays of PVK-HTs functionalized with SWCNTs at different concentrations: (**a**) 0%, (**b**) x = 1.87%, (**c**) 3.75%, (**d**) 7.5%, (**e**) 15%, (**f**) 30%, and (**g**) 60%.

6. PANI/SWCNT Nanocomposites

Following the same procedure, PANI, known by a number of oxidation degrees, is also functionalized with CNTs. The starting material, the emeraldine base (PANIEB purity 99.99%; macromolecular mass Mn > 15,000), is first doped with sulfonic acid and then functionalized with SWCNTs in the dimethylformamide (DMF) solvent [116]. The sulfonic acid doping procedure used in this case is described by MacDiarmid et al. [117].

6.1. Vibrational Study

The normalized FTIR spectra of Figure 13(1), when referring to the already published vibrational studies, reveal that concentrations lower than 2.10% cannot significantly induce apparent changes [118,119]. Starting from 2.1%, the linking of both components is evidenced by the appearance of new features at 1064, 1255, 1409, and 1438 cm^{-1}. Then, the apparent shift of both modes at 1124 and 619 cm^{-1}, the disappearance of the band at 1022 cm^{-1}, and the enhancement of bands in the range of 1350–1700 cm^{-1} reflect the new molecular arrangement induced by covalent bonding [120,121].

The weak band located at 1022 cm^{-1} ascribed to the C–H deformation disappears, and a new band appears at 1062 cm^{-1}. The bands situated in the range from 1350 to 1700 cm^{-1} show a better intensity increase, supporting the increase in aromaticity in the resulting compound. Two other new bands appear, one at 1409 cm^{-1} and one at 1438 cm^{-1}, and confirm covalent binding between both components via a C–N link [121].

A theoretical calculation including bond length, atomic charge, and spin density lead to the conclusion that nitrogen atoms are the most probable grafting sites, as in the case of similar woks [19,119]. In Figure 13(2), we illustrate the model structure that has been used for DFT calculation. Noncovalent bonding can also be present as a consequence of the strong van der Walls forces that inhibit the homogeneous dispersion process [122,123].

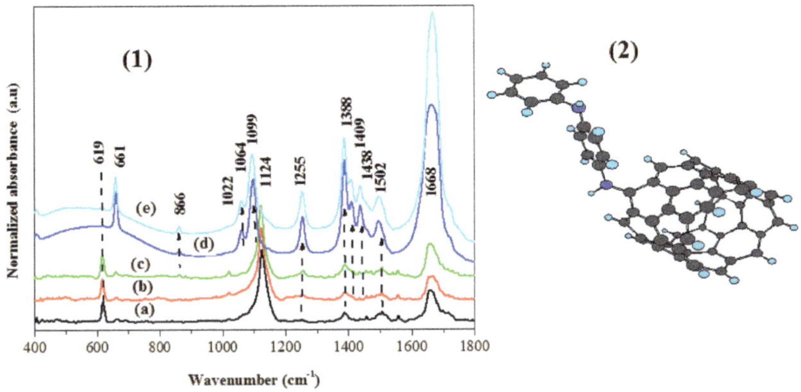

Figure 13. (**1**) FTIR spectra of PANI/SWCNT nanocomposites with various SWCNT concentrations: (**a**) 0%, (**b**) 0.71%, (**c**) 1.16%, (**d**) 2.10%, and (**e**) 5.00%. (**2**) Model structure of the nanocomposite used for DFT calculation [19].

6.2. Optical Property Changes

The modeling structure proposed above has been supported by theoretical calculation of OA spectra (Figure 14). The same transitions are found either experimentally or theoretically, for which it is noted that there is a diminution of band gap after either doping or adding SWCNTs. The energy gap of PANI in its neutral state is 3.75 eV; it decreases when doped and reaches 2.63 eV. A new band appears at −535 nm on adding SWCNTs, and the gap is 1.7 eV. This new band corresponds to the charge transfer where electron–hole pairs appear in the organic matrix, and then it is transferred to the SWCNTs by the diffusion process, leading to enhanced photovoltaic performances [18,96,124].

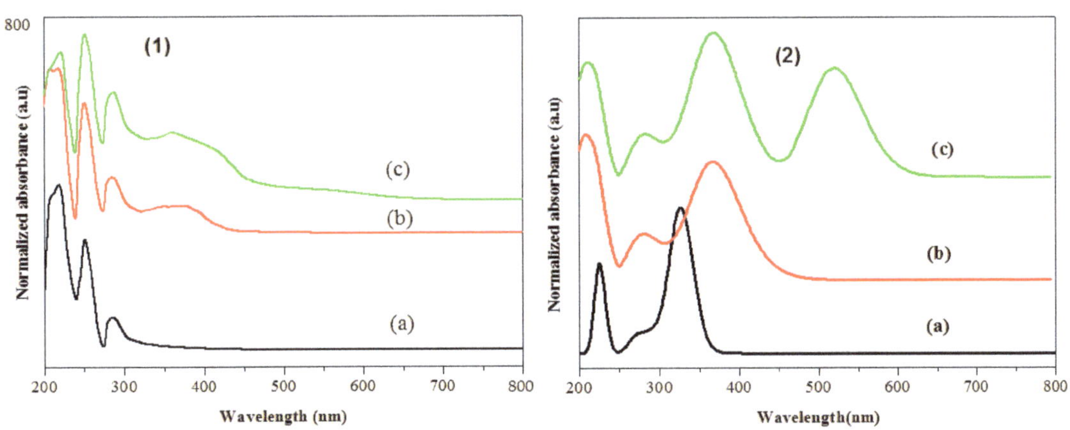

Figure 14. OA spectra of PANI in various states: (**1**) experimental and (**2**) theoretical; (**a**) neutral, (**b**) doped, and (**c**) the nanocomposite.

6.3. Electronic Study

The study on SWCNT concentration leads to the conclusion that localized states are created in the band gap with the energy of 1.62 eV and with the band width of almost 80 meV (Urbach energy) [121]. This donor–acceptor charge transfer is more elucidated by contour plots of molecular orbitals (MOs), shown in Figure 15(1) [125]. HOMO and LUMO state densities are, respectively, localized on donor and acceptor moieties [126].

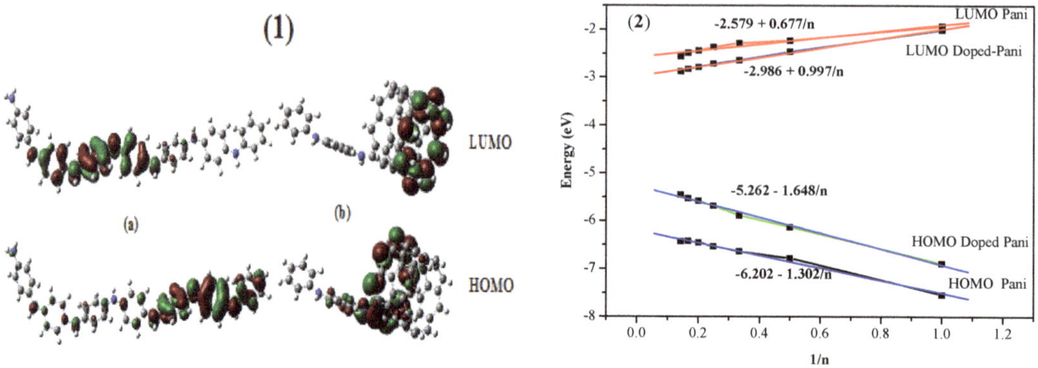

Figure 15. (**1**) HOMO and LUMO orbitals: doped PANI (**a**) and nanocomposite (**b**). (**2**) Effect of chain length on the HOMO–LUMO energy levels of PANI in pristine and doped states.

The ability of the resulting material to improve photovoltaic conversion is checked from its electronic structure (Figure 15(2)). Indeed, HOMO–LUMO energy levels are calculated from the optimization of model structures of pristine and doped states for different chain lengths [127].

The HOMO–LUMO energy levels for an infinite chain length can, therefore, be evidenced following Aleman et al. [127], where the theoretical energy gaps are 3.58 and 2.31 eV, respectively, for neutral and doped states. Other parameters can be evaluated, such as the PCE and the open-circuit voltage (V_{OC}). From the individual theoretical electronic structures (donor and acceptor), PCE can be evaluated using the Scharber diagram. Otherwise, the V_{OC} can be calculated according to Equation (1) based on the HOMO and LUMO energy levels of, respectively, donor and acceptor materials [128–131]:

$$V_{OC}(V) = \frac{1}{e}\left|E(HOMO)^{donor}(eV)\right| - \left|E(LUMO)^{acceptor}(eV)\right| - 0.3\ (V) \quad (1)$$

The choice of metal electrode is strictly based on the values of HOMO and LUMO energy levels for doped PANI and CNTs. It is found that Al and ITO are the most appropriate for the device architecture (Figure 16(1)). Using Equation (1), the V_{oc} is estimated at nearly 1.25 V, and using the Scharber diagram, PCE is estimated at 4–5% (Figure 16(2)). This is in good accordance with our previous PL results showing both the quenching effect and the total absence of luminescence in the visible region [116].

6.4. Effect of Annealing Treatment on the PANI–SWCNT Composite

Vibrational and optical behaviors of the hybrid material after annealing were analyzed in a previous study [132]. Interpretation of changes in FTIR spectra has shown the grafting and self-cross-linking of PANI moieties [133,134]. More details of the molecular structure are evidenced by the splitting of the band at 1126 cm^{-1} after annealing (Figure 17(1)). These chemical environments could not be the consequence of CNT insertion (the same line shape appears with and without CNTs). These additional bands are typically coherent with self-cross-linking (Figure 17(2)), such as benzene ring para- and ortho-link, to nitrogen atoms [132].

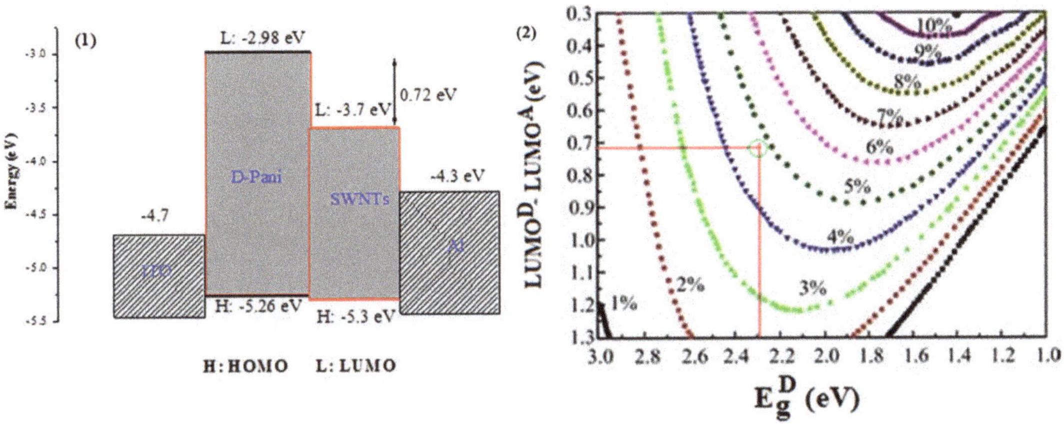

Figure 16. (**1**) Electrode choice for the photovoltaic device. (**2**) Estimation of PCE from the Scharber diagram [19].

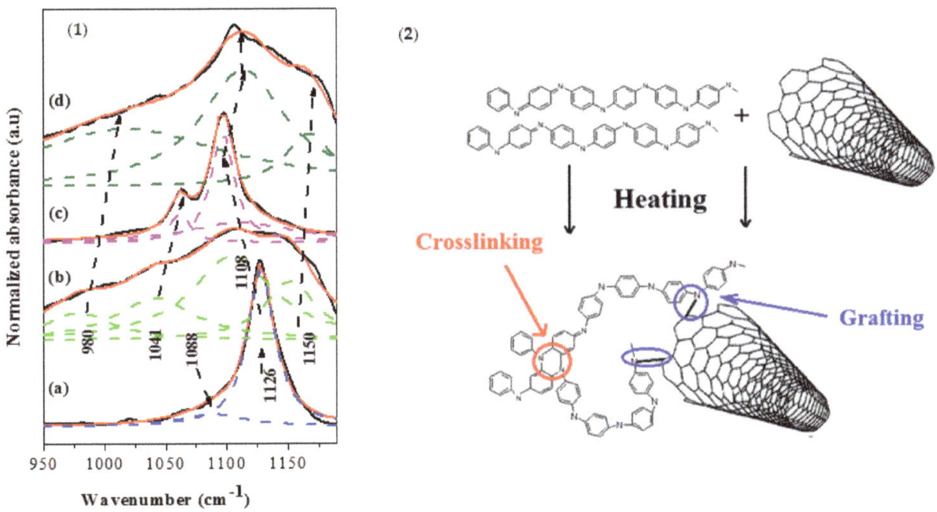

Figure 17. (**1**) Splitting of the FTIR band at 1126 cm^{-1} under annealing: pure PANI (**a**), annealed PANI (**b**), not-annealed composite (**c**), and annealed composite (**d**). (**2**) Structure of the nanocomposite obtained after annealing at 393 K.

After annealing at 393 K, new absorption bands are created at λ_{max} = 570 nm and at λ_{max} = 792 nm, respectively, for doped and functionalized states [132]. These new optical features support the presence of a quinoidal ring and agree well with the self-cross-linking process [135,136]. The direct and indirect band gaps are evaluated from absorbance coefficient data using the Tauc method [137,138]. The direct and indirect energy gaps (Eg) are calculated from the linear dependence of the plot of $(\alpha h\nu)^{1/n}$ versus energy (Figure 18) [139–141].

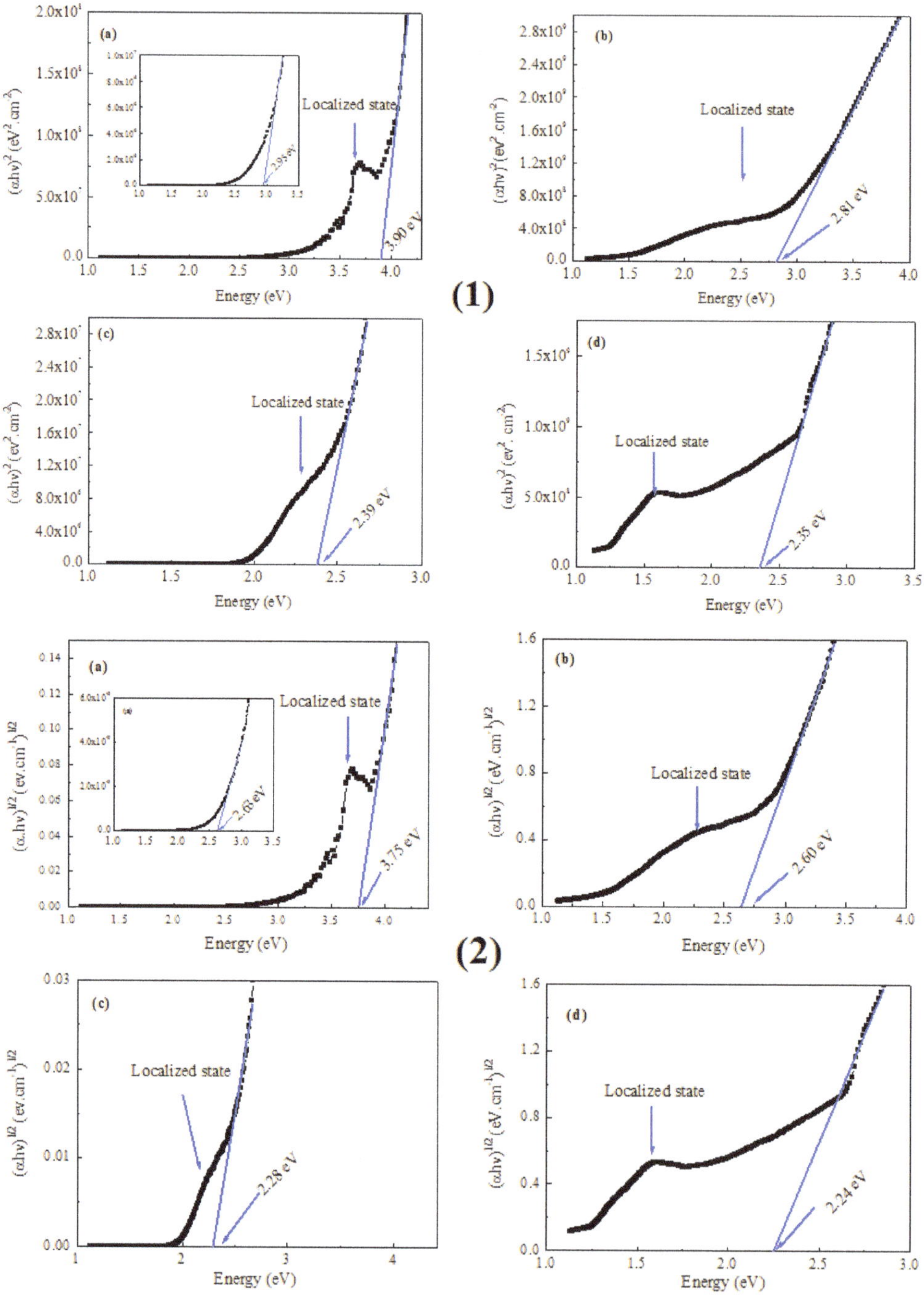

Figure 18. Evaluation of direct (**1**) and indirect (**2**) band gap for doped PANI in the not-annealed (**a**) and annealed (**b**) states and those of the doped PANI/SWCNT composite in the not-annealed (**c**) and annealed (**d**) states. The inset of (**a**) shows the threshold energy of localized states created by doping.

As previously reported, both direct and indirect transition processes dominate optical transitions since direct and indirect energy gaps are nearly similar and the variation ($\Delta E = (E_{Dir} - E_{Ind})/E_{Ind}$) is in the range from 4% to 8% [138]. The value of the transition probability index n has been also calculated to evaluate the weight of optical transition mode (values vary from 0.82 to 1.24).

The skin depth variation as a function of incident photon energy is one of the most relevant parameters reflecting the importance of adding CNTs (Figure 19).

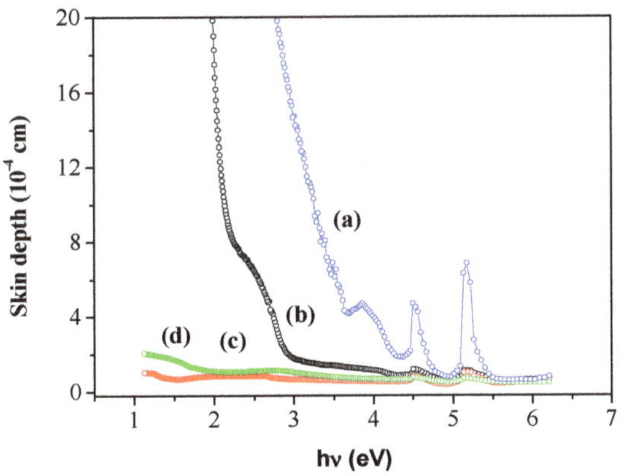

Figure 19. Skin depths of doped PANI: not annealed (**a**) and annealed (**b**); skin depths of doped PANI functionalized with CNTs: not annealed (**c**) and annealed (**d**).

When SWCNTs are added, the skin depth decreases; it decreases further after annealing, indicating the reduction in the transparency volume. This result demonstrates that the addition of CNTs leads to a better interfacial strength, as proved in a previous thermal study [142]. At a higher photon energy, the two picks of skin depth are the signature of plasma frequency, resulting from overlapping between the frequency of electron oscillation and the incident electromagnetic radiation (the wave is stagnant).

The variation in the extinction coefficient versus energy is another indicator of the change in optical properties after the addition of CNTs in both annealed and not-annealed states (Figure 20).

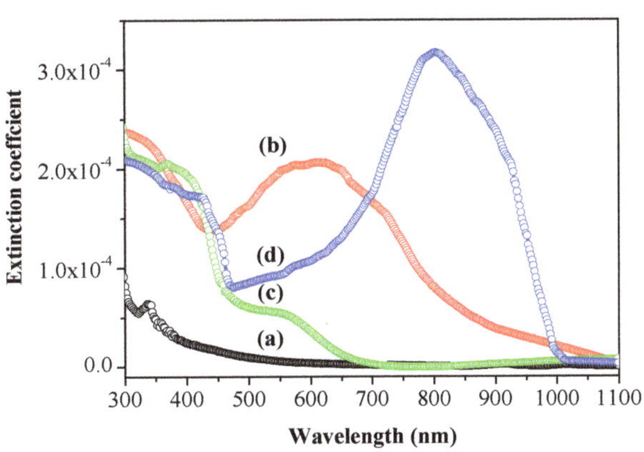

Figure 20. Extinction coefficients of doped PANI: not annealed (**a**) and annealed (**b**). Extinction coefficients of doped PANI functionalized with CNTs: not annealed (**c**) and annealed (**d**).

On the red side of the spectra, annealing improves the extinction coefficient. For the doped state of PANI, thermal treatment leads to a change in the oxidation degree and localized states can overlap with the LUMO level. The extinction coefficient may be improved as a consequence of the new energetic configuration. For the nanocomposite, a net amelioration in the spectral range from 600 to 850 nm is observed after annealing. Indeed, good SWCNT dispersion leads to a covalent bonding, as previously reported [46]. If the overlapping between diffusion length and nanoparticles repartition is satisfied, charge transfer from the doped matrix to CNTs can be established. From Tauc formalism, the calculation of band width of the corresponding localized states (E_U) demonstrates the decrease in the latter when annealing, probably due to the already discussed self-cross-linking effects [133,134,143]. Figure 21 represents all changes in the electronic structure after annealing and adding CNTs.

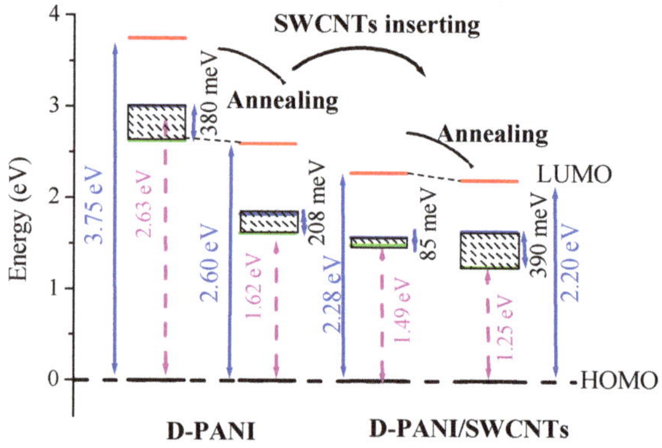

Figure 21. Change in optical parameters under doping and annealing.

For the nanocomposite, the E_U (localized state band width) is nearly 85 meV and 390 eV for not-annealed and annealed states, respectively, involving a charge transfer from the organic matrix (self-cross-linked) to CNTs, as previously reported [21]. The large value after annealing is the consequence of well-dispersed nanoparticles, involving good compatibility between the diffusion length and CNT repartition. A microscopic study mentions that bundled CNTs before annealing became relatively dispersed and homogeneously distributed, with the repartition size varying from 3 to 8 nm [132]. At the molecular scale, annealing leads to partially dispersed and oriented SWCNTs in cross-linked PANI. Electron–hole pairs, which constitute the exciton pairs, are born in the PANI matrix and then migrate by diffusion [20,144]. From the Scharber diagram, the resulting electronic structure presents a band gap suitable for a relatively higher conversion efficiency and is qualified by good compatibility with the solar spectrum [145].

7. Conclusions

A study of the structural, optical, and electronic properties of diverse polymers containing nitrogen atoms was performed. Some parameters are fundamental to achieving better dispersion of CNTs in the organic matrix, namely the choice of the solvent and thermal treatment. Short oligo-N-vinylcarbazole can be spontaneously functionalized with SWCNTs when chloroform is the medium. However, moderate thermal treatment is needed when chlorobenzene is the medium. Though the complete isolation of SWNTs can be achieved, chlorobenzene is more efficient in the SWCNT dispersion process. Correlation of experimental and theoretical study (DFT) in both cases demonstrates that there is a grafting of vinylidene groups to the nanotube side wall. For PVK-P3HT, a good dispersion state is obtained even at a higher SWCNT concentration, reaching 60% of CNTs, and the grafting is limited to the hexylthiophene sequences. The conservation of the hexylthiophene molecular structure leads to excellent properties for photovoltaic conversion, such as compatibility with the solar spectrum and the effectiveness of the dynamics of the excited states.

PANI itself is an insulating material, because of which its functionalization with CNTs was preceded by acid doping. Without thermal treatment, PANI is linked to SWCNTs via nitrogen atoms, leading to some structural and optical changes. A decrease in the optical gap of 2.28 eV gives rise to good compatibility with the solar spectrum. Moreover, donor–acceptor charge transfer creates a localized state within the band gap, with an onset energy of 1.62 eV. The modeling of a prototype bulk hetero-nanojunction photovoltaic device gives a typical open-circuit voltage of 1.25 V and a PCE of 4~5%. Annealing induces the self-cross-linking of PANI in its doped state and the dispersion of CNTs in the organic matrix. Otherwise, thermal treatment influences the electronic structure, leading to a band gap of 2.20 eV and localized states having a bandwidth of 390 meV. Moreover, it was found that skin depth and extinction coefficient variations are closely related to a molecular arrangement induced by annealing, namely self-cross-linking. As a consequence of the number of oxidation degrees of PANI, the corresponding interpenetrating network (doped PANI/CNTs) exhibits excellent compatibility with the solar spectrum and shows good coherence with the diffusion length, making it a good candidate for the active layer in organic photovoltaic devices.

As a consequence of the above-mentioned structural changes, adding CNTs to the polymer matrix leads to a broader absorption spectrum, PL quenching, and a shorter excited lifetime. In fact, these optical criteria are coherent with charge transfer from the organic matrix to the CNTs due to the interpenetrating P–N nanojunctions.

Author Contributions: Conceptualization, B.Z.; methodology, B.Z., N.S. and M.G.A.; software, B.Z.; formal analysis, B.Z.; investigation B.Z.; resources, B.Z., A.G.A. and S.D.A.; data curation, B.Z.; writing—original draft preparation, B.Z.; writing—review and editing, B.Z., N.S. and A.G.A.; visualization, N.S.; supervision, B.Z.; project administration, B.Z.; funding acquisition, B.Z., M.G.A. and N.S. All authors have read and agreed to the published version of the manuscript.

Funding: This research was funded by the Deputyship for Research & Innovation, Ministry of Education, Saudi Arabia, through project number IFP-2021-109.

Institutional Review Board Statement: Not applicable.

Informed Consent Statement: Not applicable.

Data Availability Statement: The data supporting the reported results of this study will be made available by the authors upon request.

Acknowledgments: The authors extend their appreciation to the Deputyship for Research & Innovation, Ministry of Education, Saudi Arabia, for funding this research work through project number IFP-2021-109.

Conflicts of Interest: The authors declare no conflict of interest.

References

1. Burroughes, J.H.; Bradley, D.D.C.; Brown, A.R.; Maorks, R.N.; Mackey, K.; Friend, R.H.; Burn, P.L.; Holmes, A.B. Electroluminescence on conjugated polymer. *Nature* **1990**, *397*, 547–569.
2. Saxena, V.; Malhotra, B.D. Prospects of conducting polymers in molecular electronics. *Curr. Appl. Phys.* **2003**, *3*, 293–305. [CrossRef]
3. Han, Y.-K.; Chang, M.-Y.; Ho, K.-S.; Hsieh, T.-H.; Tsai, J.-L.; Huang, P.-C. Electrochemically deposited nano polyaniline films as hole transporting layers in organic solar cells. *Solar Energy Mater. Sol. Cells* **2014**, *128*, 198–203. [CrossRef]
4. Namsheer, K.; Rout, C.S. Conducting polymers: A comprehensive review on recent advances in synthesis, properties and applications. *RSC Adv.* **2021**, *11*, 5659–5697. [CrossRef]
5. Ahllskog, M.; Menon, R.; Heeger, A.J.; Noguchi, T.; Ohnishi, T. Metal-insulator transition in oriented poly(p-phenylenevinylene). *Phys. Rev. B* **1997**, *55*, 6777–6787. [CrossRef]
6. Wang, G.; Morrin, A.; Li, M.; Liu, N.; Luo, X. Nanomaterial-doped conducting polymers for electrochemical sensors and biosensors. *J. Mater. Chem. B* **2018**, *6*, 4173–4190. [CrossRef]
7. Heeger, A.J. Nobel Prize 2000 lecture: Semiconducting and metallic polymers: The fourth generation of polymeric materials. *Curr. Appl. Phys.* **2001**, *1*, 247–267. [CrossRef]

8. Wu, W.; Li, F.; Nie, C.; Wu, J.; Chen, W.; Wu, C.; Guo, T. Improved performance of flexible white hybrid light emitting diodes by adjusting quantum dots distribution in polymer matrix. *Vacuum* **2015**, *111*, 1–4. [CrossRef]
9. Yuanhang, Y.; Hong, Z. Water-induced polymer swelling and its application in soft electronics. *Appl. Surf. Sci.* **2022**, *577*, 151895. [CrossRef]
10. Razykov, T.M.; Ferekides, C.S.; Morel, D.; Stefanakos, E.; Ullal, H.S.; Upadhyaya, H.M. Solar photovoltaic electricity: Current status and future prospects. *Solar Energy* **2011**, *85*, 1580–1608. [CrossRef]
11. Lizin, S.; Passel, S.V.; Schepper, E.D.; Vranken, L. The future of organic photovoltaic solar cells as a direct power source for consumer electronics. *Sol. Energy Mater. Sol. Cells* **2012**, *103*, 1–10. [CrossRef]
12. Deng, P.; Lei, Y.; Zheng, X.; Li, S.; Wu, J.; Zhu, F.; Ong, B.S.; Zhang, Q. Polymer based on benzothiadiazole-bridged bis-isoindigo for organic field-effect transistor applications. *Dyes Pigm.* **2016**, *125*, 407–413. [CrossRef]
13. Tehrani, Z.; Korochkina, T.; Govindarajan, S.; Thomas, D.J.; Mahony, J.O.; Kettle, J.; Claypole, T.C.; Gethin, D.T. Ultra-thin flexible screen printed rechargeable polymer battery for wearable electronic applications. *Org. Electron.* **2015**, *26*, 386–394. [CrossRef]
14. Moaseri, E.; Karimi, M.; Baniadam, M.; Maghreb, M. Improvements in mechanical properties of multi-walled carbon nanotube-reinforced epoxy composites through novel magnetic-assisted method for alignment of carbon nanotubes. *Compos. A Appl. Sci. Manuf.* **2014**, *64*, 228–233. [CrossRef]
15. Mulligan, C.J.; Bilen, C.; Zhou, X.; Belcher, W.J.; Dastoor, P.C. Levelised cost of electricity for organic photovoltaics. *Sol. Energy Mater. Sol. Cells* **2015**, *133*, 26–31. [CrossRef]
16. Subramanyam, B.V.R.S.; Mahakul, P.C.; Sa, K.; Alam, I.; Das, S.; Subudhi, S.; Mandal, M.; Patr, S.; Mahanandi, P. Applications of carbon nanotubes in different layers of P3HT: PCBM bulk heterojunction organic photovoltaic cells. *Mater. Today Proc.* **2021**, *39*, 1862–1865. [CrossRef]
17. Francis, X.P.A.; Benoy, M.D.; Stephen, S.K.; Varghese, T. Enhanced electrical properties of polyaniline carbon nanotube composites: Analysis of temperature dependence of electrical conductivity using variable range hopping and fluctuation induced tunneling models. *J. Solid State Chem.* **2021**, *300*, 122232. [CrossRef]
18. Zaidi, B.; Bouzayen, N.; Znaidia, S.; Mbarek, M.; Massuyeau, F.; Faulques, E.; Gautron, E.; Wery, J.; Duvail, J.L.; Ghedira, M.; et al. Dynamic properties of the excited states of oligo-N-vinylcarbazole functionalized with single walled carbon nanotubes. *J. Mol. Struct.* **2013**, *1039*, 46–50. [CrossRef]
19. Saoudi, M.; Ajjel, R.; Zaidi, B. Experimental and theoretical study on the charge transfer between polyaniline and single walled carbon nanotube. *J. Mater. Environ. Sci.* **2016**, *7*, 4435–4447.
20. Saaidia, A.; Saidani, M.A.; Romdhane, S.; BenFredj, A.; Egbec, D.A.M.; Tekin, E.; Bouchriha, H. Morphology-dependent exciton diffusion length in PPE-PPVs thin films as revealed by a Forster mechanism based-study. *Synth. Met.* **2017**, *226*, 177–182. [CrossRef]
21. Zaidi, B.; Bouzayen, N.; Wéry, J.; Alimi, K. Grafting of oligo-N-vinyl carbazole on single walled carbon nanotubes. *J. Mol. Str.* **2010**, *97*, 71–80. [CrossRef]
22. Ferguson, A.J.; Blackburn, J.L.; Kopidakis, N. Fullerene and carbon nanotubes as acceptor materials in organic photovoltaic. *Mater. Lett.* **2013**, *90*, 115–125. [CrossRef]
23. Gao, G.; Yu, J.; Hummelen, J.C.; Wudl, F.; Heeger, A.J. Polymer Photovoltaic Cells: Enhanced Efficiencies via a Network of Internal Donor-Acceptor Heterojunctions. *Science* **1995**, *270*, 1789–1791. [CrossRef]
24. Janssen, R.A.J.; Hummelen, J.C.; Sariciftci, N.S. Polymer–Fullerene Bulk Heterojunction Solar Cells. *MRS Bull.* **2005**, *30*, 33–36. [CrossRef]
25. Sun, Y.; Zhang, W.; Chi, H.J.; Liu, Y.; Hou, C.L.; Fang, D. Recent development of graphene materials applied in polymer solar cell, Renew. *Sust. Energ. Rev.* **2015**, *43*, 973–980. [CrossRef]
26. Neeraj, K.; Pushpendra, S. Review of next generation photovoltaic solar cell technology and comparative materialistic development. *Mater. Today Proc.* 2021; *in press*. [CrossRef]
27. Anindya, S.; Nibedita, S.; Gautam, M. Fabrication and Characterization of Flexible Semi-conducting Nanocomposite Polymer. *Ref. Modul. Mater. Sci. Mater. Eng.* 2022; *in press*. [CrossRef]
28. Huang, Y.Y.; Terentjev, E.M. Dispersion of Carbon Nanotubes: Mixing, Sonication Stabilization, and Composite Properties, Polymers. *Polymers* **2012**, *4*, 275–295. [CrossRef]
29. Alimi, K.; Zaidi, B.; Chemek, M. About Grafting of Single-walled Carbon Nanotubes on the Oligo-N-vinyl Carbazole and Copolymer Involving N-vinylcarbazole and Hexylthiophene. In *Carbon Nanotubes—Polymer Nanocomposites*; Intech: London, UK, 2011; pp. 300–330. [CrossRef]
30. Sandler, J.; Shaffer, M.S.P.; Prasse, T.; Bauhfer, W.; Schulte, K.; Windle, H. Development of a dispersion Process for carbon nanotubes in an epoxy matrix and the resulting electrical properties. *Polymer* **1999**, *40*, 5967–5971. [CrossRef]
31. Park, S.D.; Han, D.H.; Teng, D.; Kwon, Y. Rheological properties and dispersion of multi-walled Carbon nanotube (MWCNT) in polystyrene matrix. *Curr. Appl. Phys.* **2008**, *8*, 482–485. [CrossRef]
32. Chen, G.X.; Li, Y.; Shimizu, H. Ultrahigh-shear processing for the preparation of polymer/carbon nanotube composites. *Carbon* **2007**, *45*, 2334–2340. [CrossRef]
33. Moradi, O.; Yari, M.; Zare, K.; Mizra, B.; Najafi, F. A review of chemistry principles and reactions. *Fuller. Nanotub. Carbon Nanostructures* **2012**, *20*, 138–151. [CrossRef]

34. Liu, J.Q.; Xiao, T.; Liao, K.; Wu, P. Interfacial design of carbon nanotube polymer composites: A hybrid system of noncovalent and covalent functionalizations. *Nanotechnology* **2012**, *18*, 165701. [CrossRef]
35. Viswanathan, G.; Chakrapani, N.; Yang, H.; Wei, B.; Chung, H.; Cho, K.; Chang, Y.R.; Ajayan, P.M. Single-step in situ synthesis of polymer-grafted single-wall nanotube composites. *J. Am. Chem. Soc.* **2003**, *125*, 9258–9259. [CrossRef] [PubMed]
36. Holzinger, M.; Steinmetz, J.; Samaille, D.; Glerup, M.; Paillet, M.; Bernier, P.; Ley, L.; Graupner, R. Cycloaddition for cross-linking SWCNTs. *Carbon* **2004**, *42*, 941–947. [CrossRef]
37. Peng, H.; Reverdy, P.; Khabashesku, V.N.; Margrave, J.L. Side wall functionalization of single- walled carbon nanotubes with organic peroxides. *Chem. Commun.* **2003**, *3*, 362–365. [CrossRef]
38. Martin, R.; Céspedes-Guirao, F.J.; Miguel, M.D.; Lazaro, F.F.; García, H.; Sastre, S.A. Single- and multi-walled carbon nanotubes covalently linked to Synthesis, characterization and photophysical properties. *Chem. Sci.* **2012**, *3*, 470–475. [CrossRef]
39. Andreas, H. Functionalization of Single-Walled Carbon Nanotubes. *Chem. Int. Ed.* **2012**, *41*, 1853–1859. [CrossRef]
40. Shanmugharaj, A.M.; Bae, J.H.; Nayak, R.R.; Ryu, S.H. Preparation of poly(styrene-co-acrylonitrile)-grafted multiwalled carbon nanotubes via surface-initiated atom transfer radical polymerization. *J. Polym. Sci. A Polym. Chem.* **2007**, *45*, 460–470. [CrossRef]
41. Qin, S.; Qin, D.; Ford, W.T.; Resasco, D.E.; Herrera, J.E. Functionalization of single-walled carbon nanotubes with polystyrene via grafting to and grafting from method. *Macromolecules* **2004**, *37*, 752–757. [CrossRef]
42. Vijayakumar, C.; Balan, B.; Kim, M.J.; Takeuchi, M. Noncovalent Functionalization of SWNTs with Azobenzene-Containing Polymers: Solubility, Stability, and Enhancement of Photoresponsive Properties. *ACS J. Phys. Chem. C* **2011**, *115*, 4533–4539. [CrossRef]
43. Rajarajeswari, M.; Iyakutti, K.; Kawazoe, Y. Noncovalent and radicals covalent functionalization of a (5, 0) single-walled carbon nanotube with alanine and alanine. *J. Mol. Model.* **2012**, *18*, 771–781. [CrossRef] [PubMed]
44. Kong, J.; Yenilmez, E.; Tombler, T.W.; Kim, W.; Dai, H. Quantum interference and ballistic transmission in nanotube electron waveguides. *Phys. Rev. Lett.* **2001**, *87*, 106801–106807. [CrossRef] [PubMed]
45. Deng, J.; Ding, X.; Zhang, W.; Peng, Y.; Wang, J.; Long, X.; Li, P.; Chan, A. Carbon nanotube–polyaniline hybrid materials. *Eur. Polym. J.* **2002**, *38*, 2497–2501. [CrossRef]
46. Sahoo, N.G.; Jung, Y.C.; So, H.H.; Chob, J.W. Polypyrrole Coated Carbon Nanotubes: Synthesis, Characterization, and Enhanced Electrical Properties. *Synth. Met.* **2007**, *157*, 374–379. [CrossRef]
47. Kerr, C.J.; Huang, Y.Y.; Marshall, J.E.; Terentjev, E.M. Effect of filament aspect ratio on the dielectric response of multiwalled carbon nanotube composites. *J. Appl. Phys.* **2011**, *109*, 094109. [CrossRef]
48. Pietro, W.J.; Francl, M.M.; Hehre, W.J.; Defrees, D.J.; Pople, J.A.; Binkley, J.S. Self-consistent molecular orbital methods. 24. Supplemented small split-valence basis sets for second-row elements. *J. Am. Chem. Soc.* **1982**, *104*, 5039–5048. [CrossRef]
49. Pickholz, M.; dos Santos, M.C. Interchain and correlation effects in oligothiophenes. *Synh. Met.* **1999**, *101*, 528–529. [CrossRef]
50. DiCesare, N.; Belletete, M.; Marrano, C.; Leclerc, M.; Durocher, G. Conformational Analysis (ab initio HF/3-21G*) and Optical Properties of Symmetrically Disubstituted Terthiophenes. *J. Phys. Chem. A* **1998**, *102*, 5142–5149. [CrossRef]
51. Ayachi, S.; Alimi, K.; Bouachrine, M.; Hamidi, M.; Mevellec, J.Y.; Porte, J.P.L. Spectroscopic investigations of copolymers incorporating various thiophene and phenylene monomers. *Synth. Met.* **2006**, *156*, 318–326. [CrossRef]
52. Zou, L.Y.; Ren, A.M.; Feng., J.K.; Ran, X.Q.; Liu, Y.L.; Sun, C.C. Structural, electronic, and optical properties of phenol-pyridyl boron complexes for light-emitting diodes. *Int. J. Quantum Chem.* **2009**, *109*, 1419–1429. [CrossRef]
53. Khoshkholgh, M.J.; Marsusi, F.; Abolhassani, M.R. Density functional theory investigation of opto-electronic properties of thieno [3,4-b] thiophene and benzodithiophenepolymer and derivatives and their applications in solar cell. *Spectrochim. Acta Part A* **2015**, *136*, 373–380. [CrossRef] [PubMed]
54. Zhengkun, D.; Mian, C.; Li, D.; Baojin, H.; Zhong, C.; Donghong, Y.; Meng, C.L. Effect of alkylthiolated hetero-aromatic rings on the photovoltaic performance of benzodithiophene-based polymer/fullerene solar cells. *Synt. Met.* **2021**, *276*, 116756. [CrossRef]
55. Zhang, L.; Wan, M. Self-Assembly of Polyaniline—From Nanotubes to Hollow Microspheres. *Adv. Funct. Mater.* **2003**, *13*, 815–820. [CrossRef]
56. Bonard, J.M.; Stora, T.; Salvetat, J.P.; Maier, F.; Stockli, T.; Dusch, C.; Forró, L.; Châtelain, A. Purification and size selection of carbon nanotubes. *Adv. Mater.* **1997**, *9*, 827–831. [CrossRef]
57. Hu, H.; Yu, A.; Kim, E.; Zhao, B.; Itkis, M.E.; Bekyarova, E.; Haddon, R.C. Influence of the Zeta Potential on the Dispersability and Purification of Single-Walled Carbon Nanotubes. *J. Phys. Chem. B* **2005**, *109*, 11520–11524. [CrossRef] [PubMed]
58. Wang, Y.; Gao, L.; Sun, J.; Liu, Y.; Zheng, S.; Kajiura, H. An integrated route for purification, cutting and dispersion of single-walled carbon nanotubes. *Chem. Phys. Lett.* **2006**, *432*, 205–208. [CrossRef]
59. Arnold, M.S.; Green, A.A.; Hulvat, J.F.; Stupp, S.I.; Hersam, M.C. Sorting carbon nanotubes by electronic structure using density differentiation. *Nat. Nanotechnol.* **2006**, *1*, 60–65. [CrossRef]
60. Arnold, M.S.; Stupp, S.I.; Hersam, M.C. Enrichment of single-walled carbon nanotubes by diameter. *Nano Lett.* **2005**, *5*, 713–718. [CrossRef]
61. Strano, M.S.; Dyke, C.A.; Usrey, M.L.; Barone, P.W.; Allen, M.J.; Shan, H.; Kittrell, C.; Hauge, R.H.; Tour, J.M.; Smalley, R.E. Electronic Structure Control of Single-Walled Carbon Nanotube Functionalization. *Science* **2003**, *301*, 1519–1522. [CrossRef]
62. Kyoichi, O.; Masahiko, M.; Yoshihiro, T. Analysis of the dispersion state of pitch particles in polymers for nanofiber fabrication by optical microscopy and image processing. *J. Phys. Chem. Solids* **2022**, *163*, 110585. [CrossRef]

63. Becke, A.D.J. Density-functional thermochemistry. III. The role of exact exchange. *Chem. Phys.* **1993**, *98*, 5648. [CrossRef]
64. Futaba, D.N.; Mizuno, K.; Namai, T.; Yumura, M.; Iijima, S. Water-assisted highly efficient synthesis of impurity-free single-walled carbon nanotubes. *Science* **2004**, *306*, 1362–1364. [CrossRef]
65. Meyyappan, M.; Delzeit, L.; Cassell, A.; Hash, D. Carbon nanotube growth by PECVD: A review. *Plasma Sources Sci. Technol.* **2003**, *12*, 205. [CrossRef]
66. Zhang, Y.; Iijima, S. Elastic reponse of carbon nanotube bundles to visible light. *Phys. Rev. Lett.* **1999**, *82*, 3472. [CrossRef]
67. Rarvikar, N.R.; Schadler, L.S.; Vijaaraghavan, A.; Zhao, Y.; Wei, B.; Ajayan, P.M. Synthesis and characterization of thickness-aligned carbon nanotube-polymer composite films. *Chem. Mater.* **2005**, *17*, 974–983. [CrossRef]
68. Boncel, S.; Koziol, K.K.K.; Walczak, K.Z.; Windle, A.H.; Shaffer, M.S.P. Infiltration of highly aligned carbon nanotube arrays with molten polystyrene. *Mater. Lett.* **2011**, *65*, 2229–2303. [CrossRef]
69. Chen, Y.; Yu, J. Growth direction control of aligned carbon nanotubes. *Carbon* **2005**, *43*, 3181–3194. [CrossRef]
70. Chen, X.Q.; Saito, T.; Yamada, H.; Matsushige, K. Aligning singlewall carbon nanotubes with an alternating- current electric field. *Appl. Phys. Lett.* **2001**, *78*, 3714–3716. [CrossRef]
71. Domingues, D.; Logakis, E.; Skordos, A.A. The use of an electric field in the preparation of glass fibre/epoxy composites containing carbon nanotubes. *Carbon* **2012**, *50*, 2493–2503. [CrossRef]
72. Hone, J.; Llaguno, M.C.; Nemes, N.M.; Johnson, A.T.; Fischer, J.E.; Walters, D.A. Electrical and thermal transport properties of magnetically aligned single wall carbon nanotube films. *Appl. Phys. Lett.* **2000**, *77*, 666–668. [CrossRef]
73. Choi, E.S.; Brooks, J.S.; Eaton, D.L.; Al-Haik, M.S.; Hussaini, M.Y.; Garmestani, H.; Li, D.; Dahmen, K. Enhancement of thermal and electrical properties of carbon nanotube polymer composites by magnetic field processing. *J. Appl. Physiol.* **2003**, *94*, 6034–6039. [CrossRef]
74. Dimaki, M.; Bøggild, P. Dielectrophoresis of carbon nanotubes using microelectrodes: A numerical study. *Nanotechnology* **2004**, *15*, 1095–1102. [CrossRef]
75. Zhan, H.; Chen, Y.W.; Shi, Q.Q.; Zhang, Y.; Mo, R.W.; Wang, J.N. Highly aligned and densified carbon nanotube films with superior thermal conductivity and mechanical strength. *Carbon* **2022**, *186*, 205–214. [CrossRef]
76. Monti, M.; Natali, M.; Torre, L.; Kenny, J.M. The alignment of single walled carbon nanotubes in an epoxy resin by applying a DC electric field. *Carbon* **2012**, *50*, 2453–2464. [CrossRef]
77. Zhokhavets, U.; Erb, T.; Hoppe, H.; Gobsch, G.; Sariciftci, N.S. Effect of annealing of poly(3-hexylthiophene)/fullerene bulk heterojunction composites on structural and optical properties. *Thin Solid Film* **2006**, *496*, 679–682. [CrossRef]
78. Yang, X.; Loos, J.; Veenstra, S.C.; Verhees, W.J.H.; Wienk, M.M.; Kroon, J.M.; Michels, M.A.J.; Janssen, R.A.J. Nanoscale Morphology of High-Performance Polymer Solar Cells. *Nano Lett.* **2005**, *5*, 579–583. [CrossRef]
79. Savenije, T.J.; Kroeze, J.E.; Yang, X.; Loos, J. The formation of crystalline P3HT fibrils upon annealing of a PCBM:P3HT bulk heterojunction. *Thin Solid Film.* **2006**, *511–512*, 2–6. [CrossRef]
80. Zhu, H.; Wei, J.; Wang, K.; Wu, D. Applications of carbon materials in photovoltaic solar cells. *Sol. Energy Mater. Sol. Cells* **2009**, *93*, 1461–1470. [CrossRef]
81. Yun, D.; Feng, W.; Wu, H.; Li, B.; Liu, X.; Yi, W.; Qiang, J.; Gao, S.; Yan, S. Controllable functionalization of single-wall carbon nanotubes by in situ polymerization method for organic photovoltaic device. *Synt. Met.* **2008**, *158*, 977–983. [CrossRef]
82. Derbal-Habak, H.; Bergeret, C.; Cousseau, J.; Nunz, J.M. Improving the current density Jsc of organic solar cells P3HT:PCBM by structuring the photoactive layer with functionalized SWCNTs. *Sol. Energy Mater. Sola. Cells* **2011**, *95*, 553–556. [CrossRef]
83. Jun, G.H.; Jin, S.H.; Park, S.H.; Jeon, S.; Hong, S.H. Highly dispersed carbon nanotubes in organic media for polymer: Fullerene photovoltaic devices. *Carbon* **2012**, *50*, 40–46. [CrossRef]
84. Rajiv, K.; Jitendra, K.; Amit, K.; Vikram, K.; Rama, K.; Ramadhar, S. Poly(3-hexylthiophene):Functionalized single-walled carbon nanotubes: (6,6)-phenyl-C61-butyric acidmethyl ester composites for photovoltaic cell at ambient condition. *Sol. Energy Mater. Sol. Cells* **2010**, *94*, 2386–2394. [CrossRef]
85. Shaheen, S.E.; Brabec, C.J.; Sariciftci, N.S.; Padinger, F.; Fromherzet, T.; Hummelen, J.C. 2.5% efficient organic plastic solar cells. *Appl. Phys. Lett.* **2001**, *78*, 841–843. [CrossRef]
86. Peumans, P.; Forrest, S.R. Very-high-efficiency double-heterostructure copper phthalocyanine/C60 photovoltaic cells. *Appl. Phys. Lett.* **2001**, *79*, 126–128. [CrossRef]
87. Xue, M.A.; Uchida, S.; Rand, B.P.; Forrest, S.R. 4.2% efficient organic photovoltaic cells with low series resistances. *Appl. Phys. Lett.* **2004**, *84*, 3013–3015. [CrossRef]
88. Cui, Y.; Yao, H.; Hong, L.; Zhang, T.; Tang, Y.; Lin, B.; Xian, K.; Gao, B.; An, C.; Bi, P.; et al. 17% efficiency organic photovoltaic cell with superior processability. *Nat. Sci. Rev.* **2020**, *7*, 1239–1246. [CrossRef]
89. Wu, W.; Li, J.; Liu, L.; Yanga, L.; Guo, Z.-X.; Dai, L.; Zhu, D. The photoconductivity of PVK-carbon nanotube blends. *Chem. Phys. Lett.* **2002**, *364*, 196–199. [CrossRef]
90. Bernede, J.C.; Alimi, K.; Safoula, G. Influence of annealing treatment on iodine doped poly(N-vinylcarbazole) powders. *Polym. Deg. Stab.* **1994**, *46*, 269–274. [CrossRef]
91. Napo, K.; Chand, S.; Bernede, J.C.; Safoula, G.; Alimi, K. Growth and characterization of vacuum-deposited polyvinylcarbazole (PVK) films. *J. Mat. Sci.* **1992**, *27*, 6219–6222. [CrossRef]

92. Touihri, S.; Safoula, G.; Leny, R.; Bernede, J.C. Comparison of the properties of iodine-doped poly(N-vinylcarbazole)(PVK) thin films obtained by evaporation of pure powder followed by iodine post-deposition doping and iodine pre-doped powder. *Thin Solid Film.* **1997**, *304*, 16–23. [CrossRef]
93. Lee, C.; Yang, W.; Parr, R.G. Development of the Colle-Salvetti correlation-energy formula into a functional of the electron density. *Phys. Rev. B* **1988**, *37*, 785–789. [CrossRef] [PubMed]
94. Massuyeau, F.; Faulques, E.; Lefrant, S.; Majdoub, M.; Ghedira, M.; Alimi, K.; Wéry, J. Photoluminescence properties of new PPV derivatives. *J. Lumin.* **2011**, *131*, 1541–1544. [CrossRef]
95. Mbarek, M.; Zaidi, B.; Alimi, K. Theoretical study of the alkoxyls groups effect on PPV-ether excited states, a relationship with femtosecond decay. *Spectrochim. Acta A Mol. Biomol. Spectrosc.* **2012**, *88*, 23–30. [CrossRef]
96. Zaidi, B.; Bouzayen, N.; Wéry, J.; Alimi, K. Annealing treatment and carbon nanotubes concentration effects on the optical and vibrational properties of single walled carbon nanotubes functionalized with short oligo-N-vinyl carbazole. *Mter. Chem. Phys.* **2011**, *126*, 417–423. [CrossRef]
97. Baibarac, M.; Cantu, M.L.; Sol, J.O.; Baltog, I.; Pastor, N.C.; Romero, P.G. Poly(N-vinyl carbazole) and carbon nanotubes based composites and their application to rechargeable lithium batteries. *Compos. Sci. Technol.* **2007**, *67*, 2556–2563. [CrossRef]
98. Wang, C.; Guo, Z.X.; Fu, S.; Wu, W.; Zhu, D. Polymers containing fullerene or carbon nanotube structures. *Prog. Polym. Sci.* **2004**, *29*, 1079–1141. [CrossRef]
99. Yang, L.; Feng, J.-K.; Ren, A.-M.; Sun, J.-Z. The electronic structure and optical properties of carbazole-based conjugated oligomers and polymers, a theoretical investigation. *Polymers* **2006**, *47*, 1397–1404. [CrossRef]
100. Baibarac, M.; Baltog, I.; Lefrant, S. Raman spectroscopic evidence for interfacial interactions in poly(bithiophene)/single-walled carbon nanotube composites. *Carbon* **2009**, *47*, 1389–1398. [CrossRef]
101. Eklund, P.C.; Holden, J.M.; Jishi, R.A. Vibrational modes of carbon nanotubes; spectroscopy and theory. *Carbon* **1995**, *33*, 959–972. [CrossRef]
102. Huang, J.E.; Li, X.H.; Xu, J.C.; Li, H.L. Well-dispersed single-walled carbon nanotube/polyaniline composite films. *Carbon* **2003**, *41*, 2731–2736. [CrossRef]
103. Lukin, L.V. Efficiency of exciton dissociation at the interface between a conjugated polymer and an electron acceptor with consideration for a two-dimensional arrangement of interfacial dipoles. *Chem. Phys.* **2021**, *551*, 111327. [CrossRef]
104. Baibarac, M.; Baltog, I.; Lefrant, S.; Mevellec, J.Y.; Bucur, C. Vibrational and photoluminescence properties of the polystyrene functionalized single-walled carbon nanotubes. *Diam. Relat. Mater.* **2008**, *17*, 1380–1388. [CrossRef]
105. Ago, H.; Petritsch, K.; Shaffer, M.S.P.; Windle, A.H.; Friend, R.H. Composites of carbon nanotubes and conjugated polymers for photovoltaic devices. *Adv. Mater.* **1999**, *11*, 128–1285. [CrossRef]
106. Mulazzi, E.; Perego, R.; Aarab, H.; Mihut, L.; Lefrant, S.; Faulques, E.; Wéry, J. Photoconductivity and optical properties in composites of poly(paraphenylene vinylene) and single-walled carbon nanotubes. *Phys. Rev. B* **2004**, *70*, 155206–155214. [CrossRef]
107. Lin, Y.-Y.; Chen, C.-W.; Chang, T.Y.; Lin, J.; Liu, I.S.; Su, W.-F. Exciton dissociation and migration in enhanced order conjugates polymer/nanoparticlehybrid materials. *Nanotechnology* **2006**, *17*, 1260–1263. [CrossRef]
108. Lin, Y.-T.; Zeng, T.-W.; Lai, W.-Z.; Chen, C.-W. Efficient photoinduced charge transfer in tio2 nanorod/conjugated polymer hybrid materials. *Nanotechnology* **2006**, *17*, 5781–5785. [CrossRef]
109. Aarab, H.; Baitoul, M.; Wery, J.; Almairac, R.; Lefrant, S.; Faulques, E.; Duvail, J.L.; Hamedoun, M. Electrical and optical properties of PPV and single-walled carbon nanotubes composites films. *Synth. Met.* **2005**, *155*, 63–67. [CrossRef]
110. Ho, Y.B.; In, J.; Chung, H.-K.S.; Chung, Y.K. The effects of alkyls side-chain length and shape of polyfluorenes on the photoluminescence spectra and the fluorescence lifetimes of polyflorene blends with poly(n-vinyl carbazole). *Chem. Phys. Lett.* **2004**, *393*, 197–203. [CrossRef]
111. Massuyeau, F.; Aarab, H.; Mihut, L.; Lefrant, S.; Faulques, E.; Wéry, J. Optical Properties of Poly(para-phenylenevinylene) and single-walled Carbon Nanotube composite films: Effects of conversion temperature, prucursor dilution and nanotube concentrations. *Phys. Chem. C* **2007**, *111*, 15111–15118. [CrossRef]
112. Chemek, M.; Wéry, J.; Bouachrine, M.; Paris, M.; Lefrant, S.; Alimi, K. Synthesis and characterization of novel graft copolymers of Poly(N-vinylcarbazole) and Poly(3-methylthiophene) for optoelectronic applications. *Synth. Met.* **2010**, *160*, 2306–2314. [CrossRef]
113. Craley, C.R.; Zhang, R.; Kowalewski, T.; McCullough, R.D.; Stefan, M.C. Regioregular Poly(3-exylthiophene) in a Novel Conducting Amphiphilic Block Copolymer. *Macromol. Rapid Commun.* **2009**, *30*, 11–16. [CrossRef] [PubMed]
114. McCullough, R. The Chemistry of Conducting Polythiophenes. *Adv. Mater.* **1998**, *10*, 93–116. [CrossRef]
115. Wéry, J.; Aarab, H.; Lefrant, S.; Faulques, E.; Mulazzi, E.; Perego, R. Photoexcitations in compositrs of poly(paraphenylene vinylene) and single—walled carbon nanotubes. *Phys. Rev. B* **2003**, *67*, 115202–115207. [CrossRef]
116. Saoudi, M.; Zaidi, B.; Alotaibi, A.A.; Althobaiti, M.G.; Alosime, E.M.; Ajjel, R. Polyaniline: Doping and Functionalization with Single Walled Carbon Nanotubes for Photovoltaic and Photocatalytic Application. *Polymers* **2021**, *13*, 2595. [CrossRef]
117. Chiang, J.-C.; MacDiarmid, A.G. Polyaniline: Protonic acid doping of the emeraldine form to the metallic regime. *Synth. Met.* **1986**, *13*, 193. [CrossRef]
118. Mishra, A.K.; Tandon, P. A Comparative Ab Initio and DFT Study of Polyaniline Leucoemeraldine Base and Its Oligomers. *J. Phys. Chem. B* **2009**, *113*, 14629. [CrossRef]

119. Baibarac, M.; Baltog, I.; Lefrant, S.; Mevellec, J.Y.; Chauvet, O. Polyaniline and carbon nanotubes based composites containing whole units and fragments of nanotubes. *Chem. Mater.* **2003**, *15*, 4149–4156. [CrossRef]
120. Pal, G.; Kumar, S. Modeling of carbon nanotubes and carbon nanotube–polymer composites. *Prog. Aerosp. Sci.* **2016**, *80*, 33. [CrossRef]
121. Lu, X.; Hu, Y.; Wang, L.; Guo, Q.; Chen, S.; Hou, H.; Song, Y. Macroporous Carbon/Nitrogen-doped Carbon Nanotubes/Polyaniline Nanocomposites and Their Application in Supercapacitors. *Electrochim. Acta.* **2016**, *189*, 158–165. [CrossRef]
122. Byron, P.R.; Hubert, P.; Salvetat, J.-P.; Zalamea, L. Flexural deflection as a measure of van der Waals interaction forces in the CNT array. *Compos. Sci. Technol.* **2006**, *66*, 1125. [CrossRef]
123. Schroder, E.; Hyldgaard, P. The van der Waals interactions of concentric nanotubes. *Mater. Sci. Eng. C* **2003**, *23*, 721. [CrossRef]
124. He, B.; Tang, Q.; Luo, J.; Li, Q.; Chen, X.; Cai, H. Rapid charge-transfer in polypyrrole–single wall carbon nanotube complex counter electrodes: Improved photovoltaic performances of dye-sensitized solar cells. *J. Power Sources* **2014**, *256*, 170–177. [CrossRef]
125. Goswami, M.; Ghosh, R.; Maruyama, T.; Meikap, A.K. Polyaniline/carbon nanotube/CdS quantum dot composites with enhanced optical and electrical properties. *Appl. Surf. Sci.* **2016**, *364*, 176. [CrossRef]
126. El Malki, Z.; Bouachrine, M.; Hamidi, M.; Serein-Spirau, F.; Lere-Porte, J.P.; Sotiropoulos, J.M. Theoretical study of New Donor-π-Acceptor compounds based on Carbazole, Thiophene and Benzothiadiazole for Photovoltaic application as Dyesensitized solar cells. *J. Mater. Environ. Sci.* **2016**, *7*, 324–3255.
127. Aleman, C.; Ferreir, C.A.; Torras, J.; Meneguzzi, A.; Canales, M.; Rodrigues, M.A.S.; Casanovas, J. On the molecular properties of polyaniline: A comprehensive theoretical study. *J. Polym.* **2008**, *49*, 5169. [CrossRef]
128. Gadisa, A.; Svensson, M.; Andersson, M.R.; Inganas, O. Correlation between oxidation potential and open-circuit voltage of composite solar cells based on blends of polythiophenes/fullerene derivative. *Appl. Phys. Lett.* **2004**, *84*, 1609. [CrossRef]
129. Veldman, A.; Meskers, S.C.J.; Janssen, R.A.J. The Energy of Charge-Transfer States in Electron Donor–Acceptor Blends: Insight into the Energy Losses in Organic Solar Cells. *Adv. Funct. Mater.* **2009**, *19*, 1939–1948. [CrossRef]
130. Scharber, M.C.; Sariciftci, N.S. Efficiency of bulk-heterojunction organic solar cells. *Prog. Polym. Sci.* **2013**, *38*, 1929–1940. [CrossRef]
131. Azazi, A.; Mabrouk, A.; Alimi, K. Theoretical investigation on the photophysical properties of low-band-gap copolymers for photovoltaic devices Comput. *Theor. Chem.* **2011**, *978*, 7. [CrossRef]
132. Saoudi, M.; Zaidi, B.; Ajjel, R. Correlation between microstructures and optical properties of polyaniline/single-walled carbon nanotubes composites. *Polym. Compos.* **2019**, *40*, E821–E831. [CrossRef]
133. Rozlivkova, Z.; Trchova, M.; Exnerova, M.; Stejska, J. The carbonization of granular polyaniline to produce nitrogen-containing carbon. *Synt. Met.* **2011**, *161*, 11–22. [CrossRef]
134. Liu, D.; Wang, H.; Du, P.; Liu, P. Independently double-crosslinked carbon nanotubes/polyaniline composite films as flexible and robust free-standing electrodes for high-performance supercapacitors. *Carbon* **2017**, *122*, 761–774. [CrossRef]
135. Borah, R.; Banerjee, S.; Kumar, A. Surface functionalization effects on structural, conformational, and optical properties of polyaniline nanofibers. *Synth. Met.* **2014**, *197*, 225–232. [CrossRef]
136. Verma, D.; Dutt, V. Novel microstructure in spin coated polyaniline thin films. *J. Phys. Condens. Matter.* **2007**, *19*, 186–212. [CrossRef]
137. Tauc, J. *Amorphous and Liquid Semiconductors*; Plenum: New York, NY, USA, 1974.
138. Mathew, A.M.; Predeep, P. Plasma-polymerized elastomer/conducting polymer composite: Structural and optical characterization. *Polym. Compos.* **2013**, *34*, 1091–1098. [CrossRef]
139. Kabir, H.; Rahman, M.M.; Uddin, K.M.; Bhuiya, A.H. Structural, morphological, compositional and optical studies of plasma polymerized 2-furaldehyde amorphous thin films. *Appl. Surf. Sci.* **2017**, *423*, 983–994. [CrossRef]
140. Chithralekha, P.; Subramanian, E.; Padiyan, D.P. Electrodeposition of polyaniline thin films doped with dodeca tungstophosphoric acid: Effect on annealing and vapor sensing. *Sens. Actuators B Chem.* **2007**, *122*, 274. [CrossRef]
141. Banerjee, S.; Kumar, A. Swift heavy ion irradiation induced modifications in the optical band gap and Urbach's tail in polyaniline nanofibers. *Nucl. Instrum. Methods Phys. Res. Sect. B* **2011**, *269*, 2798. [CrossRef]
142. Ji, T.; Feng, Y.Y.; Qin, M.; Feng, W. Thermal conducting properties of aligned carbon nanotubes and their polymer composites. *Compos Part A Appl. Sci. Manuf.* **2016**, *91*, 351. [CrossRef]
143. Pocas, L.C.; Travain, S.A.; Duarte, J.L.; Silva, R.A.; Giacometti, J.A.; Marletta, A. Annealing effects on conductivity and optical properties of the PAni layer in ITO/PAni/PPV+DBS/Al polymer light-emitting diodes. *J. Phys. Condens. Matter.* **2007**, *19*, 436221. [CrossRef]
144. Dimitrov, S.D.; Schroeder, B.C.; Nielsen, C.B.; Bronstein, H.; Fei, Z.; McCulloch, I.; Heeney, M.; Durrant, J.R. Singlet Exciton Lifetimes in Conjugated Polymer Films for Organic Solar Cells. *Polymers* **2016**, *8*, 14. [CrossRef] [PubMed]
145. Scharber, M.C.; Mühlbacher, D.; Koppe, M.; Denk, P.; Waldauf, C.; Heeger, A.J.; Brabec, C.J. Design Rules for Donors in Bulk-Heterojunction Solar Cells—Towards 10% Energy-Conversion Efficiency. *Adv. Mater.* **2006**, *18*, 789–794. [CrossRef]

polymers

MDPI

Review

Bioactive Graphene Quantum Dots Based Polymer Composite for Biomedical Applications

Seyyed Mojtaba Mousavi [1], Seyyed Alireza Hashemi [2], Masoomeh Yari Kalashgrani [3], Navid Omidifar [4], Sonia Bahrani [3], Neralla Vijayakameswara Rao [1], Aziz Babapoor [5], Ahmad Gholami [6,*] and Wei-Hung Chiang [1,*]

1 Department of Chemical Engineering, National Taiwan University of Science and Technology, Taipei City 106335, Taiwan; kempo.smm@gmail.com (S.M.M.); vijayrao@mail.ntust.edu.tw (N.V.R.)
2 Nanomaterials and Polymer Nanocomposites Laboratory, School of Engineering, University of British Columbia, Kelowna, BC V1V 1V7, Canada; s.a.hashemi0@gmail.com
3 Biotechnology Research Center, Shiraz University of Medical Sciences, Shiraz 71468-64685, Iran; masoomeh.yari.72@gmail.com (M.Y.K.); S.bahrani22@gmail.com (S.B.)
4 Department of Pathology, Shiraz University of Medical Sciences, Shiraz 71468-64685, Iran; omidifar@gmail.com
5 Department of Chemical Engineering, University of Mohaghegh Ardabil, Ardabil 56199-11367, Iran; Babapoor@uma.ac.ir
6 Pharmaceutical Sciences Research Center, Shiraz University of Medical Sciences, Shiraz 71468-64685, Iran
* Correspondence: gholami@sums.ac.ir (A.G.); whchiang@mail.ntust.edu.tw (W.-H.C.)

Citation: Mousavi, S.M.; Hashemi, S.A.; Kalashgrani, M.Y.; Omidifar, N.; Bahrani, S.; Vijayakameswara Rao, N.; Babapoor, A.; Gholami, A.; Chiang, W.-H. Bioactive Graphene Quantum Dots Based Polymer Composite for Biomedical Applications. *Polymers* **2022**, *14*, 617. https://doi.org/10.3390/polym14030617

Academic Editors: Beata Podkościelna and Andrzej Puszka

Received: 1 January 2022
Accepted: 2 February 2022
Published: 5 February 2022

Abstract: Today, nanomedicine seeks to develop new polymer composites to overcome current problems in diagnosing and treating common diseases, especially cancer. To achieve this goal, research on polymer composites has expanded so that, in recent years, interdisciplinary collaborations between scientists have been expanding day by day. The synthesis and applications of bioactive GQD-based polymer composites have been investigated in medicine and biomedicine. Bioactive GQD-based polymer composites have a special role as drug delivery carriers. Bioactive GQDs are one of the newcomers to the list of carbon-based nanomaterials. In addition, the antibacterial and anti-diabetic potentials of bioactive GQDs are already known. Due to their highly specific surface properties, π-π aggregation, and hydrophobic interactions, bioactive GQD-based polymer composites have a high drug loading capacity, and, in case of proper correction, can be used as an excellent option for the release of anticancer drugs, gene carriers, biosensors, bioimaging, antibacterial applications, cell culture, and tissue engineering. In this paper, we summarize recent advances in using bioactive GQD-based polymer composites in drug delivery, gene delivery, thermal therapy, thermodynamic therapy, bioimaging, tissue engineering, bioactive GQD synthesis, and GQD green resuscitation, in addition to examining GQD-based polymer composites.

Keywords: polymer composites; graphene quantum dots; bioactive; biomedical; synthesis

1. Introduction

Bioactive materials play an increasingly important role in the biomaterials industry and have wide applications in nanomedicine, drug delivery systems, and biosensors. Among the medical applications of bioactive substances are the continuous development and current status of bioactive substances in medicine [1]. The term graphene was first introduced in 1986 and was created by combining the word graphite with a suffix (n) referring to polycyclic aromatic hydrocarbons. In various fields of science and technology, graphene has changed its status from an unknown substance to a bright star. This is due to the exceptional properties of graphene, including high current density, ballistic transport, chemical inertness, high thermal conductivity, optical permeability, and excellent hydrophobicity at the nanometer scale [2,3]. The first use of functionalized graphene oxide polyethylene glycol as a nan carrier of anticancer drug release was reported in Liu et al. [4]. Sanchez et al. investigated the biological properties of graphene nanomaterials and their toxicity, containing their proper interaction with biomolecules, cells, and tissues based on the number of layers and the dimensions of chemical functionalization [5]. Quantum dots are mineral semiconductor nanocrystals with a diameter of 1 to 10 nanometers that emit light after excitation and typically consist of 100 to 10,000 atoms. Their small size makes them unique compared to macrocrystalline materials [6]. In many branches of science, GDs with unique properties have been used. The use of these GDs in pharmaceutical and medical technology is also evolving [7,8]. The remarkable features of quantum dots that have made them a probe for targeted drug delivery and therapeutic fields are their rich surface area and optical properties.

The cells or tissues in question are targeted by QDs, combined with ligands. Recently, the use of QDs for in vivo applications has become more prominent with advances in the production of biocompatible QDs [9]. QDs can also lead to good solubility in various solvents, such as aqueous buffers, when in the form of a semiconductor core with a coating (as a shell) and a cap [10]. Luminescent nanocrystals, known as QDs, are used for carriers or probes in medical applications due to their rich surface chemistry, such as in drug delivery [11–14], imaging, and treatments [15]. Graphene quantum dots (GQDs) are graphene-based nanomaterials [16]. In other words, GQDs that are the product of the chemical oxidation process of carbonaceous materials can be considered extremely small derivatives of graphene oxide which contain a significant number of oxygenated functional groups, such as hydroxyl and carboxyl [17,18]. The attention of many researchers has been drawn to GQDs due to their excellent properties, such as low toxicity, good water solubility, electrical conductivity, biocompatibility, stable photoluminescence, and surface to volume ratio, and also to the application of GQDs in bioimmography, sensors, tissue engineering, photocatalysis, and energy conversion [19]. Photoluminescence is one of the main properties of GQDs that sometimes explain the diversity of their biomedical applications. However, the photoluminescence mechanisms of GQDs are not well known because their structure is complex and their surface functionalities are still unknown. Therefore, achieving optimal optical properties that affect biomedical applications is challenging.

So far, no mechanism has comprehensively explained the photoluminescence phenomenon of GQDs, limiting the regulation of their optical properties. However, several mechanisms have been proposed, including the quantum confinement effect, the emission of surface states, molecular fluorophores, the molecular diffusion of polycyclic aromatic hydrocarbons, self-trapped exciton emissions, the surface dipole emission center, the aggregate emission center, the multiple dissolution center, the relaxation of slowed solvent center, and a solvatochromic shift. Each mechanism only partially explains the photoluminescence phenomenon of GQDs. It has been suggested that bioactive compounds can specifically improve photoluminescence. The polymerization of GQD structures can also increase the biological effects of GQDs.

GQDs are typically reported to be between 3 and 20 nanometers in size, with the largest reported size being 60 nanometers. Structurally, a GQD is crystalline and has one or more layers of graphene [20,21]. GQDs can interact with biomolecules, so they can be used for drug and gene delivery by making appropriate surface changes [22]. For example, Huang et al. designed GQD nanocarriers that were able to simultaneously use specific drug delivery, imaging, and chemotherapy for cancer [23]. Apul et al. reported the synthesis of GQDs using citric acid and dicyandiamide as the starting material at 180 °C for 180 min. The synthesized GQDs showed a quantum efficiency of 36.5% and high optical stability [24]. In other studies, GQD synthesis was reported as a bottom-up method based on the degree of carbonization of citric acid [25].

The term composite (multiplayer or composite material) refers to a material that consists of two phases, a matrix, and a reinforcement, and the second phase is used at a maximum of 0%. A matrix composition with fibers (or reinforcing material) below 0% is referred to as a composite. Polymers in which fillers (in the size range from 1 to 100 nm) are homogeneously reinforced are called polymer nanocomposites (PNCs). In these nanomaterials, fillers are specified in at least one dimension less than 100 nm (<100 nm). Based on the number of dimensions that are outside the nanometer range, nanofillers can be classified into three categories: (1) Spherical (zero-dimensional), (2) layered (one-dimensional), and (3) fibrous and tubular (two-dimensional) [26–31].

There is a better interaction between the polymer matrix and the filler in polymer nanocomposites than conventional composites. The uniform distribution of nanoparticles in the polymer matrix increases the contact surface made up of the matrix and nanoparticles and improves its mechanical, thermal, and obstruction properties. The smaller the size of the nanoparticles, the more difficult it is to distribute them in the polymer matrix because the nanoparticles are very unstable and have a high tendency to aggregate or clump. The clumping of the nanoparticles is a weakness of the polymer matrix. Mineral nanoparticles with petite particle sizes make minimal changes, as in the case of polymeric materials. If properly designed and formulated, they can improve the polymer's thermal and mechanical barriers and its flammability properties [32]. Surface charge density, doping status, the degree of surface oxidation, or inks with polymers are among the properties of GQDs that can cause different interactive behaviors with biomolecules and which are dramatically different in additional others [33]. The elastomer matrix is combined with filler properties such as nanoparticles or graphene by flexible and durable polymer composites. The potential application of graphene sheets [34–36] as fillers is possible due to their outstanding optical, electrical, and mechanical properties in the context of synthesizing new nanocomposites based on polymers [37–40].

One of the most important aspects of polymer composites when it comes to obtaining lightweight materials with superior performance is the excellent mechanical and electrical properties that graphene imparts to polymer matrices. The outstanding properties that GQD-polymer composites possess with the introduction of only small amounts of nanofillers into the polymer matrix are still being considered [41–43]. Thus, research groups have made significant efforts to prepare GQD-polymer composites and study their properties [44–50]. Information on GQD-polymer composites, including the small size of GQD particles, their atomic rough surfaces, and the dispersibility in polymers have been obtained through simultaneous studies of their mechanical properties and thermal conductivity [51]. Graphene-bound biomacromolecules can improve the bioactivity and biocompatibility of advanced GQD-based biocomposites and nanoplatforms [52,53]. QD-based polymer composites on graphene base surfaces can provide significant biocompatibility and bioactivity, with many potential biomedical and medical applications [54–57]. Recently, graphene and bioactive GQD based polymer composites have been widely used in medicine for drug delivery, cancer treatment, tissue engineering, phototherapy, and stem cell and biosensor separation [58,59].

This study aimed to present advances in bioactive graphene quantum dot-based polymer composites, focusing on advances in bioactive GQD-based synthesis methods and highlighting future work areas in this field. These methods are based on techniques that rely on top-down or bottom-up processes: bioactive GQD synthesis, GQD green reduction, and GQD-based polymer composites. Finally, the applications of bioactive GQD-based polymer composites, including drug delivery, gene delivery, heat treatment, thermodynamic therapy, bioimaging, and tissue engineering, are discussed. The contents of this review are shown in Figure 1.

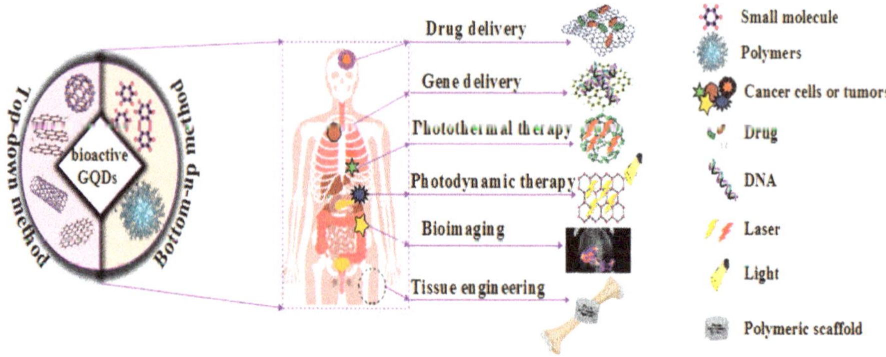

Figure 1. The applications of bioactive GQD-based polymer composites in biomedicine.

2. Bioactivation

Bioactive compound is a term commonly used to refer to secondary metabolites produced by organisms which are often not necessary for the growth and survival of organisms and, unlike the primary metabolites and vital macromolecules, are not the basis of the primary processes of the organism and do not play an important role. In general, secondary metabolites in bacteria result from specific conditions, such as limited food sources, and are produced during the idiophase period of the organism's life. Most of these compounds support the organism in specific environmental conditions (such as competition in the ecosystem). Zahner has proposed the most convincing theory for producing secondary metabolites, which he refers to as secondary metabolite evolutionary compounds. If a metabolite does not affect any stage in the differentiation of an organism (morphogenesis, movement, regulation) and is produced for specific purposes in organisms, it is called a secondary metabolite. Many of these compounds have an antifeedant function in certain organisms, sex absorbers, antibiotic agents [60,61].

The first report on the production of biologically active substances from marine bacteria was made by Zobell and Rosenfeld in the context of antibiotic production. Since then, several reports have been received in this regard. When it comes to biological compounds, 16,000 have been isolated from marine microorganisms, including antibacterial, antiviral, and antitumor compounds [62,63].

2.1. Polymers Containing Bioactivation

The stabilization of bioactive compounds to produce the functional properties of polymers has developed extensively over the past decade. This method is used in packing industries of foodstuffs and pharmaceuticals. Various bioactive compounds can be used in the stabilization process, such as proteins, peptides, antimicrobial compounds, but most are enzymes which stabilize the surface of polymers. The first step is to stabilize, design, or select a suitable polymer. The mechanical properties considered include elasticity, strength, transparency, and degradability. Stabilization operations can be performed on the surface or inside the polymer. Therefore, the polymer surface must be ready to accept biomolecules. Creating the ability to obtain and bind bioactive compounds in polymers is referred to a special operation called producing functional properties [64,65]. The stabilization of biomolecules and bioactive compounds in polymers is performed using the following methods: (1) Physical absorption, (2) entrapment, (3) covalent immobilisation, and (4) affinity immobilisation [66].

2.2. Methods of Bioactivation

2.2.1. Bioactive Carbon Sources

For biomedical applications such as MSC-based therapy, the development of bioactive materials is desirable when it comes to integrating efficient differentiation and traceable properties in a biocompatible manner. A new type of one-step hydrothermal method makes bioactive carbon dots (CD) [67]. Recently, a bottom-up synthetic strategy has been used to design and fabricate multifunctional CDs with the complex introduction of bioactive precursors, including small molecules, nucleic acids, and proteins [68,69]. Compared to traditional citric acid-based CDs, these bioactive CDs can expand their applications against cancer and pathogens and have different pharmacological activities [70–72]. In both in vitro and in vivo studies, bioactive CDs have shown improved anti-cancer performance when compared with small free molecules [73]. Therefore, dual-function precursors have not yet been developed to obtain bioactive CDs.

2.2.2. Biomass-Waste Derived GQD

When it comes to the scalable production of GQDs, biomass has been used as a source of green, natural, cheap, sustainable, and renewable carbon. Since biomass is abundant and inexpensive, and there is no published paper on the total cost of GQD materials, it is therefore expected that the price of biomass-derived precursors (waste) is much lower than other precursors (graphite, carbon fiber, CNTs, citric acid, glucose). GQD production with a product yield comparable to expensive graphene-based precursors is possible using different types of biomass, such as plant leaves, grass, rice husk, coffee grounds, and charcoal [74–78]. The quantum efficiency of GQDs derived from biomass is superior to graphene derivatives. According to reports, green production is possible through plant leaves, and GQD production is also done without inactivators, reducing agents, oxidants, or organic solvents [75].

2.2.3. Biologically Active Agents

Bioactive compounds and diagnostic probes, known as biologically active agents, effectively load drug delivery systems based on enzyme-responsive polymers via covalent or physical encapsulation. The isolation of enzyme-responsive polymers for subsequent therapeutic or diagnostic activities is performed by target enzymes to expose or activate biologically active agents. To achieve drug release/activation and physicochemical alternation, enzyme-responsive drug delivery systems can be designed to take advantage of high enzymatic properties [79]. The separation of bioactive agents after enzyme-induced cleavage is one of the direct release/activation methods, which leads to the explosive release of therapeutic cargoes or the activation of extinguished/protected functional agents [80–82].

2.3. *Roles of Bioactivation*

Metabolic activation or bioactivation is a relatively inactive organic compound produced by reactive electrophilic intermediates, including free radicals. Bioactivation plays a significant role in hepatotoxicity, dermal ADRs, blood dyscrasias, and drug-induced toxicities [83,84]. The formation of stable/inactive or reactive metabolites is accomplished by the drug's metabolism, enabling them to interact with cellular structures and functions, which indicates the role of bioactivation in drug-induced hepatotoxicity. Hepatotoxicity is induced by the reaction of electrophilic metabolites with nucleophilic sites in macromolecules either through the metabolic bioactivation of drugs in the formation of covalent compounds or through oxidative damage caused by ROS production [85].

2.4. *Design of Bioactivation*

Electrophiles have attracted the most attention concerning bioactivation in drug design because they are easier to trap and their chemistry is rationalized [86]. Thus, ambiguity in understanding subsequent covalent bonding can be linked with toxicities [87]. In contrast, the initiation of downstream cellular events associated with oxidative stress is better understood through the formation of free radicals. Most of the structural features associated with bioactivation during drug design are related to active aromatic systems (usually aniline derivatives). Assessing the risks involved in developing compounds with structural alerts and the placing bioactivation data during drug discovery are critical issues in the context of bioactivation and drug design [88].

2.5. Method of Bioactive GQD Synthesis

The synthesis method plays a significant role in the properties and application of bioactive GQDs. Therefore, the production of bioactive GQDs must be adjusted during and after their production [68]. Chemical, physical, and electrochemical techniques, including various methods, have been used to generate bioactive GQDs [89]. Bioactive GQDs are synthesized top-down and bottom-up (Figure 2). The different methods for synthesizing bioactive GQDs used in biomedicine are listed in Table 1. In the top-down synthesis method, carbon mass decomposition in nanoparticles takes place via chemical and physical processes such as acid oxide, electrochemistry, and hydrothermal processes [90–92]. Using a variety of organic compounds under carbonization conditions, GQDs are prepared in a bottom-up method [93]. Among the advantages of bottom-up synthesis are the regulation of physical properties, the use of various organic compounds as raw materials, and the structural properties of GQDs. Processes including stepwise solution chemistry, microwaves, hydrothermal processes, ultrasonic chemistry, acid oxidation, and electrochemistry make up the various physical and chemical methods used to prepare GQDs [94–96]. The restriction of access to the mass production of bioactive GQDs using this method is due to the high cost spent on raw materials in the method. Recently, due to the simplicity, cost-effectiveness, and one-step synthesis route method of the hydrothermal method, this method has been considered [25]. The hydrothermal method [97–99], microwave method [100–102], electrochemical oxidation method [103–105], the ultrasonic method [103,106], and biosynthesis methods [107,108] are among the methods used for bioactive GQDs synthesis based on top-down and bottom-up processes.

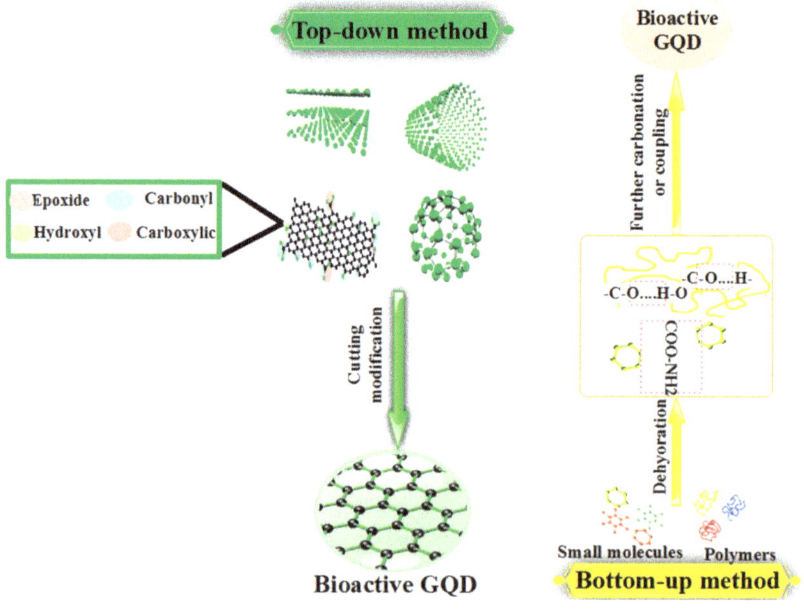

Figure 2. The top-down and bottom-up methods for the synthesis of bioactive GQDs.

Table 1. Different methods for synthesizing bioactive GQDs are used in biomedical applications.

Synthesis Methods	Application	Size	Ref.
	Bottom-Up Method		
GQD-RhB-silka	Diagnosis	3–20 nm	[109]
mango leaf extract—mGQDs	NIR-responsive fluorescence bioimaging	2–8 nm	[110]
PEGylated GQD	Fluorescence imaging of tumors	2.75 nm	[111]
GQD-PEI	Gene transfection	3–4 nm	[112]
GQDs	Drug delivery and bioimaging	~12 nm	[113]
MSN-SS-CD-DOX	Targeted and controlled drugs	2.7 nm	[114]
	Top-Down Method		
GQD	Diagnosis	5 nm	[115]
durian extract—GQDs	Bioimaging	2–6 nm	[116]
NP-GQD	Cysteine detection	10–30 nm	[117]
GQD-PEG-AG	Radiotherapy	3–4 nm	[118]
lignin—GQDs	Bioimaging	2–6 nm	[119]

2.5.1. Hydrothermal Method

A fast and straightforward way to prepare bioactive GQDs is the hydrothermal method. The hydrothermal method is a one-step method in which the organic precursor is heated in a Teflon tube to reach high temperatures and pressures. As a result of high temperatures under high pressure, the bonds between carbon materials are broken to form bioactive GQDs. The electrical-optical properties of the particles can be adjusted by using different precursors and through temperature optimization. Therefore, the hydrothermal method is low-cost and non-toxic (Figure 3. Hydrothermal methods also produce particles with a diameter of 10 nm, more prominent than other dot production methods, such as the electrochemical method [120–125]. Lianget al. easily obtained highly fluorescent quantum dots from gelatin by using hydrothermal method [126]. Liu and colleagues developed a simple, green, and inexpensive way to prepare fluorescent quantum dots using hydrothermal processing and a type of radish. They investigated the application of prepared carbon quantum dots in cell imaging and the detection of iron ions [127]. Hong Miao et al. presented a green method for synthesizing carbon quantum dots with a high quantum yield of about 14% from tomato paste and investigated the use of synthesized quantum dots in determining the type of antigen [128].

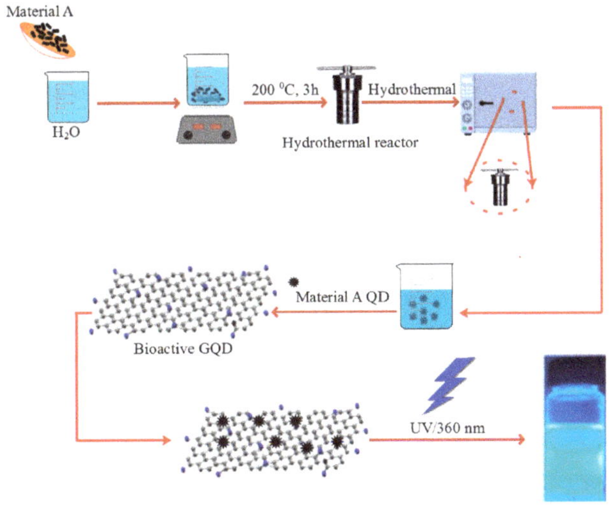

Figure 3. Synthesis of photoluminescence GQDs by using hydrothermal method from material A.

2.5.2. Microwave Method

The synthesis of nanoparticles via microwave-facilitated methods has many advantages compared to other methods. Compared to the hydrothermal method, this method has the advantage of its higher speed and lower synthesis temperature. Some of the benefits of microwave facilitated reactions are: (1) Deficient level of impurities in the products, (2) a very high product efficiency, (3) the easy control of temperature and pressure, (4) the easy control of product properties (such as size), (5) the environmental friendliness of the method, (6) the very high security of the method, (7) its reproducibility, and (8) selective heating (i.e., the reduction of energy costs) [129–131]. Ayele et al. have proposed a green method for the mass production of CdSe quantum dots by using microwaves [132]. An easy, light, and fast method for the synthesis of nitrogen-doped GQDs using microwave prepared cedar tree root without surface modification has been proposed by Shawing Sheng et al. [133]. The resulting quantum dots were used to identify Hg2 + ions selectively. Synthesized GQDs can also be used as multicolor fluorescents in bioimaging. The purification of dialysis membranes for 7 h and the preparation of GQDs via microwave irradiation for 10 min using raw materials such as aspartic acid (Asp), NH4HCO3, and solvent (DI water) were investigated by Zhang et al. The results showed that the prepared GQDs had a strong blue fluorescence and a QY of 14% (Figure 4) [134].

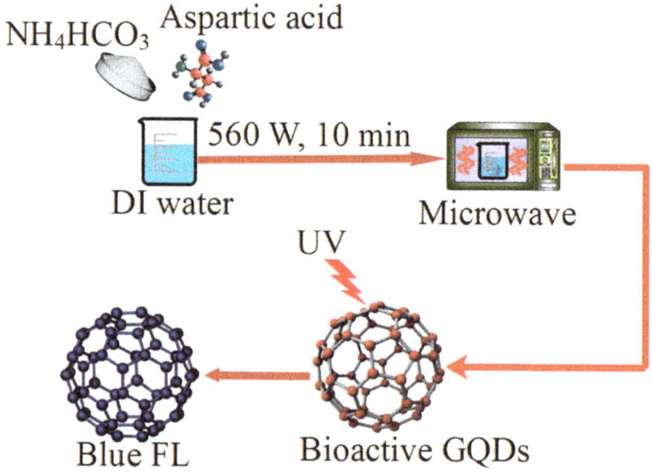

Figure 4. The purification of GQDs using microwave irradiation.

2.5.3. Electrochemical Oxidation Methods

Electrochemical methods make it possible to adjust nanostructures by controlling the current–voltage ratio. For example, applying a regulated voltage to a bulk of carbon precursors leads to the electrochemical corrosion reactions of carbon reactants and the production of carbon nanostructures. This method does not require high temperatures; it can be done quickly on a large scale with aqueous or non-aqueous solvents. This method is one of the fastest ways of producing graphene sheets (Figure 5) [103–105,135,136]. For example, Paulo et al. prepared GQDs using electrochemical and corrosion methods [137]. Wong et al. proposed a bottom-up approach for synthesizing nitrogen-doped GQDs using the electrochemical process. This method is green and simple and is suitable for large-scale synthesis with an interest rate of more than 95%. The quantum gain of the particles is 0.71. Also, the toxicity of quantum dots synthesized by this method is low [138].

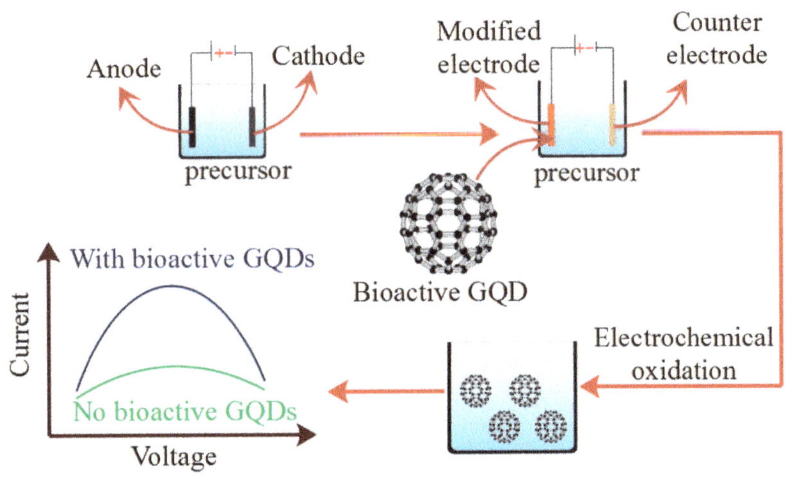

Figure 5. The synthesis method of electrochemical oxidation.

2.5.4. Ultrasonic Method

In recent years, many methods for making photoactive materials for use in solar cells and photocatalysts have been proposed. However, the benefits of preparing these materials in unusual ways are exciting from the point of view of green chemistry. The idea of producing highly efficient photocatalysts through the use of ultrasound is fascinating and essential in terms of science and technology and has excellent potential for producing photocatalysts in the near future. Ultrasonic synthesis is a promising method that allows for control over size, morphology, nanostructure, and catalytic properties. Compared to conventional energy sources, the supersonic method provides abnormal reaction conditions in the liquid phase due to the phenomenon of cavitation (very high temperatures and pressures for brief periods in liquids) [139–141]. Zhu et al. synthesized high-quantum GQDs from only graphene oxide and potassium permanganate in a one-step method using ultrasonic irradiation and used them as a test for the detection of alkaline phosphate (ALP) [142]. Oza et al. prepared GQD quickly and greenly using the precursor chemical acoustic method and a lemon [143]. The resulting quantum dots were purified using the sucrose gradient density centrifugation method, which separated the water-soluble, single-spray, photoluminescent, highly stable optical, and chemical quantum dots. Furthermore, quantum dots synthesized in this way are not toxic and can be used as optical imaging tools. The formation of tens of thousands of small bubbles in the liquid and the loss of carbon–carbon bonds are generated by mechanical force and eventually results in GQD cutting, an issue which is affected by ultrasound waves (Figure 6).

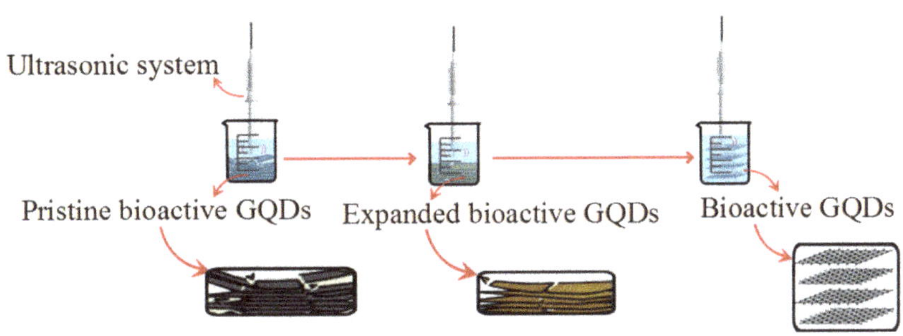

Figure 6. The ultrasonic synthesis method.

2.5.5. Biosynthesis Method

The biosynthesis method is performed in mild conditions, and the bioactive GQDs synthesized using this method have inherent biocompatibility and bio-stability without the need for encapsulation operations. Therefore, these methods provide a green way to prepare biocompatible, bioactive GQDs. As with chemical processes, biosynthesized quantum dot emissions can be controlled by their size, the instantaneous excitation of bioactive GQDs of different sizes by a single light source, and by a broad spectrum window that can be adjusted by the constructive component [107,108,144,145].

2.6. Green Reduction of CQD

Advances in nanoscience and nanotechnology have led to versatile applications and the discovery of different new nanomaterials, the properties of which depend on their shape and size. QD is known as one of the nanomaterials that has been introduced due to its variation in bandwidth size. Since there are numerous challenges involved in disposing and recycling of QDs, it can be said that they are very dangerous to process, specifically ordinary QDs such as PbS and CdSe [146]. Bioactive GQDs, an example of organic quantum dots, are not as limited as conventional QDs. Bioactive GQDs, known as nanomaterials, have advantages such as adjustable diffusion, excellent biocompatibility, resistance to light bleaching, the abundance of raw materials in nature, and low toxicity and costs [147]. Among their potential applications are their use as light-emitting diode materials [148], detection probes [149], and optical bioimaging probes. An effective way to regulate the electrical and optical properties of GQDs is to dope GQDs with heteroatoms (such as B, N, S, and F) [100]. In GQDs, the presence of N can be effective for the homogeneous distribution of metal nanoparticles. Nevertheless, N-GQDs, in addition to being generated in situ for catalytic proposals, can support metal nanoparticles such as Pd (0). N-GQDs were expected to be an adequate support for Pd (0) in the catalytic reduction of nitro compounds due to the efficient activity of graphene as a support in the green reduction reaction [150].

3. GQD Based Polymer Composite

In graphene-based polymer composites, the superior properties of graphene compared to polymers are reflected. Compared to neat polymer, graphene-based polymer composites show gas barrier, electrical, superior mechanical, flame retardant, and thermal properties [41,42,151–154]. Graphene nanofibers are used as a two-dimensional model to arrange polymers, increasing the polymers' solubility [155]. Although carbon nanotubes (CNTs) exhibit mechanical properties comparable to graphene, graphene is a better nanofiller in certain respects, such as in terms of thermal and electrical conductivity [156–161]. Interfacial bonding between graphene layers and the polymer matrix and the distribution of graphene layers in the polymer matrix affect nanocomposites' physicochemical properties. Pure graphene does not have good compatibility with organic polymers and does not form homogeneous composites. Since the use of graphene oxide (GO) sheets as nanofillers has attracted much attention in the context of polymer nanocomposites, it can be said that in contrast to graphene, GO is preferable to organic polymers [162–165]. Graphene oxide is an insulating material, so it cannot synthesize conductive nanocomposites without further resuscitation. In polymer, graphene, and solvent, factors such as polarity, molecular weight, hydrophobicity, and reactive groups play an essential role in when it comes to preparation methods [166]. For advanced hybrid nanomaterials and applications, polymer dots have been designed and used, specifically necessary and unique quantum dots. Therefore, polymer dots can be obtained by using conjugated and non-conjugated polymers [167,168]. Both QD polymer encapsulation, without changing the native surface ligand layer [169], and ligand exchange, used to replace the main hydrophobic ligands on the GQD surface with polymer molecules [170], are methods of making GQD-polymer nanocomposites. Biological imaging, fluorescence imaging, and bioassay from GQD-based nanocomposites are used [171–173]. The foundation of potentially new and innovative GQD composites [174–177] has been established through previous research on graphene-reinforced composites [178–180] and GQD synthesis [20,181–183].

One of the newest forms of graphene is GQDs. Electronic devices [166,184], energy storage [185,186], fuel cells [187], and biomedical applications [188,189] are among the wide range of applications of GQDs-polymer nanocomposites.

4. Polymerization

Polymerization is a chemical reaction in which small and simple molecules, or monomers, bond together to form a large molecule with a molecular mass several times that of the original molecule [190–192]. Bulk, emulsion, suspension, and surface polymerization are polymerization methods used to prepare microparticles. One or more monomers are heated in a bulk polymerization technique to induce polymerization in the presence of a catalyst. During the polymerization process, the drug may form. Also, monomers can be prepared or fragmented to convert the obtained polymer into a microsphere [193]. The polymerization process occurs at lower temperatures, and the polymerization of suspension is similar to bulk polymerization [194].

4.1. Methods for Polymerization

4.1.1. Suspension Polymerization

To obtain a hybrid of stable QDs of polymer particles, considering the confinement of QDs during a polymerization reaction, one of the investigated methods is the polymerization of suspension [195–197]. Hexadecylamine CdSe QDs were synthesized by O'Brien et al. The surface polymerizable QDs were then obtained via ligand exchange. As a result, the QDs were encapsulated in the synthesized polymer particles [198]. The inclusion of QDs in the polymer particles during the polymerization process of suspension results in reasonable control with respect to the position of the quantum dots or their final fluorescence properties.

4.1.2. Emulsion Polymerization

The enclosure of QDs in colloidal polymer particles is achieved through emulsion polymerization [199–201]. The coverage of CdSe quantum dots with tridactyl phosphine oxide and their dispersion in toluene were investigated by Yang and Zhan. The results show good control over the confinement of quantum dots in polymer particles to prevent the accumulation of nanocrystals and fluorescence microscope images [202]. It is not easy to completely enclose QDs in polymer particles using emulsion polymerization unless polymerization starts from the surface of QDs after their surface has been modified to combine polymerizable groups [197].

4.1.3. Mini-Emulsion Polymerization

Mini-emulsion polymerization is one of the simplest and most common methods for encapsulating QDs in hydrophobic sub-micron polymer particles. A polymerization monomer produces an aqueous dispersion of nanoparticles where the surface of the coated QDs is compatible [203–207]. The creation of polymer nanoparticles containing QDs is achieved through the polymerization of a dispersion nanodroplet. The monomer system, the type of initiator, the surfactant concentration, and the surface modification of the mineral nanoparticles are among the factors on which the successful encapsulation of the mineral nanoparticles in the polymer particles by mini-emulsion polymerization depends [208].

4.2. Physical Blending

The physical blending approach makes it easy to prepare GQD/polymer nanocomposite materials. Some physical blending methods include mixing tiny amounts of GQD with a polymer matrix, blending the solution, and processing the melt. Hydrogen bonding, electrostatic interactions, or π-π interactions between polymers and GQDs are among the non-covalent interactions that lead to nanocomposite formation. When water molecules stick together, an "electrostatic" bond is formed between an oxygen atom from one molecule (which has a negatively charged particle) and a hydrogen atom from another molecule (which has a positively charged particle), and this is a hydrogen bond. The distribution of attractive driving force interactions when using this method to prepare GQD/polymer composites led to the creation of suitable quality nanocomposites with good GQDs [209–214].

Infusion of GQD into Polymer

Injecting a bandgap through modifications that introduce quantum confinement and edge effects makes graphene nanoparticles such as GQDs and graphene nanoribbons [215]. A subset of graphene-based nanomaterials consisting of several graphene layers stacked on top of each other are called GQDs [216]. Glucose [216], carbon fibers [217], carbon nanotubes (CNTs) [218], graphite [219], and coal are among the various materials used in the synthesis of GQDs. Gobi et al. investigated the injection of graphene quantum dots to create stronger, harder, and brighter polymer composites. When comparing polymer resins with and without GQD, uniform loading with weight percentages of up to 10%, an 18% change in the maximum tensile strain, and a 2.6-fold increase in toughness, etc., are the results of GQD injection into an epoxy polymer matrix, reflecting the excellent optical properties of the composite formulation [182].

4.3. Chemical Grafting

Esterification [220], etherification epoxidation [221], methacrylation/acrylation [222], and acylation are among the abundant functional groups on the GQD surface that allow for chemical modifications to form covalent bonds with polymers due to the covalent bonds between polymer chains and GQDs. The chemical grafting of GQDs to the polymer matrix is superior to the physical blending approach in terms of promoting mechanical strength and maintaining properties over time [223–227].

4.4. In-Situ Growth

Since the secondary intermolecular interaction between GQDs and the polymer matrix is relatively weak and unstable, it can be said that the physical blending method is a practical approach that is used to fabricate GQD/polymer nanocomposites. In many cases, chemical grafting methods use organic reagents, and solvents are toxic due to their shortcomings in terms of complex reaction processes and multiple preparation steps. The physical blending method can also anchor GQDs firmly to the polymer matrix due to the formation of covalent bonds. The ensuring of a better bond strength is possible via the in-situ growth of GQDs in the polymer matrix because it is an easy and green process. Also, the bond between GQD and polymer includes both chemical and physical interactions. An in-situ growth strategy can be obtained by applying one-pot thermal treatment containing hydrothermal, pyrolysis, or low temperature heating on a mixture containing GQD precursors and polymer solution [228–232].

5. Biomedical Applications

5.1. Drug Delivery

Bioactive GQD based polymer composites can bind to various biological materials through π-π and electrostatic interactions, so the loading of most drugs can be done quickly. On the other hand, bioactive GQD based polymer composites have good membrane permeability and biocompatibility, which can improve the drug efficiency of the loaded drug and increase effectiveness in the face of drug-resistant cells [233,234]. Table 2 shows the application of bioactive GQDs in the field of drug delivery. Extensive research has been done on using bioactive GQD based polymer composites for in vivo drug delivery. Absorption by the reticuloendothelial system and renal clearance is reduced due to the average size of bioactive GQDs. The efficiency of bioactive GQDs can also be attributed to their high delivery, which increases blood circulation time [235]. Sam et al. reported the high anticancer activity of GQDs containing curcumin in both in vivo and laboratory conditions. They combined three types of curcumin-graphene composites, including GOs, DGOs, and GQDs, to form the hydrophobic cancer drug curcumin. Interactions between curcumin and the oxygen-containing functional groups DGOs, GOs, and GQDs, played a crucial role in the loading capacity of curcumin. Since the number of oxygen-containing functional groups on the surface of the graphene derivatives was pH-dependent, the effective drug loading and release behavior was pH-dependent. According to the results, the prepared GQD-curcumin composites contained the highest amounts of curcumin (40,800 mg/g). They had the best anti-cancer activity compared to other composites containing the same dose of curcumin. There is no fluorescence in the case of curcumin and its GQD composite. It was only after the release of curcumin from the composites of GQDs that the GQDs provided the remaining fluorescence signal. As a result, GQDs simultaneously act as probes for tumor imaging along with drug release [236]. Silica-coated bipolar quantum dots were investigated by Akbarzadeh et al. They reported that the synthesis of intelligent nano-aptamer was targeted by QDs coated with porous meso silica with a bimodal imaging capacity. Their results showed that the targeted hybrid system that was prepared based on their in-vivo test has MR capability and fluorescent imaging [237]. Iannazzoa et al. examined the use of graphene QDs in cancer treatment and drug delivery. Based on the results, the prepared GQDs have a remarkable ability to deliver the drug to cancer cells and are biocompatible. Similarly, it is possible to combine targeted ligand drugs in nanomaterials via a multifaceted combination. As a result, conventional chemotherapy, toxicity and side efficacy were minimized (Figure 7) [238].

Table 2. The application of bioactive GQDs in the field of drug delivery.

	Composite	Application	Ref.
1	GQD/DOX	In vitro: A549 cells	[239]
2	SiRNA/GQD/DOX	Therapy of A549 cancer cells	[240]
3	GQD/DOX	In vitro: HeLa, A549, and HEK293A cells	[241]
4	CMC/GQDDOX	Therapy of K562Leukemia cells	[242]
5	GQD/DOX	In vivo: BALB/c mice	[234]
6	GQD/CDDP	Breast cancer cells	[243]

5.2. Gene Delivery

Gene therapy is one of the new and promising ways to treat various diseases caused by genetic disorders such as cystic tissue corruption, Parkinson's disease, and cancer [244]. Figure 8 shows the treatment of cancer cells with gene therapy based on bioactive GQDs. Gene therapy for treating a wide range of disease conditions such as rare hereditary and single-gene disorders has also attracted much attention. For this purpose, nucleic acids must be delivered to human target cells and expressed. Gene therapy offers a new treatment method by offering DNA or RNA instead of chemotherapy drugs to inhibit cell pathways and induce tumor cell death. This treatment method can also remove a malfunctioning gene, replace the mutated gene, or offer new gene products to fight the disease [245–247]. Successful gene therapy requires a gene carrier that can protect DNA from nucleoside degradation and, in addition, facilitate high-yield DNA cell uptake [248]. The primary challenge facing the development of gene therapy is the lack of safe and highly efficient carriers [249]. Among the effective non-toxic carriers in gene therapy are none-viral gene delivery systems. Such systems ensure that nucleic acid payloads to cytosols and cell nuclei bypass biological barriers. Viral vectors are one of the factors influencing gene delivery due to their natural ability to invade and deliver genetic material. It is also possible to keep them safe for clinical use by blocking the immunogenicity of viral vectors [250].

Bioactive GQD-based polymer composites demonstrate the characteristics of an effective gene carrier. Therefore, many efforts have been made to develop bioactive GQD-based polymer composites as non-viral vectors in gene therapy. The development of bioactive GQD-based gene nanocarriers is possible by forming a set of bioactive GQD-based polymer composites, plasmid DNA (pDNA), and chimeric peptides. Observing the cellular uptake of GQD-peptide-pDNA complexes is possible concerning PL bioactive GQD-based polymer composites through a confocal microscope [251]. Based on previous studies, the use of bioactive GQD-based polymer composites as transmission vectors can lead to more efficient in-vitro and in-vivo gene therapy applications. The application of bioactive GQD-based polymer composite on other non-viral vectors seems to be less common than using bioactive GQD-based polymer composites as drug delivery agents. Due to the sp2 hybridized structure and π-stacking in bioactive GQDs-based polymer composite systems, the possibility of drug loading in bioactive GQD-based polymer composite systems is higher than in other nanoparticle-based drug delivery systems [252–256].

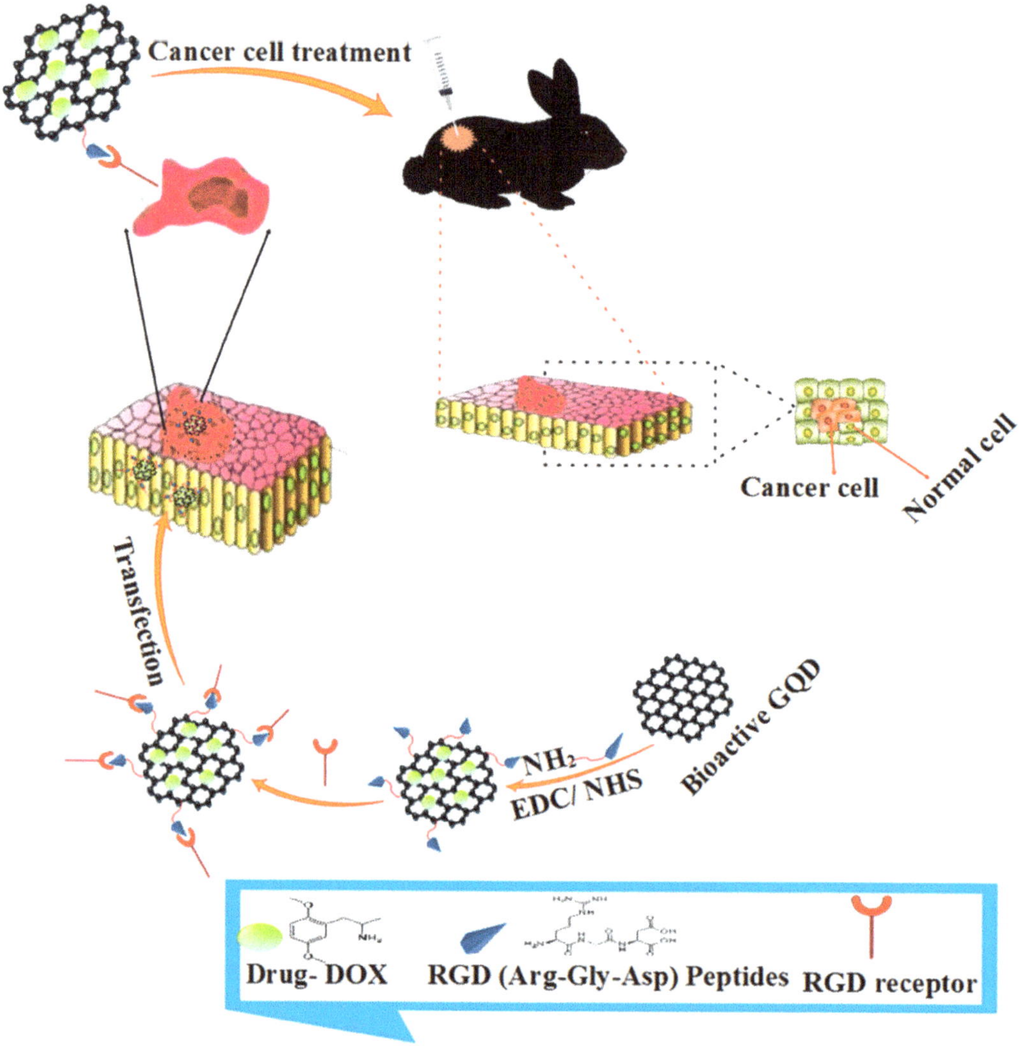

Figure 7. Treatment of cancer cells with drug delivery method based on bioactive GQDs.

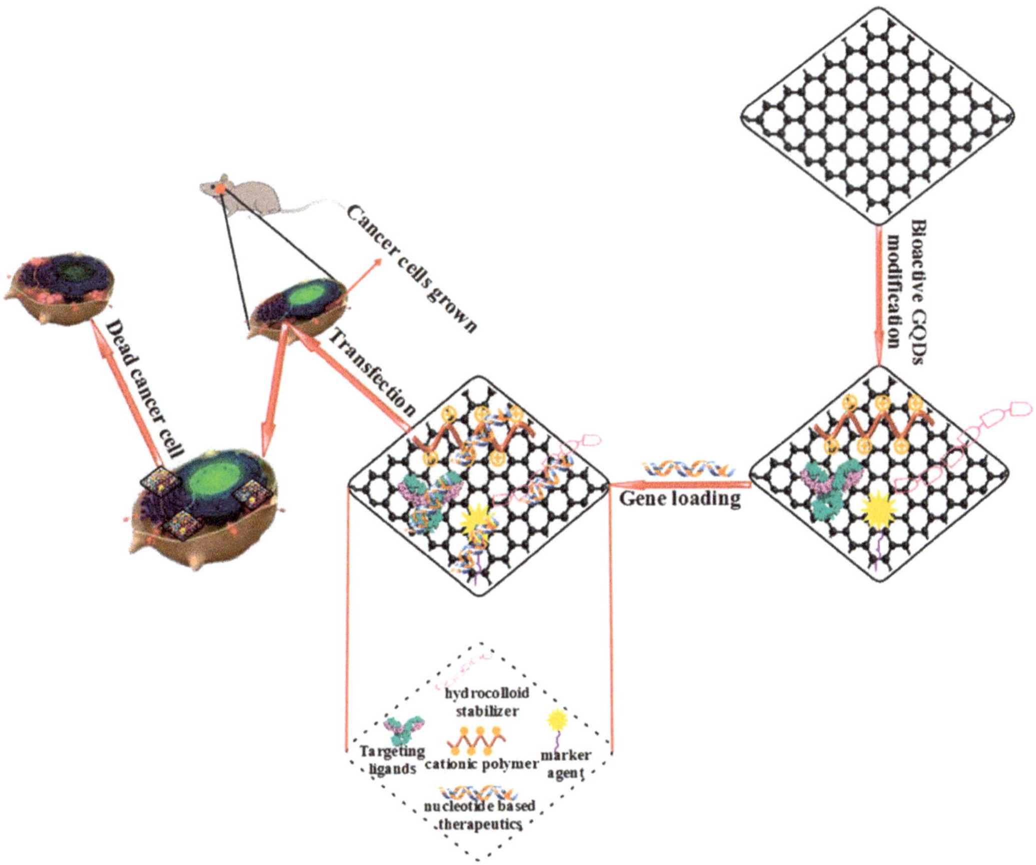

Figure 8. Treatment of cancer cells with gene therapy based on bioactive GQDs.

5.3. Photothermal Therapy (PTT)

The therapeutic window refers to the aromatic structure that provides graphene with strong light absorption in the NIR region (700–900 nm) [257,258]. The main treatments of cancer are radiology and chemotherapy, which have significant side effects. Recently, PTT has been considered due to its high efficiency in destroying malignant tumors as well as the minor amount of damage it inflicts on adjacent healthy tissues (Figure 9) [259]. In the photothermal therapy method, electromagnetic wave radiation (with wavelengths in the infrared region) and absorbent nanoparticles at these wavelengths, which have a high efficiency of light to heat, are used. Thus, by converting light energy into heat and increasing the temperature to between 42–46 °C, cell membranes and proteins are destroyed, leading to the death of cancer cells [260,261]. Many nanoparticles are identified as external agents after entering the body due to their hydrophobicity and cause poisoning by accumulating in one part of the body. Therefore, the surface of nanoparticles is coated with ligands to increase biocompatibility and reduce toxicity [262]. Nanoparticles used in cancer treatment via PTT include gold nanoparticles, carbon nanotubes, and graphene. The efficiency of light production to heat in graphene and its biocompatibility are both higher when compared to gold nanoparticles and carbon nanotubes.

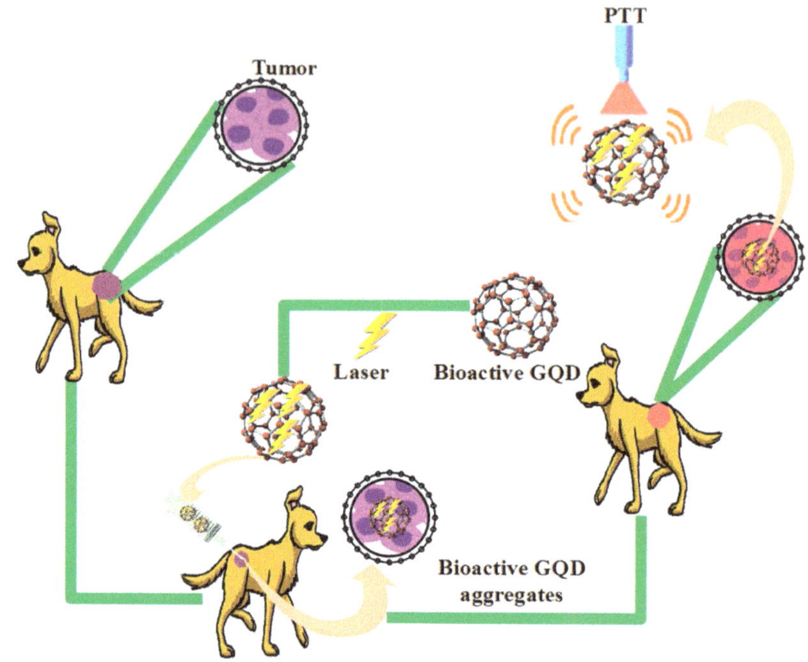

Figure 9. Tumor treatment with the PTT method based on bioactive GQDs.

On the other hand, due to the increased ability of graphene, the use of these nanoparticles has received much attention when it comes to cancer drugs and the simultaneous treatments of PTT and drug delivery. By placing graphene on the curved surface of the nanoparticle, the interaction of light with graphene increases, increasing the efficiency of PTT [263]. The features of this therapeutic method are non-invasive radiation, penetration into the skin, and its harmlessness. In recent years, graphene-based PTT nano-hybrids have been discovered to inhibit tumors in vivo. After intravenous injection, GO complexes begin to accumulate in the solid tumor severely and passively, due to the effect of EPR. The effective death of cancer cells occurs when GO complexes produce enormous heat after irradiation. Recently, PTT in vivo using GO and GO-complexes was reported to remove 100% of tumors [264]. The successful treatment of cancer cells was demonstrated by using nitrogen and boron dual doped bioactive GQD-based PTT in the near-infrared II region [265]. The effectiveness of such treatment is achieved by merging a drug in bioactive GQD-based polymer composites with PTT. The bioactive GQDs-gated hollow mesoporous carbon nanoplatform loaded with doxorubicin achieves this effect simultaneously with the controlled release of near-infrared drug delivery [266]. This combination of PTT and PDT is exciting. Thus, a multifunctional bioactive GQD is formed to treat cancer when it produces 808 nm ROS laser radiation and heats up simultaneously [267].

5.4. Photodynamic Therapy (PDT)

PDT activates a light-sensing compound by using visible light to generate samples of cytotoxic oxygen and free radicals that selectively destroy rapidly growing cells [268]. PDT is widely used to treat acne, psoriasis, and cancer using a light source and a light-sensitizing agent. Light sensitizing factors such as semiconductor quantum dots have received much attention in the context of PDT. The killing of target cells occurs via a PDT mechanism involving ROS production and oxidative stress. ROS production in cancer cells and tumors is mediated by a light-sensitizing agent such as a bioactive GQD-based polymer composite (Figure 10). The factors that make bioactive GQD-based polymer composites a good light sensitizing agent include stability in various pHs and light, their biological corrosion resistance, and their biocompatibility. To achieve PDT and the simultaneous imaging of cancer cells, bioactive GQD-based polymer composites can be used. The prevention of photobleaching and the production of a high quantum efficiency from a single oxygen molecule are among the benefits of bioactive GQD-based polymer composites. In PDT, bioactive GQD-based polymer composites are more practical than any other factor. The induction of cancer cell apoptosis and autophagy is performed by bioactive GQD-based polymer composites through oxidative stress.

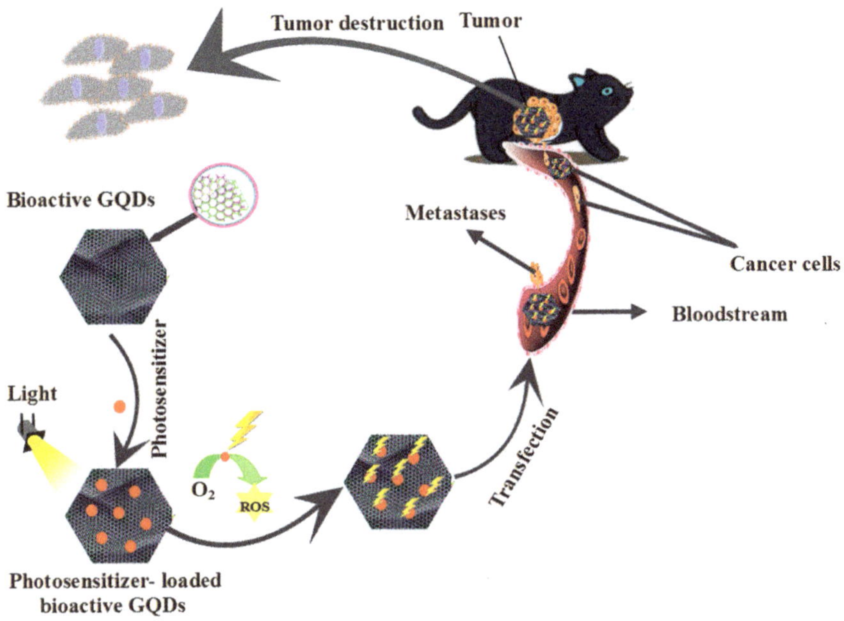

Figure 10. Tumor destruction using the photodynamic method based on bioactive GQDs.

Studies have shown that cancer treatment using bioactive GQD-based polymer composites is more effective than photodynamic therapy using HeLa cells [269–272]. Chemotherapy and photodynamic therapy guarantee a synergistic effect in the treatment of cancer. With GQDs, the targeted PDT of mitochondria is also performed [273]. Reports on the PDT of two photons were performed using a Bengal rose light sensitizer with nitrogen-doped GQDs [274]. A synergistic effect on tumor cell apoptosis was demonstrated simultaneously with drug delivery and light irradiation by silver nanoparticles of PEGylated GQDs loaded with doxorubicin (DOX) [275].

5.5. Bioimaging Application

Bioimaging is a method used in both clinical and research settings. Using different parts of the electromagnetic spectrum means that the bio-distribution of therapies separately and accurately and the observation of biological processes such as targeted delivery provides cellular uptake [22,234,276]. The application of QDs in cell imaging is multicolored and sensitive due to a significant increase in their synthesis, surface composition, and chemistry [277]. The extensive application of bioactive GQD-based polymer composite in biological imaging are stimulated by broad absorption with narrow emission spectra, strong quantum confinement, and relatively high quantum efficiencies with high molar extinction coefficients [278–280]. Bioactive GQD-based polymer composites are promising candidates for imaging deeper tissue samples because they have the property of emitting near-infrared reflection [281]. Since the detection of metastasis and the recurrence of cancer, and the early detection of tumors by sensitive imaging, are possible, it can be said that the role of imaging in the diagnosis of cancer is of particular importance. Figure 11 illustrates the diagnosis of cancer using bioimaging based on bioactive GQDs. As required with other nanoparticle platforms, the intrinsic PLs of GQDs allow them to be used as optical probes in fluorescence imaging without the further combination of fluorescent dyes. Recently, the capabilities of bioactive GQD-based polymer composites in MR imaging and NIR fluorescence imaging have led to the development of bioactive GQD-based polymer composite in bioimaging. Among the unique features of bioactive GQD-based polymer composite, we can mention their ideal imaging probes, with applications in different bioimaging methods, and their excellent biocompatibility [256]. In Sheng et al., the doping of quantum dots with nitrogen (N-GQDs) was performed using a hydrothermal process. In this study, raw materials such as citric acid, PVP K90, and glutamate were used, and the quantum efficiency was 64.2%. The results showed that a new fluorescence probe of nitrogen-doped GQDs could be used to detect chromium (VI) and mark MCF-7 cells [282].

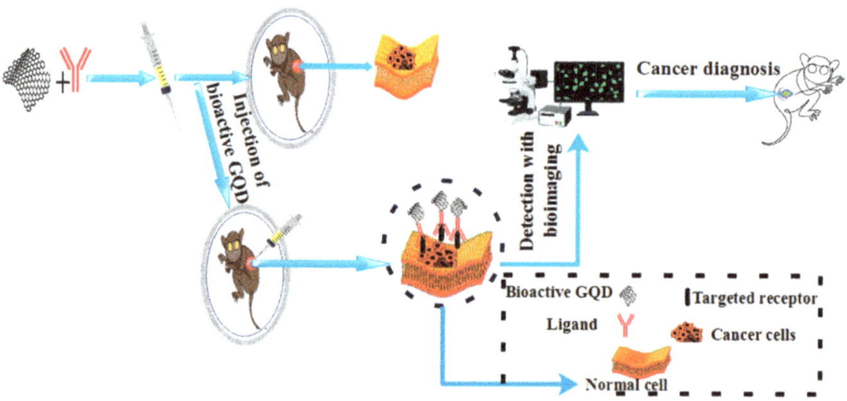

Figure 11. Diagnosis of cancer using bioimaging based on bioactive GQD.

5.6. Tissue Engineering

Tissue engineering is a method in which cells are taken from a patient, and after culturing and increasing their number, they are accepted in a scaffold. Appropriate chemical, biological, mechanical, and electrical stimuli are applied, and new tissue is formed in a short time [283]. The main components of tissue engineering are the cell, the scaffold, and the growth factor [284]. More recently, bioactive GQDs have been tested for stem cell-based tissue engineering and reconstructive medical applications. Many efforts have been made to differentiate stem cells into different cells by performing other techniques [285]. Encouraging specific stem cell differentiation under certain conditions is possible using bioactive GQD-based polymer composites. The important role of bioactive GQDs in bone differentiation was investigated by Qiu et al. [286]. The critical role of bioactive GQDs in damaged bone formation in tissue engineering is shown in Figure 12. Specifically, bioactive GQDs were found to stimulate the primary activation of ontogenesis. In medicine, bioactive GQD-based polymer composites are considered valuable in terms of regeneration because these particles have excellent mechanical properties, differentiation powers, and low toxicity. In addition, they can increase the abundance of calcium [16]. In tissue engineering, the improvement of the mechanical properties of scaffolds in which orthopedic implants can be placed is performed by bioactive GQDs, which act as reinforcing agents [287]. In addition to bone, bioactive GQDs are used for nerve and cartilage tissues [16].

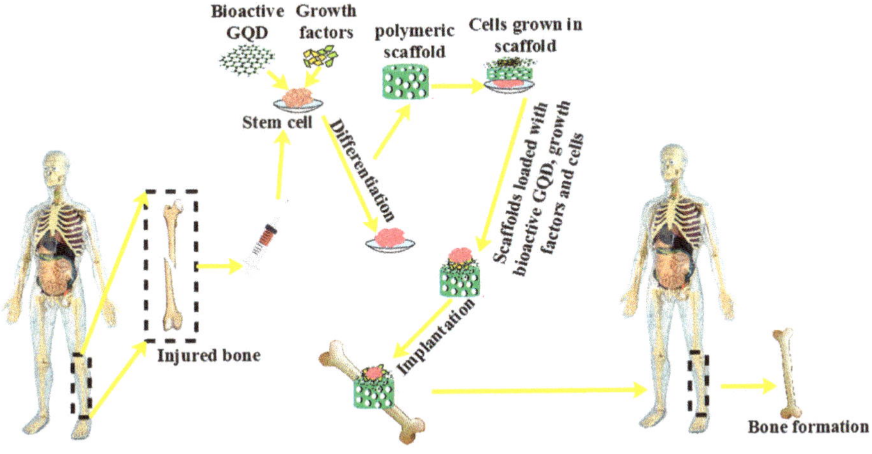

Figure 12. The important role of bioactive GQDs in damaged bone formation in tissue engineering.

6. Cytotoxicity and Biocompatibility

The size and shape of GQDs, the doping of non-metallic ions, and the modification of graphene surfaces and edges by photoluminescence (PL), one of the most prominent properties of graphene, are all easily adjustable [288,289]. A Large surface area, easy surface modification, high photostability, low cytotoxicity, and excellent biocompatibility, besides adjustable PL, are all features of GQDs [290,291]. Compared to conventional semiconductor QDs and other carbon-based luminescent nanomaterials, photoluminescent GQDs are more durable. GQDs are used in biomedical applications due to their adjustable PL, excellent biocompatibility, low cytotoxicity, high biocompatibility, and ease of functionalization, and, as imaging and labeling luminescent agents, are attractive in the context of tracking molecular targets in living cells or organs [292].

7. Conclusions and Perspective

In the last few decades, bioactive GQDs have attracted much attention due to their properties and applications in various environmental and health fields. There are problems that need to be addressed in the context of their practical biological applications, such the preparation of bioactive GQDs, their size, reproducibility, and low quantum efficiency. The size of bioactive GQDs is responsible for their fluorescence properties. The low toxicity of bioactive GQDs also makes them suitable for use in many in vivo applications. Therefore, it is possible to expand their potential applications in various fields by overcoming the problem of their low quantum efficiency by preparing bioactive GQD nanocomposites with surface factorization and band gap engineering. In this study, bioactive GQDs were introduced, and their synthesis methods were explained. The various biomedical applications of bioactive GQDs, such as bioimaging (in vivo and in vitro), drug delivery, gene delivery, photothermal therapy, photodynamic therapy, and tissue engineering, were discussed. The effortless combination of many drugs and ligands is provided through the large surface and functional groups of bioactive GQDs. Taking advantage of this, bioactive GQDs can be used as a nanocarrier for targeted drug delivery. Also, the photoluminescence of bioactive GQDs is used to develop bioimaging techniques to identify various biomolecules that offer a wide range of new strategies for disease diagnosis. They can be used to image cancer cells and to track drug delivery to cells and tissues. Bioactive GQDs present low levels of toxicity in the human body while having adjustable optical properties and fluorescence propagation in the NIR region of the light spectrum. Researchers aim to develop bioactive GQDs for biomedical applications and face a key challenge in the form of achieving high-quality products. The small-scale production of bioactive GQDs with large size distributions is possible through existing synthesis methods. Therefore, it is necessary to find new methods to achieve high performance and easy purification methods that do not require the removal of raw materials. Recent research advances in polymer composites based on bioactive GQDs focusing on their synthesis and biomedical applications, including bioimaging (in vivo and in vitro), drug delivery, gene delivery, light therapy, photodynamic therapy, and tissue engineering, were summarized in this review. Finally, we conclude that there is a promising future for further developing GQD-based polymer composites for many unresolved therapeutic barriers.

Author Contributions: S.M.M., S.B. and A.G. developed the idea and structure of the review article. S.A.H. and M.Y.K. wrote the manuscript collecting the materials from databases. N.O., S.B., N.V.R., A.B. and A.G. revised and improved the manuscript. A.G. and W.-H.C. supervised the manuscript. All authors have read and agreed to the published version of the manuscript.

Funding: This research was funded by Ministry of Science and Technology Taiwan, grant number MOST 110-2628-E-011-003.

Institutional Review Board Statement: Not applicable.

Informed Consent Statement: Not applicable.

Data Availability Statement: All data generated or analyzed during this study are included in this published article.

Conflicts of Interest: The authors declare no conflict of interest.

References

1. Zhao, X.; Courtney, J.M.; Qian, H. *Bioactive Materials in Medicine: Design and Applications*; Elsevier: Amsterdam, The Netherlands, 2011.
2. Geim, A. Graphene prehistory. *Phys. Scr.* **2012**, *2012*, 014003. [CrossRef]
3. Mousavi, S.M.; Zarei, M.; Hashemi, S.A.; Ramakrishna, S.; Chiang, W.-H.; Lai, C.W.; Gholami, A. Gold nanostars-diagnosis, bioimaging and biomedical applications. *Drug Metab. Rev.* **2020**, *52*, 299–318. [CrossRef] [PubMed]
4. Liu, Z.; Robinson, J.T.; Sun, X.; Dai, H. PEGylated nanographene oxide for delivery of water-insoluble cancer drugs. *J. Am. Chem. Soc.* **2008**, *130*, 10876–10877. [CrossRef] [PubMed]
5. Sanchez, V.C.; Jachak, A.; Hurt, R.H.; Kane, A.B. Biological interactions of graphene-family nanomaterials: An interdisciplinary review. *Chem. Res. Toxicol.* **2012**, *25*, 15–34. [CrossRef]
6. Fu, X.-W.; Shi, W.-Q.; Zhang, Q.-J.; Zhao, X.-M.; Yan, C.L.; Hou, Y.-P.; Zhou, G.-B.; Fan, Z.-Q.; Suo, L.; Wusiman, A. Positive effects of Taxol pretreatment on morphology, distribution and ultrastructure of mitochondria and lipid droplets in vitrification of in vitro matured porcine oocytes. *Anim. Reprod. Sci.* **2009**, *115*, 158–168. [CrossRef]
7. Geszke-Moritz, M.; Moritz, M. Quantum dots as versatile probes in medical sciences: Synthesis, modification and properties. *Mater. Sci. Eng. C* **2013**, *33*, 1008–1021. [CrossRef]
8. Rahimi-Aghdam, T.; Shariatinia, Z.; Hakkarainen, M.; Haddadi-Asl, V. Nitrogen and phosphorous doped graphene quantum dots: Excellent flame retardants and smoke suppressants for polyacrylonitrile nanocomposites. *J. Hazard. Mater.* **2020**, *381*, 121013. [CrossRef]
9. Yong, K.-T.; Wang, Y.; Roy, I.; Rui, H.; Swihart, M.T.; Law, W.-C.; Kwak, S.K.; Ye, L.; Liu, J.; Mahajan, S.D. Preparation of quantum dot/drug nanoparticle formulations for traceable targeted delivery and therapy. *Theranostics* **2012**, *2*, 681. [CrossRef]
10. Reshma, V.; Mohanan, P. Quantum dots: Applications and safety consequences. *J. Lumin.* **2019**, *205*, 287–298. [CrossRef]
11. Liu, L.; Jiang, H.; Dong, J.; Zhang, W.; Dang, G.; Yang, M.; Li, Y.; Chen, H.; Ji, H.; Dong, L. PEGylated MoS2 quantum dots for traceable and pH-responsive chemotherapeutic drug delivery. *Colloids Surf. B Biointerfaces* **2020**, *185*, 110590. [CrossRef]
12. Chen, F.; Gao, W.; Qiu, X.; Zhang, H.; Liu, L.; Liao, P.; Fu, W.; Luo, Y. Graphene quantum dots in biomedical applications: Recent advances and future challenges. *Front. Lab. Med.* **2017**, *1*, 192–199. [CrossRef]
13. Asadi, E.; Abdouss, M.; Leblanc, R.M.; Ezzati, N.; Wilson, J.N.; Azodi-Deilami, S. In vitro/in vivo study of novel anti-cancer, biodegradable cross-linked tannic acid for fabrication of 5-fluorouracil-targeting drug delivery nano-device based on a molecular imprinted polymer. *RSC Adv.* **2016**, *6*, 37308–37318. [CrossRef]
14. Asadi, E.; Abdouss, M.; Leblanc, R.M.; Ezzati, N.; Wilson, J.N.; Kordestani, D. Synthesis, characterization and in vivo drug delivery study of a biodegradable nano-structured molecularly imprinted polymer based on cross-linker of fructose. *Polymer* **2016**, *97*, 226–237. [CrossRef]
15. Abdouss, M.; Asadi, E.; Azodi-Deilami, S.; Beik-mohammadi, N.; Aslanzadeh, S.A. Development and characterization of molecularly imprinted polymers for controlled release of citalopram. *J. Mater. Sci. Mater. Med.* **2011**, *22*, 2273–2281. [CrossRef]
16. Lee, J.S.; Youn, Y.H.; Kwon, I.K.; Ko, N.R. Recent advances in quantum dots for biomedical applications. *J. Pharm. Investig.* **2018**, *48*, 209–214. [CrossRef]

17. Zhang, S.; Li, Y.; Song, H.; Chen, X.; Zhou, J.; Hong, S.; Huang, M. Graphene quantum dots as the electrolyte for solid state supercapacitors. *Sci. Rep.* **2016**, *6*, 1–7. [CrossRef] [PubMed]
18. Raee, M.J.; Ebrahiminezhad, A.; Gholami, A.; Ghoshoon, M.B.; Ghasemi, Y. Magnetic immobilization of recombinant E. coli producing extracellular asparaginase: An effective way to intensify downstream process. *Sep. Sci. Technol.* **2018**, *53*, 1397–1404. [CrossRef]
19. Qu, D.; Zheng, M.; Zhang, L.; Zhao, H.; Xie, Z.; Jing, X.; Haddad, R.E.; Fan, H.; Sun, Z. Formation mechanism and optimization of highly luminescent N-doped graphene quantum dots. *Sci. Rep.* **2014**, *4*, 1–11. [CrossRef]
20. Shen, J.; Zhu, Y.; Yang, X.; Li, C. Graphene quantum dots: Emergent nanolights for bioimaging, sensors, catalysis and photovoltaic devices. *Chem. Commun.* **2012**, *48*, 3686–3699. [CrossRef]
21. Zheng, X.T.; Ananthanarayanan, A.; Luo, K.Q.; Chen, P. Glowing graphene quantum dots and carbon dots: Properties, syntheses, and biological applications. *Small* **2015**, *11*, 1620–1636. [CrossRef]
22. Nurunnabi, M.; Khatun, Z.; Huh, K.M.; Park, S.Y.; Lee, D.Y.; Cho, K.J.; Lee, Y.-K. In vivo biodistribution and toxicology of carboxylated graphene quantum dots. *ACS Nano* **2013**, *7*, 6858–6867. [CrossRef] [PubMed]
23. Huang, C.-L.; Huang, C.-C.; Mai, F.-D.; Yen, C.-L.; Tzing, S.-H.; Hsieh, H.-T.; Ling, Y.-C.; Chang, J.-Y. Application of paramagnetic graphene quantum dots as a platform for simultaneous dual-modality bioimaging and tumor-targeted drug delivery. *J. Mater. Chem. B* **2015**, *3*, 651–664. [CrossRef] [PubMed]
24. Apul, O.G.; Wang, Q.; Zhou, Y.; Karanfil, T. Adsorption of aromatic organic contaminants by graphene nanosheets: Comparison with carbon nanotubes and activated carbon. *Water Res.* **2013**, *47*, 1648–1654. [CrossRef] [PubMed]
25. Ogi, T.; Iwasaki, H.; Aishima, K.; Iskandar, F.; Wang, W.-N.; Takimiya, K.; Okuyama, K. Transient nature of graphene quantum dot formation via a hydrothermal reaction. *RSC Adv.* **2014**, *4*, 55709–55715. [CrossRef]
26. Ajayan, P.M.; Schadler, L.S.; Braun, P.V. *Nanocomposite Science and Technology*; John Wiley & Sons: Hoboken, NJ, USA, 2006.
27. Mohkam, M.; Rasoul-Amini, S.; Shokri, D.; Berenjian, A.; Rahimi, F.; Sadraeian, M.; Khalvati, B.; Gholami, A.; Ghasemi, Y. Characterization and in vitro probiotic assessment of potential indigenous Bacillus strains isolated from soil rhizosphere. *Minerva Biotecnol.* **2016**, *28*, 19–28.
28. Gholami, A.; Hashemi, S.A.; Yousefi, K.; Mousavi, S.M.; Chiang, W.-H.; Ramakrishna, S.; Mazraedoost, S.; Alizadeh, A.; Omidifar, N.; Behbudi, G. 3D nanostructures for tissue engineering, cancer therapy, and gene delivery. *J. Nanomater.* **2020**, *2020*, 1852946. [CrossRef]
29. Vera, M.; Mella, C.; Urbano, B.F. Smart polymer nanocomposites: Recent advances and perspectives. *J. Chil. Chem. Soc.* **2020**, *65*, 4973–4981. [CrossRef]
30. Müller, K.; Bugnicourt, E.; Latorre, M.; Jorda, M.; Echegoyen Sanz, Y.; Lagaron, J.M.; Miesbauer, O.; Bianchin, A.; Hankin, S.; Bölz, U. Review on the processing and properties of polymer nanocomposites and nanocoatings and their applications in the packaging, automotive and solar energy fields. *Nanomaterials* **2017**, *7*, 74. [CrossRef]
31. Schadler, L.; Brinson, L.; Sawyer, W. Polymer nanocomposites: A small part of the story. *Jom* **2007**, *59*, 53–60. [CrossRef]
32. Kamigaito, O. What can be improved by nanometer composites? *J. Jpn. Soc. Powder Powder Metall.* **1991**, *38*, 315–321. [CrossRef]
33. Liang, L.; Peng, X.; Sun, F.; Kong, Z.; Shen, J.-W. A review on the cytotoxicity of graphene quantum dots: From experiment to simulation. *Nanoscale Adv.* **2021**, *3*, 904–917. [CrossRef]
34. Bonaccorso, F.; Sun, Z.; Hasan, T.; Ferrari, A. Graphene photonics and optoelectronics. *Nat. Photonics* **2010**, *4*, 611–622. [CrossRef]
35. Cooper, D.; D'Anjou, B.; Ghattamaneni, N.; Harack, B.; Hilke, M.; Horth, A. ISRN Condensed Matter. *Phys* **2012**, *2012*, 501686.
36. Kuc, A.; Heine, T.; Seifert, G. Structural and electronic properties of graphene nanoflakes. *Phys. Rev. B* **2010**, *81*, 085430. [CrossRef]
37. Chang, H.; Wang, G.; Yang, A.; Tao, X.; Liu, X.; Shen, Y.; Zheng, Z. A transparent, flexible, low-temperature, and solution-processible graphene composite electrode. *Adv. Funct. Mater.* **2010**, *20*, 2893–2902. [CrossRef]
38. Kim, K.H.; Oh, Y.; Islam, M. Graphene coating makes carbon nanotube aerogels superelastic and resistant to fatigue. *Nat. Nanotechnol.* **2012**, *7*, 562–566. [CrossRef] [PubMed]
39. Chang, C.-H.; Huang, T.-C.; Peng, C.-W.; Yeh, T.-C.; Lu, H.-I.; Hung, W.-I.; Weng, C.-J.; Yang, T.-I.; Yeh, J.-M. Novel anticorrosion coatings prepared from polyaniline/graphene composites. *Carbon* **2012**, *50*, 5044–5051. [CrossRef]
40. Mousavi, S.M.; Hashemi, S.A.; Zarei, M.; Gholami, A.; Lai, C.W.; Chiang, W.H.; Omidifar, N.; Bahrani, S.; Mazraedoost, S. Recent Progress in Chemical Composition, Production, and Pharmaceutical Effects of Kombucha Beverage: A Complementary and Alternative Medicine. *Evid.-Based Complement. Altern. Med.* **2020**, *2020*, 4397543. [CrossRef]
41. Stankovich, S.; Dikin, D.A.; Dommett, G.H.; Kohlhaas, K.M.; Zimney, E.J.; Stach, E.A.; Piner, R.D.; Nguyen, S.T.; Ruoff, R.S. Graphene-based composite materials. *Nature* **2006**, *442*, 282–286. [CrossRef]
42. Ramanathan, T.; Abdala, A.; Stankovich, S.; Dikin, D.; Herrera-Alonso, M.; Piner, R.; Adamson, D.; Schniepp, H.; Chen, X.; Ruoff, R. Functionalized graphene sheets for polymer nanocomposites. *Nat. Nanotechnol.* **2008**, *3*, 327–331. [CrossRef]
43. Eda, G.; Chhowalla, M. Graphene-based composite thin films for electronics. *Nano Lett.* **2009**, *9*, 814–818. [CrossRef]
44. Meinardi, F.; Colombo, A.; Velizhanin, K.A.; Simonutti, R.; Lorenzon, M.; Beverina, L.; Viswanatha, R.; Klimov, V.I.; Brovelli, S. Large-area luminescent solar concentrators based on 'Stokes-shift-engineered'nanocrystals in a mass-polymerized PMMA matrix. *Nat. Photonics* **2014**, *8*, 392–399. [CrossRef]
45. Cho, S.; Kwag, J.; Jeong, S.; Baek, Y.; Kim, S. Highly fluorescent and stable quantum dot-polymer-layered double hydroxide composites. *Chem. Mater.* **2013**, *25*, 1071–1077. [CrossRef]
46. Han, S.; Li, X.; Wang, Y.; Su, C. A core–shell Fe_3O_4 nanoparticle–CdTe quantum dot–molecularly imprinted polymer composite for recognition and separation of 4-nonylphenol. *Anal. Methods* **2014**, *6*, 2855–2861. [CrossRef]

47. Liu, H.; Zhou, K.; Wu, D.; Wang, J.; Sun, B. A novel quantum dots-labeled on the surface of molecularly imprinted polymer for turn-off optosensing of dicyandiamide in dairy products. *Biosens. Bioelectron.* **2016**, *77*, 512–517. [CrossRef] [PubMed]
48. Kahmann, S.; Mura, A.; Protesescu, L.; Kovalenko, M.V.; Brabec, C.J.; Loi, M.A. Opto-electronics of PbS quantum dot and narrow bandgap polymer blends. *J. Mater. Chem. C* **2015**, *3*, 5499–5505. [CrossRef]
49. Matvienko, O.O.; Savin, Y.N.; Kryzhanovska, A.S.; Vovk, O.M.; Dobrotvorska, M.V.; Pogorelova, N.V.; Vashchenko, V.V. Dispersion and aggregation of quantum dots in polymer–inorganic hybrid films. *Thin Solid Film.* **2013**, *537*, 226–230. [CrossRef]
50. Mousavi, S.M.; Hashemi, S.A.; Zarei, M.; Bahrani, S.; Savardashtaki, A.; Esmaeili, H.; Lai, C.W.; Mazraedoost, S.; Abassi, M.; Ramavandi, B. Data on cytotoxic and antibacterial activity of synthesized Fe3O4 nanoparticles using Malva sylvestris. *Data Brief* **2020**, *28*, 104929. [CrossRef]
51. Seibert, J.R.; Keleş, Ö.; Wang, J.; Erogbogbo, F. Infusion of graphene quantum dots to modulate thermal conductivity and dynamic mechanical properties of polymers. *Polymer* **2019**, *185*, 121988. [CrossRef]
52. Hu, K.; Kulkarni, D.D.; Choi, I.; Tsukruk, V.V. Graphene-polymer nanocomposites for structural and functional applications. *Prog. Polym. Sci.* **2014**, *39*, 1934–1972. [CrossRef]
53. Shi, X.; Wang, Y.; Sun, H.; Chen, Y.; Zhang, X.; Xu, J.; Zhai, G. Heparin-reduced graphene oxide nanocomposites for curcumin delivery: In vitro, in vivo and molecular dynamics simulation study. *Biomater. Sci.* **2019**, *7*, 1011–1027. [CrossRef] [PubMed]
54. Cheng, C.; Li, S.; Nie, S.; Zhao, W.; Yang, H.; Sun, S.; Zhao, C. General and biomimetic approach to biopolymer-functionalized graphene oxide nanosheet through adhesive dopamine. *Biomacromolecules* **2012**, *13*, 4236–4246. [CrossRef] [PubMed]
55. Lee, D.Y.; Khatun, Z.; Lee, J.-H.; Lee, Y.-K.; In, I. Blood compatible graphene/heparin conjugate through noncovalent chemistry. *Biomacromolecules* **2011**, *12*, 336–341. [CrossRef]
56. Cheng, C.; Nie, S.; Li, S.; Peng, H.; Yang, H.; Ma, L.; Sun, S.; Zhao, C. Biopolymer functionalized reduced graphene oxide with enhanced biocompatibility via mussel inspired coatings/anchors. *J. Mater. Chem. B* **2013**, *1*, 265–275. [CrossRef] [PubMed]
57. Shim, G.; Kim, J.-Y.; Han, J.; Chung, S.W.; Lee, S.; Byun, Y.; Oh, Y.-K. Reduced graphene oxide nanosheets coated with an anti-angiogenic anticancer low-molecular-weight heparin derivative for delivery of anticancer drugs. *J. Control. Release* **2014**, *189*, 80–89. [CrossRef] [PubMed]
58. Makar, A.; McMartin, K.; Palese, M.; Tephly, T. Formate assay in body fluids: Application in methanol poisoning. *Biochem. Med.* **1975**, *13*, 117–126. [CrossRef]
59. Heiran, R.; Sepehri, S.; Jarrahpour, A.; Digiorgio, C.; Douafer, H.; Brunel, J.M.; Gholami, A.; Riazimontazer, E.; Turos, E. Synthesis, docking and evaluation of in vitro anti-inflammatory activity of novel morpholine capped β-lactam derivatives. *Bioorganic Chem.* **2020**, *102*, 104091. [CrossRef] [PubMed]
60. Cannell, R.J. How to approach the isolation of a natural product. In *Natural Products Isolation*; Springer: Berlin/Heidelberg, Germany, 1998; pp. 1–51.
61. Zahner, H.; Drautz, H.; Weber, W. *Novel Approaches to Metabolite Screening. Bioactive Secondary Metabolites. Search and Discovery*; Bu'Lock, J.D., Nisbet, L., Winstanley, D., Eds.; Academic Press: London, UK, 1982.
62. Charyulu, E.M.; Sekaran, G.; Rajakumar, G.S.; Gnanamani, A. Antimicrobial activity of secondary metabolite from marine isolate, *Pseudomonas* sp. against Gram positive and negative bacteria including MRSA. *Indian J. Exp. Biol.* **2009**, *47*, 964–968. [PubMed]
63. Rosenfeld, W.D.; ZoBell, C.E. Antibiotic production by marine microorganisms. *J. Bacteriol.* **1947**, *54*, 393–398. [CrossRef]
64. Ahvenainen, R. *Novel Food Packaging Techniques*; Elsevier: Amsterdam, The Netherlands, 2003.
65. Le Guillou-Buffello, D.; Hélary, G.; Gindre, M.; Pavon-Djavid, G.; Laugier, P.; Migonney, V. Monitoring cell adhesion processes on bioactive polymers with the quartz crystal resonator technique. *Biomaterials* **2005**, *26*, 4197–4205. [CrossRef]
66. Aikio, S.; Grönqvist, S.; Hakola, L.; Hurme, E.; Jussila, S.; Kaukoniemi, O.-V.; Kopola, H.; Känsäkoski, M.; Leinonen, M.; Lippo, S. *Bioactive Paper and Fibre Products: Patent and Literary Survey*; VTT Technical Research Centre of Finland: Espoo, Finland, 2006.
67. Han, Y.; Zhang, F.; Zhang, J.; Shao, D.; Wang, Y.; Li, S.; Lv, S.; Chi, G.; Zhang, M.; Chen, L. Bioactive carbon dots direct the osteogenic differentiation of human bone marrow mesenchymal stem cells. *Colloids Surf. B Biointerfaces* **2019**, *179*, 1–8. [CrossRef]
68. Lim, S.Y.; Shen, W.; Gao, Z. Carbon quantum dots and their applications. *Chem. Soc. Rev.* **2015**, *44*, 362–381. [CrossRef] [PubMed]
69. Zhang, Z.; Hao, J.; Zhang, J.; Zhang, B.; Tang, J. Protein as the source for synthesizing fluorescent carbon dots by a one-pot hydrothermal route. *RSC Adv.* **2012**, *2*, 8599–8601. [CrossRef]
70. Xu, X.; Zhang, K.; Zhao, L.; Li, C.; Bu, W.; Shen, Y.; Gu, Z.; Chang, B.; Zheng, C.; Lin, C. Aspirin-based carbon dots, a good biocompatibility of material applied for bioimaging and anti-inflammation. *ACS Appl. Mater. Interfaces* **2016**, *8*, 32706–32716. [CrossRef] [PubMed]
71. Liu, J.; Lu, S.; Tang, Q.; Zhang, K.; Yu, W.; Sun, H.; Yang, B. One-step hydrothermal synthesis of photoluminescent carbon nanodots with selective antibacterial activity against Porphyromonas gingivalis. *Nanoscale* **2017**, *9*, 7135–7142. [CrossRef]
72. Horst, F.H.; da Silva Rodrigues, C.V.; Carvalho, P.H.P.R.; Leite, A.M.; Azevedo, R.B.; Neto, B.A.; Corrêa, J.R.; Garcia, M.P.; Alotaibi, S.; Henini, M. From cow manure to bioactive carbon dots: A light-up probe for bioimaging investigations, glucose detection and potential immunotherapy agent for melanoma skin cancer. *RSC Adv.* **2021**, *11*, 6346–6352. [CrossRef]
73. Zhang, F.; Zhang, M.; Zheng, X.; Tao, S.; Zhang, Z.; Sun, M.; Song, Y.; Zhang, J.; Shao, D.; He, K. Berberine-based carbon dots for selective and safe cancer theranostics. *RSC Adv.* **2018**, *8*, 1168–1173. [CrossRef]
74. Nirala, N.R.; Khandelwal, G.; Kumar, B.; Prakash, R.; Kumar, V. One step electro-oxidative preparation of graphene quantum dots from wood charcoal as a peroxidase mimetic. *Talanta* **2017**, *173*, 36–43. [CrossRef]

75. Roy, P.; Periasamy, A.P.; Chuang, C.; Liou, Y.-R.; Chen, Y.-F.; Joly, J.; Liang, C.-T.; Chang, H.-T. Plant leaf-derived graphene quantum dots and applications for white LEDs. *New J. Chem.* **2014**, *38*, 4946–4951. [CrossRef]
76. Wang, Z.; Yu, J.; Zhang, X.; Li, N.; Liu, B.; Li, Y.; Wang, Y.; Wang, W.; Li, Y.; Zhang, L. Large-scale and controllable synthesis of graphene quantum dots from rice husk biomass: A comprehensive utilization strategy. *ACS Appl. Mater. Interfaces* **2016**, *8*, 1434–1439. [CrossRef] [PubMed]
77. Suryawanshi, A.; Biswal, M.; Mhamane, D.; Gokhale, R.; Patil, S.; Guin, D.; Ogale, S. Large scale synthesis of graphene quantum dots (GQDs) from waste biomass and their use as an efficient and selective photoluminescence on–off–on probe for Ag+ ions. *Nanoscale* **2014**, *6*, 11664–11670. [CrossRef]
78. Wang, L.; Li, W.; Wu, B.; Li, Z.; Wang, S.; Liu, Y.; Pan, D.; Wu, M. Facile synthesis of fluorescent graphene quantum dots from coffee grounds for bioimaging and sensing. *Chem. Eng. J.* **2016**, *300*, 75–82. [CrossRef]
79. Wang, J.; Zhang, H.; Wang, F.; Ai, X.; Huang, D.; Liu, G.; Mi, P. Enzyme-responsive polymers for drug delivery and molecular imaging. In *Stimuli Responsive Polymeric Nanocarriers for Drug Delivery Applications, Volume 1*; Elsevier: Amsterdam, The Netherlands, 2018; pp. 101–119.
80. Maxwell, D.; Chang, Q.; Zhang, X.; Barnett, E.M.; Piwnica-Worms, D. An improved cell-penetrating, caspase-activatable, near-infrared fluorescent peptide for apoptosis imaging. *Bioconjugate Chem.* **2009**, *20*, 702–709. [CrossRef]
81. Wilson, A.N.; Salas, R.; Guiseppi-Elie, A. Bioactive hydrogels demonstrate mediated release of a chromophore by chymotrypsin. *J. Control. Release* **2012**, *160*, 41–47. [PubMed]
82. Zhao, Q.; Zhang, J.; Song, L.; Ji, Q.; Yao, Y.; Cui, Y.; Shen, J.; Wang, P.G.; Kong, D. Polysaccharide-based biomaterials with on-demand nitric oxide releasing property regulated by enzyme catalysis. *Biomaterials* **2013**, *34*, 8450–8458. [CrossRef] [PubMed]
83. Kalgutkar, A.S. Role of bioactivation in idiosyncratic drug toxicity: Structure–toxicity relationships. In *Advances in Bioactivation Research*; Springer: Berlin/Heidelberg, Germany, 2008; pp. 1–29.
84. Sanderson, J.P.; Naisbitt, D.J.; Park, B.K. Role of bioactivation in drug-induced hypersensitivity reactions. *AAPS J.* **2006**, *8*, E55–E64. [CrossRef] [PubMed]
85. Gómez-Lechón, M.J.; Tolosa, L.; Donato, M.T. Metabolic activation and drug-induced liver injury: In vitro approaches for the safety risk assessment of new drugs. *J. Appl. Toxicol.* **2016**, *36*, 752–768. [CrossRef]
86. Doss, G.A.; Baillie, T.A. Addressing metabolic activation as an integral component of drug design. *Drug Metab. Rev.* **2006**, *38*, 641–649. [CrossRef]
87. Shin, N.-Y.; Liu, Q.; Stamer, S.L.; Liebler, D.C. Protein targets of reactive electrophiles in human liver microsomes. *Chem. Res. Toxicol.* **2007**, *20*, 859–867. [CrossRef]
88. Walsh, J.S.; Miwa, G.T. Bioactivation of drugs: Risk and drug design. *Annu. Rev. Pharmacol. Toxicol.* **2011**, *51*, 145–167. [CrossRef]
89. Wang, L.; Zhou, H.S. Green synthesis of luminescent nitrogen-doped carbon dots from milk and its imaging application. *Anal. Chem.* **2014**, *86*, 8902–8905. [CrossRef] [PubMed]
90. Havrdova, M.; Hola, K.; Skopalik, J.; Tomankova, K.; Petr, M.; Cepe, K.; Polakova, K.; Tucek, J.; Bourlinos, A.B.; Zboril, R. Toxicity of carbon dots–Effect of surface functionalization on the cell viability, reactive oxygen species generation and cell cycle. *Carbon* **2016**, *99*, 238–248. [CrossRef]
91. Lu, J.; Yeo, P.S.E.; Gan, C.K.; Wu, P.; Loh, K.P. Transforming C 60 molecules into graphene quantum dots. *Nat. Nanotechnol.* **2011**, *6*, 247–252. [CrossRef]
92. Juzenas, P.; Kleinauskas, A.; George Luo, P.; Sun, Y.-P. Photoactivatable carbon nanodots for cancer therapy. *Appl. Phys. Lett.* **2013**, *103*, 063701. [CrossRef]
93. Zong, J.; Zhu, Y.; Yang, X.; Shen, J.; Li, C. Synthesis of photoluminescent carbogenic dots using mesoporous silica spheres as nanoreactors. *Chem. Commun.* **2011**, *47*, 764–766. [CrossRef] [PubMed]
94. Chua, C.K.; Sofer, Z.; Simek, P.; Jankovsky, O.; Klimova, K.; Bakardjieva, S.; Hrdličková Kučková, S.; Pumera, M. Synthesis of strongly fluorescent graphene quantum dots by cage-opening buckminsterfullerene. *ACS Nano* **2015**, *9*, 2548–2555. [CrossRef]
95. Jeong, C.J.; Roy, A.K.; Kim, S.H.; Lee, J.-E.; Jeong, J.H.; In, I.; Park, S.Y. Fluorescent carbon nanoparticles derived from natural materials of mango fruit for bio-imaging probes. *Nanoscale* **2014**, *6*, 15196–15202. [CrossRef]
96. Gholami, A.; Emadi, F.; Nazem, M.; Aghayi, R.; Khalvati, B.; Amini, A.; Ghasemi, Y. Expression of key apoptotic genes in hepatocellular carcinoma cell line treated with etoposide-loaded graphene oxide. *J. Drug Deliv. Sci. Technol.* **2020**, *57*, 101725. [CrossRef]
97. Luo, P.; Guan, X.; Yu, Y.; Li, X.; Yan, F. Hydrothermal synthesis of graphene quantum dots supported on three-dimensional graphene for supercapacitors. *Nanomaterials* **2019**, *9*, 201. [CrossRef]
98. He, J.; Li, Z.; Zhao, R.; Lu, Y.; Shi, L.; Liu, J.; Dong, X.; Xi, F. Aqueous synthesis of amphiphilic graphene quantum dots and their application as surfactants for preparing of fluorescent polymer microspheres. *Colloids Surf. A Physicochem. Eng. Asp.* **2019**, *563*, 77–83. [CrossRef]
99. Facure, M.; Schneider, R.; Mercante, L.; Correa, D. Rational hydrothermal synthesis of graphene quantum dots with optimized luminescent properties for sensing applications. *Mater. Today Chem.* **2022**, *23*, 100755. [CrossRef]
100. Kundu, S.; Yadav, R.M.; Narayanan, T.; Shelke, M.V.; Vajtai, R.; Ajayan, P.M.; Pillai, V.K. Synthesis of N, F and S co-doped graphene quantum dots. *Nanoscale* **2015**, *7*, 11515–11519. [CrossRef] [PubMed]
101. Hu, C.; Su, T.-R.; Lin, T.-J.; Chang, C.-W.; Tung, K.-L. Yellowish and blue luminescent graphene oxide quantum dots prepared via a microwave-assisted hydrothermal route using H_2O_2 and $KMnO_4$ as oxidizing agents. *New J. Chem.* **2018**, *42*, 3999–4007.

102. Li, W.; Li, M.; Liu, Y.; Pan, D.; Li, Z.; Wang, L.; Wu, M. Three minute ultrarapid microwave-assisted synthesis of bright fluorescent graphene quantum dots for live cell staining and white LEDs. *ACS Appl. Nano Mater.* **2018**, *1*, 1623–1630.
103. Li, Y.; Li, S.; Wang, Y.; Wang, J.; Liu, H.; Liu, X.; Wang, L.; Liu, X.; Xue, W.; Ma, N. Electrochemical synthesis of phosphorus-doped graphene quantum dots for free radical scavenging. *Phys. Chem. Chem. Phys.* **2017**, *19*, 11631–11638.
104. Kalita, H.; Palaparthy, V.S.; Baghini, M.S.; Aslam, M. Electrochemical synthesis of graphene quantum dots from graphene oxide at room temperature and its soil moisture sensing properties. *Carbon* **2020**, *165*, 9–17. [CrossRef]
105. Karaman, O.; Özcan, N.; Karaman, C.; Yola, B.; Atar, N.; Yola, M. Electrochemical cardiac troponin I immunosensor based on nitrogen and boron-doped graphene quantum dots electrode platform and Ce-doped SnO_2/SnS_2 signal amplification. *Mater. Today Chem.* **2022**, *23*, 100666.
106. Roeinfard, M.; Zahedifar, M.; Darroudi, M.; Zak, A.K.; Sadeghi, E. Synthesis of Graphene Quantum Dots Decorated With Se, Eu and Ag As Photosensitizer and Study of Their Potential to Use in Photodynamic Therapy. *J. Fluoresc.* **2021**, *31*, 551–557. [CrossRef]
107. Mustafa, S.M.; Barzinjy, A.A.; Hamad, A.H.; Hamad, S.M. Biosynthesis of quantum dots and their usage in solar cells: Insight from the novel researches. *Int. Nano Lett.* **2021**, *2021*, 1–13. [CrossRef]
108. Öcal, N.; Ceylan, A.; Duman, F. Intracellular biosynthesis of PbS quantum dots using Pseudomonas aeruginosa ATCC 27853: Evaluation of antibacterial effects and DNA cleavage activities. *World J. Microbiol. Biotechnol.* **2020**, *36*, 1–10.
109. Liu, Y.; Zhao, Y.; Zhang, Y. One-step green synthesized fluorescent carbon nanodots from bamboo leaves for copper (II) ion detection. *Sens. Actuators B Chem.* **2014**, *196*, 647–652. [CrossRef]
110. Kumawat, M.K.; Thakur, M.; Gurung, R.B.; Srivastava, R. Graphene quantum dots from mangifera indica: Application in near-infrared bioimaging and intracellular nanothermometry. *ACS Sustain. Chem. Eng.* **2017**, *5*, 1382–1391. [CrossRef]
111. Cao, H.; Qi, W.; Gao, X.; Wu, Q.; Tian, L.; Wu, W. Graphene Quantum Dots prepared by Electron Beam Irradiation for Safe Fluorescence Imaging of Tumor. *Nanotheranostics* **2022**, *6*, 205. [CrossRef] [PubMed]
112. Hu, L.; Sun, Y.; Li, S.; Wang, X.; Hu, K.; Wang, L.; Liang, X.-j.; Wu, Y. Multifunctional carbon dots with high quantum yield for imaging and gene delivery. *Carbon* **2014**, *67*, 508–513. [CrossRef]
113. Biswas, M.C.; Islam, M.T.; Nandy, P.K.; Hossain, M.M. Graphene Quantum Dots (GQDs) for Bioimaging and Drug Delivery Applications: A Review. *ACS Mater. Lett.* **2021**, *3*, 889–911. [CrossRef]
114. Jiao, J.; Liu, C.; Li, X.; Liu, J.; Di, D.; Zhang, Y.; Zhao, Q.; Wang, S. Fluorescent carbon dot modified mesoporous silica nanocarriers for redox-responsive controlled drug delivery and bioimaging. *J. Colloid Interface Sci.* **2016**, *483*, 343–352. [CrossRef]
115. Yu, C.; Li, X.; Zeng, F.; Zheng, F.; Wu, S. Carbon-dot-based ratiometric fluorescent sensor for detecting hydrogen sulfide in aqueous media and inside live cells. *Chem. Commun.* **2013**, *49*, 403–405. [CrossRef]
116. Wang, G.; Guo, Q.; Chen, D.; Liu, Z.; Zheng, X.; Xu, A.; Yang, S.; Ding, G. Facile and highly effective synthesis of controllable lattice sulfur-doped graphene quantum dots via hydrothermal treatment of durian. *ACS Appl. Mater. Interfaces* **2018**, *10*, 5750–5759. [CrossRef]
117. Le, T.H.; Ahn, Y.N.; Park, S.J. An effective method for cysteine determination based on fluorescence resonance energy system between co-doped graphene quantum dots and silver nanoparticles. *Korean J. Chem. Eng.* **2022**, *2022*, 1–7. [CrossRef]
118. Tao, H.; Yang, K.; Ma, Z.; Wan, J.; Zhang, Y.; Kang, Z.; Liu, Z. In vivo NIR fluorescence imaging, biodistribution, and toxicology of photoluminescent carbon dots produced from carbon nanotubes and graphite. *Small* **2012**, *8*, 281–290. [CrossRef]
119. Ding, Z.; Li, F.; Wen, J.; Wang, X.; Sun, R. Gram-scale synthesis of single-crystalline graphene quantum dots derived from lignin biomass. *Green Chem.* **2018**, *20*, 1383–1390. [CrossRef]
120. Liu, B.; Xie, J.; Ma, H.; Zhang, X.; Pan, Y.; Lv, J.; Ge, H.; Ren, N.; Su, H.; Xie, X. From graphite to graphene oxide and graphene oxide quantum dots. *Small* **2017**, *13*, 1601001. [CrossRef] [PubMed]
121. Teymourinia, H.; Salavati-Niasari, M.; Amiri, O.; Safardoust-Hojaghan, H. Synthesis of graphene quantum dots from corn powder and their application in reduce charge recombination and increase free charge carriers. *J. Mol. Liq.* **2017**, *242*, 447–455. [CrossRef]
122. Chu, X.; Dai, P.; Dong, Y.; Sun, W.; Bai, L.; Zhang, W. The acetic acid gas sensing properties of graphene quantum dots (GQDs)–ZnO nanocomposites prepared by hydrothermal method. *J. Mater. Sci. Mater. Electron.* **2017**, *28*, 19164–19173. [CrossRef]
123. Chu, X.; Dai, P.; Liang, S.; Bhattacharya, A.; Dong, Y.; Epifani, M. The acetone sensing properties of ZnFe2O4-graphene quantum dots (GQDs) nanocomposites at room temperature. *Phys. E Low-Dimens. Syst. Nanostructures* **2019**, *106*, 326–333. [CrossRef]
124. Mehrdad-Vahdati, B.; Pourhashem, S.; Sedghi, M.; Vaezi, Z.; Shojaedin-Givi, B.; Rashidi, A.; Naderi-Manesh, H. A novel aspect of functionalized graphene quantum dots in cytotoxicity studies. *Toxicol. Vitr.* **2019**, *61*, 104649. [CrossRef]
125. Gao, X.X.; Zhou, X.; Ma, Y.F.; Wang, C.P.; Chu, F.X. A fluorometric and colorimetric dual-mode sensor based on nitrogen and iron co-doped graphene quantum dots for detection of ferric ions in biological fluids and cellular imaging. *New J. Chem.* **2018**, *42*, 14751–14756. [CrossRef]
126. Liang, Q.; Ma, W.; Shi, Y.; Li, Z.; Yang, X. Easy synthesis of highly fluorescent carbon quantum dots from gelatin and their luminescent properties and applications. *Carbon* **2013**, *60*, 421–428. [CrossRef]
127. Liu, W.; Diao, H.; Chang, H.; Wang, H.; Li, T.; Wei, W. Green synthesis of carbon dots from rose-heart radish and application for Fe3+ detection and cell imaging. *Sens. Actuators B Chem.* **2017**, *241*, 190–198. [CrossRef]
128. Miao, H.; Wang, L.; Zhuo, Y.; Zhou, Z.; Yang, X. Label-free fluorimetric detection of CEA using carbon dots derived from tomato juice. *Biosens. Bioelectron.* **2016**, *86*, 83–89. [CrossRef]
129. Hou, X.; Li, Y.; Zhao, C. Microwave-assisted synthesis of nitrogen-doped multi-layer graphene quantum dots with oxygen-rich functional groups. *Aust. J. Chem.* **2015**, *69*, 357–360. [CrossRef]

130. Alves, A.K.; Frantz, A.C.S.; Berutti, F.A. Microwave-assisted oleothermal synthesis of graphene-TiO2 quantum dots for photoelectrochemical oxygen evolution reaction. *FlatChem* **2018**, *12*, 26–34. [CrossRef]
131. Fresco-Cala, B.; Soriano, M.L.; Sciortino, A.; Cannas, M.; Messina, F.; Cardenas, S. One-pot synthesis of graphene quantum dots and simultaneous nanostructured self-assembly via a novel microwave-assisted method: Impact on triazine removal and efficiency monitoring. *RSC Adv.* **2018**, *8*, 29939–29946. [CrossRef]
132. Ayele, D.W.; Chen, H.M.; Su, W.N.; Pan, C.J.; Chen, L.Y.; Chou, H.L.; Cheng, J.H.; Hwang, B.J.; Lee, J.F. Controlled Synthesis of CdSe Quantum Dots by a Microwave-Enhanced Process: A Green Approach for Mass Production. *Chem.-A Eur. J.* **2011**, *17*, 5737. [CrossRef] [PubMed]
133. Gu, D.; Shang, S.; Yu, Q.; Shen, J. Green synthesis of nitrogen-doped carbon dots from lotus root for Hg (II) ions detection and cell imaging. *Appl. Surf. Sci.* **2016**, *390*, 38–42. [CrossRef]
134. Zhang, C.; Cui, Y.; Song, L.; Liu, X.; Hu, Z. Microwave assisted one-pot synthesis of graphene quantum dots as highly sensitive fluorescent probes for detection of iron ions and pH value. *Talanta* **2016**, *150*, 54–60. [CrossRef]
135. Chen, L.; Wu, C.; Du, P.; Feng, X.; Wu, P.; Cai, C. Electrolyzing synthesis of boron-doped graphene quantum dots for fluorescence determination of Fe3+ ions in water samples. *Talanta* **2017**, *164*, 100–109. [CrossRef]
136. Huang, H.; Yang, S.; Li, Q.; Yang, Y.; Wang, G.; You, X.; Mao, B.; Wang, H.; Ma, Y.; He, P. Electrochemical cutting in weak aqueous electrolytes: The strategy for efficient and controllable preparation of graphene quantum dots. *Langmuir* **2018**, *34*, 250–258. [CrossRef] [PubMed]
137. Paulo, S.; Palomares, E.; Martinez-Ferrero, E. Graphene and carbon quantum dot-based materials in photovoltaic devices: From synthesis to applications. *Nanomaterials* **2016**, *6*, 157. [CrossRef]
138. Wong, K.K. Synthesis of Nitrogen-Doped Graphene Quantum Dots and its Antibacterial Property. Disertation Thesis, Hong Kong Polytechnic University, Hong Kong, China, 2018.
139. Huang, D.; Yin, L.; Lu, X.; Lin, S.; Niu, Z.; Niu, J. Directional electron transfer mechanisms with graphene quantum dots as the electron donor for photodecomposition of perfluorooctane sulfonate. *Chem. Eng. J.* **2017**, *323*, 406–414. [CrossRef]
140. Kaur, M.; Kaur, M.; Sharma, V.K. Nitrogen-doped graphene and graphene quantum dots: A review onsynthesis and applications in energy, sensors and environment. *Adv. Colloid Interface Sci.* **2018**, *259*, 44–64. [CrossRef]
141. Behera, A. Nanomaterials. In *Advanced Materials*; Springer: Berlin/Heidelberg, Germany, 2022; pp. 77–125.
142. Zhu, Y.; Wang, G.; Jiang, H.; Chen, L.; Zhang, X. One-step ultrasonic synthesis of graphene quantum dots with high quantum yield and their application in sensing alkaline phosphatase. *Chem. Commun.* **2015**, *51*, 948–951. [CrossRef]
143. Oza, G.; Oza, K.; Pandey, S.; Shinde, S.; Mewada, A.; Thakur, M.; Sharon, M.; Sharon, M. A green route towards highly photoluminescent and cytocompatible carbon dot synthesis and its separation using sucrose density gradient centrifugation. *J. Fluoresc.* **2015**, *25*, 9–14. [CrossRef]
144. Zhou, J.; Yang, Y.; Zhang, C.-y. Toward biocompatible semiconductor quantum dots: From biosynthesis and bioconjugation to biomedical application. *Chem. Rev.* **2015**, *115*, 11669–11717. [CrossRef]
145. Xu, G.; Zeng, S.; Zhang, B.; Swihart, M.T.; Yong, K.-T.; Prasad, P.N. New generation cadmium-free quantum dots for biophotonics and nanomedicine. *Chem. Rev.* **2016**, *116*, 12234–12327. [CrossRef]
146. Cao, L.I.; Meziani, M.J.; Sahu, S.; Sun, Y.P. Photoluminescence properties of graphene versus other carbon nanomaterials. *Acc. Chem. Res.* **2013**, *46*, 171–180. [CrossRef] [PubMed]
147. Dong, Y.; Pang, H.; Yang, B.; Guo, C.; Shao, J.; Chi, Y.; Ming, C.; Yu, T. Harbon-Based Dots Co-doped with Nitrogen and Sulfur for High Quantum Yield and Excitation-Independent Emission. *Angew. Chem. Int. Ed.* **2013**, *52*, 7800–7804. [CrossRef] [PubMed]
148. Gonçalves, H.M.; Duarte, A.J.; da Silva, J.C.E. Optical fiber sensor for Hg (II) based on carbon dots. *Biosens. Bioelectron.* **2010**, *26*, 1302–1306. [CrossRef] [PubMed]
149. Zheng, M.; Liu, S.; Li, J.; Qu, D.; Zhao, H.; Guan, X.; Hu, X.; Xie, Z.; Jing, X.; Sun, Z. Integrating Oxaliplatin with Highly Luminescent Carbon Dots: An Unprecedented Theranostic Agent for Personalized Medicine. *Adv. Mater.* **2014**, *26*, 3554–3560. [CrossRef] [PubMed]
150. Keshipour, S.; Adak, K. Pd (0) supported on N-doped graphene quantum dot modified cellulose as an efficient catalyst for the green reduction of nitroaromatics. *RSC Adv.* **2016**, *6*, 89407–89412. [CrossRef]
151. Lee, Y.R.; Raghu, A.V.; Jeong, H.M.; Kim, B.K. Properties of waterborne polyurethane/functionalized graphene sheet nanocomposites prepared by an in situ method. *Macromol. Chem. Phys.* **2009**, *210*, 1247–1254. [CrossRef]
152. Xu, Y.; Wang, Y.; Liang, J.; Huang, Y.; Ma, Y.; Wan, X.; Chen, Y. A hybrid material of graphene and poly (3, 4-ethyldioxythiophene) with high conductivity, flexibility, and transparency. *Nano Res.* **2009**, *2*, 343–348. [CrossRef]
153. Quan, H.; Zhang, B.-Q.; Zhao, Q.; Yuen, R.K.; Li, R.K. Facile preparation and thermal degradation studies of graphite nanoplatelets (GNPs) filled thermoplastic polyurethane (TPU) nanocomposites. *Compos. Part. A Appl. Sci. Manuf.* **2009**, *40*, 1506–1513. [CrossRef]
154. Emadi, F.; Emadi, A.; Gholami, A. A comprehensive insight towards pharmaceutical aspects of graphene nanosheets. *Curr. Pharm. Biotechnol.* **2020**, *21*, 1016–1027. [CrossRef]
155. Papageorgiou, D.G.; Kinloch, I.A.; Young, R.J. Mechanical properties of graphene and graphene-based nanocomposites. *Prog. Mater. Sci.* **2017**, *90*, 75–127. [CrossRef]
156. Balog, R.; Jørgensen, B.; Nilsson, L.; Andersen, M.; Rienks, E.; Bianchi, M.; Fanetti, M.; Lægsgaard, E.; Baraldi, A.; Lizzit, S. Bandgap opening in graphene induced by patterned hydrogen adsorption. *Nat. Mater.* **2010**, *9*, 315–319. [CrossRef]

157. Scarpa, F.; Adhikari, S.; Phani, A.S. Effective elastic mechanical properties of single layer graphene sheets. *Nanotechnology* **2009**, *20*, 065709. [CrossRef]
158. Van Lier, G.; Van Alsenoy, C.; Van Doren, V.; Geerlings, P. Ab initio study of the elastic properties of single-walled carbon nanotubes and graphene. *Chem. Phys. Lett.* **2000**, *326*, 181–185. [CrossRef]
159. Li, D.; Müller, M.B.; Gilje, S.; Kaner, R.B.; Wallace, G.G. Processable aqueous dispersions of graphene nanosheets. *Nat. Nanotechnol.* **2008**, *3*, 101–105. [CrossRef]
160. Lewandowska, M.; Krawczyńska, A.T.; Kulczyk, M.; Kurzydłowski, K.J. Structure and properties of nano-sized Eurofer 97 steel obtained by hydrostatic extrusion. *J. Nucl. Mater.* **2009**, *386*, 499–502. [CrossRef]
161. Gholami, A.; Mohammadi, F.; Ghasemi, Y.; Omidifar, N.; Ebrahiminezhad, A. Antibacterial activity of SPIONs versus ferrous and ferric ions under aerobic and anaerobic conditions: A preliminary mechanism study. *IET Nanobiotechnol.* **2019**, *14*, 155–160. [CrossRef] [PubMed]
162. Vickery, J.L.; Patil, A.J.; Mann, S. fabrication of graphene–polymer nanocomposites with higher-order three-dimensional architectures. *Adv. Mater.* **2009**, *21*, 2180–2184. [CrossRef]
163. Becerril, H.A.; Mao, J.; Liu, Z.; Stoltenberg, R.M.; Bao, Z.; Chen, Y. Evaluation of solution-processed reduced graphene oxide films as transparent conductors. *ACS Nano* **2008**, *2*, 463–470. [CrossRef] [PubMed]
164. Bourlinos, A.B.; Gournis, D.; Petridis, D.; Szabó, T.; Szeri, A.; Dékány, I. Graphite oxide: Chemical reduction to graphite and surface modification with primary aliphatic amines and amino acids. *Langmuir* **2003**, *19*, 6050–6055. [CrossRef]
165. Mousavi, S.M.; Hashemi, S.A.; Mazraedoost, S.; Yousefi, K.; Gholami, A.; Behbudi, G.; Ramakrishna, S.; Omidifar, N.; Alizadeh, A.; Chiang, W.-H. Multifunctional gold nanorod for therapeutic applications and pharmaceutical delivery considering cellular metabolic responses, oxidative stress and cellular longevity. *Nanomaterials* **2021**, *11*, 1868. [CrossRef]
166. Zhang, H.-B.; Zheng, W.-G.; Yan, Q.; Yang, Y.; Wang, J.-W.; Lu, Z.-H.; Ji, G.-Y.; Yu, Z.-Z. Electrically conductive polyethylene terephthalate/graphene nanocomposites prepared by melt compounding. *Polymer* **2010**, *51*, 1191–1196. [CrossRef]
167. Sun, W.; Hayden, S.; Jin, Y.; Rong, Y.; Yu, J.; Ye, F.; Chan, Y.-H.; Zeigler, M.; Wu, C.; Chiu, D.T. A versatile method for generating semiconducting polymer dot nanocomposites. *Nanoscale* **2012**, *4*, 7246–7249. [CrossRef]
168. Zhu, S.; Song, Y.; Shao, J.; Zhao, X.; Yang, B. Non-conjugated polymer dots with crosslink-enhanced emission in the absence of fluorophore units. *Angew. Chem. Int. Ed.* **2015**, *54*, 14626–14637. [CrossRef]
169. Hezinger, A.; Teßmar, J.; Göpferich, A. Polymer coating of quantum dots–a powerful tool toward diagnostics and sensorics. *Eur. J. Pharm. Biopharm.* **2008**, *68*, 138–152. [CrossRef]
170. Zrazhevskiy, P.; Sena, M.; Gao, X. Designing multifunctional quantum dots for bioimaging, detection, and drug delivery. *Chem. Soc. Rev.* **2010**, *39*, 4326–4354. [CrossRef]
171. Dou, L.; Liu, Y.; Hong, Z.; Li, G.; Yang, Y. Low-bandgap near-IR conjugated polymers/molecules for organic electronics. *Chem. Rev.* **2015**, *115*, 12633–12665. [CrossRef] [PubMed]
172. Lu, Z.S.; Li, C.M. Quantum dot-based nanocomposites for biomedical applications. *Curr. Med. Chem.* **2011**, *18*, 3516–3528. [CrossRef]
173. Asadi, E.; Chimeh, A.F.; Hosseini, S.; Rahimi, S.; Sarkhosh, B.; Bazli, L.; Bashiri, R.; Tahmorsati, A.H.V. A review of clinical applications of graphene quantum dot-based composites. *J. Compos. Compd.* **2019**, *1*, 31–40. [CrossRef]
174. Li, L.L.; Liu, K.P.; Yang, G.H.; Wang, C.M.; Zhang, J.R.; Zhu, J.J. Fabrication of graphene–quantum dots composites for sensitive electrogenerated chemiluminescence immunosensing. *Adv. Funct. Mater.* **2011**, *21*, 869–878. [CrossRef]
175. Arthisree, D.; Joshi, G.M. Influence of graphene quantum dots on electrical properties of polymer composites. *Mater. Res. Express* **2017**, *4*, 075045. [CrossRef]
176. Geng, X.; Niu, L.; Xing, Z.; Song, R.; Liu, G.; Sun, M.; Cheng, G.; Zhong, H.; Liu, Z.; Zhang, Z. Aqueous-processable noncovalent chemically converted graphene–quantum dot composites for flexible and transparent optoelectronic films. *Adv. Mater.* **2010**, *22*, 638–642. [CrossRef]
177. Xie, H.; Hou, C.; Wang, H.; Zhang, Q.; Li, Y. S, N co-doped graphene quantum dot/TiO 2 composites for efficient photocatalytic hydrogen generation. *Nanoscale Res. Lett.* **2017**, *12*, 1–8. [CrossRef]
178. Rafiee, M.A.; Rafiee, J.; Srivastava, I.; Wang, Z.; Song, H.; Yu, Z.Z.; Koratkar, N. Fracture and fatigue in graphene nanocomposites. *Small* **2010**, *6*, 179–183. [CrossRef]
179. Eksik, O.; Bartolucci, S.F.; Gupta, T.; Fard, H.; Borca-Tasciuc, T.; Koratkar, N. A novel approach to enhance the thermal conductivity of epoxy nanocomposites using graphene core–shell additives. *Carbon* **2016**, *101*, 239–244. [CrossRef]
180. Young, R.J.; Kinloch, I.A.; Gong, L.; Novoselov, K.S. The mechanics of graphene nanocomposites: A review. *Compos. Sci. Technol.* **2012**, *72*, 1459–1476. [CrossRef]
181. Zhuo, S.; Shao, M.; Lee, S.-T. Upconversion and downconversion fluorescent graphene quantum dots: Ultrasonic preparation and photocatalysis. *ACS Nano* **2012**, *6*, 1059–1064. [CrossRef] [PubMed]
182. Gobi, N.; Vijayakumar, D.; Keles, O.; Erogbogbo, F. Infusion of graphene quantum dots to create stronger, tougher, and brighter polymer composites. *ACS Omega* **2017**, *2*, 4356–4362. [CrossRef] [PubMed]
183. Jayanthi, S.C.; Kaur, R.; Erogbogbo, F. Graphene quantum dot-titania nanoparticle composite for photocatalytic water splitting. *MRS Adv.* **2016**, *1*, 2071–2077. [CrossRef]
184. Wu, J.; Becerril, H.A.; Bao, Z.; Liu, Z.; Chen, Y.; Peumans, P. Organic solar cells with solution-processed graphene transparent electrodes. *Appl. Phys. Lett.* **2008**, *92*, 237. [CrossRef]

185. Das, T.K.; Prusty, S. Graphene-based polymer composites and their applications. *Polym.-Plast. Technol. Eng.* **2013**, *52*, 319–331. [CrossRef]
186. Wang, D.; Kou, R.; Choi, D.; Yang, Z.; Nie, Z.; Li, J.; Saraf, L.V.; Hu, D.; Zhang, J.; Graff, G.L. Ternary self-assembly of ordered metal oxide−graphene nanocomposites for electrochemical energy storage. *ACS Nano* **2010**, *4*, 1587–1595. [CrossRef]
187. Zhu, J.; Liu, F.; Mahmood, N.; Hou, Y. Graphene polymer nanocomposites for fuel cells. In *Graphene-Based Polymer Nanocomposites in Electronics*; Springer: Berlin/Heidelberg, Germany, 2015; pp. 91–130.
188. Xu, C.; Wang, X.; Wang, J.; Hu, H.; Wan, L. Synthesis and photoelectrical properties of β-cyclodextrin functionalized graphene materials with high bio-recognition capability. *Chem. Phys. Lett.* **2010**, *498*, 162–167. [CrossRef]
189. Sun, X.; Liu, Z.; Welsher, K.; Robinson, J.T.; Goodwin, A.; Zaric, S.; Dai, H. Nano-graphene oxide for cellular imaging and drug delivery. *Nano Res.* **2008**, *1*, 203–212. [CrossRef] [PubMed]
190. Hashemi, S.A.; Mousavi, S.M.; Bahrani, S.; Gholami, A.; Chiang, W.H.; Yousefi, K.; Omidifar, N.; Rao, N.V.; Ramakrishna, S.; Babapoor, A.; et al. Bio-enhanced polyrhodanine/graphene Ox-ide/Fe_3O_4 nanocomposite with kombucha solvent supernatant as ultra-sensitive biosensor for detection of doxorubicin hydrochlo-ride in biological fluids. *Mater. Chem. Phys.* **2022**, *279*, 125743.
191. Staudinger, H. On Polymerization. In *A Source Book in Chemistry, 1900-1950*; Harvard University Press: Cambridge, MA, USA, 2013; pp. 259–264.
192. Clayden, J.; Greeves, N.; Warren, S.; Wothers, P. *Organic Chemistry*; Oxford University Press: Oxford, UK, 2001.
193. Surini, S.; Anggriani, V.; Anwar, E. Study of mucoadhesive microspheres based on pregelatinsed cassava starch succinate as a new carrier for drug delivery. *J. Med. Sci.* **2009**, *9*, 249–256. [CrossRef]
194. Negut, I.; Grumezescu, V.; Dorcioman, G.; Socol, G. Microscale drug delivery systems: Current perspectives and novel approaches. *Nano-Microscale Drug Deliv. Syst.* **2017**, *2017*, 1–15.
195. Lee, C.; Nam, E.; Chae, H. Photo-stable cross-linked micron bead with functionalized quantum via suspension polymerization for color conversion. *Polymer* **2019**, *177*, 19–24. [CrossRef]
196. Vaidya, S.V.; Couzis, A.; Maldarelli, C. Reduction in aggregation and energy transfer of quantum dots incorporated in polystyrene beads by kinetic entrapment due to cross-linking during polymerization. *Langmuir* **2015**, *31*, 3167–3179. [CrossRef] [PubMed]
197. Gonzalez, A.d.S.L. Nanostructured Polymeric aqueous Dispersions Containing Quantum Dots. Ph.D. Thesis, Universidad del País Vasco-Euskal Herriko Unibertsitatea, Bilbao, Spain, 2018.
198. O'Brien, P.; Cummins, S.S.; Darcy, D.; Dearden, A.; Masala, O.; Pickett, N.L.; Ryley, S.; Sutherland, A.J. Quantum dot-labelled polymer beads by suspension polymerisation. *Chem. Commun.* **2003**, *20*, 2532–2533. [CrossRef] [PubMed]
199. Chern, C. Emulsion polymerization mechanisms and kinetics. *Prog. Polym. Sci.* **2006**, *31*, 443–486. [CrossRef]
200. Das, P.; Claverie, J.P. Synthesis of single-core and multiple-core core-shell nanoparticles by RAFT emulsion polymerization: Lead sulfide-copolymer nanocomposites. *J. Polym. Sci. Part A Polym. Chem.* **2012**, *50*, 2802–2808. [CrossRef]
201. Kloust, H.; Schmidtke, C.; Merkl, J.-P.; Feld, A.; Schotten, T.; Fittschen, U.E.; Gehring, M.; Ostermann, J.; Pöselt, E.; Weller, H. Poly (ethylene oxide) and polystyrene encapsulated quantum dots: Highly fluorescent, functionalizable, and ultrastable in aqueous media. *J. Phys. Chem. C* **2013**, *117*, 23244–23250. [CrossRef]
202. Yang, X.; Zhang, Y. Encapsulation of quantum nanodots in polystyrene and silica micro-/nanoparticles. *Langmuir* **2004**, *20*, 6071–6073. [CrossRef]
203. Ugelstad, J.; El-Aasser, M.; Vanderhoff, J. Emulsion polymerization: Initiation of polymerization in monomer droplets. *J. Polym. Sci. Polym. Lett. Ed.* **1973**, *11*, 503–513. [CrossRef]
204. Fleischhaker, F.; Zentel, R. Photonic crystals from core-shell colloids with incorporated highly fluorescent quantum dots. *Chem. Mater.* **2005**, *17*, 1346–1351. [CrossRef]
205. Dinh, L.N.; Ramana, L.N.; Agarwal, V.; Zetterlund, P.B. Miniemulsion polymerization of styrene using carboxylated graphene quantum dots as surfactant. *Polym. Chem.* **2020**, *11*, 3217–3224. [CrossRef]
206. Cho, H.-H.; Yang, H.; Kang, D.J.; Kim, B.J. Surface engineering of graphene quantum dots and their applications as efficient surfactants. *ACS Appl. Mater. Interfaces* **2015**, *7*, 8615–8621. [CrossRef] [PubMed]
207. Samanta, S.; Banerjee, S.L.; Bhattacharya, K.; Singha, N.K. Graphene Quantum Dots-Ornamented Waterborne Epoxy-Based Fluorescent Adhesive via Reversible Addition–Fragmentation Chain Transfer-Mediated Miniemulsion Polymerization: A Potential Material for Art Conservation. *ACS Appl. Mater. Interfaces* **2021**, *13*, 36307–36319. [CrossRef] [PubMed]
208. Asua, J.M. Mapping the morphology of polymer–inorganic nanocomposites synthesized by miniemulsion polymerization. *Macromol. Chem. Phys.* **2014**, *215*, 458–464. [CrossRef]
209. Konwar, A.; Gogoi, N.; Majumdar, G.; Chowdhury, D. Green chitosan–carbon dots nanocomposite hydrogel film with superior properties. *Carbohydr. Polym.* **2015**, *115*, 238–245. [CrossRef]
210. Ruiz-Palomero, C.; Soriano, M.L.; Benítez-Martínez, S.; Valcarcel, M. Photoluminescent sensing hydrogel platform based on the combination of nanocellulose and S, N-codoped graphene quantum dots. *Sens. Actuators B Chem.* **2017**, *245*, 946–953. [CrossRef]
211. Malik, R.; Lata, S.; Soni, U.; Rani, P.; Malik, R.S. Carbon quantum dots intercalated in polypyrrole (PPy) thin electrodes for accelerated energy storage. *Electrochim. Acta* **2020**, *364*, 137281. [CrossRef]
212. Heidari-Maleni, A.; Gundoshmian, T.M.; Jahanbakhshi, A.; Ghobadian, B. Performance improvement and exhaust emissions reduction in diesel engine through the use of graphene quantum dot (GQD) nanoparticles and ethanol-biodiesel blends. *Fuel* **2020**, *267*, 117116. [CrossRef]

213. Cheah, C.B.; Lim, J.S.; Ramli, M.B. The mechanical strength and durability properties of ternary blended cementitious composites containing granite quarry dust (GQD) as natural sand replacement. *Constr. Build. Mater.* **2019**, *197*, 291–306. [CrossRef]
214. Kolli, V.; Mandal, P. Experimental exploration of additive NGQD as a performance and emissions improver of the CI engine with biodiesel–diesel blends. *Int. J. Environ. Sci. Technol.* **2021**, *2021*, 1–16.
215. Zhu, S.; Zhang, J.; Qiao, C.; Tang, S.; Li, Y.; Yuan, W.; Li, B.; Tian, L.; Liu, F.; Hu, R. Strongly green-photoluminescent graphene quantum dots for bioimaging applications. *Chem. Commun.* **2011**, *47*, 6858–6860. [CrossRef] [PubMed]
216. Peng, J.; Gao, W.; Gupta, B.K.; Liu, Z.; Romero-Aburto, R.; Ge, L.; Song, L.; Alemany, L.B.; Zhan, X.; Gao, G. Graphene quantum dots derived from carbon fibers. *Nano Lett.* **2012**, *12*, 844–849. [CrossRef]
217. Tang, L.; Ji, R.; Li, X.; Teng, K.S.; Lau, S.P. Size-dependent structural and optical characteristics of glucose-derived graphene quantum dots. *Part. Part. Syst. Charact.* **2013**, *30*, 523–531. [CrossRef]
218. Shinde, D.B.; Pillai, V.K. Electrochemical preparation of luminescent graphene quantum dots from multiwalled carbon nanotubes. *Chem.–A Eur. J.* **2012**, *18*, 12522–12528. [CrossRef] [PubMed]
219. Sun, Y.; Wang, S.; Li, C.; Luo, P.; Tao, L.; Wei, Y.; Shi, G. Large scale preparation of graphene quantum dots from graphite with tunable fluorescence properties. *Phys. Chem. Chem. Phys.* **2013**, *15*, 9907–9913. [CrossRef]
220. Hazarika, D.; Karak, N. Biodegradable tough waterborne hyperbranched polyester/carbon dot nanocomposite: Approach towards an eco-friendly material. *Green Chem.* **2016**, *18*, 5200–5211. [CrossRef]
221. De, B.; Kumar, M.; Mandal, B.B.; Karak, N. An in situ prepared photo-luminescent transparent biocompatible hyperbranched epoxy/carbon dot nanocomposite. *RSC Adv.* **2015**, *5*, 74692–74704. [CrossRef]
222. Li, B.; Wang, L.; Xu, F.; Gang, X.; Demirci, U.; Wei, D.; Li, Y.; Feng, Y.; Jia, D.; Zhou, Y. Hydrosoluble, UV-crosslinkable and injectable chitosan for patterned cell-laden microgel and rapid transdermal curing hydrogel in vivo. *Acta Biomater.* **2015**, *22*, 59–69. [CrossRef]
223. Islam, M.S.; Deng, Y.; Tong, L.; Roy, A.K.; Faisal, S.N.; Hassan, M.; Minett, A.I.; Gomes, V.G. In-situ direct grafting of graphene quantum dots onto carbon fibre by low temperature chemical synthesis for high performance flexible fabric supercapacitor. *Mater. Today Commun.* **2017**, *10*, 112–119. [CrossRef]
224. Park, C.H.; Yang, H.; Lee, J.; Cho, H.-H.; Kim, D.; Lee, D.C.; Kim, B.J. Multicolor emitting block copolymer-integrated graphene quantum dots for colorimetric, simultaneous sensing of temperature, pH, and metal ions. *Chem. Mater.* **2015**, *27*, 5288–5294. [CrossRef]
225. Luo, P.; Ji, Z.; Li, C.; Shi, G. Aryl-modified graphene quantum dots with enhanced photoluminescence and improved pH tolerance. *Nanoscale* **2013**, *5*, 7361–7367. [CrossRef] [PubMed]
226. Feng, Z.; Adolfsson, K.H.; Xu, Y.; Fang, H.; Hakkarainen, M.; Wu, M. Carbon dot/polymer nanocomposites: From green synthesis to energy, environmental and biomedical applications. *Sustain. Mater. Technol.* **2021**, *29*, e00304. [CrossRef]
227. De, B.; Voit, B.; Karak, N. Transparent luminescent hyperbranched epoxy/carbon oxide dot nanocomposites with outstanding toughness and ductility. *ACS Appl. Mater. Interfaces* **2013**, *5*, 10027–10034. [CrossRef]
228. Fernandes, D.; Heslop, K.; Kelarakis, A.; Krysmann, M.; Estevez, L. In situ generation of carbon dots within a polymer matrix. *Polymer* **2020**, *188*, 122159. [CrossRef]
229. Hu, S.; Zhou, Y.; Xue, C.; Yang, J.; Chang, Q. A solid reaction towards in situ hybridization of carbon dots and conjugated polymers for enhanced light absorption and conversion. *Chem. Commun.* **2017**, *53*, 9426–9429. [CrossRef]
230. Ju, J.; Chen, W. In situ growth of surfactant-free gold nanoparticles on nitrogen-doped graphene quantum dots for electrochemical detection of hydrogen peroxide in biological environments. *Anal. Chem.* **2015**, *87*, 1903–1910. [CrossRef]
231. Liu, Y.; Qin, Z.; Zhang, X.; Wang, N.; Liu, T.; Cui, S.; An, Q.-F.; Guo, H. In-situ growth of graphene quantum dots modified MoS2 membrane on tubular ceramic substrate with high permeability for both water and organic solvent. *J. Membr. Sci.* **2021**, *627*, 119247. [CrossRef]
232. Li, M.; Feng, Q.; Liu, H.; Wu, Y.; Wang, Z. In situ growth of nano-ZnO/GQDs on cellulose paper for dual repelling function against water and bacteria. *Mater. Lett.* **2021**, *283*, 128838. [CrossRef]
233. Ding, H.; Wei, J.-S.; Xiong, H.-M. Nitrogen and sulfur co-doped carbon dots with strong blue luminescence. *Nanoscale* **2014**, *6*, 13817–13823. [CrossRef]
234. Lee, J.-E.; In, I.; Lee, H.; Lee, K.D.; Jeong, J.H.; Park, S.Y. Target delivery and cell imaging using hyaluronic acid-functionalized graphene quantum dots. *Mol. Pharm.* **2013**, *10*, 3736–3744.
235. Hanifi, D.A.; Bronstein, N.D.; Koscher, B.A.; Nett, Z.; Swabeck, J.K.; Takano, K.; Schwartzberg, A.M.; Maserati, L.; Vandewal, K.; van de Burgt, Y. Redefining near-unity luminescence in quantum dots with photothermal threshold quantum yield. *Science* **2019**, *363*, 1199–1202. [CrossRef] [PubMed]
236. Wang, X.-Y.; Lei, R.; Huang, H.-D.; Wang, N.; Yuan, L.; Xiao, R.-Y.; Bai, L.-D.; Li, X.; Li, L.-M. The permeability and transport mechanism of graphene quantum dots (GQDs) across the biological barrier. *Nanoscale* **2015**, *7*, 2034–2041. [CrossRef] [PubMed]
237. Akbarzadeh, M.; Babaei, M.; Abnous, K.; Taghdisi, S.M.; Peivandi, M.T.; Ramezani, M.; Alibolandi, M. Hybrid silica-coated Gd-Zn-Cu-In-S/ZnS bimodal quantum dots as an epithelial cell adhesion molecule targeted drug delivery and imaging system. *Int. J. Pharm.* **2019**, *570*, 118645. [CrossRef]
238. Iannazzo, D.; Pistone, A.; Salamò, M.; Galvagno, S.; Romeo, R.; Giofré, S.V.; Branca, C.; Visalli, G.; Di Pietro, A. Graphene quantum dots for cancer targeted drug delivery. *Int. J. Pharm.* **2017**, *518*, 185–192. [CrossRef] [PubMed]

239. Chen, T.; Yu, H.; Yang, N.; Wang, M.; Ding, C.; Fu, J. Graphene quantum dot-capped mesoporous silica nanoparticles through an acid-cleavable acetal bond for intracellular drug delivery and imaging. *J. Mater. Chem. B* **2014**, *2*, 4979–4982. [CrossRef] [PubMed]
240. Şenel, B.; Demir, N.; Büyükköroğlu, G.; Yıldız, M. Graphene quantum dots: Synthesis, characterization, cell viability, genotoxicity for biomedical applications. *Saudi Pharm. J.* **2019**, *27*, 846–858. [CrossRef] [PubMed]
241. Wang, X.; Sun, X.; Lao, J.; He, H.; Cheng, T.; Wang, M.; Wang, S.; Huang, F. Multifunctional graphene quantum dots for simultaneous targeted cellular imaging and drug delivery. *Colloids Surf. B Biointerfaces* **2014**, *122*, 638–644. [CrossRef]
242. Javanbakht, S.; Namazi, H. Doxorubicin loaded carboxymethyl cellulose/graphene quantum dot nanocomposite hydrogel films as a potential anticancer drug delivery system. *Mater. Sci. Eng. C* **2018**, *87*, 50–59. [CrossRef]
243. Nasrollahi, F.; Koh, Y.R.; Chen, P.; Varshosaz, J.; Khodadadi, A.A.; Lim, S. Targeting graphene quantum dots to epidermal growth factor receptor for delivery of cisplatin and cellular imaging. *Mater. Sci. Eng. C* **2019**, *94*, 247–257. [CrossRef]
244. Yang, Z.; Wang, H.; Zhao, J.; Peng, Y.; Wang, J.; Guinn, B.; Huang, L. Recent developments in the use of adenoviruses and immunotoxins in cancer gene therapy. *Cancer Gene Ther.* **2007**, *14*, 599–615. [CrossRef]
245. Liu, S.; Yongsheng, G.; Sigen, A.; Zhou, D.; Gresier, U.; Guo, T.; Guo, R.; Wang, W. Biodegradable Highly Branched Poly(β-Amino Ester)s for Targeted Cancer Cell Gene Transfection. *ACS Biomater. Sci. Eng.* **2017**, *3*, 1283. [CrossRef] [PubMed]
246. Yin, F.; Hu, K.; Chen, Y.; Yu, M.; Wang, D.; Wang, Q.; Yong, K.-T.; Lu, F.; Liang, Y.; Li, Z. SiRNA delivery with PEGylated graphene oxide nanosheets for combined photothermal and genetherapy for pancreatic cancer. *Theranostics* **2017**, *7*, 1133. [CrossRef] [PubMed]
247. Dunbar, C.; High, K.; Joung, J.; Kohn, D.; Ozawa, K.; Sadelain, M. Gene therapy comes of age. *Science* **2018**, *359*, 6372. [CrossRef] [PubMed]
248. Naldini, L.; Blömer, U.; Gallay, P.; Ory, D.; Mulligan, R.; Gage, F.H.; Verma, I.M.; Trono, D. In vivo gene delivery and stable transduction of nondividing cells by a lentiviral vector. *Science* **1996**, *272*, 263–267. [CrossRef]
249. Mintzer, M.A.; Simanek, E.E. Nonviral vectors for gene delivery. *Chem. Rev.* **2009**, *109*, 259–302. [CrossRef]
250. Muro, A.F.; D'Antiga, L.; Mingozzi, F. Gene therapy in pediatric liver disease. In *Pediatric Hepatology and Liver Transplantation*; Springer: Berlin/Heidelberg, Germany, 2019; pp. 799–829.
251. Ghafary, S.M.; Nikkhah, M.; Hatamie, S.; Hosseinkhani, S. Simultaneous gene delivery and tracking through preparation of photo-luminescent nanoparticles based on graphene quantum dots and chimeric peptides. *Sci. Rep.* **2017**, *7*, 1–14. [CrossRef]
252. Zhao, C.; Song, X.; Liu, Y.; Fu, Y.; Ye, L.; Wang, N.; Wang, F.; Li, L.; Mohammadniaei, M.; Zhang, M. Synthesis of graphene quantum dots and their applications in drug delivery. *J. Nanobiotechnol.* **2020**, *18*, 1–32. [CrossRef]
253. Ahn, M.; Song, J.; Hong, B.H. Facile Synthesis of N-Doped Graphene Quantum Dots as Novel Transfection Agents for mRNA and pDNA. *Nanomaterials* **2021**, *11*, 2816. [CrossRef]
254. Lai, W.-F.; Wong, W.-T. Use of graphene-based materials as carriers of bioactive agents. *Asian J. Pharm. Sci.* **2021**, *16*, 577–588. [CrossRef]
255. Biswas, A.; Khandelwal, P.; Das, R.; Salunke, G.; Alam, A.; Ghorai, S.; Chattopadhyay, S.; Poddar, P. Oxidant mediated one-step complete conversion of multi-walled carbon nanotubes to graphene quantum dots and their bioactivity against mammalian and bacterial cells. *J. Mater. Chem. B* **2017**, *5*, 785–796. [CrossRef]
256. Chung, S.; Revia, R.A.; Zhang, M. Graphene quantum dots and their applications in bioimaging, biosensing, and therapy. *Adv. Mater.* **2021**, *33*, 1904362. [CrossRef]
257. Yang, Y.; Deng, Y.; Huang, J.; Fan, X.; Cheng, C.; Nie, C.; Ma, L.; Zhao, W.; Zhao, C. Antibacterial Nanocarbons: Size-Transformable Metal–Organic Framework–Derived Nanocarbons for Localized Chemo-Photothermal Bacterial Ablation and Wound Disinfection. *Adv. Funct. Mater.* **2019**, *29*, 1970225. [CrossRef]
258. Fan, X.; Yang, F.; Huang, J.; Yang, Y.; Nie, C.; Zhao, W.; Ma, L.; Cheng, C.; Zhao, C.; Haag, R. Metal–organic-framework-derived 2D carbon nanosheets for localized multiple bacterial eradication and augmented anti-infective therapy. *Nano Lett.* **2019**, *19*, 5885–5896. [CrossRef]
259. Huang, X.; El-Sayed, M.A. Plasmonic photo-thermal therapy (PPTT). *Alex. J. Med.* **2011**, *47*, 1–9. [CrossRef]
260. Tu, X.; Wang, L.; Cao, Y.; Ma, Y.; Shen, H.; Zhang, M.; Zhang, Z. Efficient cancer ablation by combined photothermal and enhanced chemo-therapy based on carbon nanoparticles/doxorubicin@ SiO_2 nanocomposites. *Carbon* **2016**, *97*, 35–44. [CrossRef]
261. Yang, K.; Feng, L.; Shi, X.; Liu, Z. Nano-graphene in biomedicine: Theranostic applications. *Chem. Soc. Rev.* **2013**, *42*, 530–547. [CrossRef]
262. Gonçalves, G.; Vila, M.; Portolés, M.T.; Vallet-Regi, M.; Gracio, J.; Marques, P.A.A. Nano-graphene oxide: A potential multifunctional platform for cancer therapy. *Adv. Healthc. Mater.* **2013**, *2*, 1072–1090. [CrossRef] [PubMed]
263. Li, J.-L.; Tang, B.; Yuan, B.; Sun, L.; Wang, X.-G. A review of optical imaging and therapy using nanosized graphene and graphene oxide. *Biomaterials* **2013**, *34*, 9519–9534. [CrossRef]
264. Yang, K.; Zhang, S.; Zhang, G.; Sun, X.; Lee, S.-T.; Liu, Z. Graphene in mice: Ultrahigh in vivo tumor uptake and efficient photothermal therapy. *Nano Lett.* **2010**, *10*, 3318–3323. [CrossRef]
265. Wang, H.; Mu, Q.; Wang, K.; Revia, R.A.; Yen, C.; Gu, X.; Tian, B.; Liu, J.; Zhang, M. Nitrogen and boron dual-doped graphene quantum dots for near-infrared second window imaging and photothermal therapy. *Appl. Mater. Today* **2019**, *14*, 108–117. [CrossRef]
266. Fang, J.; Liu, Y.; Chen, Y.; Ouyang, D.; Yang, G.; Yu, T. Graphene quantum dots-gated hollow mesoporous carbon nanoplatform for targeting drug delivery and synergistic chemo-photothermal therapy. *Int. J. Nanomed.* **2018**, *13*, 5991. [CrossRef]

267. Thakur, M.; Kumawat, M.K.; Srivastava, R. Multifunctional graphene quantum dots for combined photothermal and photodynamic therapy coupled with cancer cell tracking applications. *RSC Adv.* **2017**, *7*, 5251–5261. [CrossRef]
268. Choudhary, S.; Nouri, K.; Elsaie, M.L. Photodynamic therapy in dermatology: A review. *Lasers Med. Sci.* **2009**, *24*, 971–980. [CrossRef]
269. Ge, J.; Lan, M.; Zhou, B.; Liu, W.; Guo, L.; Wang, H.; Jia, Q.; Niu, G.; Huang, X.; Zhou, H. A graphene quantum dot photodynamic therapy agent with high singlet oxygen generation. *Nat. Commun.* **2014**, *5*, 1–8. [CrossRef]
270. Markovic, Z.M.; Ristic, B.Z.; Arsikin, K.M.; Klisic, D.G.; Harhaji-Trajkovic, L.M.; Todorovic-Markovic, B.M.; Kepic, D.P.; Kravic-Stevovic, T.K.; Jovanovic, S.P.; Milenkovic, M.M. Graphene quantum dots as autophagy-inducing photodynamic agents. *Biomaterials* **2012**, *33*, 7084–7092. [CrossRef]
271. Samia, A.C.; Chen, X.; Burda, C. Semiconductor quantum dots for photodynamic therapy. *J. Am. Chem. Soc.* **2003**, *125*, 15736–15737. [CrossRef]
272. Tabish, T.A.; Scotton, C.J.; Ferguson, D.C.J.; Lin, L.; der Veen, A.V.; Lowry, S.; Ali, M.; Jabeen, F.; Ali, M.; Winyard, P.G. Biocompatibility and toxicity of graphene quantum dots for potential application in photodynamic therapy. *Nanomedicine* **2018**, *13*, 1923–1937. [CrossRef]
273. Zhang, D.; Wen, L.; Huang, R.; Wang, H.; Hu, X.; Xing, D. Mitochondrial specific photodynamic therapy by rare-earth nanoparticles mediated near-infrared graphene quantum dots. *Biomaterials* **2018**, *153*, 14–26. [CrossRef]
274. Sun, J.; Xin, Q.; Yang, Y.; Shah, H.; Cao, H.; Qi, Y.; Gong, J.R.; Li, J. Nitrogen-doped graphene quantum dots coupled with photosensitizers for one-/two-photon activated photodynamic therapy based on a FRET mechanism. *Chem. Commun.* **2018**, *54*, 715–718. [CrossRef]
275. Habiba, K.; Encarnacion-Rosado, J.; Garcia-Pabon, K.; Villalobos-Santos, J.C.; Makarov, V.I.; Avalos, J.A.; Weiner, B.R.; Morell, G. Improving cytotoxicity against cancer cells by chemo-photodynamic combined modalities using silver-graphene quantum dots nanocomposites. *Int. J. Nanomed.* **2016**, *11*, 107. [CrossRef]
276. Fasbender, S.; Allani, S.; Wimmenauer, C.; Cadeddu, R.-P.; Raba, K.; Fischer, J.C.; Bulat, B.; Luysberg, M.; Seidel, C.A.; Heinzel, T. Uptake dynamics of graphene quantum dots into primary human blood cells following in vitro exposure. *RSC Adv.* **2017**, *7*, 12208–12216. [CrossRef]
277. Gao, X.; Yang, L.; Petros, J.A.; Marshall, F.F.; Simons, J.W.; Nie, S. In vivo molecular and cellular imaging with quantum dots. *Curr. Opin. Biotechnol.* **2005**, *16*, 63–72. [CrossRef]
278. Benítez-Martínez, S.; Valcárcel, M. Graphene quantum dots in analytical science. *TrAC Trends Anal. Chem.* **2015**, *72*, 93–113. [CrossRef]
279. Roy, E.; Patra, S.; Tiwari, A.; Madhuri, R.; Sharma, P.K. Introduction of selectivity and specificity to graphene using an inimitable combination of molecular imprinting and nanotechnology. *Biosens. Bioelectron.* **2017**, *89*, 234–248. [CrossRef]
280. Yao, X.; Tian, Z.; Liu, J.; Zhu, Y.; Hanagata, N. Mesoporous silica nanoparticles capped with graphene quantum dots for potential chemo–photothermal synergistic cancer therapy. *Langmuir* **2017**, *33*, 591–599. [CrossRef]
281. Miao, P.; Han, K.; Tang, Y.; Wang, B.; Lin, T.; Cheng, W. Recent advances in carbon nanodots: Synthesis, properties and biomedical applications. *Nanoscale* **2015**, *7*, 1586–1595. [CrossRef]
282. Sheng, L.; Huangfu, B.; Xu, Q.; Tian, W.; Li, Z.; Meng, A.; Tan, S. A highly selective and sensitive fluorescent probe for detecting Cr (VI) and cell imaging based on nitrogen-doped graphene quantum dots. *J. Alloy. Compd.* **2020**, *820*, 153191. [CrossRef]
283. Liu, C.; Xia, Z.; Czernuszka, J. Design and development of three-dimensional scaffolds for tissue engineering. *Chem. Eng. Res. Des.* **2007**, *85*, 1051–1064. [CrossRef]
284. Skalak, R.; Fox, C.F. *Tissue Engineering: Proceedings of a Workshop held at Granlibakken, Lake Tahoe, California, 26–29 February 1988*; Alan, R., Ed.; Liss: New York, NY, USA, 1988; Volume 107.
285. Mousavi, S.M.; Low, F.W.; Hashemi, S.A.; Samsudin, N.A.; Shakeri, M.; Yusoff, Y.; Rahsepar, M.; Lai, C.W.; Babapoor, A.; Soroshnia, S. Development of hydrophobic reduced graphene oxide as a new efficient approach for photochemotherapy. *RSC Adv.* **2020**, *10*, 12851–12863. [CrossRef]
286. Qiu, J.; Li, D.; Mou, X.; Li, J.; Guo, W.; Wang, S.; Yu, X.; Ma, B.; Zhang, S.; Tang, W. Effects of graphene quantum dots on the self-renewal and differentiation of mesenchymal stem cells. *Adv. Healthc. Mater.* **2016**, *5*, 702–710. [CrossRef]
287. Fallahinezhad, F.; Afsa, M.; Ghahramani, Y. Graphene Quantum Dots and their applications in regenerative medicine: A mini-review. *Adv. Appl. NanoBio-Technol.* **2021**, *4*, 2.
288. Li, L.; Wu, G.; Yang, G.; Peng, J.; Zhao, J.; Zhu, J.-J. Focusing on luminescent graphene quantum dots: Current status and future perspectives. *Nanoscale* **2013**, *5*, 4015–4039. [CrossRef]
289. Qu, D.; Zheng, M.; Du, P.; Zhou, Y.; Zhang, L.; Li, D.; Tan, H.; Zhao, Z.; Xie, Z.; Sun, Z. Highly luminescent S, N co-doped graphene quantum dots with broad visible absorption bands for visible light photocatalysts. *Nanoscale* **2013**, *5*, 12272–12277. [CrossRef]
290. Liu, F.; Jang, M.; Ha, H.; Kim, J.; Cho, Y.; Seo, T. Structural and optical properties of N-doped graphene quantum dots. *Adv. Mater.* **2013**, *25*, e36622013.
291. Liu, F.; Jang, M.H.; Ha, H.D.; Kim, J.H.; Cho, Y.H.; Seo, T.S. Facile synthetic method for pristine graphene quantum dots and graphene oxide quantum dots: Origin of blue and green luminescence. *Adv. Mater.* **2013**, *25*, 3657–3662. [CrossRef]
292. Wang, Z.; Zeng, H.; Sun, L. Graphene quantum dots: Versatile photoluminescence for energy, biomedical, and environmental applications. *J. Mater. Chem. C* **2015**, *3*, 1157–1165. [CrossRef]

polymers

MDPI

Article

Enhancement of the Properties of Hybridizing Epoxy and Nanoclay for Mechanical, Industrial, and Biomedical Applications

Zainab Fakhri Merzah [1], Sokina Fakhry [2], Tyser Gaaz Allami [3,*], Nor Yuliana Yuhana [3] and Ahmed Alamiery [3]

[1] Institute of Laser for Postgraduate Studies, University of Baghdad, Baghdad 10071, Iraq; zainab.fakhri@ilps.uobaghdad.edu.iq
[2] Al Furat Intermediate School for Girls, Ministry of Education, Babylon 11524, Iraq; sokinafahkry@yahoo.com
[3] Department of Chemical and Process Engineering, Faculty of Engineering and Built Environment, Universiti Kebangsaan Malaysia, Bangi 43600, Selangor Darul Ehsan, Malaysia; yuliana@ukm.edu.my (N.Y.Y.); dr.ahmed1975@ukm.edu.my (A.A.)
* Correspondence: gaaz@ukm.edu.my; Tel.: +60-1157-733-927

Abstract: The strong demand for plastic and polymeric materials continues to grow year after year, making these industries critical to address sustainability. By functioning as a filler in either a synthetic or natural starch matrix, nanoclay enables significant reductions in the impact of nonbiodegradable materials. The effect of treated nanoclay (NC) loading on the mechanical and morphological properties (EP) of epoxy is investigated in this research. The NC-EP nanocomposites were prepared via casting. The investigation begins with adding NC at concentrations of 1, 2, and 3 weight percent, followed by the effect of acid treatment on the same nanocomposites. The evaluation is focused on four mechanical tensile strength parameters: Young's modulus, maximum load, and % elongation. The addition of NC improved the mechanical properties of the four components by 27.2%, 33.38%, 46.98%, and 43.58%, respectively. The acid treatment improved 35.9%, 42.8%, 51.1%, and 83.5%, respectively. These improvements were attributed to NC's ability to alter the structural morphology as assessed by field emission scanning electron microscopy (FESEM), a tool for analysing the microstructure. FESEM images were used to visualise the interaction between the NC and EP nanocomposites. The dynamic mechanical properties of the hybrid nanocomposites were investigated using storage modulus, loss modulus, and tan(delta). The results have shown that the viscoelastic properties improved as the fraction of NC increased. The overall findings suggest that these nanocomposites could be used in various industrial and biomedical applications.

Keywords: nanoclay; nanocomposites; mechanical properties; impact properties; hardness

Citation: Merzah, Z.F.; Fakhry, S.; Allami, T.G.; Yuhana, N.Y.; Alamiery, A. Enhancement of the Properties of Hybridizing Epoxy and Nanoclay for Mechanical, Industrial, and Biomedical Applications. *Polymers* **2022**, *14*, 526. https://doi.org/10.3390/polym14030526

Academic Editors: Andrzej Puszka and Beata Podkościelna

Received: 31 December 2021
Accepted: 21 January 2022
Published: 28 January 2022

Publisher's Note: MDPI stays neutral with regard to jurisdictional claims in published maps and institutional affiliations.

1. Introduction

The nanoclay (NC) nanocomposites are filler-based materials designed to provide superior mechanical performance at a cheap cost; nevertheless, this results in the NC-EP nanocomposites being brittle, resulting in a loss of mechanical strength [1–3]. Numerous investigations have demonstrated that NC nanocomposites have an extremely high tensile strength and Young's modulus [4–6]. Clays are divided chemically and morphologically into several classes, including smectite, chlorite, kaolinite, and halloysite [7]. Another advantage of clay is that it is widely available and inexpensive and has a low environmental impact. Clay minerals are increasingly being employed as natural nanomaterials [8], since they have no adverse effect on the environment. The clay's widespread use is due to the octahedral (Al or Mg) or tetrahedral layers [9].

As a result, NC nanocomposites have found widespread applications in a range of aerospace and mechanical applications [10,11]. NC composites are frequently employed in applications that need a variety of mechanical properties along with desirable characteristics and physical traits such as lightweight [12–14]. NC is utilised as a filler in polymer composite systems, such as nanoclay–epoxy (NC-EP) systems [15], to enhance the mechanical properties of the parent polymer [16–18]. Pinto et al. [19–23] studied the mechanical characteristics of EP-based nanocomposites after dispersing nanoclay using sonication [19–23] and surfactants. NC-EP systems can fabricate solid and lightweight materials with a variety of advantageous features [24,25]. The effect of NC on microhardness, Young's modulus, and impact strength was examined in an elastic EP matrix by adding 3 wt.% of NC [26]. However, the approach they provided for producing the solution combination was ineffective in terms of distributing the NC evenly across the EP substrate. This inefficient distribution was investigated, and it was determined that the mixing process formed a substantial micrometre-sized conglomerate. As a result, the chemical nanoscale compatibility between the polymer matrix and the nanofiller was established, and homogenous nanofiller dispersion was inside the polymer matrix. This achievement demonstrates the critical nature of the interface between nanoclay fillers and the polymer matrix and the critical nature of nanoclay dispersion quality [27]. These interrelated properties dictate the morphology of polymer/nanoclay composites and, consequently, their fundamental bulk properties such as strength, elastic modulus, thermal stability, heat distortion temperature, self-healing, shape memory, and gas barrier [28].

Li et al. [29] examined the modulus of EP-nanoclay composites.

Additionally, Ma et al. [30] have shown considerable advancements in the dispersion and adhesion of EP matrix and carbon nanotubes by nanotube functionalisation. Other research [18,31] investigated whether NC is functionalised to become an EP network component, increasing polarity. However, previous research has proven that homogenous dispersion can be achieved by reinforcing EP-based nanocomposites with NC [32–35].

Epoxy–nanoclay composites are typically used to improve memory. However, the chemical structures of the thermoplastic materials and the switch temperature or glass transition temperature of the thermoset materials [36] influence the shaping of this memory.

The epoxy–nanocomposite, on the other hand, still has drawbacks such as stiffness and recovery stress. Furthermore, some of the filler percentages have become standard materials for reinforcing epoxy–nanocomposites [37]. Nanoclay-based materials are nontoxic, recyclable, environmentally friendly, and carbon neutral [38]. NC has demonstrated a wide range of advanced applications, including automotive, optically transparent materials, drug delivery, coating films, tissue technology, biomimetic materials, aerogels, sensors, three-dimensional (3D) printing, textiles, printed and flexible electronics, medical and healthcare, and scaffolding [39]. Reinforcing nanoclay with epoxy increased the composite's tensile strength at 7.5 wt.% nanoclay [40].

Previous studies primarily concentrated on thermoplastic materials such as polyurethane [41]. Thermoset is widely used in memory epoxy, which has critical applications in industry and aerospace. The unique thermochemical properties of thermoset are suitable for memory shaping recovery ratio, environmental durability, rapid response, and easy processing and stability [42]. According to Rousseau et al. [43], changing the chemistry of the epoxy could result in memory shaping by relying on changing the phase-transition temperature. They proposed that the failure strain could be attributed to a change in chemical composition or the use of nanocomposites, in which the glass temperature decreased linearly from 124 to 60 °C, depending on the contents. Aside from the effect of changing glass temperature, chemical and physical crosslinking was discovered to have excellent mechanical properties of high strain and recovery stress [36].

The filler of TiNi macrowires, for example, has demonstrated a significant increase in stiffness accompanied by a higher temperature than the neat epoxy. The benefits of the filler include reshaping memory or either accelerating the increase in temperature or decreasing the specimen's thinness. Recent research has revealed additional applications for nanoclay-added composites. These applications have suggested that the enrichment in the interfacial bonding between the matrix and reinforcement materials resulted in improved properties. Furthermore, the nanoclay increases the natural fibre's flexibility and rigidity in a single step [44].

A recent article examined the dynamic mechanical and free vibration properties of various ratios of untreated and treated jute fibre with nanoclay using compression moulding. The results indicate that NaOH concentration and the amount of nanoclay affect the storage and loss modulus, damping factor, and natural frequency. The glass's temperature increases as the natural frequency increases. Composites comprising 5% treated fibre and 5% nanoclay were shown to have low-strength structural applications in the construction and vehicle industries [45]. According to the literature review, nanoclay/natural fibres hybrid nanocomposites have piqued the interest of researchers due to their improved properties such as mechanical properties, barrier properties, thermal and fire performance. However, there has been little research on the effect of nanoclay-modified epoxy on hybrid composites reinforced with two types of natural fibre materials [46].

The viscoelastic and damping properties of polymer composite materials (sometimes referred to as dynamic mechanical analysis) are advantageous. Additionally, dynamic parameters provide information on interfacial bonding [47], crosslinking density [48], and phase transition [49]. The hybrid composite material significantly improves the dynamic mechanical properties of hybrid-reinforced composites, particularly those reinforced with natural fibres. A study of phenolic hybrid composites of kenaf/pineapple leaf fibres demonstrated an increase in storage modulus over the clean sample [50]. Another work on polyester composites reinforced with snake grass fibres demonstrated balancing storage modulus and damping. The current research evaluates several different forms of nanoclay epoxy composites. The addition of nanofiller to macromolecular polymer chains appears to change their relaxing behaviour [51]. Additionally, it was revealed that inserting nanoclay increased banana fibre's storage and loss modulus on nanoclay while reducing the loss factor, referred to as tan delta [52]. According to the literature, NC-epoxy hybrids have gained potential interest due to improvements in dynamic and mechanical properties such as barrier properties and thermal and fire performance. To date, however, there has been little research published on the effect of nanoclay-modified epoxy on hybrid composites reinforced with two types of mat form natural fibres [46].

This study aims to increase the amount of NC in the dispersion and improve its uniformity without employing a solvent. Increases in the NC content of the EP matrix resulted in noticeable improvements in the mechanical characteristics of tensile strength and hardness.

2. Materials and Methods

2.1. Materials

The primary materials used in this report are sulfuric acid and NC. The sulfuric acid was purchased with 95–98% purity from Sigma-Aldrich in Saint Louis, MO, USA. Natural Nano, Inc. provided NC (molecular weight 98.08 g/mole) (New York, NY, USA). $Al_2Si_2O_5(OH)_4.nH_2O$ (0.59) is its chemical formula, and its compositions are O:SiO_2 (61.19), Al:Al_2O_3 (18.11), and Si:SiO_2 (20.11). Surface area (65 m^2/g), pore volume (1.25 mL/g), density (2540 kg/m^3), and refractive index are among the physical properties (1.54).

2.2. Treated NC

NC was acid-treated according to a technique described by Biswas et al. [53]. The technique begins with 15 g of NC being treated with 100 mL of 3M sulfuric acid. The mixture is then divided into four pieces and held at a constant temperature of about 90 °C in a water bath. Each sample (portion) is stirred for 3 h at a speed of 200 rpm using a stirrer machine at 3000 rpm for 10 min, and a centrifugal machine separates the paste from the solution. The paste was removed four times with distilled water and dried in an oven at 70 °C for 12 h. The final step is to use a mortar to grind the dried NC.

2.3. Matrix

The method used to prepare the epoxy (EP) consists of 25% hardener and 75% transparent EP resin, with the amount of NC varying while the viscosity varies between 11,000 and 14,000 cps of clear EP resin. Furthermore, a nonmethylene dianiline aromatic amine prepared by Daemyung Chemical Technology (Gwangju, Gyeonggi Province, Korea) for use as a curing catalyst (i.e., Amicure 101) was used to obtain an accurate wetting boundary condition for the NC strengthening filler.

2.4. Preparation of Nanocomposites (NC-EP)

The sample preparation and experimental procedure were carried out following the flow chart depicted in Figure 1. The first step was to make the neat EP by combining 75% resin and 25% hardener to make 100 wt.% EP. The next step was to collect 1 gm of the neat EP and nanoclay. The first mixture was made by combining 99 wt.% (equivalent to 0.01 g) of neat EP with 1 wt.% (equivalent to 0.1 g) of NC. The 2 and 3 wt.% nanoclay were prepared using the same method. Each nanocomposite sample was kept at 25 °C for 10 min before being baked for 3 h at 40 °C to remove any moisture absorbed during the preparation. For 48 h, the reaction between NC and EP was carried out at a room temperature of 25 °C. Table 1 shows the contents of the NC and EP.

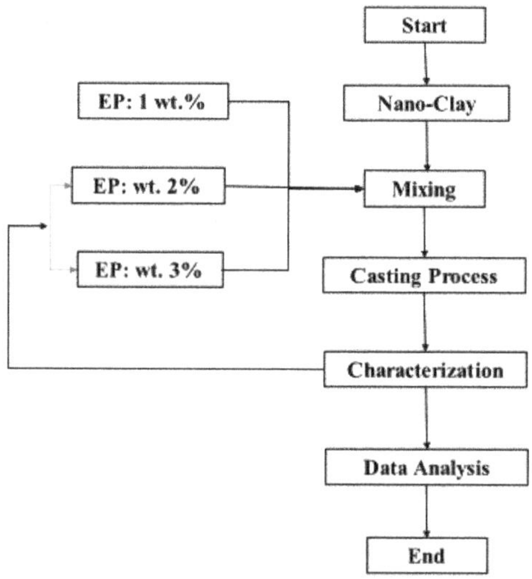

Figure 1. Sample preparation and experimental set-up.

Table 1. Ratio of the nanocomposition in terms of EP Resin, NC, and hardeners.

EP	NC (wt.%)	Nanocomposite
EP–Neat	0	NC (0 wt.%)
EP (99 wt.%)	1	NC (1 wt.%)
EP (98 wt.%)	2	NC (2 wt.%)
EP (97 wt.%)	3	NC (3 wt.%)

3. Mechanical Properties

3.1. Tensile Properties

A machine bought from Instron Company (INSTRON 5567, a product of Konigsallee, Düsseldorf, Germany) shown in Figure 2, made dog-bone-shaped specimens for the tensile tests. The American Society set up the Testing and Materials (ASTM) standard D638. The test was used to perform the mechanical ductility test at a temperature of 25 °C and a speed of 1 mm/min. An analysis of flexible and ductile materials was possible because the loading percentage ratio was kept at a low level. In addition, Bluehill software was used to measure the tensile strength, strain, modulus, and percent elongation. Five samples had NC content from 0% to 3% when they were put through the test.

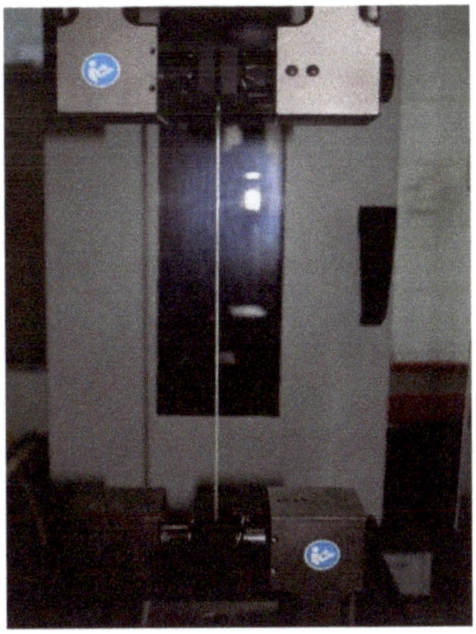

Figure 2. Universal testing machine.

3.2. Impact Properties

The Charpy impact tests were conducted with a CEAST 6545 pendulum, which was chosen because it meets the expected standard (ASTM D638). Here, 11.4 cm × 1.2 cm × 0.2 cm specimens were put together as a simple beam from both ends. This is shown in Figure 3. The scratch was made to look like a nick or cut at the centre of each sample. For a sample with a single nick in it, this equation [54] was used to figure out the total fracture energy. The calculation was performed according to Equation (1):

$$(a_c u) = \frac{w}{h.b} \times 10^{-3} \quad (1)$$

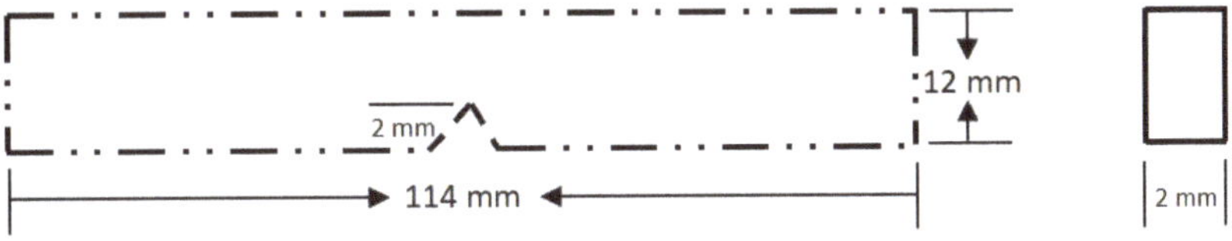

Figure 3. Charpy model impact specimen.

3.3. Hardness Properties

In this case, a microhardness tester was used to obtain average hardness readings by measuring two points simultaneously. This procedure shows the setting up of the unit. It was based on Vickers hardness, and the indentation average diagonal length was measured as shown in Figure 4. Equation (2) was used to get the Vickers hardness, as shown in [55]:

$$HV = 2Fsin136^{\circ}/2/d^2 \quad or \quad HV = 1.85F/d^2 \ (approximately) \quad (2)$$

Figure 4. Shore hardness specimen.

4. Results and Discussion

4.1. Tensile Properties NC-EP Nanocomposites

The findings of the untreated EP sample, the untreated EP mixed with 1, 2, and 3 wt.%, and the acid-treated NC mixed with 1, 2, and 3 wt.% are shown in Figure 5. All results regardless of being treated with acid or not have the untreated sample in surplus. The results indicated that the tensile strength decreased when the NC loading increased from 2 to 3 wt.% caused by a slight shift of 1% in the NC loading. The loss in tensile strength at 3% NC loading could result from NC agglomeration, which naturally results in a poor interaction between NC and EP. These findings are consistent with those of Zhang et al. [27].

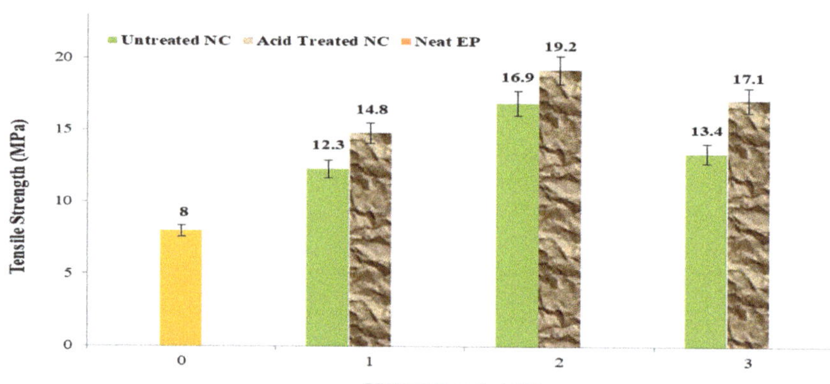

Figure 5. Tensile strength of NC-EP nanocomposites at various contents of NC.

Similarly, Figure 6 illustrates Young's modulus values for the same series of untreated EP samples, untreated EP samples loaded with 1, 2, and 3 wt.% NC, and acid-treated EP samples loaded with 1%, 2%, and 3% NC. Young's modulus values of 180 and 210 MPa for untreated and acid-treated steels were reported with 3 wt.% NC additives. The increase in Young's modulus between the two samples is 33.3% and 42.8%, respectively, compared to the untreated EP sample. The effect of acid treatment on the same sample with a 3% NC loading is visible in the 180 and 210 MPa values, representing an increase of around 14%. The findings of loadings of 1 and 2 wt.% NC were nearly identical, with a slight rise in Young's modulus. The Young's under nanoclay composite showed the same trend as in case of carbon nanotube. The interpretation relies on the effect of dispersion quality of carbon nanotube colloids [4].

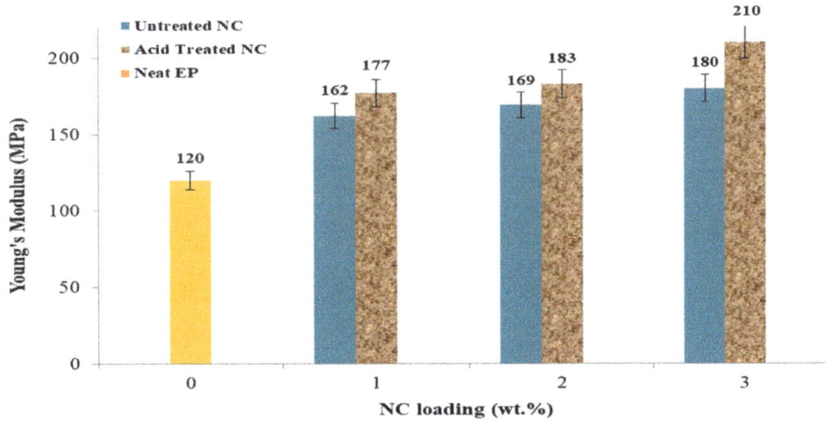

Figure 6. The effect of NC-EP of nanocomposites on Young's modulus at various NC contents.

Figure 7 shows the maximum load for the same sequence. The maximal load of untreated and acid-treated samples attained their maximum values at 2% NC loading, with 245 and 267 N, respectively. These data indicate a maximum improvement of 46.9% and 51.1%, respectively, for the nonacid and acid treatments. Around 8% of the improvement is due to acid treatment. The results indicate that adding NC at a concentration of 3 wt.% reduces the maximum load achieved by the EP-NC sample regardless of the acid treatment.

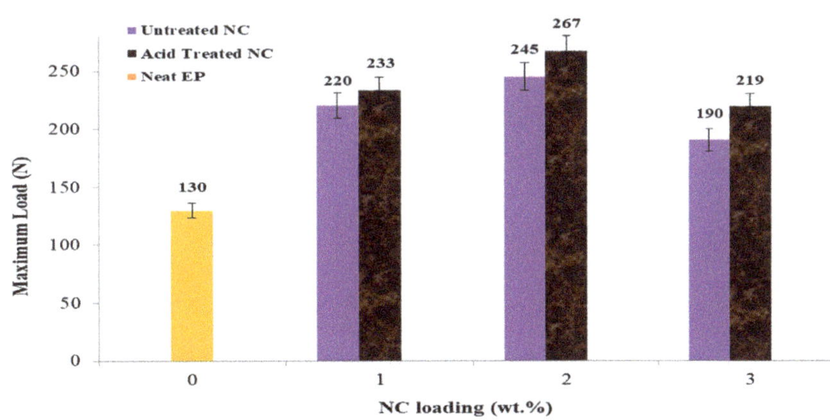

Figure 7. The effect of NC-EP nanocomposites on maximum load at various NC contents.

The final tensile attribute is the elongation percentage at the break, as seen in Figure 8. For nonacid and acid treatment, the elongation percentages are 46.1% and 68.5%, respectively, at a % -NC loading. The results indicate a growing tendency as the concentration of NC additives increases. As a result, it is impossible to determine if elongation at break hits a maximum or continues to rise following the 3 wt.% -NC addition. The author believes that additional research is necessary to study more significant NC additives.

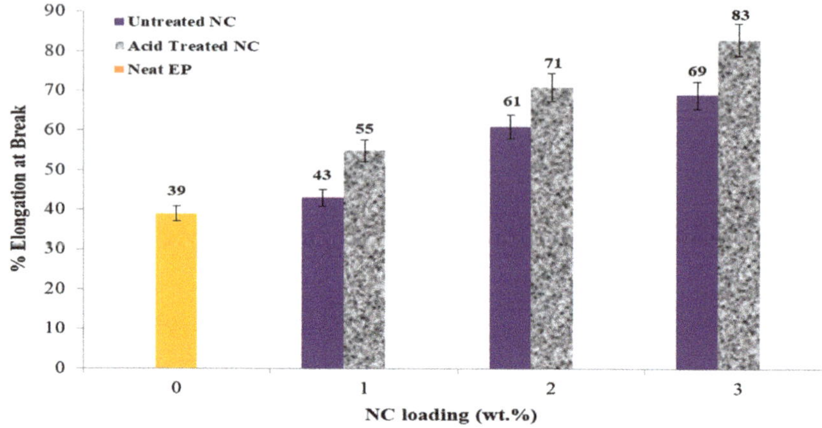

Figure 8. The effect of NC-EP nanocomposites on elongation at break at NC various contents.

4.2. Impact Properties

Figure 9 depicts the Charpy impact test results for five single-notched samples. The addition of NC at a rate of 2% increased the total gained impact energy from 2.6 to 18.7 J/mm^2. When the NC content was reduced to 3 wt.%, the impact strength decreased to 15.6 J/mm^2. The EP nanocomposites matrix combined with NC allowed for the formation of a chemical bond between EP, the filler, and the resin [56].

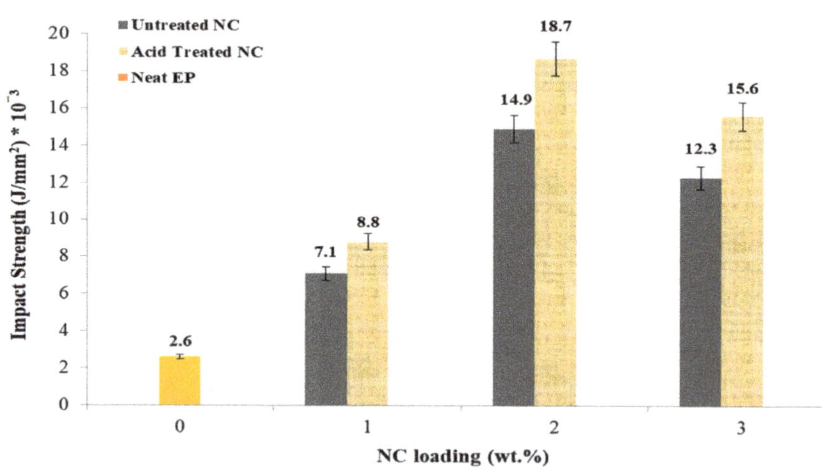

Figure 9. The variation of impact strength with NC content.

4.3. Hardness Properties

Figure 10 clearly shows the NC effect on the hardening of the EP matrix, which shows a hardness of 248.2 HV at 1 wt.% NC. By contrast, at NC-2 weight percent, the hardness increased to a maximum of 275 HV. Furthermore, it was discovered that the lowest value of Vickers hardness was found at NC of 3 wt.%. This increase can be attributed to the proposed NC ratio value. The voids produced by the fabrication increased as the NC content increased. Furthermore, the hardness test results at lower temperatures were superior to the sample at room temperature.

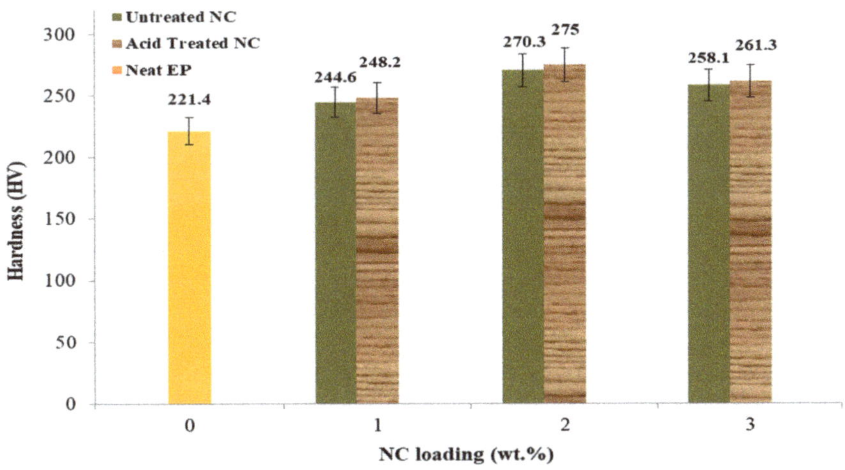

Figure 10. Effects of NC content on nanocomposites hardness.

4.4. Summary of the Mechanical Results

The tensile parameters explored include tensile strength, Young's modulus, maximum loading, and elongation at break. To completely explore the tensile properties, one sample of untreated NC was used, as well as three samples of untreated NC-EP at 1, 2, and 3 wt.% and acid-treated NC-EP at 1, 2, and 3 wt.%, as shown in Table 2. The most significant relative changes in tensile strength occurred at 2 wt.% EP of both untreated and acid-treated nanocomposites; the results indicate that the highest tensile strength occurred at 2 wt.% untreated NC-EP composites (16.9 MPa; 27.2%) and 2 wt.% sulfuric acid-treated NC-EP nanocomposites (16.9 MPa; 27.2%), (19.2 MPa; 35.9 %). Acid therapy is reported to have an 11.9% effect. However, the improvement was attributable to the NC addition for the acid-treated samples, which permits greater dispersion between the layered silicate of NC and the EP [33]. In terms of Young' modulus, adding NC at a concentration of 2% to the untreated samples resulted in relative improvements of 33.3% and 42.8%, respectively. Additionally, the acid treatment resulted in a 14.2% improvement. NC 3 wt.% addition resulted in 46.9% and 51.1% reductions in nonacid and acid treatment, respectively, for the maximum load. The effect of the acid treatment resulted in an increase of 8.1% in the maximum load. Finally, the nonacid and acid treatments increased the elongation percentage at the break by 43.5% and 53.0%, respectively, compared to the neat sample. However, the acid treatment resulted in 16.9%.

4.5. Field Emission Scanning Electron Microscopy (FESEM)

Figure 11 illustrates FESEM images of tensile fracture surfaces made of pure EP resin. As illustrated in Figure 11a, a smooth fracture surface was obtained at a magnification of 500x. The sample containing 1% NC is depicted in Figure 11b, where the fracture surface becomes cloudy and rougher as additional NC is added. Due to the wide specific surface area of the NC grid, the random formation of microfractures indicates that it is an undesirable condition. The inclusion of NC increased the strength of the nanocomposites; however, adding more than 1% of NC could result in the formation of microscale clusters via the NC network. The results indicate that the nanocomposites degrade due to strength transfer from the resin matrix to the NC network. This observation is consistent with those discussed in [57]. Figure 11c illustrates the cross-sectional produced area of the nanocomposites' fracture surface beyond a tensile test at 2% NC. The FESEM image showed that NC was frequently discontinuous in the matrix [57]. The morphology of the fractured surfaces in the impact test at 3 wt.% NC is illustrated in Figure 11d, indicating that the fracture surface was not uniform. Due to the NC agglomerates, the polymer flow was hampered. The brittleness indication, with a coarser topography, could be detected on the surface of the tidy EP nanocomposites, indicating that the NC pull-out was reduced. The images in the figures have sowed several fractures with varying numbers and sizes.

Table 2. Tensile properties of NC-EP nanocomposites.

Sample	Tensile Strength (MPa)	Young's Modulus (MPa)	Maximum Load (N)	%Elongation at Break
Neat NC	8	120	130	39
1 wt.% untreated NC-EP	12.3	162	220	43
2 wt.% untreated NC-EP	16.9 (27.2%)	169	245 (46.9%)	61
3 wt.% untreated NC-EP	13.4	180 (33.3%)	190	69 (43.5%)
1 wt.% acid-treated NC-EP	14.8	177	233	55
2 wt.% acid-treated NC-EP	19.2 (35.9%)	183	267 (51.1%)	71
3 wt.% acid-treated NC-EP	17.1	210 (42.8%)	219	83 (53.0%)
Effect of acid treatment	11.9%	14.2%	8.1%	16.9%

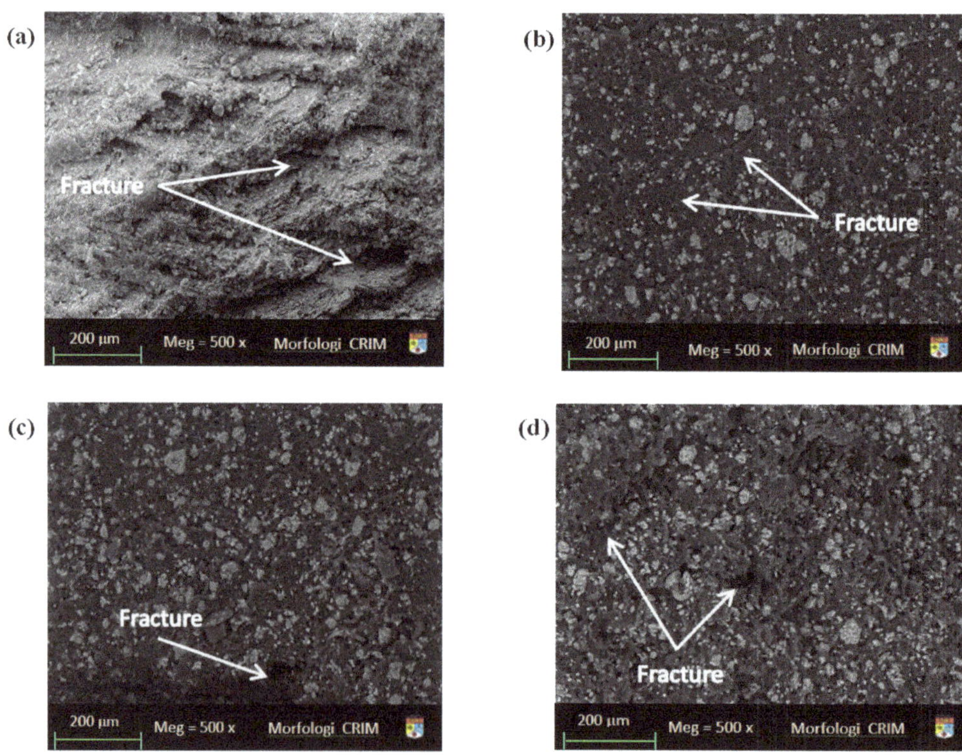

Figure 11. Tensile samples images of FESEM fracture surface (**a**) at pure EP (**b**) at NC content of 1 wt.% (**c**) at NC content of 2 wt.% (**d**) influence fracture surface at NC content of 3 wt.%.

5. Dynamic Mechanical Analysis (DMA)

5.1. Storage Modulus (E′)

The storage modulus (E') quantifies a material's elasticity under sinusoidal stress. E' provides information on a material's dynamic mechanical properties, load-bearing capacity [58], crosslink density [59], and fibre-matrix interface strength [60]. Additionally, as illustrated in Figure 12, introducing a nanoclay resulted in a substantial variance in

E'. The storage modulus of the composites was increased both before and after the glass-transition zone. The glassy region of hybrid composites was improved by 11.2%, 14.4%, and 35.3% with 1, 2, and 3 wt.% NC added to EP, respectively. Additionally, 3 wt.% NC/EP has the best storage capacity, which appears to be attributable to the fibre-reinforced epoxy, which increases the interfacial adhesion strength between fibres and matrix [61].

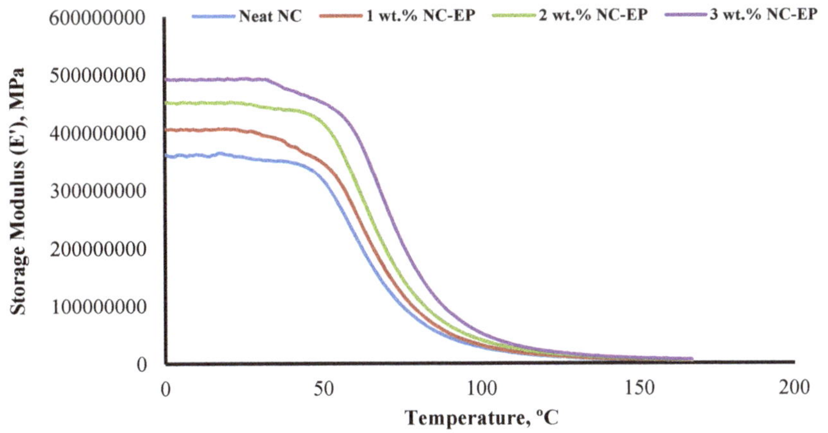

Figure 12. Effect of nanoclay on the storage modulus of the NC-EP at different weight concentrations.

The E' curves in Figure 12 indicated that the composites' glass-transition (T_g) temperatures were between 44 and 56 °C, indicating that T_g transitions from a glassy to a rubbery state. Thus, the material exhibits stiff and rigid behaviour, as indicated by the material's high modulus value throughout the mechanical property measurements. Additionally, it is projected that the epoxy matrix's free volume grows beyond T_g temperature due to the collapsing of the densely packed molecular arrangement.

5.2. Loss Modulus

The loss modulus (E'') can be used to analyse the viscous behaviour of materials under oscillating tension [62]. When a material has a large capacity for energy dissipation, it is said to have a high loss modulus, superior damping qualities, and a low damping factor. Thus, materials with a high E'' value demonstrate a strong potential for industrial applications. Figure 13 illustrates the E' behaviour of neat EP and EP composites at weight percentages of NC of 1, 2, and 3 as a function of temperature. All composites reached a maximum peak height between 64 and 72 °C, inside the T_g region. The glassy region's materials exhibit increased stiffness. However, it becomes stiff and unyielding as the material reaches the glassy temperature. As a result, its loss modulus is small. However, as it passes through the glass transition region and transitions from the glassy to the rubbery state, the material's viscosity increases dramatically. Similarly to E', when the loss modulus reaches its highest peak height, it exhibits high dissipation energy; however, when the molecules relax, the internal friction decreases, resulting in a decrease in the loss modulus.

5.3. Tan(δ)

The ratio of the loss modulus (E'') to the storage modulus (E') indicates the loss factor or damping factor. Tan delta is a considerable value, indicating a significant nonelastic strain component. On the other hand, low tan delta values indicate that the material is more elastic. The tan delta curves for EP and the composite sequence of EP-NC at 1, 2, and 3 wt.% are shown in Figure 14. The tan delta peak height of nanoclay composites is lower than that of neat EP, implying that the material has improved damping capabilities due to increased nonelastic deformation and energy dissipation. Additionally, increasing the

NC concentration decreases the peak values, indicating that the nanoclay and epoxy are interlocked [61].

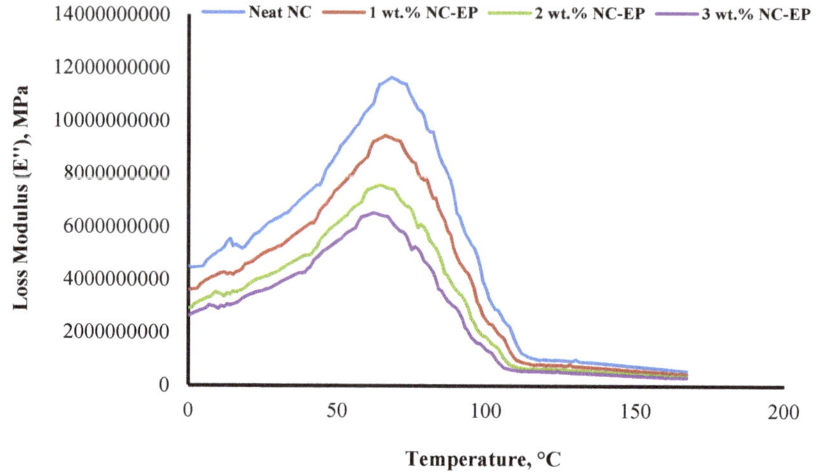

Figure 13. The loss of modulus of EP and the effect of nanoclay on the loss modulus of the hybrid composites.

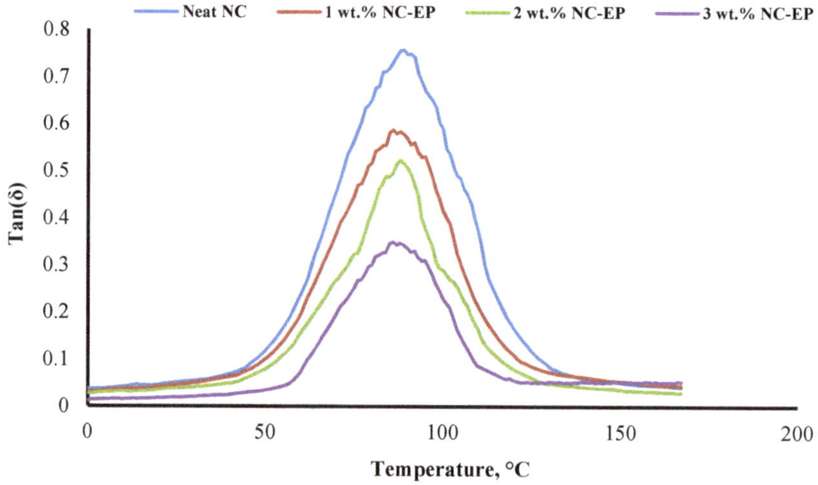

Figure 14. The values of tan delta for EP and the effect of nanoclay on EP.

5.4. Applications

Any material's application is contingent upon conformity with applicable requirements and standards. When the qualities of the material are enhanced and made modifiable, the material can be used in various applications. In this example, advancements in the qualities of epoxy materials expand their applications, with nanoclay applications divided by kind. Epoxy–NC-based materials could be employed in packaging [63], catalysis [64], and as an additive in the paper-making process [65], as well as medical applications [66]. Additionally, due to the high temperature, it was discovered that it could be utilised as thermal insulation [67]. NC is advantageous in medical applications, particularly vascular grafts, due to its small diameter [68]. NC-epoxy can also be utilised in fire extinguishers [35] and automobile components [69] due to its outstanding thermal stability and tensile strength.

Any composite's applicability is determined by its regulated durability under the conditions of use. Developing biocomposite-based nanoclay poses several obstacles because of the diverse applications of nanoclay, particularly as reinforcements [70]. The sustainability of the environment has a direct impact on economic development. By broadening the application, nanoclay sources improve, benefiting many other sectors [71].

6. Conclusions

We investigated the mechanical characteristics of an epoxy–clay nanocomposite. Modified clay boosts the adhesive yield of epoxy under high tensile conditions. The expansion percent to failure of epoxy nanocomposite-1.5% clay is nearly comparable to that of pure epoxy, with a difference of 0.047. Tension strength and failure toughness of epoxy clay nanocomposite are proportional to the amount of nanofiller in the epoxy, and improvement in these parameters is completely evident when using epoxy–clay nanocomposite rather than pure epoxy. Alternatively, adding modified clay to epoxy resin increases the material's strength and hardness simultaneously. Rack angle, development of new surfaces, and the failure of new piles are all effective methods for enhancing the toughness of an epoxy–clay nanocomposite with intercalation morphology, according to microscopic considerations.

According to reports, NC has been examined for its superior mechanical qualities and high aspect ratio. The current study examines the effects of nanocrystals on EP matrix nanocomposites. The results indicated that adding NC at a rate of 2 wt.% to the EP matrix improved the tensile strength, maximum load, impact strength, and hardness of the nanocomposites when compared to the clean EP matrix. The degree of homogeneity in nanocomposites has a significant effect on their mechanical characteristics. When the NC content was increased from 1 to wt. 3%, the Young's modulus and elongation at break both rose. Additionally, the impact and tensile strength were enhanced by incorporating NC. The results established the potential benefits of employing NC-EP nanocomposites in a number of applications, including biomedical ones.

The effect of adding nanoclay at concentrations of 1%, 2%, and 3% was investigated on dynamic mechanical characteristics. In comparison to neat EP, the NC-EP composites exhibit extremely high E′ and E″, indicating a strong interfacial connection between the fibres and matrix.

Author Contributions: Z.F.M. performed all of the experiments as part of his project and wrote the manuscript. S.F. and T.G.A. were the principal investigators and supervised the whole work. N.Y.Y. and A.A. helped with the characterisation of the manuscript. All authors have read and agreed to the published version of the manuscript.

Funding: This research received no external funding.

Acknowledgments: This work was supported by the Faculty of Engineering and Built Environment, Universiti Kebangsaan Malaysia, Malaysia and Institute of Laser for Postgraduate Studies, University of Baghdad, Iraq and Al Furat Intermediate School for Girls, Ministry of Education, Babylon, Iraq.

Conflicts of Interest: The authors declare no conflict of interest.

References

1. Chee, S.S.; Jawaid, M.; Sultan, M.T.H.; Alothman, O.Y.; Abdullah, L.C. Effects of nanoclay on physical and dimensional stability of Bamboo/Kenaf/nanoclay reinforced epoxy hybrid nanocomposites. *J. Mater. Res. Technol.* **2020**, *9*, 5871–5880. [CrossRef]
2. Yuhana, N.Y.; Ahmad, S.H.; Mebrabzadeh, M.; Bahri, A.R.S. Thermal and flexural properties of room–temperature cured PMMA grafted natural rubber toughened epoxy layered silicate nanocomposite. *Int. J. Nano Biomat.* **2014**, *5*, 45–58. [CrossRef]
3. Dzuhri, S.; Yuhana, N.Y.; Khairulazfar, M. Thermal stability and decomposition study of epoxy/clay nanocomposites. *Sains Malays.* **2015**, *44*, 441–448. [CrossRef]
4. Saba, N.; Tahir, P.M.; Jawaid, M. A review on potentiality of nano filler/natural fiber filled polymer hybrid composites. *Polymers* **2014**, *6*, 2247–2273. [CrossRef]
5. Keinänen, P. The Effect of Dispersion Quality of Carbon Nanotube Colloids on Physical Properties of Nanocomposites. Master's Thesis, Faculty of Engineering and Natural Sciences of Tampere University, Tampere, Finland, 30 October 2020.
6. Fréchette, M.F.; Reed, C.W.; Sedding, H. Progress, understanding and challenges in the field of nanodielectrics. *IEEJ Trans. Fundam. Mater.* **2006**, *126*, 1031–1043. [CrossRef]

7. Guo, F.; Aryana, S.; Han, Y.; Jiao, Y. A review of the synthesis and applications of polymer–nanoclay composites. *Appl. Sci.* **2018**, *8*, 1696. [CrossRef]
8. Rytwo, G. Clay minerals as an ancient nanotechnology: Historical uses of clay organic interactions, and future possible perspectives. *Macla* **2008**, *9*, 15–17.
9. Uddin, M.K. A review on the adsorption of heavy metals by clay minerals, with special focus on the past decade. *Chem. Eng. J.* **2017**, *308*, 438–462. [CrossRef]
10. Shahroze, R.M.; Ishak, M.R.; Salit, M.S.; Leman, Z.; Chandrasekar, M.; Munawar, N.S.; Asim, M. Sugar palm fiber/polyester nanocomposites: Influence of adding nanoclay fillers on thermal, dynamic mechanical, and physical properties. *J. Vinyl Addit. Technol.* **2020**, *26*, 236–243. [CrossRef]
11. Gaaz, T.S.; Sulong, A.B.; Ansari, M.N.M.; Kadhum, A.A.H.; Al-Amiery, A.A.; Nassir, M.H. Effect of starch loading on the thermo-mechanical and morphological properties of polyurethane composites. *Materials* **2017**, *10*, 777. [CrossRef]
12. Dutta, S.; Sengupta, S.; Chanda, J.; Das, A.; Wiessner, S.; Ray, S.S.; Bandyopadhyay, A. Distribution of nanoclay in a new TPV/nanoclay composite prepared through dynamic vulcanization. *Polym. Test.* **2020**, *83*, 106374. [CrossRef]
13. Gaharwar, A.K.; Cross, L.M.; Peak, C.W.; Gold, K.; Carrow, J.K.; Brokesh, A.; Singh, K.A. 2D nanoclay for biomedical applications: Regenerative medicine, therapeutic delivery, and additive manufacturing. *Adv. Mater.* **2019**, *31*, 1900332. [CrossRef]
14. Yuan, S.; Shen, F.; Chua, C.K.; Zhou, K. Polymeric composites for powder-based additive manufacturing: Materials and applications. *Prog. Polym. Sci.* **2019**, *91*, 141–168. [CrossRef]
15. Timmerman, J.F.; Hayes, B.S.; Seferis, J.C. Nanoclay reinforcement effects on the cryogenic microcracking of carbon fiber/epoxy composites. *Compos. Sci. Technol.* **2002**, *62*, 1249–1258. [CrossRef]
16. Simón-Herrero, C.; Gómez, L.; Romero, A.; Valverde, J.L.; Sánchez-Silva, L. Nanoclay-based PVA aerogels: Synthesis and characterization. *Ind. Eng. Chem. Res.* **2018**, *57*, 6218–6225. [CrossRef]
17. Jeyakumar, R.; Sampath, P.S.; Ramamoorthi, R.; Ramakrishnan, T. Structural, morphological and mechanical behaviour of glass fibre reinforced epoxy nanoclay composites. *Int. J. Adv. Manuf. Technol.* **2017**, *93*, 527–535. [CrossRef]
18. Zabihi, O.; Ahmadi, M.; Li, Q.; Shafei, S.; Huson, M.G.; Naebe, M. Carbon fibre surface modification using functionalized nanoclay: A hierarchical interphase for fibre-reinforced polymer composites. *Compos. Sci. Technol.* **2017**, *148*, 49–58. [CrossRef]
19. Pinto, D.; Bernardo, L.; Amaro, A.; Lopes, S. Mechanical properties of epoxy nanocomposites using titanium dioxide as reinforcement–a review. *Constr. Build. Mater.* **2015**, *95*, 506–524. [CrossRef]
20. Bakar, M.; Kostrzewa, M.; Białkowska, A.; Pawelec, Z. Effect of mixing parameters on the mechanical and thermal properties of a nanoclay-modified epoxy resin. *High Perform. Polym.* **2014**, *26*, 298–306. [CrossRef]
21. Xu, W.; Wang, X.; Wu, X.; Li, W.; Cheng, C. Organic-Inorganic dual modified graphene: Improving the dispersibility of graphene in epoxy resin and the fire safety of epoxy resin. *Polym. Degrad. Stab.* **2019**, *165*, 80–91. [CrossRef]
22. Yuhana, N.; Ahmad, S.; Bahri, A.R. The effect of ultrasonic treatment on thermal stability of the cured epoxy/layered silicate nanocomposite. *Adv. Mater. Sci. Eng.* **2012**, *2012*, 789815. [CrossRef]
23. Gaaz, T.S.; Sulong, A.B.; Kadhum, A.A.; Al-Amiery, A.A.; Nassir, M.H.; Jaaz, A.H. The Impact of Halloysite on the Thermo-Mechanical Properties of Polymer Composites. *Molecules* **2017**, *22*, 838. [CrossRef]
24. Zabihi, O.; Ahmadi, M.; Nikafshar, S.; Preyeswary, K.C.; Naebe, M. A technical review on epoxy-clay nanocomposites: Structure, properties, and their applications in fiber reinforced composites. *Compos. Part B Eng.* **2018**, *135*, 1–24. [CrossRef]
25. Khairulazfar, M.; Yuhana, N.Y.; Dzuhri, S. Curing Kinetics and Mechanical Study of Epoxy/Clay Nanocomposites. *Polym. Res. J.* **2015**, *9*, 475.
26. Suresha, B.; Varun, C.A.; Indushekhara, N.M.; Vishwanath, H.R. Effect of Nano Filler Reinforcement on Mechanical Properties of Epoxy Composites. In Proceedings of the Materials Science and Engineering, Adhi College of Engineering and Technology, Kanchipuram, Tamil Nadu, India, 8–10 May 2019; Volume 574, p. 012010.
27. Zhang, Y.; Choi, J.R.; Park, S.J. Interlayer polymerization in amine-terminated macromolecular chain-grafted expanded graphite for fabricating highly thermal conductive and physically strong thermoset composites for thermal management applications. *Compos. Part A Appl. Sci. Manuf.* **2018**, *109*, 498–506. [CrossRef]
28. Shah, R.; Kausar, A.; Muhammad, B.; Shah, S. Progression from graphene and graphene oxide to high performance polymer-based nanocomposite: A review. *Polym.-Plast. Technol. Eng.* **2015**, *54*, 173–183. [CrossRef]
29. Li, X.-F.; Lau, K.-T.; Yin, Y.-S. Mechanical properties of epoxy-based composites using coiled carbon nanotubes. *Compos. Sci. Technol.* **2008**, *68*, 2876–2881. [CrossRef]
30. Ma, P.-C.; Mo, S.-Y.; Tang, B.-Z.; Kim, J.-K. Dispersion, interfacial interaction and re-agglomeration of functionalized carbon nanotubes in epoxy composites. *Carbon* **2010**, *48*, 1824–1834. [CrossRef]
31. Azeez, A.A.; Rhee, K.Y.; Park, S.J.; Hui, D. Epoxy clay nanocomposites–processing, properties and applications: A review. *Compos. Part B Eng.* **2013**, *45*, 308–320. [CrossRef]
32. Jojibabu, P.; Zhang, Y.X.; Prusty, B.G. A review of research advances in epoxy-based nanocomposites as adhesive materials. *Intern. J. Adhes. Adhes.* **2020**, *96*, 102454. [CrossRef]
33. Esmaeili, A.; Sbarufatti, C.; Jiménez-Suárez, A.; Hamouda, A.M.S.; Rovatti, L.; Ureña, A. Synergistic effects of double-walled carbon nanotubes and nanoclays on mechanical, electrical and piezoresistive properties of epoxy based nanocomposites. *Compos. Sci. Technol.* **2020**, *200*, 108459. [CrossRef]

34. Njuguna, J.; Pielichowski, K.; Alcock, J.R. Epoxy-based fibre reinforced nanocomposites. *Adv. Eng. Mater.* **2007**, *9*, 835–847. [CrossRef]
35. Allami, T.; Alamiery, A.; Nassir, M.H.; Kadhum, A.H. Investigating Physio-Thermo-Mechanical Properties of Polyurethane and Thermoplastics Nanocomposite in Various Applications. *Polymers* **2021**, *13*, 2467. [CrossRef] [PubMed]
36. Liu, Y.; Zhao, J.; Zhao, L.; Li, W.; Zhang, H.; Yu, X.; Zhang, Z. High performance shape memory epoxy/carbon nanotube nanocomposites. *ACS Appl. Mater. Interface* **2016**, *8*, 311–320. [CrossRef]
37. Zhao, J.; Shan, J.; Van Assche, G.; Tenhu, H.; Van Mele, B. Demixing and remixing kinetics in aqueous dispersions of poly (N-isopropylacrylamide)(PNIPAM) brushes bound to gold nanoparticles studied by means of modulated temperature differential scanning calorimetry. *Macromolecules* **2009**, *42*, 5317–5327. [CrossRef]
38. Dufresne, A. Nanocellulose: A new ageless bionanomaterial. *Mater. Today* **2013**, *16*, 220–227. [CrossRef]
39. Ferreira, F.V.; Otoni, C.G.; Kevin, J.; Barud, H.S.; Lona, L.M.; Cranston, E.D.; Rojas, O.J. Porous nanocellulose gels and foams: Breakthrough status in the development of scaffolds for tissue engineering. *Mater. Today* **2020**, *37*, 126–141. [CrossRef]
40. Gabr, M.H.; Phong, N.T.; Abdelkareem, M.A.; Okubo, K.; Uzawa, K.; Kimpara, I.; Fujii, T. Mechanical, thermal, and moisture absorption properties of nano-clay reinforced nano-cellulose biocomposites. *Cellulose* **2013**, *20*, 819–826. [CrossRef]
41. Huang, W.M.; Yang, B.; Zhao, Y.; Ding, Z. Thermo-moisture responsive polyurethane shape-memory polymer and composites: A review. *J. Mater. Chem.* **2010**, *20*, 3367–3381. [CrossRef]
42. Yoonessi, M.; Shi, Y.; Scheiman, D.A.; Lebron-Colon, M.; Tigelaar, D.M.; Weiss, R.A.; Meador, M.A. Graphene polyimide nanocomposites; thermal, mechanical, and high-temperature shape memory effects. *ACS Nano* **2012**, *6*, 7644–7655. [CrossRef]
43. Rousseau, I.A.; Xie, T. Shape memory epoxy: Composition, structure, properties and shape memory performances. *J. Mater. Chem.* **2010**, *20*, 3431–3441. [CrossRef]
44. Rajeshkumar, G.; Seshadri, S.A.; Ramakrishnan, S.; Sanjay, M.R.; Siengchin, S.; Nagaraja, K.C. A comprehensive review on natural fiber/nano-clay reinforced hybrid polymeric composites: Materials and technologies. *Polym. Compos.* **2021**, *42*, 3687–3701. [CrossRef]
45. Ramakrishnan, S.; Krishnamurthy, K.; Rajeshkumar, G.; Asim, M. Dynamic mechanical properties and free vibration characteristics of surface modified jute fiber/nano-clay reinforced epoxy composites. *J. Polym. Environ.* **2021**, *29*, 1076–1088. [CrossRef]
46. Chee, S.S.; Jawaid, M.; Alothman, O.Y.; Fouad, H. Effects of nanoclay on mechanical and dynamic mechanical properties of bamboo/kenaf reinforced epoxy hybrid composites. *Polymers* **2021**, *13*, 395. [CrossRef]
47. Saba, N.; Jawaid, M.; Alothman, O.Y.; Paridah, M.T. A review on dynamic mechanical properties of natural fibre reinforced polymer composites. *Constr. Build. Mater.* **2016**, *106*, 149–159. [CrossRef]
48. Pistor, V.; Ornaghi, F.G.; Ornaghi, H.L.; Zattera, A.J. Dynamic mechanical characterization of epoxy/epoxycyclohexyl–POSS nanocomposites. *Mater. Sci. Eng. A* **2012**, *532*, 339–345. [CrossRef]
49. Chee, S.S.; Jawaid, M.; Sultan, M.T. Thermal stability and dynamic mechanical properties of kenaf/bamboo fibre reinforced epoxy composites. *BioResources* **2017**, *12*, 7118–7132.
50. Asim, M.; Paridah, M.T.; Saba, N.; Jawaid, M.; Alothman, O.Y.; Nasir, M.; Almutairi, Z. Thermal, physical properties and flammability of silane treated kenaf/pineapple leaf fibres phenolic hybrid composites. *Compos. Struct.* **2018**, *202*, 1330–1338. [CrossRef]
51. Sathishkumar, T.P. Dynamic mechanical analysis of snake grass fiber-reinforced polyester composites. *Proc. Inst. Mech. Eng. Part L J. Mater. Des. Appl.* **2016**, *230*, 160–174. [CrossRef]
52. Rajesh, M.; Jeyaraj, P.; Rajini, N. Mechanical, dynamic mechanical and vibration behavior of nanoclay dispersed natural fiber hybrid intra-ply woven fabric composite. In *Nanoclay Reinforced Polymer Composites*; Springer: Berlin/Heidelberg, Germany, 2016; pp. 281–296.
53. Biswas, B.; Warr, L.N.; Hilder, E.F.; Goswami, N.; Rahman, M.M.; Churchman, J.G.; Vasilev, K.; Pan, G.; Naidu, R. Biocompatible functionalisation of nanoclays for improved environmental remediation. *Chem. Soc. Rev.* **2019**, *48*, 3740–3770. [CrossRef]
54. Ali, M.B.; Abdullah, S.; Nuawi, M.Z.; Ariffin, A.K. Correlation of absorbed impact with calculated strain energy using an instrumented Charpy impact test. *Indian J. Eng. Mater. Sci.* **2013**, *20*, 504–514.
55. Broitman, E. Indentation hardness measurements at macro-, micro-, and nanoscale: A critical overview. *Tribol. Lett.* **2017**, *65*, 23. [CrossRef]
56. Ahmed, M.A.; Kandil, U.F.; Shaker, N.O.; Hashem, A.I. The overall effect of reactive rubber nanoparticles and nano clay on the mechanical properties of epoxy resin. *J. Radiat. Res. Appl. Sci.* **2015**, *8*, 549–561. [CrossRef]
57. Khanbabaei, G.; Aalaie, J.; Rahmatpour, A.; Khoshniyat, A.; Gharabadian, M.A. Preparation and Properties of Epoxy-Clay Nanocomposites. *J. Macromol. Sci. Part B Phys.* **2007**, *46*, 975–986. [CrossRef]
58. Anand, G.; Alagumurthi, N.; Elansezhian, R.; Venkateshwaran, N. Dynamic mechanical, thermal and wear analysis of Ni-P coated glass fiber/Al_2O_3 nanowire reinforced vinyl ester composite. *Alex. Eng. J.* **2018**, *57*, 621–631. [CrossRef]
59. Miyagawa, H.; Misra, M.; Drzal, L.T.; Mohanty, A.K. Novel biobased nanocomposites from functionalized vegetable oil and organically-modified layered silicate clay. *Polymer* **2005**, *46*, 445–453. [CrossRef]
60. Joseph, S.; Appukuttan, S.P.; Kenny, J.M.; Puglia, D.; Thomas, S.; Joseph, K. Dynamic mechanical properties of oil palm microfibril-reinforced natural rubber composites. *J. Appl. Polym. Sci.* **2010**, *117*, 1298–1308. [CrossRef]
61. Jesuarockiam, N.; Jawaid, M.; Zainudin, E.S.; Thariq, H.S.M.; Yahaya, R. Enhanced thermal and dynamic mechanical properties of synthetic/natural hybrid composites with graphene nanoplateletes. *Polymers* **2019**, *11*, 1085. [CrossRef]

62. Saba, N.; Safwan, A.; Sanyang, M.L.; Mohammad, F.; Pervaiz, M.; Jawaid, M.; Alothman, O.Y.; Sain, M. Thermal and dynamic mechanical properties of cellulose nanofibers reinforced epoxy composites. *Intern. J. Biol. Macromol.* **2017**, *102*, 822–828. [CrossRef]
63. Azeredo, H.M.; Rosa, M.F.; Mattoso, L.H.C. Nanocellulose in bio-based food packaging applications. *Ind. Crop. Prod.* **2017**, *97*, 664–671. [CrossRef]
64. Gama, M.; Dourado, F.; Bielecki, S. *Bacterial Nanocellulose: From Biotechnology to Bio-Economy*; Elsevier: Amsterdam, The Netherlands, 2016.
65. Boufi, S.; González, I.; Delgado-Aguilar, M.; Tarrès, Q.; Pèlach, M.À.; Mutjé, P. Nanofibrillated cellulose as an additive in papermaking process: A review. *Carbohydr. Polym.* **2016**, *154*, 151–166. [CrossRef]
66. Naderi, A.; Lindström, T.; Sundström, J. Carboxymethylated nanofibrillated cellulose: Rheological studies. *Cellulose* **2014**, *21*, 1561–1571. [CrossRef]
67. Gupta, P.; Singh, B.; Agrawal, A.K.; Maji, P.K. Low density and high strength nanofibrillated cellulose aerogel for thermal insulation application. *Mater. Des.* **2018**, *158*, 224–236. [CrossRef]
68. Brown, E.E.; Hu, D.; Lail, N.A.; Zhang, X. Potential of nanocrystalline cellulose–fibrin nanocomposites for artificial vascular graft applications. *Biomacromolecules* **2013**, *14*, 1063–1071. [CrossRef]
69. Ilyas, R.; Sapuan, S.; Asyraf, M.; Atikah, M.; Ibrahim, R.; Dele-Afolabi, T.; Hazrol, M.D. Introduction to biofiller-reinforced degradable polymer composites. In *Biofiller-Reinforced Biodegradable Polymer Composites*; CRC Press: Boca Raton, FL, USA, 2020; pp. 1–23.
70. Jawaid, M.; Boufi, S.; HPS, A.K. *Cellulose-Reinforced Nanofibre Composites: Production, Properties and Applications*; Woodhead Publishing: Cambridge, UK, 2017.
71. Omran, A.A.B.; Mohammed, A.A.B.A.; Sapuan, S.M.; Ilyas, R.A.; Asyraf, M.R.M.; Koloor, S.S.R.; Petrů, M. Micro-and nanocellulose in polymer composite materials: A review. *Polymers* **2021**, *13*, 231. [CrossRef]

polymers

MDPI

Article

Nanocomposites of PCL and SBA-15 Particles Prepared by Extrusion: Structural Characteristics, Confinement of PCL Chains within SBA-15 Nanometric Channels and Mechanical Behavior

Tamara M. Díez-Rodríguez [1], Enrique Blázquez-Blázquez [1], Nadine L. C. Antunes [2], M. Rosário Ribeiro [2], Ernesto Pérez [1] and María L. Cerrada [1,*]

1 Instituto de Ciencia y Tecnología de Polímeros (ICTP-CSIC), Juan de la Cierva 3, 28006 Madrid, Spain; t.diez@ictp.csic.es (T.M.D.-R.); enrique.blazquez@csic.es (E.B.-B.); ernestop@ictp.csic.es (E.P.)

2 Centro de Química Estrutural, Instituto Superior Técnico, Universidade de Lisboa, Av. Rovisco Pais 1, 1049-001 Lisboa, Portugal; nadine.leonor10@gmail.com (N.L.C.A.); rosario@tecnico.ulisboa.pt (M.R.R.)

* Correspondence: mlcerrada@ictp.csic.es; Tel.: +34-91-258-7474

Abstract: A study of different nanocomposites based on poly(ε-caprolactone) (PCL) and mesoporous SBA-15 silica that were prepared by melt extrusion was carried out by analyzing the possible effect of this filler on the crystalline details of PCL, on its mechanical behavior, and on the eventual observation of the confinement of the polymeric chains within the hollow nanometric silica channels. Thus, simultaneous Small-Angle and Wide-Angle X-ray Scattering (SAXS/WAXS) synchrotron experiments at variable temperature were performed on these PCL nanocomposites with different mesoporous silica contents. The importance of the morphological and structural features was assessed by the changes that were observed during the mechanical response of the final materials, which determined that the presence of mesoporous particles leads to a noticeable reinforcing effect.

Keywords: PCL–SBA-15 nanocomposites; real-time variable-temperature synchrotron measurements; confinement; mechanical behavior

Citation: Díez-Rodríguez, T.M.; Blázquez-Blázquez, E.; Antunes, N.L.C.; Ribeiro, M.D.R.G.; Pérez, E.; Cerrada, M.L. Nanocomposites of PCL and SBA-15 Particles Prepared by Extrusion: Structural Characteristics, Confinement of PCL Chains within SBA-15 Nanometric Channels and Mechanical Behavior. *Polymers* **2022**, *14*, 129. https://doi.org/10.3390/polym14010129

Academic Editor: Andrzej Puszka

Received: 18 November 2021
Accepted: 21 December 2021
Published: 30 December 2021

1. Introduction

Polycaprolactone (PCL) is one of the most attractive and most commonly used biodegradable polymers [1]. This material belongs to the category of polyesters that are derived from petrochemical feedstocks, and its main features include hydrophobicity, low glass transition (T_g) and melting temperatures (T_m) at about −60 °C and 60 °C, respectively, as well as important chain flexibility and outstanding processability. Furthermore, it can be spun into fibers or blown films at temperatures under 200 °C without experiencing thermal degradation. PCL degrades through the hydrolysis of its ester linkages, showing a relatively slow degradation rate [1]. This latest aspect makes PCL suitable for the fabrication of long-term implant systems and for its use as a matrix in drugs delivery systems, primarily those for controlled-release devices with long working lifetimes (1–2 years). However, PCL suffers from certain shortcomings when it is intended to be used in the tissue engineering field, including its aforementioned slow degradation rate, low cell adhesion and poor mechanical properties, which tends to be experienced under load-bearing conditions. Advanced manufacturing technologies such as electrospinning or 3D printing [2,3] and its blending it with other polymers [4–9] or incorporation of stiffer materials (fillers or fibers) [10–13] can promote the improvement of either its bioactivity or its mechanical response, enabling it to be used for bone tissue engineering applications.

Ordered mesoporous silicas emerged in the 1990s. MCM-41 and SBA-15 are the most well-known members, with both exhibiting hexagonal arrangements of uniformly sized cylindrical pores, a narrow pore size distribution, and a large surface area [14,15]. The diameter of the pores that are found in SBA-15 particles are larger than those that are found in MCM-41, with the former typically showing pore size diameters of 7–9 nm, and the latest demonstrating typical pore sizes of 3 nm. Because of their tailor-made pore shapes and sizes, these materials have become particularly important in applications that are based on molecular recognition, such as selective catalysis, molecular sieving, chemical sensing, and precise adsorption or in applications where they act as drug carriers. Furthermore, the incorporation of these ordered and hollow mesostructures as a minority component into neat polymers becomes a very attractive approach that can be used to achieve hybrid polymeric-based materials. In fact, they have been used as hosts and reactors in various polymerization reactions [16–21], where they also acted as catalyst carriers and as a reinforcing material [22–25]. In addition, these pristine mesoporous silicas are capable of chemical medication, which supposes a versatile range of methodologies that can be implemented to easily attain tailored functionalized polymeric composites [26,27] with fine-tuned performance characteristics.

Polymers may become confined in the pores during the polymerization or during their further processing if the macromolecular chains are able to be allocated in the nanometric spaces that exist in mesoporous silicas. This partial presence of a polymer within those hollow nanospaces may result in a more intimate phase affinity in the resulting composite, contributing to improved interface adhesion and leading to a significant enhancement in the ultimate properties, mainly the mechanical ones, since mesoporous silicas can act as a filler [18,22–24,26,27].

This research aims to prepare composites that are based on PCL and SBA-15 particles at different contents through extrusion in order to learn the effect that pristine mesoporous particles exert on the polymeric crystalline structure, thermal behavior, and mechanical response. Moreover, the presence of PCL chains within the nanometric pores of SBA-15 will be also evaluated. Extrusion was selected since it is a cost-effective and an environmentally friendly transformation protocol that does not involve the use of a solvent. Accordingly, the novelty of this research consists of proposing a sustainable strategy that can be implemented to improve mechanical performance of PCL-based materials at temperatures higher than room temperature by taking advantage of incorporating a more rigid component that allows the additional possibility of including the PCL chains confined within the regular nanometric arrangement of these materials. Thus, the influence of SBA-15 on the crystalline details of PCL will be assessed by X-ray diffraction, specifically with real-time variable-temperature X-ray scattering at both small angles (SAXS) and wide angles (WAXS). Differential scanning calorimetry (DSC) and SAXS experiments will be employed to verify eventual confinement occurs in the obtained materials. Finally, the ultimate mechanical performance will be examined by stress–strain tests.

2. Materials and Methods

2.1. Materials and Chemicals

A commercially available polycaprolactone (purchased from Sigma-Aldrich) with an average molar mass (M_n) of 80,000 g/mol and a density of 1.145 g/cm^3 was used as a polymeric matrix in the present research. SBA-15 particles were also purchased from Sigma-Aldrich (specific surface area, S_{BET} = 517 m^2/g; total pore volume, V_t = 0.83 cm^3/g; average mesopore diameter [28], D_p = 6.25 nm) and were used as received.

2.2. Nano Composite and Film Preparation

Composites with different contents of SBA-15 particles (3, 6, and 9% in weight, which were labeled as PCL-SBA3, PCL-SBA6, and PCL-SBA9, respectively) were processed by melt extrusion in a corotating twin-screw microextruder (Rondol). Both the polymer and SBA-15 were dried prior to extrusion. The PCL was placed in an oven at 50 °C for 20 min and was then dried under vacuum conditions at 50 °C for 20 h. The SBA-15 particles were dried under vacuum conditions at 100 °C for 24 h. In the extruder, a screw temperature profile of 100, 105, 110, 120, and 110 °C was used from the hopper to the die, where the length-to-diameter ratio was 20:1. Then, films were obtained by compression molding at 120 °C and at 30 bar for 6 minutes in a hot-plate Collin press. Afterwards, a cooling process was applied to the different materials from their molten state to room temperature for 4 min at the relatively fast rate of around 80 °C/min and at a pressure of 30 bar.

2.3. Transmission Electron Microscopy

Measurements were performed at room temperature under a 200 kV JEM-2100 JEOL microscope. The particles were dispersed in acetone in an ultrasonic bath for 5 min and were then deposited in a holder prior to observation.

2.4. Scanning Electron Microscopy

Images were attained using S-8000 Hitachi equipment at room temperature in different cryo-fractured sections of the composites with distinct mesoporous contents. These thin sections, which were around 40 nm in size, were cut by means of cryo-ultramicrotomy (Leica EM UC6) at −120 °C and were deposited in a holder.

2.5. Thermogravimetric Analysis

Thermogravimetric analysis (TGA) was performed in using the Q500 equipment from TA Instruments under either a nitrogen or air atmosphere and at a heating rate of 10 °C/min. The degradation temperatures of the distinct materials were determined as well as the exact amount of SBA-15 that had been incorporated into the composites prepared by extrusion, which was estimated as an average of the values that were obtained from the two environments.

2.6. Differential Scanning Calorimetry

Calorimetric analyses were carried out in a TA Instruments Q100 calorimeter that was connected to a cooling system and that had been calibrated with different standards. The sample weights were around 6 mg. A temperature interval ranging from −80 to 100 °C was studied at a heating rate of 20 °C/min. To determine the crystallinity of the samples, a value of 135 J/g was considered for the melting enthalpy of the 100% crystalline PCL [29,30]. Errors in the temperature determination, enthalpy calculation, and the crystallinity were estimated at ±0.5 °C, ±4 J/g, and ±0.04 units, respectively.

2.7. X-ray Experiments with Synchrotron Radiation

Simultaneous real-time variable-temperature SAXS/WAXS experiments were carried out using synchrotron radiation in the beamline BL11-NCD-SWEET at ALBA (Cerdanyola del Valles, Barcelona, Spain) at a fixed wavelength of 0.1 nm. A Pilatus detector was used for the SAXS (off beam, at a distance of 296 cm from sample) experiments, and a Rayonix one was used for the WAXS (at about 14.6 cm from sample, and a tilt angle of around 29 degrees) experiments. A Linkam Unit, which was connected to a liquid nitrogen cooling system, was employed for temperature control. The spacing calibration was determined by means of silver behenate and using Cr_2O_3 standards. The initial 2D X-ray images were converted into 1D diffractograms as function of the inverse scattering vector, $s = 1/d = 2 \sin \theta/\lambda$, using pyFAI python code (ESRF) that had been modified by the ALBA beamline staff. Film samples of around 5 × 5 × 0.2 mm were used for the synchrotron analysis.

2.8. Mechanical Behavior by Means of Stress–Strain Tests

Nominal stress–strain tests were performed in an MTS Q-Test Elite dynamometer with a load-cell of 100 N at a temperature of 25 °C and at a rate of 10 mm/min. The specimens that were used for these experiments were punched out from the polymer films. The dimensions of these strips were 15 mm in length, 1.9 mm in width, and around 0.20 mm in thickness. A minimum of six different strips were stretched until fracture at a given specimen.

3. Results and Discussion

3.1. Morphological Characteristics

Details of the mesoporous SBA-15 silica were visible through the use of transmission electron microscopy (TEM), while the dispersion of SBA-15 particles within the PCL matrix was evaluated by high-resolution field emission scanning electron microscopy (FESEM). Figure 1a shows a TEM micrograph for the particles in the mesoporous SBA-15 silica that was used in this investigation. These particles display a vermicular and elongated shape with an approximate size of 350 nm in width and 0.9 μm in length. The size that is exhibited by this commercial mesoporous silica corresponds with the one that was previously described in the literature for mesoporous SBA-15 particles that had been synthesized in a laboratory [22]. Its magnification, shown in Figure 1b, depicts the interior particle morphology, which is made up of a well-defined, uniform, and ordered channel structure with hexagonal arrangements.

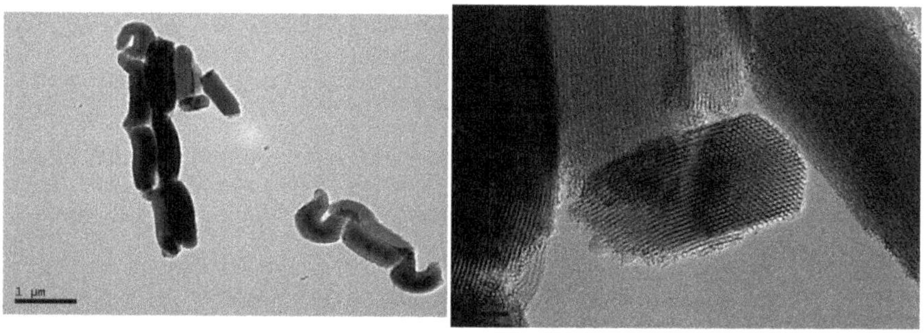

(a) (b)

Figure 1. TEM images of SBA-15 particles at scale bars of (**a**) 1 μm and (**b**) 100 nm.

Figure 2 shows FESEM micrographs for the distinct composites based on PCL and several amounts of SBA-15 particles. The SBA-15 silica demonstrates a clear uniform dispersion in the pictures, and an evident increase in the number of SBA-15 particles is also observed, as their content is increased in the final composite (see precise compositions determined from TGA results in Table 1).

Table 1. TGA results under nitrogen and air atmospheres for neat PCL and its composites prepared by melt extrusion: temperature of a loss weight of 10% ($T_{10\%}$) and temperature at the maximum (T_{max}) together with the SBA-15 wt.% at 600 °C and its average.

Sample	Average SBA-15 wt.%	*Inert Atmosphere*			*Oxidative Atmosphere*		
		$T_{10\%}$ (°C)	T_{max} (°C)	SBA-15 wt.%	$T_{10\%}$ (°C)	T_{max} (°C)	SBA-15 wt.%
PCL	0	359.0	360.0	0	344.5	356.0	0
PCL-SBA3	3.3	357.5	359.5	3.8	350.5	354.5	2.8
PCL-SBA6	6.0	358.0	359.5	6.0	353.0	357.5	6.0
PCL-SBA9	9.0	358.5	361.0	9.3	347.0	355.5	8.6

The non-existence of inorganic domains that are bulky in size is noticed across the materials that were fabricated with different contents of SBA-15. However, the formation of particles aggregates is evident, as the amount of silica in the material increases and is more apparent in the PCL-SBA9 composite, as depicted in the picture in Figure 2c.

The well-defined channel structure and the common hexagonal arrangement that is exhibited by these mesoporous SBA-15 particles are maintained when they are embedded in the PCL matrix of the different composites. Figure 2d,e undoubtedly display these features for the PCL-SBA3 and PCL-SBA6 composites. That ordering is also preserved in the composite with the highest content of mesoporous SBA-15, something that will be commented upon in more depth later. This indicates that at the mesoscale, regularity is not changed by the shear forces that are applied during extrusion, the protocol that was used for the preparation of the PCL-based materials in this study.

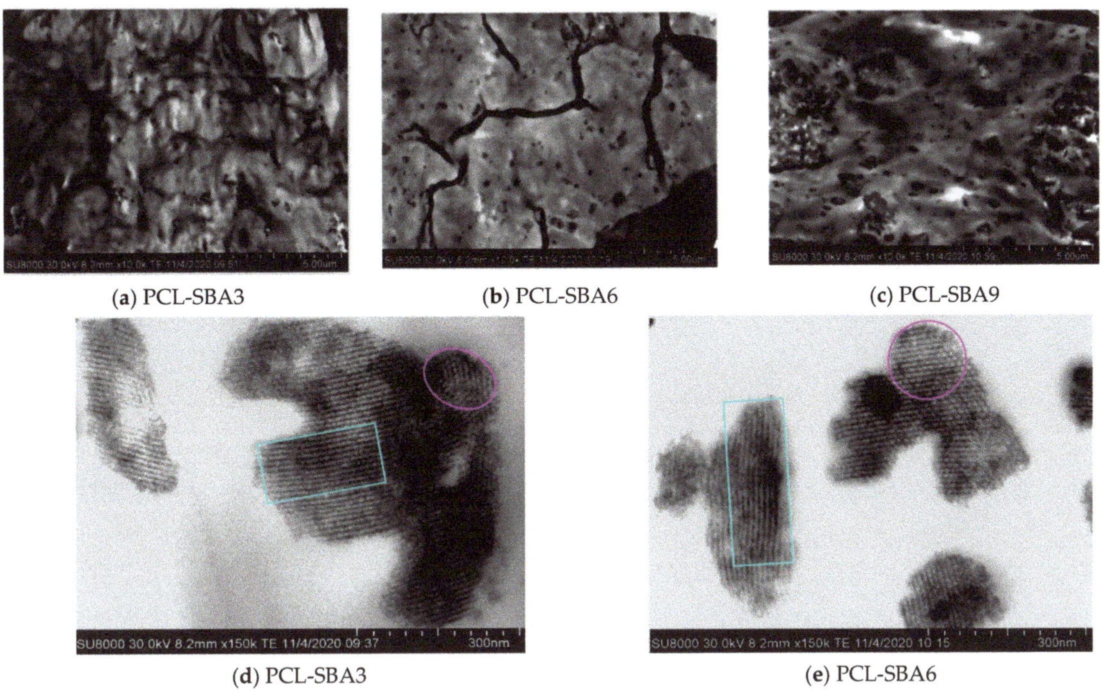

Figure 2. FESEM micrographs for the different composites: (**a**) PCL-SBA3; (**b**) PCL-SBA6; and (**c**) PCL-SBA9, at a scale bar of 5 μm. FESEM micrographs showing individual SBA-15 particles for different composites: (**d**) PCL-SBA3 and (**e**) PCL-SBA6, at a scale bar of 300 nm.

3.2. Thermal Stability

Mesoporous silicas have also sometimes been used as catalysts for thermal decomposition. This effect was found by Marcilla et al. [31] when degradation caused by TGA was studied in polyethylene (PE) under N_2 in the presence and absence of mesoporous MCM-41. Other authors [32] have shown the efficiency of mesoporous alumino-silicate MCM-41 as a promoter of the polyolefin degradation into liquid fuels. This role as a promoter of decomposition was also found when MCM-41 silica was used as a catalyst carrier and filler for in situ polymerized PE-based composites that employed either neat mesoporous particles or those decorated with undecenoic acid or silanes [18,26,33,34].

Figure 3 shows the temperature dependence of PCL weight loss under either inert or oxidative conditions in these composites. The inert TGA curve (see Figure 3a) displays one main degradation process in the pristine PCL, with an inflection point at around 360 °C (see data in Table 1), which is where most of the weight loss (approximately 95%) occurs. Nevertheless, a careful examination of the DTGA trace (see inset in Figure 3b) highlights another minor consecutive mechanism that takes place at a higher temperature. By studying the nature of the gases that were produced through the entire thermal degradation process and under the inert environment allowed a two-stage degradation mechanism to be proposed [35,36]. The first process involved a statistical rupture of the polyester chains via an ester pyrolysis reaction. The produced gases were identified as H_2O, CO_2, and 5-hexenoic acid. The second step led to the formation of ε-caprolactone (cyclic monomer) as result of an unzipping depolymerization process.

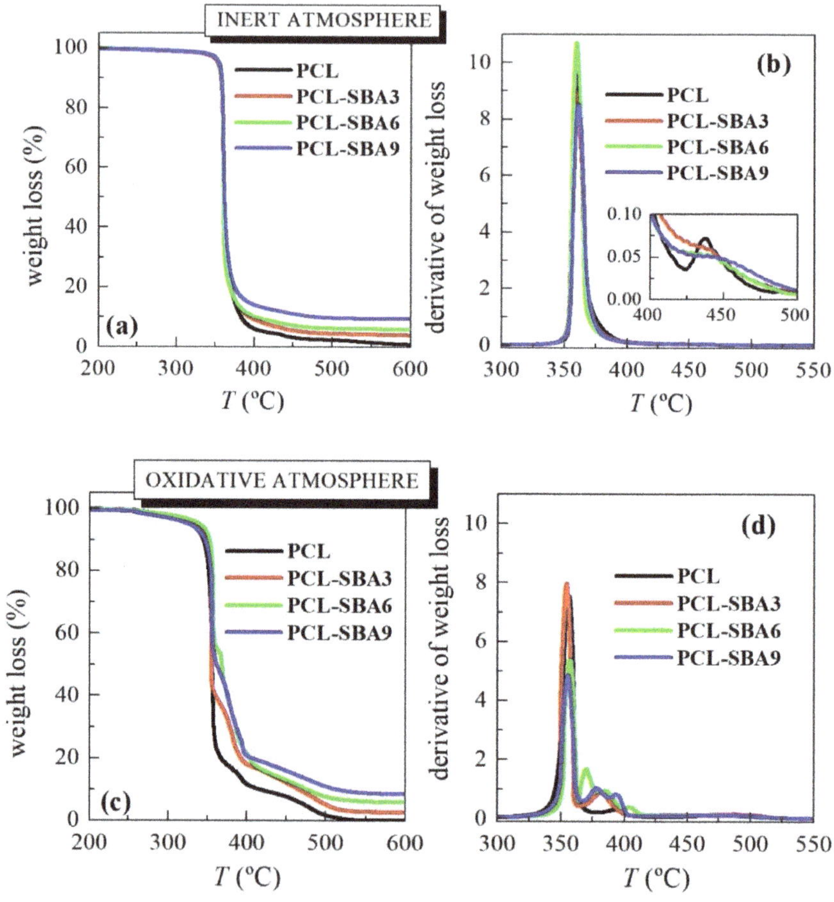

Figure 3. TGA curves either under inert atmosphere (**a**) and their derivatives (**b**) or in an oxidant environment (**c**) and their derivatives (**d**) for the neat PCL and its composites with SBA-15 at different contents prepared by melt extrusion.

If experiments are performed in an inert atmosphere, then thermal PCL degradability is not considerably affected by the incorporation of SBA-15 particles. The presence of mesoporous silica does not change the location at which most of the weight loss takes place (see values under the two atmospheres in Table 1) although the process that is related to unzipping depolymerization of PCL becomes a shoulder, and it is increased in the composites compared to in the pristine polymer. PCL degradation is affected by other parameters, such as polyester molar mass and the nature of the PCL end groups [35]. An increase in the former provokes a significant drop in the degradation rate. This behavior can be explained by the statistical chain cleavage that is triggered by the pyrolysis reaction during the first degradation process.

Figure 3c shows the temperature dependence of the PCL weight loss under an air environment. The behavior is more complex than that seen under the inert atmosphere. Thus, weight loss takes place in several stages, something that is clearly noticeable in the TGA and DTGA curves, since now PCL degradation occurs via both thermolysis and oxidation, i.e., through thermo-oxidation. The process that involves the highest amount of weight loss is not considerably affected by the presence of mesoporous SBA-15, as deduced from the T_{max} listed in Table 1. Its effect is more evident in the value of $T_{10\%}$. This is slightly shifted to higher temperatures in the composites, indicating that SBA-15 slightly postpones the beginning of PCL weight loss. All of these results indicate that the presence of SBA-15 particles in a PCL matrix does not exert a catalytic effect in these composites, contrary to the case of previous observations, where polyethylene was used as a matrix and where MCM-41 was used as the mesoporous silica [18,26,31–34]. Furthermore, the TGA curves that were obtained under both environments show that PCL degrades at an analogous range of temperatures independently of the atmosphere that is used.

3.3. Phase Transitions, Crystalline Characteristics and Confinement of PCL Chains

To analyze the effect of incorporating the mesoporous silica in the PCL transitions on the first heating run, the different transitions were evaluated separately, in the order of increasing temperatures: the glass transition temperature and the melting process temperature, as well as their corresponding characteristic temperatures, T_g and T_m. Figure 4a shows the temperature range for neat PCL and its composites in the T_g interval, with this transition being related to the amorphous regions. These curves, which have been normalized to the actual PCL amount, display that the intensity of this transition is dependent on the silica content in the different composite samples. This could indicate that the quantity of PCL that is present in the amorphous regions decreased as a result of the silica particle incorporation, i.e., the crystallinity increased due to the presence of SBA-15 particles; however, the results listed in Table 2 do not confirm this hypothesis. On the other hand, this finding could be also ascribed to a hindrance in the amorphous mobility caused by mesoporous silica incorporation. The location remains almost constant, with only slight displacement being observed at higher temperatures in the samples with the highest SBA-15 contents.

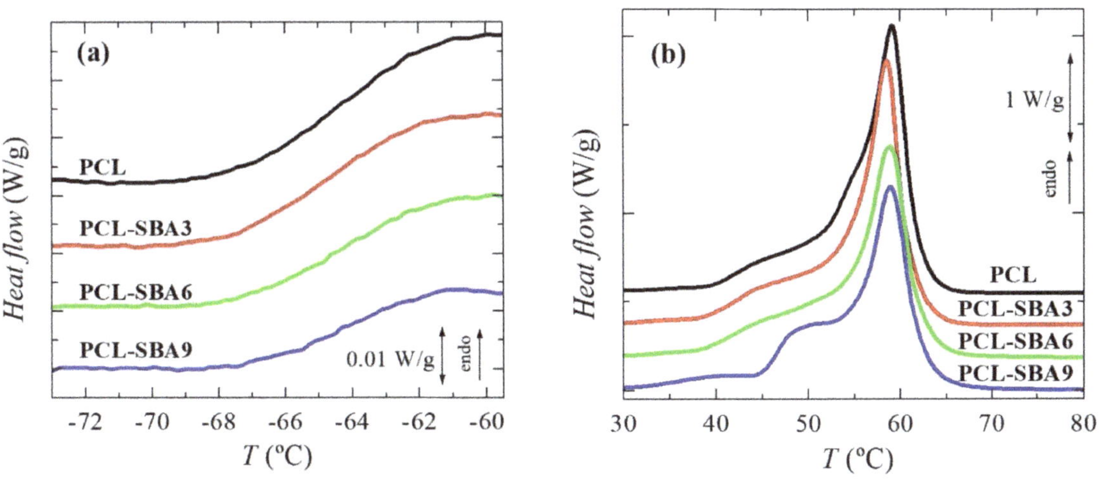

Figure 4. DSC curves (normalized to the actual PCL amount at a given material) for the neat PCL and its composites prepared by melt extrusion: (**a**) in the range of glass transition and (**b**) in the melting range. Curves have been shifted for the sake of clarity.

Table 2. Average SBA-15 wt.% estimated from TGA measurements and DSC results: glass transition (T_g)—calculated from the first heating run—, melting (T_m), and crystallization (T_c) temperatures, and overall crystallinity (normalized to the actual PCL content in the material) for the first melting ($f_c^{m1}{}_{NORM}$), crystallization ($f_c^{C}{}_{NORM}$), and second melting ($f_c^{m1}{}_{NORM}$) processes.

Sample	SBA-15 wt.%	T_g (°C)	T_{m1} (°C)	$f_c^{m1}{}_{NORM}$	T_c (°C)	$f_c^{C}{}_{NORM}$	T_{m2} (°C)	$f_c^{m2}{}_{NORM}$
PCL	0	−64.5	59.0	0.51	27.5	0.42	55.5	0.42
PCL-SBA3	3.3	−64.5	58.5	0.50	30.5	0.41	55.0	0.41
PCL-SBA6	6.0	−64.0	59.0	0.50	29.5	0.41	55.5	0.41
PCL-SBA9	9.0	−64.0	59.0	0.51	29.0	0.40	55.5	0.40

Figure 4b shows the DSC curves in the temperature range for the melting of the PCL crystallites in the neat homopolymer and the composites during the first heating stage. A complex melting process is exhibited by the PCL in the pristine polymer [9,30] and in all of the composites. Thus, distinct endothermic events are noticeable, which can be ascribed to the existence of several processes. These multiple stages could be associated with the thermal history that was imposed during their processing, implying a fast cooling rate from the melt stage that leads to thin and imperfect crystals that are able to become thicker and more perfect via annealing at room temperature during the DSC run. The main melting peak is, however, rather analogous in all of these samples, as the size and amount of the major PCL crystallites is not affected by the incorporation of SBA-15, which can be deduced from the crystallinity values after they have been normalized to the actual PCL amount in the different materials, as listed in Table 2.

Regarding the secondary melting events, important differences can be observed in the interval ranging from 40 to 53 °C. Two different processes are considered to be present in this interval. The first one is the annealing of the initially imperfect crystals at room temperature [37–40]. The as-processed films were maintained for 2 days at ambient conditions before the DSC experiments, and during that time, the imperfect crystallites were able to thicken, thus increasing their melting temperature.

The second process is the eventual melting of the PCL crystals that are confined inside the SBA-15 pores, as those crystals are much thinner than the ones located in the outer surface of the mesoporous particles. Therefore, the melting temperature of such small crystals has to be significantly lower than those for the PCL crystals outside of the channels. The reason for this can be found by considering the Gibbs–Thomson equation. A simplified equation is used [41–43] in the case of "regular" lamellar crystals, but a more general version [44,45] is necessary for the thin crystals that are confined in the SBA-15 channels, which supposedly present very low values in their lateral size because of the pore diameters. As a consequence, the corresponding melting temperatures are considerably lower.

Obviously, confinement cannot exist in neat PCL, so the melting curve for this sample only includes the secondary process that is associated with the annealing that occurs at room temperature. Therefore, the eventual confinement that takes place in the composites can be approximately determined by subtracting the DSC profile of neat PCL from those of the composites. This is just an approximation since there may be differences in the thickening of the crystals by effect of the silica. For instance, the main crystal size is slightly dependent on the SBA-15 content (see below).

Previously, DSC experiments have been proven to demonstrate the existence of crystallites that are confined inside porous materials: in organic solvents within controlled pore glasses [46], in polyethylene in cylindrical nanoporous alumina [45], and in semicrystalline polyolefins within mesoporous silica [18,22–24,26,27,47,48], through the appearance of an additional minor endotherm, indicating a decrease in the melting point (and a reduced enthalpy). The sensitivity of this technique together with the multiple small melting processes that take place in PCL processed under these conditions currently only allow the observation that something different is taking place at the highest SBA-15 content, which is the PCL-SBA9 sample. The actual demonstration of confinement for PCL crystallites will be evaluated later by real-time variable-temperature small-angle X-ray scattering (SAXS) measurements with synchrotron radiation since they have turned out to be a valuable means to reveal the presence of polypropylene (PP) chains within the SBA-15 channels in nanocomposites that have been prepared by in situ polymerization and melt extrusion [24,48,49]. A thorough analysis performed from 0.095 to 0.13 nm^{-1} in the *s* scale, where the first order of the hexagonal arrangement of the SBA-15 particles is observed, allows the determination of the existence or absence of crystallites that are growing within the nanometric pores.

The overall degree of PCL crystallinity can be also determined from the DSC curves. As mentioned before, the results that were obtained during the first melting process, which were reported in Table 2, show that incorporating mesoporous silica does not change the PCL crystallization capability and that crystallinity remains rather constant in these films, which are processed by applying a fast cooling rate after the melt and then keeping the films at room temperature.

This simultaneous existence in the first melting scan of the annealing process and the eventual melting of the confined PCL crystals does not occur in the subsequent cooling and second melting processes. Thus, the curves that were attained for the cooling process are represented in Figure 5a. An exothermic main event is clearly noticeable since these PCL-based materials are semicrystalline. Furthermore, it can be observed that the mesoporous SBA-15 particles exert an evident nucleating effect during PCL crystallization, and its location is shifted to a higher temperature in the composites. This easier capability of PCL to crystallize is more significant in the composite with the lowest silica content, this showing the highest T_c value. An increase in the SBA-15 composition leads to a progressive decrease in T_c, which can probably be ascribed to the increase in size of the inorganic domains; however, the T_c in the composites is always superior to that found in pristine PCL (see data in Table 2).

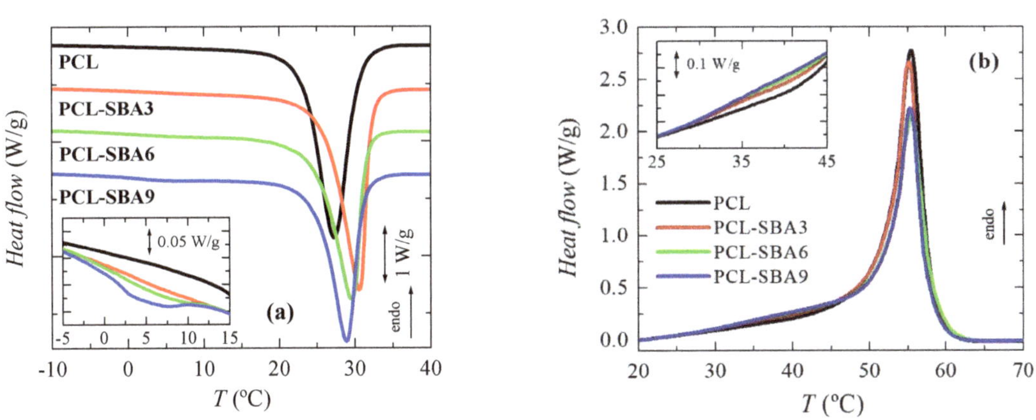

Figure 5. DSC curves corresponding to (**a**) cooling and (**b**) second heating run for the pristine PCL and its composites with SBA-15 particles. Curves have been shifted in the left plot for the sake of clarity.

In addition to this main exothermic event, another small crystallization process is observed in the composites from 0 to 10 °C, as seen in the inset of Figure 5a. This is more evident in the PCL-SBA9 material, and it almost goes undetected in PCL-SBA3. The pure PCL does not show any event in this temperature range. This process implies the formation of constrained crystallites, and it appears at lower temperatures. Since the pristine PCL does not show it, the only crystals that can be developed in these composite materials compared to in neat homopolymer are those from the PCL chains that are located within the nanometric SBA-15 channels. This point will be discussed in more detail along with results of the real-time temperature-variable SAXS experiments.

The PCL crystallinity was also estimated from these DSC curves, as listed in Table 2. A clear reduction in the crystallinity is clearly observed compared with that determined from the as-processed films. In spite of the DSC cooling rate is slower than that used for film preparation, 20 °C/min instead of at approximately 80 °C/min, and it should lead to thicker crystals, the difference is associated with the fact that the samples from the as-processed films were maintained, as mentioned before, at room temperature for 2 days before they underwent DSC testing, and during that time, the crystallites were able to undergo several melting–recrystallization processes and were able to increase in number, allowing crystallinity to increase. Under these conditions, the incorporation of mesoporous silica seems to result in a tiny diminishment in the PCL crystallinity within the experimental error.

Figure 5b depicts the DSC curves that were observed during the second heating run after PCL crystallization at 20 °C/min. The behavior is now simpler, without the presence of an annealing process being noted during the initial scan. The second heating process is initiated just after cooling finishes, and then the crystalline entities do not have enough time to develop more perfect crystallites. In addition, these crystals that are grown under 20 °C/min are thicker. Accordingly, melting–recrystallization processes are not noticeable in this second heating run, and the main T_m values are shifted to slightly lower temperatures. Its location seems to be independent of SBA-15 particle incorporation. The absence of those melting–recrystallization processes results in the crystallinity being lower than that deduced from the first melting runs. As the mesoporous silica content increases, the crystallinity is slightly reduced. The only effect of the presence of SBA-15 is seen in the interval that ranges between 30 and 45 °C, which is associated with the melting of the PCL crystals that are confined in the nanometric SBA-15 mesostructure. Evidence of confinement during DSC crystallization and during the second heating process was previously observed for high density PE (HDPE) [27] but not in the case of ultrahigh molecular weight PE (UHMWPE) [22,23,25,47] or extruded PP [48,49].

Much more detailed information can be deduced from real-time variable-temperature synchrotron experiments in the small-angle region, as aforementioned. The presence of PCL chains within the SBA-15 channel means that there is a certain amount of PCL within the constrained nanometric spaces and, accordingly, these systems could be called nanocomposites.

Figure 6 shows the WAXS profiles at room temperature for the pristine PCL, its composites, and the neat SBA-15 silica. The profiles were obtained through the use of synchrotron radiation. PCL typically crystallizes into an orthorhombic lattice [50]. There are no evident changes in the location of its characteristic diffractions ((100), (110), and (200) peaks) with the presence of mesoporous silica. The SBA-15 particles are, however, amorphous at a short-range, thus showing a wide halo, which is centered at the s value of 2.54 nm^{-1}.

The degree of crystallinity can be also determined from wide-angle X-ray diffraction by comparison of the area under the crystalline peaks to the total scattered intensity when a two-phase model is considered [51]. For that, the corresponding amount of the amorphous halo must be subtracted. This amorphous profile was obtained from the real-time variable temperature WAXS experiments (see below). The crystallinity values are listed in Table 3. They do not vary much in the composites when compared to the values that are exhibited by the pristine PCL. A slightly decreasing trend is observed, and the data are within the experimental error. Moreover, these WAXS values are rather similar to those determined by DSC (see Table 2).

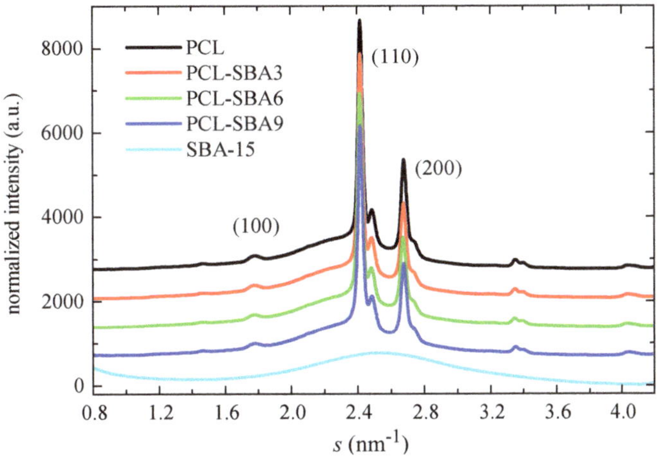

Figure 6. Wide angle profiles at room temperature, obtained with synchrotron radiation, for pristine PCL homopolymer, and SBA-15 particles together with those for the nanocomposites prepared by extrusion.

Table 3. Characteristics of the PCL crystalline phase for the pristine polymer and the different composites: $f_c^{PCL}{}_{WAXS}$ (crystallinity degree determined by WAXS at room temperature); $L^{PCL}{}_{SAXS}$ (long spacing estimated by SAXS at room temperature); and l_c (most probable crystal size calculated assuming a two-phase model: $l_c = L^{PCL}{}_{SAXS} \cdot f_c^{PCL}{}_{WAXS}$).

Sample	SBA-15 wt.%	$f_c^{PCL}{}_{WAXS}$	$L^{PCL}{}_{SAXS}$ (nm)	l_c (nm)
PCL	0	0.54	17.1	9.2
PCL-SBA3	3.3	0.53	17.6	9.3
PCL-SBA6	6.0	0.52	17.8	9.3
PCL-SBA9	9.0	0.52	18.1	9.4

Standard errors: $f_c^{NORM}{}_{WAXS}$ ± 4%; L^{SAXS} and l_c ± 0.3 nm.

Figure 7 shows the SAXS profiles for the neat PCL and mesoporous SBA-15 silica as well as its composites at room temperature. On one hand, it is clearly noticeable that the ordering of the mesoporous particles has been maintained throughout extrusion of the distinct composites, a fact that was found earlier from the FESEM pictures. Thus, the characteristic reflections of its hexagonal *p6mm* symmetry at (100), (110), (200), (210), and (300) [15,52] can be observed in the SAXS profiles. The inset in Figure 7 represents the profiles for SBA-15 and for the PLC-SBA9 composite in the logarithmic scale, making the smaller peaks more noticeable.

In addition to the diffractions ascribed to SBA-15, a broad peak is also evident in the region of low *s* values. This can be attributed to the variation in the electron density of the PCL matrix as consequence of its semicrystalline nature and its lamellar crystallites, i.e., it can be ascribed to its most probable long spacing. The values for the different samples are reported in Table 3. The presence of SBA-15 particles increases the PCL long spacing, as mesoporous content is raised in the composites. The most probable crystal size is also enlarged, increasing with the SBA-15 amount, in spite of a slightly decrease in the crystallinity values. These slightly thicker crystallites do not seem to affect the T_m of the first melting process and values remain almost constant (see Table 2). These similar T_m values can be ascribed to the existence of multiple melting–recrystallization processes during heating of the different materials.

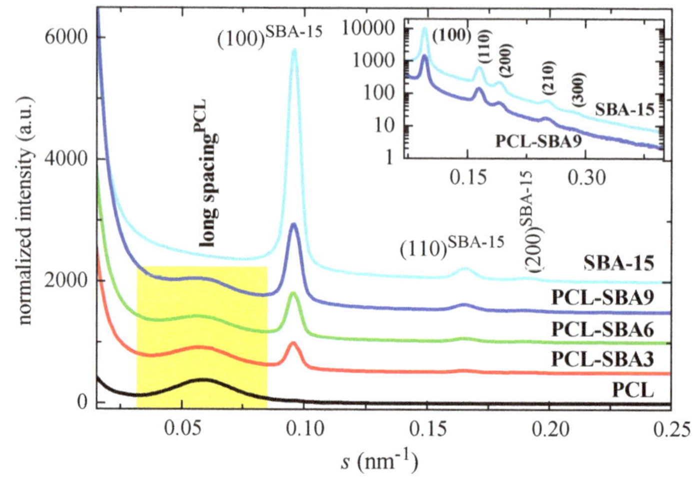

Figure 7. SAXS profiles at room temperature for the pristine PCL and the mesoporous SBA-15 as well as their composites.

It has been previously reported that the existence of the polymeric chains that are confined in the channels of SBA-15 (and other mesoporous silicas) is clearly reflected in important variations in the intensity of the SAXS diffraction peaks of the silica, as observed in different polymeric composites [48,53–56]. These variations can be interpreted by considering that the intensity of SBA-15 diffraction not only depends on the amount of pore filling but also on the scattering contrast between the walls and the inside of the SBA-15 channels [57,58]. This is also the case for the present PCL composites. For instance, Figure 8 shows the Lorentz-corrected synchrotron SAXS 1D diffractograms that were obtained during the cooling from the melt at 20 °C/min of the neat PCL homopolymer and SBA-15 particles together with those for the three composites. A decrease in the SBA-15 diffraction intensity in the composites, which is centered at approximately 11 °C, is clearly observed and is directly proportional to the silica content in the composite. Additionally, it is important to note that the pristine SBA-15 sample does not show that decrease, with the intensity being constant throughout the entire temperature interval.

Therefore, the evident changes in the intensity of the SBA-15 peaks that are observed in Figure 8 for the composites can be interpreted as arising from variations in the scattering contrast before and after the crystallization of the PCL chains that are confined inside of the SBA-15 pores. These changes appear in the same temperature range as those that are observed for the secondary exotherm in the DSC curves of Figure 5a.

Moreover, Figure 8 also shows the long spacing of the PCL crystals at lower s values when the sample crystallizes from the melt, which occurs at approximately 30 °C for PCL as well as for the composites.

That change in the intensity of the SBA-15 peak also occurs during the first and second melting processes of the composites. In fact, the derivative of that intensity has been performed and plotted against temperature, allowing it to be compared to the corresponding DSC curves. For instance, Figure 9 shows the variation in that derivative with temperature in the case of the first melting process for the different composites (neat PCL has also been included, but its value is obviously always zero).

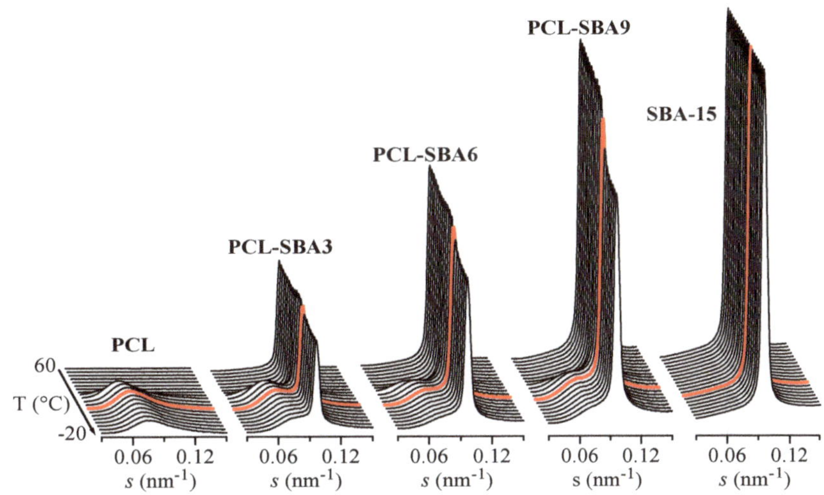

Figure 8. Lorentz-corrected synchrotron SAXS 1D diffractograms during the cooling from the melt at 20 °C/min of the different samples: neat PCL homopolymer and SBA-15 particles together with those for the three composites. Only one out every two frames were plotted for clarity. Highlighted in red: frame at T = 11 °C. The Y scale for SBA-15 is 10 times higher than for the other samples.

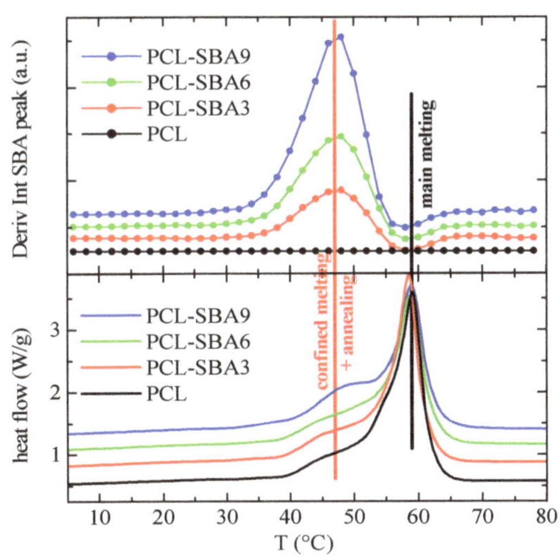

Figure 9. Variation in the temperature (**upper**) of the derivative of the intensity of the SBA-15 diffraction for the indicated samples during the first melting process, at 20 °C/min, compared to the corresponding DSC melting curves (**lower**). Derivative curves have been vertically shifted for clarity.

A comparison of the DSC melting curves allows to reach the important conclusion that the derivative of the intensity is only slightly sensitive to the main melting process, while considerably larger variations occur in the region of the secondary endotherms, which can be interpreted as being connected to the melting processes of the PCL crystals that are confined in the SBA-15 channels. Moreover, the magnitude of the derivative maxima in the composites is approximately proportional to their actual mesoporous silica content.

The corresponding results after cooling from the melt are displayed in Figure 10. As before, a comparison with the DSC cooling curves indicates that the derivative of the intensity is a little sensitive to the main crystallization intensity, which occurs at around 30 °C, while variations are considerably higher in the region of the secondary exotherms, which are associated with crystallization of the PCL chains that are confined in the SBA-15 channels.

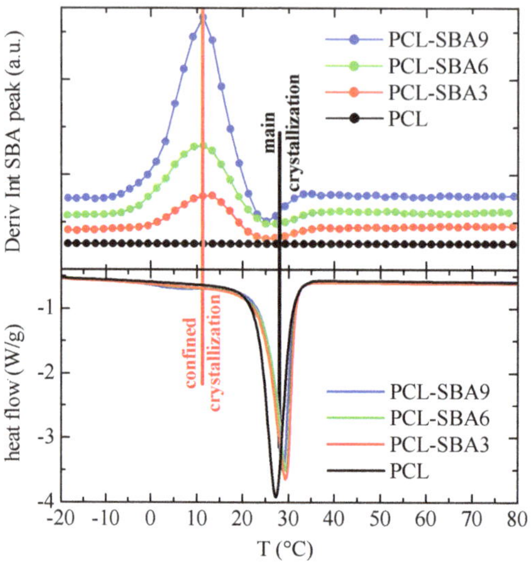

Figure 10. Variation with temperature (upper) of the derivative of the intensity of the SBA-15 diffraction for the indicated samples during the cooling from the melt, at 20 °C/min, compared with the corresponding DSC crystallization curves (lower). Derivative curves have been vertically shifted, for clarity.

Finally, Figure 11 shows the variation in the temperature of the derivative of the SBA-15 diffraction intensity for the subsequent second melting process. Now, the variations that occur in the region of the secondary endotherms (confined melting) appear to be at a temperature that is somewhat lower than it is for the first melting process, which also happens to be the case for the DSC results. In fact, the confined melting occurs at around 47 °C during the first melting process, while it appears to take place at 38 °C for the second melting process. It, therefore, seems that by keeping the materials at room temperature, not only the "regular" PCL crystals outside of and surrounding the SBA-15 particles undergo annealing processes, but the ones that are confined inside of the silica channels do as well. This leads to a drop in the melting temperature of around 12 °C in the first case (Figure 9), while it amounts to about 18 °C during the second melting process (Figure 11). Additionally, this displacement is about 18 °C for the crystallization upon cooling (Figure 10).

Therefore, the previous results show the relevance of real-time variable-temperature SAXS experiments for the determination of the presence of polymeric crystals that are confined within the channels of the SBA-15 mesoporous silica.

On the other hand, the synchrotron WAXS 1D diffractograms that were attained during the cooling process after the melt at 20 °C/min for neat PCL and for the composite PCL-SBA9 are displayed in Figure 12. No information about confinement can be deduced from these WAXS profiles. Nevertheless, these diffractograms show that the main crystallization phase occurs at around 30 °C, which is in accordance with the DSC results. The temperature for this primary ordering is around 20 degrees higher than the one for the confined crystallization process that takes place inside the silica channels, which appears at around 11 °C, as previously seen in Figure 10.

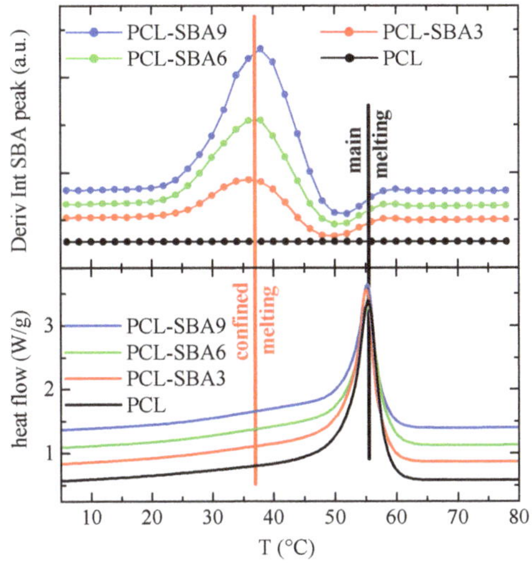

Figure 11. Variation in the temperature (upper) of the derivative of the intensity of the SBA-15 diffraction for the indicated samples during the second melting, at 20 °C/min, compared to the corresponding DSC melting curves (lower). Curves have been vertically shifted for clarity.

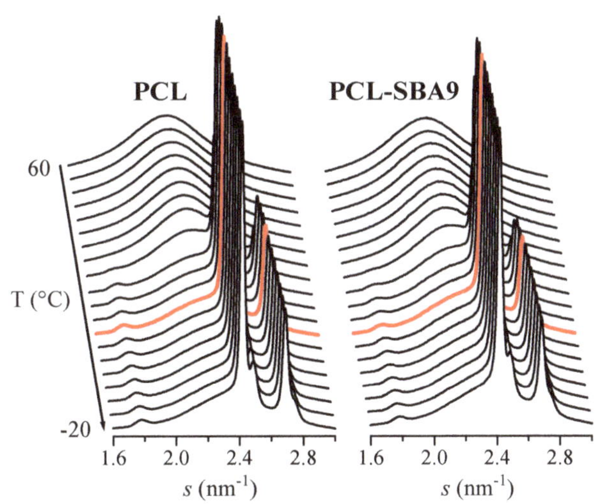

Figure 12. Synchrotron WAXS 1D diffractograms obtained during the cooling from the melt at 20 °C/min for neat PCL (left) and for the composite PCL-SBA9 (right). Only one out every two frames were plotted for clarity. Highlighted in red: frame at T = 11 °C.

3.4. Mechanical Properties

Figure 13 shows the stress–strain curves for the pure PCL and the different composites. The behavior that can be deduced from the nominal stress–strain curves for all of these specimens are characteristic of ductile polymeric materials. The engineering stress–strain curves display three distinct regions: initially, the stress rises as strain does in a linear dependence that allows Young´s modulus, E, to be determined. After this initial stage, an evident yield point is observed (see inset in Figure 13) followed by a narrow region where the stress is kept rather constant. Finally, a third region where another increase in the stress is observed, resulting in considerable strain-hardening that is related to the stress-induced orientation of the polymeric PCL chains being observed. Thus, the stress–strain curves of these samples are characterized by the formation of a neck during the deformation process, which was also confirmed during the direct observations that were made throughout the stretching process. On the other hand, the narrow necking propagation stage also indicates that these materials are relatively soft.

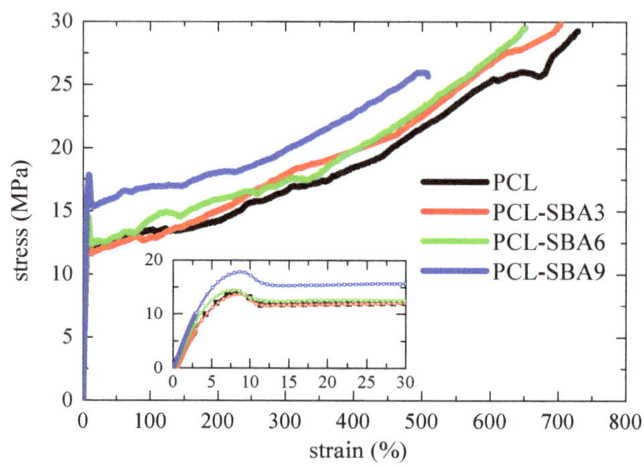

Figure 13. Stress–strain behavior of pristine PCL and its composites, which were prepared by extrusion with SBA-15 particles.

The main parameters that were achieved from the curves that are depicted in Figure 13, Young´s modulus (E) and yield stress (σ_Y), are represented in Figure 14. Regarding the E values, the incorporation of the silica particles leads to more rigid materials, and the dependence on the content is almost linear. Accordingly, the E values are at any filler content higher than that attained for the pristine PCL. In fact, PCL-SBA9 shows a value that is 40 % higher than for the neat PCL. At similar loading contents, this increase is superior to that found in nanocomposites based on polypropylene [49,56] but is analogous to the one observed when polyethylene is used as the matrix [22,59].

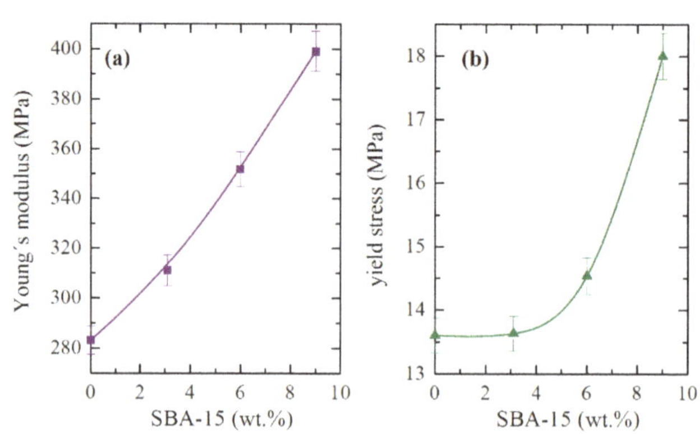

Figure 14. (**a**) Modulus values, E, and (**b**) yield stress, σ_Y, both of which were deduced from stress–strain tests for the PCL homopolymer and composites with different SBA-15 contents.

There are three possible explanations for this behavior in the studied PCL materials. On one hand, silica is stiffer than polymeric PCL. Thus, the inclusion of silica, regardless of the amount, should be expected to increase the elastic modulus proportionally to the amount of silica that is added. SBA-15 particles play a reinforcing role. Secondly, although crystallinity remained almost constant in the different materials, the crystallite size increases proportionally to the amount of mesoporous silica that is in the structure. Consequently, the PCL matrix in the nanocomposites is stiffer than it is in neat PCL. Finally, the rather good dispersion of SBA-15 particles within the PCL, the presence of PCL chains within the SBA-15 pores, and their crystallization capability in these confined nanospaces can contribute to improved interface adhesion, resulting in enhanced ultimate mechanical properties.

In relation to the yielding point, effect of the yielding stress on the SBA-15 content is somehow different to that exhibited by the E values. This could be associated with the fact that different contributions other than those that are exclusively related to rigidity (presence of fillers, polymer crystallinity and crystal size, etcetera) exert an effect when the plastic domain is initiated. Figure 14 clearly shows that no important variations in the absolute σ_Y values are observed between PCL and PCL-SBA3 in spite of an increase being noticed in the stiffness values. This could be ascribed to the fact that all of these materials behave in a ductile manner and possibly due to the fact that the amount of filler at that content is not high. However, the PCL-SBA6 and PCL-SBA9 composites exhibit considerably larger σ_Y values.

Other very interesting features are related to the strain at that break and the tensile strength. PCL shows the highest value since it is the softest material that was studied. The values decrease as the SBA-15 content increases, but PCL-SBA9 is able to be deformed up to a strain of 500 %. Its rigidity has been improved remarkably, but at rupture, these characteristics did not show a noticeable decrease. Concerning the tensile strength, the average values were determined to be at about 30 MPa for PCL, PCL-SBA3, and PCL-SBA6, and for PCL-SBA9, the tensile strength decreases to 26 MPa.

Furthermore, the stress–strain response demonstrated good reproducibility in terms of the shape of deformation process as well as in the mechanical magnitudes that were derived from these experiments for the different strips that were stretched for a given sample. This feature could be once again ascribed to an optimal SBA-15 dispersion within the PCL, as shown by the FESEM images in Figure 2.

4. Conclusions

Composites that are based on PCL and different amounts of mesoporous SBA-15 silica were prepared by means of melt extrusion. The well-defined hexagonal arrangement that is exhibited by pristine SBA-15 was found to be maintained in the particles that were embedded within the PCL matrix in the resultant composites, as shown by the FESEM pictures and the results of the SAXS experiments. Moreover, the SBA-15 particles demonstrate a rather homogenous distribution within the PCL matrix at the different contents.

The presence of an endothermic event in the first DSC melting curves, which take place at the temperature interval ranging from 40 to 53 °C, indicates the preliminary development of small crystals that are accommodated in the interior space of the nanometric SBA-15 channels. This means that there are confined PCL chains in the SBA-15 pores. This was also observed in the subsequent cooling and second melting experiments, where a secondary exotherm or endotherm, respectively, appeared. They were approximately proportional to the SBA-15 content in the composites.

The real-time variable-temperature SAXS experiments were found to be very important instrument for determining the presence of the PCL crystals that were confined within the channels of the mesoporous SBA-15 silica. Thus, the existence of those confined PCL chains can be clearly deduced from the important variations in the intensity of the main SAXS diffraction that is seen in the silica. Derivatives of this intensity are only slightly sensitive to the main melting process (or crystallization), while considerably higher variations occur in the regions that contain the confined entities.

The incorporation of SBA-15 particles does not demonstrate a large effect on the PCL thermal degradation processes under either inert or air atmospheres, and neither the location of T_g and main T_m nor the crystallinity values that were deduced from either the

DSC or WAXS experiments. Nevertheless, the PCL crystallites became slightly thicker as the amount of mesoporous silica increased in the final material. This feature, together with the higher stiffness of the SBA-15 particles, is responsible for the remarkable reinforcement effect that was observed in the composites. Accordingly, Young´s modulus increased significantly as the SBA-15 content raised. Moreover, in terms of the break behavior, the rupture strain and tensile strength were suitably maintained.

The incorporation of SBA-15 seems to be an appropriate and useful strategy that can be used to improve the mechanical performance of PCL. This could allow this type of composites to be used in a more diverse range of applications that require stiffer materials and an enhanced mechanical response.

Author Contributions: Conceptualization, M.L.C. and M.R.R.; methodology, T.M.D.-R., E.B.-B., N.L.C.A., M.R.R., E.P. and M.L.C.; software, E.P.; validation, T.M.D.-R., E.B.-B. and N.L.C.A.; formal analysis, T.M.D.-R., E.B.-B., N.L.C.A., E.P. and M.L.C.; investigation, T.M.D.-R., E.B.-B., N.L.C.A., M.R.R., E.P. and M.L.C.; resources, E.P. and M.L.C.; writing—original draft preparation, E.P. and M.L.C.; writing—review and editing, T.M.D.-R., E.B.-B., N.L.C.A., M.R.R., E.P. and M.L.C.; supervision, M.L.C.; project administration, E.P. and M.L.C.; funding acquisition, E.P. and M.L.C. All authors have read and agreed to the published version of the manuscript.

Funding: This work was supported by the Agencia Estatal de Investigación (AEI, Spain) together with the European Regional Development Fund (FEDER, UE) (grant numbers MAT2016-79869-C2-1-P and PID2020-114930GB-I00) by CSIC (grant number 2020AEP129) and ALBA Synchrotron Light Facility. TMDR is also grateful for her pre-doctoral funding (contract number BES-2017-082284).

Institutional Review Board Statement: Not applicable.

Informed Consent Statement: Not applicable.

Data Availability Statement: Data are contained within the article.

Acknowledgments: Authors are grateful to the Characterization Service at ICTP-CSIC for the FE-SEM, TGA, and mechanical facilities as well as to its personnel for their support. The synchrotron experiments were conducted at beamline BL11-NCD-SWEET at the ALBA Synchrotron Light Facility with the collaboration of ALBA staff.

Conflicts of Interest: The authors declare no conflict of interest.

References

1. Chandra, R.; Rustgi, R. Biodegradable polymers. *Prog. Polym. Sci.* **1998**, *23*, 1273–1335. [CrossRef]
2. Lim, J.S.; Ki, C.S.; Kim, J.W.; Lee, K.G.; Kang, S.W.; Kweon, H.Y.; Park, Y.H. Fabrication and Evaluation of Poly(epsilon-caprolactone)/Silk Fibroin Blend Nanofibrous Scaffold. *Biopolymers* **2012**, *97*, 265–275. [CrossRef] [PubMed]
3. Simoes, M.C.R.; Cragg, S.M.; Barbu, E.; De Sousa, F.B. The potential of electrospun poly(methyl methacrylate)/polycaprolactone core-sheath fibers for drug delivery applications. *J. Mater. Sci.* **2019**, *54*, 5712–5725. [CrossRef]
4. Rosa, D.S.; Guedes, C.G.F.; Bardi, M.A.G. Evaluation of thermal, mechanical and morphological properties of PCL/CA and PCL/CA/PE-g-GMA blends. *Polym. Test.* **2007**, *26*, 209–215. [CrossRef]
5. Arcana, M.; Bundjali, B.; Yudistira, I.; Jariah, B.; Sukria, L. Study on properties of polymer blends from polypropylene with polycaprolactone and their biodegradability. *Polym. J.* **2007**, *39*, 1337–1344. [CrossRef]
6. Newman, D.; Laredo, E.; Bello, A.; Grillo, A.; Feijoo, J.L.; Müller, A.J. Molecular Mobilities in Biodegradable Poly(DL-lactide)/Poly(ε-caprolactone) Blends. *Macromolecules* **2009**, *42*, 5219–5225. [CrossRef]
7. Ravati, S.; Favis, B.D. Interfacial coarsening of ternary polymer blends with partial and complete wetting structures. *Polymer* **2013**, *54*, 6739–6751. [CrossRef]
8. Seggiani, M.; Altieri, R.; Puccini, M.; Stefanelli, E.; Esposito, A.; Castellani, F.; Stanzione, V.; Vitolo, S. Polycaprolactone-collagen hydrolysate thermoplastic blends: Processability and biodegradability/compostability. *Polym. Degrad. Stab.* **2018**, *150*, 13–24. [CrossRef]
9. Blázquez-Blázquez, E.; Pérez, E.; Lorenzo, V.; Cerrada, M.L. Crystalline characteristics and their influence in the Mechanical Performance in Poly(ε-caprolactone)/High Density Polyethylene Blends. *Polymers* **2019**, *11*, 1874. [CrossRef]
10. Lee, S.-H.; Teramoto, Y.; Endo, T. Cellulose nanofiber-reinforced polycaprolactone/polypropylene hybrid nanocomposite. *Compos. Part A* **2011**, *42*, 151–156. [CrossRef]
11. Pan, L.; Pei, X.; He, R.; Wan, Q.; Wang, J. Multiwall carbon nanotubes/polycaprolactone composites for bone tissue engineering application. *Colloid Surface B* **2012**, *93*, 226–234. [CrossRef] [PubMed]

12. Boujemaoui, A.; Sanchez, C.C.; Engstrom, J.; Bruce, C.; Fogelstrom, L.; Carlmark, A.; Malmstrom, E. Polycaprolactone Nanocomposites Reinforced with Cellulose Nanocrystals Surface-Modified via Covalent Grafting or Physisorption: A Comparative Study. *ACS Appl. Mater. Interfaces* **2017**, *9*, 35305–35318. [CrossRef]
13. Mao, L.; Liu, Y.J.; Wu, H.Q.; Chen, J.H.; Yao, J. Poly(epsilon-caprolactone) filled with polydopamine-coated high aspect ratio layered double hydroxide: Simultaneous enhancement of mechanical and barrier properties. *Appl. Clay Sci.* **2017**, *150*, 202–209. [CrossRef]
14. Beck, J.S.; Vartuli, J.C.; Roth, W.J.; Leonowicz, M.E.; Kresge, C.T.; Schmitt, K.D.; Chu, C.T.-W.; Olson, D.H.; Sheppard, E.W.; McCullen, S.B.; et al. A new family of mesoporous molecular sieves prepared with liquid crystal templates. *J. Am. Chem. Soc.* **1992**, *114*, 10834–10843. [CrossRef]
15. Zhao, D.Y.; Feng, J.L.; Huo, Q.S.; Melosh, N.; Fredrickson, G.H.; Chmelka, B.F.; Stucky, G.D. Triblock Copolymer Syntheses of Mesoporous Silica with Periodic 50 to 300 Angstrom Pores. *Science* **1998**, *279*, 548–552. [CrossRef]
16. Kageyama, K.; Tamazawa, J.; Aida, T. Extrusion Polymerization: Catalyzed Synthesis of Crystalline Linear Polyethylene Nanofibers Within a Mesoporous Silica. *Science* **1999**, *285*, 2113–2115. [CrossRef] [PubMed]
17. Nakajima, H.; Yamada, K.; Iseki, Y.; Hosoda, S.; Hanai, A.; Oumi, Y.; Teranishi, T.; Sano, T. Preparation and characterization of polypropylene/mesoporous silica nanocomposites with confined polypropylene. *J. Polym. Sci. Part B Polym. Phys.* **2003**, *41*, 3324–3332. [CrossRef]
18. Cerrada, M.L.; Pérez, E.; Lourenço, J.P.; Campos, J.M.; Ribeiro, M.R. Hybrid HDPE/MCM-41 Nanocomposites: Crystalline Structure and Viscoelastic Behaviour. *Microporous Mesoporous Mater.* **2010**, *130*, 215–223. [CrossRef]
19. Kurek, A.; Mark, S.; Enders, M.; Kristen, M.O.; Mülhaupt, R. Mesoporous Silica Supported Multiple Single-Site Catalysts and Polyethylene Reactor Blends with Tailor-Made Trimodal and Ultra-Broad Molecular Weight Distributions. *Macromol. Rapid Commun.* **2010**, *31*, 1359–1363. [CrossRef]
20. Campos, J.M.; Lourenço, J.P.; Cramail, H.; Ribeiro, M.R. Nanostructured silica materials in olefin polymerisation: From catalytic behaviour to polymer characteristics. *Prog. Polym. Sci.* **2012**, *37*, 1764–1804. [CrossRef]
21. Dong, X.C.; Wang, L.; Jiang, G.H.; Zhao, Z.R.; Sun, T.X.; Yu, H.J.; Wang, W.Q. MCM-41 and SBA-15 supported Cp_2ZrCl_2 catalysts for the preparation of nano-polyethylene fibres via in situ ethylene extrusion polymerization. *J. Mol. Catal. A Chem.* **2005**, *240*, 239–244.
22. Ferreira, A.E.; Cerrada, M.L.; Pérez, E.; Lorenzo, V.; Cramail, H.; Lourenço, J.P.; Ribeiro, M.R. UHMWPE/SBA-15 nanocomposites synthesized by in situ polymerization. *Microporous Mesoporous Mater.* **2016**, *232*, 13–25. [CrossRef]
23. Ferreira, A.E.; Cerrada, M.L.; Pérez, E.; Lorenzo, V.; Cramail, H.; Lourenço, J.P.; Quijada, R.; Ribeiro, M.R. Hafnocene catalyst for polyethylene and its nanocomposites with SBA-15 by in situ polymerization: Immobilization approaches, catalytic behavior and properties evaluation. *Eur. Polym. J.* **2016**, *85*, 298–312. [CrossRef]
24. Barranco-García, R.; Ferreira, A.E.; Ribeiro, M.R.; Lorenzo, V.; García-Peñas, A.; Gómez-Elvira, J.M.; Pérez, E.; Cerrada, M.L. Hybrid materials obtained by in situ polymerization based on polypropylene and mesoporous SBA-15 silica particles: Catalytic aspects, crystalline details and mechanical behavior. *Polymer* **2018**, *151*, 218–230. [CrossRef]
25. Ferreira, A.E.; Ribeiro, M.R.; Cramail, H.; Lourenço, J.P.; Lorenzo, V.; Pérez, E.; Cerrada, M.L. Extraordinary Mechanical Performance in Disentangled UHMWPE Films Processed by Compression Molding. *J. Mech. Behav. Biomed. Mater.* **2019**, *90*, 202–207. [CrossRef] [PubMed]
26. Cerrada, M.L.; Pérez, E.; Lourenço, J.P.; Bento, A.; Ribeiro, M.R. Decorated MCM-41/polyethylene hybrids: Crystalline Details and Viscoelastic Behavior. *Polymer* **2013**, *54*, 2611–2620. [CrossRef]
27. Cerrada, M.L.; Bento, A.; Pérez, E.; Lorenzo, V.; Lourenço, J.P.; Ribeiro, M.R. Hybrid Materials Based on Polyethylene and MCM-41 Particles Functionalized with Silanes: Catalytic Aspects of In Situ Polymerization, Crystalline Features and Mechanical Properties. *Microporous Mesoporous Mater.* **2016**, *232*, 86–96. [CrossRef]
28. Deryło-Marczewska, A.; Zienkiewicz-Strzałka, M.; Skrzypczyńska, K.; Światkowski, A.; Kuśmierek, K. Evaluation of the SBA-15 materials ability to accumulation of 4-chlorophenol on carbon paste electrode. *Adsorption* **2016**, *22*, 801–812. [CrossRef]
29. Nakagawa, S.; Kadena, K.; Ishizone, T.; Nojima, S.; Shimizu, T.; Yamaguchi, K.; Nakahama, S. Crystallization Behavior and Crystal Orientation of Poly(ε-caprolactone) Homopolymers Confined in Nanocylinders: Effects of Nanocylinder Dimension. *Macromolecules* **2012**, *45*, 1892–1900. [CrossRef]
30. Muñoz-Bonilla, A.; Cerrada, M.L.; Fernández-García, M.; Kubacka, A.; Ferrer, M.; Fernández-García, M. Biodegradable Polycaprolactone-Titania Nanocomposites: Preparation, Characterization and Antimicrobial Properties. *Int. J. Mol. Sci.* **2013**, *14*, 9249–9266. [CrossRef]
31. Marcilla, A.; Gómez-Siurana, A.; Menargues, S.; Ruiz-Femenia, R.; García-Quesada, J.C. Oxidative degradation of EVA copolymers in the presence of MCM-41. *J. Anal. Appl. Pyrolysis* **2006**, *76*, 138–143. [CrossRef]
32. Aguado, J.; Serrano, D.P.; Romero, M.D.; Escola, J.M. Catalytic conversion of polyethylene into fuels over mesoporous MCM-41. *Chem. Commun.* **1996**, *6*, 725–726. [CrossRef]
33. Campos, J.M.; Lourenço, J.P.; Pérez, E.; Cerrada, M.L.; Ribeiro, M.R. Self-reinforced Hybrid Polyethylene/MCM-41 Nanocomposites: In-situ Polymerisation and Effect of MCM-41 Content on Rigidity. *J. Nanosci. Nanotech.* **2009**, *9*, 3966–3974. [CrossRef]
34. Bento, A.; Lourenço, J.P.; Fernandes, A.; Ribeiro, M.R.; Arranz-Andrés, J.; Lorenzo, V.; Cerrada, M.L. Gas permeability properties of decorated MCM 41/polyethylene hybrids prepared by in situ polymerization. *J. Membr. Sci.* **2012**, *415–416*, 702–711. [CrossRef]

35. Persenaire, O.; Alexandre, M.; Degée, P.; Dubois, P. Mechanisms and Kinetics of Thermal Degradation of Poly(ε-caprolactone). *Biomacromolecules* **2001**, *2*, 288–294. [CrossRef]
36. Mofokeng, J.P.; Kelnar, I.; Luyt, A.S. Effect of layered silicates on the thermal stability of PCL/PLA microfibrillar composites. *Polym. Test.* **2016**, *50*, 9–14. [CrossRef]
37. Bensason, S.; Minick, J.; Moet, A.; Chum, S.; Hiltner, A.; Baer, E. Classification of homogeneous ethylene-octene copolymers based on comonomer content. *J. Polym. Sci. Part B Polym. Phys.* **1996**, *34*, 1301–1315. [CrossRef]
38. Alizadeh, A.; Richardson, L.; Xu, J.; McCartney, S.; Marand, H.; Cheung, Y.W.; Chum, S. Influence of Structural and Topological Constraints on the Crystallization and Melting Behavior of Polymers. 1. Ethylene/1-Octene Copolymers. *Macromolecules* **1999**, *32*, 6221–6235. [CrossRef]
39. Polo-Corpa, M.J.; Benavente, R.; Velilla, T.; Quijada, R.; Pérez, E.; Cerrada, M.L. Development of mesomorphic form in propene/higher α-olefin copolymers at intermediate comonomer content and its effect on properties. *Eur. Polym. J.* **2010**, *46*, 1345–1354. [CrossRef]
40. García-Peñas, A.; Gómez-Elvira, J.M.; Blázquez-Blázquez, E.; Barranco-García, R.; Pérez, E.; Cerrada, M.L. Microstructural Details and Polymorphs in Poly(propylene-*co*-1-nonene) Copolymers Synthesized at Different Polymerization Temperatures. *Polym. Cryst.* **2021**, *4*, e10150. [CrossRef]
41. Wunderlich, B. *Macromolecular Physics*; Academic Press: New York, NY, USA, 1980; Volume 3.
42. Darras, O.; Séguéla, R. Surface free energy of the chain-folding crystal faces of ethylene-butene random copolymers. *Polymer* **1993**, *34*, 2946–2950. [CrossRef]
43. Lu, L.; Alamo, R.G.; Mandelkern, L. Lamellar thickness distribution in linear polyethylene and ethylene copolymers. *Macromolecules* **1994**, *27*, 6571–6576. [CrossRef]
44. Gedde, U.W. *Polymer Physics*; Chapman & Hall: London, UK, 1995.
45. Shin, K.; Woo, E.; Jeong, Y.G.; Kim, C.; Huh, J.; Kim, K.-W. Crystalline structures, melting, and crystallization of linear polyethylene in cylindrical nanopores. *Macromolecules* **2007**, *40*, 6617–6623. [CrossRef]
46. Jackson, C.L.; McKenna, G.B. The melting behavior of organic materials confined in porous solids. *J. Chem. Phys.* **1990**, *93*, 9002–9011. [CrossRef]
47. Ferreira, A.E.; Cerrada, M.L.; Pérez, E.; Lorenzo, V.; Vallés, E.; Ressia, J.; Cramail, H.; Lourenço, J.P.; Ribeiro, M.R. UHMWPE/HDPE in-reactor blends, prepared by in situ polymerization: Synthetic aspects and characterization. *eXPRESS Polym. Lett.* **2017**, *11*, 344–361. [CrossRef]
48. Barranco-García, R.; López-Majada, J.M.; Martínez, J.C.; Gómez-Elvira, J.M.; Pérez, E.; Cerrada, M.L. Confinement of iPP crystallites within mesoporous SBA-15 channels in extruded iPP-SBA-15 nanocomposites by Small Angle X-ray Scattering. *Microporous Mesoporous Mater.* **2018**, *272*, 209–216. [CrossRef]
49. Barranco-García, R.; López-Majada, J.M.; Lorenzo, V.; Gómez-Elvira, J.M.; Pérez, E.; Cerrada, M.L. Confinement of iPP chains in the interior of SBA-15 mesostructure ascertained by gas transport properties in iPP-SBA-15 nanocomposites prepared by extrusion. *J. Membrane Sci.* **2019**, *569*, 137–148. [CrossRef]
50. Hu, H.; Dorset, D.L. Crystal Structure of Poly(ε-caprolactone). *Macromolecules* **1990**, *23*, 4604–4607. [CrossRef]
51. Alexander, L.E. *X-ray Diffraction Methods in Polymer Science*; Wiley-Interscience: Malabar, FL, USA, 1969.
52. Zhao, D.Y.; Huo, Q.S.; Feng, J.L.; Chmelka, B.F.; Stucky, G.D. Nonionic Triblock and Star Diblock Copolymer and Oligomeric Surfactant Syntheses of Highly Ordered, Hydrothermally Stable, Mesoporous Silica Structures. *J. Am. Chem. Soc.* **1998**, *120*, 6024–6036. [CrossRef]
53. Xu, X.; Song, C.; Andrésen, J.M.; Miller, B.G.; Scaroni, A.W. Preparation and characterization of novel CO_2 "molecular basket" adsorbents based on polymer-modified mesoporous molecular sieve MCM-41. *Microporous Mesoporous Mater.* **2003**, *62*, 29–45. [CrossRef]
54. Son, W.J.; Choi, J.S.; Ahn, W.S. Adsorptive removal of carbon dioxide using polyethyleneimine-loaded mesoporous silica materials. *Microporous Mesoporous Mater.* **2008**, *113*, 31–40. [CrossRef]
55. Wang, X.; Ma, X.; Song, C.; Locke, D.R.; Siefert, S.; Winans, R.E.; Möllmer, J.; Lange, M.; Möller, A.; Gläser, R. Molecular basket sorbents polyethylenimine–SBA-15 for CO_2 capture from flue gas: Characterization and sorption properties. *Microporous Mesoporous Mater.* **2013**, *169*, 103–111. [CrossRef]
56. Barranco-García, R.; Gómez-Elvira, J.M.; Ressia, J.A.; Quinzani, L.; Vallés, E.M.; Pérez, E.; Cerrada, M.L. Effect of iPP molecular weight on its confinement within mesoporous SBA-15 silica in extruded iPP-SBA-15 nanocomposites. *Microporous Mesoporous Mater.* **2020**, *294*, 109945. [CrossRef]
57. Hammond, W.; Prouzet, E.; Mahanti, S.D.; Pinnavaia, T.J. Structure factor for the periodic walls of mesoporous MCM-41 molecular sieves. *Microporous Mesoporous Mater.* **1999**, *27*, 19–25. [CrossRef]
58. Sauer, J.; Marlow, F.; Schüth, F. Simulation of powder diffraction patterns of modified ordered mesoporous materials. *Phys. Chem. Chem. Phys.* **2001**, *3*, 5579–5584. [CrossRef]
59. Cecílio, D.M.; Cerrada, M.L.; Pérez, E.; Fernandes, A.; Lourenço, J.P.; McKenna, T.F.L.; Ribeiro, M.R. Unique stiffness-deformability features of dendrimeric silica reinforced HDPE nanocomposites obtained by an innovative route. *Microporous Mesoporous Mater.* **2022**, *331*, 111619. [CrossRef]

polymers

Review

A Review of Prestressed Fibre-Reinforced Polymer Matrix Composites

Raphael Olabanji Ogunleye * and **Sona Rusnakova**

Department of Production Engineering, Faculty of Technology, Tomas Bata University in Zlín, Vavrečkova 275, 760 01 Zlín, Czech Republic; rusnakova@utb.cz
* Correspondence: ogunleye@utb.cz

Abstract: This review examines various studies on reducing tensile stresses generated in a polymer matrix composite without increasing the mass or dimension of the material. The sources of residual stresses and their impacts on the developed composite were identified, and the different techniques used in limiting residual stresses were also discussed. Furthermore, the review elaborates on fibre-prestressing techniques based on elastically (EPPMC) and viscoelastically (VPPMC) prestressed polymer matrix composites, while advantages and limitations associated with EPPMC and VPPMC methods are also explained. The report shows that tensile residual stresses are induced in a polymer matrix composite during production as a result of unequal expansion, moisture absorption and chemical shrinkage; their manifestations have detrimental effects on the mechanical properties of the polymer composite. Both EPPMC and VPPMC have great influence in reducing residual stresses in the polymer matrix and thereby improving the mechanical properties of composite materials. The reports from this study provide some basis for selecting a suitable technique for prestressing as well as measuring residual stresses in composite materials.

Keywords: polymer composite; fibre-prestressing; residual stresses

Citation: Ogunleye, R.O.; Rusnakova, S. A Review of Prestressed Fibre-Reinforced Polymer Matrix Composites. *Polymers* **2022**, *14*, 60. https://doi.org/10.3390/polym14010060

Academic Editors: Andrzej Puszka and Beata Podkościelna

Received: 2 December 2021
Accepted: 21 December 2021
Published: 24 December 2021

1. Introduction

Composite materials are developed by combining materials offering unique properties that cannot be achieved individually by the constituent materials. Because of their excellent strength to weight ratio, they have increasingly been used as engineering materials for many applications, that include automotive and aerospace parts, construction materials, electrical parts and other consumer products [1]. The growth of composite materials for extended use, particularly in the aerospace and automobile industries, has necessitated continuous research for developing improved composites with excellent mechanical properties [2]. The constituents of composites are classified into matrix and reinforcement. While the matrix can be made of polymers, metals or ceramics, reinforcement includes fibres, whiskers and particulate fillers that exist in natural (lignocellulose, animal fibres and minerals) or synthetic form (carbon, aramid, boron, nylon, polyethylene) [3].

To produce a fibre-reinforced polymer composite, fibres of various configurations and stiffness are embedded into a polymer matrix of lower stiffness. While the fibre is responsible for carrying the load and offering much needed strength and stiffness, the polymer matrix is responsible for the mobility of the load to other parts of the fibre by providing the required binding forces [3]. The matrices also prevent the reinforced fibres from absorbing moisture, propagating micro-cracking due to microbial and chemical attacks. Polymer matrices can be either thermoplastic or thermoset. Thermoplastic polymers undergo a physical change when heated and can be softened and reformed. Some thermoplastics commonly used in composites include polyetheretherketone (PEEK), polyetherketoneketone (PEKK), polyphenylene sulphide, polyethylenimine (PEI) and polycarbonates (PC) [4]. On the other hand, thermoset polymers undergo a chemical change in the presence of a crosslinking agent (hardener) to form a three-dimensional structure. Thermosets include epoxies, phenolics, polyimides, polyester, vinyl ester, bismaleimide, melamine and silicone [4]. Thermoplastic composites offer great potential due to less processing time. However, significant drawbacks, which include a high processing temperature (the polymers must be heated to melting point to incorporate the fibre), a high tendency for the buckling of fibres in the polymer matrices, and solvent and fluid resistance features, have reduced their adoption as a substitute for thermoset in fibre-reinforced polymer composites [5].

Both tensile and compressive stresses are generated in polymer composite structures during the manufacturing process, and their occurrence in a composite can result from differences in physical and mechanical properties, a contraction of the polymer matrices before curing, manufacturing techniques and moisture absorption from the environment [6]. However, while compressive stresses have positive influences on the mechanical properties of the composite material, the tensile stresses have a detrimental effect on the manufactured polymer composite by causing distortion, dimensional instability and matrix cracking [7]. The residual stresses referred to in this study are the tensile residual stresses, and many techniques have reportedly been used in the literature to lessen the residual stresses in fibre-reinforced polymer (FRP) composites. Fibre-prestressing offers a low-cost means of reducing the influence of tensile residual stresses, thereby improving the mechanical properties without increasing the mass or dimension of the composite [8]. It is important to know that fibre-prestressing can only reduce residual stress in polymer composites if applied to a certain level [9]. Therefore, there has been continuous research on improving fibre-prestressing techniques.

This review explores the state-of-the-art fibre-prestressing techniques used in reducing residual stress in FRP composites. Moreover, the mechanism and methods used for improving the mechanical properties of FRP composites are examined.

2. Source of Residual Stresses in a Composite

In the absence of external forces, residual stresses are forms of stress that bring about the deformation of a composite material [7]. There are two primary sources of residual stresses in a composite material; (i) residual stresses generated due to differences in deformation-related properties of the materials, and (ii) production processes that induce residual stresses [9]. Moisture absorption, the chemical shrinkage of the matrix and a variation in the coefficient of thermal expansion induce residual stresses in polymer composites. Researchers have also shown that residual stresses can initiate failure in a composite material due to matrix cracking [10]. The influence of matrix cracking can lead to a dangerous failure, especially when the material is under loading conditions. Both compressive and tensile residual stresses are generated in composite materials. However, while compressive residual stresses help prevent cracks from spreading throughout the matrices, tensile residual stresses assist in the opening of micro-cracks in the polymer matrix [11]. Other defects such as fibre waviness, delamination, warping and dimensional imbalance are also due to residual stresses [12].

Many techniques have reportedly been used to reduce the effect of residual stresses in polymer composites. Process optimisation of cure cycle parameters such as dwell time, dwell temperature, dwell cycle and cooling rate can minimise residual stress during production [13–15]. Shape memory alloys (SMAs) also reduce stress concentration in FRP matrices. SMAs are novel metallic materials that can undergo reversible solid to solid change in a phase when subjected to thermal or mechanical loading that can cause sizeable inelastic strain [16]. They are generally available in the form of a wire with a diameter below 200 μm [17] and can be embedded in various configurations, as shown in Figure 1. SMAs offer other vital properties such as impact damage control, crack closure and shape morphing. Nevertheless, SMAs induce fibre misalignment and can create an uneven stress distribution in a composite [17].

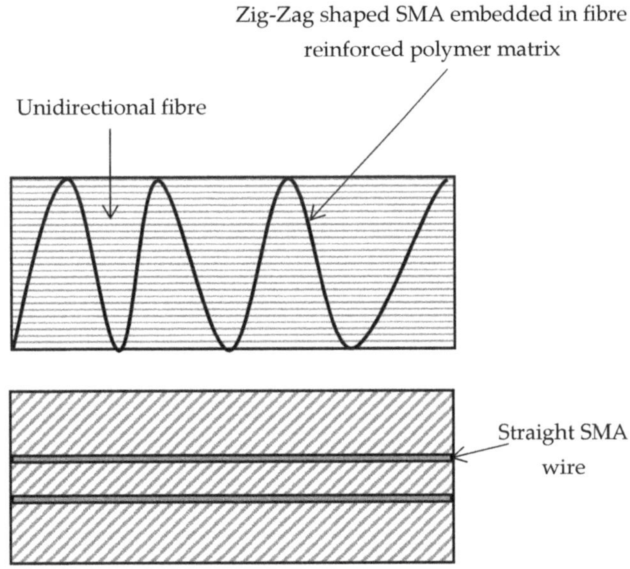

Figure 1. Integration of SMA into a Fibre-Reinforced Polymer Composite.

Similarly, the electron beam curing technique composite can also reduce stress concentration [18]. The technique involves generating free radicals from a high-energy electron accelerator that propagates the polymerisation and crosslinking reaction at room temperature. Since the curing temperature is kept below the traditional thermal curing process temperature, the residual stress generated using an electron beam will be lowered [19]. Moreover, high processing electron beam curing results in an increased production rate and low shrinkage compared to that of conventional thermally cured composites [20]. To utilise the electron beam curing process, cationic initiators are required. Hence, the initiators—that are primarily Lewis or Bronsted acids—react when exposed to high energy electron irradiation. However, a high initiator concentration can lead to a thermal degradation of the polymer matrix, while excess irradiation can damage the fibre strength [21].

Despite their positive influences, electron beam cured composites cannot replicate some of the characteristics of thermally cured composites (high fracture toughness and inter-laminar shear strength) [20]. Similarly, expanding monomers have reportedly been used as anti-shrinkage in composite materials [22]. The polymerisation reaction of the polymer matrices results in a volumetric shrinkage of the matrices during the curing stage. The addition of the expanding polymer gives rise to volumetric expansion, thereby reducing volumetric shrinkage and residual stress [23,24]. The expanding monomers can also reduce the modulus and initiate the transverse cracking of composite materials [9]. Compared to the techniques mentioned above concerning the cost and simplicity, fibre-prestressing is relatively more acceptable.

3. Fibre-Prestressing Technique

There has been growing interest in improving the mechanical performance of FRP composites by minimising the induced stress without increasing the mass or dimension of the composite [8]. Since improving the composite performance increases the production cost, improving the production technique of composites through fibre-prestressing is a possible way of mitigating the cost [9]. Fibre-prestressing has been discovered by researchers as a productive method of reducing the effect of residual stresses in composite materials during production [25–27].

Fibre-prestressing can be achieved elastically or viscoelastically. The elastically prestressed polymer matrix composite (EPPMC) is achieved by subjecting the fibre material to the applied load and maintaining the load throughout the curing cycle of the polymer composite [28,29]. The load is released after the cooling and solidification of the composite at ambient temperature. Subsequently, the prestressed fibres tend to return to their initial length, generating compressive stresses in the matrix. These compressive stresses reduce the detrimental effect of the tensile residual stresses induced during the curing process [29]. In addition, the viscoelastically prestressed polymer matrix composite (VPPMC) is formed by subjecting polymeric fibres to a tension load, which introduces creep stress, and then removing the load before moulding the fibre to the polymer matrix. After the curing stage, prestressed fibres tend to return to their initial length, thereby creating compressive stresses that counterbalance the effect of induced tensile residual stresses [28]. The advantages and shortcomings of the two prestressing techniques are discussed further in the next section.

3.1. *Elastically Prestressed Polymer Matrix Composite (EPPMC)*

The fibres are prestressed by subjecting them to a force lower than the elastic limit of the fibre and maintaining this force throughout the curing cycle of the composite. After curing, the force is removed, and the fibres tend to return to their original states. Compressive stresses are developed within the solidified matrix due to the bonding of the prestressed fibre with the polymer matrix [30]. Various methods that have reportedly been used to elastically prestressed fibres are discussed in the next section.

3.1.1. Deadweight Method

The deadweight prestressing method (Figure 2) was first presented by Jorge, Marques and De Castro [31]. They studied the influence of prestressing on the mechanical behaviour of a unidirectional polyester composite using E-glass fibre. The deadweight rig consists of steel pins arranged on a roving boundary, with the distance between the adjacent pin approximately equal to 2 mm. The fibres were rolled around the steel pins, and the two ends were subject to a load. The resin was applied, and the composite was cured at ambient temperature in an oven. Al-Dulaimy, Al-hassany and Shakir [32] utilised a similar prestressing technique using E-glass fibre and epoxy resin. The result obtained shows an increase in stiffness and a percentage elongation at break for the prestressed composite. Some of the shortcomings of using the deadweight method include the inability to obtain a uniform distribution of fibres in the polymer matrix and a fibre fracture tendency in the steel pin area when subjected to a bending force [33].

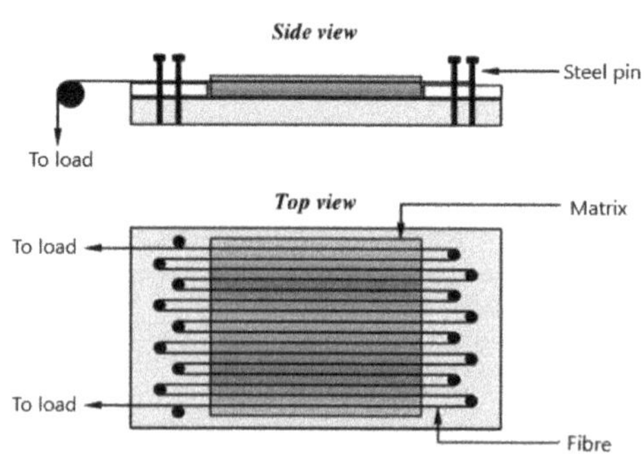

Figure 2. The deadweight prestressing method used by Jorge, Marques and De Castro [6].

3.1.2. V-Slot Fastening Method

The V-slot consists of an aluminium plate with V-shaped slots at the two ends (Figure 3). Unlike the deadweight method, the V-slot fastening method can be used to prestress laminates because it can accommodate material with a large surface area [33]. Prepreg laminates are prestressed by laying them up between the plates, and both ends of the material are fastened using the V-shaped bar to the slot and finally cure in an autoclave or hot press. The major drawback to using this method is a tendency to fracture when brittle fibres are being prestressed due to twisting when fastened and a difficulty in determining the prestressing level [9]. Moreover, fibre fracture can lead to a non-uniform distribution of stresses in the composite [34].

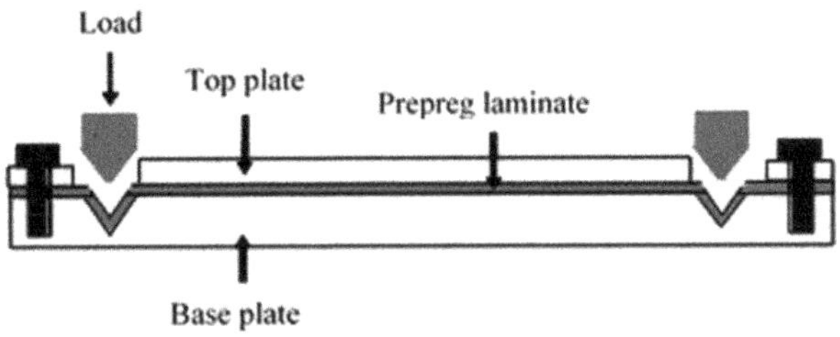

Figure 3. V-slot mechanical fastener [6].

3.1.3. Jack Prestress Rig Method

The Jack prestress rig method was presented by Abdullah and Hassan [35]. They studied the effect of varying prestressing levels on the strength of a carbon-fibre-reinforced composite laminate. The setup consists of a flat prestress surface connected to a pulling jack. The two ends of the fibre laminate were fixed to the base plate (2) and moving assembly (4), respectively (Figure 4). The tensile force needed to produce the required prestressing level was applied through the jack and the composite was cured in an oven. However, the assembly is designed to have a free movement which might cause fibre waviness during curing.

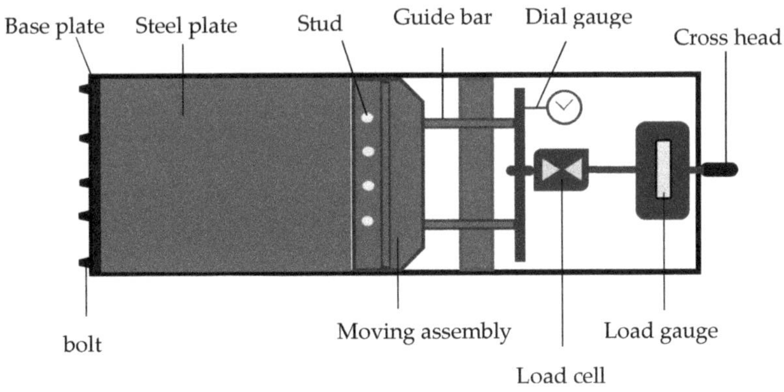

Figure 4. Jack prestress rig assembly.

3.1.4. Filament Winding

In this technique, fibres roving are pulled with a known amount of load over a liquid resin bath and then warped on a mandrel to produce a composite material (Figure 5). Composites with different prestressing levels can be made using this technique by applying varying pre-load levels [36,37]. Following the winding step, the composite is cured in an oven or autoclave or by exposing it to infrared radiation [38]. Products from this technique can vary in complexity from a simple pipe to an aeroplane fuselage [39]. The primary benefit of filament winding is the ability to use automation and robots in performing operations [40]. The greatest drawback is the difficulty in obtaining an equal fibre volume fraction due to compression and loss of resin when each layer of the composite is being warped unto the mandrel [36]. Moreover, the mandrel stiffness affects the fibre pretension; therefore, it is not easy to maintain a constant load throughout the process [9]

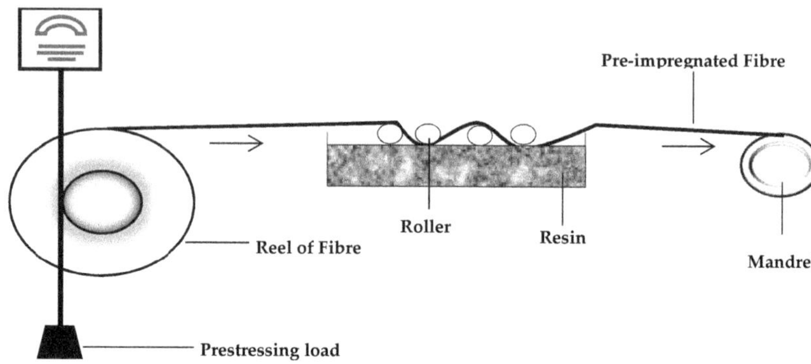

Figure 5. Schematic representation of the filament winding technique.

3.1.5. Biaxial Loading Frame and Fibre Stretching Rig

Jevons, Fernando and Kalsi [41] studied the effect of prestressing on the low-velocity impact performance of glass fibre composites using a biaxial loading frame (Figure 6). The setup is made of a C-channel consisting of four clamps joined to the frame by bolts. Prestressing levels are applied through a loading pin by making use of a mechanical test machine. The prestressed composite is clamped and tightened with a bolt to the frame, vacuum-bagged and cured in an autoclave [42]. As a result of the sharp edge of the frame, it is difficult to prepare the assembly for vacuum-bagging [43]. Moreover, maintaining a constant level of prestressing by using the tensile machine is not easy due to the limited amount of accessible space between the fixed and moveable jaws needed to place the prestressing frame inside the device [36].

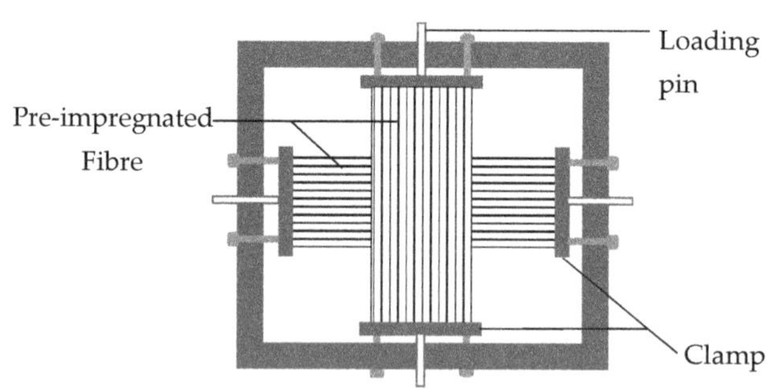

Figure 6. Schematic representation of biaxial loading frame.

Zhao and Cameron, [44] utilised a fibre-prestressing rig similar to the biaxial loading frame to prestress glass fibre-reinforced polypropylene. The fibres are wounded on the frame and preloaded using the tensile testing machine, and then the bolts are locked to achieve the required prestressing level (Figure 7).

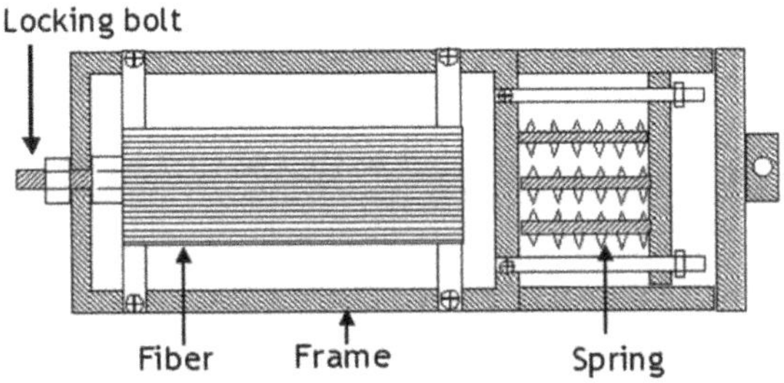

Figure 7. Schematic representation of a fibre-prestressing rig used by Zhao and Cameron.

3.1.6. Horizontal Tensile Machine

Motahhari and Cameron [45] reported a fibre-prestressing technique using a horizontal tensile machine in studying micro-residual stresses in an E-glass fibre-reinforced epoxy polymer composite. The fibres were wound on two different grips before being transferred to a horizontal tensile machine (Figure 8) that provided applied forces, and then the epoxy resin was added directly to the fibre and curing was carried out in an oven (Figure 9). The principle seems simple, but it is challenging to maintain a uniform temperature profile. Furthermore, vacuum-bagging for autoclave curing will also be difficult due to the complex assembly.

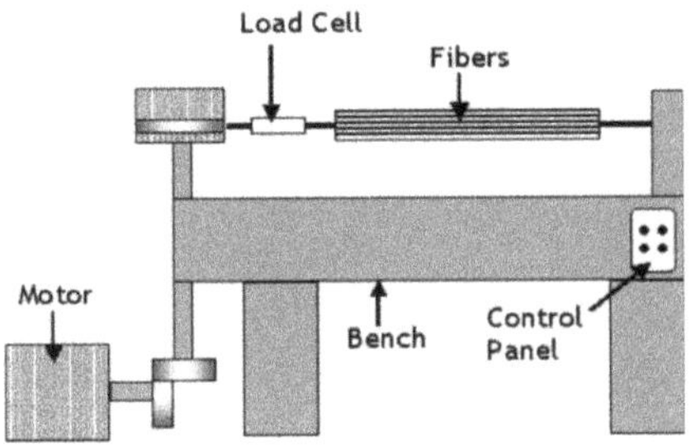

Figure 8. Schematic representation of a horizontal tensile machine.

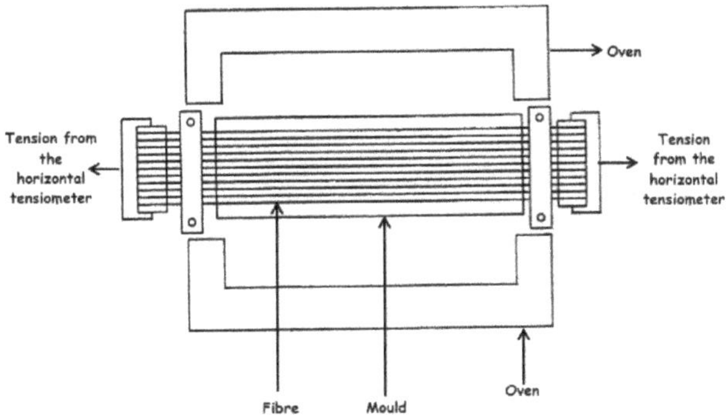

Figure 9. Schematic representation of fibre-prestressing using a horizontal tensile machine.

3.1.7. Hydraulic Prestressing Rig

Tuttle, Koehler and Keren [46] proposed a technique for fibre-prestressing using the hydraulic-cylinder-based prestressing rig shown in Figure 10. The machine consists of a movable rod connected to the hydraulic cylinder and a fixed loading rod at the other end. The prepreg was rolled around the movable rod while the tension was applied through the hydraulic cylinder that also controls the prestressing level. The technique allows the tension load to be applied only in one direction [35].

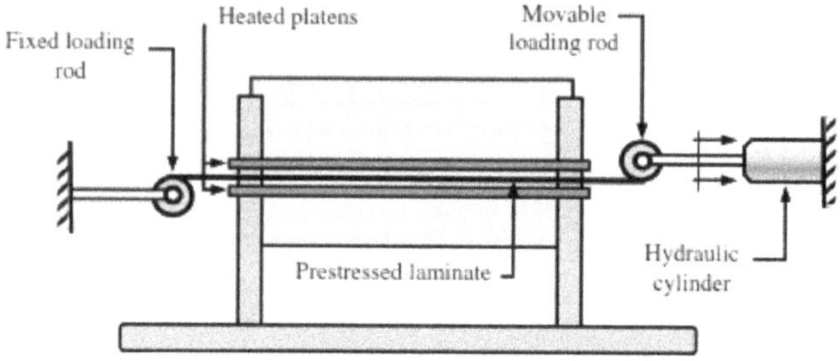

Figure 10. Schematic representation of the hydraulic prestressing rig [46].

3.1.8. Flatbed Prestressing Rig

Krishnamurthy [35] prestressed E-glass fibre/Epoxy laminates using the flatbed prestressing rig shown in Figure 11. The laminates' edges were cured by hot press and fixed to the rig by clamping with bolts. The loaded screw provided the tension while the laminate prestressing level was measured by the load cell (up to 150 MPa) and the autoclave was used for curing the laminate.

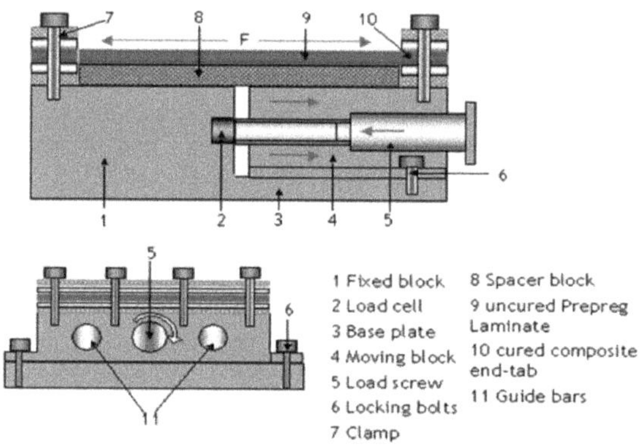

Figure 11. Schematic representation of the flatbed prestressing rig [42].

Fibre misalignments were identified close to the edge of the laminates due to the temperature variation between the hot press cure edge of the laminates and the remaining parts that were being cured in an autoclave [35]. The results of the requirement to subject the composite material to load throughout the curing process (EPPMC) provide the basis for the design of various prestressing techniques for a specific product geometry. Generally, EPPMC techniques can be applied to pre-impregnated fibre and both synthetic and natural fibres can be used. However, they are limited to continuous fibres, and there are constraints related to the product geometry [34].

3.2. Viscoelastically Prestressed Polymer Matrix Composite (VPPMC)

The technique involves using polymeric fibres to generate compressive stress in a composite material through the viscoelastic recovery process [47,48]. The fibres are prestressed by applying a load over a while to induce creep. After removal of the load, the fibres undergo a time-dependent elastic recovery. When the prestressed fibres are embedded in the polymer matrix and subjected to curing, the elastic recovery with the surrounding polymer matrix continues, thereby imparting compressive stresses that counterbalance the tensile residual stresses generated in the composite matrix [20]. Because the fibre stretching and moulding operations can be carried out independently, the technique offers flexibility in the geometry of the manufactured composite [49]. Based on previous studies, VPPMC can be classified into cellulose-fibre-based VPPMC (CFVPPMC) and synthetic polymeric fibre-based VPPMC (SPFVPPMC). CFVPPMC contain prestressed cellulose fibres predominantly from natural plants such as kenaf, sisal, flax, bamboo, jute, wheat straw, ramie and eucalyptus. Natural fibres are biodegradable and cheap and have a low density. Table 1 show physicomechnical properties of some natural and synthetic fibres commonly used in PMC.

Cui et al. [50] studied the flexural characteristics of the viscoelastically prestressed bamboo-sliver-reinforced parallel strand lumber (PSL). The flexural strength increases compared to the unprestressed counterparts. A similar result was obtained by Qin and Fancey [51] for viscoelastically prestressed cellulose yarn in a polyester casting resin matrix.

Širvaitienė et al. [52] also reported an increase in the flexural performance of viscoelastically prestressed vegetable fibre in PMC. However, due to moisture absorption, inadequate knowledge about the mechanism that controls natural fibre mechanical variations and the mode of failure, there is a limitation to their usage in polymer matrix composites [53].

Table 1. Physicomechanical properties of some commonly used fibres.

	Fibre	Density (kg/m^3)	Tensile Strength (MPa)	Elongation at Break (%)	Young Modulus (GPa)	References
Natural fibres	Kenaf	1200	295–930	2.7–6.9	53	[54,55]
	Sisal	1200	507–885	1.9–3	9.4–22	[54,55]
	Flax	1380	343–1035	1.2–3	27.6	[54,55]
	Bamboo	800–1400	391–1000	2	11–30	[54,55]
	Banana	1350	529–914	3–10	8–32	[54,55]
	Wheat straw	1600	273	2.7	4.76–6.58	[54,55]
	Hemp	1350	580–1110	1.6–4.5	70	[54,55]
	Jute	1230	187–773	1.5–3.1	13–26.5	[54,55]
	Ramie	1440	400–938	2–4	61.4–128	[54,55]
	Rice straw	1650	449	2.2	1.21–1.25	[54,55]
Synthetic fibres	E-glass	2500	2000–3000	2.5	70	[56]
	Carbon	1800	4000	1.3	300	[57]
	Kevlar	1400	3600	2.7	130	[57]
	Nylon	1100	950	18	5	[57]

Furthermore, high-performance synthetic polymer fibres are designed to resist a wide range of thermal, chemical and physical stress [58]. Nylon and ultra-high molecular weight polyethylene fibre (UHMPE) are common synthetic polymeric fibres used in polymer matrix composite development. Previous studies have shown that viscoelastically prestressed UHMPE and the Nylon 6.6 fibre enhanced the mechanical properties (strength and young modulus) of the PMC compare to their unstressed counterparts [59,60].

Unlike the EPPMC, in which the fibre-prestressing and curing operations are designed and carried out with the same assembly, VPPMC fibre stretching and moulding operations are decoupled therefore there is more flexibility in the fibre orientation (both continuous and short fibre can be used), distribution and material production. Many studies on VPPMC carried out fibre-prestressing using bespoke vertical stretching rigs (Figure 12). The creep and recovery strain for a small level of prestressing can be determined using the setup shown in Figure 12a, while higher level of fibre prestressing can be achieved using the rig shown in Figure 12b. The displacement gauge measures the strain and recovery rate by calculating the distance between two marks on the prestressed fibre yarn and the stretching rigs have the advantage of accommodating a wide variety of load levels required for prestressing operations [6]. On the contrary, the VPPMC technique is only applicable to fibres that have viscoelastic behaviour, and curing at elevated temperatures is not suitable to prevent a permanent deformation of the fibres [43].

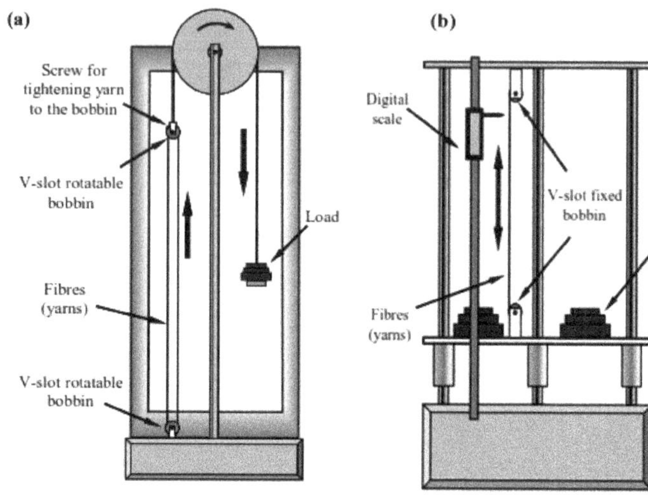

Figure 12. Vertical stretching rigs for VPPMC fibre-prestressing [6] Rig (**a**), Rig (**b**).

3.3. Mechanical Properties of Prestressed PMC

Compared to other enhancing techniques, both of the two prestressing methods have shown their significance in improving the mechanical performance of composite material without increasing the mass or section dimension. Table 2 provides up-to-date findings and results of previous studies on fibre-prestressing techniques.

Table 2. Reviews on fibre-prestressed PMC.

Material	Prestress Technique	Research Area	Results of Findings	References
Glass fibre woven into a fabric Phenol-based formaldehyde resin	Elastically prestressing of the glass fibre using tensioning rod (EPPMC).	Assessment of the compressive and tension characteristics of the composite.	Enhancement of elastic properties up to 31% was recorded due to the straightening of the warp fibres.	[61]
Unidirectional graphite/epoxy prepreg tape	Prepreg tape was subjected to tension by bending over a steel roller (EPPMC).	Tensile and elastic modulus measurement	Up to 17% increase in tensile strength Composite elastic modulus was not affected	[62]
Unidirectional carbon fibre/epoxy composite with 60% fibre volume fraction	The load was applied to fibre before curing but the nature of assembly was not reported (EPPMC).	Thermal stress analysis of the composite.	Fibre-prestresses lessen the residual stresses in the matrix.	[63]
Unidirectional E-glass fibre/polyester resin with 56% fibre volume fraction	Deadweight (EPPMC)	Tensile properties evaluation	The tensile strength increases with an increase in the level of prestressing (60–80 MPa applied load). The maximum percentage increase in tensile strength and modulus obtained were 15% and 18%, respectively.	[31]

Table 2. *Cont.*

Material	Prestress Technique	Research Area	Results of Findings	References
Carbon fibre/epoxy resin cross-ply laminate with 70% fibre volume fraction	Filament winding (EPPMC)	Modelling and experimental study of composite failure	Failure strength of the ply increased by increasing the prestress level up to 690 MPa	[64]
Graphite fibre/epoxy resin, unsymmetric cross-ply laminate with 56% fibre volume fraction	Hydraulic cylinder (EPPMC)	Examination of the tensile strength, curvature and transverse cracking	Fibre-prestressing reduced warping, curvature and transverse crack. Up to 28% increase ultimate strength	[47]
Unidirectional Nylon 6.6 fibre/polyester resin (up 3% fibre volume fraction)	Bespoke vertical stretching rig (VPPMC)	Analysis of the impact energy	Viscoelastically induced compressive stresses. Absorption of higher impact energy (25%) by the prestressed sample	[65]
E-glass fibre/epoxy resin cross-ply laminate (56% fibre volume fraction)	Biaxial loading frame (EPPMC)	Effect of low-velocity impact performance	25% increase in impact performance at low velocity due to prestressing	[41]
E-glass fibre/epoxy resin cross ply laminate (56% fibre volume fraction)	Biaxial loading frame (EPPMC)	Effect of high- and low-velocity impact performance	Improvement of impact performance at a low-level velocity	[66]
Unidirectional E-glass fibre/epoxy cross-ply laminates (58.2% fibre volume fraction)	Flatbed (EPPMC)	Tensile, fatigue life and compressive strength measurement	Improved fibre alignment, increase in resistance to onset damage due to induced compressive strength. 9% increase in tensile modulus and compressive strength at prestressing levels of 51 MPa and 80 MPa, respectively.	[35]
Unidirectional Nylon 6.6 fibre/epoxy resin (16, 28, 41) and 53% fibre volume fraction	Bespoke vertical stretching rig	Tensile strength and modulus measurement	30% and 15% tensile modulus and tensile strength, respectively.	[59]
Carbon and glass fibre/Hexcel cross-ply laminates	Flatbed (EPPMC)	Experimental and finite element analysis of bistable prestressed buckled laminate	Induction of bistable behaviour through prestressing.	[67]
Unidirectional Nylon 6.6 fibre/polyester resin (8, 12, 16% fibre volume fraction)	Bespoke vertical stretching rig (VPPMC)	Flexural properties measurement	Up to 50% increase in flexural modulus.	[60]
Unidirectional S-glass fibre/composite resins (Quixfil and Adoro) (12% fibre volume fraction)	Deadweight (EPPMC)	Flexural properties measurement	Increase in flexural strength.	[68]
Unidirectional UHMWPE fibre/polyester resin (3.6% fibre volume fraction)	Bespoke vertical stretching rig (VPPMC)	Impact properties measurement	Prestressing increases impact energy absorption (up to 40% increase in some batches).	[30]
Carbon fibre/epoxy resin (50% fibre volume fraction)	Deadweight (EPPMC)	Impact properties	Increase in strength of composite material.	[69]

Table 2. *Cont.*

Material	Prestress Technique	Research Area	Results of Findings	References
Hybrid unidirectional Nylon 6.6 and Kevlar fibres/polyester	Bespoke vertical stretching rig (for Nylon alone) (VPPMC)	Impact and flexural test	33 and 40% rise in absorption energy and flexural modulus.	[30]
Unidirectional Nylon 6.6 fibre/polymer resin (fibre volume fraction 2.2%)	Bespoke vertical stretching rig (VPPMC)	Impact assessment	Impact energy absorbed increased (40%).	[70]
Flax yarn/polyester resin	Tension frame (EPPMC)	Tensile and flexural assessment	Fibre alignment enhancement. Increased tensile strength and modulus. Increase in flexural strength and modulus.	[26]
Plain weave E-glass fabric/polyester resin (16% fibre weight fraction)	Hydraulic cylinder biaxial loading frame (EPPMC)	Flexural characteristics	Up to 16% increase in flexural strength at 50 MPa optimum prestressing level	[71]
Plain weave E-glass fabric/polyester resin (11% fibre weight fraction)	Hydraulic cylinder biaxial loading frame (EPPMC)	Tensile and fatigue characteristics	Fatigue life increased up to 43% Fatigue life improvement when under low and intermediate stress fatigue load	[43]
Unidirectional E-glass fibre mats/epoxy resin	Horizontal testing machine (EPPMC)	Flexural, tensile and compression properties	Reduction in fibre waviness Increase in flexural, tensile and compressive strength	[8]
Nylon 6.6 yarn/polyester cross-ply composite	Bespoke stretching rig (VPPMC)	Impact behaviour	Up to 29% reduction in damage depth	[28]
Unidirectional E-glass fibre/epoxy resin (10% fibre volume fraction)	Deadweight method (EPPMC)	Tensile properties	Increase in maximum strength, percentage elongation and rupture strength by 38.5%, 45.57% and 106.2%, respectively	[32]

3.4. Residual Stresses Measurement

The continuous advancement in the development of composite materials has drawn more attention to prediction and measuring the distribution and sizes of residual stresses because the behaviour of composite materials is greatly influenced by these factors [72,73]. Residual stress development in the composites cannot be negligible; therefore, these can result in failure, especially when the tensile stresses generated become greater than the critical tensile strength of the material [74]. At this stage, microcracking can occur, thereby exposing the fibres to chemical and microbial attacks. Due to the nonlinear behaviour of composite material, it is difficult to measure some residual stresses that may result in material failure [75]. However, the effective measurement will reduce material failure and further enhance the research in the development of predictive control of residual stresses in composite material.

The measurement of residual stresses in the composite material can be achieved by modelling the entire production process or through the observation and measurement of the phenomena resulting in residual stress generation in a material. The latter method can be achieved through destructive and non-destructive testing. Table 3 highlights the difference between destructive and non-destructive testing techniques. The methods that are regarded as destructive typically require cutting, slitting or drilling operations to relieve the residual stresses. Afterwards, residual stresses are estimated by considering the dimensional changes that occurred. Table 4 shows some destructive and non-destructive testing methods that are commonly used to determine residual stresses in composite materials.

Table 3. Difference between destructive and non-destructive testing.

Destructive Testing (DT)	Non-Destructive Testing (NDT)
Part of the materials is removed or damaged.	Testing can be done without removing or damaging the material.
Testing cannot be repeated on the same specimen.	Testing can be repeated on the same specimen.
Residual stress measurement is limited to a small area of the material sample.	Residual stresses can be measured within a large surface (e.g., laminate).
Global residual stresses distribution along the plies in a composite can be measured.	They cannot estimate global residual stress distributions along with composite plies.

The layer removal method is an established method for measuring residual stresses by subjecting the material to deformation on one surface while the layers of the materials are removed from the other surface [76]. The surface of a completely equilibrated stressed component is gradually removed in layers. As a result, the portion is free of stress and a force imbalance is established throughout the system. The plate is then deformed to bring it back into balance, and the strain that results from that measurement is then utilised to quantify the amount of removed residual stress. By combining with other methods such as XRD, the layer removal method can offer information about the stress profile in a material [77]. However, this method can only assess residual stresses in materials with a simple geometry and on a macro-scale level [78].

Proposed by Mathar [79], hole drilling is commonly used in measuring residual stresses in a material. It is conducted by drilling a hole into a stressed material while the stresses relax, resulting in a change in the surrounding strain field that can be determined using a strain gauge and related to the relaxed stresses [79]. The hole-drilling method was initially designed for homogenous isotropic materials such as metals, but with a few modifications it may be used for composite, inhomogeneous and anisotropic materials, existing in both crystalline and amorphous form [80]. This method provides a more accurate measurement compared to other methods such as layer removal and ring-core methods. Moreover, the hole drilling method is less destructive because only a small portion of the central material is cut away during the deformation measurement. Likewise, it can determine the biaxial residual stresses distribution because the strain gauge is capable of identifying responses due to the biaxial surface strain of the material [81]. However, the hole-drilling method is more suitable for predicting residual stresses at the macro-scale, while it is difficult to predict at the micro-scale level of the material [82].

The hole-drilling method is divided into incremental hole drilling (IHD) and deep hole drilling (DHD). Incremental hole drilling is commonly used to measure residual stresses at varying degrees of thickness through an incremental depth drilling of the material [83]. On the contrary, deep-hole drilling involves drilling a reference hole through the material and measuring the deformation at the desired region [84]. The DHD method is faster compared to IHD but does not provide adequate insight to through-thickness residual stresses in a component, because it only provides an average residual stress over the entire depth. Thus, the IHD method is widely used compared to DHD [85]. Ghasemi, Taheri-Behrooz and Shokrieh [86] predicted non-uniform residual stresses within E-glass/Epoxy laminate composites using the incremental hole-drilling method, and the result obtained revealed that the strain at the surface decreased considerably compared to the strain released in the depth when the depth of the hole is increased. This result indicates that the residual stresses at each ply of the laminates affect the released strain of the underneath ply. A similar result was obtained by Sicot et al. [80] using carbon/epoxy cross-ply laminates, by studying the effect of increasing the drilling depth and the relative position of the strain gauge with a radius of the drilled hole. A widespread use of the hole-drilling method in determining residual stresses has led to the development of standard testing methods by The American Society for Testing and Materials (ASTM E837-20) [87].

Similarly, the ring-core method works the same way as the hole-drilling method. However, as opposed to drilling a hole, an annular groove is created, with a strain gauge rosette inserted in the middle of the groove to determine the elastic behaviour of the material [88]. The ring-core approach offers several advantages over the more popular hole-drilling technique when it comes to measuring residual stress. In the ring-core approach, more stress is released throughout the hole-drilling process, which allows for a more accurate assessment of the strain. Furthermore, because the stress concentration is reduced around the machined region, greater residual stresses can be recorded without surpassing the material's yield stress. Additionally, the ring core method is tolerant of slight annular-hole diameter inaccuracies or eccentricity relative to the strain gauges. With this method, stress concentration effects are minimised, and it may measure stress levels close to the yield stress of the material. However, the diameter of the annular ring is relatively big, resulting in a significantly greater damage than the hole-drilling method. Because the diameter of the hole or ring determines the measuring region for residual stress, the results are also less localised.

One of the main drawbacks of the ring-core method is that it is more difficult to implement, especially when strain gauges are used, since the central wire gets in the way of the slot cutting process. Moreover, this technique has been limited to homogenous and isotropic materials [89]. The ring-core method has successfully been applied to quasi-isotropic FRP composites [90] and welded stainless steel [91]. Hence, it is obvious that the method continues to provide certain distinct advantages, especially at the micro-scale, and that additional study is required in this area to fully exploit its potentials.

The contour method is distinct among other methods used in measuring residual stresses by a relaxation mechanism because it offers detailed two-dimensional residual stress profiles that act on a plane of the material. To determine residual stresses using the contour method, the material is cut through the cross-sectional area by employing a wire Electro-Discharge Machining (EDM), and then the height maps of the cut surfaces are measured by using a laser profilometer or coordinate measuring machine. By cutting the material along the surface, residual stresses are released, resulting in deformation. The initially existent residual stresses parallel to the surface may be estimated using finite element analysis by evaluating the stresses necessary to restore the altered surface shape to a flat plane. The surface deformations are very small, and thus the EDM cutting and the surface height map measuring must be performed with very high accuracy. Furthermore, to prevent any asymmetry impacts, it is ideal to estimate the surfaces on both sides of the cut and to utilise the average surface height map. It is also feasible to make more cuts on perpendicular planes to acquire maps of the normal residual stresses along the planes. This method has been employed for measuring residual stresses in metal alloy melts [92], ion exchange glasses [93] and metal matrix composites [94]. However, the EDM cutting process can only be achieved if the material is electrically conductive, and therefore polymer-based and some fibre reinforcement, such as aramid, cannot be employed. Furthermore, the contour method can only measure uniaxial residual stresses, and it is difficult to apply the method to materials with a complex geometry [95].

The slitting method is another destructive testing method employed in measuring residual stresses. In several studies, this is also referred to as incremental slitting [96], crack compliance or compliance [97,98]. The method involves making a slit in a prestressed material and measuring the resulting deformation normal to the direction of the slit by using a strain gauge. The key advantage of using this method is its simple and non-complex procedure, its applicability to wide ranges of materials and its capability of measuring high-magnitude residual stresses. However, the slitting method application is limited to materials with reduced thickness. Therefore, it cannot adequately measure residual stresses in the materials at a macro-scale level [99].

It is standard practice in the microelectronics industry to employ Micro-Raman Spectroscopy to identify regions of mechanical stress in silicon circuit board components [100]. Chemical bonds in crystals may be studied using Raman spectroscopy, which employs light scattering to look at the vibrational energy of the chemical bonds. Raman peaks may be seen in the light scattered by the object. Any externally applied strain alters the location of these peaks. Consequently, the difference in Raman peak location between an unstressed and a stressed sample may be used to measure the applied strain. Moreover, the distribution of molecule orientations in the polymer matrix may be used to assess the strain in the amorphous polymer matrix [101]. Amorphous materials such as thermosetting polymers or glass have Raman peaks that are broad and irregular. Thermal plastics, on the other hand, have more clearly defined Raman peaks [89].

Furthermore, the X-ray diffraction method is another important technique used in measuring residual stresses in materials. This method can measure the residual stresses at the material surfaces up to a depth of 30 microns [102]. However, for deeper depth measurement, the removal of layers in the sample is required, which renders the process destructive. XRD determines the strain in the crystal lattice of a material through the examination of the change in crystal spacing. A diffracted X-ray beam may be generated by X-rays dispersed from a polycrystalline material; therefore, optimum diffracted intensities are recorded at certain angles. The diffraction plane's inter-planar distance may be calculated from these angles by applying Bragg's law [102]. Generally, samples with residual stresses have a spacing that is different from that of unstressed samples. Some of the advantages of using the XRD method are its simplicity and quickness in applying the process. Furthermore, the method can be applied to materials with complex shapes, and high-magnitude residual stresses at both micro and macro scales can be accurately predicted. However, the XRD method is only suitable for polycrystalline materials and the accuracy measurements can be affected by the grain size and texture [96,99]. The XRD method has reportedly been used to measure residual stresses in alloys [100,103], glass–ceramic matrix composites [104] and graphite–polymer matrix composites [105].

In addition, neutron diffraction (ND) is a method that uses neutron beam diffraction to identify deep-seated residual stresses in materials. As with X-ray diffraction, neutron diffracted beams follow Bragg's Law, allowing stress-induced variations in atomic lattice spacing to be detected. As a result, the relative spacing differences are calibrated using a stress-free material sample to compute actual stress values. In comparison to X-rays, neutrons have the benefit of measuring residual stresses in materials at high depth and precision. The Synchrotron X-ray Method operates with the same principle as the XRD and ND discussed above.

It is a common phenomenon to employ ultrasound waves in the detection of faults in engineering materials [106], but they may also be used to quantify applied and residual loads [107]. To effectively assess stress levels in a material, it is necessary to precisely measure the time-of-flight change of an ultrasonic wave travelling between the stressed and unstressed areas of that material. To obtain an absolute measurement of stress, the time-of-flight measurement in the stressed material is compared to that in the unstressed material. The method has the advantage of measuring high-magnitude residual stresses. The measuring procedure is fast and tri-axial residual stresses can be measured [107]. However, the ultrasonic method supports only crystalline materials, is extremely sensitive to changes in the microstructure and not suitable for components with complex shapes [107].

Table 4. Common destructive and non-destructive test methods used in the calculation of residual stresses in composites.

Methods	Principle	Material	Shortcomings	References
Layer Removal (DT)	It monitors the elastic response of a laminate to the release of residual stresses	Ceramics Metals Polymers Composites	Additional stresses can be imparted to the test sample due to the machining of the composite surfaces. Limited to macro-scale residual stresses	[76,78]
Hole drilling (SDT)	Drilling of a hole into the stressed object releases the stresses, leading to changes in the surrounding strain field that may be measured and related to the relaxed stresses.	Ceramics Metals Polymers Composites	It requires several assumptions to simplify the result solution. Accurate measurement around the hole, especially in the fibre direction, is very challenging. Limited to macro-scale residual stress measurement	[80,82,108]
Ring-Core Method (SDT)	It follows a principle comparable to the hole-drilling method. However, instead of discharging residual stresses by drilling a hole and measuring the elastic reaction of the surrounding material, the ring-core method discharges stress by cutting an annular groove into the surface of a component that contains residual stress.	Metals Ceramics Polymers	Limited to homogenous and isotropic material.	[109]
Contour Method (DT)	The material is sliced through by a planar surface, releasing residual stresses across the plane. As a result, the surface experiences out-of-plane deformation, which is recorded, and the underlying residual stresses across the cut are calculated using the finite element technique.	Ceramics Metals Plastics Composites	Difficulty in measuring residual stresses close to the surface of the material. Not suited for small components.	[95]
Slitting Method (DT)	A tiny slit is cut into a prestressed sample, and the resultant deformation parallel to the slot's direction induced by the restoration of force equilibrium is determined. The repetition of this procedure at increasing depths allows for the determination of residual stress across the component's thickness.	Ceramics Metals Plastics Composites	Macro-scale residual stresses cannot be fully measured. Only average stress along the transverse direction (y-axis) can be measured.	[110]
Neutron Diffraction Method (NDT)	Raman spectroscopy employs light scattering to measure the vibrational energy of crystalline chemical bonds. The dispersed light is detected, and typical Raman peaks may be detected. Any externally imposed strain alters the position of these peaks. Consequently, a stressed and unstressed sample's Raman peak position variations may be used to calculate the applied strain.	Metal Ceramics Composites	Resolution is limited, and residual stress changes smaller than 1 mm cannot be measured. Not suitable for amorphous materials	[111]
Raman Spectroscopy Method (NDT)	Stresses are determined by monitoring the frequency of certain luminescence peaks in comparison to those in an unstressed state.	Ceramics Polymers Composites	Limited to macro-scale residual stresses measurement.	[112]

Table 4. *Cont.*

Methods	Principle	Material	Shortcomings	References
X-ray Diffraction Method (NDT) (destructive if used for measuring depth)	When residual stress is determined using X-ray diffraction (XRD), the strain in the crystal lattice is determined and the related residual stress is calculated using the elastic constants, assuming that the relevant crystal lattice plane exhibits linear elastic deformation.	Metal Ceramics Composites	Applicable to polycrystalline materials only. The accuracy of this method is affected by the texture and grain size. Measurement is limited to the surface of the material	[113]
Synchrotron X-ray Method (NDT)	Similar to the X-ray diffraction method. However, X-rays are far more intense and have a much greater energy, and their tremendous energy allows them to penetrate much farther into materials.	Metal Ceramics Composites	Applicable to polycrystalline materials only	[114]
Ultrasonic Method (NDT)	The material is subjected to an ultrasonic (acoustic) wave, which is then detected by reflection, transmission or scattering. To determine the magnitude of stresses, the velocity of an ultrasonic wave in some modes is evaluated.	Metals Ceramics Composites	Not suitable for amorphous materials. Limited to macro-scale residual stress measurement.	[115]

DT: Destructive testing, SDT: Semi-destructive testing, NDT: Non-destructive testing.

4. Potential Applications and Prospects

Fibre-reinforced materials are used for many engineering applications where high mechanical properties are required. The use of carbon-fibre-reinforced polymer composites may reduce the weight of the materials up to 70% compared to steel [116]. Therefore, exploring such materials in making aeroplane and car parts provides the potential for minimising fuel consumption and carbon dioxide emission. Due to the high-impact energy absorption behaviour, prestressed fibre polymer composites can be used for ballistic protection in armours, vests and automobiles. Kevlar-based polymer matrix composites are materials used for ballistic protection [117]. It is well known that prestress induces an energy-absorbing mechanism at the fibre–polymer matrix interface [48], thereby increasing the impact absorption of the composite material.

Crashworthiness is an automobile's capacity to protect its occupants from severe injury or death in the event of a given number of collisions. It is a quantitative assessment of a structure's ability to protect its passengers in survivable crashes [118]. It is an essential factor that needs to be considered in selecting the material for vehicular structural assembly [119]. Material crashworthiness is characterised in terms of its energy absorption capacity (EAC). Generally, polymer composites have a high EAC compared to metals, and they can release deformation absorption energy during impact [120]. Prestressed fibres enhance the energy absorption capability of composite materials [70]. Moreover, the application of EPPMC and VPPMC technology in fibre-reinforced concrete development can enhance the resistance to crack propagation in a concrete structure. Bistable composite materials, which are often utilised in the aerodynamic control of aircraft and wind turbine blades, have drawn growing attention in recent years. They can be effectively manufactured utilising prestressed fibre technology [49]. The emerging developments will be toward alternate fibre materials, hybridisation, process optimisation and the use of a cost-effective fibre-prestressing technique.

5. Conclusions

The present article reviews some of the studies related to the techniques associated with fibre-prestressing in polymer matrix development. The presence of tensile residual stresses during the manufacturing process has a detrimental effect on composite materials' performance, especially when subjected to an external load. Fibre-prestressing offers the advantage of minimising residual stress without increasing the mass or section dimension in composite materials. It involves generating compressive residual stresses in a fibre–polymer matrix interface to minimise the effect of tensile residual stresses. While fibre-prestressing can be achieved both elastically (EPPMC) and viscoelastically (VPPMC), several equipment designs have reportedly been used to achieve prestressing. Despite some limitations associated with these techniques, the influence of fibre-prestressing on polymer composite structural properties has been positive. Previous studies on prestressed composites have reported up to 60% increase in the tensile modulus, 62% increase in impact strength, 140% increase in tensile strength, 50% in the flexural modulus, 9% increase in compressive strength and an increase of more than 150% in fatigue life compare to their unprestressed composite counterparts.

Moreover, various methods have been reported for measuring residual stresses in materials. These are generally grouped into destructive and non-destructive methods. The destructive methods measure residual stresses by monitoring the response of a material to deformation, while non-destructive methods monitor the change in material structures when under the influence of residual stresses. The selection of the right method depends on the type of material (metals, plastics, ceramics, composites), material structure (Crystalline or amorphous), states of stresses, speed and measurement accuracy. Out of these methods, the hole-drilling method has been widely used due to its versatility and the availability of scientific standards for measuring residual stresses in various materials.

Author Contributions: Investigation, R.O.O. and S.R.; resources, R.O.O. and S.R.; writing—original draft preparation, R.O.O.; writing—review and editing, R.O.O. and S.R.; supervision, S.R. All authors have read and agreed to the published version of the manuscript.

Funding: The work received financial support from the "Internal grant of TBU in Zlín No. IGA/FT/2022/007" funded from the resources of specific university research and was completed in connection with the project "Innovative and additive manufacturing technology—new technological solutions for 3D printing of metals and composite materials," registration no. CZ.02.1.01/0.0/0.0/17_049/0008407 financed by the Structural Funds of European Union project.

Institutional Review Board Statement: Not applicable.

Informed Consent Statement: Not applicable.

Data Availability Statement: The data presented in this study are available on request from the corresponding author.

Acknowledgments: This work and the project were realised with the financial support of the "Internal grant of TBU in Zlín No. IGA/FT/2022/007", funded by the resources of specific university research and completed in connection with the project "Innovative and additive manufacturing technology—new technological solutions for 3D printing of metals and composite materials," registration no. CZ.02.1.01/0.0/0.0/17_049/0008407, financed by the Structural Funds of the European Union project.

Conflicts of Interest: The authors declare no conflict of interest.

References

1. Grzesik, W. *Advanced Machining Processes of Metallic Materials Theory, Modelling, and Applications*, 2nd ed.; Elsevier: Amsterdam, The Netherlands, 2017.
2. Maria, M. Advanced Composite Materials of the Future in Aerospace Industry. *INCAS Bull.* **2013**, *5*, 139–150. [CrossRef]
3. Chawla, K.K. *Composite Materials*, 3rd ed.; Springer: New York, NY, USA, 2012. [CrossRef]
4. Wu, H.; Fahy, W.P.; Kim, S.; Kim, H.; Zhao, N.; Pilato, L.; Koo, J.H. Recent Developments in Polymers/Polymer Nanocomposites for Additive Manufacturing. *Prog. Mater. Sci.* **2020**, *111*, 100638. [CrossRef]
5. Campbell, F.C. Manufacturing Processes for Advanced Composites. In *Manufacturing Processes for Advanced Composites*; Elsevier: Amsterdam, The Netherlands, 2004. [CrossRef]
6. Fazal, A. Polymer Fibre Composites: Investigation into Performance Enhancement through Viscoelastically Generated Pre-Stress. Ph.D. Thesis, The University of Hull, Hull, UK, 2014.
7. Parlevliet, P.P.; Bersee, H.E.N.; Beukers, A. Residual Stresses in Thermoplastic Composites–A Study of the Literature. Part III: Effects of Thermal Residual Stresses. *Compos. Part A* **2007**, *38*, 1581–1596. [CrossRef]

8. Mohamed, M.; Selim, M.M.; Ning, H.; Pillay, S. Effect of Fibre Prestressing on Mechanical Properties of Glass Fibre Epoxy Composites Manufactured by Vacuum-Assisted Resin Transfer Moulding. *J. Reinf. Plast. Compos.* **2019**, *39*, 21–30. [CrossRef]
9. Mostafa, N.H.; Ismarrubie, Z.N.; Sapuan, S.M.; Sultan, M.T.H. Fibre Prestressed Polymer-Matrix Composites: A Review. *J. Compos. Mater.* **2016**, *51*, 39–66. [CrossRef]
10. Tan, S.C.; Nuismer, R.J. A Theory for Progressive Matrix Cracking in Composite Laminates. *J. Compos. Mater.* **1989**, *23*, 1029–1047. [CrossRef]
11. Zhou, N.; Peng, R.L.; Schönning, M.; Pettersson, R. SCC of 2304 Duplex Stainless Steel—Microstructure, Residual Stress and Surface Grinding Effects. *Materials* **2017**, *10*, 221. [CrossRef] [PubMed]
12. Russell, J.D.; Madhukar, M.S.; Genidy, M.S.; Lee, A.Y. A New Method to Reduce Cure-Induced Stresses in Thermoset Polymer Composites, Part III: Correlating Stress History to Viscosity, Degree of Cure, and Cure Shrinkage. *J. Compos. Mater.* **2000**, *34*, 1926–1947. [CrossRef]
13. Shah, P.; Halls, V.; Zheng, J.; Batra, R. Optimal Cure Cycle Parameters for Minimizing Residual Stresses in Fibre-Reinforced Polymer Composite Laminates. *J. Compos. Mater.* **2018**, *52*, 773–792. [CrossRef]
14. Ali, H.; Ghadbeigi, H.; Mumtaz, K. Processing Parameter Effects on Residual Stress and Mechanical Properties of Selective Laser Melted Ti6Al4V. *J. Mater. Eng. Perform.* **2018**, *27*, 4059–4068. [CrossRef]
15. Mann, B.; Ford, K.; Neilsen, M.; Kammler, D. Minimizing Residual Stress in Brazed Joints by Optimizing the Brazing Thermal Profile. In Volume 2A: Advanced Manufacturing. *Am. Soc. Mech. Eng.* **2020**, *84485*, V02AT02A035. [CrossRef]
16. Schrooten, J.; Michaud, V.; Parthenios, J.; Psarras, G.; Galiotis, C.; Gotthardt, R.; Humbeeck, J. Progress on Composites with Embedded Shape Memory Alloy Wires. *Mater. Trans.* **2002**, *43*, 961–973. [CrossRef]
17. Cohades, A.; Michaud, V. Shape Memory Alloys in Fibre-Reinforced Polymer Composites. *Adv. Ind. Eng. Polym. Res.* **2018**, *1*, 66–81. [CrossRef]
18. Raghavan, J. Evolution of Cure, Mechanical Properties, and Residual Stress during Electron Beam Curing of a Polymer Composite. *Compos. Part A* **2009**, *40*, 300–308. [CrossRef]
19. Razavi, S.M.; Ahmadi, S.J.; Rahmani Cherati, P.; Hadi, M.; Ahmadi, S.A.R. Effect of Electron Beam Irradiation on Mechanical Properties of Unsaturated Polyester/Nanoclay Composites Reinforced with Carbon and Glass Fibres. *Mech. Mater.* **2020**, *141*, 103265. [CrossRef]
20. Wolff-Fabris, F.; Altstadt, V.; Arnold, U.; Doring, M. *Electron Beam Curing of Composites*; Hanser Publshers: Cincinnati, OH, USA, 2010.
21. Zhang, J.; Duan, Y.; Wang, B.; Zhang, X. Interfacial Enhancement for Carbon Fibre Reinforced Electron Beam Cured Polymer Composite by Microwave Irradiation. *Polymer* **2020**, *192*, 122327. [CrossRef]
22. Orso, J.; Vizzini, J.A. Stress Effects of Inhomogeneous Expansion-Controlled Matrices in Continuous Fiber Composites. *J. Compos. Mater.* **1995**, *29*, 2003–2004. [CrossRef]
23. Marx, P.; Wiesbrock, F. Expanding Monomers as Anti-Shrinkage Additives. *Polymers* **2021**, *13*, 806. [CrossRef]
24. Fu, J.; Liu, W.; Hao, Z.; Wu, X.; Yin, J.; Panjiyar, A.; Liu, X.; Shen, J.; Wang, H. Characterization of a Low Shrinkage Dental Composite Containing Bismethylene Spiroorthocarbonate Expanding Monomer. *Int. J. Mol. Sci.* **2014**, *15*, 2400–2412. [CrossRef]
25. Sokairge, H.; Elgabbas, F.; Rashad, A.; Elshafie, H. Long-Term Creep Behaviour of Basalt Fibre Reinforced Polymer Bars. *Constr. Build. Mater.* **2020**, *260*, 120437. [CrossRef]
26. Zaidi, B.M.; Magniez, K.; Miao, M. Prestressed Natural Fibre Spun Yarn Reinforced Polymer-Matrix Composites. *Compos Part A* **2015**, *75*, 68–76. [CrossRef]
27. Dvorak, G.; Suvorov, A. The Effect of Fiber Pre-Stress on Residual Stresses and the Onset of Damage in Symmetric Laminates. *Compos. Sci. Technol.* **2000**, *60*, 1129–1139. [CrossRef]
28. Qin, Y.; Fancey, K.S. Drop Weight Impact Behaviour of Viscoelastically Prestressed Composites. *Compos Part A* **2020**, *131*, 105782. [CrossRef]
29. Ge, C.; Wang, B.; Fancey, K.S. An Evaluation of the Scanning Electron Microscope Mirror Effect to Study Viscoelastically Prestressed Polymeric Matrix Composites. *Mater. Today Commun.* **2017**, *12*, 79–87. [CrossRef]
30. Fazal, A.; Fancey, K.S. UHMWPE Fibre-Based Composites: Prestress-Induced Enhancement of Impact Properties. *Compos. Part B* **2014**, *66*, 1–6. [CrossRef]
31. Jorge, L.D.A.; Marques, A.T.; De Castro, P.M.S.T. The Influence of Prestressing on the Mechanical Behaviour of Unidirectional Composites. In *Developments in the Science and Technology of Composite Materials*; Springer: Dordrecht, The Netherlands, 1990; pp. 897–902. [CrossRef]
32. Al-Dulaimy, A.; Khalid, M.A.; Al-hassany, M.O.A.; Shakir, S.W. The Effect of Unidirectional Pre-Load on Tensile Characteristics of E-Glass Fibre and Epoxy Composite. *Mater. Today Proc.* **2021**, *42*, 2510–2515. [CrossRef]
33. Hassan, F.A.K.; Abdullah, O.A. New Methodology for Prestressing Fiber Composites. *Univers. J. Mech. Eng.* **2015**, *3*, 252–261. [CrossRef]
34. Krishnamurthy, S. Pre-Stressed Advanced Fibre Reinforced Composites Fabrication and Mechanical Performance. Ph.D. Thesis, Cranfield University, Silsoe, UK, 2006.
35. Abdullah, O.A.; Hassan, A.K.F. Effect of Prestress Level on the Strength of CFRP Composite Laminate. *J. Mech. Sci. Technol.* **2016**, *30*, 5115–5123. [CrossRef]

36. Kang, C.; Shi, Y.; Deng, B.; Yu, T.; Sun, P. Determination of Residual Stress and Design of Process Parameters for Composite Cylinder in Filament Winding. *Adv. Mater. Sci. Eng.* **2018**, *1*, 1–11. [CrossRef]
37. Toptaş, E.; Akkuş, N. Damage Detection of Carbon Fibers in Filament Winding Machines Using an Electrical Resistance Method. *Int. J. Adv. Manuf. Technol.* **2017**, *93*, 671–679. [CrossRef]
38. Korotkov, V.N.; Chekanov, Y.A.; Rozenberg, B.A. The Simultaneous Process of Filament Winding and Curing for Polymer Composites. *Compos. Sci. Technol.* **1993**, *47*, 383–388. [CrossRef]
39. Ilangovan, S.; Senthil Kumaran, S.; Naresh, K. Effect of Nanoparticles Loading on Free Vibration Response of Epoxy and Filament Winding Basalt/Epoxy and E-Glass/Epoxy Composite Tubes: Experimental, Analytical and Numerical Investigations. *Mater. Res. Express* **2020**, *7*, 025007. [CrossRef]
40. Wong, J.C.; Blanco, J.M.; Ermanni, P. Filament Winding of Aramid/PA6 Commingled Yarns with in Situ Consolidations. *J. Thermoplast. Compos. Mater.* **2018**, *31*, 465–482. [CrossRef]
41. Jevons, M.; Fernando, G.; Kalsi, G. Effect of Pre-Tensioning on the Low-Velocity Impact Performance of Glass Fibre Composites. In Proceedings of the Tenth European Conference on Composite Materials (ECCM-10), Brugge, Belgium, 3–7 June 2002; pp. 3–7.
42. Krishnamurthy, S.; Badcock, R.; Machavaram, V.; Fernando, G. Monitoring Pre-Stressed Composites Using Optical Fibre Sensors. *Sensors* **2016**, *16*, 777. [CrossRef]
43. Mostafa, N.H.; Ismarrubie, Z.N.; Sapuan, S.M.; Sultan, M.T.H. Effect of Fabric Biaxial Prestresses on the Fatigue of Woven E-glass/Polyester Composites. *Mater. Des.* **2016**, *92*, 579–589. [CrossRef]
44. Zhao, J.; Cameron, J. Polypropylene Matrix Composites Reinforced with Pre-Stressed Glass Fibres. *Polym. Compos.* **1998**, *19*, 218–224. [CrossRef]
45. Motahhari, S.; Cameron, J. Measurement of Micro-Residual Stresses in Fiber-Prestressed Composites. *J. Reinf. Plast. Compos.* **1997**, *16*, 1129–1137. [CrossRef]
46. Tuttle, M.E.; Koehler, R.T.; Keren, D. Controlling Thermal Stresses in Composites by Means of Fiber Prestress. *J. Compos. Mater.* **1996**, *30*, 486–502. [CrossRef]
47. Osman, B.; Tian, Z.; Jiang, G.; Sun, X.; Carroll, A. Experimental Study on Dynamic Properties of UHMWPE and PVA Fibers Concrete. *KSCE J. Civ. Eng.* **2020**, *24*, 2993–3011. [CrossRef]
48. Fazal, A.; Fancey, K.S. Performance Enhancement of Nylon/Kevlar Fibre Composites through Viscoelastically Generated Pre-stress. *Polym. Compos.* **2014**, *35*, 931–938. [CrossRef]
49. Wang, B.; Ge, C.; Fancey, K.S. Snap-through Behaviour of a Bistable Structure Based on Viscoelastically Generated Prestress. *Compos. Part B Eng.* **2017**, *114*, 23–33. [CrossRef]
50. Cui, H.X.; Guan, M.J.; Zhu, Y.X.; Zhang, Z.Z. The Flexural Characteristics of Prestressed Bamboo Slivers Reinforced Parallel Strand Lumber (PSL). *Key Eng. Mater.* **2012**, *517*, 96–100. [CrossRef]
51. Qin, Y.; Fancey, K.S. Towards "Green" Viscoelastically Prestressed Composites: Cellulose Fibre Reinforcement. *Compos. Part B Eng.* **2018**, *154*, 439–448. [CrossRef]
52. Širvaitienė, A.; Jankauskaitė, V.; Bekampienė, P.; Norkaitis, J. Vegetable Fibre Pre-Tensioning Influence the Composites Reinforcement. *Polym. Compos.* **2013**, *34*, 1533–1537. [CrossRef]
53. Herrera-Franco, P.J.; Valadez-González, A. Mechanical Assessment of Natural Fibre Composites. In *Interface Engineer Nature Fibre Compose Maximum Perform*; Elsevier: Amsterdam, The Netherlands, 2011; pp. 222–240. [CrossRef]
54. Djafari Petroudy, S.R. Physical and Mechanical Properties of Natural Fibres. In *Advanced High Strength Natural Fibre Composites in Construction*; Elsevier: Amsterdam, The Netherlands, 2017; pp. 59–83. [CrossRef]
55. Mohanty, A.K.; Misra, M.; Drzal, T. *Natural Fibers, Biopolymers, Biocomposites*; CRC Press: New York, NY, USA, 2005.
56. Ashori, A.; Bahreini, Z. Evaluation of Calotropis Gigantea as a Promising Raw Material for Fiber-Reinforced Composite. *J. Compos. Mater.* **2009**, *43*, 1297–1304. [CrossRef]
57. Sarkar, A.; Connor, A.J.; Koffas, M.; Zha, R.H. Chemical Synthesis of Silk-Mimetic Polymers. *Materials* **2019**, *12*, 4086. [CrossRef]
58. Bhat, G.; Kandagor, V. Synthetic Polymer Fibres and Their Processing Requirements. In *Advances in Filament Yarn Spinning of Textiles and Polymers*; Elsevier: Amsterdam, The Netherlands, 2014; pp. 3–30. [CrossRef]
59. Pang, J.W.C.; Fancey, K.S. Analysis of the Tensile Behaviour of Viscoelastically Prestressed Polymeric Matrix Composites. *Compos. Sci. Technol.* **2008**, *68*, 1903–1910. [CrossRef]
60. Pang, J.W.C.; Fancey, K.S. The Flexural Stiffness Characteristics of Viscoelastically Prestressed Polymeric Matrix Composites. *Compos. Part A Appl. Sci. Manuf.* **2009**, *40*, 784–790. [CrossRef]
61. Zhigun, I.G. Experimental Evaluation of the Effect of Prestressing the Fibres in Two Directions on Certain Elastic Characteristic of Woven-Glass Reinforced Plastics. *Polym. Mech.* **1972**, *4*, 691–695. [CrossRef]
62. Brown, G.G. *Development of Prestressed Graphite Processing Techniques*; North American Rockwell: Hampton, VA, USA, 1976.
63. Tuttle, M.E. A Mechanical/Thermal Analysis of Prestressed Composite Laminates. *J. Compos. Mater.* **1988**, *22*, 780–792. [CrossRef]
64. Rose, D.H. Effect of Prestressed Fibres Upon the Response of Composite Materials. Master's Thesis, The University of Dayton, Dayton, OH, USA, 1993.
65. Fancey, K.S. Investigation into the Feasibility of Viscoelastically Generated Pre-Stress in Polymeric Matrix Composites. *Mater. Sci. Eng. A* **2000**, *279*, 36–41. [CrossRef]
66. Jevons, M.P. The Effects of Fibre Pre-Stressing on the Impact Performance of Composite Laminates. Ph.D. Thesis, Cranfield University, Silsoe, UK, 2011.

67. Daynes, S.; Potter, K.D.; Weaver, P.M. Bistable Prestressed Buckled Laminates. *Compos. Sci. Technol.* **2008**, *68*, 3431–3437. [CrossRef]
68. Schlichting, L.H.; de Andrada, M.A.C.; Vieira, L.C.C.; de Oliveira Barra, G.M.; Magne, P. Composite Resin Reinforced with Pre-tensioned Glass Fibres. Influence of Prestressing on Flexural Properties. *Dent. Mater.* **2010**, *26*, 118–125. [CrossRef]
69. Nishi, Y.; Okada, T.; Okada, S.; Hirano, M.; Matsuda, M.; Matsuo, A.; Faudree, M.C. Effects of Tensile Prestress Level on Impact Value of 50 vol% Continuous Unidirectional 0 Degree Oriented Carbon Fiber Reinforced Epoxy Polymer (CFRP). *Mater. Trans.* **2014**, *55*, 318–322. [CrossRef]
70. Fancey, K.S.; Fazal, A. Prestressed Polymeric Matrix Composites: Longevity Aspects. *Polym. Compos.* **2015**, *37*, 2092–2097. [CrossRef]
71. Mostafa, N.H.; Ismarrubie, Z.; Sapuan, S.; Sultan, M. The Influence of Equi-Biaxially Fabric Prestressing on the Flexural Performance of Woven E-Glass/Polyester-Reinforced Composites. *J. Compos. Mater.* **2015**, *50*, 3385–3393. [CrossRef]
72. Rossini, N.; Dassisti, M.; Benyounis, K.; Olabi, A. Methods of Measuring Residual Stresses in Components. *Mater. Des.* **2012**, *35*, 572–588. [CrossRef]
73. Ersoy, N.; Vardar, O. Measurement of Residual Stresses in Layered Composites by Compliance Method. *J. Compos. Mater.* **2000**, *34*, 575–598. [CrossRef]
74. Shen, J.; Tong, Q. Prestressing Strategy for Strengthening Biocomposites: A Numerical Study. *ACS Biomater. Sci. Eng.* **2021**, *7*, 5014–5027. [CrossRef] [PubMed]
75. Perreux, D.; Lazuardi, D. The effect of Residual Stress on the Non-Linear Behaviour of Composite Laminates Part II. Layer, Laminate Non-Linear Models and the Effect of Residual Stress on the Model Parameters. *Compos. Sci. Technol.* **2001**, *61*, 177–190. [CrossRef]
76. Eijpe, M.; Powell, P. Residual Stress Evaluation in Composites Using a Modified Layer Removal Method. *Compos. Struct.* **1997**, *37*, 335–342. [CrossRef]
77. Mahmoodi, M.; Sedighi, M.; Tanner, D. Investigation of through Thickness Residual Stress Distribution in Equal Channel Angular Rolled Al 5083 Alloy by Layer Removal Technique and X-ray Diffraction. *Mater. Des.* **2012**, *40*, 516–520. [CrossRef]
78. Dreier, S.; Denkena, B. Determination of Residual Stresses in Plate Material by Layer Removal with Machine-integrated Measurement. *Procedia CIRP* **2014**, *24*, 103–107. [CrossRef]
79. Mathar, J. Determination of Initial Stresses by Measuring the Deformation around Drilled Holes. *Trans. ASME* **1934**, *56*, 249–259.
80. Sicot, O.; Gong, X.; Cherouat, A.; Lu, J. Influence of Experimental Parameters on Determination of Residual Stress Using the Incremental Hole-Drilling Method. *Compos. Sci. Technol.* **2004**, *64*, 171–180. [CrossRef]
81. Makino, A.; Nelson, D.; Fuchs, E.; Williams, D. Determination of Biaxial Residual Stresses by a Holographic-Hole Drilling Technique. *J. Eng. Mater. Technol.* **1996**, *118*, 583–588. [CrossRef]
82. Ghasemi, A.; Mohammadi, M. Residual Stress Measurement of Fiber Metal Laminates Using Incremental Hole-Drilling Technique in Consideration of the Integral Method. *Int. J. Mech. Sci.* **2016**, *114*, 246–256. [CrossRef]
83. Smit, T.; Reid, R. Residual Stress Measurement in Composite Laminates Using Incremental Hole-Drilling with Power Series. *Exp. Mech.* **2018**, *58*, 1221–1235. [CrossRef]
84. Mahmoudi, A.; Hossain, S.; Truman, C.; Smith, D.; Pavier, M. A New Procedure to Measure Near Yield Residual Stresses Using the Deep Hole Drilling Technique. *Exp. Mech.* **2008**, *49*, 595–604. [CrossRef]
85. Hossain, S.; Truman, C.; Smith, D. Benchmark Measurement of Residual Stresses in a 7449 Aluminium Alloy Using Deep-Hole and Incremental Centre-Hole Drilling Methods. *Eng. Appl. Res. Stress* **2011**, *8*, 67–74. [CrossRef]
86. Ghasemi, A.; Taheri-Behrooz, F.; Shokrieh, M. Determination of Non-Uniform Residual Stresses in Laminated Composites Using Integral Hole Drilling Method: Experimental Evaluation. *J. Compos. Mater.* **2013**, *48*, 415–425. [CrossRef]
87. ASTM E837-20*Standard Test Method for Determining Residual Stresses by the Hole-Drilling Strain-Gage Method*, ASTM International: West Conshohocken, PA, USA, 2020.
88. Ajovalasit, A.; Petrucci, G.; Zuccarello, B. Determination of Non-uniform Residual Stresses Using the Ring-Core Method. *J. Eng. Mater. Technol.* **1996**, *118*, 224–228. [CrossRef]
89. Seers, B.; Tomlinson, R.; Fairclough, P. Residual Stress in Fiber Reinforced Thermosetting Composites: A Review of Measurement Techniques. *Polym. Compos.* **2021**, *42*, 1631–1647. [CrossRef]
90. Ghaedamini, R.; Ghassemi, A.; Atrian, A. Ring-Core Method in Determining the Amount of Non-Uniform Residual Stress in Laminated Composites: Experimental, Finite Element and Theoretical Evaluation. *Arch. Appl. Mech.* **2017**, *88*, 755–767. [CrossRef]
91. Giri, A.; Mahapatra, M. On the Measurement of Sub-Surface Residual Stresses in SS 304L Welds by Dry Ring Core Technique. *Measurement* **2017**, *106*, 152–160. [CrossRef]
92. Hodek, J.; Prantl, A.; Džugan, J.; Strunz, P. Determination of Directional Residual Stresses by the Contour Method. *Metals* **2019**, *9*, 1104. [CrossRef]
93. Sun, H.; Dugnani, R. Precise Residual Stress Profile in Ion-Exchanged Silicate Glass by Modified Contour Method. *J. Eur. Ceram. Soc.* **2021**, *41*, 4355–4368. [CrossRef]
94. Araujo de Oliveira, J.; Fitzpatrick, M.; Kowal, J. Residual Stress Measurements on a Metal Matrix Composite Using the Contour Method with Brittle Fracture. *Adv. Mater. Res.* **2014**, *996*, 349–354. [CrossRef]
95. Prime, M.; Sebring, R.; Edwards, J.; Hughes, D.; Webster, P. Laser Surface-Contouring and Spline Data-Smoothing for Residual Stress Measurement. *Exp. Mech.* **2004**, *44*, 176–184. [CrossRef]

96. Salehi, S.; Rastak, M.; Shokrieh, M.; Barrallier, L.; Kubler, R. Full-Field Measurement of Residual Stresses in Composite Materials Using the Incremental Slitting and Digital Image Correlation Techniques. *Exp. Mech.* **2020**, *60*, 1239–1250. [CrossRef]
97. Nervi, S.; Szabó, B. On the Estimation of Residual Stresses by the Crack Compliance Method. *Comput. Methods Appl. Mech. Eng.* **2007**, *196*, 3577–3584. [CrossRef]
98. Zhao, L.; Santos Macías, J.; Dolimont, A.; Simar, A.; Rivière-Lorphèvre, E. Comparison of Residual Stresses Obtained by the Crack Compliance Method for Parts Produced by Different Metal Additive Manufacturing Techniques and after Friction Stir Processing. *Addit. Manuf.* **2020**, *36*, 101499. [CrossRef]
99. Shokrieh, M. *Residual Stresses in Composite Materials*; Woodhead Publishing: London, UK, 2014.
100. Ma, L.; Qiu, W.; Fan, X. Stress/Strain Characterization in Electronic Packaging by Micro-Raman Spectroscopy: A Review. *Microelectron. Reliab.* **2021**, *118*, 114045. [CrossRef]
101. Nielsen, A.; Batchelder, D.; Pyrz, R. Estimation of Crystallinity of Isotactic Polypropylene Using Raman Spectroscopy. *Polymer* **2002**, *43*, 2671–2676. [CrossRef]
102. Guo, J.; Fu, H.; Pan, B.; Kang, R. Recent Progress of Residual Stress Measurement Methods: A review. *Chin. J. Aeronaut.* **2021**, *34*, 54–78. [CrossRef]
103. Ao, S.; Li, C.; Huang, Y.; Luo, Z. Determination of Residual Stress in Resistance Spot-Welded Joint by a Novel X-ray Diffraction. *Measurement* **2020**, *161*, 107892. [CrossRef]
104. Widjaja, S. Determination of Creep-Induced Residual Stress in Fiber-Reinforced Glass–Ceramic Matrix Composites by X-ray Diffraction. *Mater. Charact.* **2001**, *47*, 47–54. [CrossRef]
105. Benedikt, B.; Lewis, M.; Rangaswamy, P.; Kumosa, M.; Predecki, P.; Kumosa, L.; Gentz, M. Residual Stress Analysis in Aged Graphite/PMR-15 Composites Using X-ray Diffraction. *Mater. Sci. Eng. A* **2006**, *421*, 1–8. [CrossRef]
106. Marquez, F.; Muñoz, C. A New Approach for Fault Detection, Location and Diagnosis by Ultrasonic Testing. *Energies* **2020**, *13*, 1192. [CrossRef]
107. Kudryavtsev, Y.; Kleiman, J. Ultrasonic Technique and Equipment for Residual Stresses Measurement. *Eng. Appl. Res. Stress* **2011**, *8*, 55–66. [CrossRef]
108. Liu, X.; Wang, X.; Guan, Z.; Jiang, T.; Geng, K.; Li, Z. Improvement and Validation of Residual Stress Measurement in Composite Laminates Using the Incremental Hole-Drilling Method. *Mech. Mater.* **2021**, *154*, 103715. [CrossRef]
109. Song, X.; Yeap, K.; Zhu, J.; Belnoue, J.; Sebastiani, M.; Bemporad, E.; Zeng, K.; Korsunsky, A. Residual Stress Measurement in Thin Films Using the Semi-Destructive Ring-Core Drilling Method Using Focused Ion Beam. *Procedia Eng.* **2011**, *10*, 2190–2195. [CrossRef]
110. Hill, M. The Slitting Method. *Pract. Residual Stress Meas. Methods* **2013**, 89–108. [CrossRef]
111. Hutchings, M.; Withers, P.; Holden, T.; Lorentzen, T. *Introduction to the Characterization of Residual Stress by Neutron Diffraction*; CRC Press: Boca Raton, FL, USA, 2005. [CrossRef]
112. Niu, X.; Zhang, H.; Pei, Z.; Shi, N.; Sun, C.; Gong, J. Measurement of Interfacial Residual Stress in SiC Fiber Reinforced Ni-Cr-Al Alloy Composites by Raman Spectroscopy. *J. Mater. Sci. Technol.* **2019**, *35*, 88–93. [CrossRef]
113. Balasingh, C.; Singh, V. Measurement of Residual Stresses in CFRP Laminates by X-ray Diffraction Method. *Bull. Mater. Sci.* **1997**, *20*, 325–332. [CrossRef]
114. Zhang, X.; Xue, Y.; Zhang, H.; Fu, H.; Wang, Z.; Nie, Z.; Wang, L. Thermal Residual Stresses in W Fibers/Zr-based Metallic Glass Composites by High-energy Synchrotron X-ray Diffraction. *J. Mater. Sci. Technol.* **2015**, *31*, 159–163. [CrossRef]
115. Xu, C.; Song, W.; Pan, Q.; Li, H.; Liu, S. Nondestructive Testing Residual Stress Using Ultrasonic Critical Refracted Longitudinal Wave. *Phys. Procedia* **2015**, *70*, 594–598. [CrossRef]
116. Lengsfeld, H.; Wolff-Fabris, F.; Krämer, J.; Lacalle, J.; Altstädt, V. *Composite Technology: Prepregs and Monolithic Part Fabrication Technologies*; Hanser Publishers: Munich, Germany, 2016.
117. Pekbey, Y.; Aslantaş, K.; Yumak, N. Ballistic Impact Response of Kevlar Composites with filled epoxy matrix. *Steel Compos. Struct.* **2017**, *24*, 191–200. [CrossRef]
118. Alkbir, M.F.M.; Sapuan, S.M.; Nuraini, A.A.; Ishak, M.R. Fibre Properties and Crashworthiness Parameters of Natural Fibre-Reinforced Composite Structure: A Literature Review. *Compos. Struct.* **2016**, *148*, 59–73. [CrossRef]
119. Jacob, G.; Starbuck, J.; Fellers, J.; Simunovic, S. Energy Absorption in Chopped Carbon Fiber Epoxy Composites for Automotive Crashworthiness. *Polym. J.* **2003**, *35*, 560–567. [CrossRef]
120. Isaac, C.W.; Ezekwem, C. A Review of the Crashworthiness Performance of Energy Absorbing Composite Structure within the Context of Materials, Manufacturing and Maintenance for Sustainability. *Compos. Struct.* **2021**, *257*, 113081. [CrossRef]

polymers

Article

Polymer-Templated Durable and Hydrophobic Nanostructures for Hydrogen Gas Sensing Applications

Mohammad Kamal Hossain [1,]* and Qasem Ahmed Drmosh [2]

[1] Interdisciplinary Research Center for Renewable Energy and Power Systems (IRC-REPS), Research Institute, King Fahd University of Petroleum & Minerals (KFUPM), Dhahran 31261, Saudi Arabia
[2] Interdisciplinary Research Center for Hydrogen and Energy Storage (IRC-HES), Research Institute, King Fahd University of Petroleum & Minerals (KFUPM), Dhahran 31261, Saudi Arabia; drmosh@kfupm.edu.sa
* Correspondence: kamalhossain@kfupm.edu.sa; Tel.: +966-138601058; Fax: +966-138607312

Abstract: A simple and hands-on one-step process has been implemented to fabricate polymer-templated hydrophobic nanostructures as hydrogen gas sensing platforms. Topographic measurements have confirmed irregular hills and dips of various dimensions that are responsible for creating air bubble pockets that satisfy the Cassie–Baxter state of hydrophobicity. High-resolution field-emission scanning electron microscopy (FESEM) has revealed double-layer structures consisting of fine microscopic flower-like structures of nanoscale petals on the top of base nanostructures. Wetting contact angle (WCA) measurements further revealed the contact angle to be ~142.0° ± 10.0°. Such hydrophobic nanostructures were expected to provide a platform for gas-sensing materials of a higher surface area. From this direction, a very thin layer of palladium, ca. 100 nm of thickness, was sputtered. Thereafter, further topographic and WCA measurements were carried out. FESEM micrographs revealed that microscopic flower-like structures of nanoscale petals remained intact. A sessile drop test reconfirmed a WCA of as high as ~130.0° ± 10.0°. Due to the inherent features of hydrophobic nanostructures, a wider surface area was expected that can be useful for higher target gas adsorption sites. In this context, a customized sensing facility was set up, and H_2 gas sensing performance was carried out. The surface nanostructures were found to be very stable and durable over the course of a year and beyond. A polymer-based hydrophobic gas-sensing platform as investigated in this study will play a dual role in hydrophobicity as well as superior gas-sensing characteristics.

Keywords: polymer; hydrogen; hydrophobic; sensing; nanostructures; palladium

Citation: Hossain, M.K.; Drmosh, Q.A. Polymer-Templated Durable and Hydrophobic Nanostructures for Hydrogen Gas Sensing Applications. *Polymers* **2021**, *13*, 4470. https://doi.org/10.3390/polym13244470

Academic Editors: Andrzej Puszka and Beata Podkościelna

Received: 22 November 2021
Accepted: 17 December 2021
Published: 20 December 2021

1. Introduction

Gas sensing, particularly hydrogen (H_2) gas sensing, has become very crucial due to its renascence as a new and alternative energy in modern life [1–3]. H_2 is an important energy carrier that is going to be complementary to current electricity very soon [4,5]. A persistence challenge is being carried out to incorporate H_2 as fuel for "zero-emissions" vehicles, to heat accommodations and workplaces and fuel aircrafts, amongst many other applications [6–8]. Therefore, not only is an efficient and sensitive sensing platform urgently needed for the safe deployment of all H_2-based applications, but also multifunctional capabilities are required to deal with extreme and critical environmental conditions. Most H_2 gas sensors, particularly used in industries and workplaces, are not suitable for advanced and sophisticated applications. The sensor needs to be smart, durable and of multitasking capacity [9–11]. The fabrication and realization of multifunctional sensing platforms have been exciting and hot areas of research in both academia and industry, including healthcare sectors and environmental protection [12,13]. There is an enticing and ever-growing interest in devising a sensing platform capable of hydrophobic and non-adhesive characteristics [14–16]. Such platforms facilitate water droplets dropped onto them, rolling them off automatically with a small tilt angle. In the process of rolling off, contaminants and dust are also carried away, and thus the sensing surfaces become ready to detect the target gas [17,18]. However, developing an artificial hydrophobic sensing surface is not that straightforward, and it requires a smooth and fine strategy so that the surface becomes stable, durable and capable of reconciling with the surrounding environment [19–22]. Most of the time, a hydrophobic surface is achieved by following the improvised Cassie–Baxter model that confirms a double-layer roughness wherein there should be nanoscale roughness on the top of the microscale structures [23–25]. Although such double-layer structures have great potential in a wide range of applications, including self-cleaning, anti-fouling, anti-corrosion and oil–water separation, the top nanostructures decay over time under extreme environmental conditions [26,27].

A wide range of organic and inorganic materials are used as base substrates to achieve hydrophobic sensing surfaces [28–30]. Polymers, particularly polycarbonate (bisphenol A polycarbonate: 2,2-bis(p-hydroxyphenyl)-propane, PC), are one of the interesting base materials that have been studied extensively and therefore used in a wide variety of applications [31–34]. Due to low-cost, high durability, low modulus of elasticity and high transparency, PC has been useful in optoelectronic and microelectronic applications [35–37]. PC is an industrially attractive soft polymer material that is being extracted as a by-product from oil and gas refineries through industrial-scale and low-cost processes [33,38]. However, one finds that such PC can be further reinforced by including hydrophobic characteristics, which are a key element in many applications that require a self-cleaning property. It is well-acknowledged that the effectiveness of hydrophobicity is demonstrated by the Cassie–Baxter model, although the same model indicates that a higher surface area is required for many surface-enhanced applications, such as molecule detection and sensing [39–41]. For multifunctional devices, particularly hydrophobic gas-sensing platforms, using PC as the base materials is industrially viable and one of the promising candidates. To the best of our knowledge, a PC-templated hydrophobic gas-sensing platform has not been reported so far. Mazen and his group have demonstrated and devised a transparent hydrophobic PC as a self-cleaning surface for PV panels installed in a dusty environment [42]. Yilbas and his team have developed a generic process to copy the micro/nanoscale structure of hydrophobic PC by polydimethylsiloxane (PDMS) that showed a higher wetting contact angle (WCA) and transparency [43]. Jhang and his group developed hydrophobic microchannels in PC that enabled the valve-free sequential injection of multiple liquids [44]. Most of the methods used in achieving such hydrophobic PC were associated with multiple treatments, apart from the requirement of skilled hands and the usage of specialized reagents.

Here in this study, we have reported a simple and inexpensive one-step process to fabricate polymer-templated hydrophobic nanostructures for a H_2 sensing application. A hydrophobic PC fabricated by a wet chemical treatment was decorated with palladium (Pd) by high vacuum sputtering technique. High-resolution field-emission scanning electron microscopy (FESEM) revealed double-layer structures consisting of fine microscopic flower-like structures of nanoscale petals on the top of base nanostructures. Sessile drop tests confirmed a WCA of the treated PC as high as $142 \pm 10.0°$, whereas the WCA of pristine PC was estimated to be ~$83 \pm 10.0°$. The as-fabricated polymer-based nanostructures were transferred to an automatic sputtering chamber for Pd sputtering and decorated the treated PC with Pd. Sessile drop tests were carried out once again to evaluate the WCA. It was noted that the Pd-decorated treated PC exhibited a relatively high WCA of ~$130 \pm 10.0°$. A customized setup was built to carry out the sensing characteristics of the Pd-decorated treated PC. Such a generic strategy is indispensable to explore new routes of multifunctional sensing platforms that are particularly important in extreme environmental conditions.

2. Materials and Methods

Commercially available PC sheet (1 mm × 1 mm × 1.6 mm) was cut into pieces (2.5 cm × 2.5 cm × 1.6 mm) and treated with 2-propanone (CH_3–CO–CH_3) under controlled lab conditions. The sample was washed copiously with deionized (DI) water. A transparent pristine PC specimen was found to turn opaque after the treatment. The treated PC was left under a fluorescent light of 30 W for 10 min. and then transferred to an automatic sputtering coater (model #NSC 4000; NANO-MASTER Inc., Austin, TX, USA) for Pd decoration, as shown in Figure 1. Inset (i)–(iii) of Figure 1a represents CCD images of pristine PC, treated PC and Pd-decorated treated PC. High purity Pd (99.999%) target was purchased from Semiconductor Wafer Inc. and used as received without any modification. Pre-sputtering for cleaning the target was carried out for 1 min. Plasma was generated by direct current (DC) magnetron power of 30 W for 20 s, keeping the chamber background pressure as low as 3.5×10^{-6} Torr in Ar gas flow of 80 SCCM. The target-to-substrate distance was fixed at 10 cm.

Initial assessment of surface topography and roughness was carried out by using a 3D optical microscope (model #Meiji Techno MX7100; Meiji, IL, USA). A Dektak profilometer (mode #BrukerXT; Bruker, MA, USA) was used to explore the microscopic surface structure of the treated specimen. Topographic confirmation and in-depth morphology of treated specimens were carried out using high-resolution FESEM (model #LYRA3; TESCAN, Brno, Czech Republic). WCA measurements were carried out using a goniometer (model #DM 501; Kyowa Interface Science Co. Ltd., Saitama, Japan) through sessile drop tests. DI water was used in the sessile drop experiments, and the droplet volume was controlled with an automatic dispensing system. The images of the droplets were taken one second after deposition of the water droplet on the surface. A customized gas chamber, Linkam stage (Model HFS-600E-PB4; Linkam Scientific Instruments, Wakefield, UK), was used to incorporate air and H_2-balanced nitrogen (1% H_2, 99% N_2) sequentially at room temperature. Two mass flow controllers (MFCs) connected with an external XPH-100 power hub supply were utilized to control the flow of H_2. The sensing measurements were performed through a resistivity measurement approach at room temperature. It is well-acknowledged that resistivity defines the ability of the materials to resist the charge flow, whereas the charge carriers available in the vicinity depend on the adsorption of target gas molecules on sensing materials.

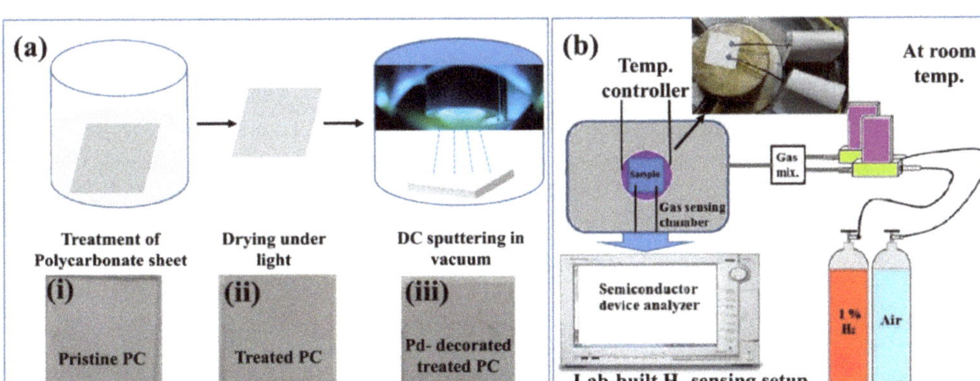

Figure 1. (**a**) A free-hand schematic for fabricating a multifunctional sensing platform; inset (**i–iii**): CCD images of pristine PC, treated PC and Pd-decorated treated PC; (**b**) customized H_2-sensing setup used in this study.

3. Results and Discussion

A quick screening of the treated sample was carried out using 3D optical microscopy, as shown in Figure 2a. Since the optical microscope was using infinity-corrected optics for reflected light observation, it facilitated the acquisition of long-range line scans, as shown in Figure 2b,c. Figure 2b,c represent long-range line scans along the horizontal and vertical axes, respectively, as shown in Figure 2a. Zoom-in views as marked by a black dashed rectangle in Figure 2b,c were shown in inset (i) and inset (ii) of Figure 2a, respectively. The insets in Figure 2a provided an impression without further details that the treated surface of PC had a microscale roughness. Further details of the surface topography were obtained by the Dektak profilometer. Figure 2d displays the 2D mapping of the treated surface, indicating the step height of the nanostructure as traced by the stylus. Line scans along the white horizontal and vertical lines, as marked in Figure 2d, were shown in Figure 2f,g, respectively. The hills and dips were visible in both line scans. The maximum and minimum step height along the horizontal line scan were found to be ~7 μm and ~−22 μm, respectively. In the case of the line scan along the vertical axis, such hills and dips were found to be ~11 μm and ~−23 μm, respectively. A 3D mapping of the same scan as shown in Figure 2d was acquired and shown in Figure 2e. The islands-like view and the abovementioned height profiles indicated that the surface topography of the treated specimen was indeed of microscopic structures.

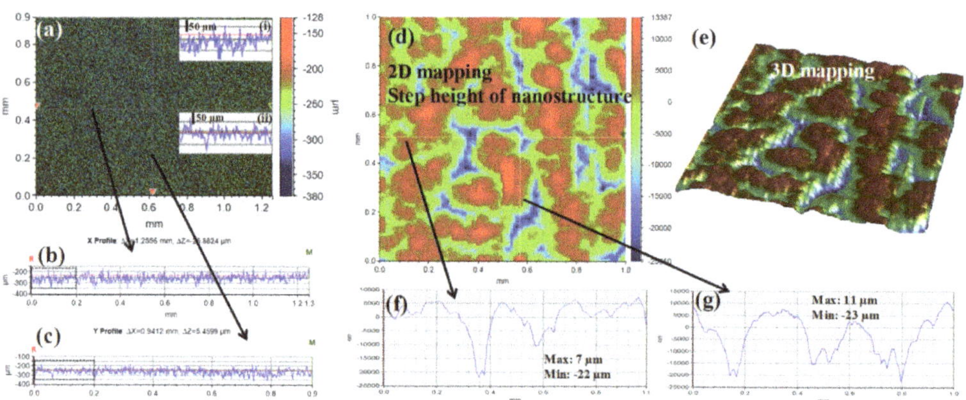

Figure 2. (**a**) Large area 3D optical microscopic image of treated PC; (**b**,**c**) line profiles along the horizontal and vertical axis as marked by black arrows in Figure 2a, respectively; (**d**) 2D image of the same extracted from surface profilometer; (**e**) 3D mapping of the same area confirming hills and dips; (**f**,**g**) line profiles along the horizontal and vertical axis as marked by black arrows in Figure 2d, respectively.

To validate and reconfirm the inherent characteristics of a hydrophobic surface, one needs to go through nanoscale micrographs similar to those captured by high-resolution FESEM as shown in Figure 3. Figure 3a represents a low-resolution FESEM micrograph confirming fine nanostructures on the top of base nanostructures. A high-resolution FESEM as shown in Figure 3b revealed that the fine nanostructures on the top of base structures were indeed something similar to nanoflowers that consisted of petals of different sizes and shapes. An individual nanoflower as marked by "A" in Figure 3b was shown in the inset of Figure 3a. A further zoomed-in view of a small area as marked by the black square in Figure 3b was displayed in Figure 3c. Seven nanoflowers of different sizes as shown in Figure 3c were observed on the top of base nanostructures. A 3D hawk-eye view of the same area as marked by the black square in Figure 3b was presented in Figure 3d. As mentioned earlier, nanoflowers were observed on the top of base nanostructures. As a result, there were two different layers of nanostructures, one at the base and the other at the top, and thus a combination of these two nanostructures indeed was responsible for making the ultimate surface hydrophobic. A line profile along the white dashed line marked as "1" across the base nanostructure was shown in Figure 3e. A further zoomed-in view of a small section of this line profile as marked by the white dashed rectangle was shown as an inset of Figure 3e. On the other hand, the line profile along the white dashed line marked as "2" across the nanoflower structure was shown in Figure 3f. The sharp hills observed therein corresponded to the petals of the nanoflowers. Hence, it was speculated that such double layers of the nanostructure would provide enough void to satisfy the Cassie and Baxter state and facilitate hydrophobicity of the treated specimen.

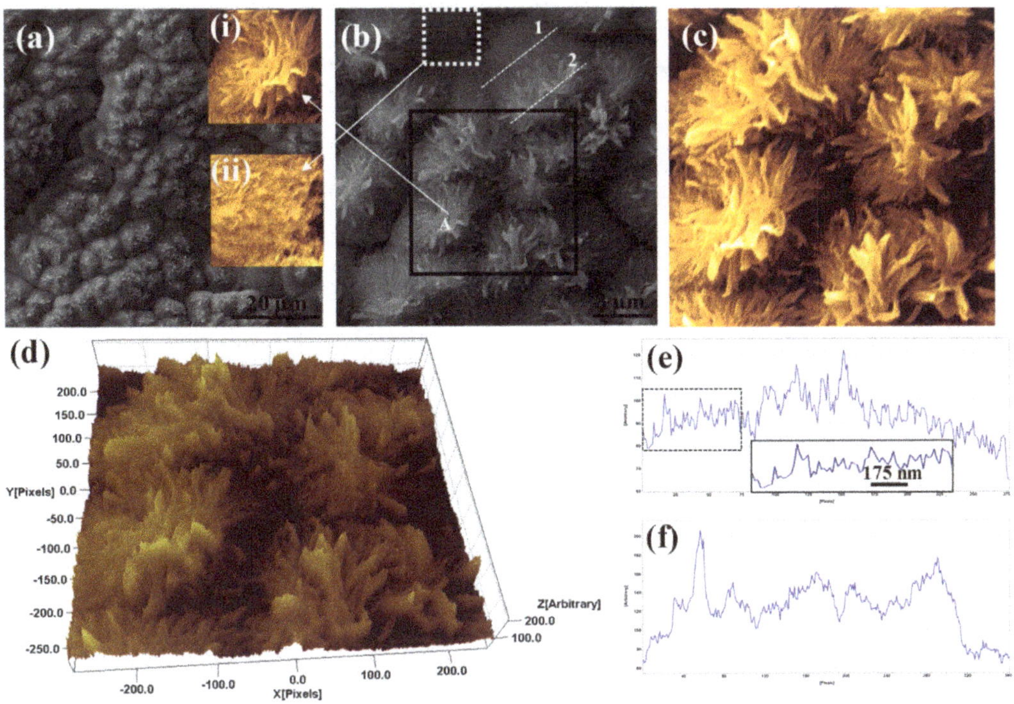

Figure 3. (**a**) FESEM micrograph of treated PC; inset (**i**): an individual flower as marked by the white arrow; inset (**ii**): zoomed-in view of the base nanostructure; (**b**) high-resolution FESEM image indicating several flowers on the top of primary PC nanostructure; (**c**) a zoomed-in view of the selected area as marked by a black square in Figure 3b; (**d**) hawk-eye view of the same area as shown in Figure 3c; (**e**,**f**) line profiles along the white dotted lines marked by "1" and "2" within Figure 3b, respectively.

It is well-acknowledged that the WCA is the measure of indication of whether the surface is hydrophobic or hydrophilic. Conventionally, a sessile drop test is used to directly measure the contact angle and determine the preferential wetting of the substrate by the reference liquid. Here in this investigation, DI water was used as the reference liquid, and the volume was controlled by an automatic dispenser. Figure 4a–c represents water droplet images of pristine PC, treated PC and Pd-decorated treated PC, respectively. The corresponding average WCA of pristine PC was estimated to be 83 $\pm$ 10.0°, as shown in Figure 4a. A further zoomed-in view of the contact region as marked by the green dashed square in Figure 4a is shown in Figure 4d. The lower part of Figure 4d represents the mirror region of the droplet on the pristine PC. It was visible clearly enough, as the pristine PC was noted to be highly transparent. Once the specimen was treated, the sessile drop test confirmed the WCA of the treated PC to be as high as ~142 $\pm$ 10.0°, as shown in Figure 4b. As explained earlier, high-resolution FESEM revealed a double layer of nanostructures wherein fine nanoflower-like structures were observed on the top of base nanostructures. Such a combination of nanostructures indeed facilitated enough voids to yield the top surface being hydrophobic. A zoomed-in view of the contact region as marked by the green dashed square in Figure 4b is shown in Figure 4e, reconfirming a high WCA of the same value. The mirror region of the droplet on the treated PC is shown in the lower part of Figure 4e. The mirror image of the treated specimen was found to be blurred, as the transparency of the same specimen dropped substantially. However, the WCA of the Pd-decorated treated PC was still higher compared to that observed in the pristine PC, but it was a bit lower than that observed in the treated PC. Figure 4c displays water droplet images of Pd-decorated treated PC along with the estimated average WCA of ~130 ± 10.0°. A zoomed-in view of the contact region as marked by the green dashed square in Figure 4c is shown in Figure 4f, and the lower part of the same image represents the mirror region of the droplet on the Pd-decorated treated PC.

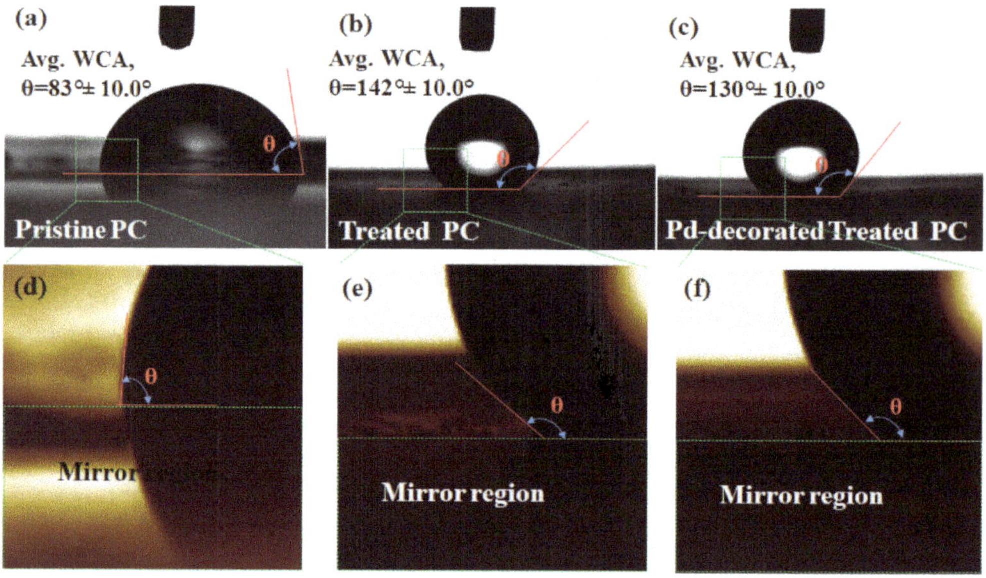

Figure 4. (**a–c**) Optical images defining the wetting contact angle of pristine PC, treated PC and Pd-decorated treated PC, respectively; (**d–f**) zoomed-in views of the contact site of pristine PC, treated PC and Pd-decorated treated PC, respectively.

As explained earlier, the H_2 sensing measurements were investigated through a special test chamber (Linkam chamber, Model HFS-600E-PB4). This gas chamber was connected to the temperature controller, cooling system and two MFCs using an external power hub supply to introduce an appropriate amount of H_2. The H_2 concentration in the Linkam chamber was varied by mixing the test gas (1% H_2 balanced with N_2) and dry air. The dynamic response of the Pd-decorated treated PC for H_2 gas concentrations of 0.05%, 0.1%, 0.2% and 0.5% was shown in Figure 5a. It was observed that for very low concentrations, such as 0.05% of H_2, the response was delayed and started to show up after ~500 s. For the case of other intakes, such as 0.1%, 0.2 | % and 0.5% of H_2 concentrations, the responses were observed starting from ~300 s, as shown in Figure 5a. However, the response characteristics from each intake of H_2 concentrations were found to be different. The rising gradients of the dynamic responses for 0.05%, 0.1%, 0.2% and 0.5% of H_2 concentrations were estimated to be 0.16 (delayed), 0.16, 0.36 and 2.05, respectively.

The response time of the gas sensor is one of the crucial parameters that defines the speed of response of a particular gas sensor. Conventionally, it is measured by the time taken for a sensor to reach 90% of the final indication of saturation. Figure 5b depicts the response time of the specimen in this investigation. It was observed that the intake of H_2 concentrations of 0.05%, 0.1%, 0.2% and 0.5% took a response time of 718, 294, 305 and 315 s, respectively.

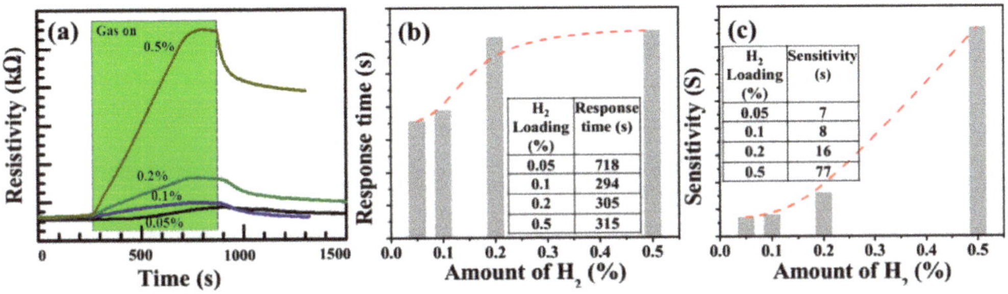

Figure 5. (**a**) Dynamic response of the sensor to different H2 gas concentrations (0.5%, 0.1%, 0.2% and 0.5%) at room temperature; (**b**) response time of the same system; (**c**) sensitivity of the same system.

Sensitivity is another crucial characteristic that indicates how efficiently the target gas can be detected by the sensor. The sensitivity of a gas sensor is defined as:

$$Sensitivity\ (\%) = \frac{R_g - R_a}{R_a} \times 100 \tag{1}$$

where R_a and R_g are the resistance of the sensor in air and target gas, respectively. The electrical resistance of the sensors with and without H_2 was measured by an Agilent B1500A semiconductor device analyzer (SDA). The sensitivity of the fabricated sensors was calculated within 0.05%–0.5 of H_2.

Figure 5c depicts the sensitivity of the specimen in this investigation for different intakes of H_2 concentrations. It was found that the sensitivities for Pd-decorated treated PC at room temperature at 0.05%, 0.1%, 0.2% and 0.5% H_2 concentrations were 7%, 8%, 16% and 77%, respectively. It is noteworthy that the change in electrical resistance (i.e., sensitivity) and the response time increased with increasing concentrations of H_2.

4. Conclusions

A one-step and low-cost process is one of the prerequisites for an industrial-scale production line. In this context, a simple and inexpensive generic route was reported to achieve polymer-templated hydrophobic nanostructures as H_2 gas sensing platforms. Double-layer micro and nanostructures of PC were generated that contributed a vital role toward satisfying the Cassie–Baxter state and yielded a hydrophobic surface with a WCA as high as ~142.0 ± 10.0°. Although preliminary topographic investigations through a 3D optical microscope and Dektak profilometer indicated irregular hills and dips of various dimensions that were responsible for creating air bubble pockets therein, an in-depth surface topography was investigated by high-resolution FESEM. Interestingly, it was noticed that fine microscopic flower-like structures of nanoscale petals were populated on the top of base nanostructures. The surface nanostructures were found to be very stable and durable over the course of a year and beyond. Such nano-flowers decorated with Pd-sensing materials exhibited a relatively high WCA and facilitated a high surface area for adsorbing the target gas simultaneously. A sessile drop test confirmed the WCA of the Pd-decorated hydrophobic PC to be as high as ~130.0 ± 10.0°. A lab-built sensing setup was employed to carry out sensing activity of the Pd-decorated hydrophobic PC for H_2 gas concentrations of 0.05%, 0.1%, 0.2% and 0.5% at room temperature, and the corresponding sensitivity and response time were evaluated.

Author Contributions: Conceptualization, M.K.H.; methodology, M.K.H.; software, M.K.H.; validation, M.K.H. and Q.A.D.; formal analysis, M.K.H. and Q.A.D.; investigation, M.K.H. and Q.A.D.; resources, M.K.H. and Q.A.D.; data curation, M.K.H. and Q.A.D.; writing—original draft preparation, M.K.H.; writing—review and editing, M.K.H. and Q.A.D.; visualization, M.K.H.; supervision, M.K.H.; project administration, M.K.H.; funding acquisition, M.K.H. All authors have read and agreed to the published version of the manuscript.

Funding: MKH would like to acknowledge the support received under the Interdisciplinary Research Center for Renewable Energy and Power Systems (IRC-REPS) (Grant #INRE2114).

Institutional Review Board Statement: Not applicable.

Informed Consent Statement: Not applicable.

Data Availability Statement: The data presented in this study are available on request from the corresponding author.

Acknowledgments: The authors would like to thank the Interdisciplinary Research Center for Renewable Energy and Power Systems (IRC-REPS) research institute and the King Fahd University of Petroleum and Minerals (KFUPM), Dhahran 31261, Saudi Arabia. QAD acknowledges the Interdisciplinary Research Center for Hydrogen and Energy Storage (IRC-HES) research institute and the King Fahd University of Petroleum and Minerals (KFUPM), Dhahran 31261, Saudi Arabia.

Conflicts of Interest: The authors declare that they have no conflict of interest.

References

1. Chauhan, P.S.; Bhattacharya, S. Hydrogen gas sensing methods, materials, and approach to achieve parts per billion level detection: A review. *Int. J. Hydrogen Energy* **2019**, *44*, 26076–26099. [CrossRef]
2. Korotcenkov, G. *Handbook of Gas Sensor Materials;* Conventional Approaches; Springer: Berlin/Heidelberg, Germany, 2013; Volume 1.
3. Padvi, M.N.; Moholkar, A.V.; Prasad, S.R. A Critical Review on Design and Development of Gas Sensing Materials. *Eng. Sci.* **2021**, *15*, 20–37. [CrossRef]
4. Kovač, A.; Paranos, M.; Marciuš, D. Hydrogen in energy transition: A review. *Int. J. Hydrogen Energy* **2021**, *46*, 10016–10035. [CrossRef]
5. Dawood, F.; Anda, M.; Shafiullah, G. Hydrogen production for energy: An overview. *Int. J. Hydrogen Energy* **2020**, *45*, 3847–3869. [CrossRef]
6. Hirscher, M.; Yartys, V.A.; Baricco, M.; von Colbe, J.B.; Blanchard, D.; Bowman, R.C., Jr.; Broom, D.P.; Buckley, C.E.; Chang, F.; Chen, P.; et al. Materials for hydrogen-based energy storage–past, recent progress and future outlook. *J. Alloy. Compd.* **2020**, *827*, 153548. [CrossRef]
7. Sheffield, J.W.; Martin, K.B.; Folkson, R. Electricity and hydrogen as energy vectors for transportation vehicles. In *Alternative Fuels and Advanced Vehicle Technologies for Improved Environmental Performance;* Woodhead Publishing: Sawston, UK, 2014; pp. 117–137.
8. Petrescu, R.V.V.; Machín, A.; Fontánez, K.; Arango, J.C.; Márquez, F.M.; Petrescu, F.I.T. Hydrogen for aircraft power and propulsion. *Int. J. Hydrogen Energy* **2020**, *45*, 20740–20764. [CrossRef]
9. Song, Z.; Ye, W.; Chen, Z.; Chen, Z.; Li, M.; Tang, W.; Wang, C.; Wan, Z.; Poddar, S.; Wen, X.; et al. Wireless Self-Powered High-Performance Integrated Nanostructured-Gas-Sensor Network for Future Smart Homes. *ACS Nano* **2021**, *15*, 7659–7667. [CrossRef] [PubMed]
10. Sazonov, E. (Ed.) *Wearable Sensors: Fundamentals, Implementation and Applications;* Academic Press: Cambridge, MA, USA, 2020.
11. Zhu, Z.; Liu, C.; Jiang, F.; Liu, J.; Liu, G.; Ma, X.; Liu, P.; Huang, R.; Xu, J.; Wang, L. Flexible fiber-shaped hydrogen gas sensor via coupling palladium with conductive polymer gel fiber. *J. Hazard. Mater.* **2021**, *411*, 125008. [CrossRef] [PubMed]
12. Zheng, X.; Cheng, H. Flexible and stretchable metal oxide gas sensors for healthcare. *Sci. China Ser. E Technol. Sci.* **2018**, *62*, 209–223. [CrossRef]
13. Soo, M.T.; Cheong, K.Y.; Noor AF, M. Advances of SiC-based MOS capacitor hydrogen sensors for harsh environment applications. *Sens. Actuators B Chem.* **2010**, *151*, 39–55. [CrossRef]
14. Li, X.; Gao, Z.; Li, B.; Zhang, X.; Li, Y.; Sun, J. Self-healing superhydrophobic conductive coatings for self-cleaning and humidity-insensitive hydrogen sensors. *Chem. Eng. J.* **2021**, *410*, 128353. [CrossRef]
15. Gao, Z.; Song, G.; Zhang, X.; Li, Q.; Yang, S.; Wang, T.; Li, Y.; Zhang, L.; Guo, L.; Fu, Y. A facile PDMS coating approach to room-temperature gas sensors with high humidity resistance and long-term stability. *Sens. Actuators B Chem.* **2020**, *325*, 128810. [CrossRef]
16. Gao, J.; Wang, L.; Guo, Z.; Li, B.; Wang, H.; Luo, J.; Huang, X.; Xue, H. Flexible, superhydrophobic, and electrically conductive polymer nanofiber composite for multifunctional sensing applications. *Chem. Eng. J.* **2020**, *381*, 122778. [CrossRef]
17. Parvate, S.; Dixit, P.; Chattopadhyay, S. Superhydrophobic Surfaces: Insights from Theory and Experiment. *J. Phys. Chem. B* **2020**, *124*, 1323–1360. [CrossRef] [PubMed]
18. Li, L.; Bai, Y.; Li, L.; Wang, S.; Zhang, T. A superhydrophobic smart coating for flexible and wearable sensing electronics. *Adv. Mater.* **2017**, *29*, 1702517. [CrossRef]
19. Kinoshita, H.; Ogasahara, A.; Fukuda, Y.; Ohmae, N. Superhydrophobic/superhydrophilic micropatterning on a carbon nanotube film using a laser plasma-type hyperthermal atom beam facility. *Carbon* **2010**, *48*, 4403–4408. [CrossRef]
20. Yilbas, B.; Khaled, M.; Abu-Dheir, N.; Al-Aqeeli, N.; Said, S.; Ahmed, A.; Varanasi, K.; Toumi, Y. Wetting and other physical characteristics of polycarbonate surface textured using laser ablation. *Appl. Surf. Sci.* **2014**, *320*, 21–29. [CrossRef]

21. Sanger, A.; Kumar, A.; Kumar, A.; Jaiswal, J.; Chandra, R. A fast response/recovery of hydrophobic Pd/V2O5 thin films for hydrogen gas sensing. *Sens. Actuators B Chem.* **2016**, *236*, 16–26. [CrossRef]
22. Hassan, K.; Chung, G.S. Fast and reversible hydrogen sensing properties of Pd-capped Mg ultra-thin films modified by hydrophobic alumina substrates. *Sens. Actuators B Chem.* **2017**, *242*, 450–460. [CrossRef]
23. Xie, H.; Huang, H. Gradient Wetting Transition from the Wenzel to Robust Cassie-Baxter States along Nanopillared Cicada Wing and Underlying Mechanism. *J. Bionic Eng.* **2020**, *17*, 1009–1018. [CrossRef]
24. Hao, J.-H.; Wang, Z.-J. Modeling Cassie–Baxter State on Superhydrophobic Surfaces. *J. Dispers. Sci. Technol.* **2015**, *37*, 1208–1213. [CrossRef]
25. Tsougeni, K.; Tserepi, A.; Boulousis, G.; Constantoudis, V.; Gogolides, E. Control of Nanotexture and Wetting Properties of Polydimethylsiloxane from Very Hydrophobic to Super-Hydrophobic by Plasma Processing. *Plasma Process. Polym.* **2007**, *4*, 398–405. [CrossRef]
26. Mortazavi, V.; Khonsari, M. On the degradation of superhydrophobic surfaces: A review. *Wear* **2017**, *372–373*, 145–157. [CrossRef]
27. Esteves, C. Self-Healing Functional Surfaces. *Adv. Mater. Interfaces* **2018**, *5*, 1800293. [CrossRef]
28. Sanjay, S.L.; Annaso, B.G.; Chavan, S.M.; Rajiv, S.V. Recent progress in preparation of superhydrophobic surfaces: A review. *J. Surf. Eng. Mater. Adv. Technol.* **2012**, *2*, 76–94.
29. Ha, C.S.; Nagappan, S. *Hydrophobic and Superhydrophobic Organic-Inorganic Nano-Hybrids*; CRC Press: Boca Raton, FL, USA, 2018.
30. Quan, Y.Y.; Chen, Z.; Lai, Y.; Huang, Z.S.; Li, H. Recent advances in fabricating durable superhydrophobic surfaces: A review in the aspects of structures and materials. *Mater. Chem. Front.* **2021**, *5*, 1655–1682. [CrossRef]
31. Hoekstra, E.J.; Simoneau, C. Release of Bisphenol A from Polycarbonate—A Review. *Crit. Rev. Food Sci. Nutr.* **2013**, *53*, 386–402. [CrossRef] [PubMed]
32. Shi, G.L.; Li, F.S.; Tian, H.B. Advances and application of polycarbonate in automobile windows and aero glass. *Mater. Rev.* **2006**, *20*, 404–407.
33. Fukuoka, S.; Fukawa, I.; Adachi, T.; Fujita, H.; Sugiyama, N.; Sawa, T. Industrialization and Expansion of Green Sustainable Chemical Process: A Review of Non-phosgene Polycarbonate from CO_2. *Org. Process Res. Dev.* **2019**, *23*, 145–169. [CrossRef]
34. Lee, S.S.; Jho, J.Y. Structure, Properties and Applications of Polycarbonate. *Polym. Sci. Technol.* **1993**, *4*, 432–438.
35. Subramani, N.K.; Shivanna, S.; Nagaraj, S.K.; Suresha, B.; Raj, B.J.; Siddaramaiah, H. Optoelectronic Behaviours of UV shielding Calcium ZirconateReinforced Polycarbonate Nanocomposite Films: An Optical View. *Mater. Today Proc.* **2018**, *5*, 16626–16632. [CrossRef]
36. Bormashenko, E.; Pogreb, R.; Stanevsky, O.; Biton, Y.; Bormashenko, Y. Self-organization in thin polycarbonate films and its optical and electro-optical applications. *J. Mater. Sci.* **2004**, *39*, 6639–6641. [CrossRef]
37. Goyal, P.K.; Kumar, V.; Gupta, R.; Mahendia, S.; Kumar, S. Modification of polycarbonate surface by Ar+ ion implantation for various opto-electronic applications. *Vacuum* **2012**, *86*, 1087–1091. [CrossRef]
38. Kamps, J.H.; Scheffler, C.; Simon, F.; Van Der Heijden, R.; Verghese, N. Functional polycarbonates for improved adhesion to carbon fibre. *Compos. Sci. Technol.* **2018**, *167*, 448–455. [CrossRef]
39. Li, G.-J.; Kawi, S. High-surface-area SnO_2: A novel semiconductor-oxide gas sensor. *Mater. Lett.* **1998**, *34*, 99–102. [CrossRef]
40. Ikram, M.; Liu, L.; Liu, Y.; Ma, L.; Lv, H.; Ullah, M.; He, L.; Wu, H.; Wang, R.; Shi, K. Fabrication and characterization of a high-surface area MoS2@WS2 heterojunction for the ultra-sensitive NO2 detection at room temperature. *J. Mater. Chem. A* **2019**, *7*, 14602–14612. [CrossRef]
41. Shen, Y.; Yamazaki, T.; Liu, Z.; Meng, D.; Kikuta, T.; Nakatani, N. Influence of effective surface area on gas sensing properties of WO_3 sputtered thin films. *Thin Solid Film.* **2009**, *517*, 2069–2072. [CrossRef]
42. Khaled, M. Directed Hierarchical Patterning of Polycarbonate Bisphenol A Glass Surface along Predictable Sites. *J. Nanomater.* **2015**, *16*, 332. [CrossRef]
43. Yilbas, B.S.; Ali, H.; Al-Aqeeli, N.; Khaled, M.; Abu-Dheir, N.; Varanasi, K.K. Solvent-induced crystallization of a polycarbonate surface and texture copying by polydimethylsiloxane for improved surface hydrophobicity. *J. Appl. Polym. Sci.* **2016**, *133*. [CrossRef]
44. Jang, M.; Park, C.K.; Lee, N.Y. Modification of polycarbonate with hydrophilic/hydrophobic coatings for the fabrication of microdevices. *Sens. Actuators B Chem.* **2014**, *193*, 599–607. [CrossRef]

polymers MDPI

Article

Evaluation of Mechanical and Tribological Properties of Corn Cob-Reinforced Epoxy-Based Composites—Theoretical and Experimental Study

Ahmed Fouly [1,2,*], Hany S. Abdo [3,4], Asiful H. Seikh [3], Khalid Alluhydan [2], Hend I. Alkhammash [5], Ibrahim A. Alnaser [2] and Mohamed S. Abdo [6]

[1] Department of Production Engineering and Mechanical Design, Faculty of Engineering, Minia University, Minia 61519, Egypt
[2] Mechanical Engineering Department, King Saud University, P.O. Box 800, Riyadh 11421, Saudi Arabia; kalluhydan@KSU.EDU.SA (K.A.); ianaser@KSU.EDU.SA (I.A.A.)
[3] Center of Excellence for Research in Engineering Materials (CEREM), King Saud University, P.O. Box 800, Riyadh 11421, Saudi Arabia; habdo@ksu.edu.sa (H.S.A.); aseikh@ksu.edu.sa (A.H.S.)
[4] Mechanical Design and Materials Department, Faculty of Energy Engineering, Aswan University, Aswan 81521, Egypt
[5] Department of Electrical Engineering, College of Engineering, Taif University, P.O. Box 11099, Taif 21944, Saudi Arabia; Khamash.h@tu.edu.sa
[6] Biomedical Engineering Department, Faculty of Engineering, Minia University, Minia 61519, Egypt; Bioengmsa@yahoo.com
* Correspondence: Ahmed.Fouly@mu.edu.eg or Amohammed7.c@ksu.edu.sa

Citation: Fouly, A.; Abdo, H.S.; Seikh, A.H.; Alluhydan, K.; Alkhammash, H.I.; Alnaser, I.A.; Abdo, M.S. Evaluation of Mechanical and Tribological Properties of Corn Cob-Reinforced Epoxy-Based Composites—Theoretical and Experimental Study. *Polymers* **2021**, *13*, 4407. https://doi.org/10.3390/polym13244407

Academic Editors: Andrzej Puszka and Beata Podkościelna

Received: 21 November 2021
Accepted: 10 December 2021
Published: 15 December 2021

Abstract: Epoxy is considered to be the most popular polymer and is widely used in various engineering applications. However, environmental considerations require natural materials-based epoxy. This necessity results in further utilization of natural materials as a natural reinforcement for different types of composites. Corn cob is an example of a natural material that can be considered as an agricultural waste. The objective of the present work is to improve the economic feasibility of corn cob by converting the original corn cob material into powder to be utilized in reinforcing epoxy-based composites. In the experiment, the corn cob was crushed and ground using a grain miller before it was characterized by scanning electron microscopy (SEM). The corn cob powder was added to the epoxy with different weight fractions (2, 4, 6, 8, 10 wt%). In order to prevent corn cob powder agglomeration and ensure homogeneous distribution of the reinforcement inside the epoxy, the ultrasonic technique and a mechanical stirrer were used. Then, the composite's chemical compositions were evaluated using X-ray diffraction (XRD). The mechanical experiments showed an improvement in the Young's modulus and compressive yield strength of the epoxy composites, increasing corn cob up to 8 wt% by 21.26% and 22.22%, respectively. Furthermore, tribological tests revealed that reinforcing epoxy with 8 wt% corn cob can decrease the coefficient of friction by 35% and increase wear resistance by 4.8%. A finite element model for the frictional process was constructed to identify different contact stresses and evaluate the load-carrying capacity of the epoxy composites. The finite element model showed agreement with the experimental results. An epoxy containing 8 wt% corn cob demonstrated the optimal mechanical and tribological properties. The rubbed surfaces were investigated by SEM to identify the wear mechanism of different composites.

Keywords: epoxy composite; green composite; corn cob

1. Introduction

Recent global challenges and improvements associated with the utilization of different materials in tribological applications and polymer composites' performance have attracted the attention of many researchers. Polymer composites are lightweight materials used in various applications such as automotive, satellites, marine, aerospace, etc. [1]. Furthermore, polymer composites have many advantages compared with metal materials, including low processing costs, corrosion resistance, lower raw material, and are non-toxic, which allow tremendous design flexibility [2]. Generally, polymer composites have been utilized in different applications and have shown outstanding results when used as a structural material for aerospace components [3]. These results motivated researchers to further investigate the tribological characteristics of polymer composites. Consequently, the tribological properties, including wear rate and friction coefficient of different polymer composites, have been studied to be used in nuts, brakes, bolts, bearing, and clutches [4–6]. To investigate the tribological characteristics, three essential matters should be studied [7]: the interfacial bonds strength; shear stress, which results in the delamination of surface contact area; and the contact area to illustrate the wear mechanism. Furthermore, additional factors can affect the composite performance during friction, such as the applied normal load and sliding velocity.

One of the most popular polymers used in different engineering applications is the epoxy thermoset, which has many advantages that encourage industrialists to use it. Epoxy-based thermosetting resin can be used for its chemical and heat resistance [8], high corrosion barrier [9], and it is an electric isolator, in addition to its high stiffness and the simplicity of its production and processing. Such advantages have nominated epoxy for various applications such as helicopter rotors, industrial flooring, pump impellers, and aircraft engines [10]. Nevertheless, epoxy has its drawbacks, such as its brittle nature, which affects its mechanical characteristics and decreases its wear resistance, preventing its utilization in frictional applications [11,12].

To enhance the properties of epoxy, researchers have tried to add some additives as a reinforcement and investigate the different characteristics of epoxy composites, including mechanical, tribological, electrical, and thermal qualities [13]. The mechanical and thermal properties of epoxy composites after adding block-copolymer (BCP) and core-shell rubber were investigated by Bajpai et al. [14]. They recorded an enhancement in the fracture toughness by 268% for 10 wt% BCP and 200% for 12 wt% CSR. Furthermore, adding 3 wt% BCP and CSR improved the fracture toughness by 100%. Kim et al. [15] investigated the effect of adding different weight fractions of graphene nanoplatelets on the mechanical properties of epoxy thermosets. They found that 4 wt% graphene nanoplatelets enhanced the fracture toughness and shear strength by 142% and 252%, respectively. Upadhyay et al. [16] sought to improve epoxy's mechanical and tribological behavior by adding different weight fractions of nano-graphene. They found that increasing the graphene weight fraction up to 5 wt% enhanced the compressive stress. Furthermore, the specific wear rate decreased by approximately 50%, and the decrease in the friction coefficient reached 60%. Xian et al. [17] investigated the tribological performance of epoxy incorporated with different volume fractions of TiO_2. They found that 4 vol% TiO_2 could considerably decrease the specific wear rate. Utilizing the same additives with varying weight fractions, TiO_2, Srivastava et al. [18] evaluated epoxy's tribological and mechanical performance. The results recorded an enhancement in the epoxy composite wear resistance at 5 wt% in addition to an improvement in both impact strength and toughness. Bazrgari et al. [19] investigated the effect of incorporating Al_2O_3 with different weight fractions into epoxy matrix. They could prove that 1 wt% Al_2O_3 can improve the mechanical and tribological characteristics of the composite. Still, with the increase in Al_2O_3 weight fraction up to 3%, all the composite properties have deteriorated due to the additives' agglomeration inside the matrix. Fouly et al. [20] studied the effect of reinforcing epoxy resin with a low loading fraction of Al_2O_3. They found that a 0.4 wt% Al_2O_3 could improve the epoxy composite's mechanical and tribological properties.

The reinforcements used with different composites can be classified as natural and synthetic additives [1]. Synthetic additives are usually nonbiodegradable, and due to environmental pollution, the utilization of natural additives has attracted the attention of many researchers. Natural additives are easily available in nature at a low cost, are easily manufactured, and their utilization is eco-friendly [21]. Venkateshwaran et al. [22]

evaluated the effect of reinforcing epoxy resin with banana fibers, and they recorded enhancement in tensile strength, flexural strength, and impact strength. Oladele et al. [23] investigated the effect of reinforcing epoxy resin with different weight fractions of eggshell powder. They recorded an enhancement in the tensile modulus, impact resistance, flexural properties, hardness, and wear resistance. Karthick et al. [24] reinforced PMMA resin with different weight fractions of seashell nanopowder and investigated its effect on the tribological properties of the polymer composite. They found that the optimal weight fraction that could enhance the tribological properties of the PMMA composite is 12 wt% seashell powder. Sim et al. [25] investigated the effect of reinforcing epoxy by fly ash on the mechanical properties of epoxy composites. Although the tensile strength was improved to a specific fly ash weight fraction, the compression strength was continuously enhanced as the fly ash weight fraction increased. Other researchers evaluated epoxy's mechanical and wettability characteristics when reinforced by natural fibers [26].

Corn cob is considered one of the most common agricultural wastes, and some researchers have tried to turn it into a useful material. Chen et al. [27] extracted corn cob fibers and reinforced polyethylene. They found that the polyethylene composites recorded sufficient mechanical and weathering performance. Zhu et al. [28] reinforced high-density polyethylene (HDPE) by corn cob powder and investigated the composite water absorption and mechanical properties. The results showed that increasing corncob content led to an increase in the water absorption of the composite. The flexural strengths and moduli were increased up to 40% with increasing corn cob powder, but the mechanical properties started to deteriorate at a high level of powder content.

According to the above literature survey, using natural materials as a reinforcement could affect polymer composites' mechanical and tribological properties. Although many researchers investigated the effect of corn cob as a reinforcement additive with polymers, there is a lack of research on the impact of a corn cob on the tribological properties of polymer composites. Consequently, the objective of the present investigation is to evaluate the performance of epoxy composite after incorporating corn cob powder with different weight fractions, 2, 4, 6, 8, and 10 wt%. The density, hardness, Young's modulus, and compressive yield strength were determined based on the corn cob content. The tribological characteristics are introduced by evaluating the friction coefficient and specific wear rate of different composites. A finite element model was constructed based on the mechanical and tribological results to assess the load-carrying capacity. The wear mechanisms of each composite were identified after evaluating the worn surfaces using scanning electron microscopy (SEM).

2. Materials and Methods

2.1. Preparation of Corn Cob Powder and Epoxy/Corn Cob Composite Samples

The matrix material utilized in the current study was the commercial-grade Kemapoxy 150 (CMB International, Wadi El Natroun City, Egypt) epoxy resin kit consisting of crystal-clear resin, a density of 1.11 $\pm$ 0.02 kg/L crystal-clear epoxy hardener. As mentioned before, corn cob is considered one of the most common agricultural wastes. Corn cob is the backbone part of the corn that the kernels grow around. After removing the kernels, corn cobs were dried under the sun for three months to decrease the moisture. Then, the corn cobs were initially crushed into smaller particles utilizing hammering before transferring the crushed powder to the mono chamber of a grain miller at a speed of 250 rpm for 3 h, with 15 min grinding and 15 min rest. Finally, the achieved powder was dried again in an oven at a temperature of 70 °C for 5 h to ensure the removal of all moisture.

The fabrication of tested samples passed through different stages. First, the corn cob powder was added to ethanol and mixed ultrasonically for 15 min to prevent powder agglomeration. Second, the liquid epoxy was added. To ensure homogenous dispersion of the corn cob powder inside the liquid epoxy, mixing using a mechanical stirrer was applied to blend the whole mixture for 30 min at 150 rpm. Thirdly, the epoxy/ethanol/corn cob mixture was inserted into a vacuum chamber to remove any voids and to fumigate the ethanol for 24 h. Fourthly, the hardener was added to the mix in a ratio of 1:2 to the epoxy weight and mixed again for 5 min. Eventually, the mixture was discharged into a mold taking the shape of needed specimens and placed again in a vacuum at 35 oC. A schematic diagram illustrating the fabrication process is shown in Figure 1. Based on the epoxy supplier instructions, the fabricated samples were left for 7 days before applying any test to achieve the recommended mechanical characteristics. The epoxy/corn cob composites were prepared by adding different weight fractions of corn cob. Table 1 illustrates the weight percentage of each material and the code of each sample.

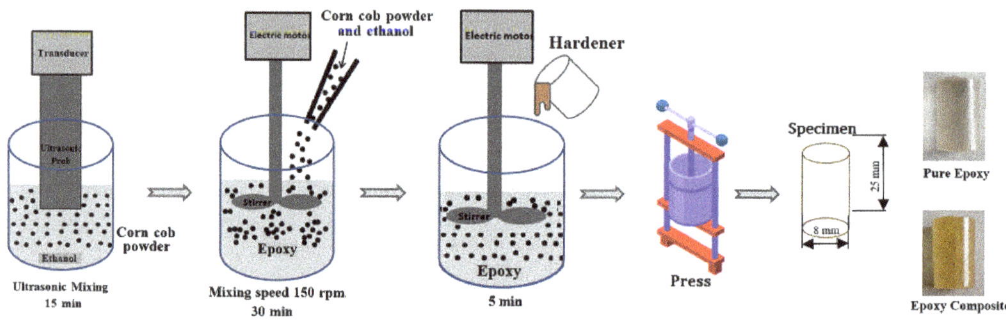

Figure 1. Schematic diagram of the epoxy composite samples' fabrication.

Table 1. Composition of epoxy/corn cob composite samples.

Code	Corn Cob wt%	Epoxy wt%	Hardener wt%
ECC0	0	66.67	33.33
ECC2	2	65.33	32.67
ECC4	4	64	32
ECC6	6	62.67	31.33
ECC8	8	61.33	30.67
ECC10	10	60	30

2.2. Mechanical Properties of Epoxy/Corn Cob Composites

The mechanical characteristics of the fabricated epoxy/corn cob composites were evaluated to investigate the effect of adding corncob powder with different weight fractions. According to ASTM D2240 [29], the D-index's shore hardness was evaluated for each composite sample. The utilized durometer specifications were 15 s dwell time with a 5 ± 0.5 kg loading. To ensure the hardness of the produced composite, measuring the hardness was conducted on different regions along the surfaces of the specimens, and the average value and standard errors were calculated. According to ISO 604:2002 Plastics [30], to apply the compression test, epoxy/corn cob samples were fabricated with a length and diameter of 16 mm and 8 mm, respectively. A 30-ton computer-controlled universal testing machine was utilized to apply the compression test with a 2 mm/min strain rate. Based on the compression test results, the elastic modulus and compressive yield strength were evaluated.

2.3. Tribological Properties of Epoxy/Corn Cob Composites

The tribological performance of epoxy/corn cob composites was evaluated in dry conditions according to ASTM G99-95 [31]. A pin-on-disc tribometer was utilized at room temperature 24:27 °C and humidity around 60%, as shown in Figure 2. The epoxy/corn cob sample was considered the tribometer pin with a cylindrical shape with a diameter and length of 8 mm and 25 mm, respectively. The composite samples were rubbed against a stainless-steel disk with a diameter of 18 cm and surface roughness of 12.5 μm. To prevent errors in the tribological results, the stainless-steel disk was cleaned and dried before every experiment to remove any remaining contaminants. The epoxy composite tribological performance was evaluated under normal loads, and sliding distances changed while the sliding speed was constant at 0.4 m/s. For each load, sliding distance, and composite composition, the friction test was repeated 6 times before calculating the specific wear rate and friction coefficient. The specific wear rate (SWR) was calculated using Equation (1), dependent on sliding distances (L), the difference in weight before and after the test (Δm), the density of each composite sample measured by Archimedes' principle (ρ), and the applied normal load (F_n).

$$SWR = \frac{\Delta m}{\rho \, L \, F_n} \tag{1}$$

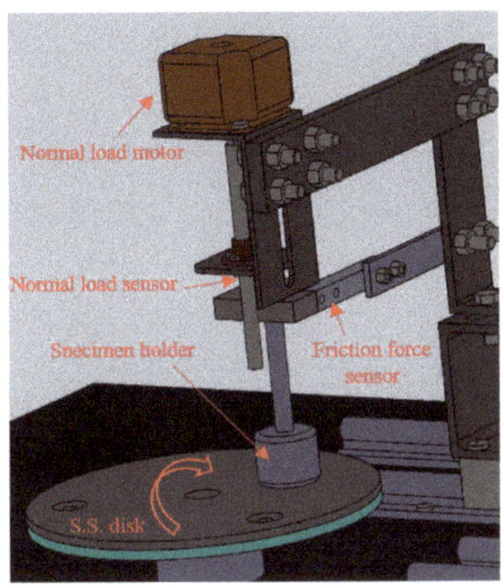

Figure 2. Pin-on-disk tribometer.

3. Results and Discussion

After the grain milling process, the corn cob particles' morphology was characterized utilizing a scanning electron microscope (SEM). As shown in Figure 3, the produced corn cob powder had an irregular, rugged shape with a rough surface and some large particles covered with agglomerates. Furthermore, some particles have much smaller sizes and are agglomerated, as shown the red circle.

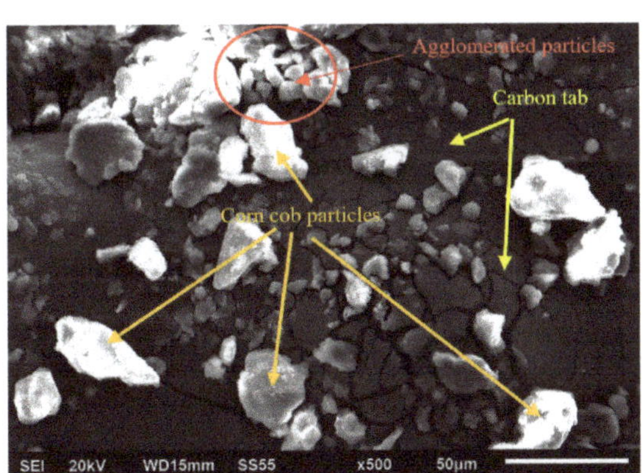

Figure 3. SEM image of corn cob particles after the grain milling process.

To investigate the effect of incorporating corn cob powder into epoxy resin on the chemical composition of the composites, X-ray diffraction (XRD) was utilized. Figure 4 shows that the carbon peak is the major peak of pure epoxy, and it obviously appears in all epoxy composites with different weight fractions of corn cob. In addition, the XRD shows that the main peak of epoxy is not sharp, which illustrates its amorphous nature [32]. There is also one major peak of corn cob powder at 21.8°, similar to the XRD pattern of alpha cellulose [33]. Incorporating corn cob powder in epoxy resin affected the main peak of epoxy, as shown in the yellow rectangle in Figure 3. However, the disappearance of new peaks indicates that corn cob incorporation did not affect the epoxy structural characteristics, and no chemical reaction occurred between the epoxy and corn cob. The increasing weight fraction of corn cob did not affect the amorphous nature of the epoxy, and the appearance of the corn cob peak with the same effect in all composites emphasizes its thorough distribution inside the epoxy resin.

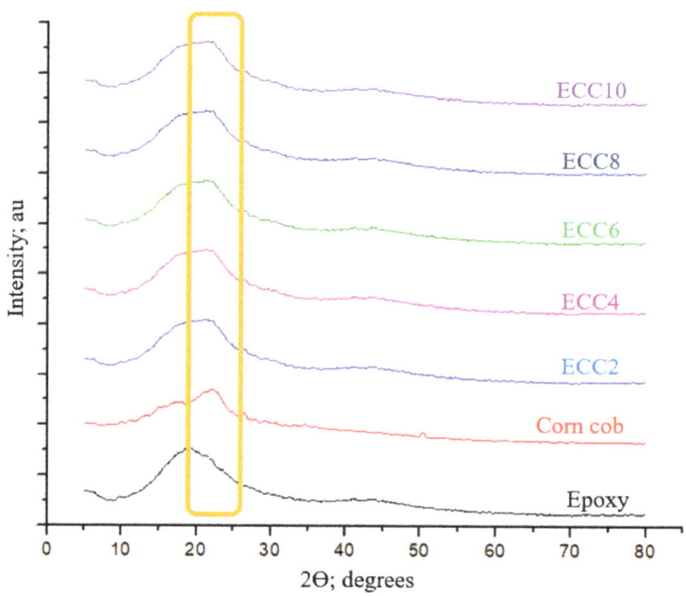

Figure 4. X-ray diffraction (XRD) traces of neat epoxy, corn cob powder, and epoxy composites.

The density of the produced composite is one of the most important properties that can affect the weight of the new generated composite. The density of composite materials usually depends on the density of both the matrix and reinforcement materials. Consequently, the current study used Archimedes' principle to calculate the density of the fabricated epoxy/corn cob composites. Figure 5 illustrates the variation in the epoxy composite density due to the corn cob weight fraction change. The density was found to be reduced as the corn cob weight fraction increased. Increasing the corn cob content to 10 wt% of the epoxy matrix decreased the composite density by 22.87%. This decrease in density can lead to a lighter polymer and support the advantages of polymer utilization in different applications. The reduction in the epoxy composite could be attributed to replacing wt% of the epoxy by a wt% of a lower density material, corn cob.

In the same figure, Figure 5, the effect of incorporating corn cob on the epoxy composites hardness is illustrated. It was evident that increasing corn cob weight fraction led to a decrease in the hardness of the composites. The average shore-D hardness of pure epoxy was 68. Increasing the corn cob up to 10 wt% decreased the hardness by 17.24% to reach 58 on the D-index scale. Due to the increased weight fraction of the corn cob, the density of the composite samples decreased; as a result, a reduction in hardness might occur [34].

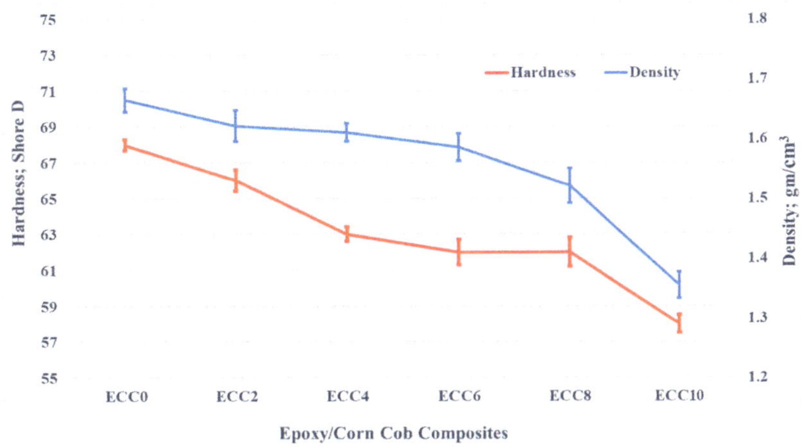

Figure 5. Density and shore-D hardness of epoxy/corn cob composites.

The experimental results of the compression test of epoxy corn cob composites are shown in Figure 6. It was noticed that increasing corn cob content contributed to the improvement of the epoxy composites' Young's modulus along with the compressive yield strength. The epoxy composite consisting of 8 wt% corn cob was the optimum one concerning the mechanical strength. Young's modulus and compressive yield strength enhancement reached 21.26% and 22.22%, respectively. The improvement in the compressive yield strength might be due to the thorough distribution of corn cob particles inside the epoxy matrix that helped the matrix transfer the compressive load to the corn cob powder, which dissipated it.

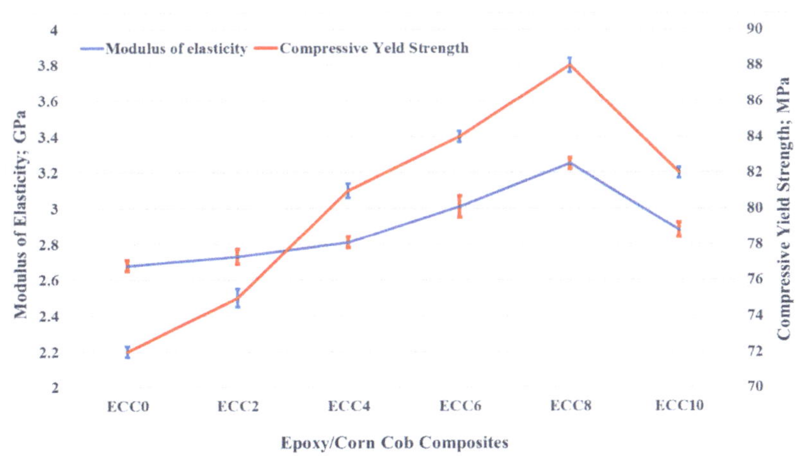

Figure 6. Young's modulus and compressive yield strength of epoxy/corn cob composites.

On the other hand, incorporating 10 wt% corn cob resulted in the deterioration of Young's modulus and the compressive yield strength. This result could be attributed to the ease of debonding corn cob particles from the epoxy matrix because of the high corn cob weight fraction. Consequently, the load transmission from the epoxy to the corn cob particles failed; hence the mechanical characteristics diminished.

The variation in the friction coefficient of neat epoxy and 2–10 wt% corn cob particles-reinforced composite under different normal loads (3–12 N) is illustrated in Figure 7.

Figure 7 shows that the incorporation of corn cob particles in the epoxy resin decreased the coefficient of friction with the variation in the normal loads. The results show that 8 wt% corn cob recorded the minimum friction coefficient (0.39), 35% less than that for pure epoxy (0.6) at 3 N normal load. However, adding more particles of corn cob to reach 10 wt% increased the friction coefficient. More analysis is offered later in the current study to understand why the friction coefficient increased when corn cob weight fraction reached 10 wt%. On the other hand, Figure 7 showed that increasing the normal load led to an increase in the friction coefficient. This might have happened because of the thermal effect caused by the high normal load [35], in which the contact area may be changed, leading to a change in the adhesion between the tested specimens and the stainless steel disk. Chang et al. [36] proved that increasing the contact temperature increases the friction coefficient.

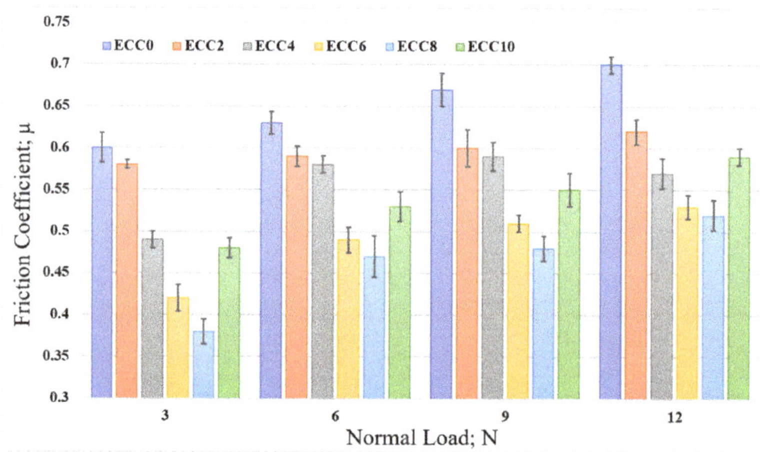

Figure 7. The friction coefficient of epoxy/corn cob composites under different normal loads.

Figure 8 shows the influence of the sliding distance on the friction coefficient of epoxy/corn cob composites. The sliding distances were varied from 120 m to 480 m under a normal load of 12 N. Generally, the friction coefficient decreased with the increase in the sliding distance, while 8 wt% corn cob still recorded the lowest coefficient of friction among other composites. The decrease in the friction coefficient could be attributed to the smoothness in the epoxy composite surface after rubbing for a long distance against the stainless steel disk. Furthermore, Dass et al. [10] reported that after rubbing polymeric samples for a long time, the generated heat could melt the surface of the composite. Consequently, a thin film of the melted polymer composite is transmitted to the counterpart. It works as a lubrication layer between the rubbed surfaces, resulting in a decrease in the friction coefficient. To conclude, the reduction in the epoxy/corn cob coefficient of friction stands for an enhancement in the epoxy composite load-carrying capacity due to the existence of corn cob particles. Kuminek et al. [37] claimed that the contact stress on the surface of rubbed samples could indicate the load-carrying capacity. Therefore, in the current study, the pin on the disk tribometer was modeled utilizing ANSYS software, as shown in Figure 9, and the stresses due to the frictional process were reported.

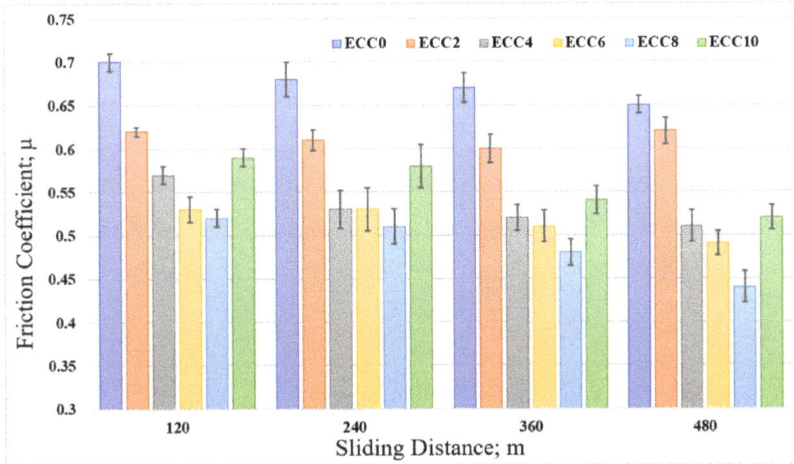

Figure 8. Friction coefficient of epoxy/corn cob composites with different sliding distances.

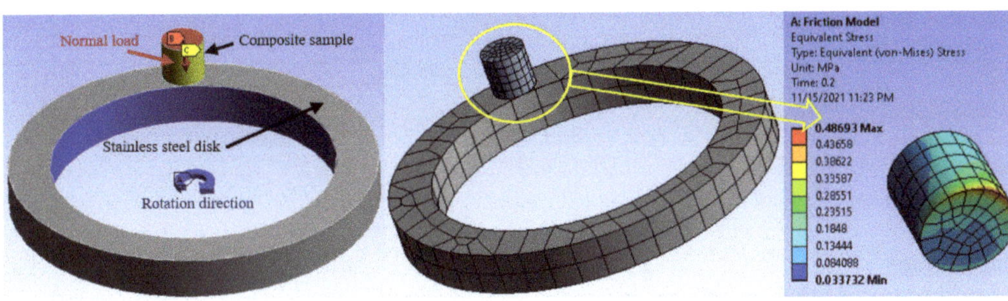

Figure 9. Finite element model of the frictional process with boundary condition, meshing, and composite sample after friction.

Figure 9 illustrates the finite element model for the frictional process. The disk was modeled as a hollow cylinder with a stainless steel specification, and the epoxy composite was modeled as a pin. The measured mechanical properties of composite samples and the measured friction coefficient between the composite samples and the stainless-steel disk were inserted into the ANSYS software. In order to measure different stresses during the rubbing process, frictional contact was defined between the pin and the disk. Normal Lagrange contact solution is commonly used when normal pressure is applied in the z-direction. In addition, output results are dependent on samples stiffness. In the present work, the normal force was solved explicitly and was added as an additional degree of freedom applied in the z-direction with 12 N. The automatic meshing provided by ANSYS software was utilized both the pin and the disk, as shown in the middle of Figure 9. After the automatic meshing, the generated grids for both the pin and the disk were a combination of tetrahedron and hexahedron shapes. The disk's elements and nodes are 214 and 1623, respectively. The pin's elements and nodes are 112 and 637, respectively. The boundary conditions for the epoxy composite pin were a fixation in the x and y direction besides a 12 N normal load on the sample surface at the z-direction. A 300 rpm speed was applied to the rotational joint of the stainless steel disk. The stress on the surface of the epoxy composite surfaces after the frictional process is illustrated in Figure 9. It was evident that the motion direction generates concentrated stress on the sample edge, where the deterioration of the sample surface may begin.

The finite element results show the generated equivalent and shear stresses variation due to the friction between different epoxy composite samples and the stainless steel disk, as shown in Figure 10. It was evident that the finite element results were affected by the mechanical properties of the fabricated composites. As the corn cob content increased, an enhancement in the samples strength was recorded, which is reflected in the finite element results as a decrease in both the equivalent and shear stresses. This decrease in the contact stresses corresponds to an improvement in the load-carrying capacity that caused a reduction in the friction coefficient [38]. Marra et al. [39] claimed that the decrease in the shear stress between contact elements could reduce the friction coefficient.

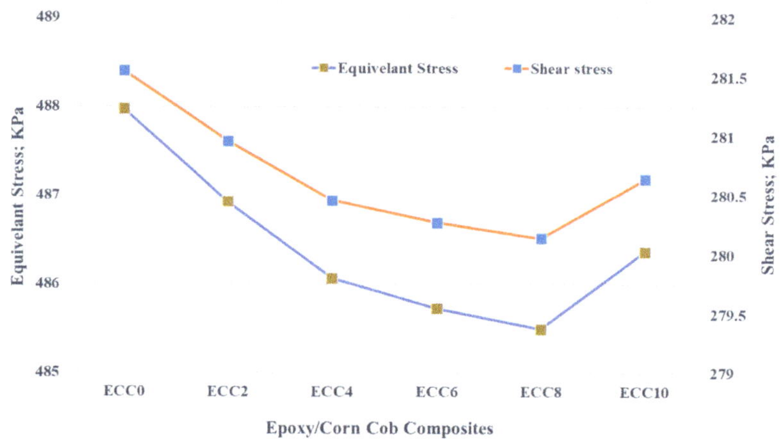

Figure 10. Contact stresses on the surface of the epoxy/corn cob composites.

Figure 11, containing finite element results, shows that the change in frictional stress generated during the rubbing process is similar to the equivalent and shear stress changes. The contact stresses can be the main reason for the deterioration of the composite samples. Therefore, wear layer thickness was recorded during the simulation of the frictional process. The results show that the wear layer thickness decreases with corn cob weight fraction. These results could also be attributed to the enhancement in the composite strength recorded during the evaluation of the mechanical characterization.

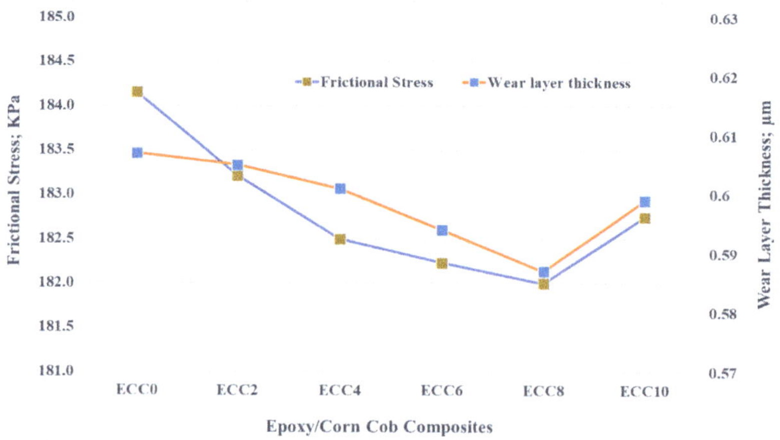

Figure 11. Frictional stress and wear layer thickness on the surface of the epoxy/corn cob composites.

To validate the finite element results and evaluate the influence of different corn cob weight fractions on the tribological performance of epoxy composites, the specific wear rate of the epoxy composites under different normal loads was calculated. Figure 12 shows that incorporating corn cob particles in the epoxy resin decreased the specific wear rate. According to the results, 8 wt% corn cob recorded the minimum specific wear rate. Nevertheless, the extra addition of corn cob to reach 10 wt% increased the specific wear rate. The increase in wear rate to more than 8 wt% corn cob could be attributed to the agglomeration of corn cob particles due to high weight percent [25]. Furthermore, the high reinforcement weight ratio and the poor interfacial bonding between the epoxy and the corn cob particles resulted in a decrease in the wear resistance of the epoxy composite. Thus, results indicate an agreement among the finite element results, wear layer thickness, and the experimental results with the specific wear rate. On the other hand, increasing the normal load led to an increase in the specific wear rate. During the rubbing process, some abrasive particles detached from the surface of the composite samples. With high loading, these particles were pushed against the sample surface and increased the frictional force. The increase in the frictional force may deteriorate the sample's surface, and, consequently, increase the specific wear rate. Figure 13 illustrates the change in the specific wear rate of epoxy/corn cob composites after sliding for different distances. A limited gradual decrease in the specific wear rate was recorded for all epoxy composites.

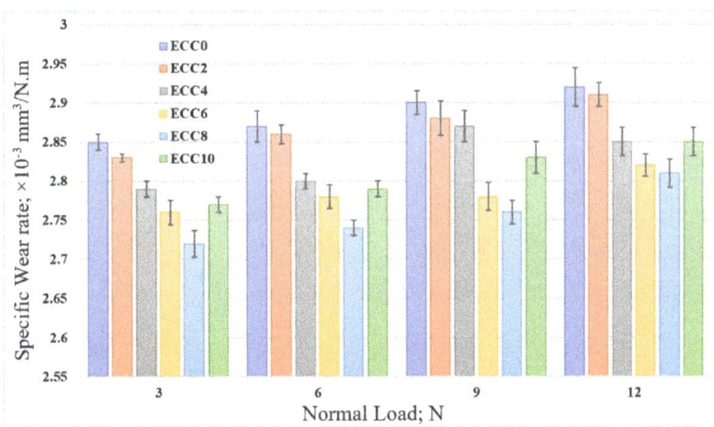

Figure 12. The specific wear rate of epoxy/corn cob composites under different normal loads.

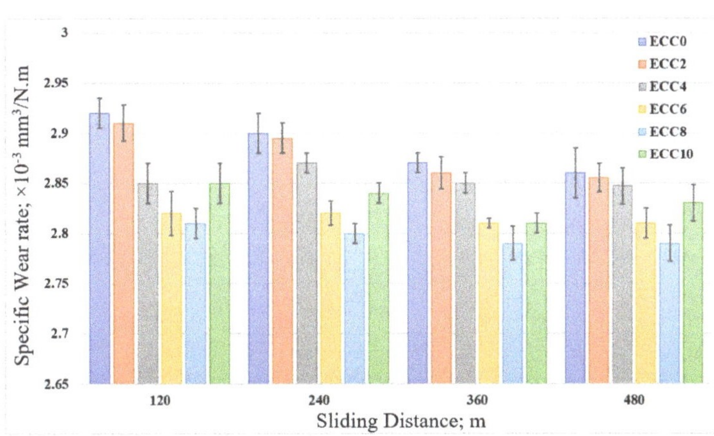

Figure 13. The specific wear rate of epoxy/corn cob composites with different sliding distances.

The microscopic analysis of epoxy composites rubbed surfaces were utilized to identify the wear mechanism during the friction process, as shown in Figure 14. The high shear stress besides the plowing action resulted in a degradation of the rubbed surface of the pure epoxy. Therefore, Figure 14a shows some eliminated layers and severe insensitive damage on the pure epoxy surface. The illustrated deterioration in the epoxy surface explains the cause of the increased specific wear rate. Furthermore, the weak separated layers increased the shear resistance; thus, the measured friction coefficient increased [40]. The rubbed surface depicts diversified wear mechanisms such as micro-cutting; its propagation resulted in the debonding of some material from the epoxy sample surface. Therefore, the primary wear mechanism could be a delamination mechanism, which usually encourages the increase in both specific wear mechanism and the friction coefficient [41]. Incorporating 2 wt% corn cob into the epoxy matrix enhanced the epoxy composite's thrust-bearing capacity, leading to fewer eliminated layers with a clear plowing surface and the initiation of microcracks, as shown in Figure 14b. However, as the corn cob content is low, still eliminated layers of the epoxy appeared on the surface of the rubbed sample. Compared to pure epoxy, adding 2 wt% corn cob enhanced the strength of the composite surface and decreased the specific wear rate and the change in the wear mechanism to be a fatigue-delamination due to the appearance of the micro cracks. The addition of 4 wt% corn cob to the epoxy matrix led to more enhancement and the eliminated layers decreased again, as shown in Figure 14c. In the case of 6 wt% corn cob, Figure 14d, the eliminated layers retracted again, indicating an enhancement in the composite sample's strength. The rubbed surface becomes smoother with micro-plowing, as a wear mechanism, and the clear view on the surface is the appearance of the microcracks. Figure 14e illustrates the worn surface of the epoxy composite after adding 8 wt% corn cob. The delaminated layers disappeared, and only some micro-cracks appeared on the surface of the composite sample. The increase in the presence of the corn cob provided improved strength to resist the generated thrust force on the epoxy composite surface during the friction process. Consequently, an enhancement in the wear resistance was achieved that indicates perfect interfacial adhesion between the corn cob particles and the epoxy, which encourages the epoxy matrix to share generated thrust force and shear stress on the sample surface. Therefore, an enhancement in the friction coefficient and specific wear rate was achieved for 8 wt%. Lastly, Figure 14f shows the worn surface of epoxy composite with 10 wt% corn cob. As mentioned before, studies proved that the high reinforcement weight ratio and the poor interfacial bonding among the composite composition reduce the capacity of the corn cob particles to support the epoxy surface against applied stress. Consequently, the wear resistance decreased, resulting in a surface debonding and severe damage. Furthermore, separated layers and particles increased plowing action, and shear stress increased the shear resistance; thus, the measured friction coefficient increased.

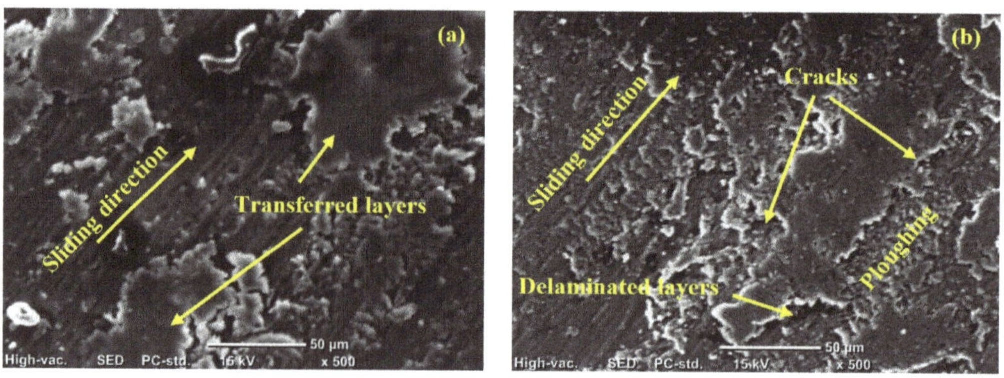

Figure 14. *Cont.*

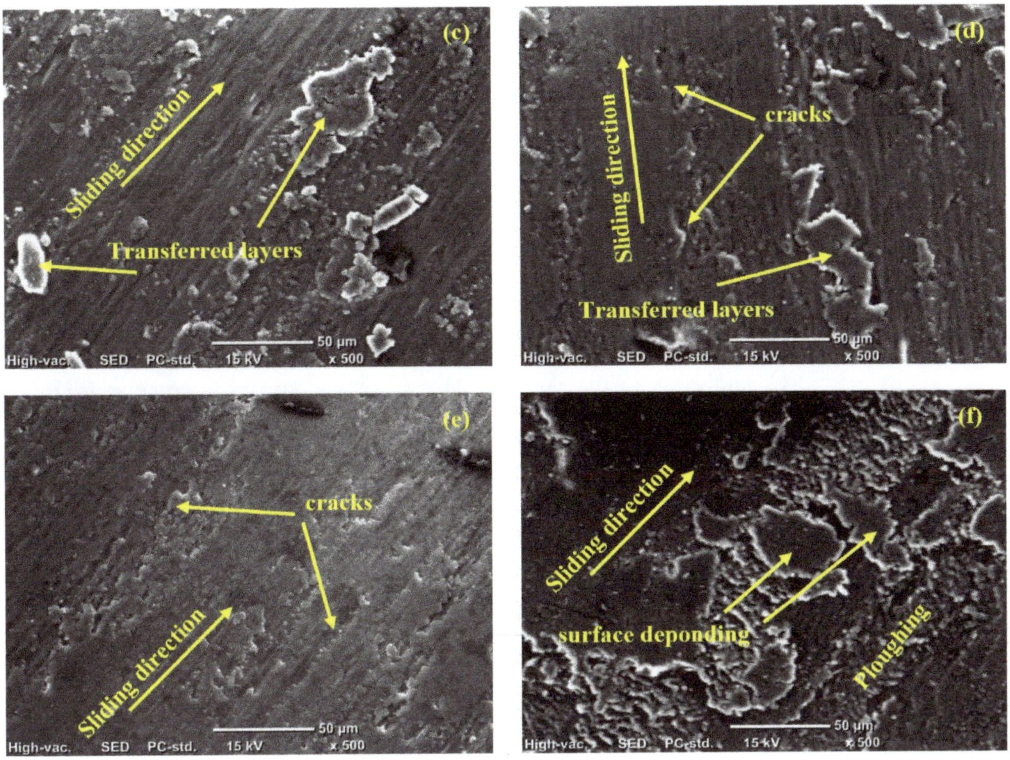

Figure 14. SEM of worn epoxy composite surfaces (**a**) ECC0, (**b**) ECC2, (**c**) ECC4, (**d**) ECC6, (**e**) ECC8, and (**f**) ECC10.

4. Conclusions

In conclusion, we conducted an experimental investigation on reinforcing epoxy-based polymer using corn cob, a natural material. The physical, mechanical, and tribological characteristics of the fabricated epoxy/corn cob composites were evaluated at different corn cob weight fractions, and the following conclusions are drawn:

- The results revealed that corn cob could be used as a natural reinforcement for polymer-based composites.
- Without using an agent to enhance the bonding between the epoxy and the corn cob, the weight fraction of the corn cob is limited to 8 wt%.
- Increasing the corn cob weight fraction to 10 wt% decreased both density and hardness of the composite by 22.87% and 17.24%, respectively.
- For 8 wt% corn cob, Young's modulus and compressive yield strength of the epoxy composite reached 21.26% and 22.22%, respectively.
- For 8 wt% epoxy/corn cob, the composite recorded a frictional coefficient of 35% less than pure epoxy, as well as enhanced wear resistance.
- The finite element results showed an enhanced load-carrying capacity due to corn cob incorporation.

Finally, corn cob particles could be modified chemically to improve the interfacial bond between the epoxy and corn cob particles and to reinvestigate the new composite characteristics.

Author Contributions: Data curing, A.F. and H.S.A.; methodology, H.S.A. and K.A.; software, A.F. and M.S.A.; formal analysis, A.F. and K.A.; investigation, A.F., H.S.A., I.A.A. and M.S.A.; resources, A.H.S. and H.I.A.; writing—original draft preparation, A.F.; writing—review and editing, K.A. and M.S.A.; visualization, A.H.S., I.A.A. and K.A.; supervision, A.F.; project administration, H.S.A.; funding acquisition, A.F. and H.I.A. All authors have read and agreed to the published version of the manuscript.

Funding: This research was funded and supported by Taif University Researchers Supporting Project number (TURSP-2020/264), Taif University, Taif, Saudi Arabia; and King Saud University, Deanship of Scientific Research, College of Engineering Research Center.

Institutional Review Board Statement: Not applicable.

Informed Consent Statement: Not applicable.

Data Availability Statement: Not applicable.

Acknowledgments: The authors would like to acknowledge the support from Taif University Researchers Supporting Project number (TURSP-2020/264), Taif University, Taif, Saudi Arabia, and they extend their sincere appreciation to the King Saud University, Deanship of Scientific Research, College of Engineering Research Center for funding this research.

Conflicts of Interest: The authors declare no conflict of interest.

References

1. Chavali, P.J.; Taru, G.B. Effect of Fiber Orientation on Mechanical and Tribological Properties of Banana-Reinforced Composites. *J. Fail. Anal. Prev.* **2021**, *21*, 1–8. [CrossRef]
2. Jaramillo, A.; Medina, C.; Flores, P.; Canales, C.; Maldonado, C.; Rivera, P.C.; Rojas, D.; Meléndrez, M.; Castaño, P. Improvement of thermomechanical properties of composite based on hydroxyapatite functionalized with alkylsilanes in epoxy matrix. *Ceram. Int.* **2020**, *46*, 8368–8378. [CrossRef]
3. Fouly, A.; Ibrahim, A.M.M.; Sherif, E.-S.M.; FathEl-Bab, A.M.R.; Badran, A.H. Effect of Low Hydroxyapatite Loading Fraction on the Mechanical and Tribological Characteristics of Poly (Methyl Methacrylate) Nanocomposites for Dentures. *Polymers* **2021**, *13*, 857. [CrossRef]
4. Yousif, B.F. Design of newly fabricated tribological machine for wear and frictional experiments under dry/wet condition. *Mater. Des.* **2013**, *48*, 2–13. [CrossRef]
5. Bajpai, P.K.; Singh, I.; Madaan, J. Tribological behavior of natural fiber reinforced PLA composites. *Wear* **2013**, *297*, 829–840. [CrossRef]
6. Ibrahim, A.M.M.; Mohamed, A.F.A.; Fathelbab, A.M.; Essa, F.A. Enhancing the tribological performance of epoxy composites utilizing carbon nano fibers additives for journal bearings. *Mater. Res. Express* **2018**, *6*, 035307. [CrossRef]
7. Farfan-Cabrera, L.I.; Tapia-Gaspar, M.; Pérez-González, J. *Tribology of Polymer Matrix Composites within the Automotive Industry*; Elsevier: Amsterdam, The Netherlands, 2021.
8. Kim, K.; Kim, M.; Hwang, Y.; Kim, J. Chemically modified boron nitride-epoxy terminated dimethylsiloxane composite for improving the thermal conductivity. *Ceram. Int.* **2014**, *40*, 2047–2056. [CrossRef]
9. Rusly SN, A.; Ismail, I.; Matori, K.A.; Abbas, Z.; Shaari, A.H.; Awang, Z.; Ibrahim, I.R.; Idris, F.M.; Zaid, M.H.M.; Mahmood, M.K.A.; et al. Influence of different BFO filler content on microwave absorption performances in BiFeO3/epoxy resin composites. *Ceram. Int.* **2020**, *46*, 737–746. [CrossRef]
10. Dass, K.; Chauhan, S.R.; Gaur, B.V. Study on the effects of nano-aluminum-oxide particulates on mechanical and tribological characteristics of chopped carbon fiber reinforced epoxy composites. *Proc. Inst. Mech. Eng. Part J. Mater. Des. Appl.* **2017**, *231*, 403–422. [CrossRef]
11. Che, Y.; Sun, Z.; Zhan, R.; Wang, S.; Zhou, S.; Huang, J. Effects of graphene oxide sheets-zirconia spheres nanohybrids on mechanical, thermal and tribological performances of epoxy composites. *Ceram. Int.* **2018**, *44*, 18067–18077. [CrossRef]
12. Fouly, A.; Ibrahim, A.M.M.; El-Bab, A.M. Promoting the tribological Properties of epoxy Composites via using Graphene Nanoplatelets as a functional Additive. *KGK-Kautsch Gummi Kunstst.* **2020**, *73*, 25–32.
13. Mostovoy, A.; Yakovlev, A.; Tseluikin, V.; Lopukhova, M. Epoxy nanocomposites reinforced with functionalized carbon nanotubes. *Polymers* **2020**, *12*, 1816. [CrossRef]
14. Bajpai, A.; Wetzel, B.; Klingler, A.; Friedrich, K. Mechanical properties and fracture behavior of high-performance epoxy nanocomposites modified with block polymer and core–shell rubber particles. *J. Appl. Polym. Sci.* **2020**, *137*, 48471. [CrossRef]
15. Kim, J.; Cha, J.; Chung, B.; Ryu, S.; Hong, S.H. Fabrication and mechanical properties of carbon fiber/epoxy nanocomposites containing high loadings of noncovalently functionalized graphene nanoplatelets. *Compos. Sci. Technol.* **2020**, *192*, 108101. [CrossRef]
16. Upadhyay, R.K.; Kumar, A. Effect of particle weight concentration on the lubrication properties of graphene based epoxy composites. *Colloid Interface Sci. Commun.* **2019**, *33*, 100206. [CrossRef]

17. Xian, G.; Walter, R.; Haupert, F. Friction and wear of epoxy/TiO2 nanocomposites: Influence of additional short carbon fibers, Aramid and PTFE particles. *Compos. Sci. Technol.* **2006**, *66*, 3199–3209. [CrossRef]
18. Srivastava, S.; Tiwari, R.K. Synthesis of epoxy-TiO2 nanocomposites: A study on sliding wear behavior, thermal and mechanical properties. *Int. J. Polym. Mater.* **2012**, *61*, 999–1010. [CrossRef]
19. Bazrgari, D.; Moztarzadeh, F.; Sabbagh-Alvani, A.A.; Rasoulianboroujeni, M.; Tahriri, M.; Tayebi, L. Mechanical properties and tribological performance of epoxy/Al2O3 nanocomposite. *Ceram. Int.* **2018**, *44*, 1220–1224. [CrossRef]
20. Fouly, A.; Alkalla, M.G. Effect of low nanosized alumina loading fraction on the physicomechanical and tribological behavior of epoxy. *Tribol. Int.* **2020**, *152*, 106550. [CrossRef]
21. Daramola, O.O.; Akinwekomi, A.D.; Adediran, A.A.; Akindote-White, O.; Sadiku, E.R. Mechanical performance and water uptake behaviour of treated bamboo fibre-reinforced high-density polyethylene composites. *Heliyon* **2019**, *5*, e02028. [CrossRef]
22. Venkateshwaran, N.; ElayaPerumal, A.; Alavudeen, A.; Thiruchitrambalam, M. Mechanical and water absorption behaviour of banana/sisal reinforced hybrid composites. *Mater. Des.* **2011**, *32*, 4017–4021. [CrossRef]
23. Oladele, I.O.; Makinde-Isola, B.A.; Adediran, A.A.; Oladejo, M.O.; Owa, A.F.; Olayanju, T.M.A. Mechanical and wear behaviour of pulverised poultry eggshell/sisal fiber hybrid reinforced epoxy composites. *Mater. Res. Express* **2020**, *7*, 045304. [CrossRef]
24. Karthick, R.; Sirisha, P.; Sankar, M.R. Mechanical and tribological properties of PMMA-sea shell based biocomposite for dental application. *Procedia Mater. Sci.* **2014**, *6*, 1989–2000. [CrossRef]
25. Sim, J.; Kang, Y.; Kim, B.J.; Park, Y.H.; Lee, Y.C. Preparation of fly ash/epoxy composites and its effects on mechanical properties. *Polymers* **2020**, *12*, 79. [CrossRef]
26. Atmakuri, A.; Palevicius, A.; Siddabathula, M.; Vilkauskas, A.; Janusas, G. Analysis of mechanical and wettability properties of natural fiber-reinforced epoxy hybrid composites. *Polymers* **2020**, *12*, 2827. [CrossRef]
27. Chen, B.; Cai, D.; Luo, Z.; Chen, C.; Zhang, C.; Qin, P.; Cao, H.; Tan, T. Corncob residual reinforced polyethylene composites considering the biorefinery process and the enhancement of performance. *J. Clean. Prod.* **2018**, *198*, 452–462. [CrossRef]
28. Zhu, S.; Guo, Y.; Tu, D.; Chen, Y.; Liu, S.; Li, W.; Wang, L. Water absorption, mechanical, and crystallization properties of high-density polyethylene filled with corncob powder. *BioResources* **2018**, *13*, 3778–3792. [CrossRef]
29. Zhao, H.; Allanson, D.; Ren, X.J. Use of shore hardness tests for in-process properties estimation/monitoring of silicone rubbers. *J. Mater. Sci. Chem. Eng.* **2015**, *3*, 142–147. [CrossRef]
30. BS EN ISO. *14126: 1999-Fibre-Reinforced Plastic Composites. Determination of Compressive Properties in the In-Plane Direction*; International Organization for Standardization: Geneva, Switzerland, 1999.
31. ASTM. *G99, Standard Test Method for Wear Testing with a Pin-on-Disk Apparatus*; ASTM International: West Conshohocken, PA, USA, 2008.
32. Sharmila, T.B.; Antony, J.V.; Jayakrishnan, M.P.; Beegum, P.S.; Thachil, E.T. Mechanical, thermal and dielectric properties of hybrid composites of epoxy and reduced graphene oxide/iron oxide. *Mater. Des.* **2016**, *90*, 66–75. [CrossRef]
33. Yu, J.; Qian, Q.; Zhao, X. Research progress on effects of structural planes of rock mass on stress wave propagation law. *Acta Armamentarii* **2009**, *30*, 308–316.
34. Guo, R.Q.; Rohatgi, P.K.; Nath, D. Preparation of aluminium-fly ash particulate composite by powder metallurgy technique. *J. Mater. Sci.* **1997**, *32*, 3971–3974. [CrossRef]
35. Khun, N.W.; Zhang, H.; Lim, L.H.; Yue, C.Y.; Hu, X.; Yang, J. Tribological properties of short carbon fibers reinforced epoxy composites. *Friction* **2014**, *2*, 226–239. [CrossRef]
36. Chang, L.; Zhang, Z.; Zhang, H.; Friedrich, K. Effect of nanoparticles on the tribological behaviour of short carbon fibre reinforced poly (etherimide) composites. *Tribol. Int.* **2005**, *38*, 966–973. [CrossRef]
37. Kuminek, T.; Aniołek, K.; Młyńczak, J. A numerical analysis of the contact stress distribution and physical modelling of abrasive wear in the tram wheel-frog system. *Wear* **2015**, *328*, 177–185. [CrossRef]
38. Tang, W.; Zhou, Y.; Zhu, H.; Yang, H. The effect of surface texturing on reducing the friction and wear of steel under lubricated sliding contact. *Appl. Surf. Sci.* **2013**, *273*, 199–204. [CrossRef]
39. Marra, K.G.; Szem, J.W.; Kumta, P.N.; DiMilla, P.A.; Weiss, L.E. In vitro analysis of biodegradable polymer blend/hydroxyapatite composites for bone tissue engineering. *J. Biomed. Mater. Res. Off. J. Soc. Biomater. Jpn. Soc. Biomater. Aust. Soc. Biomater. Korean Soc. Biomater.* **1999**, *47*, 324–335. [CrossRef]
40. Fouly, A.; Almotairy, S.M.; Aijaz, M.O.; Alharbi, H.F.; Abdo, H.S. Balanced Mechanical and Tribological Performance of High-Frequency-Sintered Al-SiC Achieved via Innovative Milling Route—Experimental and Theoretical Study. *Crystals* **2021**, *11*, 700. [CrossRef]
41. Ibrahim, A.M.M.; Shi, X.; Radwan, A.R.; Mohamed, A.F.A.; Ezzat, M.F. Enhancing the tribological properties of NiAl based nano-composites for aerospace bearing applications. *Mater. Res. Express* **2019**, *6*, 085067. [CrossRef]

polymers

MDPI

Synthesis and Characterization of Polyurethanes from Residual Palm Oil with High Poly-Unsaturated Fatty Acid Oils as Additive

Javier Chavarro Gomez [1], Rabitah Zakaria [1,*], Min Min Aung [2], Mohd Noriznan Mokhtar [1] and Robiah Yunus [3]

[1] Department of Process and Food Engineering, Faculty of Engineering, Universiti Putra Malaysia, UPM, Serdang 43400, Selangor, Malaysia; chagoja@hotmail.com (J.C.G.); noriznan@upm.edu.my (M.N.M.)
[2] Institute of Tropical Forestry and Forest Products, University Putra Malaysia, UPM, Serdang 43400, Selangor, Malaysia; minmin_aung@upm.edu.my
[3] Department of Chemical and Environmental Engineering, Faculty of Engineering, Universiti Putra Malaysia, UPM, Serdang 43400, Selangor, Malaysia; robiah@upm.edu.my
* Correspondence: rabitah@upm.edu.my; Tel.: +60-3-89464301; Fax: +60-3-89464440

Abstract: In the effort to produce renewable and biodegradable polymers, more studies are being undertaken to explore environmentally friendly sources to replace petroleum-based sources. The oil palm industry is not only the biggest vegetable-oil producer from crops but also one the biggest producers of residual oil that cannot be used for edible purposes due to its low quality. In this paper the development of biopolymers from residual palm oil, residual palm oil with 10% jatropha oil, and residual palm oil with 10% algae oil as additives were explored. Polyols from the different oils were prepared by epoxydation with peroxyacetic acid and alcoholysis under the same conditions and further reacted with poly isocyanate to form polyurethanes. Epoxidized oils, polyols and polyurethanes were analyzed by different techniques such as TGA, DSC, DMA, FTIR and H-NMR. Overall, although the IV of algae oil is slightly higher than that of jatropha oil, the usage of algae oil as additive into the residual palm oil was shown to significantly increase the hard segments and thermal stability of the bio polyurethane compared to the polymer with jatropha oil. Furthermore, when algae oil was mixed with the residual palm oil, it was possible to identify phosphate groups in the polyol which might enhance the fire-retardant properties of the final biopolymer.

Keywords: bio-based polyurethanes; jatropha oil; algae oil; recovered palm oil

Citation: Gomez, J.C.; Zakaria, R.; Aung, M.M.; Mokhtar, M.N.; Yunus, R. Synthesis and Characterization of Polyurethanes from Residual Palm Oil with High Poly-Unsaturated Fatty Acid Oils as Additive. *Polymers* **2021**, *13*, 4214. https://doi.org/10.3390/polym13234214

Academic Editors: Andrzej Puszka and Beata Podkościelna

Received: 29 September 2021
Accepted: 19 November 2021
Published: 1 December 2021

1. Introduction

Each year around the world, a large amount of polyurethanes (PU) are produced with a wide range of applications. Polyurethanes are highly versatile materials and can be used for multiple purposes such as molding, foaming, packaging, elastomers, coating, insulating, sealing etc. causing demand for them to continuously increase. In 2020, it was reported that 24 million tons of polyurethanes (PU) were produced [1], with most of it from petroleum-based feedstock [2,3]. Unfortunately, the instability in the cost of petroleum, and the non-renewability and the non-biodegradability of these polymers, have urged researchers to find other environmentally friendly alternatives.

In the past few years, several researchers have turned back their attention to the development of biopolyurethane from vegetable oils due to straightforward processing, biodegradability and renewability of the raw material. For example, drying oils such as linseed oil [4] showed great performance for the production of biopolyurethane especially for paint coating applications due to effortless epoxidation. Semi-dry oils such as soy bean [5] and sunflower [3] was shown to produce polyurethane with similar characteristics to petroleum-based rigid PU and non-drying oils such as cottonseed oil [6], castor oil [7] and refined palm oil [8] have been used for the development of foams and elastomers by different polymerization methods with the inclusion of additives. Unfortunately, the high cost of vegetable oils (especially from dry and semi-dry oils) and the competition with edible purposes limit their full implementation.

Despite this, other sources of vegetable oils with less competition for food supply can be explored as raw materials for the development of bio-polymers. In Malaysia, the extraction of crude palm oil (CPO) from *Elaeis guineensis* produces massive amounts of by-products in the form of palm pressed fiber, empty fruit bunch (EFB), palm oil mill effluent (POME), etc. which are disposed of with a high amount of oil still within. As an example, 1.13 tons of EFB was produced per 1 ton of crude palm oil (CPO) in 2012 [9] with 3–12% of CPO remaining on the OPEFB [10]. The residual palm oil (RPO) showed similar fatty acid composition to CPO but with lower oil quality due to the formation of degradation products such as free fatty acids (FFA), peroxides, etc. [11]. Therefore, with an annual production of some 19.14 million tons of crude palm oil in 2020 [12], the EFB itself can yield about 2.5 million tons of residual oil if fully recovered. Since this residual oil is not fit for consumption due to the degradation products, its utilization to other non-food products such as bio-polyurethane has potential economic and environmental benefits.

Nevertheless, the low degree of unsaturation in RPO impede the direct usage of this material for the development of polymers. To achieve a good degree of polymerization, the vegetable oil requires a high degree of unsaturation which is measured by the iodine absorption per 100 g of oil, and this value is named iodine value (IV). The functionality and the degree of unsaturation of the resulting polyol are linked and showed to have a significant effect on the mechanical properties and stability of the resulting polymer [2]. Polyols with low functionality and high molecular weight such as from palm oil (functionality of 1.7) produce soft and more fragile polymers due to relatively lesser crosslinking in the polymeric matrix [13]. This limits it uses and hampers its full implementation. Several studies have been undertaken on converting palm oil to polyol and polyurethane using technique such as epoxidation [14,15], transesterification [16] and polycondensation [17]. However, in all these studies the final polyurethane is a mixture between petrochemical polyol and palm oil polyol which is required in order to achieve the desired strength and thermal properties. This will ultimately affect the biodegradation properties of the final polymer. Tanaka et al. [18] produced rigid polyurethanes from palm oil by glycerolysis followed by hydrolysis. They converted the triglycerides (TG) in the crude palm oil into mono-glycerides using glycerol and alcohol at high temperature. This process showed to significantly increase the functionality of the polyol by adding two hydroxyl groups instead of the untreated CPO with functionality only in the fatty acid chains in the first/second or first/third position of the TG. The final polyurethane is produced by reacting the hydroxylated oil with polyethylene glycol (PEG) or diethylene glycol (DEG) and polyisocyanate (MDI). Hence, the addition of a significant amount of petrochemical compound (PEG/DEG) is needed to produce polyurethane with acceptable mechanical properties.

Cardeño et al. [13] compared the thermal stability of alkyd resins produced from refined, bleached and deodorized soy bean oil (RBD-SBO) and mixtures of RBD-SBO-recovered frying oils, palm oil and castor oil. Their research showed that the highest thermal stability was from the mixture RBD-SBO/castor oil (70/30) followed by RBD-SBO/frying oil and RBD-SBO/palm oil. Therefore, palm oil with a low a degree of unsaturation can be blended with oils of a very high degree of unsaturation, ideally from non-food sources, to increase the functionality of the oil mix. jatropha oil and algae oil are suitable candidates for such purpose.

Jatropha oil is a non-edible, highly unsaturated oil mainly from C18:1 and C18:2 fatty acids and has been utilized for the development of biopolymers. Moreover, the oil from *Jatropha curcas* has been used for the production of polymeric adhesives not only because of the high unsaturation but also because of the high content of gums that can enhance adhesive properties [19]. Algae oil is another less explored renewable source that has shown potential to be used for the development of PU [20]. With the increasing usage of microalgae technology for the capture and reduction of CO_2 emission, a huge amount of non-edible algae waste can be utilized for the production of biopolymers [21]. Algae oil contains a high degree of unsaturation (60–80% unsaturated fatty acids) basically from palmitoleic fatty acids (~37%, C16:1), oleic acid (~11.2%, C18:1) and linoleic acid (~40%, C18:2) [22]. Also, algae oil contains long polyunsaturated carbon chains (C22:3, C25:3) that add variation to the length of elastically active network chains and dangling chains in the polymer network to produce polyurethanes foams with similar thermal properties to foams from commercial petroleum polyols [23]. Roesle et al. [24] showed that oil from the diatom *Phaeodactylum tricornutum* has diglyceride structure with phosphate end-groups which can significantly enhance the thermal characteristics of the final biopolymer. However, the composition of algae oil vastly differs between species and growth condition. This variation can greatly affect the final mechanical and thermal properties of the polyurethane.

Therefore, our research focuses on the development of biodegradable polyurethanes from residual palm oil (RPO) with two highly unsaturated vegetable oils such as algae oil (AO) from *Chlorella vulgaris* and jatropha oil (JO) from *Jatropha curcas* as additive to produce polyurethanes. To the best of our knowledge, there is very limited study in the literature that explore the development of PU from recovered oil from palm oil waste such as empty fruit bunch and POME sludge. Residual palm oil has a different fatty acid composition, mono and diglyceride and free fatty acid contents from refined palm oil which may result in different macromolecular configuration after polymerization. Hence, this study aims to develop PU from RPO and compared that to the PU of RPO mixed with JO and AO. Since both JO and AO are highly unsaturated oils with a similar degree of unsaturation but with different composition and fatty acid structure, its effect on the thermal characteristic of the resulting polyurethane is evaluated.

2. Materials and Methods

2.1. Materials

RPO was recovered from palm oil mill effluent (POME) collected from the last sludge pit at Felda Sungai Tengi Palm Oil Mill located in Perak, Malaysia. AO was manually extracted from the dry algae powder of the species *Chlorella vulgaris* purchased from Purebulk (Los Angeles, California, CA, USA). JO was provided by Biofuel Bionas Sdn Bhd, Kuala Lumpur, Malaysia. Formic acid 98% was supplied by Friendemann Schimdt chemicals (Kuala Lumpur, Malaysia). Hydrogen peroxide 50% was purchased from R&M Chemicals (Selangor, Malaysia). Other chemical such as methanol reagent, hydrogen bromide, chlorobenzene, acetone, potassium hydrogen phthalate, phthalic anhydride, pyridine, sodium hydroxide and potassium hydroxide were reagent grade chemicals, locally purchased and used as received.

2.2. Oil Recovery

RPO was recovered from POME by removing water and suspended solids using a centrifuge at 7000 rpm (Centrifuge 5804, Enfield, CT, USA). The purified RPO was carefully collected from the upper layer using a pipette and was kept in the fridge at 5 °C for further usage. Dry algae powder was mixed with hexane (1:2) and homogenized using a homogenizer (Ultra turrax T25, IKA, Guangzhou, China) at 11,000 rpm for 10 min. After this, the mix was cooled down in an ice bath and was submitted to sonication for 10 min using a sonicator (Fisher Scientific, Waltham, MA, USA). The mixture of AO and hexane was separated from the solids by centrifugation at 7000 rpm for 10 min and the AO was recovered using a rotary evaporator (IKA RV10 Digital V, Guangzhou, China) at 50 °C and 150 mm of H_2O. Finally, the process was repeated three more times to recover most of the AO.

2.3. Epoxidation of the Oils

Epoxidized oils were prepared from pure RPO, 90 wt% RPO and 10 wt% JO and 90 wt% RPO and 10 wt% AO by in-situ epoxidation with peroxyformic acid and sulphuric acid in a molar ratio of 0.2:0.8:2.0:0.1 (double bound: organic acid: hydrogen peroxide:

catalyst) according to Dinda et al. [6]. The calculated amount of organic acid was added to 20 g of oil with specific amount of catalyst and the mixture was kept at a constant temperature of 40 °C with continuous stirring. The corresponding quantity of hydrogen peroxide was added drop wise to the mixture in order to prevent high temperature increase due to the exothermic reaction. Once the addition of hydrogen peroxide was completed, the temperature was increased to 65 °C and the reaction was carried out for 6 h. During this time, the formation of epoxy rings was monitored and recorded by measuring the OOC%. The final epoxidized oil was cooled down at room temperature and was washed using petroleum ether and distilled water. The solvent was removed from the epoxidized oil using a rotary evaporator at 50 °C and 180 mm of H_2O.

2.4. Hydroxylation of the Mixture

The hydroxylation was carried out by ring opening of the epoxidized oils in the presence of methanol, water and catalyst according to Hazmi et al. [25]. Approximately 10–20 g of epoxidized vegetable oil was placed into a 100 mL three-neck flask equipped with a magnetic stirrer, a thermometer and a reflux condenser. The mixture was heated to 65 °C with constant stirring and was left to react for 2 h. After this, the mixture was transferred into a 250 mL separation funnel and was carefully washed with petroleum ether and distilled water in order to remove unreacted catalyst. Finally, the excess of water, alcohol and solvent were removed by the rotary evaporator for approximately 3–4 h. The resulting hydroxyl value and acid value of the polyols were measured and the polyols were also analyzed using FTIR and NMR.

2.5. Preparation of Vegetable Oil Polyurethanes

MDI (4,4′-methylene diphenyl diisocyanate) was used to prepare polyurethane for different polyols by one shot method. The molar ratio of OH group to isocyanate (NCO) was established at 1:1.6 according to the following equation:

$$r = \frac{\frac{W_{polyol}}{EW_{polyol}}}{\frac{(W_{Pu} - W_{polyol})}{EW_{Isocyanate}}}$$

where, W_{polyol} is the weight of the polyol, EW_{polyol} is the equivalent weight of the polyol, W_{PU} the weight of the produced polyurethane and $EW_{Isocyanate}$ is the equivalent weight of the isocyanate which is 125 g/mol as supplied. The equivalent weight of the polyol was calculated as:

$$EWpolyol = \frac{56.1 \times 1000}{Acid\ number\ \left(mg\frac{KOH}{g}\right)}$$

The calculated amount of polyol was manually mixed at room temperature with MDI and molded into glass Petri dishes, no catalyst or other chemical was used. Afterwards, the mixture of polyol/isocyanate was left to be moisture-cured overnight in a vacuum desiccator. Finally, the polyurethane was removed and kept in a desiccator container for further analysis. This procedure was applied for all polyols (RPO, RPO + 10%JO and RPO + 10%AO).

2.6. Analytical Procedures

The oxirane oxygen content was measured throughout the reaction time according to AOCS official method Cd 9-57, using potassium hydrogen phthalate instead of potassium acid phthalate. Hydroxyl value (OHv) was carried out following ASTM D 4274 99 method C-reflux phthalation and titrated with NaOH solution until reaching a persisting light pink end point. The acid measurement of the vegetable polyols was determined by the MPOB method p2.5: 2004 [26] with potassium hydroxide. Finally, phospholipids from the oils (RPO, AO, CPO) and RPO(AO) polyol were recovered according to Goh et al. [27] using methanol and purified with acid-treated fluorisin. The quantification of phospholipids was conducted using a spectrophotometer at 820 nm according to Vaskovsky et al. [28] with modified Zinzadze's reagent.

2.7. Spectroscopy Analysis

The FTIR spectra was carried out using the Perkin-Elmer-spectrum with KBr plate attachment and the analysis was conducted from 4000–500 cm^{-1}. ^{1}H NMR was performed using an NMR spectrometer (Perkin-Elmer, LA, CA, USA) and the results were recorded at 400 MHz using JEOL Delta.

2.8. Gas Chromatography

Three oil samples from RPO, AO and JO were converted into the corresponding methyl esters according to the Ce 2-66 method [29], the preparation of methyl esters of fatty acids, using boron trifluoride. Then, the fatty acid composition was analyzed according to the Ce 1-62 method [30], fatty acid composition by gas chromatography, and the identification of fatty acids was undertaken using two standards from algae oil poly unsaturated fatty acid methyl ester (PUFAs) (Supelco, PA, USA) and algae oil fatty acid methyl esters (FAMEs) (Restek, PA, USA). The gas chromatography was carried out using GC-FID (Shimadzu, 2010-FID, Tokyo, Japan) and column BP-20 (SGE, Fisher Scientific, Waltham, MA, USA) with nitrogen as the carrier gas.

2.9. Morphological Analysis

The morphological analysis of the surface of different polyurethanes produced was conducted by scanning electron microscopy (SEM) (d4300, Hitachi, Tokyo, Japan) at 100 and 500 μm magnification.

2.10. Thermal Analysis of Polyurethanes

The thermal degradation of the polyurethane was verified using TGA Perkin Elmer Pyris TGA thermal analyser. Approximately, 10 mg of sample was placed into 70 μL sample holder and heated from 30 °C up to 800 °C at a heating rate of 10 °C/min under nitrogen gas flow rate (20 mL/min). DSC was undertaken using Perkin Elmer Pyris DSC thermal analyser according to ASTM E3418-03 as follows: heated from 25 °C to 120 °C at a heating rate of 10 °C/min, cooled to −40 °C at 5 °C/min, and heated again up to 120 °C at 10 C/min under nitrogen N_2. Dynamic mechanical analysis (DMA) was carried out on DMA Q800 V20.24 with liquid nitrogen. DMA analysis was conducted according to ASTM D5062-01 standard practice. The sample of 10 × 30 × 1 mm^3 was initially cooled to 100 °C and ramped to 140 °C at a rate of 5 °C/min and 1 Hz.

3. Results and Discussion

3.1. Characterization of the Vegetable Oils, Epoxidized Oils and Polyols

The free fatty acid (FFA) contents (%) of the RPO, JO and AO are 11.21–12.57, 2.12–2.23 and 11.85–11.91 respectively while the acid values based on palmitic acid (mg KOH/g) are 46.44–48.89, 12.4–12.75 and 46.46–48.83 respectively. The recovered palm oil has a higher FFA content compared to crude palm oil or refined palm oil which is usually below 5% and 0.1%, respectively [31]. This is obviously due to the hydrolysis and degradation of the recovered oil after exposure to the environment. On the other hand, algae oil also contain naturally high amount of FFA. The iodine value (I_2/100 g) of the oils are 43.15–45.26, 103.62–103.69 and 116.52–118.80 for RPO, JO and AO, respectively. In comparison, the iodine values of refined palm oil is higher at around 56 due to the removal of the saturated fraction during refining [16]. The iodine values for JO were similar to results previously reported from similar oil [19]. On the other hand, The IV value for AO reported here is higher than that reported by Petrovic et al. (85 I_2/100 g) due to the different species of algae used [32].

The fatty acids composition from the three oils used are shown in Table 1. The highest amount of unsaturated fatty acids is from AO (81.9%) followed by JO (78.91%) and RPO

(47.3%). This is in agreement with the IV values stated above. However, although algae oil content of unsaturated fatty acids is only slightly higher than that of JO, AO consists of mainly polyunsaturated fatty acids (76.9%) from C18:2, C20:5 and C22:6 whereas JO contains only 33.29% polyunsaturated fatty acids mainly from C18:2 The highest content of mono unsaturated fatty acids was found in the JO with more than 45.62% from oleic acid (C18:1); for AO there is a low amount of mono unsaturated fatty acids with approximately 4.99% mostly from oleic and palmitoleic acid. Hence, the difference in the number of double bonds in a fatty acid backbone can affect the length of the dangling chain in the resulting polyurethane which consequently will affect the mechanical and thermal properties.

Table 1. Fatty acid profile of residual palm oil (RPO), jatropha oil (JO) and algae oil (AO) obtained from gas chromatography.

Compound		Compound Name	Fatty Acid (%)		
			RPO	JO	AO
Saturated	C14:0	Methyl myristate	2.18	0.10	1.37
	C16:0	Methyl Palmitate	46.62	13.54	16.73
	C18:0	Methyl sterate	3.90	7.45	-
Mono unsaturated	C14:1	Methyl myristoleate	-	-	0.88
	C16:1	Methyl Palmitoleate	-	0.69	1.95
Polyunsaturated	C18:1n9	Methyl oleate	37.93	44.93	-
	C22:1n9	Methyl erucate	-	-	2.16
	C18:2n6	Methyl linoleate	9.37	33.29	21.05
	C20:5n3	Methyl eicosapentaenoate	-	-	52.57
	C22:6n3	Methyl docosahexaenoate	-	-	3.28

The fatty acid profile of algae oil also differs greatly for different species and different growing conditions [33]. The oil from *Chlorella vulgaris* contains a much longer hydrocarbon chain (C20:5) with five double bonds in its backbone. In comparison, algae oil from another *Chlorella* species reported by Pawar et al. [20] contains predominantly C18 hydrocarbon (64%) with one double bond (C18:1). Hence, there is a great variation in the chemical structure of the algae triglyceride which is known to impact the properties of the resulting polyurethane.

3.2. Analysis of Phospholipids

Table 2 shows the results obtained from the detection of phospholipids in RPO, AO, JO and RPO(AO) polyol. AO contains a very high concentration of phospholipids compared to JO and RPO. According to Gurr and Brawn [34], phosphate in AO from *Chlorella vulgaris* is mostly presence in the form of phosphatidyl choline from the diglycerides of C16:0, C18:1, C18:2 and C:18:3 fatty acids, which are the main reactive groups during polymerization. The addition of algae oil containing a high amount of phosphate as additive can significantly increase the thermal stability and fire-retardant properties of the final biopolymer as explained by Lewis A. [35]. However, it is apparent that the concentration of phospholipids in the polyol from RPO(AO) is significantly reduced from the original content in the AO. This reduction could be due to the formation of water-phospholipid emulsion during washing after epoxidation and hydroxylation process. The low content of phospholipids in JO and RPO indicates that only a small concentration of phospholipids will end up in the polyols and the final PU from JO and RPO.

The phospholipid content found in this study was, however, lower than that found in literatures. Nichols B. [36] showed that algae oil from *Chlorella vulgaris* contains around 30% *w/w* of phospholipid from diglycerides of phosphatidyl choline and phosphatidyl ethanolamine among others under normal light growing conditions. Similarly, Rao et al. [37] found that jatropha oil seeds contain approximately 1.4 wt% of phospholipids which is significantly higher than the value found in this study. This is possibly due to the extraction method used in this study in which hexane was used as the solvent to recover the maximum amount of oil. As hexane is a highly non-polar solvent, it is expected that less phospholipid will be extracted compared to using other more polar solvents used in the studies mentioned above [36,37].

Table 2. Phospholipid content in different oil samples and RPO(AO) polyol.

Sample	Sample (g)	λ 820 nm	ppm
AO	1.009	1.709 ± 0.16	6237 ± 582
JO	0.909	0.518 ± 0.13	92.51 ± 17
RPO	1.015	0.812 ± 0.01	98.18 ± 0.19
RPO(AO) polyol	1.005	1.832 ± 0.35	235.7 ± 42.37

3.3. H^1 Nuclear Magnetic Resonance (NMR) Spectra of the Vegetable Oils

The proton (H^1) spectrum of the RPO, AO and JO are shown in Figure 1 and the peaks assignation is shown in Table 3. The analysis of the signals was conducted according to Gustone F. [38], Hatzakis et al. [39] and Aung et al. [19]. Several similarities are shown in the structure of the different oils which are basically constituted of glyceride bone and acyl chains (i.e., fatty acid chains) with different unsaturation degrees. For example, in the spectrum of JO the strongest peak corresponds to alkenyl group (signal No. 1). However, in AO the peak found for double allycil groups (signal No. 2) is higher than signal No1. For AO, strong peaks were found in high amount between δ = 3.3 to 4 which are distinctive peaks from the hydrogen-choline compounds found in vegetable oil phospholipids. Unfortunately, the separation in this range was not completed because of the low solubility of phospholipid compounds with the solvent used during the analysis (i.e., chloroform). According to Hatzakis et al. [39], the peak found at 3.8 corresponds to protons on the fatty acyl chain of phosphtidylcholine (PC) and the peak found at 8.4 is related to the $^+NH_3$ group from phosphtidylethanolamine (PDA). Hence, the H^1 NMR for the different oils is in agreement with the analytical procedure described above in which the highest concentration of phospholipids was found in AO and zero to low amounts in RPO and JO.

Table 3. Signal number, hydrogen assignation and group of oils by H^1 NMR.

Signal	Proton	δ	Assignation of Hydrogen Atoms
1	(-CH=CH-)	5.33	All alkenyl
2	(=CHCH2CH=)	2.81	doubly allylic
3	$-CH_2CH_2COOCH_3$	2.28	Acyl chains
4	$CH_2CH{=}CHCH_2$-	2.03	Allylic chains
5	$(CH_2)_n$	1.28	Acyl chains
6	CH_2 CH_2 CH_2 $-CH_3$	1.58/0.8	Acyl chains excluding linolenyl
7	CH_2 -OCOR	4.12/4.31/4.41/5.20	Glycerol
A	CH_2 $-OP(O)_2O$	3.94	Glycerol/Choline
B	CH_2 $-N(CH_3)_3$	3.81	Choline
C	$-N(CH_3)_3$	3.36	Choline
D	CH_2 $-^+NH_3$	8.4	Anolamine

3.4. Oxirane Oxygen Content (OOC%)

The analysis of OOC% of RPO, RPO(JO) and RPO(AO) during the epoxidation process is presented in Figure 2. The graph shows the increment of epoxy groups after the inclusion of oxygen into the unsaturated bond of the different oils. The process continued until the maximum OOC was achieved and it was followed by ring opening of the cyclic ethylene oxide which led to reduction in OOC. The epoxidize RPO (ERPO) without additive showed the lowest speed of reaction, the shortest time to reach maximum OOC, and the lowest maximum OOC. This might be due to the low degree of unsaturation of the RPO. However, the addition of 10% AO and JO to the RPO significantly increase the OOC value as well as the reaction time.

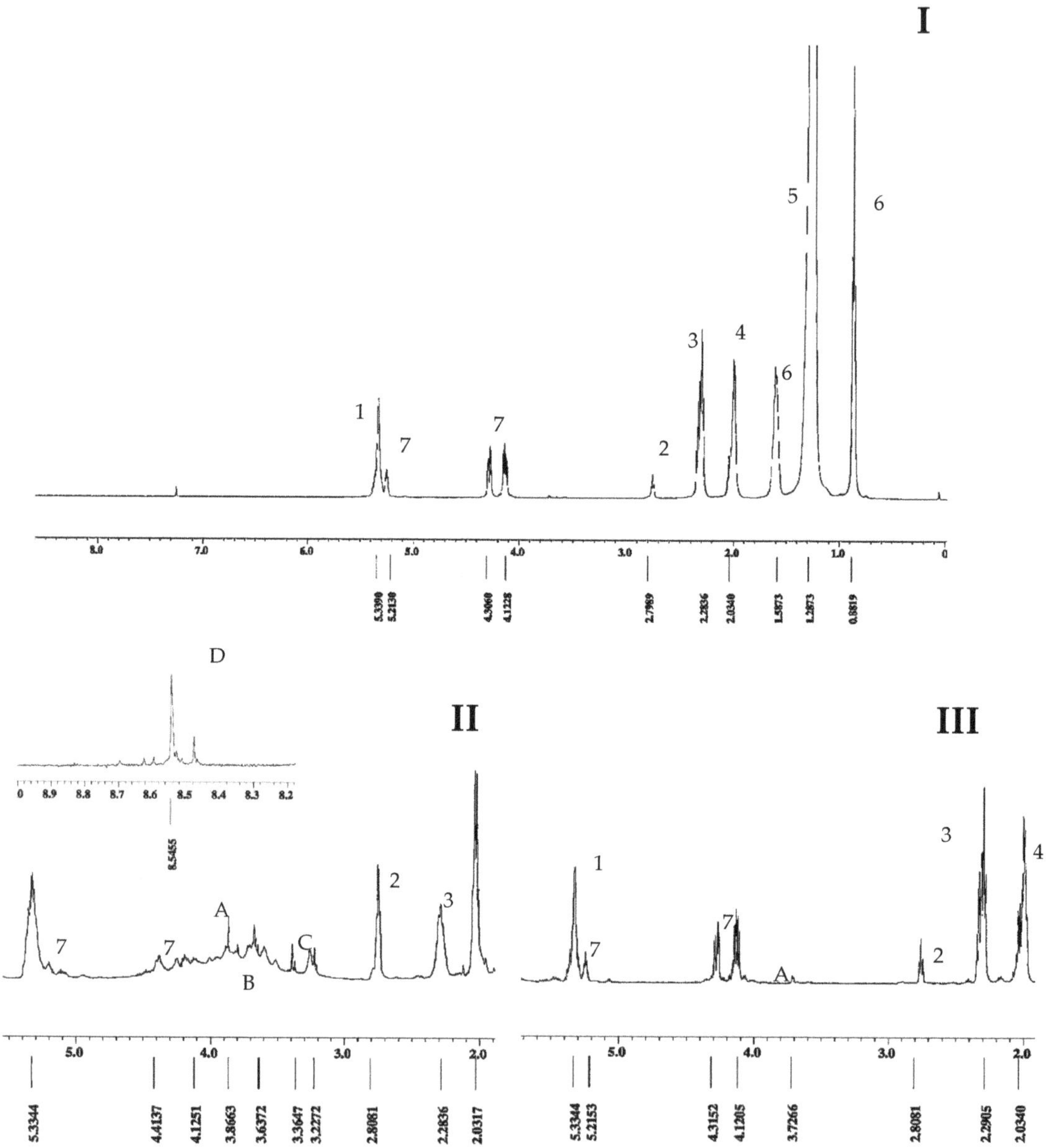

Figure 1. H^1 nuclear magnetic resonance (NMR) spectrum of RPO (**I**), AO (**II**) and JO (**III**).

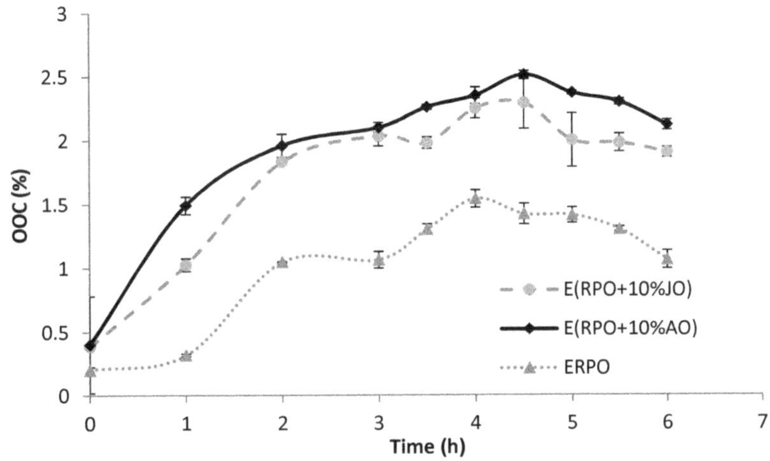

Figure 2. Analysis of OOC (oxirane oxygen content) during epoxidation of RPO, RPO + 10%JO and RPO + 10%AO.

Table 4 shows the characterization of residual palm-based epoxy and polyols with and without additive. Based on the result, the addition of only 10% AO and JO as additive increases approximately 40% and 30% the OOC of epoxidized RPO(AO) [ERPO(AO)] and epoxidized RPO(JO) [ERPO(JO)] compared to epoxidized pure RPO (ERPO), respectively. However, the maximum OOC values obtained in this study are lower compared to the theoretical OOC calculated using the IV value of each oil according to Hazmi et al. [25]. All the samples have similar oxirane conversion with a yield of 60 to 70% OOC. Finally, the value of IV in the epoxidized samples indicated that the epoxidation process was incomplete and that a certain number of ethilenic groups did not react and optimization of the reaction is required.

Table 4. Properties of the epoxidized residual palm oil (RPO), RPO with 10% algae oil (AO) and RPO with 10% jatropha oil (JO). [1] Value calculated according to Hazmi et al. [25].

Property	Epoxidized		
	RPO	RPO(JO)	RPO(AO)
	-	-	-
OOC_max, (%)	1.65 ± 0.11	2.33 ± 0.04	2.745 ± 0.23
Acid Value (mg KOH/g)	29.395 ± 0.47	6.46 ± 0.7	12.2 ± 1.14
IV,% per mole	17.6 ± 0.25	11.05 ± 0.28	12.43 ± 0.11
OOC Max. Theoretical [1]	2.71 ± 0.06	3.53 ± 0.02	3.975 ± 0.35
		hydroxylated	
Hydroxyl value (mg KOH per gram)	78.525 ± 1.29	131.34 ± 2.68	152.06 ± 0.56
Equivalent weight	60.75 ± 0.5	60.92 ± 0.4	62.40 ± 0.02

In addition, the acid value from the ERPO(AO) was shown to be higher than the ERPO(JO). Naturally, AO contains high acid value which can interfere with the polymerization due to the reaction with (N) isocyanate instead of the OH form polyol. Nevertheless, this might be minimized due to the high OH obtained from the polyunsaturated AO. Finally, the value of OH obtained from RPO was higher than values reported from untreated crude palm oil CPO (50 mg KOH/g), but, it is lower than the value reported for refined, bleached and deodorized (RBD) palm olein which was found to be 110 mg KOH/g [40]. The OH value of RPO(AO) polyol is almost double that of RPO polyol again showing that the addition of 10% algae oil increases the hydroxyl value of the polyol significantly.

3.5. Fourier Transform Infrared (FTIR) Analysis of Different Polyols

FTIR of the epoxidized oils after hydroxylation is presented in Figure 3 and the FTIR analysis was reported according to Aung et al. [19]. The formation of polyol for all the oil samples was corroborated with the characteristic hydroxyl band observed at 3475 cm^{-1}. It was found that the hydroxyl peak obtained from RPO(AO) was the highest which confirms the result of hydroxyl values in Table 3. The vibrations attributed to CH_2 bending, -CH bends, and CH_2, were found at 1464, 1407 and 1311 cm^{-1} absorption bands, respectively and could be attributed to the fatty acid chains of the triglycerides. It was found that the RPO(AO) polyol showed the highest area at these bands followed by RPO(JO) and RPO. This can be due to the longer fatty acid chains' present in AO (i.e., fatty acids from C22 and C25). In addition, characteristic bands of the hydroxyl and carbonyl groups were found in the 3373 and 1750 cm^{-1} region. Finally, no apparent peak from ethylenic bonds was found at 1630–1640 cm^{-1} which might indicate that remaining double bonds could be opened during polyol reaction due to high temperature used. (e.g., 90 °C) (Lai et al., 2013).

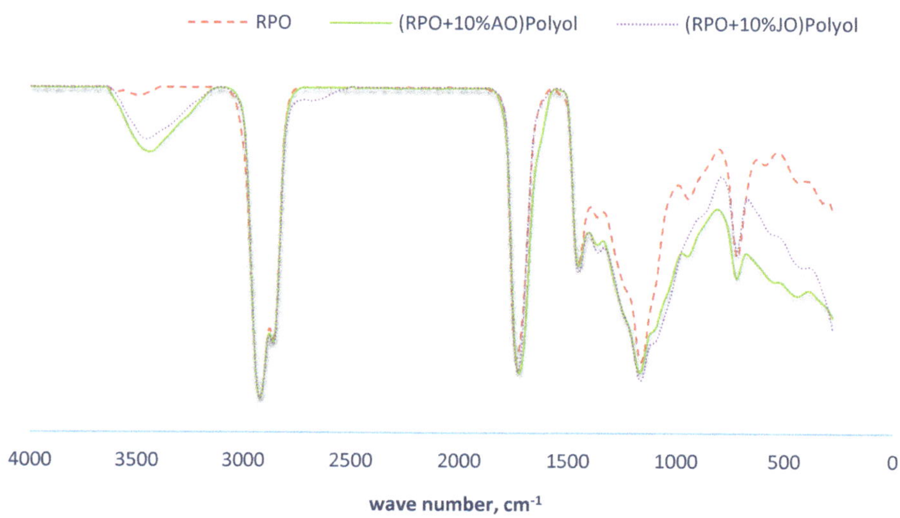

Figure 3. Hydroxylation of RPO, RPO + 10%JO and RPO + 10%AO polyols is described by Fourier transform infrared (FTIR) spectra.

3.6. H^1 NMR Analysis of RPO(AO) Polyol Formation

The hydroxylation of RPO(AO) oils was corroborated by H^1 NMR. Figure 4, shows the pathway of RPO(AO) to epoxidized oil and to hydroxylated polyol, analyzed by NMR spectra. The identification of phosphate, unsaturated, epoxy and hydroxyl groups are highlighted with less attention to other groups shown previously. As mentioned before, the structure of the different oils before processing is similar however, AO can provide phosphate groups to the final polyol and to the subsequent polyurethane. During the epoxidation and hydroxylation of the oils it is possible to identify the formation of epoxy rings at 4.3 ppm (Figure 4B) and hydroxyl carbon at 3.65 ppm (Figure 4C) which is in accordance with previous reported analysis [19]. However, in all the NMR shown it is possible to identify one small peak at $\delta = 5.33$ of alkenyl unsaturation which indicates that the epoxidation was not complete as shown in the analysis of IV (This peak was not clear in the FTIR analysis of the RPO(AO) polyol). Finally, the protons of phosphate groups were found in the final polyol from RPO(AO) which agrees with the analysis above.

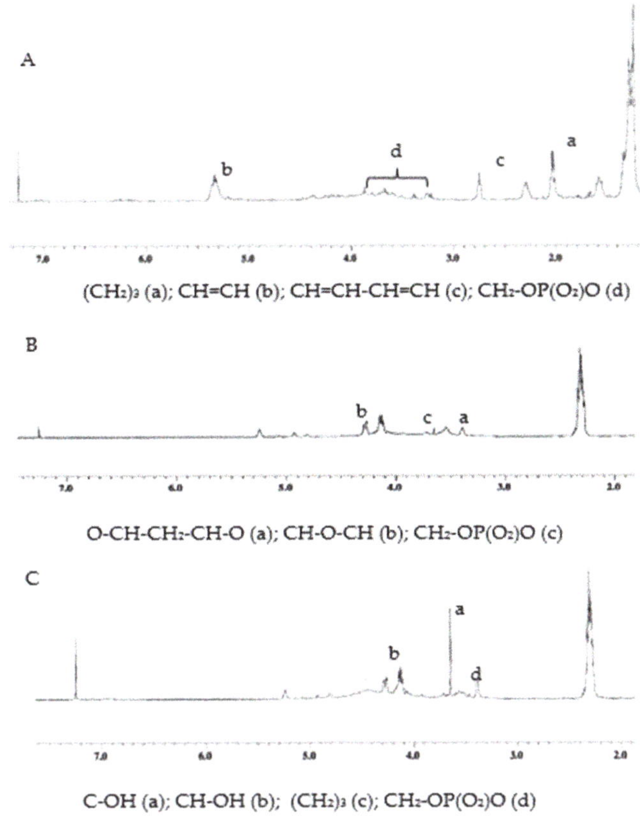

Figure 4. Pathway of the hydroxylation of RPO(AO) analyzed by H^1 NMR showing the oil RPO(AO) (**A**), the epoxidized RPO(AO) (**B**) and RPO(AO) polyol (**C**).

A polyurethane linkage is created by reacting the poly isocyanate group, -N=C=O, with the hydroxyl (-OH) group in the polyalcohol (polyol) in the presence of a catalyst (Scheme 1) [41]. Hence, when more hydroxyl groups are presented in the polyol more linkages will be obtained in the polymer. Furthermore, the hydroxyl number of the polyols has a linear effects on crosslinking density of the poly-urethanes, expressed as the number of elastically active network chains per unit volume, and inversely proportional to the average molecular weight between crosslinks which resulted in better mechanical properties [4].

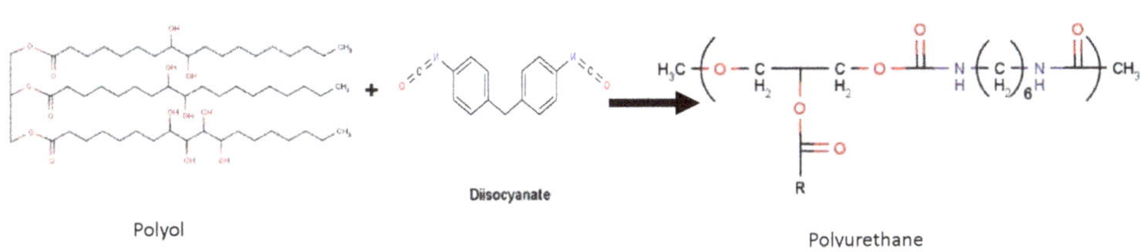

Scheme 1. Urethane linkage.

3.6.1. Spectroscopy Analysis of Polyurethanes (FTIR)

The produced polyurethanes (PU) were analyzed by FTIR spectroscopy to identify their structure and the respective polymerization. Three different PU from RPO,

RPO + 10%AO and RPO + 10%JO were prepared, analyzed by FTIR according to Velencoso et al. [42] and the results are shown in Figure 5. The stretching vibrations at 3325 cm^{-1} were identified as characteristic bands of N–H from the urethanes linkages. The PU from RPO with AO as additive showed the highest intensity compared to other PU. This indicates that a higher urethane linkage was obtained with AO at the same NCO:OH ratio. The unreacted NCO compounds were identified at 2290 cm^{-1} and shown to be similar in the three samples. Stretches at 2920 and 2853 were found for the C–H methane group as asymmetric and symmetric. The large difference between the peaks in the 3325 and 2920 region indicates that the three consist mostly of semi rigid and soft PU rather than hard sections. In addition, according to Zhang et al. [43] the stretching vibration at 1029 cm^{-1} indicates the presence of P–O–C symmetric bending vibration. This peak is clearly present in PU from RPO(AO) which can be attributed to the presence of phospholipid in the AO. No peak in the bands for the OH group (3400 cm^{-1}) was found which indicates that all the available OH groups reacted with the MDI to form the PU. Also, no stretch corresponding to urea bands (1628–1690 cm^{-1}) was found.

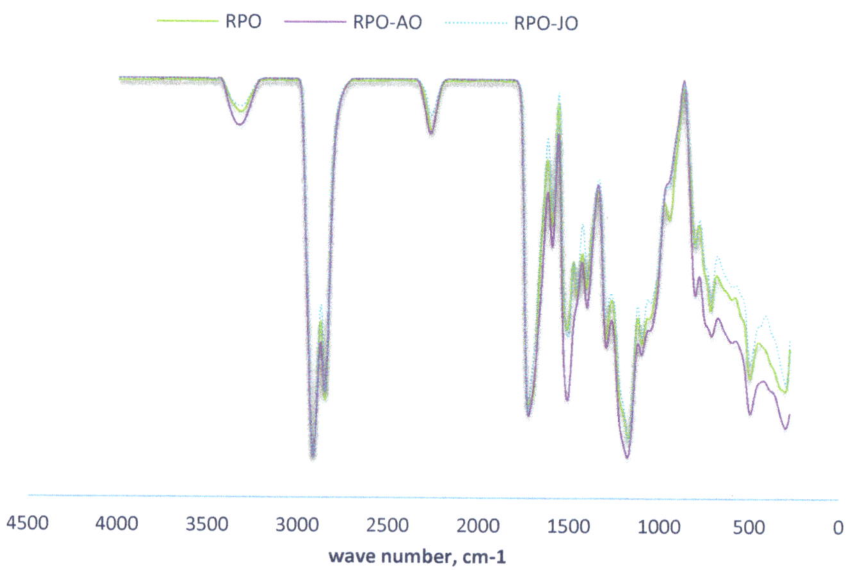

Figure 5. FTIR spectra of the polyurethanes produced by RPO, RPO + 10%JO and RPO + 10%AO polyol with isocyanate.

3.6.2. Morphological Analysis

The surface of the polyurethane obtained from hydroxylated RPO, RPO + 10%JO and RPO + 10%AO was scanned by using SEM (d4300, Hitachi, Japan) and the results are shown in Figure 6. The surface of the polyurethane obtained from RPO (Figure 6a) showed an amorphous growing of several spike shapes without specific orientation (Figure 6d). The formation of amorphous spike shapes caused cracks which weaken the polymer [44]. On the other hand, the addition of polyunsaturated oil such as JO (Figure 6b) to the original RPO significantly reduced the formation of spike shapes with better orientation. A smoother surface was obtained when the RPO was blended with 10% AO (Figure 6c).

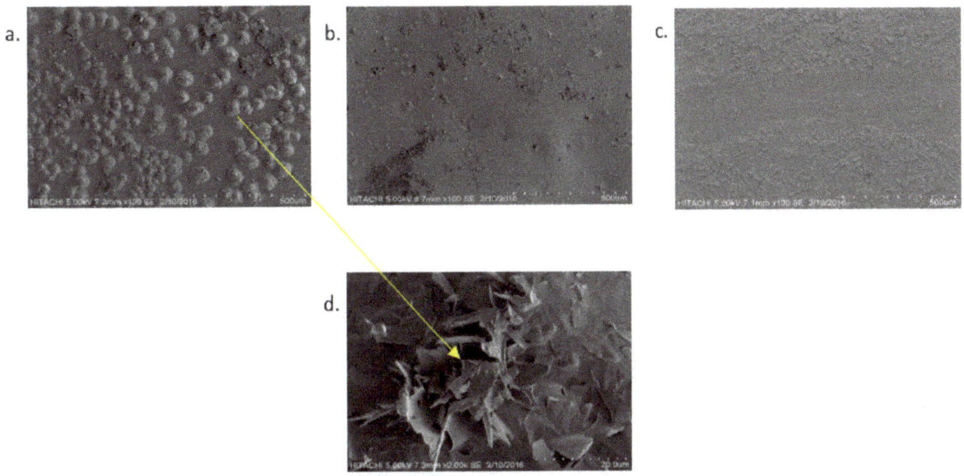

Figure 6. Surface of the polyurethanes using scanning electron microscopy (SEM, 5.0 KV:500 μm magnification) obtained from the polyols of (**a**) RPO (**b**) RPO + 10JO (**c**) RPO + 10%AO and (**d**) RPO with SEM (5.0 KV:100 μm magnification).

3.7. Polymer Thermal Analysis

3.7.1. Differential Scanning Calorimeter (DSC)

Figure 7 shows the analysis of DSC of the different polyurethanes synthesized while Table 5 summarizes the values obtained for T_{g1} and T_{g2} for the soft segment and hard segment of the different PU produced. The change of slope on the DSC shows that there is one glass transition temperature T_{g1} for soft segments and one T_{g2} for hard segments without melting peak, indicating that the samples have an amorphous structure [45]. The lowest T_{g1} was obtained from RPO(AO) at −33.3 °C, followed by RPO(JO) at −25.8 °C and RPO at around −25.6 °C. All samples exhibit similar T_{g1} because of the similar main source used (i.e., palm oil) and are similar to other palm-based bio polymers developed by other researchers (e.g., −33 to −15 °C) [46]. However, RPO(AO) exhibit slightly lower Tg1 possibly due to the longer fatty acid chains that have higher mobility at lower temperature.

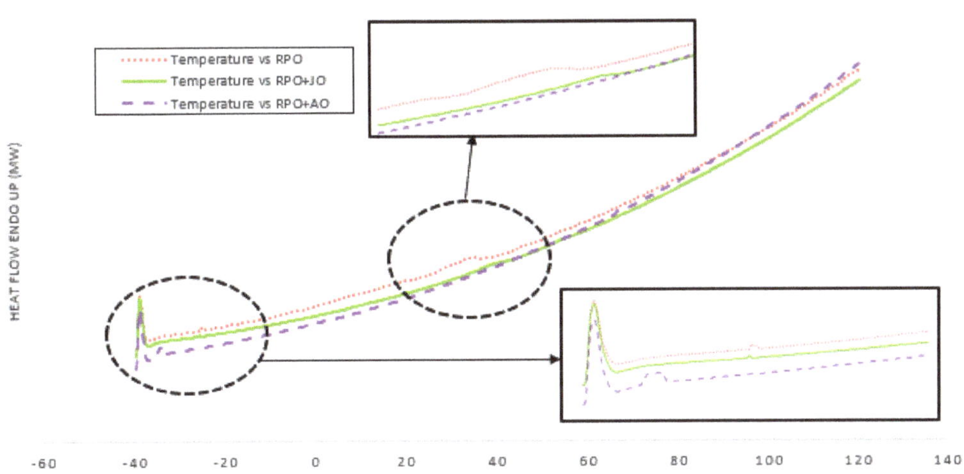

Figure 7. Differential scanning calorimetry (DSC) curve of polyurethanes from vegetable oils.

The negative values of T_{g1} suggest that all the PUs are dominant in flexible segments provided by the long fatty acid chains of the vegetable oils. When AO is added to the RPO as additive, this might add flexibility, increase elongation at break and resistance to scratching to the final PU. In addition, the lower value of T_{g1} from RPO(AO) can also be related to the inclusion of phosphate groups in final PU. Wu et al. [47] compared the T_g obtained from different PU before and after the addition of tris(chloroisopropyl) phosphate (TCPP) as a flame retardant plasticizer. They obtained significant reduction of T_g per percentage of TCPP added as plasticizer due to the swelling of the PU structure, effect of the phosphate group and dilution of aromaticity (i.e., reaction with NCO).

Table 5. Glass transition temperature (T_{g1} and T_{g2}) obtained under and over 0 °C, respectively, from residual palm oil (RPO); RPO with 10% algae oil RPO(AO); RPO with 10% jatropha oil RPO(JO); compared with jatropha oil (JO) reported by Saalah, S. (2016) [48] [a].

Sample	T_{g1}	T_{g2}
RPO	−25.6	27.8
RPO(JO)	−25.8	35.3
RPO(AO)	−33.3	37.5
JO [a]	−11.45	-

The effect of hard segments in the sample and the relation with T_{g2} can be correlated with the increasing of the OH number in the polyol. All the samples exhibited hard segments due to the crosslinking with urethane groups. However, the high OH value found in the AO leads to the production of more hard segments between the polyol and the isocyanate which restricted the chain mobility and increased T_{g2}.

3.7.2. Thermo Gravimetrical Analysis (TGA)

The degradation of the PUs with temperature was analyzed using TGA and the results are shown in Figure 8. The first degradation (TD1) is associated with the loss of moisture content and suggests that the degradation of PU from RPO(AO) is slightly faster compared to other PUs due to possible water emulsion trapped in the phospholipid groups. The second onset of degradation of the PU (TD2) from all samples started at 250 to 300 °C which is associated with the hard segments on the PU [20]. It can be seen that the PU from RPO degrades much faster possibly due to higher saturation and less urethane linkage that is formed [13] and possible free isocyanate [49]. PU from RPO/JO and RPO/AO degrades at a similar rate during this stage. The third degradation stage (TD3) started at 350 °C for RPO and at 420 °C for RPO(AO) and RPO(JO). At this stage, the decomposition of the polyol soft segment occurred. The degradation rate and the percentage of weight loss of RPO(AO) is the slowest which could be due to longer fatty acid chains of AO that are squeezed out of the PU matrix during curing and the phosphate plasticizer in the AO. Hence, the usage of AO as additive was shown to significantly improve the stability of the polymer at high temperature.

Table 6 shows the degradation at 5%, 10% weight loss ratio and max residual amount of material remaining after the thermal decomposition versus the temperature measured. The initial stage of degradation (i.e., 5% weight loss) occurred at a higher temperature for the sample with RPO compared with the other polymers possibly due to less phospholipid concentration. As mentioned before, the phospholipids might have created water-oil emulsions during the epoxidation process trapped inside the polymer and not removed during the drying process of the samples. The next 10 percentage weight loss show the highest temperature for the sample with JO followed by the sample with AO, which correlates with a higher degree of polymerization when compared to RPO. Finally, the three samples flatten the curve of weight loss (Final weight) at different temperatures and with different sample char residual. The sample with AO presented the highest char residual with 20.96% possibly due to the highest concentration of phosphatide presented in the AO.

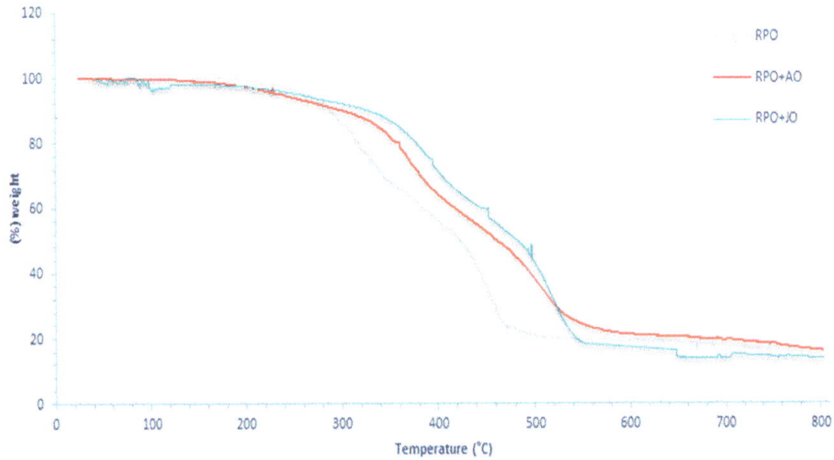

Figure 8. Overlay of thermogravimetric analysis (TGA) curves of the palm-based bio-polyurethanes produced.

Table 6. The 5, 10 and max percentage of weight loss versus temperatures for the samples with RPO, RPO(JO) and RPO(AO).

Weight Loss (%)	Temperature (°C)		
	RPO	RPO(JO)	RPO(AO)
5	261.33	239.54	234.10
10	288.02	312.05	305.56
max	516.23	550.34	669.53
Sample residual (%)	16.11	18.53	20.96

3.7.3. Dynamic Mechanical Analyzer (DMA)

DMA was conducted to analyze the mechanical and viscoelastic behavior of the three different materials under thermodynamic changes. Figure 9, depicts the dependence of the loss factor and the storage modulus (i.e., tan delta) with temperature. The continuous increment of tan delta might be due to accretion of the molecular chain mobility that soften the polymer at an elevated temperature.

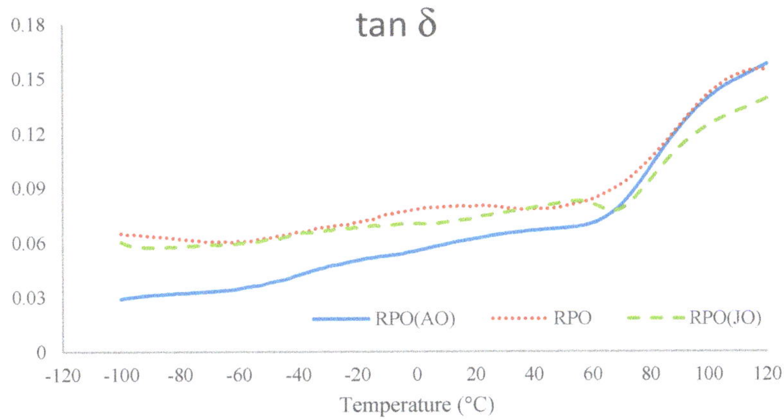

Figure 9. Variation of Tan d versus temperature for the different polymers.

The delta modulus (tan δ) has an inverse proportion with the OH number across the temperature range. Initially, the polymer developed with RPO(AO) showed the lowest tan δ value and that developed from RPO showed the highest. The lesser height of the delta peaks is associated with the crosslinking arising from the strong covalent bonds between the high functionality polyol and the hard segment component, which restrict the mobility of the dangling chains [50]. After the temperature of 67 °C the tan δ of the polymer from RPO(AO) increased significantly while the polymer from RPO(JO) displayed the lowest values. The restriction in the molecular mobility of the RPO(JO) due to stronger urethane linkages in the network under stress might increase the wear resistance of the PU. Hence, RPO(JO) can provide a better PU for coating application when compared to RPO and RPO(AO) [48].

4. Conclusions

Polyester polyols and polyurethane were prepared from recovered palm oil, recovered palm oil with 10% jatropha oil, and recovered palm oil with 10% algae oil as additive. The polymerizations were conducted via one pot epoxidation and hydrolysis followed by isocyanation of the hydroxyl groups (-OH). The characterization of the oils, epoxidized oil and polyols was conducted and the results were compared among the different samples. Residual palm oil has a much lower IV number than that of refined palm oil but a higher FFA content. Algae oil, on the other hand, has a slightly higher IV value than jatropha, but it contains predominantly C20:5 fatty acid chain which is longer and with much higher double bonds than jatropha which is predominantly C18:1. Algae oil also contains much higher phospholipid content and the phosphate groups are found to be in the polyol and the final polyurethane. The addition of 10% algae oil to RPO almost doubled the hydroxyl of value of the polyol compared to pure RPO polyol The thermal properties of the final polyurethanes were analyzed by different techniques and the results were compared. We found that the addition of polyunsaturated oil as additive into the highly saturated recovered palm oil significantly enhanced the thermal properties of the final polymer. Moreover, the polymer developed with algae oil as additive showed a higher thermal stability than that of jatropha oil due to the higher crosslinking from the polyunsaturated fatty acid chain and the inclusion of phosphate groups. Differential scanning calorimetry showed significant increment of T_g for the hard segment and a lower Tg for the soft segments of the polyurethanes produced with RPO(AO). Finally, the usage of, non-edible, biodegradable and renewable recovered palm oil with jatropha oil and algae oil as additive has a high potential for the production of environmentally friendly bio-based polyurethanes.

Author Contributions: Conceptualization, R.Z. and J.C.G.; methodology, R.Z., J.C.G. and M.M.A.; software, J.C.G.; validation, R.Z. and M.M.A.; formal analysis, J.C.G. and R.Z.; investigation, J.C.G.; resources, R.Z. and M.M.A.; data curation, J.C.G.; writing—original draft preparation, J.C.G. and R.Z.; writing—review and editing, R.Z., R.Y. and M.N.M.; visualization, J.C.G.; supervision, R.Z., M.M.A., M.N.M. and R.Y.; project administration, R.Z. and J.C.G.; funding acquisition, R.Z. and R.Y. All authors have read and agreed to the published version of the manuscript.

Funding: This research was funded by University Putra Malaysia through grant GP-IPS/2016/9503300.

Institutional Review Board Statement: Not applicable.

Informed Consent Statement: Not applicable.

Data Availability Statement: The data presented in this study are available on request from the corresponding author.

Acknowledgments: Appreciation is given to all assistant engineers in the department of process and food engineering for their guidance and laboratory assistance.

Conflicts of Interest: The authors declare no conflict of interest.

References

1. Global Polyurethane Demand 2012–2024—Statista Website1. Available online: https://www.statista.com/statistics/747004/polyurethane-demand-worldwide/ (accessed on 15 August 2021).
2. Petrović, Z.S. Polyurethanes from vegetable oils. *Polym. Rev.* **2008**, *48*, 109–155. [CrossRef]
3. Desroches, M.; Escouvois, M.; Auvergne, R.; Caillol, S.; Boutevin, B. From vegetable oils to polyurethanes: Synthetic routes to polyols and main industrial products. *Polym. Rev.* **2012**, *52*, 38–79. [CrossRef]
4. Zlatanić, A.; Lava, C.; Zhang, W.; Petrović, Z.S. Effect of structure on properties of polyols and polyurethanes based on different vegetable oils. *J. Polym. Sci. Part B Polym. Phys.* **2004**, *42*, 809–819. [CrossRef]
5. Caillol, S.; Desroches, M.; Boutevin, G.; Loubat, C.; Auvergne, R.; Boutevin, B. Synthesis of new polyester polyols from epoxidized vegetable oils and biobased acids. *Eur. J. Lipid Sci. Technol.* **2012**, *114*, 1447–1459. [CrossRef]
6. Dinda, S.; Patwardhan, A.V.; Goud, V.V.; Pradhan, N.C. Epoxidation of cottonseed oil by aqueous hydrogen peroxide catalysed by liquid inorganic acids. *Bioresour. Technol.* **2008**, *99*, 3737–3744. [CrossRef] [PubMed]
7. Beneš, H.; Vlček, T.; Černá, R.; Hromádková, J.; Walterová, Z.; Svitáková, R. Polyurethanes with bio-based and recycled components. *Eur. J. Lipid Sci. Technol.* **2012**, *114*, 71–83. [CrossRef]
8. Badri, K.H.; Ahmad, S.H.; Zakaria, S. The production of a high-functionality RBD palm kernel-based polyester polyol. *J. Appl. Polym. Sci.* **2001**, *82*, 827–832. [CrossRef]
9. MPOB. Economics, and Industry Division. MPOB Website. Available online: http://bepi.mpob.gov.my/index.php/statistics/production/71-production-2012/298-production-of-crude-oil-palm-2012.html (accessed on 10 October 2013).
10. Chavarro-Gomez, J.; Mokhtar, M.N.; Sulaiman, A.; Samsu Baharuddin, A.; Busu, Z.A. Recovery of Residual Crude Palm Oil from the Empty Fruit Bunch Spikelets Using Environmentally Friendly Processes. *Sep. Sci. Technol.* **2015**, *50*, 1677–1683. [CrossRef]
11. Gomez, J.C.; Mokhtar, M.N.; Sulaiman, A.; Zakaria, R.; Baharuddin, A.S.; Busu, Z. Study on Residual Oil Recovery from Empty Fruit Bunch by Combination of Water and Steam Process. *J. Food Process. Eng.* **2015**, *38*, 385–394. [CrossRef]
12. Malaysian Palm Oil Council Website. Available online: http://mpoc.org.my/malaysian-palm-oil-sector-performance-in-2020-and-market-opportunities/ (accessed on 15 August 2021).
13. Cardeño, F.; Rios, L.A.; Cardona, J.F.; Ocampo, D. Síntesis de Resinas Alquídicas a partir de Aceites de Higuerilla, de Palma y de Fritura, Mezclados con Aceite de Soja. *Inf. Tecnológica* **2013**, *24*, 33–42. [CrossRef]
14. Polaczek, K.; Kurańska, M.; Auguścik-Królikowska, M.; Prociak, A.; Ryszkowska, J. Open-cell polyurethane foams of very low density modified with various palm oil-based bio-polyols in accordance with cleaner production. *J. Clean. Prod.* **2021**, *290*, 125875. [CrossRef]
15. Prociak, A.; Malewska, E.; Kurańska, M.; Bąk, S.; Budny, P. Flexible polyurethane foams synthesized with palm oil-based bio-polyols obtained with the use of different oxirane ring opener. *Ind. Crop. Prod.* **2018**, *115*, 69–77. [CrossRef]
16. Arniza, M.Z.; Hoong, S.S.; Idris, Z.; Din, A.K.; Choo, Y.M. Synthesis of transesterified palm olein-based Polyol and rigid polyurethanes from this polyol. *J. Am. Oil Chem. Society* **2015**, *92*, 243–255. [CrossRef] [PubMed]
17. Ng, W.S.; Lee, C.S.; Chuah, C.H.; Cheng, S.-F. Preparation and modification of water-blown porous biodegradable polyurethane foams with palm oil-based polyester polyol. *Ind. Crop. Prod.* **2017**, *97*, 65–78. [CrossRef]
18. Tanaka, R.; Hirose, S.; Hatakeyama, H. Preparation and characterization of polyurethane foams using a palm oil-based polyol. *Bioresour. Technol.* **2008**, *99*, 3810–3816. [CrossRef] [PubMed]
19. Aung, M.M.; Yaakob, Z.; Kamarudin, S.; Abdullah, L.C. Synthesis and characterization of Jatropha (Jatropha curcas L.) oil-based polyurethane wood adhesive. *Ind. Crop. Prod.* **2014**, *60*, 177–185. [CrossRef]
20. Pawar, M.S.; Kadam, A.S.; Dawane, B.S.; Yemul, O.S. Synthesis and characterization of rigid polyurethane foams from algae oil using biobased chain extenders. *Polym. Bull.* **2016**, *73*, 727–741. [CrossRef]
21. Williams, P.J.L.B.; Laurens, L.M. Microalgae as biodiesel & biomass feedstocks: Review & analysis of the biochemistry, energetics & economics. *Energy Environ. Sci.* **2010**, *3*, 554–590.
22. Mülhaupt, R. Green Polymer Chemistry and Bio-based Plastics: Dreams and Reality. *Macromol. Chem. Phys.* **2013**, *214*, 159–174. [CrossRef]
23. D'Alessandro, D.M.; Smit, B.; Long, J.R. Carbon dioxide capture: Prospects for new materials. *Angew. Chem. Int. Ed.* **2010**, *49*, 6058–6082. [CrossRef]
24. Roesle, P.; Stempfle, F.; Hess, S.K.; Zimmerer, J.; Río Bártulos, C.; Lepetit, B.; Eckert, A.; Kroth, P.G.; Mecking, S. Synthetic polyester from algae oil. *Angew. Chem. Int. Ed.* **2014**, *53*, 6800–6804. [CrossRef]
25. Hazmi, A.S.A.; Aung, M.M.; Abdullah, L.C.; Salleh, M.Z.; Mahmood, M.H. Producing Jatropha oil-based polyol via epoxidation and ring opening. *Ind. Crop. Prod.* **2013**, *50*, 563–567. [CrossRef]
26. MPOB. *MPOB Test Method p2.5:2004, Method of Test for Palm Oil and Palm Oil Products: Determination of Acidity*; Malaysian Palm Oil Board: Selangor, Malaysia, 2004.
27. Goh, S.H.; Tong, S.L.; Gee, P.T. Inorganic phosphate in crude palm oil: Quantitative analysis and correlations with oil quality parameters. *J. Am. Oil Chem. Soc.* **1984**, *61*, 1601–1604. [CrossRef]
28. Vaskovsky, V.E.; Kostetsky, E.Y.; Vasendin, I.M. A universal reagent for phospholipid analysis. *J. Chromatogr. A* **1975**, *114*, 129–141. [CrossRef]
29. American Oil Chemists' Society. *Method Ce 2–66, Preparation of Methyl ester of Fatty Acids*, 5th ed.; Fireston, D., Ed.; Official Methods and Recommended Practices of the American Oil Chemists' Society; AOCS Press: Urbana, IL, USA, 1998.

30. American Oil Chemists' Society. *Method Ce 1–62, Fatty Acid Composition by Gas Chromatography*, 5th ed.; Fireston, D., Ed.; Official Methods and Recommended Practices of the American Oil Chemists' Society; AOCS Press: Urbana, IL, USA, 1998.
31. MPOB Official Palm Oil Source. MPOB Website. Available online: http://www.palmoilworld.org (accessed on 15 August 2021).
32. Petrović, Z.S.; Wan, X.; Bilić, O.; Milić, J.; Degruson, D. Polyols and polyurethanes from crude algal oil. *J. Am. Oil Chem. Soc.* **2013**, *90*, 1073–1078. [CrossRef]
33. Peyrton, J.; Chambaretaud, C.; Sarbu, A.; Avérous, L. Biobased Polyurethane Foams Based on New Polyol Architectures from Microalgae Oil. *ACS Sustain. Chem. Eng.* **2020**, *8*, 12187–12196. [CrossRef]
34. Gurr, M.I.; Brawn, P. The biosynthesis of polyunsaturated fatty acids by photosynthetic tissue. *Eur. J. Biochem.* **1970**, *17*, 19–22. [CrossRef]
35. Lewis, A.L. Phosphorylcholine-based polymers and their use in the prevention of biofouling. *Colloids Surf. B Biointerf.* **2000**, *18*, 261–275. [CrossRef]
36. Nichols, B.W. Light induced changes in the lipids of Chlorella vulgaris. *Biochim. Et Biophys. Acta (BBA)-Lipids Lipid Metab.* **1965**, *106*, 274–279. [CrossRef]
37. Rao, K.S.; Chakrabarti, P.P.; Rao, B.V.S.K.; Prasad, R.B.N. Phospholipid composition of Jatropha curcus seed lipids. *J. Am. Oil Chem. Soc.* **2009**, *86*, 197–200. [CrossRef]
38. Gunstone, F.D. The chemistry of oils and fats. In *Sources, Composition, Properties and Uses*; Blackwell Publishing Ltd.: Hoboken, NJ, USA, 2004; 345p.
39. Hatzakis, E.; Koidis, A.; Boskou, D.; Dais, P. Determination of phospholipids in olive oil by 31P NMR spectroscopy. J. *Agric. Food Chem.* **2008**, *56*, 6232–6240. [CrossRef]
40. Pawlik, H.; Prociak, A. Influence of palm oil-based polyol on the properties of flexible polyurethane foams. *J. Polym. Environ.* **2012**, *20*, 438–445. [CrossRef]
41. Saifuddin, N.; Chun, W.O.; Wei, Z.L.; Xin, N.K. Palm Oil Based Polyols for Polyurethane Foams Application. In Proceedings of the UNITEN-IKRAM, Selangor, Malaysia, 26 June 2010.
42. Velencoso, M.M.; Ramos, M.J.; Klein, R.; De Lucas, A.; Rodriguez, J.F. Thermal degradation and fire behaviour of novel polyurethanes based on phosphate polyols. *Polym. Degrad. Stab.* **2014**, *101*, 40–51. [CrossRef]
43. Zhang, M.; Zhang, L.; Hu, L.; Zhou, Y. Synthesis of rigid polyurethane foams with castor oil-based flame retardant polyols. *Ind. Crop. Prod.* **2014**, *52*, 380–388. [CrossRef]
44. Galià, M.; de Espinosa, L.M.; Ronda, J.C.; Lligadas, G.; Cádiz, V. Vegetable oil-based thermosetting polymers. *Eur. J. Lipid Sci. Technol.* **2010**, *112*, 87–96. [CrossRef]
45. Chen, R.; Zhang, C.; Kessler, M.R. Polyols and polyurethanes prepared from epoxidized soybean oil ring-opened by polyhydroxy fatty acids with varying OH numbers. *J. Appl. Polym. Sci.* **2015**, *132*, 41213. [CrossRef]
46. Mustafa, S.; Gan, S.N.; Yahya, R. Synthesis and Characterization of Novel Alkyds Derived from Palm Oil Based Polyester Resin. *Asian J. Chem.* **2013**, *25*, 8737–8740. [CrossRef]
47. Wu, L.; Van Gemert, J.; Camargo, R.E. Rheology Study in Polyurethane Rigid Foams. Huntsman International Technical Presentations Web Site. Available online: http://www.huntsman.com/polyurethanes/a/Products/Technical%20presentations%20overview (accessed on 10 October 2014).
48. Saalah, S. Synthesis and Characterization of Jatropha Oil-Based Waterborne Polyurethane Dispersions. Ph.D. Thesis, University Putra Malaysia, Serdang, Malaysia, 2016.
49. Guo, A.; Javni, I.; Petrovic, Z. Rigid Polyurethane Foams Based on Soybean Oil. *J. Appl. Polym. Sci.* **2000**, *77*, 467–473. [CrossRef]
50. Saba, N.; Jawaid, M.; Alothman, O.Y.; Paridah, M.T. A review on dynamic mechanical properties of natural fibre reinforced polymer composites. *Constr. Build. Mater.* **2016**, *106*, 149–159. [CrossRef]

polymers

MDPI

Article

A Study of the Strength Performance of Peat Soil: A Modified Cement-Based Stabilization Agent Using Fly Ash and Polypropylene Fiber

Mohammed K. H. Radwan [1], Foo Wei Lee [1,*], Yoke Bee Woon [1], Ming Kun Yew [1], Kim Hung Mo [2] and Soon Han Wai [3]

[1] Department of Civil Engineering, Lee Kong Chian Faculty of Engineering and Science, Universiti Tunku Abdul Rahman, Kajang 43000, Malaysia; eng.moh.rad@1utar.my (M.K.H.R.); woonyb@utar.edu.my (Y.B.W.); yewmk@utar.edu.my (M.K.Y.)

[2] Department of Civil Engineering, Faculty of Engineering, Universiti Malaya, Kuala Lumpur 50603, Malaysia; khmo@um.edu.my

[3] Department of Environmental Engineering, Faculty of Engineering and Green Technology, Universiti Tunku Abdul Rahman, Kampar 31900, Malaysia; waish@utar.edu.my

* Correspondence: leefw@utar.edu.my

Citation: Radwan, M.K.H.; Lee, F.W.; Woon, Y.B.; Yew, M.K.; Mo, K.H.; Wai, S.H. A Study of the Strength Performance of Peat Soil: A Modified Cement-Based Stabilization Agent Using Fly Ash and Polypropylene Fiber. *Polymers* **2021**, *13*, 4059. https://doi.org/10.3390/polym13234059

Academic Editors: Andrzej Puszka and Beata Podkościelna

Received: 30 September 2021
Accepted: 5 November 2021
Published: 23 November 2021

Abstract: The use of cement as a soil stabilization agent is one of the common solutions to enhancing the engineering properties of soil. However, the impact and cost of using cement have raised environmental concerns, generating much interest in the search for alternative materials to reduce the use of cement as a stabilizing agent in soil treatment. This study looked into limiting cement content in peat soil stabilization by using fly ash waste and polypropylene fiber (PPF). It focused on soil mechanical mediation for stabilization of peat with fly ash cement and PPF cement by comparing the mechanical properties, using unconfined compressive strength (UCS) and California bearing ratio (CBR) tests. The control (untreated) peat specimen and specimens with either fly ash (10%, 20% and 30%) and PPF (0.1%, 0.15% and 0.2%) were studied. Test results showed that 30% of fly ash and cement content displays the highest UCS and CBR values and gives the most reliable compressibility properties. On the other hand, UCS and CBR test results indicate optimum values of PPF–cement stabilizing agent content in the specimen of 0.15% PPF and 30% cement. Selected specimens were analyzed using scanning electron microscopy (SEM), and PPF threads were found to be well surrounded by cement-stabilized peat matrices. It was also observed that the specimen with 30% fly ash generated more hydration products when compared to the specimen with 100% cement content. It is concluded that the use of fly ash cement and PPF cement as stabilizing agents to limit the cement usage in peat soil treatment is potentially viable.

Keywords: peat soil; cement; stabilization; fly ash; polypropylene fiber; unconfined compressive strength (UCS); California bearing ratio (CBR); scanning electron microscopy (SEM)

1. Introduction

Around 8% of the total area of the world's land is covered by peat soil, and Malaysia is ranked ninth by total peat area with approximately 25,000 km^2 [1]. Peat deposits are formed from the accumulation and fossilization of partly decomposed and fragmented remains of plants [2]. It is relatively easy to differentiate peat from other soil types by its dark brown to black coloration, high organic content, high moisture content and lightweight nature [3].

Peat is categorized as a type of soft soil and is inappropriate for geotechnical applications due to its extreme compressibility (high porosity), high squeezability and low shear strength [4]. In addition, peat soil is often referred to as problematic soil due to its high acidity, high rates of creep, high water content (500–2000%) and high permeability [5].

These properties could cause differential settlement or failures in structures built on such soils, making peat soil an unsuitable material for construction.

There have been numerous studies on the feasibility of using cement as a stabilizing agent to enhance the strength properties of peat soil [6–8]. These studies have found that cement-based stabilizing agents generally improve the compressive strength of treated peat soil. It has also been proven that cement stabilization exhibits superior performance in enhancing the soil properties of peat soils with various characteristics and at different geographical locations [9]. The hydration reactions of cement and the formation of hydration products, such as calcium silicate (C–S–H) and calcium aluminate (C–A–H) hydrates, due to cement-based soil treatment, enhance the strength development of soil [10]. For example, calcium hydroxide ($Ca(OH)_2$), a cement hydration product, may react with silica or alumina sources in soil, contributing to enhancing the long-term strength of the treated soil [11]. However, cement production is known to emit a considerable amount of carbon dioxide (CO_2) into the atmosphere [12].

In order to reduce its impact on the environment, several types of industrial waste have been introduced to fully or partially replace cement as a soil stabilizer [13]. For example, Said and Taib [14] studied the use of carbide lime in peat soil stabilization. They found that the unconfined compressive strength (UCS) improved when increasing amounts of carbide lime were added to peat soil. On the other hand, Kolay and Pui [15] utilized fly ash and gypsum to stabilize peat soil, and reported that the use of fly ash gave better UCS results than gypsum. Pond ash [6], sodium bentonite [16] and clayey diatomite [5] were also used to stabilize peat soil. As efforts to search for other suitable waste materials to replace cement in soil treatment continued, it was found that soil treated with cement-waste-based stabilizer agents exhibits low flexural and tensile strength, and in some cases, even brittle behaviour [16,17]. However, very few studies have discussed and compared the use of recycled waste such as fly ash and polypropylene fiber (PPF) as peat stabilization agents in peat soil treatment.

Therefore, this study focused on investigating the mechanical properties of peat stabilized with different percentages of cement content (i.e., 10%, 20% and 30%), as the conventional peat soil stabilizer. The strength performance of cement-treated peat soil with added fly ash (10%, 20% and 30%) or PPF (0.1%, 0.15% and 0.2%) was also examined. UCS and CBR tests were conducted to evaluate the strength performance of each soil stabilizing material.

2. Materials and Methods

2.1. Peat Sample

Peat soil samples in this study were acquired from three different locations approximately 1 km apart in Johan Setia village, Klang city, located about 40 km away from Kuala Lumpur, as shown in Figure 1. The primary geotechnical properties of in-situ (untreated) peat soil are displayed in Table 1. The soil samples were obtained at depths between 0.5 and 1 m below the existing ground surface. It should be noted that the groundwater level was found to be about 0.4 m below the existing ground surface. This accounts for the high moisture content found in the peat soil samples. The physical appearance of peat soil revealed that it was fully saturated and had very dark brown coloration. Peat soil sampler was implemented in this study to determine the physical and chemical properties of peat soil. The sampler was used in a previous study by Zulkifley et al. [18]. The sampler mainly consists of a movable chamber that prevents the occurrence of disturbances, an anchored fin and a movable sampling chamber. Samples were collected using hand augers and were then oven-dried at 60 °C until a constant weight was observed (average duration of 48 h). The oven-dried samples were then stored in plastic bags after they reached room temperature.

Figure 1. Locations of peat soil sampling in this study.

Table 1. Physical properties of untreated peat soil.

Property	Standard and Specifications	Value
Depth of sampling (m)	-	0.5–1
Natural moisture content (%)	ASTM D2216	598.5
Fiber content (%)	ASTM D1997	79.33
Loss on Ignition (%)	ASTM D2974	90.84
Organic content (%)	ASTM D2974	90.47
Liquid limit (%)	BS EN 1997-2: 2006	200.2
Plastic limit (%)		Non plastic
Specific gravity (Gs)	AASHTO T 180-D	1.21
pH	BS EN 1997-2: 2006	4.5

2.2. Characteristics of Cement and Fly Ash

Ordinary Portland cement (OPC) was used in this study—cement type CEM I (Grade 42.5N/52.5N) with a specific gravity of 3.15, complying with MS EN 197-1. Coal fly ash with a specific gravity of 2.9 was used to reduce cement content, and it is classified as Class "C" in accordance with ASTM C618-17a [19]. The physical appearances of cement and fly ash are presented in Figure 2, and Table 2 lists the chemical composition, loss of ignition (LOI) and specific surface area of each of these materials.

Figure 2. Physical appearances of cement and fly ash (left to right).

Table 2. Chemical composition and specific surfaces of the binders used.

Binder	Composition (%)							LOI	Specific Surface Area (cm^2/g)
	Na_2O	Al_2O_3	SiO_2	MgO	CaO	Fe_2O_3	SO_3		
OPC	0.1	4.5	18.7	3.3	64.8	4.1	3	3.2	3110
Fly ash	3.1	16.1	35.1	8.1	15.3	17.3	2.4	2.5	3800

2.3. Characteristics of PPF

PPFs of 12 mm in length to aid the flexural behaviour of treated peat soil were used in this study (see Figure 3). The physical properties of PPF, as provided by the manufacturer, are listed in Table 3.

Figure 3. The physical appearance of PPF.

Table 3. Physical properties of PPF.

Properties	Length (mm)	Diameter (μm)	Density (g/cm^3)	Tensile Strength (MPa)	Modulus of Elasticity (GPa)	Elongation at Break (%)
PPF	12	50	0.9	350–500	3.5–3.9	12.9

2.4. Stabilizer Combinations and Sample Preparation

The proposed stabilizer combinations and proportions in each peat sample are listed in Table 4. In Phase 1, an untreated peat control specimen (labelled as U0) was prepared first. This was followed by the preparation of C10, C20 and C30 samples, where 10%, 20% and 30% cement by weight, respectively, was added to untreated peat. The cement contents could be described in weight per unit volume as 130, 140, and 150 kg/m^3. Thereafter, a series of tests were conducted. In Phase 2, the cement-treated peat was modified by adding fly ash (10%, 20% and 30% by weight, equal to 128, 137, and 146 kg/m^3, respectively). Besides that, three different levels of PPF (0.1% (1.2 kg/m^3), 0.15% (1.8 kg/m^3) and 0.2% (2.4 kg/m^3) by volume) were added to the peat samples, and similar tests were conducted on those samples.

Table 4. Mix designs of the stabilized peat specimens.

	Specimen ID	Material Content (%)		
		Cement	Fly Ash	PPF
Phase 1	U0	0	0	0
	C10	10	0	0
	C20	20	0	0
	C30	30	0	0
Phase 2	C10FA	10	(10%, 20% and 30%)	0
	C20FA	20	(10%, 20% and 30%)	0
	C30FA	30	(10%, 20% and 30%)	0
	C10PF	10	0	(0.1%, 0.15% and 0.2%)
	C20PF	20	0	(0.1%, 0.15% and 0.2%)
	C30PF	30	0	(0.1%, 0.15% and 0.2%)

The treated peat soil samples were each hand mixed in a mixing bowl for about 5–10 min until the mixture was homogeneous. The treated peat was then poured into 76 cm high cylindrical moulds of 38 mm in diameter. The moulds were demoulded the following day and placed in a water tank to be cured under laboratory conditions at 30 ± 2 °C until the desired testing days (days 7, 14 and 28).

3. Test Programme

3.1. pH Value Test

The pH values of untreated and treated peat soil were measured using a digital pH meter in accordance with ASTM D2976 (2015). First, 30 g of dry peat soil sample was sieved through a 200 μm sieve and mixed with distilled water. The sample was then stirred for 10 min to attain a homogenous mixture. The samples were left to stand overnight before pH measurements were taken.

3.2. Proctor Compaction Test

The Proctor compaction test was conducted to obtain the maximum dry density (MDD) and the optimum moisture content (OMC) values of untreated and treated peat soils in accordance with ASTM D 698 (2007). Before testing, the collected soil samples were oven-dried at 110 °C for four hours. They were then sieved using sieve no.4 (4.75 mm) and stored in sealed plastic storage boxes for further testing. Initially, 8% of water was assumed (per weight of dry peat soil), and the water content was then increased by 2% for each trial. Three layers of soil were then fed to the pre-weighted mould. Compaction was carried out using a 4.5 kg rammer by striking 27 blows with a uniform rate at about 1.5 s per drop. Each layer of compaction was about one-third of the volume of the mould. Next, the surface of the soil was trimmed off using a straight edge. The weight of the mould, together with the compacted soil, was recorded. The soil sample was then demoulded and oven-dried for 24 h. After 24 h, the sample was kept at room temperature to cool down and the dried mass was then recorded. This process was repeated five times for each sample's moisture content and the average density was recorded.

3.3. Unconfined Compressive Strength (UCS) Test

Upon the determination of OMC for each soil stabilizer, three specimens of treated and untreated peat soil were prepared according to ASTM D 2166 (2016) for UCS testing. The peat specimen was first mixed with its corresponding OMC. The mixed soil was then poured into cylindrical moulds (76 mm heigh by 38 mm diameter) and compacted into three layers. After 24 h, the specimens were extracted vertically using a hydraulic jack and cured in water before 7-day, 14-day and 28-day tests were conducted on the samples for UCS. The rate of strain was fixed at 1 mm/min.

3.4. California Bearing Ratio (CBR) Test

CBR tests to evaluate the strengths of the proposed soil stabilizers were conducted in accordance with ASTM D1883. The contents of cement, fly ash and PPF were mixed manually with the peat soil until reaching homogeneity in their OMC. The samples were then tested for UCS. The CBR moulds containing the soil samples were soaked in water to simulate actual in-situ conditions until the day of testing.

3.5. Morphology Analysis

Twenty-eight-day stabilized soil specimens were examined using a scanning electron microscope (SEM, Phenom XL Desktop, Thermo Fisher Scientific, Waltham, MA, USA). After the UCS test, small slices were collected from the cores of the samples. The sliced samples were dried using an electric laboratory oven at 60 °C for four hours to remove moisture and air vacuumed to remove dust particles. A Phenom ProX desktop microscope (Phenom XL Desktop, Thermo Fisher Scientific, MA, USA) was used to observe the morphologies of the stabilized soil samples. Scanning was conducted at an acceleration voltage of 2.0 kV and with a magnification of 2000×. An attached X-ray energy dispersive spectroscope (EDS, OmniProbe 200, Oxford Instruments, Abingdon, United Kingdom) was used to characterize and identify the elementary compositions of the soils.

4. Results and Discussions

4.1. Proctor Compaction Test

The results of OMC for the corresponding MDD for both the untreated and cement-treated peat soil are depicted in Figure 4. The results show that the untreated peat soil attained an MDD of 800 kg/m^3 when 40% moisture was utilized. One possible reason for such a high OMC is due to the presence of large voids, as organic contents and water are absorbed to fill up the large voids. This is in line with Zulkifley et al.'s description of peat soil having rather low bulk density and high water content due to the presence of high organic contents, such as branches, trunks, roots, leaves and fruits. This is also substantiated by literature showing a similar bulk density range of 800–1000 kg/m^3 in West Malaysian peat [20].

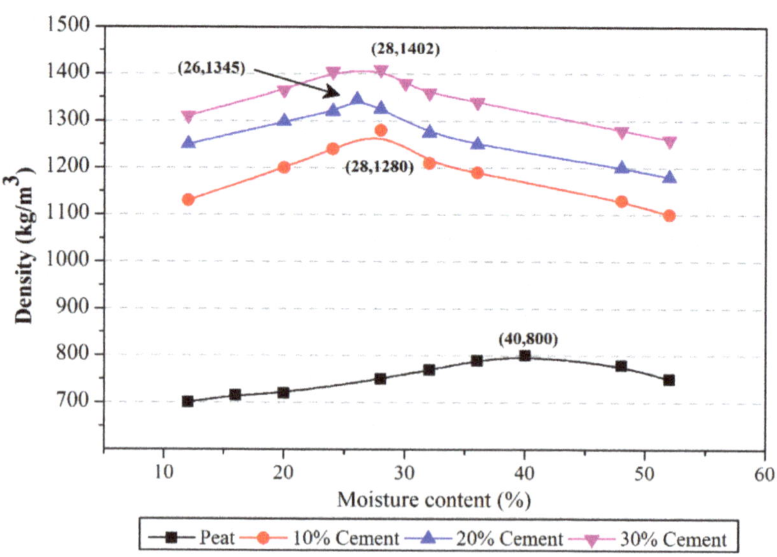

Figure 4. Optimum moisture content—maximum dry density diagram.

Values of MDD and OMC were found to be 1250 kg/m^3 and 32%, respectively, for stabilized peat soil with 10% OPC. MDD values for treated peat soil specimens were found to increase when OPC content was increased from 10% to 30%. This behaviour was a direct result of increasing the solid-to-organic content ratio and the higher specific gravity of peat soil with cement. On the other hand, OMC values were found to decrease when OPC content was increased. The low optimum moisture contents in these specimens were probably due to the voids being occupied by the presence of cement. Regardless of the amounts of cement, samples of treated peat soil showed consistent trends of increasing MDD and reducing OMC. This was due to the addition of cement, which in turn introduced further solid (non-organic) content to the peat soil. These observations are in line with the results obtained from Boobathiraja et al. [8].

4.2. pH Test

pH tests were conducted to determine the pH values of treated and untreated peat soil. The results are presented in Figure 5. The average pH value of untreated peat soil was found to be 4.5, indicating the acidic nature of peat soil. This value corresponds to the previously measured pH value of around 3.76 for local peat soil obtained from different Malaysian states [21]. The low pH value of peat soil is due to the presence of high amounts of decomposed plant fragments (organic), which influence the chemical and biological properties of peat soil [22].

The addition of cement turns the peat into an alkaline material. This was demonstrated by an increase in pH value to 11.5 when 10–30% of cement was added to peat soil. This correlates with the findings by Ghadir and Ranjbar [23], who reported the treated soil's pH values to be between 11 and 12 when cement dosages of 5% to 15% were added. A study by Moayedi and Nazir [22] also revealed that the release of OH- in remarkable amounts resulted in pH values of between 10 and 12 in stabilized peat soil.

The pH values of fly ash-treated soil were also found to increase with the addition of fly ash. Furthermore, pH increments of approximately 0.5 units per 10% increase in fly ash dosage were observed. This finding is consistent with the report from Nath et al. [24]. On the other hand, no significant changes were observed in the pH values of PPF-treated soil specimens. The slight changes in pH values in these specimens were likely due to the presence of cement in peat soil specimens.

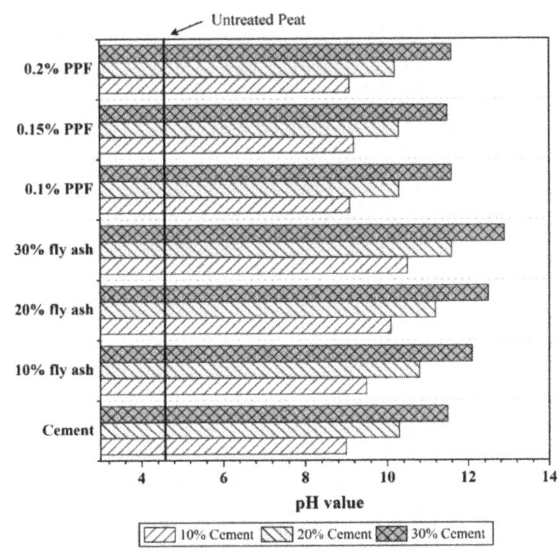

Figure 5. pH values of treated and untreated peat soil.

4.3. Unconfined Compressive Strength Test

Figure 6 shows the stress–strain diagram of untreated peat soil. The untreated peat soil had a maximum UCS of 95 kPa, corresponding to approximately 2.9% strain. This value lies within the range of reported strengths in previous research [7]. Compared to other soils (such as sand [25], clay [26,27], Laterite, Kaolin [21], etc.), peat soil exhibits lower UCS. This is mainly due to the presence of large amounts of water content, voids and organic content in peat soil [21].

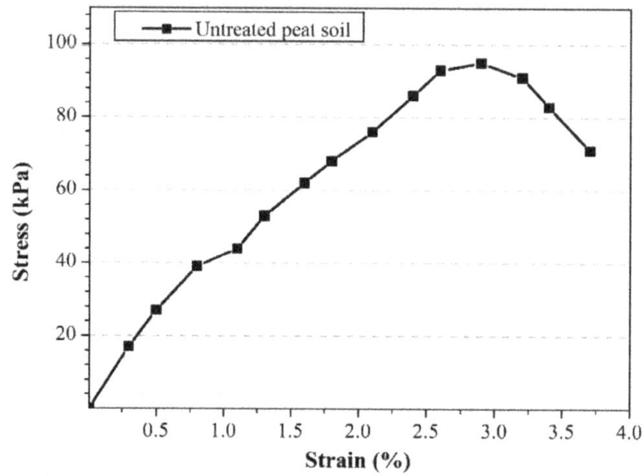

Figure 6. UCS of untreated peat soil.

4.3.1. Effect of Cement Addition on UCS

UCS test results for each proportion of cement content are presented in Figure 7. The UCS of cement-treated peat soil was found to increase with the curing period. This could be attributed to the additional hydration reactions that occur during curing [28], which in turn would have enhanced the strength characteristics of cement-treated peat soil. It was also observed that at all ages, higher UCS was attained by cement-treated peat soil as the cement content in the specimens was increased. For instance, the addition of 20% of cement resulted in about 30% higher UCS when compared to the 10% addition of cement for 7-day, 14-day and 28-day specimens. Likewise, when the cement dosage was increased up to 30%, the UCS increased by around 50% when compared to the UCS of the specimen treated with 20% cement. A similar observation was reported by Kalantari et al. [29].

4.3.2. Effect of Fly Ash Addition on UCS

Figure 8 shows UCS results for each combination of fly ash and cement. As fly ash is a cementitious material, fly ash-based test specimens also exhibited similar trends as those reported for solely cement-treated soil test specimens in Section 4.3.1. The addition of 30% fly ash resulted in higher UCS values in all mixtures considered. As expected, fly ash–cement combination specimens outperformed cement-treated specimens in terms of UCS (see Table 5). This might have been due to the hydration products produced during pozzolanic reactions that take place in the presence of fly ash. The release of calcium hydroxide ($Ca(OH)_2$), a by-product in the hydration process, activates the pozzolanic activity of fly ash, resulting in a gain in strength. The higher amount of fly ash is attributed to the increase in the filler effect of fly ash and the additional pozzolanic reaction products from fly ash.

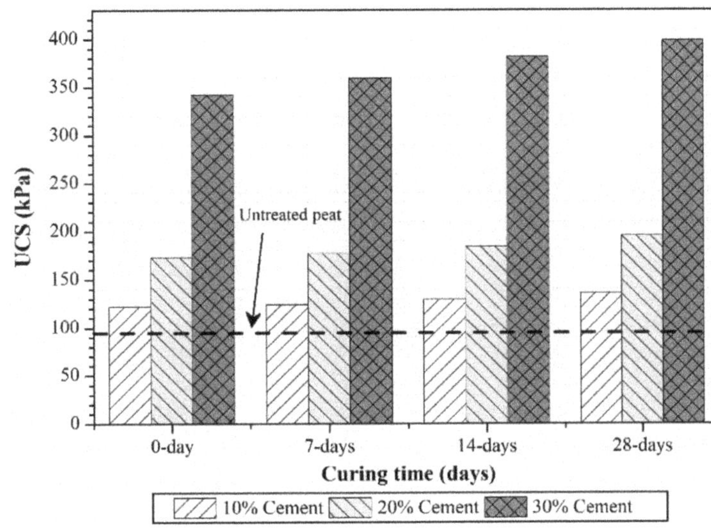

Figure 7. UCS of peat soil treated with different mix proportions of cement.

Table 5 summarizes the percentage improvements in the 28-day-curing test results of UCS for modified cement stabilizer with either fly ash or PPF. The results were calculated based on the differences between modified peat soil and untreated peat soil.

Table 5. Summary of UCS improvements (%) with added fly ash and PPF.

Untreated Peat	Cement		Fly Ash + Cement		PPF + Cement	
	Cement Content	UCS Improvement (%)	Fly Ash Content	UCS Improvement (%)	PPF Content	UCS Improvement (%)
95 KPa	10%	43.2	10%	93.7	0.10%	46.3
			20%	143.2	<u>0.15%</u>	<u>51.6</u>
			<u>30%</u>	<u>147.4</u>	0.20%	48.4
	20%	105.3	10%	155.8	0.10%	115.8
			20%	208.4	<u>0.15%</u>	<u>135.8</u>
			<u>30%</u>	<u>216.8</u>	0.20%	124.2
	<u>30%</u>	<u>319</u>	10%	352.6	0.10%	377.9
			20%	422.1	**<u>0.15%</u>**	**<u>475.8</u>**
			<u>30%</u>	**<u>429.5</u>**	0.20%	427.4

4.3.3. Effect of PPF Addition on Unconfined Compressive Strength

Figure 9a–c presents the UCS test results for 7, 14 and 28 day cures obtained from different proportions of PPF (0.1–0.2%) added to cement-treated peat soil. The addition of PPF was shown to increase the UCS of treated peat soil, particularly in samples with the 0.15% PPF and 30% cement combination. The comparison of UCS between PPF–cement samples and untreated peat soil samples is listed in Table 5. The addition of 0.1% of PPF to 10% and 20% cement-treated peat soils increased the UCS by 2.5 and 5 kPa on average, respectively. However, the UCS increased by around 55% when the same amount of PPF was added to 30% cement-treated peat soil. This might have been due to PPF being chemically inert and due to its ability to bind the cement during the hydration process. Hence, a more remarkable improvement was achieved.

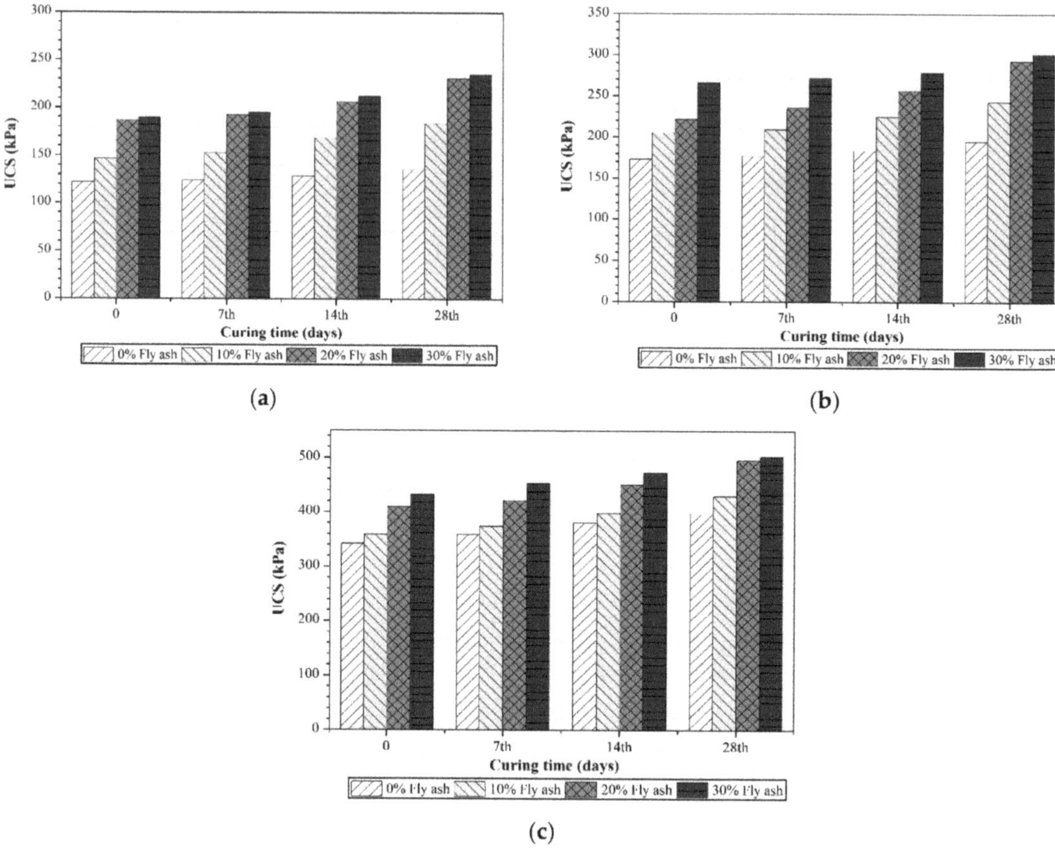

Figure 8. UCS of fly ash–cement-treated peat with (**a**) 10% cement and 0–30% fly ash, (**b**) 20% cement and 0–30% fly ash and (**c**) 30% cement and 0–30% fly ash and different curing durations.

On the other hand, the addition of 0.1% or 0.15% PPF also increased the UCS of cement-treated peat soil. This improvement was mostly due to the load-deformation behaviour through surface friction and interlocking between soil particles, the cement mixture and the PPF [30]. Soil reinforced with PPF exhibits greater toughness and ductility, increased formability and bending strength, and reduced post-peak strength loss compared to cemented-treated soil [31]. This implies that an appropriate amount of polypropylene should be considered as a stabilizing agent. The drop in UCS values of specimens with more than 2.0% of PPF content could have been due to the uneven distribution and lumping effect of PPF fragments.

4.4. California Bearing Ratio Test

The measured CBR test results for different proportions of fly ash cement and PPF-cement-treated peat soil are presented in Figures 10 and 11, respectively. The lowest CBR result observed was around 4% for untreated peat soil. Mohd Yusoff et al. [21] reported a CBR value of around 6.5% for unsoaked peat soil and assigned such a low value to the relative dry weight of peat soil. Thereafter, the addition of cement resulted in a substantial improvement in the CBR performance of peat soil. The results demonstrated that CBR values increased from 3.6% for untreated (U0) peat soil to around 10.3%, 15.2% and 19.8% in the cases of C10, C20 and C30 cement-treated peat soil specimens, respectively. In addition, the inclusion of both fly ash and PPF further enhanced the CBR performance of treated peat soil.

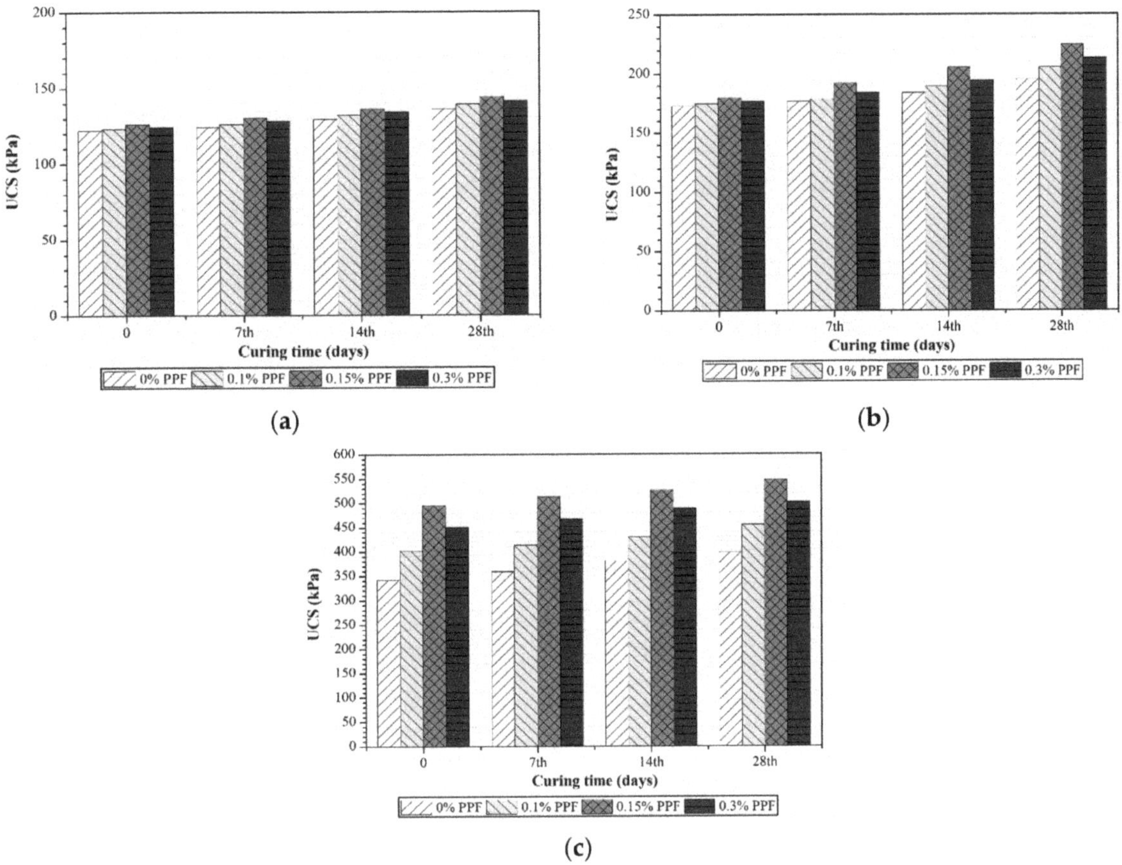

Figure 9. UCS of treated peat with (**a**) 10% cement and 0–0.2% PPF, (**b**) 20% cement and 0–0.2% PPF and (**c**) 30% cement and 0–0.2% PPF and different curing durations.

A positive correlation was observed between the CBR value and the amount of fly ash added to soil with all cement percentages, as can be seen in Figure 10. On average, the addition of 10% and 20% of fly ash to cement-treated peat soils gave about 12% and 24% higher CBR values when compared to the corresponding cement-treated peat soils (for all cement series). As an example, C20FA10 treated peat soil exhibited a CBR of 16.8%, whereas C20 (containing only 20% cement) produced a CBR of 15.2%. The largest improvement in CBR values (37% on average) was observed when 30% of fly ash was added to cement-treated peat soil. These observations are in line with UCS results, as the strength of treated peat soil samples improved with cement and fly ash contents.

Similarly, an increase in PPF content produced higher CBR results when compared to the corresponding cement-treated peat soil. Figure 11 shows that there were significant improvements in CBR values when 0.1% and 0.15% of PPF were added to cement-treated peat soils. However, slight reduced CBR values were observed when the amount of PPF was increased up to 0.2%. In all specimens, CBR values of cement-PPF treated peat soil exceeded those of untreated peat soil.

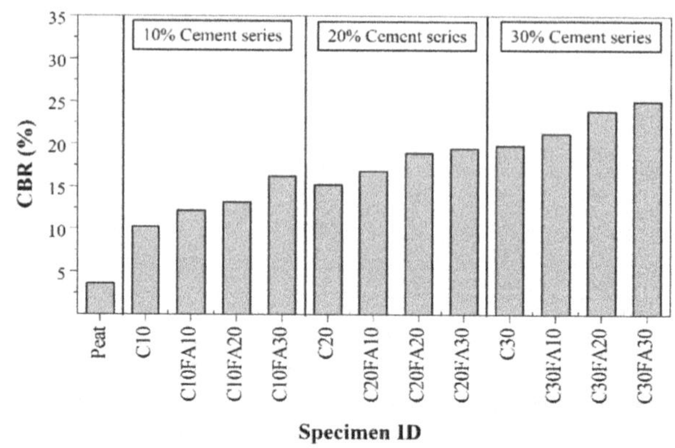

Figure 10. CBR results of fly ash cement-treated peat soil.

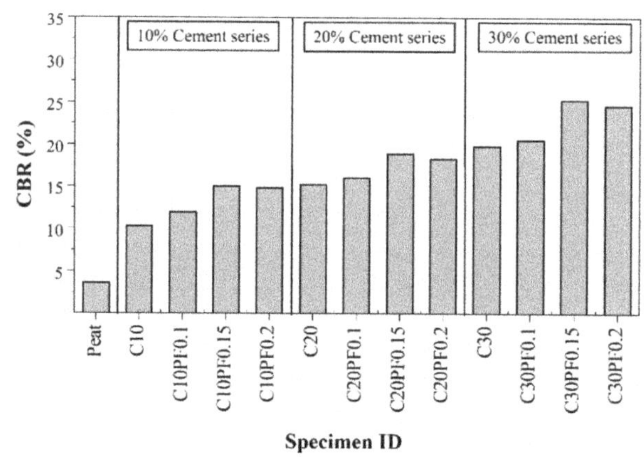

Figure 11. CBR results of PPF-cement-treated peat soil.

4.5. SEM Analysis

The results of SEM of selected stabilized soil specimens, including C30, C30FA10, C30FA30 and C30PF0.15, are presented in Figure 12. In general, all stabilized soil specimens showed a considerable number of internal microcracks accompanied by hydration products. Figure 12a shows the presence of a small, uneven distribution of microcracks in the stabilized soil with the addition of 30% cement. These microcracks could be attributed to the presence of cement particles, as exhibited by small expansion during the hydration process while the peat soil shrinks as the moisture is reduced. Hydration products are visibly shown to attach to peat particles. Figure 12b,c presents the SEM results of soil stabilized with 30% cement along with 10% or 20% fly ash, respectively. The morphologies of both soil samples demonstrate additional hydration products with fewer microcracks. The additional hydration products could be attributed to fly ash's presence, which provides additional C–S–H gel throughout the pozzolanic reactions. Unreacted fly ash particles were found to be present in both soil specimens, although it was more pronounced in the C30FA30 sample. This could be attributed to fly ash's slow pozzolanic reaction rate, which starts upon the formation of calcium hydroxide (CH) from cement hydration [28,32]. The presence of unreacted fly ash also contributed to enhancing the strength through the filler effect. This explained the enhanced UCS results when higher contents of fly ash were added to the soil. Figure 12d shows the morphology of CA30PF1.5 stabilized soil. As expected, significant amounts of hydration products were found to be distributed around the peat particles. It can be seen that the fiber threads were wrapped tightly by the cement hydration products–peat soil matrix, thereby effectively restricting the mobility of the fibre threads. The reduction in movement thus increased the reinforcement benefit of cement-stabilized soil.

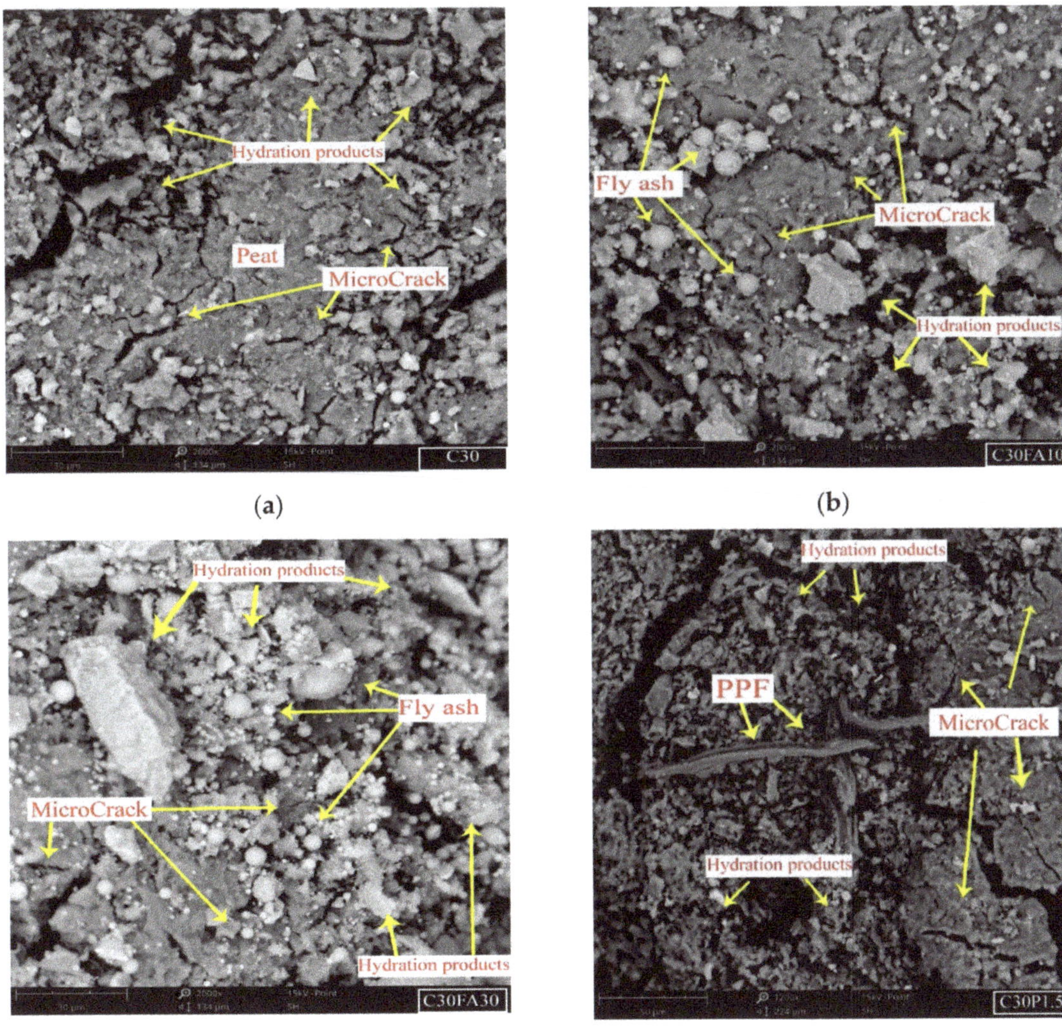

Figure 12. SEM micrographs of stabilized soil after 28 days with (**a**) C30 (30% cement), (**b**) C30FA10, (**c**) 30FA30 and (**d**) C30PF1.5%.

Figure 13 shows the clear morphology of PPF fiber thread present in the C30PF0.15 specimen under a magnification of 4500×. This clearly demonstrates the ability of the surfaces of PPF threads to bond with cement hydration products, thereby increasing the static friction between fiber and cement–soil matrix, eventually translating to an increase in the strength characteristics of fiber-added, cement-stabilized peat soil [33].

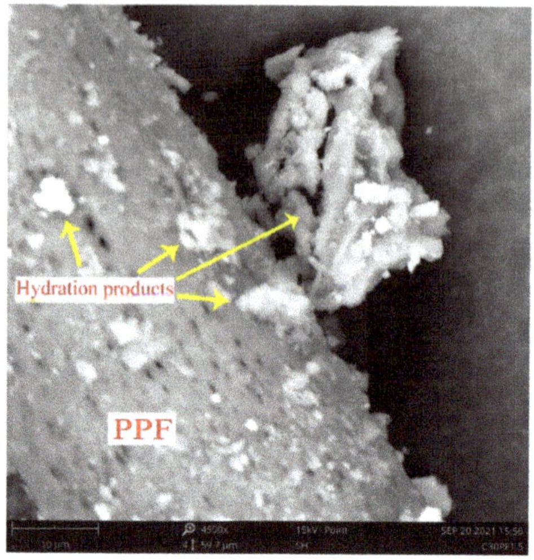

Figure 13. SEM image of C30PF1.5 with a magnification of 4500×.

5. Conclusions and Recommendations

This study evaluated the use of fly ash (10% to 30%) and polypropylene fiber (PPF) (0.1% to 0.2%) as additives to enhance cement-stabilized peat soil. MDD, OMC, pH, unconfined compressive strength (UCS) and California bearing ratio (CBR) tests were conducted on these specimens. The morphologies of stabilized soil were also examined using SEM–EDS analysis. Based on the results, the following conclusions can be drawn:

- The utilization of cement as a conventional soil stabilizing agent substantially enhanced the basic properties (pH, MDD and OMC) of peat soil, reflecting the improved geotechnical performances of treated peat soil. Higher cement content resulted in superior improvements in the UCS and CBR of cement-treated peat soil.
- Further enhancements in UCS and CBR values were attained by adding 10%, 20% and 30% of fly ash in cement-treated peat soil. The use of 30% of fly ash in the 30% cement-treated peat soil resulted in a 430% improvement in UCS and a 595% improvement in CBR.
- The addition of 0.15% PPF was found to be the optimum dosage of PPF, exhibiting a 476% improvement in UCS and a 600% improvement in CBR.
- SEM analysis depicted the presence of cement hydration products attached to the peat soil particles. The number of hydration products increased with the addition of fly ash. However, unreacted fly ash particles were presented in fly ash cement-stabilized peat soil, which enhanced UCS and CBR performances through the filler effect. A clear cement hydration product appeared on the surfaces of fiber threads—which was effectively surrounded by the cement–peat soil matrix—in PPF-cement-stabilized peat soil (i.e., C30PF1.5).

Based on the enhanced geotechnical properties obtained from this study, the use of both fly ash and polypropylene fiber can potentially limit the usage of cement in future peat soil treatments. The addition of 30% of fly ash and 0.15% of polypropylene fiber showed potential improvements in soil properties, even when no additional cement was added. For future research, it is recommended that other geotechnical soil properties, such as consistency limits, swelling parameters and undrained shear strength be further investigated. In addition, the influence of changes in flexural strength and stiffness could be considered in future investigations on the effects of PPF on cement-treated soil. The combined effects of using fly ash and PPF as cement-based stabilizer agents should also be investigated to determine the significant effects on the binding process of hydration products. It would also be interesting to study the crystalline phases presented in treated soil using X-ray diffraction analysis using different dosages of stabilizer and additives.

Author Contributions: Conceptualization, F.W.L.; methodology, M.K.H.R.; validation, K.H.M. and M.K.Y.; formal analysis, M.K.H.R.; investigation, S.H.W.; resources, K.H.M.; writing—original draft preparation, M.K.H.R.; writing—review and editing, F.W.L. and Y.B.W. All authors have read and agreed to the published version of the manuscript.

Funding: This research was funded by Universiti Tunku Abdul Rahman, grant number IPSR/RMC/UTARRF/2020-C1/L03.

Acknowledgments: The authors are grateful for the financial support provided by Universiti Tunku Abdul Rahman under the University Research Funding (IPSR/RMC/UTARRF/2020-C1/L03).

Conflicts of Interest: The authors declare no conflict of interest.

References

1. Mesri, G.; Ajlouni, M. Engineering Properties of Fibrous Peats. *J. Geotech. Geoenvironmental Eng.* **2007**, *133*, 850–866. [CrossRef]
2. Soper, E.K.; Osbon, C.C. *The Occurrence and Uses of Peat in the United States*; U.S. Government Printing Office: Washington, DC, USA, 1922.
3. Abdel-Salam, A.E. Stabilization of peat soil using locally admixture. *HBRC J.* **2018**, *14*, 294–299. [CrossRef]
4. Kazemian, S.; Huat, B.B.K.; Prasad, A.; Barghchi, M. A state of art review of peat: Geotechnical engineering perspective. *Int. J. Phys. Sci.* **2011**, *6*, 1974–1981. [CrossRef]
5. Dehghanbanadaki, A.; Sotoudeh, M.A.; Golpazir, I.; Keshtkarbanaeemoghadam, A.; Ilbeigi, M. Prediction of geotechnical properties of treated fibrous peat by artificial neural networks. *Bull. Eng. Geol. Environ.* **2019**, *78*, 1345–1358. [CrossRef]
6. Kolay, P.K.; Sii, H.Y.; Taib, S.N.L. Tropical peat soil stabilization using class F pond ash from coal fired power plant. *World Acad. Sci. Eng. Technol.* **2011**, *5*, 15–19. [CrossRef]
7. Kalantari, B.; Prasad, A. A study of the effect of various curing techniques on the strength of stabilized peat. *Transp. Geotech.* **2014**, *1*, 119–128. [CrossRef]
8. Boobathiraja, S.; Balamurugan, P.; Dhansheer, M.; Adhikari, A. Study on Strength of Peat Soil Stabilised with Cement and Other Pozzolanic Materials. *Int. J. Civ. Eng. Res.* **2014**, *5*, 2278–3652.
9. Timoney, M.J.; Mccabe, B.A.; Bell, A.L. Experiences of dry soil mixing in highly organic soils. *Proc. Inst. Civ. Eng. Gr. Improv.* **2012**, *165*, 3–14. [CrossRef]
10. Tremblay, H.; Duchesne, J.; Locat, J.; Leroueil, S. Influence of the nature of organic compounds on fine soil stabilization with cement. *Can. Geotech. J.* **2002**, *39*, 535–546. [CrossRef]
11. Paul, A.; Hussain, M. Cement Stabilization of Indian Peat: An Experimental Investigation. *J. Mater. Civ. Eng.* **2020**, *32*, 04020350. [CrossRef]
12. Onn, C.C.; Mo, K.H.; Radwan, M.K.H.; Liew, W.H.; Ng, C.G.; Yusoff, S. Strength, Carbon Footprint and Cost Considerations of Mortar Blends with High Volume Ground Granulated Blast Furnace Slag. *Sustainability* **2019**, *11*, 7194. [CrossRef]
13. Phetchuay, C.; Horpibulsuk, S.; Arulrajah, A.; Suksiripattanapong, C.; Udomchai, A. Strength development in soft marine clay stabilized by fly ash and calcium carbide residue based geopolymer. *Appl. Clay Sci.* **2016**, *127–128*, 134–142. [CrossRef]
14. Said, J.M.; Taib, S.N.L. Peat Stabilization with Carbide Lime. *J. Civ. Eng. Sci. Technol.* **2009**, *1*, 1–6. [CrossRef]
15. Kolay, P.K.; Pui, M.P. Peat Stabilization using Gypsum and Fly Ash. *J. Civ. Eng. Sci. Technol.* **2010**, *1*, 1–5. [CrossRef]
16. Hietala, M.; Koivuranta, E.; Ämmälä, A.; Oksman, K.; Illikainen, M. Moisture stability and mechanical durability of peat biocomposites. In Proceedings of the 21st International Conference on Composite Materials, Xi'an, China, 20–25 August 2017; pp. 20–25.
17. Correia, A.A.S.; Venda Oliveira, P.J.; Custódio, D.G. Effect of polypropylene fibres on the compressive and tensile strength of a soft soil, artificially stabilised with binders. *Geotext. Geomembr.* **2015**, *43*, 97–106. [CrossRef]
18. Zulkifley, M.T.M.; Fatt, N.T.; Konjing, Z.; Ashraf, M.A. Development of Tropical Lowland Peat Forest Phasic Community Zonations in the Kota. *Earth Sci. Res. J.* **2016**, *20*, O1–O10.
19. ASTM. *C618-17a Standard Specification for Coal Fly Ash and Raw or Calcined Natural Pozzolan for Use*; ASTM International: West Conshohocken, PA, USA, 2017.
20. Duraisamy, Y.; Huat, B.B.K.; Aziz, A.A. Engineering Properties and Compressibility Behavior of Tropical Peat Soil. *Am. J. Appl. Sci.* **2007**, *4*, 768–773. [CrossRef]
21. Mohd Yusoff, S.A.N.; Bakar, I.; Wijeyesekera, D.C.; Zainorabidin, A.; Madun, A. Comparison of Geotechnical Properties of Laterite, Kaolin and Peat. *Appl. Mech. Mater.* **2015**, *773–774*, 1438–1442. [CrossRef]
22. Moayedi, H.; Nazir, R. Malaysian Experiences of Peat Stabilization, State of the Art. *Geotech. Geol. Eng.* **2018**, *36*, 1–11. [CrossRef]
23. Ghadir, P.; Ranjbar, N. Clayey soil stabilization using geopolymer and Portland cement. *Constr. Build. Mater.* **2018**, *188*, 361–371. [CrossRef]

24. Nath, B.D.; Molla, M.K.A.; Sarkar, G. Study on Strength Behavior of Organic Soil Stabilized with Fly Ash. *Int. Sch. Res. Not.* **2017**, *2017*, 5786541. [CrossRef]
25. Fonseca, J.; O'Sullivan, C.; Coop, M.R.; Lee, P.D. Quantifying the evolution of soil fabric during shearing using scalar parameters. *Geotechnique* **2013**, *63*, 818–829. [CrossRef]
26. Xiao, H.; Lee, F.H.; Liu, Y. Bounding Surface Cam-Clay Model with Cohesion for Cement-Admixed Clay. *Int. J. Geomech.* **2017**, *17*, 04016026. [CrossRef]
27. Liu, Y.; He, L.Q.; Jiang, Y.J.; Sun, M.M.; Chen, E.J.; Lee, F.H. Effect of in situ water content variation on the spatial variation of strength of deep cement-mixed clay. *Geotechnique* **2019**, *69*, 391–405. [CrossRef]
28. Radwan, M.K.H.; Onn, C.C.; Mo, K.H.; Yap, S.P.; Chin, R.J.; Lai, S.H. Sustainable ternary cement blends with high-volume ground granulated blast furnace slag–fly ash. *Environ. Dev. Sustain.* **2021**, 1–35. [CrossRef]
29. Kalantari, B.; Prasad, A.; Huat, B.B.K. Stabilising peat soil with cement and silica fume. *Proc. Inst. Civ. Eng. Geotech. Eng.* **2011**, *164*, 33–39. [CrossRef]
30. Kalantari, B.; Huat, B.B.K. Peat soil stabilization, using Ordinary Portland Cement, Polypropylene fibers, and Air Curing Technique. *Electron. J. Geotech. Eng.* **2008**, *13*, 1–13.
31. Malekzadeh, M.; Bilsel, H. Effect of polypropylene fiber on mechanical behaviour of expansive soils. *Electron. J. Geotech. Eng.* **2012**, *17*, 55–63.
32. Radwan, M.K.H.; Onn, C.C.; Mo, K.H.; Yap, S.P.; Ng, C.G.; Yusoff, S. Eco-mechanical performance of binary and ternary cement blends containing fly ash and slag. *Proc. Inst. Civ. Eng.-Eng. Sustain.* **2021**, *174*, 23–36. [CrossRef]
33. Tang, C.; Shi, B.; Gao, W.; Chen, F.; Cai, Y. Strength and mechanical behavior of short polypropylene fiber reinforced and cement stabilized clayey soil. *Geotext. Geomembr.* **2007**, *25*, 194–202. [CrossRef]

polymers

Article

Preparation and Characterization of Highly Porous Polyacrylonitrile Electrospun Nanofibers Using Lignin as Soft Template via Selective Chemical Dissolution Technique

Mohd Adib Tajuddin Ahmad [1] and Norizah Abdul Rahman [1,2,*]

[1] Department of Chemistry, Faculty of Science, Universiti Putra Malaysia, Serdang 43400, Selangor, Malaysia; adibataupm@gmail.com
[2] Materials Synthesis and Characterization Laboratory, Institute of Advanced Technology, Universiti Putra Malaysia, Serdang 43400, Selangor, Malaysia
* Correspondence: a_norizah@upm.edu.my

Abstract: In this study, polyacrylonitrile (PAN) was mixed with a renewable polymer, lignin, to produce electrospun nanofibers by using an electrospinning technique. Lignin was utilized as a soft template that was removed from the nanofibers by using a selective dissolution technique to create porous PAN nanofibers. These nanofibers were characterized with Fourier transform infrared (FTIR), field emission scanning electron microscopy (FESEM), thermogravimetry analysis (TGA), X-ray diffraction (XRD), and Brunauer-Emmett-Teller (BET) to study their properties and morphology. The results showed that lignin can be homogeneously mixed into the PAN solution and successfully electrospun into nanofibers. FESEM results showed a strong relationship between the PAN: lignin ratio and the diameter of the electrospun fibers. Lignin was successfully removed from electrospun nanofibers by a selective chemical dissolution technique, which resulted in roughness and porousness on the surface of the nanofibers. Based on the BET result, the specific surface area of the PAN/lignin nanofibers was more than doubled following the removal of lignin compared to PAN nanofibers. The highest specific surface area of nanofibers after selective chemical dissolution was found at an 8:2 ratio of PAN/lignin, which was 32.42 m^2g^{-1} with an average pore diameter of 5.02 nm. The diameter of electrospun nanofibers was also slightly reduced after selective chemical dissolution. Porous PAN nanofibers can be seen as the precursors to the production of highly porous carbon nanofibers.

Keywords: polyacrylonitrile; lignin; electrospinning; selective chemical dissolution; porous nanofibers; nanofibers; soft template

Citation: Ahmad, M.A.T.; Abdul Rahman, N. Preparation and Characterization of Highly Porous Polyacrylonitrile Electrospun Nanofibers Using Lignin as Soft Template via Selective Chemical Dissolution Technique. *Polymers* **2021**, *13*, 3938. https://doi.org/10.3390/polym13223938

Academic Editor: Andrzej Puszka

Received: 15 September 2021
Accepted: 5 October 2021
Published: 15 November 2021

1. Introduction

In recent years, nanotechnology has become a promising technology for the future. There are many types of nanomaterials, such as nanoparticles, nanofibers, nanotubes, nanowires, and quantum dots, that have garnered extensive attention due to their unique properties and characteristics [1]. Among these nanomaterials, nanofibers are receiving significant attention from researchers due to their holding properties such as high porosity, a high surface area and controllable morphology, and high chemical and thermal stability [2,3]. Several techniques can be used to produce nanofibers, such as template-assisted synthesis [4], chemical vapor synthesis [5], and electrospinning.

At present, electrospinning techniques have received interest as they are economical and suited to various types of material [6], good for controlling fiber morphology [7], and easy to operate [8]. A high-voltage power supply, a conductive collector, and a spinneret are the important, basic forms of equipment that are needed for the electrospinning process. Electrospinning involves an electrohydrodynamic process, during which a high voltage is applied to a polymer droplet; this is designed to reduce the surface tension of the polymer droplet and transform it into a pointed shape. Increasing the voltage supply will trigger the formation of a Taylor cone and a jet from the tip of needle. The solvent evaporates as the jet travels to the collector, and to prevent the jet from forming into droplets (electrospraying), sufficient entanglement of the polymer occurs [9]. An electrospinning technique is a distinctive way to produce nanofibers from various types of polymers with a diameter ranging from micrometer to nanometer, and to produce a higher surface area than those obtained from conventional spinning processes [10]. Thus, electrospinning is the best tool for the production of polymer nanofibers as it is economical and the simplest method [11].

Recently, a significant number of studies that have used nanofibers as an adsorbent for the removal of pollutants from aqueous solutions, such as drugs [12,13] and heavy metals [14–17], has been reported. For the application of nanofibers as adsorbent, the adsorption process is crucial. During this process, the adsorbate molecules are transported into the surface of the absorbent. The internal mass transfer is transported to the inner surface of the porous structure from the outer surface of the adsorbent [18]. Targeted contaminant removal depends on the pore size of the adsorbent [19]. The pores of the nanofibers are made up of several categories: micropores (<2 nm), mesopores (2–50 nm), and macropores (>50 nm). The main part of the adsorption process occurs with mesopores and micropores [20].

Polyacrylonitrile (PAN) is a polymer usually used as the precursor to the formation of carbon fibers due to its excellent mechanical strength, low flammability, good thermal stability and chemical resistance [21], such as PAN-based carbon fiber electrodes for energy storage application [22,23] and for thermal materials [24]. It is a homo-polymer made up of acrylonitrile. PAN contains a nitrile group, known as a polar group, that helps it to be soluble in a highly polar solvent, and helps increase its melting points. The interaction of the nitrile group and the polymeric backbone of the polymer hinder the molecules of PAN carbon fibers to align in a particular direction. Dimethylformamide (DMF) is the most used solvent for dissolving PAN to make nanofibers; the solvent helps to reduce the interaction between the nitrile groups, and helps provide a better orientation of the produced polymer chain [25].

Lignin is among the most promising biorenewable raw materials, and it is a natural polymer that acts as a partial matrix within the structure of plants and trees. It is one of the main elements of wood and is the second most abundant natural polymer. Lignin is a highly branched macromolecular network structure, with an aromatic nature and complex compositions. Furthermore, lignin has a lack of cytotoxicity, which proves that lignin is a type of biomaterial [26]. Industries such as the pulp and paper industry are the main sources of lignin as a by-product and of second-generation bioethanol facilities derived from lignocellulosic sources [27]. The structure of lignin is composed of phenylpropene units (monolignols) of the hydroxyl group, but the plant source and its isolation process impact lignin complex configuration (Duval and Lawoko, 2014). Lignin has some advantages, such as a high carbon content, its suitability with various chemicals, good thermal stability, and its low cost and biodegradability [28]; however, there are some limitations, such as the relative heterogeneity of its structure and difficulties in its processing that still hinder its more widespread use [29]. Lignin has received interest among researchers as a potential low-cost carbon fiber precursor because it is easy to obtain and has a high carbon content.

The selective chemical dissolution technique is one of the processes to produce a porous structure and higher surface area in a polymer. For porous polymer nanofibers, this can be achieved through the electrospinning of polymer blend solutions to form a nanofiber, followed by the selective chemical dissolution process which involves the removal of one component using an appropriate solvent that only dissolves a certain polymer in the blend's nanofibers. Another alternative is the addition of additives such as salts or nanoparticles in the polymer solution that are used as a template, followed by their removal by post-electrospinning processes; the elimination of the template is another method to create a porous structure [30]. In a study by Chen et al. [31], silk fibroin/poly (L-lactic acid) (SF/PLA) fibers were produced by an electrospinning technique. The silk fibroin was removed from the fibers by dissolving the fibers in chloroform for up to 1 h.

After the selective chemical dissolution process, the fibers showed a porous structure with irregular shapes [31]. In another study, Kim et al. [23] used electrospinning to produce poly(vinyl alcohol) (PVA) and PAN blend fibers. Either PVA or PAN can be removed from the nanofibers. PVA was successfully removed from the fibers by dipping them in hot water and acetic acid, while PAN was eliminated from the fibers using DMF treatment. Both results showed the surface diameter of the electrospun fibers as decreasing, and the surface of the fibers revealed grooves along the axis and appeared rougher [32].

In this study, alkali lignin, with a low sulfonate content, was incorporated with PAN to produce nanofibers. A renewable polymer, lignin was used as a soft template to create a porous and rougher surface for the nanofibers through the selective chemical dissolution technique.

2. Materials and Methods

2.1. Reagents

The materials used in this experiment were alkali lignin (low sulfonate content) and polyacrylonitrile (PAN) (average M_w = 150,000), and were purchased from Sigma-Aldrich (Merck Group, St. Louis, MO, USA). *N,N*-dimethylformamide (DMF) was obtained from R&M Chemicals (Ever Gainful ENT., Ara Damansara, Selangor, Malaysia). De-ionized water was used throughout the experiment.

2.2. Preparation of Electrospun Nanofibers

PAN nanofibers were prepared from a PAN solution dissolved in DMF with a concentration of 7.5%. A 5 mL syringe was loaded with the PAN solution, and a 0.8 mm inner diameter needle was attached with the syringe. The distance between the tip of the needle and the aluminum foil collector was fixed at 10 cm. The electrospinning was conducted using 18 kV voltage with a 2 mL/h flow rate. The nanofibers produced were collected on the aluminum foil and characterized. Following this, the PAN/lignin nanofibers were prepared in the same way as the PAN nanofibers. The ratios of PAN/lignin were varied: 9:1, 8:2, 7:3, and 6:4.

2.3. Selective Chemical Dissolution Technique

Lignin is a water-soluble polymer, while PAN is insoluble in water. Thus, PAN/lignin nanofibers were immersed in de-ionized water at 60 °C for 30 min to eliminate lignin from the electrospun nanofibers, and then the nanofibers were dried in a convection oven for 3 h at 60 °C and sent for characterization. This procedure was repeated for the PAN nanofibers for comparison purposes.

2.4. Characterizations

FTIR analyses of the samples were conducted to determine the functional groups using an attenuated total reflectance-Fourier transform infrared spectroscopy (ATR–FTIR) spectrometer (Perkin Elmer Spectrum RXI, Waltham, MA, USA). The wavelength used to examine the functional groups ranged from 280 to 4000 cm^{-1}. The thermal properties of the samples were determined using thermal gravimetric analysis (TGA) (Mettler Toledo Thermogravimetric Model TGA/SDTA, Columbus, OH, USA). The samples were heated from 50 °C to 600 °C, with a heating rate of 10 °C/min under a nitrogen atmosphere. Field emission scanning electron microscopy (FESEM) (FEI NOVA NANOSEM 230, Hillsboro, OR, USA) was used to examine the morphologies of the nanofibers. A stub was taped with carbon tape and the samples were placed onto it. The samples were then sputtered with a thin layer of gold. The average fiber diameter was determined by using ImageJ Software, with 200 diameter readings taken. X-ray diffraction (XRD) (Shimadzu Model XRD-6000, Kyoto, Japan) analyses of the samples occurred to identify the crystalline phase, where the scan started from 2θ range and from 2° to 60°, with a scanning rate of 2°/min. The surface area and porosity of the prepared nanofiber were determined by N_2 adsorption measurements at 77K using BELSORP Mini II. The surface characterization of the nanofibers was obtained using the Brunauer-Emmet-Teller (BET Micromeritics Instrument Corporation, Model-3Flex, Norcross, GA, USA) method.

3. Results and Discussion

3.1. Fourier Transform Infrared Spectra (FTIR) Analysis

ATR-FTIR technique was used to analyze the PAN, lignin, and PAN/lignin nanofibers before and after selective chemical dissolution. Figure 1 shows the FTIR spectra of PAN, lignin, and PAN/lignin nanofibers. The ratio of PAN: lignin was varied at 9:1, 8:2, 7:3, and 6:4 to study the effect of different ratios on the fabricated PAN/lignin nanofibers. The IR spectrum of lignin shows the major lignin bands at 3400 cm^{-1} (O–H stretching), 1595 cm^{-1} (aromatic skeletal vibration), 1430 cm^{-1} (C–C stretching with C–H deformation), 1134 cm^{-1} (C–H deformation), and 1040 cm^{-1} (C–O deformation) [33]. In Figure 1, it can be seen that the higher the lignin content in the fiber, the higher the intensity of the lignin peaks at 1595 cm^{-1}, 1134 cm^{-1}, and 1040 cm^{-1}. A small peak at 2920 cm^{-1} indicates a stretching of the alkyl C–H bond and a stretching of the intermolecular hydrogen bonding of the O–H group, which is associated with the hydroxyl group in the lignin [34]. A strong peak at 2240 cm^{-1} indicates the C≡N group of PAN; the intensity of the peak was decreased as the ratio of PAN decreased. A peak at around 1450 cm^{-1} corresponds to the bending of the CH_2 scissoring of PAN [35]. The intensity of the peak at 1450 cm^{-1} was reduced when the ratio of lignin increased. As can be seen in Figure 1, the main peak position of PAN did not shift for all the nanofibers, demonstrates that the blending of PAN and lignin only involves physical blending.

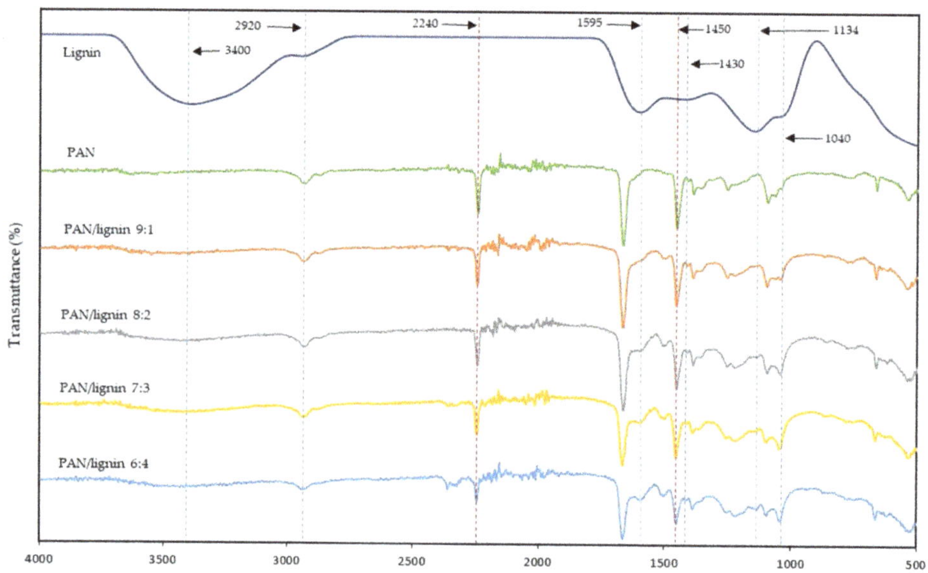

Figure 1. FTIR spectra of the PAN/lignin nanofibers after selective chemical dissolution compared to PAN and lignin.

The FTIR spectra of the PAN/lignin nanofibers after the selective chemical dissolution process when compared to PAN and lignin is shown in Figure 2. The IR spectrum of PAN shows peaks at around 2920, 2240, 1664, and 1450 cm^{-1}. These peaks indicate the alkyl C–H bond stretching, C≡N stretching, C≡N stretching, and CH_2 bending, respectively [36–38]. Selective chemical dissolution was performed to remove the lignin from the nanofibers. The results showed that, for the nanofibers after treatment, most IR peaks in PAN/lignin nanofibers exhibited mainly belong to PAN. In comparison, the peaks that belong to lignin, such as the peaks at around 1595 cm^{-1} and 1040 cm^{-1}, were reduced significantly. This is because the lignin content in the nanofibers was successfully removed through the selective chemical dissolution process.

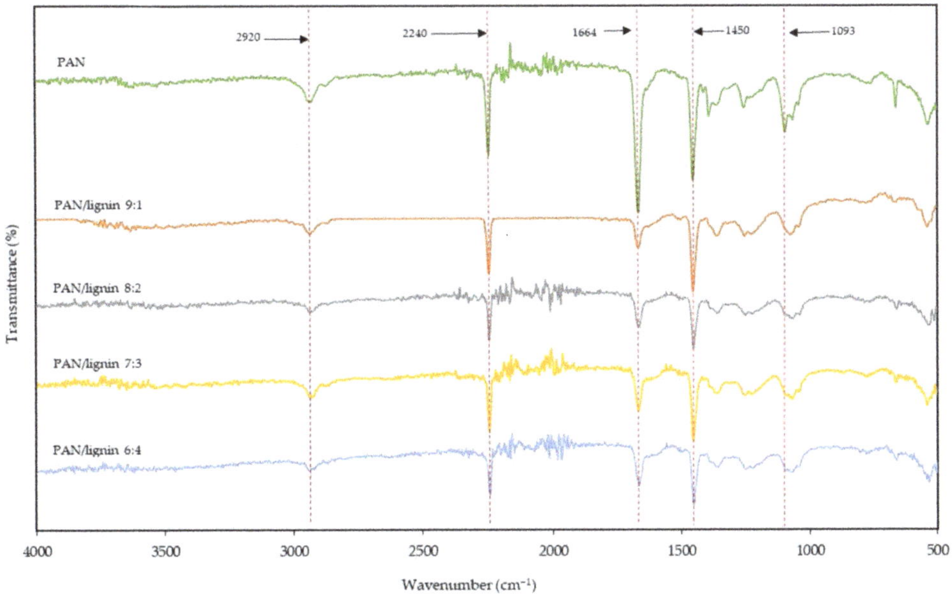

Figure 2. FTIR spectra of pure PAN and PAN/lignin nanofibers after selective chemical dissolution process.

3.2. Field Emission Scanning Electron Microscope (FESEM) Analysis

The morphology of the nanofibers was investigated using the FESEM technique. Figures 3 and 4 show FESEM images and the average diameter of the PAN/lignin nanofibers against the various ratios of PAN: lignin nanofibers, respectively. The PAN nanofibers did not contain lignin, but underwent a selective chemical dissolution process for comparison purposes. From the FESEM images in Figure 3, it can be observed that before selective chemical dissolution, the nanofibers' surface is smoother than after selective chemical dissolution. However, it can be seen that some pores and rough surfaces appear on the nanofibers before selective chemical dissolution, possibly due to the effect of rapid solvent evaporation during electrospinning and drying proses. After selective chemical dissolution, the surface nanofibers become rougher and more porous due to the removal of lignin during hot de-ionized water treatment. The average diameter of nanofibers before and after selective chemical dissolution decreased when the content of lignin increased from a ratio of 9:1 to 6:4. The average diameter of PAN and PAN/lignin nanofibers at ratios of 9:1, 8:2, 7:3, and 6:4 before selective chemical dissolution was 419 $\pm$ 49 nm, 409 $\pm$ 35 nm, 371 $\pm$ 31 nm, 362 $\pm$ 52 nm, and 358 $\pm$ 31 nm, respectively. When the content of lignin increased, the average diameter of nanofibers decreased. This is due to a decrease in the viscosity of the polymer solution that produces a lower fiber diameter, as mentioned earlier. This is due to the lower molecular weight and higher polydispersity index of lignin compared to PAN [39]. The average diameter of PAN and PAN/lignin nanofibers after dissolution was slightly decreased compared with before selective chemical dissolution; 412 $\pm$ 55 nm, 386 $\pm$ 29 nm, 349 $\pm$ 44 nm, 340 $\pm$ 40 nm, and 327 $\pm$ 35 nm for PAN and PAN/lignin with ratios of 9:1, 8:2, 7:3, and 6:4, respectively. The FESEM images show that all the nanofibers before and after dissolution were bead-free, and that the nanofibers had a non-uniform, porous, and rough surface. This proves that, by using the selective chemical dissolution technique, lignin could be successfully removed from nanofibers and could produce a rough surface and cause a high porosity of PAN/lignin nanofibers, providing a higher surface area. However, FESEM images only are insufficient to prove the higher surface area of nanofibers after selective chemical dissolution. These results can be supported by Brunauer-Emmett-Teller (BET) analysis, which will be discussed later in Section 3.5.

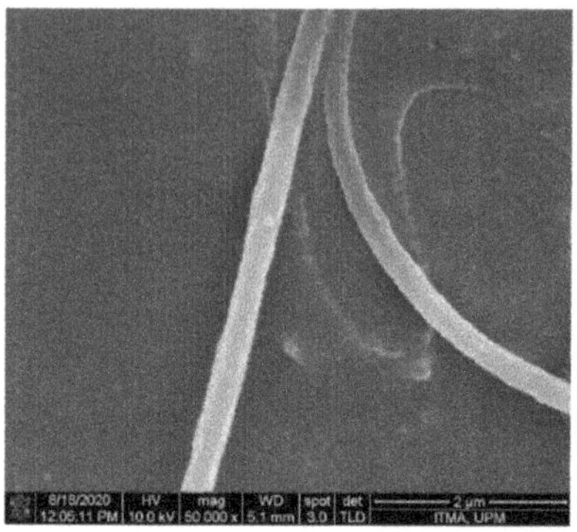

(**a**)

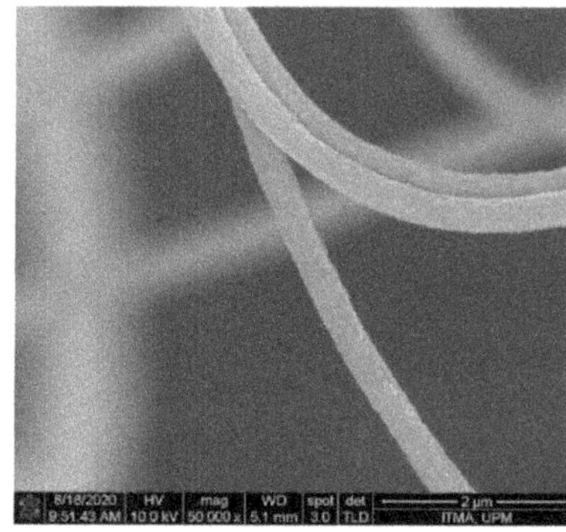

(**a′**)

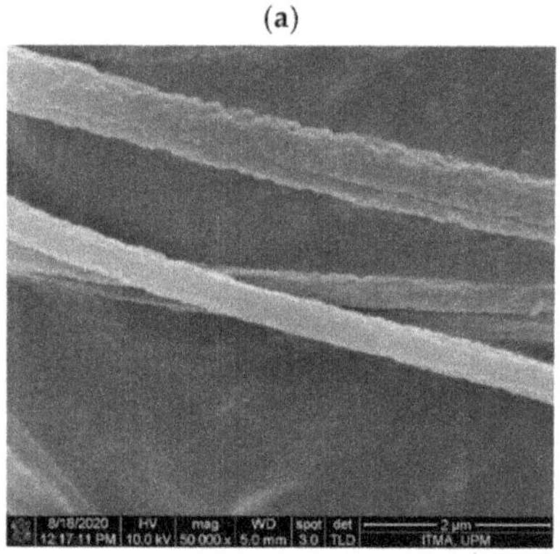

(**b**)

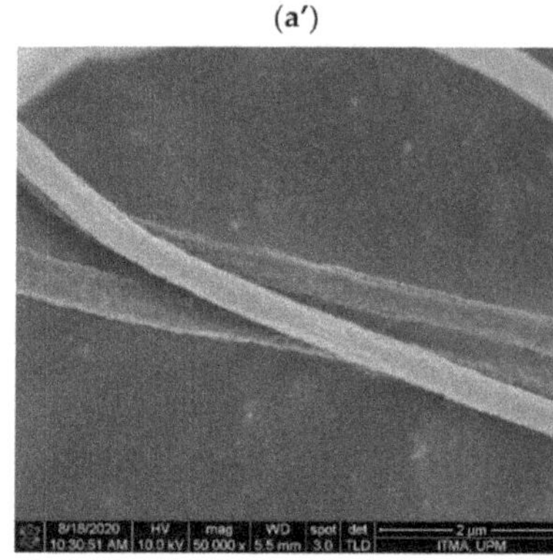

(**b′**)

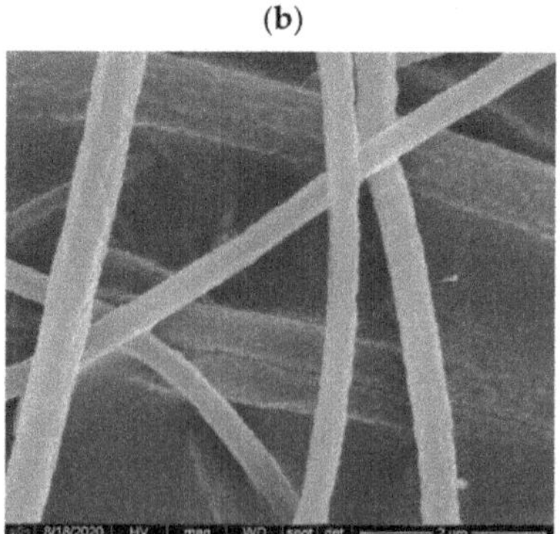

(**c**)

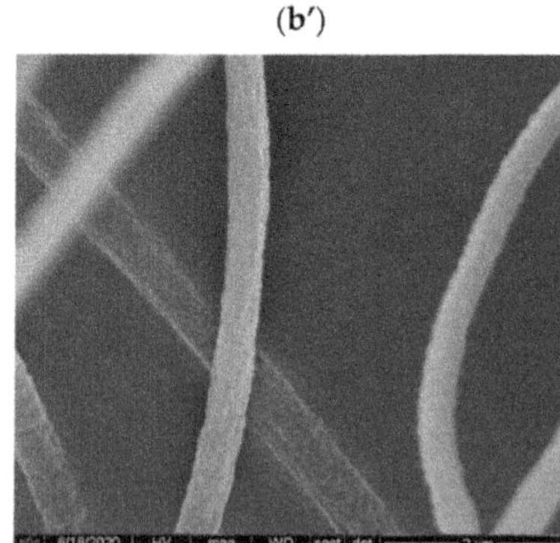

(**c′**)

Figure 3. *Cont.*

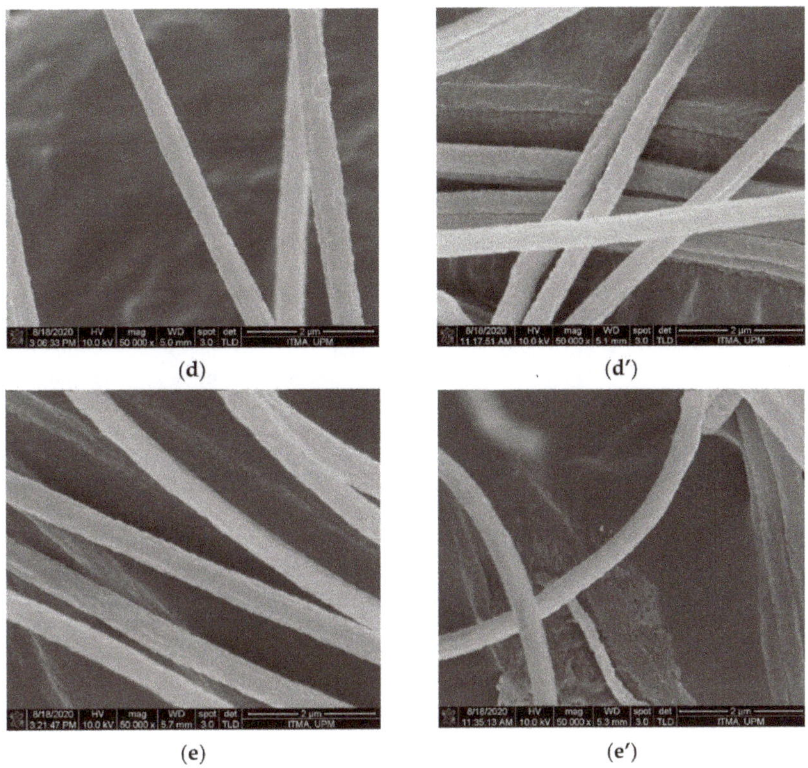

Figure 3. FESEM micrograph of nanofibers with various PAN: lignin ratios (**a**) PAN, (**b**) 9:1, (**c**) 8:2, (**d**) 7:3, (**e**) 6:4 ratio before selective chemical dissolution, and (**a′**) PAN, (**b′**) 9:1, (**c′**) 8:2, (**d′**) 7:3, and (**e′**) 6:4 ratio after selective chemical dissolution.

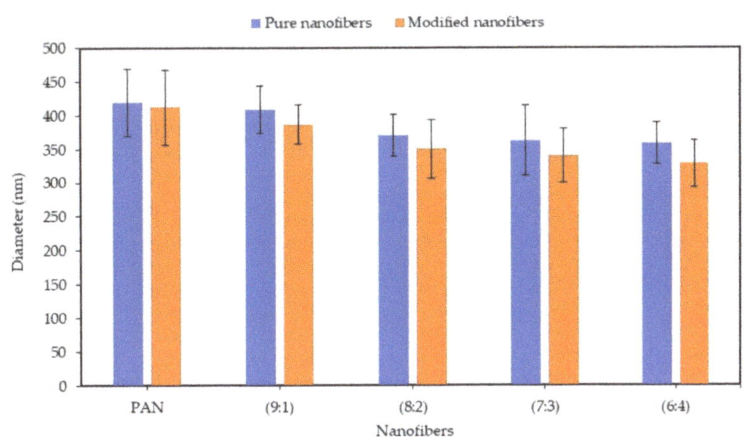

Figure 4. Graph of the diameter of nanofibers before (pure nanofibers) and after (modified nanofibers) selective chemical dissolution.

3.3. Thermal Analysis

The derivative thermogravimetry (DTG) and TGA analyses of the PAN: lignin nanofibers before and after selective chemical dissolution was carried out to study the thermal stability of the samples. Figures 5 and 6 show the DTG thermogram at different ratios of PAN: lignin nanofibers, from 50 °C to 600 °C. TGA thermograms of nanofibers before and after selective dissolution are shown in Supplementary Materials (Figures S1 and S2). The DTG and TGA thermograms of nanofibers both before and after selective chemical dissolution show a similar trend. The first weight loss can be observed at 50–120 °C and is due to the loss of water molecules that are bound to the nanofibers. TGA thermograms of the nanofibers (Figures S1 and S2) show that all the nanofibers gradually decomposed at a higher temperature range. From DTG and TGA thermograms, the main degradation began at 275 °C for PAN nanofibers, attributed to the pyrolysis of the nanofibers [40]. The addition of lignin to the nanofibers (PAN/lignin nanofibers) lowered the onset degradation temperature of the fibers (~260 °C) compared to PAN nanofibers; however, there was no significant difference between the various ratios of lignin in the fibers. The addition of lignin slightly reduced the thermal stability of the nanofibers. When comparing the results before and after selective chemical dissolution, the onset degradation temperature of the PAN/lignin nanofibers was slightly higher than before selective chemical dissolution, possibly due to the removal of lignin and because the nanofibers are mostly composed of PAN, which has a higher thermal stability than lignin. However, after selective dissolution, the onset temperature of the PAN/lignin nanofibers was still slightly lower than PAN nanofibers (Figure 6). It is noticeable that the PAN/lignin (8:2) in Figure S1 and the PAN/lignin (6:4) in Figure S2 show that weight loss exceeded 100%. The possible explanation for this is that the PAN fiber can have a reaction with nitrogen under thermal treatment [41] with a resulting weight increase of over 100%.

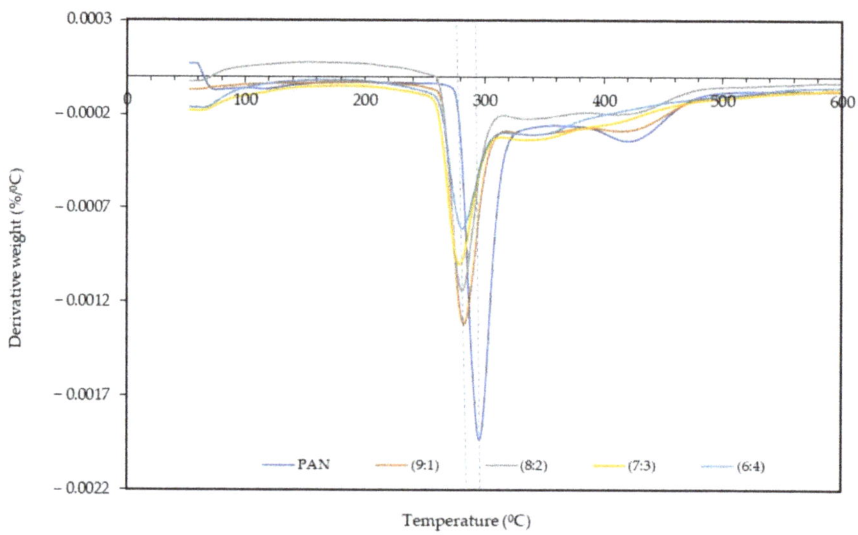

Figure 5. DTG curve of nanofibers before selective chemical dissolution.

From DTG thermograms, a sharp peak corresponding to a maximum decomposition temperature of PAN nanofibers is observed at 296 °C (Figure 5) and 299 °C (Figure 6) for before and after selective chemical dissolution, respectively. The maximum degradation temperature of the PAN/lignin nanofibers before selective chemical dissolution occurred at a temperature lower than PAN, which is 282 °C, and after the removal of lignin, the maximum degradation temperature shifted to a higher temperature in the range of 287–295 °C. This is because the removal of lignin caused the nanofibers to be composed of more PAN, and provided the nanofiber with higher thermal stability. These results also prove that the removal of lignin after de-ionized water treatment successfully occurred. It is further confirmed that the incorporation of lignin reduces the thermal stability of nanofibers.

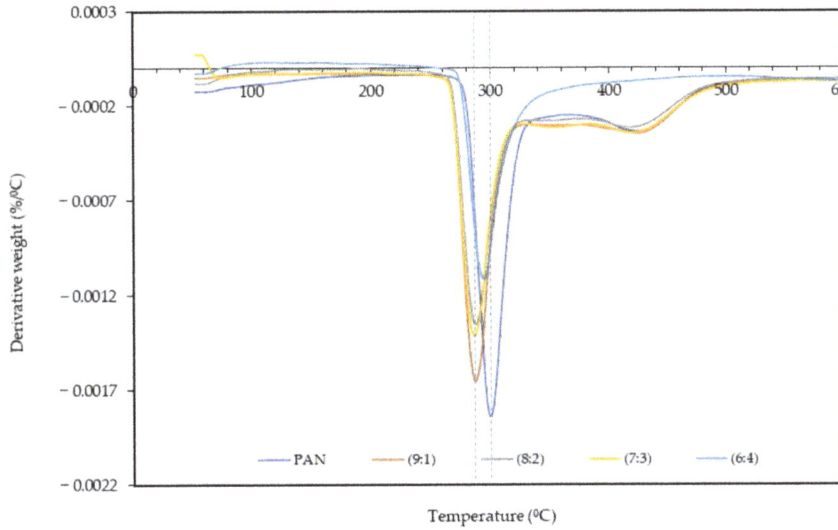

Figure 6. DTG curve of nanofibers after selective chemical dissolution.

3.4. X-ray Diffractometer (XRD) Analysis

Figure 7 shows the sample result of XRD analyses carried out to study the crystallinity of nanofibers before (pure 7:3 PAN/lignin nanofibers) and after (modified 7:3 PAN/lignin nanofibers) selective chemical dissolution. The XRD diffraction pattern of the nanofibers before and after selective chemical dissolution shows a broad peak at around 17°, which can be ascribed to (100) crystallographic planes of PAN due to their amorphous nature [14,42]. Nevertheless, the intensities of peaks at 17° were increased with the removal of lignin content in PAN-based nanofibers after selective chemical dissolution. Crystallinity decreased with the higher lignin content as lignin is a biopolymer consisting of amorphous phenyl propylene [43]. The XRD patterns of the pure 7:3 PAN/lignin nanofibers show characteristic peaks of 2 theta located at 42°, 44°,49°, and 51° that may be due to the blend of PAN with lignin. Once modified, intensity of all these peaks were reduced due to lignin removal.

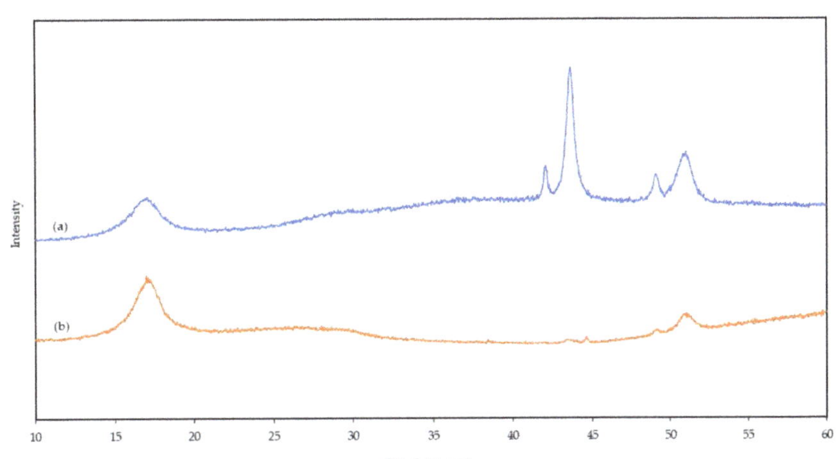

Figure 7. XRD spectra of (**a**) 7:3 PAN/lignin nanofiber and (**b**) 7:3 PAN/lignin nanofibers after selective chemical dissolution.

3.5. BET Analysis

The most frequent approach to the determination of specific surface areas of porous materials is the Brunauer-Emmett-Teller (BET) approach. The pore information of the porous nanofibers, including the BET-specific surface area and the microporous and mesoporous volumes, is summarized in Table 1. As seen in Table 1, the specific surface area of PAN: lignin nanofibers after selective chemical dissolution are 16.17 m^2g^{-1}, 20.94 m^2g^{-1}, 32.42 m^2g^{-1}, 29.13 m^2g^{-1}, and 28.15 m^2g^{-1} for PAN, with 9:1, 8:2, 7:3, and 6:4 ratios, respectively. The PAN: lignin nanofibers of 8:2 ratio showed the highest BET surface area (32.42 m^2g^{-1}), followed by 7:3, 6:4, 9:1, and PAN. This shows an increase in specific surface area of up to 50% compared to PAN nanofibers after the removal of lignin. The specific surface area ratio from 7:3 to 6:3 is a slight decrease from 8:2, possibly due to the excess lignin addition to the nanofibers, which creates a bigger pore diameter and consequently reduces the specific surface area. In this study, the increase in the specific surface area is greater than reported elsewhere in the literature. For example, Ji and co-workers [44] studied the effect of PAN and silica nanoparticles using a selective chemical dissolution process. The silica component was removed from silica/PAN composite fibers, and the surface area of the nanofibers was increased by only 20%.

Table 1. Specific surface areas (S_{BET}), pore parameters of PAN/lignin nanofibers after selective chemical dissolution with different PAN: lignin ratios.

	PAN	9:1	8:2	7:3	6:4
Specific surface area, S_{BET}/m^2g^{-1}	16.17	20.94	32.42	29.13	28.15
Pore volume, V_{total}/m^3g^{-1}	0.2360	0.2371	0.1974	0.1800	0.2970
Average pore diameter/nm	5.16	5.13	5.02	5.34	5.08

The interrelationship between the average diameter and the BET-specific surface area of nanofibers as a function of a ratio of PAN/lignin nanofibers after selective chemical dissolution is shown in Figure 8. From the graph, it can be seen that the higher the composition of lignin in the nanofibers, the lower the average diameter of the fiber and the higher the BET-specific surface area. The optimum surface area of the porous nanofibers can be obtained by only a 20% incorporation of lignin fibers with a very small pore diameter (5.02 nm) and pore volume (0.1974 m^3g^{-1}).

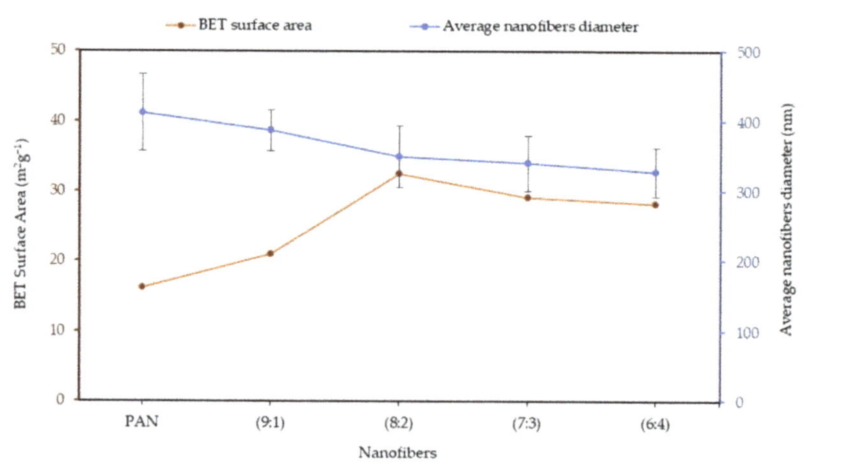

Figure 8. Graph of various ratios of PAN/lignin nanofibers, with BET surface area and average diameter.

4. Conclusions

The highly porous and specific surface area of PAN/lignin electrospun nanofibers was successfully prepared by using lignin as a soft template and through a selective chemical dissolution technique. The morphology showed a significant change in the surface of the nanofibers before and after selective chemical dissolution technique. After selective chemical dissolution, the nanofibers' surface appeared to be rougher and slightly smaller in average fiber diameter than before selective chemical dissolution. The BET analysis showed a significant increase in specific surface area after the addition of lignin to the PAN/lignin nanofibers, and the pore diameter was varied with the various ratios of lignin in the nanofibers. The optimum specific surface area of PAN/lignin nanofibers was 8:2 ratio, which is 32.42 m^2g^{-1} with a pore diameter of 5.02 nm. Additional lignin not only impacts the morphology of the nanofibers, but also the fiber diameter. The higher the ratio of lignin added, the smaller the average fiber diameter. This study showed that lignin and PAN can be homogenously blended and utilized to produce the high surface area and porosity of PAN nanofibers using a selective chemical dissolution technique. This can be used in an extensive variety of applications, for instance, as a precursor to highly porous carbon nanofibers.

Supplementary Materials: The following are available online at https://www.mdpi.com/article/10.3390/polym13223938/s1, Figure S1: TGA curve of nanofibers before selective chemical dissolution, Figure S2: TGA curve of nanofibers after selective chemical dissolution.

Author Contributions: Conceptualization, N.A.R. and M.A.T.A.; methodology, M.A.T.A.; software, M.A.T.A.; validation, N.A.R. and M.A.T.A.; formal analysis, M.A.T.A.; investigation, M.A.T.A.; resources, N.A.R.; data curation, N.A.R. and M.A.T.A.; writing—original draft preparation, M.A.T.A.; writing—review and editing, N.A.R. and M.A.T.A.; visualization, M.A.T.A.; supervision, N.A.R.; project administration, N.A.R.; funding acquisition, N.A.R. All authors have read and agreed to the published version of the manuscript.

Funding: Fundamental Research Grant Scheme (FRGS) from the Ministry of Higher Education Malaysia (FRGS/1/2019/STG01/UPM/02/7).

Institutional Review Board Statement: Not applicable.

Informed Consent Statement: Not applicable.

Data Availability Statement: Not applicable.

Acknowledgments: The authors are grateful for the financial support of the Fundamental Research Grant Scheme (FRGS) from the Ministry of Higher Education Malaysia (FRGS/1/2019/STG01/UPM/02/7).

Conflicts of Interest: The authors declare no conflict of interest.

References

1. Philip, P.; Jose, E.T.; Chacko, J.K.; Philip, K.; Thomas, P. Preparation and characterisation of surface roughened PMMA electrospun nanofibers from PEO-PMMA polymer blend nanofibers. *Polym. Test.* **2019**, *74*, 257–265. [CrossRef]
2. Yin, J.; Roso, M.; Boaretti, C.; Lorenzetti, A.; Martucci, A.; Modesti, M. PVDF-TiO_2 core-shell fibrous membranes by microwave-hydrothermal method: Preparation, characterization, and photocatalytic activity. *J. Environ. Chem. Eng.* **2021**, *9*, 106250. [CrossRef]
3. Moon, S.; Choi, J.; Farris, R.J. Highly porous polyacrylonitrile/polystyrene nanofibers by electrospinning. *Fibers Polym.* **2008**, *9*, 276–280. [CrossRef]
4. Deeney, C.; McKiernan, E.P.; Belhout, S.A.; Rodriguez, B.J.; Redmond, G.; Quinn, S.J. Template-Assisted Synthesis of Luminescent Carbon Nanofibers from Beverage-Related Precursors by Microwave Heating. *Molecules* **2019**, *24*, 1455. [CrossRef] [PubMed]
5. Manafi, S.A.; Badiee, S.H. Production of Carbon Nanofibers Using a CVD Method with Lithium Fluoride as a Supported Cobalt Catalyst. *Res. Lett. Mater. Sci.* **2008**, *2008*, 1–5. [CrossRef]
6. Yalcinkaya, F. Preparation of various nanofiber layers using wire electrospinning system. *Arab. J. Chem.* **2019**, *12*, 5162–5172. [CrossRef]
7. García-Mateos, F.J.; Ruiz-Rosas, R.; Rosas, J.M.; Rodríguez-Mirasol, J.; Cordero, T. Controlling the Composition, Morphology, Porosity, and Surface Chemistry of Lignin-Based Electrospun Carbon Materials. *Front. Mater.* **2019**, *6*, 114. [CrossRef]

8. Hoque, M.E.; Nuge, T.; Yeow, T.K.; Nordin, N. Chapter 5—Electrospun Matrices from Natural Polymers for Skin Regeneration. In *Nanostructured Polymer Composites for Biomedical Applications*; Swain, S.K., Jawaid, M., Eds.; Elsevier: Amsterdam, The Netherlands, 2019; pp. 87–104.
9. Kong, L.; Ziegler, G. Fabrication of pure starch fibers by electrospinning. *Food Hydrocoll.* **2014**, *36*, 20–25. [CrossRef]
10. Chew, S.Y.; Wen, Y.; Dzenis, Y.; Leong, K. The Role of Electrospinning in the Emerging Field of Nanomedicine. *Curr. Pharm. Des.* **2006**, *12*, 4751–4770. [CrossRef]
11. Haghi, A.K.; Zaikov, G.E. *Electrospinning Process and Nanofiber Research*; Nova Science Publishers: Hauppauge, NY, USA, 2011; pp. 1–196.
12. Kanafi, N.M.; Rahman, N.A.; Rosdi, N.H.; Bahruji, H.; Maarof, H. Hydrogel Nanofibers from Carboxymethyl Sago Pulp and Its Controlled Release Studies as a Methylene Blue Drug Carrier. *Fibers* **2019**, *7*, 56. [CrossRef]
13. Kanafi, N.M.; Rahman, N.A.; Rosdi, N.H. Citric acid cross-linking of highly porous carboxymethyl cellulose/poly(ethylene oxide) composite hydrogel films for controlled release applications. *Mater. Today Proc.* **2019**, *7*, 721–731. [CrossRef]
14. Nordin, N.A.; Rahman, N.A.; Abdullah, A.H. Effective Removal of Pb(II) Ions by Electrospun PAN/Sago Lignin-based Activated Carbon Nanofibers. *Molecules* **2020**, *25*, 3081. [CrossRef]
15. Nthumbi, R.; Ngila, J.; Kindness, A.; Moodley, B.; Petrik, L. Method Development for Flow Adsorption and Removal of Lead and Copper in Contaminated Water Using Electrospun Nanofibers of Chitosan Blend. *Anal. Lett.* **2011**, *44*, 1937–1955. [CrossRef]
16. Park, J.-A.; Kang, J.; Lee, S.-C.; Kim, S.-B. Electrospun poly(acrylic acid)/poly(vinyl alcohol) nanofibrous adsorbents for Cu(II) removal from industrial plating wastewater. *RSC Adv.* **2017**, *7*, 18075–18084. [CrossRef]
17. Bates, I.I.C.; Loranger, É; Chabot, B. Chitosan-PEO nanofiber mats for copper removal in aqueous solution using a new versatile electrospinning collector. *SN Appl. Sci.* **2020**, *2*, 1–14. [CrossRef]
18. Khulbe, K.C.; Matsuura, T. Removal of heavy metals and pollutants by membrane adsorption techniques. *Appl. Water Sci.* **2018**, *8*, 19. [CrossRef]
19. Sharma, S.; Bhattacharya, A. Drinking water contamination and treatment techniques. *Appl. Water Sci.* **2016**, *7*, 1043–1067. [CrossRef]
20. Awad, R.; Mamaghani, A.H.; Boluk, Y.; Hashisho, Z. Synthesis and characterization of electrospun PAN-based activated carbon nanofibers reinforced with cellulose nanocrystals for adsorption of VOCs. *Chem. Eng. J.* **2021**, *410*, 128412. [CrossRef]
21. Mokhtari-Shourijeh, Z.; Montazerghaem, L.; Olya, M.E. Preparation of Porous Nanofibers from Electrospun Polyacrylonitrile/Polyvinylidene Fluoride Composite Nanofibers by Inexpensive Salt Using for Dye Adsorption. *J. Polym. Environ.* **2018**, *26*, 3550–3563. [CrossRef]
22. Heo, Y.-J.; Lee, H.I.; Lee, J.W.; Park, M.; Rhee, K.Y.; Park, S.-J. Optimization of the pore structure of PAN-based carbon fibers for enhanced supercapacitor performances via electrospinning. *Compos. Part B Eng.* **2019**, *161*, 10–17. [CrossRef]
23. Zhang, H.; Tan, Y.; Luo, X.; Sun, C.; Chen, N. Polarization Effects of a Rayon and Polyacrylonitrile Based Graphite Felt for Iron-Chromium Redox Flow Batteries. *ChemElectroChem* **2019**, *6*, 3175–3188. [CrossRef]
24. Qiu, L.; Zheng, X.; Zhu, J.; Su, G.; Tang, D. The effect of grain size on the lattice thermal conductivity of an individual polyacrylonitrile-based carbon fiber. *Carbon* **2013**, *51*, 265–273. [CrossRef]
25. Ramasubramanian, G. Influence of Lignin Modification on PAN-Lignin Copolymers as Potential Carbon Fiber Precursors. Master's Thesis, Iowa State University, Ames, IA, USA, 2013.
26. Xiong, F.; Han, Y.; Wang, S.; Li, G.; Qin, T.; Chen, Y.; Chu, F. Preparation and Formation Mechanism of Renewable Lignin Hollow Nanospheres with a Single Hole by Self-Assembly. *ACS Sustain. Chem. Eng.* **2017**, *5*, 2273–2281. [CrossRef]
27. Branco, R.H.R.; Serafim, L.S.; Xavier, A.M.R.B. Second Generation Bioethanol Production: On the Use of Pulp and Paper Industry Wastes as Feedstock. *Fermentation* **2018**, *5*, 4. [CrossRef]
28. Thakur, V.K.; Thakur, M.K.; Raghavan, P.; Kessler, M.R. Progress in Green Polymer Composites from Lignin for Multifunctional Applications: A Review. *ACS Sustain. Chem. Eng.* **2014**, *2*, 1072–1092. [CrossRef]
29. Kim, J.S.; Lee, Y.Y.; Kim, T.H. A review on alkaline pretreatment technology for bioconversion of lignocellulosic biomass. *Bioresour. Technol.* **2016**, *199*, 42–48. [CrossRef] [PubMed]
30. Sabetzadeh, N.; Gharehaghaji, A.A. How Porous Nanofibers Have Enhanced Engineering of Advanced Materials: A Review. *Mat. Sci.* **2017**, *5*, 3–21.
31. Chen, X.; Huang, J.; Ling, X.; Lm, H. Preparation of porous ultrafine SF fibers via selective dissolution of electrospun SF/PLA blend fibers. *IOP Conf. Ser. Mater. Sci. Eng.* **2018**, *423*, 012153. [CrossRef]
32. Kim, M.N.; Koh, J.; Lee, Y.; Kim, H. Preparation of PVA/PAN bicomponent nanofiber via electrospinning and selective dissolution. *J. Appl. Polym. Sci.* **2009**, *113*, 274–282. [CrossRef]
33. Ibrahim, M.N.M.; Iqbal, A.; Shen, C.C.; Bhawani, S.A.; Adam, F. Synthesis of lignin based composites of TiO_2 for potential application as radical scavengers in sunscreen formulation. *BMC Chem.* **2019**, *13*, 17. [CrossRef]
34. Kubo, S.; Kadla, J.F. Lignin-based Carbon Fibers: Effect of Synthetic Polymer Blending on Fiber Properties. *J. Polym. Environ.* **2005**, *13*, 97–105. [CrossRef]
35. Seydibeyoğlu, M.Ö. A Novel Partially Biobased PAN-Lignin Blend as a Potential Carbon Fiber Precursor. *J. Biomed. Biotechnol.* **2012**, *2012*, 1–8. [CrossRef]
36. Yue, Z.; Benak, K.R.; Wang, J.; Mangun, C.L.; Economy, J. Elucidating the porous and chemical structures of $ZnCl_2$-activated polyacrylonitrile on a fiberglass substrate. *J. Mater. Chem.* **2005**, *15*, 3142–3148. [CrossRef]

37. Wangxi, Z.; Jie, L.; Gang, W. Evolution of structure and properties of PAN precursors during their conversion to carbon fibers. *Carbon* **2003**, *41*, 2805–2812. [CrossRef]
38. Jenab, A.; Roghanian, R.; Ghorbani, N.; Ghaedi, K.; Emtiazi, G. The Efficacy of Electrospun PAN/Kefiran Nanofiber and Kefir in Mammalian Cell Culture: Promotion of PC12 Cell Growth, Anti-MCF7 Breast Cancer Cells Activities, and Cytokine Production of PBMC. *Int. J. Nanomed.* **2020**, *15*, 717–728. [CrossRef]
39. Choi, D.I.; Lee, J.-N.; Song, J.; Kang, P.-H.; Park, J.-K.; Lee, Y.M. Fabrication of polyacrylonitrile/lignin-based carbon nanofibers for high-power lithium ion battery anodes. *J. Solid State Electrochem.* **2013**, *17*, 2471–2475. [CrossRef]
40. Pashaloo, F.; Bazgir, S.; Tamizifar, M.; Faghihisani, M.; Zakerifar, S. Preparation and Characterization of Carbon Nanofibers via Electrospun PAN Nanofibers. *Text. Sci. Technol. J.* **2009**, *3*, 2.
41. Dang, W.; Liu, J.; Wang, X.; Yan, K.; Zhang, A.; Yang, J.; Chen, L.; Liang, J. Structural Transformation of Polyacrylonitrile (PAN) Fibers during Rapid Thermal Pretreatment in Nitrogen Atmosphere. *Polymers* **2020**, *12*, 63. [CrossRef] [PubMed]
42. Mustafov, S.D.; Mohanty, A.K.; Misra, M.; Seydibeyoğlu, M.Ö. Fabrication of conductive Lignin/PAN carbon nanofibers with enhanced graphene for the modified electrodes. *Carbon* **2019**, *147*, 262–275. [CrossRef]
43. Goudarzi, A.; Lin, L.-T.; Ko, F.K. X-ray Diffraction Analysis of Kraft Lignins and Lignin-Derived Carbon Nanofibers. *J. Nanotechnol. Eng. Med.* **2014**, *5*, 021006. [CrossRef]
44. Ji, L.; Saquing, C.; Khan, S.A.; Zhang, X. Preparation and characterization of silica nanoparticulate-polyacrylonitrile composite and porous nanofibers. *Nanotechnology* **2008**, *19*, 085605. [CrossRef] [PubMed]

polymers

MDPI

Article

Axial Compression Behavior of Ferrocement Geopolymer HSC Columns

Taha Awadallah El-Sayed

Structural Engineering Department, Faculty of Engineering at Shoubra, Benha University, Cairo 11629, Egypt; taha.ibrahim@feng.bu.edu.eg

Abstract: Geopolymer concrete (GC) is a substantial sort that is created by utilizing metakaolin, ground granulated blast furnace slag (GGBS), silica fumes, fly ash, and other cementitious materials as binding ingredients. The current study concentrated on the structural behavior of the ferrocement geopolymer HSC-columns subjected to axial loading and produced using rice straw ash (RSA). The major goal of this research was to use the unique features of the ferrocement idea to manufacture members that function as columns bearing members. As they are more cost-effective and lower in weight, these designed elements can replace traditional RC members. The study also intended to reduce the cost of producing new parts by utilizing low-cost materials such as light weight expanded and welded wire meshes, polyethylene mesh (Tensar), and fiber glass mesh. For this purpose, an experimental plan was conducted and a finite element prototype with ANSYS2019-R1 was implemented. Nine geopolymer ferrocement columns of dimensions of 150 mm × 150 mm × 1600 mm with different volume-fraction and layers as well as a number of metallic and nonmetallic meshes were examined under axial compression loading until failure. The performance of the geopolymer columns was examined with consideration to the mid-span deflection, ultimate failure load, first crack load with various phases of loading, the cracking patterns, energy absorption and ductility index. Expanded or welded ferrocement geopolymer columns showed greater ultimate failure loads than the control column. Additionally, using expanded or welded columns had a considerable effect on ultimate failure loads, where the welded wire mesh exhibited almost 28.10% compared with the expanded wire mesh. Columns reinforced with one-layer of nonmetallic Tensar-mesh obtained a higher ultimate failure load than all tested columns without concrete cover spalling. The analytical and experimental results were in good agreement. The results displayed an accepted performance of the ferrocement geopolymer HSC-columns.

Keywords: axial behavior; geopolymer concrete (GC); ferrocement; finite element analysis (FEA)

Citation: El-Sayed, T.A. Axial Compression Behavior of Ferrocement Geopolymer HSC Columns. *Polymers* **2021**, *13*, 3789. https://doi.org/10.3390/polym13213789

Academic Editors: Andrzej Puszka and Beata Podkościelna

Received: 5 October 2021
Accepted: 28 October 2021
Published: 1 November 2021

1. Introduction

The utilization of industrial by-products in the construction area is an important means of reducing construction costs and the safe removal of manufacturing waste [1–5]. In this sense, the straight use of alkaline fly ash and GGBS is recycled to manufacture geopolymer–cement for special construction concrete manufacturing [6,7]. The energy employed in geopolymer–cement production is much less than that used in OPC, which is directly affected by greenhouse gases [7]. In some adverse environmental conditions where OPCs are not very resistant, new and alternative concretes such as geopolymer–concrete need to be developed.

Most constructions use concrete hoists where OPC is used as the main binder. As a civil engineer, a lot is known regarding the environmental issues in the cement industry. The amount of carbon dioxide produced inside the atmosphere is equal to the manufacture of cement and it is similarly known that the energy required to produce cement is high, which consumes more fossil fuels [8,9]. When the cement is partially replaced in the existence of polymers and water at room temperature, the fly ash interacts with calcium hydroxide through the hydration procedure to produce a (C–S–H) gel. The application and development of large quantities of fly-ash in concrete has made it possible to replace cement up to 60% with a concrete mass [8].

Over the course of several years, wide-ranging research has been conducted to confirm the possibility of using GC as a building material [6,7,10–14].

The use of GC is gradually rising, particularly for chemical resistant structures in industries, and research is ongoing to expand the variety of applications. In fact, significant experimental work has been performed in Australia, the United States, and Spain. Several investigators have offered suitable starting materials for the production, stiffness, mix design, mechanical properties, and durability of GC [2,15,16]. The bearing capacity and stiffness of geopolymer concrete columns are influenced by the compressive strength of the material. Larger compressive strength geopolymer concrete columns provide a higher bearing capacity, stiffness, and ductility.

Ordinary concrete columns have a lower bearing capacity and rigidity than geopolymer concrete columns. In engineering applications, geopolymer concrete columns can meet the design requirements for structural columns with sufficient load capacity, stiffness, and ductility.

Mansur and Paramasivan [17] carried out a test study on ferrocement columns under centric and eccentric compressive loads. Test findings indicated that a ferrocement column could be used as a structural column. Kaushik et al. [18] executed a study on ferrocement RC columns and realized that the ferrocement enhanced the strength and ductility of the columns for centric and eccentric compressive loads. Similarly, various studies have been undertaken on different ferrocement structural elements under centric and eccentric compressive loads [19–27].

The major goal of this research was to use the unique features of the ferrocement idea to manufacture members that function as column bearing members. As they are more cost-effective and lower in weight, these designed elements can replace traditional RC members. The study also intended to reduce the cost of producing new parts by utilizing low-cost materials such as light weight expanded and welded wire meshes, polyethylene mesh (Tensar), and fiber glass mesh. Therefore, the main reason for this research was to examine the influence of the performance of RSA based geopolymer ferrocement HSC-columns under an axial compression load with various kinds and number of layers of metallic and nonmetallic meshes. For this purpose, an experimental plan was conducted and a finite element prototype with ANSYS2019-R1 was implemented. Nine geopolymer ferrocement columns with dimensions of 150 mm × 150 mm × 1600 mm were examined under axial compression loading until failure. The variables in this investigation were the mesh types and number of layers. The performance of the geopolymer columns was examined with consideration of the mid-span deflection, ultimate failure load, first crack load with various phases of loading, the cracking patterns, energy absorption, and ductility index.

2. Experimental Study

This experimental program was conducted in the Housing and Building National Research Center-Dokki-Egypt. A 5000-kN capacity test machine capable of testing columns up to 6 m in height was used to test the columns. The main aim was to find the ultimate deflection, ultimate load, and failure mode of the GC columns.

2.1. Materials

1. Fine aggregate: Sand with 2.55 specific gravity and bulk density 1780 kg/m^3. According to the Egypt Standard Specification (ESS) 203/2020 [28], a sieve analysis was performed. The results of the sieve analysis and the physical property test results are shown in Table 1 and Figure 1.

Table 1. Physical properties of the sand.

Property	Results	ESS Acceptance Limits
Specific gravity (kg/m^3)	2.55	-
Bulk density (kg/m^3)	1780	-
Materials finer than no. 200, sieve (0.074 mm)%	1.4	Less than 4%

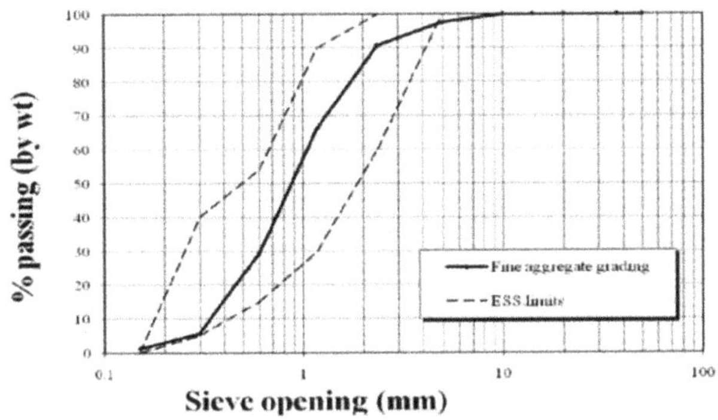

Figure 1. Fine aggregate grading curve.

2. Coarse aggregate: Crushed aggregate with size 10 mm, 2.60 specific gravity and bulk density 1750 kg/m^3. According to the Egypt Standard Specification (ESS) 203/2020 [28], sieve analysis was performed. The aggregate mechanical and physical properties are shown in Table 2, and the grading is shown in Figure 2.

Table 2. The physical and mechanical properties of the coarse aggregate.

Property	Results	ESS Acceptance Limits
Specific gravity (kg/m^3)	2.2.60	-
Unit weight (kg/m^3)	1750	-
Absorption Percentage	1.46%	No more than 2.5%

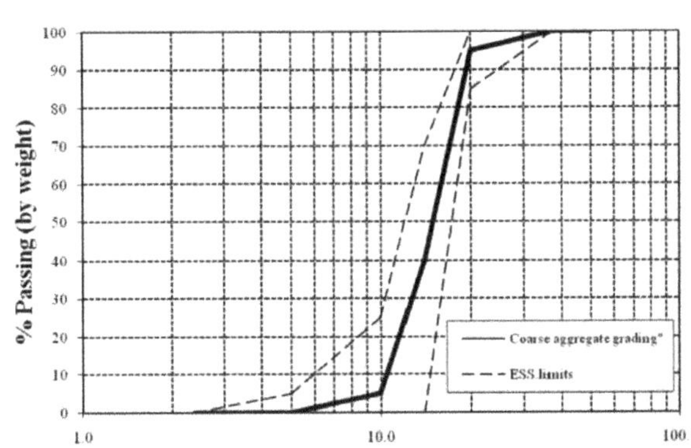

Figure 2. Coarse aggregate grading curve.

3. Recycled Rice-Straw–Ash (RSA): RSA with a specific gravity 2.91 g/cm^2, and specific surface area of 5200 cm^2/g.
4. Water: used for mixing and curing.
5. Alkaline activator: sodium meta-silicate (Na_2SiO_3) and sodium hydroxide (NaOH).
6. Steel RFT: Two types of steel were used. Plain bars (24/35) with a 6 mm diameter, and deformed bars (42/60) with a 12 mm diameter.
7. Steel wire-meshes:
   (a) Welded and expanded wire-mesh: Figure 3 shows the types of ferrocement meshes used. Table 3 shows the mechanical properties of the welded and expanded steel wire-meshes.
   (b) Polyethylene (Tensar)-mesh: This mesh is made from the high density polyethylene, "Geogrid CE 121", as shown in Figure 4, with an opening size of 6 mm × 8 mm, thickness of 3.3 mm, volume fraction of 2.04%, and weight of 725 gm/m^2.
   (c) Fiber glass mesh: Gavazzi "V3-133-A" was used with an opening dimension of 12.5 mm × 11.5 mm. The cross-section dimension 1.66 mm × 0.66 mm (longitudinal direction) and 1.0 mm × 0.5 mm (transverse direction) as shown in Figure 5. The mesh has a volume fraction of 0.535% and weight of 123 gm/m^2.

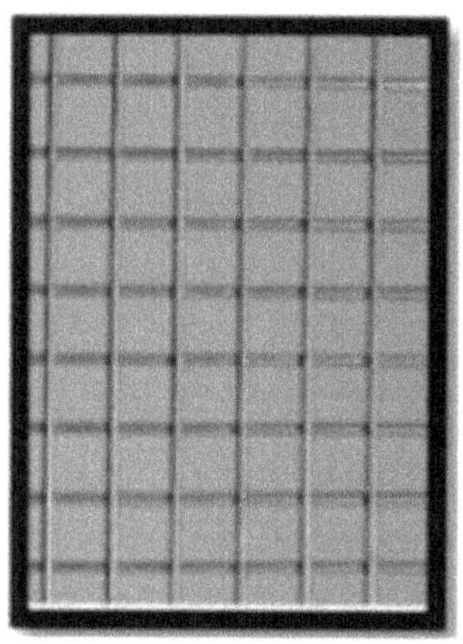

(a)

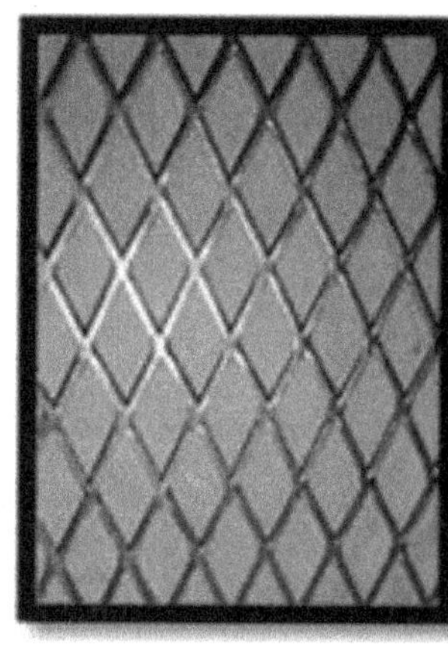

(b)

Figure 3. Types of meshes. (**a**) Welded wire-mesh, (**b**) Expanded wire-mesh [29].

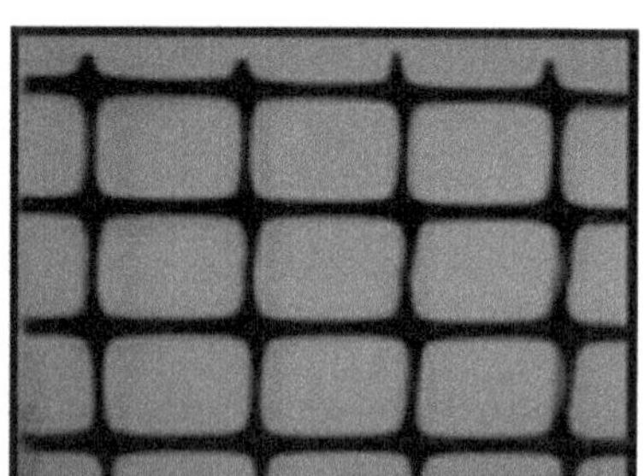

Figure 4. Polyethylene (Tensar) mesh.

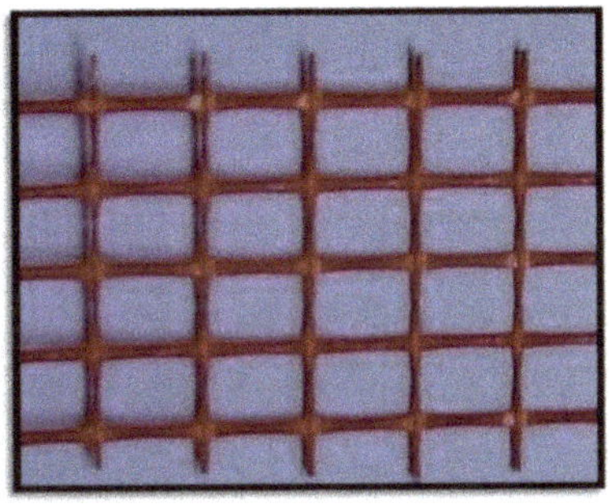

Figure 5. Fiber glass mesh.

Table 3. Mechanical properties of the welded and expanded wire-meshes [29].

Welded Wire Mesh		Expanded Wire Mesh	
Dimensions size	12.5 × 12.5 mm	Dimensions size	16.5 × 31 mm
Weight	600 gm/m^2	Weight	1660 gm/m^2
Thickness	0.7 mm	Wire Diameter	1.25 mm
Young's Modulus	17000 N/mm^2	Young's Modulus	12000 N/mm^2
Yield Stress	400 N/mm^2	Yield Stress	250 N/mm^2
Yield Strain	1.17 × 10^{-3}	Yield Strain	9.7 × 10^{-3}
Ultimate Strength	600 N/mm^2	Ultimate Strength	380 N/mm^2
Ultimate Strain	58.5 × 10^{-3}	Ultimate Strain	59.2 × 10^{-3}

2.2. Design of Mix

Table 4 shows the mix design of the high strength concrete. The mix design was used to develop HSC at 28 days with a target strength of 60 MPa.

Table 4. Mix design of HSC.

Item	RSA (kg/m^3)	Coarse Aggregate (kg/m^3)	Fine Aggregate (kg/m^3)	NaOH (kg/m^3)	Na_2SiO_3 (kg/m^3)	Water (kg/m^3)
Per m^3 of concrete	400	1150	650	50	150	47

2.3. Column Sample Description

The experimental work was made to investigate the behavior, ultimate capacity, and crack-pattern of the geopolymer HSC columns. The experimental program consisted of (nine) geopolymer HSC columns with dimensions of 150 mm × 150 mm × 1600 mm reinforced with (4 φ 12) steel bars. Columns were tested axially using a compression machine of capacity 5000 kN. The concrete dimensions of the columns and details of RFT are presented in Figure 6. All tested columns are presented in Table 5. Additionally, Table 6 shows the reinforcement configurations for all tested columns.

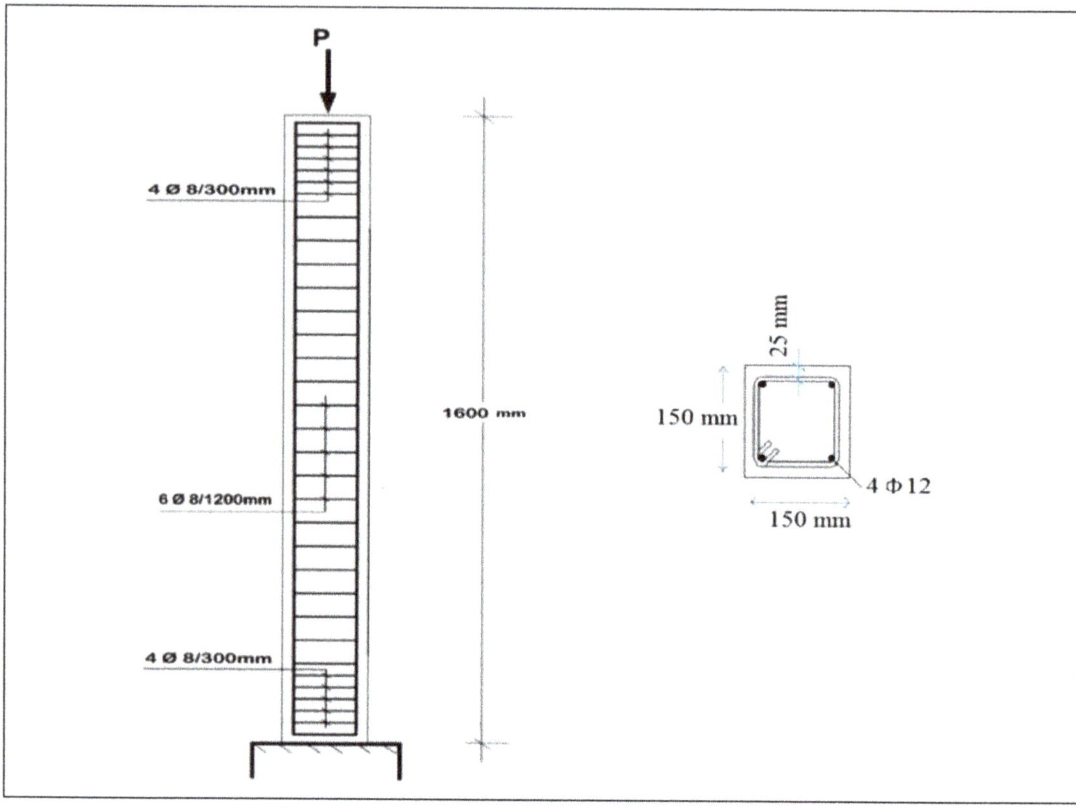

Figure 6. Concrete dimensions of columns and RFT details.

Table 5. Description of the studied columns.

Series	Sample ID	Sample Description	Volume of Fraction	RFT.	Stirrups
Control	C1	Control	——	4 φ 12	6 φ 8/′
Group A: Welded wire-mesh	C1-A	1-layer welded	0.00270	4 φ 12	——
	C2-A	2-layers welded	0.00540	4 φ 12	——
	C3-A	3-layers welded	0.00810	4 φ12	——
Group B: Expanded wire-mesh	C4-B	1-layer expanded	0.00753	4 φ 12	——
	C5-B	2-layers expanded	0.01510	4 φ 12	——
Group C: Tensar-mesh	C6-C	1-layer Tensar	0.02040	4 φ 12	——
Group D: Fiber glass-mesh	C7-D	1-layer fiber glass	0.00535	4 φ 12	——
	C8-D	2-layers fiber glass	0.01070	4 φ 12	——

Table 6. Reinforcement configurations of the tested columns.

Sample ID	Tested Columns Reinforcement Configurations
C1	
C1-A	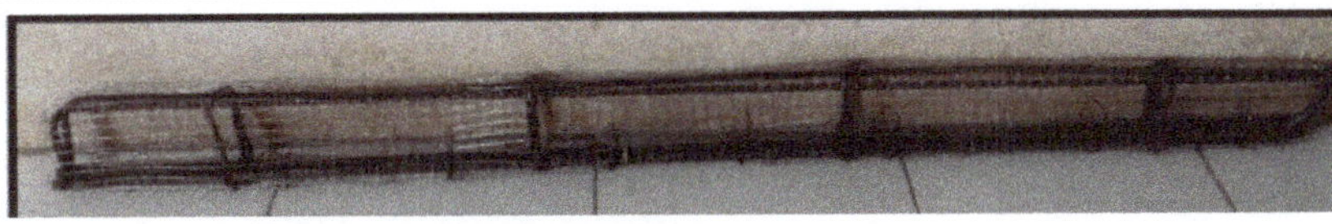
C2-A	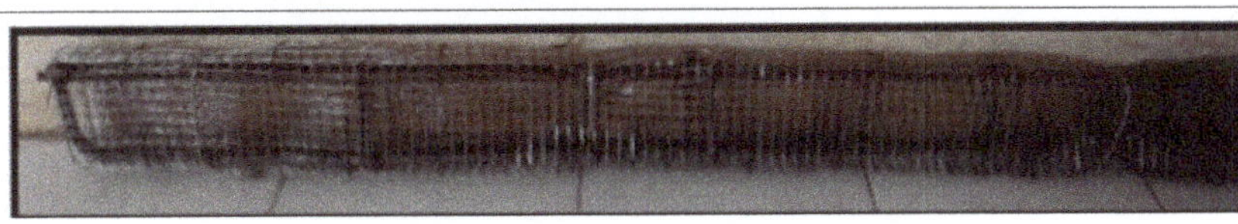
C3-A	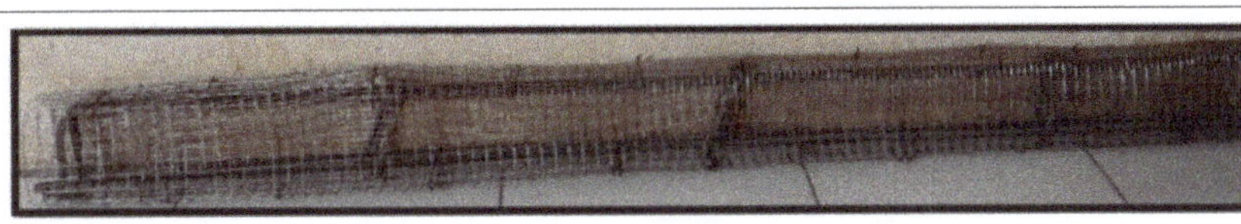
C4-B	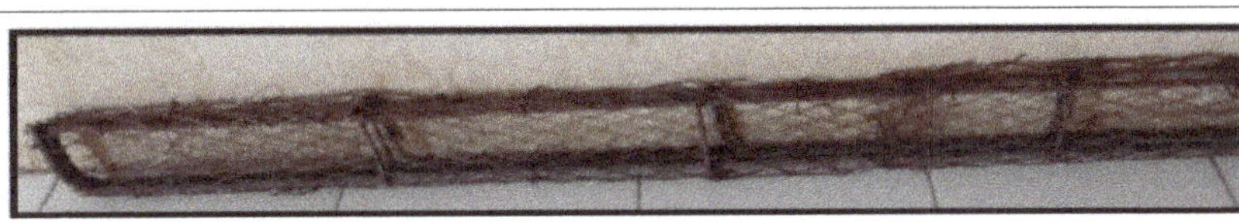
C5-B	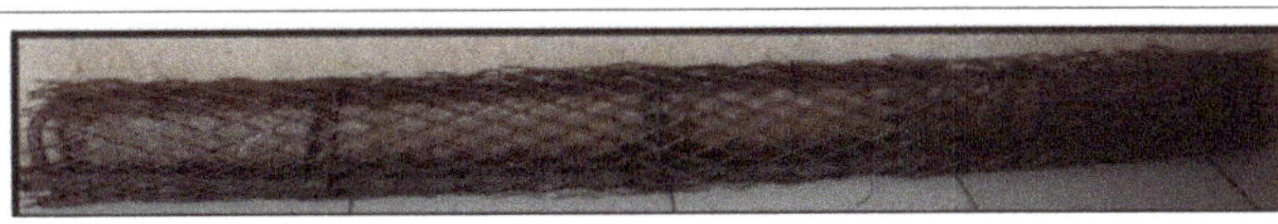
C6-C	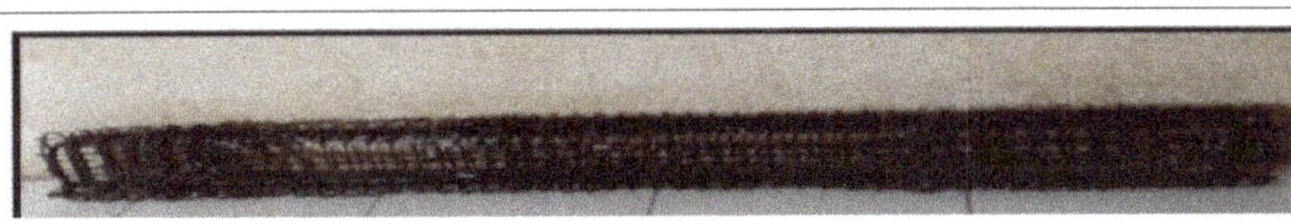
C7-D	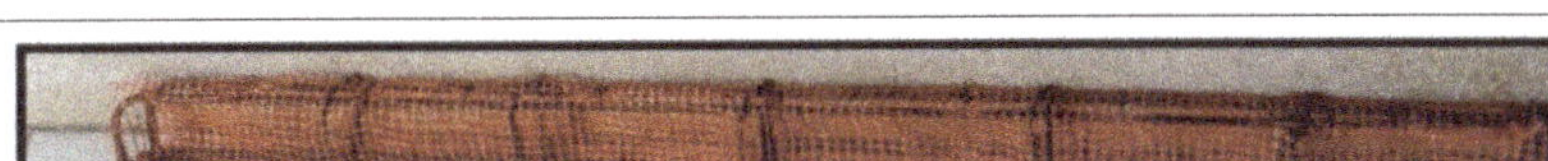
C8-D	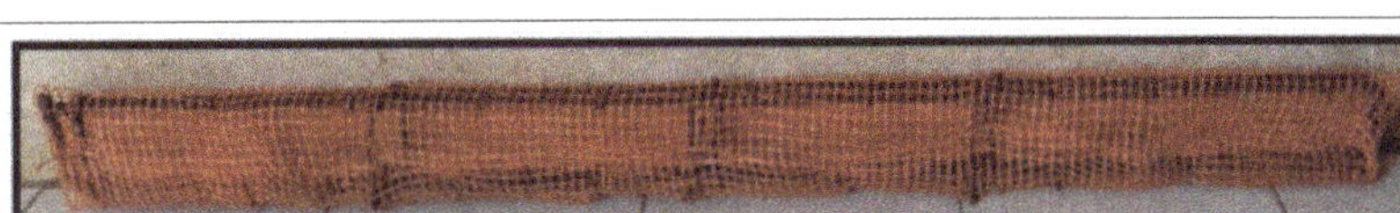

2.4. Test Setup

A compression test machine with a capacity of 5000 kN was utilized to test all column samples. Figure 7 shows a typical test setup for the columns. The deflection was measured using LVDT and all were tested until failure.

Figure 7. Test setup.

3. Discussion of Results

The behavior of the tested geopolymer HSC columns in terms of ultimate deflection, ultimate load, load–deflection relationship and failure mode, and cracking behavior are discussed as follows.

3.1. Ultimate Load

The ultimate loads for the tested columns are shown in Figure 11 and Table 7. The ultimate load of the control C1 was 738.70 kN. For group A, columns C1-A to C3-A, the ultimate loads extended between 771.80 kN and 955.60 kN. The improvement in the ultimate capacity was 4.30% to 22.70%. For group B, columns C4-B and C5-B, the ultimate loads extended between 813.70 kN and 821.70 kN. The improvement in the ultimate capacity was 9.20% and 19.90%. For group C, column C6-C, with an ultimate load of 1027.20 kN, there was a large considerable improvement of 28.10%. For group D, columns C7-D and C8-D, the ultimate loads were 793.40 kN and 841.30 kN, with an improvement of 7.20% and 12.20%, respectively.

Table 7. Experimental test results.

Column ID	First Crack Load (kN)	Serviceability Load (kN)	Ultimate Load (kN)	Def. at First Crack Load (mm)	Def. at Ult. load (mm)	Ductility Ratio —	Energy Absorption (kN·mm)
C1	296.00	460.59	738.70	3.80	12.02	3.16	6320.08
C1-A	325.00	481.27	771.80	4.30	12.72	2.96	9201.48
C2-A	380.00	525.46	842.50	4.70	14.36	3.06	9353.37
C3-A	455.00	596.15	955.60	5.20	14.50	2.79	10,380.91
C4-B	443.00	507.46	813.70	5.90	15.45	2.62	10,122.72
C5-B	449.00	574.96	921.70	6.30	16.96	2.69	12,008.52
C6-C	515.00	640.90	1027.20	4.40	13.55	3.08	10,414.93
C7-D	391.00	496.65	796.40	3.90	12.08	3.10	6983.16
C8-D	378.00	524.71	841.30	4.50	14.07	3.13	9179.36

According to the findings in Table 7, the use of fiber glass mesh is more efficient than other types of metallic and nonmetallic mesh reinforcements in increasing the ultimate capacity.

3.2. *Ultimate Deflection*

Figure 12 and Table 7 show the ultimate deflection for all tested columns. The deflection for the control C1 was 12.02 mm. For group A, the maximum deflection ranged from 12.72 mm to 14.50 mm for columns C1-A to C3-A, which were higher than that of the control C1. For group B, the maximum deflection at ultimate load was 15.45 mm and 16.96 mm for columns C4-B and C5-B respectively, which was also higher than that of the control C1. For group C, column C6-C, had an ultimate deflection of 13.55 mm. For group D, columns C7-D and C8-D, the ultimate deflections were 12.08 and 14.07 mm, respectively.

3.3. *Load–Deflection Relationship*

The relationship between the load and the deflection for the tested columns is presented in Figure 8. From this figure, it can clearly be seen that for all columns, the relationship between the load and deflection can be divided into three stages as follows:

1. Elastic behavior until the first cracking. The load–deflection relationship in this stage is linear. The slope of the load deflection curve in this stage varies with different types of test specimens. The end of this stage is marked by the deviation from linearity.
2. In the second stage, the load–deflection curve slope changed slowly as a result of the samples' stiffness reduction due to the multiple cracking.
3. In the third stage, large plastic deformation occurred as the result of the yielding of the reinforcing bars and the large extension in the reinforcing mesh of the ferrocement columns. This stage is terminated by failure of the test columns.

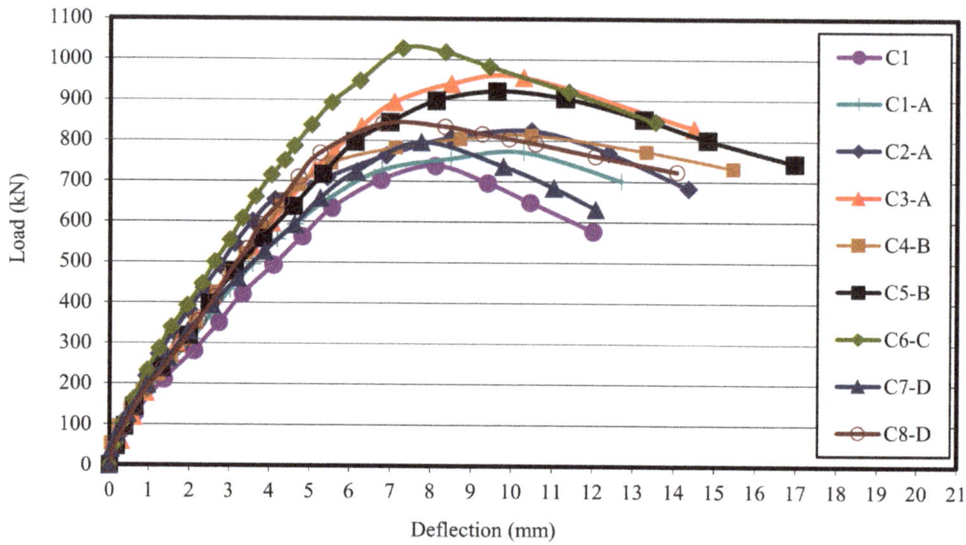

Figure 8. Comparison between the load deflection of all columns.

From Figure 9, it can be concluded that column C6-C, which was reinforced with one-layer of Tensar-mesh, had the highest first crack load and ultimate load, while the control column C1 had the lowest ultimate load carrying capacity.

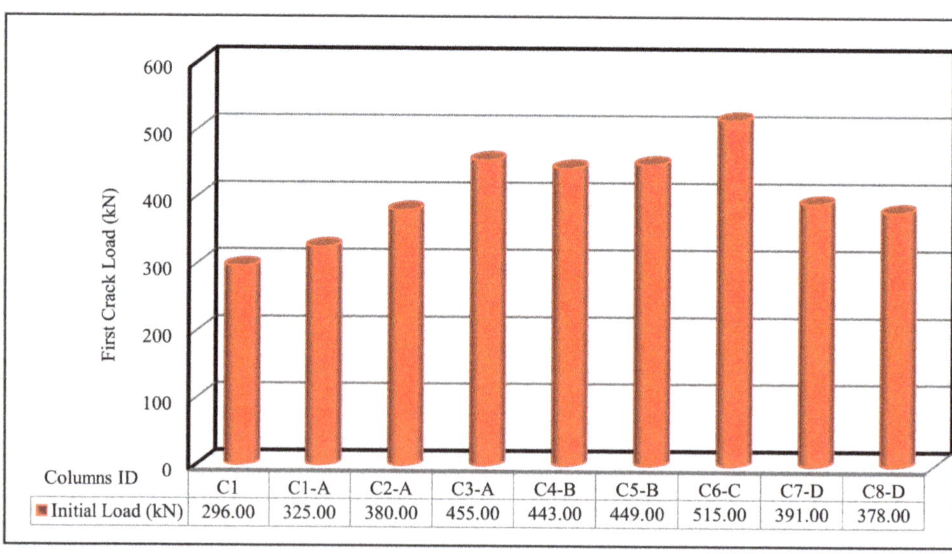

Figure 9. First crack load for the tested columns.

The results of all test specimens are listed in Table 7. This table shows the obtained results for the first crack load, service load, deflection at ultimate load, ductility ratio, and energy absorption. First crack load, ultimate load, and first crack and ultimate load deflection were gained throughout testing, but the ductility ratio, service load, and energy absorption were calculated from the load–deflection curve for each column sample. The first crack-load was obtained at the point at which the curve of load deflection began to deviate from the linear relationship. Furthermore, service load can be computed from Equation (1).

$$P_{ser} = \frac{(P_{ult} - 1.4 \times D.L.)}{1.6}, D.L. = own \quad weight \quad of \quad column. \tag{1}$$

In the current section, the comparison between the behavior up to failure of the tested column as obtained from the experimental results is illustrated. The comparisons between all tested columns are shown in Figures 9–14, which show the first crack load, serviceability load, ultimate load, deflection at ultimate load, ductility ratio, and energy absorption for all tested columns, respectively.

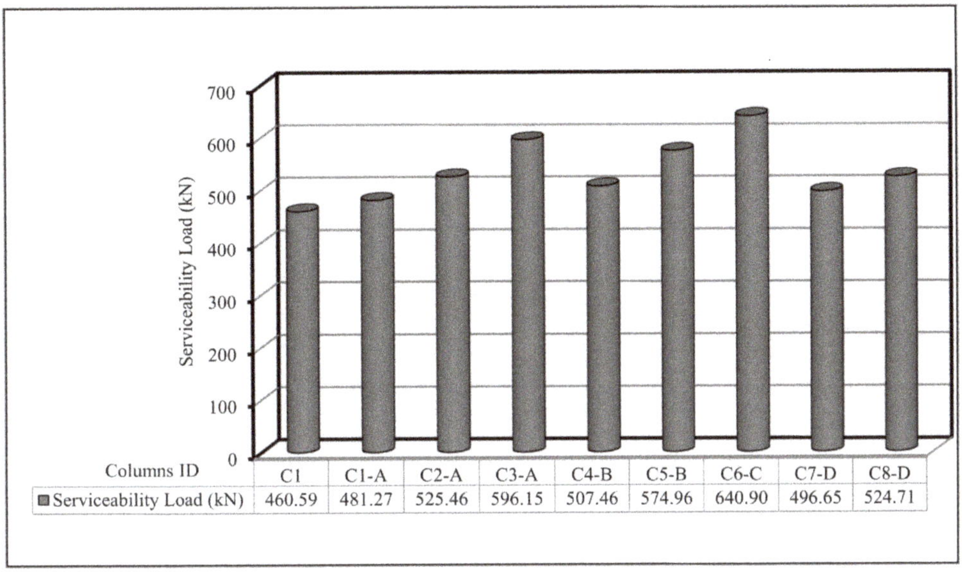

Figure 10. Serviceability load for the tested columns.

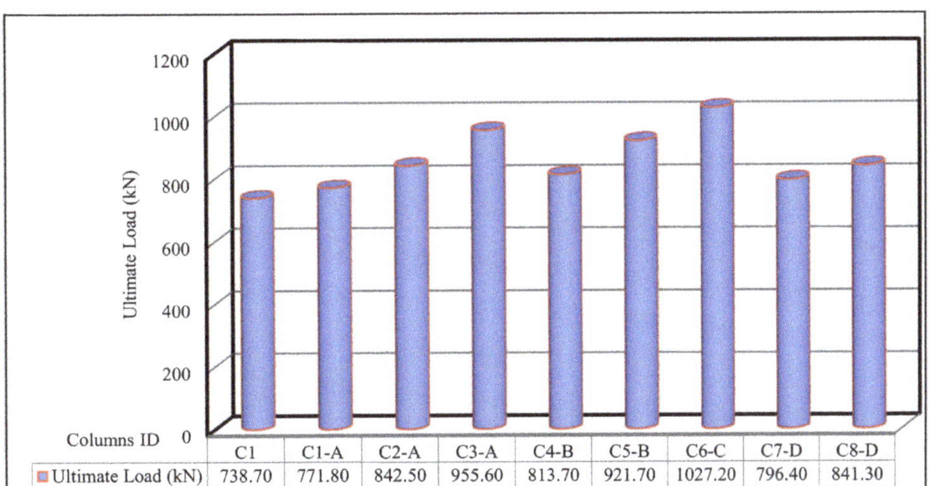

Figure 11. Ultimate load for the tested columns.

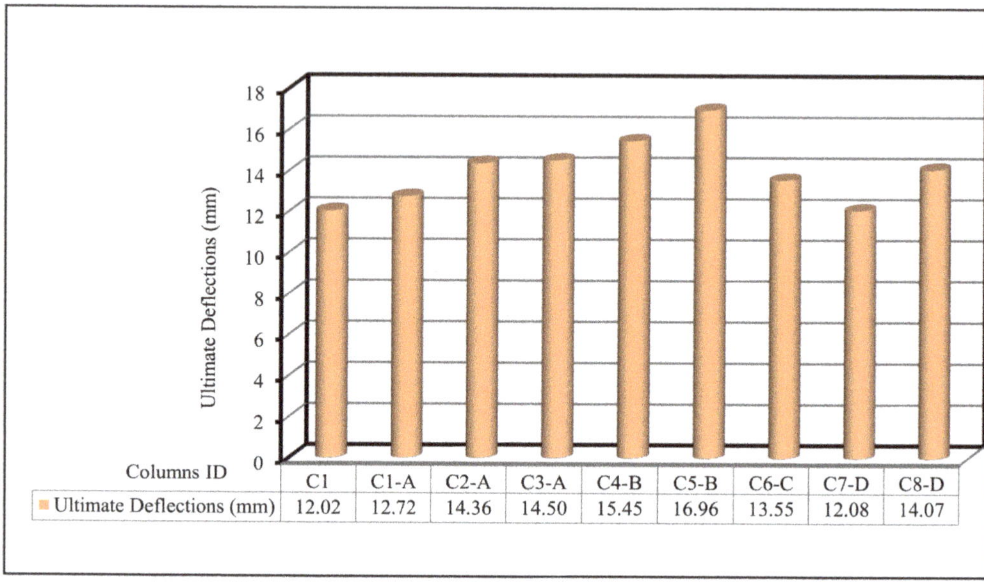

Figure 12. Ultimate deflection for the tested columns.

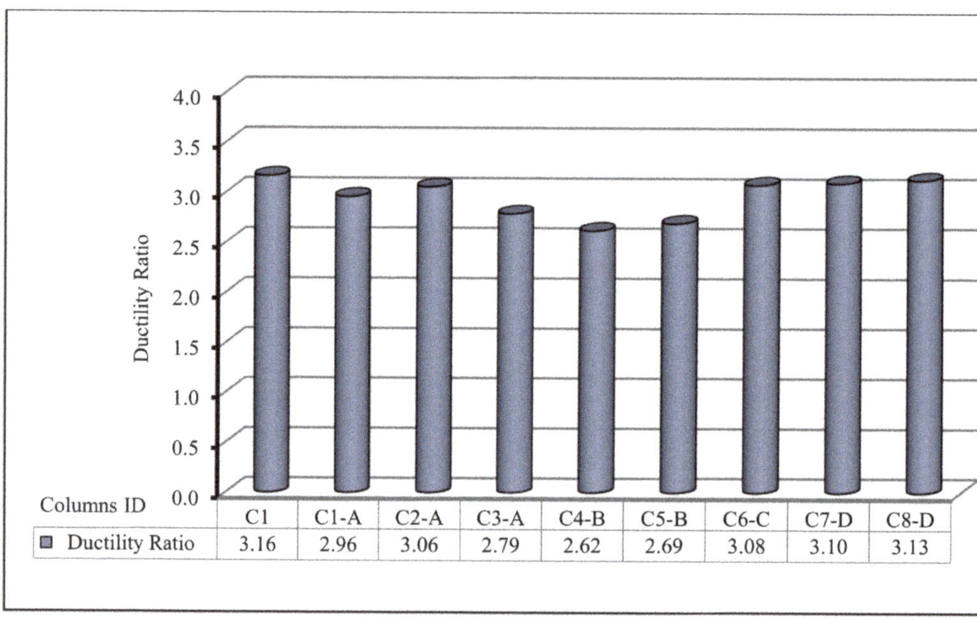

Figure 13. Ductility ratio for the tested columns.

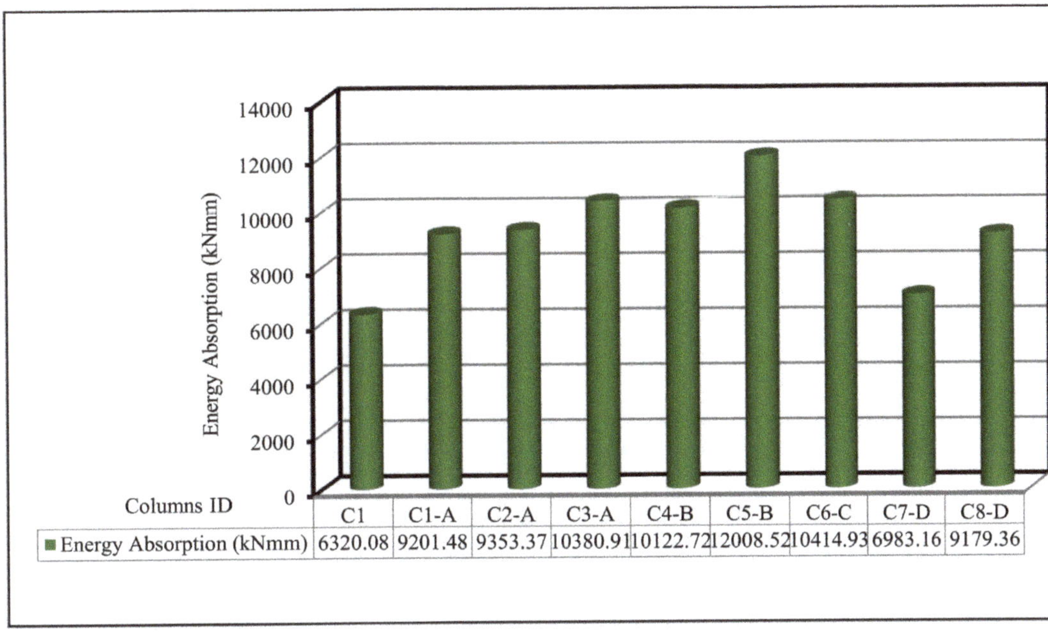

Figure 14. Energy absorption for the tested columns.

3.4. Energy Absorption

The ductility ratio was well-defined, similar to the proportion of the ultimate load deformation to the first crack load deformation, whereas the energy absorption was known as the area under the load deflection curve until collapse. Table 7 shows the ductility ratios and energy absorption values for all columns. Ongoing increase in the energy absorption as the volume fraction percentage increased, was noted. Column C8-D, with one layer of fiber glass mesh, had the highest ductility ratio when compared with the other types of meshes. Figure 14 shows the energy absorption comparison for all columns. The energy absorption for the control column C1 was 6320.08 kN·mm. For all other columns, the energy absorption was greater than the control C1. That is, it showed good enhancement with an enhancement percentage of 10% to 190%. Group D with the fiber glass mesh exhibited the smallest enhancement whereas Group C with the Tensar-mesh exhibited the highest enhancement. Column C5-B, which uses two layers of expanded wire mesh, had the highest energy absorption. It may be concluded that by improving the ductility ratio, these new composite materials improved the failure behavior.

Finally, the performance of columns was improved by employing these advanced composite materials. It may be said that it slowed the beginning of the first cracks, while also increasing the capacity of the service load. It also has high ultimate-loads, high durability, improved deformation, and improved energy absorption, all of which are advantageous in dynamic purposes.

3.5. Crack Pattern

Near collapse, the control column C1 showed a compression mode of failure with a concrete cover of local crushing and spalling. For all other tested columns, near collapse after the ultimate load value decreased up to 60% to 40% of the ultimate load. All crack patterns of the tested columns are shown in Figure 15.

C1

C1-A

C2-A

C3-A

Figure 15. *Cont.*

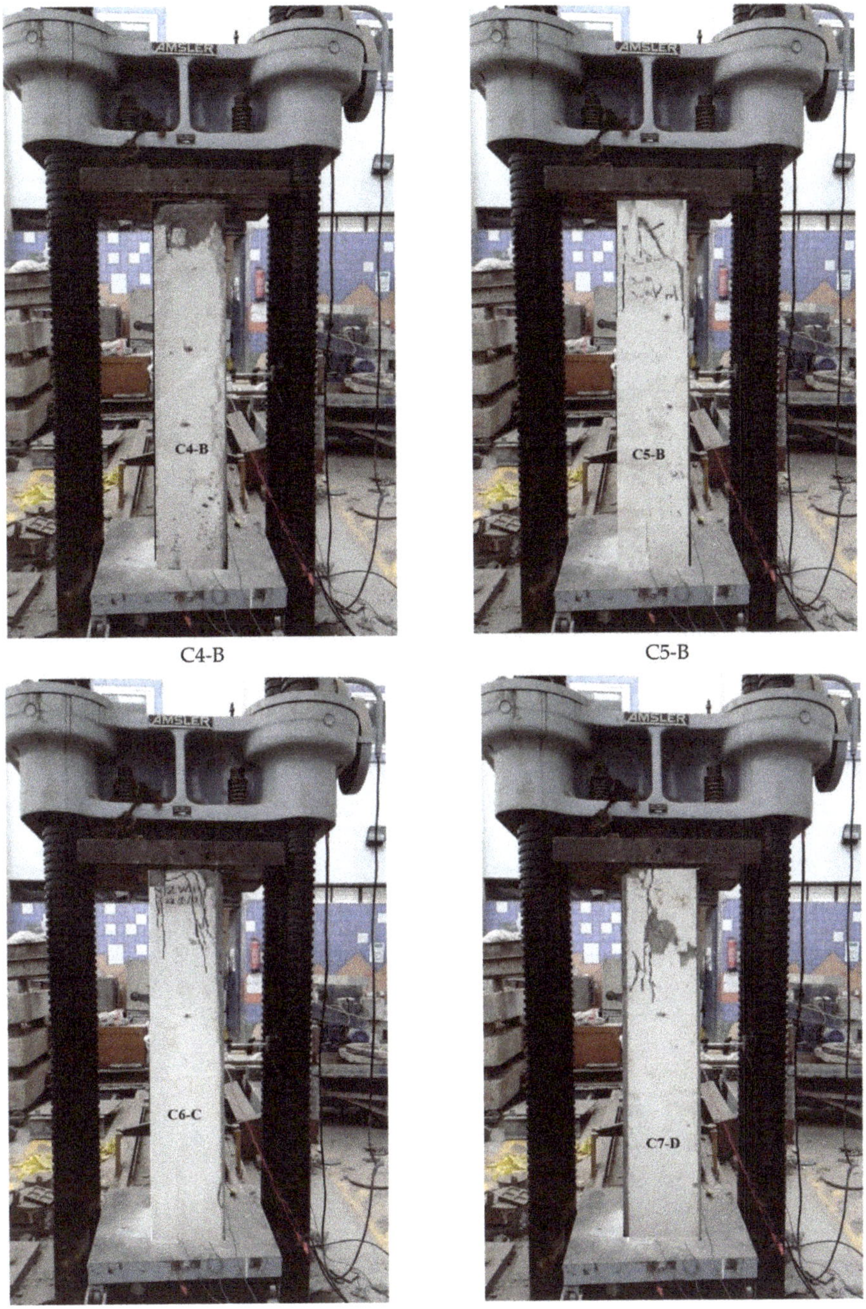

C4-B C5-B

C6-C C7-D

Figure 15. *Cont.*

C8-D

Figure 15. Crack patterns.

4. Analytical Analysis

Analytical analysis was conducted to validate the results of the experimental program. Table 8 showed the analytical results were obtained from the NLFEA software ANSYS2019-R1 [30] program.

Table 8. Analytical results.

Column ID	First Crack Load (kN)	Ultimate Load (kN)	Def. at First Crack Load (mm)	Def. at Ult. Load (mm)	Ductility Ratio —	Energy Absorption (kN·mm)
C1	260.30	812.57	2.85	8.12	2.85	4104.8310
C1-A	260.30	848.98	3.23	10.30	3.19	6200.2411
C2-A	260.30	906.95	3.53	10.50	2.98	7055.3147
C3-A	260.30	1051.16	3.90	10.31	2.64	7303.0707
C4-B	260.30	995.07	4.43	10.49	2.37	6899.9601
C5-B	260.30	1013.87	4.73	9.63	2.04	6396.8092
C6-C	260.30	1129.90	3.30	7.30	2.21	4933.7318
C7-D	260.30	876.04	2.93	7.76	2.65	4237.6949
C8-D	260.30	925.43	3.38	6.82	2.02	3782.9131

4.1. Types of Elements

For concrete, element Solid 65 (Figure 16a) was used to represent the concrete stress–strain curve, while element Link 180 3-D (Figure 16b) was used to represent the reinforcing bars and reinforcing stirrups. All ferrocement reinforcement was modeled by computing the volumetric ratio (reinforcing steel ratio to concrete) in the concrete element Solid 65. As ANSYS allows the user to enter three rebar materials in the concrete, each material corresponds to x, y, and z. The orientation angles denoted the reinforcement orientation in the smeared model. Therefore, ferrocement reinforcements were modeled as smeared layers with the volumetric ratio as indicated in Section 4.2.

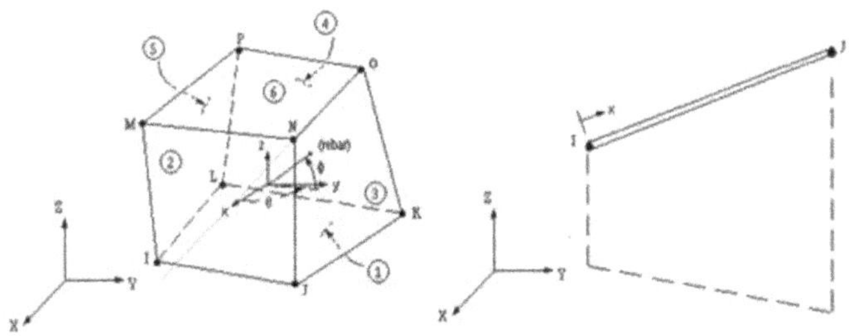

(**a**) Solid65 (**b**) Link180

Figure 16. Geometry of element types.

4.2. Properties of Modeled Materials

This section shows the material properties for concrete, reinforcing steel bars, and ferrocement wire meshes:

- The material properties for concrete:
    1. Elastic modulus of elasticity (Ec = 4400$\sqrt{}$fcu = 24,100 N/mm^2) [28].
    2. Poisson's ratio (ν = 0.3) [28].
- The material properties for reinforcing steel bars:
    1. Elastic modulus of elasticity (E_s = 200 kN/mm^2) [28].
    2. Yield stress (f_y = 400 N/mm^2 & f_{yst} = 240 N/mm^2) [28].
    3. Poisson's ratio (ν = 0.2) [28].
    4. Area of steel of φ 12 (As = 112 mm^2)
    5. Area of steel of φ 8 (As = 50.3 mm^2)
- The properties for welded mire mesh:
    1. Volumetric ratio of one layer = 0.0027
    2. Volumetric ratio of two layers = 0.0054
    3. Volumetric ratio of three layers = 0.0081
- The material properties for Expanded wire mesh:
    1. Volumetric ratio of one layer = 0.00753
    2. Volumetric ratio of two layers = 0.01510
- The material properties for Tensar mesh:
    1. Volumetric ratio of one layer = 0.02040
- The material properties for glass fiber mesh:
    1. Volumetric ratio of one layer = 0.00535
    2. Volumetric ratio of two layers = 0.01070

4.3. Specimens Modeling

A finite nonlinear analysis was conducted to evaluate the behavior of geopolymer ferrocement HSC columns, as shown in Figure 17.

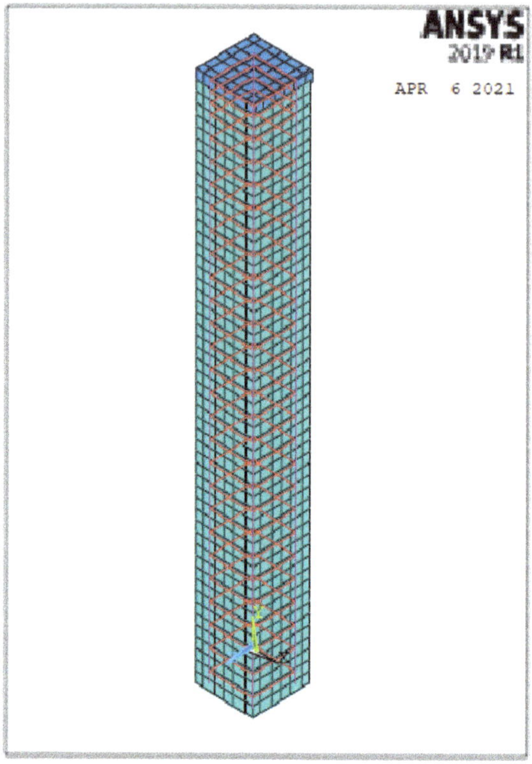

Figure 17. 3D-modeling of the columns.

4.4. Analytical Results and Discussion

Table 8 shows the analytical results such as the first crack load, ultimate load, the deflection at first crack and ultimate load, ductility ratio, and energy absorption for the modeled columns.

4.4.1. Ultimate Load

The ultimate loads for the modeled columns are shown in Table 8. The ultimate load of control C1 was 812.57 kN. For group A, columns C1-A to C3-A, the ultimate loads extended between 848.98 kN and 1051.16 kN. The improvement in the ultimate capacity was 4.40% to 22.80%. For group B, columns C4-B and C5-B, the ultimate loads extended between 995.07 kN and 1013.87 kN. The improvement in the ultimate capacity was 18.30% and 20.00%. For group C, column C6-C had an ultimate load of 1129.90 kN with a considerable improvement of 28.20%. For group D, columns C7-D and C8-D, the ultimate loads were 876.04 kN and 925.43 kN, with an improvement of 7.30% and 12.30%, respectively.

According to the findings in Table 8, the use of fiber glass mesh is more efficient than other types of metallic and nonmetallic mesh reinforcements in increasing the ultimate capacity. A good agreement was noted between the experimental and nonlinear approaches for all columns.

4.4.2. Ultimate Deflection

Table 8 shows the ultimate deflection for all modeled columns. The deflection for the control column C1 was 8.12 mm. For group A, the maximum deflection ranged from 10.30 mm to 10.50 mm for columns C1-A to C3-A, respectively, which were higher than that of the control C1. For group B, the maximum deflection at ultimate load was 10.49 mm and 9.63 mm for columns C4-B and C5-B, respectively, which was also higher than that of the control C1. For group C, column C6-C, there was an ultimate deflection of 7.30 mm. For group D, columns C7-D and C8-D, the ultimate deflections were 7.76 and 6.82 mm, respectively. It can clearly be seen that good agreement was noted between the experimental and nonlinear approaches for all columns.

4.4.3. Load–Deflection Relationship

The relationship between the load and deflection for the modeled columns are presented in Figures 21–29. From these figures, it can clearly be seen that good agreement was noted between the experimental and nonlinear approaches for all columns.

It is possible to conclude that the FE simulations produced accurate findings when compared to the experimental results. Furthermore, the analytical deflection findings outperformed the experimental results by a mean of 15%, as indicated.

4.4.4. Energy Absorption

Table 8 shows the ductility ratios and energy absorption values for all modeled columns. The energy absorption for the control column C1 was 4104.83 kN·mm. For all other columns, the energy absorption was greater than the control C1. Group D with the fiber glass mesh exhibited the smallest enhancement whereas group C with the Tensar-mesh exhibited a considerable enhancement. It may be concluded that by improving the ductility ratio and the energy absorption, these new composite materials improved the failure behavior.

4.4.5. Crack Pattern

The cracking started at an initial loading step in the molded column face near the column supports. This was due to the invisible micro-cracks in the experimental study. The cracking load was fairly lower than the experimental one. This might be acceptable as the FE analysis characterized the stage of micro-cracks. Furthermore, the wire mesh composite materials may have hidden the micro-cracks initiated in the experimental test. Conversely, the patterns of cracks at every load step showed that the crack propagation for all columns was somewhat changed compared to the experimental one due to the accuracy of the FE Ansys program in obtaining the micro-cracks. All crack patterns of the molded columns are shown in Figure 18.

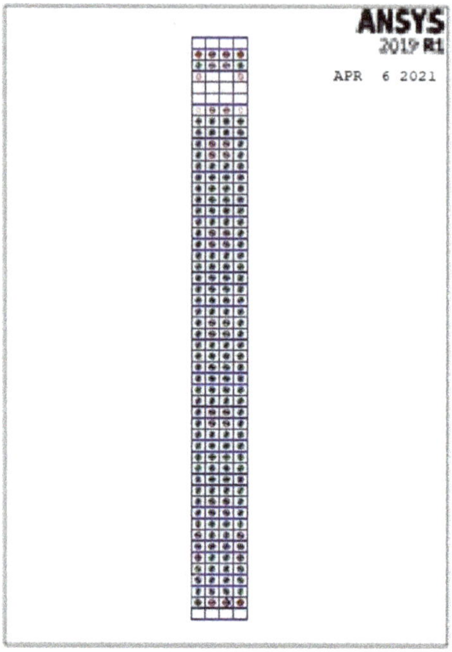

C1

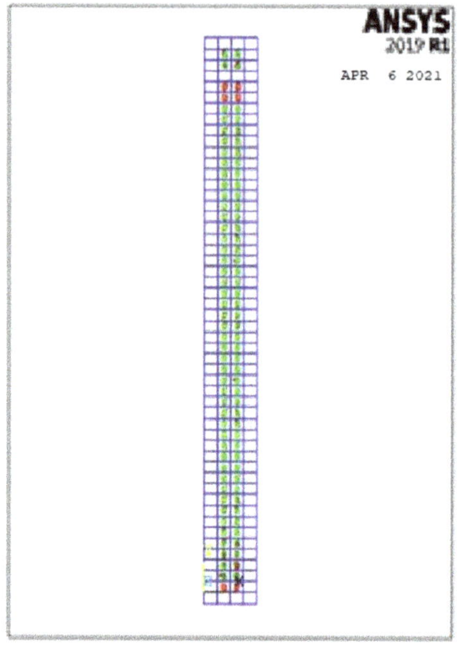

C1-A

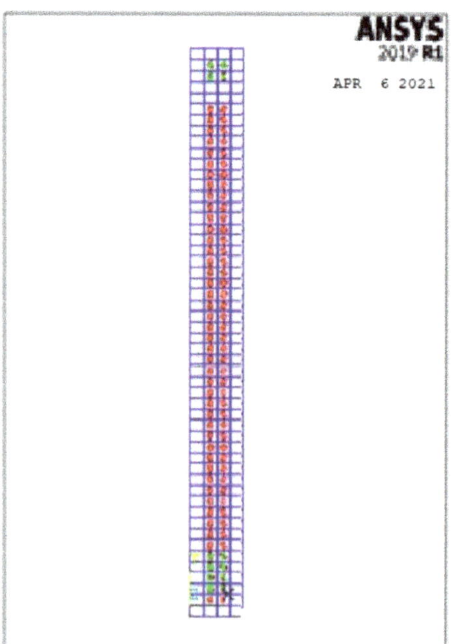

C2-A

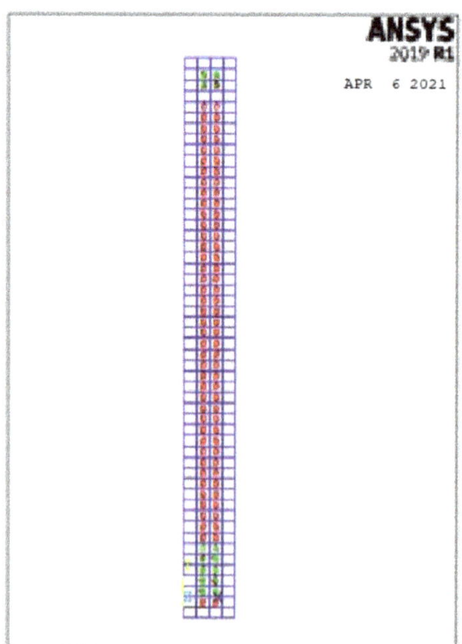

C3-A

Figure 18. *Cont.*

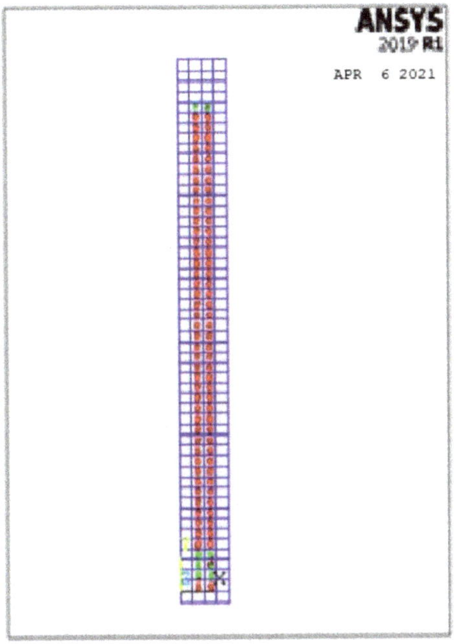

C4-B

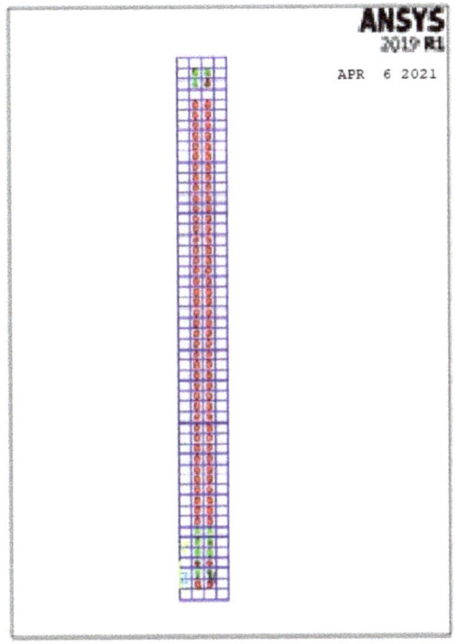

C5-B

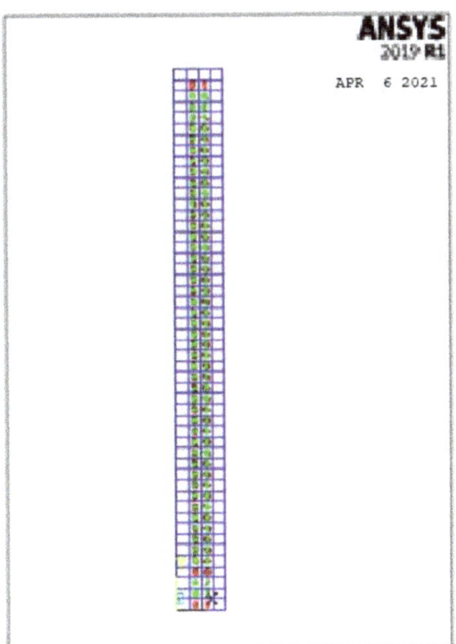

C6-C

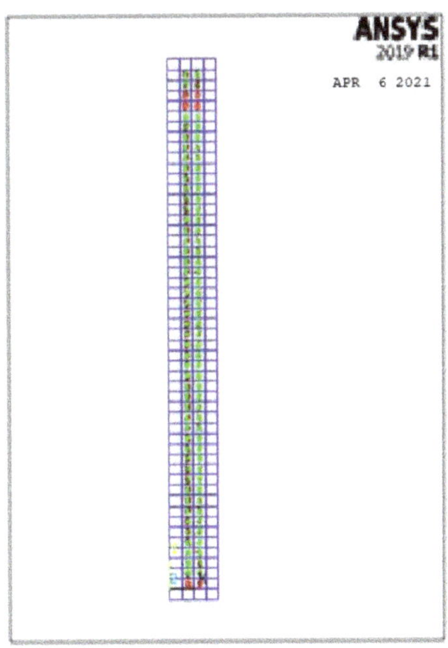

C7-D

Figure 18. *Cont.*

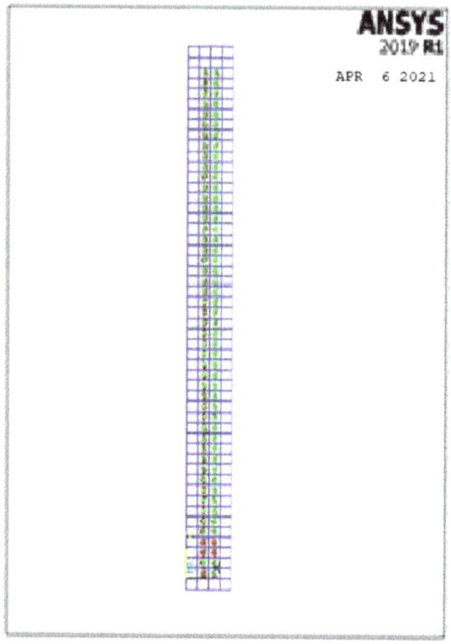

C8-D

Figure 18. Crack patterns for the modeled columns.

5. Comparisons between Analytical and Experimental Results

Comparison between analytical and experimental results confirmed an acceptable agreement in representing the geopolymer ferrocement HSC columns' performance in terms of first crack and ultimate deflection, first crack load and ultimate load, and crack pattern.

5.1. Ultimate Failure Load

Figure 19 and Table 9 show the comparison between the ultimate experimental and analytical load. There was fair agreement between the experimental and analytical ultimate loads. It is possible to conclude that the FE simulations produce accurate findings when compared to the experimental results. Furthermore, the analytical ultimate load results outperformed the experimental results by a mean of 11%.

5.2. Ultimate Deflection

Figure 20 and Table 9 show the comparison between the ultimate experimental analytical deflections. The load–defection curves as shown in Figures 21–29 for the experimental and modeled columns showed good agreement with respect to the control column deflection. Furthermore, the analytical ultimate defection results outperformed the experimental results by a mean of 15%.

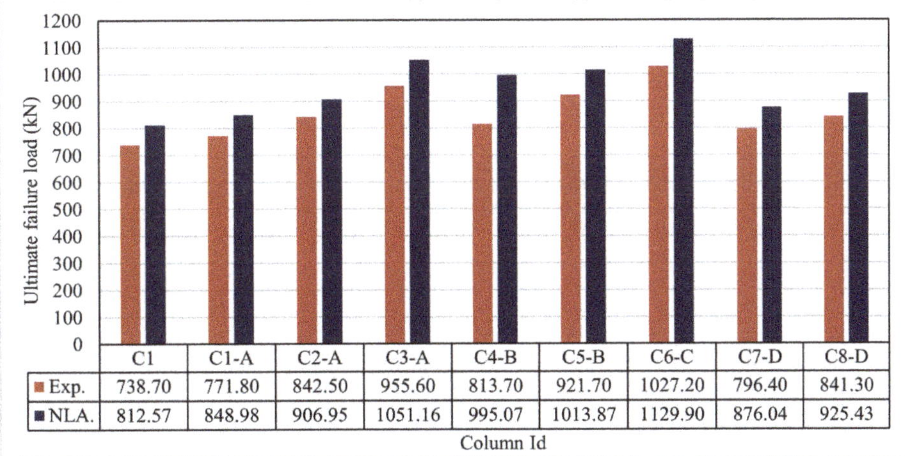

Figure 19. Comparison between Exp. and NLA ultimate loads.

Table 9. Experimental and analytical results.

Column ID	First Crack Load (kN)	First Crack Load (kN)	Ultimate Load (kN)	Ultimate Load (kN)	Def. at Ult. Load (mm)	Def. at Ult. Load (mm)
	NLA.	EXP.	NLA.	EXP.	NLA.	EXP.
C1	260.30	296.00	812.57	738.70	8.12	12.02
C1-A	260.30	325.00	848.98	771.80	10.30	12.72
C2-A	260.30	380.00	906.95	842.50	10.50	14.36
C3-A	260.30	455.00	1051.16	955.60	10.31	14.50
C4-B	260.30	443.00	995.07	813.70	10.49	15.45
C5-B	260.30	449.00	1013.87	921.70	9.63	16.96
C6-C	260.30	515.00	1129.90	1027.20	7.30	13.55
C7-D	260.30	391.00	876.04	796.40	7.76	12.08
C8-D	260.30	378.00	925.43	841.30	6.82	14.07

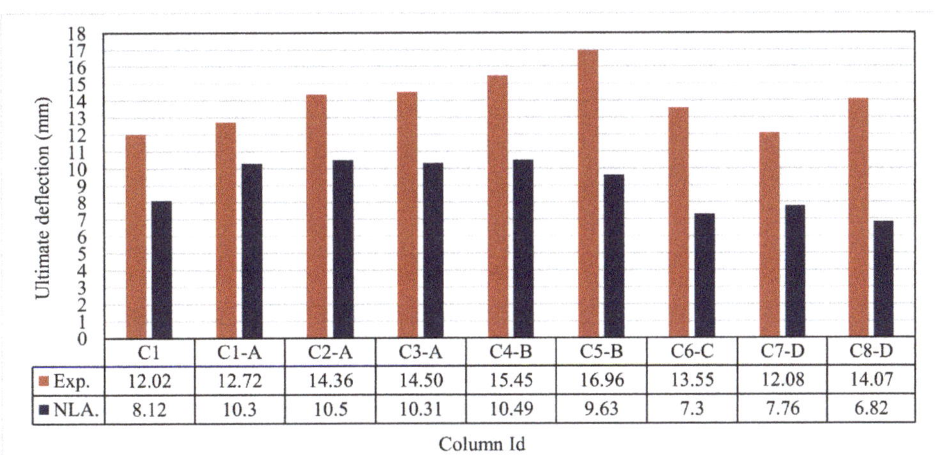

Figure 20. Comparison between Exp. and NLA ultimate deflections.

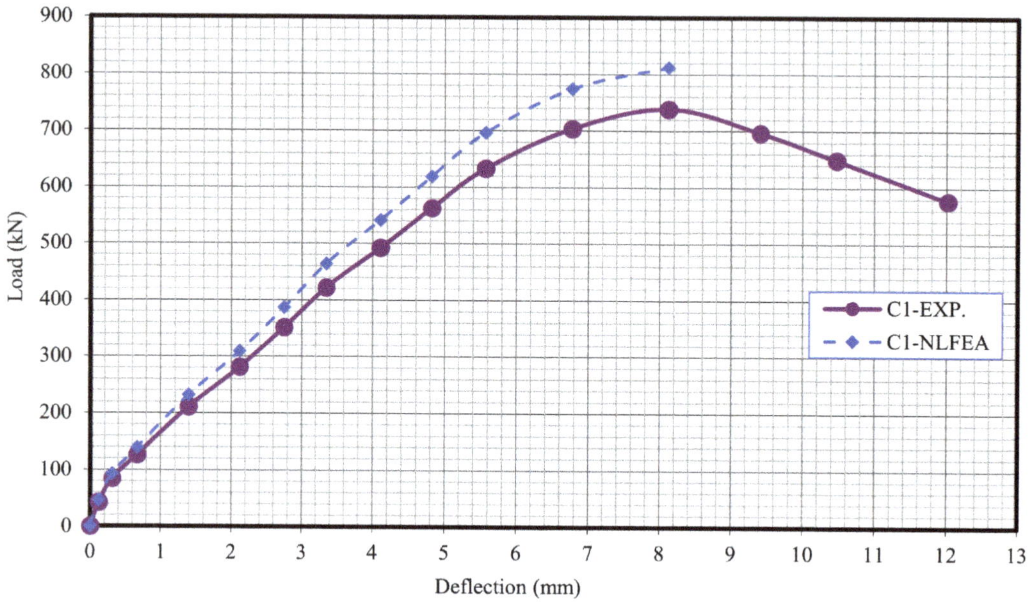

Figure 21. Load deflection of Column C1.

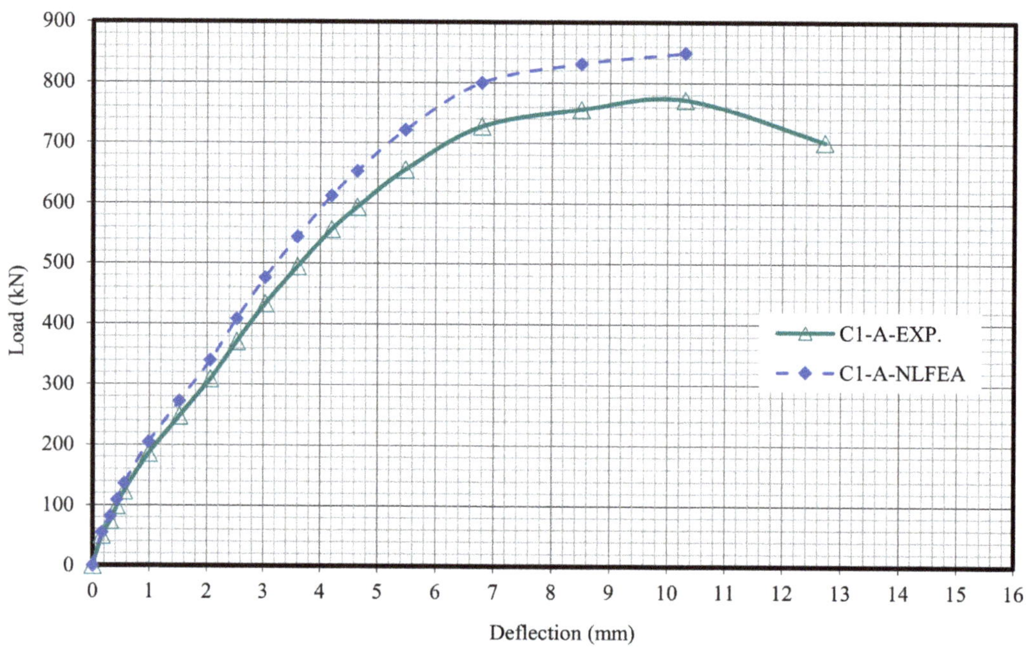

Figure 22. Load deflection of Column C1-A.

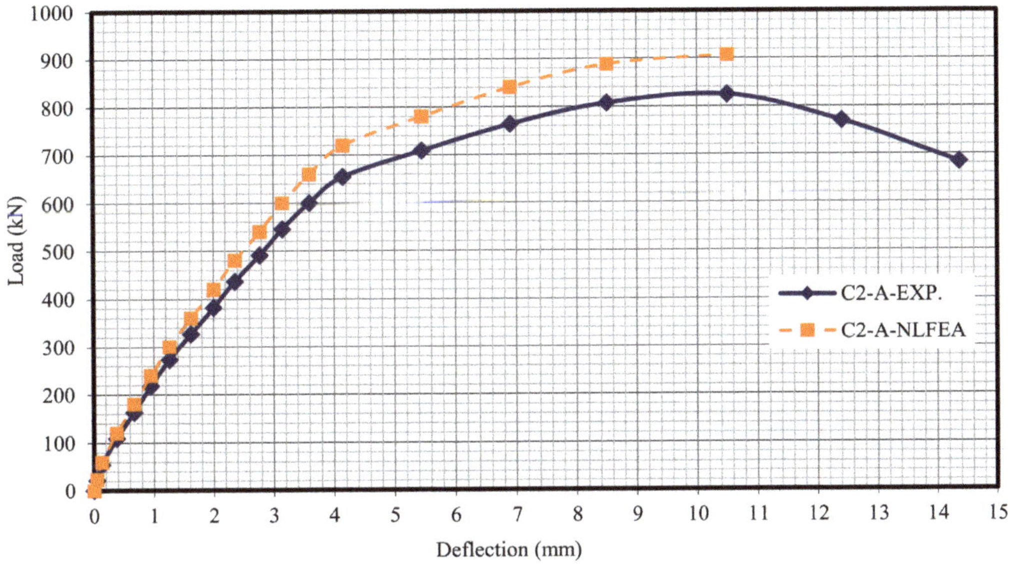

Figure 23. Load deflection of Column C2-A.

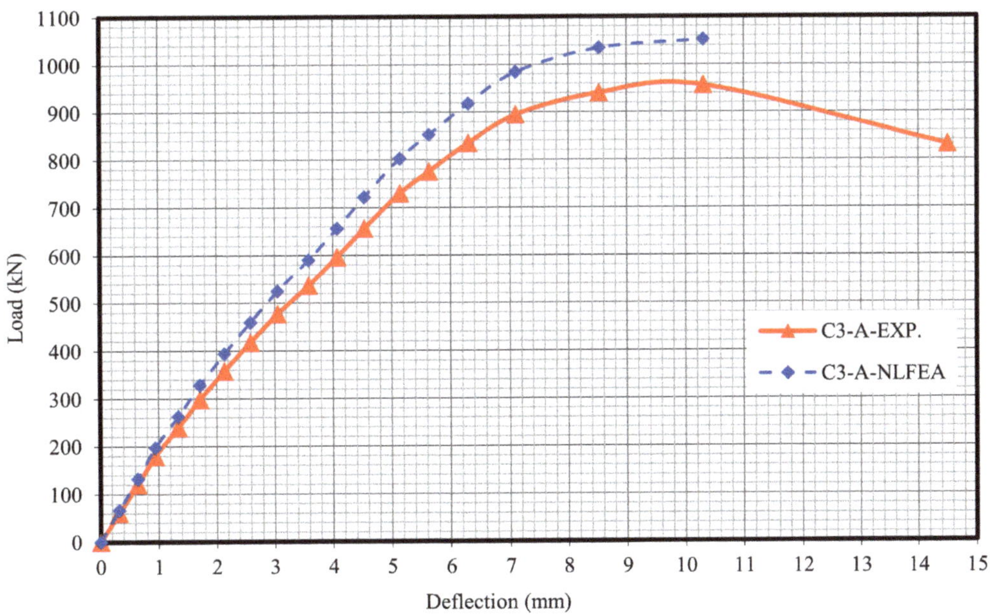

Figure 24. Load deflection of Column C3-A.

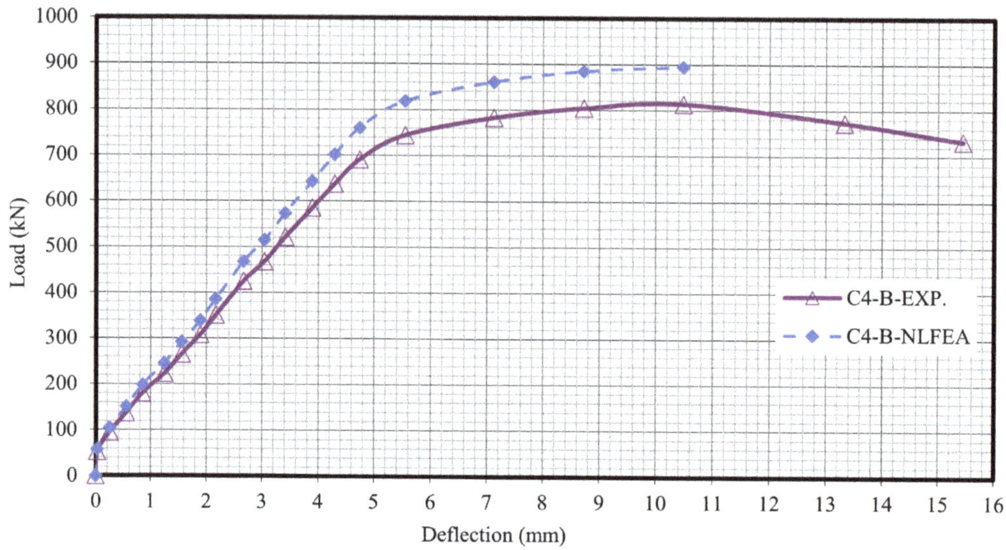

Figure 25. Load deflection of Column C4-B.

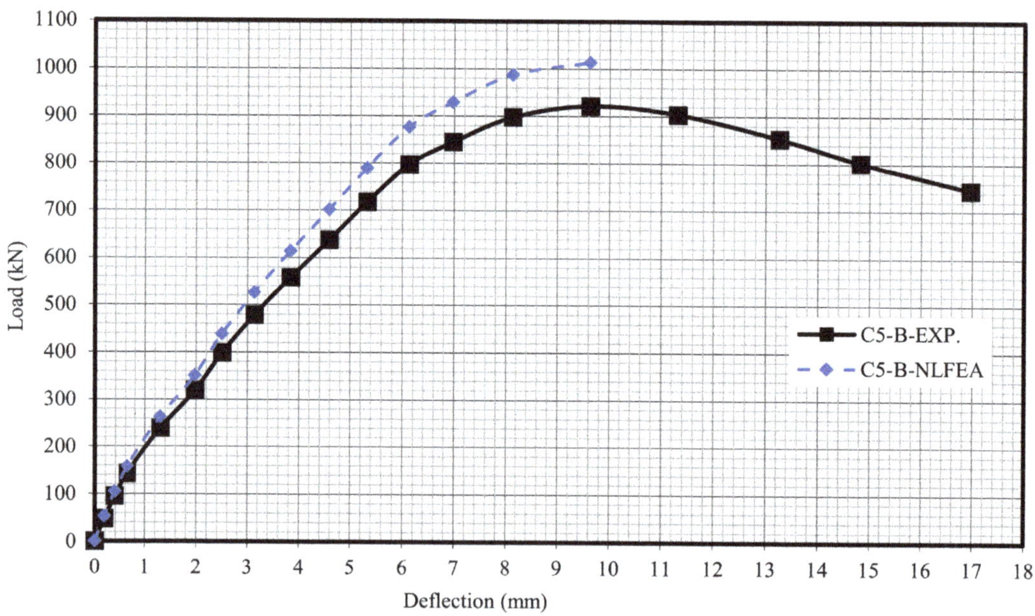

Figure 26. Load deflection of Column C5-B.

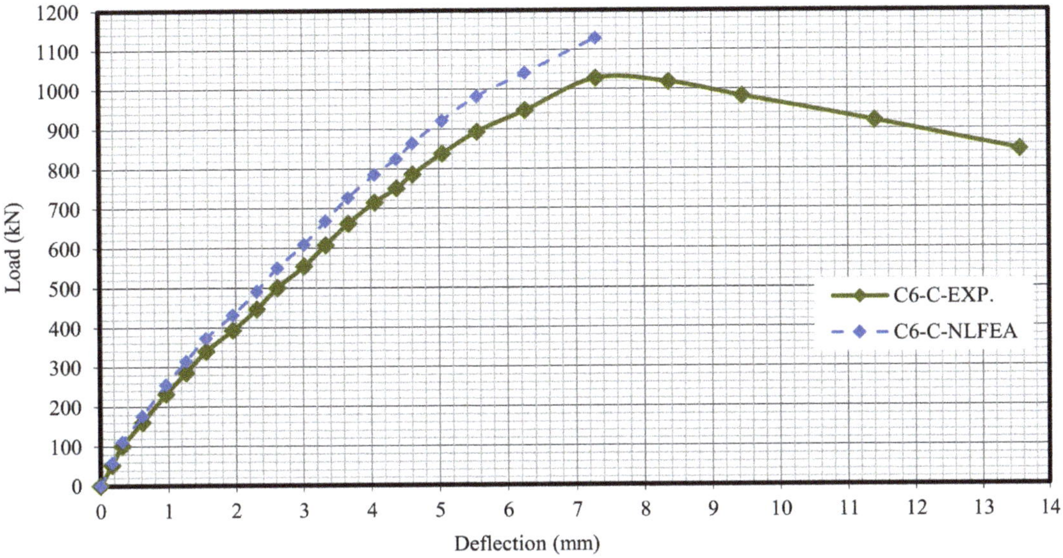

Figure 27. Load deflection of Column C6-C.

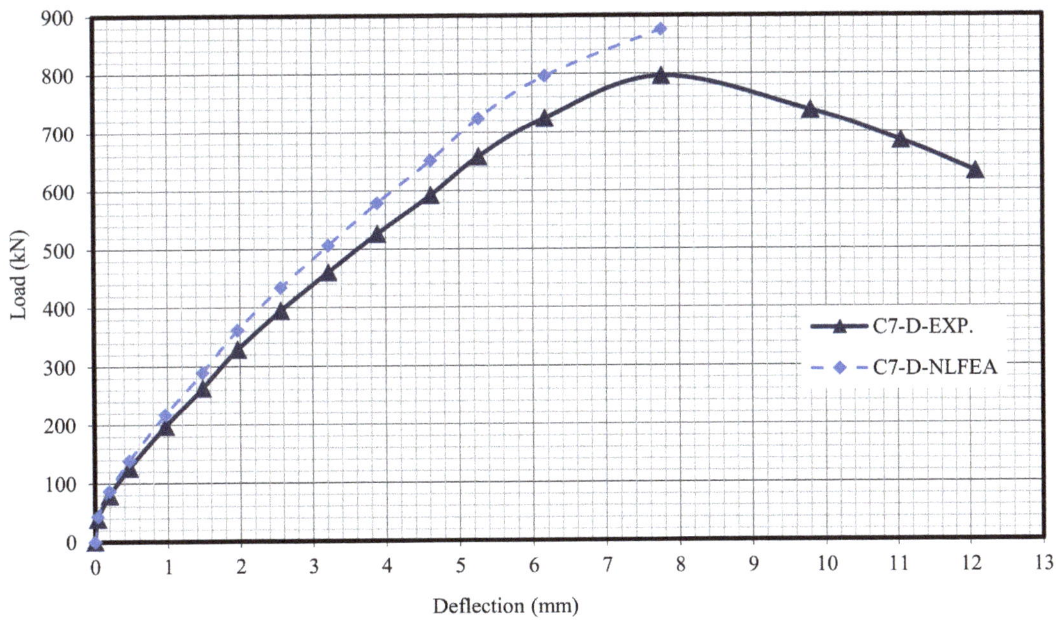

Figure 28. Load deflection of Column C7-D.

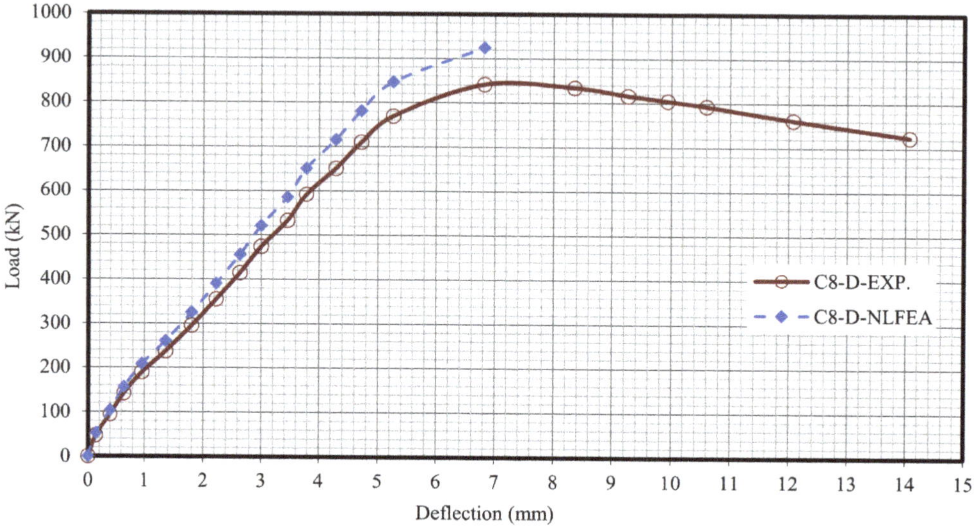

Figure 29. Load deflection of Column C8-D.

5.3. Cracking Patterns

Figures 30 and 31 show the comparison between the crack pattern for the experimental and modeled column samples. The micro-cracking stage, which occurs before observable cracking, is represented by the NLA forecasts. The cracking patterns at each load increment, on the other hand, demonstrated that the crack propagation for all columns differed somewhat from the experimental fracture pattern. This is due to the nonlinear finite element program's precision in determining micro- and large cracks as well as the impact of the reinforcing technique on cracking patterns, as shown in Table 9.

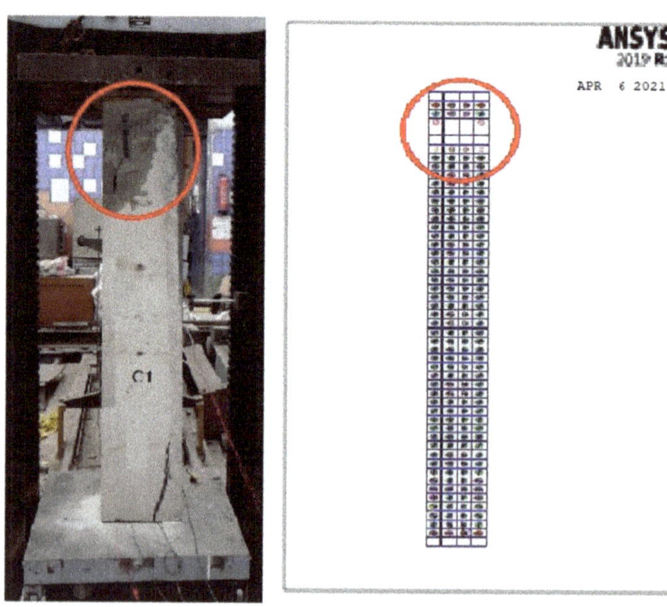

Figure 30. Crack spread for the control specimen.

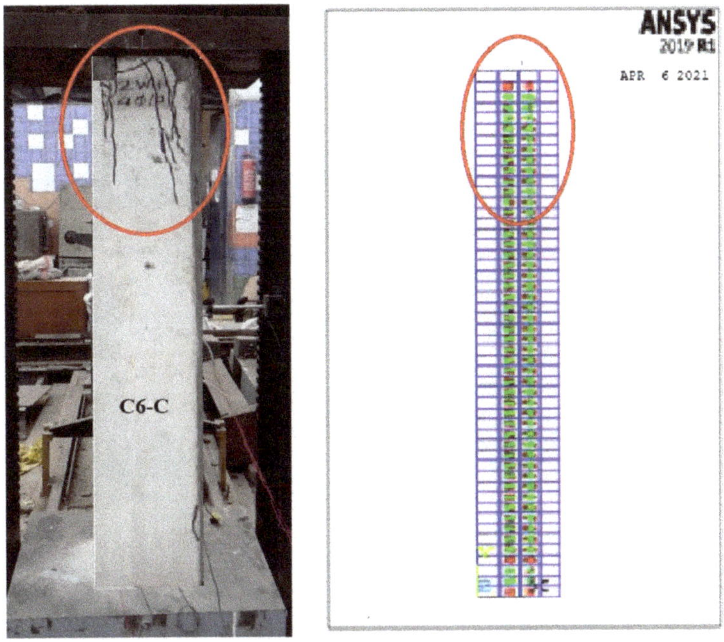

Figure 31. Cracks spread for the C6-C specimen.

6. Conclusions

Based on the experimental and analytical results, the following conclusions can be drawn:

1. Because of the lighter and easier handling of wire meshes compared with steel reinforcement, all wire meshes offer several improvements, especially for structures with complex shapes.
2. Increasing the volume fraction of the wire mesh reinforcement increased the initial cracks, ultimate loads, energy absorption, and ductility index.
3. Ferrocement geopolymer columns achieved higher ultimate load, ductility, and energy absorption compared to the steel reinforced concrete control column.
4. Cracks with greater number and narrower widths were observed for those ferrocement geopolymer columns compared with the steel geopolymer columns.
5. Expanded or welded ferrocement geopolymer columns showed greater ultimate failure loads than the control column. Additionally, using expanded or welded columns had a considerable effect on the ultimate failure loads, where welded wire mesh exhibited almost 28.10% compared with expanded wire mesh.
6. Column reinforced with one-layer of nonmetallic Tensar-mesh obtained the highest ultimate failure load out of all the tested columns without concrete cover spalling. Consequently, increasing the volume fraction had the main result of postponing the incidence of crack development with higher corrosion protection and high loading carrying capacity than columns reinforced with metallic reinforcement.
7. Column reinforced with one layer of fiber glass mesh obtained the smallest ultimate failure load compared with the control column.
8. The analytical procedures for first crack and ultimate load computations obtained good prediction for these loads and the column failure modes. Consequently, there were improved strength, deformation characteristics, and cracking behavior with great savings of reinforcement.

9. The comparison of the crack patterns obtained by the FE and experimental models led to identical crack propagation for the two approaches up to failure. The inclination of the failure surfaces and the concentration of cracks of all columns were the same in both patterns.
10. The established ferrocement geopolymer columns could be successfully used as an alternative to the traditional RC columns, which could be of true merit in both developed and developing countries aside from its anticipated economic and environmental merits.

Funding: The Scientific Research at Benha University for was funded this work thorough the Research Support Program, "Research Projects Call 2017-2019-Fifth round" project no: M/5/2/14 & "National Project Call 2018-2020-First round".

Conflicts of Interest: The author declares no conflict of interest.

References

1. Dattatreya, J.K.; Rajamane, N.P.; Sabitha, D.; Ambily, P.S.; Nataraja, M.C. Flexural behaviour of reinforced Geopolymer concrete beams. *Int. J. Civ. Struct. Eng.* **2011**, *2*, 138–159.
2. El-Sayed, T.A.; Erfan, A.M.; Abd El-Naby, R.M. Influence of Rice, Wheat Straw Ash & Rice Husk Ash on The properties of Concrete Mixes. *Jokull* **2017**, *67*, 103–119.
3. El-Sayed, T.A.; Erfan, A.M.; El-Naby, R.M.A. Recycled rice & wheat straw ash as cement replacement materials. *J. Eng. Res. Rep.* **2019**, *5*, 1–9.
4. El-Sayed, T.A.; MErfan, A.; MErfan, A.; El-Naby, R. Flexural Behavior of RC Beams by Using Agricultural Waste as a Cement Reinforcement Materials. *J. Eng. Res. Rep.* **2019**, *7*, 1–12. [CrossRef]
5. El-Sayed, T.A.; Shaheen, Y.B. Flexural performance of recycled wheat straw ash-based geopolymer RC beams and containing recycled steel fiber. *Structures* **2020**, *28*, 1713–1728. [CrossRef]
6. Davidovits, J. Geopolymers: Inorganic polymeric new materials. *J. Therm. Anal. Calorim.* **1991**, *37*, 1633–1656. [CrossRef]
7. Duxson, P.; Jimenez, A.M.F.; Provis, J.; Lukey, G.C.; Palomo, Á.; Van Deventer, J.S.J. Geopolymer technology: The current state of the art. *J. Mater. Sci.* **2006**, *42*, 2917–2933. [CrossRef]
8. Sofi, A.; Phanikumar, B. An experimental investigation on flexural behaviour of fibre-reinforced pond ash-modified concrete. *Ain Shams Eng. J.* **2015**, *6*, 1133–1142. [CrossRef]
9. Chun, L.B.; Sung, K.J.; Sang, K.T.; Chae, S.T. A study on the fundamental properties of concrete incorporating pond-ash in Korea. In Proceedings of the 3rd ACF International Conference- ACF/VCA, HoChiMinh, Vietnam, 11–13 November 2008.
10. Hardjito, D.; Rangan, B.V. *Development and Properties of Low-Calcium Fly Ash-Based Geopolymer Concrete*; Research Report GC1; Faculty of Engineering, Curtin University of Technology: Perth, Australia, 2005.
11. Bakharev, T. Geopolymeric materials prepared using Class F fly ash and elevated temperature curing. *Cem. Concr. Res.* **2005**, *35*, 1224–1232. [CrossRef]
12. Palomo, A.; Grutzeck, M.; Blanco-Varela, M.T. Alkali-activated fly ashes: A cement for the future. *Cem. Concr. Res.* **1999**, *29*, 1323–1329. [CrossRef]
13. van Jaarsveld, J.; van Deventer, J.; Lukey, G. The effect of composition and temperature on the properties of fly ash- and kaolinite-based geopolymers. *Chem. Eng. J.* **2002**, *89*, 63–73. [CrossRef]
14. Sofi, M.; van Deventer, J.; Mendis, P.; Lukey, G. Engineering properties of inorganic polymer concretes (IPCs). *Cem. Concr. Res.* **2007**, *37*, 251–257. [CrossRef]
15. Wallah, S.; Rangan, B.V. *Low-Calcium Fly Ash-Based Geopolymer Concrete: Long-Term Properties*; Research Report GC 2; Faculty of Engineering, Curtin University of Technology: Perth, Australia, 2006.
16. Bakharev, T. Resistance of geopolymer materials to acid attack. *Cem. Concr. Res.* **2005**, *35*, 658–670. [CrossRef]
17. Mansur, M.N.; Paramasivam, P. Ferro-cement Short Columns under Axial and Eccentric Compression. *ACI Struct. J.* **1990**, *84*, 523.
18. Kaushik, S.K.; Prakash, A.; Singh, K.K. Inelastic Buckling of Ferro-cement Encased Columns. In Proceedings of the Fifth International Symposium on Ferrcement, Manchester, UK, 6–9 September 1994; pp. 327–341.
19. Nedwell, P.J.; Ramesht, M.H.; Rafei-Taghanaki, S. Investigation into the Repair of Short Square Columns using Ferro-cement. In Proceedings of the Fifth International Symposium on Ferrcement, Manchester, UK, 6–9 September 1994; pp. 277–285.
20. Fahmy, E.; Shaheen, Y.B.; Korany, Y. Repairing Reinforced Concrete Columns Using Ferro-cement Laminates. *J. Ferro-Cem.* **1999**, *29*, 115–124.
21. Erfan, A.M.; Ahmed, H.H.; Mina, B.A.; El-Sayed, T.A. Structural Performance of Eccentric Ferrocement Reinforced Concrete Columns. *Nanosci. Nanotechnol. Lett.* **2019**, *11*, 1213–1225. [CrossRef]
22. Erfan, A.M.; Elnaby, R.M.A.; Elhawary, A.; El-Sayed, T.A. Improving the compressive behavior of RC walls reinforced with ferrocement composites under centric and eccentric loading. *Case Stud. Constr. Mater.* **2021**, *14*, e00541.
23. Hussin, A.A.; Erfan, A.M.; El-Sayed, T.A.; Abd El-Naby, R.M. Experimental and Analytical Analysis of Lightweight Fe rrocement Composite Slabs. *Eng. Res. J.* **2019**, *1*, 88–96.

24. Erfan, A.M.; El-Sayed, T.A. Structural Shear Behavior of Composite Box Beams Using Advanced Innovated Materials. *J. Eng. Res. Rep.* **2019**, *5*, 1–14. [CrossRef]
25. Erfan, A.M.; El-Sayed, T.A. Shear Strength of Ferrocement Composite Box Section Concrete Beams. *Int. J. Sci. Eng. Res.* **2019**, *10*, 260–279.
26. El-Sayed, T.A.; Shaheen, Y.B.; Ahmed, H.H.; Yussef, A.K. Flexural performance of GGBS-based geopolymer ferrocement beams. *Int. J. Sci. Eng. Res.* **2021**, *12*, 1–40.
27. El-Sayed, T.A. Performance of Porous Slabs Using Recycled Ash. *Polymers* **2021**, *13*, 3319. [CrossRef]
28. E.C.P. 203/2020. *Egyptian Code of Practice: Design and Construction for Reinforced Concrete Structures*; Housing and Building National Research Center: Cairo, Egypt, 2020.
29. El-Sayed, T.A.; Erfan, A.M. Improving shear strength of beams using ferrocement composite. *Constr. Build. Mater.* **2018**, *172*, 608–617. [CrossRef]
30. ANSYS. *Engineering Analysis System User's Manual, Vol. 1&2, and Theoretical Manual*; Revision 8.0; Swanson Analysis System Inc.: Houston, PA, USA, 2005.

polymers

MDPI

Review

Liquid Crystals Investigation Behavior on Azo-Based Compounds: A Review

Nurul Asma Razali and Zuhair Jamain *

Sustainable Materials and Renewable Energy (SMRE) Research Group, Faculty of Science and Natural Resources, Universiti Malaysia Sabah, Jalan UMS, Kota Kinabalu 88400, Sabah, Malaysia; nasmarazali@gmail.com
* Correspondence: zuhairjamain@ums.edu.my

Abstract: Liquid crystal is an intermediate phase between the crystalline solid and an isotropic liquid, a very common substance in our daily lives. Two major classes of liquid crystal are lyotropic, where a liquid crystal is dissolved in a specific solvent under a particular concentration and thermotropic, which can be observed under temperature difference. This review aims to understand how a structure of a certain azo compound might influence the liquid crystal properties. A few factors influence the formation of different liquid crystals: the length of the alkyl terminal chain, inter/intra-molecular interaction, presence of spacer, spacer length, polarization effects, odd-even effects, and the presence of an electron-withdrawing group or an electron-donating group. As final observations, we show the compound's different factors, the other liquid crystal is exhibited, and the structure–property relationship is explained. Liquid crystal technology is an ideal system to be applied to products to maximize their use, especially in the electronic and medical areas.

Keywords: azo compound; liquid crystal; structure–property relationship

Citation: Razali, N.A.; Jamain, Z. Liquid Crystals Investigation Behavior on Azo-Based Compounds: A Review. *Polymers* **2021**, *13*, 3462. https://doi.org/10.3390/polym13203462

Academic Editors: Beata Podkościelna and Andrzej Puszka

Received: 29 August 2021
Accepted: 7 October 2021
Published: 9 October 2021

1. Introduction

Azo compounds represent a large branch of the liquid crystal sciences. The azo compound is a compound that contains two or more organic groups, which an azo group separates by –N=N– as its linking unit [1,2]. In order to conserve natural flora and fauna sources, an alternative was taken by producing a synthetic dye using an azo compound. An azo compound is an important group used extensively in the textile industry [3], manufacture of ink [4], and the cosmetic industry [5].

Liquid crystal is a liquid that is not isotropic, has two refractive indexes, and displays interference in a polarized light [6,7]. According to Stegemeyer (1989), it was first discovered by Dr. Friedrich Reinitzer back in 1888 in a German University located in Prague, where he revealed two crucial traits of the cholesteric liquid crystal, which are the change of color with temperature and the temperature range of the cholesteric phase, and the melting point and clearing point [8,9].

It has been proven that liquid crystal can form an intermediate phase, called the mesophase. This phase is situated in the middle of a crystalline solid and an isotropic liquid [10]. Relying on a particular circumstance in which the mesophase becomes apparent, liquid crystal can be characterized into two major classes: lyotropic and thermotropic [11]. It can be suggested that lyotropic shows the liquid crystal state in a certain concentration, as Andrienko (2018) reported that the specification of this liquid crystal is its concentration [12].

However, Brightman (1954) has stated that the thermotropic phase is a phase where a liquid crystal is prepared via a heating process [7]. It can be signified as a liquid crystal formed with temperature change, in agreement with Kusabayashi and Takenaka (1984) who reported that thermotropic liquid crystals give out a few different forms with temperature variation [13]. Enantiotropic is a term to define a thermodynamically stable mesophase that appears on heating and cooling of the molecule, while a thermotropic mesophase that arises only during cooling of a molecule is termed as a monotropic mesophase. On the other hand, a molecule that exhibits liquid crystal under the same influence as a lyotropic (solvent) and a thermotropic (heat) is referred to as an amphotropic [14].

Generally, the systemic nature of the thermotropic liquid crystal is dependable on the molecular shape of the crystal, whether it is a rod-like or a disc-like shape molecule [15]. The thermotropic liquid crystal with disc-like molecules are a discotic mesogen, while a thermotropic liquid crystal with rod-like molecules is a calamitic mesogen, which can be sub-divided into nematic, cholesteric and smectic phase.

A nematic phase is the simplest liquid crystal where the molecules are positioned in a long axis across the same preferred direction [12]. This means that the rod-like molecules are arranged in a plane parallelly. In a thermotropic liquid crystal, the positional order of the molecules can be destructed if the molecules are treated with heat at a specific temperature but not the orientational order [15]. As a molecule of a nematic mesophase is aligned parallelly along its axes, this mesophase exhibits anisotropic physical properties. Typically, a nematic phase shows schlieren, marble, and pseudoisotropic textures in accordance with the structure of its surface.

Cholesteric mesophase is also known as a chiral nematic phase (N*) as it is formed via doping of a nematic liquid crystal or when the molecules of a system are chiral [16]. The director of a cholesteric mesogen tends to form a helix with pitch due to the alignment of an adjacent molecule at a trivial angle of one another. Basically, a pitch is a distance taken by the director to make one full turn in the helix. This helical structure of N* can reflect light with a wavelength uniform to the pitch length. A cholesteric mesogen commonly exhibit a schlieren textures.

Different from a nematic phase, a smectic phase is aligned in layers disclosing an association between its position along with the orientational order [12]. A different smectic phase is formed as the molecular order of the smectic phase change, which is Smectic A (SmA), Smectic B (SmB), Smectic C (SmC), Smectic F (SmF), and Smectic I (SmI). These smectic are then sub-divided into two categories depending on whether the molecules are tilted to the layer normal or not. A part of SmA and SmB, other smectic are tilted phase.

Among all other smectic, SmA or SmC are often observed. These mesogen is formed when each molecule does not have a long-range positional order [15]. The molecules in a smectic A phase are arranged in layers so that the long axis is perpendicularly aligned to the plane. Most compound that exhibits these kinds of structure is a compound that carries a terminal polar group [17]. Generally, this phase displays a focal-conic texture.

SmC differs from the SmA phase as the director of the molecule's constant tilt angle measured normally to the smectic plane [18]. The layers of the smectic C phase are closely packed, respecting the director of the phase, in a short-range. Similar to the nematic mesophase, SmC has a chiral smectic C (SmC*). The director of the layer is identical to SmC, and the only difference it makes is that the angle that rotates from a layer to another and forms a helix. According to [16], there is a sub-phase of the smectic C phase called the anticlinic smectic C. The order of this molecule is similar compared to the order of a smectic C phase. The difference between each phase is the correlation between the tilted direction in the layers.

Understanding of the structure–property relationship of the compound is required in order to understand the interconnections between the core system, linking unit, and terminal group. Currently, there is a lack research focused on this area. There has been a lack of research that focuses on the azo-based compounds, especially on the methods and the structure–property relationship. Hence, this review will help future researchers to understand the phase properties and the characterization of azo-based compounds.

2. Structure–Property Relationship

Despite being lyotropic or thermotropic, most liquid crystal can exhibit at least one liquid crystal phase. According to Safinya et al. (1986), a thermotropic liquid crystal can exhibit one to several liquid crystal phases between the crystal and isotropic states [18].

However, there is a time where a liquid crystal does not exhibit any mesophase at all. This study will explain the possibilities of liquid crystal disclosing different types of liquid crystal phase, why a structure with the same central unit gives out different liquid crystal phases, and how any substitution can affect the mesomorphic.

In accordance with a study by Collings and Hird (1997), the structure–properties relationship is crucial to understand for the synthesizing or altering of a molecule in a certain way or arrangement to get a certain mesomorphic phase [14]. This understanding is important, especially when the liquid crystal is synthesized for a particular application that requires a particular liquid crystal phase.

A study was conducted in 1997 by Parra et al. on the structure–properties relationship of azo-based compounds [19]. In this study, Parra and her co-worker synthesized two azo compounds, **1**(a–f) and **2**(a–f). All homologues of compound **1a** were claimed to display a nematic phase. In addition, homologues n = 9 and 10, compounds **1e** and **1f**, respectively, show a monotropic smectic C phase. Similar to compound **1**, all homologues of compound **2** also exhibit an enantiotropic nematic phase. However, compounds **2**(a–c) displays a monotropic smectic C while compound **2**(d–f) exhibit an enantiotropic smectic C phase. Compound **2** appears to have more extensive mesomorphic range compared to compound **1**. Parra and her co-worker stated that the low mesophase stability of compound **1** is caused by the presence of a thiophene ring which produces an additional deviation that hinders the formation of a stable mesophase.

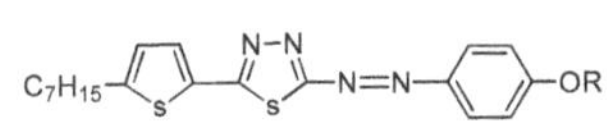

Comp	1a	1b	1c	1d	1e	1f
R	C_5H_{11}	C_6H_{13}	C_7H_{15}	C_8H_{17}	C_9H_{19}	$C_{10}H_{21}$

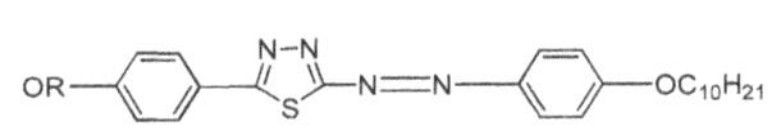

Comp	2a	2b	2c	2d	2e	2f
R	C_5H_{11}	C_6H_{13}	C_7H_{15}	C_8H_{17}	C_9H_{19}	$C_{10}H_{21}$

Two years later, Belmar et al. (1999) synthesized quite a similar azo compound, namely compounds **3**(a–h) [20]. All homologues of compound **3** give out a nematic phase showing that corresponding molecular interactions occur that give the same outcome in terms of molecular arrangement and thermal stability. However, not all homologues exhibit smectic C phase, such as compounds **3**(a–d) as a layered smectic order is not ideal due to the difference in the volume occupied by the opposite chain. Nevertheless, a tilted smectic C order is formed as the alkoxy chain lengthens starting of homologues 7 to 10, and almost the same to the opposite chain.

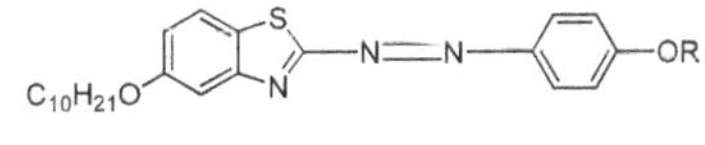

Comp	3a	3b	3c	3d	3e	3f	3g	3h
R	C_3H_7	C_4H_9	C_5H_{11}	C_6H_{13}	C_7H_{15}	C_8H_{17}	C_9H_{19}	$C_{10}H_{21}$

In 2001, Lee et al. synthesized two crystalline dyes containing a non-activated pyranylazo group, compounds **4**(a–f) and **5**(a–f) [21]. At the first heating phase, none of the homologues of compounds **4** and **5** display any mesomorphic property due to the thermo- and photochromic properties of the dye. This property causes a small ring-opened merocyanine species, which is not favored at high temperatures. Only after the first heating-cooling phase does the compound starting to exhibit a mesophase. Compounds **4a** and **4c** show an enantiotropic nematic phase, compounds **4b** and **4d** display a monotropic nematic phase, while compounds **4e** and **4f** do not form any liquid crystal phase. In contrast, all homologues of compound **5** form a nematic phase, and some even display a smectic A phase, except compounds **5d** and **5e**.

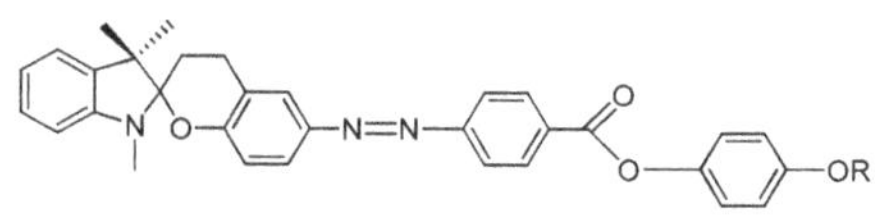

Comp	4a	4b	4c	4d	4e	4f
R	$(CH_2)_4CH_3$	$(CH_2)_5CH_3$	$(CH_2)_6CH_3$	$(CH_2)_7CH_3$	$(CH_2)_8CH_3$	$(CH_2)_9CH_3$

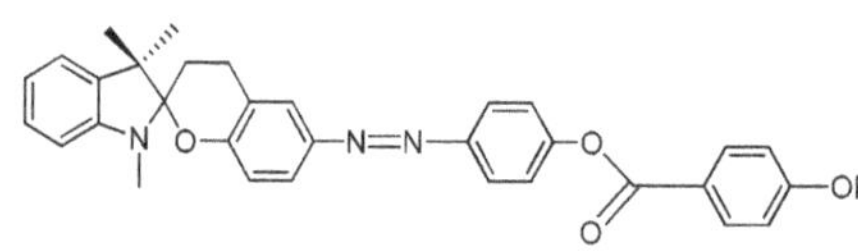

Comp	5a	5b	5c	5d	5e	5f
R	$(CH_2)_4CH_3$	$(CH_2)_5CH_3$	$(CH_2)_6CH_3$	$(CH_2)_7CH_3$	$(CH_2)_8CH_3$	$(CH_2)_9CH_3$

In the same year, two azo compounds were synthesized containing a similar structural unit, comprised of pyridine in compounds **6**(a–f) and 1,3,4-thiadiazole rings in both compounds **6**(a–f) and **7**(a–f) [22]. All homologues of both compounds exhibit a liquid crystal phase. As for compound **6**, all homologues display a crystal to isotropic transition, and only two of the highest homologues, compounds **6e** and **6f** exhibit a monotropic nematic phase. Compound **7** shows a dimorphism of a nematic phase and a smectic C phase. However, compounds **7**(a–c) exhibit a monotropic smectic C phase. Nevertheless, the thermal stability of compound **7** is higher compared to the thermal stability of compound **6**. The most prominent difference between compounds **6** and **7** can be seen from the structure. Compound **6** comprises a pyridine unit at the end of the central core and only one lateral chain, while compound **7** has a benzene ring instead of a pyridine, two lateral units, and a greater molecular length. To briefly summarize, compound **6** contains a pyridine unit, and is not long enough to be polarized enough to exhibit a stable liquid crystal phase compared to compound **7**.

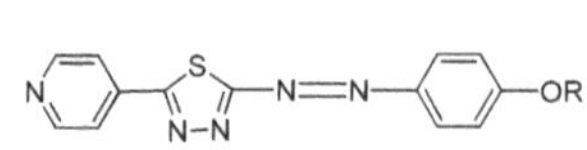

Comp	6a	6b	6c	6d	6e	6f
R	C_5H_{11}	C_6H_{13}	C_7H_{15}	C_8H_{17}	C_9H_{19}	$C_{10}H_{21}$

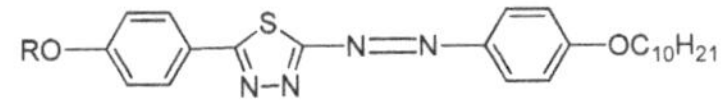

Comp	7a	7b	7c	7d	7e	7f
R	C_5H_{11}	C_6H_{13}	C_7H_{15}	C_8H_{17}	C_9H_{19}	$C_{10}H_{21}$

So et al. (2003) reported in his study about compounds **8**(a–d), where this compound has two two-ring mesogenic units that are connected by a spacer, namely 2-hydroxy-1,3-dioxypropylene [23]. All homologues of compound **8** have display an enantiotropic schlieren and/or a broken fan textured smectic C phase except for compound **8**a. Figure 1 shows a micrograph of compound **8**c. This phenomenon happens due to the temperature range of mesophase increase with the increasing length of the terminal alkyl chain.

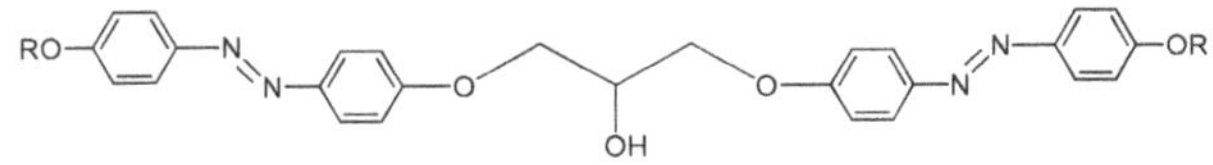

Comp	8a	8b	8c	8d
R	C_6H_{13}	C_8H_{17}	$C_{10}H_{21}$	$C_{12}H_{25}$

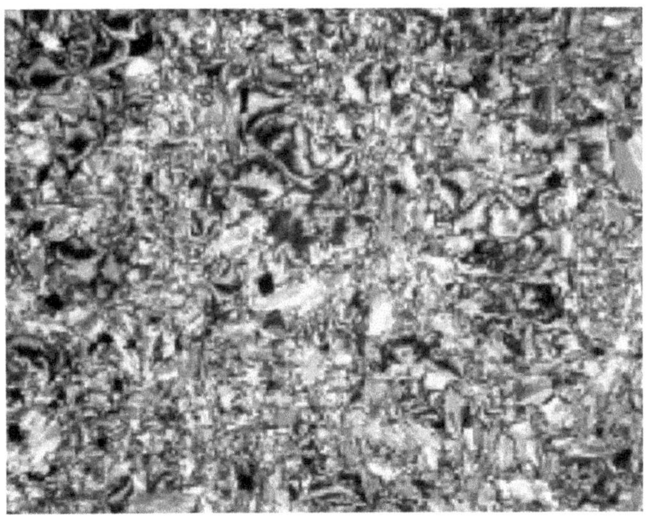

Figure 1. Polarizing optical micrographs upon cooling of isotropic liquid of compound **8**c at 168.0 °C [23].

Compounds **9**(a–c) was synthesized by Abbasi et al. (2006) [24]. Compounds **9**(a–b) display a monotropic liquid crystal phase during cooling from the isotropic liquid. However, compound **9**c formed an enantiotropic mesophase behavior and showed a liquid crystalline characteristic on heating and cooling from the isotropic liquid. Abbasi et al. declared that the stability of mesophase is influenced by the alkyl length (R). Compound **9**c with the longest alkyl chain length exhibit enantiotropic mesophase which is thermodynamically stable compound, and compound with a short chain length (**9**a and **9**b) exhibit unstable mesomorphic behavior.

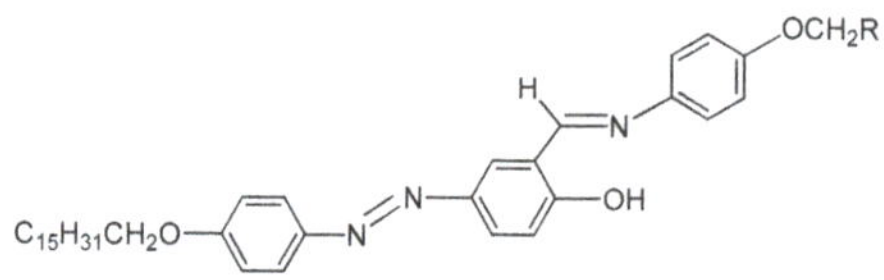

Comp	9a	9b	9c
R	C_7H_{15}	$C_{11}H_{23}$	$C_{15}H_{31}$

In the same year, compounds **10**(a–c) and **11**(a–c) were synthesized by Rezvani et al. (2006) [25]. Compounds **10**(a–c) did not show any liquid crystal phase as this compound directly melted into an isotropic liquid. It was claimed that the melting point of the ligands decreases with increasing alkyl chain length value. Compound **11**a exhibit four endothermic transitions, where the first two transitions are equivalent to the crystal-to-crystal transition. The third transition peak at 204.3 °C with enthalpy values of 532.43 kJ mol^{-1} is corresponding to a crystal phase to a mesophase. A Schlieren texture typically for the smectic C phase was detected, as reported in Figure 2. The fourth transition peak at 240.3 °C with a low enthalpy value of 5.25 kJ mol^{-1} is responsible for the transition from mesophase to the isotropic liquid. The high clearing enthalpy corresponding to the mesophase to isotropic transition indicates that the mesophase structure is highly in order.

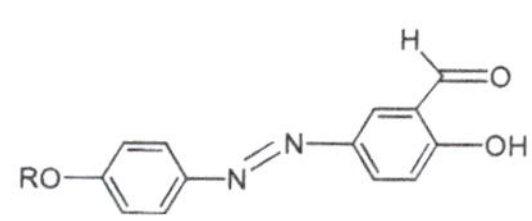

Comp	10a	10b	10c
R	C_9H_{19}	$C_{11}H_{23}$	$C_{13}H_{27}$

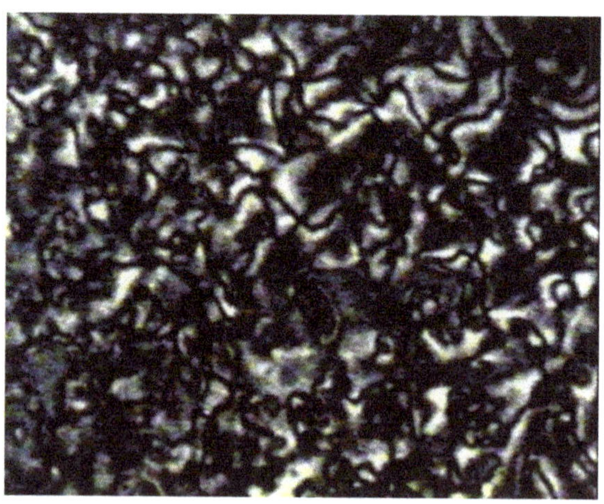

Figure 2. Optical texture of compound **11**a at 210.0 °C [25].

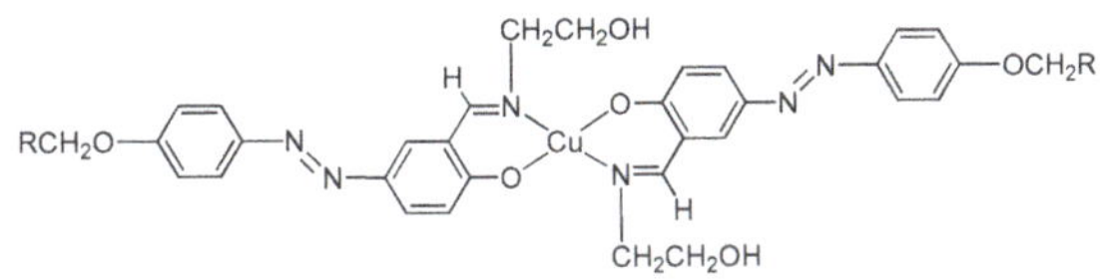

Comp	11a	11b	11c
R	C_9H_{19}	$C_{11}H_{23}$	$C_{13}H_{27}$

Compounds **12**(a–c) were first synthesized by So et al. in 2001. In 2006, he conducted another study using the same compound but with a different alkyl chain [26]. During the heating phase, compound with the highest homologues, compound **12**c, exhibits a crystal to smectic C phase at 163.0 °C, and smectic C to isotropic liquid at 168.0 °C and compounds **12**a and **12**b does not show any mesomorphic phase. However, during the cooling scan, compounds **12**b and **12**c show a smectic C mesophase (Figure 3) but not compound **12**a. Compound **12**b form a monotropic smectic C liquid crystal phase with a schlieren or broken-fan texture, whereas compound **12**c exhibit an enantiotropic smectic C liquid crystal phase with a schlieren texture and compound **12**a shows no mesomorphic behavior. So et al. (2006) reported that increasing the alkyl chain is expected to increase the length-to-breadth, resulting in a liquid crystal phase, especially the smectic phase [27]. He also claimed that as the length of the terminal chain increase, the temperature range of the smectic phase also increases. It was also discussed in the study that the entropy change keeps increasing as the terminal chain increase, hence, a smectic mesophase is prone to show, and as the terminal chain increases, the thermal stability of a tilted smectic C also increases.

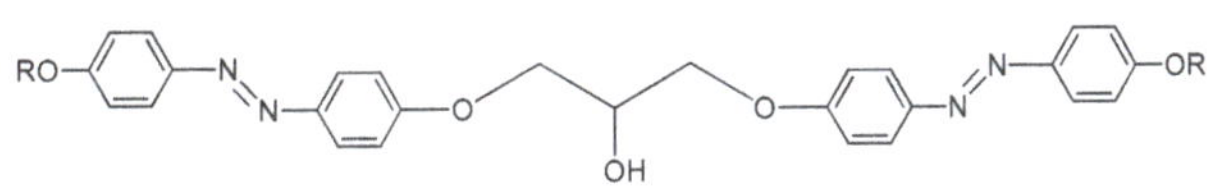

Comp	12a	12b	12c
R	C_5H_{11}	C_7H_{15}	C_9H_{19}

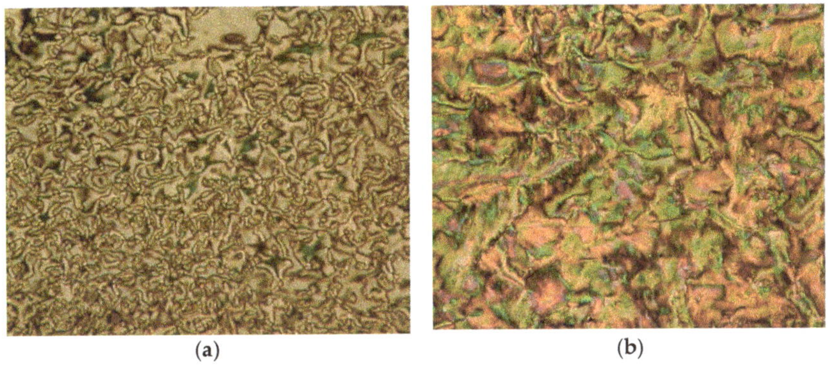

Figure 3. (**a**) Schlieren textured smectic C of compound **12**b at 160.0 °C, and (**b**) Schlieren textured smectic C of compound **12**c at 160.2 °C [26].

Lutfor et al. (2009) have synthesized compounds **13**(a–f) with six different homologues [28]. As for raw compound **13**a, two transition peaks were observed during cooling of compound at 175.2 and 123.1 °C, which are assigned for the isotropic to smectic and smectic to crystal transition, respectively. Chloro-substituted compound **13**b also shows two peaks on cooling, 144.3 and 126.0 °C, which are corresponded to isotropic to smectic and smectic to crystal transition, respectively. Compound **13**c and **13**f do not display any liquid crystal phase, and it melts at 168.0 and 165.0 °C and crystallized at 162.0 °C and 160.0 °C, respectively. Compound **13**d is a fluoro-substituted compound that displays two peaks on cooling at exactly 150.5 and 127.1°C, corresponding to isotropic to smectic and smectic to crystal transition, respectively. Compound **13**e is a fluoro, and chloro-substituted compound that shows two peaks on cooling at 142.9 and 126.3 °C, correspond to isotropic to smectic phase and smectic to crystal phase. In summary, the substitution of F atom to the aromatic core in compound **13**d and chloro-substituted of compound **13**b decreases the mesophase-isotropic transition temperature compared to compound **13**a. Compound **13**e possessing both F and Cl atoms have the lowest transition temperature compared to the other compounds.

When cooling from the isotropic liquid, compounds **13**d and **13**e exhibit a fan-like texture which is typical for a smectic phase. Both compounds form a smectic phase at a lower temperature due to the addition of F atoms in compound **13**d and F and Cl atoms in compound **13**e. Compound **13**(a, b, d, and e) were found to be monotropic in nature (Figure 4), and only compound **13a** was studied to be thermodynamically more stable compared to the other compound. However, it was expected that compound **13a** has the highest mesomorphic stability as it contains no lateral substitution that may disturb the aromatic core packing.

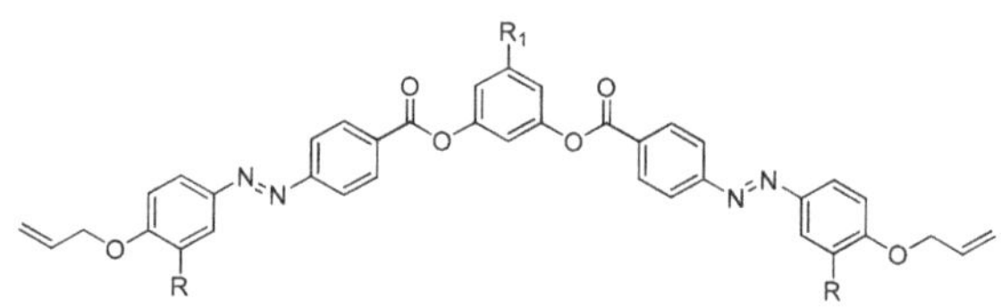

Comp	13a	13b	13c	13d	13e	13f
R	H	H	H	F	F	F
R′	H	Cl	COOH	H	Cl	COOH

There are several possible reasons for the absence of a smectic mesophase. Firstly, the dipole of the substituent is partly canceled by the dipole of the ester group. Next, no other dipole across the long axis of molecules. Lastly, the strength of intermolecular lateral attractions is reduced as a result of a long narrow to be pushed further apart due to an increase in molecular breadth [29]. Compounds **14**(a–b) consists of CH_3 and F substituted compounds, respectively. Compound **14**a is prone to show a nematic phase while compound **14**b tend to exhibit both nematic and smectic phase. Briefly summarize, the ratio of lateral to the terminal attraction of compound **14**b is higher than compound **14**a. This is because the F atom does not increase the molecule width due to the small atomic size compared to the adjacent substituent. Although the C–F bond has a high dipole moment, the bond itself cannot be entirely canceled by the dipole moment because of the ester group that presents in the compound. The dipole moment of the long axis will approach the dipole moment of the lateral axis, causing the attractive terminal forces is almost the same as the attractive lateral forces. Hence, two phases were formed at low temperatures.

Figure 4. Optical micrograph of (**a**) compound **13**a at 164.0 °C, (**b**) compound **13**b at 130.0 °C, (**c**) compound **13**d at 129.0 °C, and (**d**) compound **13**e at 134.0 °C [28].

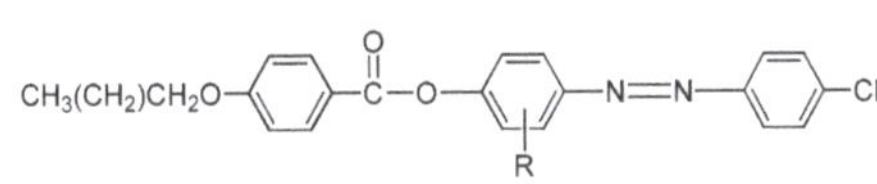

Comp	14a	14b
R	CH_3	F

In 2011, two ligands containing Cu and Ni atoms were synthesized by Yeap et al. (2011), namely compounds **15**(a–e) and **16**(a–e) [30]. All compound **15** first exhibits a schlieren textured nematic phase followed by a focal conic fan shape textured typically for smectic A. However, all compound **16** does not exhibit a mesomorphic behavior due to direct isotropization during the heating and cooling phase without exhibiting any liquid crystal phase. The layer spacing of compound **15**(a–e) is 1.04 Å. This layer spacing corresponds to the molecular length, and the layer thickness of compound **15** is slightly lower than its molecular length. This explains the presence of interdigitation of the alkoxy chain and its neighboring layer.

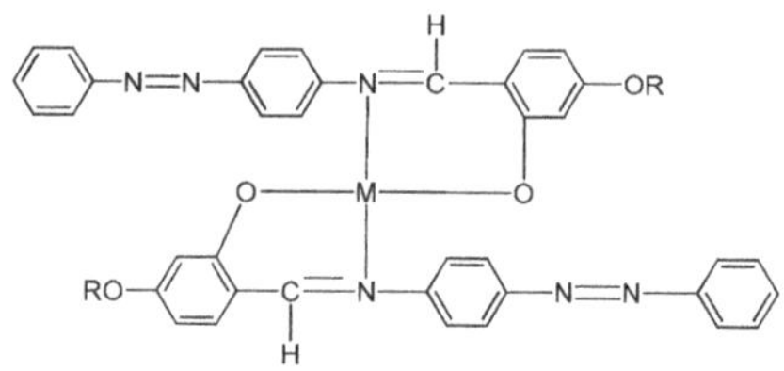

Comp	15a	15b	15c	15d	15e
M	Cu	Cu	Cu	Cu	Cu
R	C_8H_{17}	$C_{10}H_{21}$	$C_{12}H_{25}$	$C_{14}H_{29}$	$C_{16}H_{33}$
Comp	**16a**	**16b**	**16c**	**16d**	**16e**
M	Ni	Ni	Ni	Ni	Ni
R	C_8H_{17}	$C_{10}H_{21}$	$C_{12}H_{25}$	$C_{14}H_{29}$	$C_{16}H_{33}$

Yeap et al. (2011) also focused on the study of a structure–property relationship of compound **17**, which was synthesized by Reddy et al. (1991) [30,31]. Compound **17** exhibit a smectic A phase and is thermodynamically enantiotropic up to the clearing temperature of 33.0–261.0 °C. Such a thing happens due to the introduction of azobenzene cored that may induce mesophase formation in compound **17** and lower the temperature of complexes.

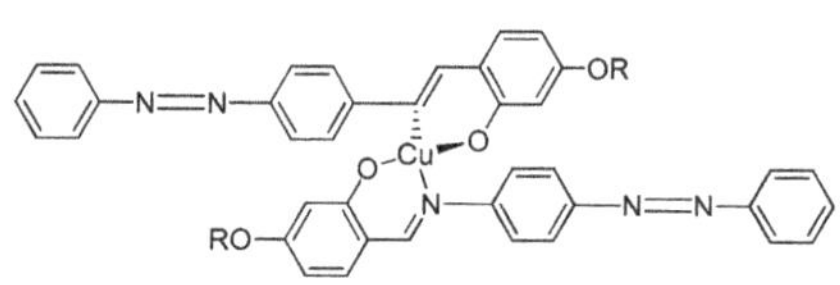

Comp	17
R	C_5H_{11}

A year later, compounds **18**(a–d) were synthesized by Yang et al. (2012) [32]. In this study, all compounds **18** exhibit mesomorphic phases. This is due to the six or eleven segments (n = 6 or 11) present in the compound that acts as a spacer. In his study, Yang et al. (2012) argued that in order for a compound to exhibit a mesomorphic phase, a liquid crystal must contain a suitable spacer length between the center and the terminal chain [32]. This flexible spacer is also crucial as it affects the dipole and molecular interaction between the compound, which will play an important role in mesophase formation. During the heating scan, compound **18a** melted, and a cholesteric liquid crystal was exhibited between 138.7 and 189.1 °C. On the other hand, compound **18c** was observed to show an enantiotropic mesophase at between 160.0–185.4 °C and 135.6–183.1 °C, upon heating and cooling scan, respectively. Compound **18a** displayed a broader phase transition temperature range compared to **18c**, which shows that the electron-withdrawing nitro moiety intensifies the head-to-tail molecular interactions. However, in compound **18b**, the longer methylene spacer decreases the head-to-tail interactions, resulting in a phase transition temperature range. In contrast, compounds **18c** and **18d** that contain a methoxy terminal chain exhibit different behavior. Compound **18d** has a broader phase transition temperature change compared to compound **18c**. Based on this observation, the rigidity of the mesomorphic core, length of the flexible spacer, and the type of terminal chain play an important role in the dipole–dipole interaction that leads to the variability of the phase transition temperature. Hence, it will later influence the formation of different types of liquid crystal properties and textures.

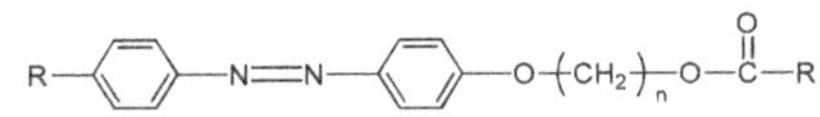

Comp	18a	18b	18c	18d
n	6	11	6	11
R	NO_2	NO_2	OCH_3	OCH_3

Two azo compounds bearing two different moieties used at the end terminal were synthesized: compounds **19**(a–i) and **20**(a–i). Lower homologues of both compounds only exhibit a nematic phase, while compounds with higher homologues, which are compounds **19**h, **19**i, **20**f, **20**g, **20**h, and **20**i, possess a longer terminal chain exhibit an additional phase transition directing to a layered mesophase. During the cooling scan, all homologues of both compounds are enantiotropic in nature and exhibit a schlieren and marbled texture typically for a nematic mesophase. As observed, the nematic phase range decreases with an increasing chain length in both compounds [33].

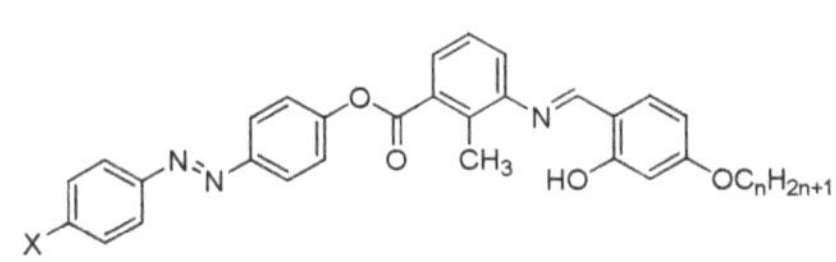

Comp	19a	19b	19c	19d	19e	19f	19g	19h	19i
n	4	5	6	7	8	9	10	11	12
X	CN	CN	CN	CN	CN	CN	CN	CN	CN
Comp	**20a**	**20b**	**20c**	**20d**	**20e**	**20f**	**20g**	**20h**	**20i**
n	4	5	6	7	8	9	10	11	12
X	NO_2	NO_2	NO_2	NO_2	NO_2	NO_2	NO_2	NO_2	NO_2

Compounds **21**(a–d) have a similar molecular structure except for the R group that was substituted on the benzothiazole ring. Hence, different types of mesophase were displayed by each compound during the heating and cooling cycles. Generally, the terminal methacrylate unit and the substituent on the benzothiazole ring are crucial to forming a liquid crystal. Compound **21**a shows only a smectic mesophase, while compounds **21**b, **21**c, and **21**d display both nematic and smectic phases. The terminal methacrylate unit is responsible for forming the smectic phase as this unit elevates the polarizability anisotropy, which is ideal for the lateral attraction of molecules to form a strong smectic phase. As mentioned earlier, compounds **21**b, **21**c, and **21**d exhibited both smectic and nematic liquid crystal phases. Nematic liquid crystal is favored as the replacement of hydrogen atom on compound **21**a by a methyl group on compound **21**b, methoxy group on compound **21**c, and ethoxy group on compound **21**d, to the sixth position of the benzothiazole ring was facilitated by conjugation of the high core polarizability with the short terminal chain [34]. The polarization of benzothiazole moiety is directly affected by the electron distribution in the electron-donating substituent. The substituent size also influences the mesophase temperature transition in the sixth position of the benzothiazole ring. The ethoxy substituent has the greatest mesophase stability compared to the methoxy or methyl substituent. To conclude, the terminal methacrylate unit and the sixth position substitution play a vital role in confirming the liquid crystal phase formation.

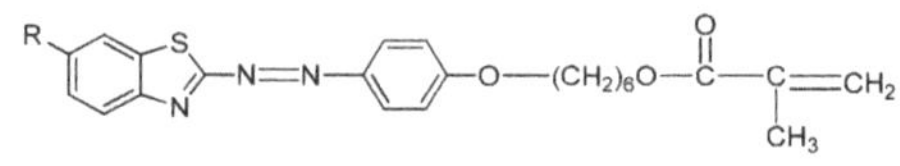

Comp	21a	21b	21c	21d
R	H	CH3	OCH3	OC2H5

Compounds **22** and **23** are azobenzene chromophores with a fluoro substituent. Both compounds are similar compared to the number of fluorine atom each compound possess, compound **22** contain only one fluorine atom while compound **23** contain two fluorine atoms. Compound **22** shows a nematic to isotropic transition at 158.9 °C, whereas compound **23** exhibit a smectic A to isotropic transition at exactly 157.6 °C. The number of fluorine substituent present in the compound influence the transition temperature by increasing the number of fluorine atom, the transition temperature will decrease [35].

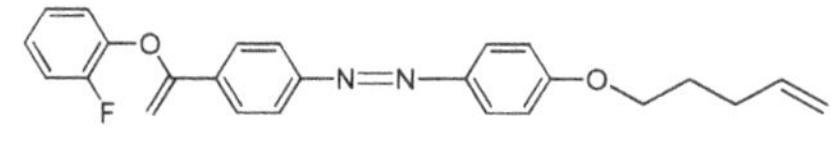

Compound 22

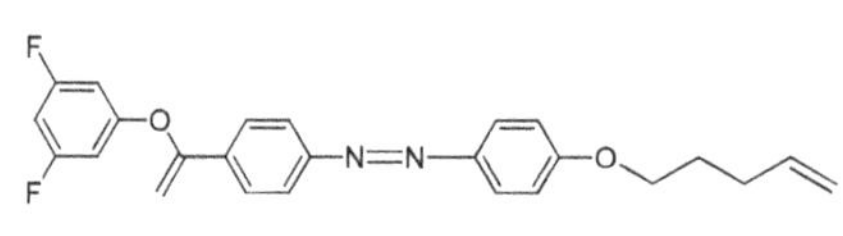

Compound 23

All compounds **24**(a–f) exhibited a schlieren texture in an enantiotropic liquid crystal phase which is typical for a nematic phase. The thermal transition of the nematic phase of methoxy homologues (**24**d, **24**e, and **24**f) is higher than the methyl homologues compounds (**24**a, **24**b, and **24**c). The presence of an oxygen atom on the methoxy group is vital domination in the mesophase thermal stability and the thermal phase range. Upon heating, the enthalpy values of the crystal to the nematic phase transition of compounds **24**a, **24**b, and **24**c are 38.8, 46.6, and 50.3 °C, respectively. As for compounds **24**d, **24**e, and **24**f are 99.0, 77.8, and 72.4 °C, respectively. These large enthalpy values suggest a strong Van der Waals interaction between the end unit of an adjacent molecule. However, upon cooling, the enthalpy values recorded for the crystal to the nematic phase transition are 26.2, 27.1, 25.8, 46.6, 43.7, and 47.0 °C for each compound, respectively. These lower enthalpy values upon cooling are lower compared to the enthalpy values upon heating. This is due to the fact that the strong Van der Waals interactions between the end of the chain of a neighboring molecule keep them static in the crystalline state before the gradation to liquid crystal starts in the heating cycle [36].

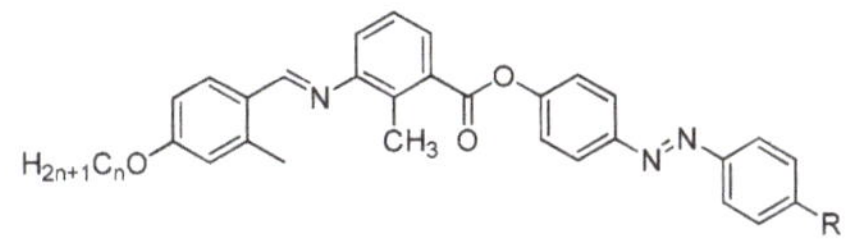

Comp	24a	24b	24c	24d	24e	24f
n	4	5	6	4	5	6
R	CH3	CH3	CH3	OCH3	OCH3	OCH3

Compounds **25**(a,b) are novel benzoates that contain three phenyl rings and a trimethylsilyl group in the terminal position. Both compounds possess a similar structural formula having a central azo compound with an eminently polar nitro group and a pentyl chain except the R group at the end of the chain. Despite having a similar structural formula, only one compound exhibits a liquid crystal property, which is compound **25**a. The optical texture of compound **25**a is illustrated in Figure 5. The polar nitro terminal favored a smectic A phase, while compound **25**b possessing an alkyl chain does not display any liquid crystalline phase property. To conclude, the property of a liquid crystal is reduced as the chain length lengthen as per the disruption of the packing layers. Hence, the only molecule with a short-chain compound can display any liquid crystalline property [37].

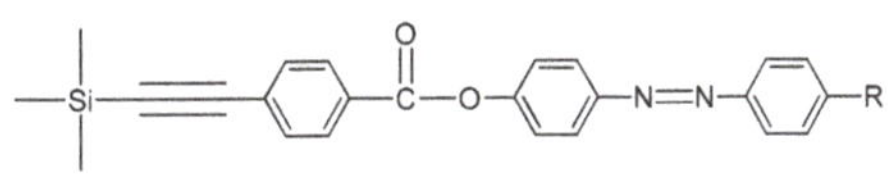

Comp	25a	25b
R	NO_2	C_5H_{11}

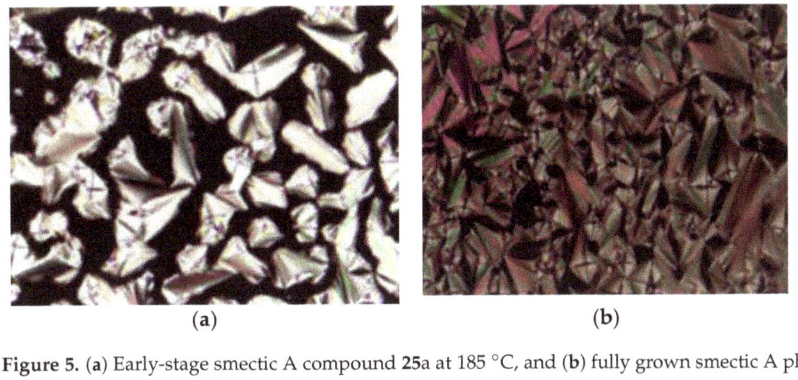

Figure 5. (**a**) Early-stage smectic A compound **25**a at 185 °C, and (**b**) fully grown smectic A phase of compound **25**b at 100.0 °C [37].

Compounds **26**(a–e) were synthesized by Selvarasu and Kannan, in 2015 [38]. Generally, the transition temperature of compounds **26**a and **26**b is larger compared to the transition temperature range of compounds **26**d and **26**e, as the electron-withdrawing substituent in compounds **26**a and **26**b have a strong Π to Π* interaction that increase the head to tail molecular structure. The nematic to the isotropic thermal stability of compounds **26**a and **26**b are higher compared to compounds **26**d and **26**e. The reason is that the –CN and –Cl groups as well as the benzene ring located in the terminal position will cause both compounds to have high polarity and thermal attraction that later will causes the thermal stability to increase. The only difference that compounds **26**a and **26**e show is on the substitution unit at the terminal end. Both of these compounds share the same central unit, a cinnamoyloxy group (–C_6H_5–CH=CH–COO–) which is directly connected to the alkyloxyphenylester group. The presence of the double bond in the cinnamoyloxy group lengthen the length of polarizability and elevate the thermal stability.

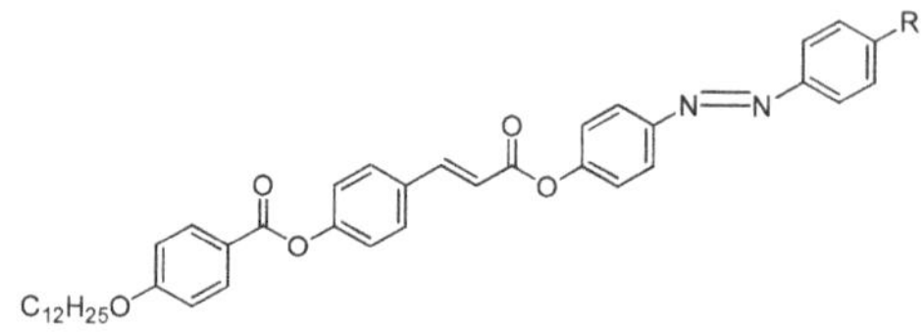

Comp	26a	26b	26c	26d	26e
R	CN	Cl	H	CH_3	OCH_3

A year later, Selvarasu and Kannan (2016) synthesized a new azobenzene compound, compound **27** [39]. Compounds **27**a, **27**b, and **27**c exhibit a nematic phase with the lowest homologues of n: 6 forming small droplets that affiliate into a classic schlieren nematic and compound of homologues n: 8 and 10 display a focal-conic textured nematic phase while the compound with the highest homologues exhibits a nematic phase and an addition of a focal conic textured smectic C phase upon cooling from the isotropic liquid. The formation of a nematic phase is a result of the alkoxy chain terminal being more noticeable. Compound **27** contains a cinnamoyloxy group connected directly to the azo compound and the ester group that acts as the central unit of the molecule, and a double bond that increase the polarizability. The core of compound **27** is directly bonded to an ester group, causing the molecule to lose stability. This phenomenon is due to the ability of the oxygen atom of the carbonyl group to bump into the non-bonded sides of the adjacent hydrogen in the aromatic ring, which attributed to the strain in the molecules. An ester linkage contains a ketonic bond, where the electron is being pulled by the ester group which causes the transition temperature to decline. To sum everything up, the attributes responsible for mesophase formation are the effect of terminal alkoxy containing azobenzene moiety and the central linkage with or without the spacer. The presence of an olefinic unit reinforces the length of polarizability, and an additional olefinic unit would proliferate the thermal stability with a high polarizability.

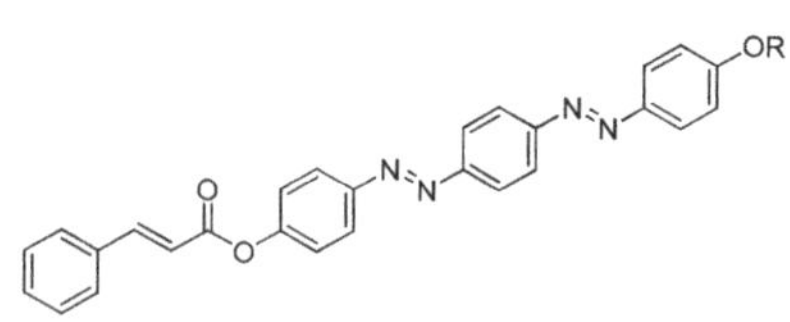

Comp	27a	27b	27c	27d
R	C_6H_{13}	C_8H_{17}	C_{10}H21	$C_{12}H_{25}$

Compounds **28**(a–f) are azo-bridged compounds with triphenylene cores. The only difference between the compounds **28(a–f)** is the length of the methylene units. The length of the spacer has a vital role in the fluctuation of the transition temperature. Upon cooling, the melting point of compound **28**a is higher compared to compound **28**b. However, it gradually elevates from compounds **28**b to **28**e. According to Yeap et al. (2016), it has something to do with the spacer length [40]. As the length of spacer increase, the melting point continue to decrease from compounds **28**e to **28**f. Nonetheless, the clearing temperature of compound increase from compounds **28**b to **28**c and keep on descending as the length of alkyl chain increase. By reason, it may be due to the dilution of the mesogenic core, which causes the spacer to be flexible.

Compound **28**f exhibits a smectic C phase, where it has been claimed that the formation of a smectic phase results from a side-by-side organization of azobenzene base peripheral units that are adjacent in parallel to each other. Many flexible spacers link all homologues of compound **28**. This flexible spacer enables a conformational change so that a rod-shaped unit could align in a parallel arrangement resulting in a layered structure of a smectic phase [40].

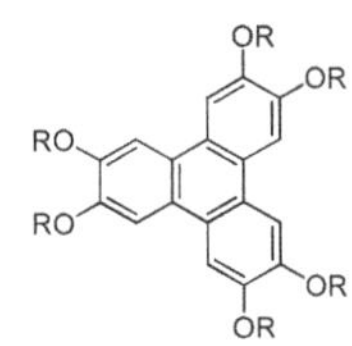

Comp	28a	28b	28c	28d	28e	28f
n	5	6	7	8	9	10
R	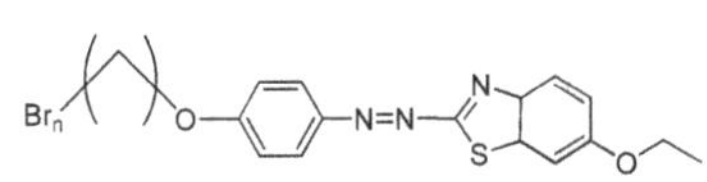					

Compounds **29**(a–c) were synthesized by Sen et al. (2016) [41]. All of compounds **29**(a–c) exhibit only one mesophase transition (Figure 6). Uncommonly for a rod-shaped molecule, the enthalpy values for the liquid crystalline phase of the isotropic liquid are higher than the enthalpy values of a solid–liquid crystalline phase. This unusual phenomenon may be due to the molecule's arrangement within each layer, which the minimized steric forces have driven.

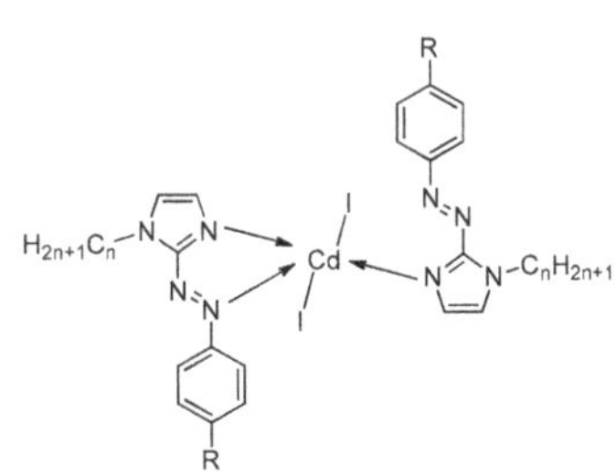

Comp	29a	29b	29c
n	10	12	18
R	H	H	H

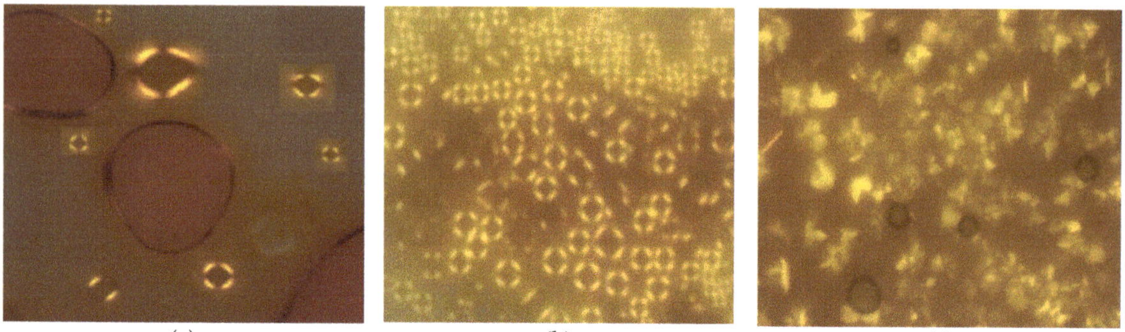

(**a**) (**b**) (**c**)

Figure 6. Polarizing micrograph ring-like texture of (**a**) compound **29**a at 123.0 °C, (**b**) compound **29**b at 191.0 °C, and (**c**) cone-like texture of hexatic phase of compound **29**c at 123.0 °C [41].

Jaworska et al. (2017) synthesized compounds **30**(a–n) [42]. Out of 14 compounds with different homologues, he showed that two of them do not exhibit any liquid crystalline property. Compounds **30**a and **30**b do not possess any liquid crystal phase; by reason both compounds have high melting points compared to other homologues. Other compounds studied exhibit at least one enantiotropic mesophase, which was identified as a focal-conic textured smectic A. Compound **30**c is the first homologue displaying smectic A and the width is 30.0 °C, as it can be seen in Figure 7. Starting from compound **30**d until compound **30**k, the width of the smectic A phase is broader, ranging from 40.0 to 75.0 °C. The smectic range keeps narrowing corresponding to increasing the alkoxy chain to n: 14 and above.

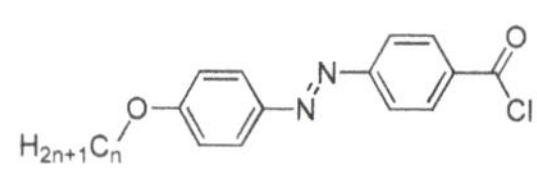

Comp	30a	30b	30c	30d	30e	30f	30g
n	4	5	6	7	8	9	10
Comp	**30h**	**30i**	**30j**	**30k**	**30l**	**30m**	**30n**
n	4	5	6	7	8	9	10

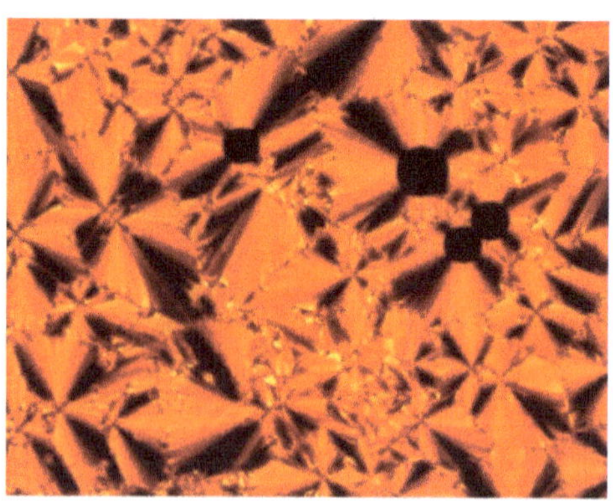

Figure 7. Focal-conic textured smectic A liquid crystalline phase present in compound **30**c [42].

All compounds **31**(a–e) possess at least one liquid crystal phase except compound **31**e, where it exhibits an extra liquid crystalline phase familiar as a smectic phase. This is a result of the length of the alkoxy chain in compound **31**e is long enough to stabilize the lamellar molecular arrangement ideal for the molecule to display a smectic phase. The variation of the terminal group by the methylene spacer located in 1,4-disubstituted triazole derivatives, the molecular alignment of mesophase is stabilized [43]. Imine as a central unit causes the polarity to increase through retaining of linear patter. Minor changes in the alkyl chain may cause major changes in the transition temperature and types of mesophase formed.

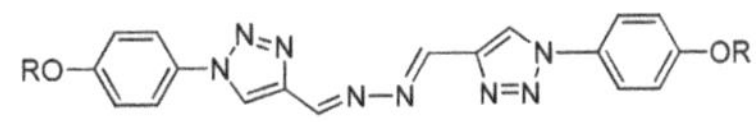

Comp	**31a**	**31b**	**31c**	**31d**	**31e**
R	C_6H_{13}	C_7H_{15}	C_8H_{17}	C_9H_{19}	$C_{10}H_{21}$

In the same year, two identical compounds, **32** and **33**, were synthesized. The only difference between these two compounds is the central unit, where it exchanges place with the adjacent group. The melting point and the clearing point of compounds **32** and **33** are 104.8, 130.2, and 166.5, 162.2 °C, respectively. These temperature changes may be due to the length to diameter ratio, which eventually increases the transition temperature [44].

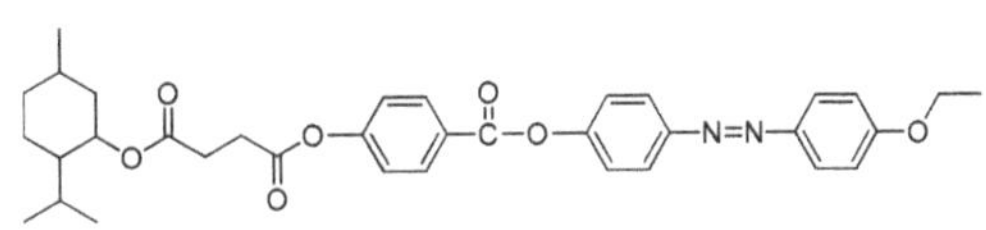

Compound 32

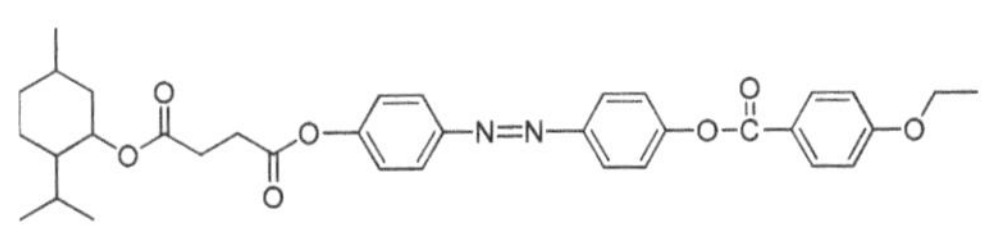

Compound 33

Each homologue of compounds **34**(a–c) exhibit a different liquid crystal phase despite having similar compound. The formation of different mesophase within this compound is influenced by the substitution terminal (–R). Compound **33**a contains an end non-polar methyl group, and it shows a nematic phase at 31.7 °C upon cooling. Compound **33**b comprises a relatively polar chloro moiety, revealing a nematic phase about 3.0 °C above the isotropic liquid and a smectic A phase at 47.0 °C upon cooling. A polar nitro moiety is present in compound **33**c, a long-range smectic A phase is observed at 106.3 °C in the cooling scan. The polarity of substituent not only affects the formation of different mesophase, but also caused the clearing temperature to change. The mesophase range increase as the polarity of substituent increase from **33**a to **33**c [45].

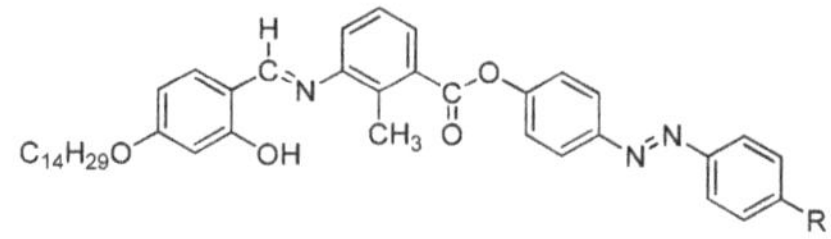

Comp	34a	34b	34c
R	CH_3	Cl	NO_2

Compounds **35**(a–e) were synthesized in 2019 by Madiahlagan et al., and are a series of azo-coumarin compounds with different aliphatic chain lengths [46]. Lower homologues of compound **35** displayed a nematic phase at 120.0 °C for compound **35**b, while compound **35**e exhibited a broken fan-shaped smectic A at 170.0 °C upon cooling from the isotropic liquid. The difference in mesomorphic formation and the transition temperature is greatly influenced by the increasing carbon atom of the aliphatic substituent.

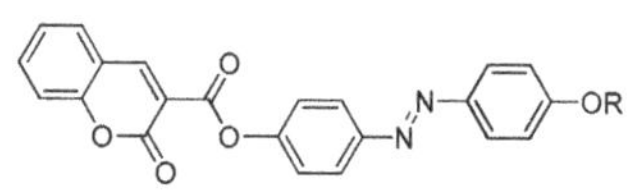

Comp	35a	35b	35c	35d	35e
R	C_6H_{13}	C_8H_{17}	$C_{10}H_{21}$	$C_{12}H_{25}$	$C_{14}H_{29}$

The shortest terminal chain, compounds **36a** and **36e**, exhibited three liquid crystal phases which are nematic phase, smectic A phase, and smectic C phase, while compound **36c** containing a butyl terminal chain exhibit a twist grain boundary smectic A (TGBA) rather than a smectic A itself and a smectic C phase. However, the broadest smectic S phase was observed to be of compound **36**b. The TGBA phase does not present in **36**b, and **36**d is due to the odd-even effect. Compound **36i** with the longest aliphatic group displayed a smectic A to smectic C mesophase transition with the temperature narrowed. It was affected by the elongation of the non-chiral alkyl chain, which eventually decreases the mesophase range.

As for compounds **37**(a–i), the TGBA phase appeared in compound **37**a and disappeared from compounds **37**b to **37**d and re-appear in compound **37e** implying that a change has happened for compounds with the middle length alkyl chain. Further studies were done to identify the cause, however, further increase of the carboxylate group leads to a reduction of mesophase property [47].

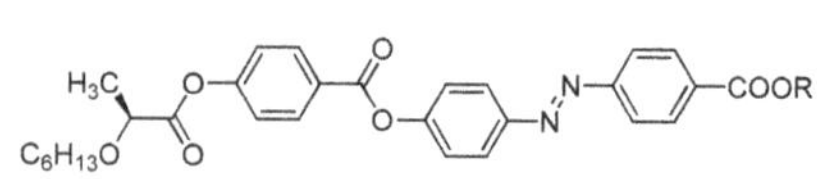

Comp	36a	36b	36c	36d	36e	36f	36g	36h	36i
R	C_2H_5	C_3H_7	C_4H_9	C_5H_{11}	C_6H_{13}	C7H15	C_8H_{17}	$C_{10}H_{21}$	$C_{12}H_{25}$

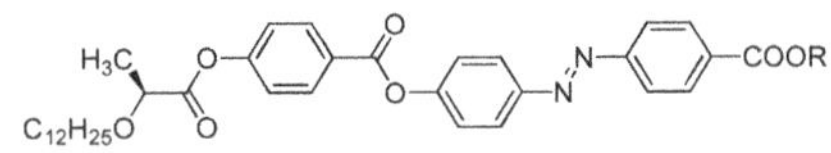

Comp	37a	37b	37c	37d	37e	37f	37g	37h	37i
R	C_2H_5	C_3H_7	C_4H_9	C_5H_{11}	C_6H_{13}	C_7H_{15}	C_8H_{17}	$C_{10}H_{21}$	$C_{12}H_{25}$

In the following year, compounds **38**(a–e) and **39**(a–e) were synthesized, where all compounds studied exhibit a liquid crystalline property. Within compound **38** and all higher homologues were observed to be enantiotropic except compounds **38**a, and **38**b, while compounds **39**a and **39**d are enantiotropic in nature. Compounds **38**b and **38**c display an enantiotropic smectic B1 phase, while compound **38**d exhibits an enantiotropic smectic B1 phase and an addition of a monotropic smectic B2 phase. Theoretically, a change in the alkyl chain will cause the mesophase property to be varied, and the formation of smectic B proves the point [48].

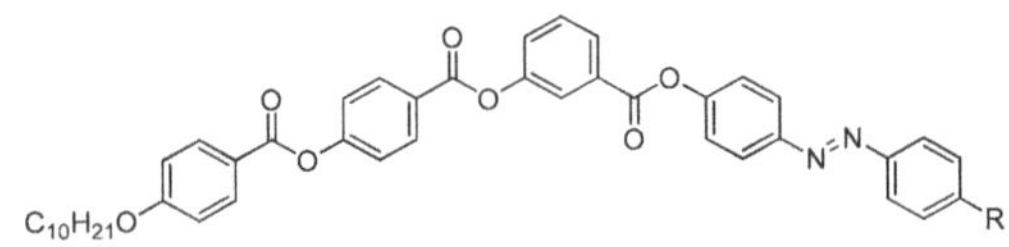

Comp	38a	38b	38c	38d	38e
R	C4H9	C6H13	C8H17	C10H21	C12H25
Comp	**39a**	**39b**	**39c**	**39d**	**39e**
R	OC4H9	OC6H13	OC8H17	OC10H21	OC12H25

Sardon et al. (2021) synthesized compounds **40**(a–d) and compared them with a compound studied back in 2016 by Karim et al., labeled as compounds **41**(b–c) [49,50]. The result of this observation is illustrated in Table 1.

Table 1. Phase transition temperature and enthalpy changes of **40**(a–d) and **41**(b–c).

Compound	Scan	Transition Temperature (°C) (Enthalpy Changes (kJ mol^{-1}))	Mesophase Range (°C)	
			SmA	N
40a	Heat Cool	Cr_1 61 (4.5) Cr_2 88 (16.1) N 98 (3.7) I 57 (−0.4) N 18 (−3.8) Cr	-	10
40b	Heat Cool	Cr 108 (34.6) N 140 (0.5) I I 138 (−0.3) N 73 (−33.3) Cr	-	32
40c	Heat Cool	Cr_1 93 (0.9) Cr_2 115 (31.8) N 136 (0.5) I I 134 (−0.4) N 33 (−15.7) Cr	-	21
40d	Heat Cool	Cr 127 (58.4) N 159 (0.9) I I 156 (−1.2) N 74 (−49.3) Cr	-	32
41b	Heat Cool	Cr 116 (33.1) SmA 185 (0.9) N 203 (0.3) I I 197 (−0.3) N 173 (−0.5) SmA 101 (−15.7)	69	18
41c	Heat Cool	Cr 124 (15.7) SmA 192 (1.1) N 204 (0.3) I I 192 (−0.2) N 174 (−0.3) SmA 112 (−13.0)	68	12

Abbreviations: Cr = crystalline phase; SmA = smectic A phase; N = nematic phase; I = isotropic phase.

Based on Table 1, it can be observed that the mesophase stability is inconsistent. This inconsistency is caused by the substituent (–H, –Cl, –Br, –CN). However, the nematic mesophase range of compounds **40**(b–d) are wider. The substitution of hydrogen atom on the para-position of benzene ring affects the polarizability of mesogens that aid the formation of the liquid crystal phase. Compound **40**d presented the highest nematic to isotropic clearing transition temperature because of higher polarization effects caused by the cyano substitution that gives high temperature to break the association within the molecule.

Compounds **40**b and **40**c only reveal the formation of a nematic phase, while compounds **41**b and **41**c displayed both nematic and smectic A phases. The methyl side chain substituent in compounds **40**b and **40**c truncate the melting and clearing temperature compared to compounds **41**b and **41**c that possess no methyl side chain. Generally, any lateral group attached to the core unit of a compound may cause some changes in the mesomorphic behavior. The presence of methyl side-chain group widens the core unit where the intermolecular separation is eventually increased and leads to lower mesophase stability.

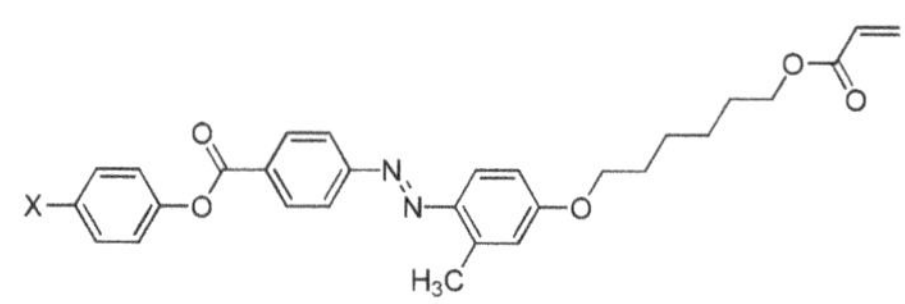

Comp	40a	40b	40c	40d
R	H	Cl	Br	CN

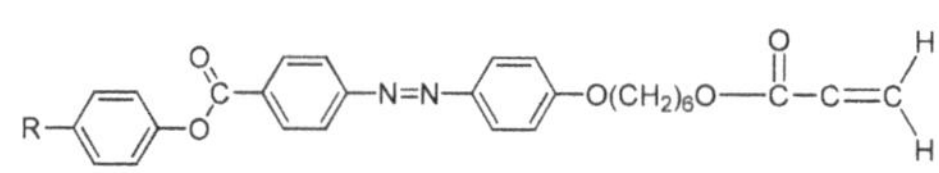

Comp	41a	41b
R	Cl	Br

3. Application

In this era, everyone uses at least one liquid crystal in their daily life. Over the years, the use of electronic devices has increased to meet the demand of easing our lives. The most common liquid crystal used is the liquid crystal display (LCD) that is present on our mobile phones, computers, laptops, and many more [51,52]. LCD is an electronic display device with a flat panel that uses a liquid crystal as its main operation medium [53]. An LCD possess both nematic and smectic phase. However, a smectic LCD is more demanded compared to the nematic LCD as the response is faster. The display on an LCD is not emitted by the liquid crystal, instead of liquid crystal acts as a reflector that produces the image in either monochrome or in color [27].

Another application is the liquid crystal thermometer. As informed, a thermometer detects the temperature of something. Liquid crystal is applied to this system so that when it detects any temperature, a specific color persists. Recalling from a study by Kitzerow and Bahr (2001), a chiral nematic liquid crystal reflects light of a wavelength of the same length of the pitch [16]. As the pitch is temperature-dependent, the color that is reflected is also temperature-dependent. This set-up is applied to various applications, such as the mood ring that has recently gone viral, where the ring will conjure different color according to our body temperature. In addition to that, this application is ideal for use in the medical area where it can detect different temperature distributions on a tumor patient as the tumor, and its surrounding tissues develop different temperatures [14].

4. Conclusions

There are several factors observed that might influence the formation of a different liquid crystal phase. The most common factor that can be scrutinized in most studies conducted by most authors is the increase in the alkyl terminal chain. These factors may have several consequences, which are the main influences on mesophase development. The effects of the increase in the terminal alkyl chain are that it regulates the increase in molecular length-to-breadth, amplifies the entropy chain, enhances thermal stability, and diminishes the clearing temperature. Moreover, the azo linking unit increase the linearity of the molecule and this characteristic able to induce the mesophase transition.

Author Contributions: Conceptualization, N.A.R. and Z.J.; validation, N.A.R. and Z.J.; formal analysis, N.A.R.; investigation, N.A.R.; data curation, N.A.R. and Z.J.; writing—original draft preparation, N.A.R. and Z.J.; writing—review and editing, N.A.R. and Z.J.; visualization, N.A.R.; supervision, Z.J.; project administration, Z.J.; funding acquisition, Z.J. Both authors have read and agreed to the published version of the manuscript.

Funding: This research was funded by Universiti Malaysia Sabah (UMS), grant number SBK0488-2021 and The APC was funded by Universiti Malaysia Sabah (UMS).

Institutional Review Board Statement: Not applicable.

Informed Consent Statement: Not applicable.

Data Availability Statement: Not applicable.

Acknowledgments: The authors would like to thanks Universiti Malaysia Sabah (UMS) for the technical and financial support.

Conflicts of Interest: The authors declare no conflict of interest.

References

1. Jamain, Z.; Habil, S.; Makmud, M.Z.H.; Khairuddean, M. Synthesis, structural and dielectric characteristics of liquid crystalline azo-based compounds with different terminal length. In Proceedings of the 2021 IEEE International Conference on the Properties and Applications of Dielectric Materials (ICPADM), Johor Bahru, Malaysia, 12–14 July 2021; pp. 49–52.
2. Jamain, Z.; Khairuddean, M.; Loh, M.L.; Manaff, N.L.A.; Makmud, M.Z.H. Synthesis and characterization of hexasubstituted cyclotriphosphazene derivatives with azo linking units. *Malays. J. Chem.* **2020**, *22*, 125–140.
3. Benkhaya, S.; Rabet, S.M.; Harfi, A.E. A review on classifications, recent synthesis and applications of textile dyes. *Inorg. Chem. Commun.* **2020**, *115*, 107891. [CrossRef]
4. Selvaraj, V.; Karthika, T.S.; Mansiya, C.; Alagar, M. An over review on recently developed techniques, mechanisms and intermediate involved in the advanced azo dye degradation for industrial applications. *J. Mol. Struct.* **2021**, *1224*, 129195. [CrossRef]
5. Kyei, S.K.; Akaranta, O.; Darko, G. Synthesis, characterization and antimicrobial activity of peanut skin extract-azo-compounds. *Sci. Afr.* **2020**, *8*, e00406. [CrossRef]
6. Jamain, Z.; Khairuddean, M.; Guan-Seng, T.; Rahman, A.B.A. Synthesis, characterization and mesophase transition of hexasubstituted cyclotriphosphazene molecules with Schiff base and azo linking units and determination of their fire retardant properties. *Macromol. Res.* **2021**, *29*, 331–341. [CrossRef]
7. Brightman, R. *Dictionary of Chemistry in Nature*; Mc-Graw Hill: New York, NY, USA, 1954; pp. 219–376.
8. Stegemeyer, H. Centenary of the discovery of liquid crystals. *Liq. Cryst.* **1989**, *5*, 5–6. [CrossRef]
9. Lejcek, L.; Glogarova, M.; Tesinska, E. Commemorative plaque of Prof. Friedrich Reinitzer installed in Prague. *Liq. Cryst. Today* **2017**, *26*, 66–68. [CrossRef]
10. Yildiz, S.; Cetinkaya, M.C.; Ozbek, H.; Tzitzios, V.; Nounesis, G. High-resolution birefringence investigation on the effect of surface-functionalized CdSe nanoparticles on the phase transitions of a smectic—A liquid crystal. *J. Mol. Liq.* **2020**, *298*, 112029. [CrossRef]
11. Carlescu, I. Introductory Chapter: Liquid Crystals. In *Liquid Crystals—Self-Organized Soft Functional Materials for Advanced Applications*; IntechOpen: Bucharest, Romania, 2018.
12. Andrienko, D. Introduction to liquid crystals. *J. Mol. Liq.* **2018**, *267*, 520–541. [CrossRef]
13. Kusabayashi, S.; Takenaka, S. Structure and Properties of Liquid Crystals. *J. Syn. Org. Chem.* **1984**, *42*, 2–12. [CrossRef]
14. Lehn, J.M. *Supramolecular Chemistry*; Wiley-VCH: Weinheim, Germany, 1995; pp. 177–219.

15. Collings, T.D.; Hird, M. *Introduction to Liquid Crystal*; Taylor & Francis: London, UK, 1997.
16. Kitzerow, H.S.; Bahr, C. *Chirality in Liquid Crystals*; Springer: Berlin/Heidelberg, Germany, 2001.
17. Goodby, J.W. *Materials and Phase Structure of Calamatic and Discotic Liquid Crystals*; Springer: Cham, Switzerland, 2016.
18. Safinya, C.R.; Varady, W.A. Calamitic liquid crystals-nematic and smectic mesophases. *Liq. Cryst.* **1986**, *57*, 43–77.
19. Parra, M.; Villouta, S.; Vera, V.; Belmar, J.; Zúñiga, C.; Zunza, H. Synthesis and mesomorphic properties of azo compounds derived from phenyl- and thienyl-1,3,4-thiadiazole. *Z. Fur. Naturforsch.-B J. Chem. Sci.* **1997**, *52*, 1533–1538. [CrossRef]
20. Belmar, J.; Parra, M.; Zuniga, C.; Perez, C.; Munoz, C. New liquid crystals containing the benzothiazol unit: Amides and azo compounds. *Liq. Cryst.* **1999**, *26*, 389–396. [CrossRef]
21. Lee, M.J.; Yoo, B.W.; Shin, S.T.; Keum, S.R. Synthesis and properties of new liquid crystalline compounds containing an indolinobenzospiropyranylazo group. *Dyes Pigm.* **2001**, *51*, 15–24. [CrossRef]
22. Parra, M.; Alderete, J.; Zuñiga, C.; Gallardo, H.; Hidalgo, P.; Vergara, J.; Hernandez, S. Azo compounds and Schiff's bases derived from 5-(4-pyridyl)-2-amino-1,3,4-thiadiazole: Synthesis, mesomorphic properties and structural study by semi-empirical calculations. *Liq. Cryst.* **2001**, *28*, 1659–1666. [CrossRef]
23. So, B.K.; Jang, M.C.; Park, J.H.; Lee, K.S.; Song, H.H.; Lee, S.M. Novel bent-shaped liquid crystalline compounds: 1. Synthesis and structure analysis of dimesogenic compounds with azo units. *Opt. Mater.* **2003**, *21*, 685–689. [CrossRef]
24. Abbasi, A.R.; Rezvani, Z.; Nejati, K. Synthesis and properties of new liquid crystalline compounds containing an alkoxyphenylazo group. *Dyes Pigm.* **2006**, *70*, 71–75. [CrossRef]
25. Rezvani, Z.; Divband, B.; Abbasi, A.R.; Nejati, K. Liquid crystalline properties of copper (II) complexes derived from azo-containing salicylaldimine ligands. *Polyhedron* **2006**, *25*, 1915–1920. [CrossRef]
26. So, B.K.; Kim, H.J.; Lee, S.M.; Song, H.H.; Park, J.H. Novel bent-shaped liquid crystalline compounds: IV. Dimesogenic compounds containing 2-hydroxy-1,3-dioxypropylene and azobenzene mesogens. *Dyes Pigm.* **2006**, *70*, 38–42. [CrossRef]
27. Dunmur, D.; Walton, H.G. "Liquid Crystal Display". Encyclopedia Britannica, 3 August 2015. Available online: https://www.britannica.com/technology/liquid-crystal-display (accessed on 10 August 2021).
28. Lutfor, M.R.; Hegde, G.; Kumar, S.; Tschierske, C.; Chigrinov, V.G. Synthesis and characterization of bent-shaped azobenzene monomers: Guest-host effects in liquid crystals with azo dyes for optical image storage devices. *Opt. Mater.* **2009**, *32*, 176–183. [CrossRef]
29. Al-hamdani, U.J.; Gassim, T.E.; Radhy, H.H. Synthesis and Characterization of Azo Compounds and Study of the Effect of Substituents on Their Liquid Crystalline Behavior. *Molecules* **2010**, *15*, 5620–5628. [CrossRef] [PubMed]
30. Yeap, G.Y.; Heng, B.T.; Kakeya, M.; Takeuchi, D.; Gorecka, E.; Ito, M.M. Synthesis, 2D NMR and X-ray diffraction studies on Cu(II) and Ni(II) complexes with ligands derived from azobenzene-cored Schiff base: Mesomorphic behaviors of Cu(II)-phenolates and crystal structure of bis [4-(4-alkoxy-2-hydroxybenzylideneamino)azobenzen. *J. Mol. Struct.* **2011**, *999*, 68–82. [CrossRef]
31. Reddy, K.P.; Brown, T.L. Synthesis and thermal behaviour of mesomorphic Cu(II) Schiff's base complexes. *J. Mater. Chem.* **1991**, *1*, 757–764. [CrossRef]
32. Yang, P.C.; Li, C.Y.; Wu, H.; Chiang, J.C. Synthesis and mesomorphic properties of photoresponsive azobenzene-containing chromophores with various terminal groups. *J. Taiwan Inst. Chem. Eng.* **2012**, *43*, 480–490. [CrossRef]
33. Debnath, S.; Mohiuddin, G.; Turlapati, S.; Begum, N.; Sarkar, D.D.; Rao, V.S.N. Nematic phases in achiral unsymmetrical four-ring bent-core azo compounds possessing strongly polar cyano and nitro moieties as end substituents: Synthesis and characterization. *Dyes Pigm.* **2013**, *99*, 447–455. [CrossRef]
34. Karim, R.; Salleh, N.M.; Yahya, R.; Hassan, A.; Hoque, A. Synthesis and characterization of azo benzothiazole chromophore based liquid crystal macromers: Effects of substituents on benzothiazole ring and terminal group on mesomorphic, thermal and optical properties. *Mater. Chem. Phys.* **2013**, *140*, 543–552. [CrossRef]
35. Rahman, M.L.; Biswas, T.K.; Sarkar, S.M.; Yusoff, M.M.; Yuvaraj, A.R.; Kumar, S. Synthesis of new liquid crystals embedded gold nanoparticles for photoswitching properties. *J. Colloid Interface Sci.* **2016**, *478*, 384–393. [CrossRef]
36. Paul, M.K.; Kalita, G.; Laskar, A.R.; Debnath, S.; Gude, V.; Sarkar, D.D.; Mohiuddin, G.; Varshney, S.K.; Nandiraju, R.V.S. Synthesis and mesomorphic behaviour of achiral four-ring unsymmetrical bent-core liquid crystals: Nematic phases. *J. Mol. Struct.* **2013**, *1049*, 78–89. [CrossRef]
37. Srinivasa, H.T.; Hariprasad, S. 4-[2-(Trimethylsilyl)ethynyl]benzoates: Synthesis and evaluation for mesomorphic properties of some novel calamitic molecules. *J. Organomet. Chem.* **2014**, *774*, 19–25. [CrossRef]
38. Selvarasu, C.; Kannan, P. Synthesis, characterization of azobenzene and cinnamate ester based calamitic liquid crystalline compounds and their photoresponsive properties. *J. Mol. Struct.* **2015**, *1092*, 176–186. [CrossRef]
39. Selvarasu, C.; Kannan, P. Effect of azo and ester linkages on rod shaped Schiff base liquid crystals and their photophysical investigations. *J. Mol. Struct.* **2016**, *1125*, 234–240. [CrossRef]
40. Yeap, G.; Abdul, M.; Lin, C.; Lin, H.; Maeta, N.; Ito, M.M. Synthesis and liquid crystalline studies of disc-shaped molecule on azo-bridged benzothiazole-phenyl ethers. *J. Mol. Liq.* **2016**, *223*, 734–740. [CrossRef]
41. Sen, C.; Mallick, D.; Patra, C.; Roy, S.; Sinha, R.K.; Ghosh, R.; Mondal, J.A.; Palit, D.K.; Sinha, C. Spectroscopic characterization, photochromism and mesomorphism of cadmium(II)-1-alkyl-2-(arylazo)imidazole complexes and DFT correlative studies. *Polyhedron* **2016**, *117*, 463–477. [CrossRef]
42. Jaworska, J.; Korbecka, I.; Kochel, A.; Galewski, Z. 4-(4′-Alkyloxyphenylazo)benzoyl chlorides not only intermediates but photosensitive mesogens for optical switching. *J. Mol. Liq.* **2017**, *225*, 877–882. [CrossRef]

43. Boopathi, T.S.; Senthil, S. Synthesis and characterization of heterocyclic nonlinear mesogen. *Mater. Today Proc.* **2018**, *5*, 16172–16180. [CrossRef]
44. Luo, C.C.; Sun, S.L.; Wang, Y.S.; Meng, F.B.; Hu, J.S.; Jia, Y.G. The effect of various functional groups on mesophase behavior and optical property of blue phase liquid crystal compounds based on (−) menthol. *J. Mol. Liq.* **2018**, *269*, 755–765. [CrossRef]
45. Patranabish, S.; Mohiuddin, G.; Begum, N.; Laskar, A.R.; Pal, S.K.; Rao, N.V.S.; Sinha, A. Cybotactic nematic phase of achiral unsymmetrical bent-core liquid crystals—Quelling of polar ordering and the influence of terminal substituent moiety. *J. Mol. Liq.* **2018**, *257*, 144–154. [CrossRef]
46. Madiahlagan, E.; Sunil, B.N.; Ngaini, Z.; Hegde, G. Synthesis, liquid crystalline properties and photo switching properties of coumarin-azo bearing aliphatic chains: Application in optical storage devices. *J. Mol. Liq.* **2019**, *292*, 111328. [CrossRef]
47. Poryvai, A.; Bubnov, A.; Pociecha, D.; Svoboda, J.; Kohout, M. The effect of the length of terminal n-alkyl carboxylate chain on self-assembling and photosensitive properties of chiral lactic acid derivatives. *J. Mol. Liq.* **2019**, *275*, 829–838. [CrossRef]
48. Hegde, R.S.; Sunil, B.N.; Hegde, G.; Prasad, V. Influence of alkyl and alkoxy groups on photoresponsive behaviour of bent-core azo mesogens: Synthesis, mesomorphic and photoswitching properties. *J. Mol. Liq.* **2020**, *309*, 113091. [CrossRef]
49. Karim, M.R.; Sheikh, M.R.K.; Yahya, R.; Mohamad, S.N.; Lo, K.M.; Mahmud, H.N.M.E. The effect of terminal substituents on crystal structure, mesophase behaviour and optical property of azo ester linked materials. *Liq. Cryst.* **2016**, *43*, 1862–1874. [CrossRef]
50. Sardon, S.N.F.; Rahman, N.M.M.A.; Karim, M.R.; Zahid, N.I.; Salleh, N.M. Effects of lateral methyl and terminal substituents on thermal, mesomorphic and optical properties of azo ester mesogens. *J. Mol. Struct.* **2021**, *1225*, 129112. [CrossRef]
51. Jamain, Z.; Khairuddean, M.; Guan-Seng, T. Synthesis of new star-shaped liquid crystalline cyclotriphosphazene derivatives with fire retardancy bearing amide-azo and azo-azo linking units. *Int. J. Mol. Sci.* **2020**, *21*, 4267. [CrossRef] [PubMed]
52. Jamain, Z.; Faizawani, N.; Khairuddean, M. Synthesis and determination of thermotropic liquid crystalline behaviour of cinnamaldehyde-based molecules with two Schiff base linking units. *Molecules* **2020**, *25*, 3780. [CrossRef] [PubMed]
53. Smith, C. A review of liquid crystal display technologies, electronic interconnection and failure analysis. *Circuit World* **2008**, *34*, 35–41. [CrossRef]

polymers

MDPI

Review

A Review on Synthesis, Structural, Flame Retardancy and Dielectric Properties of Hexasubstituted Cyclotriphosphazene

Siti Nur Khalidah Usri, Zuhair Jamain * and Mohamad Zul Hilmey Makmud

Sustainable Materials and Renewable Energy (SMRE) Research Group, Faculty of Science and Natural Resources, Universiti Malaysia Sabah (UMS), Kota Kinabalu 88400, Sabah, Malaysia; sitinurkhalidahusri07@gmail.com (S.N.K.U.); mzhilmey@ums.edu.my (M.Z.H.M.)
* Correspondence: zuhairjamain@ums.edu.my

Abstract: Hexachlorocyclotriphosphazene is a ring compound consisting of an alternating phosphorus and nitrogen atom with two chlorine substituents attached to the phosphorus atom. The six chlorine atoms attached to this cyclo compound can be substituted with any different nucleophile that leads to changes in different chemical and physical properties. The major topics that were investigated in this research are the flame retardancy and dielectric properties of cyclotriphosphazene compounds. Cyclotriphosphazene compounds have high potential to act as a flame retardant, and this compound consists of two active elements attributed to its high flame-retardant character. This compound also demonstrated good ability as a flame retardant due to its low toxicity and less smoke produced. In addition, cyclotriphosphazene compounds were also investigated for their dielectric properties. Cyclotriphosphazene has high potential in the electrical field since it has dielectric properties that can be widely studied in the investigation of any potential application. This review presented literature studies focused on recent research development and studies in the field of cyclotriphosphazene that focused on synthesis, structural, flame retardancy, and dielectric properties of hexachlorocyclotriphosphazene compounds.

Keywords: cyclotriphosphazane; flame retardancy; dielectric properties

Citation: Usri, S.N.K.; Jamain, Z.; Makmud, M.Z.H. A Review on Synthesis, Structural, Flame Retardancy and Dielectric Properties of Hexasubstituted Cyclotriphosphazene. *Polymers* **2021**, *13*, 2916. https://doi.org/10.3390/polym13172916

Academic Editor: Andrzej Puszka

Received: 12 August 2021
Accepted: 25 August 2021
Published: 30 August 2021

1. Introduction

The discovery of the area of organic chemistry led to a new era of material science. In the early years, the cyclotriphosphazene compound was extensively explored and had a high impact on innovation in the chemistry field. The cyclotriphosphazene compound was first investigated in the mid-1950's by Allock and his co-workers. To date, this compound has been used in many fields due to its fascinating properties [1,2], and this compound is still being explored to make it more useful in modern technology.

Cyclotriphosphazene (Figure 1a) is a ring compound that consists of alternating phosphorus and nitrogen atoms with two substituents attached to phosphorus atoms [3]. Cyclotriphosphazane, also known as hexachlorophosphazene, is a hexa-membered cyclic ring of organophosphazene with the molecular formula $(NPCl_2)_6$. This compound is easily modified by replacing the chloro atom existing on the phosphorus atom with an appropriate reagent [4]. Due to the high reactivity of the P–Cl bond, the corresponding substitution method allows for the introduction of a wide range of substituents [5] by nucleophilic reaction (Figure 1b) to obtain new cyclotriphosphazene derivatives having different chemical and physical properties depending on the characteristics of the substituted groups [6]. Many studies have focused attention on cyclotriphopshazene derivatives [7,8]. Previous research has been carried out by changing the side group -R of phosphazene, such as $-NH_2$, -OR, $-OC_6H_5$, or fluorinated derivatives that are able to improve oxidative and thermal stabilities [9]. Furthermore, the properties of polyphosphazenes change as a result of the different side groups, such as elasticity, thermal stability, solvent resistance, biological utility, electro-optical properties, and so on, indicating that the polyphosphazenes can be adjusted and controlled depending on the materials they are applied with [10–14].

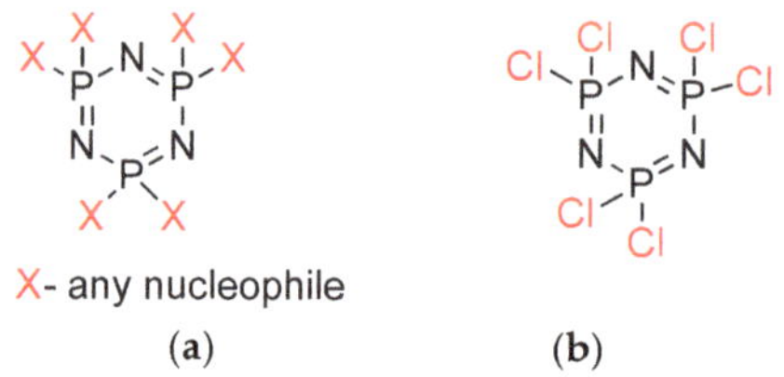

Figure 1. Chemical structure of (**a**) cycloytriphosphazene and (**b**) HCCP.

Dogan et al. also reported that cyclotriphosphazene derivatives containing a bis-aryl Schiff base having a different terminal group (H, F, Cl, and Br) have good thermal properties, which makes them suitable as a flame retardant additive to slow down the spread of the flame ignition or slow down the spread of flame after ignition of materials [15].

Cyclotriphosphazene compounds are extensively investigated, as they are reported to have potential in some industrial applications, such as in flame retardant [16,17]. Two active elements found in flame retardants are phosphorus and nitrogen, and these give a synergistic effect greater than a simple addictive. Their flame retardant effect combines good flame resistance and self-extinguishing ability in the polymer [18–21]. These materials also offer exciting technological prospects, as they can be applied as starting materials to synthesize industrially valuable substituted cyclotriphosphazenes. The derivatives of cyclotriphosphazene have been used as an additive in various fields such as special rubber [22] and biomedicine [23], the most common being in flame retardant materials [24].

Several cyclotriphosphazene derivatives with excellent properties have been synthesized by replacing Cl atoms of HCCP [25,26]. Before this, halogen was used as a flame retardant substance, but it is reported that these halides release toxic and corrosive substances during their decomposition and cause environmental problems [27]. Therefore, it is necessary to develop halogen-free and environment-friendly flame retardants to improve flame retardancy. Thus, cyclotriphosphazene compounds that contain phosphorus atoms as the main compound have shown high potential to replace halogen as an environmentally friendly flame retardant due to this compound possessing excellent properties such as low smoke emission, low toxicity, and the formation of a stable carbonized layer after burning [28].

Furthermore, related research constitutes a relatively new area that has emerged from the flame retardancy of cyclotriphosphazene to studies in the dielectric properties of this compound. Dielectric materials are non-conductors of electricity or can be defined as electrical insulators that can be highly polarized by an electrical field that is the material's dielectric constant. Dielectric materials are non-conductors of electricity that can be substantially polarized by an electrical field, which is the material's dielectric constant. On removal of the electric field, the material returns to its original state, and the time taken to do this is referred to as the relaxation period, which is a characteristic of the dielectric material.

In addition to cyclotriphosphazene and their dielectric properties, polymer compound are well known to be used in electrical applications, especially as an insulator, due to its properties of the superior service [29]. Thus, the cyclotriphosphazene compound is the most potent compound that can be used in advanced electrical devices. It has good flame retardant properties, which are also important in the build-up of the electronic application, and has a dielectric property. To conclude, an extensive study about these two related properties with high-potential compounds, which is cyclotriphosphazene, needs to be carried out to enhance multifunctional properties that might be beneficial in future production in an electrical field.

This article presented a review of the recent synthesis of hexasubstituted cyclotriphosphazene compounds. The recent studies related to the flame retardancy and dielectric properties of this compound were also discussed. The paper is divided into four significant sections related to the synthesis and properties of cyclotriphosphazene compounds. The first section deals with the study of synthesis hexasubstituted cyclotriphosphazene, and its flame retardancy properties in recent years. The second section discusses the effect of the structural cyclotriphosphazene compounds, including the linking unit and enhancements achieved through the linking unit of the new hexasubstituted compounds. The third section deals with the dielectric properties of cyclotriphosphazene compounds and, finally, the fourth section details the combination of two properties which are flame retardancy and dielectric properties of cyclotriphosphazene that can be used as a reference since this compound has high potential in a high flame-resistant electrical application or other electrical applications, such as insulation, or as a capacitor in an electrical component responsible for the energy-storage properties of the devices. In addition, the use of dielectric materials in sensors for a multitude of applications, such as self-driving cars, has made dielectric science and technology research even more significant than before.

2. Synthesis of Hexasubstituted Cyclotriphosphazene and Flame Retardant Properties

Various approaches are available for cyclotriphosphazene synthesis methods. The starting material to synthesize this compound is hexachlorocyclotriphosphazene (HCCP). An important task in preparing hexasubstituted cyclotriphosphazene is to react HCCP with other compounds to form a new hexasubstituted cyclotriphosphazene, where the chlorine atoms in HCCP are fully substituted with a new nucleophile to form a halogen-free cyclotriphosphazene compound. Over time, extensive literature has developed on cyclotriphosphazene as a flame retardant. Incorporation of HCCP into compounds increases the resistance of the material towards ignition. The flame retardant function minimizes the flame risk and prevents a small fire from becoming a major catastrophe [30]. Flame-resistant materials play a critical role in preventing property damage. The removal of heat from the materials that can burn and the creation of char during the fire results in flame retardation, interrupting the contact from combustion. In addition, some studies reported cyclotriphosphazene with an alkyl chain increases the thermal properties and flame retardancy because of phosphorus and nitrogen-flame-retardant synergy [17]. Recently, a number of novel synthesis approaches have been proposed to synthesize a new hexasubstituted cyclotriphosphazene with flame retardancy properties. In this work, different synthesis approaches were reviewed to synthesize a cyclotriphosphazene compound with flame retardancy properties. It provided an up-to-date summary of the synthesis method and also the flame retardancy of the compound. Limiting Oxygen Index (LOI) is a common technique used to evaluate the flame retardancy of textile materials and films. Materials exhibiting LOI values of 25% and greater are considered self-extinguishing. Usually, inherently flame retardant materials exhibit LOI values greater than 30%. Generally, UL-94 rating tests are widely used for the flammability of plastic standards, which are used to evaluate the material's ability to extinguish after being ignited [31,32]. In particular, the horizontal or vertical sample is ignited numerous times according to a specific flame height and flame application angle, and the evaluation is based on the sample's ignition duration and burning performance. The UL-94 rating performance of polymers can be multiply appraised into many distinct levels depending on the burning speed, burning time, anti-dripping ability, and if the bead is burning [33]. Table 1 summarizes the studies on the synthesis of hexasubstituted cyclotriphosphazene and their flame retardancy properties.

Table 1. Summary of the synthesis of hexasubstituted cyclotriphosphazene and their flame retardancy properties.

Title of Journal	Structure and Name of Compound	Flame Retardant Properties	Ref.
Flame retardant properties of cyclotriphosphazene derivatives for ABS	BPCP 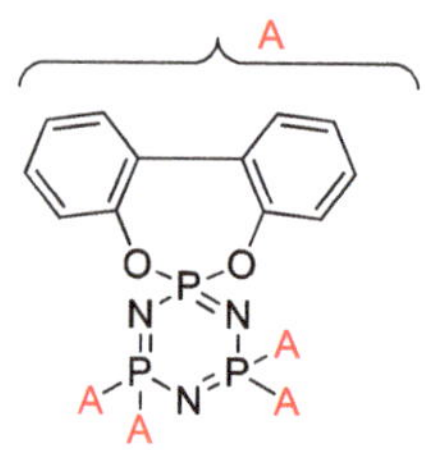 NTCP 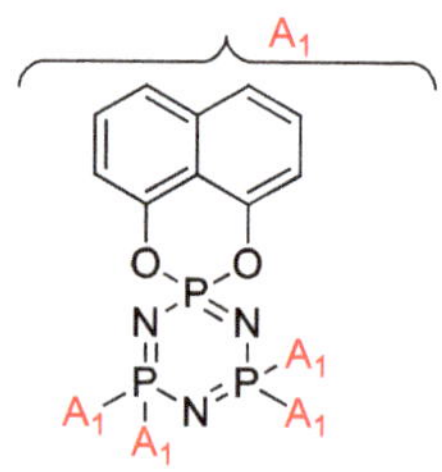 PTCP 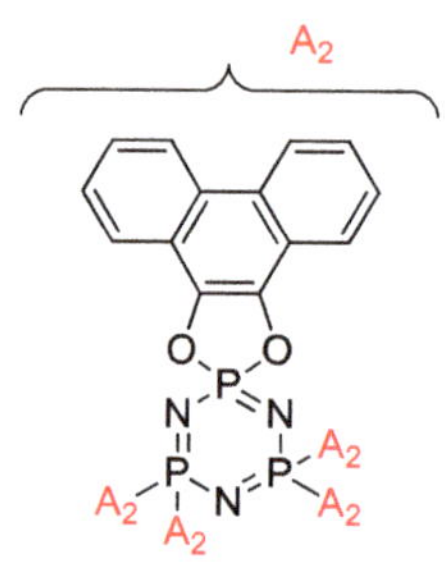	- LOI value: BPCP (21.2%, 22.1%, and 22.73% respectively), NTCP (22.3%, 23.1%, and 24.5% respectively, PTCP (23.1%, 24.1%, and 25.2%, respectively) - Additive used: 10 wt%, 15 wt%, and 20 wt% of each of BPCP, NTCP, and PTCP, respectively - Molding: ABS resin UL-94: Achieved V-1 rating	[34]
The flame retardancy and thermal stability properties of the poly(ethylene terephthalate)/hexakis(4-nitrophenoxy) cyclotriphosphazene system.	Hexakis(4-nitrophenoxy) cyclotriphosphazene (HNCP) 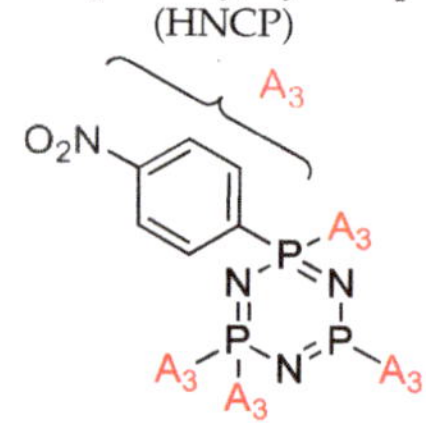	- LOI value: 35.1% - Additive used: 10 wt% of HNCP - Molding: Polyethylene terephthalate) PET UL-94: Achieved V-0 rating	[35]
Preparation and properties of halogen free flame-retardant blending modification polyester	Hexa(phenylamino)cyclotriphosphazene (HPACP) 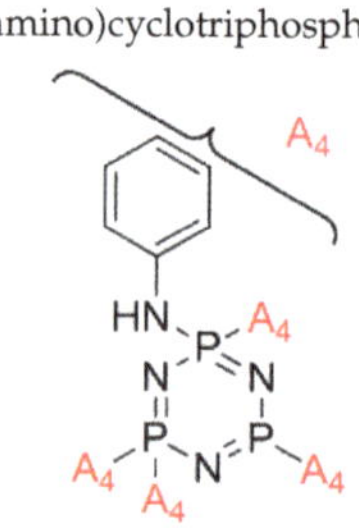	- LOI value: 28.3% - Additive used: 10 wt% of HPACP - Molding: Poly(ethylene terephthalate) PET UL-94: Not mentioned in this study	[36]

Table 1. *Cont.*

Title of Journal	Structure and Name of Compound	Flame Retardant Properties	Ref.
Synthesis of hexa-allylamino cyclotriphosphazene as a reactive flame retardant for unsaturated polyester.	Hexa-Allylamino Cyclotriphosphazene (HAC) 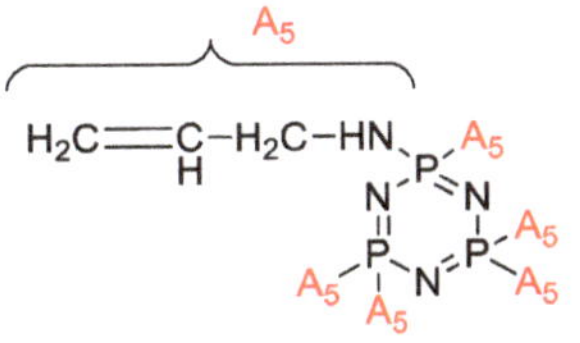	- LOI value: 36.5% - Additive used: Hexa-Allylamino Cyclotriphosphazene (HAC) - Molding: Diglycidyl ether bisphenol-A (DGEBA) UL-94: Not mentioned in this study	[37]
Synthesis of a phosphorus/nitrogen-containing compound based on maleimide and cyclotriphosphazene and its flame-retardant mechanism on epoxy resin.	Hexa(4-maleimido-phenoxyl)-cyclotriphosphazene (HMCP) 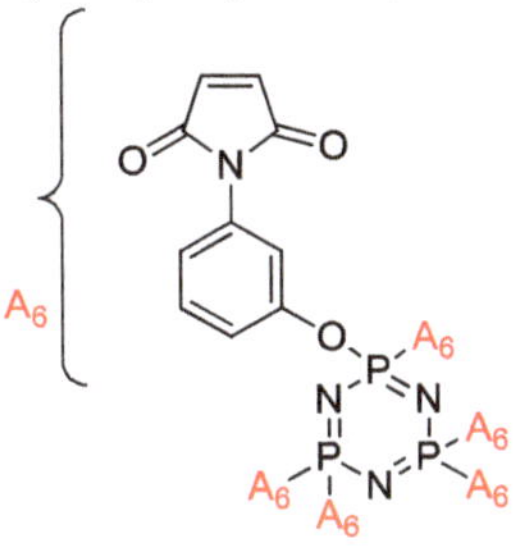	- LOI value: 36.5% - Additive used: Hexa(4-maleimido-phenoxyl)-cyclotriphosphazene (HMCP) - Molding: Epoxy Resin UL-94: Achieved V-0 rating	[38]
Synthesis of a novel flame retardant based on cyclotriphosphazene and DOPO groups and its application in epoxy resins	Hexa-[4-p-hydroxyanilino-phosphaphenanthrene-methyl)phenoxyl]-cyclotriphosphazene (HPMPC) 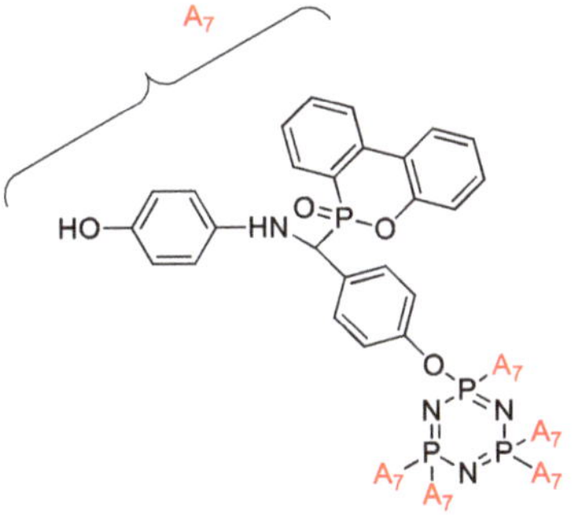	- LOI value: 36.6% - Additive used: 10.6 wt% of HPMPC - Molding: Diglycidyl ether bisphenol-A (DGEBA) UL-94: Achieved V-0 rating	[39]
Preparation of hexakis (4-aldehyde phenoxy) cyclotriphosphazene grafted kalonite and its synergistic fire resistance in poly(butylene succinate)	Hexakis (4-aldehyde phenoxy) cyclotriphosphazene (HAPC) 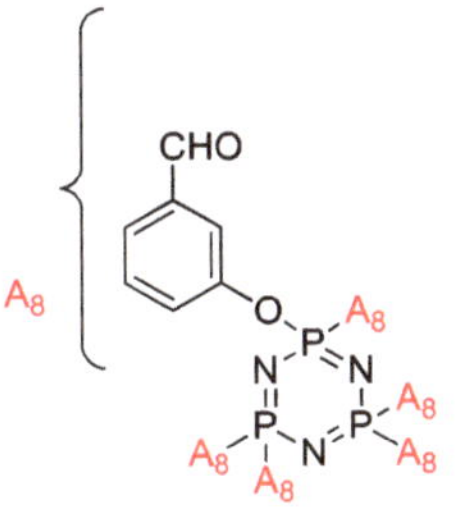	- LOI value: 40.3% - Additive used: 3 wt% of HAPC - Molding: Poly(butylene succinate) UL-94: Achieved V-0 rating	[40]

Table 1. *Cont.*

Title of Journal	Structure and Name of Compound	Flame Retardant Properties	Ref.
Synthesis, mechanical properties, and fire behavior of rigid polyurethane foam with a reactive flame retardant containing phosphazene and phosphate.	Hexa-(phosphite-hydroxyl-methyl-phenoxyl)-cyclotriphosphazene (HPHPCP) 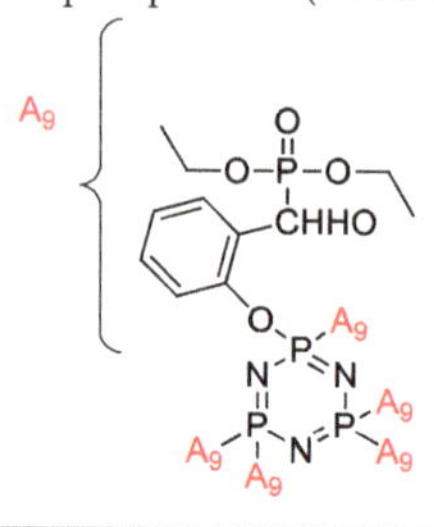	- LOI value: 25% - Additive used: 15 wt% of HPHPCP - Molding: Rigid polyurethane foams (FR-RFUFs) UL-94: Achieved HF-1 rating	[20]
Synthesis and characterization of flame-retardant rigid polyurethane foam based on reactive flame retardant containing phosphazene and cyclophosphonate	Hexa-(5,5-dimethyl-1,3,2-dioxaphosphinane-hydroxyl-methyl-phenoxyl-cycyclotriphosphazene (HDPCP) 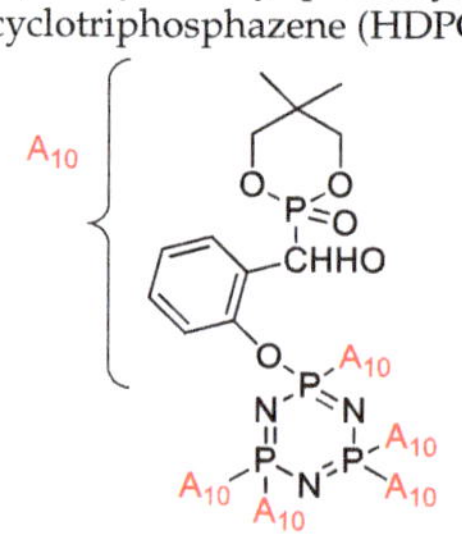	- LOI value: 25% - Additive used: 25 wt% of HDPCP - Molding: Rigid polyurethane foams (FR-RFUFs) UL-94: Not mentioned in this study	[41]
Application of cyclophosphazene derivatives as flame retardant for ABS	PNCP 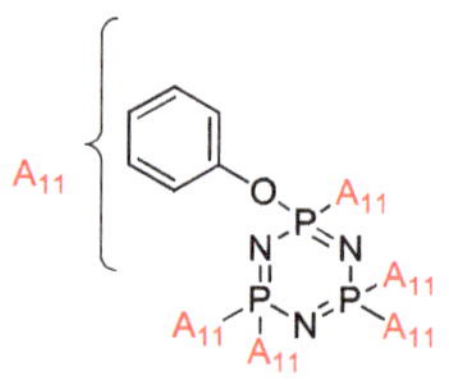 CTCP 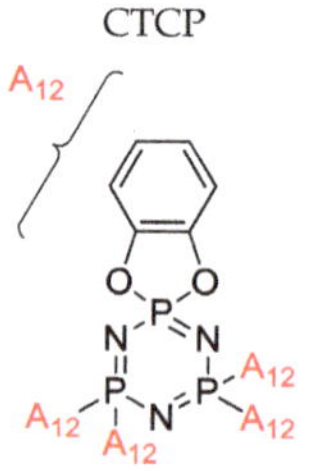 HNCP 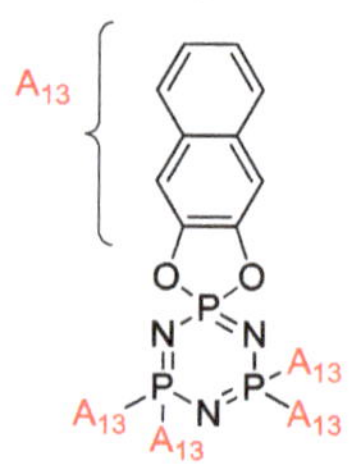	- LOI value: PNCP (20.8%, 21.5% and 22.1%, respectively) - CTCP (22.0%, 22.7% and 23.1%, respectively) - HNCP (22.9%, 23.6% and 24.9%, respectively) - Additive used: 10 wt%, 15 wt% and 20 wt% of PNCP, CTCP and HNCP, respectively - Molding: ABS resin UL-94: 15 wt% of the flame retardants were classified as the V-2 class, and 20 wt% of the flame retardants using PNCP were classified as V-2 but using HNCP or CTCP were classed as V-1	[42]

Table 1. *Cont.*

Title of Journal	Structure and Name of Compound	Flame Retardant Properties	Ref.
The non-halogen flame retardant epoxy resin based on novel compound with phosphenanthrene and cyclotriphosphazene double functional groups	Hexa-(phosphaphenanthrene-hydroxyl-methyl-phenoxyl)-cyclotriphosphazene (HAP-DOPO) 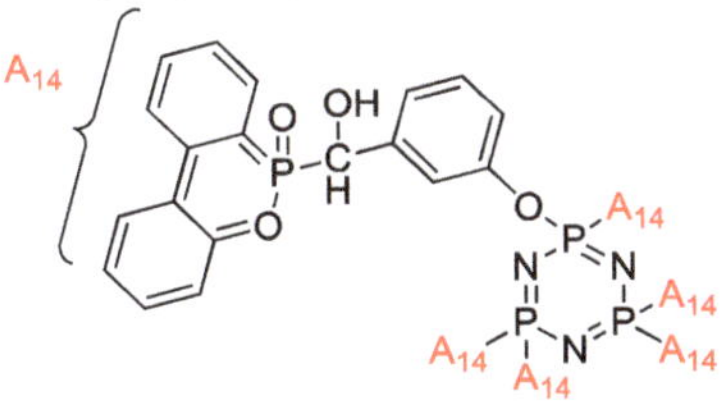	- LOI value: 35.2% - Additive used: 1.5 wt% of HAP-DOPO - Molding: epoxy resin UL-94: Achieved V-0 rating	[43]
Computer simulation study on the compatibility of cyclotriphosphazene containing the aminopropylsilicone functional group in flame retardant polypropylene/ammonium polyphosphate composites	APESP 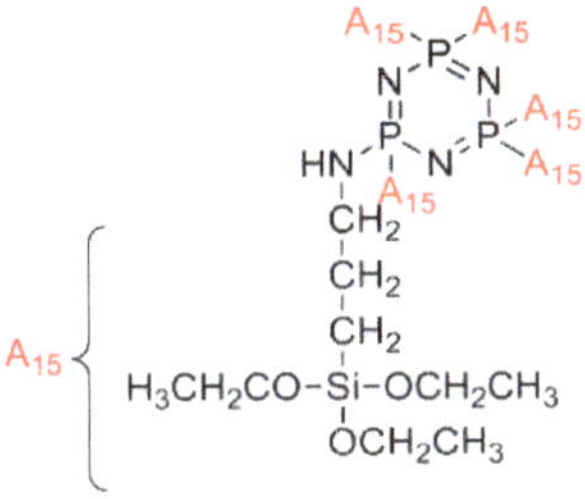	- LOI value: 26.5% - Additive used: APESP - Molding: Polypropylene (PP)/ammonium polyphospate (APP) UL-94: Achieved rating V-2	[44]
Hexa (eugenol) cyclotriphosphazene modified bismaleimide resins with unique thermal stability and flame retardancy	Hexa(eugenol)cyclotriphosphazene (HEC) 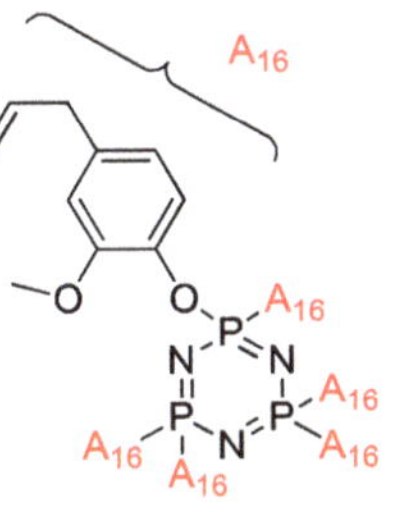	- LOI value: 39%, 48.4%, 50.1%, 49.8%, and 48.9%, respectively. - Additive used: Hexa(eugenol)cyclotriphosphazene (HEC) - Molding: 4,4′-bismaleimidodiphenylmethane (BMI) UL-94: Achieved rating V-0	[45]
Aminobenzothiazole-substituted cyclotriphosphazene derivatives as reactive fire retardant for epoxy resin.	Aminobenzothiazole-substituted cyclotriphosphazene derivative (ABCP) 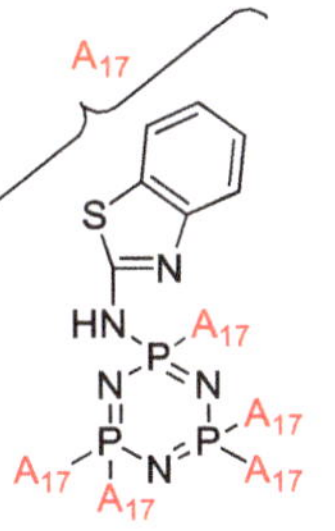	- LOI value: 31.2% - Additive used: Aminobenzothiazole-substituted cyclotriphosphazene (ABCP) - Molding: Epoxy Resin UL-94: Achieved V-0 rating	[46]

Table 1. *Cont.*

Title of Journal	Structure and Name of Compound	Flame Retardant Properties	Ref.
Effect of cyclotriphosphazene-based curing agents on the fire retardant of epoxy resin	Hexacyclohexylamino-cyclotriphosphazene (HCACTP) 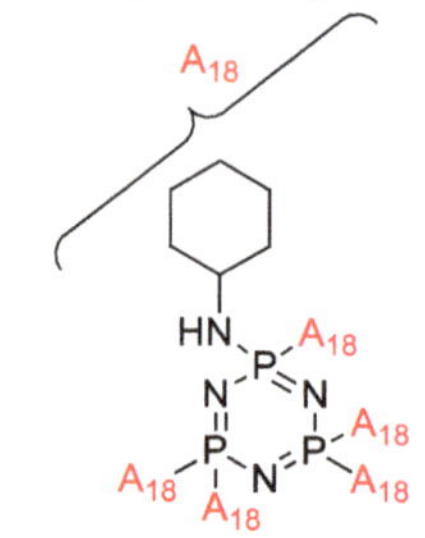 Diaminotetracyclohexylamino-cyclotriphosphazene (DTCATP) 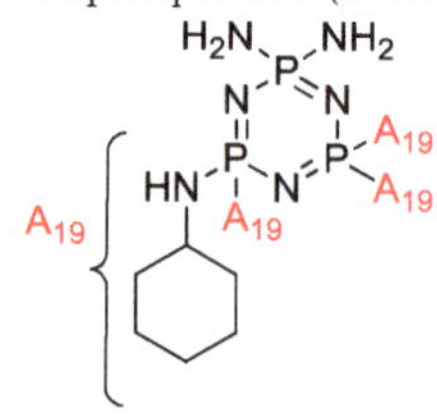	- LOI value: 25.3% and 22.2% respectively. - Additive used: Hexacyclohexylamino-cyclotriphosphazene (HCACTP) and Diaminotetracyclohexylamino-cyclotriphosphazene (DTCATP). - Molding: Epoxy Resin UL-94: Not mentioned in this study	[47]
The flame retardant properties and mechanism of poly(ethylene terephalate)/hexakis (para-allyloxyphenoxy) cyclotriphosphazene systems.	Para-allyl ether phenol derivative of cyclophosphazene (PACP) 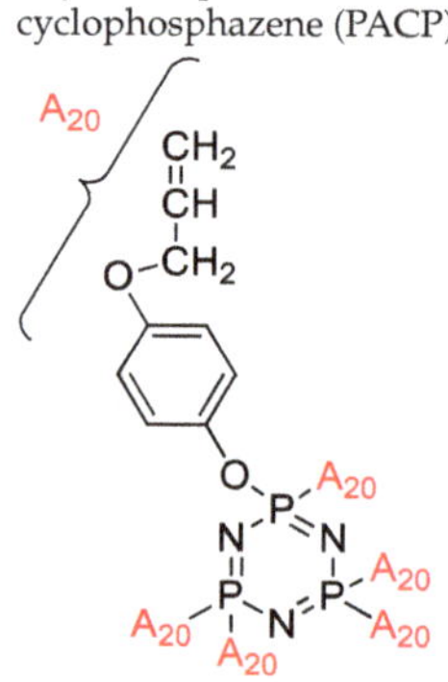	- LOI value: 33.5%, 33%, and 32.1%, respectively. - Additive: 5 wt%, 10 wt% and 15 wt% of PACP - Molding: Poy(ethylene terephalate) PET. UL-94: Achieved V-0 rating	[48]
Benzimidazolyl-substituted cyclotriphosphazene derivatives as a latent flame-retardant curing agent for a one-component epoxy resin system with excellent comprehensive performance.	Benzimidazolyl- substituted cyclotriphosphazene (BICP) 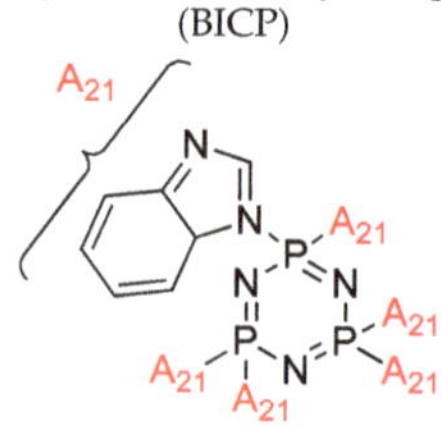	- LOI value: 33.5% - Additive: Benzimidazolyl-substituted cyclotriphosphazene (BICP) - Molding: Epoxy Resin UL-94: Achieved V-0 rating	[49]

Table 1. *Cont.*

Title of Journal	Structure and Name of Compound	Flame Retardant Properties	Ref.
Synthesis of melamine-cyclotriphosphazene derivatives and its application as flame retardant on cotton gauze	Melamine-Cyclotriphosphazene (MCP). 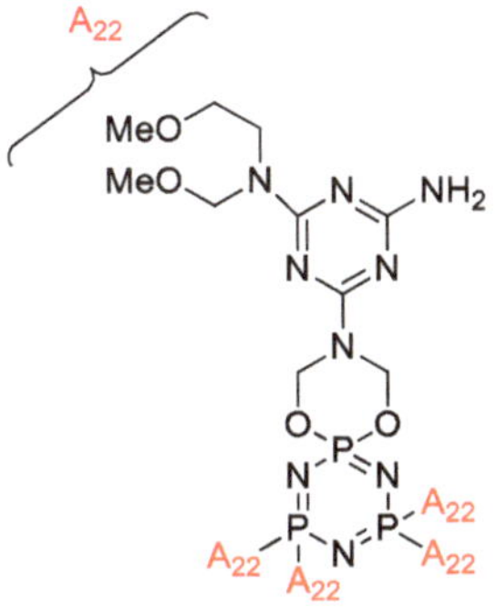	- LOI value: 16.9%, 23.5%, 24.6%, 25.4% 26.5%, and 25.8% respectively - Additive: Melamine-Cyclotriphosphazene (MCP). - Molding: Cotton Gauze UL-94: Not mentioned in this study	[50]
The flame retardancy and thermal stability properties of flame-retardant epoxy resin based on a-hydroxyphosponate cyclotriphosphazene	a-hydroxyphosponate cyclotriphos-phazene 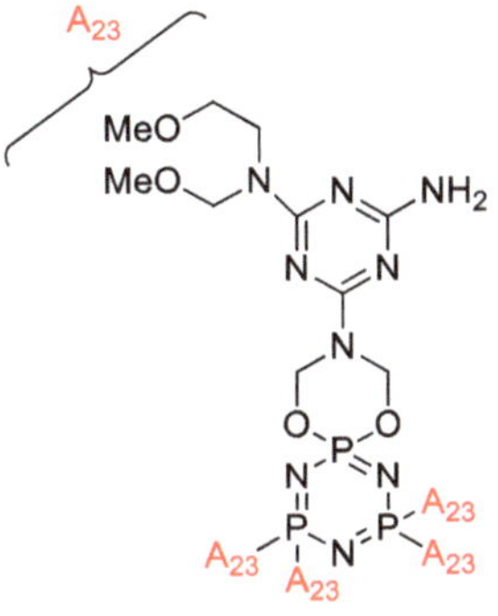	- LOI value: 30.7% - Additive: a hydroxyphosponate cyclotriphosphazene - Molding: Epoxy Resin UL-94: V-0 rating	[51]
Synthesis and flame retardant properties of cyclotriphosphazene derivatives containing Boron.	Cyclotriphosphazene derivatives containing Boron (CP-6B) 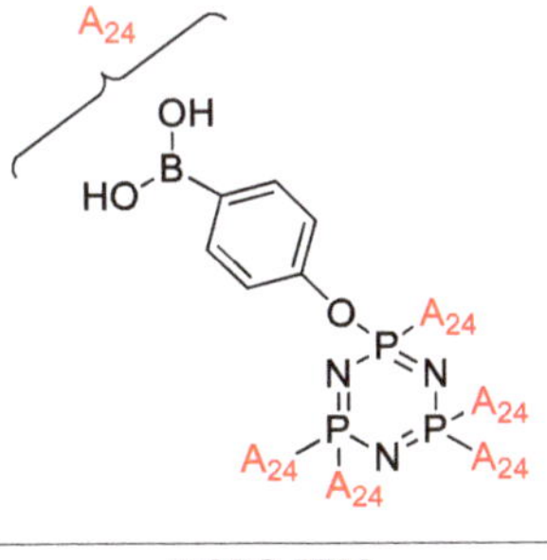	- LOI value: 32.3% - Additive: CP-6B - Molding: Epoxy Resin UL-94: Achieved V-0 rating.	[52]
An effective flame retardant for Poly(ethylene terephthalate) synthesis by phosphaphenanthrene and cyclotriphosphazene.	DOPO-TPN 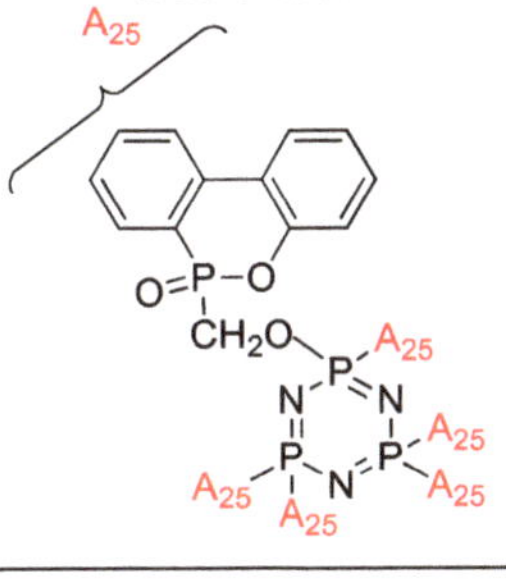	- LOI value: 34% - Additive: DOPO-TPN - Molding: Poly(ethylene terephtthalate) PET UL-94: Achieved V-0 rating	[53]

Table 1. *Cont.*

Title of Journal	Structure and Name of Compound	Flame Retardant Properties	Ref.
The influence of synergistic effects of hexakis (4-nitrophenoxy)cyclo-triphosphazene and POE-g-MA on anti-dripping and flame retardancy of PET	Hexakis (4-nitrophenoxy)cyclotriphosphazene (HNCP) 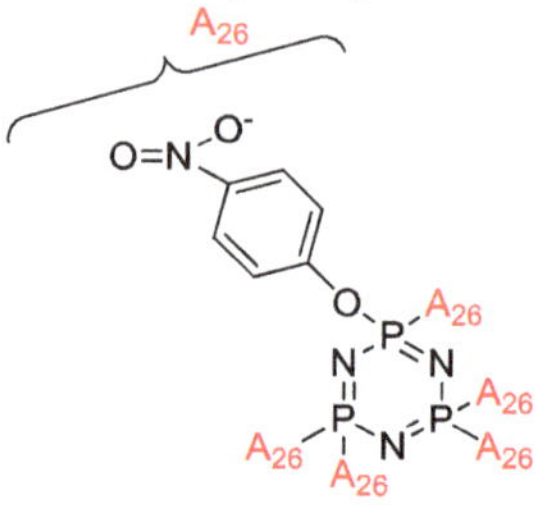	- LOI value: 28.3% - Additive used: 10 wt% - Molding: Poly(etyhylene terepththalate PET - UL-94: PET/10 wt% HNCP with 0.5 wt% POE-g-MA achieved a rating of V-0 and PET/10 wt% HNCP with 3 wt% POE-g-MA achieved rating V-0	[54]
Design of a self-healing and flame-retardant cyclotriphosphazene-based epoxy vitrimer	Cyloliner cylotriphosphazene-based epoxy resin (CTP-EP) 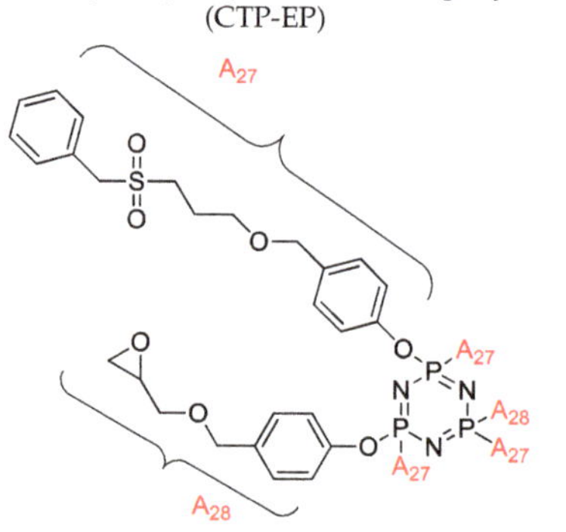	- LOI value: 30.5% - Additive used: CTP-EP/DTDA - Molding: Epoxy Resin UL-94: Achieved V-0 rating.	[55]
Synthesis and flame retardant efficacy of hexakis(3-(triethoxysilyl)propyloxy)cyclo-triphosphazene/silica coatings for cotton fabrics	Hexakis(3-(triethoxysilyl)propyloxy)cyclotriphosphazene (HTPC) 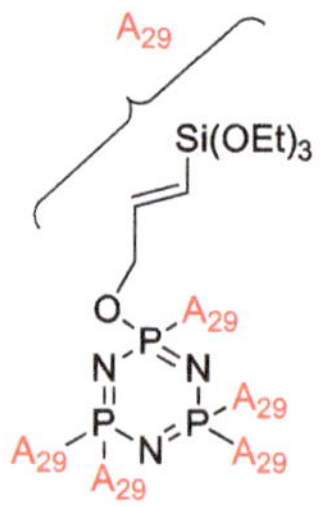	- LOI value: 27.7% - Additive used: Hexakis(3-(triethoxysilyl)propyloxy)cyclo-triphosphazene (HTPC) - Molding: Cotton Fibre UL-94: Not mentioned in this study	[56]
Synthesis and characterization of thermally stable and flame retardant of hexakis(4-aminophenoxy)cyclo-triphosphazene-based polyimide matrices	Hexakis(4-aminophenoxy)cyclotriphosphazene-based polyimide matrices (HACTP) 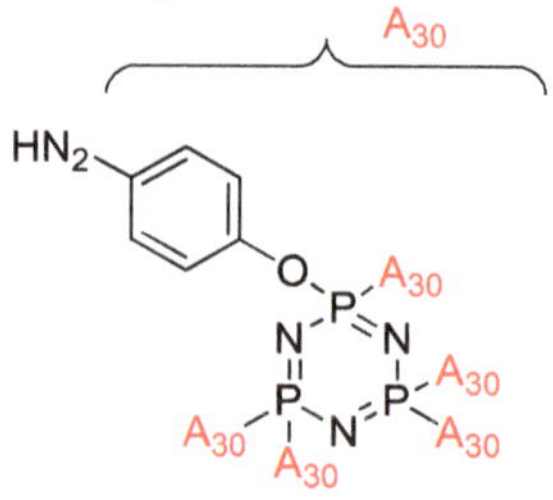	- LOI value: Range from 44.3% until 46.7%, respectively - Additive: HACTP - Molding: PI-PMDA and PI-BPDA UL-94: Achieved V-0 rating	[57]

Table 1. *Cont.*

Title of Journal	Structure and Name of Compound	Flame Retardant Properties	Ref.
The synthesis, curing kinetics, thermal properties, and flame retardancy of cyclotriphosphazene-containing multifunctional epoxy resin	Hexa-[4-(glycidyloxymethyl)phenoxy]-cyclotriphosphazene (HGPCP) 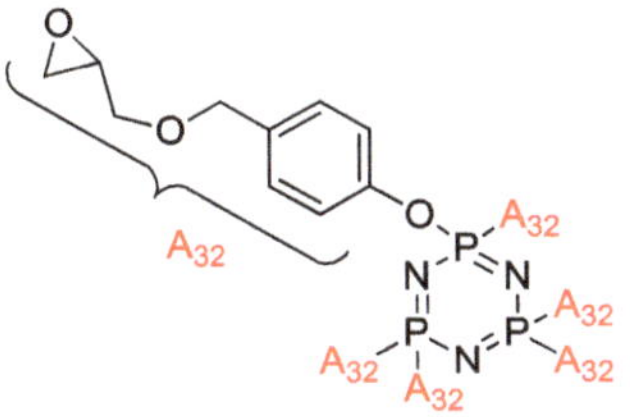	- LOI value: 32.5% and 35.4%, respectively - Additive: HGPCP/DDS and HGPCP/DDM - Molding: Epoxy Resin UL-94: Achieved V-0 rating	[58]
Hexa-[4-(glycidloxycarbonyl)phenoxy]cyclo-triphosphazene chain extender for preparing high-performance flame retardant polyamide-six composites	Hexa-[4-(glycidloxycarbonyl)phenoxy]cyclo-triphosphazene (CTP-EP) 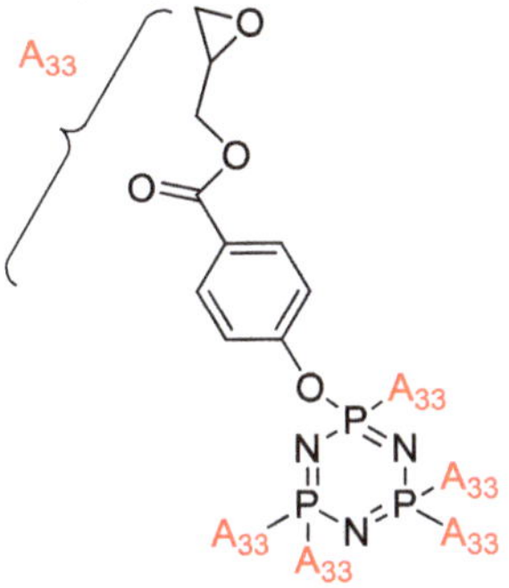	- LOI value: 31.3% - Additive: 11 wt% of ALPi/CTP-EP - Molding: Polyamide 6/aliminium diethylphpsphinate (PA6/AlPi) UL-94: Achieved V-0 rating	[59]
Synergistic effect of the intumescent flame-retardant system consisting of hexophenoxy cyclotriphosphazene and ammonium polyphosphate on methyl ethyl silicon rubber	Hexophenoxy cyclotriphosphazene (HPCP) 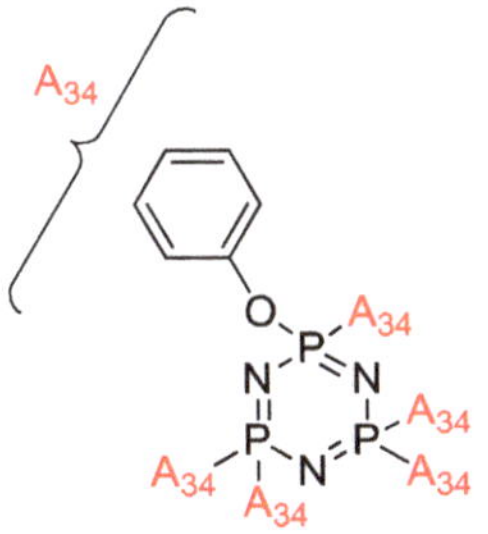	- LOI value: 30.6% - Additive: Hexophenoxy cyclotriphosphazene (HPCP) - Molding: Methyl ethyl silicone rubber (VMQ) UL-94: Achieved V-0 rating	[60]

Table 1. *Cont.*

Title of Journal	Structure and Name of Compound	Flame Retardant Properties	Ref.
High transmittance and environmentally friendly flame-retardant optical resin based on poly(methyl methacrylate) and cyclotriphosphazene derivatives	Ethyl p-hydroxybenzoate derivatives of cyclotriphosphazene 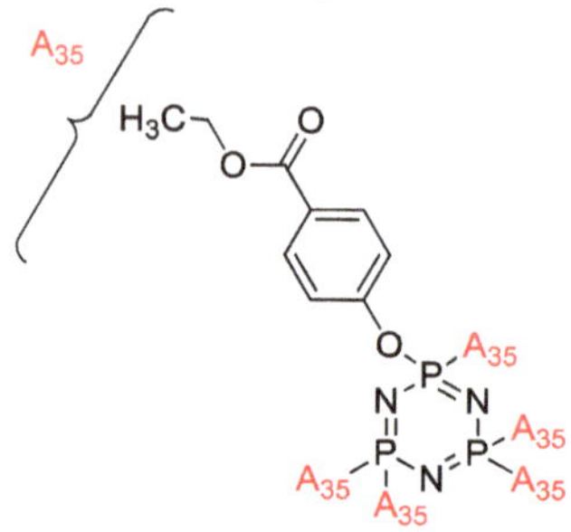 Phenol derivatives of cyclotriphosphazene 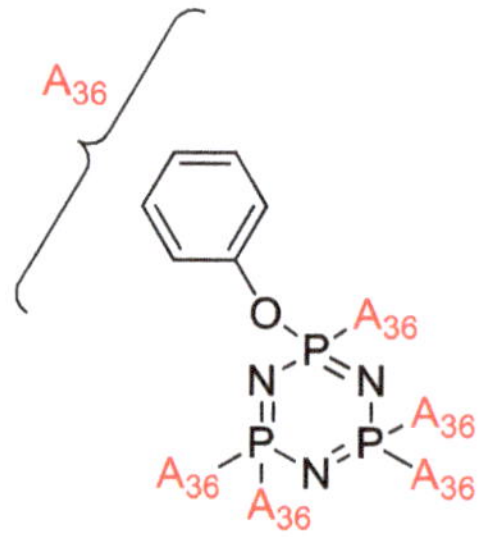	- LOI value: 26% and 22%, respectively - Additive: Two cyclotriphosphazene derivatives - Molding: High transparent optical resin based on methacrylate (PMMA) UL-94: Not mentioned in this study	[61]
Study of thermal properties of flame-retardant epoxy resin treated with hexakis[*p*-(hydroxymethyl)phenoxy] cyclotriphosphazene	Hexakis[*p*-(hydroxymethyl)phenoxy] cyclotriphosphazene (HHPCP) 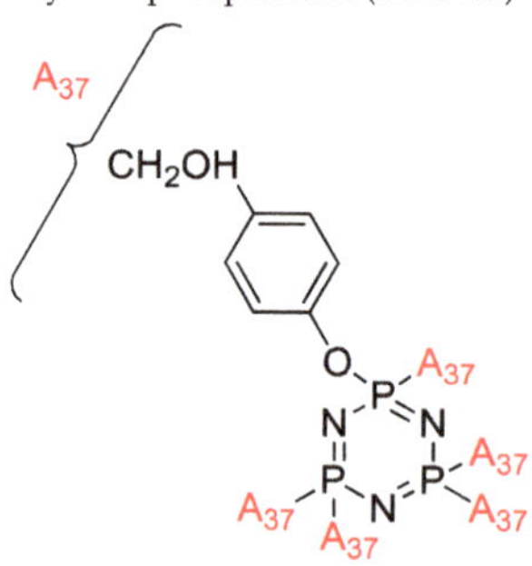	- LOI value: 26.5% - Additive: Hexakis[p-(hydroxymethyl)phenoxy] cyclotriphosphazene (HHPCP) - Molding: Epoxy Rein UL-94: Not mentioned in this study	[62]
The flame retardancy and thermal properties of poly (ethylene terephthalate)/cyclotriphosphazene modified by the montmorillonite system	Hexachlorocyclotriphosphazene mofidies by montmorillonite (HCCP-OMMT) 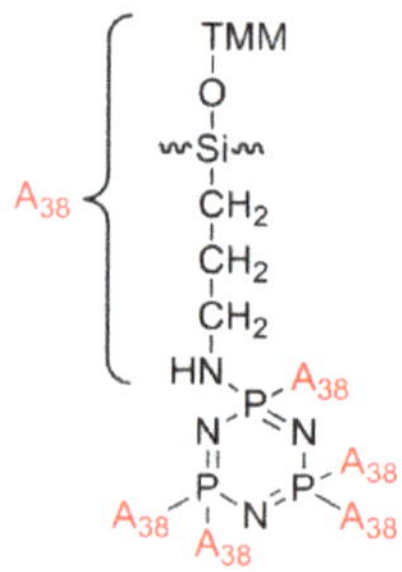	- LOI value: 31.5% - Additive: 3% of hexachlorocyclotriphosphazene mofidied by montmorillonite (HCCP-OMMT) - Molding: Poly(ethylene terephthalate) UL-94: Achieved V-0 rating	[63]

Table 1. *Cont.*

Title of Journal	Structure and Name of Compound	Flame Retardant Properties	Ref.
Synthesis, characterization, and utilization of novel phosphorus/nitrogen-containing flame retardant	Hexa(phosphaphenanthrene aminophenoxyl)-cyclotriphosphazene (HPAPC) 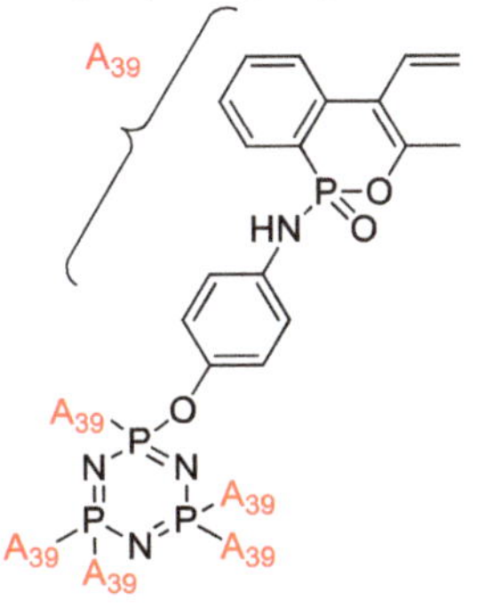	- LOI value: 34.7% - Additive: Hexa(phosphaphenanthrene aminophenoxyl)-cyclotriphosphazene (HPAPC) - Molding: Poly(lactic acid) (PLA) UL-94: Achieved V-0 rating	[25]
Effect of surface chemical modification for aluminum hypophosphite with hexa-(4-aldehyde-phenoxy)-cyclotriphosphazene on flame retardancy, water resistance, and thermal properties for polyamide 6	Hexa-(4-aldehyde-phenoxy)-cyclotriphosphazene (AHP) 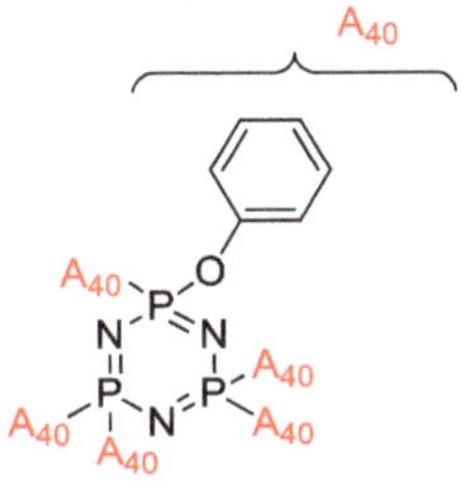	- LOI value: 27.6% - Additive: Hexa-(4-aldehyde-phenoxy)-cyclotriphosphazene (AHP) - Molding: Polyamide 6 (PA6) - UL-94: Achieved V-0 rating	[64]
Synthesis of tris(phenoxy)triflorocyclo-triphosphazene and study of its effect on the flammable, thermal, optical, and mechanical properties of bisphenol-A polycarbonate	Tris(phenoxy)trifluorocyclotriphosphazene (TCTP) 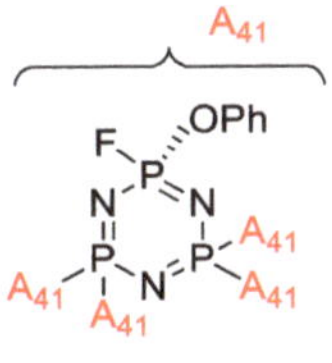	- LOI value: 40% - Additive: Tris(phenoxy)trifluorocyclo-triphosphazene (TCTP) - Molding: Polycarbonate - UL-94: Achieved V-0 rating	[65]
Effect of trisilanolphenyl-POSS on rheological, mechanical, and flame retardancy properties of the poly(ethylene terephthalate)/cyclotriphosphazene system	Hexakis (para-alloxyphenoxy) cyclotriphosphazene (PACP) 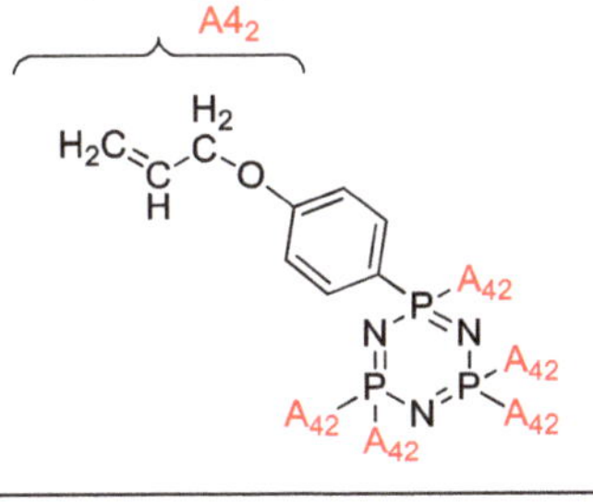	- LOI value: Not mentioned in this study - Additive: Hexakis (para-alloxyphenoxy) cyclotriphosphazene (PACP) - Molding: Poly(ethylene terepthtalate) (PET) UL-94: Achieved V-0 rating	[66]

Table 1. *Cont.*

Title of Journal	Structure and Name of Compound	Flame Retardant Properties	Ref.
Preparation and properties of novel inherent flame-retardant cyclotriphophazene containing epoxy resin	Bis-(4-hydroxyphenylsulfonylphenoxy) tereohenoxycyclotriphosphazene (HSPPZ) 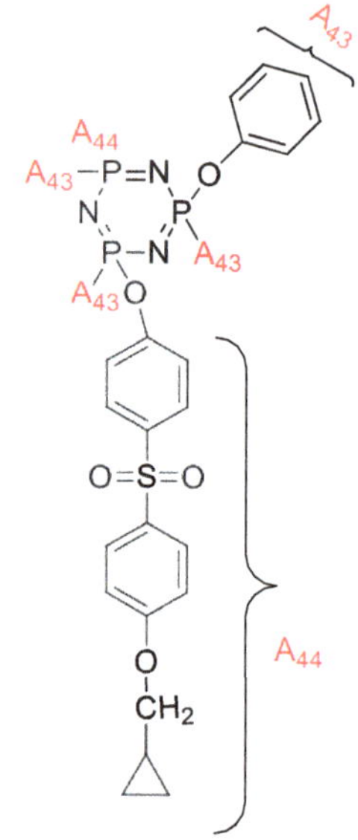	- LOI value: Not mentioned in this study - Additive: Bis-(4-hydroxyphenylsulfonylphenoxy) tereohenoxycyclotriphosphazene (HSPPZ) - Molding: Epoxy Resin UL-94: Achieved V-0 rating	[67]

3. Effect of Linking Unit in Flame Retardancy Properties of Cyclotriphosphazene

Linking units are usually structural units that connect one core to another, which maintain the linearity of the core while being compatible with the rest of the structure [3]. As the connecting units give a point of link up in synthesis, compounds with linking groups are more accessible to synthesis than those with direct bonds. There are several types of linking units, which are azo (-N=N-), Schiff base (-CH=N-), acetylene (-C≡C-), diacetylene (-C≡C-C≡C-), and stilbene (-CH=CH-). Based on the reports from previous research, a comparison of two linking units, which are Schiff base and azo linking units, has been made. This study concluded that Schiff base linkage showed higher thermal stability than the azo due to rigidity in the Schiff base's linkage [68]. Furthermore, strong molecular interaction in Schiff base derivatives results in high temperature compared with azo derivatives. Until now, few linking units have been explored related to the enhancement of properties of the cyclotriphosphazene compound, which are the azo and Schiff base linking units. Generally, the azo linking unit can be classified into two types which are monoazo and bisazo linking unit (Figure 2), while the Schiff base functional group consists of a carbon-nitrogen double bond (C=N) with the nitrogen atom connected to the aryl or alkyl group but not with the hydrogen. The general formula of Schiff base is $R_1R_2C=NR_3$ with R as a side chain [69]. The Schiff base is an intriguing linking unit that provides a stepped-core structure that allows molecules to maintain linearity, provide excellent stability, and change the physical properties contributed by the linking unit itself [70].

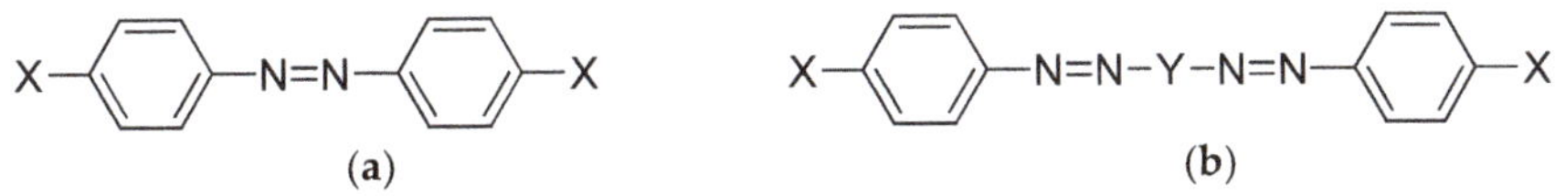

Figure 2. General chemical structure of (**a**) the monoazo linking unit and (**b**) the bisazo linking unit.

Some authors have driven the further development of studies in linking unit effects in flame retardancy as the contribution of several types of linking units in the enhancement of flame retardancy properties in cyclotriphosphazene compounds. Recently, Jamain et al. worked on synthesizing a new hexasubstituted cyclotriphosphazene compound with two Schiff base linking units and containing different terminal substituents as shown in Figure 3 [71]. The Schiff base linking unit was also reported to improve the flame retardant properties due to its thermal stability [72]. This research aimed to study the flame retardant properties of hexasubstitutd cyclotriphosphazene compounds containing two Schiff base linking units using the limiting oxygen index (LOI). All the samples were prepared with 1 wt.%, and polyester resin was used as molding. The result obtained from this study reported that the LOI value increased from 22.53% of pure polyester resin to 24.71% when hexasubstituted cyclotriphosphazene compound was incorporated with polyester resin. The best results achieved from this study were from the compounds with nitro and chloro terminal groups with LOI values of 28.37% and 27.90%, respectively. The phenomenon was due to the electron-withdrawing properties of both nitro and chlorine groups, which induced flame retardancy properties.

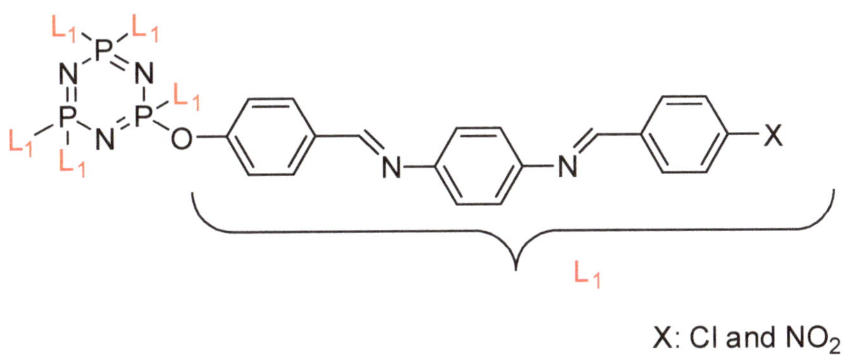

Figure 3. Hexasubstituted cyclotriphosphazene with two Schiff base linking units.

Other linking units were studied, azo and amide linking units, and their contribution to the flame retardancy properties of cyclotriphosphazene is summarized in Table 2.

Table 2. Synthesis of the hexasubstituted cyclotriphosphazene compound with several types of linking units.

Title of Journal	Types of Linking Unit	Description	Ref.
Synthesis of new star-shaped liquid crystalline cyclotriphosphazene derivatives with flame-retardant bearing amide-azo and azo-azo linking units. 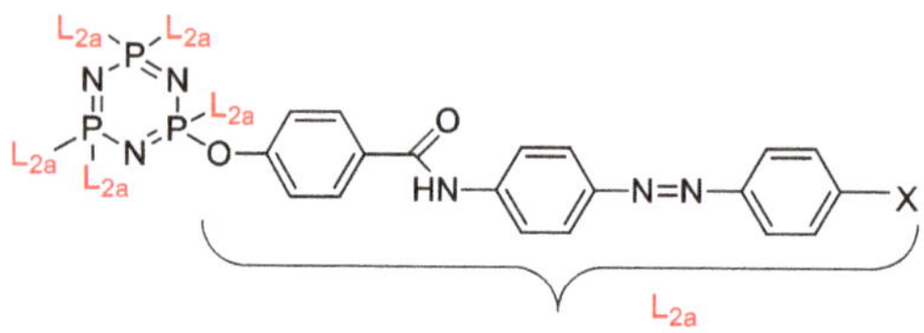 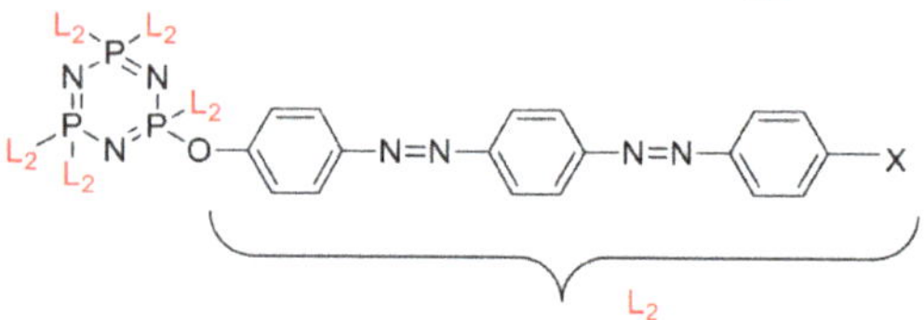 X: OC_7H_{15};OC_9H_{19};OC_{10}.H_{21}.$OC_{12}H_{25}$.$OC_{14}H_{29}$;OH;COOH;Cl;NO_2	Amide azo and azo-azo linking units	- This study aimed to synthesize two series of new hexasubstituted cyclotriphosphazene containing two types of linking units: amide-azo and azo-azo. - The homologues of the same series contain different terminal substituents such as heptyl, nonyl, decyl, dodecyl, tetradecyl, hydroxyl, carboxyl, chloro, nitro, and amino. - The flame retardancy of synthesized compounds is measured using the Limiting Oxygen Index. - From the result obtained, compounds containing heptyl have a higher LOI value compared with other compounds. The LOI values for alkylated compounds decrease as the aliphatic chain length increases. This study also compared the flame retardancy properties of two compounds containing two different linking units. The compound containing amide-azo was reported to have a high LOI value. The phenomenon indicates the attribute to the electron withdrawing properties of amide moiety.	[73]
Synthesis of novel liquid crystalline and flame-retardant molecules based on six armed cyclotriphosphazene cores containing Schiff base and amide linking units. 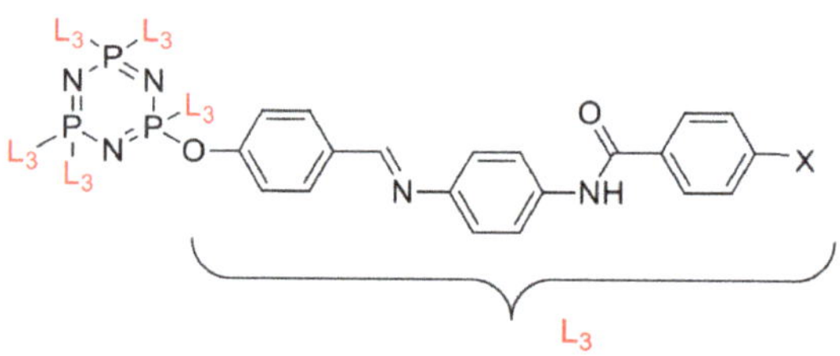 X:OC_7H_{15};OC_9H_{19};$OC_{10}H_{21}$;$OC_{12}H_{25}$;$OC_{12}H_{29}$;OH;COOH;Cl;NO_2	Schiff base and amide linking units	- This study aimed to synthesize hexa substituted cyclotriphosphazene containing Schiff base and amide linking units, characterization, and chemical testing (flame retardancy using LOI value) were also measured. - From the result obtained, the LOI value indicating the flame retardancy properties of the compound showed that the compound with the nitro group had a high LOI value due to the electron withdrawing group that enhances the synergistic effect of P-N bonds. - This study also reported that cyclotriphosphazene compounds could enhance the flame retardancy of polyester resin. The LOI value of polyester resin increased from 22.53% to 24.71%. Schiff base linking unit was found to enhance these properties due to char formation in the condensed phase. - Other than that, the amide linking unit also contributed to the increase of LOI value. This is because the electron withdrawing of the amide bond enhances the flammability of these compounds.	[74]
Synthesis, characterization, and mesophase transition of Hexasubstituted cyclotriphosphazene molecules with Schiff base and Azo linking units and determination of their flame-retardant properties. 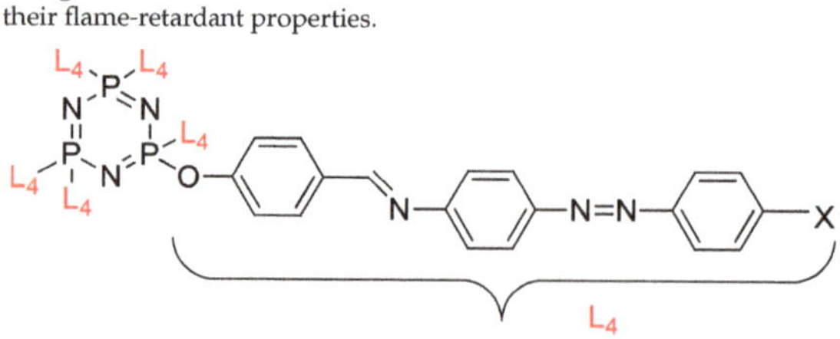 X:OC_7H_{15};OC_9H_{19};$OC_{10}H_{21}$;$OC_{12}H_{25}$;$OC_{14}H_{29}$.OH;COOH;Cl;NO_2	Schiff base and azo linking units	- This study aimed to synthesis hexasubstituted cyclotriphosphazene containing Schiff base and Azo linking units. The characterization and chemical testing (flame Retardancy using LOI value) were also measured. - From the result obtained, the LOI valued increased when incorporated with hexasubstituted cyclotriphosphazene. The LOI value of the compound containing the nitro group was recorded as the highest, which was 27.90%. This is due to the nitro group's electron withdrawing, which releases the electron from their resonance effects to their corresponding P-N bonds. As a result, the P-N synergistic effect was enhanced, and they exhibited both the condensed and gas phase action, which caused the compound to have a high LOI value.	[75,76]

4. Dielectric Properties in Cyclotriphosphazene

A non-metallic substance with high specific resistance, a negative temperature coefficient of resistance, and a substantial insulating resistance is referred as a dielectric material. The dielectric material can also be defined as a non-conducting material that stores electrical charges. When a dielectric is placed in an electric field, the electric charges do not flow through the materials. Electric charges slightly shift from their average equilibrium positions, causing dielectric polarization. Positive charges flow in the direction of the field, while negative charges shift in the opposite direction of the field due to dielectric polarization. This phenomenon yields an internal electric field, which in turn reduces the overall electric field within the dielectric materials.

Cyclotriphosphazene compounds and their versatility of substituent placed in cyclic and macromolecules backbones make this compound have significant potential in applications such as dielectric [77–79], additives of flame retardant [26,80], fluorescence materials [81,82], and liquid crystals [83]. Recently, there have been many studies reported related to the dielectric properties of cyclotriphosphazene compounds. Dielectric properties such as dielectric loss, dielectric constant, and conductivity are the most common properties extensively studied to evaluate solid materials. In addition, dielectric constant measurements are among the most popular methods of evaluating solid materials such as electric insulators and polymers, in which dielectric constant measurements can be performed easier than with chemical analysis techniques. Impedance spectroscopy is a relatively recent and powerful technique for determining several electrical properties of electrolyte materials and their interactions with electronically conducting electrodes. The permittivity of a material describes its ability to absorb, transmit, and reflect electromagnetic energy. These also are important properties required to design electronic devices and have numerous potential application areas in multifunctional electronic and optoelectronic devices [84–87].

Koran et al. recently worked on synthesizing cyclotriphosphazene derivatives from the reaction of substituted chalcone compounds containing different organic side groups at the para position with cyclotriphosphazene containing bearing dioxypheneyl [88]. Dielectric properties of the final compound, which is the dielectric constant, were measured using an impedance analyzer. From this study, the result obtained reported that the cyclotriphosphazene compound containing methoxy was observed to have a high dielectric constant. This is due to the increase in polarization with a conjunction. Meanwhile, compounds containing methyl groups showed the lowest dielectric constant. Thus, compounds containing methyl groups were chosen to determine the influence of Eu3+ doping on the dielectric properties of phosphazene. In addition, the dielectric behavior of new oxime-cyclotriphopshazene derivatives was also explored [79]. Cyclotriphosphazene compounds bearing oxime ether and ester as side groups were synthesized in this study (Table 3). The dielectric properties, which are dielectric loss and dielectric constant of this compound, were measured using an impedance analyzer. The results reported that decreases in the dielectric constants were obvious at low-frequency values. On the other hand, when closing the higher-frequency levels, the decline of the dielectric constants was reduced. The reason behind the results is the existence of polar groups on the structure, which lead to an increased dielectric constant. The carbonyl group which is bonded to oxime-ester groups made the dielectric constant higher compared with other derivatives. Both studies also reported that phosphazene derivatives are promising candidate materials in multifunctional and optoelectronic devices.

Table 3. Cyclotriphosphazene compound bearing oxime ether and ester as the side group.

Compound	Name
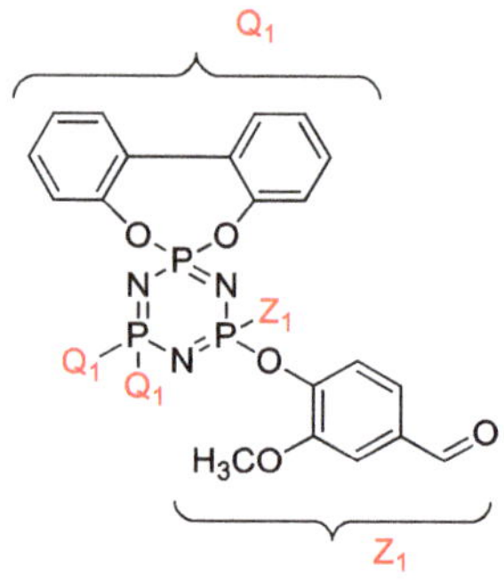	2,2-Bis(4-formyl-2-methoxyphenoxy)-4,4,6,6-bis[spiro(2′,2″-dioxy-1′,1″-biphenylyl)] cyclotriphosphazene
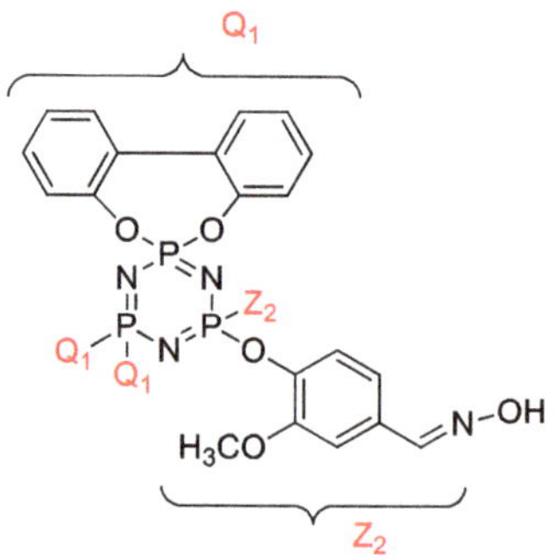	2,2-di[(4-(hydroxyimino)-2-methoxy)phenoxy]-4,4,6,6-bis[spiro(2′,2″-dioxy-1′1″-biphenylyl)]cyclotriphosphazene
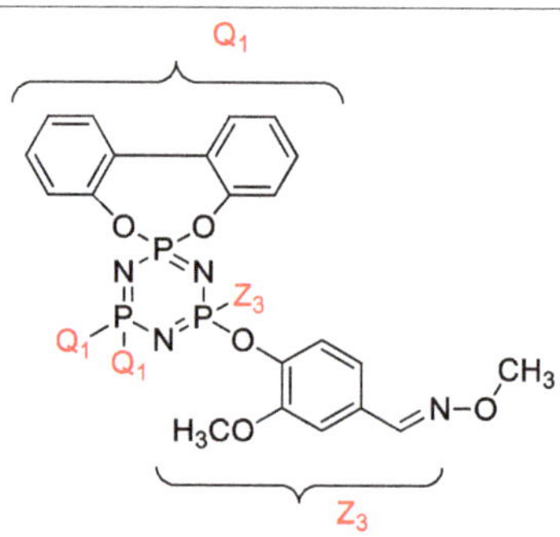	2,2-di[{(4-(methyloxy)imino)-2-methoxy} phenoxy]-4,4,6,6-bis[spiro(2′2″-dioxy-1′,1″-biphenylyl)] cyclotriphosphazene
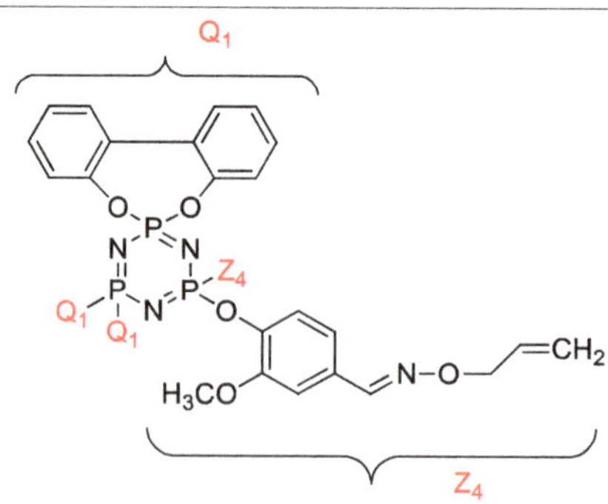	2,2-di[{(4-(allyloxy)imino)-2-methoxy} phenoxy]-4,4,6,6-bis[spiro(2′,2″-dioxy-1′,1″-biphenylyl)] cyclotriphosphazene

Table 3. *Cont.*

Compound	Name
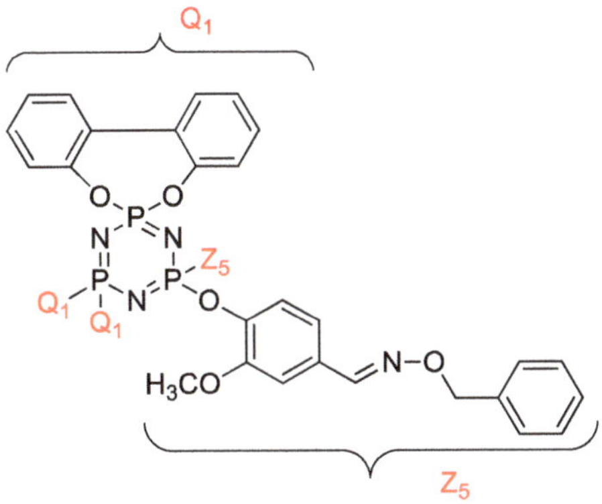	2,2-di[{(4-(benzyloxy)imino)-2-methoxy} phenoxy]-4,4,6,6-bis[spiro(2′,2″-dioxy-1′,1″-biphenylyl)] cyclotriphosphazene
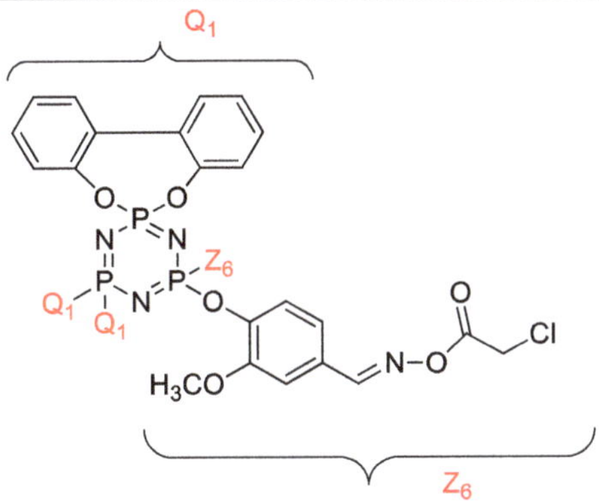	2,2-di[{(4-(chloroacetyloxy)imino)-2-methoxy} phenoxy]-4,4,6,6-bis[spiro(2′,2″-dioxy-1′,1″-biphenylyl)] cyclotriphosphazene
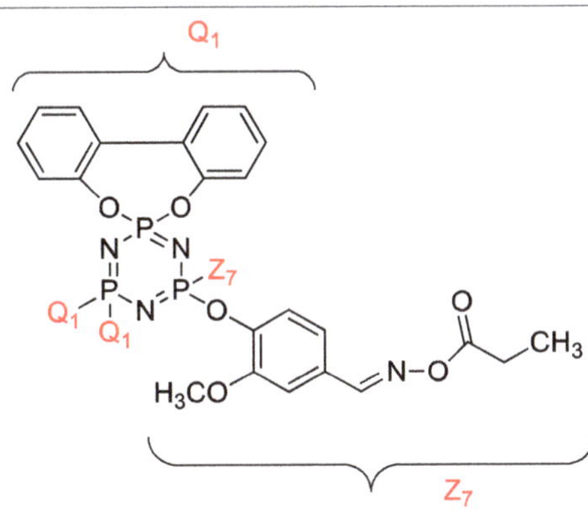	2,2-di[{(4-(propanoyloxy)imino)-2-methoxy} phenoxy]-4,4,6,6-bis[spiro(2′,2″-dioxy-1′,1″-biphenylyl)] cyclotriphosphazene

Table 3. *Cont.*

Compound	Name
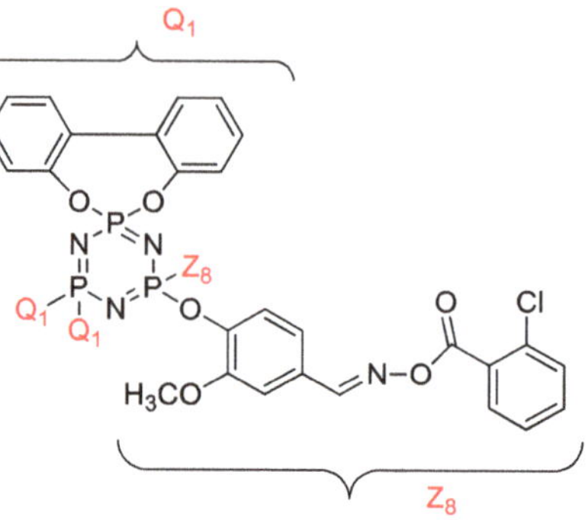	2,2-di[{(4-(o-chlorobenzoyloxy)imino)-2-methoxy}phenoxy]-4,4,6,6-bis[spiro(2′,2″-dioxy-1′,1″-biphenylyl)] cyclotriphosphaz
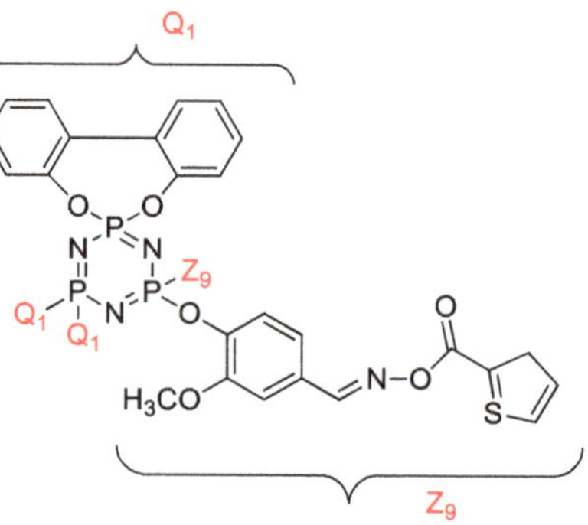	2,2-di[{(4-(thiophene-2-carbonyloxy)imino)-2-methoxy}phenoxy]-4,4,6,6-bis[spiro(2′,2″-dioxy-1′,1″-biphenylyl)] cyclotriphosphazene

In addition, there are studies that reported on the development of phosphazene imine-modified epoxy composited for low dielectric. One study aimed to synthesize the phosphazene imine (PZ-imine), and the dielectric properties of this compound were measured [89]. Based on the result obtained, the dielectric constant for neat epoxy were 1, 3, 5, and 7 wt.% of PZ-imine, reinforced epoxy nanocomposites were 3.42, 3.10, 2.93, 2.75, and 2.41, respectively, the dielectric loss of neat epoxy were 1, 3, 5, and 7 wt.% of PZ-imine, and reinforced epoxy nanocomposites were 0.25, 0.20, 0.18, 0.14, and 0.12 at 1 MHz, respectively. When the frequency increases with a decrease in the orientation of polarization and its dipole moments need a longer time than the electronic and ionic polarization, this decreased the dielectric constant. In this regard, the addition of PZ-imine to the epoxy helped the reduction in the dielectric constant and dielectric loss values. In modified epoxy resins, the bulky group of PZ-imine efficiently inhibits dipole orientation and relaxation. The PZ groups minimize the inter-phase contact between the inorganic reinforcement and the organic domain, significantly lowering the dielectric constant of the resultant nanocomposites [90]. The influencing action of PZ causes a drop in the dielectric constant value due to an increased free volume to some extent. The dielectric loss of PZ-imine nanocomposites reduces as the amount of PZ-imine in the epoxy matrix increases. Direct current conduction, space charge migration, and the movement of molecular dipoles are all phenomena that contribute to the value of dielectric loss in materials. Polymer reinforced with PZ-imine results in an insulating layer outside the dielectric cores that control the migration and build-up of space charges within the nanocomposites, resulting in a lower dielectric loss.

The information provided in Table 4 reveals that hexasubstituted cyclotriphosphazene was studied for its dielectric properties. It has fascinating properties that may contribute to the electrical field and details of explanation on how the structure affects the properties.

Table 4. Reported study of cyclotriphosphazene and their dielectric properties.

Compound Name	Dielectric Constant	Dielectric Loss	Ref.
2,2,4,4-tetra(4′-oxy-substituted-chalcone)- 6,6-diphenyl cyclotriphosphazene derivatives 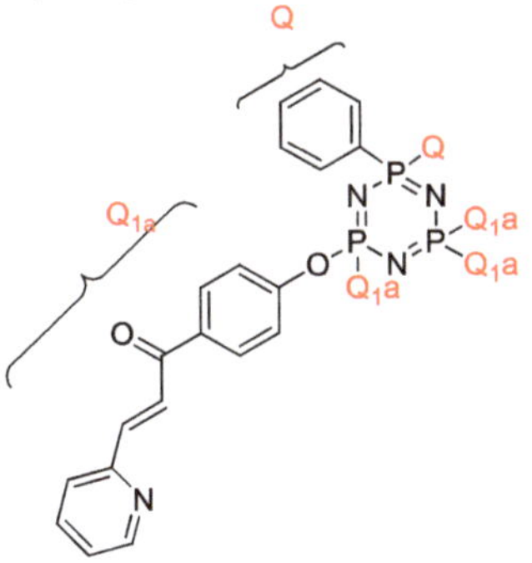 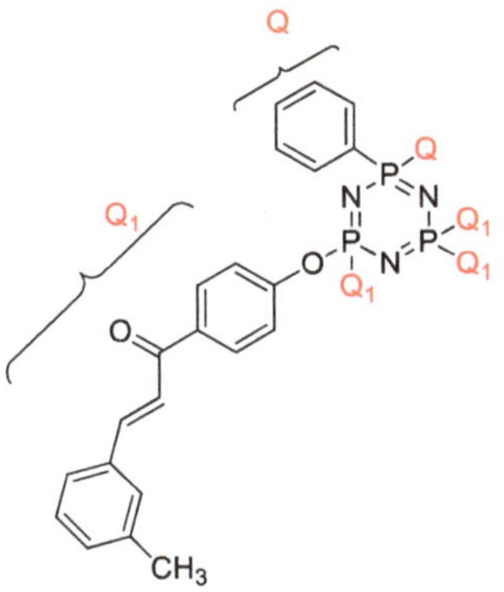	- Dielectric constants decreased with the increasing frequency, while they remained constant at high frequencies. **Structural Reasons**: This can be thought to be a polarization effect. Polarization occurs since the effect of dipole increases the frequencies. - Dielectric constant hexasubstituted cyclotrphosphazene compound containing Cl in ortho position was found as higher compared with the synthesized compound of hexasubstituted cyclotriphosphazene containing F atom in ortho position. **Structural Reasons**: The chlorine atom in the structure increased the polarity, and hence the dipole moment increased. - Dielectric constant of the synthesis compound containing pyridine was recorded to be the highest. **Structural Reasons**: This is due to the presence of the hetero atom in a ring, which can contribute to higher polarization.	Dielectric loss values of the compounds decreased along with the increasing frequency and remained unchanged after some point.	[4]
Mono(4-fluorobenzyl)cyclotriphosphazene derivatives with (dimethylamino)ethoxy and (dimethylamino)propoxy chains. 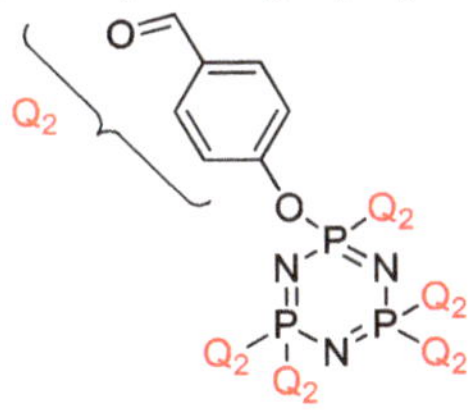	- Sharp decrease of the dielectric constant in the low region of frequency. **Structural Reasons**: This is due to the relaxation process of the diffusion ion. - Compounds containing ethyl groups as an alkyl chain had a higher dielectric constant than those containing methyl groups. **Structural Reasons**: Increasing the side chain resulted in a deteriorated charge transport and unfavorable intermolecular interactions, leading to a decreasing dielectric constant.	-	[91–93]

Table 4. *Cont.*

Compound Name	Dielectric Constant	Dielectric Loss	Ref.
Hexasubstitued cyclotriphosphazene compounds containing chalcone derivatives 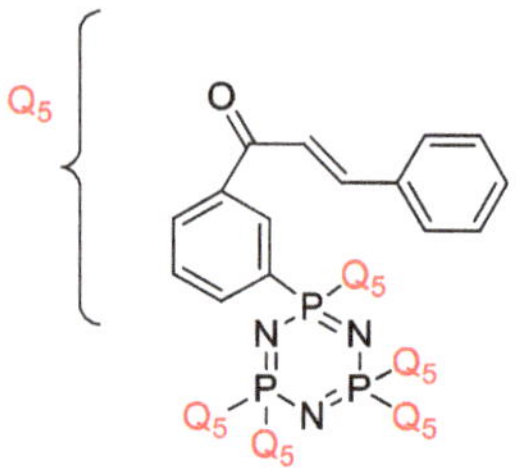 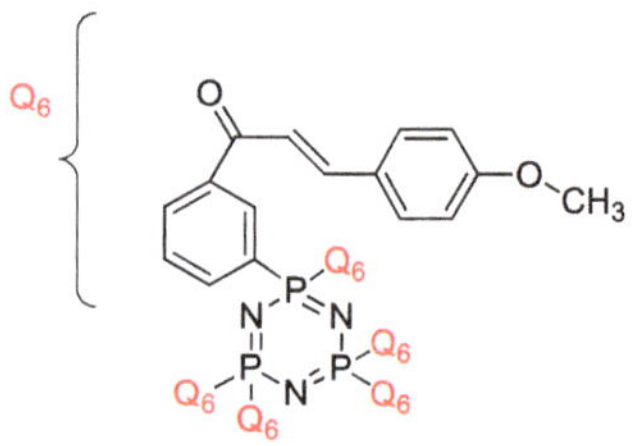 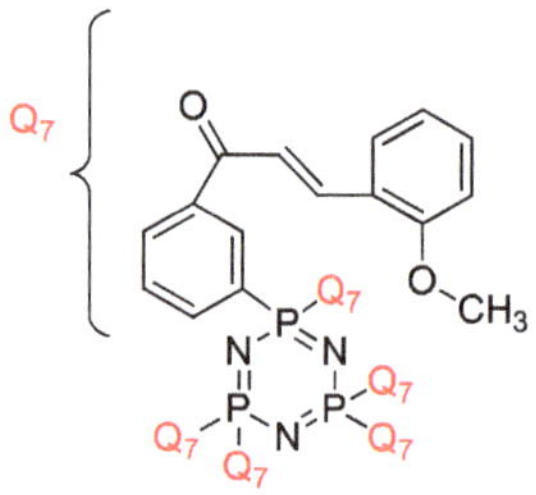 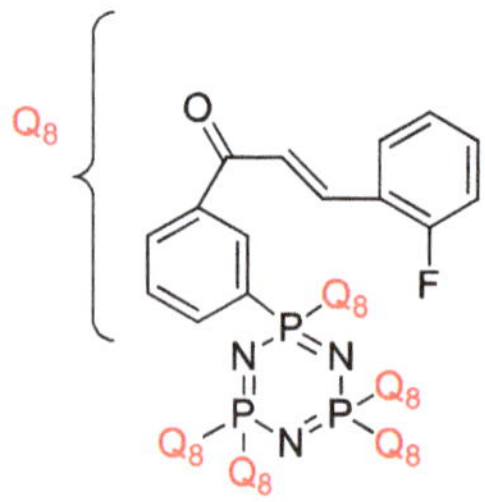 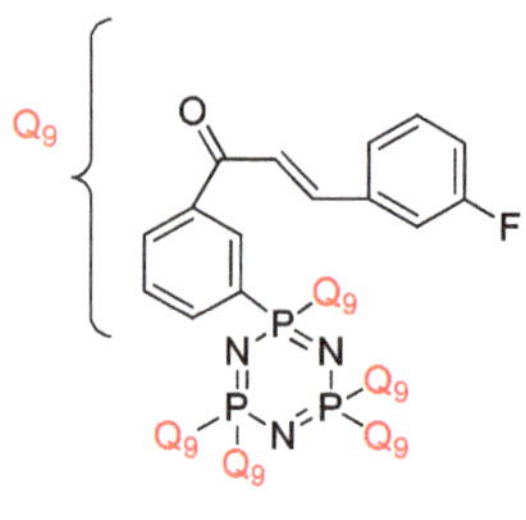	- Dielectric constant decreased with increasing frequency. **Structural Reasons:** This is because the atom in the compound was consistent with the direct charge of the alternating electric field (AEF) at low frequencies that reduce the dielectric constant value at high frequencies.	- Dielectric loss of the sample decreased with increasing frequency. **Structural Reasons**: Interfacial dipoles had less time to orient themselves in the direction of the alternate field. As a result, the low-frequency region is attributed to the contribution of charge accumulation at the interface.	[94]

Table 4. *Cont.*

Compound Name	Dielectric Constant	Dielectric Loss	Ref.
Cyclotri(trifluoroethoxy, acryloyloxy-ethyleneoxy)phosphazene 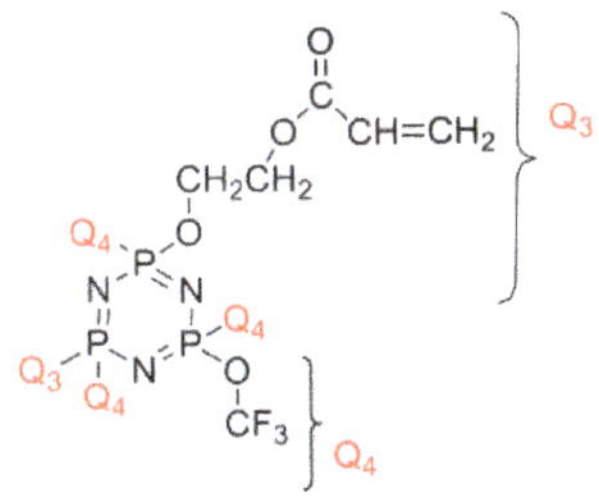	- Dielectric constant increased with increasing temperature and with decreasing frequency. **Structural Reasons:** Due to the rigidity of the polymer matrix.	- Dielectric loss increased with increasing temperature and with decreasing frequency. **Structural Reasons**: Due to the viscosity of the PNF polymer decrease, which favors dipole orientation.	[95]

5. Flame Retardant and Dielectric Properties of Cyclotriphosphazene

The research on cyclotriphosphazene compounds usually only focuses on one property, either flame retardant or dielectric properties. Regarding the development of cyclotriphosphazene, a few articles were published related to studies of a combination of two properties: flame retardant and dielectric properties on cyclotriphosphazene.

Krishbaderi conducted an experiment to synthesize the hexa(aminophenyl)cyclotriphosphazene-modified cyanate ester composite for high temperatures as illustrated in Figure 4 [96]. This study aimed to enhance CEs thermal stability and flame-retardant properties since CE has excellent thermomechanical properties that make it useful in many microelectronic applications. Hexa(aminophenyl)cyclotriphosphazene was synthesized from the reaction between hexachlorocyclotriphosphazene with a mixture of 4-acetamidophenol and calcium carbonate. The CPA/CE composite was prepared, and the flame retardancy and dielectric properties of this compound were measured using the Limiting Oxygen Index (LOI) and an impedance analyzer.

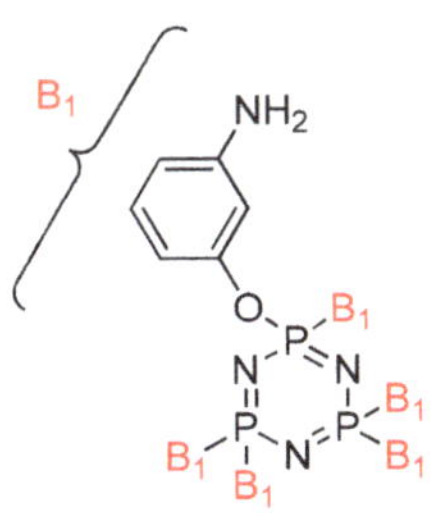

Figure 4. Structure of hexa(aminophenyl)cyclotriphosphazene.

The results obtained showed LOI values of 5%, 10%, and 15% and CPA/CE were 38%, 41%, and 44%, respectively. It showed that the LOI value improved with the increase of both phosphorus and nitrogen content. For the dielectric properties measured, the CPA/CE dielectric constant and dielectric loss were lower than the neat CE matrix. It also concluded that it is quite apparent that incorporating units with low polarity and large molar volume is essential for any molecule to have a low dielectric constant. Data obtained also indicate that CE containing cyclotriphosphazene can be found in micro-electrical applications since it has good thermal properties, flame retardancy, and dielectric behavior.

In addition, Lin and Chang also studied the development of flame retardancy and dielectric properties of cyclotriphosphazene [97]. Their study aimed to synthesize a thermally stable, flame-retardant, and dielectric polymer from cyclotriphosphazene compounds. The hexachlorocyclotriphosphazene was synthesized to form cyclotriphosphazene containing acetylene, phenyl phenoxy, and also cyclotriphosphazene-containing styrene. The LOI and dielectric properties were referred to as the dielectric constant and dielectric losses of these three compounds were determined. From this study, the hexachlorocyclotriphosphazene compound with phenyl phenoxy had a high LOI value and char formation percentage, while for dielectric properties test, the cyclotriphosphazene compound containing styrene showed the higher dielectric properties with 2.40 of dielectric constant and 0.0014 at 1gHz of dielectric losses, respectively. This compound showed excellent dielectric properties and may be useful as a future low dielectric.

There is a study reported on the synthesis of halogen free flame-retardant cyclotriphosphazene nanofibre-reinforced polybenzoxazine/epoxy [98]. Flame retardancy, thermal stability, and dielectric properties (dielectrics constant and dielectric loss) of this compound were measured in this study. The flame retardancy of this compound was measured using LOI and UL-94. Based on the result obtained, the LOI value of this compound increased when the concentration of compound increased, and 0.5, 1.0, and 1.5 wt.% showed 37%, 38%, and 41%, respectively. For dielectric properties, the dielectric constant showed 4.4, 4.0, and 3.4 with the increase of the PZT-fiber loading level of 0.5, 1.0, and 1.5 wt.%, respectively, for 1MHz at 30 degrees. These results indicated that when temperature increases, the dielectric constant will reduce significantly. The dielectric loss will reduce significantly with the increasing temperature, and the reverse trend in this case of dielectric loss was observed at 170. This study also reported that nanocomposites-based PZT fiber inhibits the lower value of the dielectric constant and dielectric loss over a wide range of temperatures, indicating its suitability for effective and thermally stable, flame-retardant dielectric material for micro electric applications.

Aromatic polyimides are usually used as a matrix material in printed circuit boards. When conventional polyimides are exposed to UV, they suffer from severe thermal breakdown due to internal heat generation [99]. Continuous usage and continuous exposure to high energy radiation even cause fire to the host materials. Namely, flame-retardant compounds such as phosphorus and halogen insertion provide heat resistance and preserve the host matrix materials. However, there are few publications available on polymer matrix precursors based on phosphorus and halogen compounds. It was recently discovered that when phosphorus and nitrogen-containing precursors are reinforced or co-polymerized, they improve the host polymers' thermal performance and flame retardancy [100–102]. Revathi et al. worked to develop phosphazene-core-based polyamide and polyhedral oligomeric silsesquioxane (POSS)-reinforced phosphazene polyamide nanocomposites [103]. This study succeeded in developing new POSS, and cyclotriphosphazene nanocomposites with improved thermal stability, antibacterial activity, and low dielectric constants may find high-performance applications in coatings and microelectronics. This polyamide was prepared in three steps, which were the synthesis of hexakis (4-acetomidophenoxy) cyclotriphosphazene, synthesis of hexa (aminophenol) cyclotriphosphazene (PZI), and also the preparation of a neat PZI matrix and POSS/PZI nanocomposites. The developed polyamide and its composites were studied for optical, thermal, and dielectric properties. From the result obtained, the 10 wt.% of POSS/PZI showed higher flame retardancy. This is due to the silica moiety in addition to the flame retardancy of phosphorus and nitrogen atoms. For dielectric behavior, the dielectric constant was measured using an impedance analyzer in the frequency range 100 Hz to 1Mhz. From the result obtained, it was observed that the higher percentage reinforcement of POSS (10 wt.%) into the PZI matrix showed a lower value of the dielectric constant ($k = 2.1$).

In addition, a seminal contribution was made by Cheng et al. in studying the hexasubstituted cyclotriphopshazene compound and its potential properties [49]. This study focused on the synthesized benzimidazoyl-substituted cyclotriphosphazene (BICP) (Figure 5) as a latent flame-retardant curing agent for one-component epoxy resin systems with excellent comprehensive performance. The flame retardancy, dielectric, and also mechanical properties of these compounds were also studied. From the result obtained, the LOI value of these compounds was recorded highest with the value of 33.5%, and the UL-94 test achieved a V-0 rating. This indicated that the BIMP enhanced the flame-retardant properties of epoxy resin without altering other properties when BICP cured EP. For the dielectric properties, the comparison of EP/BICP and EP/BIM was performed to indicate which of these two combinations have potential in the electric and electronic fields. This comparison concluded that EP/BICP has a lower dielectric loss and constant compared with the EP/BIM. The result of this research found clear support for increasing the free volume and/or decreasing the polarization of the materials, and the dielectric constant can be lowered. As a result, it was determined that the lower dielectric constant of EP/BICP thermosets could be attributed to the increased free volume of the EP matrix due to the integration of bulky, inflexible BICP molecules [104]. This simple strategy for preparing one-component flame retardant of epoxy resin with many potential properties could be useful in many fields.

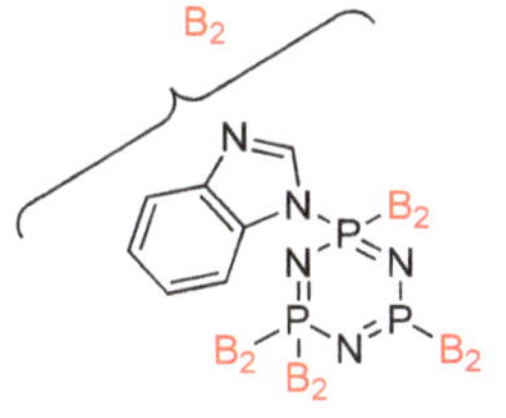

Figure 5. Benzimidazoyl substituted cyclotriphosphazene (BICP).

Table 5 reveals the structural contribution of the hexasubstituted cyclotriphosphazene compound that enhances the flame-retardant and dielectric properties.

Table 5. Structural contribution of the enhancement of flame-retardant and dielectric properties.

Addictive	Polymer Matrix	Properties Studies	Finding(s)	Ref.
Hexa(aminophenyl)cyclo-triphosphazene	Cyanate ester (CE)	Thermal properties Dielectric properties	Limiting oxygen index (LOI) increased. **Structural reasons**: CPA contains phosphorus, and nitrogen compounds exhibit a good flame retardancy by forming phosphorus char that will act as a protecting layer of heat transfer that will reduce combustible gas production. Nitrogen in CPA will enhance the formation of phosphorus char and a nitrogen compound, which is also efficient in preventing oxygen from burning materials [105]. Dielectric loss and dielectric constant of CPA/CE lower than neat CE. **Structural reasons**: CPA has bulky sextet moiety with flexible structure; hence incorporated CE composites with CPA contribute to low dielectric properties.	[96]

Table 5. *Cont.*

Addictive	Polymer Matrix	Properties Studies	Finding(s)	Ref.
Cyclotriphosphazene reinforced polybenzoxazine (PBZ)	EP/PZT	Flame-retardant properties Dielectric properties (Temperature and Volume)	Limiting oxygen index (LOI) increased. **Structural reasons**: Incorporated phosphorus and nitrogen-containing phosphazene fiber into the FBZ/EP matrix contributed to flame retardancy since the phosphorus can act in a condensed phase to promote char formation, shielding the polymer from heat and air. Dielectric loss and dielectric constant increased with decreasing the temperature. **Structural reasons**: It can be ascertained that the composite samples exhibit stable dielectric behavior with regard to variation in temperature. Dielectric constant decreased when the volume of the loaded composite increases. **Structural reasons:** The dielectric of polymer was determined by the three factors in this study: volume, charge distribution, and statically thermal motion of the polar group. The reduction in the dielectric constant may be explained due to the enhancement of the free volume and, to a certain extent, due to the influencing effect of PZT fiber.	[98]
Hexa(aminophenyl)cyclo-triphosphazene (PZA)	PZI matrix	Flame-retardant properties Dielectric constant	Limiting oxygen index (LOI) of PZA/PZI matrix was higher than neat PZI. **Structural Reasons:** Due to the nitrogen and phosphorus in skeletal moiety [105]. In addition, the network structure of PZI matrix also offered an alternative to phosphazene cores. Low dielectric constant obtained from this compound. **Structural Reasons:** Introduction of such porous materials creates a free volume and contributes to obtaining materials that have a low dielectric constant.	[103]
BICP	Epoxy Resin	Fire retardant Dielectric constant	Limiting oxygen index (LOI) of curing EP with BICP improved. **Structural Reasons:** These organophosphorus compounds facilitated the formation of more residual chars with better thermo-oxidative stability. Dielectric constant and loss of EP/BICP thermosets were lower in comparison with those of the EP/BIM10 thermoset over a wide frequency range. **Structural Reasons**: Increased free volume of the EP matrix due to the incorporation of a bulky, rigid BICP molecule might be responsible for the decreased dielectric constant of EP/BICP thermosets. Moreover, in the case of thermosets with the homogeneous network, the dielectric properties of polymers generally depend on the orientation and relaxation of dipoles which are closely related to the movement of polymer-chain segments [106].	[104]

Table 5. *Cont.*

Addictive	Polymer Matrix	Properties Studies	Finding(s)	Ref.
HSSPZ	Epoxy Resin	Flame-retardant Electrical properties	The UL-94 achieved a V-0 rating. **Structural reasons**: This is attributed to the high nitrogen and phosphorus content in the resins. These are believed to have a synergistic effect during burning and yield the higher char amount that can retard the flammability of the materials. The resistivity of the compound increased. **Structural Reasons:** Inorganic ring of alternating phosphorus and nitrogen atoms of cyclotriphosphazene skeleton in the molecular backbone. The skeleton of cyclotriphosphazene is not a conjugated structure [107], and no electron can be supplied. Thus, the addition of CPEP could retain the electrical resistance of EP.	[67]

6. Conclusions

The hexasubstituted cyclotriphosphazene compound revealed that this compound is an excellent compound with fascinating properties, and it has been extensively studied as addictive for flame retardant. This is due to the presence of phosphorus and nitrogen atoms in this compound and the contribution of the linking unit and side chain that helps enhance the flame retardancy of this compound. In addition, hexasubstituted cyclotriphosphazene compounds also have their place in the electrical field, since the dielectric properties of these compounds are starting to receive attention from researchers. This compound has promising potential in the electrical field since it has flame-retardant and dielectric properties that could be valuable in manufacturing electronic devices.

Author Contributions: Conceptualization, S.N.K.U., Z.J. and M.Z.H.M.; writing—original draft preparation, S.N.K.U. and Z.J.; writing—review and editing, S.N.K.U., Z.J. and M.Z.H.M.; visualization, S.N.K.U.; supervision, Z.J. and M.Z.H.M.; project administration, S.N.K.U. and Z.J.; funding acquisition, Z.J. All authors have read and agreed to the published version of the manuscript.

Funding: This research was funded by Universiti Malaysia Sabah (UMS), grant number SGI0133-2020 and the APC was funded by Universiti Malaysia Sabah (UMS).

Institutional Review Board Statement: Not applicable.

Informed Consent Statement: Not applicable.

Data Availability Statement: Not applicable.

Acknowledgments: The authors gratefully acknowledge the technical and financial support from the Universiti Malaysia Sabah (UMS).

Conflicts of Interest: The authors declare no conflict of interest.

References

1. Barberá, J.; Jiménez, J.; Laguna, A.; Oriol, L.; Pérez, S.; Serrano, J.L. Cyclotriphosphazene as a dendritic core for the preparation of columnar supermolecular liquid crystals. *Chem. Mater.* **2006**, *18*, 5437–5445. [CrossRef]
2. Khurana, J.M.; Chauhan, S.; Bansal, G. Facile hydrolysis of esters with KOH-Methanol at ambient temperature. *Monatshefte Chem.* **2004**, *135*, 83–87. [CrossRef]
3. Jamain, Z.; Khairuddean, M.; Saidin, S.A. Synthesis and characterization of 1,4-phenylenediamine derivatives containing hydroxyl and cyclotriphosphazene as terminal group. *J. Mol. Struct.* **2019**, *1186*, 293–302. [CrossRef]
4. Koran, K.; Tekin, Ç.; Biryan, F.; Tekin, S.; Sandal, S.; Görgülü, A.O. Synthesis, structural and thermal characterizations, die-lectric properties and in vitro cytotoxic activities of new 2,2,4,4-tetra(4′-oxy-substituted-chalcone)-6,6-diphenyl cyclotri-phosphazene derivatives. *Med. Chem. Res.* **2017**, *26*, 962–974. [CrossRef]

5. Moriya, K.; Suzuki, T.; Yano, S.; Kajiwara, M. Liquid crystalline phase transitions in cyclotriphosphazenes with different mesogenic moieties in the side chains. *Liq. Cryst.* **1995**, *19*, 711–713. [CrossRef]
6. Allcock, H.R. The crucial role of inorganic ring chemistry in the development of new polymers. *Phosphorus Sulfur Silicon Relat. Elem.* **2004**, *179*, 661–671. [CrossRef]
7. Bose, S.; Mukherjee, M.; Das, C.K.; Saxena, A.K. Effect of polyphosphazene elastomer on the compatibility and properties of PES/TLCP composites. *Polym. Compos.* **2009**, *31*, 543–552. [CrossRef]
8. Chaubal, M.V.; Gupta, A.S.; Lopina, S.T.; Bruley, D.F. Polyphosphates and other phosphorus-containing polymers for drug delivery applications. *Crit. Rev. Ther. Drug Carr. Syst.* **2003**, *20*, 295–315. [CrossRef] [PubMed]
9. Allcock, H. *Phosphorus-Nitrogen Compounds: Cyclic, Linear, and High Polymeric Systems*; Elsevier: Saint Louis, MO, USA, 2012.
10. Allcock, H.R.; Steely, L.B.; Singh, A. Hydrophobic and superhydrophobic surfaces from polyphosphazenes. *Polym. Int.* **2006**, *55*, 621–625. [CrossRef]
11. Allcock, H.R.; Wood, R.M. Design and synthesis of ion-conductive polyphosphazenes for fuel cell applications: Review. *J. Polym. Sci. Part B Polym. Phys.* **2006**, *44*, 2358–2368. [CrossRef]
12. Gunatillake, P.; Mayadunne, R.; Adhikari, R. Recent developments in biodegradable synthetic polymers. *Biotechnol. Annu. Rev.* **2006**, *12*, 301–347. [CrossRef] [PubMed]
13. Honarkar, H.; Rahimi, A. Applications of Inorganic Polymeric Materials, III: Polyphosphazenes. *Monatshefte Chemie Chem. Mon.* **2007**, *138*, 923–933. [CrossRef]
14. Wycisk, R.; Pintauro, P.N. Polyphosphazene membranes for fuel cells. *Adv. Polym. Sci.* **2008**, 157–183. [CrossRef]
15. Doğan, S.; Tümay, S.O.; Balci, C.M.; Yeşilot, S.; Beşli, S. Synthesis of new cyclotriphosphazene derivatives bearing Schiff bases and their thermal and absorbance properties. *Turk. J. Chem.* **2020**, *44*, 31–47. [CrossRef] [PubMed]
16. Zhao, B.; Liang, W.-J.; Wang, J.-S.; Li, F.; Liu, Y.-Q. Synthesis of a novel bridged-cyclotriphosphazene flame retardant and its application in epoxy resin. *Polym. Degrad. Stab.* **2016**, *133*, 162–173. [CrossRef]
17. Levchik, G.F.; Grigoriev, Y.V.; Balabanovich, A.I.; Levchik, S.V.; Klatt, M. Phosphorus–nitrogen containing fire retardants for poly(butylene terephthalate). *Polym. Int.* **2000**, *49*, 1095–1100. [CrossRef]
18. Fontenot, K.R.; Nguyen, M.M.; Al-Abdul-Wahid, M.S.; Easson, M.W.; Chang, S.; Lorigan, G.A.; Condon, B.D. The thermal degradation pathway studies of a phosphazene derivative on cotton fabric. *Polym. Degrad. Stab.* **2015**, *120*, 32–41. [CrossRef]
19. Mao, Z.; Li, J.; Pan, F.; Zeng, X.; Zhang, L.; Zhong, Y.; Sui, X.; Xu, H. High-temperature auto-cross-linking cyclotriphosphaznene: Synthesis and application in flame retardance and antidripping poly(ethylene terephthalate). *Ind. Eng. Chem. Res.* **2015**, *54*, 3788–3799. [CrossRef]
20. Yang, R.; Hu, W.; Xu, L.; Song, Y.; Li, J. Synthesis, mechanical properties and fire behaviors of rigid polyurethane foam with a reactive flame retardant containing phosphazene and phosphate. *Polym. Degrad. Stab.* **2015**, *122*, 102–109. [CrossRef]
21. Allcock, H.R. New approaches to hybrid polymers that contain phosphazene rings. *J. Inorg. Organomet. Polym. Mater.* **2007**, *17*, 349–359. [CrossRef]
22. Kumar, D.; Fohlen, G.M.; Parker, J.A. High-strength fire- and heat-resistant imide resins containing cyclotriphosphazene and hexafluoroisopropylidene groups. *J. Polym. Sci. Polym. Chem. Ed.* **1984**, *22*, 927–943. [CrossRef]
23. El Gouri, M.; El Bachiri, A.; Hegazi, S.E.; Rafik, M.; Elharfi, A. Thermal degradation of a reactive flame retardant based on cyclotriphosphazene and its blend with DGEBA epoxy resin. *Polym. Degrad. Stab.* **2009**, *94*, 2101–2106. [CrossRef]
24. El Gouri, M.; El Bachiri, A.; Hegazi, S.E.; Ziraoui, R.; Rafik, M.; El Harfi, A. A phosphazene compound multipurpose application -Composite material precursor and reactive flame retardant for epoxy resin materials. *J. Mater. Environ. Sci.* **2011**, *2*, 319–334.
25. Jiang, P.; Gu, X.; Zhang, S.; Wu, S.; Zhao, Q.; Hu, Z. Synthesis, characterization, and utilization of a novel phosphorus/nitrogen-containing flame retardant. *Ind. Eng. Chem. Res.* **2015**, *54*, 2974–2982. [CrossRef]
26. Allen, C.W. The use of phosphazenes as fire resistant materials. *J. Fire Sci.* **1993**, *11*, 320–328. [CrossRef]
27. Jiang, J.; Wang, Y.; Luo, Z.; Qi, T.; Qiao, Y.; Zou, M.; Wang, B. Design and application of highly efficient flame retardants for polycarbonate combining the advantages of cyclotriphosphazene and silicone oil. *Polymers* **2019**, *11*, 1155. [CrossRef] [PubMed]
28. Liu, Q.; Wang, D.; Li, Z.; Li, Z.; Peng, X.; Liu, C.; Zhang, Y.; Zheng, P. Recent developments in the flame-retardant system of epoxy resin. *Materials* **2020**, *13*, 2145. [CrossRef] [PubMed]
29. Ehsani, M.; Borsi, H.; Gockenbach, E.; Morshedian, J.; Bakhshandeh, G. An investigation of dynamic mechanical, thermal, and electrical properties of housing materials for outdoor polymeric insulators. *Eur. Polym. J.* **2004**, *40*, 2495–2503. [CrossRef]
30. Liu, Y.L.; Chiu, Y.C.; Chen, T.Y. Phosphorus-containing polyaryloxydiphenylsilanes with high flame retardance arising from a phosphorus-silicon synergic effect. *Polym. Int.* **2003**, *52*, 1256–1261. [CrossRef]
31. Huang, G.; Huo, S.; Xu, X.; Chen, W.; Jin, Y.; Li, R.; Song, P.; Wang, H. Realizing simultaneous improvements in mechanical strength, flame retardancy and smoke suppression of ABS nanocomposites from multifunctional graphene. *Compos. Part B Eng.* **2019**, *177*, 107377. [CrossRef]
32. Mu, X.; Wang, D.; Pan, Y.; Cai, W.; Song, L.; Hu, Y. A facile approach to prepare phosphorus and nitrogen containing macromolecular covalent organic nanosheets for enhancing flame retardancy and mechanical property of epoxy resin. *Compos. Part B Eng.* **2019**, *164*, 390–399. [CrossRef]
33. Zhou, X.; Qiu, S.; Mu, X.; Zhou, M.; Cai, W.; Song, L.; Xing, W.; Hu, Y. Polyphosphazenes-based flame retardants: A review. *Compos. Part B Eng.* **2020**, *202*, 108397. [CrossRef]

34. Shin, Y.J.; Shin, M.J.; Shin, J.S. Flame retardant properties of cyclotriphosphazene derivatives for ABS. *Polym. Polym. Compos.* **2018**, *26*, 309–314. [CrossRef]
35. Zhang, X.; Zhong, Y.; Mao, Z.-P. The flame retardancy and thermal stability properties of poly (ethylene terephthalate)/hexakis (4-nitrophenoxy) cyclotriphosphazene systems. *Polym. Degrad. Stab.* **2012**, *97*, 1504–1510. [CrossRef]
36. Chen, S.; Su, M.; Lan, J.W.; Zheng, Q.K. Preparation and Properties of halogen-free flame retardant blending modification polyester. *Adv. Mater. Res.* **2012**, *463–464*, 515–518. [CrossRef]
37. Kuan, J.-F.; Lin, K.-F. Synthesis of hexa-allylamino-cyclotriphosphazene as a reactive fire retardant for unsaturated polyesters. *J. Appl. Polym. Sci.* **2003**, *91*, 697–702. [CrossRef]
38. Yang, S.; Wang, J.; Huo, S.; Wang, J.; Tang, Y. Synthesis of a phosphorus/nitrogen-containing compound based on maleimide and cyclotriphosphazene and its flame-retardant mechanism on epoxy resin. *Polym. Degrad. Stab.* **2016**, *126*, 9–16. [CrossRef]
39. Xu, M.-J.; Xu, G.-R.; Leng, Y.; Li, B. Synthesis of a novel flame retardant based on cyclotriphosphazene and DOPO groups and its application in epoxy resins. *Polym. Degrad. Stab.* **2016**, *123*, 105–114. [CrossRef]
40. Zhang, S.; Li, Y.; Guo, J.; Gu, L.; Li, H.; Fei, B.; Sun, J.; Gu, X. Preparation of hexakis (4-aldehyde phenoxy) cyclotriphosphazene grafted kaolinite and its synergistic fire resistance in poly (butylene succinate). *Polym. Compos.* **2019**, *41*, 1024–1035. [CrossRef]
41. Yang, R.; Wang, B.; Han, X.; Ma, B.; Li, J. Synthesis and characterization of flame retardant rigid polyurethane foam based on a reactive flame retardant containing phosphazene and cyclophosphonate. *Polym. Degrad. Stab.* **2017**, *144*, 62–69. [CrossRef]
42. Shin, Y.J.; Ham, Y.R.; Kim, S.H.; Lee, D.H.; Kim, S.B.; Park, C.S.; Yoo, Y.M.; Kim, J.G.; Kwon, S.H.; Shin, J.S. Application of cyclophosphazene derivatives as flame retardants for ABS. *J. Ind. Eng. Chem.* **2010**, *16*, 364–367. [CrossRef]
43. Qian, L.-J.; Ye, L.-J.; Xu, G.-Z.; Liu, J.; Guo, J.-Q. The non-halogen flame retardant epoxy resin based on a novel compound with phosphaphenanthrene and cyclotriphosphazene double functional groups. *Polym. Degrad. Stab.* **2011**, *96*, 1118–1124. [CrossRef]
44. Lan, Y.; Li, D.; Yang, R.; Liang, W.; Zhou, L.; Chen, Z. Computer simulation study on the compatibility of cyclotriphosphazene containing aminopropylsilicone functional group in flame retarded polypropylene/ammonium polyphosphate composites. *Compos. Sci. Technol.* **2013**, *88*, 9–15. [CrossRef]
45. Zhang, X.; Akram, R.; Zhang, S.; Ma, H.; Wu, Z.; Wu, D. Hexa(eugenol)cyclotriphosphazene modified bismaleimide resins with unique thermal stability and flame retardancy. *React. Funct. Polym.* **2017**, *113*, 77–84. [CrossRef]
46. Cheng, J.; Wang, J.; Yang, S.; Zhang, Q.; Hu, Y.; Ding, G.; Huo, S. Aminobenzothiazole-substituted cyclotriphosphazene de-rivative as reactive flame retardant for epoxy resin. *React. Funct. Polym.* **2020**, *146*, 104412. [CrossRef]
47. Zarybnicka, L.; Machotova, J.; Kopecka, R.; Sevcik, R.; Hudakova, M.; Pokorny, J.; Sal, J. Effect of Cyclotriphosphazene-Based Curing Agents on the Flame Resistance of Epoxy Resins. *Polymers* **2020**, *13*, 8. [CrossRef]
48. Li, J.; Pan, F.; Zeng, X.; Xu, H.; Zhang, L.; Zhong, Y.; Sui, X.; Mao, Z. The flame-retardant properties and mechanisms of poly(ethylene terephthalate)/hexakis (para-allyloxyphenoxy) cyclotriphosphazene systems. *J. Appl. Polym. Sci.* **2015**, *132*, 42711. [CrossRef]
49. Cheng, J.; Wang, J.; Yang, S.; Zhang, Q.; Huo, S.; Zhang, Q.; Hu, Y.; Ding, G. Benzimidazolyl-substituted cyclotriphosphazene derivative as latent flame-retardant curing agent for one-component epoxy resin system with excellent comprehensive performance. *Compos. Part B Eng.* **2019**, *177*, 107440. [CrossRef]
50. Lv, M.; Yao, C.; Yang, D.; Zeng, H. Synthesis of a melamine-cyclotriphosphazene derivative and its application as flame retardant on cotton gauze. *J. Appl. Polym. Sci.* **2016**, *133*. [CrossRef]
51. Zhou, L.; Zhang, G.; Li, J.; Jing, Z.; Qin, J.; Feng, Y. The flame retardancy and thermal stability properties of flame-retarded epoxy resins based on α-hydroxyphosphonate cyclotriphosphazene. *J. Therm. Anal. Calorim.* **2017**, *129*, 1667–1678. [CrossRef]
52. Ai, L.; Chen, S.; Zeng, J.; Liu, P.; Liu, W.; Pan, Y.; Liu, D. Synthesis and flame retardant properties of cyclophosphazene de-rivatives containing boron. *Polym. Degrad. Stab.* **2018**, *155*, 250–261. [CrossRef]
53. Fang, Y.; Zhou, X.; Xing, Z.; Wu, Y. An effective flame retardant for poly(ethylene terephthalate) synthesized by phosphaphenanthrene and cyclotriphosphazene. *J. Appl. Polym. Sci.* **2017**, *134*, 45246. [CrossRef]
54. Zhang, X.; Zhang, L.; Wu, Q.; Mao, Z. The influence of synergistic effects of hexakis (4-nitrophenoxy) cyclotriphosphazene and POE-g-MA on anti-dripping and flame retardancy of PET. *J. Ind. Eng. Chem.* **2012**, *19*, 993–999. [CrossRef]
55. Zhou, L.; Zhang, G.; Feng, Y.; Zhang, H.; Li, J.; Shi, X. Design of a self-healing and flame-retardant cyclotriphosphazene-based epoxy vitrimer. *J. Mater. Sci.* **2018**, *53*, 7030–7047. [CrossRef]
56. Dutkiewicz, M.; Przybylak, M.; Januszewski, R.; Maciejewski, H. Synthesis and flame retardant efficacy of hex-akis(3-(triethoxysilyl)propyloxy)cyclotriphosphazene/silica coatings for cotton fabrics. *Polym. Degrad. Stab.* **2018**, *148*, 10–18. [CrossRef]
57. Devaraju, S.; Selvi, M.; Alagar, M. Synthesis and characterization of thermally stable and flame retardant hexakis(4-aminophenoxy)cyclotriphosphazene-based polyimide matrices. *Int. J. Polym. Anal. Charact.* **2018**, *23*, 29–37. [CrossRef]
58. Zhou, L.; Zhang, G.; Yang, S.; Yang, L.; Cao, J.; Yang, K. The synthesis, curing kinetics, thermal properties and flame rertardancy of cyclotriphosphazene-containing multifunctional epoxy resin. *Thermochim. Acta* **2019**, *680*, 178348. [CrossRef]
59. MJ, X.; Ma, K.; DW, J.; JX, Z.; Zhao, M.; XK, G.; Shao, Q.; Wujcik, E.; Li, B.; ZH, G. Hexa-[4-(glycidyloxycarbonyl) phenoxy]cyclotriphosphazene chain extender for preparing high-performance flame retardant polyamide 6 composites. *Polymer* **2018**, *146*, 63–72.
60. Qi, J.; Wen, Q.; Zhu, J. Synergistic effect of intumescent flame retardant system consisting of hexophenoxy cyclotriphosphazene and ammonium polyphosphate on methyl ethyl silicone rubber. *Mater. Lett.* **2019**, *249*, 62–65. [CrossRef]

61. Guo, Y.; Qiu, J.-J.; Tang, H.-Q.; Liu, C.-M. High transmittance and environment-friendly flame-resistant optical resins based on poly(methyl methacrylate) and cyclotriphosphazene derivatives. *J. Appl. Polym. Sci.* **2011**, *121*, 727–734. [CrossRef]
62. Xu, J.; He, Z.; Wu, W.; Ma, H.; Xie, J.; Qu, H.; Jiao, Y. Study of thermal properties of flame retardant epoxy resin treated with hexakis[p-(hydroxymethyl)phenoxy]cyclotriphosphazene. *J. Therm. Anal. Calorim.* **2013**, *114*, 1341–1350. [CrossRef]
63. Wang, J.; Su, X.; Mao, Z. The flame retardancy and thermal property of poly (ethylene terephthalate)/cyclotriphosphazene modified by montmorillonite system. *Polym. Degrad. Stab.* **2014**, *109*, 154–161. [CrossRef]
64. Xu, M.-J.; Liu, C.; Ma, K.; Leng, Y.; Li, B. Effect of surface chemical modification for aluminum hypophosphite with hexa-(4-aldehyde-phenoxy)-cyclotriphosphazene on the fire retardancy, water resistance, and thermal properties for polyamide 6. *Polym. Adv. Technol.* **2017**, *28*, 1382–1395. [CrossRef]
65. Yang, Y.; Kong, W.; Wang, Y.; Cai, X. Synthesis of tris(phenoxy)trifluorocyclotriphosphazenes and study of its effects on the flammable, thermal, optical, and mechanical properties of bisphenol-A polycarbonate. *J. Therm. Anal. Calorim.* **2015**, *122*, 805–816. [CrossRef]
66. Li, J.; Yan, X.; Zeng, X.; Yi, L.; Xu, H.; Mao, Z. Effect of trisilanolphenyl-POSS on rheological, mechanical, and flame-retardant properties of poly(ethylene terephthalate)/cyclotriphosphazene systems. *J. Appl. Polym. Sci.* **2018**, *135*, 45912. [CrossRef]
67. Liu, F.; Wei, H.; Huang, X.; Zhang, J.; Zhou, Y.; Tang, X. Preparation and Properties of Novel Inherent Flame-Retardant Cyclotriphosphazene-Containing Epoxy Resins. *J. Macromol. Sci. Part B* **2010**, *49*, 1002–1011. [CrossRef]
68. Parra, M.; Alderete, J.B.; Zuñﬁiga, C.; Gallardo, H.; Hidalgo, P.; Vergara, J.; Ndez, S.H. Azo compounds and Schiff 's bases derived from 5-(4-pyridyl)-2-amino-1,3,4-thiadiazole: Synthesis, mesomorphic properties and structural study by semi-empirical calculations. *Liq. Cryst.* **2001**, *28*, 1659–1666. [CrossRef]
69. Jamain, Z.; Omar, N.F.; Khairuddean, M. Synthesis and Determination of Thermotropic Liquid Crystalline Behavior of Cinnamaldehyde-Based Molecules with Two Schiff Base Linking Units. *Molecules* **2020**, *25*, 3780. [CrossRef]
70. Jamain, Z.; Khairuddean, M.; Zulbaharen, N.N.; Chung, T.K. Synthesis, characterization and determination of mesophase transition of azo-azomethine derivatives with different terminal chain lengths. *Malaysian J. Chem.* **2019**, *21*, 73–85.
71. Jamain, Z.; Khairuddean, M.; Guan-Seng, T. Liquid-crystal and fire-retardant properties of new hexasubstituted cyclotriphosphazene compounds with two Schiff base linking units. *Molecules* **2020**, *25*, 2122. [CrossRef]
72. Abd El-Wahab, H.; Abd El-Fattah, M.; El-alfy, H.M.Z.; Owda, M.E.; Lin, L.; Hamdy, I. Synthesis and characterisation of sulphonamide (Schiff base) ligand and its copper metal complex and their efficiency in polyurethane varnish as flame retardant and antimicrobial surface coating additives. *Prog. Org. Coat.* **2020**, *142*, 105577. [CrossRef]
73. Jamain, Z.; Khairuddean, M.; Guan-Seng, T. Synthesis of New Star-Shaped Liquid Crystalline Cyclotriphosphazene Derivatives with Fire Retardancy Bearing Amide-Azo and Azo-Azo Linking Units. *Int. J. Mol. Sci.* **2020**, *21*, 4267. [CrossRef] [PubMed]
74. Jamain, Z.; Khairuddean, M.; Guan-Seng, T. Synthesis of novel liquid crystalline and fire retardant molecules based on six-armed cyclotriphosphazene core containing Schiff base and amide linking units. *RSC Adv.* **2020**, *10*, 28918–28934. [CrossRef]
75. Jamain, Z.; Khairuddean, M.; Guan-Seng, T.; Rahman, A.B.A. Synthesis, characterization and mesophase transition of hexasubstituted cyclotriphosphazene molecules with Schiff base and azo linking units and determination of their fire retardant properties. *Macromol. Res.* **2021**, *29*, 331–341. [CrossRef]
76. Shariatinia, Z.; Javeri, N.; Shekarriz, S. Flame retardant cotton fibers produced using novel synthesized halogen-free phosphoramide nanoparticles. *Carbohydr. Polym.* **2015**, *118*, 183–198. [CrossRef]
77. Alidağı, H.A.; Gırgıç, Ö.M.; Zorlu, Y.; Hacıvelioğlu, F.; Çelik, S.Ü.; Bozkurt, A.; Kılıç, A.; Yeşilot, S. Synthesis and proton conductivity of azole-substituted cyclic and polymeric phosphazenes. *Polymer* **2013**, *54*, 2250–2256. [CrossRef]
78. Inoue, K.; Yamauchi, T.; Itoh, T.; Ihara, E. Ionic conductivity of cross-linked polymethacrylate derivatives/cyclophosphazenes/Li+ Salt Complexes. *J. Inorg. Organomet. Polym. Mater.* **2007**, *17*, 367–375. [CrossRef]
79. Koran, K.; Özen, F.; Biryan, F.; Görgülü, A.O. Synthesis, structural characterization and dielectric behavior of new oxime-cyclotriphosphazene derivatives. *J. Mol. Struct.* **2016**, *1105*, 135–141. [CrossRef]
80. Jiang, P.; Gu, X.; Zhang, S.; Sun, J.; Wu, S.; Zhao, Q. Syntheses and characterization of four phosphaphenanthrene and phosphazene-based flame retardants. *Phosphorus Sulfur Silicon Relat. Elem.* **2014**, *189*, 1811–1822. [CrossRef]
81. Çoşut, B. Highly efficient energy transfer in BODIPY–pyrene decorated cyclotriphosphazene. *Dye. Pigment.* **2014**, *100*, 11–16. [CrossRef]
82. Bolink, H.J.; Barea, E.M.; Costa, R.D.; Coronado, E.; Sudhakar, S.; Zhen, C.; Sellinger, A. Efficient blue emitting organic light emitting diodes based on fluorescent solution processable cyclic phosphazenes. *Org. Electron.* **2008**, *9*, 155–163. [CrossRef]
83. Jamain, Z.; Khairuddean, M.; Loh, M.L.; Manaff, N.L.A.; Makmud, M.Z.H. Synthesis and characterization of hexasubstituted cyclotriphosphazene derivatives with azo linking units. *Malaysian J. Chem.* **2020**, *22*, 125–140.
84. Fratini, S.; Morpurgo, A.F.; Ciuchi, S. Electron–phonon and electron–electron interactions in organic field effect transistors. *J. Phys. Chem. Solids* **2008**, *69*, 2195–2198. [CrossRef]
85. Haughey, A.-M.; Guilhabert, B.; Kanibolotsky, A.L.; Skabara, P.J.; Burley, G.A.; Dawson, M.D.; Laurand, N. An organic semiconductor laser based on star-shaped truxene-core oligomers for refractive index sensing. *Sens. Actuat. B Chem.* **2013**, *185*, 132–139. [CrossRef]
86. Tsai, Y.; Juang, F.; Yang, T.; Yokoyama, M.; Ji, L.; Su, Y. Effects of different buffer layers in flexible organic light-emitting diodes. *J. Phys. Chem. Solids* **2008**, *69*, 764–768. [CrossRef]

87. Bokolia, R.; Thakur, O.; Rai, V.K.; Sharma, S.; Sreenivas, K. Dielectric, ferroelectric and photoluminescence properties of Er3+ doped Bi4Ti3O12 ferroelectric ceramics. *Ceram. Int.* **2015**, *41*, 6055–6066. [CrossRef]
88. Koran, K.; Özen, F.; Biryan, F.; Demirelli, K.; Görgülü, A.O. Eu+3-doped chalcone substituted cyclotriphosphazenes: Synthesis, characterizations, thermal and dielectrical properties. *Inorg. Chim. Acta* **2016**, *450*, 162–169. [CrossRef]
89. Lakshmikandhan, T.; Sethuraman, K.; Chandramohan, A.; Alagar, M. Development of phosphazene imine modified epoxy nanocomposites for low dielectric, antibacterial activity and UV shielding applications. *Polym. Compos.* **2015**, *38*, E24–E33. [CrossRef]
90. Fan, J.; Hu, X.; Yue, C.Y. Dielectric properties of self-catalytic interpenetrating polymer network based on modified bismaleimide and cyanate ester resins. *J. Polym. Sci. Part B Polym. Phys.* **2003**, *41*, 1123–1134. [CrossRef]
91. Akbaş, H.; Karadağ, A.; Destegül, A.; Çakırlar, Ç.; Yerli, Y.; Tekin, K.C.; Malayoglu, U.; Kılıç, Z. Synthesis, and spectroscopic, thermal and dielectric properties of phosphazene based ionic liquids: OFET application and tribological behavior. *New J. Chem.* **2019**, *43*, 2098–2110. [CrossRef]
92. Fu, J.Y. On the theory of the universal dielectric relaxation. *Philos. Mag.* **2014**, *94*, 1788–1815. [CrossRef]
93. Sangoro, J.; Iacob, C.; Naumov, S.; Valiullin, R.; Rexhausen, H.; Hunger, J.; Buchner, R.; Strehmel, V.; Kärger, J.; Kremer, F. Diffusion in ionic liquids: The interplay between molecular structure and dynamics. *Soft Matter* **2011**, *7*, 1678–1681. [CrossRef]
94. Koran, K.; Özen, F.; Torğut, G.; Pıhtılı, G.; Çil, E.; Orhan Görgülü, A.; Arslan, M. Synthesis, characterization and dielectric properties of phosphazenes containing chalcones. *Polyhedron* **2014**, *79*, 213–220. [CrossRef]
95. Kimura, T.; Kajiwara, M. Electrical properties of a cyclotriphosphazene matrix polymer: Cyclotri(trifluoroethoxy, acryloyloxy-ethyleneoxy)phosphazene matrix polymer. *J. Inorg. Organomet. Polym. Mater.* **1992**, *2*, 431–441. [CrossRef]
96. Krishnadevi, K.; Grace, A.N.; Alagar, M.; Selvaraj, V. Development of hexa (aminophenyl)cyclotriphosphazene-modified cyanate ester composites for high-temperature applications. *High Perform. Polym.* **2013**, *26*, 89–96. [CrossRef]
97. Lim, H.; Chang, J.Y. Thermally stable and flame retardant low dielectric polymers based on cyclotriphosphazenes. *J. Mater. Chem.* **2010**, *20*, 749–754. [CrossRef]
98. Selvi, M.; Devaraju, S.; Alagar, M. Cyclotriphosphazene nanofiber-reinforced polybenzoxazine/epoxy nanocomposites for low dielectric and flame-retardant applications. *Polym. Bull.* **2018**, *76*, 3785–3801. [CrossRef]
99. Wang, S.; Ahmad, Z.; Mark, J.E. Polyimide-Silica Hybrid Materials Modified by Incorporation of an Organically Substituted Alkoxysilane. *Chem. Mater.* **1994**, *6*, 943–946. [CrossRef]
100. Lu, S.-Y.; Hamerton, I. Recent developments in the chemistry of halogen-free flame retardant polymers. *Prog. Polym. Sci.* **2002**, *27*, 1661–1712. [CrossRef]
101. Choi, J.; Yee, A.F.; Laine, R.M. Organic/inorganic hybrid composites from cubic silsesquioxanes. Epoxy resins of octa(dimethylsiloxyethylcyclohexylepoxide) silsesquioxane. *Macromolecules* **2003**, *36*, 5666–5682. [CrossRef]
102. Chen, D.-Q.; Wang, Y.-Z.; Hu, X.-P.; Wang, D.-Y.; Qu, M.-H.; Yang, B. Flame-retardant and anti-dripping effects of a novel char-forming flame retardant for the treatment of poly(ethylene terephthalate) fabrics. *Polym. Degrad. Stab.* **2005**, *88*, 349–356. [CrossRef]
103. Revathi, R.; Prabunathan, P.; Alagar, M. Synthesis and studies on phosphazene core-based POSS-reinforced polyimide nanocomposites. *Polym. Bull.* **2018**, *76*, 387–407. [CrossRef]
104. Chang, H.C.; Lin, H.T.; Lin, C.H. Benzoxazine-based phosphinated bisphenols and their application in preparing flame-retardant, low dielectric cyanate ester thermosets. *Polym. Chem.* **2012**, *3*, 970–978. [CrossRef]
105. Xiong, Y.; Jiang, Z.; Xie, Y.; Zhang, X.; Xu, W. Development of a DOPO-containing melamine epoxy hardeners and its thermal and flame-retardant properties of cured products. *J. Appl. Polym. Sci.* **2012**, *127*, 4352–4358. [CrossRef]
106. Chen, X.; Gu, A.; Liang, G.; Yuan, L.; Zhuo, D.; Hu, J. Novel low phosphorus-content bismaleimide resin system with outstanding flame retardancy and low dielectric loss. *Polym. Degrad. Stab.* **2012**, *97*, 698–706. [CrossRef]
107. Lund, L.G.; Paddock, N.L.; Proctor, J.E.; Searle, H.T. 514. Phosphonitrilic derivatives. Part I. The preparation of cyclic and linear phosphonitrilic chlorides. *J. Chem. Soc.* **1960**, 2542–2547. [CrossRef]

polymers

MDPI

Article

Failure Investigation of Layered LFT SB1plus Package after Ballistic Tests for Level IIA

Cătălin Pîrvu [1,2,*] and Lorena Deleanu [2,*]

1 National Institute of Aerospace Research "Elie Carafoli" INCAS, 061126 Bucharest, Romania
2 Faculty of Engineering, "Dunărea de Jos" University of Galati, 800008 Galati, Romania
* Correspondence: pirvu.catalin@incas.ro (C.P.); lorena.deleanu@ugal.ro (L.D.)

Abstract: The main objective of this study focuses on designing and testing body protection systems using advanced materials based on aramid fibers, for high impact speeds of up to 420 ± 10 m/s. Ballistic applications of aramid fiber-based composites mostly include soft body armors. The investigation of the failure mechanisms identifies issues of protective fabrics, major challenges and technological problems for efficient development of these systems. The authors present an investigation on the failure processes and destructive stages of a ballistic package made of successive layers of LFT SB1plus, a trade name for a multiaxial fabric by Twaron Laminated Fabric Technology (LFT), taking into account the particular test conditions from NIJ Standard-0101.06 Ballistic Resistance of Body Armor. The main parameter of interest was the backface signature (BFS), but also details of projectile arrest and SEM investigation could offer arguments for using this material for individual protection. For the reported tests, the maximum and minimum values for BFS were 12 mm and 24 mm, the mean value being 18.66 mm and the standard deviation being 3.8 mm.

Keywords: aramid fiber; ballistic test; failure mechanism

Citation: Pîrvu, C.; Deleanu, L. Failure Investigation of Layered LFT SB1plus Package after Ballistic Tests for Level IIA. *Polymers* **2021**, *13*, 2912. https://doi.org/10.3390/polym13172912

Academic Editors: Beata Podkościelna and Andrzej Puszka

Received: 8 August 2021
Accepted: 26 August 2021
Published: 29 August 2021

1. Introduction

For hundreds of years, metallic materials have been used not only for body armor, but also for the protection of larger objects, such as vehicles, which is called "heavy protection". However, just a few decades ago, at the end of World War II, lighter solutions emerged, especially for military personnel, in the form of nylon ballistic vests. However, those did not come close to the current level of ballistic protection offered by aramid fibers, yarns and fabrics, included in personal armors. Another advantage of ultra-fine polymer filaments (not only aramid [1]) is that they offer extraordinarily flexible fabrics, which favors a high level of comfort for the wearer.

Ballistic applications of aramid fiber-based composites mostly include soft body armors. The mechanical properties of the aramids and the ballistic effects on its fabrics and composites have been investigated in several studies [2–5] that involved both experimental testing and FEM (finite element method) [6–9] and established the effectiveness of the ballistic protecting system and the protection level of bulletproof vests. Recent reviews on ballistic protection [10,11] point out interesting comments on failure mechanisms and very particular solutions in combining different materials for facing very different threats. Research has involved aramid fibers with different architectures, from simple or treated woven fabrics [12] to 3D fabrics [13] and unidirectional or multiaxial non-crimp fabrics, each solution being simulated with particular model conditions and tested for a specific threat, the conclusion of the documentation conducted by the authors being that each solution has to be experimentally investigated and the failure mechanisms understood in order to offer a design with very strong statistical reliability before use in combat.

Bajya et al. [7] recently presented an experimental report on ballistic protection and failure mechanisms of soft armor packages made of different structures, including woven and unidirectional fabrics. Packages made of para-aramid woven fabrics and unidirectional laminates and hybrid ones were developed and subjected to ballistic impact against 9 mm lead core bullets, at an impact velocity of 430 m/s. The ballistic performance of each structure was evaluated in terms of backface signature (BFS), perforation ratio (PR) and the expansion of the bullet. The unidirectional aramid packages failed mainly due to fiber break, including fibrillation, debonding and delamination. The fiber orientation was only $(0^\circ/90^\circ) \times 2$ and the surface density of this type of fabric was 230 g/m^2, with this inducing enough flexibility to allow for a too high BFS as compared to other solutions offered by these authors.

In general, ballistic threats are either from bullets or fragments. Bullets can be defined as projectiles of different shapes and consistencies, fired from weapons such as pistols, revolvers and rifles. Fragments, on the other hand, may also result from explosions (for example, detonations of a grenade) or from the target the projectiles had already hit.

The various kinetic characteristics and deformation behaviors of such a wide range of bullets and fragments must require a customized study for each of this multitude of different projectiles and for the design required to face an impact process, in order to have an effective ballistic protection.

Even if, nowadays, many models and simulations have been reported, from micro [14,15], meso [13,16], to macro [17,18] scales, the tests remain the ultimate and restrictive step in adopting a certain type of individual armor, generally based on aramid fabrics.

An active field of research concerning the impact resistance of aramid fabrics involves the study of ballistic performance, from destructive investigation based on standard structures of packages [19] or structures involving new entries such as aerogels [20] and methodologies accepted by the interested parties. The standard structure of a ballistic package involves a certain number of layers of material or semi-finished product (as fabrics) for ballistic protection, which can vary within fairly wide limits between 8 and 40 layers (or folds), depending on its architecture and the material characteristics for the ballistic protection.

The aim of this study is to evaluate a particular multiaxial fabric, LSF SB1plus, produced by Teijin [21], for designing body armor for the threat level IIA as classified in Ballistic Resistance of Personal Body Armor NIJ Standard–0101.04 [22], to identify failure mechanisms and to measure a well-accepted characteristic, backface signature (BFS), for the designed ballistic protection package. The authors measured BFS and investigated failure mechanisms of this particular package in order to evaluate it as a future application in body armor.

2. Material and Testing Campaign

Even though fabrics based on aramid fibers are lightweight and offer better protection, the designer will be forced to reduce the number of fabric layers required without compromising the effectiveness of the final protecting armor to reduce the cost. In general, aramid fibers are 43% lighter than glass fibers, as strong as E-Glass, ten times stronger than aluminum and are close in strength to carbon fibers (in tensile tests). They also have good dimensional stability with a slightly negative coefficient of thermal expansion (-2.4×10^{-6} 1/°C) and could resist chemicals with the exception of a few strong acids and alkalis. Aramid fibers exhibit excellent stability over a wide range of temperatures for prolonged periods with no strength loss at temperatures as low as −196 °C and do not melt but will start to carbonize at approximately 427 °C [10].

The layers are available in the form of fabrics in multidirectional yarns (0°, 90°, −45°, 45°) (Figure 1), namely LTF SB1plus (a trade name for a multiaxial fabric by Twaron Laminated Fabric Technology) (Arnhem production facility, Arnhem, Netherlands), and they are laminated together with a very thin foil of resin [21]. These fabrics can be combined into various structures and composites, each type having unique protection performance, forming ballistic packages with different uses and providing protection for a wide range of threats.

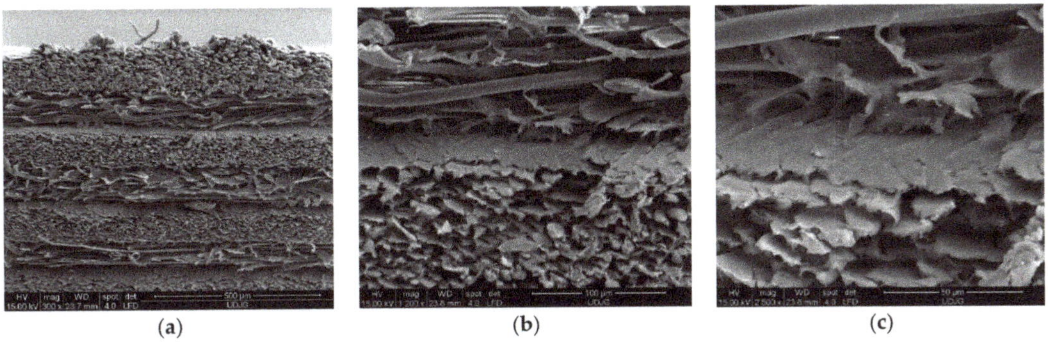

Figure 1. SEM images of the LFT SB1plus layers after being cut by the help of a power guillotine. (**a**) Magnification ×300; (**b**) magnification ×1200; (**c**) magnification ×2500.

A recent tendency of research in trying to improve the response of aramid fabrics to ballistic impact is 3D architecture [2], but fabrics with unidirectional layers with different angle orientations in sublayers are also intensively used for body armor, due to the fact that the strength of the yarns is not "consumed" in bending them in a 2D or 3D architecture. Of course, technology is obliged to find the adequate film and yarns for maintaining these multiaxial orientations in a single layer.

The architecture of the ballistic package is usually made of the same type of layer or combinations of two or more materials, the designer being interested in comparing ballistic packages with different numbers of layers in their structure [6]. For individual armor, fabrics could be used as delivered, without adding adhesive among layers. No influence of stitching the layers on the ballistic resistance properties of the package has been noticed.

Taking into account the stated criteria, the development of a type of soft armor, from the same aramid textile product, is quite numerous and technologically advantageous.

Designing a new individual protective armor requires adequate layers and yarn architecture, taking into account the particular response of the system to particular threats and the protection degree, including here the limitative request as an acceptable value of backface signature (BFS).

The packages were obtained as successive layers of Twaron LFT SB1plus, as delivered in rolls by the producer, Teijin Aramid [21]. Twaron LFT SB1plus is a fabric of four layers of unidirectional Twaron yarns, with orientations (0°, 90°, +45°, −45°), laminated together with a very small amount of polymer [21]. It has a surface density of 430 g/m^2 and the theoretical mass of the 12-layer package of 500 mm × 500 mm is 1290 g. The manufacturing of the packages includes the cutting and coupling of the fabrics (in order to have a narrow variation in their weight), sewing and control. The package has the dimensions 500 mm × 500 mm, with an area of 0.25 m^2, between recommended values by NIJ-C-4 (0.23 m^2) and NIJ-C-5 (0.3 m^2) for high and very high surfaces (Figure 2).

Figure 2. Cutting process of the layers for the ballistic package.

After being cut, 12 layers formed a package, this number being selected by the authors after preliminary tests on 8 layers that gave both results as total and partial perforation. In previous works [6,17], macro simulation on equivalent layers gave similar results to the tested package with the same number of layers (bullet was stopped on the last broken layer for a virtual package of 8 layers and simulation on a 12-layer package gave satisfactory results; thus, this number of layers (12) was used in forming a protective package).

The projectile was an FMJ (Full Metal Jacket) 9 mm bullet, as required for protection level IIA as given in NIJ Standard-0101.06/2008 [23]. The layers were laterally fixed by sewing on two sides, on a central location for a length of approx. 90 mm, in order to maintain its integrity and the layers' order. The packages were tested in a controlled environment enclosure (temperature kept in the range of 20 ± 5 °C and relative humidity of 50 ... 65%, atmospheric pressure of 760 ± 15 mm Hg), using a ballistic smooth barrel, the measured velocity of the projectiles being 420 ± 10 m/s. There were used a chronograph Oehler 43 (Oehler Research, Inc., Austin, Texas, USA) for measuring the impact velocity and a firing table with compensated rebond.

3. Ballistic Impact Testing

In practice, in order to assess the impact resistance of a protection system, there are reference standards that propose methods of assessment, such as NIJ Standard-0101.06/2008 [23] and NIJ 0101.04/2004 [22], the results giving the possibility to include the system in a level of protection [10].

The assessment of the ballistic resistance of the studied protection packages was performed according to NIJ 0101.04/2004 (Figure 3) [22], by firing a 9 mm caliber bullet, with an initial speed of 420 ± 10 m/s, from a distance of 3 m (normal conditions), with shots fired in a laboratory facility.

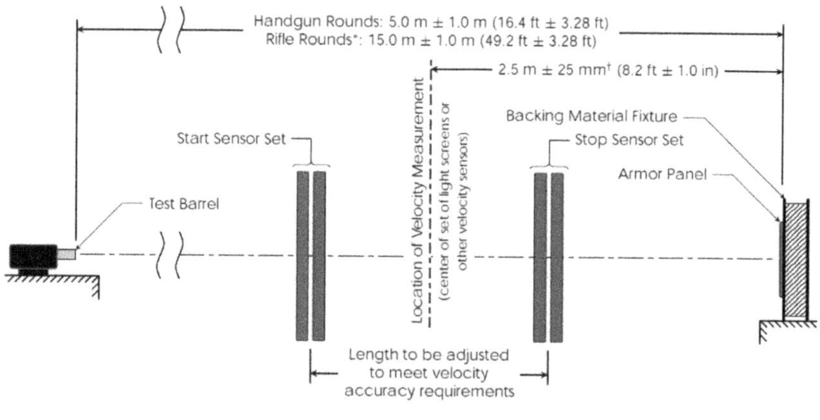

Figure 3. Firing arrangement equipment, Reprinted with permission from ref. [23], Copyright Year Copyright Owner's Name. * For rife rounds the length may be further adjusted to minimize yaw at impact; however, in such cases the yaw at impact must be experimentally shown to be less than 5° and reasonably close to minimal. † Tolerance for 0° shots. For 30° and 45° shots the tolerance shall be +25 mm/−190 mm (+1.0 in/−7.5 in).

The testing of the realized protection packages in order to evaluate their resistance to the 9 mm FMJ bullet was discussed, taking into account the resistance to penetration through the depth of the trace left in the support material (ballistic clay), namely backface signature (BFS). The assessment of the total penetration of a package is in many cases simple, when it is found that there is a hole with a diameter at least equal to the caliber of the bullet and that the entire bullet passes through it.

When testing personal ballistic protection equipment, the trauma on the human body is assessed by the depth of the BFS imprint that is formed in the ballistic clay on which the sample is fixed. The standard admits materials and protection systems as satisfactory when they produce an imprint in the support clay that does not exceed 44 mm [23].

The following steps are performed when testing personal armors or parts of it:

- The test equipment is positioned in the clamping support, at the distance imposed for each piece of equipment from the mouth of the barrel; the types of weapons and ammunition required for levels II and IIA are verified [23];
- The bullet speed measurement system is positioned, starting with the distance of 2 m from the mouth of the barrel, so that the frames of the system are in planes perpendicular to the firing direction; the distance between the frames of the system is 0.5 m and the distances are measured with an accuracy of 1 mm;
- Firing is executed on the test package.

4. Results and Discussion of Failure Mechanisms

Factors influencing ballistic protection against a specified threat include those characterizing the target material(s), in terms of fiber and yarn properties, weave architecture, surface density and the number of plies, boundary conditions and those characterizing the impact (impact velocity, impact angle, projectile shape and materials, etc.) [10].

How do we evaluate the ballistic response of body armor? The investigation is done at macro scale and the tests produce clear results: if the target resists, the backface signature could reveal the degree of protection and, if the target does not resist, the residual velocity would indicate how far away the protective solution is from the requested parameters, but in a qualitative way. However, this scale does not offer answers to improve the ballistic resistance, and the study of how projectiles and targets are destroyed at lower scale is of great interest. Thus, researchers will investigate the target and projectile failure mechanisms using optical and scanning electron microscopes, high speed cameras and spectrometers, etc.

A synthesis of the damage mechanisms under impact is given in Figure 4.

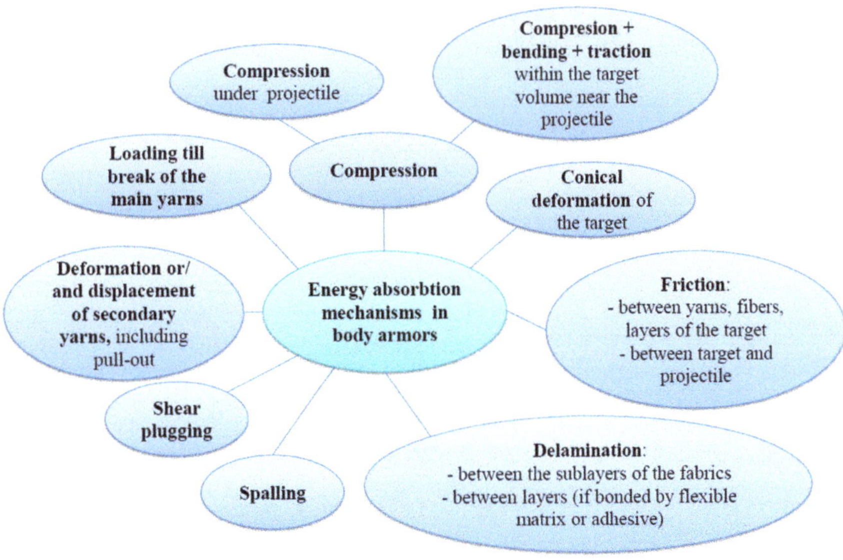

Figure 4. Damage mechanisms for packages based on fabrics.

Based on a general overview [10,24] on how the kinetic energy of the projectile is absorbed, this study aims to identify damage mechanisms for a particular package of 12 layers of LFT SB1plus when impacted with a 9 mm FMJ:

- Compression of the package directly under the bullet;
- Compression in the target volume surrounding the impacted zone;
- Conical deformation on the package backface;
- Loading the main yarns that face the higher strain; these yarns tend to fail when the induced tensile strain of these yarns exceeds the ultimate strain;
- Deformation and/or displacement of secondary yarns;
- Delamination, especially on the layers situated towards the back of the target;
- Shear plugging;
- Spalling of the already broken layers (yarns);
- Friction between the projectile and the target and friction between yarns and layers.

The secondary yarns are close to the main ones, absorbing energy by their strain distribution within the yarns, and the highest values are found near the top face of the deformed cone.

As the friction coefficient is difficult to measure for high impact velocities, simulation offers a fair evaluation by covering a large range of values for friction coefficient. The friction coefficient could have lower values, being more intense when the bullet is forced to be deformed inside the target.

When the bullet hits the target, because of the sudden drop in contact force, only the upper few layers fail by shear as the shear wave propagates along the thickness direction. Then, the undamaged layer absorbs the residual kinetic energy of the projectile through creating a cone-shaped deformation. For woven fabrics, this deformation could be rhomboidal, the rhombus diagonals being almost oriented along the warp and weft yarns. Multiaxial unidirectional fabrics have an intermediate behavior, that is, the shape of deformation could be almost conical.

Shear plugging is one of the major damage mechanisms during impact for energy absorptions by the targets. This process occurs due to the initial impact contact force between the projectile and the target, resulting shear stress through its thickness, around the periphery of the projectile. If shear plugging stress exceeds the ultimate shear plugging strength, the target will fail. The failure by shear plugging is not symmetrical as if the first yarns will be broken by shear, these yarns are de-stressed and able to be displaced when the projectile advances.

During an experimental campaign, the investigation of failure mechanisms points out the ballistic performance of a specific target configuration when hit by different projectiles, with different impact velocities.

The test results from nine fires on three packages made of 12 layers of SB1plus are plotted in Figure 5, on a web chart, with the red line being the accepted limit of 44 m [23] and the blue points being the values measured as the depth of the imprint in the ballistic clay.

Figure 6 presents the shape, as taken with a mold resin, produced by a fire, in the ballistic clay behind the package in order to highlight the importance of this characteristic for body armor. From this study, the authors want to suggest a more detailed evaluation of the entire shape of the backface signature, not only its depth in the ballistic clay, with BFS being of 18 mm here.

An investigation of the destruction of flexible packages was carried out with the help of a professional camera to highlight characteristics dependent on fabrics' architecture and the number of layers. A detailed, photographic study on each layer of the tested ballistic packages was performed to point out the failure processes in stages of the layers.

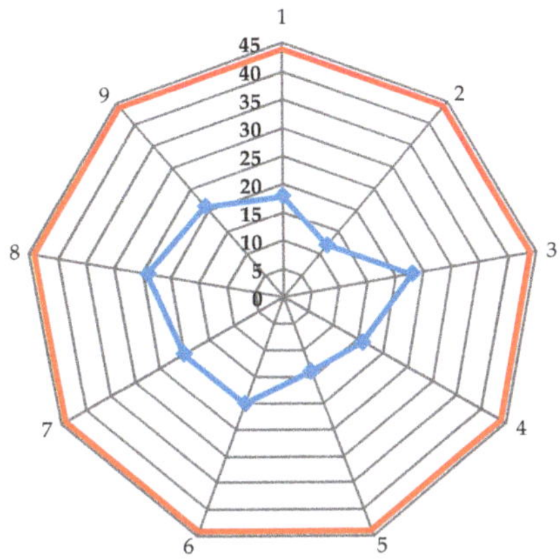

Figure 5. Data for BFS measured for 9 fires on three packages made of 12 layers of SB1plus fabric.

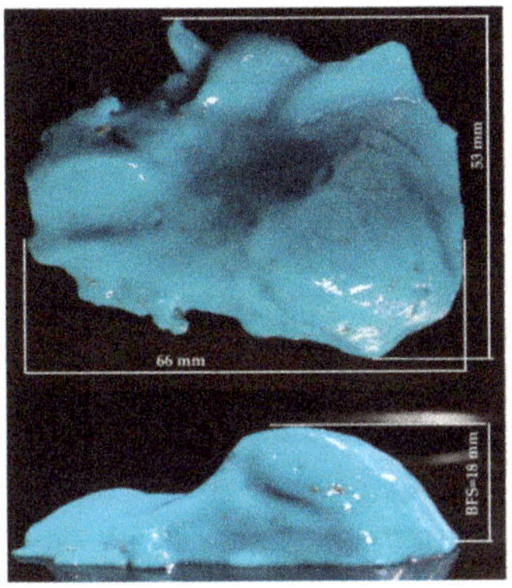

Figure 6. Cavity shape in the ballistic clay, obtained with a molding resin.

Figures 7–11 show photographs of LFT SB1plus layers from a 12-layer package. It should be noted that the tests were performed under conditions of a small variation for the bullet velocity (420 m/s ± 10 m/s). The angle of impact is normal on the target surface, the deviations being 5% at the mouth of the barrel.

In the details of the photos, for each fire the main yarns (broken under the projectile) and the secondary yarns (those affected by the penetration of the projectile, by twisting, firing and scattering) may be counted.

One may see the total penetration of the first layer in Figure 7: (a) general view from the impact direction and (b), (c) and (d) details of each of the three fires on the same target. The second layer is presented in Figure 8: (a) general view of the designed ballistic package, and the localized ruptures of the yarns is highlighted, for each fire, in (b), (c) and (d). This type of failure occurs when all the fibers of the thread break in almost the same place. The two causes of the breaking of the yarns in layer 1 are their stretching along the length next to the contact with the projectile and the shearing on their thickness, observed in Figure 7 (detail b). Theoretically, the fibers in the yarn will fail when the induced strain exceeds the strain at break, but the process is dynamic (depending on strain rate) and statistic, given the large number of fibers in a yarn.

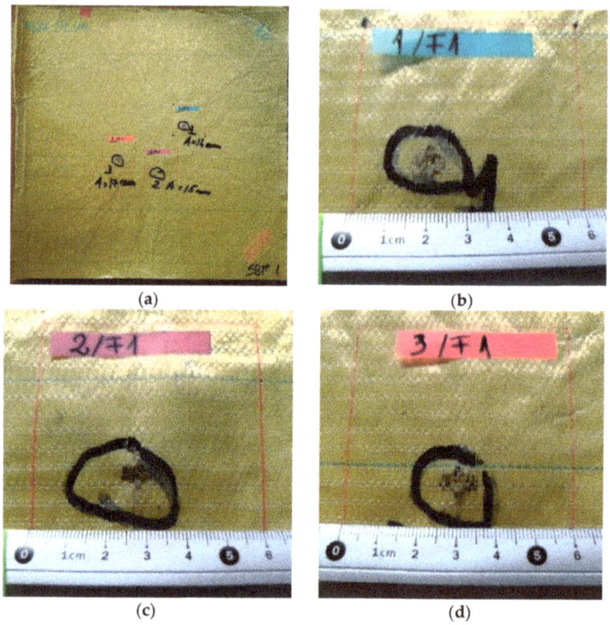

Figure 7. Frontal view (**a**) of a 12-layer LFT SB1plus package: (**a**) layer 1 (face) and details of the fires on the first layer of the package: (**b**) fire 1, (**c**) fire 2 and (**d**) fire 3. BFS was noted as A on the photo.

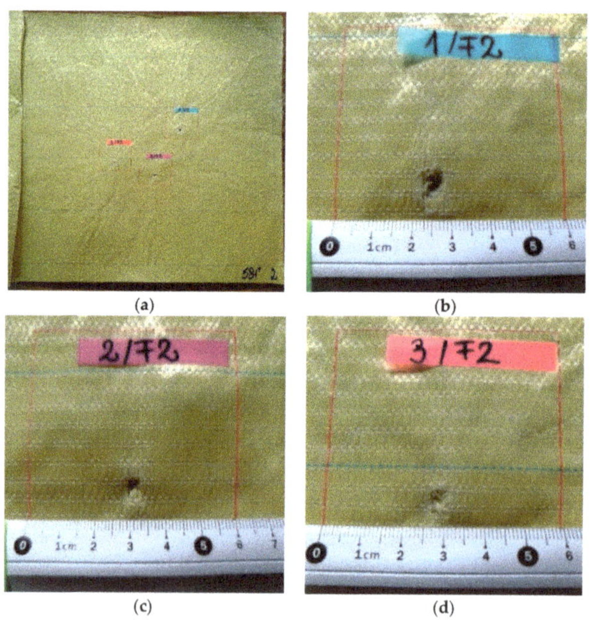

Figure 8. Frontal view of the sample with 12 layers of LFT SB1plus: (**a**) layer 2 (face), with details of the bullet holes of (**b**) fire 1, (**c**) fire 2 and (**d**) fire 3.

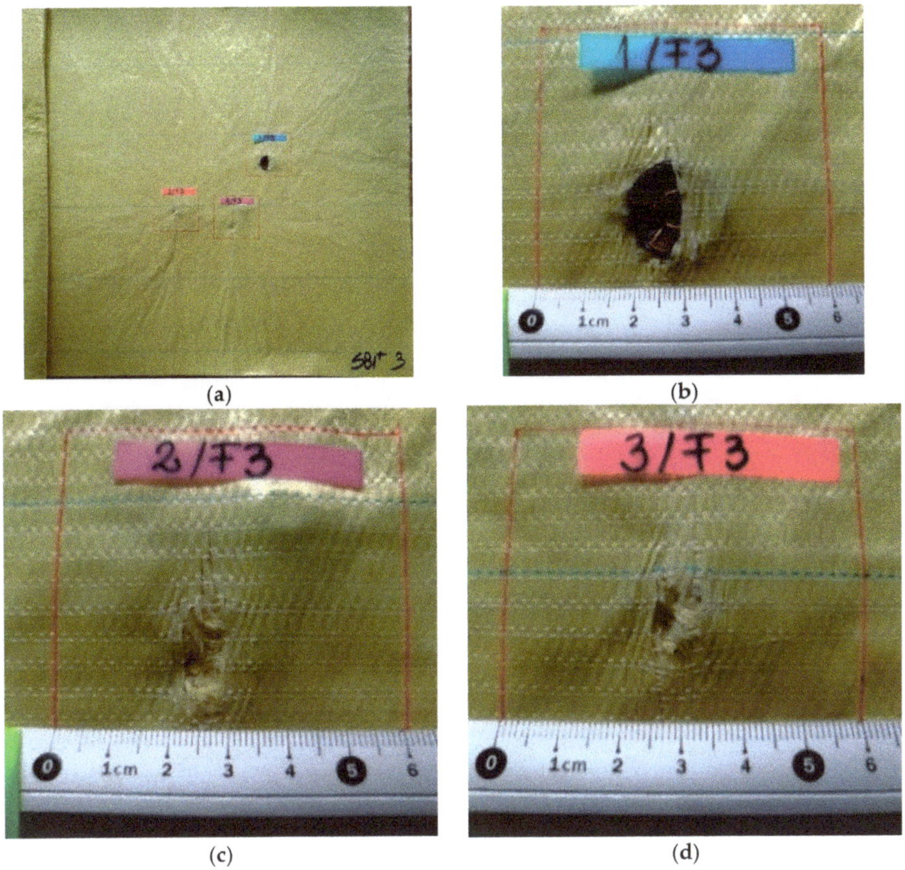

Figure 9. Frontal view of the sample with 12 layers of LFT SB1plus: (**a**) layer 3 (face), with details of the bullet holes of (**b**) fires 1, (**c**) fires 2 and (**d**) fires 3.

Starting from layer 3 (Figure 9) and in layer 4 (Figure 10), the enlargement of the holes is observed (due to the lateral plastic deformation of the projectile when it starts to be stopped), and also the process of pulling the threads. The tear-off of the yarn (even partially) from its fabric architecture occurs when the yarns are not broken, sometimes only on one side of the projectile [6].

Layer 4 is the last layer in the SB1plus LFT package through which the bullets passed or stopped. In the current case of failure, the first four layers have the role of slowing down and even arresting and deforming the projectile.

Starting from layer 3 (Figure 9) and in layer 4 (Figure 10), the widening of the holes is observed (due to the lateral deformation of the projectile when it starts to be stopped, and the process of pulling the yarns). The first fired bullet (Figure 9a) was found in the slot produced in the third layer, the second was found between the third layer and the fourth one (see the mark on Figure 10c) and the third bullet fired on the package had a similar position and shape, but advanced one layer (it was arrested between the fourth layer and fifth one). For body armor, it is very important that successive fires do not create failures with very different and progressive damages on the package. The authors consider this response promising in facing multiple and localized fires.

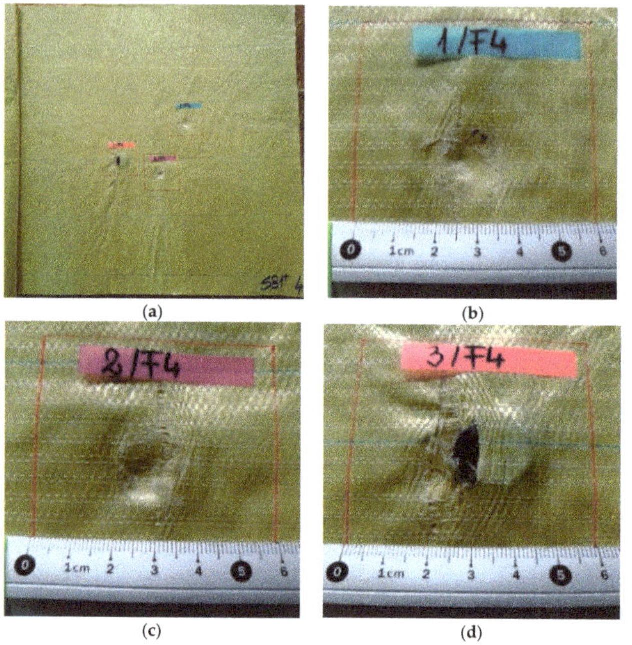

(a) (b) (c) (d)

Figure 10. Front view of sample made of LFT SB1plus: (**a**) layer 4 (face), with details of the penetration: (**b**) fire 1, (**c**) fire 2 and (**d**) fire 3.

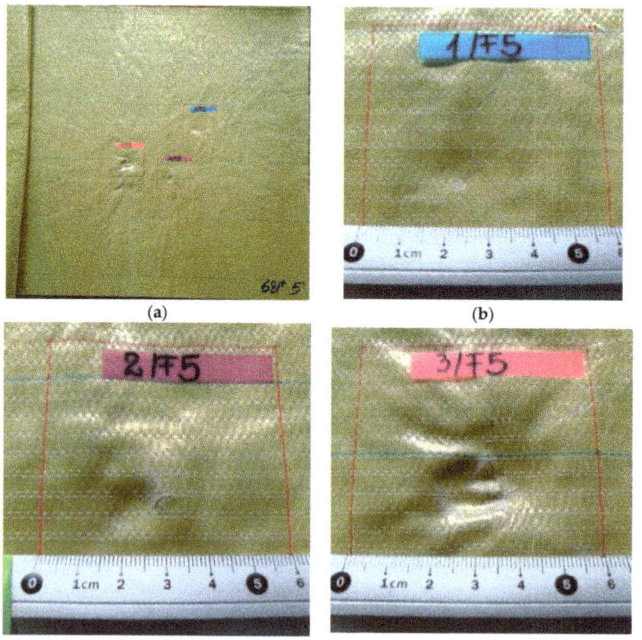

(a) (b) (c) (d)

Figure 11. Frontal view of a package made of layers of LFT SB1plus: (**a**) layer 5 (face), with details of (**b**) fire 1, (**c**) fire 2 and (**d**) fire 3.

The quality of the material SB1plus is noticed by the high degree of resemblance of the penetration zones and even by the size of zones with broken and torn-off yarns.

Layer 5 (Figure 11) shows traces of crushing (compression), circular in shape, which would be justified by the architecture of the LFT SB1plus semi-finished product (with the orientation of the yarns 0°/90°/45°/−45°), inducing a tendency to uniformity of the properties of the fabrics in four sub-layers and, therefore, of its answer.

Layers 6 and 7, theoretically, have the role of retaining (arresting) the projectile and dumping its tendency to advance together with the fabrics. These layers and the following ones are also responsible for reducing the BFS.

Figure 12 presents the last layer of the package: (a) the general view being taken from the opposite direction of the impact (named back face in this text), (b) the print on the last layer of fire 1, where aramid fibers are not broken, only the auxiliary (white) yarns that keep the resistant aramid yarns in their aligned position, (c) for fire 2 the yarns are not broken but (d) for the last fire, there are several bundles of fibers broken due to the large deformation on this layer.

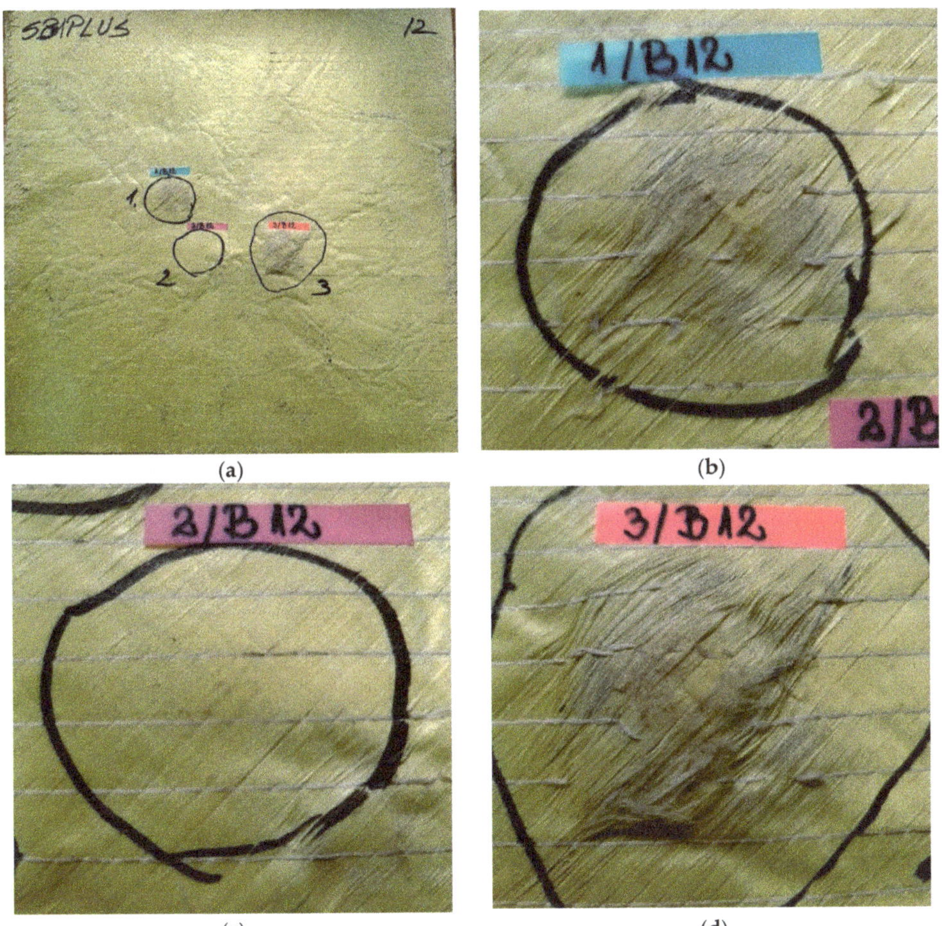

(a) (b) (c) (d)

Figure 12. Back view of package made of 12 layers of LFT SB1plus: (**a**) layer 12 (back), with details of (**b**) fire 1, (**c**) fire 2 and (**d**) fire 3.

After testing, it can be seen that the arching in the orthogonal yarns is more dominant in the back layers of the multilayer package, where the projectile tries to enter through an edge-like approach after being considerably slowed down by the initial layers [8]. The existence of the phenomenon of passing through the fabrics usually produces a hole smaller than the diameter of the projectile, with a smaller number of yarns being broken as compared to the number of yarns that intersect the projectile.

In Figure 13a, two broken main yarns can be observed on the first layer of unidirectional yarns. The ends of the broken yarns relax after being tensioned, hence the appearance of scattered fibers. Figure 13b shows the deviation of the secondary yarns from layer 2 (from the four substrates of a layer) and the more disordered aspect of the fibers, as compared to layer 1.

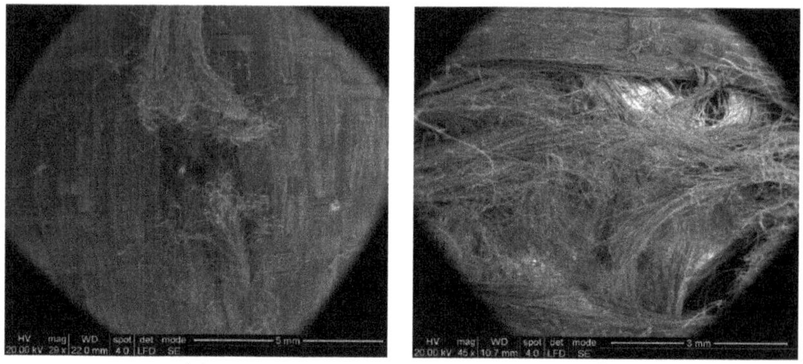

(**a**) (**b**)

Figure 13. Aspect of broken yarns from the ballistic protection package made of LFT SB1plus: (**a**) face of layer 1 (fire 1); (**b**) face of layer 2 (fire 1).

Some particular aspects of the failure of aramid fibers could be highlighted with the help of SEM images for the tested ballistic package of 12 layers. According to SEM images, breaking a fiber can be performed by fibrillation (Figure 14), shearing (Figure 15), necking (obviously, by tensile stress, as one may see in Figure 15 details (c) and (d) and twisting).

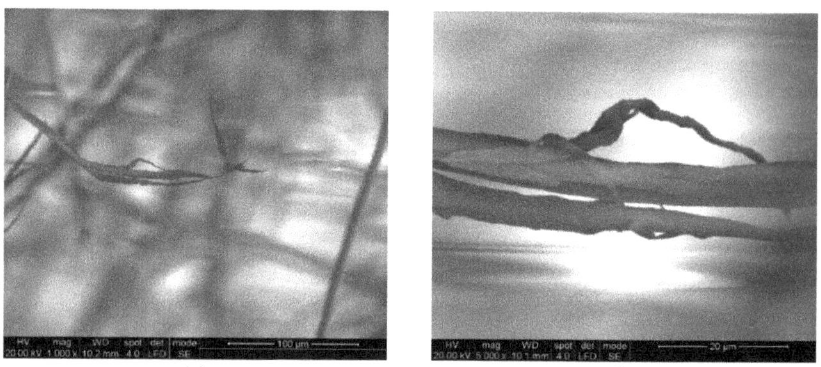

(**a**) (**b**)

Figure 14. Details of fibers from the ballistic package LFT SB1plus: (**a**) fibril from front layer 2; (**b**) detail from (**a**).

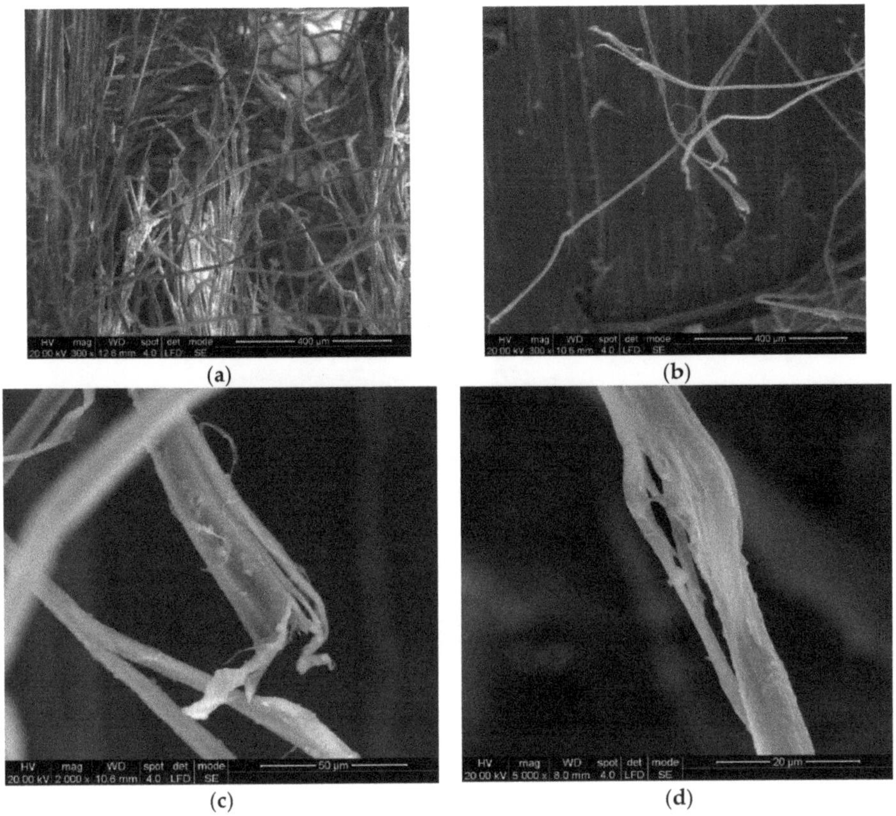

Figure 15. Details of the broken fibrils from the 2nd layer of the LFT SB1plus ballistic package. (**a**) layer 1, front view of the broken fibers, from the impact direction; (**b**) details of layer 1, front view of the broken fibers; (**c**) details of a higher magnification of a fiber; (**d**) details of (**b**) with non-uniform fibrillation and necking.

In Figure 14, by the help of scanning electron microscopy, one may see the fibrillation, thinning and rupture of the aramid fibrils, resulting from the process of destroying a thread on the 2nd layer.

The fibrillation at the broken ends of the yarns (Figures 14a,b and 15c) or along the yarn (Figure 15d) is a type of failure favored by the abrasive action of the projectile on the length of the fiber, but it is also dependent on the local traction characteristics of the yarn fibers. Of course, the location of the fibrillation is a defect of the molecular chains or their arrangement in the fiber. In layered systems, friction between layers results in abrasion friction among fibers, inducing local scratches that diminish the fiber resistance. All projectiles that have the ability to penetrate through the fabric also cause the scratching and/or splitting of the fibers, which favors the appearance of fibrillation, a specific failure of aramid fibers.

Figure 15 shows the details of some fibers damaged by fibrillation; there are noticed discontinuities in the fiber and the thinning of the fibrils that supported the load and ruptures of fibrils resulting from tensile and shearing loading (deformed breaks with less deformations, as a cut) during the destruction process.

The values for BFS (backface signature) for the LFT SB1plus package are given in Table 1 and are measured according to Ballistic Resistance of Body Armor, NIJ Standard-0101.06 [2] (Figure 16).

Table 1. BFS of a ballistic package made of LFT SB1plus layers.

LFT SB1plus Package of 12 Layers	
Fires	**BSF (mm)**
1	14
2	15
3	17

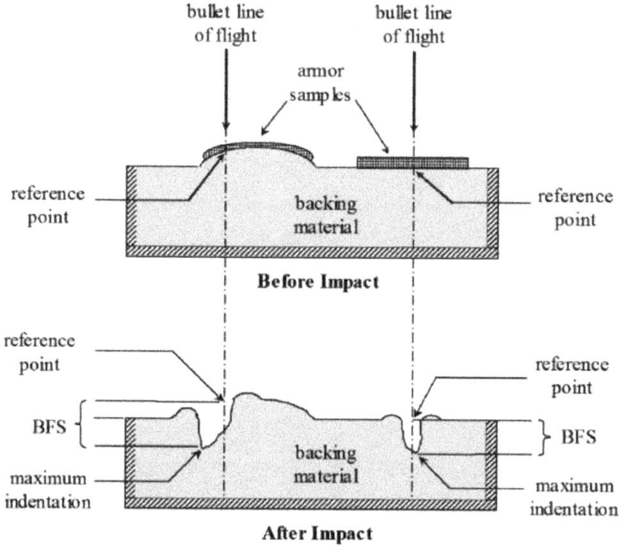

Figure 16. Measuring BFS in agreement with Ballistic Resistance of Body Armor, NIJ Standard-0101.06 [23].

Imprints in the clay support material, characterized by the BFS, for a new armor package, were analyzed to determine whether the armor will provide adequate protection against trauma (without perforation/penetration of the package). The requirements of the NIJ [23] state that all depths of the measured traces in the support material obtained from fires falling within the firing requirements must not exceed 44 mm, or if the BFS exceeds 44 mm then there must be a coefficient of 95% confidence that 80% of the depths of the traces in the support material will be 44 mm or less. Under no circumstances should a trace in the support material exceed 50 mm.

Measured BFSs for a new armor package were analyzed to determine whether the armor will provide adequate protection against trauma (without perforation/penetration of the package). The requirements of the NIJ [23] state that all depths of BFS in the support clay, obtained from fires obeying the firing requirements, must not exceed 44 mm, or if the BFS exceeds 44 mm then there must be a coefficient of 95% confidence that 80% of the depths of the imprints in the support material will be 44 mm or less. Under no circumstances should a BFS in the support material exceed 50 mm.

Following the extraction of the bullets from a package of 12 LFT SB1plus layers, the following characteristics of the recovered projectiles could be highlighted based on the analyses of the photographs in Figure 17:

- Bullets were flatted like a mushroom cap;
- Bullets had initiated fragmentation on their edges, but no fragments seemed to be separated, with the jacket and core torn off the second bullet;
- Their faces in contact with the target had trapped aramid fibers, especially from the first and second hit yarns.

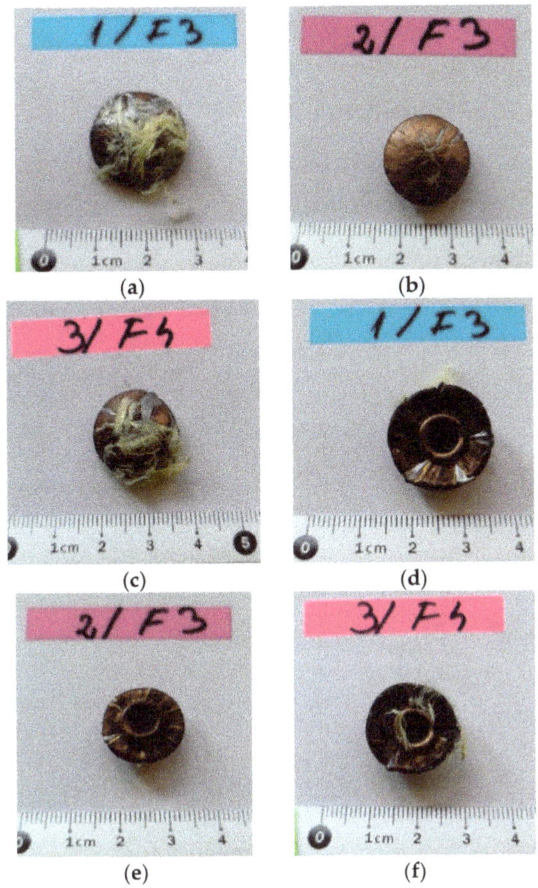

Figure 17. Bullets (9 mm FMJ) as recovered after testing 12-layer ballistic package: front view of the bullets (**a**–**c**); back view of the same bullets (**d**–**f**).

The number of layers penetrated in the package influences the way the projectile is damaged, and the bullets retained within the package (Figure 17) had a very flattened shape, with peripheral ruptures of the jacket and core.

Figure 18 presents details of the impacted face of the projectile after testing. One may notice the embedding of fragmented aramid yarns on the projectile front. The first yarn contacting the bullet was broken near its contacting length with the bullet due to high tensile stress; this central fragment was so compressed that its fibers were forced into the projectile material and remained fixed on the bullet, even if this one was intensively deformed and cracked. The bullet from the 3rd fire kept a similar aspect of the bullet failure (local fragmentation and plastic deformation), but only several fibers were "stuck" to the bullet. Several fibers were trapped in the cracks of the projectile, both in the lead core and copper alloy jacket.

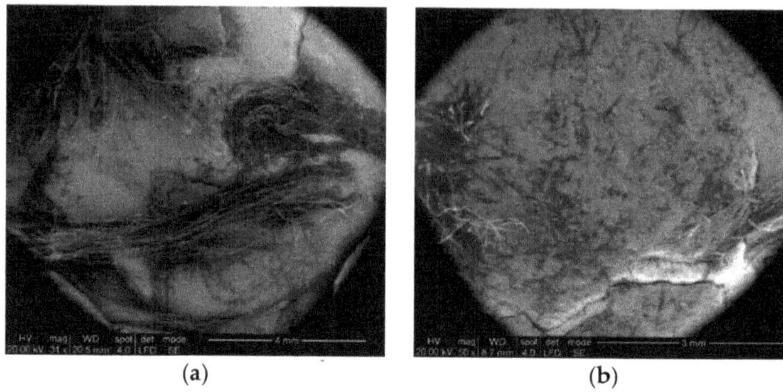

Figure 18. SEM details of bullet of 9 mm FMJ, as retrieved after firing. (a) Bullet of 2nd fire; (b) bullet of 3rd fire.

5. Conclusions

The main objective of this study focuses on designing and testing protection systems using advanced materials based on aramid fibers for high-impact speeds of up to 420 ± 10 m/s. The investigation of the failure mechanisms identifies current issues of protective materials, major challenges and technological problems for developing these types of systems, including the fulfilling of ballistic impact requirements according to NIJ Standard-0101.06 [23].

The purpose of the paper was to conduct an investigation on the failure processes and destructive stages of a ballistic package made of successive layers of LFT SB1plus, taking into account the particular test conditions from NIJ Standard-0101.06 [23]. The main parameter of interest was the backface signature (BFS), but also details of projectile arrest and SEM investigation could offer an argument for using this material for individual protection. For the reported tests, the maximum and minimum values for BFS were 12 mm and 24 mm, the mean value being 18.66 mm and the standard deviation being 3.8 mm. These values recommend the package for use as a protective panel for body armor for threat level IIA.

Author Contributions: Conceptualization, C.P. and L.D.; methodology, C.P. and L.D.; software, C.P.; validation, L.D.; formal analysis, C.P.; investigation, C.P.; resources, C.P.; data curation, L.D. and C.P.; writing—original draft preparation, C.P.; writing—review and editing, L.D.; visualization, C.P.; supervision, L.D.; project administration, L.D. and C.P.; funding acquisition, C.P. All authors have read and agreed to the published version of the manuscript.

Funding: This research was funded by the National Institute of Aerospace Research "Elie Carafoli" INCAS, Bucharest.

Institutional Review Board Statement: Not applicable.

Informed Consent Statement: Not applicable.

Data Availability Statement: The data presented in this study are available on request from the corresponding author.

Acknowledgments: We acknowledge the support in the test campaign given by the CBRN Defense and Ecology Scientific Research Centre, Bucharest, Romania.

Conflicts of Interest: The authors declare no conflict of interest.

References

1. Claus, J.; Santos, R.A.M.; Gorbatikh, L.; Swolfs, Y. Effect of matrix and fibre type on the impact resistance of woven composites. *Compos. Part B Eng.* **2020**, *183*, 107736. [CrossRef]
2. Abtew, M.A.; Boussu, F.; Bruniaux, P.; Liu, H. Fabrication and Mechanical Characterization of Dry Three-Dimensional Warp Interlock Para-Aramid Woven Fabrics: Experimental Methods toward Applications in Composite Reinforcement and Soft Body Armor. *Materials* **2020**, *13*, 4233. [CrossRef]
3. Bhatnagar, A. *Lightweight Ballistic Composites*; CRC Press: Boca Raton, FL, USA; Boston, MA, USA; New York, NY, USA, 2006.
4. Termonia, Y. Impact Resistance of Woven Fabrics. *Text. Res. J.* **2004**, *748*, 723–729. [CrossRef]
5. Tam, D.K.Y.; Ruan, S.; Gao, P.; Yu, T. *High-Performance Ballistic Protection Using Polymer Nanocomposites (Chapter 10) in Advances in Military Textiles and Personal Equipment*; Woodhead Publishing Limited: Sawston, UK, 2012.
6. Pirvu, C. Contribution on Experimental and Numerical Study of Ballistic Protection Packages Made of Aramid Fabrics. Ph.D. thesis, "Dunarea de Jos" University, Galați, Romania, 2015.
7. Bajya, M.; Majumdar, A.; Butola, B.S.; Arora, S.; Bhattacharjee, D. Ballistic performance and failure modes of woven and unidirectional fabric based soft armour panels. *Compos. Struct.* **2021**, *255*, 112941. [CrossRef]
8. Guo, Z.; Chen, W. A merit parameter to determine the stacking order of heterogeneous diphasic soft armor systems. *Compos. Struct.* **2020**, *241*, 112086. [CrossRef]
9. Nunes, S.G.; Scazzosi, R.; Manes, A.; Amico, S.C.; de Amorim Júnior, W.F.; Giglio, M. Influence of projectile and thickness on the ballistic behavior of aramid composites: Experimental and numerical study. *Int. J. Impact Eng.* **2019**, *132*, 103307. [CrossRef]
10. Abtew, M.A.; Boussu, F.; Bruniaux, P.; Loghin, C.; Cristian, I. Ballistic impact mechanisms—A review on textiles and fibre-reinforced composites impact responses. *Compos. Struct.* **2019**, *223*, 110966. [CrossRef]
11. Naik, S.; Dandagwhal, R.D.; Loharkar, P.K. A review on various aspects of Kevlar composites used in ballistic applications. *Mater. Today Proc.* **2020**, *21*, 1366–1374. [CrossRef]
12. Gürgen, S. Numerical modeling of fabrics treated with multi-phase shear thickening fluids under high velocity impacts. *Thin-Walled Struct.* **2020**, *148*, 106573. [CrossRef]
13. Yang, Y.; Zhang, X.; Chen, X.; Min, S. Numerical Study on the Effect of Z-Warps on the Ballistic Responses of Para-Aramid 3D Angle-Interlock Fabrics. *Materials* **2021**, *14*, 479. [CrossRef] [PubMed]
14. Sockalingam, S.; Gillespie, J.W.; Keefe, M. Role of Inelastic Transverse Compressive Behavior and Multiaxial Loading on the Transverse Impact of Kevlar KM2 Single Fiber. *Fibers* **2017**, *5*, 9. [CrossRef]
15. Chiper, T.L.; Ojoc, G.G.; Deleanu, L.; Pîrvu, C. Simulation of impact behavior of a glass yarn. *Mech. Test. Diagn.* **2020**, *1*, 10–17. [CrossRef]
16. Feito, N.; Loya, J.A.; Muñoz-Sánchez, A.; Das, R. Numerical Modelling of Ballistic Impact Response at Low Velocity in Aramid Fabrics. *Materials* **2019**, *12*, 2087. [CrossRef] [PubMed]
17. Ionescu, T.F.; Pirvu, C.; Badea, S.; Georgescu, C.; Deleanu, L. The Influence of Friction Characteristics in Simulating the Impact Bullet-Stratified Materials. In Proceedings of the 15th International Conference on Tribology, Kragujevac, Serbia, 17–19 May 2017.
18. Kedzierski, P.; Popławski, A.; Gieleta, R.; Morka, A.; Sławinski, G. Experimental and numerical investigation of fabric impact behavior. *Compos. Part B* **2015**, *69*, 452–459. [CrossRef]
19. Mou, H.; Jiang, X.; Hui, P.; Feng, Z.; Hongzhang, G. Ballistic impact tests and stacked shell simulation analysis of aramid fabric containment system. *Aerosp. Sci. Technol.* **2020**, *107*, 106344. [CrossRef]
20. Ayten, A.İ.; Taşdelen, M.A.; Ekici, B. An experimental investigation on ballistic efficiency of silica-based crosslinked aerogels in aramid fabric. *Ceram. Int.* **2020**, *46*, 26724–26730. [CrossRef]
21. Teijin Aramid. *Ballistics Material Handbook 38-1405/2012*; Teijin Aramid: Arnhem, The Netherlands, 2012.
22. *Ballistic Resistance of Personal Body Armor NIJ Standard–0101.04*; U.S. Department of Justice Office of Justice Programs National Institute of Justice: Rockville, MD, USA, 2004.
23. *Ballistic Resistance of Body Armor, NIJ Standard-0101.06*; U.S. Department of Justice Office of Justice Programs National Institute of Justice: Washington, DC, USA, 2008.
24. Attwood, J.P.; Russell, B.P.; Wadley, H.N.G.; Deshpande, V.S. Mechanisms of the penetration of ultra-high molecular weight polyethylene composite beams. *Int. J. Impact Eng.* **2016**, *93*, 153–165. [CrossRef]

polymers

MDPI

Article

Radiation-Based Crosslinking Technique for Enhanced Thermal and Mechanical Properties of HDPE/EVA/PU Blends

Jang-Gun Lee †, Jin-Oh Jeong †, Sung-In Jeong and Jong-Seok Park *

Advanced Radiation Technology Institute, Korea Atomic Energy Research Institute, Jeongeup-si 56212, Korea; dlwkdrjs@kaeri.re.kr (J.-G.L.); nanjinoh@gmail.com (J.-O.J.); sijeong@kaeri.re.kr (S.-I.J.)

* Correspondence: jspark75@kaeri.re.kr; Tel.: +82-63-570-3067

† These two authors contributed equally to this work.

Abstract: Crosslinking of polyolefin-based polymers can improve their thermal and mechanical properties, which can then be used in various applications. Radiation-induced crosslinking can be done easily and usefully by irradiation without a crosslinking agent. In addition, polymer blending can improve thermal and mechanical properties, and chemical resistance, compared to conventional single polymers. In this study, high-density polyethylene (HDPE)/ethylene vinyl acetate (EVA)/polyurethane (PU) blends were prepared by radiation crosslinking to improve the thermal and mechanical properties of HDPE. This is because HDPE, a polyolefin-based polymer, has the weaknesses of low thermal resistance and flexibility, even though it has good mechanical strength and machinability. In contrast, EVA has good flexibility and PU has excellent thermal properties and wear resistance. The morphology and mechanical properties (e.g., tensile and flexure strength) were characterized using scanning electron microscopy (SEM) and a universal testing machine (UTM). The gel fraction, thermal shrinkage, and abrasion resistance of samples were confirmed. In particular, after storing at 180 °C for 1 h, the crosslinked HDPE-PU-EVA blends exhibited ~4-times better thermal stability compared to non-crosslinked HDPE. When subjected to a radiation dose of 100 kGy, the strength of HDPE increased, but the elongation sharply decreased (80%). On the other hand, the strength of the HDPE-PU-EVA blends was very similar to that of HDPE, and the elongation was more than 3-times better (320%). Finally, the abrasion resistance of crosslinked HDPE-PU-EVA was ~9-times better than the crosslinked HDPE. Therefore, this technology can be applied to various polymer products requiring high heat resistance and flexibility, such as electric cables and industrial pipes.

Keywords: polymer blending; radiation crosslinking; polyethylene; polyurethane; heat resistance; mechanical property

Citation: Lee, J.-G.; Jeong, J.-O.; Jeong, S.-I.; Park, J.-S. Radiation-Based Crosslinking Technique for Enhanced Thermal and Mechanical Properties of HDPE/EVA/PU Blends. *Polymers* **2021**, *13*, 2832. https://doi.org/10.3390/polym13162832

Academic Editors: Beata Podkościelna and Andrzej Puszka

Received: 5 August 2021
Accepted: 21 August 2021
Published: 23 August 2021

1. Introduction

Polyethylene, in particular, is the most representative general-purpose polymer among the polyolefin-based polymers. It exhibits properties such as crystallinity and a decrease in crystal size depending on the number and length of branches connected to the main chain, and variations can be divided into HDPE and low-density polyethylene (LDPE). Among them, HDPE has a small number of short and/or long molecular chains, has a high crystallinity (60 to 80%), and has a melting point of 135 °C [1–4]. In addition, HDPE is used in diverse ways, such as film, blow molding, injection molding, pipe processing, and wires due to its many advantages (high tensile strength, excellent processability, excellent low-temperature resistance, and electrical insulation) [5–7]. However, HDPE has problems with rapid mechanical degradation and thermal shrinkage at above 130 °C near the melting point [8,9]. To overcome these disadvantages, polymer blends are prepared by blending and/or crosslinking to improve their mechanical properties [10–13]. In our previous study, improved thermal and mechanical properties were confirmed after blending styrene-grafted polyurethane (PU) into polypropylene (PP), a polyolefin-based polymer. PU is a non-toxic polymer with excellent elasticity, abrasion resistance, and workability. It was confirmed that the original mechanical properties (disadvantageous low-temperature characteristics) of PP were improved. In this way, mechanical properties can be improved through polymer blending [14].

In addition, it is possible to improve further or maintain the thermal and mechanical properties of polymer blends through polymer crosslinking [15]. Crosslinking refers to the connection by physical and chemical interactions between polymer chains. Polymer crosslinking can be divided into physical and chemical crosslinking [16]. Physical crosslinking is a crosslinking method involving interactions such as the entanglement of chains or ionic bonds between chains. Chemical crosslinking is provided by chemical bonds (such as covalent bonding) between molecular chains [17–23]. In particular, among the chemical crosslinking methods, the radiation crosslinking method has the advantage that a reaction is induced without the use of a chemical additive, such as a crosslinking agent or initiator. Moreover, a reaction can be induced at various temperatures and in various states (e.g., solid, liquid, and gas) [24–27]. The crosslinking of HDPE using radiation crosslinking technology is useful for generating radicals in the HDPE polymer chain. HDPE has the characteristic that when tertiary carbon atoms in the branch chain lose hydrogen, they generate radicals that are easily attacked by external free radicals and stabilized. Thus, crosslinking occurs readily due to easy radical generation [28,29].

In this study, an HDPE blend was prepared by blending ethylene vinyl acetate (EVA) and PU to improve their thermal and mechanical properties. EVA is a copolymer of ethylene and vinyl acetate (VA), and its density, flexibility, elasticity, and durability vary depending on the VA content. In addition, EVA is widely used as an over wire cable and as a solar sheet for its excellent low-temperature characteristics and impact resistance [30,31]. EVA and PU, which have these advantages, were blended with HDPE to prepare a blend and then crosslinked using electron beam irradiation. To determine the properties of the crosslinked HDPE/EVA/PU (H/V/U) blend, the gel fraction, shrinkage rate, tensile strength, flexural strength, and wear resistance were confirmed. We intend to confirm the improvement of the thermal and mechanical properties of this blend after electron beam crosslinking.

2. Materials and Methods

2.1. Materials

High-density polyethylene (HDPE) and ethylene-vinyl acetate (EVA) were purchased from the Lotte Chemical Corporation (Seoul, Korea). Polyurethane (PU) was obtained from Songwon Industrial Co., Ltd. (Ulsan, Korea). Polyethylene-graft-maleic anhydride (PE-g-MA) (viscosity of 500 cP) was purchased from Sigma-Aldrich (St. Louis, MI, USA). All other reagents and solvents were of analytical grade and used as received.

2.2. Preparation of the HDPE-EVA-PU Blends

HDPE, EVA, and PU were blended using a Brabender mixer (Brabender D-47055, Brabender, Duisburg, Germany) at 45 rpm and 190 °C for 20 min. HDPE-EVA-PU blend sheets were prepared using a hot press at 190 °C. To achieve the crosslinking, the sheets were exposed to electron beam irradiation (2.5. MeV, UELV-10-105, Korea Atomic Energy Research Institute, Jeongeup, Korea) to achieve doses of 50, 75, and 100 kGy (25 kGy/cycle).

2.3. Characterization of the HDPE-EVA-PU Blends

The surface morphology of the samples was observed using scanning electron microscopy (SEM, TM3030, HITACHI, Tokyo, Japan). To acquire high-resolution images, the samples were coated with gold for 60 s using sputter coating to conduct SEM with a 15 kV electron beam and a working distance of 8.1 mm. In addition, chemical component and quantitative analyses were performed using energy-dispersive X-ray spectroscopy (EDS, TM3030, HITACHI, Tokyo, Japan).

To analyze the gel fraction of the HDPE-EVA-PU blends using Soxhlet extraction, samples (0.5 × 0.5 cm^2) were prepared, and their initial weights were recorded prior to immersion in xylene for 8 h at 140 °C to remove unreacted polymers. After that, the samples were dried in an oven at 80 °C for 8 h and then dried naturally for 4 h after which the weight was recorded. The gel fraction was calculated from the following equation:

$$\text{Gel fraction (\%)} = (W_f/W_i) \times 100$$

where W_i and W_f represent the initial and final weights of the dried samples.

Prior to testing for tensile strength, the samples were prepared according to ASTMD638 and measured using a universal testing machine (UTM, Instron 5982, Norfolk, MA, USA) with a 100 kN range of load and crosshead speed of 50 mm/min. In addition, the flexural strength of samples (8 × 1 cm^2) was measured using the UTM with a 10 kN range of load, 48 mm of span distance, and 50 mm/min of crosshead speed.

To confirm the thermal shrinkage, samples were prepared (4 × 4 cm^2) and placed in an oven at 150 or 180 °C for 1 h before the result was recorded. The thermal shrinkage of samples was calculated using the following equation:

$$\text{Thermal shrinkage (\%)} = [1 - (S_a/S_b)] \times 100$$

where S_a and S_b represent the thermal shrinkage after and before tested area, respectively.

The abrasion resistance of the samples was measured according to KSD8314 (1000 cycles, speed 60 rpm, test load of 1 kg, stroke of 50 ± 1 mm at room temperature). The abrasive material was sandpaper (Koptri, #220). The weight reduction rate of the samples was calculated using the following equation:

$$\text{Weight reduction rate (\%)} = [(T_a - T_b)/T_a] \times 100$$

T_a is the weight before the test and T_b is the weight after the test.

3. Results and Discussion

3.1. Preparation of HDPE/EVA/PU Blends

Polymer blending is one of the methods that can be used to supplement the weaknesses of a polymer, and properties can be enhanced, and desired properties can be manufactured through blending [32–34]. In this study, HDPE, EVA, and PU were blended using a Brabender mixer to prepare H/V/U blend sheets using a hot press. To crosslink each sample, the H/V/U blends were irradiated using electron beam irradiation at different doses (50, 75, and 100 kGy). The overall schematic illustration and chemical composition of the H/V/U blends are shown in Figure 1 and Table 1.

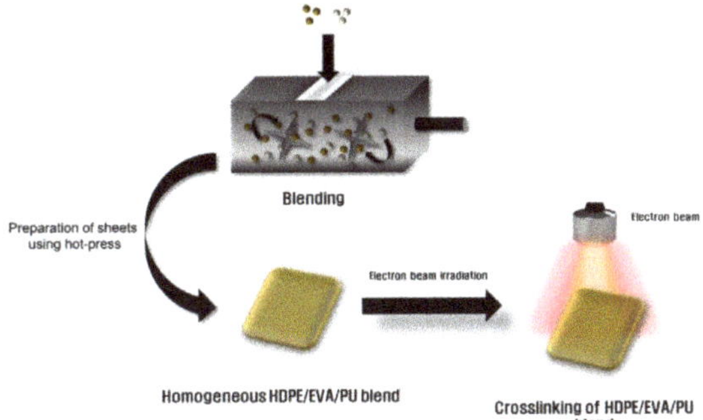

Figure 1. Schematic illustration of the HDPE/EVA/PU blends after using electron beam irradiation.

Table 1. Chemical composition of the HDPE/EVA/PU blends.

	H	H/U	H/V	H/V/U
HDPE	200 g	200 g	200 g	200 g
EVA	-	-	20 phr	20 phr
PU	-	5 phr	-	5 phr
PE-g-MA	-	3 phr	3 phr	3 phr

3.2. Characterization of HDPE/EVA/PU Blends

HDPE is composed of pure carbon and hydrogen and exhibits non-polar properties [14]. EVA is a copolymer of ethylene and VA and is also non-polar, whereas VA contains oxygen atoms and exhibits polar properties [30,31]. Therefore, EVA mixes well with non-polar polymers and polar polymers and has the property that it can easily be mixed with various kinds of chemical additives. In addition, because PU also exhibits non-polar properties, HDPE, EVA, and PU are considered to blend effectively. Figure 2 shows the results of the surface morphology and elemental composition of H, H/U, H/V, and H/V/U determined using SEM/EDS. As shown in Figure 2a, as a result of the surface resulting after preparing a sheet by blending each polymer using a Brabender mixer and a hot press, a flat surface was confirmed in all samples. In addition, it was confirmed that the form of separation between polymers was not significant. In addition, Figure 2b shows the EDS results for each element (green: carbon, blue: nitrogen, and red: oxygen). The Nitrogen (blue) of PU was confirmed in H/U and H/V/U, and the oxygen (red) of EVA and PU was confirmed in H/U, H/V, and H/V/U. It was also confirmed that each element was uniformly distributed in the blends, which showed effective blending.

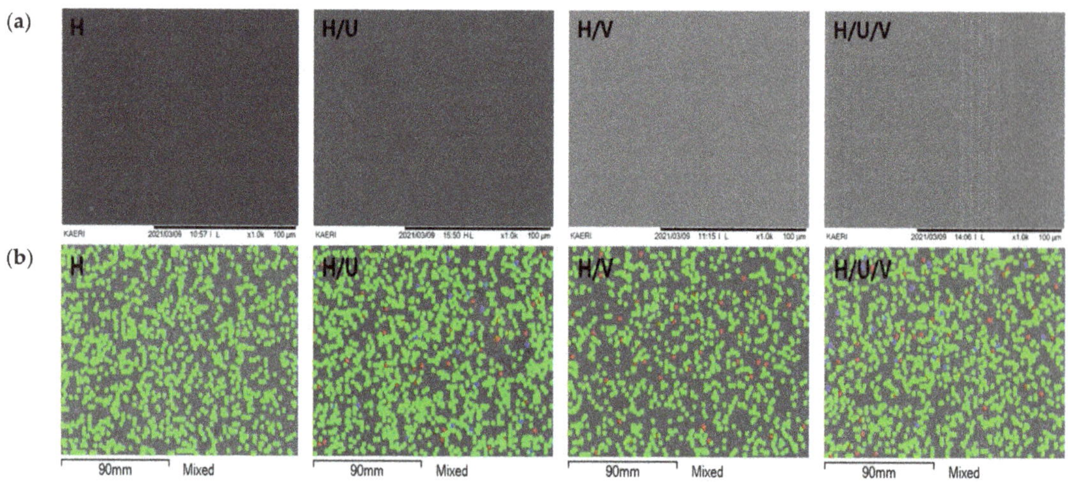

Figure 2. Surface morphology and elemental composition of the HDPE/EVA/PU blends by (**a**) scanning electron microscopy and (**b**) energy-dispersive X-ray spectroscopy.

Figure 3 shows the gel fraction of H and the H/V and H/V/U blends after electron beam irradiation. As shown in Figure 3a, the gel fraction of each sample was confirmed according to the radiation dose of 100 kGy. In the case of H, a 58% gel fraction was confirmed, and the gel fraction of the H/V and H/V/U blends was higher (66 and 65%, respectively). This result was believed to be due to crosslinking of HDPE as well as EVA. In general, the mechanical properties of EVA could be improved by crosslinking using peroxide (e.g., dicummyl peroxide or perbutyl peroxide) because EVA has weaker mechanical properties than PE as applied to polyethylene and polyester [35]. In this study, it was confirmed that the gel fraction was increased through crosslinking of HDPE and EVA using only electron beam irradiation (that is, without the use of a crosslinking agent). In the case of H/U blends containing PU, the gel fraction was not significant compared to that of H/V. This result is believed to be because EVA was crosslinked by electron beam irradiation more efficiently than PU. In addition, the gel fraction of each sample was confirmed by the dose of electron beam irradiation. Figure 3b shows that the gel fraction result was confirmed by the different radiation doses of HDPE. The gel fraction increase in relation to the radiation dose of 50, 75, and 100 kGy is 50, 56, and 58%, respectively. The gel fraction of the H/V blend increases to 55, 61, and 66%, respectively, as the radiation dose increases to 50, 75, and 100 kGy (shown in Figure 3c). In addition, the increase in the gel fraction of the H/V/U blend with the dose of 50, 75, and 100 kGy is 54, 65, and 65%, respectively.

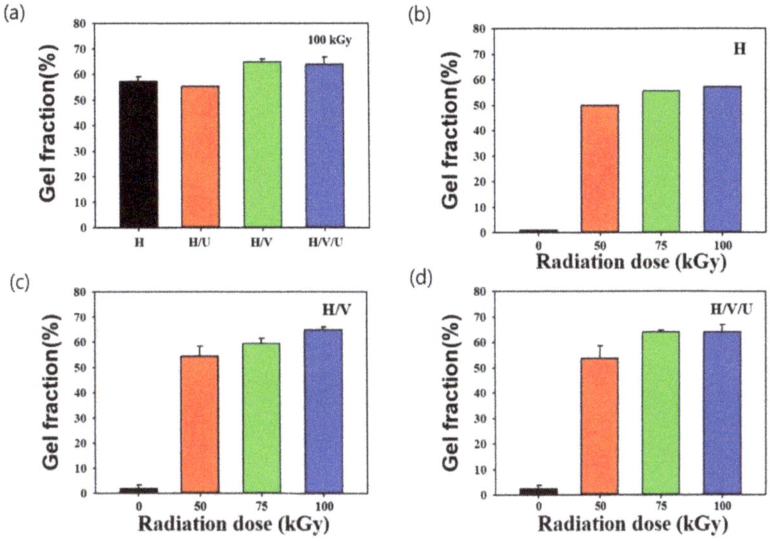

Figure 3. Gel fraction of the HDPE/EVA/PU blends: (**a**) radiation dose at 100 kGy, (**b**) different radiation doses at 0, 50, 75, and 100 kGy in HDPE, (**c**) different radiation doses at 0, 50, 75, and 100 kGy in HDPE/EVA blends, and (**d**) different radiation doses at 0, 50, 75, and 100 kGy in HDPE/EVA/PU blends.

As shown in Figure 4, the mechanical properties of the samples were confirmed through crosslinking by electron beam irradiation, and Figure 4a shows the tensile stress of each sample. In the case of the tensile stress of H and the H/U blends, it was confirmed that the tensile stress was only slightly changed, even if the radiation dose was increased from 50 to 100 kGy. In particular, it was confirmed that the tensile stress of the H/V and H/V/U blends decreased when the radiation dose was 100 kGy. Although the tensile stress of the H/V blends increased with an increase in the radiation dose from 0 kGy (25 MPa) to 50 kGy (25.3 MPa) and 75 kGy (27 MPa), at the dose of 100 kGy, the tensile stress decreased to 23 MPa. In addition, the tensile stress of the H/V/U blends increased with an increase in the radiation dose but decreased to 23 MPa in the case of 10 kGy (like for the H/V blends). Figure 4b shows the result of the tensile strain of each sample. With the increase in radiation dose from 0 to 50, 75, and 100 kGy, the HDPE strain decreases to 510%, 370%, 100%, and 80%. In particular, it was confirmed that the strain decreased rapidly at 75 and 100 kGy. This result is believed to be because the tensile strain decreases as the amount of crosslinking in HDPE was increased with an increase in the radiation dose. In the case of the tensile strain of the H/U blend, the tensile strain decreased slightly (compared to

HDPE) because PU has excellent mechanical properties. The H/V and H/V/U blends with EVA exhibited higher tensile strain than with HDPE at 100 kGy. Moreover, it was confirmed that the mechanical properties were improved. The H/V blends decreased to 570%, 500%, 450%, and 390% as the radiation dose increased from 0 to 100 kGy. In the case of the H/V/U blends, the strain in specimens with radiation doses of 0, 50, 75, and 100 kGy was decreased to 530%, 480%, 410%, and 320%. The tensile strains of H and the H/V/U blends with a radiation dose of 100 kGy were 80% and 320%, confirming that the strain of the H/V/U blends with EVA and PU increased approximately four times. This result shows that the H/V/U blend exhibits excellent tensile strength.

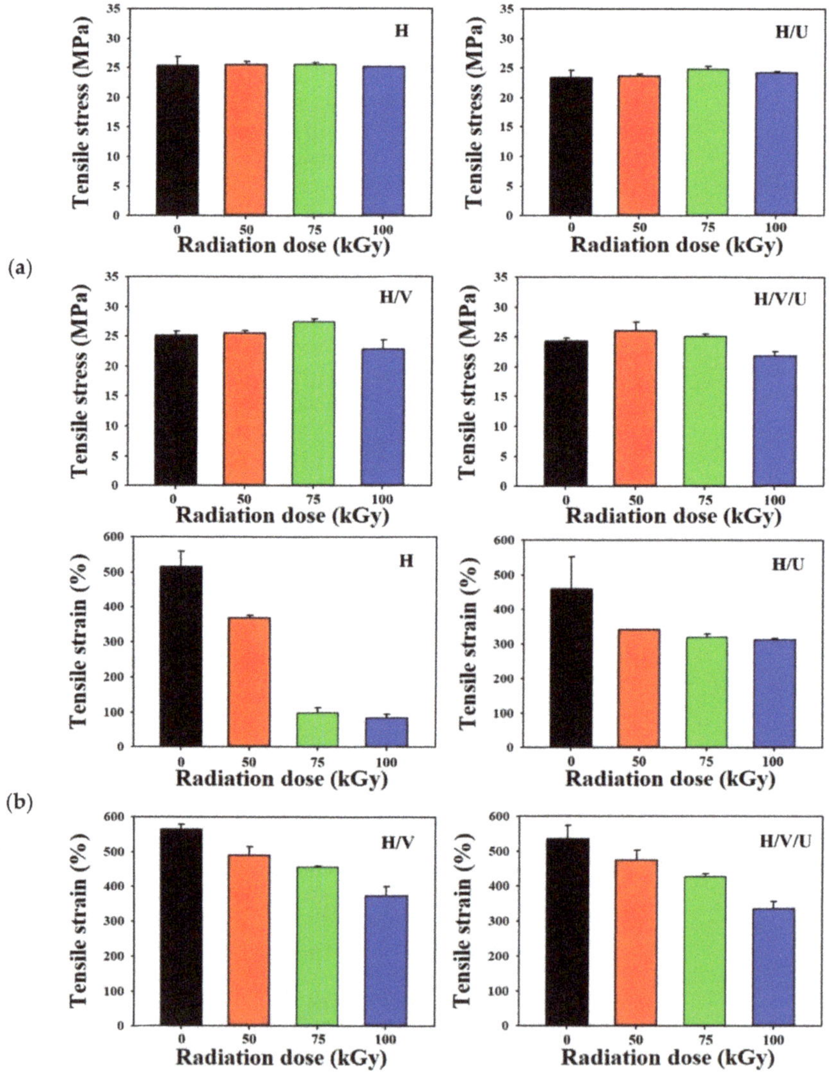

Figure 4. Tensile properties of HDPE/EVA/PU blends: (**a**) tensile stress of each sample with different radiation dose, (**b**) tensile strain of each sample with different radiation dose.

Flexure strength is the resistance to the bending of various materials such as plastic, ceramic, and rubber samples. It is defined as the maximum value at which the load does not increase anymore when applying a bending force. Figure 5a shows the flexure stress of each sample. The flexure stress of HDPE is 27, 30, 29, and 29 MPa at the radiation dose of 0, 50, 75, and 100 kGy, and the flexure stress of the H/U blend is 33, 29.7, 30.2, and 29 MPa as the radiation dose increases to 0, 50, 75, and 100 kGy. Even after crosslinking by electron beam irradiation, there is only a slight change in the flexure stress. On the other hand, the flexure stress of the H/V and H/V/U blends is lower than that of the H and the H/U blend. It was confirmed that there was only a slight change, even when the radiation dose increased. The flexure stress of the H/V blend was 22.5, 22.3, 23, and 22.9 MPa according to the radiation dose of 0, 50, 75, and 100 kGy. In addition, the flexure stress of the H/V/U blend was 21.5, 22, 21.3, and 21.4 MPa at the radiation dose of 0, 50, 75, and 100 kGy. This result shows that the flexure stress was lowered by blending with EVA. Figure 5b shows the flexure strain of each sample. The flexure strain of HDPE decreased after crosslinking by electron beam irradiation. The flexure strain of HDPE was 19.8, 14.7, 14, and 13.8% according to the radiation dose of 0, 50, 75, and 100 kGy. On the other hand, it was confirmed that the flexure strain of the H/U blend blended with PU increased to 19.5, 20.7, 22.5, and 24% at the radiation dose of 0, 50, 75, and 100 kGy. This result is believed to be because the flexure strain increased due to the excellent tensile properties of PU. In addition, the flexure strains of the H/V and H/V/U blends in which EVA was blended were also confirmed to have similar tendencies by the inclusion of PU (as in the result for H and the H/U blend). The flexure strain of the H/V blend was 19.9, 17.3, 16.1, and 16.8% according to the radiation dose of 0, 50, 75, and 100 kGy, and the flexure stain of the H/V/U blend was determined by the radiation dose of 0, 50, 75, and 100 kGy (with the results of 21%, 24%, 24.8%, and 25.1%).

(a)

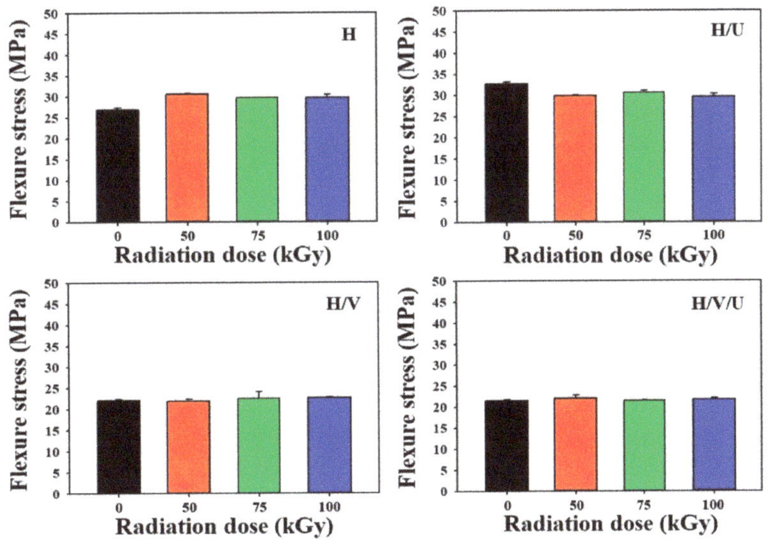

Figure 5. *Cont.*

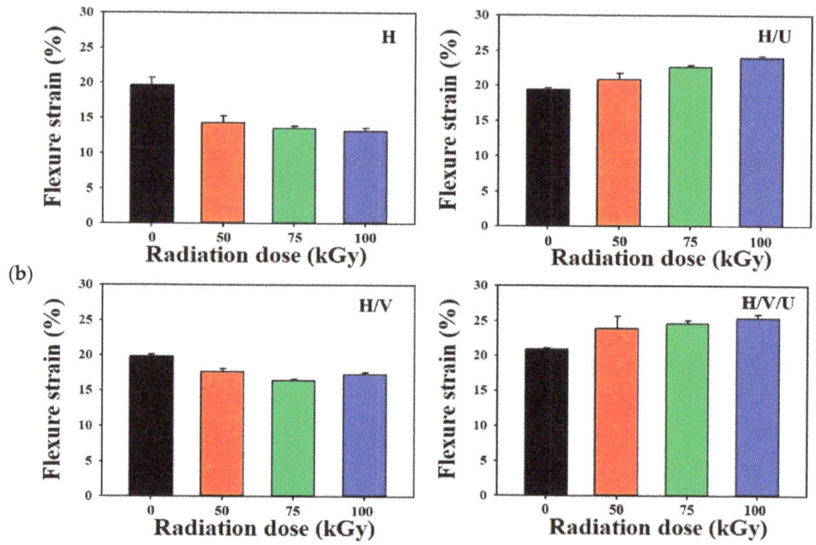

Figure 5. Flexure properties of the HDPE/EVA/PU blends: (**a**) flexure stress of each sample with different radiation dose, (**b**) flexure strain of each sample with different radiation dose.

The physical properties (gel fraction, tensile, and flexure strength) were also confirmed by the ANOVA one-way (Table 2).

Table 2. ANOVA table of physical properties.

Physical Property	F-Value	*p*-Value	F Crit
Gel fraction	0.04	0.96	4.26
Tensile stress	1.97	0.17	3.49
Tensile strain	2.21	0.14	3.49
Flexure stress	112.43	0.00	3.49
Flexure strain	9.46	0.00	3.49

Polyethylene (PE) begins to collapse when the crystal structure is above 70 °C, which increases the amorphous state. The melting point is where it becomes 100% amorphous. The melting point of HDPE is 132–138 °C, the heat of fusion is 55–66 cal/g, and the melting point changes depending on the copolymerized branches and molecular weight. In addition, the glass transition phenomenon in HDPE, a semi-crystalline polymer, occurs by the thermal behavior of the amorphous region between the crystalline lamellae. The glass transition phenomenon determines the fragility of the polymer according to the thermal deformation temperature. The glass transition temperature (Tg) is lowered due to the cooling rate of HDPE and having more chain branches and is increased by the higher molecular weight and intermolecular crosslinking [36–38]. In particular, the thermal conductivity of HDPE is caused by the complex vibration of atoms in the lattice structure. For this reason, the thermal conductivity of HDPE with high crystallinity is very high compared to other polyolefins with low crystallinity, such as LDPE or PP. The thermal conductivity of HDPE is strongly proportional to its density, decreases with increasing temperature, and increases with increasing pressure [39–41]. In particular, it is very important to consider the thermal conductivity of HDPE when preparing thick HDPE moldings. Figure 6 shows the thermal shrinkage results at 150 °C and 180 °C for H, and the H/U, H/V, and H/V/U blends with radiation doses at 0 and 75 kGy. As shown in Figure 6a, in the case of non-irradiated samples, the confirmed reduction rate of HDPE is 9.8%, and the H/U, H/V, and H/V/U blends are 6.7%, 9.3%, and 6.2%, respectively. On the other hand, in the case of the sample irradiated with 75 kGy, it was confirmed that the reduction rate was reduced by 3.8%, 0.1%, 2.9%, and 1.8%, respectively, compared to non-irradiated samples. In particular, in the case of the H/U blend, it was confirmed that thermal shrinkage hardly occurred after electron beam irradiation. This result suggests that thermal shrinkage did not occur due to the excellent thermal properties of PU. These results were the same as the thermal shrinkage results of PP blended and styrene-grafted PU blends in previous studies. To improve the thermal properties of PP, after blending PU with grafted styrene, it was left at 150 °C for 1 h to confirm from the shape of samples that there was almost no change. In addition, in the case of the H/V/U blends, it was possible to confirm the result of crosslinking of HDPE by electron beam irradiation.

(a)

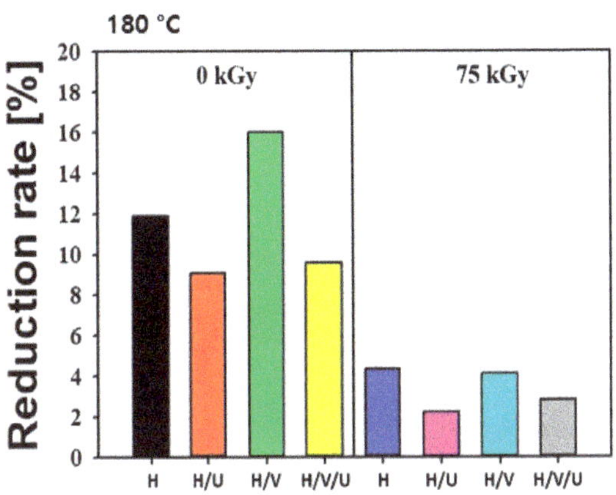

(b)

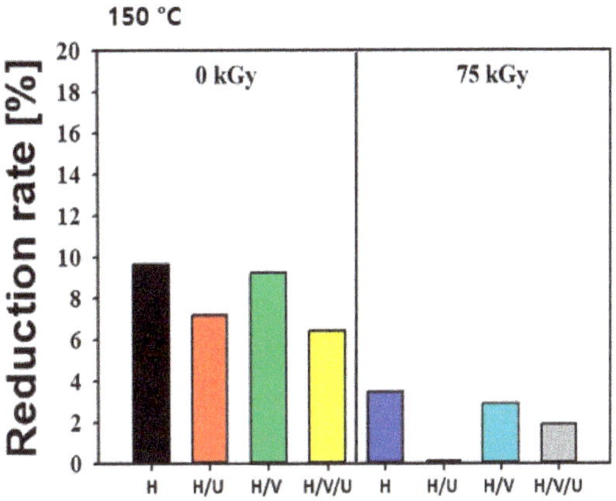

(c)

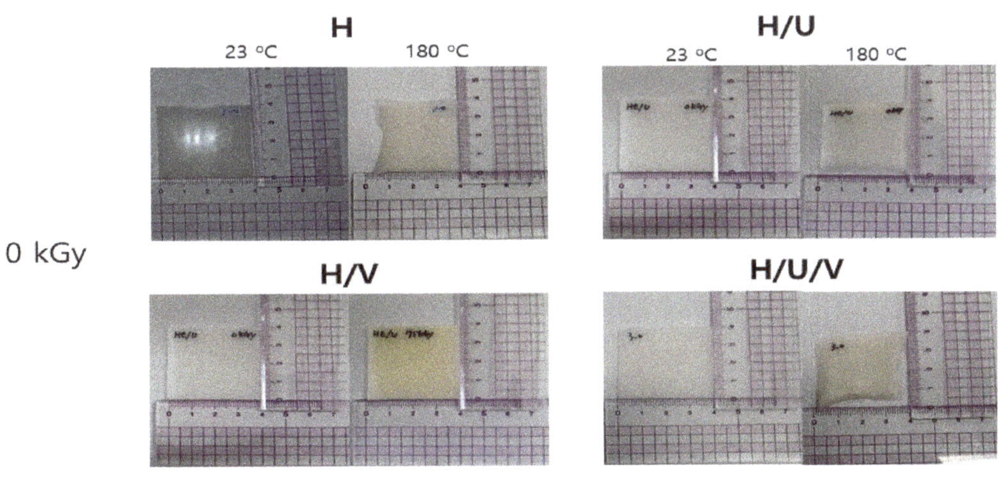

Figure 6. *Cont.*

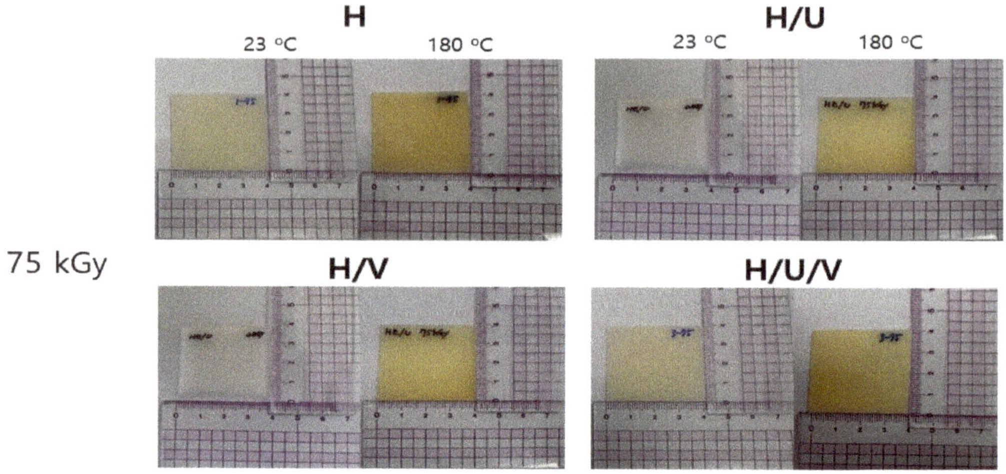

Figure 6. Thermal shrinkage of HDPE/EVA/PU blends: (**a**) reduction rate of non-irradiated sample and samples irradiated at 75 kGy and 150 °C for 1 h, (**b**) reduction rate of non-irradiated sample and samples irradiated at 75 kGy and 180 °C for 1 h, and (**c**) optical images of the HDPE/EVA/PU blends with non-irradiated and samples irradiated at 23 °C and 180 °C for 1 h.

After efficient blending of EVA and PU, a reduction rate of 1.8% was confirmed after irradiation at 75 kGy. In Figure 6b, the result of thermal shrinkage showed in samples after storage at 180 °C for 1 h. In cases of non-irradiated samples, the reduction rate of HDPE was confirmed to be 11.9%, and reduction rates of 9.1%, 16%, and 9.7% were confirmed for the H/U, H/V, and H/V/U blends, respectively. In addition, in the case of the sample irradiated at 75 kGy, it was confirmed that the shrinkage rate for the H, H/U, H/V, and H/V/U blends was reduced by 4.2%, 2.1%, 4.1%, and 2.8%, respectively. With non-irradiated H/V/U blends, in particular, it was confirmed that the thermal shrinkage phenomenon rapidly increased as the temperature increased from 150 to 180 °C. It is considered that the poor thermal properties of HDPE at high temperatures affect the thermal shrinkage results. Based on these results, the possibility of improving the thermal properties of plastic polymers was confirmed through the easy and useful blending of polymers and crosslinking using electron beam irradiation. Figure 6c shows an optical image of the thermal shrinkage of each sheet-type sample.

Abrasion resistance refers to the property of resisting abrasion well, and it is applied by using abrasion-resistant materials for accessory parts for such as electric wires, automobiles, and airplanes. For example, regarding wires, if its cover is not abrasion-resistant, it could cause a major fire due to the peeling of the insulation off the wire over time. Thus, if the stability of the material decreases and causes a serious risk of personal injury or accident, this is a big disadvantage [42,43]. Therefore, abrasion resistance is an important factor for plastic materials. Figure 7 shows the results of tests pf abrasion resistance of HDPE and the crosslinked samples irradiated at 100 kGy using electron beam irradiation. With non-irradiated HDPE (H 0), a weight reduction rate of 0.08% was confirmed, but in the case of HDPE irradiated at 100 kGy, a high weight reduction rate of 0.27% was shown. In particular, it was confirmed that crosslinked HDPE had weak abrasion resistance. On the other hand, when the radiation dose was 100 kGy, in the H/U (H/U 100), H/V (H/V 100), and H/V/U (H/V/U 100) blends with EVA and PU, excellent abrasion resistance properties were indicated by the low weight reduction rates of 0.02, 0.04, and 0.03%.

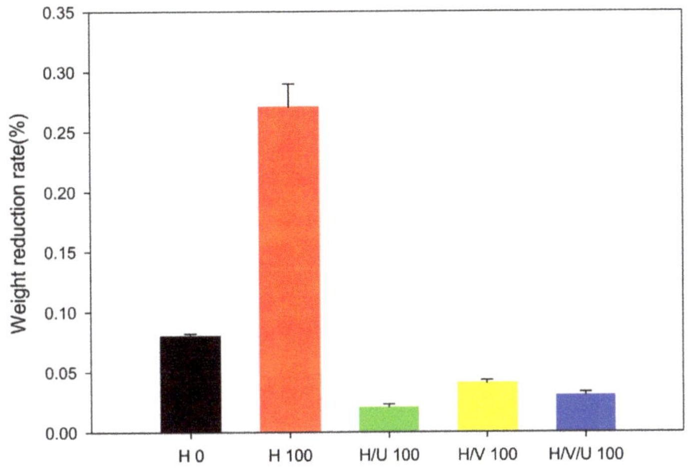

Figure 7. Abrasion resistance of HDPE/EVA/PU blends.

4. Conclusions

In this study, H/V/U blends were prepared by radiation crosslinking to improve the thermal and mechanical properties of HDPE. In the case of H/V/U blends with radiation doses of 75 and 100 kGy, the mechanical properties, and thermal shrinkage were sharply enhanced compared to non-irradiated HDPE. After storage at 180 °C for 1 h, the crosslinked HDPE/PU/EVA blends exhibited about 4-times better thermal stability compared to non-crosslinked HDPE. In addition, when the radiation dose was 100 kGy, the strength of HDPE increased, but the elongation sharply decreased (80%). On the other hand, the strength of the HDPE-PU-EVA blends was similar to that of HDPE, and the elongation was >3-times better (320%). Finally, the abrasion resistance of the crosslinked HDPE/PU/EVA was ~9-times better than with crosslinked HDPE. These excellent properties of the H/V/U blends can be used not only for insulated cable, thermal shrinkage tube, and pipes but also in automobile components.

Author Contributions: J.-S.P. conceived and designed the experiments; J.-O.J. and J.-G.L. performed the experiments; S.-I.J. analyzed the data; J.-S.P., J.-G.L. and J.-O.J. wrote the manuscript. The manuscript was reviewed by all authors. All authors have read and agreed to the published version of the manuscript.

Funding: This work was supported by the Nuclear R&D program through the National Research Foundation (NRF) of Korea grant funded by the Ministry of Science and ICT (Information and Communication Technology) and Future Planning, Korea government (NRF- 2020M2D8A1045967).

Institutional Review Board Statement: Not applicable.

Informed Consent Statement: Not applicable.

Data Availability Statement: The data presented in this study are available on request from the corresponding author.

Conflicts of Interest: The authors declare no conflict of interest.

References

1. Li, D.; Zhou, L.; Wang, X.; He, L.; Yang, X. Effect of Crystallinity of Polyethylene with Different Densities on Breakdown Strength and Conductance Property. *Materials* **2019**, *12*, 1746. [CrossRef] [PubMed]
2. Ishii, R.; Cho, D.C.; Mori, T.; Mizu, T.; Lshi, M. Electrical Properties of Low-density Polyethylene Prepared by Different Manufactur-ing Process. In Proceedings of the 1999 Annual Report Conference on Electrical Insulation and Dielectric Phenomena, Austin, TX, USA, 17–20 October1999; pp. 642–645.

3. De Faria, I.P.; Barreira Martinez, M.L.; de Queiroz, A.A.A. Electrical Performance Evaluation of Plasticized Polyolefin Formulation Developed for Manufacturing Surge Arresters Housings. *IEEE Trans. Dielectr. Electr. Insul.* **2015**, *22*, 3429–3441. [CrossRef]
4. Li, T. Structural characteristics and application development of polyethylene chemical materials. *Chem. Ind.* **2014**, *32*, 20–22.
5. Yilmaz, G.; Ellingham, T.; Turng, L.S. Improved Processability and the Processing-Structure-Properties Relationship of Ultra-High Molecular Weight Polyethylene via Supercritical Nitrogen and Carbon Dioxide in Injection Molding. *Polymers* **2018**, *10*, 36. [CrossRef]
6. Houben, S.J.A.; Verpaalen, R.C.P.; Engels, T.A.P. Processing and Properties of Melt Processable UHMW-PE Based Fibers Using Low Molecular Weight Linear Polyethylene's. *Macromol. Mater. Eng.* **2020**, *305*, 2000360. [CrossRef]
7. Oblak, P.; Gonzalez-Gutierrez, J.; Zupančič, B.; Aulova, A.; Emri, I. Processability and mechanical properties of extensively recycled high density polyethylene. *Polym. Degrad. Stab.* **2015**, *114*, 133–145. [CrossRef]
8. Kosciuszko, A.; Marciniak, D.; Sykutera, D. Post-Processing Time Dependence of Shrinkage and Mechanical Properties of Injection-Molded Polypropylene. *Materials* **2021**, *14*, 22. [CrossRef] [PubMed]
9. Khonakdar, H.A.; Morshedian, J.; Mehrabzadeh, M.; Wagenknecht, U.; Jafari, S.H. Thermal and shrinkage behaviour of stretched peroxide-crosslinked high-density polyethylene. *Eur. Polym. J.* **2003**, *39*, 1729–1734. [CrossRef]
10. Kikugawa, G.; Desai, T.G.; Keblinski, P.; Ohara, T. Effect of crosslink formation on heat conduction in amorphous polymers. *J. Appl. Phys.* **2013**, *114*, 034302. [CrossRef]
11. Morgan, V.T. The thermal conductivity of crosslinked polyethylene insulation in aerial bundled cables. *IEEE Trans. Elect. Insul.* **1991**, *26*, 1153–1158. [CrossRef]
12. Kang, S.H.; Kim, K.W.; Kim, B.J. Carbon Fibers from High-Density Polyethylene Using a Hybrid Cross-Linking Technique. *Polymers* **2021**, *13*, 2157. [CrossRef]
13. Bradler, P.R.; Fischer, J.; Wallner, G.M.; Lang, R.W. Characterization of Irradiation Crosslinked Polyamides for Solar Thermal Applications—Basic Thermo-Analytical and Mechanical Properties. *Polymers* **2018**, *10*, 969. [CrossRef] [PubMed]
14. Jeong, J.O.; Lim, Y.M.; Park, J.S. Improving thermal stability and mechanical performance of polypropylene/polyurethane blend prepared by radiation-based techniques. *Eur. Polym. J.* **2017**, *94*, 366–375. [CrossRef]
15. Manas, D.; Manas, M.; Mizera, A.; Stoklasek, P.; Navratil, J.; Sehnalek, S.; Drabek, P. The High Density Polyethylene Composite with Recycled Radiation Cross-Linked Filler of rHDPE. *Polymers* **2018**, *10*, 1361. [CrossRef]
16. Tolinski, M. Chapter 15-Crosslinking. In *Additives for Polyolefins, Getting the Most Out of Polypropylene, Polyethylene and TPO*, 2nd ed.; Elsevier Inc.: Amsterdam, Netherlands, 2015; pp. 215–220.
17. Kontou, E.; Spathis, G.; Niaounakis, M.; Kefalas, V. Physical and chemical cross-linking effects in polyurethane elastomers. *Colloid Polym. Sci.* **1990**, *268*, 636–644. [CrossRef]
18. Kaltenegger-Uray, A.; Rieß, G.; Lucyshyn, T.; Holzer, C.; Kern, W. Physical Foaming and Crosslinking of Polyethylene with Modified Talcum. *Polymers* **2019**, *11*, 1472. [CrossRef] [PubMed]
19. Bhattacharjee, P.; Ahearne, M. Significance of Crosslinking Approaches in the Development of Next Generation Hydrogels for Corneal Tissue Engineering. *Pharmaceutics* **2021**, *13*, 319. [CrossRef] [PubMed]
20. Bialik-Wąs, K.; Królicka, E.; Malina, D. Impact of the Type of Crosslinking Agents on the Properties of Modified Sodium Alginate/Poly(vinyl Alcohol) Hydrogels. *Molecules* **2021**, *26*, 2381. [CrossRef]
21. Khunmanee, S.; Jeong, Y.; Park, H. Crosslinking method of hyaluronic-based hydrogel for biomedical applications. *J. Tissue Eng.* **2017**, *8*, 1–16. [CrossRef]
22. Rozbeský, D.; Rosůlek, M.; Kukačka, Z.; Chmelík, J.; Man, P.; Novák, P. Impact of Chemical Cross-Linking on Protein Structure and Function. *Anal. Chem.* **2018**, *90*, 1104–1113. [CrossRef]
23. Hoffman, E.A.; Frey, B.L.; Smith, L.M.; Auble, D.T. Formaldehyde Crosslinking: A Tool for the Study of Chromatin Complexes. *J. Biol. Chem.* **2015**, *290*, 26404–26411. [CrossRef] [PubMed]
24. Yasin, T.; Khan, S.; Shafiq, M.; Gill, R. Radiation crosslinking of styrene–butadiene rubber containing waste tire rubber and polyfunctional monomers. *Radiat. Phys. Chem.* **2015**, *106*, 343–347. [CrossRef]
25. Manas, D.; Ovsik, M.; Mizera, A.; Manas, M.; Hylova, L.; Bednarik, M.; Stanek, M. The Effect of Irradiation on Mechanical and Thermal Properties of Selected Types of Polymers. *Polymers* **2018**, *10*, 158. [CrossRef] [PubMed]
26. Ghobashy, M.M.; Abdeen, Z.I. Radiation Crosslinking of Polyurethanes: Characterization by FTIR, TGA, SEM, XRD, and Raman Spectroscopy. *J. Polym.* **2016**, *2016*, 9802514. [CrossRef]
27. Park, J.S.; Lim, Y.M.; Nho, Y.C. Radiation-Induced Grafting with One-Step Process of Waste Polyurethane onto High-Density Polyethylene. *Materials* **2016**, *9*, 13. [CrossRef]
28. Burgstaller, C.; Höftberger, T.; Gallnböck-Wagner, B.; Stadlbauer, W. Effects of radiation type and dose on the properties of selected polymers. *Polym. Eng. Sci.* **2021**, *61*, 39–54. [CrossRef]
29. Ren, Y.; Sun, X.; Chen, L.; Li, Y.; Sun, M.; Duan, X.; Liang, W. Structures and impact strength variation of chemically crosslinked high-density polyethylene: Effect of crosslinking density. *RSC Adv.* **2021**, *11*, 6791–6797. [CrossRef]
30. Son, C.E.; Choi, S.S. Characterization of Poly(ethylene-co-vinyl acetate) (EVA) Using Thermal Analytical Techniques. *Elastomers Compos.* **2019**, *54*, 61–69.
31. Kakkar, D.; Maiti, S.N. Effect of Flexibility of Ethylene Vinyl Acetate and Crystallization of Polypropylene on the Mechanical Properties of i-PP/EVA Blends. *J. Appl. Polym. Sci.* **2012**, *123*, 1905–1912. [CrossRef]

32. Barick, A.K.; Tripathy, D.K. Preparation and characterization of thermoplastic polyurethane/organoclay nanocomposites by melt intercalation technique: Effect of nanoclay on morphology, mechanical, thermal, and rheological properties. *J. Appl. Polym. Sci.* **2010**, *117*, 639–654. [CrossRef]
33. Jia, S.; Zhu, Y.; Wang, Z.; Chen, J.; Fu, L. Influences of PP-g-MA on the surface free energy, morphologies and mechanical properties of thermoplastic polyurethane/polypropylene blends. *J. Polym. Res.* **2015**, *22*, 1–10. [CrossRef]
34. Xu, Y.; Chen, M.; Ning, X.; Chen, X.L. Influences of coupling agent on thermal properties, flammability and mechanical properties of polypropylene/thermoplastic polyurethanes composites filled with expanded graphite. *Therm. Anal. Calorim.* **2014**, *115*, 689–695. [CrossRef]
35. Chae, E.; Cho, S.S. Analysis of Peroxide Curing Agents in EVA Compounds and Vulcanizates. *Elastomers Compos.* **2020**, *55*, 6–12.
36. Salakhov, I.I.; Shaidullin, N.M.; Chalykh, A.E.; Matsko, M.A.; Shapagin, A.V.; Batyrshin, A.Z.; Shandryuk, G.A.; Nifant'ev, I.E. Low-Temperature Mechanical Properties of High-Density and Low-Density Polyethylene and Their Blends. *Polymers* **2021**, *13*, 1821. [CrossRef] [PubMed]
37. Depan, D.; Chirdon, W.; Khattab, A. Morphological and Chemical Analysis of Low-Density Polyethylene Crystallized on Carbon and Clay Nanofillers. *Polymers* **2021**, *13*, 1558. [CrossRef] [PubMed]
38. Khanam, P.N.; AlMaadeed, M.A.A. Processing and characterization of polyethylene based composites. *Adv. Manuf.: Polym. Compos. Sci.* **2015**, *1*, 63–79. [CrossRef]
39. Tarani, E.; Terzopoulou, Z.; Bikiaris, D.N.; Kyratsi, T.; Chrissafis, K.; Vourlias, G. Thermal conductivity and degradation behavior of HDPE/graphene nanocomposites. *J. Therm. Anal. Calorim.* **2017**, *129*, 1715–1726. [CrossRef]
40. Sobolciak, P.; Abdulgader, A.; Mrlik, M.; Popelka, A.; Abdala, A.A.; Aboukhlewa, A.A.; Karkri, M.; Kiepfer, H.; Bart, H.J.; Krupa, I. Thermally Conductive Polyethylene/Expanded Graphite Composites as Heat Transfer Surface: Mechanical, Thermo-Physical and Surface Behavior. *Polymers* **2020**, *12*, 2863. [CrossRef]
41. Mysiukiewicz, O.; Kosmela, P.; Barczewski, M.; Hejna, A. Mechanical, Thermal and Rheological Properties of Polyethylene-Based Composites Filled with Micrometric Aluminum Powder. *Materials* **2020**, *13*, 1242. [CrossRef]
42. Amjadi, M.; Fatemi, A. Tensile Behavior of High-Density Polyethylene Including the Effects of Processing Technique, Thickness, Temperature, and Strain Rate. *Polymers* **2020**, *12*, 1857. [CrossRef]
43. Ferreira, E.H.C.; Vieira, A.A.; Vieira, L.; Fechine, G.J.M. High-Tribological-Performance Polymer Nanocomposites: An Approach Based on the Superlubricity State of the Graphene Oxide Agglomerates. *Polymers* **2021**, *13*, 2237. [CrossRef] [PubMed]

polymers

MDPI

Article

Biocomposites of Epoxidized Natural Rubber/Poly(Lactic Acid) Modified with Natural Substances: Influence of Biomolecules on the Aging Properties (Part II)

Anna Masek * and **Stefan Cichosz**

Institute of Polymer and Dye Technology, Faculty of Chemistry, Lodz University of Technology, Stefanowskiego 12/16, 90-924 Lodz, Poland; stefan.cichosz@dokt.p.lodz.pl
* Correspondence: anna.masek@p.lodz.pl

Abstract: The aim of this study is to present the possible influence of natural substances on the aging properties of epoxidized natural rubber (ENR) and poly(lactic acid) (PLA) eco-friendly elastic blends. Therefore, the ENR/PLA blends were filled with natural pro-health substances of potentially antioxidative behavior, namely, δ-tocopherol (vitamin E), curcumin, β-carotene and quercetin. In this way, the material biodeterioration potential was maintained and the material's lifespan was prolonged while subjected to increased temperatures or high-energy UVA irradiation (340 nm). The investigation of the samples' properties indicated that curcumin and quercetin are the most promising natural additives that may contribute to the delay of ENR/PLA degradation under the above-mentioned conditions. The efficiency of the proposed new natural anti-aging additives was proven with static mechanical analysis, color change investigation, as well as mass loss during a certain aging. The aging coefficient, which compares the mechanical properties before and after the aging process, indicated that the ENR/PLA performance after 200 h of accelerated aging might decrease only by approximately 30% with the blend loaded with quercetin. This finding paves new opportunities for bio-based and green anti-aging systems employed in polymer technology.

Keywords: poly(lactic acid); epoxidized natural rubber; polymer blend; natural additives; antioxidant

Citation: Masek, A.; Cichosz, S. Biocomposites of Epoxidized Natural Rubber/Poly(Lactic Acid) Modified with Natural Substances: Influence of Biomolecules on the Aging Properties (Part II). *Polymers* **2021**, *13*, 1677. https://doi.org/10.3390/polym13111677

Academic Editors: Beata Podkościelna and Andrzej Puszka

Received: 12 May 2021
Accepted: 19 May 2021
Published: 21 May 2021

1. Introduction

Hybrid plastic materials are referred as blends of two types of polymers: bio- and petro-based. Due to substituting a part of the petro-based resin with polymers and additives from renewable resources, they are considered to be more sustainable in comparison with their pure synthetic analogues [1–6]. Hybrid plastic materials are being created for many reasons. Yet, some major factors might be highlighted: rising environmental awareness, trial of incorporation of less noxious nature-derived substances into the polymer technology and economic reasons (the need to balance the costs of biodegradable polymeric matrices by manufacturers) [7–13].

As it was mentioned in the previous part [14] of this research, Hamad et al. [15] have gathered a broad range of precise information concerning blends containing poly(lactic acid) (PLA), their properties and applications. An additional subsection of the mentioned review debated on the future perspectives. The authors underlined the importance of PLA-based materials in 3D-printing and the development of porous PLA materials for biomedical applications, but they also drew attention to the importance of the recyclability and reformability of the PLA-containing blends. The last problem was also underlined in the recently published work by La Mantia et al. [16], namely, the lack of sufficient scientific background regarding the recycling of biopolymer blends. Contrary to synthetic polymer blends, PLA-containing materials exhibit a higher degradation rate, which is obviously attributed to the poly(lactic acid) content [17–19].

Clearly, it is more desirable to create a material which fully undergoes biodeterioration. Yet, since PLA blending is a promising modification technique enabling us to overcome some noticeable drawbacks of neat poly(lactic acid) [20–26] and hybrid bioplastic material creation helps to balance the production costs, the stabilization of PLA-based blends requires significant attention. Fortunately, similarly to pure bio- and synthetic polymers, the recycling of their blends might be carried out via mechanical or chemical recycling methods [1,16].

Nevertheless, research published by La Mantia et al. [27] confirmed that the presence of even small amounts of PLA in the poly(ethylene terephthalate) (PET) waste can noticeably affect the rheological properties of a recycled material. Moreover, scientists proved that the mechanical properties of the reproduced PET product were significantly changed.

Another example may be another work of Hamad et al. [28], which describes the properties of the PLA and polystyrene (PS) blend (PLA/PS = 50/50 wt%). According to the presented results, multiple extrusion and injection of a PLA/PS blend resulted in a drop in the values of stress and strain at break even after two processing cycles. Yet, according to the information reported by the authors, Young's modulus was not significantly affected at this time. Unfortunately, while four processing cycles were performed, the Young's modulus showed a reduction by 26%. The most noticeable property degradation was observed regarding the strain at break with 73% loss and stress at break with 79% drop. The authors attributed this phenomenon to the reduction in the polymers' molecular weight with the multiple processing cycles.

The examples given above reveal how important it is to stabilize the PLA-containing blends. Hamad et al. [15] claim that it is possible to introduce antioxidative agents through the creation of composites containing some nanoparticles, e.g., nanoclays, silver nanoparticles, metal oxides or functional biopolymers.

According to Li et al. [29,30], the stabilization of rubber-like materials may be obtained with the addition of sulfur. Nonetheless, this research is a proposal of an alternative way to achieve this goal with some eco-friendly plant-derived substances, e.g., vitamins, flavonoids and carotenes. According to the information gathered in the literature, these plant-derived substances may play the role of antioxidants and anti-aging factors [31–33]. Lately, it was found that, among others, eugenol [34], rosmarinic acid [35], phytic acid [36] or catechol [37] might be successfully applied in polymeric materials in order to prevent their properties' loss during aging. Additionally, the significant aspect of the research performed by our group is the broadening of the knowledge on the antioxidative potential of natural substances [38–43].

Recently, our team has proven that apart from the above-mentioned plant-derived compounds, hesperidin, which can be found in various citric fruits, may play the role of the effective anti-aging factor in silica-filled ethylene-norbornene copolymer (EN) based materials. Not only did it prohibit the carbonyl groups' formation during the performed 400-h-long weathering aging, but also hesperidin prevented the loss of the mechanical properties of the polymer composite during weathering (initial tensile strength of the blends at the level of 40 MPa; after 400 h of aging: approximately 10 MPa for EN + silica and 30 MPa while hesperidin added) [44].

Therefore, the aim of this study is to investigate the effect of different plant-derived substances, whose anti-oxidant properties have been primarily assessed by our team [45–48]. δ-tocopherol (vitamin E; δ-TF), curcumin (CM), β-carotene (β-CT) and quercetin (QU) were chosen regarding the elastic blends of epoxidized natural rubber (ENR) and poly(lactic acid) stabilization. According to the previous research, the above-mentioned biomolecules may play the role of effective natural antioxidants, simultaneously being relatively cheap [45–48].

Thanks to the incorporation of natural additives, material biodeterioration potential described in the previous part of the research [14] might be maintained and the composites' lifespan prolonged while subjected to increased temperatures or UV irradiation. The stabilization effect was tested within the 200 h-long thermo-oxidative and UV accelerated aging processes. Thus, it was proven that some of the mentioned natural additives may successfully play the role of antioxidants in the investigated elastic and eco-friendly ENR/PLA blends.

2. Materials and Methods

2.1. Materials

Polymer matrixes employed in this research: epoxidized natural rubber (ENR) (Epoxyprene ENR-50; 50 mol% epoxidation) obtained from Kumpulan Guthrie Berhad (Kuala Lumpur, Malaysia) and poly(lactic acid) (PLA), Ingeo 4043D purchased from Nature Works (Minnetonka, MN, USA). Lauric acid (97% purity), 1,2-dimethylimidazole (highest available purity) provided by Sigma-Aldrich (Darmstadt, Germany) and elastin hydrolysate purchased from Proteina (Lodz, Poland) were used as a cross-linking system. Natural additives, namely, δ-tocopherol (δ-TF), CM—curcumin (CM), β-carotene (β-CT), quercetin (QU), of the highest available purities were purchased from Sigma-Aldrich (Darmstadt, Germany). Structures of natural additives are shown in Figure 1.

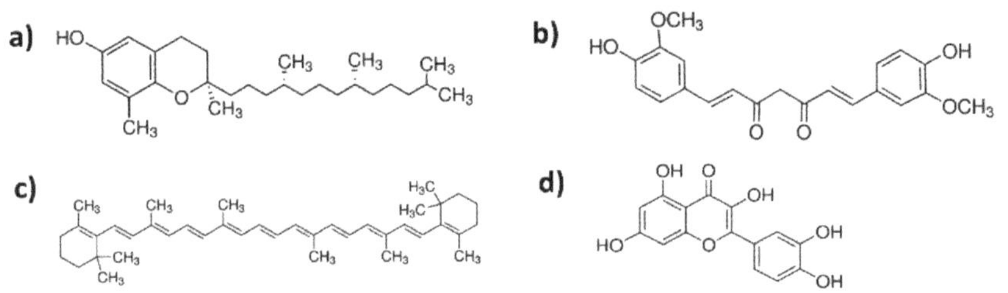

Figure 1. Structures of natural additives employed in this research: (**a**) δ-tocopherol (δ-TF), (**b**) curcumin (CM), (**c**) β-carotene (β-CT), (**d**) quercetin (QU).

2.2. Preparation of ENR/PLA Samples

Poly(lactic acid) was conditioned for 24 h at 70 °C in a laboratory oven (Binder, Tuttlingen, Germany) before being incorporated into epoxidized natural rubber. All mixture components (Table 1) were mixed in a micromixer (Brabender Lab-Station from Plasti-Corder with Julabo cooling system) at 160 °C for 30 min (50 rpm). Next, such prepared material was put between two roll mills (100 mm × 200 mm)—roll's temperature: 20–25 °C, friction: 1:1.1, time: 2 min. The last step was to form the plate-like samples between two steel molds (with the use of Teflon sheets separating the mixture from the mold) in a hydraulic press—temperature: 160 °C, time: 60 min, pressure: 125 bar.

Table 1. Composition of the polymer blend mixtures prepared for analysis in this research. Abbreviations: ENR—epoxidized natural rubber, PLA—poly(lactic acid), LA—lauric acid, DMI—1,2-dimethylimidazole, EH—elastin hydrolysate, δ-TF—δ-tocopherol, CM—curcumin, β-CT—β-carotene, QU—quercetin, phr—per hundred rubber (it means: for one hundred parts by weight of rubber there are x parts by weight of the substance).

Sample	Polymer Mixture Composition [phr]								
	ENR	PLA	LA	DMI	EH	δ-TF	CM	β-CT	QU
ENR/PLA	100	75	3	0.6	0.6	----	----	----	----
+ δ-tocopherol	100	75	3	0.6	0.6	3	----	----	----
+ curcumin	100	75	3	0.6	0.6	----	3	----	----
+ β-carotene	100	75	3	0.6	0.6	----	----	3	----
+ quercetin	100	75	3	0.6	0.6	----	----	----	3

2.3. Accelerated Aging of the Materials

2.3.1. Thermo-Oxidative Aging

Thermo-oxidative aging was performed in a laboratory oven (Binder, Tuttlingen, Germany) at 70 °C for 200 h. Four samples of each ENR/PLA blend were placed in an oven and were taken out individually after, respectively, 50 h, 100 h, 150 h and 200 h.

2.3.2. UV Irradiation

ENR/PLA blend samples were mounted in the special holders in the Atlas UV 2000 apparatus (Duisburg, Germany). Again, specimens were taken out individually after, respectively, 50 h, 100 h, 150 h and 200 h. The aging cycle consisted of two alternating segments: day segment (240 min, 60 °C, UV irradiation: 0.7 W/m^2) and night segment (120 min, 50 °C, no UV radiation).

2.4. Methods of Polymer Blend Sample Characterization

2.4.1. Swelling in Toluene

The experiment aim was to measure the weight of the sample immersed in toluene before and after the swelling process. Thus, 4 samples of different shapes of each material were prepared (30–40 mg). The samples were weighed before the measurement (m_1) and then put into the vessel with toluene. When 48 h were over, they were taken out and weighed in the swollen state (m_2). The excess toluene was cleaned with the filter paper and each sample was immersed in ethyl ether for 1–2 s. Subsequently, the samples were dried to constant weight in an oven at 50 °C for 96 h and weighed when the time was over (m_3). In order to analyze the potential cross-linking density, two parameters were calculated: $m_{rise} = m_2 - m_1$ (m_{rise} reveals how much the sample is swollen in the solvent), $m_{loss} = m_3 - m_1$ (m_{loss} informs about the low molecular weight compounds washed out during the swelling in toluene).

2.4.2. Contact Angle Measurement

Contact angle measurements were carried out for: distilled water and 1,4-diiodomethane (droplet volume set for: 1 µL). Before the measurement, surfaces of polymer composite samples were cleaned with the use of acetone. OCA 15EC goniometer by DataPhysics Instruments GmbH (Filderstadt, Germany) equipped with single direct dosing system (0.01–1 mL B. Braun syringe, Hassen, Germany) was employed. Surface free energy is calculated with the Owens–Wendt–Rabel–Kaelble (OWRK) method.

2.4.3. Tensile Tests

Mechanical properties, namely, tensile strength (TS) and elongation at break (Eb), were determined based on the ISO-37 with the use of a Zwick-Roell 1435 device (Ulm, Germany). Tests were carried out on a "dumbbell" shape (thickness: 1 mm, width of measured fragment: 4 mm, length of measured fragment: 25 mm, total length: 75 mm, width at the ends: 12.5 mm). The samples were stretched at a speed of 500 mm/min. On the basis of obtained results, the aging coefficient K was calculated according to the Equation (1) as a quotient of the product of TS and Eb after and before the performed aging process [49]:

$$K = \frac{(TS \cdot Eb)_{after\ ageing}}{(TS \cdot Eb)_{before\ ageing}} \tag{1}$$

2.4.4. Color Change

Spectrophotometer UV-VIS CM-36001 from Konica Minolta. Sample color was described with the CIE-Lab system (L—lightness, a—red-green, b—yellow-blue) employed in this research. Then, color difference (ΔE), whiteness index (W_i), chroma (C_{ab}) and hue angle (h_{ab}) values were calculated according to the equations given below (2–4):

$$\Delta E = \sqrt{(\Delta a)^2 + (\Delta b)^2 + (\Delta L)^2} \quad (2)$$

$$W_i = 100 - \sqrt{a^2 + b^2 + (100 - L)^2} \quad (3)$$

$$C_{ab} = \sqrt{a^2 + b^2} \quad (4)$$

$$h_{ab} = \begin{cases} arctg\left(\frac{b}{a}\right),\ when\ a > 0 \wedge b > 0 \\ 180^\circ + arctg\left(\frac{b}{a}\right),\ when\ (a < 0 \wedge b > 0) \vee (a < 0 \wedge b < 0) \\ 360^\circ + arctg\left(\frac{b}{a}\right),\ when\ a > 0 \wedge b < 0 \end{cases} \quad (5)$$

2.4.5. Mass Loss during the Degradation Process

In order to observe some mass changes, while samples were subjected to elevated temperature and UV irradiation, specially prepared samples were being weighed at the following times during the accelerated aging processes: 0 h, 50 h, 100 h, 150 h and 200 h.

2.4.6. Scanning Electron Microscopy (SEM) Analysis

Scanning electron microscopy (Zeiss, ULTRA Plus, Oberchoken, Germany) was employed in order to examine the morphology of prepared ENR/PLA blend samples. Magnification was 250 and 1000 times.

3. Results and Discussion

The previous part [14] of this research study showed that it is possible to control the degradation rate of ENR/PLA blends with plant-derived fibers while subjected to increased temperature or UV irradiation, simultaneously causing their easier biodeterioration. Moreover, it was revealed that thanks to the organic–inorganic phase ratio, the control over the material degradation rate during thermo-oxidation and UV aging could be enabled. However, the incorporation of natural fibers into ENR/PLA blends without any additional anti-aging substances was the reason for their lower resistance to the two above-mentioned factors (temperature, UV), and only MMT was found to be able to prolong the lifespan of analyzed ENR/PLA blends.

Therefore, in this part of the research, different substances leading to the ENR/PLA blends' stabilization are investigated. Some natural compounds of the potentially antioxidative properties [31–33] have been chosen: δ-tocopherol (δ-TF), curcumin (CM), β-carotene (β-CT), quercetin (QU). Their effect on the ENR/PLA blend properties before and after the accelerated aging processes is shown below.

Similarly to the previous part [14] of the study, analyzed polymer blends were investigated regarding the properties which may indicate some information about the degradation of the ENR/PLA blends during the accelerated aging processes. A swelling experiment was caried out in order to assess the cross-linking density as the crosslinks may stabilize the structure of the polymer network and prevent quick material degradation [50]. Moreover, the polymer composites exhibiting a higher polar component of surface free energy are known to be more prone to degradation during the aging processes regarding the ongoing radical reactions [51], e.g., oxidation. Thus, surface free energy was examined with the employment of contact angle technique. Finally, it was examined if the natural additives proposed in this research influence the mechanical performance of the analyzed ENR/PLA blends. Then, the effect of thermo-oxidative and UV accelerated aging was tested.

3.1. Characterization of Specimens before the Accelerated Aging Process

The swelling experiment, tensile tests and contact angle measurements were carried out in order to, respectively, assess the cross-linking density of the prepared material, investigate the influence of natural additives on the blends' mechanical performance and predict the degradation potential of the prepared specimens.

In Figure 2a, two parameters are presented: mass rise after the swelling process, which is proportional to the number of bonds created between different macromolecules, and mass loss during the swelling process, which is proportional to the amount of low molecular weight particles washed out in the swollen state [52]. Thus, it may be concluded that δ-TF and QU did not affect cross-linking density in a significant way as the samples are swollen similarly to the reference ENR/PLA blend. At the same time, in the case of the ENR/PLA samples with the addition of CM and β-CT, a great drop in mass rise after the swelling process was detected. This may indicate that these two compounds (CM, β-CT) led to the significant improvement in cross-linking density.

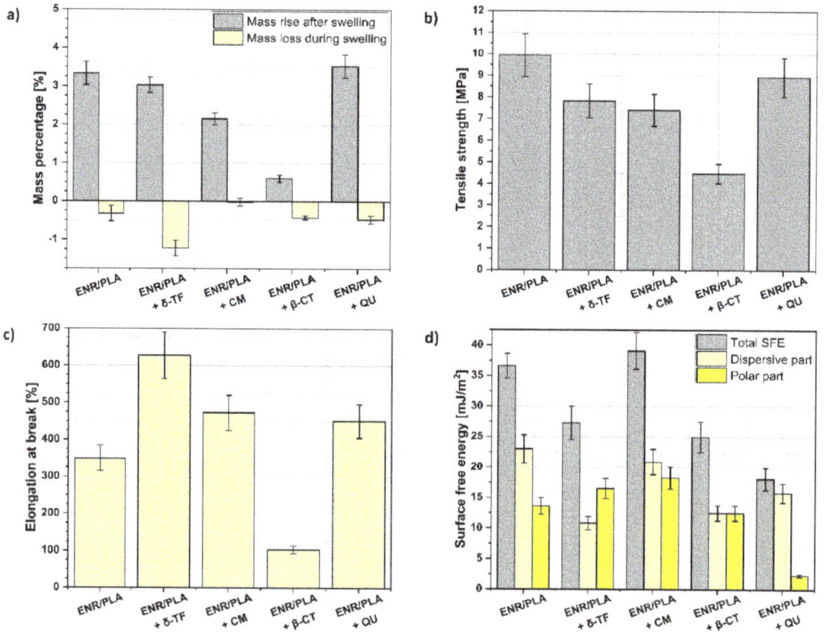

Figure 2. ENR/PLA blend properties before the accelerated aging processes: (**a**) analysis of the swelling experiment, (**b**) tensile strength of analyzed specimens, (**c**) elongation at break of investigated samples, (**d**) surface free energy and its components analysis.

Furthermore, analyzing the data concerning the mass loss during the swelling process, it is visible that different amounts of low-molecular weight substances were washed out during the swelling of ENR/PLA. The reason for the observed differences could be the varied rate in the migration of the applied compounds to the surface of the polymer blend [52], e.g., the mass loss of the CM-loaded sample is the lowest, while for δ-TF addition the highest rinsing effect is detected.

Moving forward and analyzing the mechanical properties of prepared ENR/PLA blends (Figure 2b,c), some more variations between the samples are revealed. The specimen which was expected to have cross-linked the most and exhibited the lowest swelling degree, namely, ENR/PLA + β-CT, was revealed to have the lowest tensile strength and elongation at break. This could be explained with the fact that β-CT possesses in its structure many C=C bonds that can easily take part in cross-linking during the vulcanization process, therefore stiffening the material and preventing its elongation [53]. At the same time, the rest of the prepared samples revealed similar values of tensile strength on the level of the ENR/PLA reference blend or slightly lower.

More variations can be found regarding the elongation at break (Figure 2c). Each natural additive, apart from β-CT, enabled obtaining the material which can elongate more than the reference ENR/PLA blend. The highest elongation is achieved in the case of the sample filled with δ-TF. This could be explained with the fact that its molecule possesses in its structure a long carbon chain that may work as a plasticizing agent [18].

The plasticization effect of the analyzed ENR/PLA blend specimens with an addition of δ-TF and CM may also be visible regarding the tensile stress values for the elongations of 100%, 200% and 300% (Table 2). In the case of the samples mentioned above, the tensile stress at any point is lower than for the reference ENR/PLA polymer blend.

Table 2. Tensile stress at elongation of 100%, 200%, 300% for the unfilled and filled ENR/PLA blend samples.

Sample	Tensile Stress [MPa] at Elongation of:		
	100%	200%	300%
ENR/PLA	4.5 ± 0.2	6.7 ± 0.3	8.8 ± 0.2
ENR/PLA + δ-tocopherol	3.3 ± 0.1	4.7 ± 0.2	5.8 ± 0.2
ENR/PLA + curcumin	3.8 ± 0.4	5.2 ± 0.5	6.3 ± 0.3
ENR/PLA + β-carotene	4.0 ± 0.3	—	—
ENR/PLA + quercetin	5.2 ± 0.3	6.8 ± 0.4	8.0 ± 0.5

Moreover, according to the data gathered in Figure 2d, some variations regarding the surface free energy and both its components are revealed. Taking into consideration the presented results, specimens loaded with δ-TF and β-CT seem to be more hydrophilic. This is evidenced with contact angle measurements (Table 3) as these materials exhibit easier wetting with water (lower value of contact angle). On the other hand, QU incorporation into the ENR/PLA blend leads to the surface hydrophobization and the contact angle with water rises up to (101 ± 4)°, which is the highest of observed values.

Table 3. Water and diiodomethane contact angles for the unfilled and filled ENR/PLA blend samples.

Sample	Contact Angle [°]	
	Water	Diiodomethane
ENR/PLA	70 ± 3	57 ± 2
ENR/PLA + δ-tocopherol	67 ± 3	82 ± 1
ENR/PLA + curcumin	65 ± 2	60 ± 2
ENR/PLA + β-carotene	82 ± 1	80 ± 2
ENR/PLA + quercetin	101 ± 4	80 ± 3

Samples with the addition of δ-TF, β-CT and QU exhibit lower total surface free energy in comparison with the reference ENR/PLA blend. The only material containing natural additives and exhibiting surface properties similar to the reference sample is ENR/PLA + CM. The lowest value for the polar component of surface free energy is detected for the sample loaded with QU, which may indicate the highest stabilization potential regarding further aging tests (potentially less polar active centers prone to oxidation) [50].

Finally, the morphology of the prepared ENR/PLA blend samples was examined with the use of the SEM method. Images captured during the measurement are presented in Figure 3. The magnifications of 250 and 1000 times were applied.

Importantly, no signs of the blend components' separation can be denoted. This means that the mixing process of ENR and PLA was sufficient and effective.

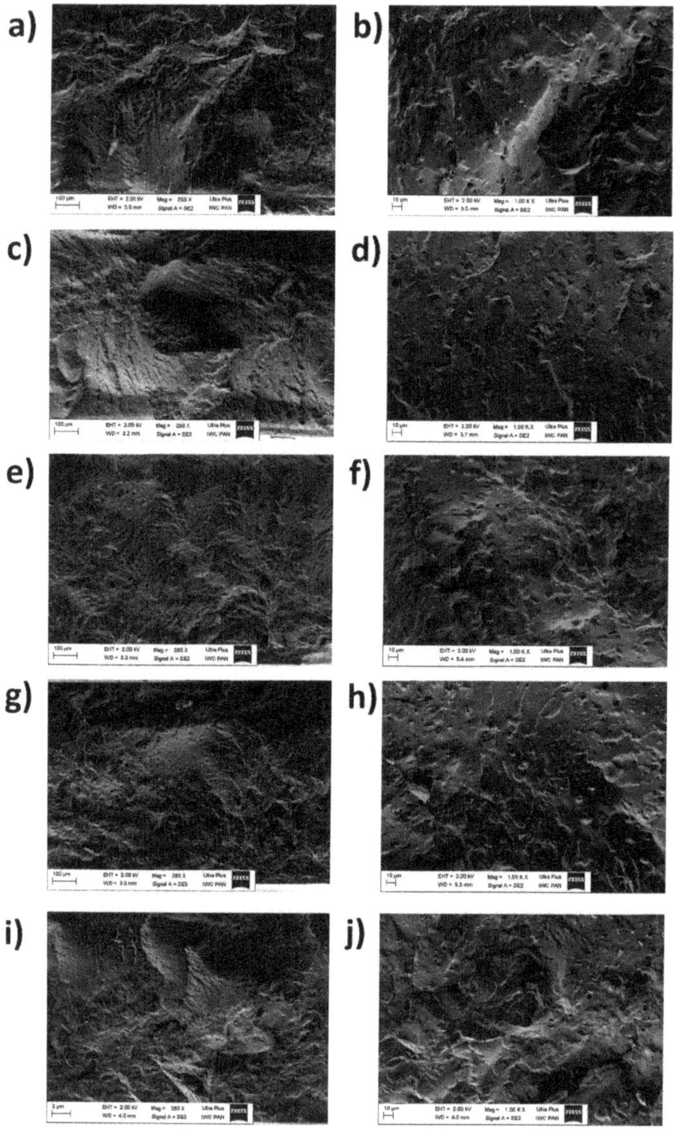

Figure 3. SEM images for: reference sample of ENR/PLA blend (**a**,**b**), ENR/PLA + δ-tocopherol (**c**,**d**), ENR/PLA + curcumin (**e**,**f**), ENR/PLA + β-carotene (**g**,**h**), ENR/PLA + quercetin (**i**,**j**) with magnifications, respectively, 250 and 1000.

Moreover, performed morphology analysis proved that the natural additives employed in this study are fairly well dispersed in the polymer matrix and no big aggregates are present. Nonetheless, some inclusions might be denoted, e.g., specimens filled with β-CT and QU.

It is also visible that the blends' surfaces are characterized by different regularity. The reference sample of ENR/PLA blend has an uneven and ragged surface, while specimens prepared with the addition of natural substances seem to be smoother. The only exception is the sample of QU-modified ENR/PLA blend.

3.2. Characterization of the Aging Impact

ENR/PLA blend samples were subjected to accelerated aging of two types: thermo-oxidative (increased temperature in the presence of air) and UV (chamber with UV lamps; 340 nm). During the aging processes, radical reactions might be initiated via temperature or UV irradiation and contribute to polymer chains' crosslinking, oxidation or scission. This leads to some irreversible changes in the properties of the material subjected to the aging process [54].

The performed accelerated aging lasted for 200 h, and during the processes, after each 50 h, the tensile properties, mass of the polymer blend and color parameters were tested in order to assess the rate of the degradation process (Figure 4) and if the accelerated aging caused crosslinking resulting in the material stiffening or oxidation/scission that may be evidenced with, e.g., lowering mechanical performance of the investigated polymer blend [55,56].

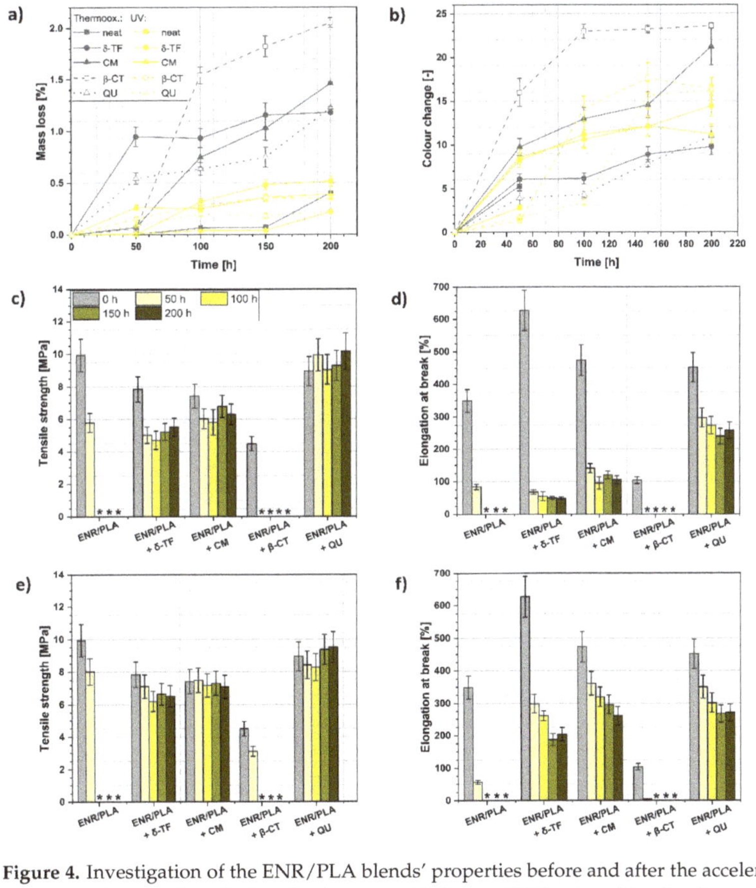

Figure 4. Investigation of the ENR/PLA blends' properties before and after the accelerated aging processes: (**a**) mass loss during the thermo-oxidative and UV aging, (**b**) color change during the thermo-oxidative and UV aging, (**c**) tensile strength changes and (**d**) elongation at break variations during the thermo-oxidative aging, (**e**) tensile strength changes and (**f**) elongation at break variations during the UV aging. *—sample too brittle to be examined with the selected method; the measurement was impossible.

Regarding the data gathered in Figure 4a, some differences between thermo-oxidative and UV aging could be found. Firstly, higher mass loss might be detected in the case of the first type of aging. It might be explained with the moisture content in the hydrophilic components of the loaded ENR/PLA blend [57,58], which is lost while the sample is subjected to the elevated temperature.

For thermo-oxidative aging, the highest mass variations could be evidenced in the case of β-CT and δ-TF-loaded ENR/PLA blends. In turn, CM incorporation seems to cause a slower mass decrease. Yet, after 200 h of aging, the mass loss of the CM-filled specimen achieved a value higher than that observed for δ-TF addition.

On the contrary, regarding the UV aging, weight loss occurs considerably slower and steadier. Moreover, the mass decrease values are the highest in the case of the reference ENR/PLA polymer blend, which means that the natural substances added to the composition could have possibly acted as successful stabilizers during the UV aging.

Moving forward to the graphs presented in Figures 4b and 5, the detected color changes are quite significant for some polymer blend samples, and in some cases they reach the value of color change of approximately 20, e.g., δ-TF and CM-loaded specimens. Moreover, they become darker or lighter depending on the aging.

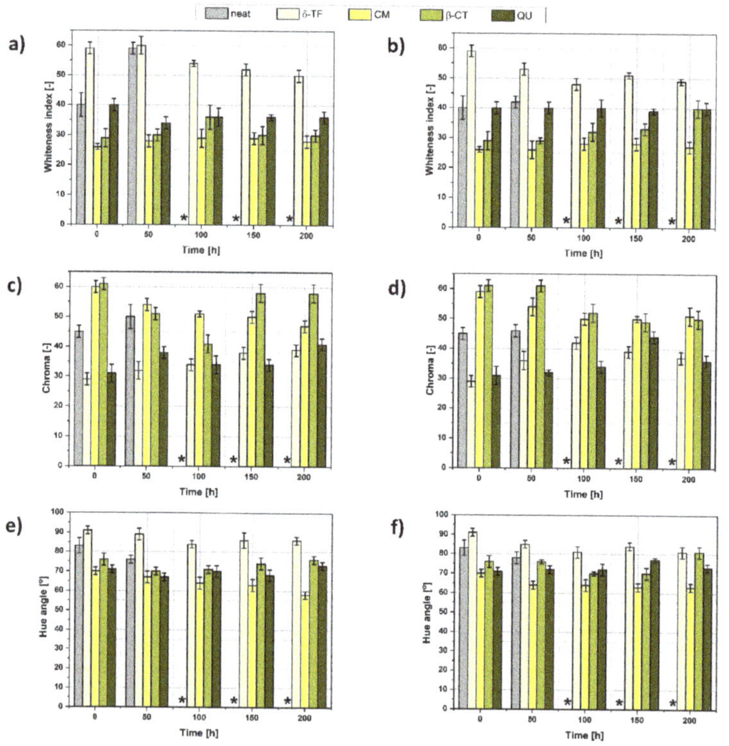

Figure 5. Additional parameters, namely, whiteness index, chroma, hue angle, contributing to color change of ENR/PLA blends and their variations during the thermo-oxidative (respectively: (**a**,**c**,**e**)) and UV (respectively: (**b**,**d**,**f**)) accelerated aging processes. *—sample too brittle to be examined with the selected method; the measurement was impossible.

A color change higher than 3 means that variations evidenced for these samples could be easily visible with the human eye. Furthermore, it was observed that all substances steadily and constantly change their color upon the carried out aging processes (Figures 4b and 6). Thus, the applied natural substances, apart from their application as natural antioxidants, might be very promising regarding the bio-based aging indicators for, e.g., food packaging [48,59].

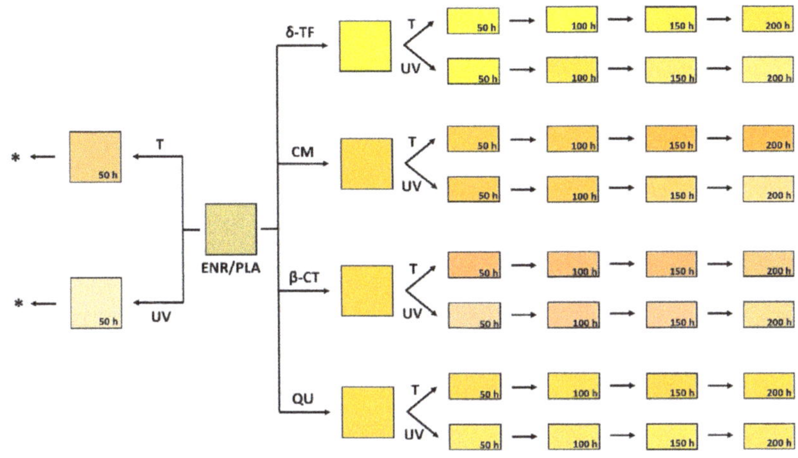

Figure 6. Color change representation for the prepared ENR/PLA blends before and after the accelerated thermo-oxidative (T), as well as UV-initiated (UV) aging processes. *—sample destroyed significantly.

It is worth noticing that samples of the reference ENR/PLA polymer blend after 100 h, 150 h, and 200 h of thermo-oxidative and UV aging were degraded noticeably. Thus, it was impossible to investigate the properties of the reference ENR/PLA polymer blend as it became too brittle to carry out the measurement.

Additionally, more data regarding the color change described with the parameters such as whiteness index, chroma and hue angle are presented in Figure 5. Based on the gathered information, some interesting conclusions might be made, e.g., the ENR/PLA + δ-TF specimen subjected to thermo-oxidation becomes darker and the color is more intense with the aging time. On the other hand, the color of the same sample subjected to UV irradiation becomes lighter. The color change representation is shown in Figure 6.

Moving forward and analyzing the mechanical properties of the prepared ENR/PLA blends during the aging process, it is visible that the reference material and the sample with the addition of β-CT are the most affected by the performed accelerated aging. Fortunately, δ-TF, CM and QU exhibit a stabilizing effect to a certain extent, depending on the used natural additive (Figure 4c–f).

According to the gathered results, the most effective stabilization is visible for the QU-filled ENR/PLA blend. The sample's performance slightly rose during the aging process and the elongation at break decreased the least in comparison with different samples.

The improvement in tensile strength upon the aging process observed for the QU-loaded sample might be explained with the previously evidenced ability of quercetin to cross-link ENR-containing blends [60]. On the other hand, the negative effect of β-CT on the ENR/PLA blend's properties during the aging process might be explained with two factors: the above-mentioned ability to take part in the crosslinking process (and, thus, over-crosslinking of the polymer blend) and possible prooxidative behavior related to the not properly adjusted amount of β-CT in the polymeric mixture [61–63].

Considering Figure 4c,d, which refer to the aging performed in the elevated temperature, samples filled with δ-TF and CM become stiffer during the aging process as the tensile strength stays almost at the same level and elongation at break drops significantly.

This phenomenon may be explained with the ability of quercetin reported in the literature to crosslink the ENR [60]. It is possible that increased temperature or UV irradiation during the performed accelerated aging processes could have initiated the crosslinking of the polymer matrix and, thus, further stabilized the ENR/PLA blend.

In turn, Figure 4e,f refer to the UV aging. It is visible that QU, again, exhibits a stabilizing effect and prevents the mechanical properties' loss. However, CM also effectively delays the material degradation, leading to keeping the tensile strength at the same level during the aging process.

A perfect confirmation for the stabilizing effect of these two compounds is presented in Figure 7. The values of the *K* coefficient, which compares the mechanical properties before and after the aging process, are presented for the samples prepared with and without potential natural anti-aging substances. Therefore, according to the data gathered in Figure 7, the positive effect of some among the applied compounds might be found (certain natural anti-aging substances are able to prevent a material's significant performance drop during the aging processes). Additionally, information gathered in Table 4 helps to compare the results presented in this research with the previously carried out study, which debates on plant fibers' incorporation into ENR/PLA blends.

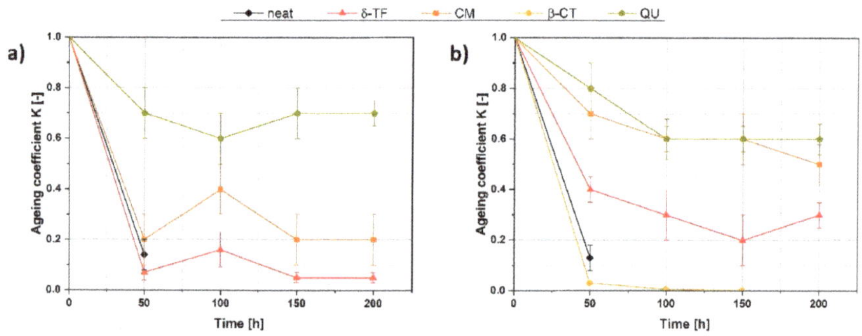

Figure 7. Aging coefficients attributed to the filled and unfilled ENR/PLA specimens at a certain aging time during thermo-oxidative (**a**) and UV (**b**) aging.

Table 4. Comparison of aging coefficient values with the results presented in the previous part of the study [14]; CF—cellulose fibers, FF—flax fibers, MMT—montmorillonite. *—sample too brittle to be examined with the selected method; the measurement was impossible.

Sample ENR/PLA	Aging Coefficient *K* [-]—Thermo-Oxidation			
	50 h	**100 h**	**150 h**	**200 h**
+ CF [14]	*	*	*	*
+ FF [14]	0.004 ± 0.002	0.002 ± 0.001	0.002 ± 0.001	*
+ FF + MMT [14]	0.3 ± 0.1	0.3 ± 0.1	0.20 ± 0,08	0.20 ± 0,07
Sample ENR/PLA	**Aging Coefficient *K* [-]—UV Irradiation**			
	50 h	**100 h**	**150 h**	**200 h**
+ CF [14]	0.03 ± 0.01	0.06 ± 0.03	0.02 ± 0.01	*
+ FF [14]	0.02 ± 0.01	0.06 ± 0.02	0.06 ± 0.02	0.01 ± 0.01
+ FF + MMT [14]	0.2 ± 0.1	0.18 ± 0.07	0.15 ± 0.05	0.21 ± 0.08

It is clearly visible that QU is the substance of the highest stabilizing potential and it can be applied while the material is subjected to elevated temperature or UV irradiation. The maximum shift of the aging coefficient detected for the QU-loaded sample is from 1 to 0.6 ± 0.3 for 200 h-lasting UV aging and to 0.7 ± 0.3 for thermo-oxidative aging, which means a mechanical performance loss by, respectively, 40% and 30%. QU stabilizing activity may be explained as a synergic effect of antioxidative properties and the ability to crosslink ENR [47]. Moreover, according to the gathered data, CM might also be a quite efficient anti-aging factor for ENR/PLA blends while considering UV-resistance (the performance drop by approximately 50%).

3.3. Possible Stabilization Effect of the Natural Substances Employed in This Research

The aim of this subsection is to clarify the possible stabilization behavior of the natural substances employed in this research and provide the reader with some useful information regarding the polymer blends' aging process analysis and the probable stabilization behavior of plant-derived substances employed in this research (Figure 8).

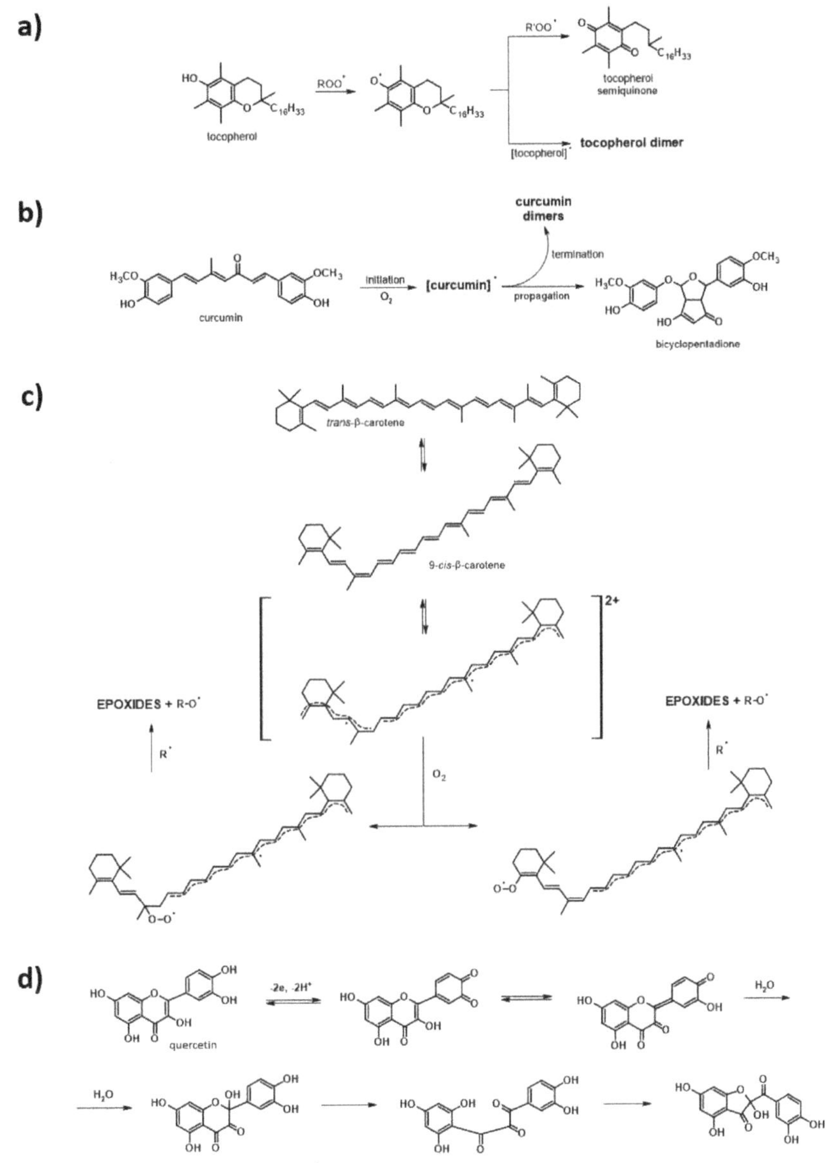

Figure 8. Possible mechanisms of free radical scavenging and oxidation behavior according to the available literature data: (**a**) δ-tocopherol (δ-TF), (**b**) curcumin (CM), (**c**) β-carotene (β-CT), (**d**) quercetin (QU).

The overall mechanism of polymer material degradation might be divided into four stages, as follows:

i. Initiation of the degradation reaction with external stimuli, e.g., temperature, irradiation, with simultaneous chemical bonds' cleavage and formation of free alkyl radicals;
ii. Initial propagation—reaction of alkyl radicals with oxygen and formation of peroxy/hydroxy radicals;
iii. Further propagation—chain branching and transferring radical activity to another chain;
iv. Termination and inactivated form creation with possible disproportionation reactions.

In general, polymer degradation leads to the shortening of the polymer chains' length and, therefore, it provides a material with different properties [32,64].

Thermal degradation starts while the interatomic vibration energy is equal to or exceeds the energy of interatomic bond dissociation. Then, the bond cleavage occurs and two active macroradicals, which may further take part in the propagation processes, are formed. In turn, UV light leads to the polymer chain photodegradation.

Similarly, it is initiated when the energy of absorbed radiation is greater than or at least equal to the dissociation energy of the individual bonds in the macromolecule. However, this type of degradation may also be initiated due to the presence of chromophore moieties which could shift their forms from basic singlet state to the excited singlet/triplet state, hence becoming incredibly reactive. In the case of PLA, the chromophore moieties are C=O carbonyl moieties [32].

On the other hand, regarding the structure of ENR, C=C bonds and unreacted oxirane rings are expected to be the most fragile to the free radicals' presence, thus being prone to oxidation or further crosslinking during aging. Another problem could be the presence of some proteins which may affect the material's resistance to elevated temperatures or UV irradiation [65,66].

Additionally, the above-described degradation processes are the perfect examples of avalanche reactions as the propagation step provides various different free radical forms, e.g., oxygen-incorporated (hydroxy/peroxy) and alkyl radicals. Thus, once initiated, the degradation process is highly challenging to stop.

The plant-derived substances employed in this research, namely, δ-tocopherol (δ-TF), curcumin (CM), β-carotene (β-CT), and quercetin (QU), exhibit strong antioxidant properties, hence they are promising natural antioxidants for polymer-based materials. According to various studies presented in the literature, they may easily react with free radicals generated during the degradation of the polymer matrix, acting as free radical scavengers and simultaneously being oxidized [64,67–72].

In this way, the chemical compounds that promote ENR/PLA degradation are eliminated and the deterioration of the polymer matrix is prevented or delayed (only the natural additive is degrading at that time and not the polymer matrix itself) [64,67–69]. Some possible stabilization mechanisms are presented in the works performed by Choe et al. [64] (δ-TF), Schneider et al. [67] (CM), Penicaud et al. [68] (β-CT), and Sokolova et al. [69] (QU). Some possible stabilization mechanisms, created on the basis of the available literature data presented in the works mentioned above, are given in Figure 8.

4. Conclusions

Taking into consideration the gathered data, some natural substances potentially able to stabilize elastic blends of epoxidized natural rubber (ENR) with poly(lactic acid) (PLA) have been found. Nonetheless, further analysis is highly advised. Among the tested natural additives, quercetin was proven to be a substance of the highest stabilizing potential. Thus, it can be applied while the material is subjected to elevated temperature or UV irradiation. The maximum shift of the aging coefficient detected for the QU-loaded sample is from 1 to 0.6 ± 0.3 for 200 h-lasting UV aging and to 0.7 ± 0.3 for thermo-oxidative aging, which means a mechanical performance loss by only, respectively, 40% and 30%. Moreover, curcumin seems to also be a quite efficient anti-aging additive for ENR/PLA blends while considering UV-resistance (the performance drop by approximately 50%). Curcumin is also very promising as a future natural aging indicator. The hue angle of CM-filled samples might be shifted significantly from (70 ± 2)° to (58 ± 2)° for thermo-oxidative aging and up to (63 ± 2)° for UV aging. Additionally, it was evidenced that these two substances, namely, quercetin and curcumin, do not lead to a deterioration of the mechanical performance of the ENR/PLA blends while added to the polymer matrix. The presented research indicates that the anti-aging effect in PLA-containing blends could be obtained with certain plant-derived additives. This provides new opportunities for the creation of the materials characterized by facilitated recycling and controllable lifespan.

Author Contributions: Conceptualization, formal analysis, data analysis, investigation, methodology, review and editing, A.M.; conceptualization, data analysis, investigation, methodology and writing, S.C. All authors have read and agreed to the published version of the manuscript.

Funding: Non-competition conceptual Ministry of Science and Higher Education project The best of the best! 4.0 implemented under the Priority Axis III of the Operational Program Knowledge Education Development 2014–2020 co-financed by the European Social Fund (application number for funding POWR.03.03.00-00-P019/18; contract with the Ministry of Science and Higher Education–No. MNiSW/2020/169/DIR/NN4 signed on 4 June 2020).

Institutional Review Board Statement: Not applicable.

Informed Consent Statement: Not applicable.

Data Availability Statement: No data available.

Acknowledgments: This work has been completed while the second author was the Lodz University of Technology, Poland.

Conflicts of Interest: The authors declare no conflict of interest.

References

1. Soroudi, A.; Jakubowicz, I. Recycling of bioplastics, their blends and biocomposites: A review. *Eur. Polym. J.* **2013**, *49*, 2839–2858. [CrossRef]
2. Andrzejewski, J.; Szostak, M.; Barczewski, M.; Łuczak, P. Cork-wood hybrid filler system for polypropylene and poly(lactic acid) based injection molded composites. Structure evaluation and mechanical performance. *Compos. Part B Eng.* **2019**, *163*, 655–668. [CrossRef]
3. Barczewski, M.; Sałasińska, K.; Szulc, J. Application of sunflower husk, hazelnut shell and walnut shell as waste agricultural fillers for epoxy-based composites: A study into mechanical behavior related to structural and rheological properties. *Polym. Test.* **2019**, *75*, 1–11. [CrossRef]
4. Aliotta, L.; Gigante, V.; Coltelli, M.B.; Cinelli, P.; Lazzeri, A. Evaluation of Mechanical and Interfacial Properties of Bio-Composites Based on Poly(Lactic Acid) with Natural Cellulose Fibers. *Int. J. Mol. Sci.* **2019**, *20*, 960. [CrossRef]
5. Rayung, M.; Ibrahim, N.A.; Zainuddin, N.; Saad, W.Z.; Razak, N.I.A.; Chieng, B.W. The Effect of Fiber Bleaching Treatment on the Properties of Poly(lactic acid)/Oil Palm Empty Fruit Bunch Fiber Composites. *Int. J. Mol. Sci.* **2014**, *15*, 14728–14742. [CrossRef] [PubMed]
6. Blessing, B.; Trout, C.; Morales, A.; Rybacki, K.; Love, S.A.; Lamoureux, G.; O'Malley, S.M.; Hu, X.; La Cruz, D.S.-D. The Impact of Composition and Morphology on Ionic Conductivity of Silk/Cellulose Bio-Composites Fabricated from Ionic Liquid and Varying Percentages of Coagulation Agents. *Int. J. Mol. Sci.* **2020**, *21*, 4695. [CrossRef]
7. KAIRYTĖ, A.; Kremensas, A.; Vaitkus, S.; Członka, S.; Strąkowska, A. Fire Suppression and Thermal Behavior of Biobased Rigid Polyurethane Foam Filled with Biomass Incineration Waste Ash. *Polymers* **2020**, *12*, 683. [CrossRef]
8. Członka, S.; Strąkowska, A. Rigid Polyurethane Foams Based on BioPolyol and Additionally Reinforced with Silanized and Acetylated Walnut Shells for the Synthesis of Environmentally Friendly Insulating Materials. *Materials* **2020**, *13*, 3245. [CrossRef] [PubMed]
9. Członka, S.; Kairytė, A.; Miedzińska, K.; Strąkowska, A. Polyurethane Hybrid Composites Reinforced with Lavender Residue Functionalized with Kaolinite and Hydroxyapatite. *Materials* **2021**, *14*, 415. [CrossRef]
10. Bartos, A.; Anggono, J.; Farkas, Á.E.; Kun, D.; Soetaredjo, F.E.; Móczó, J.; Purwaningsih, H.; Pukánszky, B. Alkali treatment of lignocellulosic fibers extracted from sugarcane bagasse: Composition, structure, properties. *Polym. Test.* **2020**, *88*, 106549. [CrossRef]
11. Barczewski, M.; Mysiukiewicz, O.; Kloziński, A. Complex modification effect of linseed cake as an agricultural waste filler used in high density polyethylene composites. *Iran. Polym. J.* **2018**, *27*, 677–688. [CrossRef]
12. Szadkowski, B.; Kuśmierek, M.; Rybiński, P.; Żukowski, W.; Marzec, A. Application of Earth Pigments in Cycloolefin Copolymer: Protection against Combustion and Accelerated Aging in the Full Sunlight Spectrum. *Materials* **2020**, *13*, 3381. [CrossRef] [PubMed]

13. Plota, A.; Masek, A. Lifetime Prediction Methods for Degradable Polymeric Materials—A Short Review. *Materials* **2020**, *13*, 4507. [CrossRef]
14. Masek, A.; Cichosz, S.; Piotrowska, M. Biocomposites of Epoxidized Natural Rubber/Poly(lactic acid) Modified with Natural Fillers (Part I). *Int. J. Mol. Sci.* **2021**, *22*, 3150. [CrossRef]
15. Hamad, K.; Kaseem, M.; Ayyoob, M.; Joo, J.; Deri, F. Polylactic acid blends: The future of green, light and tough. *Prog. Polym. Sci.* **2018**, *85*, 83–127. [CrossRef]
16. La Mantia, F.P.; Botta, L.; Mistretta, M.C.; Di Fiore, A.; Titone, V. Recycling of a Biodegradable Polymer Blend. *Polymers* **2020**, *12*, 2297. [CrossRef]
17. Piemonte, V.; Gironi, F. Kinetics of Hydrolytic Degradation of PLA. *J. Polym. Environ.* **2013**, *21*, 313–318. [CrossRef]
18. Elsawy, M.A.; Kim, K.-H.; Park, J.-W.; Deep, A. Hydrolytic degradation of polylactic acid (PLA) and its composites. *Renew. Sustain. Energy Rev.* **2017**, *79*, 1346–1352. [CrossRef]
19. Najafi, N.; Heuzey, M.; Carreau, P.; Wood-Adams, P.M. Control of thermal degradation of polylactide (PLA)-clay nanocomposites using chain extenders. *Polym. Degrad. Stab.* **2012**, *97*, 554–565. [CrossRef]
20. Bijarimi, M.; Ahmad, S.; Rasid, R. Mechanical, thermal and morphological properties of poly(lactic acid)/epoxidized natural rubber blends. *J. Elastomers Plast.* **2012**, *46*, 338–354. [CrossRef]
21. Andrzejewski, J.; Skórczewska, K.; Kloziński, A. Improving the Toughness and Thermal Resistance of Polyoxymethylene/Poly(lactic acid) Blends: Evaluation of Structure–Properties Correlation for Reactive Processing. *Polymers* **2020**, *12*, 307. [CrossRef] [PubMed]
22. Formela, K.; Zedler, Ł.; Hejna, A.; Tercjak, A. Reactive extrusion of bio-based polymer blends and composites—Current trends and future developments. *Express Polym. Lett.* **2018**, *12*, 24–57. [CrossRef]
23. Yao, Q.; Cosme, J.G.; Xu, T.; Miszuk, J.M.; Picciani, P.H.; Fong, H.; Sun, H. Three dimensional electrospun PCL/PLA blend nanofibrous scaffolds with significantly improved stem cells osteogenic differentiation and cranial bone formation. *Biomaterials* **2017**, *115*, 115–127. [CrossRef]
24. Ahmed, M.F.; Li, Y.; Yao, Z.; Cao, K.; Zeng, C. TPU/PLA blend foams: Enhanced foamability, structural stability, and implications for shape memory foams. *J. Appl. Polym. Sci.* **2019**, *136*, 47416. [CrossRef]
25. Siracusa, V.; Karpova, S.; Olkhov, A.; Zhulkina, A.; Kosenko, R.; Iordanskii, A. Gas Transport Phenomena and Polymer Dynamics in PHB/PLA Blend Films as Potential Packaging Materials. *Polymers* **2020**, *12*, 647. [CrossRef] [PubMed]
26. Sookprasert, P.; Hinchiranan, N. Morphology, mechanical and thermal properties of poly(lactic acid) (PLA)/natural rubber (NR) blends compatibilized by NR-graft-PLA. *J. Mater. Res.* **2017**, *32*, 788–800. [CrossRef]
27. La Mantia, F.; Botta, L.; Morreale, M.; Scaffaro, R. Effect of small amounts of poly(lactic acid) on the recycling of poly(ethylene terephthalate) bottles. *Polym. Degrad. Stab.* **2011**, *97*, 21–24. [CrossRef]
28. Hamad, K.; Kaseem, M.; Deri, F. Effect of recycling on rheological and mechanical properties of poly(lactic acid)/polystyrene polymer blend. *J. Mater. Sci.* **2010**, *46*, 3013–3019. [CrossRef]
29. Li, S.; Leng, D.; Li, W.; Qie, L.; Dong, Z.; Cheng, Z.; Fan, Z. Recent progress in developing Li2S cathodes for Li–S batteries. *Energy Storage Mater.* **2020**, *27*, 279–296. [CrossRef]
30. Li, S.; Fan, Z. Encapsulation methods of sulfur particles for lithium-sulfur batteries: A review. *Energy Storage Mater.* **2021**, *34*, 107–127. [CrossRef]
31. Domínguez, R.; Barba, F.J.; Gómez, B.; Putnik, P.; Kovačević, D.B.; Pateiro, M.; Santos, E.M.; Lorenzo, J.M. Active packaging films with natural antioxidants to be used in meat industry: A review. *Food Res. Int.* **2018**, *113*, 93–101. [CrossRef] [PubMed]
32. Kirschweng, B.; Tátraaljai, D.; Földes, E.; Pukánszky, B. Natural antioxidants as stabilizers for polymers. *Polym. Degrad. Stab.* **2017**, *145*, 25–40. [CrossRef]
33. Sanches-Silva, A.; Costa, D.; Albuquerque, T.; Buonocore, G.G.; Ramos, F.; Castilho, M.C.; Machado, A.V.; Costa, H.S. Trends in the use of natural antioxidants in active food packaging: A review. *Food Addit. Contam. Part A* **2014**, *31*, 374–395. [CrossRef] [PubMed]
34. Goñi, M.L.; Gañan, N.; Strumia, M.C.; Martini, R.E. Eugenol-loaded LLDPE films with antioxidant activity by supercritical carbon dioxide impregnation. *J. Supercrit. Fluids* **2016**, *111*, 28–35. [CrossRef]
35. Doudin, K.; Al-Malaika, S.; Sheena, H.; Tverezovskiy, V.; Fowler, P. New genre of antioxidants from renewable natural resources: Synthesis and characterisation of rosemary plant-derived antioxidants and their performance in polyolefins. *Polym. Degrad. Stab.* **2016**, *130*, 126–134. [CrossRef]
36. Diouf-Lewis, A.; Commereuc, S.; Verney, V. Toward greener polyolefins: Antioxidant effect of phytic acid from cereal waste. *Eur. Polym. J.* **2017**, *96*, 190–199. [CrossRef]
37. Zhan, K.; Ejima, H.; Yoshie, N. Antioxidant and Adsorption Properties of Bioinspired Phenolic Polymers: A Comparative Study of Catechol and Gallol. *ACS Sustain. Chem. Eng.* **2016**, *4*, 3857–3863. [CrossRef]
38. Masek, A.; Latos-Brozio, M.; Kałużna-Czaplińska, J.; Rosiak, A.; Chrzescijanska, E. Antioxidant Properties of Green Coffee Extract. *Forests* **2020**, *11*, 557. [CrossRef]
39. Masek, A.; Latos-Brozio, M.; Chrzescijanska, E.; Podsedek, A. Polyphenolic Profile and Antioxidant Activity of Juglans regia L. Leaves and Husk Extracts. *Forests* **2019**, *10*, 988. [CrossRef]
40. Masek, A.; Chrzescijanska, E.; Latos-Brozio, M.; Zaborski, M. Characteristics of juglone (5-hydroxy-1,4,-naphthoquinone) using voltammetry and spectrophotometric methods. *Food Chem.* **2019**, *301*, 125279. [CrossRef] [PubMed]

41. Masek, A.; Chrzescijanska, E.; Latos, M.; Zaborski, M. Influence of hydroxyl substitution on flavanone antioxidants properties. *Food Chem.* **2017**, *215*, 501–507. [CrossRef] [PubMed]
42. Olejnik, O.; Masek, A.; Kiersnowski, A. Thermal Analysis of Aliphatic Polyester Blends with Natural Antioxidants. *Polymers* **2020**, *12*, 74. [CrossRef] [PubMed]
43. Olejnik, O.; Masek, A. Bio-Based Packaging Materials Containing Substances Derived from Coffee and Tea Plants. *Materials* **2020**, *13*, 5719. [CrossRef] [PubMed]
44. Masek, A.; Plota, A. Influence of a Natural Plant Antioxidant on the Ageing Process of Ethylene-norbornene Copolymer (Topas). *Int. J. Mol. Sci.* **2021**, *22*, 4018. [CrossRef] [PubMed]
45. Masek, A.; Chrzescijanska, E.; Zaborski, M. Characteristics of curcumin using cyclic voltammetry, UV–vis, fluorescence and thermogravimetric analysis. *Electrochim. Acta* **2013**, *107*, 441–447. [CrossRef]
46. Masek, A.; Chrzescijanska, E.; Diakowska, K.; Zaborski, M. Application of β-carotene, a natural flavonoid dye, to polymeric materials as a natural antioxidant and determination of its characteristics using cyclic voltammetry and FTIR spectroscopy. *Int. J. Electrochem. Sci.* **2015**, *10*, 3372–3386.
47. Masek, A.; Latos, M.; Piotrowska, M.; Zaborski, M. The potential of quercetin as an effective natural antioxidant and indicator for packaging materials. *Food Packag. Shelf Life* **2018**, *16*, 51–58. [CrossRef]
48. Latos-Brozio, M.; Masek, A. The application of natural food colorants as indicator substances in intelligent biodegradable packaging materials. *Food Chem. Toxicol.* **2020**, *135*, 110975. [CrossRef]
49. Masek, A.; Zaborski, M.; Kosmalska, A.; Chrzescijanska, E. Eco-friendly elastomeric composites containing Sencha and Gun Powder green tea extracts. *Comptes Rendus Chim.* **2012**, *15*, 331–335. [CrossRef]
50. Liu, H.; Fang, Z.; Peng, M.; Shen, L.; Wang, Y. The effects of irradiation cross-linking on the thermal degradation and flame-retardant properties of the HDPE/EVA/magnesium hydroxide composites. *Radiat. Phys. Chem.* **2009**, *78*, 922–926. [CrossRef]
51. Żołek-Tryznowska, Z.; Prica, M.; Pavlović, Ž.; Cveticanin, L.; Annusik, T. The influence of aging on surface free energy of corona treated packaging films. *Polym. Test.* **2020**, *89*, 106629. [CrossRef]
52. Homma, H.; Kuroyagi, T.; Mirley, C.L.; Ronzello, J.; Boggs, S.A. Diffusion of low molecular weight siloxane from bulk to surface [outdoor insulators]. In Proceedings of the Conference Record of the 1996 IEEE International Symposium on Electrical Insulation, Montreal, QC, Canada, 16–19 June 1996; Volume 1, pp. 279–282.
53. Atta, E.M.; Mohamed, N.H.; Abdel Gawad, A.A.M. Antioxidants: An Overview on the Natural and Synthetic Types. *Eur. Chem. Bull.* **2017**, *6*, 365–375. [CrossRef]
54. Seguchi, T.; Tamura, K.; Shimada, A.; Sugimoto, M.; Kudoh, H. Mechanism of antioxidant interaction on polymer oxidation by thermal and radiation ageing. *Radiat. Phys. Chem.* **2012**, *81*, 1747–1751. [CrossRef]
55. Wu, J.; Dong, J.; Wang, Y.; Gond, B.K. Thermal oxidation ageing effects on silicone rubber sealing performance. *Polym. Degrad. Stab.* **2017**, *135*, 43–53. [CrossRef]
56. Gardette, M.; Perthue, A.; Gardette, J.-L.; Janecska, T.; Földes, E.; Pukánszky, B.; Therias, S. Photo- and thermal-oxidation of polyethylene: Comparison of mechanisms and influence of unsaturation content. *Polym. Degrad. Stab.* **2013**, *98*, 2383–2390. [CrossRef]
57. Kakanuru, P.; Pochiraju, K. Moisture Ingress and Degradation of Additively Manufactured PLA, ABS and PLA/SiC Composite Parts. *Addit. Manuf.* **2020**, *36*, 101529. [CrossRef]
58. Bayart, M.; Gauvin, F.; Foruzanmehr, M.R.; Elkoun, S.; Robert, M. Mechanical and moisture absorption characterization of PLA composites reinforced with nano-coated flax fibers. *Fibers Polym.* **2017**, *18*, 1288–1295. [CrossRef]
59. Anetta, B.; Joanna, W. Innovations in the food packaging market—Intelligent packaging—A review. *Czech J. Food Sci.* **2017**, *35*, 1–6. [CrossRef]
60. Masek, A.; Zaborski, M. ENR/PCL Polymer biocomposites from renewable resources. *Comptes Rendus Chim.* **2014**, *17*, 944–951. [CrossRef]
61. Zeb, A.; Murkovic, M. Pro-Oxidant Effects of β-Carotene During Thermal Oxidation of Edible Oils. *J. Am. Oil Chem. Soc.* **2013**, *90*, 881–889. [CrossRef]
62. Ha, D.-O.; Park, C.U.; Kim, M.-J.; Lee, J. Antioxidant and prooxidant activities of β-carotene in accelerated autoxidation and photosensitized model systems. *Food Sci. Biotechnol.* **2012**, *21*, 607–611. [CrossRef]
63. Dintcheva, N.T.; Arrigo, R.; Baiamonte, M.; Rizzarelli, P.; Curcuruto, G. Concentration-dependent anti-/pro-oxidant activity of natural phenolic compounds in bio-polyesters. *Polym. Degrad. Stab.* **2017**, *142*, 21–28. [CrossRef]
64. Choe, E.; Min, D.B. Mechanisms of Antioxidants in the Oxidation of Foods. *Compr. Rev. Food Sci. Food Saf.* **2009**, *8*, 345–358. [CrossRef]
65. Yasin, S.; Hussain, M.; Zheng, Q.; Song, Y. Effects of ionic liquid on cellulosic nanofiller filled natural rubber bionanocomposites. *J. Colloid Interface Sci.* **2021**, *591*, 409–417. [CrossRef]
66. Rolere, S.; Liengprayoon, S.; Vaysse, L.; Sainte-Beuve, J.; Bonfils, F. Investigating natural rubber composition with Fourier Transform Infrared (FT-IR) spectroscopy: A rapid and non-destructive method to determine both protein and lipid contents simultaneously. *Polym. Test.* **2015**, *43*, 83–93. [CrossRef]
67. Schneider, C.; Gordon, O.N.; Edwards, R.L.; Luis, P.B. Degradation of Curcumin: From Mechanism to Biological Implications. *J. Agric. Food Chem.* **2015**, *63*, 7606–7614. [CrossRef]
68. Pénicaud, C.; Achir, N.; Dhuique-Mayer, C.; Dornier, M.; Bohuon, P. Degradation of β-carotene during fruit and vegetable processing or storage: Reaction mechanisms and kinetic aspects: A review. *Fruits* **2011**, *66*, 417–440. [CrossRef]

69. Sokolová, R.; Ramešová, Š.; Degano, I.; Hromadová, M.; Gál, M.; Žabka, J. The oxidation of natural flavonoid quercetin. *Chem. Commun.* **2012**, *48*, 3433–3435. [CrossRef]
70. Abourashed, E.A. Bioavailability of Plant-Derived Antioxidants. *Antioxidants* **2013**, *2*, 309–325. [CrossRef]
71. Roleira, F.M.; Tavares-Da-Silva, E.J.; Varela, C.L.; Costa, S.C.; Silva, T.; Garrido, J.; Borges, F. Plant derived and dietary phenolic antioxidants: Anticancer properties. *Food Chem.* **2015**, *183*, 235–258. [CrossRef]
72. Hernández, I.; Alegre, L.; Van Breusegem, F.; Munné-Bosch, S. How relevant are flavonoids as antioxidants in plants? *Trends Plant Sci.* **2009**, *14*, 125–132. [CrossRef] [PubMed]

MDPI
St. Alban-Anlage 66
4052 Basel
Switzerland
Tel. +41 61 683 77 34
Fax +41 61 302 89 18
www.mdpi.com

Polymers Editorial Office
E-mail: polymers@mdpi.com
www.mdpi.com/journal/polymers

www.ingramcontent.com/pod-product-compliance
Lightning Source LLC
Chambersburg PA
CBHW061450220926
51850CB00004B/57
9783036558981